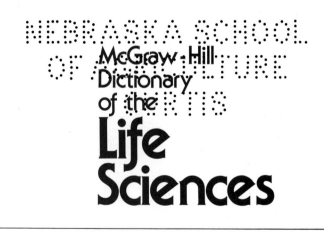

McGraw-Hill
Dictionary
of the
Life
Sciences

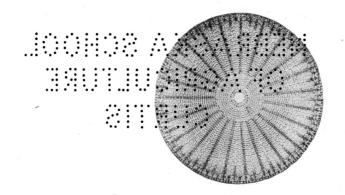

McGraw-Hill Book Company

New York	Mexico
St. Louis	Montreal
San Francisco	New Delhi
	Panama
Auckland	Paris
Bogota	São Paulo
Düsseldorf	Singapore
Johannesburg	Sydney
London	Tokoyo
Madrid	Toronto

McGraw-Hill
Dictionary
of the
Life
Sciences

Daniel N. Lapedes Editor in Chief

Library of Congress Cataloging in Publication Data

McGraw-Hill dictionary of the life sciences.

1. Life sciences—Dictionaries. 2. Biology—Dictionaries.
3. Science—Dictionaries. I. Lapedes, Daniel N.
QH302.5.M3   570'.3   76-17817
ISBN 0-07-045262-8

Contents

Preface vii

Editorial Staff vii

Consulting Editors viii

How to Use the Dictionary .... ix—x

Field Abbreviations xi

Scope of Fields .... xii—xiv

Dictionary of the Life Sciences 1–907

Appendix A1–A38

    U.S. Customary System and the metric system A1

    The International System, or SI A2–A7

    Conversion factors for the measurement systems A7–A11

    Symbols and atomic numbers for the chemical elements ... A12

    Fundamental constants A13–A15

    Spectrum of antibiotic activity A16–A19

    Normal clinical chemistry and cytology values A20–A23

    Animal taxonomy A24–A30

    Bacterial taxonomy A31–A37

    Plant taxonomy A37–A38

Preface

The *McGraw-Hill Dictionary of the Life Sciences* is intended to provide the student, researcher, teacher, librarian, and the general public with the vocabulary of the biological sciences and of related disciplines such as chemistry, statistics, and physics. The concept of the Dictionary is based on the awareness that the life sciences embrace so many specialized areas that a specialist in one aspect is a nonspecialist in another.

The more than 20,000 terms and definitions in the *McGraw-Hill Dictionary of the Life Sciences* are, in the opinion of the Board of Consulting Editors, fundamental to understanding the life sciences. The definitions either were written especially for this work or were drawn from the broader *McGraw-Hill Dictionary of Scientific and Technical Terms* (1974). The present Dictionary is thus a product of data-base operations, for the terms and definitions selected from the larger work were extracted from a master file stored on magnetic tape. As additional terms were written and reviewed by the consulting editors, they were alphabetically collated with the original terms on tape. The present Dictionary was generated from this tape and set by computer composition.

Each definition is preceded by an abbreviation identifying the field in which it is primarily used. Some of the fields covered are zoology, microbiology, genetics, anatomy, ecology, physical chemistry, spectroscopy, thermodynamics, and nuclear physics. When a definition applies equally to more than one field, it is identified by a more general field. For example, a definition that applies to both botany and zoology is assigned to biology.

The usefulness of this Dictionary is enhanced by illustrations, cross-references, and the Appendix. There are approximately 800 illustrations to amplify the definitions. Synonyms are given in the alphabetical sequence and are cross-referenced to the term where the definition appears. The Appendix has an explanation of the International System of Units, with conversion tables; a list of the chemical elements; taxonomy of animals, plants, and bacteria; and a table which gives the spectrum of activity of various antibiotics and antimicrobial agents.

An explanation of the alphabetization, cross-referencing, format, field abbreviations, and other information on how to use the Dictionary begins on page ix.

The *McGraw-Hill Dictionary of the Life Sciences* is the result of the ideas and efforts of the editorial staff and the consulting editors. It is a reference tool in which the editors have tried to achieve a clear and simple style, with the hope that the information and communication needs of the community in the field of life sciences will be served.

DANIEL N. LAPEDES
Editor in Chief

Editorial Staff

Daniel N. Lapedes, *Editor in Chief*

Sybil P. Parker, *Senior editor*
Jonathan Weil, *Staff editor*

Joe Faulk, *Editing manager*
Ellen Okin, *Editing supervisor*
Patricia Albers, *Editing assistant*

Edward J. Fox, *Art director*

Ann D. Bonardi, *Art production supervisor*
Richard A. Roth, *Art editor*
Cecilia M. Giunta, *Art/traffic*

Consulting Editors

How to Use the Dictionary

I. ALPHABETIZATION

The terms in the *McGraw-Hill Dictionary of the Life Sciences* are alphabetized on a letter-by-letter basis; word spacing, hyphen, comma, solidus, and apostrophe in a term are ignored in the sequencing. For example, an ordering of terms would be:

agar
Agarbacterium
agar-gel reaction
A horizon
Al

Also ignored in the sequencing of terms (usually, chemistry terms) are italic elements, numbers, small capitals, and Greek letters. For example, the following terms appear within alphabet letter "A":

para-**aminobenzoic acid**
D-**aminobenzyl penicillin**
4-aminofolic acid

II. FORMAT

The basic format for a defining entry provides the term in boldface, the field in small capitals, and the single definition in lightface:

term [FIELD] Definition.

A field may be followed by multiple definitions, each introduced by a boldface number:

term [FIELD] **1.** Definition. **2.** Definition. **3.** Definition.

A term may have definitions in two or more fields:

term [BOT] Definition. [GEOL] Definition.

A simple cross-reference entry appears as:

term *See* another term.

A cross-reference may also appear in combination with definitions:

term [BOT] Definition. [GEOL] *See* another term.

III. CROSS-REFERENCING

A cross-reference entry directs the user to the defining entry. For example, the user looking up "backbone" finds:

backbone *See* spine.

The user then turns to the "S" terms for the definition. Cross-references are also made from variant spellings, acronyms, abbreviations, and symbols.

aesthacyte *See* esthacyte.
ATP *See* adenosinetriphosphate.
at wt *See* atomic weight.
Au *See* gold.

IV. ALSO KNOWN AS . . . , etc.

A definition may conclude with a mention of a synonym of the term, a variant spelling, an abbreviation for the term, or other such information, introduced by "Also known as . . . ," "Also spelled . . . ," "Abbreviated . . . ," "Symbolized . . . ," "Derived from . . . ." When a term has more than one definition, the positioning of any of these phrases conveys the extent of applicability. For example:

> **term**   [BOT] **1.** Definition. Also known as synonym. **2.** Definition. Symbolized T.

In the above arrangement, "Also known as . . ." applies only to the first definition: "Symbolized . . ." applies only to the second definition.

> **term**   [BOT] **1.** Definition. **2.** Definition. [GEOL] Definition. Also known as synonym.

In the above arrangement "Also known as . . ." applies only to the second field.

> **term**   [BOT] Also known as synonym. **1.** Definition. **2.** Definition. [GEOL] Definition.

In the above arrangement, "Also known as . . ." applies to both definitions in the first field.

> **term**   Also known as synonym. [BOT] **1.** Definition. **2.** Definition. [GEOL] Definition.

In the above arrangement, "Also known as . . ." applies to all definitions in both fields.

V. CHEMICAL FORMULAS

Chemistry definitions may include either an empirical formula (say, for acetaldehyde, C_2H_2O) or a line formula (for acetone, CH_3COCH_3), whichever is appropriate.

Field Abbreviations

AGR	agriculture	MATH	mathematics
ANALY CHEM	analytical chemistry	MECH	mechanics
ANAT	anatomy	MED	medicine
ANTHRO	anthropology	MICROBIO	microbiology
ARCHEO	archeology	MOL BIO	molecular biology
ATOM PHYS	atomic physics	MYCOL	mycology
BIOCHEM	biochemistry	NUC PHYS	nuclear physics
BIOL	biology	OPTICS	optics
BIOPHYS	biophysics	ORG CHEM	organic chemistry
BOT	botany	PALEOBOT	paleobotany
CHEM	chemistry	PALEON	paleontology
CRYO	cryogenics	PATH	pathology
CRYSTAL	crystallography	PHARM	pharmacology
CYTOL	cytology	PHYS	physics
ECOL	ecology	PHYS CHEM	physical chemistry
EMBRYO	embryology	PHYSIO	physiology
ENG	engineering	PL PATH	plant pathology
EVOL	evolution	PL PHYS	plasma physics
FL MECH	fluid mechanics	PSYCH	psychology
FOOD ENG	food engineering	SCI TECH	science and technology
FOR	forestry	SPECT	spectroscopy
GEN	genetics	STAT	statistics
GEOL	geology	SYST	systematics
HISTOL	histology	THERMO	thermodynamics
IMMUNOL	immunology	VERT ZOO	vertebrate zoology
INORG CHEM	inorganic chemistry	VET MED	veterinary medicine
INV ZOO	invertebrate zoology	VIROL	virology
MATER	materials	ZOO	zoology

Scope of Fields

agriculture — The production of plants and animals useful to humans, involving soil cultivation and the breeding and management of crops and livestock.

analytical chemistry — Science and art of determining composition of materials in terms of elements and compounds which they contain.

anatomy — The branch of morphology concerned with the gross and microscopic structure of animals, especially humans.

anthropology — The study of the interrelations of biological, cultural, geographical, and historical aspects of the human race.

archeology — The scientific study of the material remains of the cultures of historical and prehistorical peoples.

atomic physics — A branch of physics concerned with the structures of the atom, the characteristics of the electrons and other elementary particles of which the atom is composed, the arrangement of the atom's energy states, and the processes involved in the radiation of light and x-rays.

biochemistry — The study of the chemical substances that occur in living organisms, the processes by which these substances enter into or are formed in the organisms and react with each other and the environment, and the methods by which the substances and processes are identified, characterized, and measured.

biology — The science of living organisms, concerned with the study of embryology, anatomy, physiology, cytology, morphology, taxonomy, genetics, evolution, and ecology.

biophysics — The hybrid science involving the methods and ideas of physics and chemistry to study and explain the structures of living organisms and the mechanics of life processes.

botany — That branch of biological science which embraces the study of plants and plant life, including algae; deals with taxonomy, morphology, physiology, and other aspects.

chemistry — The scientific study of the properties, composition, and structure of matter, the changes in structure and composition of matter, and accompanying energy changes.

cryogenics — The science of producing and maintaining very low temperatures, of phenomena at those temperatures, and of technical operations performed at very low temperatures.

crystallography — The branch of science that deals with the geometric description of crystals, their internal arrangement, and their properties.

cytology — The branch of biological science which deals with the structure, behavior, growth, and reproduction of cells and the function and chemistry of cells and cell components.

ecology — The study of the interrelationships between organisms and their environment.

embryology — The study of the development of the organism from the zygote, or fertilized egg.

engineering—The science by which the properties of matter and the sources of power in nature are made useful to humans in structures, machines, and products.

evolution—The processes of biological and organic change in organisms by which descendants come to differ from their ancestors, and a history of the sequence of such change.

fluid mechanics—The science concerned with fluids, either at rest or in motion, and dealing with pressures, velocities, and accelerations in the fluid, including fluid deformation and compression or expansion.

food engineering—Technical discipline involved in food manufacturing and processing.

forestry—The science of developing, cultivating, and managing forest lands for wood, forage, water, wildlife, and recreation; the management of growing timber.

genetics—The science concerned with biological inheritance, that is, with the causes of the resemblances and differences among related individuals.

geology—The study or science of earth, its history, and its life as recorded in the rocks; includes the study of the geologic features of an area, such as the geometry of rock formations, weathering and erosion, and sedimentation.

histology—The study of the structure and chemical composition of animal tissues as related to their function.

immunology—The division of biological science concerned with the native or acquired resistance of higher animal forms and humans to infection with microorganisms.

inorganic chemistry—A branch of chemistry that deals with reactions and properties of all chemical elements and their compounds, excluding hydrocarbons but usually including carbides and other simple carbon compounds (such as CO_2, CO, and HCN).

invertebrate zoology—A branch of zoology concerned with the taxonomy, behavior, and morphology of invertebrate animals.

materials—The study of admixtures of matter or the basic matter from which products are made; includes adhesives, building materials, fuels, paints, leathers, and so on.

mathematics—The deductive study of shape, quantity, and dependence; the two main areas are applied mathematics and pure mathematics, the former arising from the study of physical phenomena, the latter involving the intrinsic study of mathematical structures.

mechanics—The branch of physics which seeks to formulate general rules for predicting the behavior of a physical system under the influence of any type of interaction with its environment.

medicine—The study of cause and treatment of human disease, including the healing arts dealing with diseases which are treated by a physician or a surgeon.

microbiology—The science and study of microorganisms, especially bacteria and rickettsiae, and of antibiotic substances.

molecular biology—That branch of biology which attempts to interpret biological events in terms of the molecules in the cell.

mycology—A branch of biological science concerned with the study of fungi.

nuclear physics—The study of the characteristics, behavior, and internal structure of the atomic nucleus.

optics—The study of phenomena associated with the generation, transmission, and detection of electromagnetic radiation in the spectral range extending from the long-wave edge of the x-ray region to the short-wave edge of the radio region; and the science of light.

organic chemistry—The study of the composition, reactions, and properties of carbon compounds except CO_2, CO, and certain ionic compounds such as Na_2CO_3 and $NaCN$.

paleobotany—The study of fossil plants and vegetation of the geologic past.

paleontology—The study of life in the geologic past as recorded by fossil remains.

pathology—The study of the nature of disease, through study of its causes, its processes, and its effects, together with the associated alterations of structure and function; and the laboratory findings of disease, as distinguished from clinical signs and symptoms.

pharmacology—The science of detection and measurement of the effects of drugs or other chemicals on biological systems; includes all chemicals used as drugs.

physical chemistry—The description and prediction of chemical behavior by means of physical theory; main subject areas are structure, thermodynamics, and kinetics.

physics—The science concerned with those aspects of nature which can be understood in terms of elementary principles and laws.

physiology—The branch of biological science concerned with the basic activities that occur in cells and tissues of living organisms and involving physical and chemical studies of these organisms.

plant pathology—The branch of botany concerned with diseases of plants.

psychology—The science of the function of the mind and the behavior of an organism, both animal and human, in relation to its environment.

science and technology—The study of the natural sciences and the application of this knowledge for practical purposes.

spectroscopy—The branch of physics concerned with the production, measurement, and interpretation of electromagnetic spectra arising from either emission or absorption of radiant energy by various substances.

statistics—The science dealing with the collection, analysis, interpretation, and presentation of masses of numerical data.

systematics—The science of animal and plant classification.

thermodynamics—The branch of physics which seeks to derive, from a few basic postulates, relations between properties of substances, especially those which are affected by changes in temperature, and a description of the conversion of energy from one form to another.

vertebrate zoology—A branch of zoology concerned with the taxonomy, behavior, and morphology of vertebrate animals.

veterinary medicine—That branch of medical practice which treats of the diseases and injuries of animals.

virology—The science that deals with the study of viruses.

zoology—The science that deals with the taxonomy, behavior, and morphology of animal life.

A

A *See* angstrom.

aaa disease [MED] An endemic hookworm disease accompanied by anemia that occurred in ancient Egypt and is designated in the Ebers Papyrus.

Aalenian [GEOL] Lowermost Middle or uppermost Lower Jurassic geologic time.

aapamoor [ECOL] A moor with elevated areas or mounds supporting dwarf shrubs and sphagnum, interspersed with low areas containing sedges and sphagnum, thus forming a mosaic.

aardvark [VERT ZOO] A nocturnal, burrowing, insectivorous mammal of the genus *Orycteropus* in the order Tubulidentata. Also known as earth pig.

aardwolf [VERT ZOO] *Proteles cristatus.* A hyenalike African mammal of the family Hyaenidae.

abaca [BOT] *Musa textilis.* A plant of the banana family native to Borneo and the Philippines, valuable for its hard fiber. Also known as Manila hemp.

abactinal [INV ZOO] In radially symmetrical animals, pertaining to the surface opposite the side where the mouth is located.

abalienation [PSYCH] Mental deterioration or derangement.

abalone [INV ZOO] A gastropod mollusk composing the single genus, *Haliotis,* of the family Haliotidae. Also known as ear shell; ormer; paua.

A band [HISTOL] The region between two adjacent I bands in a sarcomere; characterized by partial overlapping of actin and myosin filaments.

abapertural [INV ZOO] Away from the shell aperture, referring to mollusks.

abapical [BIOL] On the opposite side to, or directed away from, the apex.

abarognosis [MED] Lack of ability to estimate the weight of an object one is holding.

abasia [MED] Lack of muscular coordination in walking.

abaxial [BIOL] On the opposite side to, or facing away from, the axis of an organ or organism.

abaxile [BOT] Referring to an embryo whose axis has a different direction from that of the seed.

Abbe condenser [OPTICS] A variable large-aperture lens system arranged substage to image a light source into the focal plane of a microscope objective.

abcauline [BOT] Positioned away from the stem.

Abderhalden reaction [PATH] A chemical blood test for the identification of certain enzymes associated with pregnancy and a few diseases.

abdomen [ANAT] **1.** The portion of the vertebrate body between the thorax and the pelvis. **2.** The cavity of this part of the body. [INV ZOO] The elongate region posterior to the thorax in arthropods.

AARDVARK

The aardvark *(Orycteropus afer),* a nocturnal, burrowing animal ranging from Ethiopia to southern Africa.

ABALONE

Typical abalone ear-shaped shell perforated by pores.

abdominal apoplexy [MED] Vascular occlusion and hemorrhage in an abdominal organ, usually the small intestine, or in the peritoneal cavity.

abdominal depth [ANTHRO] Maximum horizontal contact dimension, measured front to back.

abdominal gestation [MED] Development of a fetus outside the uterus in the abdominal cavity.

abdominal hernia *See* ventral hernia.

abdominal hysterectomy [MED] Surgical removal of all or part of the uterus through an incision in the abdomen.

abdominal pore [VERT ZOO] Any of the single or paired pores leading from the coelom to the exterior in cyclostomes and certain fishes.

abdominal reflex [PHYSIO] A superficial or cutaneous reflex involving contraction of the abdominal muscles, induced by stroking the overlying skin.

abdominal regions [ANAT] Nine theoretical areas delineated on the abdomen by two horizontal and two parasagittal lines: above, the right hypochondriac, epigastric, and left hypochondriac; in the middle, the right lateral, umbilical, and left lateral; and below, the right inguinal, hypogastric, and left inguinal.

abdominal rib [VERT] One of the ossifications that occur in fibrous tissue between the skin and muscles of certain reptiles.

abducens [ANAT] The sixth cranial nerve in vertebrates; a paired, somatic motor nerve arising from the floor of the fourth ventricle of the brain and supplying the lateral rectus eye muscles.

abduction [PHYSIO] Movement of an extremity or other body part away from the axis of the body.

abductor [PHYSIO] Any muscle that draws a part of the body or an extremity away from the body axis.

abenteric [MED] Involving abdominal organs and structures outside the intestine.

aberrant [BIOL] An atypical group, individual, or structure, especially one with an aberrant chromosome number.

aberration [OPTICS] Any deviation from perfect reproduction so that a point is not imaged as a point, a straight line as straight, or an angle as an equal angle.

Abies [BOT] The firs, a genus of trees in the pine family characterized by erect cones, absence of resin canals in the wood, and flattened needlelike leaves.

abiocoen [ECOL] A nonbiotic habitat.

abiogenesis [BIOL] The obsolete concept that plant and animal life arise from nonliving organic matter. Also known as autogenesis; spontaneous generation.

abiotic [BIOL] Referring to the absence of living organisms.

abiotic environment [ECOL] All physical and nonliving chemical factors, such as soil, water, and atmosphere, which influence living organisms.

abiotrophy [MED] Disordered functioning of an organ or system, as in Huntington's chorea, due to an inherited pathologic trait, which trait, however, may remain latent in the individual rather than becoming apparent; this mechanism is still conceptual.

abjection [MYCOL] The discharge or casting off of spores by the spore-bearing structure of a fungus.

abjunction [BOT] Spore formation by hyphal septation.

ablactation [PHYSIO] Termination of the period of mammary secretion.

ablastin [IMMUNOL] An antibodylike substance elicited by *Trypanosoma lewisi* in the blood serum of infected rats that inhibits reproduction of the parasite.

ablation [MED] The removal of tissue or a part of the body by surgery, such as by excision or amputation.

abnormal behavior [PSYCH] Personality functioning that is socially undesirable or that renders the individual unable to cope with day-to-day living. Also known as behavior disorder.

abnormal psychology [PSYCH] A branch of psychology that deals with behavior disorders and internal psychic conflict in addition to certain normal phenomena such as dreams, motivations, and anxiety.

ABO blood group [IMMUNOL] An immunologically distinct, genetically determined group of human erythrocyte antigens represented by two blood factors (A and B) and four blood types (A, B, AB, and O).

abomasitis [VET MED] Inflammation of the abomasum in ruminants.

abomasum [VERT ZOO] The final chamber of the complex stomach of ruminants; has a glandular wall and corresponds to a true stomach.

aboral [INV ZOO] Opposite to the mouth.

abortifacient [MED] Any agent that induces abortion.

abortion [MED] The spontaneous or induced expulsion of the fetus prior to the time of viability, most often during the first 20 weeks of the human gestation period.

abortive [BIOL] Imperfectly formed or developed.

abortus [MED] An aborted fetus.

abrachiocephalia [MED] Congenital lack of arms and head. Also known as acephalobrachia.

abrasion [MED] A spot denuded of skin, mucous membrane, or superficial epithelium by rubbing or scraping.

abruptly acuminate [BOT] Of leaves, having an apex or base which is suddenly sharply pointed.

abruptly pinnate [BOT] Of leaves, having parts arranged like a feather. [INV ZOO] Referring to an epipodium having no wings on the main axis but having a number of secondary axes which bear wings.

abscess [MED] A localized collection of pus surrounded by inflamed tissue.

abscisic acid [BIOCHEM] $C_{16}H_{20}O_4$ A plant hormone produced by fruits and leaves that promotes abscission and dormancy and retards vegetative growth. Formerly known as abscisin.

abscisin See abscisic acid.

abscissa [MATH] One of the coordinates of a two-dimensional coordinate system, usually the horizontal coordinate, denoted by x.

abscission [BOT] A physiological process promoted by abscisic acid whereby plants shed a part, such as a leaf, flower, seed, or fruit.

absolute boiling point [CHEM] The boiling point of a substance expressed in the unit of an absolute temperature scale.

absolute deviation [STAT] The difference, without regard to sign, between a variate value and a given value.

absolute gravity [CHEM] Density or specific gravity of a fluid reduced to standard conditions; for example, with gases, to 760 mm Hg pressure and 0°C temperature. Also known as absolute density.

absolute pressure [PHYS] The pressure above the absolute zero value of pressure that theoretically obtains in empty space or at the absolute zero of temperature, as distinguished from gage pressure.

absolute refractory period [PHYSIO] The brief time during discharge of a nerve impulse when the neuron cannot be fired again.

absolute temperature [THERMO] **1.** The temperature measurable in theory on the thermodynamic temperature scale. **2.** The temperature in Celsius degrees relative to the absolute

ABSCISIC ACID

Structural formula for S-abscisic acid, the naturally occurring form.

ABSOLUTE TEMPERATURE

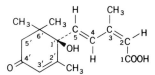

Comparisons of Kelvin, Celsius, Rankine, and Fahrenheit temperature scales. Temperatures are rounded off to nearest degree. (*From M. W. Zemansky, Temperatures Very Low and Very High, Van Nostrand, 1964*)

zero at $-273.16°C$ (the Kelvin scale) or in Fahrenheit degrees relative to the absolute zero at $-459.69°F$ (the Rankine scale).

absolute temperature scale [THERMO] A scale with which temperatures are measured relative to absolute zero. Also known as absolute scale.

absolute threshold [PHYSIO] The minimum stimulus energy that an organism can detect.

absolute zero [THERMO] The temperature of $-273.16°C$, or $-459.69°F$, or 0 K, thought to be the temperature at which molecular motion vanishes and a body would have no heat energy.

absorb [CHEM] To take up matter in bulk. [ELECTROMAG] To take up energy from radiation. [PHYS] To take up matter or radiation.

absorbance [PHYS CHEM] The common logarithm of the reciprocal of the transmittance of a pure solvent. Also known as absorbancy; extinction.

absorbancy *See* absorbance.

absorbed dose [NUCLEO] The amount of energy imparted by ionizing particles to a unit mass of irradiated material at a place of interest. Also known as dosage; dose.

absorption [CHEM] The taking up of matter in bulk by other matter, as in dissolving of a gas by a liquid. [PHYSIO] Passage of a chemical substance through a body membrane.

absorption atelectasis *See* obstructive atelectasis.

absorption band [PHYS] A range of wavelengths or frequencies in the electromagnetic spectrum within which radiant energy is absorbed by a substance.

absorption curve [PHYS] A graph showing the curvilinear relationship of the variation in absorbed radiation as a function of wavelength.

absorption edge [SPECT] The wavelength corresponding to a discontinuity in the variation of the absorption coefficient of a substance with the wavelength of the radiation. Also known as absorption limit.

absorption limit *See* absorption edge.

absorption line [SPECT] A minute range of wavelength or frequency in the electromagnetic spectrum within which radiant energy is absorbed by the medium through which it is passing.

absorption spectrophotometer [SPECT] An instrument used to measure the relative intensity of absorption spectral lines and bands. Also known as difference spectrophotometer.

absorption spectroscopy [SPECT] The study of spectra obtained by the passage of radiant energy from a continuous source through a cooler, selectively absorbing medium.

absorption spectrum [SPECT] The array of absorption lines and absorption bands which results from the passage of radiant energy from a continuous source through a cooler, selectively absorbing medium.

absorption test [IMMUNOL] Analysis of the antigenic components of bacterial cells and large macromolecules by a series of precipitation or agglutination reactions with specific antibodies.

absorptive power *See* absorptivity.

absorptivity [ANALY CHEM] The constant a in the Beer's law relation $A = abc$, where A is the absorbance, b the path length, and c the concentration of solution. Also known as absorptive power.

abstinence syndrome [MED] A disturbance of metabolic equilibrium that occurs when a narcotic drug is withdrawn from the user.

abstriction [MYCOL] In fungi, the cutting off of spores in hyphae by formation of septa followed by abscission of the spores, especially by constriction.

abterminal [BIOL] Referring to movement from the end toward the middle; specifically, describing the mode of electric current flow in a muscle.

abulia [PSYCH] Loss of ability to make decisions.

abyssal-benthic [OCEANOGR] Pertaining to the bottom of the abyssal zone.

abyssal floor [GEOL] The ocean floor, or bottom of the abyssal zone.

abyssal zone [OCEANOGR] The biogeographic realm of the great depths of the ocean beyond the limits of the continental shelf, generally below 1000 meters.

abyssobenthic [BIOL] Pertaining to, or found on, the bottom of the ocean.

Ac *See* actinium.

acalyculate [BOT] Lacking a calyx.

Acalyptratae [INV ZOO] A large group of small, two-winged flies in the suborder Cyclorrhapha characterized by small or rudimentary calypters.

acantha [BIOL] A sharp spine; a spiny process, as on vertebrae.

Acanthaceae [BOT] A family of dicotyledonous plants in the order Scrophulariales distinguished by their usually herbaceous habit, irregular flowers, axile placentation, and dry, dehiscent fruits.

acanthaceous [BOT] Armed with prickles or spines.

Acantharia [INV ZOO] A subclass of essentially pelagic protozoans in the class Actinopodea characterized by skeletal rods constructed of strontium sulfate (celestite).

Acanthaster [INV ZOO] A genus of Indo-Pacific starfishes, including the crown-of-thorns, of the family Asteriidae; economically important as a destroyer of oysters in fisheries.

acanthella [INV ZOO] A transitional larva of the phylum Acanthocephala in which rudiments of reproductive organs, lemnisci, a proboscis, and a proboscis receptacle are formed.

acanthion [ANAT] A point at the tip of the anterior nasal spine.

acanthocarpous [BOT] Having spiny fruit.

Acanthocephala [INV ZOO] The spiny-headed worms, a phylum of helminths; adults are parasitic in the alimentary canal of vertebrates.

acanthocephalous [INV ZOO] Having a proboscis that is hooked.

acanthocladous [BOT] Having spiny branches.

acanthocyst [INV ZOO] In Nemertea, a sac that contains lateral or reserve stylets.

acanthocytosis [MED] A disorder of erythrocytes in which spiny projections appear on the blood cells.

Acanthodidae [PALEON] A family of extinct acanthodian fishes in the order Acanthodiformes.

Acanthodiformes [PALEON] An order of extinct fishes in the class Acanthodii having scales of acellular bone and dentine, one dorsal fin, and no teeth.

Acanthodii [PALEON] A class of extinct fusiform fishes, the first jaw-bearing vertebrates in the fossil record.

acanthodion [INV ZOO] In Acarina, a seta on the tarsus that contains an extension of a sensory basal cell.

acanthoid [BIOL] Shaped like a spine.

Acanthometrida [INV ZOO] An order of marine protozoans in the subclass Acantharia with 20 or less skeletal rods.

acanthophore [INV ZOO] In Nemertea, a conical mass that forms the basis of a median stylet.

Acanthophractida [INV ZOO] An order of marine protozoans

ACANTHELLA

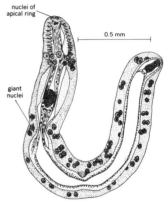

A stage in the life history of *Moniliformis dubius*, a helminth, with the acanthella dissected from its enveloping sheath.

ACANTHOMETRIDA

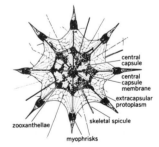

A drawing of *Acanthometra* showing the characteristic pattern of the radially arranged rods (skeletal spicules). (*From L. H. Hyman, The Invertebrates, vol. 1, McGraw-Hill, 1940*)

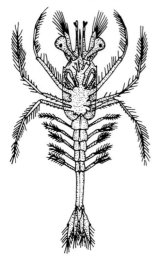

1 mm

The mysis (acanthosoma) larva
of sergestid shrimp.
(Smithsonian Institution)

in the subclass Acantharia; skeleton includes a latticework shell and skeletal rods.

acanthopodous [BOT] Having a spiny or prickly petiole or peduncle.

acanthopore [PALEON] A tubular spine in some fossil bryozoans.

Acanthopteri [VERT ZOO] An equivalent name for the Perciformes.

Acanthopterygii [VERT ZOO] An equivalent name for the Perciformes.

acanthosis [MED] Any thickening of the prickle-cell layer of the epidermis; associated with many skin diseases.

acanthosoma [INV ZOO] The last primitive larval stage, the mysis, in the family Sergestidae.

Acanthosomatidae [INV ZOO] A small family of insects in the order Hemiptera.

acanthosphenote [INV ZOO] In echinoids, a type of spine composed of solid wedges that are separated from one another by porous tissue.

acanthostegous [INV ZOO] Being overlaid with two series of spines, as the ovicell or ooecium of certain bryozoans.

acanthozooid [INV ZOO] A specialized individual in a bryozoan colony that secretes tubules which project as spines above the colony's outer surface.

Acanthuridae [VERT ZOO] The surgeonfishes, a family of perciform fishes in the suborder Acanthuroidei.

Acanthuroidei [VERT ZOO] A suborder of chiefly herbivorous fishes in the order Perciformes.

acapnia [MED] Absence of carbon dioxide in the blood and tissues.

Acari [INV ZOO] The equivalent name for Acarina.

acariasis [MED] Any skin disease resulting from infestation with acarids or mites.

acaricide [MATER] A pesticide used to destroy mites on domestic animals, crops, and man. Also known as miticide.

Acaridiae [INV ZOO] A group of pale, weakly sclerotized mites in the suborder Sarcoptiformes, including serious pests of stored food products and skin parasites of warm-blooded vertebrates.

Acarina [INV ZOO] The ticks and mites, a large order of the class Arachnida, characterized by lack of body demarcation into cephalothorax and abdomen.

acarocecidium [PL PATH] A gall that is caused by gall mites.

acarology [INV ZOO] A branch of zoology dealing with the mites and ticks.

acarophily [ECOL] A symbiotic relationship between plants and mites.

acarpellous [BOT] Lacking carpels.

acarpous [BOT] Not producing fruit.

acatalasia [MED] Congenital absence of the enzyme catalase.

acaulous [BOT] 1. Lacking a stem. 2. Being apparently stemless but having a short underground stem.

accelerated hypertension *See* malignant hypertension.

acceleration globulin [BIOCHEM] A globulin that acts to accelerate the conversion of prothrombin to thrombin in blood clotting; found in blood plasma in an inactive form.

acceleration tolerance [PHYSIO] The maximum *g* forces an individual can withstand without losing control or consciousness.

acceleratory reflex [PHYSIO] Any reflex originating in the labyrinth of the inner ear in response to a change in the rate of movement of the head.

accessorius [ANAT] Any muscle that reinforces the action of another.

accessory bodies [CYTOL] Small argyrophil particles that originate from the Golgi apparatus in spermatocytes.

accessory bud [BOT] An embryonic shoot occurring above or to the side of an axillary bud. Also known as supernumerary bud.

accessory cell [BOT] A morphologically distinct epidermal cell adjacent to, and apparently functionally associated with, guard cells on the leaves of many plants.

accessory chromosome *See* supernumerary chromosome.

accessory gland [ANAT] A mass of glandular tissue separate from the main body of a gland. [INV ZOO] A gland associated with the male reproductive organs in insects.

accessory movement *See* synkinesia.

accessory nerve [ANAT] The eleventh cranial nerve in tetrapods, a paired visceral motor nerve; the bulbar part innervates the larynx and pharynx, and the spinal part innervates the trapezius and sternocleidomastoid muscles.

accessory pulsatory organ [INV ZOO] In certain insects, one of several saclike structures that pulsate independently.

accidental whorl [ANAT] A type of whorl fingerprint pattern which is a combination of two different types of pattern, with the exception of the plain arch, with two or more deltas; or a pattern which possesses some of the requirements for two or more different types; or a pattern which conforms to none of the definitions; in accidental whorl tracing three types appear: an outer (O), inner (I), or meeting (M).

Accipitridae [VERT ZOO] The diurnal birds of prey, the largest and most diverse family of the order Falconiformes, including hawks, eagles, and kites.

acclimation *See* acclimatization.

acclimatization [EVOL] Adaptation of a species or population to a changed environment over several generations. Also known as acclimation.

accommodation [PHYSIO] A process in most vertebrates whereby the focal length of the eye is changed by automatic adjustment of the lens to bring images of objects from various distances into focus on the retina.

accommodation reflex [PHYSIO] Changes occurring in the eyes when vision is focused from a distant to a near object; involves pupil contraction, increased lens convexity, and convergence of the eyes.

accretion line [HISTOL] A microscopic line on a tooth, marking the addition of a layer of enamel or dentin.

acellular [BIOL] Not composed of cells.

acellular gland [PHYSIO] A gland, such as intestinal glands, the pancreas, and the parotid gland, that secretes a noncellular product.

acellular slime mold [MYCOL] The common name for members of the Myxomycetes.

acentric [BIOL] Not oriented around a middle point. [GEN] A chromosome or chromosome fragment lacking a centromere.

acentrous [VERT ZOO] Lacking vertebral centra and having the notochord persistent throughout life, as in certain primitive fishes.

Acephalina [INV ZOO] A suborder of invertebrate parasites in the protozoan order Eugregarinida characterized by nonseptate trophozoites.

acephalobrachia *See* abrachiocephalia.

acephalocyst [INV ZOO] An abnormal cyst of the *Echinococcus granulosus* larva, lacking a head and brood capsules, found in human organs.

acephalous [BOT] Having the style originate at the base instead of at the apex of the ovary. [ZOO] Lacking a head.

Acer [BOT] A genus of broad-leaved, deciduous trees of the

ACCIDENTAL WHORL

type lines

A reproduction of an I tracing of an accidental whorl. *(Federal Bureau of Investigation)*

order Sapindales, commonly known as the maples; the sugar or rock maple (*A. saccharum*) is the most important commercial species.

acerate [BOT] Needle-shaped, specifically referring to leaves.

Acerentomidae [INV ZOO] A family of wingless insects belonging to the order Protura; the body lacks tracheae and spiracles.

acerose [BOT] Of leaves, being shaped like a needle.

acerous [ZOO] 1. Lacking horns. 2. Being without antennae. 3. Being without tentacles.

acervate [BIOL] Growing in heaps or dense clusters.

acervulus [MYCOL] A cushion- or disk-shaped mass of hyphae, peculiar to the Melanconiales, on which there are dense aggregates of conidiophores.

acetabulum [ANAT] A cup-shaped socket on the hipbone that receives the head of the femur. [INV ZOO] 1. A cavity on an insect body into which a leg inserts for articulation. 2. The sucker of certain invertebrates such as trematodes and tapeworms.

acetal [ORG CHEM] 1. $CH_3CH(OC_2H_5)_2$ A colorless, flammable, volatile liquid used as a solvent and in manufacture of perfumes. Also known as 1,1-diethoxyethane. 2. Any of a class of compounds, stable ethers, formed from aldehydes and 1,1-dihydroxy alcohols.

acetaldehydase [BIOCHEM] An enzyme that catalyzes the oxidation of acetaldehyde to acetic acid.

acetaldehyde [ORG CHEM] C_2H_4O A colorless, flammable liquid used chiefly to manufacture acetic acid. Also known as ethanal.

acetamide [ORG CHEM] CH_3CONH_2 The crystalline, colorless amide of acetic acid, used in organic synthesis and as a solvent. Also known as ethanamide.

acetaminophen [PHARM] $C_8H_9O_2N$ A drug used as an analgesic and antipyretic.

acetanilide [PHARM] $C_6H_5NHCOCH_3$ A white, crystalline compound used medicinally to relieve pain and reduce fever.

acetic acid [ORG CHEM] CH_3COOH 1. A clear, colorless liquid or crystalline mass with a pungent odor, miscible with water or alcohol; crystallizes in deliquescent needles; a component of vinegar. Also known as ethanoic acid. 2. A mixture of the normal and acetic salts; used as a mordant in the dyeing of wool.

acetic acid bacteria *See* Acetobacter.

acetic fermentation [MICROBIO] Oxidation of alcohol to produce acetic acid by the action of bacteria of the genus *Acetobacter.*

acetic thiokinase [BIOCHEM] An enzyme that catalyzes the formation of acetyl coenzyme A from acetate and adenosinetriphosphate.

acetoacetyl coenzyme A [BIOCHEM] $C_{25}H_{41}O_{18}N_7P_3S$ An intermediate product in the oxidation of fatty acids.

Acetobacter [MICROBIO] A genus of aerobic, peritrichously flagellated bacteria in the family Achromobacteraceae. Also known as acetic acid bacteria; vinegar bacteria.

acetolactic acid [BIOCHEM] $C_5H_8O_4$ A monocarboxylic acid formed as an intermediate in the synthesis of the amino acid valine.

Acetomonas [MICROBIO] A genus of aerobic, polarly flagellated vinegar bacteria in the family Pseudomonadaceae; used industrially to produce vinegar, gluconic acid, and L-sorbose.

acetone [ORG CHEM] CH_3COCH_3 A colorless, volatile, extremely flammable liquid, miscible with water; used as a solvent and reagent. Also known as 2-propanone.

acetone body *See* ketone body.

acetonemia [MED] A condition characterized by large amounts of acetone bodies in the blood. Also known as ketonemia.

acetyl [ORG CHEM] CH_3CO- A two-carbon organic radical containing a methyl group and a carbonyl group.

acetylase [BIOCHEM] Any enzyme that catalyzes the formation of acetyl esters.

acetylation [ORG CHEM] The process of bonding an acetyl group onto an organic molecule.

acetylcholine [BIOCHEM] $C_7H_{17}O_3N$ A compound released from certain autonomic nerve endings which acts in the transmission of nerve impulses to excitable membranes.

acetylcholinesterase [BIOCHEM] An enzyme found in excitable membranes that inactivates acetylcholine.

acetyl coenzyme A [BIOCHEM] $C_{23}H_{39}O_{17}N_7P_3S$ A coenzyme, derived principally from the metabolism of glucose and fatty acids, that takes part in many biological acetylation reactions; oxidized in the Krebs cycle.

acetyl phosphate [BIOCHEM] $C_2H_5O_5P$ The anhydride of acetic and phosphoric acids occurring in the metabolism of pyruvic acid by some bacteria; phosphate is used by some microorganisms, in place of ATP, for the phosphorylation of hexose sugars.

acetylsalicylic acid [ORG CHEM] $CH_3COOC_6H_4COOH$ A white, crystalline, weakly acidic substance, with melting point 137°C; slightly soluble in water; used medicinally as an antipyretic. Also known by trade name aspirin.

acetylurea [ORG CHEM] $CH_3CONHCONH_2$ Crystals that are colorless and are slightly soluble in water.

Achaenodontidae [PALEON] A family of Eocene dichobunoids, piglike mammals belonging to the suborder Palaeodonta.

achalasia [MED] Inability of a hollow muscular organ or ring of muscle (sphincter) to relax.

ache [MED] A constant dull or throbbing pain.

acheb [ECOL] Short-lived vegetation regions of the Sahara composed principally of mustards (Cruciferae) and grasses (Gramineae).

acheilary [BOT] Having an undeveloped labellum, as in certain orchids.

achene [BOT] A small, dry, indehiscent fruit formed from a simple ovary bearing a single seed.

achievement age [PSYCH] Accomplishment, or actual level of scholastic performance, expressed as equivalent to the age in years of the average child showing similar attainments.

achievement quotient [PSYCH] The ratio, usually multiplied by 100, between the achievement age, or actual scholastic level, and the mental age.

achilary [BOT] In flowers, having the lip (labellum) undeveloped or lacking.

Achilles jerk [PHYSIO] A reflex action seen as plantar flection in response to a blow to the Achilles tendon. Also known as Achilles tendon reflex.

Achilles tendon [ANAT] The tendon formed by union of the tendons of the calf muscles, the soleus and gastrocnemius, and inserted into the heel bone.

Achilles tendon reflex See Achilles jerk.

achlamydeous [BOT] Lacking a perianth.

achlorhydria [MED] Absence of hydrochloric acid in the stomach.

achondroplasia [MED] A hereditary deforming disease of the skeletal system, inherited in man as an autosomal dominant trait and characterized by insufficient growth of the long bones, resulting in reduced length. Also known as chondrodystrophy fetalis.

achondroplastic dwarf [MED] A human with short legs and arms due to achondroplasia.

achordate [VERT ZOO] Lacking a notochord.

achroglobin [BIOCHEM] A colorless respiratory pigment present in some mollusks and urochordates.

achromat *See* achromatic lens.

Achromatiaceae [MICROBIO] A family of large, spherical to ovoid bacteria in the order Beggiatoales.

achromatic [OPTICS] Capable of transmitting light without decomposing it into its constituent colors.

achromatic lens [OPTICS] A combination of two or more lenses having a focal length that is the same for two quite different wavelengths, thereby removing a major portion of chromatic aberration. Also known as achromat.

achromatin [CYTOL] The portion of the cell nucleus which does not stain easily with basic dyes.

achromatism [MED] **1.** Total color blindness. **2.** Absence of chromatic aberration.

achromic [BIOL] Colorless; lacking normal pigmentation.

Achromobacter [MICROBIO] A genus of motile and nonmotile, gram-negative, rod-shaped bacteria in the family Achromobacteraceae.

Achromobacteraceae [MICROBIO] A family of true bacteria, order Eubacteriales, characterized by aerobic metabolism.

Achromycin [MICROBIO] Trade name for the antibiotic tetracycline.

acicular [SCI TECH] Needlelike; slender and pointed.

aciculum [INV ZOO] A stiff basal seta found in the parapodium of chaetopods.

acid [CHEM] **1.** Any of a class of chemical compounds whose aqueous solutions turn with blue litmus paper red, react with and dissolve certain metals to form salts, and react with bases to form salts. **2.** A compound capable of transferring a hydrogen ion in solution. **3.** By extension of the term, a substance that ionizes in solution to yield the positive ion of the solvent. **4.** A molecule or ion that combines with another molecule or ion by forming a covalent bond with two electrons from the other species.

acid alcohol [ORG CHEM] A compound containing both a carboxyl group ($-COOH$) and an alcohol group ($-CH_2OH$, $=CHOH$, or $\equiv COH$).

acid-base balance [PHYSIO] Physiologically maintained equilibrium of acids and bases in the body.

acid-base equilibrium [CHEM] The condition when acidic and basic ions in a solution exactly neutralize each other; that is, the pH is 7.

acid-base indicator [ANALY CHEM] A substance that reveals, through characteristic color changes, the degree of acidity or basicity of solutions.

acid-base titration [ANALY CHEM] A titration in which an acid of known concentration is added to a solution of base of unknown concentration, or the converse.

acidemia [MED] A condition in which the pH of the blood falls below normal.

acid-fast bacteria [MICROBIO] Bacteria, especially mycobacteria, that stain with basic dyes and fluorochromes and resist decoloration by acid solutions.

acid-fast stain [MICROBIO] A differential stain used in identifying species of *Mycobacterium* and one species of *Nocardia*.

acidity [CHEM] The state of being acid.

acid number *See* acid value.

acidolysis [ORG CHEM] A chemical reaction involving the decomposition of a molecule, with the addition of the elements of an acid to the molecule; the reaction is comparable to hydrolysis or alcoholysis, in which water or alcohol, respectively, is used in place of the acid. Also known as acyl exchange.

acidophil [BIOL] **1.** Any substance, tissue, or organism having an affinity for acid stains. **2.** An organism having a preference for an acid environment. [HISTOL] **1.** An alpha cell of the adenohypophysis. **2.** *See* eosinophil.

acidophilia *See* eosinophilia.

acidophilic erythroblast *See* normoblast.

acidosis [MED] A condition of decreased alkali reserve of the blood and other body fluids.

acid phosphatase [BIOCHEM] Any nonspecific phosphatase requiring an acid medium for optimum activity.

aciduric [BIOL] Tolerating a greater than optimum acid environment.

acid value [CHEM] The acidity of a solution expressed in terms of normality. Also known as acid number.

acinaciform [BOT] Of leaves, being shaped like a sword.

acinar [ANAT] Pertaining to an acinus.

acinar cell [ANAT] Any of the cells lining an acinous gland.

acinarious [BOT] Having vesicles that are globose; applied to certain algae.

Acinetobacter [MICROBIO] A term suggested as the genus name for some species of *Achromobacter.*

aciniform [ZOO] Shaped like a berry or a bunch of grapes.

acinotubular gland *See* tubuloalveolar gland.

acinous [BIOL] Of or pertaining to acini.

acinous gland [ANAT] A multicellular gland with sac-shaped secreting units. Also known as alveolar gland.

acinus [ANAT] The small terminal sac of an acinous gland, lined with secreting cells. [BOT] An individual drupelet of a multiple fruit.

Acipenser [VERT ZOO] A genus of actinopterygian fishes in the sturgeon family, Acipenseridae.

Acipenseridae [VERT ZOO] The sturgeons, a family of actinopterygian fishes in the order Acipenseriformes.

Acipenseriformes [VERT ZOO] An order of the subclass Actinopterygii represented by the sturgeons and paddlefishes.

Acmaeidae [INV ZOO] A family of gastropod mollusks in the order Archaeogastropoda; includes many limpets.

acme harrow [AGR] A type of harrow having a transverse horizontal frame with stiff curved blades. Also known as blade harrow; curved knife-tooth harrow; pulverizer.

acne [MED] A pleomorphic, inflammatory skin disease involving sebaceous follicles of the face, back, and chest and characterized by blackheads, whiteheads, papules, pustules, and nodules.

acne rosacea [MED] A form of acne occurring in older persons and seen as reddened inflamed areas on the forehead, nose, and cheeks.

Acnidosporidia [INV ZOO] An equivalent name for the Haplosporea.

Acoela [INV ZOO] An order of marine flatworms in the class Turbellaria characterized by the lack of a digestive tract and coelomic cavity.

Acoelea [INV ZOO] An order of gastropod mollusks in the subclass Opistobranchia; includes many sea slugs.

Acoelomata [INV ZOO] A subdivision of the animal kingdom; individuals are characterized by lack of a true body cavity.

acoelomate [INV ZOO] Any animal that lacks a true coelom.

acoelous [ZOO] **1.** Lacking a true body cavity or coelom. **2.** Lacking a true stomach or digestive tract.

Aconchulinida [INV ZOO] An order of protozoans in the subclass Filosia comprising a small group of naked amebas having filopodia.

acondylous [BOT] Lacking nodes or joints.

ACINOUS GLAND

An example of a typical compound type of acinous gland.

acone [INV ZOO] Describing a compound eye that lacks a crystalline or liquid secretion in cone cells.

aconitase [BIOCHEM] An enzyme involved in the Krebs cycle in muscles that catalyzes the breakdown of citric acid to *cis*-aconitic and isocitric acids.

aconite [BOT] Any plant of the genus *Aconitum*. Also known as friar's cowl; monkshood; mousebane; wolfsbane. [PHARM] A toxic drug obtained from the dried tuberous root of *Aconitum napellus;* the principal alkaloid is aconitine.

aconitine [PHARM] $C_{34}H_{47}O_{11}N$ A poisonous, white, crystalline alkaloid compound obtained from aconites such as monkshood (*Aconitum napellus*).

acontia [INV ZOO] In actinians, threadlike processes composed of mesenteric filaments that are armed with stinging cells.

acorn [BOT] The nut of the oak tree, usually surrounded at the base by a woody involucre.

acorn barnacle [INV ZOO] Any of the sessile barnacles that are enclosed in conical, flat-bottomed shells and attach to ships and near-shore rocks and piles.

acorn disease [PL PATH] A virus disease of citrus plants characterized by malformation of the fruit, which is somewhat acorn-shaped.

acorn worm [INV ZOO] Any member of the class Enteropneusta, free-living animals that usually burrow in sand or mud. Also known as tongue worm.

acotyledon [BOT] A plant without cotyledons.

acoustic nerve *See* auditory nerve.

acoustic reflex [PHYSIO] Brief, involuntary closure of the eyes due to stimulation of the acoustic nerve by a sudden sound.

acquired [BIOL] Not present at birth, but developed by an individual in response to the environment and not subject to hereditary transmission.

acquired immunity [IMMUNOL] Resistance to a microbial or other antigenic substance taken on by a naturally susceptible individual; may be either active or passive.

acquired immunological tolerance [IMMUNOL] Failure of immunological responsiveness, that is, inability of antigen-sensitive cells to synthesize antibodies; induced by exposure to large amounts of an antigen. Also known as immunological paralysis.

acquisition [PSYCH] Strengthening of a response that is learned.

acrania [MED] Partial or complete absence of the cranium at birth.

Acrania [ZOO] A group of lower chordates with no cranium, jaws, vertebrae, or paired appendages; includes the Tunicata and Cephalochordata.

Acrasiales [BIOL] A group of microorganisms that have plant and animal characteristics; included in the phylum Myxomycophyta by botanists and Mycetozoia by zoologists.

Acrasida [MYCOL] An order of Mycetozoia containing cellular slime molds.

acrasin [BIOCHEM] The chemotactic substance thought to be secreted by, and to effect aggregation of, myxamebas during their fruiting phase.

acraspedote [INV ZOO] Describing tapeworm segments which are not overlapping.

acridine [ORG CHEM] $(C_6H_4)_2NCH$ A typical member of a group of organic heterocyclic compounds containing benzene rings fused to the 2,3 and 5,6 positions of pyridene; derivatives include dyes and medicines.

acridine orange [ORG CHEM] A dye with an affinity for nucleic acids; the complexes of nucleic acid and dye fluoresce

ACORN WORM

proboscis genital ridge hepatic ceca
collar
mouth gills trunk anus

A drawing of the acorn worm, *Saccoglossus*, showing anatomical features. (From T. I. Storer and R. L. Usinger, *General Zoology*, 4th ed., McGraw-Hill, 1965)

orange with RNA and green with DNA when observed in the fluorescence microscope.

acriflavine   [ORG CHEM] $C_{14}H_{14}N_3Cl$ A yellow acridine dye obtained from proflavine by methylation in the form of red crystals; used as an antiseptic in solution.

acroblast   [CYTOL] A vesicular structure in the spermatid formed from Golgi material.

acrobryous   [BOT] Growing only from the tip.

acrocarpous   [BOT] In some mosses of the subclass Eubrya, having the sporophyte at the end of a stem and therefore exhibiting the erect habit.

acrocentric chromosome   [CYTOL] A chromosome having the centromere close to one end.

acrocephalosyndactylism   [MED] A congenital malformation consisting of an enlarged, pointed skull and defective separation of fingers and toes. Also known as Apert's syndrome.

acrocephaly   *See* oxycephaly.

Acroceridae   [INV ZOO] The humpbacked flies, a family of orthorrhaphous dipteran insects in the series Brachycera.

acrochordal   [VERT ZOO] In birds, an unpaired chondrocranial frontal cartilage.

acrochroic   [MYCOL] Of the terminus of hyphae, having pigmentation.

acrocoracoid   [VERT ZOO] In birds, a process located at the dorsal end of the coracoid.

acrocyst   [INV ZOO] A round, gelatinous cyst formed by gonophores when the generative cells mature in coelenterates.

acrodont   [ANAT] Having teeth fused to the edge of the supporting bone.

acrodromous   [BOT] Of leaves, having the veins converge at the point.

acrodynia   [MED] A childhood syndrome associated with mercury ingestion and characterized by periods of irritability alternating with apathy, anorexia, pink itching hands and feet, photophobia, sweating, tachycardia, hypertension, and hypotonia.

acrogenous   [BOT] Growing from the summit or apex.

acrogynous   [BOT] In certain liverworts, having archegonia that arise from the apical cell.

acrolein test   [ANALY CHEM] A test for the presence of glycerin or fats; a sample is heated with potassium bisulfate, and acrolein is released if the test is positive.

acromegaly   [MED] A chronic condition in adults caused by hypersecretion of the growth hormone and marked by enlarged jaws, extremities, and viscera, accompanied by certain physiological changes.

acromelalgia   *See* erythromelalgia.

acromere   [HISTOL] The distal portion of a rod or cone in the retina.

acromion   [ANAT] The flat process on the outer end of the scapular spine that articulates with the clavicle and forms the outer angle of the shoulder.

acron   [EVOL] Unsegmented head of the ancestral arthropod. [INV ZOO] **1.** The preoral, nonsegmented portion of an arthropod embryo. **2.** The prostomial region of the trochophore larva of some mollusks.

acroparesthesia   [MED] A chronic self-limited symptom complex associated with a variety of systemic diseases, characterized by tingling, pins-and-needles sensations, numbness or stiffness, and occasionally pains in the hands and feet.

acropetal   [BOT] From the base toward the apex, as seen in the formation of certain organs or the spread of a pathogen.

acrophyte   [ECOL] A plant that grows at a high altitude.

acroplasm [MYCOL] The cytoplasm in the apex of an ascus.

acropodium [ZOO] **1.** A digit. **2.** A finger or toe.

acrorhagus [INV ZOO] In certain Actiniaria, a tubercle which is located near the margin and which contains specialized nematocysts.

Acrosaleniidae [PALEON] A family of Jurassic and Cretaceous echinoderms in the order Salenoida.

acrosarc [BOT] A pulpy berry that arises from the union of the ovary and the calyx.

acroscopic [BOT] Facing, or on the side toward, the apex.

acrosome [CYTOL] The anterior, crescent-shaped body of spermatozoon, formed from Golgi material of the spermatid. Also known as perforatorium.

acrospore [MYCOL] In fungi, a spore formed at the outer tip of a hypha.

acroteric [ZOO] Referring to the outermost points of certain structures, such as digits, nose, ears, or tail.

Acrothoracica [INV ZOO] A small order of burrowing barnacles in the subclass Cirripedia that inhabit corals and the shells of mollusks and barnacles.

acrotonic [BOT] Having the anther united with the rostellum at its apex.

Acrotretacea [PALEON] A family of Cambrian and Ordovician inarticulate brachipods of the suborder Acrotretidina.

Acrotretida [INV ZOO] An order of brachiopods in the class Inarticulata; representatives are known from Lower Cambrian to the present.

Acrotretidina [INV ZOO] A suborder of inarticulate brachiopods of the order Acrotretida; includes only species with shells composed of calcium phosphate.

acrotrophic [INV ZOO] Referring to an ovariole that has nutritive cells at its apex which are joined to oocytes by nutritive cords.

Actaeonidae [INV ZOO] A family of gastropod mollusks in the order Tectibranchia.

Actaletidae [INV ZOO] A family of insects belonging to the order Collembola characterized by simple tracheal systems.

ACTH *See* adrenocorticotropic hormone.

Actidione [MICROBIO] Trade name for the antibiotic cyclohexamide.

actin [BIOCHEM] A muscle protein that is the chief constituent of the Z-band myofilaments of each sarcomere.

actinal [INV ZOO] **1.** Pertaining to the area of an echinoderm body where tube feet are located. **2.** In Actiniaria, referring to the oral area bearing tentacles.

actine [INV ZOO] A star-shaped spicule.

Actiniaria [INV ZOO] The sea anemones, an order of coelenterates in the subclass Zoantharia.

actinic [PHYS] Pertaining to electromagnetic radiation capable of initiating photochemical reactions, as in photography or the fading of pigments.

actinium [CHEM] A radioactive element, symbol Ac, atomic number 89; its longest-lived isotope is Ac^{227} with a half-life of 21.7 years; the element is trivalent; chief use is, in equilibrium with its decay products, as a source of alpha rays.

Actinobacillus [MICROBIO] A genus of aerobic, gram-negative bacteria in the family Brucellaceae; species are pathogenic for animals, occasionally for man.

actinobiology [BIOL] The study of radiation effects on organisms.

actinoblast [INV ZOO] In Porifera, the mother cell from which a spicule is formed.

actinocarpous [BOT] Having flowers and fruit radiating from one point.

actinochemistry [CHEM] A branch of chemistry concerned

ACTINIUM

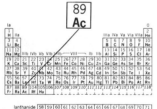

Periodic table of the chemical elements showing the position of actinium.

with chemical reactions produced by light or other radiation.

actinochitin [BIOCHEM] A form of birefringent or anisotropic chitin found in the seta of certain mites.

Actinochitinosi [INV ZOO] A group name for two closely related suborders of mites, Trombidiformes and Sarcoptiformes.

actinodrome [BOT] Having a palmate arrangement of veins.

actinogonidial [INV ZOO] Having the genital organs arranged in a radial pattern.

actinoid [INV ZOO] **1.** Rayed. **2.** Star-shaped. **3.** Stellate.

actinomere [INV ZOO] One of the segments composing the body of a radially symmetrical animal.

actinomorphic [BIOL] Descriptive of an organism, organ, or part that is radially symmetrical.

Actinomyces [MICROBIO] A genus of anaerobic and microaerophilic pathogenic bacteria belonging to the family Actinomycetaceae.

Actinomycetaceae [MICROBIO] A family of bacteria belonging to the order Actinomycetales characterized by the formation of a true mycelium.

Actinomycetales [MICROBIO] An order of bacteria in the class Schizomycetes; individuals produce filamentous cells or hyphae.

actinomycete [MICROBIO] Any member of the bacterial family Actinomycetaceae.

actinomycin [MICROBIO] The collective name for a large number of red chromoprotein antibiotics elaborated by various strains of *Streptomyces.*

actinomycosis [MED] An infectious bacterial disease caused by *Actinomyces bovis* in cattle, hogs, and occasionally in man. Also known as lumpy jaw.

Actinomyxida [INV ZOO] An order of protozoan invertebrate parasites of the class Myxosporidea characterized by trivalved spores with three polar capsules.

actinophage [MICROBIO] A bacteriophage that infects and lyses members of the order Actinomycetales.

actinopharynx [INV ZOO] The gullet in sea anemones.

Actinophryida [INV ZOO] An order of protozoans in the subclass Heliozoia; individuals lack an organized test, a centroplast, and a capsule.

Actinoplanaceae [MICROBIO] A family of bacteria in the order Actinomycetales with well-developed mycelia and spores formed on sporangia.

Actinoplanes [MICROBIO] A genus of bacteria in the family Actinoplanaceae having aerial mycelia and spherical to subspherical sporangiospores with a tuft of flagella.

Actinopodea [INV ZOO] A class of protozoans belonging to the superclass Sarcodina; most are free-floating, with highly specialized pseudopodia.

Actinopteri [VERT ZOO] An equivalent name for the Actinopterygii.

Actinopterygii [VERT ZOO] The ray-fin fishes, a subclass of the Osteichthyes distinguished by the structure of the paired fins, which are supported by dermal rays.

actinospore [MICROBIO] The spore type produced by actinomycetes.

actinost [VERT ZOO] In teleost fishes, the basal bone of a fin ray.

actinostele [BOT] A protostele characterized by xylem that is either star-shaped in cross section or has ribs radiating from the center.

actinostome [BIOL] **1.** The mouth of a radiate animal. **2.** The peristome of an echinoderm.

Actinostromariidae [PALEON] A sphaeractinoid family of extinct marine hydrozoans.

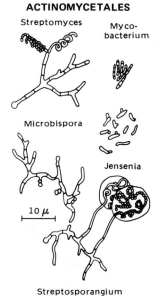

ACTINOMYCETALES

Streptomyces

Mycobacterium

Microbispora

Jensenia

10 μ

Streptosporangium

Some genera of Actinomycetales.

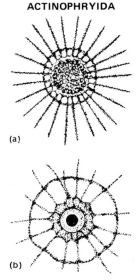

ACTINOPHRYIDA

(a)

(b)

Examples of Actinophryida. *(a)* Single specimen of *Actinosphaerium eichorni (after Pernard). (b)* Single specimen of *Actinophrys pontica. (From R. P. Hall, Protozoology, Prentice-Hall, 1953)*

actinotrichia [VERT ZOO] Horny rays lacking joints which are located at the edge of fins in many fishes.

actinula [INV ZOO] A larval stage of some hydrozoans that has tentacles and a mouth; attaches and develops into a hydroid in some species, or metamorphoses into a medusa.

action current [PHYSIO] The electric current accompanying membrane depolarization and repolarization in an excitable cell.

action potential [PHYSIO] A transient change in electric potential at the surface of a nerve or muscle cell occurring at the moment of excitation.

activated complex [PHYS CHEM] An energetically excited state which is intermediate between reactants and products in a chemical reaction.

activation [CHEM] Treatment of a substance by heat, radiation, or activating reagent to produce a more complete or rapid chemical or physical change. [PHYSIO] The designation for all changes in the ovum during fertilization, from sperm contact to the dissolution of nuclear membranes.

activation energy [PHYS CHEM] The energy, in excess over the ground state, which must be added to an atomic or molecular system to allow a particular process to take place.

activator [PHARM] A drug that increases the activity level of a depressed or withdrawn person.

activator RNA [GEN] Ribonucleic acid molecules which form a sequence-specific complex with receptor genes linked to producer genes.

active anaphylaxis [IMMUNOL] The allergic response following reintroduction of an antigen into a hypersensitive individual.

active avoidance conditioning [PSYCH] Learning to respond to a warning signal in a particular way, such as moving away from a noxious stimulus.

active immunity [IMMUNOL] Disease resistance in an individual due to antibody production after exposure to a microbial antigen following disease, inapparent infection, or inoculation.

active transport [PHYSIO] The pumping of ions or other substances across a cell membrane against an osmotic gradient, that is, from a lower to a higher concentration.

activity [NUC PHYS] The intensity of a radioactive source. Also known as radioactivity. [PHYS CHEM] A thermodynamic function that correlates changes in the chemical potential with changes in experimentally measurable quantities, such as concentrations or partial pressures, through relations formally equivalent to those for ideal systems.

activity coefficient [PHYS CHEM] A characteristic of a quantity expressing the deviation of a solution from ideal thermodynamic behavior; often used in connection with electrolytes.

actomyosin [BIOCHEM] A protein complex consisting of myosin and actin; the major constituent of a contracting muscle fibril.

actophilous [ECOL] Having a seashore growing habit.

acuate [BIOL] **1.** Having a sharp point. **2.** Needle-shaped.

acuity [BIOL] Sharpness of sense perception, as of vision or hearing.

Aculeata [INV ZOO] A group of seven superfamilies that constitute the stinging forms of hymenopteran insects in the suborder Apocrita.

aculeate [BIOL] **1.** Having prickles. **2.** Having sharp points. **3.** Having a sting.

aculeiform [BOT] Having the shape of a prickle or thorn.

aculeus [INV ZOO] **1.** A sharp, hairlike spine, as on the wings of certain lepidopterans. **2.** An insect stinger modified from an ovipositor.

Aculognathidae [INV ZOO] The ant-sucking beetles, a family of coleopteran insects in the superfamily Cucujoidea.

acuminate [BOT] Tapered to a slender point, especially referring to leaves.

acuminiferous [INV ZOO] Having tubercles that are pointed.

acuminulate [BOT] Having a sharp, tapered point.

acupuncture [MED] The ancient Chinese art of puncturing the body with long, fine gold or silver needles to relieve pain and cure disease.

acute [BIOL] Ending in a sharp point. [MED] Referring to a disease or disorder of rapid onset, short duration, and pronounced symptoms.

acute alcoholism [MED] Drunkenness accompanied by an acute, transient disturbance of physiological and mental functions.

acute appendicitis [MED] A sudden, severe attack of appendicitis characterized by abdominal pain, usually localized in the lower right quadrant, with nausea, vomiting, and constipation.

acute arthritis [MED] A severe joint inflammation with a short course.

acute ascending myelitis [MED] Severe inflammation of the spinal cord beginning in the lower segments and progressing toward the head.

acute bacterial endocarditis [MED] Fulminant, rapidly progressive endocarditis, usually associated with a significant systemic infection.

acute benign lymphoblastosis *See* infectious mononucleosis.

acute berylliosis [MED] Severe chemical pneumoconiosis caused by inhalation of beryllium salts.

acute cerebellar ataxia [MED] A severe childhood syndrome of sudden onset characterized by muscular incoordination, impaired articulation, oscillations of the eyeballs, and decreased intraocular pressure.

acute dermatitis [MED] Any severe inflammation of the skin.

acute glomerulonephritis [MED] Severe kidney inflammation, usually following infection with group A hemolytic streptococci, particularly type 12.

acute granulocytic leukemia [MED] A severe blood disorder in which the abnormal white cells are immature forms of granulocytes. Also known as myeloblastic leukemia.

acute infective encephalomyelitis *See* epidemic neuromyasthenia.

acute inflammation [MED] Severe inflammation with rapid progress and pronounced symptoms.

acute leukemia [MED] A severe blood disorder characterized by rapid onset and progress, with anemia and hemorrhagic manifestations; immature forms of leukocytes are predominant.

acute lymphocytic leukemia [MED] A severe blood disorder in which abnormal leukocytes are identified as immature forms of lymphocytes. Also known as lymphoblastic leukemia.

acute monocytic leukemia [MED] A severe blood disorder in which abnormal leukocytes are identified as immature forms of monocytes. Also known as monoblastic leukemia.

acute necrotizing hemorrhagic encephalomyelitis [MED] A sudden, severe central nervous system disease with variable symptoms; pathology includes hemorrhages and necrosis of the white matter.

acute nonsuppurative hepatitis *See* interstitial hepatitis.

acute radiation syndrome [MED] A complex of symptoms involving the intestinal tract, blood-forming organs, and skin following whole-body irradiation.

acute respiratory disease [MED] Severe adenovirus infection of the respiratory tract characterized by fever, sore throat, and cough.

acute rheumatic fever [MED] A severe infectious process caused by beta hemolytic streptococci; characterized by fever and frequently accompanied by painful inflamed joints, endocarditis, chorea, or glomerulonephritis.

acute rhinitis [MED] Inflammation of the nasal mucous membrane due to either infection or allergy.

acute toxic encephalopathy [MED] A severe childhood syndrome characterized by sudden onset of coma or stupor, fever, convulsions, and impaired respiratory and cardiovascular functioning.

acute tubular necrosis *See* lower nephron nephrosis.

acute yellow atrophy [MED] Rapid liver destruction following viral hepatitis, toxic chemicals, or other agents.

acutifoliate [BOT] Having sharply pointed leaves.

acutilobate [BOT] Having sharply pointed lobes.

acyclic [BOT] Having flowers arranged in a spiral instead of a whorl.

acyl [ORG CHEM] A radical formed from an organic acid by removal of a hydroxyl group; the general formula is RCO, where R may be aliphatic, alicyclic, or aromatic.

acylation [ORG CHEM] Any process (with the exception of the Friedel-Crafts method) whereby the acyl group is incorporated into a molecule by substitution.

acyl carrier protein [BIOCHEM] A protein in fatty acid synthesis that picks up acetyl and malonyl groups from acetyl coenzyme A and malonyl coenzyme A and links them by condensation to form β-keto acid acyl carrier protein, releasing carbon dioxide and the sulfhydryl form of acyl carrier protein. Abbreviated ACP.

acyl exchange *See* acidolysis.

adamantinoma *See* ameloblastoma.

adapical [BOT] Near or toward the apex or tip.

adaptation [GEN] The occurrence of genetic changes in a population or species as the result of natural selection so that it adjusts to new or altered environmental conditions. [PHYSIO] The occurrence of physiological changes in an individual exposed to changed conditions; for example, tanning of the skin in sunshine, or increased red blood cell counts at high altitudes.

adaptation syndrome [MED] Endocrine-mediated stress reaction of the body in response to systemic injury; involves an initial stage of shock, followed by resistance or adaptation and then healing or exhaustion.

adaptive behavior [PSYCH] Any behavior that helps the organism adjust to its environment.

adaptive colitis *See* irritable colon.

adaptive divergence [EVOL] Divergence of new forms from a common ancestral form due to adaptation to different environmental conditions.

adaptive enzyme [MICROBIO] Any bacterial enzyme formed in response to the presence of a substrate specific for that enzyme.

adaptive radiation [EVOL] Diversification of a dominant evolutionary group into a large number of subsidiary types adapted to more restrictive modes of life (different adaptive zones) within the range of the larger group.

adaptive value [GEN] The probability of one of two alleles of the same gene being transmitted to the next generation, as compared to its partner; it is dependent on the phenotypic characteristics that the gene imparts to its carriers in homozygous and heterozygous conditions. Also known as fitness.

adaxial [BIOL] On the same side as or facing toward the axis of an organ or organism.

adcauline [BOT] Located toward or being nearest to the stem.

adder [VERT ZOO] Any of the venomous viperine snakes included in the family Viperidae.

addiction [MED] Habituation to a specific practice, such as drinking alcoholic beverages or using drugs.

Addis count [PATH] A renal function test which estimates the blood cell count in a 12-hour urine specimen.

Addison's disease [MED] A primary failure or insufficiency of the adrenal cortex to secrete hormones.

adduction [PHYSIO] Movement of one part of the body toward another or toward the median axis of the body.

adductor [ANAT] Any muscle that draws a part of the body toward the median axis.

adeciduate [BOT] Not falling off or separating from; applied to evergreens or to a placenta.

adecticous [INV ZOO] Pertaining to a pupa that lacks functional mandibles.

Adeleina [INV ZOO] A suborder of protozoan invertebrate parasites in the order Eucoccida in which the sexual and asexual stages are in different hosts.

adelocodonic [INV ZOO] In certain Gymnoblastea, pertaining to an undetached medusa which degenerates after ripe sexual cells are discharged.

adelomorphic [HISTOL] Pertaining to certain poorly defined cells in the gastric glands.

adelomycete [MYCOL] A fungus that lacks the sexual spore stage; an imperfect fungus.

adelophycean [BOT] Pertaining to the stage or generation of certain seaweeds when they have the appearance of prostrate microthalli.

adelphous [BOT] Having stamens fused together by their filaments.

adenase [BIOCHEM] An enzyme that catalyzes the hydrolysis of adenine to hypoxanthine and ammonia.

adendritic [HISTOL] Referring to certain spinal ganglion cells that lack dendrites.

adenine [BIOCHEM] $C_5H_5N_5$ A purine base, 6-aminopurine; one of the fundamental components of all nucleic acids.

adenitis [MED] Inflammation of a gland or lymph node.

adenoacanthoma [MED] An adenocarcinoma, common in the endometrium, in which squamous cells replace the cylindrical epithelium.

adeno-associated satellite virus [VIROL] A defective virus that is unable to reproduce without the help of an adenovirus.

adenocarcinoma [MED] A malignant tumor originating in glandular or ductal epithelium and tending to produce acinic structures.

adenocheiri [INV ZOO] In Turbellaria, elaborate accessory copulatory organs which are outgrowths of the atrial walls.

adenohypophysis [ANAT] The glandular part of the pituitary gland, composing the anterior and intermediate lobes.

adenoid [ANAT] **1.** A mass of lymphoid tissue. **2.** Lymphoid tissue of the nasopharynx. Also known as pharyngeal tonsil.

adenoma [MED] A benign tumor of glandular origin and structure.

adenomatoid tumor [MED] A benign genital-tract tumor composed of stroma whose spaces are lined by cells that resemble epithelium, endothelium, and mesothelium.

adenomatosis [MED] A condition characterized by multiple adenomas within an organ or in several related organs.

adenomatous goiter [MED] An asymmetric goiter due to

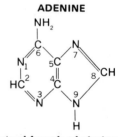

ADENINE

Structural formula of adenine; the carbon atoms are numbered.

isolated nodular masses of thyroid tissue. Also known as multiple colloid goiter; nodular goiter.

adenomere [EMBRYO] The embryonic structure which will become the functional portion of a gland.

adenomyoma [MED] A benign tumor of glandular and muscular elements occurring principally in the uterus and rectum.

adenomyosis [MED] **1.** The invasion of muscular tissue, such as of the uterine wall or Fallopian tubes, by endometrial tissue. **2.** Any abnormal growth of muscle or glandular tissues.

adenopathy [MED] Any glandular disease; common usage limits the term to any abnormal swelling or enlargement of lymph nodes.

adenophore [BOT] The stalk of a nectary.

Adenophorea [INV ZOO] A class of unsegmented worms in the phylum Nematoda.

adenophyllous [BOT] Having leaves with glands.

adenopodous [BOT] Having glands on peduncles or on petioles.

adenose [BOT] Having a glandular function.

adenosine [BIOCHEM] $C_{10}H_{13}N_5O_4$ A nucleoside composed of adenine and D-ribose.

adenosinediphosphatase [BIOCHEM] An enzyme that catalyzes the hydrolysis of adenosinediphosphate. Abbreviated ADPase.

adenosinediphosphate [BIOCHEM] $C_{10}H_{15}N_5O_{10}P_2$ A coenzyme composed of adenosine and two molecules of phosphoric acid that is important in intermediate cellular metabolism. Abbreviated ADP.

adenosinemonophosphate *See* adenylic acid.

adenosinetriphosphatase [BIOCHEM] An enzyme that catalyzes the hydrolysis of adenosinetriphosphate. Abbreviated ATPase.

adenosinetriphosphate [BIOCHEM] $C_{10}H_{16}N_5O_{12}P_3$ A coenzyme composed of adenosinediphosphate with an additional phosphate group; an important energy compound in metabolism. Abbreviated ATP.

adenosis [MED] Any nonneoplastic glandular disease, especially one involving the lymph nodes.

adenostemonous [BOT] Having glands located on the stamens.

adeno-SV40 hybrid virus [VIROL] A defective virus particle in which part of the genetic material of papovavirus SV40 is encased in an adenovirus protein coat.

adenovirus [VIROL] A group of animal viruses which cause febrile catarrhs and other respiratory diseases.

adenylic acid [BIOCHEM] **1.** A generic term for a group of isomeric nucleotides. **2.** The phosphoric acid ester of adenosine. Also known as adenosinemonophosphate (AMP).

adeoniform [INV ZOO] **1.** A lobate, bilamellar zooarium. **2.** Resembling the fossil bryozoan *Adeona.*

Adephaga [INV ZOO] A suborder of insects in the order Coleoptera characterized by fused hind coxae that are immovable.

adequate contact [MED] The degree of contact required between an infectious and a susceptible individual to cause infection of the latter.

adequate stimulus [PHYSIO] The energy of any specific mode that is sufficient to elicit a response in an excitable tissue.

adetopneustic [INV ZOO] In certain Stelleroidea, having the dermal gills located beyond the abactinal surface.

adfrontal [INV ZOO] Pertaining to the oblique plates located beside the frons of certain insect larvae.

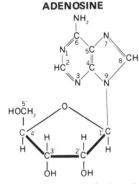

ADENOSINE

Structural formula of adenosine.

ADH *See* vasopressin.

adhesion [BOT] Growing together of members of different and distinct whorls. [MED] The abnormal union of an organ or part with some other part by formation of fibrous tissue. [PHYS] The tendency, due to intermolecular forces, for matter to cling to other matter.

adhesive cell [INV ZOO] Any of various glandular cells in ctenophores, turbellarians, and hydras used for adhesion to a substrate and for capture of prey. Also known as colloblast; glue cell; lasso cell.

adiabatic law [PHYS] The relationship which states that, for adiabatic expansion of gases, $P\rho^{-\gamma}$ = constant, where P = pressure, ρ = density, and γ = ratio of specific heats C_P/C_V.

adiabatic system [SCI TECH] A body or system whose condition is altered without gaining heat from or losing heat to the surroundings.

adiadochokinesis [MED] A type of motor incoordination associated with cerebellar damage in which repetitive movements controlled by antagonistic muscles cannot be performed without severe muscular incoordination.

Adimeridae [INV ZOO] An equivalent name for the Colydiidae.

adipocere [MED] A light-colored, waxy material formed by postmortem conversion of body fats to higher fatty acids.

adipocyte [INV ZOO] One of the cells which give rise to the fat body in insects.

adipogenesis [PHYSIO] The formation of fat or fatty tissue.

adipolysis [PHYSIO] Hydrolysis of fats by fat-splitting enzymes during digestion.

adiponecrosis neonatorum [MED] Localized fatty-tissue necrosis occurring in large, healthy infants born after difficult labor.

adipose [BIOL] Fatty; of or relating to fat.

adipose fin [VERT ZOO] A modified posterior dorsal fin that is fleshy and lacks rays; found in salmon and typical catfishes.

adipose tissue [HISTOL] A type of connective tissue specialized for lipid storage.

adiposis dolorosa [MED] An uncommon type of obesity in which the excess fat deposits are tender and painful.

adiposogenital dystrophy [MED] A syndrome involving obesity, retarded gonad development, and sometimes diabetes insipidus resulting from impaired functioning of the pituitary and hypothalamus. Also known as Froehlich's syndrome.

aditus [ANAT] An entrance or inlet.

adjustment [PSYCH] The process of mental change to allow an individual to function harmoniously with the environment.

adjustment reaction [PSYCH] A transient, situational personality disorder occurring in reaction to some significant person, immediate event, or internal emotional conflict.

adjustor [ANAT] The ganglionic portion of a reflex arc; connects the receptor and the effector. [INV ZOO] In brachiopods, a muscle that connects the stalk and the valve.

adjustor neuron [ANAT] Any of the interconnecting nerve cells between sensory and motor neurons of the central nervous system.

adjuvant [PHARM] A material that enhances the action of a drug or antigen.

adlacrimal [VERT ZOO] The lacrimal bone in reptiles.

adlittoral [OCEANOGR] Of, pertaining to, or occurring in shallow waters adjacent to a shore.

admedial [ANAT] In proximity to or approaching the median plane or central axis.

adminiculum [ANAT] The posterior fibers of the linea alba which are attached to the pubis. [INV ZOO] In certain pupae, a locomotory spine.

adnasal [VERT ZOO] In certain fishes, a small bone located in front of each nasal bone.

adnate [BIOL] United through growth; used especially for unlike parts. [BOT] Pertaining to growth with one side adherent to a stem.

adnexa [BIOL] Subordinate or accessory parts, such as eyelids, Fallopian tubes, and extraembryonic membranes.

adolescence [PSYCH] The period of life from puberty to maturity.

adont hinge [INV ZOO] A type of ostracod hinge articulation which either lacks teeth and has overlapping valves or has a ridge and groove.

adoral [ZOO] Near the mouth.

ADP *See* adenosinediphosphate.

ADPase *See* adenosinediphosphatase.

adradius [INV ZOO] In coelenterates, a third-order radius located midway between the perradius and the interradius.

adrenal cortex [ANAT] The cortical moitie of the suprarenal glands which secretes glucocorticoids, mineralocorticoids, androgens, estrogens, and progestagens.

adrenal cortex hormone [BIOCHEM] Any of the steroids produced by the adrenal cortex. Also known as adrenocortical hormone; corticoid.

adrenal cortical insufficiency [MED] Failure of the adrenal cortex to secrete adequate hormones.

adrenalectomy [MED] Surgical removal of an adrenal gland.

adrenal gland [ANAT] An endocrine organ located close to the kidneys of vertebrates and consisting of two morphologically distinct components, the cortex and medulla. Also known as suprarenal gland.

adrenaline *See* epinephrine.

adrenal medulla [ANAT] The hormone-secreting chromaffin cells of the adrenal gland that produce epinephrine and norepinephrine.

adrenal virilism [MED] **1.** The development of male characteristics in the female resulting from excessive production of adrenal hormones with androgenic activity. **2.** A rare form of pseudohermaphroditism.

adrenergic [PHYSIO] Describing the chemical activity of epinephrine or epinephrine-like substances.

adrenochrome [BIOCHEM] $C_9H_9O_3N$ A brick-red oxidation product of epinephrine which can convert hemoglobin into methemoglobin.

adrenocortical hormone *See* adrenal cortex hormone.

adrenocorticotropic hormone [BIOCHEM] The chemical secretion of the adenohypophysis that stimulates the adrenal cortex. Abbreviated ACTH. Also known as adrenotropic hormone.

adrenogenital syndrome [MED] A group of symptoms associated with hypersecretion of adrenal cortex hormones; effects vary with sex and time of development.

adrenotropic [PHYSIO] Of or pertaining to an effect on the adrenal cortex.

adrenotropic hormone *See* adrenocorticotropic hormone.

adrostral [BIOL] Located near to or being closely connected with the rostrum.

adsorbent [CHEM] A solid or liquid that adsorbs other substances; for example, charcoal, silica, metals, water, and mercury.

adsorption [CHEM] The surface retention of solid, liquid, or gas molecules, atoms, or ions by a solid or liquid, as opposed to absorbtion, the penetration of substances into the bulk of the solid or liquid.

adsorption chromatography [ANALY CHEM] Separation of a chemical mixture (gas or liquid) by passing it over an adsorbent bed which adsorbs different compounds at different rates.

adsorption isobar [PHYS CHEM] A graph showing how adsorption varies with some parameter, such as temperature, while holding pressure constant.

adsorption isotherm [PHYS CHEM] The relationship between the gas pressure p and the amount w, in grams, of a gas or vapor taken up per gram of solid at a constant temperature.

adtidal [BIOL] Pertaining to organisms that live just below the low-tide mark.

adultoid [INV ZOO] Pertaining to a nymph whose imaginal characters are more highly differentiated than in a normal nymph.

adult rickets *See* osteomalacia.

aduncate [BOT] Having the form of a hook.

adventitia [ANAT] The external, connective-tissue covering of an organ or blood vessel. Also known as tunica adventitia.

adventitious [BIOL] Also known as adventive. **1.** Acquired spontaneously or accidentally, not by heredity. **2.** Arising, as a tissue or organ, in an unusual or abnormal place.

adventitious bud [BOT] A bud that arises at points on the plant other than at the stem apex or a leaf axil.

adventitious root [BOT] A root that arises from any plant part other than the primary root (radicle) or its branches.

adventitious vein [INV ZOO] The vessel between the intercalary and accessory veins on certain insect wings.

adventive [BIOL] **1.** An organism that is introduced accidentally and is imperfectly naturalized; not native. **2.** *See* adventitious.

Aechminidae [PALEON] A family of extinct ostracods in the order Paleocopa in which the hollow central spine is larger than the valve.

aeciospore [MYCOL] A spore produced by an aecium.

aecium [MYCOL] The fruiting body or sporocarp of rust fungi.

Aedes [INV ZOO] A genus of the dipterous subfamily Culicinae in the family Culicidae, with species that are vectors for many diseases of man.

Aedes aegypti [INV ZOO] A cosmopolitan mosquito that transmits the etiological agents of yellow fever, dengue, equine encephalomyelitis, and Bancroft's filariasis.

Aeduellidae [PALEON] A family of Lower Permian palaeoniscoid fishes in the order Palaeonisciformes.

Aegeriidae [INV ZOO] The clearwing moths, a family of lepidopteran insects in the suborder Heteroneura characterized by the lack of wing scales.

Aegialitidae [INV ZOO] An equivalent name for the Salpingidae.

Aegidae [INV ZOO] A family of isopod crustaceans in the suborder Flabellifera whose members are economically important as fish parasites.

aegithognathous [VERT ZOO] Referring to a bird palate in which the vomers are completely fused and truncate in appearance.

Aegothelidae [VERT ZOO] A family of small Australo-Papuan owlet-nightjars in the avian order Caprimulgiformes.

Aegypiinae [VERT ZOO] The Old World vultures, a subfamily of diurnal carrion feeders of the family Accipitridae.

Aegyptopithecus [PALEON] A primitive primate that is thought to represent the common ancestor of both the family of man and that of the apes.

aelophilous [BOT] Describing a plant whose disseminules are dispersed by wind.

Aelosomatidae [INV ZOO] A family of microscopic fresh-water annelid worms in the class Oligochaeta characterized by a ventrally ciliated prostomium.

Aepophilidae [INV ZOO] A family of bugs in the hemipteran superfamily Saldoidea.

Aepyornis [PALEON] A genus of extinct ratite birds representing the family Aepyornithidae.

Aepyornithidae [PALEON] The single family of the extinct avian order Aepyornithiformes.

Aepyornithiformes [PALEON] The elephant birds, an extinct order of ratite birds in the superorder Neognathae.

aerial [BIOL] Of, in, or belonging to the air or atmosphere.

aerial mycelium [MYCOL] A mass of hyphae that occurs above the surface of a substrate.

aerial root [BOT] A root exposed to the air, usually anchoring the plant to a tree, and often functioning in photosynthesis.

aerial stem [BOT] A stem with an erect or vertical growth habit above the ground.

aeroallergen [MED] Any airborne particulate matter that can induce allergic responses in sensitive persons.

Aerobacter [MICROBIO] A genus of nonpathogenic, rod-shaped, gram-negative bacteria of the family Enterobacteriaceae.

aerobe [BIOL] An organism that requires air or free oxygen to maintain its life processes.

aerobic bacteria [MICROBIO] Any bacteria requiring free oxygen for the metabolic breakdown of materials.

aerobic process [BIOL] A process requiring the presence of oxygen.

aerobiology [BIOL] The study of the atmospheric dispersal of airborne fungus spores, pollen grains, and microorganisms; and, more broadly, of airborne propagules of algae and protozoans, minute insects such as aphids, and pollution gases and particles which exert specific biologic effects.

aeroembolism [MED] A condition marked by the presence of nitrogen bubbles in the blood and other body tissues resulting from a sudden fall in atmospheric pressure. Also known as air embolism.

aeromedicine *See* aerospace medicine.

Aeromonas [MICROBIO] A genus of short, gram-negative, rod-shaped, motile, heterotrophic bacteria in the family Pseudomonadaceae; the majority are pathogenic to marine and fresh-water animals.

aerootitis *See* barotitis.

aeroplankton [BIOL] Organisms or structures that drift in the air, such as spores, pollen, or bacteria.

aerosinusitis *See* barosinusitis.

aerosol [CHEM] A gaseous suspension of ultramicroscopic particles of a liquid or a solid.

aerospace medicine [MED] The branch of medicine dealing with the effects of flight in the atmosphere or space upon the human body and with the prevention or cure of physiological or psychological malfunctions arising from these effects. Also known as aeromedicine; aviation medicine.

Aerosporin [MICROBIO] Trade name for the antibiotic polymyxin B.

aerostat [INV ZOO] An air sac in the body of an insect. [VERT ZOO] One of the air sacs in the bones of a bird.

aerostatic [BIOL] Containing air sacs or cavities.

aerotaxis [BIOL] The movement of an organism, especially aerobic and anaerobic bacteria, with reference to the direction of oxygen or air.

aerotropism [BOT] A response in which the growth direction of a plant component changes due to modifications in oxygen tension.

aeschynomenous [BOT] Having sensitive leaves that droop when touched, such as members of the Leguminosae.

Aesculus [BOT] A genus of deciduous trees or shrubs belonging to the order Sapindales. Commonly known as buckeye.

Aeshnidae [INV ZOO] A family of odonatan insects in the suborder Anisoptera distinguished by partially fused eyes.

aesthacyte *See* esthacyte.

aesthesia *See* esthesia.

aesthesiometer *See* esthesiometer.

aestidurilignosa [ECOL] A mixed woodland of evergreen and deciduous hardwoods.

aestilignosa [ECOL] A woodland of trophophytic vegetation in temperate regions.

aestivation [BOT] The arrangement of floral parts in a bud. [PHYSIO] The condition of dormancy or torpidity.

aethalium [MYCOL] In slime molds, an aggregation of plasmodia or of sporangia that form a compound fruit.

Aetosauria [PALEON] A suborder of Triassic archosaurian quadrupedal reptiles in the order Thecodontia armored by rings of thick, bony plates.

affect [PSYCH] Conscious awareness of feelings; mood.

affection [MED] Any pathology or diseased state of the body. [PSYCH] The feeling aspect of consciousness.

affectional drive [PSYCH] An unlearned drive to be in close contact with another person.

affective psychosis [PSYCH] A severe disturbance marked by extremes of mood, such as depression or manic elation.

afferent [PHYSIO] Conducting or conveying inward or toward the center, specifically in reference to nerves and blood vessels.

afferent code [PHYSIO] The pattern of neural input to the central nervous system that corresponds to various aspects of the external stimulating environment.

afferent neuron [ANAT] A nerve cell that conducts impulses toward a nerve center, such as the central nervous system.

affiliation need [PSYCH] The need to associate with people. Also known as affiliative need.

affinity [BIOL] A relationship between species or higher taxonomic groups that depends on anatomical similarities and indicates a common origin. [CHEM] The selective tendency of an element to combine with one rather than another element when conditions are equal.

affinity chromatography [ANALY CHEM] A chromatographic technique that utilizes the ability of biological molecules to bend to certain ligands specifically and reversibly; used in protein biochemistry.

afforestation [FOR] Establishment of a new forest by seeding or planting on nonforested land.

afibrinogenemia [MED] Complete absence of fibrinogen in the blood.

aflagellar [BIOL] Lacking a flagellum.

aflatoxin [BIOCHEM] The toxin produced by some strains of the fungus *Aspergillus flavus*, the most potent carcinogen yet discovered.

African sleeping sickness [MED] A disease of man confined to tropical Africa, caused by the protozoans *Trypanosoma gambiense* or *T. rhodesiense;* symptoms include local reaction at the site of the bite, fever, enlargement of adjacent lymph nodes, skin rash, edema, and during the late phase, somnolence and emaciation. Also known as African trypanosomiasis; maladie du sommeil; sleeping sickness.

African trypanosomiasis *See* African sleeping sickness.

afterbirth [EMBRYO] The placenta and fetal membranes expelled from the uterus following birth of offspring in viviparous mammals.

AETOSAURIA

Restoration of *Desmatosuchus*, a quadrupedal thecodont, length to 3 meters. *(From E. H. Colbert, Evolution of the Vertebrates, Wiley, 1956)*

afterimage [PHYSIO] A visual sensation occurring after the stimulus to which it is a response has been removed.

afterpotential [PHYSIO] A small positive or negative wave that follows and is dependent on the main spike potential, seen in the oscillograph tracing of an action potential passing along a nerve.

afterripening [BOT] A period of dormancy after a seed is shed during which the synthetic machinery of the seed is prepared for germination and growth.

aftershaft [VERT ZOO] An accessory, plumelike feather near the upper umbilicus on the feathers of some birds.

agalactia [MED] Nonsecretion or imperfect secretion of milk after childbirth.

agameon [BIOL] An organism which reproduces only by asexual means. Also known as agamospecies.

agamete [BIOL] An asexual reproductive cell that develops into an adult individual.

Agamidae [VERT ZOO] A family of Old World lizards in the suborder Sauria that have acrodont dentition.

agammaglobulinemia [MED] The condition characterized by lack of or extremely low levels of gamma globulin in the blood, together with defective antibody production and frequent infections; primary agammaglobulinemia occurs in three clinical forms: congenital, acquired, and transient.

agamont [INV ZOO] A stage in certain Sporozoa which gives rise to agametes.

agamospecies *See* agameon.

agamospermy [BOT] Apogamy in which sexual union is incomplete because of abnormal development of the pollen and the embryo sac.

Agaontidae [INV ZOO] A family of small hymenopteran insects in the superfamily Chalcidoidea; commonly called fig insects for their role in cross-pollination of figs.

agar [MATER] A gelatinous product extracted from certain red algae and used chiefly as a gelling agent in culture media.

Agarbacterium [MICROBIO] A genus of motile and nonmotile, rod-shaped, gram-negative bacteria in the family Achromobacteraceae that digest agar.

agar-gel reaction [IMMUNOL] A precipitin type of antigen-antibody reaction in which the reactants are introduced into different regions of an agar gel and allowed to diffuse toward each other.

Agaricales [MYCOL] An order of fungi in the class Basidiomycetes containing all forms of fleshy, gilled mushrooms.

Agavaceae [BOT] A family of flowering plants in the order Liliales characterized by parallel, narrow-veined leaves, a more or less corolloid perianth, and an agavaceous habit.

age [BIOL] Period of time from origin or birth to a later time designated or understood; length of existence. [GEOL] **1.** Any one of the named epochs in the history of the earth marked by specific phases of physical conditions or organic evolution, such as the Age of Mammals. **2.** One of the smaller subdivisions of the epoch as geologic time, corresponding to the stage or the formation, such as the Lockport Age in the Niagara Epoch.

age structure [ANTHRO] Categorization of the population of communities or countries by age groups, allowing demographers to make projections of the growth or decline of the particular population.

agglomerate [INV ZOO] **1.** Describing an adhering mass of Protozoa. **2.** The rosette grouping of trypanosomes.

agglutinate [CHEM] **1.** To fuse, cohere, or adhere. **2.** Of suspended particles, to aggregate, thus forming clumps. **3.** To cause to aggregate or form clumps.

agglutination reaction [IMMUNOL] Clumping of a particulate

AGGLUTINATION REACTION

Reagents

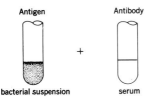

Antigen   Antibody

bacterial suspension   serum

Results: macroscopic test

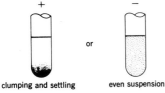

clumping and settling   even suspension

Results: microscopic test

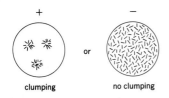

clumping   no clumping

Macroscopic or microscopic results of agglutination reaction. (From C. J. Witton, Microbiology with Application to Nursing, 2d ed., McGraw-Hill, 1956)

suspension of antigen by a reagent, usually an antibody.

agglutinin [IMMUNOL] An antibody from normal or immune serum that causes clumping of its complementary particulate antigen, such as bacteria or erythrocytes.

agglutinogen [IMMUNOL] An antigen that stimulates production of a specific antibody (agglutinin) when introduced into an animal body.

agglutinoid [IMMUNOL] An agglutin that lacks the power to agglutinate but has the ability to unite with its agglutinogen.

aggregate [BOT] Of fruit, formed in a cluster; that is, the fruit forms from the apocarpous gynoecium of a single flower, as a raspberry.

aggregation [BIOL] A grouping of individual organisms. [BOT] The movement of protoplasm in tentacle or tendril cells of sensitive plants; causes bending of the tentacle or tendril toward the point of stimulation.

aggression [PSYCH] Feelings and behavior of anger and hostility usually manifested by punitive or destructive actions; often associated with frustration.

aging [BIOL] Growing older.

aging-lung emphysema [MED] An asymptomatic pulmonary disease associated with aging, characterized by alveolar dilation due to loss of tissue elasticity.

Aglaspida [PALEON] An order of Cambrian and Ordovician merostome arthropods in the subclass Xiphosurida characterized by a phosphatic exoskeleton and vaguely trilobed body form.

aglomerular [HISTOL] Lacking glomeruli.

Aglossa [VERT ZOO] A suborder of anuran amphibians represented by the single family Pipidea and characterized by the absence of a tongue.

aglossate [ZOO] Lacking a tongue.

aglycon [BIOCHEM] The nonsugar compound resulting from the hydrolysis of glycosides; an example is 3,5,7,3',4'-pentahydroxyflavylium, or cyanidin.

aglyphous [VERT ZOO] Having solid teeth.

Agnatha [VERT ZOO] The most primitive class of vertebrates, characterized by the lack of true jaws.

agnathia [MED] Lack of the jaws.

agnathostomatous [VERT ZOO] Having a mouth without jaws, as the lamprey.

agnosia [MED] Loss of the ability to recognize persons or objects and their meaning.

Agonidae [VERT ZOO] The poachers, a small family of marine perciform fishes in the suborder Cottoidei.

agonist [PHARM] A drug that can combine with a receptor and initiate drug action. [PHYSIO] A contracting muscle involved in moving a part and opposed by an antagonistic muscle. Also known as protagonist.

agonistic behavior [PSYCH] In social animals, fighting and escape behavior common in males during the rutting season.

agouti [VERT ZOO] A hystricomorph rodent, *Dasyprocta*, in the family Dasyproctidae, represented by 13 species.

agranular [SCI TECH] **1.** Not granular. **2.** Lacking granules.

agranular leukocyte [HISTOL] A type of white blood cell, including lymphocytes and monocytes, characterized by the absence of cytoplasmic granules and by a relatively large spherical or indented nucleus.

agranulocytosis [MED] An acute febrile illness, usually resulting from drug hypersensitivity, manifested as severe leukopenia, often with complete disappearance of granulocytes.

agraphia [MED] Loss of the ability to write.

agrestal [ECOL] Growing wild in the fields.

agricultural chemistry [AGR] The science of chemical com-

AGLYCON

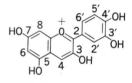

Cyanidin, the commonest aglycon.

AGOUTI

The agouti *(Dasyprocta aguti),* a rodent found in Mexico, South America, and the West Indies.

positions and changes involved in the production, protection, and use of crops and livestock; includes all the life processes through which food and fiber are obtained for man and his animals, and control of these processes to increase yields, improve quality, and reduce costs.

agricultural climatology [AGR] In general, the study of climate as to its effect on crops; it includes, for example, the relation of growth rate and crop yields to the various climatic factors and hence the optimum and limiting climates for any given crop. Also known as agroclimatology.

agricultural engineering [AGR] A discipline concerned with developing and improving the means for providing food and fiber for mankind's needs.

agricultural machinery [AGR] Machines utilized for tillage, planting, cultivation, and harvesting of crops.

agricultural meteorology [AGR] The study and application of relationships between meteorology and agriculture, involving problems such as timing the planting of crops. Also known as agrometeorology.

agricultural science [AGR] A discipline dealing with the selection, breeding, and management of crops and domestic animals for more economical production.

agricultural wastes [AGR] Those liquid or solid wastes that result from agricultural practices, such as cattle manure, crop residue (for example, corn stalks), pesticides, and fertilizers.

agriculture [BIOL] The production of plants and animals useful to man, involving soil cultivation and the breeding and management of crops and livestock.

Agriochoeridae [PALEON] A family of extinct tylopod ruminants in the superfamily Merycoidodontoidea.

Agrionidae [INV ZOO] A family of odonatan insects in the suborder Zygoptera characterized by black or red markings on the wings.

Agrobacterium [MICROBIO] A genus of rod-shaped, gram-negative soil bacteria composed of four species, three of which are plant pathogens.

Agromyzidae [INV ZOO] A family of myodarian cyclorrhaphous dipteran insects of the subsection Acalypteratae; commonly called leaf-miner flies because the larvae cut channels in leaves.

agronomy [AGR] The principles and procedures of soil management and of field crop and special-purpose plant improvement, management, and production.

agrophilous [ECOL] Having a natural habitat in grain fields.

agrostology [BOT] A division of systematic botany concerned with the study of grasses.

ahaptoglobinemia [MED] An inherited lack of haptoglobin, a blood serum protein.

A horizon [GEOL] The upper, or leached, layers of soil; the topsoil.

ainhum [MED] A tropical disease of unknown etiology that is peculiar to male Negroes, in which a toe is slowly and spontaneously amputated by a fibrous ring.

aiophyllous *See* evergreen.

air cell [ZOO] A cavity or receptacle for air such as an alveolus, an air sac in birds, or a dilation of the trachea in insects.

air chamber [INV ZOO] A gas-filled cavity of a nautilus shell that was previously occupied by the animal body.

air duct [VERT ZOO] The channel connecting the swim bladder with the gut in some fishes.

air embolism *See* aeroembolism.

air layering [BOT] A method of vegetative propagation, usually of a wounded part, in which the branch or shoot is enclosed in a moist medium until roots develop, and then it is severed and cultivated as an independent plant.

air pollution [ECOL] The presence in the outdoor atmosphere of one or more contaminants such as dust, fumes, gas, mist, odor, smoke, or vapor in quantities and of characteristics and duration such as to be injurious to human, plant, or animal life or to property, or to interfere unreasonably with the comfortable enjoyment of life and property.

air pressure [PHYS] The force per unit area that the air exerts on any surface in contact with it, arising from the collisions of the air molecules with the surface.

air sac [INV ZOO] One of large, thin-walled structures associated with the tracheal system of some insects. [VERT ZOO] In birds, any of the small vesicles that are connected with the respiratory system and located in bones and muscles to increase buoyancy.

airsickness [MED] Motion sickness associated with flying due to the effects of acceleration.

air sinus [ANAT] A cavity containing air within a bone, especially one communicating with the nasal passages.

Aistopoda [PALEON] An order of Upper Carboniferous amphibians in the subclass Lepospondyli characterized by reduced or absent limbs and an elongate, snakelike body.

aitionastic [BOT] Pertaining to the curvature of a plant part as induced by a diffuse stimulus.

Aizoaceae [BOT] A family of flowering plants in the order Caryophyllales; members are unarmed leaf-succulents, chiefly of Africa.

akaryote [CYTOL] A cell that lacks a nucleus.

akimbo span [ANTHRO] The distance measured between elbow points when the subject stands with arms flexed in a horizontal plane, with the wrists straight, palms down, fingers straight and together, and the thumbs touching the chest; or the distance measured with upper arms horizontal, forearms vertical.

akinesia [MED] **1.** Loss of or impaired motor function. **2.** Immobility from any cause.

akinete [BOT] A thick-walled resting cell of unicellular and filamentous green algae.

akureyri disease *See* epidemic neuromyasthenia.

Al *See* aluminum.

ala [BIOL] A wing or winglike structure.

alalia [MED] Loss of speech.

alang-alang *See* cogon.

alanine [BIOCHEM] $C_3H_7NO_2$ A white, crystalline, nonessential amino acid of the pyruvic acid family.

alanyl [ORG CHEM] The radical CH_3CHNH_2CO-; occurs in, for example, alanyl alanine, a dipeptide.

alarm reaction [BIOL] The sum of all nonspecific phenomena which are elicited by sudden exposure to stimuli, which affect large portions of the body, and to which the organism is quantitatively or qualitatively not adapted.

alate [BIOL] Winged. [BOT] Pertaining to a petiole or stem having a winglike expansion.

Alaudidae [VERT ZOO] The larks, a family of Oscine birds in the order Passeriformes.

Albamycin [MICROBIO] A trade name for the antibiotic novobiocin.

albatross [VERT ZOO] Any of the large, long-winged oceanic birds composing the family Diomedeidae of the order Procellariiformes.

albescent [BOT] Developing a whitish appearance.

albinism [MED] A hereditary, metabolic disorder transmitted as an autosomal recessive and characterized by the inability

ALANINE

Structural formula of the amino acid alanine.

ALBATROSS

The laysan albatross, with the characteristic hooked bill and long, tubular nostrils of oceanic birds.

to form melanin in the skin, hair, and eyes due to tyrosinase deficiency.

albino [BIOL] An individual exhibiting albinism.

albomycin [MICROBIO] An antibiotic produced by *Actinomyces subtropicus*; effective against penicillin-resistant pneumococci and staphylococci.

albumen [CYTOL] The white of an egg, composed principally of albumin.

albumin [BIOCHEM] Any of a group of plant and animal proteins which are soluble in water, dilute salt solutions, and 50% saturated ammonium sulfate.

albumin-globulin ratio [BIOCHEM] The ratio of the concentrations of albumin to globulin in blood serum.

albuminous cells [BOT] In pteridophytes and gymnosperms, parenchymal cells that are associated with the sieve cells.

albuminuria [MED] The presence of albumin in the urine; usually indicating renal disease.

Alcaligenes [MICROBIO] A genus of motile and nonmotile, gram-negative, rod-shaped bacteria of the family Achromobacteraceae, found generally in the intestinal tract of vertebrates.

alcaptonuria *See* alkaptonuria.

Alcedinidae [VERT ZOO] The kingfishers, a worldwide family of colorful birds in the order Coraciiformes; characterized by large heads, short necks, and heavy, pointed bills.

Alcidae [VERT ZOO] A family of shorebirds, predominantly of northern coasts, in the order Charadriiformes, including auks, puffins, murres, and guillemots.

Alciopidae [INV ZOO] A pelagic family of errantian annelid worms in the class Polychaeta.

alcohol [ORG CHEM] **1.** C_2H_5OH A colorless, volatile liquid; boiling point of pure liquid is 78.3°C; it is soluble in water, chloroform, and methyl alcohol; used as solvent and in manufacture of many chemicals and medicines. Also known as ethanol; ethyl alcohol. **2.** Any of a class of organic compounds containing the hydroxyl group, OH.

alcohol dehydrogenase [BIOCHEM] The enzyme that catalyzes the oxidation of ethanol to acetaldehyde.

alcoholic [MED] An individual who consumes excess amounts of alcoholic beverages to the extent of being addicted, habituated, or dependent.

alcoholic fermentation [MICROBIO] The process by which certain yeasts decompose sugars in the absence of oxygen to form alcohol and carbon dioxide; method for production of ethanol, wine, and beer.

alcoholimeter *See* alcoholometer.

alcoholism [MED] Compulsive consumption of and dependence on alcoholic beverages, usually leading to pathology of the digestive and nervous systems.

alcoholmeter *See* alcoholometer.

alcoholometer [ENG] A device, such as a form of hydrometer, that measures the quantity of an alcohol contained in a liquid. Also known as alcoholimeter; alcoholmeter.

alcoholysis [ORG CHEM] The breaking of a carbon-to-carbon bond by addition of an alcohol.

Alcyonacea [INV ZOO] The soft corals, an order of littoral anthozoans of the subclass Alcyonaria.

Alcyonaria [INV ZOO] A subclass of the Anthozoa; members are colonial coelenterates, most of which are sedentary and littoral.

aldehyde [ORG CHEM] One of a class of organic compounds containing the CHO radical.

aldehyde lyase [BIOCHEM] Any enzyme that catalyzes the nonhydrolytic cleavage of an aldehyde.

ALCYONACEA

3 cm

Representative alcyonacean *Alcyonium palmatum (after Y. Delage).*

alder [BOT] The common name for several trees of the genus Alnus.

aldohexose [ORG CHEM] A hexose, such as glucose or mannose, containing the aldehyde group.

aldolase [BIOCHEM] An enzyme in anaerobic glycolysis that catalyzes the cleavage of fructose 1,6-diphosphate to glyceraldehyde 3-phosphate; used also in the reverse reaction.

aldose [ORG CHEM] A class of monosaccharide sugars; the molecule contains an aldehyde group.

aldosterone [BIOCHEM] $C_{21}H_{28}O_5$ A steroid hormone extracted from the adrenal cortex that functions chiefly in regulating sodium and potassium metabolism.

aldosteronism [MED] Hypertension induced by excessive secretion of aldosterone.

Aldrich syndrome [MED] A recessive, sex-linked disease characterized by a complex of symptoms, including eczematoid dermatitis, thrombocytopenia, black stool, and a deficiency of immune globulins.

alecithal [CYTOL] Referring to an egg without yolk, such as the eggs of placental mammals.

alepidote [ZOO] Lacking scales.

Alepocephaloidei [VERT ZOO] The slickheads, a suborder of deap-sea teleostean fishes of the order Salmoniformes.

aletophyte [ECOL] A weedy plant growing on the roadside or in fields where natural vegetation has been disrupted by man.

aleukemia [MED] Leukemia in which the white blood cell count is normal or low.

aleuriospore [MYCOL] **1.** A lateral conidium on certain dermatophytes. **2.** A spore or the tip of a hypha that is separated by a septum or by contraction of the protoplasm early in development. **3.** A simple terminal or lateral thick-walled nondeciduous spore produced by some fungi of the order Moniliales.

aleuron [BOT] Protein in the form of grains stored in the embryo, endosperm, or perisperm of many seeds.

Aleutian disease [VET MED] A disease of mink characterized by accumulations of plasma cells in several organs, hyaline changes in the walls of small arteries, and interstitial fibrosis of the kidneys.

alewife [VERT ZOO] *Pomolobus pseudoharengus.* A food fish of the herring family that is very abundant on the Atlantic coast.

alexia [MED] Loss of the ability to read.

Aleyrodidae [INV ZOO] The whiteflies, a family of homopteran insects included in the series Sternorrhyncha; economically important as plant pests.

alfalfa [BOT] *Medicago sativa.* A herbaceous perennial legume in the order Rosales, characterized by a deep taproot. Also known as lucerne.

algae [BOT] General name for the chlorophyll-bearing organisms in the plant subkingdom Thallobionta.

algae bloom [ECOL] A heavy growth of algae in and on a body of water as a result of high phosphate concentration from farm fertilizers and detergents.

alged malaria *See* falciparum malaria.

algesia [PHYSIO] Sensitivity to pain.

algesiroreceptor [PHYSIO] A pain-sensitive cutaneous sense organ.

algicide [MATER] A chemical used to kill algae.

algicolous [BIOL] Living on algae.

Alginobacter [MICROBIO] A genus of gram-negative, peritrichously flagellated, rod-shaped soil bacteria in the family Enterobacteriaceae that ferment alginic acid and glucose, with the production of acid and gas.

ALEURIOSPORE

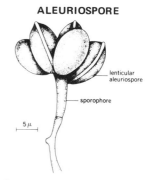

Papularia sphaerosperma (Coniosporium arundinis) with dark aleuriospores. *(After G. Goidanich, 1938)*

ALFALFA

Buffalo alfalfa, showing crown and root in 8-month-old plants. *(Iowa State University Photo Service)*

Alginomonas [MICROBIO] A genus of gram-negative, non-sporeforming, rod-shaped bacteria in the family Pseudo-monadaceae; species decompose alginic acid.

algology [BOT] The study of algae.

algometer [MED] An instrument for measuring pressure stimuli which produce pain.

Algonkian [GEOL] Geologic time between the Archean and Paleozoic. Also known as Proterozoic.

algor mortis [PATH] Postmortem cooling of the body.

alicyclic [ORG CHEM] **1.** Having the properties of both aliphatic and cyclic substances. **2.** Referring to a class of organic compounds containing only carbon and hydrogen atoms joined to form one or more rings.

aliform [INV ZOO] Wing-shaped, referring to insect muscles.

alimentary [BIOL] Of or relating to food, nutrition, or diet.

alimentary canal [ANAT] The tube through which food passes; in man, includes the mouth, pharynx, esophagus, stomach, and intestine.

aliphatic [ORG CHEM] Of or pertaining to any organic compound of hydrogen and carbon characterized by a straight chain of the carbon atoms; three subgroups of such compounds are alkanes, alkenes, and alkynes.

Alismataceae [BOT] A family of flowering plants belonging to the order Alismatales characterized by schizogenous secretory cells, a horseshoe-shaped embryo, and one of two ovules.

Alismatales [BOT] A small order of flowering plants in the subclass Alismatidae, including aquatic and semiaquatic herbs.

Alismatidae [BOT] A relatively primitive subclass of aquatic or semiaquatic herbaceous flowering plants in the class Liliopsida, generally having apocarpous flowers, and trinucleate pollen and lacking endosperm.

alisphenoid [ANAT] **1.** The bone forming the greater wing of the sphenoid in adults. **2.** Of or pertaining to the sphenoid wing.

alitrunk [INV ZOO] In insects, the thorax when fused with the first abdominal segment.

alivincular [INV ZOO] In some bivalves, having the long axis of the short ligament transverse to the hinge line.

alkalemia [MED] An increase in blood pH above normal levels.

alkali [CHEM] Any compound having highly basic qualities.

alkali chlorosis [PL PATH] Yellowing of plant foliage due to excess amounts of soluble salts in the soil.

alkali metal [CHEM] Any of the elements of group Ia in the periodic table: lithium, sodium, potassium, rubidium, cesium, and francium.

alkaline [CHEM] **1.** Having properties of an alkali. **2.** Having a pH greater than 7.

alkaline-earth metals [CHEM] The heaviest members of group IIa in the periodic table; usually calcium, strontium, magnesium, and barium.

alkaline phosphatase [BIOCHEM] A phosphatase active in alkaline media.

alkaline tide [PHYSIO] The temporary decrease in acidity of urine and body fluids after eating, attributed by some to the withdrawal of acid from the body due to gastric digestion.

alkaloid [ORG CHEM] One of a group of nitrogenous bases of plant origin, such as nicotine, cocaine, and morphine.

alkalosis [MED] A condition of high blood alkalinity caused either by high intake of sodium bicarbonate or by loss of hydrochloric acid or blood carbon dioxide.

alkane [ORG CHEM] A member of a series of saturated aliphatic hydrocarbons having the empirical formula C_nH_{2n+2}.

alkaptonuria [MED] A hereditary metabolic disorder transmitted as an autosomal recessive in man in which large amounts of homogentisic acid (alkapton) are excreted in the urine due to a deficiency of homogentisic acid oxidase. Also spelled alcaptonuria.

alkene [ORG CHEM] One of a class of unsaturated aliphatic hydrocarbons containing one or more carbon-to-carbon double bonds.

alkyl [ORG CHEM] A monovalent radical, C_nH_{2n+1}, which may be considered to be formed by loss of a hydrogen atom from an alkane; usually designated by R.

alkylation [ORG CHEM] A chemical process in which an alkyl radical is introduced into an organic compound by substitution or addition.

alkyne [ORG CHEM] One of a group of organic compounds containing a carbon-to-carbon triple bond.

allachesthesia [MED] A tactile sensation experienced remote from the point of stimulation but on the same side of the body.

allantochorion [EMBRYO] The extraembryonic membrane formed by fusion of the outer wall of the allantois with the primitive chorion.

allantoic acid [BIOCHEM] $C_4H_8N_4O_4$ A crystalline acid obtained by hydrolysis of allantoin; intermediate product in nucleic acid metabolism.

allantoid [BIOL] Having the shape of a sausage.

allantoin [BIOCHEM] $C_4H_6N_4O_3$ A crystallizable oxidation product of uric acid found in allantoic and amniotic fluids and in fetal urine.

allantoinase [BIOCHEM] An enzyme, occurring in nonmammalian vertebrates, that catalyzes the hydrolysis of allantoin.

allantois [EMBRYO] A fluid-filled, saclike, extraembryonic membrane lying between the chorion and amnion of reptilian, bird, and mammalian embryos.

allantoxanic acid [BIOCHEM] $C_4H_3N_3O_4$ An acid formed by oxidation of uric acid or allantoin.

allanturic acid [BIOCHEM] $C_3H_4N_2O_3$ An acid formed principally by the oxidation of allantoin.

allassotonic [BOT] Pertaining to movements of mature plants that are induced by stimulus.

Alleculidae [INV ZOO] The comb claw beetles, a family of mostly tropical coleopteran insects in the superfamily Tenebrionoidea.

Allee's principle [GEN] The concept of an intermediate optimal population density by which groups of organisms often flourish best if neither too few nor too many individuals are present.

Alleghenian life zone [ECOL] A biome that includes the eastern mixed coniferous and deciduous forests of New England.

allele [GEN] One of a pair of genes, or of multiple forms of a gene, located at the same locus of homologous chromosomes. Also known as allelomorph.

allelic mutant [GEN] A cell or organism with characters different from those of the parent due to alterations of one or more alleles.

allelomimetic behavior [PSYCH] Behavior in social animals in which each animal does the same thing as those nearby.

allelomorph *See* allele.

allelopathy [BOT] The harmful influence on a plant by another living plant that secretes a toxic substance.

allelotropism [BIOL] A mutual attraction between two cells or organisms.

allergen [IMMUNOL] Any antigen, such as pollen, a drug, or food, that induces an allergic state in humans or animals.

allergic arteritis [MED] Inflammation of the arterial walls resulting from an allergic state.

The American alligator (*Alligator mississipiensis*).

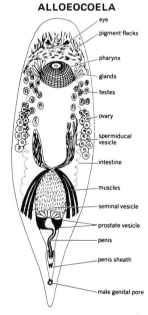

eye
pigment flecks
pharynx
glands
testes
ovary
spermiducal vesicle
intestine
muscles
seminal vesicle
prostate vesicle
penis
penis sheath
male genital pore

Plagiostomum morgani, one of the two largest and best-known genera of the Alloeocoela. (*After L. Graff, 1911*)

allergic dermatitis [MED] Inflammation of the skin following contact of an allergen with sensitized tissue.

allergic reaction *See* allergy.

allergic rhinitis *See* hay fever.

allergic vasculitis syndrome [MED] A skin disease, possibly immunologic, characterized by ulcers which result from destructive inflammation of underlying blood vessels.

allergy [MED] A type of antigen-antibody reaction marked by an exaggerated physiologic response to a substance that causes no symptoms in nonsensitive individuals. Also known as allergic reaction.

alliance [SYST] A group of related families ranking between an order and a class.

alligator [VERT ZOO] Either of two species of archosaurian reptiles in the genus *Alligator* of the family Crocodylidae.

Alligatorinae [VERT ZOO] A subgroup of the crocodilian family Crocodylidae that includes alligators, caimans, *Melanosuchus*, and *Paleosuchus*.

Allium [BOT] A genus of bulbous herbs in the family Liliaceae including leeks, onions, and chives.

allo- [CHEM] Prefix applied to the stabler form of two isomers.

allobiosis [BIOL] Altered reactivity of an organism in a changed environment.

allocarpy [BOT] Formation of fruit as a result of cross fertilization.

allocheiral [BOT] Pertaining to reversed symmetry, that is, having the left and right sides reversed.

allocheiria [MED] A form of allachesthesia in which the tactile sensation is experienced on the side opposite the one to which the stimulus was applied; seen in tabes dorsalis and other conditions.

allochoric [BOT] Describing a species that inhabits two or more closely related communities, such as forest and grassland, in the same region.

allochronic [EVOL] Pertaining to species or other taxonomic groups that are not contemporary in evolution.

Alloeocoela [INV ZOO] An order of platyhelminthic worms of the class Turbellaria distinguished by a simple pharynx and a diverticulated intestine.

allogamous [BOT] Having reproduction by cross fertilization.

allogenic [ECOL] Caused by external factors, as in reference to the change in habitat of a natural community resulting from draught.

Allogromiidae [INV ZOO] A little-known family of protozoans in the order Foraminiferida; adults are characterized by a chitinous test.

Allogromiina [INV ZOO] A suborder of marine and freshwater protozoans in the order Foraminiferida characterized by an organic test of protein and acid mucopolysaccharide.

alloheteroploidy [GEN] Heteroploidy resulting from the union of specifically distinct genomes.

allometry [BIOL] **1.** The quantitative relation between a part and the whole or another part as the organism increases in size. Also known as heterauxesis; heterogony. **2.** The quantitative relation between the size of a part and the whole or another part, in a series of related organisms that differ in size.

Allomyces [MYCOL] A genus of aquatic phycomycetous fungi in the order Blastocladiales characterized by basal rhizoids, terminal hyphae, and zoospores with a single posterior flagellum.

alloparalectotype [SYST] A nomenclatural type for a specimen obtained from the same collection as the holotype but being of opposite sex and described subsequently.

allopatric [ECOL] Referring to organisms that inhabit separate and mutually exclusive geographic regions. [EVOL] Originating in or occupying different geographical areas.

allopatric speciation [EVOL] Differentiation of populations in geographical isolation to the point where they are recognized as separate species.

allophore [HISTOL] A chromatophore which contains a red pigment that is soluble in alcohol; found in the skin of fishes, amphibians, and reptiles.

allophytoid [BOT] A propagative bud.

alloplasm [CYTOL] The differentiated portion of cytoplasm that does not form organelles.

allopolyploid [GEN] An organism or strain arising from a combination of genetically distinct chromosome sets from two diploid organisms.

allorhizal [BOT] Having root and shoot poles of opposite direction.

all-or-none law [PHYSIO] The principle that transmission of a nerve impulse is either at full strength or not at all.

alloscutum [INV ZOO] In tick larvae, the dorsal area or sclerite posterior to the scutum.

allosome [GEN] **1.** Sex chromosome. **2.** Any atypical chromosome.

allosteric enzyme [BIOCHEM] Any of the regulatory bacterial enzymes, such as those involved in end-product inhibition.

allostery [BIOCHEM] The property of an enzyme able to shift reversibly between an active and an inactive configuration.

Allotheria [PALEON] A subclass of Mammalia that appeared in the Upper Jurassic and became extinct in the Cenozoic.

allotrioblast *See* xenoblast.

Allotriognathi [VERT ZOO] An equivalent name for the Lampridiformes.

allotropous [INV ZOO] Referring to certain insects which are not limited to visiting only certain flower species.

allotype [SYST] A paratype of the opposite sex to the holotype.

alloxan [BIOCHEM] $C_4H_2N_2O_4$ Crystalline oxidation product of uric acid; induces diabetes experimentally by selective destruction of pancreatic beta cells. Also known as mesoxalyurea.

allspice [BOT] The dried, unripe berries of a small, tropical evergreen tree, *Pimenta officinalis,* of the myrtle family; yields a pungent, aromatic spice.

allulose [ORG CHEM] $CH_2OHCHCO(CHOH)_3CH_2OH$ A constituent of cane sugar molasses; it is nonfermentable. Also known as D-piscose; d-ribo-2-ketohexose.

allyl [ORG CHEM] C_3H_5- An unsaturated radical found in compounds such as allylbromide (3-bromopropene).

alm [ECOL] A meadow in alpine or subalpine mountain regions.

almond [BOT] *Prunus amygdalus.* A small deciduous tree of the order Rosales; it produces a drupaceous edible fruit with an ellipsoidal, slightly compressed nutlike seed.

alopecia [MED] Loss of hair; baldness.

Alopiidae [VERT ZOO] A family of pelagic isurid elasmobranchs commonly known as thresher sharks because of their long, whiplike tail.

alpaca [VERT ZOO] *Lama pacos.* An artiodactyl of the camel family (Camelidae); economically important for its long, fine wool.

alpage [ECOL] A summer grazing area composed of natural plant pasturage in upland or mountainous regions.

alpha cell [HISTOL] Any of the acidophilic chromophiles in the anterior lobe of the adenohypophysis.

alpha globulin [BIOCHEM] A heterogeneous fraction of serum

ALLSPICE

A branch of *Pimenta officinalis* with the berry fruit.

ALMOND

Almond twig with leaves and fruit.

ALPACA

The alpaca (*Lama pacos*), sheared every 2 years, to yield 18–24 pounds of wool in its lifetime.

globulins containing the proteins of greatest electrophoretic mobility.

alpha granules [BOT] Metachromatic granules found in the central area of the protoplast; occurring, for example, in blue-green algae.

alpha helix [MOL BIO] A spatial configuration of the polypeptide chains of proteins in which the chain assumes a helical form, 0.54 nanometer in pitch, 3.6 amino acids per turn, presenting the appearance of a hollow cylinder with radiating side groups.

alpha hemolysis [MICROBIO] Partial hemolysis of red blood cells with green discoloration in a blood agar medium by certain hemolytic streptococci.

alpha particle [ATOM PHYS] A positively charged particle consisting of two protons and two neutrons, identical with the nucleus of the helium atom; emitted by several radioactive substances.

alpha rhythm [PHYSIO] An electric current from the occipital region of the brain cortex having a pulse frequency of 8 to 13 per second; associated with a relaxed state in normal human adults.

alpha taxonomy [SYST] The initial or descriptive phase in the history of animal systematics.

Alpheidae [INV ZOO] The snapping shrimp, a family of decapod crustaceans included in the section Caridea.

alpine [ECOL] Any plant native to mountain peaks or boreal regions.

alsinaceous [BOT] Pertaining to a polypetalous corolla having intervals between the petals.

alter ego [PSYCH] A close friend who represents a second self to an individual.

alternate [BOT] **1.** Of the arrangement of leaves on opposite sides of the stem at different levels. **2.** Of the arrangement of the parts of one whorl between members of another whorl.

alternation of generations *See* metagenesis.

alterne [ECOL] A community exhibiting alternating dominance with other communities in the same area.

alternipinnate [BOT] Referring to leaflets which arise at alternating points on each side of the midrib.

altherbosa [ECOL] Communities of tall herbs, usually succeeding where forests have been destroyed.

altitudinal vegetation zone [ECOL] A geographical band of physiognomically similar vegetation correlated with vertical and horizontal gradients of environmental conditions.

altricial [VERT ZOO] Pertaining to young that are born or hatched immature and helpless, thus requiring extended development and parental care.

alula [ZOO] **1.** Digit of a bird wing homologous to the thumb. **2.** *See* calypter.

aluminium *See* aluminum.

aluminosis [MED] A lung disorder caused by inhalation of alumina dust.

aluminum [CHEM] A chemical element, symbol Al, atomic number 13, and atomic weight 26.9815. Also spelled aluminium.

alveator [INV ZOO] A type of pedicellaria in echinoderms.

alveolar [BIOL] Of or relating to an alveolus.

alveolar-capillary block syndrome [MED] Arterial oxygen deficiency due to improper functioning of the membranes between the alveoli and capillaries.

alveolar gland *See* acinous gland.

alveolate [BIOL] Being pitted or honeycombed in appearance.

alveolated cell *See* epithelioid cell.

alveolitoid [INV ZOO] A type of tabulate coral having a

ALUMINUM

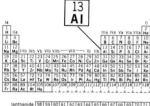

Periodic table of the chemical elements showing the position of aluminum.

vaulted upper wall and a lower wall parallel to the surface of attachment.

alveolus [ANAT] **1.** A tiny air sac of the lung. **2.** A tooth socket. **3.** A sac of a compound gland.

Alydidae [INV ZOO] A family of hemipteran insects in the superfamily Coreoidea.

alymphocytosis [MED] Absence or deficiency of blood lymphocytes.

Alzheimer's disease [MED] A type of presenile dementia associated with sclerosis of the cerebral cortex.

Am *See* americium; ammonium.

amacrine [HISTOL] Having no apparent axon, as cells in the inner nuclear layer of the retina.

amantadine [PHARM] $C_{10}H_{17}N$ A symmetrical amine used as a viral chemoprophylactic because it selectively inhibits certain myxoviruses; also of value in the treatment of parkinsonism. Also known as 1-aminoadamantane.

amanthophilous [BOT] Of plants having a habitat in sandy plains or hills.

Amaranthaceae [BOT] The characteristic family of flowering plants in the order Caryophyllales; they have a syncarpous gynoecium, a monochlamydeous perianth that is more or less scarious, and mostly perfect flowers.

Amaryllidaceae [BOT] The former designation for a family of plants now included in the Liliaceae.

amaurosis [MED] Total or partial blindness.

amaurotic familial idiocy [MED] A hereditary condition, transmitted as an autosomal recessive, predominantly in Jewish children, characterized by blindness, muscular weakness, and subnormal mental development; when onset is in infancy, the disease is known commonly as Tay-Sachs disease.

amber [MINERAL] A transparent yellow, orange, or reddish-brown fossil resin derived from a coniferous tree; used for ornamental purposes; it is amorphous, has a specific gravity of 1.05-1.10, and a hardness of 2-2.5 on Mohs scale.

ambergris [PHYSIO] A fatty substance formed in the intestinal tract of the sperm whale; used in the manufacture of perfume.

amber mutation [GEN] Alteration of another triplet to *UAG*, a terminator codon controlling termination of polypepide chain synthesis in bacteria.

ambidextrous [PHYSIO] Capable of using both hands with equal skill.

ambiens [VERT ZOO] In certain birds, a thigh muscle which acts to flex the toes, allowing a firm grasp on the perch.

ambigenous [BOT] Of a perianth whose outer leaves resemble the calyx while the inner leaves resemble the corolla.

ambilateral [ANAT] Relating to or affecting both sides of the body.

ambiparous [BOT] Referring to buds which contain the rudimentary parts of both flowers and leaves.

ambisexual [MED] An individual having undifferentiated primordia of both sexes. [PSYCH] Having feelings and exhibiting behavior common to both sexes.

ambitus [BIOL] The periphery or external edge, as of a mollusk shell or leaf.

ambivalence [PSYCH] The coexistence of conflicting reactions toward a person or object.

amblyopia [MED] Dimness of vision, especially that not due to refractive errors or organic disease of the eye; may be congenital or acquired.

Amblyopsidae [VERT ZOO] The cave fishes, a family of actinopterygian fishes in the order Percopsiformes.

Amblyopsiformes [VERT ZOO] An equivalent name for the Percopsiformes.

Amblypygi [INV ZOO] An order of chelicerate arthropods in the class Arachnida, commonly known as the tailless whip scorpions.

amboceptor [IMMUNOL] According to P. Ehrlich, an antibody present in the blood of immunized animals which contains two specialized elements: a cytophil group that unites with a cellular antigen, and a complementophil group that joins with the complement.

ambon [ANAT] The fibrocartilaginous ring that surrounds the glenoid cavity of the scapula.

ambulacriform [BIOL] Having the form of, or resembling, an ambulacrum.

ambulacrum [INV ZOO] In echinoderms, any of the radial series of plates along which the tube feet are arranged.

ambulatorial [ZOO] **1.** Capable of walking. **2.** In reference to a forest animal, having adapted to walking, as opposed to running, crawling, or leaping.

ambulatory schizophrenia [PSYCH] A condition in which a person exhibits symptoms of both manic-depressive and schizophrenic psychosis but is not considered to require institutionalization.

Ambystoma [VERT ZOO] A genus of common salamanders; the type genus of the family Ambystomatidae.

Ambystomatidae [VERT ZOO] A family of urodele amphibians in the suborder Salamandroidea; neoteny occurs frequently in this group.

Ambystomoidea [VERT ZOO] A suborder to which the family Ambystomatidae is sometimes elevated.

ameba [INV ZOO] The common name for a number of species of naked unicellular protozoans of the order Amoebida. An example is a member of the genus *Amoeba*.

Amebelodontinae [PALEON] A subfamily of extinct elephantoid proboscideans in the family Gomphotheriidae.

amebiasis [MED] A parasitic disease of man caused by the ameba *Entamoeba histolytica,* characterized by clinical-pathological intestinal manifestations, including an acute dysentery phase. Also known as amebic dysentery.

amebic abscess [MED] Liquefactive necrosis of the brain and liver, without suppuration, caused by amebas, usually *Entamoeba histolytica.*

amebic dysentery *See* amebiasis.

amebicide [MATER] A chemical used to kill amebas, especially parasitic species.

amebocyte [CYTOL] Any cell similar in shape or properties to an ameba. [INV ZOO] One of a specialized cell type found in the coelom of many invertebrates.

ameboid [BIOL] Being similar to an ameba in shape or properties.

ameboid movement [CYTOL] A type of cellular locomotion involving the formation of pseudopodia.

ameiosis [GEN] Nonreduction of chromosome number due to suppression of one of the miotic divisions, as in parthenogenesis.

Ameiuridae [VERT ZOO] A family of North American catfishes belonging to the suborder Siluroidei.

amelification [PHYSIO] The formation of enamel on the teeth.

ameloblast [EMBRYO] One of the columnar cells of the enamel organ that form dental enamel in developing teeth.

ameloblastic odontoma [MED] A neoplasm of epithelial and mesenchymal odontogenic tissue. Also known as odontoblastoma.

ameloblastoma [MED] An epithelial tumor associated with

AMEBA

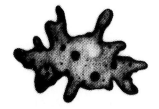

Typical species of ameba showing numerous pseudopodia, which are used for locomotion.

the enamel organ; cells of basal layers resemble the ameloblast. Also known as adamantinoma.

amenorrhea [MED] Absence of menstruation due to either normal or abnormal conditions.

ament [BOT] A catkin. [MED] A person with congenital mental deficiency; an idiot.

amentaceous [BOT] Pertaining to plants that bear amenta or catkins.

amentia [MED] Congenital subnormal intellectual development.

Amera [INV ZOO] One of the three divisions of the phylum Vermes proposed by O. Bütschli in 1910 and given the rank of a subphylum.

American boreal faunal region [ECOL] A zoogeographic region comprising marine littoral animal communities of the coastal waters off east-central North America.

American lion *See* puma.

American mucocutaneous leishmaniasis [MED] A form of leishmaniasis caused by *Leishmania braziliensis*, transmitted by sandflies of the genus *Phlebotomus,* and characterized by skin ulcers and ulceration and necrosis of the mucosa of the mouth and nose. Also known as South American leishmaniasis.

American spotted fever *See* Rocky Mountain spotted fever.

americium [CHEM] A chemical element, symbol Am, atomic number 95; the mass number of the isotope with the longest half-life is 243.

ameristic [BIOL] Not divided into segments.

Amerosporae [MYCOL] A spore group of the Fungi Imperfecti characterized by one-celled or threadlike spores.

ametabolic [INV ZOO] **1.** Referring to ciliates that do not change form. **2.** Referring to insects that do not undergo a distinct metamorphosis.

ametabolous metamorphosis [INV ZOO] A growth stage of certain insects characterized by an increase in size without distinct external changes.

amethopterin [PHARM] $C_{20}H_{22}N_8O_5 \cdot H_2O$ An antimetabolite effective as a folic acid antagonist and used for treatment of acute and subacute leukemia. Also known as methotrexate.

ametropia [MED] Any deficiency in the refractive ability of the eye that causes an unfocused image to fall on the retina.

amictic [INV ZOO] **1.** In rotifers, producing diploid eggs that are incapable of being fertilized. **2.** Pertaining to the egg produced by the amictic female.

amidase [BIOCHEM] Any enzyme that catalyzes the hydrolysis of nonpeptide $C = N$ linkages.

amide [ORG CHEM] One of a class of organic compounds containing the $CONH_2$ radical.

amidohydrolase [BIOCHEM] An enzyme that catalyzes deamination.

Amiidae [VERT ZOO] A family of actinopterygian fishes in the order Amiiformes represented by a single living species, the bowfin *(Amia calva).*

Amiiformes [VERT ZOO] An order of actinopterygian fishes characterized by an abbreviate heterocercal tail, fusiform body, and median fin rays.

amin- *See* amino-.

amine [ORG CHEM] One of a class of organic compounds which can be considered to be derived from ammonia by replacement of one or more hydrogens by organic radicals.

amine oxidase [BIOCHEM] An enzyme that catalyzes the oxidation of tyramine and tryptamine to aldehyde.

amino- [CHEM] Having the property of a compound in which

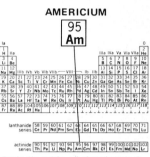

AMERICIUM

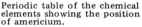

Periodic table of the chemical elements showing the position of americium.

the group NH_2 is attached to a radical other than an acid radical. Also spelled amin-.

aminoacetic acid *See* glycine.

amino acid [BIOCHEM] Any of the organic compounds that contain one or more basic amino groups and one or more acidic carboxyl groups and that are polymerized to form peptides and proteins; only 20 of the more than 80 amino acids found in nature serve as building blocks for proteins; examples are tyrosine and lysine.

amino aciduria [MED] A group of disorders in which excess amounts of amino acids are excreted in the urine; caused by abnormal protein metabolism.

***para*-aminobenzoic acid** [BIOCHEM] $C_7H_7O_2N$ A yellow-red, crystalline compound that is part of the folic acid molecule; essential in metabolism of certain bacteria. Abbreviated PABA.

D-aminobenzyl penicillin *See* Ampicillin.

amino diabetes [MED] A congenital disorder characterized by excessive quantities of amino acids, glucose, and phosphate in the urine, resulting from deficient resorption in the proximal convoluted tubules of the kidney. Also known as Fanconi's syndrome.

4-aminofolic acid *See* aminopterin.

aminopeptidase [BIOCHEM] An enzyme which catalyzes the liberation of an amino acid from the end of a peptide having a free amino group.

aminophylline [PHARM] $C_{16}H_{24}N_{10}O_4$ A drug in the form of white or slightly yellowish, water-soluble granules or powder, used as a smooth-muscle relaxant, myocardial stimulant, and diuretic.

aminopolypeptidase [BIOCHEM] A proteolytic enzyme that cleaves polypeptides containing either a free amino group or a basic nitrogen atom having at least one hydrogen atom.

aminoprotease [BIOCHEM] An enzyme that hydrolyzes a protein and unites with its free amino group.

aminopterin [PHARM] $C_{19}H_{20}N_8O_5 \cdot 2H_2O$ A yellow crystalline acid which is similar to folic acid and is used clinically as an antagonist of folic acid. Trade name for 4-aminofolic acid; 4-aminopteroylglutamic acid.

4-aminopteroylglutamic acid *See* aminopterin.

amino sugar [BIOCHEM] A monosaccharide in which a nonglycosidic hydroxyl group is replaced by an amino or substituted amino group; an example is D-glucosamine.

aminotransferase *See* transaminase.

amitosis [CYTOL] Cell division by simple fission of the nucleus and cytoplasm without chromosome differentiation.

Ammanian [GEOL] Middle Upper Cretaceous geologic time.

ammochaeta [INV ZOO] One of the bristles arranged in groups on the head of a desert ant and used for removing sand from the forelegs.

ammocoete [ZOO] A protracted larval stage of lampreys.

ammocolous [ECOL] Describing plants having a habitat in dry sand.

Ammodiscacea [INV ZOO] A superfamily of foraminiferal protozoans in the suborder Textulariina, characterized by a simple to labyrinthic test wall.

Ammodytoidei [VERT ZOO] The sand lances, a suborder of marine actinopterygian fishes in the order Perciformes, characterized by slender, eel-shaped bodies.

ammonation [INORG CHEM] A reaction in which ammonia is added to other molecules or ions by covalent bond formation utilizing the unshared pair of electrons on the nitrogen atom, or through ion-dipole electrostatic interactions.

ammonia [INORG CHEM] NH_3 A colorless gaseous alkaline compound that is very soluble in water, has a characteristic

AMINO SUGAR

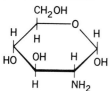

The Haworth formula of D-glucosamine.

pungent odor, is lighter than air, and is formed as a result of the decomposition of most nitrogenous organic material; used as a fertilizer and as a chemical intermediate.

ammonification [CHEM] Addition of ammonia or ammonia compounds, especially to the soil.

ammonifiers [ECOL] Fungi, or actinomycetous bacteria, that participate in the ammonification part of the nitrogen cycle and release ammonia (NH₃) by decomposition of organic matter.

ammonium [CHEM] The radical NH_4^+. Abbreviated Am.

Ammonoidea [PALEON] An order of extinct cephalopod mollusks in the subclass Tetrabranchia; important as index fossils.

ammonolysis [CHEM] **1.** A dissociation reaction of the ammonia molecule producing H^+ and NH_2^- species. **2.** Breaking of a bond by addition of ammonia.

Ammon's law [ANTHRO] The law stating that cephalic index and stature vary inversely.

Ammotheidae [INV ZOO] A family of marine arthropods in the subphylum Pycnogonida.

amnesia [MED] The pathological loss or impairment of memory brought about by psychogenic or physiological disturbances.

amnesic aphasia [MED] Loss of memory for the appropriate names of objects, conditions, or relations, accompanied by fragmented or hesitant speech.

amnicolous [ECOL] Describing plants having a habitat on sandy riverbanks.

amniocentesis [MED] A procedure during pregnancy by which the abdominal wall and fetal membranes are punctured with a cannula to withdraw amniotic fluid.

amnion [EMBRYO] A thin extraembryonic membrane forming a closed sac around the embryo in birds, reptiles, and mammals.

Amniota [VERT ZOO] A collective term for the Reptilia, Aves, and Mammalia, all of which have an amnion during development.

amniote [VERT ZOO] Any animal characterized by the presence of an amnion during embryonic development.

amniotic fluid [PHYSIO] A substance that fills the amnion to protect the embryo from desiccation and shock.

Amoeba [INV ZOO] A genus of naked, rhizopod protozoans in the order Amoebida characterized by a thin pellicle and thick, irregular pseudopodia.

Amoebida [INV ZOO] An order of rhizopod protozoans in the subclass Lobosia characterized by the absence of a protective covering (test).

Amoebobacter [MICROBIO] A sulfur purple bacteria, a genus of spherical to rod-shaped photosynthetic bacteria in the family Thiorhodaceae.

amoebula [INV ZOO] A type of spore in protists that has pseudopodia.

amorphous [PHYS] Pertaining to a solid which is noncrystalline, having neither definite form nor structure.

AMP *See* adenylic acid.

Ampeliscidae [INV ZOO] A family of tube-dwelling amphipod crustaceans in the suborder Gammaridea.

Ampharetidae [INV ZOO] A large, deep-water family of polychaete annelids belonging to the Sedentaria.

Ampharetinae [INV ZOO] A subfamily of annelids belonging to the family Ampharetidae.

amphetamine [PHARM] $C_6H_5CH_2CHNHCH_3$ A volatile, colorless liquid used as a central nervous system stimulant. Also known as racemic 1-phenyl-2-aminopropane and by the trade name Benzedrine.

amphiarthrosis [ANAT] An articulation of limited movement in which bones are connected by fibrocartilage, such as that between vertebrae or that at the tibiofibular junction.

Amphibia [VERT ZOO] A class of vertebrate animals in the superclass Tetrapoda characterized by a moist, glandular skin, gills at some stage of development, and no amnion during the embryonic stage.

Amphibicorisae [INV ZOO] A subdivision of the insect order Hemiptera containing surface water bugs with exposed antennae.

Amphibioidei [INV ZOO] A family of tapeworms in the order Cyclophyllidea.

amphibiotic [BIOL] Living in water as a larva or nymph and on land as an adult.

amphibious [BIOL] Capable of living both on dry or moist land and in water.

amphiblastic cleavage [EMBRYO] The unequal but complete cleavage of telolecithal eggs.

amphiblastula [EMBRYO] A blastula resulting from amphiblastic cleavage. [INV ZOO] The free-swimming flagellated larva of many sponges.

amphibolic pathway [BIOCHEM] A microbial biosynthetic and energy-producing pathway, such as the glycolytic pathway.

Amphibolidae [INV ZOO] A family of gastropod mollusks in the order Basommatophora.

amphicarpic [BOT] Having two types of fruit, differing either in form or ripening time.

Amphichelydia [PALEON] A suborder of Triassic to Eocene anapsid reptiles in the order Chelonia; these turtles did not have a retractable neck.

amphichrome [BOT] A plant that produces flowers of different colors on the same stalk.

Amphicoela [VERT ZOO] A small suborder of amphibians in the order Anura characterized by amphicoelous vertebrae.

amphicoelous [VERT ZOO] Describing vertebrae that have biconcave centra.

amphicondylous [VERT ZOO] Having two occipital condyles.

amphicone [PALEON] A molar cusp in extinct mammals.

amphicribral [BOT] Having the phloem surrounded by the xylem, as seen in certain vascular bundles.

Amphicyonidae [PALEON] A family of extinct giant predatory carnivores placed in the infraorder Miacoidea by some authorities.

amphicytula [EMBRYO] A zygote that is capable of holoblastic unequal cleavage.

amphid [INV ZOO] Either of a pair of sensory receptors in nematodes, believed to be chemoreceptors and situated laterally on the anterior end of the body.

amphidetic [INV ZOO] Of a bivalve ligament, extending both before and behind the beak.

amphidiploid [GEN] An organism having a diploid set of chromosomes from each parent.

amphidisc [INV ZOO] A spicule in the shape of a grapnel; peculiar to certain fresh-water poriferans.

Amphidiscophora [INV ZOO] A subclass of sponges in the class Hexactinellida characterized by an anchoring tuft of spicules and no hexasters.

Amphidiscosa [INV ZOO] An order of hexactinellid sponges in the subclass Amphidiscophora characterized by amphidisc spicules, that is, spicules having a stellate disk at each end.

amphigean [ECOL] An organism that is native to both Old and New Worlds.

amphigenesis [BIOL] Reproduction by sexual means.

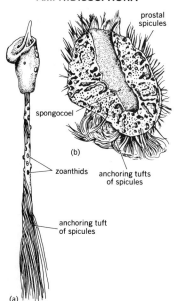

AMPHIDISCOPHORA

prostal spicules

spongocoel

(b)

zoanthids   anchoring tufts of spicules

anchoring tuft of spicules

(a)

Representative amphidiscophorans. (a) *Hyalonema*, with zoanthids encrusting root tuft of spicules (after Hyman, 1940); (b) *Pheronema*, sectioned longitudinally (from Hyman, 1940, after Schulze, 1887)

amphigenous [BOT] Located or growing on both sides of a structure, as certain leaves.

amphigynous [BOT] Pertaining to an antheridium that surrounds the base of the oogonium; occurs in certain Peronosporales.

Amphilestidae [PALEON] A family of Jurassic triconodont mammals whose subclass is uncertain.

Amphilinidea [INV ZOO] An order of tapeworms in the subclass Cestodaria characterized by a protrusible proboscis, anterior frontal glands, and no holdfast organ; they inhabit the coelom of sturgeon and other fishes.

Amphimerycidae [PALEON] A family of late Eocene to early Oligocene tylopod ruminants in the superfamily Amphimerycoidea.

Amphimerycoidea [PALEON] A superfamily of extinct ruminant artiodactyls in the infraorder Tylopoda.

amphimixis [PHYSIO] The union of egg and sperm in sexual reproduction.

Amphimonadidae [INV ZOO] A family of zoomastigophorean protozoans in the order Kinetoplastida.

Amphineura [INV ZOO] A class of the phylum Mollusca; members are bilaterally symmetrical, elongate marine animals, such as the chitons.

Amphinomidae [INV ZOO] The stinging or fire worms, a family of amphinomorphan polychaetes belonging to the Errantia.

Amphinomorpha [INV ZOO] Group name for three families of errantian polychaetes: Amphenomidae, Euphrosinidae, and Spintheridae.

amphinucleolus [CYTOL] A double nucleolus composed of both basiphil and oxyphil components.

amphiodont [INV ZOO] In stag beetles, referring to an intermediate stage in development of the mandible.

amphiont [BIOL] A zygote or a sporont formed by union of two individuals.

amphioxus [ZOO] Former designation for the lancelet, *Branchiostoma.*

amphiphloic [BOT] Pertaining to the central vascular cylinder of stems having phloem on both sides of the xylem.

amphiplatyan [ANAT] Describing vertebrae having centra that are flat both anteriorly and posteriorly.

amphipneustic [VERT ZOO] Having both gills and lungs through all life stages, as in some amphibians.

Amphipoda [INV ZOO] An order of crustaceans in the subclass Malacostraca; individuals lack a carapace, bear unstalked eyes, and respire through thoracic branchiae or gills.

amphiprotic *See* amphoteric.

amphirhinal [VERT ZOO] Pertaining to or having two nostrils.

amphisarca [BOT] An indehiscent fruit characterized by many cells and seeds, pulpy flesh, and a hard rind; melon is an example.

Amphisbaenidae [VERT ZOO] A family of tropical snakelike lizards in the suborder Sauria.

Amphisopidae [INV ZOO] A family of isopod crustaceans in the suborder Phreactoicoidea.

amphispermous [BOT] Having the pericarp closely surrounding the seed.

amphisporangiate [BOT] Having both megasporangia and microsporangia on the sporophylls.

amphispore [MYCOL] A specialized urediospore with a thick, colorful wall; a resting spore.

Amphissitidae [PALEON] A family of extinct ostracods in the suborder Beyrichicopina.

Amphistaenidae [VERT ZOO] The worm lizards, a family of

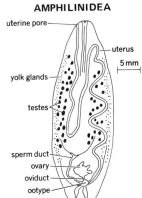

AMPHILINIDEA

uterine pore

uterus

5 mm

yolk glands

testes

sperm duct

ovary

oviduct

ootype

Amphilina, the only tapeworm in the order Amphilinidea whose life history is completely known.

reptiles in the suborder Sauria; structural features are greatly reduced, particularly the limbs.

amphistomatic [BOT] Referring to certain leaves that have stomata on both the upper and lower surface.

amphistome [INV ZOO] An adult type of digenetic trematode having a well-developed ventral sucker (acetabulum) on the posterior end.

amphistylic [VERT ZOO] Having the jaw suspended from the brain case and the hyomandibular cartilage, as in some sharks.

amphitene *See* zygotene.

amphithecium [BOT] The external cell layer during development of the sporangium in mosses.

Amphitheriidae [PALEON] A family of Jurassic therian mammals in the infraclass Pantotheria.

amphitoky [BIOL] Reproduction of both males and females by parthenogenesis in annelids.

amphitriaene [INV ZOO] A poriferan spicule having three divergent rays at each end.

amphitrichous [BIOL] Having flagella at both ends, as in certain bacteria.

Amphitritinae [INV ZOO] A subfamily of sedentary polychaete worms in the family Terebellidae.

amphitrocha [INV ZOO] In Annelida, a free-swimming larva having two bands of cilia.

amphitropous [BOT] Having a half-inverted ovule with the funiculus attached near the middle.

Amphiumidae [VERT ZOO] A small family of urodele amphibians in the suborder Salamandroidea composed of three species of large, eellike salamanders with tiny limbs.

amphivasal [BOT] Having the xylem surrounding the phloem, as seen in certain vascular bundles.

Amphizoidae [INV ZOO] The trout stream beetles, a small family of coleopteran insects in the suborder Adephaga.

amphogenic [BIOL] Producing both male and female offspring.

Amphoriscidae [INV ZOO] A family of calcareous sponges in the order Sycettida.

amphoteric [CHEM] Having both acidic and basic characteristics. Also known as amphiprotic.

amphotericin [MICROBIO] An amphoteric antifungal antibiotic produced by *Streptomyces nodosus* and having of two components, A and B.

amphotericin A [MICROBIO] The relatively inactive component of amphotericin.

amphotericin B [MICROBIO] $C_{46}H_{73}O_{20}N$ The active component of amphotericin, suitable for systemic therapy of deep or superficial mycotic infections.

Ampicillin [MICROBIO] $C_{16}H_{19}N_3O_4S$ Semisynthetic broad-spectrum penicillin produced initially by *Penicillium chrysogenum* and then chemically modified; used as an antibiotic. Trade name for D-aminobenzyl penicillin.

amplectant [BOT] Clasping tightly or winding around some support, such as tendrils.

amplexicaul [BOT] Pertaining to a sessile leaf with the base or stipules embracing the stem.

amplexus [BOT] Having the edges of a leaf overlap the edges of a leaf above it in vernation. [VERT ZOO] The copulatory embrace of frogs and toads.

ampuliaceous [ZOO] Pertaining to flask-shaped sensillae.

Ampulicidae [INV ZOO] A small family of hymenopteran insects in the superfamily Sphecoidea.

ampulla [ANAT] A dilated segment of a gland or tubule. [BOT] A small air bladder in some aquatic plants. [INV ZOO] The sac at the base of a tube foot in certain echinoderms.

ampulla of Lorenzini [VERT ZOO] Any of the cutaneous receptors in the head region of elasmobranchs; thought to have a thermosensory function.

ampulla of Vater [ANAT] Dilation at the junction of the bile and pancreatic ducts and the duodenum in man. Also known as papilla of Vater.

amputation [MED] The surgical, congenital, or spontaneous removal of a limb or projecting body part.

amu *See* atomic mass unit.

amygdala [ANAT] A mass of gray matter in the lateral wall and roof of the lateral ventricle; it is connected with the hypothalamus and is concerned with emotion.

amygdalin [BIOCHEM] $C_6H_5CH(CN)OC_{12}H_{21}O_{10}$ A glucoside occurring in the kernels of certain plants of the genus *Prunus.*

amyl [ORG CHEM] Any of the eight isomeric arrangements of the radical C_5H_{11} or a mixture of them. Also known as pentyl.

amylase [BIOCHEM] An enzyme that hydrolyzes reserve carbohydrates, starch in plants and glycogen in animals.

amyloid [PATH] An abnormal protein deposited in tissues, formed from the infiltration of an unknown substance, probably a carbohydrate.

amyloid body [PATH] Any of the microscopic hyaline bodies that stain like amyloid with metachromatic aniline dyes.

amyloidosis [MED] Deposition of amyloid in one or more organs of the body.

amylolysis [BIOCHEM] The enzyme-catalyzed hydrolysis of starch to soluble products.

amylolytic enzyme [BIOCHEM] A type of enzyme capable of denaturing starch molecules; used in textile manufacture to remove starch added to slash sizing agents.

amylopectin [BIOCHEM] A highly branched, high-molecular-weight carbohydrate polymer composed of about 80% corn starch.

amyloplast [BOT] A colorless cell plastid packed with starch grains and occurring in cells of plant storage tissue.

amylopsin [BIOCHEM] An enzyme in pancreatic juice that acts to hydrolyze starch into maltose.

amylose [BIOCHEM] A linear starch polymer.

amylostatolith [BIOL] A statolith that is a starch grain.

Amynodontidae [PALEON] A family of extinct hippopotamuslike perissodactyl mammals in the superfamily Rhinoceratoidea.

amyotonia [MED] Absence of muscle tone.

amyotonia congenita [MED] A congenital disease of the central nervous system characterized by absence of voluntary muscle tone and reflexes. Also known as Oppenheim's disease.

amyotrophic lateral sclerosis [MED] A degenerative disease of the pyramidal tracts and lower motor neurons characterized by motor weakness and a spastic condition of the limbs associated with muscular atrophy, fibrillary twitching, and final involvement of nuclei in the medulla. Also known as lateral sclerosis.

Anabantidae [VERT ZOO] A fresh-water family of actinopterygian fishes in the order Perciformes, including climbing perches and gourami.

Anabantoidei [VERT ZOO] A suborder of fresh-water labyrinth fishes in the order Perciformes.

anabiosis [BIOL] State of suspended animation induced by desiccation and reversed by addition of moisture; can be achieved in rotifers.

anabolism [BIOCHEM] A part of metabolism involving the

union of smaller molecules into larger molecules; the method of synthesis of tissue structure.

anaboly [EVOL] The addition, through evolutionary differentiation, of a new terminal stage to the morphogenetic pattern.

Anacanthini [VERT ZOO] An equivalent name for the Gadiformes.

anacanthous [BOT] Lacking spines or thorns.

Anacardiaceae [BOT] A family of flowering plants, the sumacs, in the order Sapindales; many species are allergenic to man.

anaclitic depression [PSYCH] Seriously impaired physical, social, and intellectual development in some infants following sudden separation from a loving mother or mother figure.

anaclitic object choice [PSYCH] The choosing of a love object resembling the person upon whom the individual was emotionally dependent during infancy.

anaconda [VERT ZOO] *Eunectes murinus.* The largest living snake, an arboreal-aquatic member of the boa family (Boidae).

anacrogynous [BOT] In certain liverworts, having female reproductive bodies that do not arise at or near the apex of the shoot.

anacromyoidian [VERT ZOO] In birds, having the syringeal muscles attached to the dorsal end of bronchial semirings.

Anactinochitinosi [INV ZOO] A group name for three closely related suborders of mites and ticks: Onychopalpida, Mesostigmata, and Ixodides.

anadromous [VERT ZOO] Said of a fish, such as the salmon and shad, that ascends fresh-water streams from the sea to spawn.

Anadyomenaceae [BOT] A family of green marine algae in the order Siphonocladales characterized by the expanded blades of the thallus.

anaerobe [BIOL] An organism that does not require air or free oxygen to maintain its life processes.

anaerobic bacteria [MICROBIO] Any bacteria that can survive in the partial or complete absence of air; two types are facultative and obligate.

anaerobic glycolysis [BIOCHEM] A metabolic pathway in plants by which, in the absence of oxygen, hexose is broken down to lactic acid and ethanol with some adenosinetriphosphate synthesis.

anaerobic petri dish [MICROBIO] A glass laboratory dish for plate cultures of anaerobic bacteria; a thioglycollate agar medium and restricted air space give proper conditions.

anaerobiosis [BIOL] A mode of life carried on in the absence of molecular oxygen.

anakinesis [BIOCHEM] A process in living organisms by which energy-rich molecules, such as adenosinetriphosphate, are formed.

anal [ANAT] Relating to or located near the anus.

analbuminemia [MED] A disorder transmitted as an autosomal recessive, characterized by drastic reduction or absence of serum albumin.

anal character [PSYCH] A type of personality in which anal erotic traits dominate beyond the period of childhood.

analeptic [PHARM] Any drug used to restore respiration and a wakeful state.

anal fin [VERT ZOO] An unpaired fin located medially on the posterior ventral part of the fish body.

analgesia [PHYSIO] Insensibility to pain with no loss of consciousness.

analgesic [PHARM] Any drug, such as salicylates, morphine, or opiates, used primarily for the relief of pain.

anal gland [INV ZOO] A gland in certain mollusks that se-

ANADYOMENACEAE

Anadyomene, a genus in Anadyomenaceae, with expanded blades.

ANAEROBIC PETRI DISH

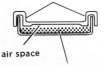

Brewer anaerobic petri dish. *(Courtesy BioQuest, Division of Becton, Dickinson and Co.)*

cretes a purple substance. [VERT ZOO] A gland located near the anus or opening into the rectum in many vertebrates.

analogous [BIOL] Referring to structures that are similar in function and general appearance but not in origin, such as the wing of an insect and the wing of a bird.

anal plate [EMBRYO] An embryonic plate formed of endoderm and ectoderm through which the anus later ruptures. [VERT ZOO] **1.** One of the plates on the posterior portion of the plastron in turtles. **2.** A large scale anterior to the anus of most snakes.

anal sphincter [ANAT] Either of two muscles, one voluntary and the other involuntary, controlling closing of the anus in vertebrates.

anal stage [PSYCH] The stage, in psychoanalytic theory, during which a child's interest is focused on anal activities.

analytical chemistry [CHEM] The branch of chemistry dealing with techniques which yield any type of information about chemical systems.

analytic psychology [PSYCH] The school of psychology that regards the libido not as an expression of the sex instinct, but of the will to live; the unconscious mind is thought to express certain archaic memories of race. Also known as Jungian psychology.

anamestic [VERT ZOO] Pertaining to small bones of varying shapes that fill spaces between larger bones that are more fixed in their position; these occur in fish skulls.

anamnestic response [IMMUNOL] A rapidly increased antibody level following renewed contact with a specific antigen, even after several years. Also known as booster response.

Anamnia [VERT ZOO] Vertebrate animals which lack an amnion in development, including Agnatha, Chondrichthyes, Osteichthyes, and Amphibia.

Anamniota [VERT ZOO] The equivalent name for Anamnia.

anamniote [VERT ZOO] Any animal in which the amnion is absent during embryonic development.

anamorphism *See* anamorphosis.

anamorphosis [EVOL] Gradual increase in complexity of form and function during evolution of a group of animals or plants. Also known as anamorphism. [OPTICS] The production of a distorted image by an optical system.

Anancinae [PALEON] A subfamily of extinct proboscidean placental mammals in the family Gomphotheriidae.

anandrous [BOT] Lacking stamens.

anantherous [BOT] Lacking anthers.

ananthous [BOT] **1.** Not producing flowers. **2.** Lacking an inflorescence.

anaphase [CYTOL] **1.** The stage in mitosis and in the second meiotic division when the centromere splits and the chromatids separate and move to opposite poles. **2.** The stage of the first meiotic division when the two halves of a bivalent chromosome separate and move to opposite poles.

anaphylactic shock [MED] A syndrome seen as one of the clinical manifestations of anaphylaxis.

anaphylactoid reaction [MED] A nonallergic reaction resembling anaphylaxis and depending on the toxicity of the inductant.

anaphylaxis [MED] Hypersensitivity following parenteral injection of an antigen; local or systemic allergenic reaction occurs when the antigen is reintroduced after a time lapse.

anaphylotoxin [IMMUNOL] The vasodilator principle, a toxic substance released by tissues of sensitized animals when antigen and antibody react.

anaphysis [BOT] A filamentous outgrowth resembling a sterigma in the apothecium of certain lichens.

ANASPIDA

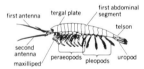

Pharyngolepis oblongus Kiaer of the Anaspida, reconstruction. *(From A. Ritchie)*

ANASPIDIDAE

A male *Anaspides tasmaniae* Thompson about 5 centimeters long, found in freshwater habitats of Tasmania. *(From R. E. Snodgrass, A Textbook of Arthropod Anatomy, Comstock, 1952)*

anaphyte [BOT] **1.** A transverse shoot segment. **2.** An internode.

anaplasia [MED] Reversion of cells to an embryonic, immature, or undifferentiated state; degree usually corresponds to malignancy of a tumor.

Anaplasmataceae [MICROBIO] A family of small bacteria in the order Rickettsiales, parasitic in red blood cells of domestic cloven-hoofed animals.

Anaplotheriidae [PALEON] A family of extinct tylopod ruminants in the superfamily Anaplotherioidea.

Anaplotherioidea [PALEON] A superfamily of extinct ruminant artiodactyls in the infraorder Tylopoda.

anapophysis [ANAT] An accessory process on the dorsal side of the transverse process of the lumbar vertebrae in man and other mammals.

anapsid [VERT ZOO] Having an imperforate skull or one that is completely roofed over.

Anapsida [VERT ZOO] A subclass of reptiles characterized by a roofed temporal region in which there are no temporal openings.

anaptychus [PALEON] An aptychus or an operculum that consists of a single plate; found in certain ammonites.

Anasca [PALEON] A suborder of extinct bryozoans in the order Cheilostomata.

Anaspida [PALEON] An order of extinct fresh- or brackish-water vertebrates in the class Agnatha.

Anaspidacea [INV ZOO] An order of the crustacean superorder Syncarida.

Anaspididae [INV ZOO] A family of crustaceans in the order Anaspidacea.

anastomosis [SCI TECH] The union or intercommunication of branched systems in either two or three dimensions. Also known as inosculation.

Anatidae [VERT ZOO] A family of waterfowl, including ducks, geese, mergansers, pochards, and swans, in the order Anseriformes.

anatomical dead space *See* dead space.

anatomy [BIOL] A branch of morphology dealing with the structure of animals and plants.

anatropous [BOT] Having the ovule fully inverted so that the micropyle adjoins the funiculus.

anaxiai [BIOL] Lacking an axis, therefore being irregular in form.

ancestroecium [INV ZOO] The tube that encloses an ancestrula.

ancestrula [INV ZOO] The first polyp of a bryozoan colony.

anchovy [VERT ZOO] Any member of the Engraulidae, a family of herringlike fishes harvested commercially for human consumption.

anchylosis *See* ankylosis.

ancipital [BOT] Having two edges, specifically referring to flattened stems, as of certain grasses.

anconeal [ANAT] Pertaining to the elbow.

anconeus [ANAT] The small extensor muscle that covers the elbow.

ancylopoda [PALEON] A suborder of extinct herbivorous mammals in the order Perissodactyla.

Ancylostoma [INV ZOO] A genus of roundworms, commonly known as hookworms, in the order Ancylostomidae; parasites of man, dogs, and cats.

Ancylostomidae [INV ZOO] A family of nematodes belonging to the group Strongyloidea.

Andreaeales [BOT] The single order of mosses of the subclass Andreaeobrya.

Andreaeceae [BOT] The single family of the Andreaeales, an order of mosses.

Andreaeobrya [BOT] The granite mosses, a subclass of the class Bryopsida.

Andrenidae [INV ZOO] The mining or burrower bees, a family of hymenopteran insects in the superfamily Apoidea.

andric [BIOL] Male.

androconium [INV ZOO] One of the modified wing scales that produce a sexual attractant in certain male butterflies.

androcyte [BOT] A cell that gives rise to an antherozoid.

androdioecious [BOT] Having the male and hermaphroditic reproductive organs on different individuals.

androecium [BOT] The aggregate of stamens in a flower.

androgen [BIOCHEM] A class of steroid hormones produced in the testis and adrenal cortex which act to regulate masculine secondary sexual characteristics.

androgenesis [EMBRYO] Development of an embryo from a fertilized irradiated egg, involving only the male nucleus.

androgenetic merogony [EMBRYO] The fertilization of egg fragments that lack a nucleus.

androgenic gland [INV ZOO] Any of the accessory glands associated with the sperm duct in male crustaceans and required for differentiation of a functional male.

androgenous [BIOL] Generating only male offspring.

androgynary [BOT] Having flowers with petals developed from stamens and pistils.

androgynophore [BOT] A common stalk which bears both the stamens and the carpels.

androgyny [MED] A form of pseudohermaphroditism in humans in which the individual has female external sexual characteristics, but has undescended testes. Also known as male pseudohermaphroditism.

android pelvis *See* masculine pelvis.

andromerogony [EMBRYO] Development of an egg fragment following cutting, shaking, or centrifugation of a fertilized or unfertilized egg.

andromonoecious [BOT] Having the male and hermaphroditic reproductive organs on the same individual.

andropetalous [BOT] Having stamens that resemble flower petals.

androphore [BOT] A stalk that supports stamens or antheridia. [INV ZOO] A gonophore in coelenterates in which only male elements develop.

androsin [BIOCHEM] $C_{15}H_{20}O_8$ A glucoside found in the herb *Apocynum androsaemifolium;* yields glucose and acetovanillone on hydrolysis.

androstane [BIOCHEM] $C_{19}H_{32}$ The parent steroid hydrocarbon for all androgen hormones. Also known as etioallocholane.

androstenedione [BIOCHEM] $C_{19}H_{26}O_2$ Any one of three isomeric androgens produced by the adrenal cortex.

androsterone [BIOCHEM] $C_{19}H_{30}O_2$ An androgenic hormone occurring as a hydroxy ketone in the urine of men and women.

androtype [SYST] Holotype of the male of a species.

anelectrotonus [PHYSIO] A positive electric potential generated by the passage of a current on the surface of a nerve or muscle in the region of the anode.

Anelytropsidae [VERT ZOO] A family of lizards represented by a single Mexican species.

anemia [MED] A condition marked by significant decreases in hemoglobin concentration and in the number of circulating red blood cells. Also known as oligochromemia.

anemic necrosis [MED] Tissue death following a critical decrease in blood flow or oxygen levels.

ANDREAECEAE

Andreaea petrophila. (From H. E. Jaques, Plant Families, How to Know Them, 2d ed., Brown, 1949)

anemochory [ECOL] Wind dispersal of plant and animal disseminules.

anemophilous [BOT] Pollinated by wind-carried pollen.

anemoplankton [BIOL] Organisms that are carried by the wind.

anemotaxis [BIOL] Orientation movement of a free-living organism in response to wind.

anemotropism [BIOL] Orientation response of a sessile organism to air currents and wind.

anencephalia [MED] A congenital malformation in which all or most of the brain and flat skull bones are absent.

anenterous [ZOO] Having no intestine, as a tapeworm.

aner [INV ZOO] A male ant.

anergy [IMMUNOL] The condition of exhibiting no response to an antigen or antibody. [MED] The condition of exhibiting a lack of energy.

anesthesia [PHYSIO] **1.** Insensibility, general or local, induced by anesthetic agents. **2.** Loss of sensation, of neurogenic or psychogenic origin.

anesthesiology [MED] A branch of medicine dealing with the administration of anesthetics.

anesthetic [PHARM] A drug, such as ether, that produces loss of sensibility.

anestrus [VERT ZOO] A prolonged period of inactivity between two periods of heat in cyclically breeding female mammals.

aneuploidy [GEN] Deviation from a normal haploid, diploid, or polyploid chromosome complement by the presence in excess of, or in defect of, one or more individual chromosomes.

aneurine *See* thiamine.

aneuronic [PHYSIO] Pertaining to chromatophores that lack innervation and are controlled by hormones.

aneurysm [MED] Localized abnormal dilation of an artery due to weakening of the vessel wall.

angelica [FOOD ENG] **1.** A spice from the perennial herb *Angelica archangelica* of the ginger family. **2.** An amber or a yellow sweet wine without muscat flavor.

angiectasis [MED] Abnormal blood vessel dilation.

angina [MED] **1.** A sore throat. **2.** Any tense, constricting pain.

angina pectoris [MED] Constricting chest pain which may be accompanied by pain radiating down the arms, up into the jaw, or to other sites.

angioblast [EMBRYO] A mesenchyme cell derived from extraembryonic endoderm that differentiates into embryonic blood cells and endothelium.

angiocardiography [MED] Roentgenographic visualization of the heart chambers and thoracic vessels following injection of a radiopaque material.

angiocarpic [MYCOL] **1.** Having the spores enclosed. **2.** Having a fruiting body which is closed until spore maturation.

Angiococcus [MICROBIO] A genus of bacteria in the family Myxococcaceae characterized by a fruiting body consisting of numerous round, thin-walled cysts.

angiogenesis [EMBRYO] The origin and development of blood vessels.

angiogram [MED] An x-ray photograph of blood vessels following injection of a radiopaque material.

angiography [MED] Roentgenographic visualization of blood vessels following injection of a radiopaque material.

angiology [MED] The branch of medicine concerned with the blood vessels and the lymphatic system.

angioma [MED] A tumor composed of lymphatic vessels or blood.

angioneurotic edema [MED] Acute, localized accumulations of tissue fluid causing swellings around the face; condition may be due either to heredity or a food allergy.

angiosarcoma [MED] A malignant soft-tissue tumor arising from vascular elements.

angioscotoma [MED] A visual-field disturbance caused by dilated blood-vessels in the retina.

angiosperm [BOT] The common name for members of the plant division Magnoliophyta.

Angiospermae [BOT] An equivalent name for Magnoliophyta.

angiostomatous [ZOO] Having a narrow mouth, referring to certain mollusks and to some snakes.

angiotensin [BIOCHEM] A decapeptide hormone that influences blood vessel constriction and aldosterone secretion by the adrenal cortex. Also known as hypertensin.

anglerfish [VERT ZOO] Any of several species of the order Lophiiformes characterized by remnants of a dorsal fin seen as a few rays on top of the head that are modified to bear a terminal bulb.

Angoumian [GEOL] Upper middle Upper Cretaceous (Upper Turonian) geologic time.

angstrom [MECH] A unit of length, 10^{-10} meter, used primarily to express wavelengths of optical spectra. Abbreviated A. Also known as tenthmeter.

Anguidae [VERT ZOO] A family of limbless, snakelike lizards in the suborder Sauria, commonly known as slowworms or glass snakes.

Anguilliformes [VERT ZOO] A large order of actinopterygian fishes containing the true eels.

Anguilloidei [VERT ZOO] The typical eels, a suborder of actinopterygian fishes in the order Anguilliformes.

angular leaf spot [PL PATH] A bacterial disease of plants characterized by angular spots on leaves and caused by *Pseudomonas lachrymans* in cucumbers and *Xanthomonas malvacearum* in cotton.

angular momentum [MECH] **1.** The cross product of a vector from a specified reference point to a particle, with the particle's linear momentum. Also known as moment of momentum. **2.** For a system of particles, the vector sum of the angular momenta (first definition) of the particles.

angulosplenial [VERT ZOO] In Amphibia, the bone that forms most of the lower and inner portion of the mandible.

angulus [ANAT] The angle formed by the junction of the manubrium and the body of the sternum.

angustifoliate [BOT] Having narrow leaves.

angustirostrate [VERT ZOO] Having a narrow beak or snout.

anhalonium alkaloid [PHARM] Any of the alkaloids found in mescal buttons, including hordenine and mescaline; used as a cerebral stimulant and motor depressant. Also known as cactus alkaloid.

anhidrosis [MED] Absent or deficient secretion of sweat.

Anhimidae [VERT ZOO] The screamers, a family of birds in the order Anseriformes characterized by stout bills, webbed feet, and spurred wings.

Anhingidae [VERT ZOO] The anhingas or snakebirds, a family of swimming birds in the order Pelecaniformes.

anhydrase [BIOCHEM] Any enzyme that catalyzes the removal of water from a substrate.

anhydremia [MED] A decreased amount of water in the plasma.

anhydride [CHEM] A compound formed from another by removal of water.

Aniliidae [VERT ZOO] A small family of nonvenomous, burrowing snakes in the order Squamata.

ANGLERFISH

Cryptopsaras couesi, the anglerfish, length up to 18 inches (45 centimeters). *(After G. B. Goode and T. H. Bean, Oceanic Ichthyology, U.S. Nat. Mus. Spec. Bull. no. 2, 1895)*

aniline [ORG CHEM] $C_6H_5NH_2$ An aromatic amine compound that is a pale brown liquid at room temperature; used in the dye, pharmaceutical, and rubber industries.

animal [ZOO] Any living organism distinguished from plants by the lack of chlorophyll, the requirement for complex organic nutrients, the lack of a cell wall, limited growth, mobility, and greater irritability.

animal communication [PSYCH] The discipline within the field of animal behavior that deals with the receipt and use of signals by animals.

animal community [ECOL] An aggregation of animal species held together in a continuous or discontinuous geographic area by ties to the same physical environment, mainly vegetation.

animal ecology [ECOL] A study of the relationships of animals to their environment.

animal fiber [TEXT] A natural textile fiber of animal origin; wool and silk are the most important.

animal husbandry [AGR] A branch of agriculture concerned with the breeding and feeding of domestic animals.

Animalia [SYST] The animal kingdom.

animal kingdom [SYST] One of the two generally accepted major divisions of living organisms which live or have lived on earth (the other division being the plant kingdom).

animal pole [CYTOL] The region of an ovum which contains the least yolk and where the nucleus gives off polar bodies during meiosis.

animal virus [VIROL] A small infectious agent able to propagate only within living animal cells.

Animikean [GEOL] The middle subdivision of Proterozoic geologic time. Also known as Penokean; Upper Huronian.

anion [CHEM] An ion that is negatively charged.

Anisakidae [INV ZOO] A family of parasitic roundworms in the superfamily Ascaridoidea.

anise [BOT] The small fruit of the annual herb *Pimpinella anisum* in the family Umbelliferae; fruit is used for food flavoring, and oil is used in medicines, soaps, and cosmetics.

Anisean [GEOL] Lower Middle Triassic geologic time.

anisocarpous [BOT] Referring to a flower whose number of carpels is different from the number of stamens, petals, and sepals.

anisocercal [VERT ZOO] Having unequal lobes on the tail fin.

anisochela [INV ZOO] A chelate sponge spicule with dissimilar ends.

anisocytosis [MED] A condition in which the erythrocytes show a considerable variation in size due to excessive quantities of hemoglobin.

anisodactylous [VERT ZOO] Having unequal digits, especially referring to birds with three toes forward and one backward.

anisogamete *See* heterogamete.

anisogamy *See* heterogamy.

anisognathous [VERT ZOO] **1.** Having jaws of unequal width. **2.** Having different types of teeth in the upper and lower jaws.

anisomerous [BOT] Referring to flowers that do not have the same number of parts in each whorl.

anisometric particle [VIROL] Any unsymmetrical, rod-shaped plant virus.

Anisomyaria [INV ZOO] An order of mollusks in the class Bivalvia containing the oysters, scallops, and mussels.

anisophyllous [BOT] Having leaves of two or more shapes and sizes.

Anisoptera [INV ZOO] The true dragonflies, a suborder of insects in the order Odonata.

anisostemonous [BOT] Referring to a flower whose number

ANISOPTERA

pterostigma

An adult dragonfly, showing the thickened spot, pterostigma, on the costal margin of the wing.

of stamens is different from the number of carpels, petals, and sepals.

Anisotomidae [INV ZOO] An equivalent name for Leiodidae.

anisotropy [BOT] The property of a plant that assumes a certain position in response to an external stimulus. [ZOO] The property of an egg that has a definite axis or axes.

ankle [ANAT] The joint formed by the articulation of the leg bones with the talus, one of the tarsal bones.

ankle breadth [ANTHRO] The distance measured between projections at lower ends of the tibia and fibula.

ankle thickness [ANTHRO] Distance measured perpendicular to ankle breadth.

Ankylosauria [PALEON] A suborder of Cretaceous dinosaurs in the reptilian order Ornithischia characterized by short legs and flattened, heavily armored bodies.

ankylosis Also spelled anchylosis. [MED] Stiffness or immobilization of a joint due to a surgical or pathologic process.

ankyroid [BOT] Having the shape of a hook.

anlage [EMBRYO] Any group of embryonic cells when first identifiable as a future organ or body part. Also known as blastema; primordium.

Annedidae [VERT ZOO] A small family of limbless, snakelike, burrowing lizards of the suborder Sauria.

Annelida [INV ZOO] A diverse phylum comprising the multisegmented wormlike animals.

Anniellidae [VERT ZOO] A family of limbless, snakelike lizards in the order Squamata.

Annonaceae [BOT] A large family of woody flowering plants in the order Magnoliales, characterized by hypogynous flowers, exstipulate leaves, a trimerous perianth, and distinct stamens with a short, thick filament.

annual plant [BOT] A plant that completes its growth in one growing season and therefore must be planted annually.

annual ring [BOT] A line appearing on tree cross sections marking the end of a growing season and showing the volume of wood added during the year.

annular [BIOL] Ringlike.

annular budding [BOT] Budding by replacement of a ring of bark on a stock with a ring bearing a bud from a selected species or variety.

annular hernia *See* umbilical hernia.

annulate [BIOL] **1.** Having the shape of a ring. **2.** Composed of annular segments. **3.** Having color arranged in annular bands.

annulus [ANAT] Any ringlike anatomical part. [BOT] **1.** An elastic ring of cells between the operculum and the mouth of the capsule in mosses. **2.** A line of cells, partly or entirely surrounding the sporangium in ferns, which constricts, thus causing rupture of the sporangium to release spores. **3.** A whorl resembling a calyx at the base of the strobilus in certain horsetails. [MYCOL] A ring of tissue representing the remnant of the veil around the stipe of some agarics.

Anobiidae [INV ZOO] The deathwatch beetles, a family of coleopteran insects of the superfamily Bostrichoidea.

anococcygeal [ANAT] Pertaining to the region between coccyx and anus.

anode [PHYS CHEM] The positive terminal of an electrolytic cell.

Anomalinacea [INV ZOO] A superfamily of marine and benthic sarcodinian protozoans in the order Foraminiferida.

anomalous [SCI TECH] Deviating from the normal; irregular.

anomalous color defect [PHYSIO] A condition in which a person is able to discriminate vivid colors but is color-blind when colors are poorly saturated.

Anomaluridae [VERT ZOO] The African flying squirrels, a

ANKYLOSAURIA

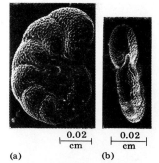

Restoration of the armored Cretaceous dinosaur *Ankylosaurus* (about 20 feet or 6 meters long).

ANOMALINACEA

0.02		0.02
cm		cm
(a)		(b)

Scanning electron micrograph of the foraminiferan *Holmanella*, from the Miocene of California. (*a*) Spiral view and (*b*) edge view of bievolute planispiral test, with a coarsely perforate granular margin, and a slitlike aperture extending up the terminal face. (*R. B. MacAdam, Chevron Oil Field Research Co.*)

small family in the order Rodentia characterized by the climbing organ, a series of scales at the root of the tail.

anomaly [BIOL] An abnormal deviation from the characteristic form of a group. [MED] Any part of the body that is abnormal in position, form, or structure.

Anomocoela [VERT ZOO] A suborder of toadlike amphibians in the order Anura characterized by a lack of free ribs.

anomophyllous [BOT] Having irregularly positioned leaves.

Anomphalacea [PALEON] A superfamily of extinct gastropod mollusks in the order Aspidobranchia.

Anomura [VERT ZOO] A section of the crustacean order Decapoda that includes lobsterlike and crablike forms.

anoopsia [MED] Strabismus in which the eye is turned upward.

Anopheles [INV ZOO] A genus of mosquitoes in the family Culicidae; members are vectors of malaria, dengue, and filariasis.

Anopla [INV ZOO] A class or subclass of the phylum Rhynchocoela characterized by a simple tubular proboscis and by having the mouth opening posterior to the brain.

Anoplocephalidae [INV ZOO] A family of tapeworms in the order Cyclophyllidea.

Anoplura [INV ZOO] The sucking lice, a small group of mammalian ectoparasites usually considered to constitute an order in the class Insecta.

anorexia [MED] Loss of appetite.

anorthopia [MED] A defect of vision in which straight lines do not seem straight, and parallelism or symmetry is not properly perceived.

anoscope [MED] An instrument for examining the lower rectum and anal canal.

anosmia [MED] Absence of the sense of smell.

Anostraca [INV ZOO] An order of shrimplike crustaceans generally referred to the subclass Branchiopoda.

anoxia [MED] The failure of oxygen to gain access to, or to be utilized by, the body tissues.

Anser [VERT ZOO] A genus of birds in the family Anatidae comprising the typical geese.

Anseranatini [VERT ZOO] A subfamily of aquatic birds in the family Anatidae represented by a single species, the magpie goose.

Anseriformes [VERT ZOO] An order of birds, including ducks, geese, swans, and screamers, characterized by a broad, flat bill and webbed feet.

ant [INV ZOO] The common name for insects in the hymenopteran family Formicidae; all are social, and colonies exhibit a highly complex organization.

antacid [CHEM] Any substance that counteracts or neutralizes acidity.

antagonism [BIOL] **1.** Mutual opposition as seen between organisms, muscles, physiologic actions, and drugs. **2.** Opposing action between drugs and disease or drugs and functions.

antarctic faunal region [ECOL] A zoogeographic region describing both the marine littoral and terrestrial animal communities on and around Antarctica.

anteater [VERT ZOO] Any of several mammals, in five orders, which live on a diet of ants and termites.

antebrachium *See* forearm.

antecosta [INV ZOO] The internal, anterior ridge of the tergum or sternum of many insects that provides a surface for attachment of the longitudinal muscles.

antecubital [ANAT] Pertaining to the region anterior to the elbow.

ANT

The female black ant
(Monororium minimum),
about ¹/₁₆ inch (1.6 millimeters).

antedorsal [VERT ZOO] Located anterior to the dorsal fin in fishes.

anteflexion [MED] Forward bending of an organ on itself.

antefurca [INV ZOO] In insects, a forked process on the anterior segment of the thorax.

antelabrum [INV ZOO] In insects, the anterior portion of the labrum when differentiated from the other portions.

antelope [VERT ZOO] Any of the hollow-horned, hoofed ruminants assigned to the artiodactyl subfamily Antilopinae; confined to Africa and Asia.

antemarginal [BOT] Referring to sori which lie within the margin of a fern frond.

antemortem [MED] Before death.

antenatal [MED] Occurring or existing before birth.

antenna [INV ZOO] Any one of the paired, segmented, and movable sensory appendages on the heads of many arthropods.

antennal gland [INV ZOO] An excretory organ in the cephalon of adult crustaceans and best developed in the Malacostraca. Also known as green gland.

Antennata [INV ZOO] An equivalent name for the Mandibulata.

antennifer [INV ZOO] **1.** In arthropods, the antennal socket. **2.** In myriapods, a projection on the rim of the antennal socket.

antennule [INV ZOO] **1.** A small antenna. **2.** In Crustacea, the first pair of antennae.

antepartum [MED] Pertaining to the period before delivery or birth.

anter [INV ZOO] Part of a bryozoan operculum which serves to close off a portion of the operculum.

anterior [ZOO] Situated near or toward the front or head of an animal body.

anterior arm reach [ANTHRO] The distance measured from the wall to the tip of the right middle finger when the subject makes the maximum forward reach with both arms, standing with heels, buttocks, middle of back (in lateral sense), and occiput against the wall.

anterograde amnesia [MED] Loss of memory for the period subsequent to a sudden trauma or a seizure.

antesternite [INV ZOO] In insects, the anterior sclerite of the sternum.

anthelminthic [PHARM] A chemical substance used to destroy tapeworms in domestic animals.

anther [BOT] The pollen-producing structure of a flower.

antheraxanthin [BIOCHEM] A neutral yellow plant pigment unique to the Euglenophyta.

antheridiophore [BOT] A specialized stemlike structure that supports an antheridium in some mosses and liverworts.

antheridium [BOT] **1.** The sex organ that produces male gametes in cryptogams. **2.** A minute structure within the pollen grain of seed plants.

antherophore [BOT] A filament that bears several anthers.

antherozoid [BOT] A motile male gamete produced by plants.

anther smut [MYCOL] *Ustilago violacea*. A smut fungus that attacks certain plants and forms spores in the anthers.

anthesis [BOT] The flowering period in plants.

Anthicidae [INV ZOO] The antlike flower beetles, a family of coleopteran insects in the superfamily Tenebrionoidea.

anthoblast [INV ZOO] A developmental stage of some corals; produced by budding.

Anthocerotae [BOT] A small class of the plant division Bryophyta, commonly known as hornworts or horned liverworts.

ANTELOPE

The black buck or Indian antelope (*Antilope cervicapra*). The male is black with long, spiral horns, while the female is beige and hornless.

anthocodium [INV ZOO] The free, oral end of an anthozoan polyp.

Anthocoridae [INV ZOO] The flower bugs, a family of hemipteran insects in the superfamily Cimicimorpha.

anthocyanidin [BIOCHEM] Any of the colored aglycone plant pigments obtained by hydrolysis of anthocyanins.

anthocyanin [BIOCHEM] Any of the intensely colored, sap-soluble glycoside plant pigments responsible for most scarlet, purple, mauve, and blue coloring in higher plants.

Anthocyathea [PALEON] A class of extinct marine organisms in the phylum Archaeocyatha characterized by skeletal tissue in the central cavity.

anthocyathus [INV ZOO] The oral portion of the anthocaulus of some solitary corals that becomes a new zooid.

Anthomedusae [INV ZOO] A suborder of hydrozoan coelenterates in the order Hydroida characterized by athecate polyps.

Anthomyzidae [INV ZOO] A family of cyclorrhaphous myodarian dipteran insects belonging to the subsection Acalypteratae.

anthophagous [ZOO] Feeding on flowers.

anthophilous [BIOL] **1.** Attracted by flowers. **2.** Feeding on flowers.

anthophore [BOT] A stalklike extension of the receptacle bearing the pistil and corolla in certain plants.

Anthosomidae [INV ZOO] A family of fish ectoparasites in the crustacean suborder Caligoida.

anthostele [INV ZOO] A thick-walled, nonretractile aboral region of certain coelenterates.

anthotaxis [BOT] The floral arrangement on an axis.

anthoxanthin [BIOCHEM] Any of a group of flavonoid pigments ranging in color from yellow to orange-red; they differ chemically from anthocyanins principally in having a more oxidized heterocyclic ring.

Anthozoa [INV ZOO] A class of marine organisms in the phylum Coelenterata including the soft, horny, stony, and black corals, the sea pens, and the sea anemones.

anthozooid [INV ZOO] Any of the individual zooids of a compound anthozoan.

anthracnose [PL PATH] A fungus disease of plants caused by members of the Melanconiales and characterized by dark or black limited stem lesions.

Anthracosauria [PALEON] An order of Carboniferous and Permian labyrinthodont amphibians that includes the ancestors of living reptiles.

anthracosilicosis [MED] Chronic lung inflammation caused by inhalation of carbon and silicon particles.

anthracosis [MED] The accumulation of inhaled black coal dust particles in the lung accompanied by chronic inflammation. Also known as blacklung.

Anthracotheriidae [PALEON] A family of middle Eocene and early Pleistocene artiodactyl mammals in the superfamily Anthracotherioidea.

Anthracotherioidea [PALEON] A superfamily of extinct artiodactyl mammals in the suborder Paleodonta.

anthraquinone [ORG CHEM] $C_6H_4(CO)_2C_6H_4$ Yellow crystalline diketone that is insoluble in water; used in the manufacture of dyes. Also known as dihydrodiketoanthracene.

anthraquinone pigments [BIOCHEM] Coloring materials which occur in plants, fungi, lichens, and insects; consists of about 50 derivatives of the parent compound, anthraquinone.

anthrax [VET MED] An acute, infectious bacterial disease of sheep and cattle caused by *Bacillus anthracis;* transmissible to man. Also known as splenic fever; wool-sorter's disease.

ANTHRAQUINONE

Structural formula of anthraquinone.

Anthribidae [INV ZOO] The fungus weevils, a family of coleopteran insects in the superfamily Curculionoidea.

anthropocentric [PSYCH] **1.** Regarding man as the most important factor in the universe. **2.** Evaluating all occurrences
solely by human values.

anthropochory [ECOL] Dispersal of plant and animal disseminules by man.

anthropogenesis [BIOL] Development of humans, ontoge-
netically and phylogenetically.

anthropography [ANTHRO] A branch of anthropology that
deals with the geographic distribution of divisions of humans
based on physical character, language, customs, and institutions.

anthropoid [VERT ZOO] Pertaining to or resembling the
Anthropoidea.

Anthropoidea [VERT ZOO] A suborder of mammals in the
order Primates including New and Old World monkeys.

anthropology [BIOL] The study of the interrelations of biological, cultural, geographical, and historical aspects of man.

anthropometry [ANTHRO] Description of the physical variation in man by measurement; a basic technique of physical
anthropology.

anthroposcopy [ANTHRO] The description of physical variation in man by visual inspection; a basic technique of
physical anthropology.

anthroposphere [ECOL] A term sometimes used to refer to
the biosphere of the great geological activities of man. Also
known as noosphere.

Anthuridea [INV ZOO] A suborder of crustaceans in the order
Isopoda characterized by slender, elongate, subcylindrical
bodies, and by the fact that the outer branch of the paired tail
appendage (uropod) arches over the base of the terminal
abdominal segment, the telson.

antiae [VERT ZOO] In certain birds, feathers at the base of the
bill ridge.

antiagglutinin [IMMUNOL] A substance that neutralizes a
corresponding agglutinin.

antiaggressin [IMMUNOL] An antibody that neutralizes ag-
gressin, a substance produced by pathogenic microorganisms
to enhance virulence.

Antiarchi [PALEON] A division of highly specialized placoderms restricted to fresh-water sediments of the Middle and
Upper Devonian.

antibacterial agent [MICROBIO] A synthetic or natural compound which inhibits the growth and division of bacteria.

antibiosis [ECOL] Antagonistic association between two organisms in which one is adversely affected.

antibiotic [MICROBIO] A chemical substance, produced by
microorganisms and synthetically, that has the capacity in
dilute solutions to inhibit the growth of, and even to destroy,
bacteria and other microorganisms.

antibiotic assay [MICROBIO] A method for quantitatively determining the concentration of an antibiotic by its effect in
inhibiting the growth of a susceptible microorganism.

antibody [IMMUNOL] A protein, found principally in blood
serum, originating either normally or in response to an
antigen and characterized by a specific reactivity with its
complementary antigen. Also known as immune body.

antibody-deficiency syndrome [MED] Any of the human defects of antibody production, such as hypogammaglobulinemia, agammaglobulinemia, and dysgammaglobulinemia, usually associated with reduced serum concentrations of
immunoglobulins.

antiboreal faunal region [ECOL] A zoogeographic region in-

ANTHURIDEA

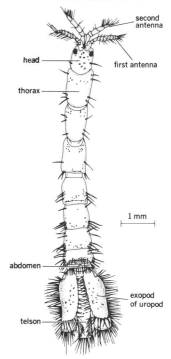

A drawing of *Paranthura
infundibulata* Richardson,
showing some of the
characteristics of the
Anthuridea.

ANTIARCHI

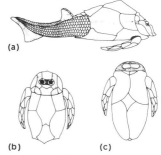

Pterichthyodes (Pterichthys),
a Middle Devonian antiarch.
(a) Lateral view showing scale-
covered tail; about 15 centimeters
in length. *(b)* Dorsal view and
(c) ventral view of the armor.
(After Traquair)

cluding marine littoral faunal communities at the southern end of South America.

anticarcinogen [PHARM] Any substance which is antagonistic to the action of a carcinogen.

anticholinesterase [BIOCHEM] Any agent, such as a nerve gas, that inhibits the action of cholinesterase and thereby destroys or interferes with nerve conduction.

anticoagulant [PHARM] An agent, such as sodium citrate, that prevents coagulation of a colloid, especially blood.

anticodon [GEN] A three-nucleotide sequence of transfer RNA that complements the codon in messenger RNA.

anticonvulsant [PHARM] An agent, such as Dilantin, that prevents or arrests a convulsion.

anticryptic [ECOL] Pertaining to protective coloration that makes an animal resemble its surroundings so that it is inconspicuous to its prey.

antidepressant [PHARM] A drug, such as imipramine and tranylcypromine, that relieves depression by increasing central sympathetic activity.

antidiabetic [PHARM] An agent, such as insulin, that is effective in controlling diabetes.

antidiarrheal [PHARM] An agent, such as Kaopectate, that prevents or arrests diarrhea.

antidiuretic [PHARM] An agent, such as vasopressin, that prevents the excretion of urine.

antidiuretic hormone *See* vasopressin.

antidote [PHARM] An agent that relieves or counteracts the action of a poison.

antienzyme [BIOCHEM] An agent that selectively inhibits the action of an enzyme.

antifertilizin [BIOCHEM] An immunologically specific substance produced by animal sperm to implement attraction by the egg before fertilization.

antifibrinolysin [BIOCHEM] Any substance that inhibits the proteolytic action of fibrinolysin.

antigen [IMMUNOL] A substance which reacts with the products of specific humoral or cellular immunity, even those induced by related heterologous immunogens.

antigen-antibody reaction [IMMUNOL] The combination of an antigen with its antibody.

antihelix [ANAT] The curved ridge of the pinna anterior to, and following most of the course of, the helix.

antihemophilic factor [BIOCHEM] A soluble protein clotting factor in mammalian blood. Also known as factor VIII; thromboplastinogen.

antihemorrhagic vitamin *See* vitamin K.

antihistamine [PHARM] A drug that prevents or diminishes the effect of histamine; used in treating allergic reactions and common-cold symptoms.

antihypertensive agent [PHARM] A substance, such as reserpine, that reduces hypertension.

anti-infective vitamin *See* vitamin A.

anti-inflammatory agent [PHARM] A substance, such as cortisone, that counteracts inflammation.

Antilocapridae [VERT ZOO] A family of artiodactyl mammals in the superfamily Bovoidea; the pronghorn is the single living species.

Antilopinae [VERT ZOO] The antelopes, a subfamily of artiodactyl mammals in the family Bovidae.

antilymphocyte serum [IMMUNOL] An immunosuppressive agent effective in prolonging the lives of homografts in experimental animals by reducing the circulating lymphocytes.

antimalarial [PHARM] **1.** A drug, such as quinacrine, that prevents or suppresses malaria. **2.** Acting against malaria.

antimere [INV ZOO] Any one of the equivalent parts into which a radially symmetrical animal may be divided.

antimetabolite [PHARM] A substance, such as sulfanilamide or amethopterin, that inhibits utilization of an essential metabolite because it is an analog of the metabolite.

antimicrobial agent [MICROBIO] A chemical compound that either destroys or inhibits the growth of microscopic and submicroscopic organisms.

antimitotic drug [PHARM] A substance, such as colchicine, vincristine, or vinblastine, that interferes with mitotic cellular division; used in the chemotherapy of leukemia.

antimony [CHEM] A chemical element, symbol Sb, atomic number 51, atomic weight 121.75.

antimycin A [MICROBIO] $C_{28}H_{40}O_9N_2$ An antifungal antibiotic produced by *Streptomyces kitazawaensis;* the principal fraction of antimycin.

antineoplastic drug [PHARM] An agent, such as mercaptopurine compounds, that is antagonistic to the growth of a neoplasm.

antioxidant [CHEM] An inhibitor, such as ascorbic acid, effective in preventing oxidation by molecular oxygen.

antiparasitic agent [PHARM] An agent, such as emetine or quinine, that destroys or suppresses human and animal parasites.

Antipatharia [INV ZOO] The black or horny corals, an order of tropical and subtropical coelenterates in the subclass Zoantharia.

antipetalous [BOT] Having stamens positioned opposite to, rather than alternating with, the petals.

antipodal [BOT] Any of three cells grouped at the base of the embryo sac, that is, at the end farthest from the micropyle, in most angiosperms.

antipruritic [PHARM] An agent, such as camphor, that relieves itching.

antipygidial [INV ZOO] In fleas, pertaining to bristles on the seventh abdominal segment which extend to the pygidium.

antipyretic [PHARM] Any agent, such as aspirin, that reduces or prevents fever.

antirachitic vitamin *See* vitamin D.

anti-Rh agglutinin [IMMUNOL] An antibody against any Rh antigen; it must be acquired and is never natural.

anti-Rh immunoglobulin [IMMUNOL] A serum protein that destroys Rh-positive fetal erythrocytes in an Rh-negative mother when administered after delivery.

anti-Rh serum [IMMUNOL] A blood serum containing anti-Rh antibodies.

antirostrum [INV ZOO] The terminal segment of the appendages of certain mites.

antisepalous [BOT] Having the stamens located opposite the sepals.

antiseptic [PHARM] A substance used to destroy or prevent the growth of infectious microorganisms on or in the body.

antiserum [IMMUNOL] Any immune serum that contains antibodies active chiefly in destroying a specific infecting virus or bacterium.

antismallpox vaccine *See* smallpox vaccine.

antispadix [INV ZOO] A group of the four modified tentacles opposite the spadix in internal lateral lobes of mollusks of the genus *Nautilus.*

antispasmodic [PHARM] An agent, such as benzyl benzoate, that relieves convulsions and the pain of muscular spasms.

antistreptolysin [IMMUNOL] The antibody that neutralizes the streptolysin of group A hemolytic streptococci.

antithetic variable [STAT] One of two random variables having high negative correlation, used in the antithetic variate

ANTIMONY

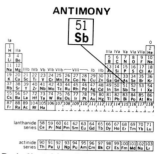

Periodic table of the chemical elements showing the position of antimony.

ANTIPATHARIA

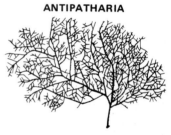

Irregularly branching plantlike colony, *Antipathes rhipidion.* (After F. Pax)

method of estimating the mean of a series of observations.

antithrombin [BIOCHEM] A substance in blood plasma that inactivates thrombin.

antitoxin [IMMUNOL] An antibody elaborated by the body in response to a bacterial toxin that will combine with and generally neutralize the toxin.

antitrochanter [VERT ZOO] In birds, the surface of the ilium which articulates with the femoral trochanter.

antitropous [BOT] Pertaining to embryos that are inverted, that is, with the radicle directed away from hilum.

antitubercular agent [PHARM] A substance, such as streptomycin or isoniazid, used in the treatment of tuberculosis.

antitumor antibiotic [MICROBIO] A substance, such as actinomycin, luteomycin, or mitomycin C, which is produced by microorganisms and is effective against some forms of cancer.

antitussive [PHARM] An agent, such as benylin expectorants, that relieves coughing.

antitype [SYST] A specimen of the same type as the holotype and collected at the same time and place.

antivenin [IMMUNOL] An immune serum that neutralizes the venoms of certain poisonous snakes and black widow spiders.

antivernalization [BOT] Delayed flowering in plants due to treatment with heat.

antiviral agent [PHARM] A substance, such as interferon or amantadine, that decreases virus multiplication in the body.

antivitamin [BIOCHEM] Any substance that prevents a vitamin from normal metabolic functioning.

antixerophthalmic vitamin *See* vitamin A.

antler [VERT ZOO] One of a pair of solid bony, usually branched outgrowths on the head of members of the deer family (Cervidae); shed annually.

antlia [INV ZOO] In Lepidoptera, the spiral suctorial proboscis.

ant lion [INV ZOO] The common name for insects of the family Myrmeleontidae in the order Neuroptera; larvae are commonly called doodlebugs.

antrorse [BIOL] Turned or directed forward or upward.

antrum [ANAT] A cavity of a hollow organ or a sinus.

Anura [VERT ZOO] An order of the class Amphibia comprising the frogs and toads.

anuresis [MED] Retention of urine in the urinary bladder due to inability to void.

anuria [MED] Complete absence of urinary output.

anurous [VERT ZOO] Lacking a tail.

anus [ANAT] The posterior orifice of the alimentary canal.

anvil *See* incus.

anxiety [PSYCH] A physiological and mental state of apprehension and fear of something unknown to the conscious.

anxiety neurosis [PSYCH] A psychoneurotic disorder characterized by diffuse anxious expectation not restricted to definite situations, persons, or objects, and by emotional instability, irritability, apprehensiveness, and a sense of fatigue; caused by incomplete resolution of repressed emotional problems, and frequently associated with somatic symptoms.

aorta [ANAT] The main vessel of systemic arterial circulation arising from the heart in vertebrates. [INV ZOO] The large dorsal or anterior vessel in many invertebrates.

aortic aneurysm [MED] Dilation of the wall of the aorta, usually the ascending portion.

aortic arch [ANAT] The portion of the aorta extending from the heart to the third thoracic vertebra; single in warm-blooded vertebrates and paired in fishes, amphibians, and reptiles.

aortic body *See* aortic paraganglion.

aortic paraganglion [ANAT] A structure in vertebrates be-

ANT LION

The ant lion
(Myrmeleon immaculatus).

AORTIC ARCH

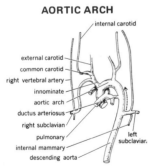

Ventral view of the aortic arch in the human, showing the blood vessels that arise from it. *(Modified from B. M. Patten)*

longing to the chromaffin system and found on the front of the abdominal aorta near the mesenteric arteries. Also known as aortic body; organs of Zuckerkandl.

aortic stenosis [MED] Abnormal narrowing of the aortic valve orifice; may be either congenital or acquired.

aortitis [MED] Inflammation of the aorta.

apandrous [BOT] Lacking male organs or having nonfunctional male organs.

Apatemyidae [PALEON] A family of extinct rodentlike insectivorous mammals belonging to the Proteutheria.

apatetic [ECOL] Pertaining to the imitative protective coloration of an animal subject to being preyed upon.

Apathornithidae [PALEON] A family of Cretaceous birds, with two species, belonging to the order Ichthyornithiformes.

ape [VERT ZOO] Any of the tailless primates of the families Hylobatidae and Pongidae in the same superfamily as man.

aperturate [BIOL] Having one or more apertures. [BOT] Referring to pollen grains having one or more thin spots in the wall or gaps in the outer layer of a spore wall.

aperture aberration [OPTICS] Errors in optical imaging which occur because rays of different distances from the axis do not come to the same focus.

apetalous [BOT] Lacking petals.

apex [ANAT] 1. The upper portion of a lung extending into the root. 2. The pointed end of the heart. 3. The tip of the root of a tooth. [BOT] The pointed tip of a leaf.

apex impulse [PHYSIO] The point of maximum outward movement of the left ventricle of the heart during systole, normally localized in the fifth left intercostal space in the midclavicular line. Also known as left ventricular thrust.

aphagia [MED] Inability to swallow; may be organic or psychic in origin.

aphakia [MED] Absence of the lens of the eye.

Aphanomyces [MYCOL] A genus of fungi in the phycomycetous order Saprolegniales; species cause root rot in plants.

aphasia [MED] Impairment in the use or comprehension of language caused by lesions of the cerebral cortex.

aphasic seizure [MED] A transient inability to speak due to an abnormal electrical discharge from the speech areas of the brain.

Aphasmidea [INV ZOO] An equivalent name for the Adenophorea.

Aphelenchoidea [INV ZOO] A superfamily of plant and insect-associated nematodes in the order Tylenchida.

Aphelocheiridae [INV ZOO] A family of hemipteran insects belonging to the superfamily Naucoroidea.

aphid [INV ZOO] The common name applied to the soft-bodied insects of the family Aphididae; they are phytophagous plant pests and vectors for plant viruses and fungal parasites.

Aphididae [INV ZOO] The true aphids, a family of homopteran insects in the superfamily Aphidoidea.

Aphidoidea [INV ZOO] A superfamily of sternorrhynchan insects in the order Homoptera.

aphlebia [BOT] A lateral outgrowth from the base of a frond stalk in certain ferns.

aphodus [INV ZOO] In certain Porifera, the tube that leads from the flagellate chamber to the excurrent canal.

aphonia [MED] Loss of voice and power of speech.

Aphredoderidae [VERT ZOO] A family of actinopterygian fishes in the order Percopsiformes containing one species, the pirate perch.

aphrodisiac [PHYSIO] Any chemical agent or odor that stimulates sexual desires.

APHID

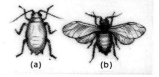

(a)   (b)

Two forms during the life cycle of an aphid: *(a)* wingless female, *(b)* winged female.

APHRODITIDAE

The sea mouse, *Aphrodita*, of the Aphroditidae.

APLANOSPORE

aplanospores

sporangium

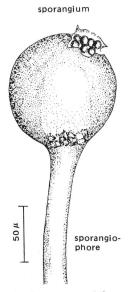

sporangiophore

Several aplanospores and the sporangium that contains them.

Aphroditidae [INV ZOO] A family of scale-bearing polychaete worms belonging to the Errantia.

Aphrosalpingoidea [PALEON] A group of middle Paleozoic invertebrates classified with the calcareous sponges.

aphtha [MED] A white, painful oral ulcer of unknown cause.

Aphylidae [INV ZOO] An Australian family of hemipteran insects composed of two species; not placed in any higher toxonomic group.

aphyllous [BOT] Lacking foliage leaves.

apiary [AGR] A place where bees are kept, especially for breeding and honey making.

apical [BOT] Relating to the apex or tip.

apical bud *See* terminal bud.

apical dominance [BOT] Inhibition of lateral bud growth by the apical bud of a shoot, believed to be a response to auxins produced by the apical bud.

apical meristem [BOT] A region of embryonic tissue occurring at the tips of roots and stems.

apical plate [INV ZOO] A group of cells at the anterior end of certain trochophore larvae; believed to have nervous and sensory functions.

apiculate [BOT] Ending abruptly in a short, sharp point.

apiculture [AGR] Large-scale commercial beekeeping.

Apidae [INV ZOO] A family of hymenopteran insects in the superfamily Apoidea including the honeybees, bumblebees, and carpenter bees.

Apioceridae [INV ZOO] A family of orthorrhaphous dipteran insects in the series Brachycera.

Apis [INV ZOO] A genus of bees, the type genus of the Apidae.

Apistobranchidae [INV ZOO] A family of spioniform annelid worms belonging to the Sedentaria.

Aplacophora [INV ZOO] A subclass of vermiform mollusks in the class Amphineura characterized by no shell and calcareous integumentary spicules.

aplanetic [BIOL] Pertaining to nonmotile spores.

aplanogamete [BIOL] A gamete that lacks motility.

aplanosporangium [MYCOL] A sporangium that produces aplanospores.

aplanospore [MYCOL] A nonmotile, asexual spore, usually a sporangiospore, common in the Phycomycetes.

aplasia [MED] Defective development resulting in the virtual absence of a tissue or organ; only a remnant appears.

aplastic anemia [MED] A blood disorder in which lymphocytes predominate while there is a deficiency of erythrocytes, hemoglobin, and granulocytes.

aplerotic [BOT] Indicating a condition in which oospores do not fill the oogonium.

aploperistomatous [BOT] Pertaining to a peristome with one row of teeth.

aplostemonous [BOT] Having only one row of stamens.

apnea [MED] A transient cessation of respiration.

Apneumonomorphae [INV ZOO] A suborder of arachnid arthropods in the order Araneida characterized by the lack of book lungs.

apneusis [PHYSIO] In certain lower vertebrates, sustained tonic contraction of the respiratory muscles to allow prolonged inspiration.

apobasidium [MYCOL] A basidium characterized by sterigmata having terminal spores.

apocarpous [BOT] Having carpels separate from each other.

apocrine gland [PHYSIO] A multicellular gland, such as a mammary gland or an axillary sweat gland, that extrudes part of the cytoplasm with the secretory product.

Apocynaceae [BOT] A family of tropical and subtropical flowering trees, shrubs, and vines in the order Gentianales, characterized by a well-developed latex system, granular pollen, a poorly developed corona, and the carpels often united by the style and stigma; well-known members are oleander and periwinkle.

Apoda [VERT ZOO] The caecilians, a small order of wormlike, legless animals in the class Amphibia.

Apodacea [INV ZOO] A subclass of echinoderms in the class Holothuroidea characterized by simple or pinnate tentacles and reduced or absent tube feet.

apodal [ZOO] **1.** Lacking feet. **2.** Not having a ventral fin.

apodeme [INV ZOO] An internal ridge or process on an arthropod exoskeleton to which organs and muscles attach.

apoderma [INV ZOO] In some Acarina, an enveloping membrane that is secreted during the resting stage between instars.

Apodes [VERT ZOO] An equivalent name for the Anguilliformes.

Apodi [VERT ZOO] The swifts, a suborder of birds in the order Apodiformes.

Apodida [INV ZOO] An order of worm-shaped holothurian echinoderms in the subclass Apodacea.

Apodidae [VERT ZOO] The true swifts, a family of apodiform birds belonging to the suborder Apodi.

Apodiformes [VERT ZOO] An order of birds containing the hummingbirds and swifts.

apoenzyme [BIOCHEM] The protein moiety of an enzyme; determines the specificity of the enzyme reaction.

apogamy [BIOL] Asexual, parthenogenetic development of diploid cells, such as the development of a sporophyte from a gametophyte without fertilization.

apogeny [BOT] Loss of the function of reproduction.

apogeotropism [BOT] Negative geotropism; growth up or away from the soil.

Apogonidae [VERT ZOO] The cardinal fishes, a family of tropical marine fishes in the order Perciformes; males incubate eggs in the mouth.

Apoidea [INV ZOO] The bees, a superfamily of hymenopteran insects in the suborder Apocrita.

apomeiosis [CYTOL] Meiosis that is either suppressed or imperfect.

apomixis [EMBRYO] Parthenogenetic development of sex cells without fertilization.

apomorph [SYST] Any derived character occurring at a branching point and carried through one descending group in a phyletic lineage.

apomorphine [PHARM] $C_{17}H_{17}NO_2$ A crystalline alkaloid obtained by dehydration of morphine; acts as a powerful emetic.

aponeurosis [ANAT] A broad sheet of regularly arranged connective tissue that covers a muscle or serves to connect a flat muscle to a bone.

apopetalous [BOT] Having free petals.

apophyllous [BOT] Having the parts of the perianth distinct.

apophysis [ANAT] An outgrowth or process on an organ or bone. [MYCOL] A swollen filament in fungi.

apoplastid [CYTOL] A plastid that lacks chromatophores.

apoplexy [MED] **1.** A symptom complex caused by an acute vascular lesion of the brain and characterized by unconsciousness with various degrees of paralysis and sensory impairment. **2.** Sudden, severe hemorrhage into any organ.

apopyle [INV ZOO] Any one of the large pores in a sponge by which water leaves a flagellated chamber to enter the exhalant system.

Aporidea [INV ZOO] An order of tapeworms of uncertain composition and affinities; parasites of anseriform birds.

APODIDA

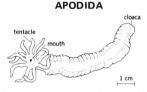

Typical appearance of an apodous holothurian.

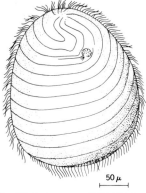

50 μ

Foettingeria, an example of an apostomatid.

aporogamy [BOT] Entry of the pollen tube into the embryo sac through an opening other than the micropyle.

aporrhysa [INV ZOO] Exhalent canals in Porifera.

apospory [MYCOL] Suppression of spore formation with development of the haploid (sexual) generation directly from the diploid (asexual) generation.

apostasis [BOT] Abnormal axial growth which causes separation of the perianth whorls from one another.

Apostomatida [INV ZOO] An order of ciliated protozoans in the subclass Holotrichia; majority are commensals on marine crustaceans.

apotele [ANAT] A scalloped ridge around the edge of an otolith.

apothecaries' measure [PHARM] A system of units of volume, usually of liquid drugs, in which 16 fluid ounces equals 1 pint.

apothecium [MYCOL] A spore-bearing structure in some Ascomycetes and lichens in which the fruiting surface or hymenium is exposed during spore maturation.

apotracheal [BOT] Having parenchyma and vessels independent or dispersed in the xylem.

apotropous [BOT] **1.** Referring to an erect ovule having the raphe in a ventral position. **2.** Referring to a pendulous ovule having the raphe in a dorsal position.

apozymase [BIOCHEM] The protein component of a zymase.

apparent motion [PHYSIO] Perceived motion when there is actual movement of the stimulus pattern over the receptor.

appendage [BIOL] Any subordinate or nonessential structure associated with a major body part. [ZOO] Any jointed, peripheral extension, especially limbs, of arthropod and vertebrate bodies.

appendectomy [MED] Surgical removal of the vermiform appendix.

appendicitis [MED] Inflammation of the vermiform appendix.

appendicular skeleton [ANAT] The bones of the pectoral and pelvic girdles and the paired appendages in vertebrates.

appendiculate [BIOL] Having or forming appendages.

appendiculum [MYCOL] Remnants of the veil on the rim of the pileus.

appendix [ANAT] **1.** Any appendage. **2.** *See* vermiform appendix.

appendix testis [MED] A remnant of the cranial part of the paramesonephric or Müllers duct, attached to the testis. Also known as hydatid of Morgagni.

appersonification [PSYCH] Unconscious identification with and acquisition of another person's characteristics.

appestat [PHYSIO] The center for appetite regulation in the hypothalamus.

applanate [BOT] Flattened or horizontally expanded.

Apple *(Malus domestica);* mature fruit.

apple [BOT] *Malus domestica.* A deciduous tree in the order Rosales which produces an edible, simple, fleshy, pome-type fruit.

apple of Peru *See* jimsonweed.

apple pox *See* blister canker.

applied ecology [ECOL] Activities involved in the management of natural resources.

apposition [BOT] Formation of successive layers during growth of a cell wall.

apposition eye [INV ZOO] A compound eye in diurnal insects and crustaceans in which each ommatidium focuses on a small part of the whole field of light, producing a mosaic image.

appressed [BIOL] Pressed close to or lying flat against something.

appressorium [MYCOL] A modified hyphal tip which may form haustoria or infection hyphae in certain parasitic fungi.

approach-approach conflict [PSYCH] A conflict in which a person is motivated toward two incompatible goals.

approach-avoidance conflict [PSYCH] A conflict in which a person is motivated both toward and away from the same goal.

apraxia [MED] The inability to perform purposeful acts as a result of brain lesions; characteristically, paralysis is absent and kinesthesia is unimpaired.

apricot [BOT] *Prunus armeniaca.* A deciduous tree in the order Rosales which produces a simple fleshy stone fruit.

aproctous [MED] Having an imperforate anus. [ZOO] Lacking an anus.

aproterodont [VERT ZOO] Lacking premaxillary teeth.

Apsidospondyli [VERT ZOO] A term used to include, as a subclass, amphibians in which the vertebral centra are formed from cartilaginous arches.

apterous [BIOL] Lacking wings, as in certain insects, or winglike expansions, as in certain seeds.

Apterygidae [VERT ZOO] The kiwis, a family of nocturnal ratite birds in the order Apterygiformes.

Apterygiformes [VERT ZOO] An order of ratite birds containing three living species, the kiwis, characterized by small eyes, limited eyesight, and nostrils at the tip of the bill.

Apterygota [INV ZOO] A subclass of the Insecta characterized by being primitively wingless.

aptitude [PSYCH] The natural inclination or capacity for skillful performance of an as yet unlearned task.

aptitude test [PSYCH] Any standardized examination used to evaluate a person'a ability to learn a particular skill.

aptyalism [MED] Deficiency or absence of saliva.

aptychus [PALEON] In ammonites, a horny or calcareous plate regarded as an operculum.

Apus [VERT ZOO] A genus of birds comprising the Old World swifts.

apyrase [BIOCHEM] Any enzyme that hydrolyzes adenosine-triphosphate, with liberation of phosphate and energy, and that is believed to be associated with actomyosin activity.

apyrexia [MED] Absence of fever.

aquaculture *See* aquiculture.

aqualung [ENG] A self-contained underwater breathing apparatus (scuba) of the demand or open-circuit type developed by J.Y. Cousteau.

aquatic [BIOL] Living or growing in, on, or near water; having a water habitat.

aqueduct of Sylvius [ANAT] The cavity of the midbrain connecting the third and fourth ventricles.

aqueous humor [PHYSIO] The transparent fluid filling the anterior chamber of the eye.

aquiculture [BIOL] Cultivation of natural faunal resources of water. Also spelled aquaculture.

Aquifoliaceae [BOT] A family of woody flowering plants in the order Celastrales characterized by pendulous ovules, alternate leaves, imbricate petals, and drupaceous fruit; common members include various species of holly (*Ilex*).

aquiprata [ECOL] Communities of plants which are found in areas such as wet meadows where groundwater is a factor.

Ar *See* argon.

Arabellidae [INV ZOO] A family of polychaete worms belonging to the Errantia.

arabinose [BIOCHEM] $C_5H_{10}O_5$ A pentose sugar obtained in crystalline form from plant polysaccharides such as gums, hemicelluloses, and some glycosides.

Araceae [BOT] A family of herbaceous flowering plants in

APRICOT

Apricot *(Prunus armeniaca)* branch showing fruit cluster.

ARABINOSE

The structural formula of β-L-arabinose, obtained from the cherry gum plant.

the order Arales; plants have stems, roots, and leaves, the inflorescence is a spadix, and the growth habit is terrestrial or sometimes more or less aquatic; well-known members include dumb cane (*Dieffenbachia*), jack-in-the-pulpit (*Arisaema*), and *Philodendron*.

Arachnida [INV ZOO] A class of arthropods in the subphylum Chelicerata characterized by four pairs of thoracic appendages.

arachnidium [INV ZOO] The spinning apparatus of a spider, consisting of spinning glands and spinnerets.

arachnodactyly [MED] A rare congenital defect of the skeletal system marked by abnormally long hand and foot bones.

arachnoid [ANAT] A membrane that covers the brain and spinal cord and lies between the pia mater and dura mater. [BOT] Of cobweblike appearance, caused by fine white hairs. Also known as araneose. [INV ZOO] Any invertebrate related to or resembling the Arachnida.

arachnoidal granulations [ANAT] Projections of the arachnoid layer of the cerebral meninges through the dura mater. Also known as arachnoid villi; Pacchionian bodies.

Arachnoidea [INV ZOO] The name used in some classification schemes to describe a class of primitive arthropods.

arachnology [INV ZOO] The study of arachnids.

Aradidae [INV ZOO] The flat bugs, a family of hemipteran insects in the superfamily Aradoidea.

Aradoidea [INV ZOO] A small superfamily of hemipteran insects belonging to the subdivision Geocorisae.

Araeoscelidia [PALEON] A provisional order of extinct reptiles in the subclass Euryapsida.

Arales [BOT] An order of monocotyledonous plants in the subclass Arecidae.

Araliaceae [BOT] A family of dicotyledonous trees and shrubs in the order Umbellales; there are typically five carpels and the fruit, usually a berry, is fleshy or dry; well-known members are ginseng (*Panax*) and English ivy (*Hedera helix*).

Aramidae [VERT ZOO] The limpkins, a family of birds in the order Gruiformes.

Aran-Duchenne atrophy [MED] A muscular system disorder of adults involving progressive spinal muscular atrophy.

Araneae [INV ZOO] An equivalent name for Araneida.

Araneida [INV ZOO] The spiders, an order of arthropods in the class Arachnida.

araneology [INV ZOO] The study of spiders.

araneose *See* arachnoid.

Arbacioida [INV ZOO] An order of echinoderms in the superorder Echinacea.

arboreal Also known as arboreous. [BOT] Relating to or resembling a tree. [ZOO] Living in trees.

arboreous [BOT] **1.** Wooded. **2.** *See* arboreal.

arborescence [BIOL] The state of being treelike in form and appearance.

arboretum [BOT] An area where trees and shrubs are cultivated for educational and scientific purposes.

arboriculture [BOT] The cultivation of ornamental trees and shrubs.

arbor vitae [ANAT] The treelike arrangement of white nerve tissue seen in a median section of the cerebellum.

arborvitae [BOT] Any of the ornamental trees, sometimes called the tree of life, in the genus *Thuja* of the order Pinales.

arboviral encephalitides [MED] Diseases which are caused by arthropod-borne viruses (arboviruses), such as the encephalitis infections.

arbovirus [VIROL] Small, arthropod-borne animal viruses that are unstable at room temperature and inactivated by

ARBORVITAE

Eastern aborvitae
(Thuja occidentalis).

sodium deoxycholate; cause several types of encephalitis. Also known as arthropod-borne virus.

arbutin [ORG CHEM] $C_{12}H_{16}O_7$ A bitter glycoside from the bearberry and certain other plants; sometimes used as a urinary antiseptic.

Arcellinida [INV ZOO] An order of rhizopodous protozoans in the subclass Lobosia characterized by lobopodia and a well-defined aperture in the test.

Archaeoceti [PALEON] The zeuglodonts, a suborder of aquatic Eocene mammals in the order Cetacea; the oldest known cetaceans.

Archaeocidaridae [PALEON] A family of Carboniferous echinoderms in the order Cidaroida characterized by a flexible test and more than two columns of interambulacral plates.

Archaeocopida [PALEON] An order of Cambrian crustaceans in the subclass Ostracoda characterized by only slight calcification of the carapace.

Archaeogastropoda [INV ZOO] An order of gastropod mollusks that includes the most primitive snails.

Archaeopteridales [PALEOBOT] An order of Upper Devonian sporebearing plants in the class Polypodiopsida characterized by woody trunks and simple leaves.

Archaeopterygiformes [PALEON] The single order of the extinct avian subclass Archaeornithes.

Archaeopteryx [PALEON] The earliest known bird; a genus of fossil birds in the order Archaeopterygiformes characterized by flight feathers like those of modern birds.

Archaeornithes [PALEON] A subclass of Upper Jurassic birds comprising the oldest fossil birds.

archaic [PSYCH] Designating elements, largely unconscious, in the psyche which are remnants of man's prehistoric past and which reappear in dreams and other symbolic manifestations.

archallaxis [BIOL] Deviation from an ancestral pattern early in development, eliminating duplication of the phylogenetic history.

Archangiaceae [MICROBIO] A family of slime bacteria in the order Myxobacterales, characterized by fruiting bodies and nonspherical microcysts surrounded by a wall or found in cysts.

archedictyon [INV ZOO] A fine network of veins in the wings of some primitive insects.

archegonium [BOT] The multicellular female sex organ in all plants of the Embryobionta except the Pinophyta and Magnoliophyta.

archencephalon [EMBRYO] The primitive embryonic forebrain from which the forebrain and midbrain develop.

archenteron [EMBRYO] The cavity of the gastrula formed by ingrowth of cells in vertebrate embryos. Also known as gastrocoele; primordial gut.

archeocyte [INV ZOO] A type of ovoid amebocyte in sponges, characterized by large nucleolate nuclei and blunt pseudopodia; gives rise to germ cells.

archeology [SCI TECH] The scientific study of the material remains of the cultures of historical or prehistorical peoples.

Archeozoic [GEOL] **1.** The era during which, or during the latter part of which, the oldest system of rocks was made. **2.** The last of three subdivisions of Archean time, when the lowest forms of life probably existed; as more physical measurements of geologic time are made, the usage is changing; it is now considered part of the Early Precambrian.

archerfish [VERT ZOO] The common name for any member of the fresh-water family Toxotidae in the order Perciformes; individuals eject a stream of water from the mouth to capture insects.

ARCHAEOPTERYX

Archaeopteryx lithographica. (From W. E. Swinton, *Fossil Birds*, British Museum-Natural History, 1958)

ARCHERFISH

The archerfish *(Toxotes jaculator)*, maximum length 7 inches (18 centimeters).

ARCHINEPHROS

archinephric duct

nephrostome

glomerulus

archinephric tubule

Diagram showing hypothetical structure of archinephros. *(From C. K. Weichert, Elements of Chordate Anatomy, 3d ed., McGraw-Hill, 1967)*

archesporium [BOT] **1.** A cell or mass of cells which give rise to archespores in angiosperms. **2.** In liverworts, spore mother cells and elater-forming cells.

archetype [BIOL] An original type or form from which others are derived. [EVOL] The original forerunner of a group of animals or plants.

Archiacanthocephala [INV ZOO] An order of worms in the phylum Acanthocephala; adults are endoparasites of terrestrial vertebrates.

archiamphiaster [CYTOL] The amphiaster which gives rise to the first or second polar body in developing ova.

Archiannelida [INV ZOO] A group name applied to three families of unrelated annelid worms: Nerillidae, Protodrilidae, and Dinophilidae.

archibenthic zone [OCEANOGR] The biogeographic realm of the ocean extending from a depth of about 200 meters to 800–1100 meters (665 feet to 2625–3610 feet).

archiblast [CYTOL] Protoplasm of an ovum.

archiblastula [EMBRYO] Hollow sphere of cells derived by total and equal segmentation of a fertilized ovum.

archicarp [MYCOL] In certain fungi, a spirally coiled portion of the thallus, or of a stalk that bears an oogonium.

archicerebrum [INV ZOO] **1.** The supraesophageal ganglia of more advanced invertebrates. **2.** The primary brain of arthropods.

Archichlamydeae [BOT] An artificial group of flowering plants, in the Englerian system of classification, consisting of those families of dicotyledons that lack petals or have petals separate from each other.

archichlamydeous [BOT] **1.** Lacking petals. **2.** Having the petals separate from one another.

archicoel [ZOO] The segmentation cavity persisting between the ectoderm and endoderm as a body cavity in certain lower forms.

archigastrula [EMBRYO] A gastrula formed by invagination, as opposed to ingrowth of cells.

Archigregarinida [INV ZOO] An order of telosporean protozoans in the subclass Gregarinia; endoparasites of invertebrates and lower chordates.

archinephridium [INV ZOO] One of a pair of primitive nephridia found in each segment of some annelid larvae.

archinephros [VERT ZOO] The paired excretory organ of primitive vertebrates and the larvae of hagfishes and caecilians.

archipallium [PHYSIO] The olfactory pallium or the olfactory cerebral cortex; phylogenetically, the oldest part of the cerebral cortex.

archipterygium [VERT ZOO] A fin with a skeleton consisting of an elongated segmented central axis and two rows of jointed rays.

architomy [INV ZOO] In some annelids, reproduction by fission followed by regeneration.

Archosauria [VERT ZOO] A subclass of reptiles composed of five orders: Thecodontia, Saurischia, Ornithschia, Pterosauria, and Crocodilia.

Archostemata [INV ZOO] A suborder of insects in the order Coleoptera.

arch pattern [ANAT] A fingerprint pattern in which ridges enter on one side of the impression, form a wave or angular upthrust, and flow out the other side.

arcocentrum [ANAT] A centrum formed of modified, fused mesial parts of the neural or hemal arches.

Arctiidae [INV ZOO] The tiger moths, a family of lepidopteran insects in the suborder Heteroneura.

Arctocyonidae [PALEON] A family of extinct carnivorelike mammals in the order Condylarthra.

Arctolepiformes [PALEON] A group of the extinct joint-necked fishes belonging to the Arthrodira.

Arcturidae [INV ZOO] A family of isopod crustaceans in the suborder Valvifera characterized by an almost cylindrical body and extremely long antennae.

arcuale [EMBRYO] Any of the four pairs of primitive cartilages from which a vertebra is formed.

arcuate [ANAT] Arched or curved; bow-shaped.

arculus [INV ZOO] An arc formed by two veins in the wings of certain insects.

arc-welder's disease *See* siderosis.

Arcyzonidae [PALEON] A family of Devonian paleocopan ostracods in the superfamily Kirkbyacea characterized by valves with a large central pit.

Ardeidae [VERT ZOO] The herons, a family of wading birds in the order Ciconiiformes.

ardellae [BOT] Small dustlike apothecia of some lichens.

area amniotica [EMBRYO] The transparent part of the blastodisc in mammals.

area opaca [EMBRYO] The opaque peripheral area of the blastoderm of birds and reptiles, continuous with the yolk.

area pellucida [EMBRYO] The central transparent area of the blastoderm of birds and reptiles, overlying the subgerminal cavity.

area placentalis [EMBRYO] The part of the trophoblast in immediate contact with the uterine mucosa in the embryos of early placental vertebrates.

area vitellina [EMBRYO] The outer nonvascular zone of the area opaca; consists of ectoderm and endoderm.

Arecaceae [BOT] The palms, the single family of the order Arecales.

Arecales [BOT] An order of flowering plants in the subclass Arecidae composed of the palms.

Arecidae [BOT] A subclass of flowering plants in the class Liliopsida characterized by numerous, small flowers subtended by a prominent spathe and often aggregated into a spadix, and broad, petiolate leaves without typical parallel venation.

Arenicolidae [INV ZOO] The lugworms, a family of mud-swallowing worms belonging to the Sedentaria.

arenicolous [ZOO] Living or burrowing in sand.

areola [ANAT] **1.** The portion of the iris bordering the pupil of the eye. **2.** A pigmented ring surrounding a nipple, vesicle, or pustule. **3.** A small space, interval, or pore in a tissue.

areola mammae [ANAT] The circular pigmented area surrounding the nipple of the breast. Also known as areola papillaris; mammary areola.

areola papillaris *See* areola mammae.

Argasidae [INV ZOO] The soft ticks, a family of arachnids in the suborder Ixodides; several species are important as ectoparasites and disease vectors for man and domestic animals.

argentaffin cell [HISTOL] Any of the cells of the gastrointestinal tract that are thought to secrete serotonin.

argentaffin fiber *See* reticular fiber.

argenteum [VERT ZOO] In fishes, a layer of iridocytes in the dermis.

Argentinoidei [VERT ZOO] A family of marine deepwater teleostean fishes, including deep-sea smelts, in the order Salmoniformes.

argentophil [BIOL] Of cells, tissues, or other structures, having an affinity for silver.

Argidae [INV ZOO] A small family of hymenopteran insects in the superfamily Tenthredinoidea.

ARCTURIDAE

Arcturella, a member of the family Arcturidae. *(From G. O. Sars, An Account of the Crustacea of Norway, vol. 2, 1899)*

ARGASIDAE

Argasid tick, *Ornithodoros coriaceus*, enlarged to about 4 times natural size.

ARGON

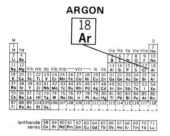

Periodic table of the chemical elements showing the position of argon.

ARMADILLO

Nine-banded armadillo *(Dasypus novemcinctus)*, the only edentate inhabiting the United States.

arginase [BIOCHEM] An enzyme that catalyzes the splitting of urea from the amino acid arginine.

arginine [BIOCHEM] $C_6H_{14}N_4O_2$ A colorless, crystalline, water-soluble, essential amino acid of the α-ketoglutaric acid family.

argon [CHEM] A chemical element, symbol Ar, atomic number 18, atomic weight 39.998.

Argovian [GEOL] Upper Jurassic (lower Lusitanian), a substage of geologic time in Great Britain.

Arguloida [INV ZOO] A group of crustaceans known as the fish lice; taxonomic status is uncertain.

argyria [MED] A dusky-gray or bluish discoloration of the skin and mucous membranes produced by the prolonged administration or application of silver preparations.

argyrophil lattice fiber *See* reticular fiber.

Arhynchobdellae [INV ZOO] An order of annelids in the class Hirudinea characterized by the lack of an eversible proboscis; includes most of the important leech parasites of man and warm-blooded animals.

Arhynchodina [INV ZOO] A suborder of ciliophoran protozoans in the order Thigmotrichida.

arhythmicity [BIOL] A condition characterized by the absence of an expected behavioral or physiologic rhythm.

ariboflavinosis [MED] Dietary deficiency of riboflavin, associated with the syndrome of angular cheilosis and stomatitis, corneal vascularity, nasolabial seborrhea, and genitorectal dermatitis.

arid biogeographic zone [ECOL] Any region of the world that supports relatively little vegetation due to lack of water.

Arid Transition life zone [ECOL] The zone of climate and biotic communities occurring in the chaparrals and steppes from the Rocky Mountain forest margin to California.

Ariidae [VERT ZOO] A family of tropical salt-water catfishes in the order Siluriformes.

Arikareean [GEOL] Lower Miocene geologic time.

aril [BOT] An outgrowth of the funiculus in certain seeds that either remains as an appendage or envelops the seed.

arilode [BOT] An aril originating from tissues in the micropyle region; a false aril.

Arionidae [INV ZOO] A family of mollusks in the order Stylommatophora, including some of the pulmonate slugs.

arista [INV ZOO] The bristlelike or hairlike structure in many organisms, especially at or near the tip of the antenna of many Diptera.

Aristolochiaceae [BOT] The single family of the plant order Aristolochiales.

Aristolochiales [BOT] An order of dicotyledonous plants in the subclass Magnoliidae; species are herbaceous to woody, often climbing, with perigynous to epigynous, apetalous flowers, uniaperturate or nonaperturate pollen, and seeds with a small embryo and copious endosperm.

aristopedia [INV ZOO] Replacement of the arista by a nearly perfect leg.

Aristotle's lantern [INV ZOO] A five-sided feeding and locomotor apparatus surrounding the esophagus of most sea urchins.

arm [ANAT] The upper or superior limb in man, consisting of the upper arm with one bone and the forearm with two bones.

armadillo [VERT ZOO] Any of 21 species of edentate mammals in the family Dasypodidae.

Armilliferidae [INV ZOO] A family of pentastomid arthropods belonging to the suborder Porocephaloidea.

arm length [ANTHRO] The length of the arm measured from top of the clavicle to the tip of the middle finger, with the arm straight down at the side of the body.

armyworm [INV ZOO] Any of the larvae of certain species of noctuid moths composing the family Phalaenidae; economically important pests of corn and other grasses.

Arneth's classification *See* Arneth's index.

Arneth's count *See* Arneth's index.

Arneth's formula *See* Arneth's index.

Arneth's index [HISTOL] A system for dividing peripheral blood granulocytes into five classes according to the number of nuclear lobes, the least mature cells being tabulated on the left, giving rise to the terms "shift to left" and "shift to right" as an indication of granulocytic immaturity or hypermaturity respectively. Also known as Arneth's classification; Arneth's count; Arneth's formula.

Arnold sterilizer [MICROBIO] An apparatus that employs steam under pressure at 212°F (100°C) for fractional sterilization of specialized bacteriological culture media.

Arodoidea [INV ZOO] A superfamily of hemipteran insects belonging to the subdivision Geocorisae.

arolium [INV ZOO] A pad projecting between the tarsal claws of some insects and arachnids.

aromatic [ORG CHEM] Pertaining to or characterized by the presence of at least one benzene ring.

aromatic amino acid [BIOCHEM] An organic acid containing at least one amino group and one or more aromatic groups; for example, phenylalanine, one of the essential amino acids.

aromatic spirit of ammonia [PHARM] A flavored, hydroalcoholic solution of ammonia and ammonium carbonate having an aromatic, pungent odor; used as a reflex stimulant.

aromatic vinegar [PHARM] A flavored solution of acetic acid used as smelling salts.

Arrhenius equation [PHYS CHEM] The relationship that the specific reaction rate constant k equals the frequency factor constant s times $\exp(-\Delta H_{act}/RT)$, where ΔH_{act} is the heat of activation, R the gas constant, and T the absolute temperature.

arrhenoblastoma [MED] A solid, sometimes malignant, tumor of the ovary that usually produces male sex hormones, inducing virilism.

arrhenotoky [BIOL] Production of only male offspring by a parthenogenetic female.

arrhizal [BOT] Lacking true roots; refers to certain parasitic plants.

arrhythmia [MED] Absence of rhythm, especially of heart beat or respiration.

Arridae [VERT ZOO] A family of catfishes in the suborder Siluroidei found from Cape Cod to Panama.

arrowroot [BOT] Any of the tropical American plants belonging to the genus *Maranta* in the family Marantaceae.

arrowworm [INV ZOO] Any member of the phylum Chaetognatha; useful indicator organism for identifying displaced masses of water.

arsenic [CHEM] A chemical element, symbol As, atomic number 33, atomic weight 74.9216.

arsenotherapy [MED] Treatment of disease by means of arsenical drugs.

arsphenamine [PHARM] $C_{12}H_{12}As_2N_2O_2 \cdot 2HCl \cdot 2H_2O$ The antisyphilitic diaminodihydroxyarsenobenzene dihydrochloride, effective also on protozoan infections, first prepared by P. Ehrlich in 1909. Also known as Ehrlich's 606.

Artacaminae [INV ZOO] A subfamily of polychaete annelids in the family Terebellidae of the Sedentaria.

artefact [BIOL] An apparent structure made visible in tissue preparation.

arteriogram [MED] A roentgenogram of an artery after injection with radiopaque material.

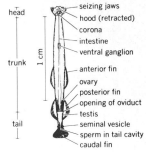

ARROWWORM

Dorsal view of *Sagitta enflata*, an arrowworm.

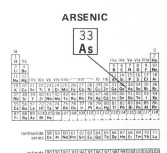

ARSENIC

Periodic table of the chemical elements showing the position of arsenic.

ARTACAMINAE

Dorsal view of *Artacamella*, a genus of the Artacaminae.

arteriography [MED] **1.** Graphic presentation of the pulse. **2.** Roentgenography of the arteries after the intravascular injection of a radiopaque substance.

arteriole [ANAT] An artery of small diameter that terminates in capillaries.

arteriolopathy [MED] Disease of the arterioles.

arteriolosclerosis [MED] Thickening of the lining of arterioles, usually due to hyalinization or fibromuscular hyperplasia.

arteriosclerosis [MED] A degenerative arterial disease marked by hardening and thickening of the vessel walls.

arteriosclerosis obliterans [MED] Hardening of the artery walls with obstruction of the lumen due to proliferation of the innermost vessel layer.

arteriotomy [MED] Incision or opening of an artery.

arteriovenous anastomosis [ANAT] A blood vessel that connects an arteriole directly to a venule without capillary intervention.

arteriovenous aneurysm [MED] **1.** Dilation of the walls of an artery and a vein via an abnormal canal (fistula) between the vessels. **2.** Dilation of an arteriovenous fistula.

arteritis [MED] Inflammation of an artery.

artery [ANAT] A vascular tube that carries blood away from the heart.

arthochromatic erythroblast *See* normoblast.

Arthoniaceae [BOT] A family of lichens in the order Hysteriales.

arthritis [MED] Any inflammatory process affecting joints or their component tissues.

arthritis deformans [MED] A chronic rheumatoid arthritis marked by deformation of affected joints.

arthritis urethritica *See* Reiter's syndrome.

Arthrobacter [MICROBIO] A genus of aerobic, cellulolytic rod-shaped soil bacteria.

arthrodesis [MED] Fusion of a joint by removing the articular surfaces and securing bony union. Also known as operative ankylosis.

arthrodia [ANAT] A diarthrosis permitting only restricted motion between a concave and a convex surface, as in some wrist and ankle articulations. Also known as gliding joint.

Arthrodira [PALEON] The joint-necked fishes, an Upper Silurian and Devonian order of the Placodermi.

Arthrodonteae [BOT] A family of mosses in the subclass Eubrya characterized by thin, membranous peristome teeth composed of cell walls.

arthrogram [MED] A roentgenogram of a joint space after injection of radiopaque material.

arthrography [MED] Roentgenography of a joint space after the injection of radiopaque material.

arthrogryposis [MED] Permanent fixation of a joint in a flexed position.

arthromere [INV ZOO] Somite of an arthropod.

Arthromitaceae [MICROBIO] A family of nonmotile bacteria in the order Caryophanales found in the intestine of millipedes, cockroaches, and toads.

arthropathy [MED] **1.** Any joint disease. **2.** A neurotrophic disorder of a joint, usually due to lack of pain sensation, found in association with tabes dorsalis, leprosy, syringomyelia, diabetic polyneuropathy, and occasionally multiple sclerosis and myelodysplasias.

arthroplasty [MED] **1.** The making of an artificial joint. **2.** Reconstruction of a new and functioning joint from an ankylosed one; a plastic operation upon a joint.

Arthropoda [INV ZOO] The largest phylum in the animal

kingdom; adults typically have a segmented body, a sclerotized integument, and many-jointed segmental limbs.

arthropod-borne virus *See* arbovirus.

arthropodin [BIOCHEM] A water-soluble protein which forms part of the endocuticle of insects.

arthropterous [VERT ZOO] In fishes, having jointed fin rays.

arthrosis [ANAT] An articulation or suture uniting two bones. [MED] Any degenerative joint disease.

arthrospore [BOT] A jointed, vegetative resting spore resulting from filament segmentation in some blue-green algae and hypha segmentation in many Basidiomycetes.

Arthrotardigrada [INV ZOO] A suborder of microscopic invertebrates in the order Heterotardigrada characterized by toelike terminations on the legs.

Arthus reaction [IMMUNOL] An allergic reaction of the immediate hypersensitive type that results from the union of antigen and antibody, with complement present, in blood vessel walls.

artichoke [BOT] *Cynara scolymus.* A herbaceous perennial plant belonging to the order Asterales; the flower head is edible.

articulamentum [INV ZOO] The innermost layer of a calcareous plate in a chiton.

articular [ANAT] Pertaining to or located at or near a joint.

articular cartilage [ANAT] Cartilage that covers the articular surfaces of bones.

articular disk [ANAT] A disk of fibrocartilage, dividing the cavity of certain joints.

articular membrane [INV ZOO] A flexible region of the cuticle between sclerotized areas of the exoskeleton of an arthropod; functions as a joint.

Articulata [INV ZOO] **1.** A class of the Brachiopoda having hinged valves that usually bear teeth. **2.** The only surviving subclass of the echinoderm class Crinoidea.

articulation [ANAT] *See* joint. [BOT] A joint between two parts of a plant that can separate spontaneously. [INV ZOO] A joint between rigid parts of an animal body, as the segments of an appendage in insects. [PHYSIO] The act of enunciating speech.

artifact [ARCHEO] Any man-made object of common use that reflects the skills of man in past cultures. [HISTOL] A structure in a fixed cell or tissue formed by manipulation or by the reagent.

artificial heart [MED] An endoprosthetic device used to replace or assist the heart.

artificial insemination [MED] A process by which spermatozoa are collected from males and deposited in female genitalia by instruments rather than by natural service.

artificial kidney [MED] An apparatus that performs the work of the kidney in purifying blood; used only in cases of renal failure or shutdown.

artificial parthenogenesis [PHYSIO] Activation of an egg by chemical and physical stimuli in the absence of sperm.

artificial respiration [MED] The maintenance of breathing by artificial ventilation, in the absence of normal spontaneous respiration; effective methods include mouth-to-mouth breathing and the use of a respirator.

Artinskian [GEOL] A European stage of geologic time including Lower Permian (above Sakmarian, below Kungurian).

artiodactyl [VERT ZOO] Characterized by an even number of digits.

Artiodactyla [VERT ZOO] An order of terrestrial, herbivorous mammals characterized by having an even number of toes and by having the main limb axes pass between the third and fourth toes.

ARTHROTARDIGRADA

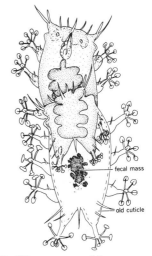

Batillipes, a genus of the Arthrotardigrada; defecation during molt.

ARTICHOKE

Edible artichoke flower heads. The bottom one is shown in cross section.

ASCIDIACEA

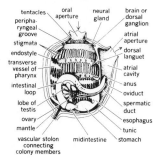

Left side of a zooid of the colonial ascidian *Perophora*. The tunic, mantle, and anterior wall of the pharynx have been removed from the left. The two arrows indicate direction of water intake and expulsion.

ASCOCARP

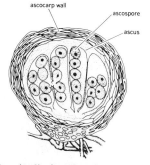

Longitudinal section showing several asci in ascocarp of *Erysiphe aggregata*.

arytenoid [ANAT] Relating to either of the paired, pyramid-shaped, pivoting cartilages on the dorsal aspect of the larynx, in man and most other mammals, to which the vocal cords and arytenoid muscles are attached.

As *See* arsenic.

asbestosis [MED] A chronic lung inflammation caused by inhalation of asbestos dust.

Ascaphidae [VERT ZOO] A family of amphicoelous frogs in the order Anura, represented by four living species.

ascariasis [MED] Any parasitic infection of man and other domestic mammals caused by species of *Ascaris.*

ascarid [INV ZOO] The common name for any roundworm belonging to the superfamily Ascaridoidea.

Ascaridata [INV ZOO] An equivalent name for the Ascaridina.

Ascaridida [INV ZOO] An order of parasitic nematodes in the subclass Phasmidia.

Ascarididae [INV ZOO] A family of parasitic nematodes in the superfamily Ascaridoidea.

Ascaridina [INV ZOO] A suborder of parasitic nematodes in the order Ascaridida.

Ascaridoidea [INV ZOO] A large superfamily of parasitic nematodes of the suborder Ascaridina.

Ascaris [INV ZOO] A genus of roundworms that are intestinal parasites in mammals, including man.

Ascaroidea [INV ZOO] An equivalent name for Ascaridoidea.

ascending aorta [ANAT] The first part of the aorta, extending from its origin in the heart to the aortic arch.

ascending chromatography [ANALY CHEM] A technique for the analysis of mixtures of two or more compounds in which the mobile phase (sample and carrier) rises through the fixed phase.

Ascheim-Zondek test [PATH] A human pregnancy test that uses the reaction of ovaries in immature white mice to an injection of urine from a woman.

Aschelminthes [INV ZOO] A heterogeneous phylum of small to microscopic wormlike animals; individuals are pseudo-coelomate and mostly unsegmented and are covered with a cuticle.

Aschoff body [MED] The lesion of rheumatic fever found around blood vessels in the myocardium.

Ascidiacea [INV ZOO] A large class of the phylum Tunicata; adults are sessile and may be solitary or colonial.

ascidiform [BOT] Pitcher-shaped, as certain leaves.

ascidium [BOT] A pitcher-shaped plant organ or part.

ascigerous [MYCOL] Bearing asci.

ascites [MED] An abnormal accumulation of serous fluid in the abdominal cavity.

Asclepiadaceae [BOT] A family of tropical and subtropical flowering plants in the order Gentianales characterized by a well-developed latex system; milkweed (*Asclepias*) is a well-known member.

ascocarp [MYCOL] The mature fruiting body bearing asci with ascospores in higher Ascomycetes.

ascogonium [MYCOL] The specialized female sexual organ in higher Ascomycetes.

Ascolichenes [BOT] A class of the lichens characterized by the production of asci similar to those produced by Ascomycetes.

ascoma [MYCOL] 1. A disc-shaped ascocarp. 2. A sporocarp with asci.

Ascomycetes [MYCOL] A class of fungi in the subdivision Eumycetes, distinguished by the ascus.

ascon [INV ZOO] A sponge or sponge larva having incurrent canals leading directly to the paragaster.

asconoid [INV ZOO] Saclike as referring to the architecture of a sponge in which the canals are straight pathways.

ascoplasm [MYCOL] The cytoplasm of an ascus undergoing formation of spores.

ascorbic acid [BIOCHEM] $C_6H_8O_6$ A white, crystalline, water-soluble vitamin found in many plant materials, especially citrus fruit. Also known as vitamin C.

ascospore [MYCOL] An asexual spore representing the final product of the sexual process, borne on an ascus in Ascomycetes.

ascostome [MYCOL] The apical pore of an ascus.

Ascothoracica [INV ZOO] An order of marine crustaceans in the subclass Cirripedia occurring as endo- and ectoparasites of echinoderms and coelenterates.

ascus [MYCOL] An oval or tubular spore sac bearing ascospores in members of the Ascomycetes.

Aselloidea [INV ZOO] A group of free-living, fresh-water isopod crustaceans in the suborder Asellota.

Asellota [INV ZOO] A suborder of morphologically and ecologically diverse aquatic crustaceans in the order Isopoda.

asepsis [MED] The state of being free from pathogenic microorganisms.

aseptate [BIOL] Lacking a septum.

asexual [BIOL] 1. Not involving sex. 2. Exhibiting absence of sex or of functional sex organs.

asexual reproduction [BIOL] Formation of new individuals from a single individual without the involvement of gametes.

ash [BOT] 1. A tree of the genus *Fraxinus*, deciduous trees of the olive family (Oleaceae) characterized by opposite, pinnate leaflets. 2. Any of various Australian trees having wood of great toughness and strength; used for tool handles and in work requiring flexibility. [CHEM] The incombustible matter remaining after a substance has been incinerated.

Ashgillian [GEOL] A European stage of geologic time in the Upper Orodovician (above Upper Caradocian, below Llandoverian of Silurian).

Asian flu [MED] An acute viral respiratory infection of man caused by influenza A-2 virus.

Asilidae [INV ZOO] The robber flies, a family of predatory, orthorrhaphous, dipteran insects in the series Brachycera.

asiphonate [INV ZOO] Referring to larvae having respiratory tubes that open directly to the exterior.

asomatognosia [MED] Lacking awareness of paralysis because the brain is damaged.

Asopinae [INV ZOO] A family of hemipteran insects in the superfamily Pentatomoidea including some predators of caterpillars.

asparaginase [BIOCHEM] An enzyme that catalyzes the hydrolysis of asparagine to asparaginic acid and ammonia.

asparagine [BIOCHEM] $C_4H_8N_2O_3$ A white, crystalline amino acid found in many plant seeds.

asparagus [BOT] *Asparagus officinalis.* A dioecious, perennial monocot belonging to the order Liliales; the shoot of the plant is edible.

aspartase [BIOCHEM] A bacterial enzyme that catalyzes the deamination of aspartic acid to fumaric acid and ammonia.

aspartate [BIOCHEM] A compound that is an ester or salt of aspartic acid.

aspartic acid [BIOCHEM] $C_4H_7NO_4$ A nonessential, crystalline dicarboxylic amino acid found in plants and animals, especially in molasses from young sugarcane and sugarbeet.

aspen [BOT] Any of several species of poplars (*Populus*) characterized by weak, flattened leaf stalks which cause the leaves to flutter in the slightest breeze.

ASCORBIC ACID

The structural formula of ascorbic acid.

ASCOTHORACICA

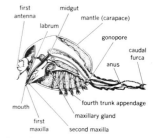

Ascothorax ophioctenis, a parasite in the bursae of brittle stars. (*After Wagin from A Kaestner; from R. D. Barnes, Invertebrate Zoology, 2d ed., Saunders, 1968*)

ASPEN

A leaf scar with axial bud, leaf, and twig of the quaking aspen (*Populus tremuloides*).

asperate [BOT] Having a surface that is somewhat rough to the touch.

Aspergillaceae [MYCOL] Former name for the Eurotiaceae.

Aspergillales [MYCOL] Former name for the Eurotiales.

aspergillic acid [BIOCHEM] $C_{12}H_{20}O_2N_2$ A diketopiperazine-like antifungal antibiotic produced by certain strains of *Aspergillus flavus.*

aspergillin [MYCOL] **1.** A black pigment found in spores of some molds of the genus *Aspergillus.* **2.** A broad-spectrum antibacterial antibiotic produced by the molds *Aspergillus flavus* and *A. fumigatus.*

aspergillosis [MED] A rare fungus infection of man and animals caused by several species of *Aspergillus.*

Aspergillus [MYCOL] A genus of fungi including several species of common molds and some human and plant pathogens.

asperifoliate [BOT] Rough-leaved.

aspermatism [MED] **1.** Failure to ejaculate or secrete semen. **2.** Defective or absent sperm formation.

asperulate [BOT] Delicately roughened.

asphyxia [MED] Suffocation due to oxygen deprivation, resulting in anoxia and carbon dioxide accumulation in the body.

aspiculate [INV ZOO] Lacking spicules, referring to Porifera.

Aspidiotinae [INV ZOO] A subfamily of homopteran insects in the superfamily Coccoidea.

Aspidiphoridae [INV ZOO] An equivalent name for the Sphindidae.

Aspidobothria [INV ZOO] An equivalent name for the Aspidogastrea.

Aspidobranchia [INV ZOO] An equivalent name for the Archaeogastropoda.

Aspidochirotacea [INV ZOO] A subclass of bilaterally symmetrical echinoderms in the class Holothuroidea characterized by tube feet and 10–30 shield-shaped tentacles.

Aspidochirotida [INV ZOO] An order of holothurioid echinoderms in the subclass Aspidochirotacea characterized by respiratory trees and dorsal tube feet converted into tactile warts.

Aspidocotylea [INV ZOO] An equivalent name for the Aspidogastrea.

Aspidodiadematidae [INV ZOO] A small family of deep-sea echinoderms in the order Diadematoida.

Aspidogastrea [INV ZOO] An order of endoparasitic worms in the class Trematoda having strongly developed ventral holdfasts.

Aspidogastridae [INV ZOO] A family of trematode worms in the order Aspidogastrea occurring as endoparasites of mollusks.

Aspidorhynchidae [PALEON] The single family of the Aspidorhynchiformes, an extinct order of holostean fishes.

Aspidorhynchiformes [PALEON] A small, extinct order of specialized holostean fishes.

Aspinothoracida [PALEON] The equivalent name for Brachythoraci.

aspiration [SCI TECH] Act or the result of removing, carrying along, or drawing by suction. [MED] The removal of fluids from a cavity by suction. [MICROBIOL] The use of suction to draw up a sample in a pipette.

aspirin *See* acetylsalicylic acid.

asporogenous [BOT] Not producing spores, especially of certain yeasts.

asporous [BOT] Lacking spores.

Aspredinidae [VERT ZOO] A family of salt-water catfishes in the order Siluriformes found off the coast of South America.

ass [VERT ZOO] Any of several perissodactyl mammals in the

ASPIDORHYNCHIFORMES

Aspidorhynchus acutirostris (Blainville), Upper Jurassic, Bavaria, length to 3 feet (91 centimeters); a typical fish of the order Aspidorhynchiformes. *(After Assmann)*

family Equidae belonging to the genus *Equus*, especially *E. hemionus* and *E. asinus*.

assay [ANALY CHEM] Qualitative or quantitative determination of the components of a material, as an ore or a drug.

assemblage [ARCHEO] All related cultural traits and artifacts associated with one archeological manifestation. [ECOL] A group of organisms sharing a common habitant by chance.

assimilation [PHYSIO] Conversion of nutritive materials into protoplasm.

associate [PSYCH] An item or event that is linked to another in the mind of an individual.

associated automatic movement *See* synkinesia.

association [CHEM] Combination or correlation of substances or functions. [ECOL] Major segment of a biome formed by a climax community, such as an oak-hickory forest of the deciduous forest biome. [PSYCH] A connection formed through learning.

association area [PHYSIO] An area of the cerebral cortex that is thought to link and coordinate activities of the projection areas.

association center [INV ZOO] In invertebrates, a nervous center coordinating and distributing stimuli from sensory receptors.

association fiber [ANAT] One of the white nerve fibers situated just beneath the cortical substance and connecting the adjacent cerebral gyri.

association neuron [ANAT] A neuron, usually within the central nervous system, between sensory and motor neurons.

association test [PSYCH] Any test designed to determine the nature of the mental or emotional link between a stimulus and a response.

associative facilitation [PSYCH] Ease in establishing a new association because of previous associations.

associative inhibition [PSYCH] Difficulty in establishing a new association because of previous associations.

associative learning [PSYCH] The principle that items experienced together are mentally linked so that they tend to reinforce one another.

associative memory [PSYCH] Recalling a previously experienced item by thinking of something that is linked with it, thus invoking the association.

associative thinking [PSYCH] **1.** The mental process of making associations between a given subject and all pertinent present factors without drawing on past experience. **2.** Free association.

assortative mating [GEN] Nonrandom mating with respect to phenotypes.

Astacidae [INV ZOO] A family of fresh-water crayfishes belonging to the section Macrura in the order Decapoda, occurring in the temperate regions of the Northern Hemisphere.

astacin [BIOCHEM] $C_{40}H_{48}O_4$ A red carotenoid ketone pigment found in crustaceans, as in the shell of a boiled lobster.

Astacinae [INV ZOO] A subfamily of crayfishes in the family Astacidae including all North American species west of the Rocky Mountains.

astasia [MED] Lack of muscular coordination in standing.

astatine [CHEM] A radioactive chemical element, symbol At, atomic number 85, the heaviest of the halogen elements.

astaxanthin [BIOCHEM] $C_{40}H_{52}O_4$ A violet carotenoid pigment found in combined form in certain crustacean shells and bird feathers.

Asteidae [INV ZOO] A small, obscure family of cyclorrha-

ASTATINE

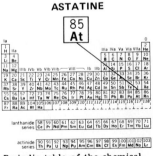

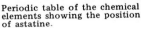

Periodic table of the chemical elements showing the position of astatine.

ASTER

New England aster *(Aster novae-angliae)*, showing ray and disk flowers. *(Courtesy of Alvin E. Staffan, from National Audubon Society)*

ASTOMATIDA

40 μ

Anoplophyra, an example of an astomatid.

phous myodarian dipteran insects in the subsection Acalypteratae.

astelic [BOT] Lacking a stele or having a discontinuous arrangement of vascular bundles.

aster [BOT] Any of the herbaceous ornamental plants of the genus *Aster* belonging to the family Compositae. [CYTOL] The star-shaped structure that encloses the centrosome at the end of the spindle during mitosis.

Asteraceae [BOT] An equivalent name for the Compositae.

Asterales [BOT] An order of dicotyledonous plants in the subclass Asteridae, including aster, sunflower, zinnia, lettuce, artichoke, and dandelion.

astereognosis [MED] Loss of recognition of objects by touch, although recognition occurs through another sense, usually vision. Also known as tactile agnosia.

Asteridae [BOT] A large subclass of dicotyledonous plants in the class Magnoliopsida; plants are sympetalous, with unitegmic, tenuinucellate ovules and with the stamens usually as many as, or fewer than, the corolla lobes and alternate with them.

asterigmate [BOT] Referring to spores that are not born on sterigmata.

Asteriidae [INV ZOO] A large family of echinoderms in the order Forcipulatida, including many predatory sea stars.

Asterinidae [INV ZOO] The starlets, a family of echinoderms in the order Spinulosida.

asteriscus [VERT ZOO] In teleosts, a small otolith in the cochlea.

asternal [ANAT] **1.** Not attached to the sternum. **2.** Without a sternum.

asteroid [INV ZOO] Pertaining to star-shaped starfish.

Asteroidea [INV ZOO] The starfishes, a subclass of echinoderms in the subphylum Asterozoa characterized by five radial arms.

Asteroschematidae [INV ZOO] A family of ophiuroid echinoderms in the order Phrynophiurida with individuals having a small disk and stout arms.

asterospondylous [VERT ZOO] Having radiating calcified cartilage on the centrum.

Asterozoa [INV ZOO] A subphylum of echinoderms characterized by a star-shaped body and radially divergent axes of symmetry.

aster wilt [PL PATH] A fungus disease of asters caused by *Fusarium oxysporum* f. *callistephi.*

aster yellows [PL PATH] A widespread virus disease of asters, other ornamental plants, and many vegetables, characterized by yellowing and dwarfing; leafhoppers are vectors.

asthenia [MED] Loss or lack of strength.

asthenopia [MED] Weakness of the eye muscles or of visual acuity, sometimes accompanied by pain and headache.

asthma [MED] A pulmonary disease marked by labored breathing, wheezing, and coughing; cause may be emotional stress, chemical irritation, or exposure to an allergen.

Astian [GEOL] A European stage of geologic time: upper Pliocene, above Plaisancian, below the Pleistocene stage known as Villafranchian, Calabrian, or Günz.

astichous [BOT] Not arranged in rows.

astigmatism [MED] A defect of vision due to irregular curvatures of the refractive surfaces of the eye so that focal points of light are distorted.

astigmatous [INV ZOO] Lacking stigmata.

astomatal [BOT] Lacking stomata. Also known as astomous.

Astomatida [INV ZOO] An order of mouthless protozoans in the subclass Holotrichia; all species are invertebrate parasites, typically in oligochaete annelids.

astomatous [INV ZOO] Lacking a mouth, especially a cytostome, as in certain ciliates.

astomocnidae nematocyst [INV ZOO] A stinging cell whose thread has a closed end and either is adhesive or acts as a lasso to entangle prey.

astomous [BOT] **1.** Having a capsule that bursts irregularly and is not dehiscent by an operculum. **2.** *See* astomatal.

Astrapotheria [PALEON] A relatively small order of large, extinct South American mammals in the infraclass Eutheria.

Astrapotheroidea [PALEON] A suborder of extinct mammals in the order Astrapotheria, ranging from early Eocene to late Miocene.

astroblast [HISTOL] A cell that gives rise to astrocytes.

astrocyte [HISTOL] A star-shaped cell; specifically, a neuroglial cell.

astroglia [HISTOL] Neuroglia composed of astrocytes.

Astropectinidae [INV ZOO] A family of echinoderms in the suborder Paxillosina occurring in all seas from tidal level downward.

astropyle [INV ZOO] A small, rounded projection from the central capsule of some radiolarians.

astrosclereid [BOT] A sclereid or stone cell of multiradiate form.

astrosphere [CYTOL] The center of the aster exclusive of the rays.

asty [INV ZOO] A bryozoan colony.

asymbolia [MED] An aphasia in which there is an inability to understand or use acquired symbols, such as speech, writing, or gestures, as a means of communication.

asynergia [MED] Faulty coordination of groups of organs or muscles normally acting in unison; particularly, the abnormal state of muscle antagonism in cerebellar disease.

asystole [MED] The absence of cardiac contraction; cardiac arrest.

At *See* astatine.

Atabrine [PHARM] A trade name for quinacrine.

atactostele [BOT] A type of monocotyledonous siphonostele in which the vascular bundles are dispersed irregularly throughout the center of the stem.

atavism [EVOL] Appearance of a distant ancestral form of an organism or one of its parts due to reactivation of ancestral genes.

atelectasis [MED] **1.** Total or partial collapsed state of the lung. **2.** Failure of the lung to expand at birth.

Ateleopoidei [VERT ZOO] A family of oceanic fishes in the order Cetomimiformes characterized by an elongate body, lack of a dorsal fin, and an anal fin continuous with the caudal fin.

Atelopodidae [VERT ZOO] A family of small, brilliantly colored South and Central American frogs in the suborder Procoela.

Atelostomata [INV ZOO] A superorder of echinoderms in the subclass Euechinoidea characterized by a rigid, exocylic test and lacking a lantern, or jaw, apparatus.

Athalamida [INV ZOO] An order of naked amebas of the subclass Granuloreticulosia in which pseudopodia are branched and threadlike (reticulopodia).

Athecanephria [INV ZOO] An order of tube-dwelling, tentaculate animals in the class Pogonophora characterized by a saclike anterior coelom.

Atherinidae [VERT ZOO] The silversides, a family of actinopterygian fishes of the order Atheriniformes.

Atheriniformes [VERT ZOO] An order of actinopterygian fishes in the infraclass Teleostei, including flyingfishes, needlefishes, killifishes, silversides, and allied species.

ASTRAPOTHERIA

Astrapotherium magnum **skeleton. Genus is known from late Oligocene to late Miocene.** *(From Riggs)*

ATHALAMIDA

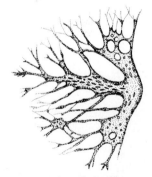

A representative athalamid, *Biomyxa vagans (after Leidy)*. *(From R. P. Hall, Protozoology, Prentice-Hall, 1953)*

ATHYRIDIDINA

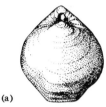

(a)

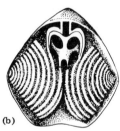

(b)

Composita, a genus in Athyrididina: (*a*) dorsal view, (*b*) ventral view with pedicle cut away to show spire. *(From R. C. Moore, ed., Treatise on Invertebrate Palentology, pt. H, Geological Society of America, Inc., and University of Kansas Press, 1965)*

atheroma [MED] **1.** Fatty degeneration of the inner arterial walls. **2.** A fatty cyst.

atherosclerosis [MED] Deposition of lipid with proliferation of fibrous connective tissue cells in the inner walls of the arteries.

athetosis [MED] Slow, recurrent, involuntary wormlike movements of various parts of the body associated with lesions of the basal ganglia.

athetotic speech [MED] Disorder of articulation rhythm involving a general jerkiness in speech production that interferes with the normal rate of speech; associated with athetosis.

Athiorhodaceae [MICROBIO] The nonsulfur photosynthetic bacteria, a family of small, gram-negative, nonsporeforming, motile bacteria in the suborder Rhodobacteriineae.

athrocyte [HISTOL] A cell that engulfs extraneous material and stores it as granules in the cytoplasm.

Athyrididina [PALEON] A suborder of fossil articulate brachiopods in the order Spiriferida characterized by laterally or, more rarely, ventrally directed spires.

Atlantacea [INV ZOO] A superfamily of mollusks in the subclass Prosobranchia.

atlas [ANAT] The first cervical vertebra.

atm *See* atmosphere.

atmosphere [MECH] A unit of pressure equal to 1.013250×10^6 dynes/cm^2, which is the air pressure at mean sea level. Abbreviated atm. Also known as standard atmosphere.

atmospheric pressure [PHYS] The pressure at any point in an atmosphere due solely to the weight of the atmospheric gases above the point concerned. Also known as barometric pressure.

atom [CHEM] The individual structure which constitutes the basic unit of any chemical element.

atomic fission *See* fission.

atomic mass [PHYS] The mass of a neutral atom usually expressed in atomic mass units.

atomic mass unit [PHYS] An arbitrarily defined unit in terms of which the masses of individual atoms are expressed; the standard is the unit of mass equal to $\frac{1}{12}$ the mass of the carbon atom, having as nucleus the isotope with mass number 12. Abbreviated amu. Also known as dalton.

atomic nucleus *See* nucleus.

atomic number [NUC PHYS] The number of protons in an atomic nucleus.

atomic weight [CHEM] The relative mass of an atom based on a scale in which a specific carbon atom (carbon-12) is assigned a mass value of 12. Abbreviated at. wt.

atony [MED] Absence or extremely low degree of tonus.

atopic allergy [MED] A type of immediate hypersensitivity in humans resulting from spontaneous sensitization, usually by inhaled or ingested antigens; for example, asthma, hay fever, or hives.

atopy [MED] Clinically evident hypersensitivity.

ATP *See* adenosinetriphosphate.

ATPase *See* adenosinetriphosphatase.

Atractidae [INV ZOO] A family of parasitic nematodes in the superfamily Oxyuroidea.

atresia [MED] Imperforation or closure of a natural orifice or passage of the body.

atrial flutter [MED] A cardiac arrhythmia characterized by rapid, irregular atrial impulses and ineffective atrial contractions; the heartbeat varies from 60 to 180 per minute and is grossly irregular in intensity and rhythm. Also known as auricular flutter.

atrial septum [ANAT] The muscular wall between the atria of the heart. Also known as interatrial septum.

atrichia [MED] Any congenital or acquired condition in which hair is essentially absent.

atrichic [BIOL] Lacking flagella.

Atrichornithidae [VERT ZOO] The scrubbirds, a family of suboscine perching birds in the suborder Menurae.

atrichous [CYTOL] Lacking flagella.

atriopore [ZOO] The opening of an atrium as seen in lancelets and tunicates.

atrioventricular canal [EMBRYO] The common passage between the atria and ventricles in the heart of the mammalian embryo before division of the organ into right and left sides.

atrioventricularis communis [MED] A congenital malformation of the heart in which partitioning has not occurred.

atrioventricular valve [ANAT] A structure located at the orifice between the atrium and ventricle which maintains a unidirectional blood flow through the heart.

atrium [ANAT] **1.** The heart chamber that receives blood from the veins. **2.** The main part of the tympanic cavity, below the malleus. **3.** The external chamber to receive water from the gills in lancelets and tunicates.

atrochal [INV ZOO] Referring to a trochophore lacking a preoral circle of cilia but having a uniformly ciliated surface.

atrophic arthritis *See* rheumatoid arthritis.

atrophic gastritis [MED] Chronic inflammation of the stomach with atrophy of the mucosa.

atrophy [MED] Diminution in the size of a cell, tissue, or organ that was once fully developed and of normal size.

atropine [PHARM] $C_{17}H_{23}O_3N$ An alkaloid extracted from *Atropa belladonna* and related plants of the family Solanaceae; used to relieve muscle spasms and pain, and to dilate the pupil of the eye.

atropous *See* orthotropous.

Atrypidina [PALEON] A suborder of fossil articulate brachiopods in the order Spiriferida.

attached X chromosome [CYTOL] A V-shaped chromosome formed by the fusion at or near the centromeres of two rod-shaped X chromosomes or by misdivision of the centromere of one rod-shaped X chromosome.

attention hypothesis [PSYCH] The theory that a person's attention to objects and other persons is selectively drawn toward areas of great personal interest.

attention span [PSYCH] The period of time a person is able to concentrate his attentions on a given item, usually with respect to learning.

attenuated [BIOL] Having reduced density, strength, or pathogenicity.

attenuated vaccine [IMMUNOL] A suspension of weakened bacteria, viruses, or fractions thereof used to produce active immunity.

attenuation [BOT] Tapering, sometimes to a long point. [MICROBIO] Weakening or reduction of the virulence of a microorganism.

at wt *See* atomic weight.

Atyidae [INV ZOO] A family of decapod crustaceans belonging to the section Caridea.

Au *See* gold.

Auberger blood group system [IMMUNOL] An immunologically distinct, genetically determined human erythrocyte antigen, demonstrated by reaction with anti-Au^a (anti-Auberger) antibody.

Auchenorrhyncha [INV ZOO] A group of homopteran families and one superfamily, in which the beak arises at the anteroventral extremity of the face and is not sheathed by the propleura.

audibility curve [ACOUS] **1.** The limits of hearing represented

ATRIOVENTRICULAR CANAL

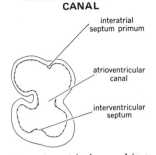

The atrioventricular canal in a 4–5 millimeter human embryo. (*Slightly modified from B. M Patten, Developmental defects at the foramen ovale, Amer. J. Pathol., 14(2):135–161, 1938*)

ATROPINE

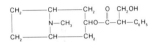

Structural formula of atropine.

ATRYPIDINA

Spiralium of *Zygospira*.

AUDIBILITY CURVE

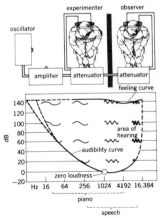

The audibility curve plotted by measuring range of sound intensities in decibels (dB) against the sound frequencies in hertz (Hz).

graphically as an area by plotting the minimum audible intensity of a sine wave sound versus frequency. **2.** *See* equal loudness contour.

audiogenic seizure [MED] A transient episode of muscular, sensory, or psychic dysfunction induced by sound.

audiogram [ACOUS] A graph showing hearing loss, percent hearing loss, or percent hearing as a function of frequency.

audiology [ACOUS] The science of hearing.

audiometer [ENG] An instrument composed of an oscillator, amplifier, and attenuator and used to measure hearing acuity for pure tones, speech, and bone conduction.

audition [PHYSIO] Ability to hear.

auditory [PHYSIO] Pertaining to the act or the organs of hearing.

auditory association area [PHYSIO] The cortical association area in the brain just inferior to the auditory projection area, related to it anatomically and functionally by association fibers.

auditory canal *See* external auditory meatus.

auditory nerve [ANAT] The eighth cranial nerve in vertebrates; either of a pair of sensory nerves composed of two sets of nerve fibers, the cochlear nerve and the vestibular nerve. Also known as acoustic nerve; vestibulocochlear nerve.

auditory placode [EMBRYO] An ectodermal thickening from which the inner ear develops in vertebrates.

Auerbach's plexus *See* myenteric plexus.

aufwuch [ECOL] A plant or animal organism which is attached or clings to surfaces of leaves or stems of rooted plants above the bottom stratum.

auk [VERT ZOO] Any of several large, short-necked diving birds (*Alca*) of the family Alcidae found along North Atlantic coasts.

aulodont dentition [INV ZOO] In echinoderms, grooved teeth with epiphyses that do not meet so that the foramen magnum of the jaw is open.

Aulolepidae [PALEON] A family of marine fossil teleostean fishes in the order Ctenothrissiformes.

aulophyte [ECOL] A nonparasitic plant that lives in the cavity of another plant for shelter.

Auloporidae [PALEON] A family of Paleozoic corals in the order Tabulata.

aulostomatous [VERT ZOO] Having a tube-shaped mouth or snout.

aural [BIOL] Pertaining to the ear or the sense of hearing.

aureofacin [MICROBIO] An antifungal antibiotic produced by a strain of *Streptomyces aureofaciens*.

Aureomycin [MICROBIO] The trade name for the antibiotic chlortetracycline.

aureothricin [MICROBIO] $C_9H_{10}O_2N_2S_2$ An antibacterial antibiotic produced by a strain of *Actinomyces*.

auricle [ANAT] **1.** An ear-shaped appendage to an atrium of the heart. **2.** Any ear-shaped structure. **3.** *See* pinna.

auricular fibrillation [MED] Arrhythmic contractions of the auricles.

auricular flutter *See* atrial flutter.

auricular height [ANTHRO] Height of the cranium, measured from the auditory point to the vertex.

auricularia larva [INV ZOO] A barrel-shaped, food-gathering larval form with a winding ciliated band, common to holothurians and asteroids.

auricularis [ANAT] Any of the three muscles attached to the cartilage of the external ear.

auriculate [BOT] Of leaves, heart-shaped, such as having an expanded base surrounding the stem or having lobes separate from the rest of the blade.

AUDIOGRAM

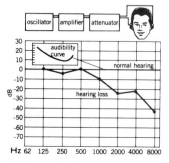

Audiogram for determining the audibility curve for pure-tone hearing loss at various frequency levels.

AULODONT DENTITION

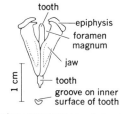

Lantern structure in aulodont dentition; a single interradial jaw is shown and below it a cross section of the tooth.

AURICULARIA LARVA

Auricularia larva, showing the ciliated band.

aurophore [INV ZOO] A bell-shaped structure which is part of the float of certain coelenterates.

aurothioglucose [PHARM] $C_6H_{11}AuO_5S$ A compound of gold with thioglucose, used for the treatment of rheumatoid arthritis and nondisseminated lupus erythematosus; administered in oil suspension.

auscultation [MED] The act of listening to sounds from internal organs, especially the heart and lungs, to aid in diagnosing their physical state.

Australia antigen [IMMUNOL] An infectious agent that causes hepatitis in some people; similar to an inherited serum protein in being polymorphic.

Australian faunal region [ECOL] A zoogeographic region that includes the terrestrial animal communities of Australia and all surrounding islands except those of Asia.

Australian X disease *See* Murray Valley encephalitis.

Australopithecinae [PALEON] The near-men, a subfamily of the family Hominidae composed of the single genus *Australopithecus.*

Australopithecus [PALEON] A genus of near-men in the subfamily Australopithecinae representing a side branch of human evolution.

austral region [ECOL] A North American biogeographic region including the region between transitional and tropical zones.

Austroastacidae [INV ZOO] A family of crayfish in the order Decapoda found in temperate regions of the Southern Hemisphere.

Austrodecidae [INV ZOO] A monogeneric family of marine arthropods in the subphylum Pycnogonida.

Austroriparian life zone [ECOL] The zone in which occurs the climate and biotic communities of the southeastern coniferous forests of North America.

autecology [ECOL] The ecological relations between a plant species and its environment.

authoritarian character [PSYCH] A personality that asks unquestioning subordination and obedience but is intolerant of weakness in others, rejects members of groups other than his own, is very rigid, and requires all issues to be decided in black and white, yet may accept superior authority in a servile way.

autism [PSYCH] A schizophrenic symptom characterized by absorption in fantasy to the exclusion of perceptual reality.

autistic [PSYCH] **1.** Pertaining to or characterized by autism. **2.** Behavior, most commonly in children, characterized by aphonia, disregard for reality, self-manipulation, and sometimes a preoccupation with pointed or shiny objects.

autoagglutination [IMMUNOL] Agglutination of an individual's erythrocytes by his own serum. Also known as autohemagglutination.

autoagglutinin [IMMUNOL] An antibody in an individual's blood serum that causes agglutination of his own erythrocytes.

autoanalysis [PSYCH] Self-analysis, a technique of psychotherapy.

autoantibody [IMMUNOL] An antibody formed by an individual against his own tissues; common in hemolytic anemias.

autoantigen [IMMUNOL] A tissue within the body which acquires the ability to incite the formation of complementary antibodies.

autoasphyxiation [PHYSIO] Asphyxiation by the products of metabolic activity.

autobasidium [MYCOL] An undivided basidium typically found in higher Basidiomycetes.

autocarp [BOT] **1.** A fruit formed as the result of self-fertilization. **2.** A fruit consisting of the ripened pericarp without adnate parts.

autocarpy [BOT] Production of fruit by self-fertilization.

autocatalysis [CYTOL] Dissolution of a cell due to the effect of its own secretion.

autochory [ECOL] Active self-dispersal of individuals or their disseminules.

autochthon [GEOL] A succession of rock beds that have been moved comparatively little from their original site of formation, although they may be folded and faulted extensively. [PALEON] A fossil occurring where the organism once lived.

autochthonous microorganism [MICROBIO] An indigenous form of soil microorganisms, responsible for chemical processes that occur in the soil under normal conditions.

autoclave [ENG] An airtight vessel for heating and sometimes agitating its contents under high steam pressure; used for industrial processing, sterilizing, and cooking with moist or dry heat at high temperatures.

autocopulation [INV ZOO] Self-copulation; sometimes occurs in certain hermaphroditic worms.

autodont [VERT ZOO] Pertaining to teeth that are indirectly attached to the jaws; found in cartilaginous fishes.

autoecious *See* autoicous.

autogamy [BIOL] A process of self-fertilization that results in homozygosis; occurs in some flowering plants, fungi, and protozoans.

autogenesis *See* abiogenesis.

autogenous [SCI TECH] Self-generated; produced without external influence.

autogenous vaccine [IMMUNOL] A vaccine prepared from a culture of microorganisms taken directly from the infected person.

autograft [BIOL] A tissue transplanted from one part to another part of an individual's body.

autohemagglutination *See* autoagglutination.

autohemorrhage [INV ZOO] Voluntary exudation or ejection of nauseous or poisonous blood by certain insects as a defense against predators.

autohemotherapy [MED] Treatment of disease with the patient's own blood, withdrawn by venipuncture and then injected intramuscularly.

autoicous [BOT] Having male and female organs on the same plant but on different branches. Also spelled autoecious.

autoimmune disease [IMMUNOL] An illness involving the formation of autoantibodies which appear to cause pathological damage to the host.

autoimmunity [IMMUNOL] An immune state in which antibodies are formed against the person's own body tissues.

autoinfection [MED] Reinfection by an organism existing within the body or transferred from one part of the body to another.

autoinoculation [MED] **1.** Spread of a disease from one part of the body to another. **2.** Injection of an autovaccine.

autointoxication [MED] Poisoning by metabolic products elaborated within the body; generally, toxemia of pathologic states.

autokinetic effect [PHYSIO] Apparent motion of a small spot of light against a dark background in a room without light.

autolysis [PATH] Self-digestion by body cells following somatic or organ death or ischemic injury.

autolysosome *See* autophagic vacuole.

autolytic enzyme [BIOCHEM] A bacterial enzyme, located in the cell wall, that causes disintegration of the cell following injury or death.

Autolytinae [INV ZOO] A subfamily of errantian polychaetes in the family Syllidae.

automatism [BIOL] Spontaneous activity of tissues or cells. [MED] An act performed with no apparent exercise of will, as in sleepwalking and certain hysterical and epileptic states.

autonomic agent [MED] A compound that reduces or enhances nerve-impulse transmission across synaptic junctions, especially in the autonomic nervous system.

autonomic conditioning [PHYSIO] The conditioning of responses such as salivation, heart rate, or dilation or constriction of blood vessels, which are controlled by the autonomic nervous system.

autonomic nervous system [ANAT] The visceral or involuntary division of the nervous system in vertebrates, which enervates glands, viscera, and smooth, cardiac, and some striated muscles.

autopalatine [VERT ZOO] In some teleosts, a bony structure formed at the anterior end of the pterygoquadrate.

autophagic vacuole [CYTOL] A membrane-bound cellular organelle that engulfs pieces of the substance of the cell itself. Also known as autolysosome.

autophagocytosis [CYTOL] The cellular process of phagocytizing a portion of protoplasm by a vacuole within the cell.

autophagy [CYTOL] The cellular process of self-digestion.

autophyllogeny [BOT] Growth of a leaf on or from another leaf.

autophyte [BOT] A plant that obtains nourishment directly from inorganic matter.

autopolyploid [GEN] A cell or organism having three or more sets of chromosomes all derived from the same species.

autopsy [PATH] A postmortem examination of the body to determine cause of death.

autoradiography [ENG] A technique for detecting radioactivity in a specimen by producing an image on a photographic film or plate. Also known as radioautography.

autoserum [IMMUNOL] A serum obtained from a patient used for treatment of that patient.

autosexing [BIOL] Displaying differential sex characters at birth, noted particularly in fowl bred for sex-specific colors and patterns.

autoskeleton [INV ZOO] The endoskeleton of a sponge.

autosome [GEN] Any chromosome other than a sex chromosome.

autostylic [VERT ZOO] Having the jaws attached directly to the cranium, as in chimeras, amphibians, and higher vertebrates.

autosyndesis [CYTOL] The act of pairing of homologous chromosomes from the same parent during meiosis in polyploids.

autotomy [MED] Surgical removal of a part of one's own body. [ZOO] The process of self-amputation of appendages in crabs and other crustaceans and tails in some salamanders and lizards under stress.

autotroph [BIOL] An organism capable of synthesizing organic nutrients directly from simple inorganic substances, such as carbon dioxide and inorganic nitrogen.

autotropism [BIOL] **1.** Tendency of plants to remain in a normal orientation and to be unaffected by a stimulus. **2.** Tendency to resume original orientation after movement in response to a stimulus.

autozooecium [INV ZOO] The tube enclosing an autozooid.

autozooid [INV ZOO] An unspecialized feeding individual in a bryozoan colony, possessing fully developed organs and exoskeleton.

Autunian [GEOL] A European stage of geologic time: Lower

AUTOLYTINAE

Procerastea of the Syllidae (Autolytinae); epitokous male, dorsal view.

AUTORADIOGRAPHY

A reproduction of a photograph produced by autoradiography of metatarsus of a calf. The dense areas indicate highest concentration of the radioisotope. (*From C. L. Comar, Radioisotopes in Biology and Agriculture, McGraw-Hill, 1955*)

Permian (above Stephanian of Carboniferous, below Saxonian).

auxanogram [MICROBIO] A plate culture provided with variable growth conditions to determine the effects of specific environmental factors.

auxanography [MICROBIO] The study of growth-inhibiting or growth-promoting agents by means of auxanograms.

auxanometer [ENG] An instrument used to detect and measure plant growth rate.

auxesis [PHYSIO] Growth resulting from increase in cell size.

auxilia [INV ZOO] In insects, two small sclerites between the unguitractor and the claws.

auxin [BIOCHEM] Any organic compound which promotes plant growth along the longitudinal axis when applied to shoots free from indigenous growth-promoting substances.

auxoautotrophic [BIOL] Requiring no exogenous growth factors.

auxocyte [BIOL] A gamete-forming cell, such as an oocyte or spermatocyte, or a sporocyte during its growth period.

auxograph [ENG] An automatic device that records changes in the volume of a body.

auxoheterotrophic [BIOL] Requiring exogenous growth factors.

auxospore [INV ZOO] A reproductive cell in diatoms formed in association with rejuvenescence by the union of two cells that have diminished in size through repeated divisions.

auxotonic [BOT] Induced by growth rather than by exogenous stimuli.

auxotrophic mutant [GEN] An organism that requires a specific growth factor, such as an amino acid, for its growth.

available-chlorine method [MICROBIO] A technique for the standardization of chlorine disinfectants intended for use as germicidal rinses on cleaned surfaces; increments of bacterial inoculum are added to different disinfectant concentrations, and after incubation the results indicate the capacity of the disinfectant to handle an increasing bacterial load before exhaustion of available chlorine, the germicidal principle.

avalanche conduction [PHYSIO] Conduction of a nerve impulse through several neurons which converge, increasing the discharge intensity by summation.

Avena [BOT] A genus of grasses (family Gramineae), including oats, characterized by an inflorescence that is loosely paniculate, two-toothed lemmas, and deeply furrowed grains.

avenin [BIOCHEM] The glutelin of oats.

aversion therapy [PSYCH] A method of behavior modification in which stimuli eliciting the undesirable behavior are paired with unpleasant conditions; the stimuli will tend to be avoided in time.

aversive behavior [PSYCH] Avoidance behavior.

Aves [VERT ZOO] A class of animals composed of the birds, which are warm-blooded, egg-laying vertebrates primarily adapted for flying.

avianize [VIROL] To attenuate a virus by repeated culture on chick embryos.

avian leukosis [VET MED] A disease complex in fowl probably caused by viruses and characterized by autonomous proliferation of blood-forming cells.

avian pneumoencephalitis *See* Newcastle disease.

avian pseudoplague *See* Newcastle disease.

avian tuberculosis [VET MED] A tuberculosis-like mycobacterial disease of fowl caused by *Mycobacterium avium.*

aviation medicine *See* aerospace medicine.

avicolous [ECOL] Living on birds, as of certain insects.

avicularium [INV ZOO] A specialized individual in a bryozoan colony with a beak that keeps other animals from settling on the colony.

avifauna [VERT ZOO] **1.** Birds, collectively. **2.** Birds characterizing a period, region, or environment.

avitaminosis [MED] Any vitamin-deficiency disease.

avocado [BOT] *Persea americana.* A subtropical evergreen tree of the order Magnoliales that bears a pulpy pear-shaped edible fruit.

Avogadro's number [PHYS] The number (6.02×10^{23}) of molecules in a gram-molecular weight of a substance.

Avogadro's hypothesis *See* Avogadro's law.

Avogadro's law [PHYS] The law which states that under the same conditions of pressure and temperature, equal volumes of all gases contain equal numbers of molecules; for example, 359 cubic feet at 32°F and 1 atmosphere for a perfect gas. Also known as Avogadro's hypothesis.

avoidance-avoidance conflict [PSYCH] A conflict in which a person is caught between two negative goals; as one goal is avoided, the person is brought closer to the other goal, and vice versa.

avulsion [MED] Tearing one part away from the other, either by trauma or surgery.

awn [BOT] Any of the bristles at the ends of glumes or bracts on the spikelets of oats, barley, and some wheat and grasses. Also known as beard.

axenic culture [BIOL] The growth of organisms of a single species in the absence of cells or living organisms of any other species.

axial filament [CYTOL] The central microtubule elements of a cilium or flagellum.

axial gland [INV ZOO] A structure enclosing the stone canal in certain echinoderms; its function is uncertain.

axial gradient [EMBRYO] In some invertebrates, a graded difference in metobolic activity along the anterior-posterior, dorsal-ventral, and medial-lateral embryonic axes.

axial musculature [ANAT] The muscles that lie along the longitudinal axis of the vertebrate body.

axial sinus [INV ZOO] In echinoderms, a vertical canal that opens into the internal region of the oral ring sinus and communicates with the stone canal.

axial skeleton [ANAT] The bones composing the skull, vertebral column, and associated structures of the vertebrate body.

axiation [EMBRYO] The formation or development of axial structures, such as the neural tube.

Axiidae [INV ZOO] A family of decapod crustaceans, including the hermit crabs, in the suborder Reptantia.

axil [BIOL] The angle between a structure and the axis from which it arises, especially for branches and leaves.

axile [BOT] Pertaining to, located in, or belonging to the axis, especially with regard to placentation.

axile placenta [BOT] **1.** A placenta located on the vertical midline of the septum of an ovary that has two or more locules. **2.** The surface of a central axis of an ovary supporting the ovules or seeds.

axilla [ANAT] The depression between the arm and the thoracic wall; the armpit. [BOT] An axil.

axillary [ANAT] Of, pertaining to, or near the axilla or armpit. [BOT] Placed or growing in the axis of a branch or leaf.

axillary bud [BOT] A lateral bud borne in the axil of a leaf.

axillary sweat gland [ANAT] An apocrine gland located in the axilla.

Axinellina [INV ZOO] A suborder of sponges in the order Clavaxinellida.

axis [ANAT] **1.** The second cervical vertebra in higher verte-

AVOCADO

Avocado foliage and fruit.

AXILLARY BUD

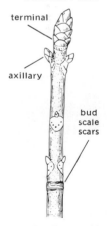

Position of axillary bud in the buckeye.

brates; the first vertebra of amphibians. **2.** The center line of an organism, organ, or other body part.

axis cylinder [CYTOL] **1.** The central mass of a nerve fiber. **2.** The core of protoplasm in a medullated nerve fiber.

axoblast [INV ZOO] **1.** The germ cell in mesozoans; cells are linearly arranged in the longitudinal axis and produce the primary nematogens. **2.** The individual scleroblast of the axis epithelium which produces spicules in octocorals.

axogamy [BOT] Having sex organs on a leafy stem.

axolemma [CYTOL] The plasma membrane of an axon.

axolotl [VERT ZOO] The neotenous larva of some salamanders in the family Ambystomidae.

axon [ANAT] The process or nerve fiber of a neuron that carries the unidirectional nerve impulse away from the cell body. Also known as neuraxon; neurite.

axoneme [CYTOL] The axial filament of a flagellum.

axon hillock [HISTOL] The conically shaped portion of a neuron from which the axon originates.

AYE–AYE

Aye-aye *(Daubentonia madagascariensis)*, nocturnal, arboreal primate found only in eastern Madagascar.

Axonolaimoidea [INV ZOO] A superfamily of free-living nematodes with species inhabiting marine and brackish-water environments.

axoplasm [CYTOL] The protoplasm of an axon.

axoplast [INV ZOO] In trypanosomes, a filament that extends from the kinetoplast to the end of the body.

axopodium [INV ZOO] A semipermanent pseudopodium composed of axial filaments surrounded by a cytoplasmic envelope.

axospermous [BOT] Having axile placentation.

axostyle [INV ZOO] In many flagellates, a thin, flexible, organic rod that provides the body with a supporting axis.

aye-aye [VERT ZOO] *Daubentonia madagascariensis.* A rare prosimian primate indigenous to eastern Madagascar; the single species of the family Daubentoniidae.

Azoic [GEOL] That portion of the earlier Precambrian time in which there is no trace of life.

azomycin [MICROBIO] $C_3H_3O_2N_3$ An antimicrobial antibiotic produced by a strain of *Nocardia mesenterica.*

azotemia [MED] The presence of excessive amounts of nitrogenous compounds in the blood.

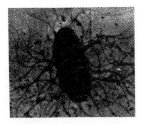

AZOTOBACTER

Azotobacter vinelandii with peritrichous flagella. *(From A. W. Hofer, Flagellation of Azotobacter, J. Bacteriol., 48:697-70l, 1944)*

Azotobacter [MICROBIO] A genus of large, usually motile, rod-shaped, oval, or spherical bacteria of the family Azotobacteraceae, found mostly in soils; the fixation of elemental nitrogen is their most important property.

Azotobacteraceae [MICROBIO] A family of aerobic bacteria in the order Eubacteriales capable of vigorous assimilation of elemental nitrogen.

Azotomonas [MICROBIO] A genus of soil bacteria in the family Pseudomonadaceae able to fix atmospheric nitrogen.

azoturia [MED] A condition characterized by excess amounts of urea or other nitrogenous substances in the urine. Also known as lumbago.

azurophil [CYTOL] Subject to easy staining with blue aniline dyes.

azygos [ANAT] An unpaired anatomic structure.

azygospore [MYCOL] A spore which is morphologically similar to a zygospore but is formed parthenogenetically.

azygos vein [ANAT] A branch of the right precava which drains the intercostal muscles and empties into the superior vena cava.

azygote [BIOL] An individual produced by haploid parthenogenesis.

B

B   *See* boron.

Ba   *See* barium.

babesiasis   [VET MED] A tick-borne protozoan disease of mammals other than man caused by species of *Babesia.*

Babesiidae   [INV ZOO] A family of protozoans in the suborder Haemosporina containing parasites of vertebrate red blood cells.

Babinski reflex   [MED] An abnormal reflex after infancy associated with a disturbance of the pyramidal tract, characterized by extension of the great toe with fanning of the other toes on sharply stroking the lateral aspect of the sole.

baboon   [VERT ZOO] Any of five species of large African and Asian terrestrial primates of the genus *Papio,* distinguished by a doglike muzzle, a short tail, and naked callosities on the buttocks.

babuina   [VERT ZOO] A female baboon.

bacca   [BOT] **1.** A pulpy fruit. **2.** A berry.

baccate   [BOT] **1.** Bearing berries. **2.** Having pulp like a berry.

bacciferous   [BOT] Bearing berries.

Bacillaceae   [MICROBIO] A family of rod-shaped, sporeforming bacteria in the order Eubacteriales.

Bacillariophyceae   [BOT] The diatoms, a class of algae in the division Chrysophyta.

Bacillariophyta   [BOT] An equivalent name for Bacillariophyceae.

bacillary   [MICROBIO] **1.** Rod-shaped. **2.** Produced by, pertaining to, or resembling bacilli.

bacillary dysentery   [MED] A highly infectious bacterial disease of man, localized in the bowels; caused by *Shigella.*

bacillary white diarrhea   *See* pullorum disease.

bacilluria   [MED] The presence of bacilli in the urine.

bacillus   [MICROBIO] Any rod-shaped bacterium.

Bacillus   [MICROBIO] The type genus of rod-shaped, aerobic and sometimes anaerobic bacteria in the family Bacillaceae.

Bacillus anthracis   [MICROBIO] A nonmotile pathogenic species of bacteria that causes anthrax.

Bacillus Calmette-Guerin vaccine   [IMMUNOL] A vaccine prepared from attenuated human tubercle bacilli and used to immunize humans against tuberculosis. Abbreviated BCG vaccine.

Bacillus diphtheriae   *See* Corynebacterium diphtheriae.

bacitracin   [MICROBIO] A group of polypeptide antibiotics produced by *Bacillus licheniformis.*

back   [ANAT] The part of the human body extending from the neck to the base of the spine.

backbone   *See* spine.

back bulb   [BOT] A pseudobulb on certain orchid plants that remains on the plant after removal of the terminal growth, and that is used for propagation.

BABOON

A baboon, a representative of Old World cercopithecoid monkeys.

BACILLARIOPHYCEAE

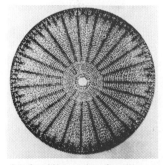

Arachnoidiscus ehrenbergii, a concentric diatom with radial symmetry. *(From H. J. Fuller and O. Tippo, College Botany, rev. ed., Holt, 1954)*

backcross [GEN] A cross between an F_1 heterozygote and an individual of P_1 genotype.

bacteremia [MED] Presence of bacteria in the blood.

bacteremic shock [MED] A state of shock occurring during the course of bacteremia, especially if caused by gram-negative bacteria.

bacteria [MICROBIO] Extremely small, relatively simple prokaryotic microorganisms traditionally classified with the fungi as Schizomycetes.

Bacteriaceae [MICROBIO] Former designation for Brevibacteriaceae.

bacterial blight [PL PATH] Any blight disease of plants caused by bacteria, including common bacterial blight, halo blight, and fuscous blight.

bacterial brown spot [PL PATH] A bacterial blight disease of plants caused by *Pseudomonas syringae;* marked by water-soaked reddish-brown spots or cankers. Also known as bacterial canker.

bacterial canker *See* bacterial brown spot.

bacterial capsule [MICROBIO] A thick, mucous envelope, composed of polypeptides or carbohydrates, surrounding some bacteria.

bacterial coenzyme [MICROBIO] Organic molecules that participate directly in a bacterial enzymatic reaction and may be chemically altered during the reaction.

bacterial encephalitis [MED] Inflammation of the brain caused by primary or secondary bacterial infection.

bacterial endocarditis [MED] Inflammation of the endocardium due to bacterial invasion. Also known as subacute bacterial endocarditis.

bacterial endoenzyme [MICROBIO] An enzyme produced and active within the bacterial cell.

bacterial endospore [MICROBIO] A body, resistant to extremes of temperature and to dehydration, produced within the cells of gram-positive, sporeforming rods of *Bacillus* and *Clostridium* and by the coccus *Sporosarcina.*

bacterial enzyme [MICROBIO] Any of the metabolic catalysts produced by bacteria.

bacterial genetics [GEN] The study of inheritance and variation patterns in bacteria.

bacterial infection [MED] Establishment of an infective bacterial agent in or on the body of a host.

bacterial leaf spot [PL PATH] A bacterial disease of plants characterized by spotty discolorations on the leaves; examples are angular leaf spot and leaf blotch.

bacterial luminescence [MICROBIO] A light-producing phenomenon exhibited by certain bacteria.

bacterial metabolism [MICROBIO] Total chemical changes carried out by living bacteria.

bacterial motility [MICROBIO] Self-propulsion in bacteria, either by gliding on a solid surface or by moving the flagella.

bacterial photosynthesis [MICROBIO] Use of light energy to synthesize organic compounds in green and purple bacteria.

bacterial pigmentation [MICROBIO] The organic compounds produced by certain bacteria which give color to both liquid cultures and colonies.

bacterial pneumonia [MED] Consolidation of the lung caused by inflammatory exudation due to bacterial infection.

bacterial pustule [PL PATH] A bacterial blight of plants caused by *Xanthomonas phaseoli;* characterized by blisters on the leaves.

bacterial soft rot [PL PATH] A bacterial disease of plants marked by disintegration of tissues.

bacterial speck [PL PATH] A bacterial disease of plants characterized by small lesions on plant parts.

BACTERIAL CAPSULE

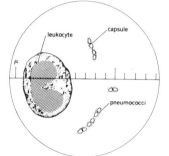

Drawing of pneumococci in sputum showing capsule surrounding the bacteria. *(From A. B. Sabin, J. Amer. Med. Assoc., 100(20):1585, 1933)*

bacterial spot [PL PATH] Any bacterial disease of plants marked by spotting of the infected part.

bacterial vaccine [IMMUNOL] A preparation of living, attenuated, or killed bacteria used to enhance the immune reaction in an individual already infected with the same bacteria.

bacterial wilt disease [PL PATH] A common bacterial disease of cucumber and muskmelon, caused by *Erwinia tracheiphila,* characterized by wilting and shriveling of the leaves and stems.

bactericide [MATER] An agent that destroys bacteria.

bactericidin [IMMUNOL] An antibody that kills bacteria in the presence of complement.

bacteriochlorophyll [BIOCHEM] $C_{52}H_{70}O_6N_4Mg$ A tetrahydroporphyrin chlorophyll compound occurring in the forms *a* and *b* in photosynthetic bacteria; there is no evidence that *b* has the empirical formula given.

bacteriocyte [INV ZOO] A modified fat cell found in certain insects that contains bacterium-shaped rods believed to be symbiotic bacteria.

bacteriogenic [MICROBIO] Caused by bacteria.

bacteriologist [MICROBIO] A specialist in the study of bacteria.

bacteriology [MICROBIO] The science and study of bacteria; a specialized branch of microbiology.

bacteriolysin [MICROBIO] An antibody that is active against and causes lysis of specific bacterial cells.

bacteriolysis [MICROBIO] Dissolution of bacterial cells.

bacteriophage [VIROL] Any of the viruses that infect bacterial cells; each has a narrow host range. Also known as phage.

bacteriosis [PL PATH] Any bacterial disease of plants.

bacteriostasis [MICROBIO] Inhibition of bacterial growth and metabolism.

bacteriotoxin [MICROBIO] **1.** Any toxin that destroys or inhibits growth of bacteria. **2.** A toxin produced by bacteria.

bacteriotropin [IMMUNOL] An antibody that is increased in amount during specific immunization and that renders the corresponding bacterium more susceptible to phagocytosis.

bacterioviridin *See* chlorobium chlorophyll.

bacteriuria [MED] The occurrence of bacteria in the urine.

Bacteroides [MICROBIO] A genus of gram-negative, anaerobic, rod-shaped bacteria in the family Bacteroidaceae, found in the alimentary and urogenital tracts of man and animals.

Bactrian camel [VERT ZOO] *Camelus bactrianus.* The two-humped camel.

baculum [VERT ZOO] The penis bone, or os priapi, in lower mammals.

badger [VERT ZOO] Any of eight species of carnivorous mammals in six genera comprising the subfamily Melinae of the weasel family (Mustelidae).

bagasse disease *See* bagassosis.

bagassosis [MED] A pneumoconiosis caused by the inhalation of bagasse dust, a dry sugarcane residue. Also known as bagasse disease.

Bagridae [VERT ZOO] A family of semitropical catfishes in the suborder Siluroidei.

Bainbridge reflex [PHYSIO] A poorly understood reflex acceleration of the heart rate due to rise of pressure in the right atrium and vena cavae, possibly mediated through afferent vagal fibers.

Bairdiacea [INV ZOO] A superfamily of ostracod crustaceans in the suborder Podocopa.

Bajocian [GEOL] A European stage: the middle Middle or

BACTERIOPHAGE

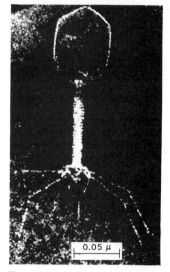

0.05 μ

Electron micrograph of negatively stained T2 bacteriophage showing head and tail components.
(Courtesy of H. Fernandez-Moran)

BACULUM

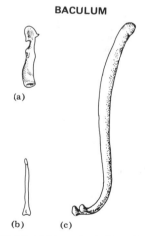

(a)

(b)   (c)

Bacula of *(a)* squirrel, *(b)* cotton mouse, *(c)* otter.

lower Middle Jurassic geologic time; above Toarcian, below Bathonian.

Bakanae disease [PL PATH] A fungus disease of rice in Japan, caused by *Gibberella fujikurae;* a foot rot disease.

Balaenicipitidae [VERT ZOO] A family of wading birds composed of a single species, the shoebill stork (*Balaeniceps rex*), in the order Ciconiiformes.

Balaenidae [VERT ZOO] The right whales, a family of cetacean mammals composed of five species in the suborder Mysticeti.

balanced anesthesia [MED] Anesthesia produced by safe doses of two or more agents or methods of anesthesia, each of which contributes to the total desired effect.

balanced fertilizer [MATER] A material of varying composition added to soil so as to provide essential mineral elements at required levels, improve soil structure, or enhance microbial activity.

balanced lethal [GEN] A strain heterozygous in trans for at least two linked recessive lethal mutants, between which crossing-over is prevented, that appears to breed true because neither of the two homozygotes is viable.

balanced polymorphism [GEN] Maintenance in a population of two or more alleles in equilibrium at frequencies too high to be explained, particularly for the rarer of them, by mutation balanced by selection; for example, the selective advantage of heterozygotes over both homozygotes.

balanced translocation [GEN] Positional change of one or more chromosome segments in cells or gametes without alteration of the normal diploid or haploid complement of genetic material.

balancer [INV ZOO] *See* haltere. [VERT ZOO] Either of a pair of rodlike lateral appendages on the heads of some larval salamanders.

Balanidae [INV ZOO] A family of littoral, sessile barnacles in the suborder Balanomorpha.

balanitis [MED] Inflammation of the glans of the penis or of the clitoris.

Balanomorpha [INV ZOO] The symmetrical barnacles, a suborder of sessile crustaceans in the order Thoracica.

Balanopaceae [BOT] A small family of dioecious dicotyledonous plants in the order Fagales characterized by exstipulate leaves, seeds with endosperm, and the pistillate flower solitary in a multibracteate involucre.

Balanopales [BOT] An ordinal name suggested for the Balanopaceae in some classifications.

Balanophoraceae [BOT] A family of dicotyledonous terrestrial plants in the order Santalales characterized by dry nutlike fruit, one to five ovules, unisexual flowers, attachment to the stem of the host, and the lack of chlorophyll.

Balanopsidales [BOT] An order in some systems of classification which includes only the Balanopaceae of the Fagales.

balantidiasis [MED] An intestinal infection of man caused by the protozoan *Balantidium coli.*

Balanus [INV ZOO] A genus of barnacles composed of sessile acorn barnacles; the type genus of the family Balanidae.

balausta [BOT] An indehiscent fruit characterized by many cells, many seeds, and a tough pericarp; an example is a pomegranate.

Balbiani rings [CYTOL] Localized swellings of a polytene chromosome.

baldness [MED] Loss or absence of hair.

baleen [VERT ZOO] A horny substance, growing as fringed filter plates suspended from the upper jaws of whalebone whales. Also known as whalebone.

Balfour's law [EMBRYO] The law that the speed with which

BALANUS

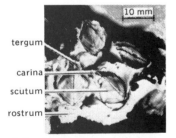

tergum

carina

scutum

rostrum

Balanus balanoides, apical view. (From D. P. Henry, Studies on the sessile Cirripedia of the Pacific coast of North America, Univ. Wash. Publ. Oceanogr., 4(3):99–131, 1942)

any part of the ovum segments is roughly proportional to the protoplasm's concentration in that area; the segment's size is inversely proportional to the protoplasm's concentration.

ball-and-socket joint *See* enarthrosis

ballistocardiogram [MED] The recording made by a ballistocardiograph.

ballistocardiograph [MED] A device to measure the volume of blood passing through the heart in a given period of time.

ballistospore [MYCOL] A type of fungal spore that is forcibly discharged at maturity.

balsa [BOT] *Ochroma lagopus.* A tropical American tree in the order Malvales; its wood is strong and lighter than cork.

Balsaminaceae [BOT] A family of flowering plants in the order Geraniales, including touch-me-not (*Impatiens*); flowers are irregular with five stamens and five carpels, leaves are simple with pinnate venation, and the fruit is an elastically dehiscent capsule.

bamboo [BOT] The common name of various tropical and subtropical, perennial, ornamental grasses in five genera of the family Gramineae characterized by hollow woody stems up to 6 inches in diameter.

Bambusoideae [BOT] A subfamily of grasses, composed of bamboo species, in the family Gramineae.

banana [BOT] Any of the treelike, perennial plants of the genus *Musa* in the family Musaceae characterized by soft, pulpy flesh and a thin rind.

banana freckle [PL PATH] A fungus disease of the banana caused by *Macrophoma musae*, producing brown or black spots on the fruit and leaves.

Bancroft's filariasis *See* wuchereriasis.

band *See* band spectrum.

bandage [MED] A strip of gauze, muslin, flannel, or other material, usually in the form of a roll, but sometimes triangular or tailed, used to hold dressing in place, to apply pressure, to immobilize a part, to support a dependent or injured part, to obliterate tissue cavities, or to check hemorrhage.

bandicoot [VERT ZOO] **1.** Any of several large Indian rats of the genus *Nesokia* and related genera. **2.** Any of several small insectivorous and herbivorous marsupials comprising the family Peramelidae and found in Tasmania, Australia, and New Guinea.

band spectrum [SPECT] A spectrum consisting of groups or bands of closely spaced lines in emission or absorption, characteristic of molecular gases and chemical compounds. Also known as band.

Bangiophyceae [BOT] A class of red algae in the plant division Rhodophyta.

Bang's disease *See* contagious abortion.

Banti's disease [MED] Portal hypertension, congestive splenomegaly, and hypersplenism due to an obstructive lesion in the splenic vein, portal vein, or intrahepatic veins.

barbel [VERT ZOO] **1.** A slender, tactile process near the mouth in certain fishes, such as catfishes. **2.** Any European fresh-water fish in the genus *Barbus.*

barbellate [BIOL] Having short, stiff, hooked bristles.

barbicel [VERT ZOO] One of the small, hook-bearing processes on a barbule of the distal side of a barb or a feather.

barbital [ORG CHEM] $C_8H_{12}N_2O_3$ A compound crystallizing in needlelike form from water; has a faintly bitter taste; melting point 188–192°C; used to make sodium barbital, a long-duration hypnotic and sedative. Also known as diethylbarbituric acid; diethylmalonylurea.

barbiturate [PHARM] Any of a group of ureides, such as

BANANA

Pseudostem of commercial banana plant *(Musa sapientum)*, showing characteristic foliage and single stem of bananas. Plant grows to height of 30 feet (9 meters) or more.

BARIUM

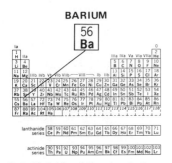

Periodic table of the chemical elements showing the position of barium.

BARK

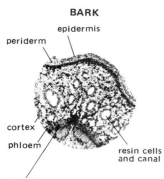

Tranverse section of young twig of balsam fir (*Abies balsamae* L.) showing tissues often considered to compose the bark. (*Forest Products Laboratory, USDA*)

BARNACLE

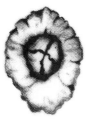

Top view of adult stage of *Balanus*, the acorn barnacle.

phenobarbital, Amytal, or Seconal, that act as central nervous system depressants.

barbituric acid [ORG CHEM] $C_4H_4O_3N_2$ 2,4,6-Trioxypyrimidine, the parent compound of the barbiturates; colorless crystals melting at 245°C, slightly soluble in water. Also known as malonyl urea.

barbiturism [MED] Intoxication following an overdose of barbiturates; characterized by delirium, coma, and sometimes death.

barbule [VERT ZOO] **1.** One of the hooked processes on either side of a barb of a feather. **2.** An appendage on the lower jaw in certain teleosts.

bar chart *See* bar graph.

bar graph [STAT] A diagram of frequency-table data in which a rectangle with height proportional to the frequency is located at each value of a variate that takes only certain discrete values. Also known as bar chart.

barium [CHEM] A chemical element, symbol Ba, with atomic number 56 and atomic weight of 137.34.

barium enema [MED] A suspension of barium sulfate administered as an enema into the lower bowel to render it radiopaque.

barium meal [MED] A suspension of barium sulfate taken orally to render the upper gastrointestinal tract radiopaque.

barium sulfate [INORG CHEM] $BaSO_4$ A salt occurring in the form of white, rhombic crystals, insoluble in water; used as a white pigment, as an opaque contrast medium for roentgenographic processes, and as an antidiarrheal.

bark [BOT] The tissues external to the cambium in a stem or root.

bark graft [BOT] A graft made by slipping the scion beneath a slit in the bark of the stock.

barley [BOT] A plant of the genus *Hordeum* in the order Cyperales that is cultivated as a grain crop; the seed is used to manufacture malt beverages and as a cereal.

barley scald [PL PATH] A fungus disease of barley caused by *Rhynchosporium secalis* and characterized by bluish-green to yellow blotches and blighting of the foliage.

barley smut [PL PATH] **1.** A loose smut disease of barley caused by *Ustilago nuda.* **2.** A covered smut disease of barley caused by *U. hordei.*

barley stripe [PL PATH] A fungus disease of barley characterized by light green or yellow stripes on the leaves; incited by the diffusible toxin of *Helminthosporium gramineum.*

barn [AGR] A farm building used for storage of agricultural products and equipment or for housing farm animals.

barnacle [INV ZOO] The common name for a number of species of crustaceans which compose the subclass Cirripedia.

baroceptor [HISTOL] A nerve ending, located principally in the walls of the carotid sinus and of the aortic arch, which is sensitive to stretching of the walls due to changes in blood pressure or pressure from outside the vessel. Also known as pressoreceptor.

baroduric bacteria [MICROBIO] Bacteria that can tolerate conditions of high hydrostatic pressure.

barometric pressure *See* atmospheric pressure.

barophile [MICROBIO] An organism that thrives under conditions of high hydrostatic pressure.

barosinusitis [MED] Inflammation of the sinuses, characterized by edema and hemorrhage, due to expansion of air within the sinuses at decreased barometric pressure. Also known as aerosinusitis.

barotaxis [BIOL] Orientation movement of an organism in response to pressure changes.

barotitis [MED] Inflammation of the ear, or a part of it, caused by changes in atmospheric pressure. Also known as aerootitis.

barotrauma [MED] Injury to air-containing structures, such as the middle ears, sinuses, lungs, and gastrointestinal tract, due to unequal pressure differences across their walls.

barracuda [VERT ZOO] The common name for about 20 species of fishes belonging to the genus *Sphyraena* in the order Perciformes.

Barr body [CYTOL] A condensed, inactivated X chromosome inside the nuclear membrane in interphase somatic cells of women and most female mammals.

barrel chest *See* emphysematous chest.

Barremian [GEOL] Lower Cretaceous geologic age, between Hauterivian and Aptian.

barrier [ECOL] Any physical or biological factor that restricts the migration or free movement of individuals or populations.

Barstovian [GEOL] Upper Miocene geologic time.

Bartholin's duct [ANAT] **1.** The duct of a large mucus-secreting vestibular gland at the lower end of the vagina. **2.** The duct of a major sublingual gland.

Bartholin's gland [ANAT] One of a pair of large mucus-secreting vestibular glands in the lower end of the vagina.

Bartonellaceae [MICROBIO] A family of the order Rickettsiales including parasites of red blood cells of man and animals.

bartonellosis *See* Carrion's disease.

Bartonian [GEOL] A European stage: Eocene geologic time above Auversian, below Ludian. Also known as Marinesian.

Barychilinidae [PALEON] A family of Paleozoic crustaceans in the suborder Platycopa.

Barylambdidae [PALEON] A family of late Paleocene and early Eocene aquatic mammals in the order Pantodonta.

Barytheriidae [PALEON] A family of extinct proboscidean mammals in the suborder Barytherioidea.

Barytherioidea [PALEON] A suborder of extinct mammals of the order Proboscidea, in some systems of classification.

basal [BIOL] Pertaining to or located at the base. [PHYSIO] Being the minimal level for, or essential for maintenance of, vital activities of an organism, such as basal metabolism.

basalar [INV ZOO] In insects, pertaining to sclerites below the base of the wing.

basal body [CYTOL] A cellular organelle that induces the formation of cilia and flagella and is similar to and sometimes derived from a centriole. Also known as kinetosome.

basal-cell carcinoma [MED] A locally invasive, rarely metastatic nevoid tumor of the epidermis. Also known as basal-cell epithelioma.

basal-cell epithelioma *See* basal-cell carcinoma.

basal disc [BIOL] The expanded basal portion of the stalk of certain sessile organisms, used for attachment to the substrate.

basale [VERT ZOO] In fish, the proximal bone of a series forming the axis of a paired fin.

basal ganglia [ANAT] The corpus striatum, or the corpus striatum and the thalamus considered together as the important subcortical centers.

basal granule [INV ZOO] In some protozoans, an enlarged body at the base of a flagellum.

basalia [VERT ZOO] The cartilaginous rods that support the base of the pectoral and pelvic fins in elasmobranchs.

basalis [HISTOL] The basal portion of the endometrium; it is not shed during menstruation.

basal lamina [EMBRYO] The portion of the gray matter of the

BARRACUDA

The great or predatory barracuda
(Sphyraena barracuda), with a
long jaw and numerous sharp
teeth.

embryonic neural tube from which motor nerve roots develop.

basal membrane [ANAT] The tissue beneath the pigment layer of the retina that forms the outer layer of the choroid.

basal metabolic rate [PHYSIO] The amount of energy utilized per unit time under conditions of basal metabolism; expressed as calories per square meter of body surface or per kilogram of body weight per hour. Abbreviated BMR.

basal metabolism [PHYSIO] The sum total of anabolic and catabolic activities of an organism in the resting state providing just enough energy to maintain vital functions.

basal placenta [BOT] A placenta arising from the proximal end of the ovary.

basal plates [EMBRYO] The portion of the decidua that is fused with the placenta. [INV ZOO] **1.** In crinoids, a series of plates at or near the top of the stalk. **2.** In echinoids, a series of plates that form part of the apical disc.

basal wall [BOT] The wall dividing the oospore into an anterior and a posterior half in plants bearing archegonia.

base [CHEM] Any chemical species, ionic or molecular, capable of accepting or receiving a proton (hydrogen ion) from another substance; the other substance acts as an acid in giving of the proton; the hydroxyl ion is a base.

basement membrane [HISTOL] A delicate connective-tissue layer underlying the epithelium of many organs.

basendite [INV ZOO] In crustaceans, either of a pair of lobes at the end of each specialized paired appendage.

base pairing [MOL BIO] The hydrogen bonding of complementary purine and pyrimidine bases—adenine with thymine, guanine with cytosine—in double-stranded deoxyribonucleic acids (DNA) or ribonucleic acids (RNA) or in DNA/RNA hybrid molecules.

basic dye [MATER] Any of the dyes which are salts of colored organic bases containing amino and imino groups, combined with a colorless acid, such as hydrochloric or sulfuric.

basic group [CHEM] A chemical group (for example, OH⁻) which, when freed by ionization in solution, produces a pH greater than 7.

basichromatic [BIOL] Staining readily with basic dyes.

basicoxite [INV ZOO] A ring at the base of the coxa.

basicranial [ANAT] Pertaining to the base of the skull.

basidiocarp [MYCOL] The fruiting body of a fungus in the class Basidiomycetes.

Basidiolichenes [BOT] A class of the Lichenes characterized by the production of basidia.

Basidiomycetes [MYCOL] A class of fungi in the subdivision Eumycetes; important as food and as causal agents of plant diseases.

basidiophore [MYCOL] A basidia-bearing sporophore.

basidiospore [MYCOL] A spore produced by a basidium.

basidium [MYCOL] A cell, usually terminal, of Basidiomycetes that produces spores (basidiospores) by nuclear fusion followed by meiosis.

basifemur [INV ZOO] In some Acarina, the proximal segment of the femur, between the trochanter and the telofemur.

basifixed [BOT] Attached at or near the base.

basil [BOT] The common name for any of the aromatic plants in the genus *Ocimum* of the mint family; leaves of the plant are used for food flavoring.

basilabium [INV ZOO] In insects, a sclerite formed by fusion of the basal portions of the labium.

basilar [BIOL] Of, pertaining to, or situated at the base.

basilar groove [ANAT] The cavity which is located on the upper surface of the basilar process of the brain and upon which the medulla rests.

BASIDIUM

A drawing of the basidiospores and basidium in fungi.

basilar index [ANTHRO] The ratio of the distance from the basion to the alveolar point to the total skull length multiplied by 100.

basilar membrane [ANAT] A membrane of the mammalian inner ear supporting the organ of Corti and separating two cochlear channels, the scala media and scala tympani.

basilar meningitis [MED] Inflammation of the meninges which affects chiefly the base of the brain, or in which exudate collects predominantly at the basal cisterns.

basilar papilla [ANAT] **1.** A sensory structure in the lagenar portion of an amphibian's membranous labyrinth between the oval and round windows. **2.** The organ of Corti in mammals.

basilar plate [EMBRYO] An embryonic cartilaginous plate in vertebrates that is formed from the parachordals and anterior notochord and gives rise to the ethmoid and other bones of the skull.

basilar process [ANAT] A strong, quadrilateral plate of bone forming the anterior portion of the occipital bone, in front of the foramen magnum.

basilic vein [ANAT] The large superficial vein of the arm on the medial side of the biceps brachii muscle.

basilingual [VERT ZOO] A broad cartilaginous plate composing the major portion of the hyoid in crocodiles, turtles, and amphibians.

basimandibula [INV ZOO] In insects, small sclerite on the head at the base of the mandible.

basimaxilla [INV ZOO] In insects, a sclerite at the base of the maxilla.

basioccipital [ANAT] Pertaining to the basilar part of the occipital bone.

basion [ANAT] In craniometry, the point on the anterior margin of the foramen magnum where the midsagittal plane of the skull intersects the plane of the foramen magnum.

basion-bregma height [ANTHRO] Distance between basion and bregma.

basipetal [BIOL] Movement or growth from the apex toward the base.

basipharynx [INV ZOO] In insects, the epipharynx and the hypopharynx fused.

basiphthalmite [INV ZOO] In crustaceans, the proximal joint of the eyestalk.

basipodite [INV ZOO] The distal segment of the protopodite of a biramous appendage in arthropods.

basiproboscis [INV ZOO] The membranous part of the proboscis in some insects.

basipterygium [VERT ZOO] A basal bone or cartilage supporting one of the paired fins in fishes.

basisphenoid [ANAT] The lower portion of the sphenoid bone; it develops in the embryo as a separate bone.

basisternum [INV ZOO] In insects, the anterior one of the two sternal skeletal plates.

basistyle [INV ZOO] Either of a pair of flexible processes on the hypopygium of certain male Diptera.

basitarsus [INV ZOO] The basal segment of the tarsus in arthropods.

basitemporal [ANAT] Pertaining to the lower part of the temporal bone.

basitonic [BOT] In orchids, having the base of the anther united with the apex of the gynoecium.

basket cell [HISTOL] A type of cell in the cerebellum whose axis-cylinder processes terminate in a basketlike network around the cells of Purkinje.

basket nerve ending [HISTOL] A specialized structure at the root of a hair; thought to be a pressure or touch receptor.

BASSWOOD

American basswood (*Tilia americana*).

BAT

Epauletted fruit bat.
(*Épomophorus wahlbergi*).

BATHYNELLACEA

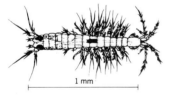

1 mm

Bathynella natans, female.

BATRACHOIDIFORMES

Atlantic midshipman (*Porichthys porosissimus*), about 8 inches (20 centimeters) long. (*After D. S. Jordan*)

basket star [INV ZOO] The common name for ophiuroid echinoderms belonging to the family Gorgonocephalidae.

Basommatophora [INV ZOO] An order of mollusks in the subclass Pulmonata containing many aquatic snails.

basonym [SYST] The original, validly published name of a taxon.

basophil [HISTOL] A white blood cell with granules that stain with basic dyes and are water-soluble.

basophilia [BIOL] An affinity for basic dyes. [MED] An increase in the number of basophils in the circulating blood. [PATH] Stippling of the red cells with basic staining granules, representing a degenerative condition as seen in severe anemia, leukemia, malaria, lead poisoning, and other toxic states.

basophilous [BIOL] Staining readily with basic dyes. [ECOL] Of plants, growing best in alkaline soils.

bass [VERT ZOO] The common name for a number of fishes assigned to two families, Centrarchidae and Serranidae, in the order Perciformes.

basswood [BOT] A common name for trees of the genus *Tilia* in the linden family of the order Malvales. Also known as linden.

bast fiber [BOT] Any fiber stripped from the inner bark of plants, such as flax, hemp, jute, and ramie; used in textile and paper manufacturing.

bat [VERT ZOO] The common name for all members of the mammalian order Chiroptera.

Batales [BOT] A small order of dicotyledonous plants in the subclass Caryophillidae of the class Magnoliopsida containing a single family with only one genus, *Batis*.

Batesian mimicry [ECOL] Resemblance of an innocuous species to one that is distasteful to predators.

Bathornithidae [PALEON] A family of Oligocene birds in the order Gruiformes.

bathyal zone [OCEANOGR] The biogeographic realm of the ocean depths between 100 and 1000 fathoms (180 and 1800 meters).

Bathyctenidae [INV ZOO] A family of bathypelagic coelenterates in the phylum Ctenophora.

Bathyergidae [VERT ZOO] A family of mammals, including the South African mole rats, in the order Rodentia.

Bathylaconoidei [VERT ZOO] A suborder of deep-sea fishes in the order Salmoniformes.

Bathynellacea [INV ZOO] An order of crustaceans in the superorder Syncarida found in subterranean waters in England and central Europe.

Bathynellidae [INV ZOO] The single family of the crustacean order Bathynellacea.

bathypelagic zone [OCEANOGR] The biogeographic realm of the ocean lying between depths of 900 and 3700 meters.

Bathypteroidae [VERT ZOO] A family of benthic, deep-sea fishes in the order Salmoniformes.

Bathysquillidae [INV ZOO] A family of mantis shrimps, with one genus (*Bathysquilla*) and two species, in the order Stomatopoda.

Batoidea [VERT ZOO] The skates and rays, an order of the subclass Elasmobranchii.

Batrachoididae [VERT ZOO] The single family of the order Batrachoidiformes.

Batrachoidiformes [VERT ZOO] The toadfishes, an order of teleostean fishes in the subclass Actinopterygii.

battered-child syndrome [MED] A clinical condition in young children due to serious physical abuse, generally from a parent or foster parent.

Baumé hydrometer scale [PHYS CHEM] A calibration scale

for liquids that is reducible to specific gravity by the following formulas: for liquids heavier than water, specific gravity = 145 ÷ 145 − *n* (at 60°F); for liquids lighter than water, specific gravity = 140 ÷ 130 + *n* (at 60°F); *n* is the reading on the Baumé scale, in degrees Baumé. Baumé is abbreviated Bé.

bay [BOT] *Laurus nobilis.* An evergreen tree of the laurel family.

bayberry [BOT] **1.** *Pimenta acris.* A West Indian tree related to the allspice; a source of bay oil. Also known as bay-rum tree; Jamaica bayberry; wild cinnamon. **2.** Any tree of the genus *Myrica.*

B cell [IMMUNOL] One of a heterogeneous population of bone-marrow-derived lymphocytes which participates in the immune responses.

BCG vaccine *See* Bacillus Calmette-Guerin vaccine.

Bdelloidea [INV ZOO] An order of the class Rotifera comprising animals which resemble leeches in body shape and manner of locomotion.

Bdellomorpha [INV ZOO] An order of ribbonlike worms in the class Enopla containing the single genus *Malacobdella.*

Bdellonemertini [INV ZOO] An equivalent name for the Bdellomorpha.

Bé *See* Baumé hydrometer scale.

Be *See* beryllium.

beak [BOT] Any pointed projection, as on some fruits, that resembles a bird bill. [INV ZOO] The tip of the umbo in bivalves. [VERT ZOO] **1.** The bill of a bird or some other animal, such as the turtle. **2.** A projecting jawbone element of certain fishes, such as the sawfish and pike.

beaker [SCI TECH] A deep, open-mouthed, cylindrical vessel with thin walls, which usually has a projecting lip for pouring.

bean [BOT] The common name for various leguminous plants used as food for man and livestock; important commercial beans are true beans (*Phaseolus*) and California blackeye (*Vigna sinensis*).

bean anthracnose [PL PATH] A fungus disease of the bean caused by *Colletotrichum lindemuthianum*, producing pink to brown lesions on the pod and seed and dark discolorations on the veins on the lower surface of the leaf.

bean blight [PL PATH] A bacterial disease of the bean caused by *Xanthomonas phaseoli*, producing water-soaked lesions that become yellowish-brown spots on all plant parts.

bear [VERT ZOO] The common name for a few species of mammals in the family Ursidae.

beard *See* awn.

beaver [VERT ZOO] The common name for two different and unrelated species of rodents, the mountain beaver (*Aplodontia rufa*) and the true or common beaver (*Castor canadensis*).

bedbug [INV ZOO] The common name for a number of species of household pests in the insect family Cimicidae that infest bedding, and by biting humans obtain blood for nutrition.

Bedoulian [GEOL] Lower Cretaceous (lower Aptian) geologic time in Switzerland.

Bedsonia [MICROBIO] The psittacosis-lymphogranulomatrachoma (PLT) group of bacteria belonging to the Chlamydozoaceae; all are obligatory intracellular parasites. Also known as Chlamydia.

bee [INV ZOO] Any of the membranous-winged insects which compose the superfamily Apoidea in the order Hymenoptera characterized by a hairy body and by sucking and chewing mouthparts.

beech [BOT] Any of various deciduous trees of the genus

BDELLOIDEA

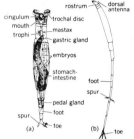

Bdelloid rotifers, with elongated, segmented bodies. *(a) Rotaria. (b) R. neptunia. (After Weber, 1898; from L. Hyman, The Invertebrates, vol. 3, McGraw-Hill, 1951)*

BEAVER

The common beaver *(Castor canadensis)*, showing characteristic webbed hindfeet and broad, flat tail.

BEECH

American beech *(Fagus grandifolia)*, showing slender scaly winter buds, a leaf, an open spiny involucre containing a pair of edible three-sided nuts, and a three-sided nut.

Fagus in the beech family (Fagaceae) characterized by smooth gray bark, triangular nuts enclosed in burs, and hard wood with a fine grain.

beech bark disease [PL PATH] A disease of beech caused by the beech scale (*Cryptococcus fagi*) and a fungus (*Nectria coccinea faginata*) acting together; bark is destroyed, foliage wilts, and the tree eventually dies.

beef [AGR] The flesh of a bovine animal, such as a cow or steer, used as food.

beehive Also known as hive. [AGR] A container that is constructed to house a colony of honeybees. [INV ZOO] A colony of bees.

beekeeping [AGR] The management and maintenance of colonies of honeybees.

beer drinkers' cardiomyopathy [MED] Congestive heart failure and nonspecific cardiomyopathy presumed due to cobalt added to beer.

Beer-Lambert-Bouguer law *See* Bouguer-Lambert-Beer law.

Beer's law [PHYS CHEM] The law which states that the absorption of light by a solution changes exponentially with the concentration, all else remaining the same.

beet [BOT] *Beta vulgaris.* The red or garden beet, a cool-season biennial of the order Caryophyllales grown for its edible, enlarged fleshy root.

beetle [INV ZOO] The common name given to members of the insect order Coleoptera.

Beggiatoaceae [MICROBIO] A family of bacteria in the order Beggiatoales that belong to the physiological group known as the sulfur bacteria.

Beggiatoales [MICROBIO] An order of motile, filamentous, and unicellular bacteria in the class Schizomycetes.

Begoniaceae [BOT] A family of dicotyledonous plants in the order Violales characterized by an inferior ovary, unisexual flowers, stipulate leaves, and two to five carpels.

behavior [PSYCH] Any overt activity of an organism.

behavioral psychophysics [PSYCH] A branch of psychology concerned primarily with the measurement of sensory capacities of normal, intact animals.

behavioral sciences [SCI TECH] The sciences concerned with human and animal behavior; they include psychology, sociology, social anthropology, and certain aspects of history, economics, political science, and zoology.

behavior disorder *See* abnormal behavior.

behaviorism [PSYCH] A school of psychology concerned with observable, tangible, and measurable data regarding behavior and human activities, but excluding ideas and emotions as purely subjective phenomena.

behavior modification [PSYCH] Psychotherapy which centers on altering behavioral problems by means of techniques of classical conditioning, operant conditioning, and perceptual learning.

bejel [MED] An infectious nonvenereal treponemal disease occurring principally in children in the Middle East.

Belemnoidea [PALEON] An order of extinct dibranchiate mollusks in the class Cephalopoda.

belladonna [BOT] *Atropa belladonna.* A perennial poisonous herb that belongs to the family Solanaceae; atropine is produced from the roots and leaves; used as an antispasmodic, as a cardiac and respiratory stimulant, and to check secretions. Also known as deadly nightshade.

Bellerophontacea [PALEON] A superfamily of extinct gastropod mollusks in the order Aspidobranchia.

Bell's law [PHYSIO] **1.** The law that in the spinal cord the ventral roots are motor and the dorsal roots sensory in

BELLADONNA

Belladonna *(Atropa belladonna),* flowering branch and isolated flowers (two views).

function. **2.** The law that in a reflex arc the nerve impulse can be conducted in one direction only.

belly [ANAT] **1.** The abdominal cavity or abdomen. **2.** The most prominent, fleshy, central portion of a muscle.

Beloniformes [VERT ZOO] The former ordinal name for a group of fishes now included in the order Atheriniformes.

Belostomatidae [INV ZOO] The giant water bugs, a family of hemipteran insects in the subdivision Hydrocorisae.

belt [ECOL] **1.** Any altitudinal vegetation zone or band from the base to the summit of a mountain. **2.** Any benthic vegetation zone or band from sea level to the ocean depths. **3.** Any of the concentric vegetation zones around bodies of fresh water.

Bence-Jones protein [PATH] An abnormal group of globulins appearing in the serum and urine, usually in association with multiple myeloma and characterized by coagulation at 50-60°C.

Bence-Jones proteinuria [MED] The presence of Bence-Jones protein in the urine.

bends *See* caisson disease.

benign [MED] Of no danger to life or health.

benign lymphoreticulosis *See* cat scratch disease.

benign myalgic encephalomyelitis *See* neuromyasthenia.

benign tumor [MED] A nonmalignant neoplasm.

benjamin gum *See* benzoin gum.

Bennettitales [PALEOBOT] An equivalent name for the Cycadeoidales.

Bennettitatae [PALEOBOT] A class of fossil gymnosperms in the order Cycadeoidales.

benthic [OCEANOGR] Of, pertaining to, or living on the bottom or at the greatest depths of a large body of water. Also known as benthonic.

benthonic *See* benthic.

benthos [ECOL] Bottom-dwelling forms of marine life. Also known as bottom fauna. [OCEANOGR] The floor or deepest part of a sea or ocean.

Benzedrine *See* amphetamine.

benzene [ORG CHEM] C_6H_6 A colorless, liquid, flammable, aromatic hydrocarbon that boils at 80.1°C and freezes at 5.4–5.5°C; used to manufacture styrene and phenol. Also known as benzol.

benzene ring [ORG CHEM] The six-carbon ring structure found in benzene, C_6H_6, and in organic compounds formed from benzene by replacement of one or more hydrogen atoms by other chemical atoms or radicals.

benzene series [ORG CHEM] A series of carbon-hydrogen compounds based on the benzene ring, with the general formula C_nH_{2n-6}, where n is 6 or more; examples are benzene, C_6H_6; toluene, C_7H_8; and xylene, C_8H_{10}.

benzoic acid [ORG CHEM] C_6H_5COOH An aromatic carboxylic acid that melts at 122.4°C, boils at 250°C, and is slightly soluble in water and relatively soluble in alcohol and ether; derivatives are valuable in industry, commerce, and medicine.

benzoin [ORG CHEM] $C_{14}H_{12}O_2$ An optically active compound; white or yellowish crystals, melting point 137°C; soluble in acetone, slightly soluble in water; used in organic synthesis. Also known as benzoylphenyl carbinol; 2-hydroxy-2-phenyl acetophenone; phenyl benzoyl carbinol.

benzyl [ORG CHEM] The radical $C_6H_5CH_2-$ found, for example in benzyl alcohol, $C_6H_5CH_2OH$.

benzyl benzoate [ORG CHEM] $C_6H_5COOCH_2C_6H_5$ An oily, colorless liquid ester; used as an antispasmodic drug and as a scabicide.

benzyl penicillinic acid [ORG CHEM] $C_{16}H_{18}N_2O_4S$ An

BERKELIUM

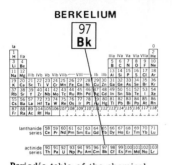

**Periodic table of the chemical
elements** showing the position
of berkelium.

BERYCIFORMES

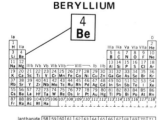

Squirrelfish (*Holocentrus
ascensionis*) of the family
Holocentridae that are nocturnal
fishes found in shallow tropical
and subtropical reefs. (*After G. B.
Goode, Fishery Industries of the
United States, 1884*)

BERYLLIUM

**Periodic table of the chemical
elements** showing the position
of beryllium.

amorphous white powder extracted with ether or chloroform
from an acidified aqueous solution of benzyl penicillin.

benzyl penicillin potassium [MICROBIO] $C_{16}H_{17}KN_2O_4S$
Moderately hygroscopic crystals; soluble in water; inactivated by acids and alkalies; obtained from fermentation of
Penicillium chrysogenum; used as an antimicrobial drug in
human and animal disease. Also known as penicillin G_1
potassium; potassium benzyl penicillinate; potassium penicillin G_1.

benzyl penicillin sodium [MICROBIO] $C_{16}H_{17}N_2NaO_4S$
Crystals obtained from a methanol-ethyl acetate acidified
extract of fermentation broth of *Penicillium chrysogenum*; used
as an antimicrobial in human and animal disease. Also
known as penicillin; sodium benzyl penicillinate; sodium
penicillin G_1.

Berberidaceae [BOT] A family of dicotyledonous herbs and
shrubs in the order Ranunculales characterized by alternate
leaves, perfect, well-developed flowers, and a seemingly
solitary carpel.

Bergmann's rule [ECOL] The principle that in a polytypic
wide-ranging species of warm-blooded animals the average
body size of members of each geographic race varies with the
mean environmental temperature.

beriberi [MED] A disorder resulting from the deficiency of
vitamin B_1 and characterized by neurologic symptoms, cardiovascular abnormalities, edema, and cerebral manifestations.

Berkefeld filter [MICROBIO] A diatomaceous-earth filter used
for sterilization of heat-labile liquids, such as blood serum,
enzyme solutions, and antibiotics.

berkelium [CHEM] A radioactive element, symbol Bk, atomic
number 97, the eighth member of the actinide series; properties resemble those of the rare-earth cerium.

Bermuda grass [BOT] *Cynodon dactylon.* A long-lived perennial in the order Cyperales.

Beroida [INV ZOO] The single order of the class Nuda in the
phylum Ctenophora.

Berriasian [GEOL] Part of or the underlying stage of the
Valanginian at the base of the Cretaceous.

berry [BOT] A usually small, simple, fleshy or pulpy fruit,
such as a strawberry, grape, tomato, or banana.

Berthelot equation [PHYS CHEM] A form of the equation of
state which relates the temperature, pressure, and volume of
a gas with the gas constant.

Bertrand's rule [MICROBIO] The rule stating that in those
compounds, and only in those compounds, having cis secondary alcoholic groups containing at least one carbon atom of
D configuration which is subtended by a primary alcohol
group, or having a methyl-substituted primary alcohol group
of D configuration, the D-carbon atom will be dehydrogenated
by the vinegar bacteria *Acetobacter suboxydans,* yielding a
ketone.

Beryciformes [VERT ZOO] An order of actinopterygian fishes
in the infraclass Teleostei.

Berycomorphi [VERT ZOO] An equivalent name for the Beryciformes.

berylliosis [MED] Chronic lung inflammation due to inhalation of beryllium oxide dust.

beryllium [CHEM] A chemical element, symbol Be, atomic
number 4, atomic weight 9.0122.

Berytidae [INV ZOO] The stilt bugs, a small family of hemipteran insects in the superfamily Pentatomorpha.

beta carotene [BIOCHEM] $C_{40}H_{56}$ A carotenoid hydrocarbon pigment found widely in nature, always associated with

chlorophylls; converted to vitamin A in the liver of many animals.

beta cell [HISTOL] **1.** Any of the basophilic chromophiles in the anterior lobe of the adenohypophysis. **2.** One of the cells of the islets of Langerhans which produce insulin.

betacyanin [BIOCHEM] A group of purple plant pigments found in leaves, flowers, and roots of members of the order Caryophyllales.

beta globulin [BIOCHEM] A heterogeneous fraction of serum globulins containing transferrin and various complement components.

beta hemolysis [MICROBIO] A sharply defined, clear, colorless zone of hemolysis surrounding certain streptococci colonies growing on blood agar.

betanin [BIOCHEM] An anthocyanin that contains nitrogen and constitutes the principal pigment of garden beets.

beta particle [NUC PHYS] An electron or positron emitted from a nucleus during beta decay.

beta ray [NUC PHYS] A stream of beta particles.

beta rhythm [PHYSIO] An electric current of low voltage from the brain, with a pulse frequency of 13–30/sec, encountered in a person who is aroused and anxious.

beta taxonomy [SYST] The phylogenic phase in the history of animal systematics.

betaxanthin [BIOCHEM] A class of plant pigments similar to betacyanins, but ranging in color from yellow to orange-red.

betel nut [BOT] A dried, ripe seed of the palm tree *Areca catechu* in the family Palmae; contains a narcotic.

Bethylidae [INV ZOO] A small family of hymenopteran insects in the superfamily Bethyloidea.

Bethyloidea [INV ZOO] A superfamily of hymenopteran insects in the suborder Apocrita.

Betula [BOT] The birches, a genus of deciduous trees composing the family Betulaceae.

Betulaceae [BOT] A small family of dicotyledonous plants in the order Fagales characterized by stipulate leaves, seeds without endosperm, and by being monoecious with female flowers mostly in catkins.

Betz cell [HISTOL] Any of the large conical cells composing the major histological feature of the precentral motor cortex in man.

Beyrichacea [PALEON] A superfamily of extinct ostracods in the suborder Beyrichicopina.

Beyrichicopina [PALEON] A suborder of extinct ostracods in the order Paleocopa.

Beyrichiidae [PALEON] A family of extinct ostracods in the superfamily Beyrichacea.

Bi *See* bismuth.

biacuminate [BOT] Being tapered to two slender points, especially referring to leaves.

biarticulate [BIOL] Having two joints.

Bial's test [PATH] A test for the presence of a pentose in urine, utilizing oracin, hydrochloric acid, and ferric chloride; a green color or green precipitate indicates pentose.

biased statistic [STAT] A statistic whose expected value, as obtained from a random sampling, does not equal the parameter or quantity being estimated.

bias error [STAT] A measurement error that remains constant in magnitude for all observations; a kind of systematic error.

Bibionidae [INV ZOO] The March flies, a family of orthorrhaphous dipteran insects in the series Nematocera.

bicameral [BIOL] Having two chambers, as the heart of a fish.

bicapsular [BIOL] **1.** Having two capsules. **2.** Having a capsule with two locules.

bicarinate [BIOL] Having two keellike projections.

bicarpellate [BOT] Having two carpels.

bicaudal [ZOO] Having two tails.

bicellular [BIOL] Having two cells.

bicephalous [ZOO] Having two heads.

biceps [ANAT] **1.** A bicipital muscle. **2.** The large muscle of the front of the upper arm that flexes the forearm; biceps brachii. **3.** The thigh muscle that flexes the knee joint and extends the hip joint; biceps femoris.

biciliate [BIOL] Having two cilia.

bicipital [ANAT] **1.** Pertaining to muscles having two origins. **2.** Pertaining to ribs having double articulation with the vertebrae. [BOT] Having two heads or two supports.

bicipital tuberosity [ANAT] An eminence on the anterior inner aspect of the neck of the radius; the tendon of the biceps muscle is inserted here.

bicolligate [VERT ZOO] In birds, having feet with two stretches of webbing.

bicornuate uterus [ANAT] A uterus with two horn-shaped processes on the superior aspect.

Bicosoecida [INV ZOO] An order of colorless, free-living protozoans, each having two flagella, in the class Zoomastigophorea.

bicostate [BOT] Of a leaf, having two principal longitudinal ribs.

bicrenate [BOT] Of leaves, having a notched crenate margin.

bicuspid [ANAT] Any of the four double-pointed premolar teeth in man. [BIOL] Having two points or prominences.

bicyclic [BOT] Consisting of or arranged in two whorls.

Bidder's organ [VERT ZOO] A structure in the males of some toad species that may develop into an ovary in older individuals.

bidentate [BIOL] Having two teeth or teethlike processes.

biennial plant [BOT] A plant that requires two growing seasons to complete its life cycle.

bifanged [ANAT] Of a tooth, having two roots.

bifarious [BOT] Consisting of or arranged in two rows, one on each side of the axis.

bifid [BIOL] Divided into two equal parts by a median cleft.

biflabellate [INV ZOO] The shape of certain insect antennae, characterized by short joints with long, flattened processes on opposite sides.

biflagellate [BIOL] Having two flagella.

biflorate [BOT] Bearing two flowers.

bifoliate [BOT] Two-leaved.

biforate [BIOL] Having two perforations.

big bud [PL PATH] **1.** A parasitic disease of currants caused by a gall mite (*Eriophyes ribis*) and characterized by abnormal swelling of the buds. **2.** A virus disease of the tomato characterized by swelling of the buds.

bigeminal [BIOL] Being doubled or paired.

Bignoniaceae [BOT] A family of dicotyledonous trees or shrubs in the order Scrophulariales characterized by a corolla with mostly five lobes, mature seeds with little or no endosperm and with wings, and opposite or whorled leaves.

big vein [PL PATH] A soil-borne virus disease of lettuce characterized by enlargement and yellowing of the leaf veins.

bijugate [BOT] Of a pinnate leaf, having two pairs of leaflets.

bilabial [ANTHRO] Distance between the highest point on the upper lip and lowest point on the lower lip.

bilabiate [BOT] Having two lips, such as certain corollas.

bilaminar [BIOL] Having or being arranged in two layers.

BICOSOECIDA

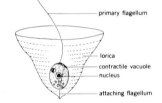

primary flagellum

lorica

contractile vacuole

nucleus

attaching flagellum

A bicosoecid, *Codomonas annulata.*

bilateral [BIOL] Of or relating to both right and left sides of an area, organ, or organism.

bilateral cleavage [EMBRYO] The division pattern of a zygote that results in a bilaterally symmetrical embryo.

bilateral hermaphroditism [ZOO] The presence of an ovary and a testis on each side of the animal body.

bilateral symmetry [BIOL] Symmetry such that the body can be divided by one median, or sagittal, dorsoventral plane into equivalent right and left halves, each a mirror image of the other.

Bilateria [ZOO] A major division of the animal kingdom embracing all forms with bilateral symmetry.

bile [PHYSIO] An alkaline fluid secreted by the liver and delivered to the duodenum to aid in the emulsification, digestion, and absorption of fats. Also known as gall.

bile acid [BIOCHEM] Any of the liver-produced steroid acids, such as taurocholic acid and glycocholic acid, that appear in the bile as sodium salts.

bile duct [ANAT] Any of the major channels in the liver through which bile flows toward the hepatic duct.

bile pigment [BIOCHEM] Either of two colored organic compounds found in bile: bilirubin and biliverdin.

bile salt [BIOCHEM] The sodium salt of glycocholic and taurocholic acids found in bile.

biliary atresia [MED] Failure of the bile ducts to develop in the embryo.

biliary cirrhosis [MED] A progressive inflammatory disease of the liver due to obstruction of bile ducts.

biliary colic [MED] Severe abdominal pain caused by passage of a gallstone through the bile ducts into the duodenum.

biliary diskinesia [MED] A functional spasticity of the sphincter of Oddi with disturbances in the speed of evacuation of the biliary tract.

biliary system [ANAT] The complex of canaliculi, or microscopic bile ducts, that empty into the larger intrahepatic bile ducts.

bilicyanin [BIOCHEM] A blue pigment found in gallstones; an oxidation product of biliverdin or bilirubin.

bilification [PHYSIO] Formation and excretion of bile.

biliprotein [BIOCHEM] The generic name for the organic compounds in certain algae that are composed of phycobilin and a conjugated protein.

bilirubin [BIOCHEM] $C_{33}H_{36}N_4O_6$ An orange, crystalline pigment occurring in bile; the major metabolic breakdown product of heme.

bilirubinemia [MED] The presence of bilirubin in the blood.

biliverdin [BIOCHEM] $C_{33}H_{34}N_4O_6$ A green, crystalline pigment occurring in the bile of amphibians, birds, and man; oxidation product of bilirubin in man.

bill [INV ZOO] A flattened portion of the shell margin of the broad end of an oyster. [VERT ZOO] The jaws, together with the horny covering, of a bird. [ZOO] Any jawlike mouthpart.

Billingsellacea [PALEON] A group of extinct articulate brachiopods in the order Orthida.

bilobate [BIOL] Divided into two lobes.

bilobular [BIOL] Having two lobules.

bilocular [BIOL] Having two cells or compartments.

bilophodont [ZOO] Having two transverse ridges, as the molar teeth of certain animals.

bimanous [VERT ZOO] Having two hands, especially referring to Primates.

bimaxillary [ANTHRO] Pertaining to the distance between the lower margins of the sutures of the maxilla and malar bones. [ZOO] Pertaining to the two halves of the maxilla.

binary fission [BIOL] A method of asexual reproduction

accomplished by the splitting of a parent cell into two equal, or nearly equal, parts, each of which grows to parental size and form.

binate [BOT] Growing in pairs.

binaural hearing [PHYSIO] The perception of sound by stimulation of two ears.

Binet age [PSYCH] An individual's mental age as determined by the Binet-Simon intelligence scale.

Binet's formula [PSYCH] The premise that children under 9 years of age whose mental development is retarded by 2 years are probably mentally deficient, and children of 9 years or older retarded by 3 years are definitely deficient.

Binet-Simon intelligence scale [PSYCH] Test for determining the relative mental development of children between 3 and 12 years of age; results are expressed as an intelligence quotient, the ratio of mental to chronological age.

binocular [BIOL] **1.** Of, pertaining to, or used by both eyes. **2.** Of a type of visual perception providing depth-of-field focus due to angular difference between the two retinal images.

binocular accommodation [PHYSIO] Automatic lens adjustment by both eyes simultaneously for focusing on distant objects.

binocular microscope [OPTICS] A microscope having two oculars, allowing the use of both eyes at once.

binodal [BIOL] Having two nodes.

binomen [SYST] A binomial name assigned to species, as *Canis familiaris* for the dog.

binomial nomenclature [SYST] The Linnean system of classification requiring the designation of a binomen, the genus and species name, for every species of plant and animal.

binomial trials [STAT] A sequence of trials, on each trial of which a certain result may or may not happen.

binuclear [CYTOL] Having two nuclei.

binucleolate [CYTOL] Having two nucleoli.

bioacoustics [BIOL] The study of the relation between living organisms and sound.

bioassay [ANALY CHEM] A method for quantitatively determining the concentration of a substance by its effect on the growth of a suitable animal, plant, or microorganism under controlled conditions.

biobubble [ECOL] A model concept of the ecosphere in which all living things are considered as particles held together by nonliving forces.

biocatalyst [BIOCHEM] A biochemical catalyst, especially an enzyme.

biocenology [ECOL] The study of natural communities and of interactions among the members of these communities.

biocenose *See* biotic community.

biochemical genetics [GEN] The study of inherited variation in respect of biochemical features.

biochemical oxygen demand [MICROBIO] The amount of dissolved oxygen required to meet the metabolic needs of anaerobic microorganisms in water rich in organic matter, such as sewage. Abbreviated BOD. Also known as biological oxygen demand.

biochemical oxygen demand test [MICROBIO] A standard laboratory procedure for measuring biochemical oxygen demand; standard measurement is made for 5 days at 20°C. Abbreviated BOD test.

biochemistry [CHEM] The study of chemical substances occurring in living organisms and the reactions and methods for identifying these substances.

biochemorphology [BIOCHEM] The science dealing with the chemical structure of foods and drugs and their reactions on living organisms.

biochore [ECOL] A group of similar biotopes.

biochrome [BIOCHEM] Any naturally occurring plant or animal pigment.

biochron [PALEON] A fossil of relatively short range of time.

biochronology [PALEON] The study of the fauna and flora of specific geologic time ranges.

biociation [ECOL] A subdivision of a biome distinguished by the predominant animal species.

biocide *See* pesticide.

bioclimatograph [ECOL] A climatograph showing the relation between climatic conditions and some living organisms.

bioclimatology [ECOL] The study of the effects of the natural environment on living organisms.

biocycle [ECOL] A group of similar biotopes composing a major division of the biosphere; there are three biocycles: terrestrial, marine, and fresh-water.

biocytin [BIOCHEM] $C_{16}H_{28}N_4O_4S$ Crystals with a melting point of $241-243°C$; obtained from dilute methanol or acetone solutions; characterized by its utilization by *Lactobacillus casei* and *L. delbrückii* LD5 as a biotin source, and by its unavailability as a biotin source to *L. arabinosus*. Also known as biotin complex of yeast.

biodynamic [AGR] Of or pertaining to a system of organic farming. [BIOPHYS] Of or pertaining to the dynamic relation between an organism and its environment.

bioelectric current [PHYSIO] A self-propagating electric current generated on the surface of nerve and muscle cells by potential differences across excitable cell membranes.

bioelectric model [PHYSIO] A conceptual model for the study of animal electricity in terms of physical principles.

bioelectrochemistry [PHYSIO] The study of the control of biological growth and repair processes by electrical stimulation.

bioenergetics [BIOCHEM] The branch of biology dealing with energy transformations in living organisms.

bioengineering [ENG] The application of engineering knowledge to the fields of medicine and biology.

bioflavonoid [BIOCHEM] A group of compounds obtained from the rinds of citrus fruits and involved with the homeostasis of the walls of small blood vessels; in guinea pigs a marked reduction of bioflavonoids results in increased fragility and permeability of the capillaries; used to decrease permeability and fragility in capillaries in certain conditions. Also known as citrus flavonoid compound; vitamin P complex.

biogenesis [BIOL] Development of a living organism from a similar living organism.

biogenetic law *See* recapitulation theory.

biogenic [BIOL] **1.** Essential to the maintenance of life. **2.** Produced by actions of living organisms.

biogenic reef [GEOL] A mass consisting of the hard parts of organisms, or of a biogenically constructed frame enclosing detrital particles, in a body of water; most biogenic reefs are made of corals or associated organisms.

biogenic sediment [GEOL] A deposit resulting from the physiological activities of organisms.

biogeography [ECOL] The science concerned with the geographical distribution of animal and plant life.

bioinstrumentation [ENG] The use of instruments attached to animals and man to record biological parameters such as breathing rate, pulse rate, body temperature, or oxygen in the blood.

biological [BIOL] Of or pertaining to life or living organisms. [IMMUNOL] A biological product used to induce immunity to

BIOGENIC REEF

A biogenic reef, the Ine Anchorage Reef, Arno Atoll, Marshall Islands.

various infectious diseases or noxious substances of biological origin.

biological balance [ECOL] Dynamic equilibrium that exists among members of a stable natural community.

biological clock [PHYSIO] Any physiologic factor that functions in regulating body rhythms.

biological control [ECOL] Natural or applied regulation of populations of pest organisms, especially insects, through the role or use of natural enemies.

biological equilibrium [BIOPHYS] A state of body balance for an actively moving animal, when internal and external forces are in equilibrium.

biological half-life [PHYSIO] The time required by the body to eliminate half of the amount of an administered substance through normal channels of elimination.

biological oxidation [BIOCHEM] Energy-producing reactions in living cells involving the transfer of hydrogen atoms or electrons from one molecule to another.

biological oxygen demand *See* biochemical oxygen demand.

biological productivity [ECOL] The quantity of organic matter or its equivalent in dry matter, carbon, or energy content which is accumulated during a given period of time.

biological specificity [BIOL] The principle that defines the orderly patterns of metabolic and developmental reactions giving rise to the unique characteristics of the individual and of its species.

biological standardization [PHARM] The standardization of drugs or biological products that cannot be chemically analyzed by studying the drugs' pharmacologic action on animals.

biological value [BIOCHEM] A measurement of the efficiency of the protein content in a food for the maintenance and growth of the body tissues of an individual.

biology [SCI TECH] A division of the natural sciences concerned with the study of life and living organisms.

bioluminescence [BIOL] The emission of visible light by living organisms.

biolysis [BIOL] **1.** Death and the following tissue disintegration. **2.** Decomposition of organic materials, such as sewage, by living organisms.

biomass [ECOL] The dry weight of living matter, including stored food, present in a species population and expressed in terms of a given area or volume of the habitat.

biome [ECOL] A complex biotic community covering a large geographic area and characterized by the distinctive life-forms of important climax species.

biomechanics [BIOPHYS] The study of the mechanics of living things.

biomedical engineering [ENG] The application of engineering technology to the solution of medical problems; examples are the development of prostheses such as artificial valves for the heart, various types of sensors for the blind, and automated artificial limbs.

biomedicine [MED] The science concerned with the study of the environment required for astronauts in space vehicles.

biomere [ECOL] A biostratigraphic unit bounded by abrupt nonevolutionary changes in the dominant elements of a single phylum.

biometeorology [BIOL] The study of the relationship between living organisms and atmospheric phenomena.

biometrics [STAT] The use of statistics to analyze observations of biological phenomena.

bion [ECOL] An independent, individual organism.

bionavigation [VERT ZOO] The ability of animals such as birds to find their way back to their roost, even if the landmarks on the outward-bound trip were effectively concealed from them.

bionics [ENG] The study of systems, particularly electronic systems, which function after the manner of living systems.

biophage *See* macroconsumer.

biophagous [ZOO] Feeding on living organisms.

biophysics [SCI TECH] The hybrid science involving the application of physical principles and methods to study and explain the structures of living organisms and the mechanics of life processes.

biophyte [BOT] A plant which derives nourishment from living organisms.

biopotency [BIOCHEM] Capacity of a chemical substance, as a hormone, to function in a biological system.

biopotential [PHYSIO] Voltage difference measured between points in living cells, tissues, and organisms.

biopsy [PATH] The removal and examination of tissues, cells, or fluids from the living body for the purposes of diagnosis.

biorheology [BIOPHYS] The application of the principles of rheology to biological systems; involves study of biological fluids such as blood, mucus, and synovial fluid.

biosonar [PHYSIO] A guidance system in certain animals, such as bats, utilizing the reflection of sounds that they produce as they move about.

biosphere [ECOL] The life zone of the earth, including the lower part of the atmosphere, the hydrosphere, soil, and the lithosphere to a depth of about 2 kilometers.

biosynthesis [BIOCHEM] Production, by synthesis or degradation, of a chemical compound by a living organism.

biota [BIOL] **1.** Animal and plant life characterizing a given region. **2.** Flora and fauna, collectively.

biotelemetry [ENG] The use of telemetry techniques, especially radio waves, to study behavior and physiology of living things.

biotherapy [MED] Treatment of disease with biologicals, that is, materials produced by living organisms.

biotic [BIOL] **1.** Of or pertaining to life and living organisms. **2.** Induced by the actions of living organisms.

biotic community [ECOL] An aggregation of organisms characterized by a distinctive combination of both animal and plant species in a particular habitat. Also known as biocenose.

biotic district [ECOL] A subdivision of a biotic province.

biotic environment [ECOL] That environment comprising living organisms, which interact with each other and their abiotic environment.

biotic isolation [ECOL] The occurrence of organisms in isolation from others of their species.

biotic province [ECOL] A community, according to some systems of classification, occupying an area where similarity of climate, physiography, and soils leads to the recurrence of similar combinations of organisms.

biotin [BIOCHEM] $C_{10}H_{16}N_2O_3S$ A colorless, crystalline vitamin of the vitamin B complex occurring widely in nature, mainly in bound form.

biotin complex of yeast *See* biocytin.

biotope [ECOL] An area of uniform environmental conditions and biota.

biotype [GEN] A group of organisms having the same genotype.

biovulate [BOT] Having two ovules.

biozone [PALEON] The range of a single taxonomic entity in geologic time as reflected by its occurrence in fossiliferous rocks.

biparasitic [ECOL] Parasitic upon or in a parasite.

BIOTIN

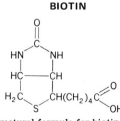

Structural formula for biotin.

BIPARTITE UTERUS

The bipartite uterus of the pig.
(From C. K. Wichert, Elements
of Chordate Anatomy, 3d. ed.,
McGraw-Hill, 1967)

BIPINNARIA

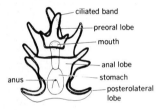

Bipinnaria larva showing
component parts.

biparietal [ANTHRO] Distance between the most distant opposite points of the parietal bone.

biparous [BOT] Having branches on dichotomous axes. [VERT ZOO] Bringing forth two young at a birth.

bipartite uterus [ANAT] A uterus divided into two parts almost to the base.

bipectinate [INV ZOO] Of the antennae of certain moths, having two margins with comblike teeth. [ZOO] Branching like a feather on both sides of a main shaft.

biped [VERT ZOO] **1.** A two-footed animal. **2.** Any two legs of a quadruped.

bipedal [BIOL] Having two feet.

bipeltate [BOT] Having two shield-shaped parts. [ZOO] Having a shell or other covering resembling a double shield.

bipenniform [ANAT] Of the arrangement of muscle fibers, resembling a feather barbed on both sides.

bipetalous [BOT] Having two petals.

biphasic [BOT] Having both a sporophyte and a gametophyte generation in the life cycle.

biphyletic [EVOL] Descended in two branches from a common ancestry.

Biphyllidae [INV ZOO] The false skin beetles, a family of coleopteran insects in the superfamily Cucujoidea.

bipinnaria [INV ZOO] The complex, bilaterally symmetrical, free-swimming larval stage of most asteroid echinoderms.

bipinnate [BOT] Pertaining to a leaf that is pinnate for both its primary and secondary divisions.

biplicate [BIOL] Having two folds.

bipocillus [INV ZOO] A microsclere characterized by a curved shaft and a cup-shaped enlargement at each end.

bipolar [SCI TECH] Having two poles.

bipolar cell [HISTOL] **1.** A neuron with only one axon and one dendrite. **2.** In the eye, a cell that connects the rods and cones with ganglion cells.

Bipolarina [INV ZOO] A suborder of protozoan parasites in the order Myxosporida.

bipotential [BIOL] Having the potential to develop in either of two mutually exclusive directions.

bipotentiality [BIOL] **1.** Capacity to function either as male or female. **2.** Hermaphroditism.

biradial symmetry [BIOL] Symmetry both radial and bilateral. Also known as disymmetry.

biramous [BIOL] Having two branches, such as an arthropod appendage.

birch [BOT] The common name for all deciduous trees of the genus *Betula* that compose the family Betulaceae in the order Fagales.

bird [VERT ZOO] Any of the warm-blooded vertebrates which make up the class Aves.

bird louse [INV ZOO] The common name for any insect of the order Mallophaga. Also known as biting louse.

bird of prey [VERT ZOO] Any of various carnivorous birds of the orders Falconiformes and Strigiformes which feed on meat taken by hunting.

bird's eye rot [PL PATH] A fungus disease of the grape caused by *Elsinoe ampelina* and characterized by small, dark, sunken spots with light centers on the fruit.

bird's eye spot [PL PATH] **1.** A plant disease characterized by dark, round spots surrounded by lighter tissue. **2.** A fungus disease of tea leaves caused by *Orcospora theae*. **3.** A leaf spot of the hevea rubber tree caused by the fungus *Helminthosporium heveae*.

birimose [BOT] Opening by two slits, as an anther.

birotulate [INV ZOO] A sponge spicule characterized by two wheel-shaped ends.

birth [BIOL] The emergence of a new individual from the body of its parent.

birth canal [ANAT] The channel in mammals through which the fetus is expelled during parturition; consists of the cervix, vagina, and vulva.

birth control [MED] Limitation of the number of children born by preventing or reducing the frequency of impregnation.

birth defect *See* congenital anomaly.

birthmark [MED] Any abnormal cellular or vascular benign nevus that is present at birth or that appears sometime later.

birth rate [BIOL] The ratio between the number of live births and a specified number of people in a population over a given period of time.

biseptate [BIOL] Having two partitions.

biserial [BIOL] Arranged in two rows or series.

biserrate [BIOL] **1.** Having serrated serrations. **2.** Serrate on both sides.

bisexual [BIOL] Of or relating to two sexes. [PSYCH] **1.** Having mental and behavioral characteristics of both sexes. **2.** Having sexual desires for members of both sexes.

bismuth [CHEM] A metallic element, symbol Bi, of atomic number 83 and atomic weight 208.980. [MINERAL] The brittle, rhombohedral mineral form of the native element bismuth.

bismuth citrate [ORG CHEM] BiC_6H_5O A salt of citric acid that forms white crystals, insoluble in water; used as an astringent.

bismuth oleate [ORG CHEM] $Bi(C_{17}H_{33}COO)_3$ A salt of oleic acid obtained as yellow granules; used in medicines to treat skin diseases.

bismuth phenate [ORG CHEM] $C_6H_5O \cdot Bi(OH)_2$ An odorless, tasteless, gray-white powder; used in medicine. Also known as bismuth phenolate; bismuth phenylate.

bismuth phenolate *See* bismuth phenate.

bismuth phenylate *See* bismuth phenate.

bismuth pyrogallate [ORG CHEM] $Bi(OH)C_6H_3(OH)O_2$ An odorless, tasteless, yellowish-green, amorphous powder; used in medicine as intestinal antiseptic and dusting powder. Also known as basic bismuth pyrogallate; helcosol.

bismuth subsalicylate [INORG CHEM] $Bi(C_7H_5O)_3Bi_2O_3$ A white powder that is insoluble in ethanol and water; used in medicine. Also known as basic bismuth salicylate.

bison [VERT ZOO] The common name for two species of the family Bovidae in the order Artiodactyla; the wisent or European bison (*Bison bonasus*), and the American species (*B. bison*).

bisporangiate [BOT] Having two different types of sporangia.

bispore [BOT] In certain red algae, an asexual spore that is produced in pairs.

bisporic embryo sac [BOT] An embryo sac that arises from two megaspores.

bistipulate [BOT] Having two stipules.

bisulcate [BOT] Having two grooves.

bite [BIOL] **1.** To seize with the teeth. **2.** Closure of the lower teeth against the upper teeth. [MED] Skin injury produced by an animal's teeth or the mouthparts of an insect.

bitegmic [BOT] Having two integuments, especially in reference to ovules.

biternate [BOT] Of a ternate leaf, having each division ternate.

biting louse *See* bird louse.

bittern [VERT ZOO] Any of various herons of the genus *Botaurus* characterized by streaked and speckled plumage.

BISMUTH

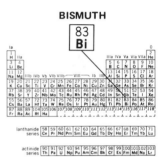

Periodic table of the chemical elements showing the position of bismuth.

BISON

North American bison (*Bison bison*), with enormous forequarter development and hump behind the head.

bitter pit [PL PATH] A disease of uncertain etiology affecting apple, pear, and quince; spots of dead brown tissue appear in the flesh of the fruit and discolored depressions are seen on its surface. Also known as Baldwin spot; stippen.

bivalent chromosome [CYTOL] The structure formed following synapsis of a pair of homologous chromosomes from the zygotene stage of meiosis up to the beginning of anaphase.

bivalve [BOT] Referring to a seed capsule consisting of two parts or plates. [INV ZOO] **1.** Having two valves, as a clam. **2.** The common name for a number of diverse, bilaterally symmetrical animals, including mollusks, ostracod crustaceans, and brachiopods, having a soft body enclosed in a calcareous two-part shell.

Bivalvia [INV ZOO] A large class of the phylum Mollusca containing the clams, oysters, and other bivalves.

biventer [ANAT] A muscle having two bellies.

bivittate [ZOO] Having a pair of longitudinal stripes.

bivium [INV ZOO] The pair of starfish rays that extend on either side of the madreporite.

bivoltine [INV ZOO] **1.** Having two broods in a season, used especially of silkworms. **2.** Of insects, producing two generations a year.

Bk *See* berkelium.

blackberry [BOT] Any of the upright or trailing shrubs of the genus *Rubus* in the order Rosales; an edible berry is produced by the plant.

blackbird [VERT ZOO] Any bird species in the family Icteridae, of which the males are predominantly or totally black.

black blight [PL PATH] Any of several diseases of tropical plants caused by superficial sooty molds.

black canker *See* ink disease.

black chaff [PL PATH] A bacterial disease of wheat caused by *Xanthomonas translucens undulosa* and characterized by dark, longitudinal stripes on the chaff.

black coral [INV ZOO] The common name for antipatharian coelenterates having black, horny axial skeletons.

black disease [VET MED] Necrotic hepatitis of sheep, resulting from infection with *Clostridium novyi* type B, with the necessary conditions for the growth of the clostridia provided by the damaged liver tissue produced by the fluke *Fasciola hepatica*.

black end [PL PATH] **1.** A disease of the pear marked by blackening of the epidermis and flesh in the region of the calyx; believed to be a result of a disturbed water relation. **2.** A fungus disease of the banana caused by several species, especially *Gloesporium musarum*, characterized by discoloration of the stem of the fruit.

blackeye bean *See* cowpea.

blackfire [PL PATH] A bacterial disease of tobacco caused by *Pseudomonas angulata* and characterized by angular leaf spots which gradually darken and may fall out, leaving ragged holes.

blackhead disease [PL PATH] **1.** A parasitic disease of the banana caused by eelworms of the family Tylenchidae. **2.** A rot disease of the banana rootstock caused by the fungus *Thielaviopsis paradoxa*.

black kernel [PL PATH] A fungus disease of rice caused by *Curvularia lunata* and characterized by dark discoloration of the kernels.

black knot [PL PATH] A fungus disease of certain fruit and nut trees that is characterized by black excrescences on the branches; destructive to plum, cherry, gooseberry, filbert, and hazel.

blackleg [VET MED] An acute, usually fatal bacterial disease

BLACKBERRY

Thorny, biennial stem of blackberry shrub.

of cattle, and occasionally of sheep, goats, and swine, caused by *Clostridium chauvoei.*

black line [PL PATH] A disease of walnuts, especially of English varieties grafted to black walnuts, characterized by a black line of dead tissue at the graft union, eventually leading to death of the tree.

blacklung *See* anthracosis.

black measles [MED] *See* hemorrhagic measles. [PL PATH] A disease of California grapevines of uncertain etiology; characterized by black spotting of the skin of the fruit, browning and dropping of the leaves, and dying back of the canes from the tip.

black mold [MYCOL] Any dark fungus belonging to the order Mucorales. [PL PATH] A fungus disease of rose grafts and onion bulbs marked by black appearance due to the mold.

blacknose [PL PATH] A physiological disease of the date, with the distal end of the fruit becoming dark, cracking, and shriveling.

black patch [PL PATH] A fungus disease of red clover characterized by simultaneous blackening of plants in groups.

black pepper [FOOD ENG] A spice; the dried unripe berries of *Piper nigrum,* a vine of the pepper family (Piperaceae) in the order Piperales.

black pod [PL PATH] A pod rot of cacao caused by the fungus *Phytophthora faveri.*

black point [PL PATH] A disease of wheat and other cereals characterized by blackening of the embryo ends of the grains

black ring [PL PATH] **1.** A virus disease of cabbage and other members of the family Cruciferae characterized by dark necrotic and often sunken rings on the surface of the leaf. **2.** A virus disease of the tomato characterized in the early stage by small black rings on young leaves.

black root [PL PATH] Any plant disease characterized by black discolorations of the roots.

black root rot [PL PATH] **1.** Any of several plant diseases characterized by dark lesions of the root. **2.** A fungus disease of the apple caused by *Xylaria mali.* **3.** A fungus disease of tobacco and other plants caused by *Thielaviopsis basicola.*

black rot [PL PATH] Any fungal or bacterial disease of plants characterized by dark brown discoloration and decay of a plant part.

black scour [VET MED] Hemorrhagic enteritis of sheep, swine, and cattle, usually associated with a heavy worm burden but sometimes caused by bacterial infection.

black shank [PL PATH] A black rot disease of tobacco caused by *Phytophthora parasitica* var. *nicotianae.*

black spot [PL PATH] Any bacterial or fungal disease of plants characterized by black spots on a plant part.

black stem [PL PATH] Any of several fungal diseases of plants characterized by blackening of the stem.

black thread [PL PATH] A fungus disease of the para rubber tree at the point where the tree is tapped, caused by *Phytophthora meadii* and characterized by black stripes extending through the exposed bast into the cambium or wood. Also known as black stripe; stripe canker.

black tip [PL PATH] Any of several plant diseases characterized by dark necrotic areas at the tip of the seed or fruit.

blacktongue [VET MED] A niacin-deficiency disease of dogs characterized by black discoloration of the tongue.

black vomit [MED] Dark vomited matter, consisting of digested blood and gastric contents.

blackwater [MED] Any disease characterized by dark-colored urine.

blackwater fever [MED] A complication of falciparum ma-

BLACK PEPPER

Berries on a branch of *Piper nigrum.*

laria, characterized by intravascular hemolysis, hemoglobinuria, tachycardia, high fever, and poor prognosis.

bladder [ANAT] Any saclike structure in man and animals, such as a swimbladder or urinary bladder, that contains a gas or functions as a receptacle for fluid. [BIOL] *See* vesicle.

bladder cell [INV ZOO] Any of the large vacuolated cells in the outer layers of the tunic in some tunicates.

blade [BOT] The broad, flat portion of a leaf. [VERT ZOO] A single plate of baleen.

Blancan [GEOL] Upper Pliocene or lowermost Pleistocene geologic time.

blast cell [HISTOL] An undifferentiated precursor of a human blood cell in the reticuloendothelial tissue.

blastema [EMBRYO] **1.** A mass of undifferentiated protoplasm capable of growth and differentiation. **2.** *See* anlage.

Blastobasidae [INV ZOO] A family of lepidopteran insects in the superfamily Tineoidea.

blastocarpous [BOT] Germinating in the pericarp.

blastochyle [EMBRYO] The fluid filling the blastocoele.

Blastocladiales [MYCOL] An order of aquatic fungi in the class Phycomycetes.

blastocoele [EMBRYO] The cavity of a blastula. Also known as segmentation cavity.

blastocone [EMBRYO] An incomplete blastomere.

blastocyst [EMBRYO] A modified blastula characteristic of placental mammals.

blastocyte [EMBRYO] An embryonic cell that is undifferentiated. [INV ZOO] An undifferentiated cell capable of replacing damaged tissue in certain lower animals.

blastoderm [EMBRYO] The blastodisk of a fully formed vertebrate blastula.

blastodisk [EMBRYO] The embryo-forming, protoplasmic disk on the surface of a yolk-filled egg, such as in reptiles, birds, and some fish.

Blastoidea [PALEON] A class of extinct pelmatozoan echinoderms in the subphylum Crinozoa.

blastokinesis [INV ZOO] Movement of the embryo into the yolk in some insect eggs.

blastoma [MED] **1.** A tumor whose parenchymal cells have certain embryonal characteristics. **2.** A true tumor.

blastomere [EMBRYO] A cell of a blastula.

blastomycosis [MED] A term for two infectious, yeastlike fungus diseases of man: North American blastomycosis, caused by *Blastomyces dermatitidis*, and South American (paracoccidioidomycosis) caused by *Blastomyces brasiliensis.*

blastophore [CYTOL] The cytoplasm that is detached from a spermatid during its transformation to a spermatozoon. [INV ZOO] An amorphose core of cytoplasm connecting cells of the male morula of developing germ cells in oligochaetes.

blastopore [EMBRYO] The opening of the archenteron.

blastospore [MYCOL] A fungal resting spore that arises by budding.

blastostyle [INV ZOO] A zooid on certain hydroids that lacks a mouth and tentacles and functions to produce medusoid buds.

blastotomy [EMBRYO] Separation of cleavage cells during early embryogenesis.

blastozooid [INV ZOO] A zooid produced by budding.

blastula [EMBRYO] A hollow sphere of cells characteristic of the early metazoan embryo.

blastulation [EMBRYO] Formation of a blastula from a solid ball of cleaving cells.

Blattidae [INV ZOO] The cockroaches, a family of insects in the order Orthoptera.

BLASTOIDEA

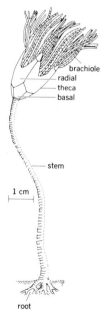

brachiole
radial
theca
basal

stem

1 cm

root

A drawing of a blastoid, *Pentremites.*

bleb [MED] A localized collection of fluid, as serum or blood, in the epidermis.

bleeder [MED] **1.** A person subject to frequent hemorrhages, as a hemophiliac. **2.** A blood vessel from which there is persistent uncontrolled bleeding. **3.** A blood vessel which has escaped closure by cautery or ligature during a surgical procedure.

bleeding canker [PL PATH] A fungus disease of hardwoods caused by *Phytophthora cactorum* and characterized by cankers which exude a reddish ooze on the trunk and branches.

bleeding disease [PL PATH] A fungus disease of the coconut palm caused by *Ceratostomella paradoxa*, characterized by a rust-colored exudation from cracks on the stem.

bleeding time [PHYSIO] The time required for bleeding to stop after a small puncture wound.

blending inheritance [GEN] Inheritance in which the character of the offspring is a blend of those in the parents; a common feature for quantitative characters, such as stature, determined by large numbers of genes and affected by environmental variation.

Blenniidae [VERT ZOO] The blennies, a family of carnivorous marine fishes in the suborder Blennioidei.

Blennioidei [VERT ZOO] A large suborder of small marine fishes in the order Perciformes that live principally in coral and rock reefs.

blennorrhagia [MED] Excessive discharge of mucus. Also known as blennorrhea.

blennorrhea *See* blennorrhagia.

blephara [BOT] In mosses, a peristome tooth.

Blephariceridae [INV ZOO] A family of dipteran insects in the suborder Orthorrhapha.

blepharism [MED] Spasm of the eyelids causing rapid, repetitive involuntary winking.

blepharitis [MED] Inflammation of the eyelids.

blepharoconjunctivitis [MED] Inflammation of the eyelids and the conjunctiva.

blepharoplast [BIOL] A basal granule or kinetoplast of uncertain nature.

blepharospasm [MED] Spasmodic winking due to spasms of the orbicular muscle of the eyelid.

blight [PL PATH] Any plant disease or injury that results in general withering and death of the plant without rotting.

blight canker [PL PATH] A phase of fire blight characterized by cankers.

blindness [MED] **1.** Loss or absence of the ability to perceive visual images. **2.** The condition of a person having less than 1/10 (20/200 on the Snellen test) normal vision.

blind seed [PL PATH] A fungus disease of forage grasses caused by *Phealea temulenta*, resulting in abortion of the seed.

blind spot [PHYSIO] A place on the retina of the eye that is insensitive to light, where the optic nerve passes through the eyeball's inner surface.

blind trial [STAT] A trial in which experimenters and subjects are kept uninformed as to whether or not treatment has been given.

blister [MED] A local swelling of the skin resulting from the accumulation of serous fluid between the epidermis and true skin.

blister blight [PL PATH] **1.** A fungus disease of the tea plant caused by *Exobasidium vexans* and characterized by blister-like lesions on the leaves. **2.** A rust disease of Scotch pine caused by *Cronartium asclepiadeum* and characterized by blisterlike lesions on the twigs.

blister canker [PL PATH] A fungus disease of the apple tree caused by *Nummularia discreta* and characterized by rough, black cankers on the trunk and large branches. Also known as apple pox.

blister rust [PL PATH] Any of several diseases of pines caused by rust fungi of the genus *Cronartium;* the sapwood and inner bark are affected and blisters are produced externally.

blister spot [PL PATH] A bacterial disease of the apple caused by *Pseudomonas papulans* and characterized by dark-brown blisters on the fruit and cankers on the branches.

bloat [VET MED] Distension of the rumen in cattle and other ruminants due to excessive gas formation following heavy fermentation of legumes eaten wet.

blood [HISTOL] A fluid connective tissue consisting of the plasma and cells that circulate in the blood vessels.

blood agar [MICROBIO] A nutrient microbiologic culture medium enriched with whole blood and used to detect hemolytic strains of bacteria.

blood bank [ENG] A place for storing whole blood or plasma under refrigeration.

blood blister [MED] A blister that is filled with blood.

blood cell [HISTOL] An erythrocyte or a leukocyte.

blood chimerism [GEN] Having red blood cells of two genetic types.

blood count [PATH] Determination of the number of white and red blood cells in a definite volume of blood.

blood crisis [MED] The sudden appearance of large numbers of nucleated erythrocytes in the circulating blood.

blood disease [PL PATH] A bacterial disease affecting the vascular tissues of the banana in the Celebes; believed to be caused by *Xanthomonas celebensis* and characterized by blighting of the leaves and reddish-brown rot of the fruit.

blood dyscrasia [MED] Any abnormal condition of the formed elements of blood or of the constituents required for clotting.

blood gills [INV ZOO] In certain aquatic insects, thin-walled sacs filled with blood that may function in respiration or osmotic activity.

blood group [IMMUNOL] An immunologically distinct, genetically determined class of human erythrocyte antigens, identified as A, B, AB, and O.

blood island [EMBRYO] One of the areas in the yolk sac of vertebrate embryos allocated to the production of the first blood cells.

blood-plate hemolysis [MICROBIO] Destruction of red blood cells in a blood agar medium by a bacterial toxin.

blood platelet *See* thrombocyte.

blood poisoning *See* septicemia.

blood pressure [PHYSIO] Pressure exerted by blood on the walls of the blood vessels.

bloodstream [PHYSIO] The flow of blood in its circulation through the body.

blood sugar [BIOCHEM] The carbohydrate, principally glucose, of the blood.

blood test [PATH] **1.** A serologic test for syphilis. **2.** A blood count. **3.** A test for detection of blood, usually one based on the peroxidase activity of blood, such as the benzidine test or guaiac test.

blood typing [IMMUNOL] Determination of an individual's blood group.

blood vessel [ANAT] A tubular channel for blood transport.

bloom [BOT] **1.** An individual flower. Also known as blossom. **2.** To yield blossoms. **3.** The waxy coating that appears as a powder on certain fruits, such as plums, and leaves, such as cabbage. [ECOL] Colored area on the surface of bodies of water caused by heavy planktonic growth.

blossom *See* bloom.

blossom-end rot [PL PATH] **1.** Any rot disease of fruit that originates at the blossom end. **2.** A physiological disease of tomato believed to be caused by extreme fluctuations in available moisture; characterized by shallow leathery depressions with a water-soaked appearance around the tip end of the fruit.

blowball [BOT] A fluffy seed ball, as of the dandelion.

blowhole [VERT ZOO] The nostril on top of the head of cetacean mammals.

blubber [INV ZOO] A large sea nettle or medusa. [VERT ZOO] A thick insulating layer of fat beneath the skin of whales and other marine mammals.

blue baby [MED] An infant with congenital cyanosis due to cardiac or pulmonary defect, causing shunting of unoxygenated blood into the systemic circulation.

blueberry [BOT] Any of several species of plants in the genus *Vaccinium* of the order Erecales; the fruit, a berry, occurs in clusters on the plant.

bluefish [VERT ZOO] *Pomatomus saltatrix.* A predatory fish in the order Perciformes. Also known as skipjack.

bluegrass [BOT] The common name for several species of perennial pasture and lawn grasses in the genus *Poa* of the order Cyperales.

blue-green algae [BOT] The common name for members of the Cyanophyceae.

blue nevus [MED] A nevus composed of spindle-shaped pigmented melanocytes in the middle and lower two-thirds of the dermis.

blue rot [PL PATH] A fungus disease of conifers caused by members of the genus *Ceratostomella,* characterized by blue discoloration of the wood. Also known as bluing.

blue stain [PL PATH] A bluish stain of the sapwood of many trees caused by several fungi, especially of the genera *Fusarium, Ceratostomella,* and *Penicillium.*

bluestem grass [BOT] The common name for several species of tall, perennial grasses in the genus *Andropogon* of the order Cyperales.

bluetongue [VET MED] **1.** A disease of African horses characterized by lesions that are most pronounced about the head. Also known as thickhead. **2.** A virus disease of African sheep characterized by hyperemia, cyanosis, and punctate hemorrhages and by swelling and sloughing of the epithelium about the mouth and tongue.

bluing *See* blue rot.

blunt dissection [MED] In surgery, the exposure of structures or separation of tissues without cutting.

BMR *See* basal metabolic rate.

boa [VERT ZOO] Any large, nonvenomous snake of the family Boidae in the order Squamata.

boar *See* wild boar.

Bobasatranidae [PALEON] A family of extinct palaeonisciform fishes in the suborder Platysomoidei.

BOD *See* biochemical oxygen demand.

Bodonidae [INV ZOO] A family of protozoans in the order Kinetoplastida characterized by two unequally long flagella, one of them trailing.

BOD test *See* biochemical oxygen demand test.

body cavity [ANAT] The peritoneal, pleural, or pericardial cavities, or the cavity of the tunica vaginalis testis.

body rhythm [PHYSIO] Any bodily process having some degree of regular periodicity.

body-righting reflex [PHYSIO] A postural reflex, initiated by the asymmetric stimulation of the body surface by the weight of the body, so that the head tends to assume a horizontal position.

BLUEBERRY

A cluster of blueberries.

BLUEFISH

The bluefish, an excellent food and game fish.

BODONIDAE

20 μ

Bodo, a genus in the family Bodonidae, showing the two unequally long flagella.

bog [ECOL] A plant community that develops and grows in areas with permanently waterlogged peat substrates. Also known as moor; quagmire.

bog harrow [AGR] A type of disk harrow with extra-large notched disks.

Boidae [VERT ZOO] The boas, a family of nonvenomous reptiles of the order Squamata, having teeth on both jaws and hindlimb rudiments.

boil *See* furuncle.

boil disease [VET MED] A protozoan disease of fish caused by *Myxobolus pfeiffer* that forms large tumorous masses in the muscles and connective tissues, finally causing death.

boiling point [PHYS CHEM] Abbreviated bp. **1.** The temperature at which the transition from the liquid to the gaseous phase occurs in a pure substance at fixed pressure. **2.** *See* bubble point.

boiling point elevation [CHEM] The raising of the normal boiling point of a pure liquid compound by the presence of a dissolved substance, the elevation being in direct relation to the dissolved substance's molecular weight.

boil smut [PL PATH] A fungus disease of corn caused by *Ustilago maydis*, characterized by galls containing black spores.

boll [BOT] A pod or capsule (pericarp), as of cotton and flax.

boll rot [PL PATH] A fungus rot of cotton bolls caused by *Glomerella gossypii* and *Xanthomonas malvacearum*.

boll weevil [INV ZOO] A beetle, *Anthonomus grandis*, of the order Coleoptera; larvae destroy cotton plants and are the most important pests in agriculture.

Boltzmann constant [STAT MECH] The ratio of the universal gas constant to the Avogadro number.

bolus [PHARM] A large pill. [PHYSIO] The mass of food prepared by the mouth for swallowing.

Bombacaceae [BOT] A family of dicotyledonous tropical trees in the order Malvales with dry or fleshy fruit usually having woolly seeds.

bomb calorimeter [ENG] A calorimeter designed with a strong-walled container constructed of a corrosion-resistant alloy, called the bomb, immersed in about 2.5 liters of water in a metal container; the sample, usually an organic compound, is ignited by electricity, and the heat generated is measured.

Bombidae [INV ZOO] A family of relatively large, hairy, black and yellow bumblebees in the hymenopteran superfamily Apoidea.

Bombycidae [INV ZOO] A family of lepidopteran insects of the superorder Heteroneura that includes only the silkworms.

Bombyliidae [INV ZOO] The bee flies, a family of dipteran insects in the suborder Orthorrhapha.

bond [CHEM] The strong attractive force that holds together atoms in molecules and crystalline salts. Also known as chemical bond.

bond angle [PHYS CHEM] The angle between bonds sharing a common atom. Also known as valence angle.

bond distance [PHYS CHEM] The distance separating the two nuclei of two atoms bonded to each other in a molecule. Also known as bond length.

bond energy [PHYS CHEM] The heat of formation of a molecule from its constituent atoms.

bond length *See* bond distance.

bone [ANAT] One of the parts of a vertebrate skeleton. [HISTOL] A hard connective tissue that forms the major portion of the vertebrate skeleton.

Bone Age [ARCHEO] A prehistoric period of human culture

BOMBIDAE

A drawing of a hairy, black and yellow bumblebee. *(From T. I. Storer and R. L. Usinger, General Zoology, 3d ed., McGraw-Hill, 1957)*

characterized by the use of implements made of bone and antler.

bone conduction [BIOPHYS] Transmission of sound vibrations to the internal ear via the bones of the skull.

Bonellidae [INV ZOO] A family of wormlike animals belonging to the order Echiuroinea.

bone marrow [HISTOL] A vascular modified connective tissue occurring in the long bones and certain flat bones of vertebrates.

Bononian [GEOL] Upper Jurassic (lower Portlandian) geologic time.

bony fish [VERT ZOO] The name applied to all members of the class Osteichthyes.

bony labyrinth [ANAT] The system of canals within the otic bones of vertebrates that houses the membranous labyrinth of the inner ear.

Boodleaceae [BOT] A family of green marine algae in the order Siphonocladales.

book gill [INV ZOO] A type of gill in king crabs consisting of folds of membranous tissue arranged like the leaves of a book.

book lung [INV ZOO] A saccular respiratory organ in many arachnids consisting of numerous membranous folds arranged like the pages of a book.

Boopidae [INV ZOO] A family of lice in the order Mallophaga, parasitic on Australian marsupials.

booster [IMMUNOL] The dose of an immunizing agent given to stimulate the effects of a previous dose of the same agent.

booster response *See* anamnestic response.

Bopyridae [INV ZOO] A family of epicaridean isopods in the tribe Bopyrina known to parasitize decapod crustaceans.

Bopyrina [INV ZOO] A tribe of dioecious isopods in the suborder Epicaridea.

Bopyroidea [INV ZOO] An equivalent name for Epicaridea.

Boraginaceae [BOT] A family of flowering plants in the order Lamiales comprising mainly herbs and some tropical trees.

Bordeaux mixture [MATER] A fungicide made from a mixture of lime, copper sulfate, and water.

bordered [BOT] Having a margin with a distinctive color or texture; used especially of a leaf.

bordered pit [BOT] A wood-cell pit having the secondary cell wall arched over the cavity of the pit.

borderline mental retardation [PSYCH] Below-normal intellectual functioning associated with an intelligence quotient of about 68–85.

borderline psychosis [PSYCH] The psychiatric diagnosis of an individual whose symptoms are severe but are not clearly psychotic or neurotic.

borderline syndrome [PSYCH] An impairment of ego function noted in individuals who have been angry most of their lives and who cannot relate meaningfully to or love other people, and who in consequence develop nonpsychotic but deviant mechanisms, such as withdrawal while expressing angry concern, or showing passive compliance at the price of real involvement, to maintain a precarious mental equilibrium.

Bordetella [MICROBIO] A genus of minute, gram-negative, parasitic coccobacilli in the family Brucellaceae.

Bordetella parapertussis [MICROBIO] A small, nonmotile coccobacillus that causes parapertussis.

Bordetella pertussis [MICROBIO] A small, nonmotile coccobacillus that causes whooping cough.

Bordet-Gengou bacillus *See* Hemophilus pertussis.

boreal [ECOL] Of or relating to northern geographic regions.

Boreal life zone [ECOL] The zone comprising the climate and

BONELLIDAE

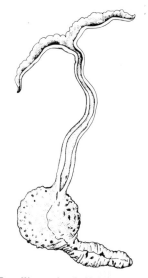

Bonellia species, half size, showing the long and cleft (at the top) prostomium characteristic of the Bonellidae.

BORON

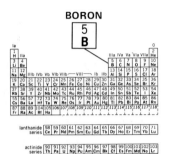

Periodic table of the chemical elements showing the position of boron.

BOTHRIDIUM

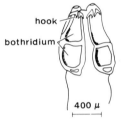

hook

bothridium

400 μ

A drawing of the scolex of *Acanthobothrium* species showing the hooks on the bothridium.

BOTHRIOCIDAROIDA

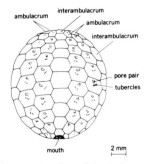

interambulacrum

ambulacrum

ambulacrum

interambulacrum

pore pair

tubercles

mouth   2 mm

A reconstruction of the test of *Bothriocidaris*.

biotic communities between the Arctic and Transitional zones.

borer [INV ZOO] Any insect or other invertebrate that burrows into wood, rock, or other substances.

Borhyaenidae [VERT ZOO] A family of carnivorous mammals in the superfamily Borhyaenoidea.

Borhyaenoidea [VERT ZOO] A superfamily of carnivorous mammals in the order Marsupialia.

boring sponge [INV ZOO] Marine sponge of the family Clionidae represented by species which excavate galleries in mollusks, shells, corals, limestone, and other calcareous matter.

boron [CHEM] A chemical element, symbol B, atomic number 5, atomic weight 10.811; it has three valence electrons and is nonmetallic.

bosque *See* temperate and cold scrub.

bosset [VERT ZOO] The rudimentary antler of male red deer, formed during the first year.

Bostrichidae [INV ZOO] The powder-post beetles, a family of coleopteran insects in the superfamily Bostrichoidea.

Bostrichoidea [INV ZOO] A superfamily of beetles in the coleopteran suborder Polyphaga.

botanical garden [BOT] An institution for the culture of plants collected chiefly for scientific and educational purposes.

botany [BIOL] A branch of the biological sciences which embraces the study of plants and plant life.

bothridium [INV ZOO] A muscular holdfast organ, often with hooks, on the scolex of tetraphyllidean tapeworms.

Bothriocephaloidea [INV ZOO] The equivalent name for the Pseudophyllidea.

Bothriocidaroida [PALEON] An order of extinct echinoderms in the subclass Perischoechinoidea in which the ambulacra consist of two columns of plates, the interambulacra of one column, and the madreporite is placed radially.

bothrium [INV ZOO] A suction groove on the scolex of pseudophyllidean tapeworms.

botryomycosis [VET MED] A chronic infectious bacterial disease of horses caused by *Staphylococcus aureus* and characterized by localized fibromatous tumors.

Botrytis disease [PL PATH] Any of various fungus diseases of plants caused by fungi of the genus *Botrytis;* characterized by soft rotting.

bottle graft [BOT] A plant graft in which the scion is a detached branch and is protected from wilting by keeping the base of the branch in a bottle of water until union with the stock.

bottom fauna *See* benthos.

bottom rot [PL PATH] **1.** A fungus disease of lettuce, caused by *Pellicularia filamentosa*, that spreads from the base upward. **2.** A fungus disease of tree trunks caused by pore fungi.

botulin [MICROBIO] The neurogenic toxin which is produced by *Clostridium botulinum* and *C. parabotulinum* and causes botulism. Also known as botulinus toxin.

botulinus [MICROBIO] A bacterium that causes botulism.

botulinus toxin *See* botulin.

botulism [MED] Food poisoning due to intoxication by the exotoxin of *Clostridium botulinum* and *C. parabotulinum.*

Bouguer-Lambert-Beer law [ANALY CHEM] The intensity of a beam of monochromatic radiation in an absorbing medium decreases exponentially with penetration distance. Also known as Beer-Lambert-Bouguer law; Lambert-Beer law.

Bouin's solution [MATER] A picric acid–acetic acid–formaldehyde fixative and preserving fluid for contractile forms.

bouton [ANAT] A club-shaped enlargement at the end of a nerve fiber. Also known as end bulb.

Bovidae [VERT ZOO] A family of pecoran ruminants in the superfamily Bovoidea containing the true antelopes, sheep, and goats.

bovine [VERT ZOO] **1.** Any member of the genus *Bos.* **2.** Resembling or pertaining to a cow or ox.

bovine mastitis [VET MED] Inflammation of the udder of a cow; may result from injury or bacterial infection.

bovine staggers [VET MED] A disease of cattle in southern Africa caused by eating the poisonous herb *Matricaria nigellaefolia* and characterized by staggering, emaciation, and finally paralysis.

Bovoidea [VERT ZOO] A superfamily of pecoran ruminants in the order Artiodactyla comprising the pronghorns and bovids.

bowel [ANAT] The intestine.

bowfin [VERT ZOO] *Amia calva.* A fish recognized as the only living species of the family Amiidae. Also known as dogfish; grindle; mudfish.

Bowman's capsule [ANAT] A two-layered membranous sac surrounding the glomerulus and constituting the closed end of a nephron in the kidneys of all higher vertebrates.

Bowman's gland [ANAT] Any one of the tubular serous glands located in the olfactory mucous membrane.

Boyle's law [PHYS] The law that the product of the volume of a gas times its pressure is a constant at fixed temperature. Also known as Mariotte's law.

bp *See* boiling point.

Br *See* bromine.

braccate [VERT ZOO] In birds having feathers on the legs or feet.

brachial [ZOO] Of or relating to an arm or armlike process.

brachial artery [ANAT] An artery which originates at the axillary artery and branches into the radial and ulnar arteries; it distributes blood to the various muscles of the arm, the shaft of the humerus, the elbow joint, the forearm, and the hand.

brachial cavity [INV ZOO] The anterior cavity which is located inside the valves of brachiopods and into which the brachia are withdrawn.

brachialis [ANAT] A muscle that lies beneath the biceps brachii and covers the anterior surface of the elbow joint.

brachial plexus [ANAT] A plexus of nerves located in the neck and axilla and composed of the anterior rami of the lower four cervical and first thoracic nerves.

Brachiata [INV ZOO] A phylum of deuterostomous, sedentary bottom-dwelling marine animals that live encased in tubes.

brachiate [BOT] Having widely divergent branches. [ZOO] Having arms.

brachidia [INV ZOO] In some brachiopods, the calcareous skeleton supporting a brachium.

brachiocephalic [ANAT] Pertaining to the arm and the head.

brachiolaria [INV ZOO] A transitional larva in the development of certain starfishes that is distinguished by three anterior processes homologous with those of the adult.

Brachiopoda [INV ZOO] A phylum of solitary, marine, bivalved coelomate animals.

brachium [ANAT] **1.** A peduncle of the cerebellum. **2.** An arm. [INV ZOO] **1.** In Brachiopoda, either of two spirally coiled structures at the sides of the mouth. **2.** A ray of a crinoid. **3.** A tentacle of a cephalopod.

brachyblast [BOT] A short shoot often bearing clusters of leaves.

brachycephalic [ANTHRO] Being short- or broad-headed, that is, having a cephalic index of over 80.

BOWFIN

Bowfin *(Amia calva)*, which may attain 2½ feet (76 centimeters) and is found only in some parts of the United States.

BRACHYGNATHA

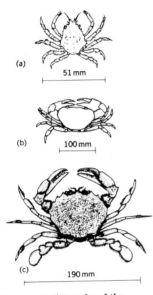

(a)

|— 51 mm —|

(b)

|— 100 mm —|

(c)

|— 190 mm —|

Representative crabs of the Brachygnatha: *(a)* spider crab, *Mithrax acuticornis;* *(b)* fresh-water crab, *Epilobocera sinuatifrons;* *(c)* swimming crab, *Ovalipes ocellatus. (Smithsonian Institution)*

brachycephaly [ANTHRO] The state of being brachycephalic.

brachycerebral [ANTHRO] Having a round or short brain.

brachycranial [ANTHRO] Being short- or broad-skulled, that is, with a cephalic index of over 80.

brachydactylia [MED] Abnormal shortening of fingers or toes.

brachydont [ANAT] Of teeth, having short crowns, well-developed roots, and narrow root canals; characteristic of man.

Brachygnatha [INV ZOO] A subsection of brachyuran crustaceans to which most of the crabs are assigned.

Brachypsectridae [INV ZOO] A family of coleopteran insects in the superfamily Cantharoidea represented by a single species.

Brachypteraciidae [VERT ZOO] The ground rollers, a family of colorful Madagascan birds in the order Coraciiformes.

brachypterous [INV ZOO] Having rudimentary or abnormally small wings, referring to certain insects.

brachysclereid [BOT] A sclereid that is more or less isodiametric and is found in certain fruits and in the pith, cortex, and bark of many stems. Also known as stone cell.

brachyskelic [ANTHRO] Having legs short in proportion to trunk length, with a skelic index of 75–80.

brachysm [BOT] Plant dwarfing in which there is shortening of the internodes only.

brachystomatous [INV ZOO] Having a short proboscis, especially referring to insects.

brachytherapy [MED] Radiation treatment using a solid or enclosed radioisotopic source on the surface of the body or at a short distance from the area to be treated.

Brachythoraci [PALEON] An order of the extinct joint-necked fishes.

Brachyura [INV ZOO] The section of the crustacean order Decapoda containing the true crabs.

brachyural [INV ZOO] Having a short abdomen that is usually tucked under the thorax; refers to certain crabs.

bracket fungus [MYCOL] A basidiomycete characterized by shelflike sporophores, sometimes seen on tree trunks.

Braconidae [INV ZOO] The braconid wasps, a family of hymenopteran insects in the superfamily Ichneumonoidea.

bract [BOT] A modified leaf associated with plant reproductive structures.

bracteate [BOT] Having bracts.

bracteolate [BOT] Having bracteoles.

Bradfordian [GEOL] Uppermost Devonian geologic time.

bradyauxesis [BIOL] Allometric growth in which a part lags behind the body as a whole in development.

bradycardia [MED] Slow heart rate.

bradykinin [BIOCHEM] $C_{50}H_{73}N_{15}O_{11}$ A polypeptide kinin; forms an amorphous precipitate in glacial acetic acid; released from plasma precursors by plasmin. Also known as callideic I; kallidin I.

Bradyodonti [PALEON] An order of Paleozoic cartilaginous fishes (Chondrichthyes), presumably derived from primitive sharks.

Bradypodidae [VERT ZOO] A family of mammals in the order Edentata comprising the true sloths.

bradytely [EVOL] Evolutionary change that is either arrested or occurring at a very slow rate over long geologic periods.

Bragg diffraction *See* Bragg scattering.

Bragg reflection *See* Bragg scattering.

Bragg scattering [SOLID STATE] Scattering of x-rays or neutrons by the regularly spaced atoms in a crystal, for which constructive interference occurs only at definite angles called

Bragg angles. Also known as Bragg diffraction; Bragg reflection.

Bragg's equation *See* Bragg's law.

Bragg's law [SOLID STATE] A statement of the conditions under which a crystal will reflect a beam of x-rays with maximum intensity. Also known as Bragg's equation; Bravais' law.

Bragg spectrometer [SOLID STATE] An instrument for x-ray analysis of crystal structure, in which a homogeneous beam of x-rays is directed on the known face of a crystal and the reflected beam is detected in a suitably placed ionization chamber. Also known as crystal spectrometer; crystal diffraction spectrometer; ionization spectrometer.

brain [ANAT] The portion of the vertebrate central nervous system enclosed in the skull. [ZOO] The enlarged anterior portion of the central nervous system in most bilaterally symmetrical animals.

braincase *See* cranium.

brain coral [INV ZOO] A reef-building coral resembling the human cerebrum in appearance.

brainstem [ANAT] The portion of the brain remaining after the cerebral hemispheres and cerebellum have been removed.

brain wave [PHYSIO] A rhythmic fluctuation of voltage between parts of the brain, ranging from about 1 to 60 hertz and 10 to 100 microvolts.

bramble [BOT] **1.** A plant of the genus *Rubus*. **2.** A rough, prickly vine or shrub.

branch [BOT] A shoot or secondary stem on the trunk or a limb of a tree. [ORG CHEM] A carbon side chain attached to a molecule's main carbon chain. [SCI TECH] An area of study representing an independent offshoot of a related basic discipline.

branched acinous gland [ANAT] A multicellular structure with saclike glandular portions connected to the surface of the containing organ or structure by a common duct.

branched tubular gland [ANAT] A multicellular structure with tube-shaped glandular portions connected to the surface of the containing organ or structure by a common secreting duct.

branchial [ZOO] Of or pertaining to gills.

branchial arch [VERT ZOO] One of the series of paired arches on the sides of the pharynx which support the gills in fishes and amphibians.

branchial basket [ZOO] A cartilaginous structure that supports the gills in protochordates and certain lower vertebrates such as cyclostomes.

branchial cleft [EMBRYO] A rudimentary groove in the neck region of air-breathing vertebrate embryos. [VERT ZOO] One of the openings between the branchial arches in fishes and amphibians.

branchial heart [INV ZOO] A muscular enlarged portion of a vein of a cephalopod that contracts and forces the blood into the gills.

branchial plume [INV ZOO] An accessory respiratory organ that extends out under the mantle in certain Gastropoda.

branchial pouch [ZOO] In cyclostomes and some sharks, one of the respiratory cavities occurring in the branchial clefts.

branchial sac [INV ZOO] In tunicates, the dilated pharyngeal portion of the alimentary canal that has vascular walls pierced with clefts and serves as a gill.

branchial segment [EMBRYO] Any of the paired pharyngeal segments indicating the visceral arches and clefts posterior to and including the third pair in air-breathing vertebrate embryos.

branchiate [VERT ZOO] Having gills.

BRAGG SPECTROMETER

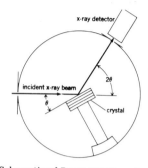

Schematic of Bragg spectrometer. θ is angle between incident beam and crystallographic planes; 2θ is angle between incident and diffracted beams.

BRANCHED TUBULAR GLAND

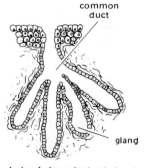

A simple branched tubular gland.

BRANCHIURA

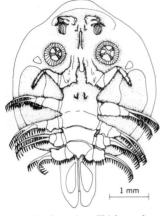

Argulus japonicus Thiele, male fish louse, of the Branchiura. *(After O. L. Meehean, 1940)*

BREADFRUIT

Breadfruit *(Artocarpus altilis)*; end of branch with multiple fruits.

branchicolous [ECOL] Being parasitic on the gills of fish; refers specifically to certain crustaceans.

branchiocranium [VERT ZOO] The division of the fish skull constituting the mandibular and hyal regions and the branchial arches.

branchiomere [EMBRYO] An embryonic metamere that will differentiate into a visceral arch and cleft; a branchial segment.

branchiomeric musculature [VERT ZOO] Those muscles derived from branchial segments in vertebrates.

Branchiopoda [INV ZOO] A subclass of crustaceans containing small or moderate-sized animals commonly called fairy shrimps, clam shrimps, and water fleas.

branchiostegite [INV ZOO] A gill cover and chamber in certain malacostracan crustaceans, formed by lateral expansion of the carapace.

Branchiotremata [INV ZOO] The hemichordates, a branch of the subphylum Oligomera.

Branchiura [INV ZOO] The fish lice, a subclass of fish ectoparasites in the class Crustacea.

Brassicaceae [BOT] An equivalent name for the Cruciferae.

brassin [BIOCHEM] Any of a class of plant hormones characterized as long-chain fatty-acid esters; brassins act to induce both cell elongation and cell division in leaves and stems.

Brathinidae [INV ZOO] The grass root beetles, a small family of coleopteran insects in the superfamily Staphylinoidea.

Braulidae [INV ZOO] The bee lice, a family of cyclorrhaphous dipteran insects in the section Pupipara.

Bravais lattice [CRYSTAL] One of the 14 possible arrangements of lattice points in space such that the arrangement of points about any chosen point is identical with that about any other point.

Bravais' law *See* Bragg's law.

Brazil nut [BOT] *Bertholettia excelsa.* A large broad-leafed evergreen tree of the order Lecythedales; an edible seed is produced by the tree fruit.

breadfruit [BOT] *Artocarpus altilis.* An Indo-Malaysian tree, a species of the mulberry family (Moraceae). The tree produces a multiple fruit which is edible.

bread mold [MYCOL] Any fungus belonging to the family Mucoraceae in the order Mucorales.

breadth-height index [ANTHRO] Ratio of the maximum breadth of the skull to its maximum height multiplied by 100.

breakage and reunion [CYTOL] The classical model of crossing over by means of physical breakage and crossways reunion of completed chromatids during the process of meiosis.

breakbone fever *See* dengue.

breast [ANAT] The human mammary gland.

breeding [AGR] The application of genetic principles to the improvement of farm animals and cultivated plants. [GEN] Controlled mating and selection, or hybridization of plants and animals in order to improve the species.

bregma [ANAT] The point at which the coronal and sagittal sutures of the skull meet.

Brentidae [INV ZOO] The straight-snouted weevils, a family of coleopteran insects in the superfamily Curculionoidea.

Brevibacteriaceae [MICROBIO] A family of gram-positive, rod-shaped, schizomycetous bacteria in the order Eubacteriales.

Brewer anaerobic jar [MICROBIO] A glass container in which petri dish cultures are stacked and maintained under anaerobic conditions.

bridge graft [BOT] A plant graft in which each of several scions is grafted in two positions on the stock, one above and the other below an injury.

brightness constancy [PHYSIO] A phenomenon of perception in which an object is perceived as having a constant brightness despite marked differences in the physical energy emitted by the object and stimulating the eye.

Bright's disease [MED] Any of several kidney diseases attended by glomerulonephritis.

Brill's disease [MED] A mild recurrence of typhus some years after the initial infection.

Brisingidae [INV ZOO] A family of deep-water echinoderms with as many as 44 arms, belonging to the order Forciculatida.

bristle [BIOL] A short stiff hair or hairlike structure on an animal or plant.

brittle star [INV ZOO] The common name for all members of the echinoderm class Ophiuroidea.

broadleaf tree [BOT] Any deciduous or evergreen tree having broad, flat leaves.

broad-spectrum antibiotic [MICROBIO] An antibiotic that is effective against both gram-negative and gram-positive bacterial species.

Broca's area [PHYSIO] The anterior portion of the opercular part of the inferior frontal gyrus; development of this area is more pronounced on the left side in right-handed persons and is designated as the area of articulate speeh. Also known as Brodmann's area 44.

broccoli [BOT] *Brassica oleracea* var. *italica*. A biennial crucifer of the order Capparales which is grown for its edible stalks and buds.

brochidodrome [BOT] Referring to leaf veins that form loops.

Brodmann's areas [PHYSIO] Numbered regions of the cerebral cortex used to identify cortical functions.

broken wind *See* heaves.

bromatium [ECOL] A swollen hyphal tip on fungi growing in ants nests that is eaten by the ants.

bromegrass [BOT] The common name for a number of forage grasses of the genus *Bromus* in the order Cyperales.

bromelain [BIOCHEM] An enzyme that digests protein and clots milk; prepared by precipitation by acetone from pineapple juice; used to tenderize meat, to chill-proof beer, and to make protein hydrolysates. Also spelled bromelin.

Bromeliaceae [BOT] The single family of the flowering plant order Bromeliales.

Bromeliales [BOT] An order of monocotyledonous plants in the subclass Commelinidae, including terrestrial xerophytes and some epiphytes.

bromelin *See* bromelain.

bromine [CHEM] A chemical element, symbol Br, atomic number 35, atomic weight 79.904; used to make dibromide ethylene and in organic synthesis and plastics.

bromism [MED] A disease state produced by prolonged usage or overdosage of bromide compounds.

bromouracil [BIOCHEM] $C_4H_3N_2O_2Br$ 5-Bromouracil, an analog of thymine that can react with deoxyribonucleic acid to produce a polymer with increased susceptibility to mutation.

bronchial adenoma [MED] A low-grade malignant or potentially malignant tumor of bronchi.

bronchial asthma [MED] Asthma usually due to hypersensitivity to an inhaled or ingested allergen.

bronchial tree [ANAT] The arborization of the bronchi of the lung, considered as a structural and functional unit.

bronchiectasis [MED] Dilation of the bronchi and bronchi-

BROMINE

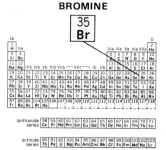

Periodic table of the chemical elements showing the position of bromine.

oles following a chronic inflammatory process or an infection attended by pus formation.

bronchiolar carcinoma [MED] Adenocarcinoma of the lung characterized by mucus-producing cells which spread over the alveoli.

bronchiole [ANAT] A small, thin-walled branch of a bronchus, usually terminating in alveoli.

bronchiolitis [MED] Inflammation of the bronchioles.

bronchiolitis obliterans [MED] Inflammation of the bronchioles with the formation of an exudate and fibrous tissue that obliterate the lumen.

bronchitis [MED] An inflammation of the bronchial tubes.

bronchodilator [MED] An instrument used to increase the caliber of the pulmonary air passages. [PHARM] Any drug which has the property of increasing the caliber of the pulmonary air passages.

bronchogram [MED] Radiography of the bronchial tree made after the introduction of a radiopaque substance.

bronchography [MED] Roentgenographic visualization of the bronchial tree following injection of a radiopaque material.

bronchopneumonia [MED] Inflammation of the lungs which has spread from infected bronchi. Also known as lobular pneumonia.

bronchopulmonary [ANAT] Pertaining to the bronchi and the lungs.

bronchorrhea [MED] Excessive discharge of mucus from the bronchial mucous membranes.

bronchoscope [MED] An instrument for the visual examination of the interior of the bronchi.

bronchospasm [MED] Temporary narrowing of the bronchi due to violent, involuntary contraction of the smooth muscle of the bronchi.

bronchospirometry [MED] The determination of various aspects of the functional capacity of a single lung or lung segment.

bronchus [ANAT] Either of the two primary branches of the trachea or any of the bronchi's pulmonary branches having cartilage in their walls.

Brönsted-Lowry theory [CHEM] A theory that all acid-base reactions consist simply of the transfer of a proton from one base to another. Also known as Brönsted theory.

Brönsted theory *See* Brönsted-Lowry theory.

Brontotheriidae [PALEON] The single family of the extinct mammalian superfamily Brontotherioidea.

Brontotherioidea [PALEON] The titanotheres, a superfamily of large, extinct perissodactyl mammals in the suborder Hippomorpha.

brood [BOT] Heavily infested by insects. [ZOO] **1.** The young of animals. **2.** To incubate eggs or cover the young for warmth. **3.** An animal kept for breeding.

brood capsule [INV ZOO] A secondary scolex-containing cyst constituting the infective agent of a tapeworm.

brood parasitism [ECOL] A type of social parasitism among birds characterized by a bird of one species laying and abandoning its eggs in the nest of a bird of another species.

brood pouch [VERT ZOO] A pouch of an animal body where eggs or embryos undergo certain stages of development.

Brotulidae [VERT ZOO] A family of benthic teleosts in the order Perciformes.

brown algae [BOT] The common name for members of the Phaeophyta.

brown bast [PL PATH] A physiological disease of the para rubber tree characterized by grayish- or greenish-brown discolorations of the inner bark near the tapping cut.

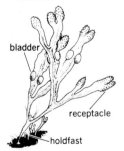

BROWN ALGAE

bladder

receptacle

holdfast

Fucus, a brown alga. *(From H. J. Fuller and O. Tippo, College Botany, rev. ed., Holt, 1954)*

brown blight [PL PATH] A virus disease of lettuce characterized by spots and streaks on the leaves, reduction in leaf size, and gradual browning of the foliage, beginning at the base.

brown blotch [PL PATH] **1.** A bacterial disease of mushrooms caused by *Pseudomonas tolaasi* and characterized by brown blotchy discolorations. **2.** A fungus disease of the pear characterized by brown blotches on the fruit.

brown canker [PL PATH] A fungus disease of roses caused by *Cryptosporella embrina* and characterized by lesions that are initially purple and gradually become buff.

brown fat cell [HISTOL] A moderately large, generally spherical cell in adipose tissue that has small fat droplets scattered in the cytoplasm.

brown felt blight [PL PATH] A fungus disease of conifers caused by several Ascomycetes, especially *Herpotrichia nigra* and *Neopeckia coulteri;* a dense felty growth of brown or black mycelia forms on the branches.

brown induration [MED] A pathologic condition marked by acute pulmonary congestion and edema with leakage of blood into the alveoli.

browning [PL PATH] Any plant disorder or disease marked by brown discoloration of a part. Also known as stem break.

brown leaf rust [PL PATH] A fungus disease of rye caused by *Puccinia dispersa.*

brown patch [PL PATH] A fungus disease of grasses in golf greens and lawns caused by various soil-inhabiting species, typically producing brown circular areas surrounded by a band of grayish-black mycelia.

brown root [PL PATH] A fungus disease of numerous tropical plants, such as coconut and rubber, caused by *Hymenochaete noxia* and characterized by defoliation and by incrustation of the roots with earth and stones held together by brown mycelia.

brown root rot [PL PATH] **1.** A fungus disease of plants of the pea, cucumber, and potato families caused by *Thielavia basicola* and characterized by blackish discoloration and decay of the roots and stem base. **2.** A disease of tobacco and other plants comparable to the fungus disease but believed to be caused by nematodes.

brown rot [PL PATH] Any fungus or bacterial plant disease characterized by browning and tissue decay.

brown seaweed [BOT] A common name for the larger algae of the division Phaeophyta.

brown spot [PL PATH] Any fungus disease of plants, especially Indian corn, characterized by brown leaf spots.

brown stem rot [PL PATH] A fungus disease of soybeans caused by *Cephalosporium gregatum* in which there is brownish internal stem rot followed by discoloration and withering of leaves.

brown stringy rot [PL PATH] A disease of conifers caused by the Indian point fungus; rusty or brown fibrous streaks appear in the heartwood.

browse [BIOL] **1.** Twigs, shoots, and leaves eaten by livestock and other grazing animals. **2.** To feed on this vegetation.

Brucella [MICROBIO] A genus of short, rod-shaped to coccoid, encapsulated, gram-negative, parasitic and pathogenic bacteria of the family Brucellaceae.

Brucella abortus [MICROBIO] The etiological agent of contagious abortion in cattle, brucellosis in man, and a wasting disease in chickens.

Brucellaceae [MICROBIO] A family of small, coccoid to rod-shaped, gram-negative bacteria in the order Eubacteriales.

brucellergen [BIOCHEM] A nucleoprotein fraction of brucellae used in skin tests to detect the presence of *Brucella* infections.

brucellergen test [IMMUNOL] A diagnostic skin test for detection of *Brucella* infections.

brucellosis [MED] An infectious bacterial disease of man caused by *Brucella* species acquired by contact with diseased animals. Also known as Malta fever; Mediterranean fever; undulant fever. [VET MED] *See* contagious abortion.

Bruchidae [INV ZOO] The pea and bean weevils, a family of coleopteran insects in the superfamily Chrysomeloidea.

Bruch's membrane [ANAT] The membrane of the retina that separates the pigmented layer of the retina from the choroid coat of the eye.

Brunner's glands [ANAT] Simple, branched, tubular mucus-secreting glands in the submucosa of the duodenum in mammals. Also known as duodenal glands; glands of Brunner.

brush *See* tropical scrub.

brush border [CYTOL] A superficial protoplasic modification in the form of filiform processes or microvilli; present on certain absorptive cells in the intestinal epithelium and the proximal convolutions of nephrons.

brush fire [FOR] A fire involving growth that is heavier than grass but less than full tree size.

brussels sprouts [BOT] *Brassica oleracea* var. *gemmifera*. A biennial crucifer of the order Capparales cultivated for its small, edible, headlike buds.

Bruxellian [GEOL] Lower middle Eocene geologic time.

bryology [BOT] The study of bryophytes.

Bryophyta [BOT] A small phylum of the plant kingdom, including mosses, liverworts, and hornworts, characterized by the lack of true roots, stems, and leaves.

Bryopsida [BOT] The mosses, a class of small green plants in the phylum Bryophyta. Formerly known as Musci.

Bryopsidaceae [BOT] A family of green algae in the order Siphonales.

Bryozoa [INV ZOO] The moss animals, a major phylum of sessile aquatic invertebrates occurring in colonies with hardened exoskeleton.

bubble point [PHYS CHEM] In a solution of two or more components, the temperature at which the first bubbles of gas appear. Also known as boiling point.

bubo [MED] An inflammatory enlargement of lymph nodes, usually of the groin or axilla; commonly associated with chancroid, lymphogranuloma venereum, and plague.

bubonic plague *See* plague.

buccal cavity [ANAT] The space anterior to the teeth and gums in the mouths of all vertebrates having lips and cheeks. Also known as vestibule.

buccal gland [ANAT] Any of the mucous glands in the membrane lining the cheeks of mammals, except aquatic forms.

Buccinacea [INV ZOO] A superfamily of gastropod mollusks in the order Prosobranchia.

buccinator [ANAT] The principal muscle of the cheek.

Buccinidae [INV ZOO] A family of marine gastropod mollusks in the order Neogastropoda containing the whelks in the genus *Buccinum*.

buccolabial [ANAT] Pertaining to the cheek and the lip.

buccolingual [ANAT] Pertaining to the cheek and the tongue.

bucconasal [ANAT] Pertaining to the cheek and the nose.

Bucconidae [VERT ZOO] The puffbirds, a family of neotropical birds in the order Piciformes.

buccopharyngeal [ANAT] Pertaining to the cheek and the pharynx.

Bucerotidae [VERT ZOO] The hornbills, a family of Old World tropical birds in the order Coraciiformes.

BRUSSELS SPROUTS

Brussels sprouts *(Brassica oleracea* var. *gemmifera),* Jade Cross. *(Joseph Harris Co., Rochester, N.Y.)*

bucket *See* calyx.

buckeye [BOT] The common name for deciduous trees composing the genus *Aesculus* in the order Sapindales; leaves are opposite and palmately compound, and the seed is large with a firm outer coat.

buckwheat [BOT] The common name for several species of annual herbs in the genus *Fagopyrum* of the order Polygonales; used for the starchy seed.

bud [BOT] An embryonic shoot containing the growing stem tip surrounded by young leaves or flowers or both and frequently enclosed by bud scales.

budbreak [BOT] Initiation of growth from a bud.

budding [BIOL] A form of asexual reproduction in which a new individual arises as an outgrowth of an older individual. Also known as gemmation. [BOT] A method of vegetative propagation in which a single bud is grafted laterally onto a stock.

bud grafting [BOT] Grafting a plant by budding.

budling [BOT] The shoot that develops from the bud which was the scion of a bud graft.

bud rot [PL PATH] Any plant disease or symptom involving bud decay.

bud scale [BOT] One of the modified leaves enclosing and protecting buds in perennial plants.

Buerger's disease *See* thromboangitis obliterans.

buffalo [VERT ZOO] The common name for several species of artiodactyl mammals in the family Bovidae, including the water buffalo and bison.

buffer [CHEM] A solution selected or prepared to minimize changes in hydrogen ion concentration which would otherwise occur as a result of a chemical reaction. Also known as buffer solution. [ECOL] An animal that is introduced to serve as food for other animals to reduce the losses of more desirable animals.

buffer solution *See* buffer.

Bufonidae [VERT ZOO] A family of toothless frogs in the suborder Procoela including the true toads (*Bufo*).

bug [INV ZOO] Any insect in the order Hemiptera.

bulb [BOT] A short, subterranean stem with many overlapping fleshy leaf bases or scales, such as in the onion and tulip.

bulbar paralysis [MED] A clinical syndrome due to involvement of the nuclei of the last four or five cranial nerves, characterized principally by paralysis or weakness of the muscles which control swallowing, talking, movement of the tongue and lips, and sometimes respiratory paralysis.

bulbocavernosus *See* bulbospongiosus.

bulb of percussion [ARCHEO] A cone-shaped bulge on a fractured flint surface that was made by a blow striking at an angle.

bulb of the penis [ANAT] The expanded proximal portion of the corpus spongiosum of the penis.

bulbospongiosus [ANAT] A muscle encircling the bulb and adjacent proximal parts of the penis in the male, and encircling the orifice of the vagina and covering the lateral parts of the vestibular bulbs in the female. Also known as bulbocavernosus.

bulbourethral gland [ANAT] Either of a pair of compound tubular glands, anterior to the prostate gland in males, which discharge into the urethra. Also known as Cowper's gland.

bulimia [MED] Excessive, insatiable appetite, seen in psychotic states; a symptom of diabetes mellitus and of certain cerebral lesions. Also known as hyperphagia.

Buliminacea [INV ZOO] A superfamily of benthic, marine foraminiferans in the suborder Rotaliina.

bulla [ANAT] A rounded bony prominence having a thin

BUCKEYE

Leaf of horse chestnut
(Aesculus hippocastanum).

BUFFALO

The Indian or water buffalo
(Bubalus bubalis), domesticated for draft purposes as well as milk production.

BULIMINACEA

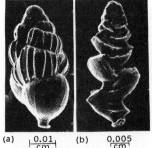

(a) |_0.01_| (b) 0.005
      cm   cm

Scanning electron micrographs of foraminiferans, suborder Rotaliina, superfamily Buliminacea, *(a) Eouvigerina,* from Upper Cretaceous of Texas. *(b) Uvigerina,* from Miocene of California. *(R. B. MacAdam, Chevron Oil Field Research Co.)*

BULLIFORM CELL

bulliform cells

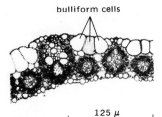

125 μ

Transection of leaf of *Andropogon* showing large bulliform cells in the upper epidermis.

BUPRESTIDAE

An example of Buprestidae. *(From T. I. Storer and R. L. Usinger, General Zoology, 3d ed., McGraw-Hill, 1957)*

wall. [MED] A large bleb or blister filled with lymph or serum and found either within or beneath the epidermis.

bulliform [BOT] Type of plant cell involved in tissue contraction or water storage, or of uncertain function.

bulliform cell [BOT] One of the large, highly vacuolated cells occurring in the epidermis of grass leaves. Also known as motor cell.

bullous emphysema [MED] Acute overinflation of the lungs due to extreme efforts to inhale to overcome a bronchial obstruction.

bull's-eye rot [PL PATH] A fungus disease of apples caused by either *Neofabraea malicorticis* or *Gloeosporium perennans* and characterized by spots resembling eyes on the fruit.

bumblebee [INV ZOO] The common name for several large, hairy social bees of the genus *Bombus* in the family Apidae.

bunchy top [PL PATH] Any viral disease of plants in which there is a shortening of the internodes with crowding of leaves and shoots at the stem apex.

bundle branch [ANAT] Either of the components of the atrioventricular bundle passing to the right and left ventricles of the heart.

bundle scar [BOT] A mark within a leaf scar that shows the point of an abscised vascular bundle.

bundle sheath [BOT] A sheath around a vascular bundle that consists of a layer of parenchyma.

Bundsandstein *See* Bunter.

bunion [MED] A swelling of a bursa of the foot, especially of the metatarsophalangeal joint of the great toe; associated with thickening of the adjacent skin and a forcing of the great toe into adduction.

bunodont [ANAT] Having tubercles or rounded cusps on the molar teeth, as in man.

bunolophodont [VERT ZOO] **1.** Of teeth, having the outer cusps in the form of blunt cones and the inner cusps as transverse ridges. **2.** Having such teeth, as in tapirs.

bunoselenodont [VERT ZOO] **1.** Of teeth, having the inner cusps in the form of blunt cones and the outer ones as longitudinal crescents. **2.** Having such teeth, as in the extinct titanotheres.

bunt [PL PATH] A fungus disease of wheat caused by two *Tilletia* species and characterized by grain replacement with fishy-smelling smut spores.

Bunter [GEOL] Lower Triassic geologic time. Also known as Bundsandstein.

buoyant density [PHYS] A technique that uses the sedimentation equilibrium in a density gradient to characterize a solute.

Buprestidae [INV ZOO] The metallic wood-boring beetles, the large, single family of the coleopteran superfamily Buprestoidea.

Buprestoidea [INV ZOO] A superfamily of coleopteran insects in the suborder Polyphaga including many serious pests of fruit trees.

Burdigalian [GEOL] Upper lower Miocene geologic time.

buret [CHEM] A graduated glass tube used to deliver variable volumes of liquid; usually equipped with a stopcock to control the liquid flow.

Burhinidae [VERT ZOO] The thick-knees or stone curlews, a family of the avian order Charadriiformes.

Burkett's lymphoma [MED] A malignant lymphoma of children, typically involving the retroperitoneal area and the mandible, but sparing the peripheral lymph nodes, bone marrow, and spleen.

Burmanniaceae [BOT] A family of monocotyledonous plants in the order Orchidales characterized by regular flowers, three or six stamens opposite the petals, and ovarian nectaries.

burn [MED] An injury to tissues caused by heat, chemicals, electricity, or irradiation effects.

Burnett's syndrome *See* milk-alkali syndrome.

burr [BOT] **1.** A rough or prickly envelope on a fruit. **2.** A fruit so characterized.

burro [VERT ZOO] A small donkey used as a pack animal.

bursa [ANAT] A simple sac or cavity with smooth walls containing a clear, slightly sticky fluid and interposed between two moving surfaces of the body to reduce friction.

bursa of Fabricius [VERT ZOO] A thymuslike organ in the form of a diverticulum at the lower end of the alimentary canal in birds.

Burseraceae [BOT] A family of dicotyledonous plants in the order Sapindales characterized by an ovary of two to five cells, prominent resin ducts in the bark and wood, and an intrastaminal disk.

bursitis [MED] Inflammation of a bursa.

burster [BOT] An abnormally double flower having the calyx split or fragmented.

bushbaby [VERT ZOO] Any of six species of African arboreal primates in two genera (*Galago* and *Euoticus*) of the family Lorisidae. Also known as galago; night ape.

butane dicarboxylic acid *See* adipic acid.

butanol fermentation [MICROBIO] Butanol production as a result of the fermentation of corn and molasses by the anaerobic bacterium *Clostridium acetobutylicum*.

Butomaceae [BOT] A family of monocotyledonous plants in the order Alismatales characterized by secretory canals, linear leaves, and a straight embryo.

butterfly [INV ZOO] Any insect belonging to the lepidopteran suborder Rhopalocera, characterized by a slender body, broad colorful wings, and club-shaped antennae.

buttocks [ANAT] The two fleshy parts of the body posterior to the hip joints.

butt rot [PL PATH] A fungus decay of the base of a tree trunk caused by polypores (such as *Fomea* species).

butyric acid [ORG CHEM] $CH_3CH_2CH_2COOH$ A colorless, combustible liquid with boiling point 163.5°C (757 mm Hg); soluble in water, alcohol, and ether; used in synthesis of flavors, in pharmaceuticals, and in emulsifying agents. Also known as butanoic acid; *n*-butyric acid; ethyl acetic acid; propylformic acid.

butyric fermentation [BIOCHEM] Fermentation in which butyric acid is produced by certain anaerobic bacteria acting on organic substances, such as butter; occurs in putrefaction and in digestion in herbivorous mammals.

butyrinase [BIOCHEM] An enzyme that hydrolyzes butyrin, found in the blood serum.

Buxbaumiales [BOT] An order of very small, atypical mosses (Bryopsida) composed of three genera and found on soil, rock, and rotten wood.

B virus [VIROL] An animal virus belonging to subgroup A of the herpesvirus group.

Byrrhidae [INV ZOO] The pill beetles, the single family of the coleopteran insect superfamily Byrrhoidea.

Byrrhoidea [INV ZOO] A superfamily of coleopteran insects in the superorder Polyphaga.

byssinosis [MED] A pneumoconiosis caused by the inhalation of cotton dust.

byssus [INV ZOO] A tuft of filaments by which certain bivalve mollusks become attached. [MYCOL] The stalk of certain fungi.

Byturidae [INV ZOO] The raspberry fruitworms, a small family of coleopteran insects in the superfamily Cucujoidea.

BUTTERFLY

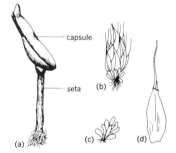

Adult of the marsh (meadow) fritillary.

BUXBAUMIALES

capsule

seta (b)

(a)   (c)   (d)

Morphological features of Buxbaumiales. (*a*) Sporophyte of *Buxbaumia aphylla* (from W. H. Welch, *Mosses of Indiana, Indiana Department of Conservation, 1957*). (*b*) *Diphyscium foliosum*, habit sketch, (*c*) leaves, and (*d*) perichaetial leaf (from H. S. Conard, *How to Know the Mosses and Liverworts*, William C. Brown, 1956).

C

C *See* carbon.

Ca *See* calcium.

caatinga [ECOL] A sparse, stunted forest in areas of little rainfall in northeastern Brazil; trees are leafless in the dry season.

cabbage [BOT] *Brassica oleracea* var. *capitata*. A biennial crucifer of the order Capparales grown for its head of edible leaves.

cabbage yellows [PL PATH] A fungus disease of cabbage caused by *Fusarium conglutinans* and characterized by yellowing and dwarfing.

cabezon [VERT ZOO] *Scorpaenichthys marmoratus*. A fish that is the largest of the sculpin species, weighing as much as 25 pounds (11.3 kilograms) and reaching a length of 30 inches (76 centimeters).

cable skidding [FOR] The use of a chain or cable choker to move tree lengths or tree segments for a short distance over unimproved terrain.

Cabot's ring [CYTOL] A ringlike body in immature erythrocytes that may represent the remains of the nuclear membrane.

cacao [BOT] *Theobroma cacao*. A small tropical tree of the order Theales that bears capsular fruits which are a source of cocoa powder and chocolate. Also known as chocolate tree.

cachexia [MED] Weight loss, weakness, and wasting of the body encountered in certain diseases or in terminal illnesses.

cacomistle [VERT ZOO] *Bassariscus astutus*. A raccoonlike mammal that inhabits the southern and southwestern United States; distinguished by a bushy black-and-white ringed tail. Also known as civet cat; ringtail.

caconym [SYST] A taxonomic name that is linguistically unacceptable.

Cactaceae [BOT] The cactus family of the order Caryophyllales; represented by the American stem succulents, which are mostly spiny with reduced leaves.

cactus [BOT] The common name for any member of the family Cactaceae, a group characterized by a fleshy habit, spines and bristles, and large, brightly colored, solitary flowers.

cadang-cadang [PL PATH] An infectious virus disease of the coconut palm characterized by yellow-bronzing of the leaves.

cadaver [MED] A dead animal or human body to be studied by dissection.

cadaverine [BIOCHEM] $C_5H_{14}N_2$ A nontoxic, organic base produced as a result of the decarboxylation of lysine by the action of putrefactive bacteria on flesh.

caddis fly [INV ZOO] The common name for all members of the insect order Trichoptera; adults are mothlike and the immature stages are aquatic.

CABBAGE

Cabbage (*Brassica oleracea* var. *capitata*), cultivar Golden Acre 84. *(Joseph Harris Co., Inc., Rochester, N.Y.)*

CADDIS FLY

Adult caddis fly.

CADMIUM

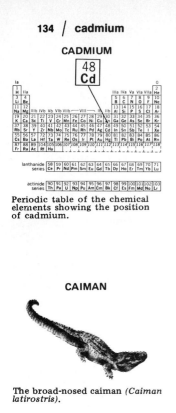

Periodic table of the chemical elements showing the position of cadmium.

CAIMAN

The broad-nosed caiman *(Caiman latirostris).*

CALAMITALES

Scouring rush *(Calamites)* restored with whorls of branches at each joint of articulated stems. *(From R. C. Hussey, Historical Geology, 2d ed., McGraw-Hill, 1947)*

cadmium [CHEM] A chemical element, symbol Cd, atomic number 48, atomic weight 112.40.

cadophore [INV ZOO] In some tunicates, dorsal outgrowth which bears buds.

caducicorn [VERT ZOO] Having deciduous horns, as certain deer.

caducous [BOT] Lasting on a plant only a short time before falling off.

caecilian [VERT ZOO] The common name for members of the amphibian order Apoda.

caecum *See* cecum.

Caenolestidae [PALEON] A family of extinct insectivorous mammals in the order Marsupialia.

Caenolestoidea [VERT ZOO] A superfamily of marsupial mammals represented by the single living family Caenolestidae.

caenostylic [VERT ZOO] Having the first two visceral arches attached to the cranium and functioning in food intake; a condition found in sharks, amphibians, and chimaeras.

Caesalpinoidea [BOT] A subfamily of dicotyledonous plants in the legume family, Leguminosae.

caiman [VERT ZOO] Any of five species of reptiles of the genus *Caiman* in the family Alligatoridae, differing from alligators principally in having ventral armor and a sharper snout.

Cainotheriidae [PALEON] The single family of the extinct artiodactyl superfamily Cainotherioidea.

Cainotherioidea [PALEON] A superfamily of extinct, rabbit-sized tylopod ruminants in the mammalian order Artiodactyla.

caisson disease [MED] A condition resulting from a rapid change in atmospheric pressure from high to normal, causing nitrogen bubbles to form in the blood and body tissues. Also known as bends; compressed-air illness.

cal *See* calorie.

Cal *See* kilocalorie.

Calabar swelling [MED] Edematous, painful, subcutaneous swelling occurring in the body of natives of Calabar and of other parts of West Africa, probably due to an allergic reaction to *Loa loa* infection.

calamine [PHARM] A powder mixture of zinc oxide and ferric oxide, used in skin lotions and ointments.

calamistrum [INV ZOO] A comblike projection on the metatarsus of certain spiders.

Calamitales [PALEOBOT] An extinct group of reedlike plants of the subphylum Sphenopsida characterized by horizontal rhizomes and tall, upright, grooved, articulated stems.

calamus [BOT] A hollow stem that has no nodes. [VERT ZOO] The quill of a feather.

Calanoida [INV ZOO] A suborder of the crustacean order Copepoda, including the larger and more abundant of the pelagic species.

Calappidae [INV ZOO] The box crabs, a family of reptantian decapods in the subsection Oxystomata of the section Brachyura.

calathiform [BIOL] Cup-shaped, being almost hemispherical.

calcaneocuboid ligament [ANAT] The ligament that joins the calcaneus and the cuboid bones.

calcaneum [ANAT] 1. A bony projection of the metatarsus in birds. 2. *See* calcaneus.

calcaneus [ANAT] A bone of the tarsus, forming the heel bone in man. Also known as calcaneum.

calcar [ZOO] A spur or spurlike process, especially on an appendage or digit.

calcarate [BOT] Having spurs.

Calcarea [INV ZOO] A class of the phylum Porifera, including sponges with a skeleton composed of calcium carbonate spicules.

calcareous [SCI TECH] Resembling, containing, or composed of calcium carbonate.

calcareous algae [BOT] Algae that grow on limestone or in soil impregnated with lime.

calcarine fissure *See* calcarine sulcus.

calcarine sulcus [ANAT] A sulcus on the medial aspect of the occipital lobe of the cerebrum, between the lingual gyrus and the cuneus. Also known as calcarine fissure.

Calcaronea [INV ZOO] A subclass of sponges in the class Calcarea in which the larva are amphiblastulae.

calcemia *See* hypercalcemia.

calcicole [BOT] Requiring soil rich in calcium carbonate for optimum growth.

calcicosis [MED] A form of pneumoconiosis caused by the inhalation of marble (calcium carbonate) dust.

calciferous [BIOL] Containing or producing calcium or calcium carbonate.

calciferous gland [INV ZOO] One of a series of glands that secrete calcium carbonate into the esophagus of certain oligochaetes.

calcification [PHYSIO] The deposit of calcareous matter within the tissues of the body.

calcifuge [ECOL] A plant that grows in an acid medium that is poor in calcareous matter.

Calcinea [INV ZOO] A subclass of sponges in the class Calcarea in which the larvae are parenchymulae.

calcinosis [MED] Deposition of calcium salts in the skin, subcutaneous tissue, or other part of the body in certain pathologic conditions.

calcitonin [BIOCHEM] A polypeptide, calcium-regulating hormone produced by the ultimobranchial bodies in vertebrates. Also known as thyrocalcitonin.

calcium [CHEM] A chemical element, symbol Ca, atomic number 20, atomic weight 40.08; used in metallurgy as an alloying agent for aluminum-bearing metal, as an aid in removing bismuth from lead, and as a deoxidizer in steel manufacture, and also used as a cathode coating in some types of photo tubes.

calcium lactate [ORG CHEM] $Ca(C_3H_5O_3)_2 \cdot 5H_2O$ A salt of lactic acid in the form of white crystals that are soluble in water; used in calcium therapy and as a blood coagulant.

calcium metabolism [BIOCHEM] Biochemical and physiological processes involved in maintaining the concentration of calcium in plasma at a constant level and providing a sufficient supply of calcium for bone mineralization.

Calclamnidae [PALEON] A family of Paleozoic echinoderms of the order Dendrochirotida.

calculus [ANAT] A small cuplike structure. [PATH] An abnormal, solid concretion of minerals and salts formed around organic materials and found chiefly in ducts, hollow organs, and cysts.

calf [VERT ZOO] The young of the domestic cow, elephant, rhinoceros, hippopotamus, moose, whale, and others.

calf circumference [ANTHRO] An average of three measurements taken at maximum horizontal distance around the left calf as the subject stands with weight distributed evenly on both feet.

Caliciaceae [BOT] A family of lichens in the order Caliciales in which the disk of the apothecium is borne on a short stalk.

Caliciales [BOT] An order of lichens in the class Ascolichenes characterized by an unusual apothecium.

californium [CHEM] A chemical element, symbol Cf, atomic number 98; all isotopes are radioactive.

CALCINEA

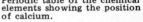

Clathrina clathrus, a calcinean sponge. *(After Minchin, 1900)*

CALCIUM

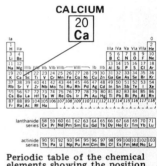

Periodic table of the chemical elements showing the position of calcium.

CALIFORNIUM

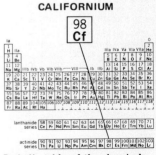

Periodic table of the chemical elements showing the position of californium.

CALIGOIDA

1 mm

Pandarus satyrus Dana, male, a caligoidan.

Caligidae [INV ZOO] A family of fish ectoparasites belonging to the crustacean suborder Caligoida.

Caligoida [INV ZOO] A suborder of the crustacean order Copepoda, including only fish ectoparasites and characterized by a sucking mouth with styliform mandibles.

caliper splint [MED] A splint designed for the leg, consisting of two metal rods from a posterior-thigh band or a padded ischial ring to a metal plate attached to the sole of the shoe at the instep.

Callichthyidae [VERT ZOO] A family of tropical catfishes in the suborder Siluroidei.

callideic I *See* bradykinin.

Callionymoidei [VERT ZOO] A suborder of fishes in the order Perciformes, including two families of colorful marine bottom fishes known as dragonets.

Callipallenidae [INV ZOO] A family of marine arthropods in the subphylum Pycnogonida lacking palpi and having chelifores and 10-jointed ovigers.

Calliphoridae [INV ZOO] The blow flies, a family of myodarian cyclorrhaphous dipteran insects in the subsection Calypteratae.

Callithricidae [VERT ZOO] The marmosets, a family of South American mammals in the order Primates.

Callorhinchidae [VERT ZOO] A family of ratfishes of the chondrichthyan order Chimaeriformes.

callose [BIOCHEM] A carbohydrate component of plant cell walls; associated with sieve plates where calluses are formed. [BIOL] Having hardened protuberances, as on the skin or on leaves and stems.

callosity [BIOL] A hardened or thickened area on the skin, or on the bark of a stem.

callow [VERT ZOO] Of a bird, not having feathers.

callus [BOT] **1.** A thickened callose deposit on sieve plates. **2.** Hard tissue that forms over a damaged plant surface. [MED] Hard, thick area on the surface of the skin.

caloreceptor [PHYSIO] A cutaneous sense organ that is stimulated by heat.

calorie [THERMO] Abbreviated cal; often designated c. **1.** A unit of heat energy, equal to 4.1868 joules. Also known as international table calorie (IT calorie). **2.** A unit of energy, equal to the heat required to raise the temperature of 1 gram of water from 14.5° to 15.5°C at a constant pressure of 1 standard atmosphere; equal to 4.1855 ± 0.0005 joules. Also known as fifteen-degrees calorie; gram-calorie (g-cal); small calorie. **3.** A unit of heat energy equal to 4.184 joules; used in thermochemistry. Also known as thermochemical calorie.

calorimeter [ENG] An apparatus for measuring heat quantities generated in or emitted by materials in processes such as chemical reactions, changes of state, or formation of solutions.

calorimetry [ENG] The measurement of the quantity of heat involved in various processes, such as chemical reactions, changes of state, and formations of solutions, or in the determination of the heat capacities of substances; fundamental unit of measurement is the joule or the calorie (4.184 joules).

calotte [BIOL] A cap or caplike structure. [INV ZOO] **1.** The four-celled polar cap in Dicyemidae. **2.** A ciliated, retractile disc in certain bryozoan larva. **3.** A dark-colored anterior area in certain nematomorphs.

calthrop [INV ZOO] A sponge spicule having four axes in which the rays are equal or almost equal in length.

calvarium [ANAT] A skull lacking facial parts and the lower jaw.

Calvé's disease *See* osteochondrosis.

calving [VERT ZOO] Giving birth to a calf.

calycanthemy [PL PATH] Abnormal development of calyx structures into petals or petaloid structures.

calyciflorous [BOT] Of flowers, having stamens and petals adnate to the calyx.

calyculate [BOT] Having bracts that imitate a second, external calyx.

calymma [INV ZOO] The outer, vacuolated protoplasmic layer of certain radiolarians.

Calymnidae [INV ZOO] A family of echinoderms in the order Holasteroida characterized by an ovoid test with a marginal fasciole.

calypter [INV ZOO] A scalelike or lobelike structure above the haltere of certain two-winged flies. Also known as alula; squama.

Calypteratae [INV ZOO] A subsection of dipteran insects in the suborder Cyclorrhapha characterized by calypters associated with the wings.

Calyptoblastea [INV ZOO] A suborder of coelenterates in the order Hydroida, including the hydroids with protective cups around the hydranths and gonozooids.

calyptra [BOT] **1.** A membranous cap or hoodlike covering, especially the remains of the archegonium over the capsule of a moss. **2.** Tissue surrounding the archegonium of a liverwort. **3.** Root cap.

calyptrate [BOT] Having a calyptra.

calyptrogen [BOT] The specialized cell layer from which a root cap originates.

Calyssozoa [INV ZOO] The single class of the bryozoan subphylum Entoprocta.

calyx [BOT] The outermost whorl of a flower; composed of sepals. [INV ZOO] A cup-shaped structure to which the arms are attached in crinoids.

Camallanida [INV ZOO] An order of phasmid nematodes in the subclass Spiruria, including parasites of domestic animals.

camarodont dentition [INV ZOO] In echinoderms, keeled teeth meeting the epiphyses so that the foramen magnum of the jaw is closed.

Cambaridae [INV ZOO] A family of crayfishes belonging to the section Macrura in the crustacean order Decapoda.

Cambarinae [INV ZOO] A subfamily of crayfishes in the family Astacidae, including all North American species east of the Rocky Mountains.

cambium [BOT] A layer of cells between the phloem and xylem of most vascular plants that is responsible for secondary growth and for generating new cells.

Cambrian [GEOL] The lowest geologic system that contains abundant fossils of animals, and the first (earliest) geologic period of the Paleozoic era from 570 to 500 million years ago.

camel [VERT ZOO] The common name for two species of artiodactyl mammals, the bactrian camel (*Camelus bactrianus*) and the dromedary camel (*C. dromedarius*), in the family Camelidae.

Camelidae [VERT ZOO] A family of tylopod ruminants in the superfamily Cameloidea of the order Artiodactyla, including four species of camels and llamas.

Cameloidea [VERT ZOO] A superfamily of tylopod ruminants in the order Artiodiodactyla.

Camerata [PALEON] A subclass of extinct stalked echinoderms of the class Crinoidea.

Campanian [GEOL] European stage of Upper Cretaceous.

Campanulaceae [BOT] A family of dicotyledonous plants in the order Campanulales characterized by a style without an

CAMARODONT DENTITION

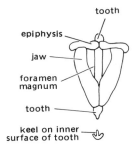

Jaw and tooth arrangement.

CAMBRIAN

PRECAMBRIAN		
CAMBRIAN		
ORDOVICIAN		
SILURIAN		PALEOZOIC
DEVONIAN		
Mississippian	CARBON-IFEROUS	
Pennsylvanian		
PERMIAN		
TRIASSIC		
JURASSIC		MESOZOIC
CRETACEOUS		
TERTIARY		CENOZOIC
QUATERNARY		

Position of the Cambrian system and its relationship to other periods.

indusium but with well-developed collecting hairs below the stigmas, and by a well-developed latex system.

Campanulales [BOT] An order of dicotyledonous plants in the subclass Asteridae distinguished by a chiefly herbaceous habit, alternate leaves, and inferior ovary.

campanulate [BOT] Bell-shaped; applied particularly to the corolla.

campestrian [ECOL] Of or pertaining to the northern Great Plains area.

camphor tree [BOT] *Cinnamomum camphora.* A plant of the laurel family (Lauraceae) in the order Magnoliales from which camphor is extracted.

Camp-Meidell condition [STAT] For determining the distribution of a set of numbers, the guideline stating that if the distribution has only one mode, if the mode is the same as the arithmetic mean, and if the frequencies decline continuously on both sides of the mode, then more than $1 - (1/2.25t^2)$ of any distribution will fall within the closed range $\bar{X} \pm t\sigma$, where t = number of items in a set, $\bar{X}$ = average, and σ = standard deviation.

Campodeidae [INV ZOO] A family of primarily wingless insects in the order Diplura which are most numerous in the Temperate Zone of the Northern Hemisphere.

campos [ECOL] The savanna of South America.

camptotrichia [VERT ZOO] Jointed dermal fin rays that occur in certain primitive fishes.

camptrodrome [BOT] Pertaining to a leaf having secondary veins which anastomose before reaching the margin.

campylodrome [BOT] Pertaining to a leaf having veins that converge at its tip.

campylotropous [BOT] Having the ovule symmetrical but half inverted, with the micropyle and funiculus at right angles to each other.

Canaceidae [INV ZOO] The seashore flies, a family of myodarian cyclorrhaphous dipteran insects in the subsection Acalypteratae.

Canadian life zone [ECOL] The zone comprising the climate and biotic communities of the portion of the Boreal life zone exclusive of the Hudsonian and Arctic-Alpine zones.

canal [BIOL] A tubular duct or passage in bone or soft tissues.

canal cell [BOT] One of the row of cells that make up the axial row within the neck of an archegonium.

canaliculate [BIOL] Having small channels, canals, or grooves.

canaliculus [HISTOL] **1.** One of the minute channels in bone radiating from a Haversian canal and connecting lacunnae with each other and with the canal. **2.** A passage between the cells of the cell cords in the liver.

canalization [MED] Surgical method of wound drainage without tubes by forming channels. [PHYSIO] The formation of new channels in tissues, such as the formation of new blood vessels through a thrombus.

canal of Schlemm [ANAT] An irregular channel at the junction of the sclera and cornea in the eye that drains aqueous humor from the anterior chamber.

canal valve [ANAT] The semilunar valve in the right atrium of the heart between the orifice of the inferior vena cava and the right atrioventricular orifice. Also known as eustachian valve.

Canastotan [GEOL] Lower Upper Silurian geologic time.

canavanine [BIOCHEM] $C_5H_{12}O_3N_4$ An amino acid found in the jack bean.

cancellate [BIOL] Lattice-shaped. Also known as clathrate.

cancellous [BIOL] Having a reticular or spongy structure.

cancellous bone [HISTOL] A form of bone near the ends of

CAMPODEIDAE

cerci

Campodea folsomi Silvestri, in the family Campodeidae.

long bones having a cancellous matrix composed of rods, plates, or tubes; spaces are filled with marrow.

cancer [MED] Any malignant neoplasm, including carcinoma and sarcoma.

cancer eye [VET MED] A malignant epithelioma of the eye of cattle, common in regions of intense sunlight.

cancroid [MED] A squamous-cell carcinoma.

Candida [MYCOL] A genus of yeastlike, pathogenic imperfect fungi that produce very small mycelia.

candidiasis [MED] A fungus infection of the skin, lungs, mucous membranes, and viscera of man caused by a species of *Candida,* usually *C. albicans.* Also known as moniliasis.

cane [BOT] **1.** A hollow, usually slender, jointed stem, such as in sugarcane or the bamboo grasses. **2.** A stem growing directly from the base of the plant, as in most Rosaceae, such as blackberry and roses.

cane blight [PL PATH] A fungus disease affecting the canes of several bush fruits, such as currants and raspberries; caused by several species of fungi.

canescent [BOT] Having a grayish epidermal covering of short hairs.

cane sugar [ORG CHEM] Sucrose derived from sugarcane.

Canidae [VERT ZOO] A family of carnivorous mammals in the superfamily Canoidea, including dogs and their allies.

canine [ANAT] A conical tooth, such as one located between the lateral incisor and first premolar in man and many other mammals. Also known as cuspid. [VERT ZOO] Pertaining or related to dogs or to the family Canidae.

canine distemper [VET MED] A pantropic virus disease occurring among animals of the family Canidae.

caninus [ANAT] A muscle attached to the skin at the angle of the mouth; functions in facial expression.

canker [PL PATH] An area of necrosis on a woody stem resulting in shrinkage and cracking followed by the formation of callus tissue around the area, ultimately killing the stem. [VET MED] A localized chronic inflammation of the ear in cats, dogs, foxes, ferrets, and others caused by the mite *Otodectes cynotis.*

canker sore [MED] Small ulceration of the mucous membrane of the mouth, sometimes caused by a food allergy.

canker stain [PL PATH] A fungus disease of plane trees caused by *Endoconidiophora fimbriata platani* and characterized by bluish-black or reddish-brown discolorations beneath blackened cankers on the trunk and sometimes the branches.

cankerworm [INV ZOO] Any of several lepidopteran insect larvae in the family Geometridae which cause severe plant damage by feeding on buds and foliage.

Cannabaceae [BOT] A family of dicotyledonous herbs in the order Urticales, including Indian hemp (*Cannabis sativa*) and characterized by erect anthers, two styles or style branches, and the lack of milky juice.

cannabidiol [ORG CHEM] $C_{21}H_{28}(OH)_2$ A constituent of cannabis which, upon isomerization to a tetrahydrocannabinol, has some of the physiologic activity of marijuana.

Cannabis [BOT] A genus of tall annual herbs in the family Cannabaceae having erect stems, leaves with three to seven elongate leaflets, and pistillate flowers in spikes along the stem.

cannabism [MED] Poisoning resulting from excessive or habitual use of cannabis.

Cannaceae [BOT] A family of monocotyledonous plants in the order Zingiberales characterized by one functional stamen, a single functional pollen sac in the stamen, mucilage canals in the stem, and numerous ovules in each of the one to three locules.

CANNACEAE

Flowers of a cultivated *Canna* of the order Zingiberales. These plants have been so extensively hybridized that most of them can no longer be identified with any of the wild species. (*Photograph by Luoma Photos, from National Audubon Society*)

Canoidea [VERT ZOO] A superfamily belonging to the mammalian order Carnivora, including all dogs and doglike species such as seals, bears, and weasels.

canonical correlation [STAT] The maximum correlation between linear functions of two sets of random variables when specific restrictions are imposed upon the coefficients of the linear functions of the two sets.

canopy [FOR] Topmost layer of leaves, twigs, and branches on the trees in a forest.

cantaloupe [BOT] The fruit (pepo) of *Cucumis malo*, a small, distinctly netted, round to oval muskmelon of the family Cucurbitaceae in the order Violales.

canthariasis [MED] Infection or disease caused by coleopteran insects or their larvae.

Cantharidae [INV ZOO] The soldier beetles, a family of coleopteran insects in the superfamily Cantharoidea.

Cantharoidea [INV ZOO] A superfamily of coleopteran insects in the suborder Polyphaga.

canthus [ANAT] Either of the two angles formed by the junction of the eyelids, designated outer or lateral, and inner or medial.

capillarity [FL MECH] The action by which the surface of a liquid where it contacts a solid is elevated or depressed, because of the relative attraction of the molecules of the liquid for each other and for those of the solid.

capillaroscope [MED] A microscope used for diagnostic examination of the cutaneous capillaries, as in the nail beds and conjunctiva.

capillary [ANAT] The smallest vessel of both the circulatory and lymphatic systems; the walls are composed of a single cell layer.

capillary angioma *See* hemangioma.

capillary bed [ANAT] The capillaries, collectively, of a given area or organ.

capillary pressure [PHYSIO] Pressure exerted by blood against capillary walls.

capillary tube [ENG] A tube sufficiently fine so that capillary attraction of a liquid into the tube is significant.

capillitium [MYCOL] A network of threadlike tubes or filaments in which spores are embedded within sporangia of certain fungi, such as the slime molds.

capitate [BIOL] Enlarged and swollen at the tip. [BOT] Forming a head, as certain flowers of the Compositae.

capitellate [BOT] 1. Having a small knoblike termination. 2. Grouped to form a capitulum.

Capitellidae [INV ZOO] A family of mud-swallowing annelid worms, sometimes called bloodworms, belonging to the Sedentaria.

capitellum [ANAT] A small head or rounded process of a bone.

Capitonidae [VERT ZOO] The barbets, a family of pantropical birds in the order Piciformes.

capitulum [BIOL] A rounded, knoblike, usually terminal protuberance on a structure. [BOT] One of the rounded cells on the manubrium in the antheridia of lichens belonging to the Caliciales.

capon [AGR] A castrated male chicken.

Caponidae [INV ZOO] A family of arachnid arthropods in the order Araneida characterized by having tracheae instead of book lungs.

Capparaceae [BOT] A family of dicotyledonous herbs, shrubs, and trees in the order Capparales characterized by parietal placentation; hypogynous, mostly regular flowers; four to many stamens; and simple to trifoliate or palmately compound leaves.

CAPITELLIDAE

Left lateral view of *Capitomastus*, a genus of Capitellidae.

Capparales [BOT] An order of dicotyledonous plants in the subclass Dilleniidae.

Caprellidae [INV ZOO] The skeleton shrimps, a family of slender, cylindrical amphipod crustaceans in the suborder Caprellidea.

Caprellidea [INV ZOO] A suborder of marine and brackish-water animals of the crustacean order Amphipoda.

capreolate [BOT] **1.** Having tendrils. **2.** Tendril-shaped.

Caprifoliaceae [BOT] A family of dicotyledonous, mostly woody plants in the order Dipsacales, including elderberry and honeysuckle; characterized by distinct filaments and anthers, typically five stamens and five corolla lobes, more than one ovule per locule, and well-developed endosperm.

Caprimulgidae [VERT ZOO] A family of birds in the order Caprimulgiformes, including the nightjars, or goatsuckers.

Caprimulgiformes [VERT ZOO] An order of nocturnal and crepuscular birds, including nightjars, potoos, and frogmouths.

caprylic acid [ORG CHEM] $C_8H_{16}O_2$ A liquid fatty acid occurring in butter, coconut oil, and other fats and oils. Also known as hexylacetic acid; n-octanoic acid; octylic acid.

Capsaloidea [INV ZOO] A superfamily of ectoparasitic trematodes in the subclass Monogenea characterized by a sucker-shaped holdfast with anchors and hooks.

capsanthin [BIOCHEM] $C_{40}H_{58}O_3$ Carmine-red carotenoid pigment occurring in paprika.

capsicum [BOT] The fruit of a plant of the genus *Capsicum*, especially *C. frutescens*, cultivated in southern India and the tropics; a strong irritant to mucous membranes and eyes.

capsid [INV ZOO] The name applied to all members of the family Miridae. [VIROL] The protein coat of a virus particle.

capsomere [VIROL] An individual protein subunit of a capsid.

capsular ligament [ANAT] A saclike ligament surrounding the articular cavity of a freely movable joint and attached to the bones.

capsulate [BIOL] Enclosed in a capsule.

capsule [ANAT] A membranous structure enclosing a body part or organ. [BOT] A closed structure bearing seeds or spores; it is dehiscent at maturity. [MICROBIO] A thick, mucous envelope, composed of polypeptide or carbohydrate, surrounding certain microorganisms. [PHARM] A soluble shell in which drugs are enclosed for oral administration.

captaculum [INV ZOO] A tentacular growth on the head of a tooth-shell mollusk; functions in food capture.

capybara [VERT ZOO] *Hydrochoerus capybara.* An aquatic rodent (largest rodent in existence) found in South America and characterized by partly webbed feet, no tail, and coarse hair.

Carabidae [INV ZOO] The ground beetles, a family of predatory coleopteran insects in the suborder Adephaga.

Caracarinae [VERT ZOO] The caracaras, a subfamily of carrion-feeding birds in the order Falconiformes.

Caradocian [GEOL] Lower Upper Ordovician geologic time.

Carangidae [VERT ZOO] A family of perciform fishes in the suborder Percoidei, including jacks, scads, and pompanos.

carapace [INV ZOO] A dorsolateral, chitonous case covering the cephalothorax of many arthropods. [VERT ZOO] The bony, dorsal part of a turtle shell.

Carapidae [VERT ZOO] The pearlfishes, a family of sinuous, marine shore fishes in the order Gadiformes that live as commensals in the body cavity of holothurians.

carate *See* pinta.

caraway [BOT] *Carum carvi.* A white-flowered perennial

CAPRELLIDEA

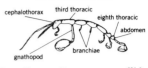

Caprella grandimana, a caprellid.

CAPSALOIDEA

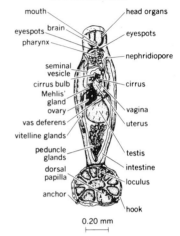

Heterocotyle acetobactis Hargis, from the spotted eagle ray, ventral view.

CARABIDAE

A drawing of a representative of the Carabidae. *(From T. I. Storer and R. L. Usinger, General Zoology, 3d ed., McGraw-Hill, 1957)*

herb of the family Umbelliferae; the fruit is used as a spice and flavoring agent.

carbachol [PHARM] $C_6H_{15}ClN_2O_2$ The choline ester carbamoylcholine chloride, used principally as a miotic in the local treatment of glaucoma.

carbamic acid [BIOCHEM] NH_2COOH An amino acid known for its salts, such as urea and carbamide. Also known as amidocarbonic acid; aminoformic acid.

carbamic acid β-hydroxyphenethyl ester *See* styramate.

carbamidine *See* guanidine.

carbamino [BIOCHEM] A compound formed by the combination of carbon dioxide with a free amino group in an amino acid or a protein.

carbanion [CHEM] One of the charged fragments which arise on heterolytic cleavage of a covalent bond involving carbon; the fragment carries an unshared pair of electrons and bears a negative charge.

carbohydrase [BIOCHEM] Any enzyme that catalyzes the hydrolysis of disaccharides and more complex carbohydrates.

carbohydrate [BIOCHEM] Any of the group of organic compounds composed of carbon, hydrogen, and oxygen, including sugars, starches, and celluloses.

carbohydrate metabolism [BIOCHEM] The sum of the biochemical and physiological processes involved in the breakdown and synthesis of simple sugars, oligosaccharides, and polysaccharides and in the transport of sugar across cell membranes.

carbolfuchsin [MATER] A solution of fuchsin, phenol, alcohol, and water; used as a stain in the identification of bacteria.

carbomycin [MICROBIO] $C_{42}H_{67}O_{16}N$ A colorless, crystalline antibiotic produced by *Streptomyces halstedii;* principally active against gram-positive bacteria.

carbomycin B [MICROBIO] $C_{42}H_{67}O_{15}N$ A colorless, crystalline antibiotic differing from carbomycin only in having one less oxygen atom in its molecule.

carbon [CHEM] A nonmetallic chemical element, symbol C, atomic number 6, atomic weight 12.01115; occurs freely as diamond, graphite, and coal.

carbon-14 [NUC PHYS] A naturally occurring radioisotope of carbon having a mass number of 14 and half-life of 5780 years; used in radiocarbon dating and in the elucidation of the metabolic path of carbon in photosynthesis. Also known as radiocarbon.

carbonaceous [SCI TECH] Relating to or composed of carbon.

carbonate [CHEM] **1.** An ester or salt of carbonic acid. **2.** A compound containing the carbonate (CO_3^{--}) radical. **3.** Containing carbonates.

carbon dioxide [INORG CHEM] CO_2 A colorless, odorless, tasteless gas about 1.5 times as dense as air.

carbonic anhydrase [BIOCHEM] An enzyme which aids carbon dioxide transport and release by catalyzing the synthesis, and the dehydration, of carbonic acid from, and to, carbon dioxide and water.

Carboniferous [GEOL] A division of late Paleozoic rocks and geologic time including the Mississippian and Pennsylvanian periods.

carbon monoxide [INORG CHEM] CO A colorless, odorless gas resulting from the incomplete oxidation of carbon; found, for example, in mines and automobile exhaust; poisonous to animals.

carbonmonoxyhemoglobin [BIOCHEM] A stable combination of carbon monoxide and hemoglobin formed in the blood when carbon monoxide is inhaled. Also known as carbonylhemoglobin and carboxyhemoglobin.

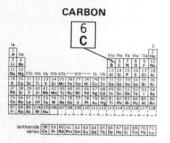

CARBON

Periodic table of the chemical elements showing the position of carbon.

CARBONIFEROUS

PRECAMBRIAN		
CAMBRIAN		
ORDOVICIAN		
SILURIAN	PALEOZOIC	
DEVONIAN		
Mississippian	CARBONIFEROUS	
Pennsylvanian		
PERMIAN		
TRIASSIC		
JURASSIC	MESOZOIC	
CRETACEOUS		
TERTIARY	CENOZOIC	
QUATERNARY		

Position of the Carboniferous and its relationship to the eras and periods of geologic time.

carbon-nitrogen cycle [NUC PHYS] A series of thermonuclear reactions, with release of energy, which presumably occurs in the sun and other stars; the net accomplishment is the synthesis of four hydrogen atoms into a helium atom, the emission of two positrons and much energy, and restoration of a carbon-12 atom with which the cycle began. Also known as carbon cycle; nitrogen cycle.

carbon-nitrogen-phosphorus ratio [OCEANOGR] The relatively constant relationship between the concentrations of carbon (C), nitrogen (N), and phosphorus (P) in plankton, and N and P in sea water, owing to removal of the elements by the organisms in the same proportions in which the elements occur and their return upon decomposition of the dead organisms.

carbonyl [ORG CHEM] A radical (CO) that is made up of one atom of carbon and one atom of oxygen connected by a double bond; found, for example, in aldehydes and ketones. Also known as carbonyl group.

carbonyl group *See* carbonyl.

carboxy *See* carboxyl.

carboxyhemoglobin *See* carbonmonoxyhemoglobin.

carboxyl [ORG CHEM] COOH The radical which determines the basicity of an organic acid. Also known as carboxy; oxatyl.

carboxylase [BIOCHEM] Any enzyme that catalyzes a carboxylation or decarboxylation reaction.

carboxylation [ORG CHEM] Addition of a carboxyl group into a molecule.

carboxylic [CHEM] Having chemical properties resembling those of carboxylic acid.

carboxypeptidase [BIOCHEM] Any enzyme that catalyzes the hydrolysis of a peptide at the end containing the free carboxyl group.

carbuncle [MED] A bacterial infection of subcutaneous tissue caused by *Staphylococcus aureus;* multiple sinuses are created by tissue destruction.

carcerulus [BOT] A superior, dry fruit, characterized by many cells and by indehiscent carpels having one or a few seeds cohering to a central axis by means of united styles.

Carcharhinidae [VERT ZOO] A large family of sharks belonging to the charcharinid group of galeoids, including the tiger sharks and blue sharks.

Carchariidae [VERT ZOO] A family of shallow-water predatory sharks belonging to the isurid group of galeoids.

carcharodont [VERT ZOO] Having sharp, flat, triangular teeth with serrated margins, like those of the man-eating sharks.

carcinogen [MED] Any agent that incites development of a carcinoma or any other sort of malignancy.

carcinoma [MED] A malignant epithelial tumor.

carcinoma in situ [MED] A malignant tumor in the premetastatic stage, when cells are at the site of origin.

carcinomatosis [MED] Metastasis of a primary carcinoma to many sites throughout the body.

cardamom *See* cardamon.

cardamon [BOT] *Elettaria cardamomum.* A perennial herbaceous plant in the family Zingiberaceae; the seed of the plant is used as a spice. Also spelled cardamom.

cardia [ANAT] **1.** The orifice where the esophagus enters the stomach. **2.** The large, blind diverticulum of the stomach adjoining the orifice. [INV ZOO] Anterior enlargement of the ventriculus in some insects.

cardiac [ANAT] **1.** Of, pertaining to, or situated near the heart. **2.** Of or pertaining to the cardia of the stomach.

cardiac arrest [MED] Cessation of the heartbeat.

cardiac cirrhosis [MED] Progressive fibrosis of the liver due

CARDAMON

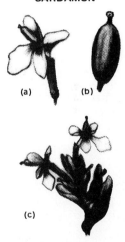

(a)   (b)

(c)

Cardamon *(Elettaria cardamomum).* (a) Flower. (b) Fruit. (c) Branch tip with flowers and fruits.

to prolonged venous blood retention as a result of prolonged and severe heart failure.

cardiac cycle [PHYSIO] The sequence of events in the heart between the start of one contraction and the start of the next.

cardiac electrophysiology [PHYSIO] The science that is concerned with the mechanism, spread, and interpretation of the electric currents which arise within heart muscle tissue and initiate each heart muscle contraction.

cardiac failure [MED] A complex of symptoms resulting from failure of the heart to pump sufficient quantities of blood. Also known as heart failure.

cardiac gland [ANAT] Any of the mucus-secreting, compound tubular structures near the esophagus or in the cardia of the stomach of vertebrates; capable of secreting digestive enzymes.

cardiac massage [MED] Rhythmic compression of the heart by a physician or other person in the effort to maintain effective circulation following heart failure.

cardiac murmur [MED] Any adventitious sound heard in the region of the heart.

cardiac muscle [HISTOL] The principal tissue of the vertebrate heart; composed of a syncytium of striated muscle fibers.

cardiac plexus [ANAT] A network of visceral nerves situated at the base of the heart; contains both sympathetic and vagal nerve fibers.

cardiac sphincter [ANAT] The muscular ring at the orifice between the esophagus and stomach.

cardiac tamponade [MED] Cardiac compression caused by an accumulation of fluid within the pericardium.

cardiac valve [ANAT] Any of the structures located within the orifices of the heart that maintain unidirectional blood flow.

cardiectomy [MED] Excision of the cardiac end of the stomach.

cardinal teeth [INV ZOO] Ridges and grooves on the inner surfaces of both valves of a bivalve mollusk near the anterior end of the hinge.

cardinal vein [EMBRYO] Any of four veins in the vertebrate embryo which run along each side of the vertebral column; the paired veins on each side discharge blood to the heart through the duct of Cuvier.

cardioblast [INV ZOO] Any of certain early embryonic cells in insects from which the heart develops.

cardiogenic plate [EMBRYO] An area of splanchnic mesoderm in the early mammalian embryo from which the heart develops.

cardiogenic shock [MED] Shock due to inadequate arterial blood flow following left-ventricular failure or pulmonary embolism.

cardiography [MED] Analysis of heart movements in the cardiac cycle by means of electronic instruments, especially by tracings.

cardioid condenser [OPTICS] A substage condenser that cuts off the direct light and allows only the light diffracted or dispersed from the object to enter the microscope; used in dark-field microscopes.

cardiolipin [BIOCHEM] A complex phospholipid found in the ether alcohol extract of powdered beef heart; mixed with leutin and cholesterol, it functions as the antigen in the Wassermann complement-fixation test for syphilis. Also known as diphosphatidyl glycerol.

cardiology [MED] The study of the heart.

cardiomyopathy *See* myocardiopathy.

cardiorrhaphy [MED] Suturing of the heart muscle.

cardioscope [MED] **1.** An instrument for the examination or visualization of the interior of the cardiac chambers. **2.** An instrument which, by means of a cathode-ray oscillograph, projects an electrocardiographic record on a luminous screen.

cardiospasm [MED] Failure of the cardiac sphincter to relax; associated with spasm of the cardiac portion of the stomach and dilation of the esophagus.

cardiotomy [MED] Dissection or incision of the heart or the cardia of the stomach.

cardiotonic drug [PHARM] Any agent, such as digitalis, that increases cardiac muscle tonus.

cardiovascular system [ANAT] Those structures, including the heart and blood vessels, which provide channels for the flow of blood.

carditis [MED] Inflammation of the heart tissues.

cardo [INV ZOO] **1.** The hinge connecting the valves of a bivalve mollusk. **2.** In insects, a basal sclerite of the maxilla.

Carettochelyidae [VERT ZOO] A family of reptiles in the order Chelonia containing only one species, the New Guinea plateless turtle (*Carettochelys insculpta*).

Cariamidae [VERT ZOO] The long-legged cariamas, a family of birds in the order Gruiformes.

Caribosireninae [VERT ZOO] A subfamily of trichechiform sirenean mammals in the family Dugongidae.

Caridea [INV ZOO] A large section of decapod crustaceans in the suborder Natantia including many diverse forms of shrimps and prawns.

caries [MED] **1.** Bone decay. **2.** Tooth decay. Also known as dental caries.

carina [BIOL] A ridge or a keel-shaped anatomical structure. [VERT ZOO] *See* keel.

carinate [BIOL] Having a ridge or keel, as the breastbone of certain birds.

Carinomidae [INV ZOO] A monogeneric family of littoral ribbonlike worms in the order Palaeonemertini.

carnassial [ANAT] Of or pertaining to molar or premolar teeth specialized for cutting and shearing.

Carnian [GEOL] Lower Upper Triassic geologic time. Also spelled Karnian.

carnitine [BIOCHEM] $C_7H_{15}NO_3$ α-Amino-β-hydroxybutyric acid trimethylbetaine; a constituent of striated muscle and liver, identical with vitamin B_T.

Carnivora [VERT ZOO] A large order of placental mammals, including dogs, bears, and cats, that is primarily adapted for predation as evidenced by dentition and jaw articulation.

carnivorous [BIOL] Eating flesh or, as in plants, subsisting on nutrients obtained from animal protoplasm.

carnivorous plant *See* insectivorous plant.

Carnosauria [PALEON] A group of large, predacious saurischian dinosaurs in the suborder Theropoda having short necks and large heads.

carnosine [BIOCHEM] $C_9H_{14}N_4O_3$ A colorless, crystalline dipeptide occurring in the muscle tissue of vertebrates.

Carolinian life zone [ECOL] A zone comprising the climate and biotic communities of the oak savannas of eastern North America.

carotenase [BIOCHEM] An enzyme that effects the hydrolysis of carotenoid compounds, used in bleaching of flour.

carotene [BIOCHEM] $C_{40}H_{56}$ Any of several red, crystalline, carotenoid hydrocarbon pigments occurring widely in nature, convertible in the animal body to vitamin A, and characterized by preferential solubility in petroleum ether. Also known as carotin.

carotenemia [MED] The presence of carotene in the blood; may cause yellowing of the skin.

CARP

Carp (*Cyprinus carpio*).

CARPOGONIUM

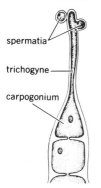

spermatia

trichogyne

carpogonium

Filament with terminal carpogonium (containing an egg) and bearing an elongated trichogyne. (*From H. J. Fuller and O. Tippo, College Botany, rev. ed., Holt, 1954*)

carotenoid [BIOCHEM] A class of labile, easily oxidizable, yellow, orange, red, or purple pigments that are widely distributed in plants and animals and are preferentially soluble in fats and fat solvents.

carotid artery [ANAT] Either of the two principal arteries on both sides of the neck that supply blood to the head and neck. Also known as common carotid artery.

carotid body [ANAT] Either of two chemoreceptors sensitive to changes in blood chemistry which lie near the bifurcations of the carotid arteries. Also known as glomus caroticum.

carotid ganglion [ANAT] A group of nerve cell bodies associated with each carotid artery.

carotid sinus [ANAT] An enlargement at the bifurcation of each carotid artery that is supplied with sensory nerve endings and plays a role in reflex control of blood pressure.

carotol [BIOCHEM] $C_{15}H_{25}OH$ A sesquiterpenoid alcohol in carrots.

carp [VERT ZOO] The common name for a number of freshwater, cypriniform fishes in the family Cyprinidae, characterized by soft fins, pharyngeal teeth, and a suckerlike mouth.

carpal [ANAT] Pertaining to the wrist or carpus.

carpel [BOT] The basic specialized leaf of the female reproductive structure in angiosperms; a megasporophyll.

carpellate [BOT] Having carpels.

carpogonium [BOT] The basal, egg-bearing portion of the female reproductive organ in some thallophytes, especially red algae.

Carpoidea [PALEON] Former designation for a class of extinct homalozoan echinoderms.

carpoids [PALEON] An assemblage of three classes of enigmatic, rare Paleozoic echinoderms formerly grouped together as the class Carpoidea.

carpolith [PALEOBOT] A fossil fruit.

carpology [BOT] The study of the morphology of fruit and seeds.

carpometacarpus [VERT ZOO] In birds, the fused carpal and metacarpal bones of the distal portion of the wing skeleton.

carpophagous [ZOO] Feeding on fruits.

carpophore [BOT] The portion of a flower receptacle that extends between and attaches to the carpels. [MYCOL] The stalk of a fruiting body in fungi.

carpophyte [BOT] A thallophyte that forms a sporocarp following fertilization.

carpopodite [INV ZOO] 1. In some crustaceans, the third joint of the endopodite. 2. The patella in spiders.

carposporangium [BOT] In red algae, a sporangium that forms the cystocarp and contains carpospores.

carpospore [BOT] In red algae, a diploid spore produced terminally by a gonimoblast, giving rise to the diploid tetrasporic plant.

carposporophyte [BOT] The diploid generation of red algae.

carpus [ANAT] 1. The wrist in man or the corresponding part in other vertebrates. 2. The eight bones of the human wrist. [INV ZOO] The fifth segment from the base of a generalized crustacean appendage.

carrageen [BOT] *Chondrus crispus.* A cartilaginous red algae harvested in the Northern Atlantic as a source of carrageenin.

carrageenin [ORG CHEM] A colloid extracted from carrageen and used chiefly as an emulsifying and stabilizing agent in foods, pharmaceuticals, and cosmetics.

carrier [GEN] An individual who is heterozygous for a recessive gene. [MED] A person who harbors and eliminates an infectious agent and so transmits it to others, but who may not show signs of the disease.

Carrion's disease [MED] A bacterial infection of man endemic in the Andes caused by *Bartonella bacilliformis,* which attacks red blood cells and blood-forming organs. Also known as bartonellosis.

carrot [BOT] *Daucus carota.* A biennial umbellifer of the order Umbellales with a yellow or orange-red edible root.

carsickness [MED] Motion sickness resulting from acceleratory movements of a train or automobile.

Carterinacea [INV ZOO] A monogeneric superfamily of marine, benthic foraminiferans in the suborder Rotaliina characterized by a test with monocrystal calcite spicules in a granular groundmass.

cartilage [HISTOL] A specialized connective tissue which is bluish, translucent, and hard but yielding.

cartilage bone [HISTOL] Bone formed by ossification of cartilage.

cartilaginous fish [VERT ZOO] The common name for all members of the class Chondrichthyes.

caruncle [ANAT] Any normal or abnormal fleshy outgrowth, such as the comb and wattles of fowl or the mass in the inner canthus of the eye. [BOT] A fleshy outgrowth developed from the seed coat near the hilum in some seeds, such as the castor bean.

Caryophanaceae [MICROBIO] A family of large, gram-negative bacteria belonging to the order Caryophanales and having disklike cells arranged in chains.

Caryophanales [MICROBIO] An order of bacteria in the class Schizomycetes occurring as trichomes; produce short structures that function as reproductive units.

Caryophyllaceae [BOT] A family of dicotyledonous plants in the order Caryophyllales differing from the other families in lacking betalains.

Caryophyllales [BOT] An order of dicotyledonous plants in the subclass Caryophyllidae characterized by free-central or basal placentation.

Caryophyllidae [BOT] A relatively small subclass of plants in the class Magnoliopsida characterized by trinucleate pollen, ovules with two integuments, and a multilayered nucellus.

caryopsis [BOT] A small, dry, indehiscent fruit having a single seed with such a thin, closely adherent pericarp that a single body, a grain, is formed.

casaba melon [BOT] *Cucumis melo.* A winter muskmelon with a yellow rind and sweet flesh belonging to the family Cucurbitaceae of the order Violales.

caseation necrosis [PATH] Tissue death involving loss of cellular integrity with the consequent conversion to a cheeselike substance; typical in tuberculosis.

cashew [BOT] *Anacardium occidentale.* An evergreen tree of the order Sapindales grown for its kidney-shaped edible nuts and resinous oil.

Casparian strip [BOT] A thin band of suberin- or lignin-like deposition in the radial and transverse walls of certain plant cells during the primary development phase of the endodermis.

Cassadagan [GEOL] Middle Upper Devonian geologic time, above Chemungian.

cassava [BOT] *Manihot esculenta.* A shrubby perennial plant grown for its starchy, edible tuberous roots. Also known as manihot; manioc.

Casselian *See* Chattian.

Cassidulinacea [INV ZOO] A superfamily of marine, benthic foraminiferans in the suborder Rotaliina, characterized by a test of granular calcite with monolamellar septa.

Cassiduloida [INV ZOO] An order of exocyclic Euechinoidea possessing five similar ambulacra which form petal-shaped areas (phyllodes) around the mouth.

CASHEW

Cashew *(Anacardium occidentale).* Branch with leaves, and fruits with nuts attached.

CASPARIAN STRIP

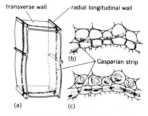

Details of endodermal structure. *(a)* Diagram of cell showing location of Casparian strip. *(b)* Cells of endodermis and of ordinary parenchyma before treatment with alcohol. *(c)* Cells after treatment. Casparian strip is seen only in sectional views in *b* and *c. (From K. Esau, Plant Anatomy, Wiley, 2d ed., 1967)*

CASTOR BEAN

(a)

(b)

Ricinus communis, the castor bean plant. *(a)* Plant grows 3–40 feet (0.9–12 meters) high. *(b)* Pods of the fruit cluster liberate smooth, distinctively marked seeds.

CASUARINACEAE

5 cm

Casuarina equisetifolia, typical member of the family Casuarinaceae and order Casuarinales. *(Photograph by W. D. Brush)*

cassowary [VERT ZOO] Any of three species of large, heavy, flightless birds composing the family Casuariidae in the order Casuariiformes.

cast [MED] **1.** A rigid dressing used to immobilize a part of the body. **2.** *See* strabismus. [PALEON] A fossil reproduction of a natural object formed by infiltration of a mold of the object by waterborne minerals. [PHYSIO] A mass of fibrous material or exudate having the form of the body cavity in which it has been molded; classified from its source, such as bronchial, renal, or tracheal.

caste [INV ZOO] One of the levels of mature social insects in a colony that carry out a specific function; examples are workers and soldiers.

Castle's intrinsic factor *See* intrinsic factor.

Castniidae [INV ZOO] The castniids; large diurnal, butterfly-like moths composing the single, small family of the lepidopteran superfamily Castnioidea.

Castnioidea [INV ZOO] A superfamily of neotropical and Indo-Australian lepidopteran insects in the suborder Heteroneura.

castor bean [BOT] The seed of the castor oil plant (*Ricinus communis*), a coarse, erect annual herb in the spurge family (Euphorbiaceae) of the order Geraniales.

castor oil [MATER] A colorless or greenish nondrying oil extracted from the castor bean; used as a cathartic, in soap, and after processing as a lubricant, and as a leather preservative. Also known as ricinus oil.

castration [MED] Removing, or inhibiting the function or development of, the ovaries or testes.

castration anxiety [PSYCH] Anxiety due to the fear of loss of the genitals or injury to them.

casual carrier [MED] A person who carries an infectious microorganism but never manifests the disease.

Casuariidae [VERT ZOO] The cassowaries, a family of flightless birds in the order Casuariiformes lacking head and neck feathers and having bony casques on the head.

Casuariiformes [VERT ZOO] An order of large, flightless, ostrichlike birds of Australia and New Guinea.

Casuarinaceae [BOT] The single, monogeneric family of the plant order Casuarinales characterized by reduced flowers and green twigs bearing whorls of scalelike, reduced leaves.

Casuarinales [BOT] A monofamilial order of dicotyledonous plants in the subclass Hamamelidae.

cat [VERT ZOO] The common name for all members of the carnivoran mammalian family Felidae, especially breeds of the domestic species, *Felis domestica.*

catabiosis [PHYSIO] Degenerative changes accompanying cellular senescence.

catabolism [BIOCHEM] That part of metabolism concerned with the breakdown of large protoplasmic molecules and tissues, often with the liberation of energy.

catabolite [BIOCHEM] Any product of catabolism.

catabolite repression [BIOCHEM] An intracellular regulatory mechanism in bacteria whereby glucose, or any other carbon source that is an intermediate in catabolism, prevents formation of inducible enzymes.

catadromous [VERT ZOO] Pertaining to fishes which live in fresh water and migrate to spawn in salt water.

catalase [BIOCHEM] An enzyme that catalyzes the decomposition of hydrogen peroxide into oxygen and water and the oxidation by hydrogen peroxide of alcohols to aldehydes.

catalectrotonus [PHYSIO] The negative electric potential during the passage of a current on the surface of a nerve or muscle in the region of the cathode.

catalepsy [PSYCH] Suspended animation with loss of voluntary motion associated with hysteria and the schizophrenic reactions in man, and with organic nervous system disease in animals.

catalysis [CHEM] A phenomenon in which a relatively small amount of substance augments the rate of a chemical reaction without itself being consumed.

catalyst [CHEM] Substance that alters the velocity of a chemical reaction and may be recovered essentially unaltered in form and amount at the end of the reaction.

catamount *See* puma.

catapetalous [BOT] Characterized by petals that are united with the base of monadelphous stamens.

catapleurite [INV ZOO] In certain Thysanura, the pleurite of the thorax between the anapleurite and the trochantin.

cataplexy [MED] **1.** Sudden loss of muscle tone provoked by exaggerated emotion, such as excessive anger or laughter, often associated with a profound desire for sleep. **2.** Prostration by the sudden onset of disease. **3.** Hypnotic sleep.

Catapochrotidae [INV ZOO] A monospecific family of coleopteran insects in the superfamily Cucujoidea.

cataract [MED] An opacity in the crystalline lens or the lens capsule of the eye.

catarrh [MED] An old term for an inflammation of mucous membranes, particularly of the respiratory tract.

catarrhal jaundice *See* infectious hepatitis.

catastrophism [PALEON] The theory that the differences between fossils in successive stratigraphic horizons resulted from a general catastrophe followed by creation of the different organisms found in the next-younger beds.

catatonia [PSYCH] A type of schizophrenic reaction in which the individual remains speechless and motionless, assumes fixed postures, and lacks the will and resists attempts to activate speech and movement.

catechol [ORG CHEM] One of a group of three isomeric dihydroxy benzenes in which the two hydroxyl groups are ortho to each other. Also known as catechin; pyrocatechol; pyrocatechuic acid.

catecholamine [BIOCHEM] Any one of a group of sympathomimetic amines containing a catechol moiety, including especially epinephrine, norepinephrine (levarterenol), and dopamine.

catenulate [BIOL] Having a chainlike form.

Catenulida [INV ZOO] An order of threadlike, colorless freshwater rhabdocoeles with a simple pharynx and a single, median protonephridium.

caterpillar [INV ZOO] **1.** The wormlike larval stage of a butterfly or moth. **2.** The larva of certain insects, such as scorpion flies and sowflies.

catfish [VERT ZOO] The common name for a number of fishes which constitute the suborder Siluroidei in the order Cypriniformes, all of which have barbels around the mouth.

catharsis [PSYCH] Release of tension by releasing deepseated emotions or reliving a traumatic experience.

cathartic [PHARM] Any drug, such as castor oil, mineral oil, or a laxative, that causes defecation.

Cathartidae [VERT ZOO] The New World vultures, a family of large, diurnal predatory birds in the order Falconiformes that lack a voice and have slightly webbed feet.

cathepsin [BIOCHEM] Any of several proteolytic enzymes occurring in animal tissue that hydrolyze high-molecular-weight proteins to proteoses and peptones.

catheter [MED] A hollow, tubular device for insertion into a cavity, duct, or vessel to permit injection or withdrawal of fluids or to establish patency of the passageway.

catheterization [MED] Insertion or use of a catheter.

CATENULIDA

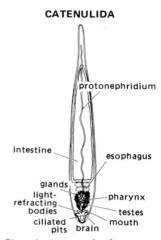

Stenostomum grande, the commonest and most widely distributed of all freshwater rhabdocoeles, in the order Catenulida. *(After Nuttycombe and Waters, 1938)*

CATERPILLAR

Forest tent caterpillar *(Malacosoma disstria).*

CATFISH

Brown bullhead *(Ictaturus nebulosus),* a catfish found in warm mud-bottomed waters across the United States.

cathode [PHYS CHEM] The electrode at which reduction takes place in an electrochemical cell, that is, a cell through which electrons are being forced.

cation exchange resin [ORG CHEM] A highly polymerized synthetic organic compound consisting of a large, nondiffusible anion and a simple, diffusible cation, which later can be exchanged for a cation in the medium in which the resin is placed.

catkin [BOT] An indeterminate type of inflorescence that resembles a scaly spike and sometimes is pendant.

Catostomidae [VERT ZOO] The suckers, a family of cypriniform fishes in the suborder Cyprinoidei.

cat scratch disease [MED] A benign systematic illness in man characterized by malaise, fever, and a granulomatous lymphadenitis; the causative organism has not been identified. Also known as benign lymphoreticulosis.

Cattell infant intelligence scale [PSYCH] A modification of the revised Stanford-Binet test adapted for children from 2 to 30 months of age.

cattle [AGR] Domesticated bovine animals, including cows, steers, and bulls, raised and bred on a ranch or farm.

Cauchy distribution [STAT] A distribution function having the form $M/[\pi M^2 + (x - a)^2]$, where x is the variable and M and a are constants. Also known as Cauchy frequency distribution.

Cauchy frequency distribution *See* Cauchy distribution.

cauda [ANAT] The posterior portion of an organ. [INV ZOO] A posterior abdominal tube in certain insects. [VERT ZOO] **1.** A tail. **2.** A taillike appendage.

cauda equina [ANAT] The roots of the sacral and coccygeal nerves, collectively; so called because of their resemblance to a horse's tail.

caudal [ZOO] Toward, belonging to, or pertaining to the tail or posterior end.

caudal artery [VERT ZOO] The extension of the dorsal aorta in the tail of a vertebrate.

caudal vertebra [ANAT] Any of the small bones of the vertebral column that support the tail in vertebrates; in man, three to five are fused to form the coccyx.

Caudata [VERT ZOO] An equivalent name for Urodela.

caudate [ZOO] **1.** Having a tail or taillike appendage. **2.** Any member of the Caudata.

caudate lobe [ANAT] The tailed lobe of the liver that separates the right extremity of the transverse fissure from the commencement of the fissure for the inferior vena cava.

caudate nucleus [ANAT] An elongated arched gray mass which projects into and forms part of the lateral wall of the lateral ventricle.

caudex [BOT] The main axis of a plant, including stem and roots.

caudicle [BOT] A slender appendage attaching pollen masses to the stigma in orchids.

Caulerpaceae [BOT] A family of green algae in the order Siphonales.

caulescent [BOT] Having an aboveground stem.

caulicle [BOT] **1.** A rudimentary stem. **2.** The axis of a young seedling.

cauliflorous [BOT] Producing flowers on the older branches or main stem.

cauliflower [BOT] *Brassica oleracea* var. *botrytis.* A biennial crucifer of the order Capparales grown for its edible white head or curd, which is a tight mass of flower stalks.

cauliflower disease [PL PATH] **1.** A disease of the strawberry plant caused by the eelworm and manifested as clustered, puckered, and malformed leaves. **2.** A bacterial disease of the

CAULERPACEAE

Caulerpa **showing the stolonlike branches, rhizoidal branches, and erect featherlike frond typical of the Caulerpaceae.**

strawberry and some other plants caused by *Corynebacterium fascians.*

cauliform [BIOL] Stemlike.

cauline [BOT] Belonging to or arising from the stem, particularly if on the upper portion.

Caulobacteraceae [MICROBIO] A family of stalked, aquatic, gram-negative bacteria in the order Pseudomonadales.

caulocarpic [BOT] Having stems that bear flowers and fruit every year.

caulome [BOT] The stem structure or stem axis of a plant as a whole.

cauterization [MED] Use of a device or chemical agent to coagulate or destroy tissue.

Cavellinidae [PALEON] A family of Paleozoic ostracods in the suborder Platycopa.

cavernicolous [BIOL] Inhabiting caverns.

cavernous sinus [ANAT] Either of a pair of venous sinuses of the dura mater located on the side of the body of the sphenoid bone.

cavicorn [VERT ZOO] Having hollow horns, referring especially to certain ruminants.

Caviidae [VERT ZOO] A family of large, hystricomorph rodents distinguished by a reduced number of toes and a rudimentary tail.

cavitation [PATH] The formation of one or more cavities in an organ or tissue, especially as the result of disease.

cavity [BIOL] A hole or hollow space in an organ, tissue, or other body part. [ELECTROMAG] *See* cavity resonator.

cavy [VERT ZOO] Any of the rodents composing the family Caviidae, which includes the guinea pig, rock cavies, mountain cavies, capybara, salt desert cavy, and mara.

Caytoniales [PALEOBOT] An order of Mesozoic plants.

Cayugan [GEOL] Upper Silurian geologic time.

Cazenovian [GEOL] Lower Middle Devonian geologic time.

cc *See* cubic centimeter.

Cd *See* cadmium.

CD *See* circular dichroism.

Ce *See* cerium.

Cebidae [VERT ZOO] The New World monkeys, a family of primates in the suborder Anthropoidea including the capuchins and howler monkeys.

Cebochoeridae [PALEON] A family of extinct palaeodont artiodactyls in the superfamily Entelodontoidae.

Cebrionidae [INV ZOO] The robust click beetles, a family of cosmopolitan coleopteran insects in the superfamily Elateroidea.

cecidium [PL PATH] Plant gall produced either by insects in ovipositing or by fungi as a consequence of infection.

Cecropidae [INV ZOO] A family of crustaceans in the suborder Caligoida which are external parasites on fish.

cecum [ANAT] The blind end of a cavity, duct, or tube, especially the sac at the beginning of the large intestine. Also spelled caecum.

cedar [BOT] The common name for a large number of evergreen trees in the order Pinales having fragrant, durable wood.

Celastraceae [BOT] A family of dicotyledonous plants in the order Celastrales characterized by erect and basal ovules, a flower disk that surrounds the ovary at the base, and opposite or sometimes alternate leaves.

Celastrales [BOT] An order of dicotyledonous plants in the subclass Rosidae marked by simple leaves and regular flowers.

celery [BOT] *Apium graveolens* var. *dulce.* A biennial umbellifer of the order Umbellales with edible petioles or leaf stalks.

CAULOBACTERACEAE

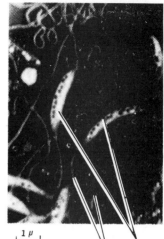

1 μ

cells

stalks

Electron micrograph of *Caulobacter* species in the Caulobacteraceae, showing cells and stalks. *(Courtesy of J. Wachsman)*

CEDAR

Leaf arrangement for the Atlas cedar *(Cedrus atlantica).*

CELLOBIOSE

CH$_2$OH·CH·CHOH·CHOH·CHOH·CH---O
|_____O_____|
----CH·CHOH·CHOH·CH·OH
CH------O----|
CH$_2$OH

Structural formula of cellobiose.

celiac [ANAT] Of, in, or pertaining to the abdominal cavity.

celiac plexus [ANAT] A nerve plexus located in front of the aorta around the origin of the celiac trunk.

celiac syndrome [MED] A complex of symptoms produced by intestinal malabsorption of fat and marked by bulky, loose, foul-smelling stools, high in fatty-acid content. Also known as idiopathic steatorrhea.

cell [BIOL] The microscopic functional and structural unit of all living organisms, consisting of a nucleus, cytoplasm, and a limiting membrane. [PHYS CHEM] A cup, jar, or vessel containing electrolyte solutions and metal electrodes to produce an electric current (conductiometric or potentiometric) or for electrolysis (electrolytic). Also known as electric cell.

cell constancy [BIOL] The condition in which the entire body, or a part thereof, consists of a fixed number of cells that is the same for all adults of the species.

cell differentiation [CYTOL] The series of events involved in the development of a specialized cell having specific structural, functional, and biochemical properties.

cell division [CYTOL] The process by which living cells multiply; may be mitotic or a amitotic.

Cellfalcicula [MICROBIO] Genus of gram-negative, spindle-shaped soil bacteria belonging to the family Spirillaceae.

cell-free extract [CYTOL] A fluid obtained by breaking open cells; contains most of the soluble molecules of a cell.

cell inclusion [CYTOL] A small, nonliving intracellular particle, usually representing a form of stored food, not immediately vital to life processes.

cell lineage [EMBRYO] The developmental history of individual blastomeres from their first cleavage division to their ultimate differentiation into cells of tissues and organs.

cell membrane [CYTOL] A thin layer of protoplasm, consisting mainly of lipids and proteins, which is present on the surface of all cells. Also known as plasmalemma; plasma membrane.

cell movement [CYTOL] **1.** Intracellular movement of cellular components. **2.** Movement of a cell relative to its environment.

cellobiose [ORG CHEM] C$_{12}$H$_{22}$O$_{11}$ A disaccharide, not known to exist in nature, obtained by partial hydrolysis of cellulose.

celloidin [MATER] A concentrated solution of pyroxylin used principally in microscopy for embedding specimens or for section-cutting.

cell pathology [PATH] Abnormalities of the events taking place within cells.

cell permeability [CYTOL] The permitting or activating of the passage of substances into, out of, or through cells.

cell plate [CYTOL] A membrane-bound disk formed during cytokinesis in plant cells which eventually becomes the middle lamella of the wall formed between daughter cells.

cell sap [CYTOL] The liquid content of a plant cell vacuole.

cells of Paneth [HISTOL] Coarsely granular secretory cells found in the crypts of Lieberkühn in the small intestine. Also known as Paneth cells.

cell theory [BIOL] **1.** A principle that describes the cell as the fundamental unit of all living organisms. **2.** A principle that describes the properties of an organism as the sum of the properties of its component cells.

cellular [BIOL] Characterized by, consisting of, or pertaining to cells.

cellular affinity [BIOL] The phenomenon of selective adhesiveness observed among the cells of certain sponges, slime molds, and vertebrates.

cellular endosperm [BOT] Endosperm in which cell wall

formation occurs immediately after mitosis so that there is no initial free-nuclear stage during early ontogeny.

cellular infiltration [MED] **1.** Passage of cells into tissues in the course of inflammation. **2.** Migration of or invasion by cells of neoplasms.

cellulase [BIOCHEM] Any of a group of extracellular enzymes, produced by various fungi, bacteria, insects, and other lower animals, that hydrolyze cellulose.

cellulitis [MED] Diffuse inflammation of a connective tissue.

cellulolytic [BIOL] Having the ability to hydrolyze cellulose; applied to certain bacteria and protozoans.

cellulose [BIOCHEM] $(C_6H_{10}O_5)_n$ The main polysaccharide in living plants, forming the skeletal structure of the plant cell wall; a polymer of β-D-glucose units linked together, with the elimination of water, to form chains of 2000–4000 units.

cell wall [CYTOL] A semirigid, permeable structure that is composed of cellulose, lignin, or other substances and that envelops most plant cells.

Celsius degree [THERMO] Unit of temperature interval or difference equal to the kelvin.

Celsius temperature scale [THERMO] Temperature scale in which the temperature Θ_c in degrees Celsius (°C) is related to the temperature T_k in kelvins by the formula $\Theta_c = T_k - 273.15$; the freezing point of water at standard atmospheric pressure is very nearly 0°C and the corresponding boiling point is very nearly 100°C. Formerly known as centigrade temperature scale.

Celyphidae [INV ZOO] A family of myodarian cyclorrhaphous dipteran insects in the subsection Acalypteratae.

cement [HISTOL] A calcified tissue which fastens the roots of teeth to the alveolus. Also known as cementum. [INV ZOO] Any of the various adhesive secretions, produced by certain invertebrates, that harden on exposure to air or water and are used to bind objects.

cement gland [INV ZOO] A structure in many invertebrates that produces cement.

cementum *See* cement.

Cenomanian [GEOL] Lower Upper Cretaceous geologic time.

Cenozoic [GEOL] The youngest of the eras, or major subdivisions of geologic time, extending from the end of the Mesozoic Era to the present, or Recent.

center of a distribution [STAT] The expected value of any random variable which has the given distribution.

centesis [MED] Surgical puncture, or perforation as of a tumor or membrane.

centi- [SCI TECH] A prefix representing 10^{-2}, which is 0.01 or one-hundredth. Abbreviated c.

centigrade temperature scale *See* Celsius temperature scale.

centimeter-gram-second system [PHYS] An absolute system of metric units in which the centimeter, gram mass, and the second are the basic units. Abbreviated cgs system.

centimorgan [GEN] A unit of genetic map distance, equal to the distance along a chromosome that gives a recombination frequency of 1%.

centipede [INV ZOO] The common name for an arthropod of the class Chilopoda.

central apparatus [CYTOL] The centrosome or centrosomes together with the surrounding cytoplasm. Also known as cytocentrum.

central canal [ANAT] The small canal running through the center of the spinal cord from the conus medullaris to the lower part of the fourth ventricle; represents the embryonic neural tube.

Centrales [BOT] An order of diatoms (Bacillariophyceae) in

CELLULOSE

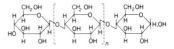

Structure of the cellulose polymer.

CENOZOIC

PRECAMBRIAN		
CAMBRIAN		PALEOZOIC
ORDOVICIAN		
SILURIAN		
DEVONIAN		
Mississippian	CARBONIFEROUS	
Pennsylvanian		
PERMIAN		
TRIASSIC		MESOZOIC
JURASSIC		
CRETACEOUS		
TERTIARY		CENOZOIC
QUATERNARY		

Chart showing the relationship of the Cenozoic to other eras.

CENTIPEDE

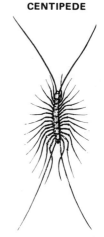

Scutigera coleoptrata (L.), the house centipede; body length about 25 millimeters. *(From R. E. Snodgrass, A Textbook of Arthropod Anatomy, Cornell University Press, 1952)*

CENTRAL PLACENTATION

Free central placentation,
ovules shown in black.

CENTRAL POCKET LOOP

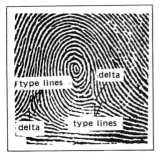

type lines

delta

delta   type lines

Central pocket loop, a type of
whorl pattern. *(Federal Bureau
of Investigation)*

CENTROHELIDA

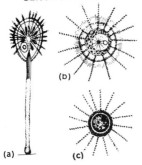

(a)   (c)

(b)

Some Centrohelida species.
(a) Actinolophus pedunculatus,
sessile on Bryozoa *(after
Villeneuve), (b) Heterophrys
myriopoda (after Penard),
(c) Pompholyxophrys punicea
(after Penard). (From R. P.
Hall, Protozoology, Prentice-
Hall, 1953)*

which the form is often circular and the markings on the valves are radial.

central nervous system [ANAT] The division of the vertebrate nervous system comprising the brain and spinal cord.

central placentation [BOT] Having the ovules located in the center of the ovary.

central pocket loop [ANAT] A whorl type of fingerprint pattern having two deltas and at least one ridge that make a complete circuit.

central sulcus [ANAT] A groove situated about the middle of the lateral surface of the cerebral hemisphere, separating the frontal from the parietal lobe.

Centrarchidae [VERT ZOO] A family of fishes in the order Perciformes, including the fresh-water or black basses and several sunfishes.

centric [ANAT] Having all teeth of both jaws meet normally with perfect distribution of forces in the dental arch.

centrifuge [MECH ENG] **1.** A rotating device for separating liquids of different specific gravities or for separating suspended colloidal particles, such as clay particles in an aqueous suspension, according to particle-size fractions by centrifugal force. **2.** A large motor-driven apparatus with a long arm, at the end of which human and animal subjects or equipment can be revolved and rotated at various speeds to simulate the prolonged accelerations encountered in rockets and spacecraft.

centrifuge microscope [OPTICS] An instrument which permits magnification and observation of living cells being centrifuged; image of the material magnified by the objective which rotates near the periphery of the centrifuge head is brought to the axis of rotation where it is observed in a stationary occular.

centrilobular emphysema [MED] A disorder marked by pulmonary inflation, primarily affecting the respiratory bronchioles and usually more severe in the upper lobes.

centriole [CYTOL] A complex cellular organelle forming the center of the centrosome in most cells; usually found near the nucleus in interphase cells and at the spindle poles during mitosis.

centripetal canals [INV ZOO] Blind canals extending from the circular canal posterior toward the apex of the bell in certain Trachomedusae.

Centrohelida [INV ZOO] An order of protozoans in the subclass Heliozoia lacking a central capsule and having axopodia or filopodia, and siliceous scales and spines.

centrolecithal ovum [CYTOL] An egg cell having the yolk centrally located; occurs in arthropods.

Centrolenidae [VERT ZOO] A family of arboreal frogs in the suborder Procoela characterized by green bones.

centromere [CYTOL] A specialized chromomere to which the spindle fibers are attached during mitosis. Also known as kinetochore; kinomere; primary constriction.

Centronellidina [PALEON] A suborder of extinct articulate brachiopods in the order Terebratulida.

centrosome [CYTOL] A spherical hyaline region of the cytoplasm surrounding the centriole in many cells; plays a dynamic part in mitosis as the focus of the spindle pole.

Centrospermae [BOT] An equivalent name for the Caryophyllales.

Centrospermales [BOT] An equivalent name for the Caryophyllales.

centrosphere [CYTOL] The differentiated layer of cytoplasm immediately surrounding the centriole.

centrum [ANAT] The main body of a vertebra. [BOT] The central space in hollow-stemmed plants.

Cephalaspida [PALEON] An equivalent name for the Osteostraci.

Cephalaspidomorphi [VERT ZOO] An equivalent name for Monorhina.

cephalic [ZOO] Of or pertaining to the head or anterior end.

cephalic index [ANTHRO] The ratio of maximum breadth to maximum length of the head multiplied by 100.

cephalic module [ANTHRO] A measure of absolute head size derived by averaging the length, breadth, and auricular height of the head.

cephalic vein [ANAT] A superficial vein located on the lateral side of the arm which drains blood from the radial side of the hand and forearm into the axillary vein.

cephalin [BIOCHEM] Any of several acidic phosphatides whose composition is similar to that of lecithin but having ethanolamine, serine, and inositol instead of choline; found in many living tissues, especially nervous tissue of the brain.

Cephalina [INV ZOO] A suborder of protozoans in the order Eugregarinida that are parasites of certain invertebrates.

Cephalobaenida [INV ZOO] An order of the arthropod class Pentastomida composed of primitive forms with six-legged larvae.

Cephalocarida [INV ZOO] A subclass of Crustacea erected to include the primitive crustacean *Hutchinsoniella macracantha*.

Cephalochordata [VERT ZOO] A subphylum of the Chordata comprising the lancelets, including *Branchiostoma*.

Cephaloidae [INV ZOO] The false longhorn beetles, a small family of coleopteran insects in the superfamily Tenebrionoidea.

cephalomere [INV ZOO] One of the somites that make up the head of an arthropod.

cephalometry [ANTHRO] The science of measuring the head, especially for determining the characteristics of a particular race, sex, or somatotype.

cephalont [INV ZOO] A sporozoan just prior to spore formation.

Cephalopoda [INV ZOO] Exclusively marine animals constituting the most advanced class of the Mollusca, including squids, octopuses, and *Nautilus*.

cephalopodium [INV ZOO] The head region in cephalopods, consisting of the head and arms.

cephalosporin [MICROBIO] Any of a group of antibiotics produced by strains of the imperfect fungus *Cephalosporium*.

cephalostyle [VERT ZOO] The sheathed anterior end of the notochord.

cephalotheca [INV ZOO] Integument of the head of an insect pupa.

cephalothin [MICROBIO] An antibiotic derived from the fungus *Cephalosporium*, resembling penicillin units in structure and activity, and effective against many gram-positive cocci that are resistant to penicillin.

cephalothorax [INV ZOO] The body division comprising the united head and thorax of arachnids and higher crustaceans.

Cephalothrididae [INV ZOO] A family of ribbonlike worms in the order Palaeonemertini.

cephalotrichous flagellation [CYTOL] Insertion of flagella in polar tufts.

cephalotrocha [INV ZOO] A turbellarian larva having eight processes in the oral region.

Cephidae [INV ZOO] The stem sawflies, composing the single family of the hymenopteran superfamily Cephoidea.

Cephoidea [INV ZOO] A superfamily of hymenopteran insects in the suborder Symphyta.

Ceractinomorpha [INV ZOO] A subclass of sponges belonging to the class Demospongiae.

CEPHALOCARIDA

Hutchinsoniella macracantha of the Cephalocarida. *(After Scourfield)*

CERAMBYCIDAE

A longhorn beetle of the Cerambycidae. (From T. I. Storer and R. L. Usinger, General Zoology, 3d ed., McGraw-Hill, 1957

Cerambycidae [INV ZOO] The longhorn beetles, a family of coleopteran insects in the superfamily Chrysomeloidea.

Ceramoporidae [PALEON] A family of extinct, marine bryozoans in the order Cystoporata.

Cerapachyinae [INV ZOO] A subfamily of predacious ants in the family Formicidae, including the army ant.

Ceraphronidae [INV ZOO] A superfamily of hymenopteran insects in the superfamily Proctotrupoidea.

cerata [INV ZOO] Respiratory papillae of the mantle in certain nudibranchs.

ceratine [INV ZOO] A hornlike material secreted by some anthozoans.

Ceratiomyxaceae [MYCOL] The single family of the fungal order Ceratiomyxales.

Ceratiomyxales [MYCOL] An order of myxomycetous fungi in the subclass Ceratiomyxomycetidae.

Ceratiomyxomycetidae [MYCOL] A subclass of fungi belonging to the Myxomycetes.

ceratite [PALEON] A fossil ammonoid of the genus *Ceratites* distinguished by a type of suture in which the lobes are further divided into subordinate crenulations while the saddles are not divided and are smoothly rounded.

Ceratodontidae [PALEON] A family of Mesozoic lungfishes in the order Dipteriformes.

Ceratomorpha [VERT ZOO] A suborder of the mammalian order Perissodactyla including the tapiroids and the rhinoceratoids.

Ceratophyllaceae [BOT] A family of rootless, free-floating dicotyledons in the order Nymphaeales characterized by unisexual flowers and whorled, cleft leaves with slender segments.

Ceratopsia [PALEON] The horned dinosaurs, a suborder of Upper Cretaceous reptiles in the order Ornithischia.

ceratotheca [INV ZOO] In insect pupae, the part of the casing protecting the antennae.

cercaria [INV ZOO] The larval generation which terminates development of a digenetic trematode in the intermediate host.

cercid [INV ZOO] In Porifera, a minute wandering cell formed by division of an archaeocyte.

Cercopidae [INV ZOO] A family of homopteran insects belonging to the series Auchenorrhyncha.

Cercopithecidae [VERT ZOO] The Old World monkeys, a family of primates in the suborder Anthropoidea.

cercopod [INV ZOO] **1.** Either of two filamentous projections on the posterior end of notostracan crustaceans. **2.** *See* cercus.

Cercospora leaf spot [PL PATH] Any of several fungus diseases of plants caused by *Cercospora* species and characterized by areas of discoloration on the leaves.

cercus [INV ZOO] Either of a pair of segmented sensory appendages on the last abdominal segment of many insects and certain other arthropods. Also known as cercopod.

cere [VERT ZOO] A soft, swollen mass of tissue at the base of the upper mandible through which the nostrils open in certain birds, such as parrots and birds of prey.

cereal [BOT] Any member of the grass family (Graminae) which produces edible, starchy grains usable as food by man and his livestock. Also known as grain.

cerebellar ataxia [MED] Incoordination of muscles due to disease of the cerebellum.

cerebellum [ANAT] The part of the vertebrate brain lying below the cerebrum and above the pons, consisting of three lobes and concerned with muscular coordination and the maintenance of equilibrium.

cerebral cortex [ANAT] The superficial layer of the cerebral hemispheres, composed of gray matter and concerned with coordination of higher nervous activity.

cerebral hemisphere [ANAT] Either of the two lateral halves of the cerebrum.

cerebral localization [ANAT] Designation of a specific region of the brain as the area controlling a specific physiologic function or as the site of a lesion.

cerebral palsy [MED] Any nonprogressive motor disorder in man caused by brain damage incurred during fetal development.

cerebral peduncle [ANAT] One of two large bands of white matter (containing descending axons of upper motor neurons) which emerge from the underside of the cerebral hemispheres and approach each other as they enter the rostral border of the pons.

cerebroganglion [INV ZOO] A supraesophageal ganglion of invertebrates.

cerebrose *See* galactose.

cerebroside [BIOCHEM] Any of a complex group of glycosides found in nerve tissue, consisting of a hexose, a nitrogenous base, and a fatty acid. Also known as galactolipid.

cerebroside lipoidosis *See* Gaucher's disease.

cerebrospinal fluid [PHYSIO] A clear liquid that fills the ventricles of the brain and the spaces between the arachnoid mater and pia mater.

cerebrospinal meningitis [MED] Inflammation of the meninges of the brain and spinal cord.

cerebrovascular accident [MED] A symptom complex resulting from cerebral hemorrhage, embolism, or thrombosis of the cerebral vessels, characterized by sudden loss of consciousness.

cerebrum [ANAT] The enlarged anterior or upper part of the vertebrate brain consisting of two lateral hemispheres.

Ceriantharia [INV ZOO] An order of the Zoantharia distinguished by the elongate form of the anemone-like body.

Ceriantipatharia [INV ZOO] A subclass proposed by some authorities to include the anthozoan orders Antipatharia and Ceriantharia.

Cerithiacea [INV ZOO] A superfamily of gastropod mollusks in the order Prosobranchia.

cerium [CHEM] A chemical element, symbol Ce, atomic number 58, atomic weight 140.12; a rare-earth metal, used as a getter in the metal industry, as an opacifier and polisher in the glass industry, in Welsbach gas mantles, in cored carbon arcs, and as a liquid-liquid extraction agent to remove fission products from spent uranium fuel.

cernuous [BOT] Drooping or inclining.

Cerophytidae [INV ZOO] A small family of coleopteran insects in the superfamily Elateroidea.

ceruloplasmin [BIOCHEM] The copper-binding serum protein in human blood.

cerumen [PHYSIO] The waxy secretion of the ceruminous glands of the external ear. Also known as earwax.

ceruminous gland [ANAT] A modified sweat gland in the external ear that produces earwax.

cervical [ANAT] Of or relating to the neck, a necklike part, or the cervix of an organ.

cervical canal [ANAT] Canal of the cervix of the uterus.

cervical flexure [EMBRYO] A ventrally concave flexure of the embryonic brain occurring at the junction of hindbrain and spinal cord.

cervical ganglion [ANAT] Any ganglion of the sympathetic nervous system located in the neck.

CERIANTHARIA

Cerianthus solitarius **lives in sandy marine substrata. (After Y. Delage)**

CERIUM

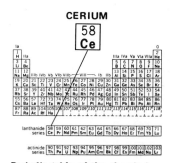

Periodic table of the chemical elements showing the position of cerium.

CERVICAL VERTEBRA

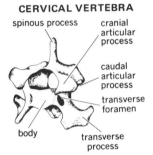

spinous process

cranial
articular
process

caudal
articular
process

transverse
foramen

body

transverse
process

Cervical vertebra of a dog. (From
M. E. Miller, G. C. Christensen,
and H. E. Evans, Anatomy of the
Dog, Saunders, 1964)

CESIUM

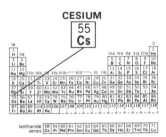

Periodic table of the chemical
elements showing the position
of cesium.

CESTIDA

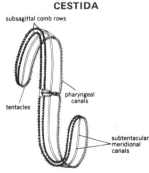

subsagittal comb rows

pharyngeal
canals

tentacles

subtentacular
meridional
canals

Velamen species.

cervical plexus [ANAT] A plexus in the neck formed by the anterior branches of the upper four cervical nerves.

cervical sinus [EMBRYO] A triangular depression caudal to the hyoid arch containing the posterior visceral arches and grooves.

cervical vertebra [ANAT] Any of the bones in the neck region of the vertebral column; the transverse process has a characteristic perforation by a transverse foramen.

cervicitis [MED] Inflammation of the cervix uteri.

Cervidae [VERT ZOO] A family of pecoran ruminants in the superfamily Cervoidea, characterized by solid, deciduous antlers; includes deer and elk.

cervix [ANAT] A constricted portion or neck of an organ or part.

cervix uteri [ANAT] The narrow outer end of the uterus.

Cervoidea [VERT ZOO] A superfamily of tylopod ruminants in infraorder Pecora, including deer, giraffes, and related species.

cesarean section [MED] Delivery of the fetus through an abdominal incision.

cesium [CHEM] A chemical element, symbol Cs, atomic number 55, atomic weight 132.905.

cesium-137 [NUC PHYS] An isotope of cesium with atomic mass number of 137; emits negative beta particles and has a half-life of 30 years; offers promise as an encapsulated radiation source for therapeutic and other purposes. Also known as radiocesium.

cespitose [BOT] **1.** Tufted; growing in tufts, as grass. **2.** Having short stems forming a dense turf.

Cestida [INV ZOO] An order of ribbon-shaped ctenophores having a very short tentacular axis and an elongated pharyngeal axis.

Cestoda [INV ZOO] A subclass of tapeworms including most members of the class Cestoidea; all are endoparasites of vertebrates.

Cestodaria [INV ZOO] A small subclass of worms belonging to the class Cestoidea; all are endoparasites of primitive fishes.

Cestoidea [INV ZOO] The tapeworms, endoparasites composing a class of the phylum Platyhelminthes.

Cetacea [VERT ZOO] An order of aquatic mammals, including the whales, dolphins, and porpoises.

cetology [VERT ZOO] The study of whales.

Cetomimiformes [VERT ZOO] An order of rare oceanic, deepwater, soft-rayed fishes that are structurally diverse.

Cetomimoidei [VERT ZOO] The whalefishes, a suborder of the Cetomimiformes, including bioluminescent, deep-sea species.

Cetorhinidae [VERT ZOO] The basking sharks, a family of large, galeoid elasmobranchs of the isurid line.

cgs system See centimeter-gram-second system.

chaeta See seta.

Chaetetidae [PALEON] A family of Paleozoic corals of the order Tabulata.

Chaetodontidae [VERT ZOO] The butterflyfishes, a family of perciform fishes in the suborder Percoidei.

Chaetognatha [INV ZOO] A phylum of abundant planktonic arrowworms.

Chaetonotoidea [INV ZOO] An order of the class Gastrotricha characterized by two adhesive tubes connected with the distinctive paired, posterior tail forks.

Chaetophoraceae [BOT] A family of algae in the order Ulotrichales characterized as branched filaments which taper toward the apices, sometimes bearing terminal setae.

Chaetopteridea [INV ZOO] A family of spioniform polychaete annelids belonging to the Sedentaria.

Chagas' disease [MED] An acute and chronic protozoan disease of man caused by the hemoflagellate *Trypanosoma* (*Schizotrypanum*) *cruzi*. Also known as South American trypanosomiasis.

chain balance [ANALY CHEM] An analytical balance with one end of a fine gold chain suspended from the beam and the other fastened to a device which moves over a graduated vernier scale.

chain isomerism [ORG CHEM] A type of molecular isomerism seen in carbon compounds; as the number of carbon atoms in the molecule increases, the linkage between the atoms may be a straight chain or branched chains producing isomers that differ from each other by possessing different carbon skeletons.

chair form [PHYS CHEM] A particular nonplanar conformation of a cyclic molecule with more than five atoms in the ring; for example, in the chair form of cyclohexane, the hydrogens are staggered and directed perpendicularly to the mean plane of the carbons (axial conformation, *a*) or equatorially to the center of the mean plane (equatorial conformation, *e*).

chalaza [BOT] The region at the base of the nucellus of an ovule; gives rise to the integuments. [CYTOL] One of the paired, spiral, albuminous bands in a bird's egg that attach the yolk to the shell lining membrane at the ends of the egg.

chalazion [MED] A small tumor of the eyelid formed by retention of tarsal gland secretions. Also known as a Meibomian cyst.

chalazogamy [BOT] A process of fertilization in which the pollen tube passes through the chalaza to reach the embryo sac.

Chalcididae [INV ZOO] The chalcids, a family of hymenopteran insects in the superfamily Chalcidoidea.

Chalcidoidea [INV ZOO] A superfamily of hymenopteran insects in the suborder Apocrita, including primarily insect parasites.

chalice cell *See* goblet cell.

chalicosis [MED] A pulmonary affection caused by inhalation of stone dust.

Chalicotheriidae [PALEON] A family of extinct perissodactyl mammals in the superfamily Chalicotherioidea.

Chalicotherioidea [PALEON] A superfamily of extinct, specialized perissodactyls having claws rather than hooves.

chalkstone [PATH] A gouty deposit, usually of sodium urate, in the hands or feet.

chalones [BIOCHEM] Substances thought to be molecules of the protein-polypeptide class that are produced as part of the growth-control systems of tissues; known to inhibit cell division by acting on several phases in the mitotic cycle.

Chamaeleontidae [VERT ZOO] The chameleons, a family of reptiles in the suborder Sauria.

Chamaemyidae [INV ZOO] The aphid flies, a family of myodarian cyclorrhaphous dipteran insects of the subsection Acalypteratae.

chamaephyte [ECOL] Any perennial plant whose winter buds are within 25 centimeters of the soil surface.

chamaerrhine [ANTHRO] Having a short broad nose with a nasal index of 51-57.9 for a skull, and 85-99.9 for a person.

Chamaesiphonales [BOT] An order of blue-green algae of the class Cyanophyceae; reproduce by cell division, colony fragmentation, and endospores.

chamecephalic [ANTHRO] Having a flattened backward-slanting head with a length-height index of 70 or less.

chamecranial [ANTHRO] Having a flat low skull with a length-height index of less than 70.

CHALCIDIDAE

A member of the Chalcididae (*From T. I. Storer and R. L. Usinger, General Zoology, 3d ed., McGraw-Hill, 1957*)

CHAMELEON

Flap-necked chameleon *(Chamaeleo dilepis)*, with opposable toes, a prehensile tail; the neck flap is raised.

CHANIDAE

The milkfish *(Chanos chanos)*, length up to 5 feet (1.5 meters), lives in marine and estuarine waters of the tropical Indo-Pacific. *(After Jordan and Evermann, The Fishes of North and Middle America, U.S. Nat. Mus. Bull., no. 47, pt. 4, 1900)*

CHARADRIIFORMES

The common or Wilson's snipe *(Gallinago gallinago)*, a typical wader. It frequents wet woodlands and marshes of North America, France, and the British Isles.

chameleon [VERT ZOO] The common name for about 80 species of small-to-medium-sized lizards composing the family Chamaeleontidae.

chameprosopic [ANTHRO] Having a broad low face with a facial index of 90 or less.

chamois [VERT ZOO] *Rupicapra rupicapra.* A goatlike mammal included in the tribe Rupicaprini of the family Bovidae.

chancre [MED] **1.** A lesion or ulcer at the site of primary inoculation by an infecting organism. **2.** The initial lesion of syphilis.

chancroid [MED] A lesion of the genitalia, usually of venereal origin, caused by *Hemophilus ducreyi.* Also known as soft chancre.

Chanidae [VERT ZOO] A monospecific family of teleost fishes in the order Gonorynchiformes containing the milkfish *(Chanos chanos),* distinguished by the lack of teeth.

Channidae [VERT ZOO] The snakeheads, a family of freshwater perciform fishes in the suborder Anabantoidei.

Chaoboridae [INV ZOO] The phantom midges, a family of dipteran insects in the suborder Orthorrhapha.

chaparral [ECOL] A vegetation formation characterized by woody plants of low stature, impenetrable because of tough, rigid, interlacing branches, which have simple, waxy, evergreen, thick leaves.

Characeae [BOT] The single family of the order Charales.

Characidae [VERT ZOO] The characins, the single family of the suborder Characoidei.

Characoidei [VERT ZOO] A suborder of the order Cypriniformes including fresh-water fishes with toothed jaws and an adipose fin.

character disorder [PSYCH] A pattern of behavior and emotional response that is socially disapproved or unacceptable, with little evidence of anxiety or other symptoms seen in neuroses.

Charadrii [VERT ZOO] The shore birds, a suborder of the order Charadriiformes.

Charadriidae [VERT ZOO] The plovers, a family of birds in the superfamily Charadrioidea.

Charadriiformes [VERT ZOO] An order of cosmopolitan birds, most of which live near water.

Charadrioidea [VERT ZOO] A superfamily of the suborder Charadrii, including plovers, sandpipers, and phalaropes.

Charales [BOT] Green algae composing the single order of the class Charophyceae.

charcoal rot [PL PATH] A fungus disease of potato, corn, and other plants caused by *Macrophomina phaseoli;* tissues of the root and lower stem are destroyed and blackened.

Chareae [BOT] A tribe of green algae belonging to the family Characeae.

Charles' law [PHYS] The law that at constant pressure the volume of a fixed mass or quantity of gas varies directly with the pressure; a close approximation. Also known as Gay-Lussac law.

Charmouthian [GEOL] Middle Lower Jurassic geologic time.

Charophyceae [BOT] A class of green algae, division Chlorophyta.

Charophyta [BOT] A group of aquatic plants, ranging in size from a few inches to several feet in height, that live entirely submerged in water.

chartaceous [BOT] Resembling paper.

chartreusin [MICROBIO] $C_{18}H_{18}O_{18}$ Crystalline, greenish-yellow antibiotic produced by a strain of *Streptomyces chartreusis;* active against gram-positive microorganisms, acid-fast bacilli, and phage of *Staphylococcus pyogenes.*

chasmochomophyte [ECOL] A plant that grows on detritus in rock crevices.

chasmophyte [ECOL] A plant that grows in rock crevices.

Chattian [GEOL] Upper Oligocene geologic time. Also known as Casselian.

Chauffard-Still disease *See* Still's disease.

chaulmoogra oil [MATER] Any of several fixed oils extracted from seeds of trees in the family Flacourtiaceae; widely used at one time to treat leprosy and other diseases.

Chautauquan [GEOL] Upper Devonian geologic time, below Bradfordian.

Cheadle's disease *See* infantile scurvy.

Chebyshev's inequality [STAT] Given a nonnegative random variable $f(x)$, and $k > 0$, the probability that $f(x) \geq k$ is less than or equal to the expected value of f divided by k. Also spelled Tchebycheff's inequality.

check ligament [ANAT] A thickening of the orbital fascia running from the insertion of the lateral rectus muscle to the medial orbital wall (medial check ligament) or from the insertion of the lateral rectus muscle to the lateral orbital wall (lateral check ligament).

cheek [ANAT] The wall of the mouth in man and other mammals. [ZOO] The lateral side of the head in submammalian vertebrates and in invertebrates.

cheek pouch [VERT ZOO] A saclike dilation of the cheeks in certain animals, such as rodents, in which food is held.

cheetah [VERT ZOO] *Acinonyx jubatus.* A doglike carnivoran mammal belonging to the cat family, having nonretractile claws and long legs.

cheiloplasty [MED] Any plastic operation upon the lip.

cheilosis [MED] Cracking at the corners of the mouth and scaling of the lips, usually associated with riboflavin deficiency.

Cheilostomata [INV ZOO] An order of ectoproct bryozoans in the class Gymnolaemata possessing delicate erect or encrusting colonies composed of loosely grouped zooecia.

Cheiracanthidae [PALEON] A family of extinct acanthodian fishes in the order Acanthodiformes.

cheiromegaly [MED] Enlargement of one or both hands that is not attributable to disease of the hypophysis. Also spelled chiromegaly.

cheiroplasty [MED] Any plastic operation on the hand. Also spelled chiroplasty.

chela [INV ZOO] **1.** A claw or pincer on the limbs of certain crustaceans and arachnids. **2.** A sponge spicule with talonlike terminal processes.

chelate [ORG CHEM] A molecular structure in which a heterocyclic ring can be formed by the unshared electrons of neighboring atoms.

chelicera [INV ZOO] Either appendage of the first pair in arachnids, usually modified for seizing, crushing, or piercing.

Chelicerata [INV ZOO] A subphylum of the phylum Arthropoda; chelicerae are characteristically modified as pincers.

Chelidae [VERT ZOO] The side-necked turtles, a family of reptiles in the suborder Pleurodira.

chelifore [INV ZOO] Either of the first pair of appendages on the cephalic segment of pycnogonids.

cheliform [INV ZOO] Having a forcepslike organ formed by a movable joint closing against an adjacent segment; referring especially to a crab's claw.

cheliped [INV ZOO] Either of the paired appendages bearing chelae in decapod crustaceans.

Chelonariidae [INV ZOO] A family of coleopteran insects in the superfamily Dryopoidea.

CHEETAH

The cheetah *(Acinonyx jubatus),* which has been trained for hunting, especially in India.

CHEILOSTOMATA

Anascan cheilostome, showing the surface of encrusting sheetlike colony of *Membranipora,* Miocene-Recent. *(T. Hincks, from L. H. Hyman, 1959)*

Chelonethida [INV ZOO] An equivalent name for the Pseudo-scorpionida.

Chelonia [VERT ZOO] An order of the Reptilia, subclass Anapsida, including the turtles, terrapins, and tortoises.

Cheloniidae [VERT ZOO] A family of reptiles in the order Chelonia including the hawksbill, loggerhead, and green sea turtles.

Cheluridae [INV ZOO] A family of amphipod crustaceans in the suborder Gammaridea.

Chelydridae [VERT ZOO] The snapping turtles, a small family of reptiles in the order Chelonia.

chemical dating [ANALY CHEM] The determination of the relative or absolute age of minerals and of ancient objects and materials by measurement of their chemical compositions.

chemical energy [PHYS CHEM] Energy of a chemical compound which, by the law of conservation of energy, must undergo a change equal and opposite to the change of heat energy in a reaction; the rearrangement of the atoms in reacting compounds to produce new compounds causes a change in chemical energy.

chemical equilibrium [CHEM] A condition in which a chemical reaction is occurring at equal rates in its forward and reverse directions, so that the concentrations of the reacting substances do not change with time.

chemical formula [CHEM] A notation utilizing chemical symbols and numbers to indicate the chemical composition of a pure substance; examples are CH_4 for methane and HCl for hydrogen chloride.

chemical indicator [ANALY CHEM] A substance whose physical appearance is altered at or near the end point of a chemical titration.

chemical kinetics [PHYS CHEM] That branch of physical chemistry concerned with the mechanisms and rates of chemical reactions. Also known as reaction kinetics.

chemical polarity [PHYS CHEM] Tendency of a molecule, or compound, to be attracted or repelled by electrical charges because of an asymmetrical arrangement of atoms around the nucleus.

chemical potential [PHYS CHEM] In a thermodynamic system of several constituents, the rate of change of the Gibbs function of the system with respect to the change in the number of moles of a particular constituent.

chemical reaction [CHEM] A change in which a substance (or substances) is changed into one or more new substances; there is only a minute change, Δm, in the mass of the system, given by $\Delta E = \Delta mc^2$, where ΔE is the energy emitted or absorbed and c is the speed of light.

chemical shift [PHYS CHEM] Shift in a nuclear magnetic-resonance spectrum resulting from diamagnetic shielding of the nuclei by the surrounding electrons.

chemical symbol [CHEM] A notation for one of the chemical elements, consisting of letters; for example Ne, O, C, and Na represent neon, oxygen, carbon, and sodium.

chemiluminescence [PHYS CHEM] Emission of light as a result of a chemical reaction without an apparent change in temperature.

chemistry [SCI TECH] The scientific study of the properties, composition, and structure of matter, the changes in structure and composition of matter, and accompanying energy changes.

chemoautotroph [MICROBIO] Any of a number of autotrophic bacteria and protozoans which do not carry out photosynthesis.

chemocline [HYD] The transition in a meromictic lake between the mixolimnion layer (at the top) and the monimolimnion layer (at the bottom).

chemodectoma [MED] A benign tumor of the carotid body.

chemodifferentiation [EMBRYO] The process of cellular differentiation at the molecular level by which embryonic cells become specialized as tissues and organs.

chemoprophylaxis [MED] Use of drugs to prevent the development of infectious diseases.

chemoreception [PHYSIO] Reception of a chemical stimulus by an organism.

chemoreceptor [PHYSIO] Any sense organ that responds to chemical stimuli.

chemostat [MICROBIO] An apparatus, and a principle, for the continuous culture of bacterial populations in a steady state.

chemosynthesis [BIOCHEM] The synthesis of organic compounds from carbon dioxide by microorganisms using energy derived from chemical reactions.

chemotaxis [BIOL] The orientation movement of a motile organism with reference to a chemical agent.

chemotherapeutic [PHARM] Any agent used for chemotherapy.

chemotherapy [MED] Administering chemical substances for treatment of disease, especially cancer and diseases caused by parasites.

chemotropism [BIOL] Orientation response of a sessile organism with reference to chemical stimuli.

Chemungian [GEOL] Middle Upper Devonian geologic time, below Cassodagan.

Chenopodiaceae [BOT] A family of dicotyledonous plants in the order Caryophyllales having reduced, mostly greenish flowers.

Chermidae [INV ZOO] A small family of minute homopteran insects in the superfamily Aphidoidea.

Cherminae [INV ZOO] A subfamily of homopteran insects in the family Chermidae; all forms have a beak and an open digestive tract.

cherry [BOT] **1.** Any trees or shrub of the genus *Prunus* in the order Rosales. **2.** The simple, fleshy, edible drupe or stone fruit of the plant.

cherry leaf spot [PL PATH] A fungus disease of the cherry caused by *Coccomyces hiemalis;* spotting and chlorosis of the leaves occurs, with consequent retardation of tree and fruit development.

chersophyte [ECOL] A plant that grows in dry waste lands.

chest breadth [ANTHRO] The measurement across the chest at nipple level.

chest circumference [ANTHRO] The horizontal circumference taken just above the nipples during a period of quiet breathing.

chest depth [ANTHRO] A measurement of the chest taken front to back from the sternum to the spinal groove.

Chesterian [GEOL] Upper Mississippian geologic time.

chestnut [BOT] The common name for several species of large, deciduous trees of the genus *Castanea* in the order Fagales, which bear sweet, edible nuts.

chestnut blight [PL PATH] A fungus disease of the chestnut caused by *Endothia parasitica,* which attacks the bark and cambium, causing cankers that girdle the stem and kill the plant. Also known as chestnut canker.

chestnut canker *See* chestnut blight.

chevron [ANAT] The bone forming the hemal arch of a caudal vertebra.

chevrotain [VERT ZOO] The common name for four species of mammals constituting the family Tragulidae in the order Artiodactyla. Also known as mouse deer.

CHESTNUT

Twig, leaf, and bud of American chestnut *(Castanea dentata).*

CHEVROTAIN

Indian chevrotain, distinguished by white spots on coat.

CHICLE

Fruit of *Achras zapota*. (From
*L. H. Bailey, The Standard
Cyclopedia of Horticulture,
vol. 3, Macmillan, 1937*)

CHIMAERIFORMES

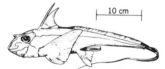

Modern chimaeriform of genus
*Chimaera. (From H. B. Bigelow
and W. C. Schroeder, Fishes of
the Western North Atlantic,
pt. 2, Sears Foundation for
Marine Research, 1953)*

CHIMPANZEE

The chimpanzee, with a highly
developed brain, and an
anatomical resemblance to man.

Cheyne-Stokes respiration [MED] Breathing characterized
by periods of hyperpnea alternating with periods of apnea;
rhythmic waxing and waning of respiration; occurs most
commonly in older patients with heart failure and cerebrovascular disease.

chiasma [ANAT] A cross-shaped point of intersection of two
parts, especially of the optic nerves. [CYTOL] The point of
junction and fusion between paired chromatids or chromosomes, first seen during diplotene of meiosis.

Chiasmodontidae [VERT ZOO] A family of deep-sea fishes in
the order Perciformes.

chicken [VERT ZOO] *Galus galus.* The common domestic fowl
belonging to the order Galliformes.

chickenpox [MED] A mild, highly infectious viral disease of
man caused by a herpesvirus and characterized by vesicular
rash. Also known as varicella.

chicle [MATER] A gummy exudate obtained from the bark of
Achras zapota, an evergreen tree belonging to the sapodilla
family (Sapotaceae); used as the principal ingredient of
chewing gum.

chicory [BOT] *Cichorium intybus.* A perennial herb of the
order Campanulales grown for its edible green leaves.

chief cell [HISTOL] **1.** A parenchymal, secretory cell of the
parathyroid gland. **2.** A cell in the lumen of the gastric fundic
glands.

chigger [INV ZOO] The common name for bloodsucking larval mites of the Trombiculidae which parasitize vertebrates.

chilarium [INV ZOO] One of a pair of processes between the
bases of the fourth pair of walking legs in the king crab.

chilidium [INV ZOO] A plate that overlies the deltidial fissure
in the dorsal valve of certain brachiopods.

Chilobolbinidae [PALEON] A family of extinct ostracods in
the superfamily Hollinacea showing dimorphism of the velar
structure.

Chilopoda [INV ZOO] The centipedes, a class of the Myriapoda that is exclusively carnivorous and predatory.

Chimaeridae [VERT ZOO] A family of the order Chimaeriformes.

Chimaeriformes [VERT ZOO] The single order of the chondrichthyan subclass Holocephali comprising the ratfishes,
marine bottom-feeders of the Atlantic and Pacific oceans.

chimera [BIOL] An organism or a part made up of tissues or
cells exhibiting chimerism.

chimerism [BIOL] The admixture of cell populations from
more than one zygote.

chimopelagic [ECOL] Pertaining to, belonging to, or being
marine organisms living at great depths throughout most of
the year; during the winter they move to the surface.

chimpanzee [VERT ZOO] Either of two species of Primates of
the genus *Pan* indigenous to central-west Africa.

chin [ANAT] The lower part of the face, at or near the
symphysis of the lower jaw.

chin breadth [ANTHRO] Contact measurement of the maximum width of the chin, taken between the points of intersection of the mandible and the menton.

chinchilla [VERT ZOO] The common name for two species of
rodents in the genus *Chinchilla* belonging to the family
Chinchillidae.

Chinchillidae [VERT ZOO] A family of rodents comprising the
chinchillas and viscachas.

chin-neck projection [ANTHRO] Measurement from the tip of
the thyroid cartilage to the midpoint of the menton.

chin projection [ANTHRO] Measurement of the distance from
the gonion to the most forward point on the vertical midline
of the menton.

Chionididae [VERT ZOO] The white sheathbills, a family of birds in the order Charadriiformes.

chionophile [ECOL] Having a preference for snow.

chip blower [MED] A dental instrument used to blow drilling debris from a tooth cavity that is being prepared for filling.

chipmunk [VERT ZOO] The common name for 18 species of rodents belonging to the tribe Marmotini in the family Sciuridae.

Chiridotidae [INV ZOO] A family of holothurians in the order Apodida having six-spoked, wheel-shaped spicules.

Chirodidae [PALEON] A family of extinct chondrostean fishes in the suborder Platysomoidei.

Chirognathidae [PALEON] A family of conodonts in the suborder Neurodontiformes.

chiromegaly *See* cheiromegaly.

chiroplasty *See* cheiroplasty.

chiropodist [MED] One who treats minor ailments of the feet. Also known as podiatrist.

chiropractic [MED] A system of therapeutics based upon the theory that disease is caused by abnormal function of the nervous system; attempts to restore normal function are made through manipulation and treatment of the structures of the body, especially those of the spinal column.

chiropractor [MED] One who practices the chiropractic arts.

Chiroptera [VERT ZOO] The bats, an order of mammals having the front limbs modified as wings.

chiropterophilous [BIOL] Pollinated by bats.

chiropterygium [VERT ZOO] A typical vertebrate limb, thought to have evolved from a finlike appendage.

Chirotheuthidae [INV ZOO] A family of mollusks comprising several deep-sea species of squids.

chirotype [SYST] A nomenclatural type for the specimen of a species designated by a manuscript name or chironym, and validated as being the type specimen on publication.

chi-square distribution [STAT] The distribution of the sample variances indicated by

$$S^2 = \sum_{i=1}^{n} (x_i - \bar{x})^2/(n - 1),$$

where $x_1, x_2, \ldots x_n$ are observations of a random sample n drawn from a normal population.

chi-square test [STAT] A generalization, and an extension, of a test for significant differences between a binomial population and a multinomial population, wherein each observation may fall into one of several classes and furnishes a comparison among several samples instead of just two.

chitin [BIOCHEM] A white or colorless amorphous polysaccharide that forms a base for the hard outer integuments of crustaceans, insects, and other invertebrates.

chitinase [BIOCHEM] An externally secreted digestive enzyme produced by certain microorganisms and invertebrates that hydrolyzes chitin.

Chitinozoa [PALEON] An extinct group of unicellular microfossils of the kingdom Protista.

chiton [INV ZOO] The common name for over 600 extant species of mollusks which are members of the class Polyplacophora.

Chitral fever *See* phlebotomus fever.

chlamydeous [BOT] **1.** Pertaining to the floral envelope. **2.** Having a perianth.

Chlamydia *See* Bedsonia.

Chlamydiaceae [MICROBIO] A family of parasitic bacteria in the order Rickettsiales.

Chlamydobacteriaceae [MICROBIO] A family of gram-nega-

CHIPMUNK

The eastern chipmunk *(Tamias striatus)* with characteristic longitudinal black and yellow stripes on its back.

CHIROGNATHIDAE

$\vdash$ 0.5 mm $\dashv$

Chirognathus, an example of the conodont Chirognathidae —minute, toothlike microfossils.

CHITIN

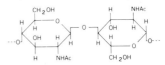

β-*N*-acetyl-D-glucosamine unit of chitin.

CHITON

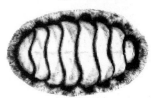

Mossy chiton with characteristic shell of eight plates and surrounding girdle bearing spicules on upper surface.

CHLAMYDOSPORE

|─10 μ─|

A chlamydospore, an asexual
spore type produced by
Ascomycetes and Basidiomycetes.

CHLORINE

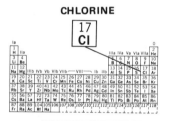

Periodic table of the chemical
elements showing the position
of chlorine.

CHLOROCOCCALES

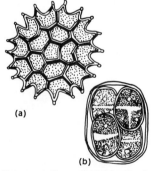

(a)

(b)

Representative colonial forms of
chlorococcales. (a) *Pediastrum
boryanum.* (b) *Oöcystis borgei.*
(*From G. M. Smith, Fresh-water
Algae of the United States, 2d
ed., McGraw-Hill, 1950*)

tive bacteria in the order Chlamydobacteriales possessing
trichomes in which false branching may occur.

Chlamydobacteriales [MICROBIO] An order comprising colorless, gram-negative, algae-like bacteria of the class Schizomycetes which occur in trichomes.

Chlamydomonadidae [INV ZOO] A family of colorless, flagellated protozoans in the order Volvocida considered to be
close relatives of protozoans that possess chloroplasts.

Chlamydoselachidae [VERT ZOO] The frilled sharks, a family
of rare deep-water forms having a combination of hybodont
and modern shark characteristics.

chlamydospore [MYCOL] A thick-walled, unicellular resting
spore developed from vegetative hyphae in almost all parasitic fungi.

Chlamydozoaceae [MICROBIO] A family of small, gram-
negative, coccoid bacteria in the order Rickettsiales; members are obligate intracytoplasmic parasites, or saprophytes.

chloasma [MED] Patchy tan, brown, and black hyperpigmen-
tation, especially on the brow and cheek; of unknown cause,
but may be due to the action of sunshine upon perfume or to
endocrinopathy.

Chloracea [MICROBIO] The green sulfur bacteria, a family of
photosynthetic bacteria in the suborder Rhodobacteriineae.

chloracne [MED] An acnelike eruption caused by chlorinated hydrocarbons.

chloragogen [INV ZOO] Of or pertaining to certain specialized cells forming the outer layer of the alimentary tract in
earthworms and other annelids.

chloramphenicol [MICROBIO] $C_{11}H_{12}O_2N_2Cl_2$ A colorless,
crystalline, broad-spectrum antibiotic produced by *Streptomyces venezuelae;* industrial production is by chemical synthesis. Also known as chloromycetin.

Chlorangiaceae [BOT] A primitive family of colonial green
algae belonging to the Tetrasporales in which the cells are
directly attached to each other.

chloranthy [BOT] A reverting of normally colored floral
leaves or bracts to green foliage leaves.

chlordiazepoxide hydrochloride [PHARM] $C_{16}H_{14}ON_3Cl$ A
white crystalline conpound, soluble in water; the hydrochloride salt is used as a tranquilizer.

chlorenchyma [BOT] Chlorophyll-containing tissue in parts
of higher plants, as in leaves.

chloride shift [PHYSIO] The reversible exchange of chloride
and bicarbonate ions between erythrocytes and plasma to
effect transport of carbon dioxide and maintain ionic equilibrium during respiration.

chlorine [CHEM] A chemical element, symbol Cl, atomic
number 17, atomic weight 35.453; used in manufacture of
solvents, insecticides, and many non-chlorine-containing
compounds, and to bleach paper and pulp.

Chlorobacteriaceae [MICROBIO] The equivalent name for
Chlorobiaceae.

Chlorobiaceae [MICROBIO] A family of green sulfur bacteria
and a few chocolate-brown bacteria in the suborder Rhodobacteriineae; all members contain bacteriochlorophyll *c* or *d*
as the predominant green pigment.

Chlorobium [MICROBIO] A genus of gram-negative, nonmotile, sulfur green bacteria in the family Chlorobiaceae occurring singly and in chains.

chlorobium chlorophyll [BIOCHEM] $C_{51}H_{67}O_4N_4Mg$ Either
of two spectral forms of chlorophyll occurring as esters of
farnesol in certain (*Chlorobium*) photosynthetic bacteria.
Also known as bacterioviridin.

Chlorococcales [BOT] A large, highly diverse order of unicellular or colonial, mostly fresh-water green algae in the class Chlorophyceae.

chlorocruorin [BIOCHEM] A green respiratory pigment found in the plasma of some species of annelids.

Chlorodendrineae [BOT] A suborder of colonial green algae in the order Volvocales comprising some genera with individuals capable of detachment and motility.

chloroform [ORG CHEM] $CHCl_3$ A colorless, sweet-smelling, nonflammable liquid; used at one time as an anesthetic. Also known as trichloromethane.

chloroguanide hydrochloride [PHARM] $C_{11}H_{16}N_5Cl$ Very effective suppressive drug in low doses, against the three kinds of malaria. Also known as chloroguanide.

chloroma [MED] A focal tumorous proliferation of granulocytes, with or without the blood findings of granulocytic leukemia; the sectioned surfaces of the mass are green.

Chloromonadida [INV ZOO] An order of flattened, grass-green or colorless, flagellated protozoans of the class Phytamastigophorea.

Chloromonadina [INV ZOO] The equivalent name for Chloromonadida.

Chloromonadophyceae [BOT] A group of algae considered by some to be a class of the division Chrysophyta.

Chloromonadophyta [BOT] A division of algae in the plant kingdom considered by some to be a class, Chloromondophyceae.

chloromycetin *See* chloramphenicol.

Chlorophyceae [BOT] A class of microscopic or macroscopic green algae, division Chlorophyta, composed of fresh- or salt-water, unicellular or multicellular, colonial, filamentous or sheetlike forms.

chlorophyll [BIOCHEM] The generic name for any of several oil-soluble green tetrapyrrole plant pigments which function as photoreceptors of light energy for photosynthesis.

chlorophyll a [BIOCHEM] $C_{55}H_{72}O_5N_4Mg$ A magnesium chelate of dihydroporphyrin that is esterified with phytol and has a cyclopentanone ring; occurs in all higher plants and algae.

chlorophyllase [BIOCHEM] An enzyme that splits or hydrolyzes chlorophyll.

chlorophyll b [BIOCHEM] $C_{55}H_{70}O_6N_4Mg$ An ester similar to cholorphyll *a* but with a $-CHO$ substituted for a $-CH_3$; occurs in small amounts in all green plants and algae.

Chlorophyta [BOT] The green algae, a highly diversified plant division characterized by chloroplasts, having chlorophyll *a* and *b* as the predominating pigments.

Chloropidae [INV ZOO] The chloropid flies, a family of myodarian cyclorrhaphous dipteran insects in the subsection Acalypteratae.

chloroplast [BOT] A type of cell plastid occurring in the green parts of plants, containing chlorophyll pigments, and functioning in photosynthesis and protein synthesis.

chloropsia [MED] A defect of vision in which all objects appear green.

chlorosis [MED] A form of macrocytic anemia in young females characterized by marked reduction in hemoglobin and a greenish skin color. [PL PATH] A disease condition of green plants seen as yellowing of green parts of the plant.

chlorotic streak [PL PATH] A systemic virus disease of sugarcane characterized by yellow or white streaks on the foliage.

chlorpromazine [PHARM] $C_{17}H_{19}ClN_2S$ A gray-white, crystalline compound used as a sedative and in preventing or relieving nausea and vomiting.

chlortetracycline [MICROBIO] $C_{22}H_{23}O_8N_2Cl$ Yellow, crys-

CHLOROFORM

Structural formula of chloroform.

CHLOROMONADIDA

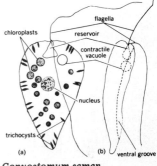

Gonyostomum semen.
(a) **Dorsal view.** *(b)* **Side view.**

talline, broad-spectrum antibiotic produced by a strain of *Streptomyces aureofaciens.*

choana [ANAT] A funnel-shaped opening, especially the posterior nares. [INV ZOO] A protoplasmic collar surrounding the basal ends of the flagella in certain flagellates and in the choanocytes of sponges.

choanate fish [PALEON] Any of the lobefins composing the subclass Crossopterygii.

Choanichthyes [VERT ZOO] An equivalent name for the Sarcopterygii.

choanocyte [INV ZOO] Any of the choanate, flagellate cells lining the cavities of a sponge. Also known as collar cell.

Choanoflagellida [INV ZOO] An order of single-celled or colonial, colorless flagellates in the class Zoomastigophorea; distinguished by a thin protoplasmic collar at the anterior end.

choanosome [INV ZOO] The inner layer of a sponge; composed of choanocytes.

chocolate spot [PL PATH] A fungus disease of legumes caused by species of *Botrytis* and characterized by brown spots on leaves and stems, with withering of shoots.

chocolate tree *See* cacao.

Choeropotamidae [PALEON] A family of extinct palaeodont artiodactyls in the superfamily Entelodontoidae.

choked disk *See* papilledema.

cholagogic [PHYSIO] Inducing the flow of bile.

cholane [BIOCHEM] $C_{24}H_{42}$ A tetracyclic hydrocarbon which may be considered as the parent substance of sterols, hormones, bile acids, and digitalis aglycons.

cholangiogram [MED] The x-ray film produced by means of cholangiography.

cholangiography [MED] Roentgenography of the bile ducts.

cholangiole [ANAT] A terminal or interlobular bile duct.

cholangiolitis [MED] Inflammation of the bile capillaries.

cholangioma [MED] Adenocarcinoma of the bile ducts.

cholangitis [MED] Inflammation of the bile ducts.

cholate [BIOCHEM] Any salt of cholic acid.

cholecystectomy [MED] Surgical removal of the gallbladder and cystic duct.

cholecystitis [MED] Inflammation of the gallbladder.

cholecystography [MED] Radiography of the gallbladder following injection or ingestion of a radiopaque substance excreted in bile. Also known as Graham-Cole test.

cholecystokinin [BIOCHEM] A hormone produced by the mucosa of the upper intestine which stimulates contraction of the gallbladder.

cholecystostomy [MED] The establishment of an opening into the gallbladder, usually for external drainage of its contents.

choledochoduodenal junction [ANAT] The point where the common bile duct enters the duodenum.

choledocholithiasis [MED] The presence of calculi in the common bile duct.

choledochostomy [MED] The draining of the common bile duct through the abdominal wall.

choleglobin [BIOCHEM] Combined native protein (globin) and open-ring iron-porphyrin, which is bile pigment hemoglobin; a precursor of biliverdin.

cholelithiasis [MED] The production of or the condition associated with gallstones in the gallbladder or bile ducts.

cholera [MED] **1.** An acute, infectious bacterial disease of man caused by *Vibrio comma;* characterized by diarrhea, delirium, stupor, and coma. **2.** Any condition characterized by profuse vomiting and diarrhea.

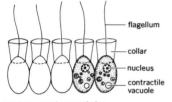

CHOANOFLAGELLIDA

flagellum

collar

nucleus

contractile vacuole

A linear colony of the choanoflagellate *Desmarella moniliformis.* *(After McCracken)*

cholera vibrio [MICROBIO] *Vibrio comma,* the bacterium that causes cholera.

cholesteatoma [MED] An epidermal inclusion cyst of the middle ear, or mastoid bone, sometimes in the external ear canal, brain, or spinal cord. Also known as pearly tumor.

cholesterol [BIOCHEM] $C_{27}H_{46}O$ A sterol produced by all vertebrate cells, particularly in the liver, skin, and intestine, and found most abundantly in nerve tissue.

cholic acid [BIOCHEM] $C_{24}H_{40}O_5$ An unconjugated, crystalline bile acid.

choline [BIOCHEM] $C_5H_{15}O_2N$ A basic hygroscopic substance constituting a vitamin of the B complex; used by most animals as a precursor of acetylcholine and a source of methyl groups.

cholinergic [PHYSIO] Liberating, activated by, or resembling the physiologic action of acetylcholine.

cholinergic nerve [PHYSIO] Any nerve, such as autonomic preganglionic nerves and somatic motor nerves, that releases a cholinergic substance at its terminal points.

cholinesterase [BIOCHEM] An enzyme found in blood and in various other tissues that catalyzes hydrolysis of choline esters, including acetylcholine. Abbreviated chE.

choluria [MED] The presence of bile in the urine.

chomophyte [ECOL] A plant that grows in detritus on rocks.

Chondrichthyes [VERT ZOO] A class of vertebrates comprising the cartilaginous, jawed fishes characterized by the absence of true bone.

chondrification [PHYSIO] Formation of or conversion into cartilage.

chondrin [BIOCHEM] A horny gelatinous protein substance obtainable from the collagen component of cartilage.

chondriome [CYTOL] Referring collectively to the chondriosomes (mitochondria) of a cell as a functional unit.

chondriosome [CYTOL] Any of a class of self-perpetuating lipoprotein complexes in the form of grains, rods, or threads in the cytoplasm of most cells; thought to function in cellular metabolism and secretion.

chondroblast [HISTOL] A cell that produces cartilage.

Chondrobrachii [VERT ZOO] The equivalent name for Ateleopoidei.

chondroclast [HISTOL] A cell that absorbs cartilage.

chondrocranium [ANAT] The part of the adult cranium derived from the cartilaginous cranium. [EMBRYO] The cartilaginous, embryonic cranium of vertebrates.

chondrocyte [HISTOL] A cartilage cell.

chondrodysplasia *See* enchondromatosis.

chondrodystrophy fetalis *See* achondroplasia.

chondrogenesis [EMBRYO] The development of cartilage.

chondroglossus [ANAT] Pertaining to the cartilaginous tip of the hyoid bone and the tongue.

chondroitin [BIOCHEM] A nitrogenous polysaccharide occurring in cartilage in the form of condroitinsulfuric acid.

chondrology [ANAT] The anatomical study of cartilage.

chondroma [MED] A benign tumor of bone, cartilage, or other tissue which simulates the structure of cartilage in its growth.

chondromucoid [BIOCHEM] A mucoid found in cartilage; a glycoprotein in which chondroitinsulfuric acid is the prosthetic group.

Chondromyces [MICROBIO] A genus of myxobacteria in the family Polyangiaceae having numerous cysts grouped at the end of a colored stalk.

chondrophone [INV ZOO] In bivalve mollusks, a structure or cavity supporting the internal hinge cartilage.

CHONETIDINA

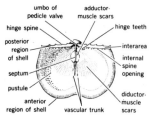

umbo of pedicle valve · adductor muscle scars · hinge spine · hinge teeth · posterior region of shell · interarea · internal spine opening · septum · pustule · diductor muscle scars · anterior region of shell · vascular trunk

Neochonetes **showing features of the pedicle valve.**

CHONOTRICHIDA

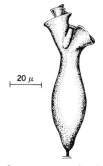

20 μ

Spirochona, **an example of a chonotrichid.**

Chondrophora [INV ZOO] A suborder of polymorphic, colonial, free-floating coelenterates of the class Hydrozoa.

chondrophore [INV ZOO] In bivalves, a structure supporting the inner hinge cartilage.

chondroprotein [BIOCHEM] A protein (glycoprotein) occurring normally in cartilage.

chondrosarcoma [MED] A malignant tumor of cartilage.

chondroseptum [ANAT] The cartilaginous portion of the nasal septum.

chondroskeleton [ANAT] **1.** The parts of the bony skeleton formed from cartilage. **2.** Cartilaginous parts of a skeleton. [VERT ZOO] A cartilaginous skeleton, as in Chondrostei.

Chondrostei [PALEON] The most archaic infraclass of the subclass Actinopterygii, or rayfin fishes.

Chondrosteidae [PALEON] A family of extinct actinopterygian fishes in the order Acipenseriformes.

chone [INV ZOO] In Porifera, a passage through the cortex having one or more external openings and one internal opening.

Chonetidina [PALEON] A suborder of extinct articulate brachiopods in the order Strophomenida.

Chonotrichida [INV ZOO] A small order of vase-shaped ciliates in the subclass Holotrichia; commonly found as ectocommensals on marine crustaceans.

chordae tendineae [ANAT] The tendons of the papillary muscles attached to the atrioventricular valves in the heart ventricles. Also known as tendinous cords.

chordae Willisii [ANAT] Fibrous bands which cross through the sinuses of the dura mater.

chordamesoderm [EMBRYO] The portion of the mesoderm in the chordate embryo from which the notochord and related structures arise, and which induces formation of ectodermal neural structures.

Chordata [ZOO] The highest phylum in the animal kingdom, characterized by a notochord, nerve cord, and gill slits; includes the urochordates, lancelets, and vertebrates.

chordate [VERT ZOO] Having a notochord.

Chordodidae [INV ZOO] A family of worms in the order Gordioidea distinguished by a rough cuticle containing thickenings called areoles.

chordotomy [MED] Surgical division of a spinal nerve tract to relieve severe intractable pain.

chordotonal [INV ZOO] Referring to certain bristlelike sense organs in various parts of an insect body thought to be receptors for mechanical and sound stimuli.

chorea [MED] A nervous disorder seen as part of a syndrome following an organic dysfunction or an infection and characterized by irregular, involuntary movements of the body, especially of the face and extremities.

choreiform syndrome [MED] A complex of symptoms representing a form or component of minimal brain dysfunction in children, manifested by twitching movements of the face, trunk, and extremities.

chorioadenoma [MED] A tumor intermediate in malignancy between a hydatidiform mole and choriocarcinoma.

chorioallantois [EMBRYO] A vascular fetal membrane that is formed by the close association or fusion of the chorion and allantois.

choriocapillary lamina [ANAT] The inner layer of the choroid; consists of a capillary plexus.

choriocarcinoma [MED] A highly malignant tumor derived from chorionic tissue; found most commonly in the uterus and testis. Also known as chorioepithelioma.

chorioepithelioma *See* choriocarcinoma.

chorion [EMBRYO] The outermost of the extraembryonic

membranes of amniotes, enclosing the embryo and all of its other membranes.

chorionic gonadotropin   *See* human chorionic gonadotropin.

chorionitis   *See* scleroderma.

chorioretinitis   [MED] Inflammation of the choroid and retina of the eye.

choripetalous   *See* polypetalous.

chorisepalous   *See* polysepalous.

chorisis   [BOT] Separation of a leaf or floral part into two or more parts during development.

Choristida   [INV ZOO] An order of sponges in the class Demospongiae in which at least some of the megascleres are tetraxons.

Choristodera   [PALEON] A suborder of extinct reptiles of the order Eosuchia composed of a single genus, *Champsosaurus.*

choroid   [ANAT] The highly vascular layer of the vertebrate eye, lying between the sclera and the retina.

choroiditis   [MED] Inflammation of the choroid.

choroid plexus   [ANAT] Any of the highly vascular, folded processes that project into the third, fourth, and lateral ventricles of the brain.

Christmas disease   [MED] A hereditary, sex-linked, hemophilia-like disease involving failure of the clotting mechanism due to a deficiency of Christmas factor.

Christmas factor   [BIOCHEM] A soluble protein blood factor involved in blood coagulation. Also known as factor IX; plasma thromboplastin component (PTC).

Chromadoria   [INV ZOO] A subclass of nematode worms in the class Adenophorea.

Chromadorida   [INV ZOO] An order of principally aquatic nematode worms in the subclass Chromadoria.

Chromadoridae   [INV ZOO] A family of soil and fresh-water, free-living nematodes in the superfamily Chromadoroidea; generally associated with algal substances.

Chromadoroidea   [INV ZOO] A superfamily of small to moderate-sized, free-living nematodes with spiral, transversely ellipsoidal amphids and a striated cuticle.

chromaffin   [BIOL] Staining with chromium salts.

chromaffin body   *See* paraganglion.

chromaffin cell   [HISTOL] Any cell of the suprarenal organs in lower vertebrates, of the adrenal medulla in mammals, of the paraganglia, or of the carotid bodies that stains with chromium salts.

chromaffin system   [PHYSIO] The endocrine organs and tissues of the body that secrete epinephrine; characterized by an affinity for chromium salts.

chromatic aberration   [OPTICS] An optical lens defect causing color fringes, because the lens material brings different colors of light to focus at different points. Also known as color aberration.

chromatic vision   [PHYSIO] Vision pertaining to the color sense, that is, the perception and evaluation of the colors of the spectrum.

chromatid   [CYTOL] **1.** One of the pair of strands formed by longitudinal splitting of a chromosome which are joined by a single centromere in somatic cells during mitosis. **2.** One of a tetrad of strands formed by longitudinal splitting of paired chromosomes during diplotene of meiosis.

chromatin   [BIOCHEM] The deoxyribonucleoprotein complex forming the major portion of the nuclear material and of the chromosomes.

chromatogram   [ANALY CHEM] The pattern formed by zones of separated pigments and of colorless substance in chromatographic procedures.

chromatographic adsorption   [ANALY CHEM] Preferential ad-

CHROMATOGRAPHIC ADSORPTION

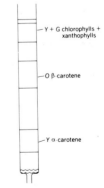

Separation of carotenes by adsorption on activated magnesium oxide. The solvent consists of petroleum ether plus 1% acetone. *G* signifies green; *O*, orange; *Y*, yellow.

sorption of chemical compounds (gases or liquids) in an ascending molecular-weight sequence onto a solid adsorbent material, such as activated carbon, alumina, or silica gel; used for analysis and separation of chemical mixtures.

chromatography [ANALY CHEM] A method of separating and analyzing mixtures of chemical substances by chromatographic adsorption.

chromatophore [HISTOL] A type of pigment cell found in the integument and certain deeper tissues of lower animals that contains color granules capable of being dispersed and concentrated.

chromatophorotrophin [BIOCHEM] Any crustacean neurohormone which controls the movement of pigment granules within chromatophores.

chromatoplasm [BOT] The peripheral protoplasm in blue-green algae containing chlorophyll, accessory pigments, and stored materials.

chromatopsia [MED] A disorder of visual sensation in which color impressions are disturbed or arise subjectively, with objects appearing as unnaturally colored or colorless objects as colored; may be caused by a disturbance of the optic centers, psychic disturbance, or drugs.

chromatosis [MED] A pathologic process or pigmentary disease in which there is a deposit of coloring matter in a normally unpigmented site, or an excessive deposit in a normally pigmented area.

chromium [CHEM] A metallic chemical element, symbol Cr, atomic number 24, atomic weight 51.996. [MET] A blue-white, hard, brittle metal used in chrome plating, in chromizing, and in many alloys.

chromoblastomycosis [MED] A granulomatous skin disease caused by any of several fungi, usually *Hormodendrum pedrosoi,* and characterized by warty nodules which may ulcerate. Also known as chromomycosis.

chromocenter [CYTOL] An irregular, densely staining mass of heterochromatin in the chromosomes, with six armlike extensions of euchromatin, in the salivary glands of *Drosophila.*

chromocyte [HISTOL] A pigmented cell.

chromogen [BIOCHEM] A pigment precursor. [MICROBIO] A microorganism capable of producing color under suitable conditions.

chromogenesis [BIOCHEM] Production of colored substances as a result of metabolic activity; characteristic of certain bacteria and fungi.

chromolipid *See* lipochrome.

chromomere [CYTOL] Any of the linearly arranged chromatin granules in leptotene and pachytene chromosomes and in polytene nuclei.

chromometer *See* colorimeter.

chromomycin [MICROBIO] Any of five components of an antibiotic complex produced by a strain of *Streptomyces griseus;* components are designated A_1 to A_5, of which A_3 ($C_{51}H_{72}O_{32}$) is biologically active.

chromomycosis *See* chromoblastomycosis.

chromonema [CYTOL] The coiled core of a chromatid; it is thought to contain the genes.

chromophile [BIOL] Staining readily.

chromophobe [BIOL] Not readily absorbing a stain.

chromophyll [BIOCHEM] Any plant pigment.

chromoplasm [BOT] The pigmented, peripheral protoplasm of blue-green algae cells; contains chlorophyll, carotenoids, and phycobilins.

chromoplast [CYTOL] Any colored cell plastid, excluding chloroplasts.

CHROMIUM

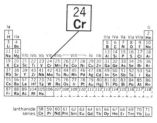

Periodic table of the chemical elements showing the position of chromium.

CHROMOCYTE

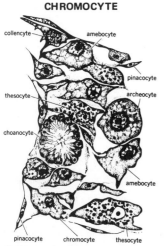

Cell types found in a fresh-water sponge as seen in a cross section through the interior of the sponge. Note chromocyte at bottom of tissue. *(After Meewis, 1936)*

chromoprotein [BIOCHEM] Any protein, such as hemoglobin, with a metal-containing pigment.

chromosomal hybrid sterility [GEN] Sterility caused by inability of homologous chromosomes to pair during meiosis due to a chromosome aberration.

chromosome [CYTOL] Any of the complex, threadlike structures seen in animal and plant nuclei during karyokinesis which carry the linearly arranged genetic units.

chromosome aberration [GEN] Modification of the normal chromosome complement due to deletion, duplication, or rearrangement of genetic material.

chromosome banding [CYTOL] Transverse banded structure of chromosomes revealed by a number of staining techniques which permits the identification of individual chromosome pairs, for instance in humans, even if indistinguishable by gross morphology.

chromosome complement [GEN] The species-specific, normal diploid set of chromosomes in somatic cells.

chromosome map *See* genetic map.

chromosome puff [CYTOL] Chromatic material accumulating at a restricted site on a chromosome; thought to reflect functional activity of the gene at that site during differentiation.

chronaxie [PHYSIO] The time interval required to excite a tissue by an electric current of twice the galvanic threshold.

chronic [MED] Long-continued; of long duration.

chronic alcoholism [MED] Excessive consumption of alcohol over a prolonged period of time.

chronic carrier [MED] A person who harbors and transmits an infectious agent for an indefinite period.

chronic hyperplastic perihepatitis *See* polyserositis.

chronic infectious arthritis *See* rheumatoid arthritis.

chronic leukemia [MED] A leukemia in which the life expectancy is prolonged; leukemias are classified as to acute or chronic, and according to the predominant cell type; the life expectancy is highly variable depending on the latter.

chronocline [PALEON] A cline shown by successive morphological changes in the members of a related group, such as a species, in successive fossiliferous strata.

Chroococcales [BOT] An order of blue-green algae (Cyanophyceae) that reproduce by cell division and colony fragmentation only.

Chryomyidae [INV ZOO] A family of myodarian cyclorrhaphous dipteran insects in the subsection Acalypteratae.

Chrysididae [INV ZOO] The cuckoo wasps, a family of hymenopteran insects in the superfamily Bethyloidea having brilliant metallic blue and green bodies.

chrysocarpous [BOT] Bearing yellow fruits.

Chrysochloridae [PALEON] The golden moles, a family of extinct lipotyphlan mammals in the order Insectivora.

Chrysomelidae [INV ZOO] The leaf beetles, a family of coleopteran insects in the superfamily Chrysomeloidea.

Chrysomeloidea [INV ZOO] A superfamily of coleopteran insects in the suborder Polyphaga.

Chrysomonadida [INV ZOO] An order of yellow to brown, flagellated colonial protozoans of the class Phytamastigophorea.

Chrysomonadina [INV ZOO] The equivalent name for the Chrysomonadida.

Chrysopetalidae [INV ZOO] A small family of scale-bearing polychaete worms belonging to the Errantia.

Chrysophyceae [BOT] Golden-brown algae making up a class of fresh- and salt-water unicellular forms in the division Chrysophyta.

Chrysophyta [BOT] The golden-brown algae, a division of

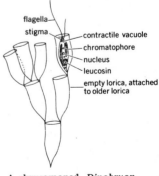

CICADELLIDAE

A cicadellid, dorsal view.

CICINDELIDAE

A tiger beetle. *(From T. I. Storer and R. L. Usinger, General Zoology, 3d ed., McGraw-Hill, 1957)*

CILIA

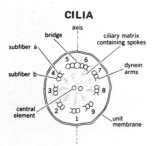

Cross section of a cilium with the peripheral filaments numbered. *(From P. Satir, Studies on Cilia, II: Examination of the distal region of the ciliary shaft and the role of the filaments in motility, J. Cell Biol, 26:805–834, 1965)*

plants with a predominance of carotene and xanthophyll pigments in addition to chlorophyll.

chrysotherapy [MED] The use of gold compounds in the treatment of disease.

Chthamalidae [INV ZOO] A small family of barnacles in the suborder Thoracica.

chyle [PHYSIO] Lymph containing emulsified fat, present in the lacteals of the intestine during digestion of ingested fats.

chylomicron [BIOCHEM] One of the extremely small lipid droplets, consisting chiefly of triglycerides, found in blood after ingestion of fat.

chylophyllous [BOT] Having succulent or fleshy leaves.

chylothorax [MED] An accumulation of chyle in the pleural cavity.

chyluria [MED] The presence of chyle or lymph in the urine, usually caused by a fistulous communication between the urinary and lymphatic tracts or by lymphatic obstruction.

chyme [PHYSIO] The semifluid, partially digested food mass that is expelled into the duodenum by the stomach.

chymosin *See* rennin.

chymotrypsin [BIOCHEM] A proteinase in the pancreatic juice that clots milk and hydrolyzes casein and gelatin.

chymotrypsinogen [BIOCHEM] An inactive proteolytic enzyme of pancreatic juice; converted to the active form, chymotrypsin, by trypsin.

Chytridiales [MYCOL] An order of mainly aquatic fungi of the class Phycomycetes having a saclike to rhizoidal thallus and zoospores with a single posterior flagellum.

Chytridiomycetes [MYCOL] A class of true fungi.

cibarium [INV ZOO] In insects, the space anterior to the mouth cavity in which food is chewed.

Cicadellidae [INV ZOO] A large family of homopteran insects belonging to the series Auchenorrhyncha; includes leaf hoppers.

Cicadidae [INV ZOO] A family of large homopteran insects belonging to the series Auchenorrhyncha; includes the cicadas.

cicatrix [BIOL] A scarlike mark, usually caused by previous attachment of a part or organ. [MED] The connective-tissue scar formed at the site of a healing wound.

Cichlidae [VERT ZOO] The cichlids, a family of perciform fishes in the suborder Percoidei.

Cicindelidae [INV ZOO] The tiger beetles, a family of coleopteran insects in the suborder Adephaga.

Ciconiidae [VERT ZOO] The tree storks, a family of wading birds in the order Ciconiiformes.

Ciconiiformes [VERT ZOO] An order of predominantly long-legged, long-necked birds, including herons, storks, ibises, spoonbills, and their relatives.

Cidaroida [INV ZOO] An order of echinoderms in the subclass Perischoechinoidea in which the ambulacra comprise two columns of simple plates.

CID disease *See* combined immunological deficiency disease.

Ciidae [INV ZOO] The minute, tree-fungus beetles, a family of coleopteran insects in the superfamily Cucujoidea.

cilia [ANAT] Eyelashes. [CYTOL] Relatively short, centriole-based, hairlike processes on certain anatomical cells and motile organisms.

ciliary body [ANAT] A ring of tissue lying just anterior to the retinal margin of the eye.

ciliary movement [BIOL] A type of cellular locomotion accomplished by the rhythmical beat of cilia.

ciliary muscle [ANAT] The smooth muscle of the ciliary body.

ciliary process [ANAT] Circularly arranged choroid folds continuous with the iris in front.

Ciliatea [INV ZOO] The single class of the protozoan subphylum Ciliophora.

ciliolate [BIOL] Ciliated to a very minute degree.

Ciliophora [INV ZOO] The ciliated protozoans, a homogeneous subphylum of the Protozoa distinguished principally by a mouth, ciliation, and infraciliature.

Cimbicidae [INV ZOO] The cimbicid sawflies, a family of hymenopteran insects in the superfamily Tenthredinoidea.

Cimicidae [INV ZOO] The bat, bed, and bird bugs, a family of flattened, wingless, parasitic hemipteran insects in the superfamily Cimicimorpha.

Cimicimorpha [INV ZOO] A superfamily, or group according to some authorities, of hemipteran insects in the subdivision Geocorisae.

Cimicoidea [INV ZOO] A superfamily of the Cimicimorpha in some systems of classification.

Cincinnatian [GEOL] Upper Ordovician geologic time.

Cinclidae [VERT ZOO] The dippers, a family of insect-eating songbirds in the order Passeriformes.

cinclides [INV ZOO] Pores in the body wall of some sea anemones for the release of water and stinging cells.

cineplasty *See* kineplasty.

cinereous [BIOL] 1. Ashen in color. 2. Having the inert and powdery quality of ashes.

Cingulata [VERT ZOO] A group of xenarthran mammals in the order Edentata, including the armadillos.

cingulate [BIOL] Having a girdle of bands or markings.

cingulum [ANAT] 1. The ridge around the base of the crown of a tooth. 2. The tract of association nerve fibers in the brain, connecting the callosal and hippocampal convolutions. [BOT] The part of a plant between stem and root. [INV ZOO] 1. Any girdlelike structure. 2. A band of color or a raised line on certain bivalve shells. 3. The outer zone of cilia on discs of certain rotifers. 4. The clitellum in annelids.

cinnamon [BOT] *Cinnamomum zeylanicum.* An evergreen shrub of the laurel family (Lauraceae) in the order Magnoliales; a spice is made from the bark.

circadian rhythm [PHYSIO] A rhythmic process within an organism occurring independently of external synchronizing signals.

circinate [BIOL] Having the form of a flat coil with the apex at the center.

circinate vernation [BOT] Uncoiling of new leaves from the base toward the apex, as in ferns.

circle of Willis [ANAT] A ring of arteries at the base of the cerebrum.

circular birefringence [OPTICS] The phenomenon in which an optically active substance transmits right circularly polarized light with a different velocity from left circularly polarized light.

circular deoxyribonucleic acid [BIOCHEM] A single- or double-stranded ring of deoxyribonucleic acid found in certain bacteriophages and in human wart virus. Also known as ring deoxyribonucleic acid.

circular dichroism [OPTICS] A change from planar to elliptic polarization when an initially plane-polarized light wave traverses an optically active medium. Abbreviated CD.

circular paper chromatography [ANALY CHEM] Paper chromatographic technique in which migration from a spot in the sheet takes place in 360° so that zones separate as a series of concentric rings.

circulation [PHYSIO] The movement of blood through de-

CINNAMON

Cinnamomum zeylanicum. (USDA)

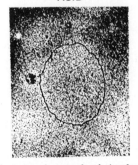

CIRCULAR DEOXYRIBONUCLEIC ACID

Electron micrograph of circular deoxyribonucleic acid extracted from the human wart virus. *(Courtesy of E. A. C. Follett)*

fined channels and tissue spaces; movement is through a closed circuit in vertebrates and certain invertebrates.

circulatory system [ANAT] The vessels and organs composing the lymphatic and cardiovascular systems.

circulin [MICROBIO] Any of a group of peptide antibiotics produced by *Bacillus circulans* which are related to polymixin and are active against both gram-negative and gram-positive bacteria.

circulus [BIOL] Any of various ringlike structures, such as the vascular circle of Willis or the concentric ridges on fish scales.

circumcision [MED] Surgical excision of the foreskin.

circumduction [ANAT] Movement of the distal end of a body part in the form of an arc; performed at ball-and-socket and saddle joints. [PHYSIO] Movement of a limb having its proximal end fixed and its distal part describing a circle.

circumflex artery [ANAT] Any artery that follows a curving or winding course.

circumfluence [INV ZOO] In Protozoa, an ingestion mechanism by which protoplasm flows toward and surrounds food.

circummutation [BOT] The irregular spiral growth movement of the apex of a stem, shoot, or tendril.

circumpharyngeal connective [INV ZOO] One of a pair of nerve strands passing around the esophagus in annelids and anthropods, connecting the brain and subesophageal ganglia.

circumscissile [BOT] Dehiscing along the line of a circumference, as exhibited by a pyxidium.

circumvallate papilla *See* vallate papilla.

circumvallation [CYTOL] An ingestion mechanism in protozoans and phagocytes by which food is engulfed by extruded pseudopodia.

Cirolanidae [INV ZOO] A family of isopod crustaceans in the suborder Flabellifera composed of actively swimming predators and scavengers with biting mouthparts.

Cirratulidae [INV ZOO] A family of fringe worms belonging to the Sedentaria which are important detritus feeders in coastal waters.

cirrhosis [MED] A progressive, inflammatory disease of the liver characterized by a real or apparent increase in the proportion of hepatic connective tissue.

cirriform [ZOO] Having the form of a cirrus; generally applied to a prolonged, slender process.

Cirripedia [INV ZOO] A subclass of the Crustacea, including the barnacles and goose barnacles; individuals are free-swimming in the larval stages but permanently fixed in the adult stage.

Cirromorpha [INV ZOO] A suborder of cephalopod mollusks in the order Octopoda.

cirrose [BIOL] Having cirri or tendrils.

cirrus [INV ZOO] **1.** The conical locomotor structure composed of fused cilia in hypotrich protozoans. **2.** Any of the jointed thoracic appendages of barnacles. **3.** Any hairlike tuft on insect appendages. **4.** The male copulatory organ in some mollusks and trematodes. [VERT ZOO] Any of the tactile barbels of certain fishes. [ZOO] A tendrillike animal appendage.

cirrus sac [INV ZOO] A pouch or channel containing the copulatory organ (cirrus) in certain invertebrates.

cis arrangement [GEN] The occurrence, in a heterozygote for two or more linked mutants, of all of them on the same one of the two homologous chromosomes. Also known as coupling.

cistern [ANAT] A closed, fluid-filled sac or vesicle, such as the subarachnoid spaces or the vesicles comprising the dictyosomes of a Golgi apparatus.

CIRRATULIDAE

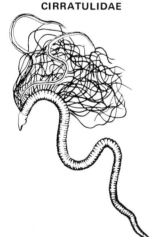

Chaetozone in left lateral view.

cis-trans isomerism [ORG CHEM] A type of geometrical isomerism found in alkenic systems in which it is possible for each of the doubly bonded carbons to carry two different atoms or groups; two similar atoms or groups maybe on the same side (cis) or on opposite sides (trans) of a plane bisecting the alkenic carbons and perpendicular to the plane of the alkenic system.

cis-trans test [GEN] A procedure used to determine whether mutants affecting the same characteristic are in one gene, that is, are alleles, or are in different adjacent genes.

cistron [MOL BIO] The genetic unit within which mutants do not complement each other; determined by the cis/trans complementation test. Also known as structural gene.

Citheroniinae [INV ZOO] A subfamily of lepidopteran insects in the family Saturniidae, including the regal moth and the imperial moth.

citramalase [BIOCHEM] An enzyme that is involved in fermentation of glutamate by *Clostridium tetanomorphum;* catalyzes the breakdown of citramalic acid to acetate and pyruvate.

citrate test [MICROBIO] A differential cultural test to identify genera within the bacterial family Enterobacteriaceae that are able to utilize sodium citrate as a sole source of carbon.

citric acid [BIOCHEM] $C_6H_8O_7 \cdot H_2O$ A colorless crystalline or white powdery organic, tricarboxylic acid occurring in plants, especially citrus fruits, and used as a flavoring agent, as an antioxidant in foods, and as a sequestering agent; the commercially produced form melts at 153°C.

citric acid cycle *See* Krebs cycle.

citriculture [BOT] The cultivation of citrus fruits.

citron [BOT] *Citrus medica.* A shrubby, evergreen citrus tree in the order Sapindales cultivated for its edible, large, lemonlike fruit.

citronella [BOT] *Cymbopogon nardus.* A tropical grass; the source of citronella oil.

citrulline [BIOCHEM] $C_6H_{13}O_3N_3$ An amino acid formed in the synthesis of arginine from ornithine.

citrus anthracnose [PL PATH] A fungus disease of citrus plants caused by *Colletotrichum gloeosporioides* and characterized by tip blight, stains on the leaves, and spots, stains, or rot on the fruit.

citrus blast [PL PATH] A bacterial disease of citrus trees caused by *Pseudomonas syringae* and marked by drying and browning of foliage and twigs and black pitting of the fruit.

citrus canker [PL PATH] A bacterial disease of citrus plants caused by *Xanthomonas citri* and producing lesions on twigs, foliage, and fruit.

citrus flavanoid compound *See* bioflavanoid.

citrus fruit [BOT] Any of the edible fruits having a pulpy endocarp and a firm exocarp that are produced by plants of the genus *Citrus* and related genera.

citrus scab [PL PATH] A fungus disease of citrus plants caused by *Sphaceloma rosarum,* producing scablike lesions on all plant parts.

civet [PHYSIO] A fatty substance secreted by the civet gland; used as a fixative in perfumes. [VERT ZOO] Any of 18 species of catlike, nocturnal carnivores assigned to the family Viverridae, having a long head, pointed muzzle, and short limbs with nonretractile claws.

civet cat *See* cacomistle.

civet gland [VERT ZOO] A large anal scent gland in civet cats that secretes civet.

civetone [BIOCHEM] $C_{17}H_{30}O$ 9-Cycloheptadecen-1-one, a macrocyclic ketone component of civet used in perfumes

CITRIC ACID

Structural formula of citric acid.

CITRON

Commercial citron *(Citrus medica),* foliage and fruit. Inset shows cross section of fruit. *(From J. Horace McFarland Co.)*

CLADOSELACHII

Cladoselache, a Late Devonian sharklike fish. The original specimens range from about 1½ to 4 feet (0.48–1.2 meters) in length. *(After Harris and Dean)*

CLAM WORM

Anterior end of a clam worm showing chitinous jaws, head appendages, and parapodia on each segment.

CLASPER

The clasper of the dogfish *(Squalus).*

because of its pleasant odor and lasting quality; believed to function as a sex attractant among civet cats.

cladanthous [BOT] Having archegonia positioned terminally on short lateral branches.

cladautoicous [BOT] Having antheridia on a specialized stalk, as in mosses.

Cladistia [VERT ZOO] The equivalent name for Polypteriformes.

Cladocera [INV ZOO] An order of small, fresh-water branchiopod crustaceans, commonly known as water fleas, characterized by a transparent bivalve shell.

Cladocopa [INV ZOO] A suborder of the order Myodocopida including marine animals having a carapace that lacks a permanent aperture when the two valves are closed.

Cladocopina [INV ZOO] The equivalent name for Cladocopa.

cladode *See* cladophyll.

cladodont [PALEON] Pertaining to sharks of the most primitive evolutionary level.

cladogenesis [EVOL] Evolution associated with altered habit and habitat, usually in isolated species populations.

cladogenic adaptation *See* divergent adaptation.

Cladoniaceae [BOT] A family of lichens in the order Lecanorales, including the reindeer mosses and cup lichens, in which the main thallus is hollow.

Cladophorales [BOT] An order of coarse, wiry, filamentous, branched and unbranched algae in the class Chlorophyceae.

cladophyll [BOT] A branch arising from the axil of a true leaf and resembling a foliage leaf. Also known as cladode.

cladoptosis [BOT] The annual abscission of twigs or branches instead of leaves.

Cladoselachii [PALEON] An order of extinct elasmobranch fishes including the oldest and most primitive of sharks.

cladus [BOT] A branch of a ramose spicule.

clairvoyance [PSYCH] A form of extrasensory perception in which there is a receiver and an extant event of which he has knowledge that has not been conveyed to him through his sensory channels.

clam [INV ZOO] The common name for a number of species of bivalve mollusks, many of which are important as food.

Clambidae [INV ZOO] The minute beetles, a family of coleopteran insects in the superfamily Dascilloidea.

clammy [BIOL] Moist and sticky, as the skin or a stem.

clam worm [INV ZOO] The common name for a number of species of dorsoventrally flattened annelid worms composing the large family Nereidae in the class Polychaeta; all have a distinct head, with numerous appendages.

Clapeyron-Clausius equation *See* Clausius-Clapeyron equation.

Clapeyron equation *See* Clausius-Clapeyron equation.

Clariidae [VERT ZOO] A family of Asian and African catfishes in the suborder Siluroidei.

Clarkecarididae [PALEON] A family of extinct crustaceans in the order Anaspidacea.

clasper [VERT ZOO] A modified pelvic fin of male elasmobranchs and holocephalians used for the transmission of sperm.

class [SYST] A taxonomic category ranking above the order and below the phylum or division.

classic epidemic typhus [MED] An epidemic disease caused by *Rickettsia prowazeki* var. *prowazekii,* and characterized by violent headache, a rash, neurological symptoms, and high fever. Also known as epidemic typhus.

classification [SYST] A systematic arrangement of plants and animals into categories based on a definite plan, considering evolutionary, physiologic, cytogenetic, and other relationships.

class interval [STAT] One of several convenient intervals into which the values of the variate of a frequency distribution may be grouped.

class mark [STAT] The mid-value of a class interval, or the integral value nearest the midpoint of the interval.

clathrate *See* cancellate.

Clathrinida [INV ZOO] A monofamilial order of sponges in the subclass Calcinea having an asconoid structure and lacking a true dermal membrane or cortex.

Clathrinidae [INV ZOO] The single family of the order Clathrinida.

Clausius-Clapeyron equation [THERMO] An equation governing phase transitions of a substance, $dp/dT = \Delta H/(T\Delta V)$, in which p is the pressure, T is the temperature at which the phase transition occurs, ΔH is the change in heat content (enthalpy), and ΔV is the change in volume during the transition. Also known as Clapeyron-Clausius equation; Clapeyron equation.

claustrophobia [PSYCH] An abnormal fear of confined spaces.

claustrum [ANAT] A thin layer of gray matter in each cerebral hemisphere between the lenticular nucleus and the island of Reil.

clava [BIOL] A club-shaped structure, as the tip on the antennae of certain insects or the fruiting body of certain fungi.

clavate [BIOL] Club-shaped. Also known as claviform.

Clavatoraceae [PALEOBOT] A group of middle Mesozoic algae belonging to the Charophyta.

Clavaxinellida [INV ZOO] An order of sponges in the class Demospongiae; members have monaxonid megascleres arranged in radial or plumose tracts.

clavicle [ANAT] A bone in the pectoral girdle of vertebrates with articulation occurring at the sternum and scapula.

claviculate [ANAT] Having a clavicle.

claviform *See* clavate.

clavola [INV ZOO] In insects, the terminal joints of an antenna.

clavus [INV ZOO] Any of several rounded or fingerlike processes, such as the club of an insect antenna or the pointed anal portion of the hemelytron in hemipteran insects.

claw [ANAT] A sharp, slender, curved nail on the toe of an animal, such as a bird. [INV ZOO] A sharp-curved process on the tip of the limb of an insect.

clearance test [PATH] The use of a substance such as urea or creatinine, or an injected foreign substance such as inulin to measure renal excretory activity; the ratio of the amount of these excreted substances in two 1-hour periods contrasted with the level of these substances in the blood is calculated in terms of the amount of blood cleared of these substances in a given unit time.

clear-cell carcinoma *See* renal-cell carcinoma.

cleavage [EMBRYO] The subdivision of activated eggs into blastomeres.

cleavage nucleus [EMBRYO] The nucleus of a zygote formed by fusion of male and female pronuclei.

cleft grafting [BOT] A top-grafting method in which the scion is inserted into a cleft cut into the top of the stock.

cleft lip *See* harelip.

cleft palate [MED] A birth defect resulting from incomplete closure of the palate during embryogenesis.

cleidocranial dysostosis [MED] A congenital defect in which

CLAVATORACEAE

250 μ

A calcified part of the unfertilized oogonia of a Clavatoraceae fossil.

CLEFT GRAFTING

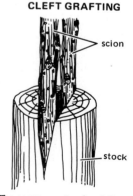

scion

stock

The components of a cleft graft. (*From H. J. Fuller and Z. B. Carothers, The Plant World, 4th ed., Holt, Rinehart, and Winston, 1963*)

there is deficient formation of bone in the skull and clavicle.

cleistocarp *See* cleistothecium.

cleistocarpous [BOT] Of mosses, having the capsule opening irregularly without an operculum. [MYCOL] Forming or having cleistothecia.

cleistogamy [BOT] Production of small closed flowers that are self-pollinating and contain numerous seeds.

cleistothecium [MYCOL] A closed sporebearing structure in Ascomycetes; asci and spores are freed of the fruiting body by decay or desiccation. Also known as cleistocarp.

cleithrum [VERT ZOO] A bone external and adjacent to the clavicle in certain fishes, stegocephalians, and primitive reptiles.

Cleridae [INV ZOO] The checkered beetles, a family of coleopteran insects in the superfamily Cleroidea.

Cleroidea [INV ZOO] A superfamily of coleopteran insects in the suborder Polyphaga.

Climatiidae [PALEON] A family of archaic tooth-bearing fishes in the suborder Climatioidei.

Climatiiformes [PALEON] An order of extinct fishes in the class Acanthodii having two dorsal fins and large plates on the head and ventral shoulder.

Climatioidei [PALEON] A suborder of extinct fishes in the order Climatiiformes.

climatopathology [MED] The study of disease in relation to the effects of the natural environment.

climatotherapy [MED] Placing a person in a suitable climate to treat a certain disease.

climax [ECOL] A mature, relatively stable community in an area, which community will undergo no further change under the prevailing climate; represents the culmination of ecological succession.

climax plant formation [ECOL] A mature, stable plant population in a climax community.

climbing stem [BOT] A long, slender stem that climbs up a support or along the tops of other plants by using spines, adventitious roots, or tendrils for attachment.

clinandrium [BOT] In orchids, a chamber in the column between the anthers.

cline [BIOL] A graded series of morphological or physiological characters exhibited by a natural group (as a species) of related organisms, generally along a line of environmental or geographic transition.

clinical genetics [GEN] The study of biological inheritance by direct observation of the living patient.

clinical pathology [PATH] A medical specialty encompassing the diagnostic study of disease by means of laboratory tests of material from the living patient.

clinical psychology [PSYCH] A branch of psychology concerned with recognizing, treating, and determining causes of behavior disorders.

clinidium [MYCOL] A sporogenous filament in a pycnidium.

Clionidae [INV ZOO] The boring sponges, a family of marine sponges in the class Demospongiae.

clisere [ECOL] The succession of ecological communities, especially climax formations, as a consequence of intense climatic changes.

clitellum [INV ZOO] The thickened, glandular, saddlelike portion of the body wall of some annelid worms.

clitoris [ANAT] The homolog of the penis in females, located in the anterior portion of the vulva.

clivus [ANAT] A shallow depression behind the dorsum sellae in the sphenoid bone.

cloaca [INV ZOO] The chamber functioning as a respiratory, excretory, and reproductive duct in certain invertebrates.

[VERT ZOO] The chamber which receives the discharges of the intestine, urinary tract, and reproductive canals in monotremes, amphibians, birds, reptiles, and many fish.

cloacal bladder [VERT ZOO] A diverticulum of the cloacal wall in monotremes, amphibians, and some fish, into which urine is forced from the cloaca.

cloacal gland [VERT ZOO] Any of the sweat glands in the cloaca of lower vertebrates, as snakes or amphibians.

clone [BIOL] All individuals, considered collectively, produced asexually or by parthenogenesis from a single individual.

clonorchiasis [MED] A parasitic infection of man and other fish-eating mammals caused by the trematode *Opisthorchis* (*Clonorchis*) *sinensis,* which is usually found in the bile ducts.

clonus [PHYSIO] Irregular, alternating muscular contractions and relaxations.

closed ecological system [ECOL] A community into which a new species cannot enter due to crowding and competition.

closed reduction [MED] Reduction of fractures or dislocations by manipulation without surgical intervention.

clot [PHYSIO] A semisolid coagulum of blood or lymph.

clot retraction time [PATH] The length of time required for the appearance or completion of the contraction or shrinkage of a blood clot, resulting in the extrusion of serum.

clotting time [PHYSIO] The length of time required for shed blood to coagulate under standard conditions. Also known as coagulation time.

cloud forest *See* temperate rainforest.

clove [BOT] 1. The unopened flower bud of a small, conical, symmetrical evergreen tree, *Eugenia caryophyllata,* of the myrtle family (Myrtaceae); the dried buds are used as a pungent, strongly aromatic spice. 2. A small bulb developed within a larger bulb, as in garlic.

clover [BOT] 1. A common name designating the true clovers, sweet clovers, and other members of the Leguminosa. 2. A herb of the genus *Trifolium.*

clubfoot [MED] Congenital malpositioning of a foot such that the forefoot is inverted and rotated with a shortened Achilles tendon.

club fungi [MYCOL] The common name for members of the class Basidiomycetes.

club moss [BOT] The common name for members of the class Lycopodiatae.

clubroot [PL PATH] A disease principally of crucifers, such as cabbage, caused by the slime mold *Plasmodiophora brassicae* in which roots become enlarged and deformed, leading to plant death.

Clupeidae [VERT ZOO] The herrings, a family of fishes in the suborder Clupoidea composing the most primitive group of higher bony fishes.

Clupeiformes [VERT ZOO] An order of teleost fishes in the subclass Actinopterygii, generally having a silvery, compressed body.

clupeine [BIOCHEM] A protamine found in salmon sperm, mainly composed of arginine (74.1%) and small percentages of threonine, serine, proline, alanine, valine, and isoleucine.

Clupoidea [VERT ZOO] A suborder of fishes in the order Clupeiformes comprising the herrings and anchovies.

Clusiidae [INV ZOO] A family of myodarian cyclorrhaphous dipteran insects in the subsection Acalypteratae.

clustering [PSYCH] In free recall, the tendency for items to be recalled in groups because of their similarity in meaningfulness, hierarchy, or conceptual category, or because of their simultaneity of occurrence.

Clypeasteroida [INV ZOO] An order of exocyclic Euechinoi-

CLOVE

Closed flower buds (cloves) and open buds on a branch of the evergreen tree *Eugenia caryophyllata.*

CLYPEASTEROIDA

Typical clypeasteroids. *Laganum,* (*a*) monobasal apical system, (*b*) aboral aspect, and (*c*) posterior aspect. *Arachnoides,* (*d*) aboral aspect, and (*e*) posterior aspect.

dea having a monobasal apical system in which all the genital plates fuse together.

clypeus [INV ZOO] An anterior medial plate on the head of an insect, commonly bearing the labrum on its anterior margin. [MYCOL] A disk of black tissue about the mouth of the perithecia in certain ascomycetes.

clysis [MED] **1.** Administration of an enema. **2.** Subcutaneous or intravenous administration of fluids.

Clythiidae [INV ZOO] The flat-footed flies, a family of cyclorrhaphous dipteran insects in the series Aschiza characterized by a flattened distal end on the hind tarsus.

cnemidium [VERT ZOO] The lower, generally scaly, featherless part of bird's leg.

Cnidaria [INV ZOO] The equivalent name for Coelenterata.

cnidoblast [INV ZOO] A cell that produces nematocysts. Also known as nettle cell; stinging cell.

cnidocil [INV ZOO] The trigger on a cnidoblast that activates discharge of the nematocyst when touched.

cnidophore [INV ZOO] A modified structure bearing nematocysts in certain coelenterates.

cnidosac [INV ZOO] A kidney-shaped swelling on dactylozooids of Siphonophora.

Cnidospora [INV ZOO] A subphylum of spore-producing protozoans that are parasites in cells and tissues of invertebrates, fishes, a few amphibians, and turtles.

Co *See* cobalt.

Co^{60} *See* cobalt-60.

CoA *See* coenzyme A.

coacervation [CHEM] The separation, by addition of a third component, of an aqueous solution of a macromolecule colloid (polymer) into two liquid phases, one of which is colloid-rich (the coacervate) and the other an aqueous solution of the coacervating agent (the equilibrium liquid).

coaction [ECOL] Interaction of living organisms on one another in a community.

coagulability test [PATH] Any of several clinical tests of the ability of blood to coagulate, such as clot retraction time and quantification, prothrombin time, partial thromboplastin time, and platelet enumeration.

coagulant [CHEM] An agent that causes coagulation.

coagulase [BIOCHEM] Any enzyme that causes coagulation of blood plasma.

coagulation [CHEM] A separation or precipitation from a dispersed state of suspensoid particles resulting from their growth; may result from prolonged heating, addition of an electrolyte, or from a condensation reaction between solute and solvent; an example is the setting of a gel.

coagulation time *See* clotting time.

coagulum [PHYSIO] A mass of coagulated material.

coalification [GEOL] Formation of coal from plant material by the processes of diagenesis and metamorphism. Also known as bituminization; carbonification; incarbonization; incoalation.

coal paleobotany [PALEOBOT] A branch of the paleobotanical sciences concerned with the origin, composition, mode of occurrence, and significance of fossil plant materials that occur in or are associated with coal seams.

coarctate [BIOL] **1.** Compressed. **2.** Closely connected. [INV ZOO] Referring to insect pupae enclosed in a hard case formed from the skin of the last larval stage.

coati [VERT ZOO] The common name for three species of carnivorous mammals assigned to the raccoon family (Procyonidae) characterized by their elongated snout, body, and tail.

cobalamin *See* vitamin B$_{12}$.

COATI

Common coati *(Nasua nasua)* with a long snout and striped bushy tail, which is held erect.

cobalt [CHEM] A metallic element, symbol Co, atomic number 27, atomic weight 58.93; used chiefly in alloys.

cobalt-60 [NUC PHYS] A radioisotope of cobalt, symbol Co^{60}, having a mass number of 60; emits gamma rays and has many medical and industrial uses; the most commonly used isotope for encapsulated radiation sources.

cobalt-beam therapy [MED] Therapy involving the use of gamma radiation from a cobalt-60 source mounted in a cobalt bomb. Also known as cobalt therapy.

Cobb's disease [PL PATH] A bacterial disease of sugarcane caused by *Xanthomonas vascularum* and characterized by a slime in the vascular bundles, dwarfing, streaking of leaves, and decay. Also known as sugarcane gummosis.

Cobitidae [VERT ZOO] The loaches, a family of small fishes, many eel-shaped, in the suborder Cyprinoidei, characterized by barbels around the mouth.

Coblentzian [GEOL] Upper Lower Devonian geologic time.

cobra [VERT ZOO] Any of several species of venomous snakes in the reptilian family Elaphidae characterized by a hoodlike expansion of skin on the anterior neck that is supported by a series of ribs.

coca [BOT] *Erythroxylon coca*. A shrub in the family Erythroxylaceae; its leaves are the source of cocaine.

cocaine [PHARM] $C_{17}H_{21}O_4N$ An alkaloid obtained from coca leaves that is used for local anesthesia and as a tonic in digestive and nervous disorders. Also known as erythroxylon; methylbenzoylecgonine.

cocarboxylase *See* thiamine pyrophosphate.

cocarcinogen [MED] A noncarcinogenic agent which augments the carcinogenic process.

Coccidia [INV ZOO] A subclass of protozoans in the class Telosporea; typically intracellular parasites of epithelial tissue in vertebrates and invertebrates.

coccidioidomycosis [MED] An infectious fungus disease of man and animals of either a pulmonary or a cutaneous nature; caused by *Coccidioides immitis.* Also known as San Joaquin Valley fever.

coccidiosis [MED] The state of or the conditions associated with being infected by coccidia.

coccine [INV ZOO] Pertaining to a sessile vegetative condition in protists during which reproduction does not take place.

Coccinellidae [INV ZOO] The ladybird beetles, a family of coleopteran insects in the superfamily Cucujoidea.

coccobacillus [MICROBIO] A short, thick, oval bacillus, midway between the coccus and the bacillus in appearance.

Coccoidea [INV ZOO] A superfamily of homopteran insects belonging to the Sternorrhyncha; includes scale insects and mealy bugs.

coccolith [BOT] One of the small, interlocking calcite plates covering members of the Coccolithophorida.

Coccolithophora [INV ZOO] An order of phytoflagellates in the protozoan class Phytamastigophorea.

Coccolithophorida [BOT] A group of unicellular, biflagellate, golden-brown algae characterized by a covering of coccoliths.

Coccomyxaceae [BOT] A family of algae belonging to the Tetrasporales composed of elongate cells which reproduce only by vegetative means.

Coccosteomorphi [PALEON] An aberrant lineage of the joint-necked fishes.

coccus [MICROBIO] A form of eubacteria which are more or less spherical in shape.

coccygeal body [ANAT] A small mass of vascular tissue near the tip of the coccyx.

coccygectomy [MED] Surgical excision of the coccyx.

COBALT

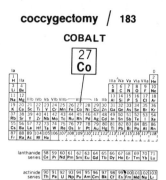

Periodic table of the chemical elements showing the position of cobalt.

COCCINELLIDAE

A ladybird beetle. *(From T. I. Storer and R. L. Usinger, General Zoology, 3d ed., McGraw-Hill, 1957)*

COCKLE

The shell of the rock cockle
(Protothaca laciniata), covered
with radiating ribs crossed with
concentric ribs.

COD

Gadus morrhua, a codfish
of the northern seas.

CODIACEAE

A portion of a dichotomously
branched thallus of *Codium*.

COELACANTH

Living coelacanth, *Latimeria
chalumnae*; 5 feet (1.5 meters).
*(After P. P. Grassé, ed., Traité
de Zoologie, tome 13, fasc. 3,
1958)*

coccyx [ANAT] The fused vestige of caudal vertebrae forming the last bone of the vertebral column in man and certain other primates.

cochlea [ANAT] The snail-shaped canal of the mammalian inner ear; it is divided into three channels and contains the essential organs of hearing.

cochlear duct *See* scala media.

Cochleariidae [VERT ZOO] A family of birds in the order Ciconiiformes composed of a single species, the boatbill.

cochlear nerve [ANAT] A sensory branch of the auditory nerve which receives impulses from the organ of Corti.

cochlear nucleus [ANAT] One of the two nuclear masses in which the fibers of the cochlear nerve terminate; located ventrad and dorsad to the inferior cerebellar peduncle.

cochleate [BIOL] Spiral; shaped like a snail shell.

Cochliodontidae [PALEON] A family of extinct chondrichthian fishes in the order Bradyodonti.

cockle [INV ZOO] The common name for a number of species of marine mollusks in the class Bivalvia characterized by a shell having convex radial ribs.

coconut [BOT] *Cocos nucifera.* A large palm in the order Arecales grown for its fiber and fruit, a large, ovoid, edible drupe with a fibrous exocarp and a hard, bony endocarp containing fleshy meat (endosperm).

coconut bud rot [PL PATH] A fungus disease of the coconut palm caused by *Phytophthora palmivora* and characterized by destruction of the terminal bud and adjacent leaves.

cocoon [INV ZOO] **1.** A protective case formed by the larvae of many insects, in which they pass the pupa stage. **2.** Any of the various protective egg cases formed by invertebrates.

cod [VERT ZOO] The common name for fishes of the subfamily Gadidae, especially the Atlantic cod (*Gadus morrhua*).

codecarboxylase [BIOCHEM] The prosthetic component of the enzyme carboxylase which catalyzes decarboxylation of L-amino acids. Also known as pyridoxal phosphate.

codehydrogenase I *See* diphosphopyridine nucleotide.

codehydrogenase II *See* triphosphopyridine nucleotide.

codeine [PHARM] $C_{18}H_{21}NO_3$ An alkaloid prepared from morphine; used as mild analgesic and cough suppressant.

Codiaceae [BOT] A family of green algae in the order Siphonales having macroscopic thalli composed of aggregates of tubes.

codon [GEN] The basic unit of the genetic code, comprising three-nucleotide sequences of messenger ribonucleic acid, each of which is translated into one amino acid in protein synthesis.

coefficient of absorption *See* absorption coefficient.

coefficient of contingency [STAT] A measure of the strength of dependence between two statistical variables, based on a contingency table.

coefficient of correlation [STAT] A number between $+1.00$ and -1.00 that expresses the degree of relationship between two sets of measurements arranged in pairs; perfect correlation is expressed by a coefficient of $+1.00$ (or -1.00); no correlation is expressed by .00.

coefficient of variation [STAT] The ratio of the standard deviation of a distribution to its arithmetic mean.

coelacanth [VERT ZOO] Any member of the Coelacanthiformes, an order of lobefin fishes represented by a single living genus, *Latimeria.*

Coelacanthidae [PALEON] A family of extinct lobefin fishes in the order Coelacanthiformes.

Coelacanthiformes [VERT ZOO] An order of lobefin fishes in the subclass Crossopterygii which were common fresh-water animals of the Carboniferous and Permian; one genus, *Latimeria*, exists today.

Coelacanthini [VERT ZOO] The equivalent name for Coelacanthiformes.

Coelenterata [INV ZOO] A phylum of the Radiata whose members typically bear tentacles and possess intrinsic nematocysts.

coelenteron [INV ZOO] The internal cavity of coelenterates.

coeloblastula [EMBRYO] A simple, hollow blastula with a single-layered wall.

Coelolepida [PALEON] An order of extinct jawless vertebrates (Agnatha) distinguished by skin set with minute, close-fitting scales of dentine, similar to placoid scales of sharks.

coelom [ZOO] The mesodermally lined body cavity of most animals higher on the evolutionary scale than flatworms and nonsegmented roundworms.

Coelomata [ZOO] The equivalent name for Eucoelomata.

coelomocyte [INV ZOO] A corpuscle, including amebocytes and eleocytes, in the coelom of certain animals, especially annelids.

coelomoduct [INV ZOO] Either of a pair of ciliated excretory and reproductive channels passing from the coelom to the exterior in certain invertebrates, including annelids and mollusks.

Coelomomycetaceae [MYCOL] A family of entomophilic fungi in the order Blastocladiales which parasitize primarily mosquito larvae.

coelomopores [INV ZOO] In *Nautilus*, ducts which lead directly from the pericardial cavity to the exterior.

coelomostome [INV ZOO] The opening of a coelomoduct into the coelom.

Coelomycetes [MYCOL] A group set up by some authorities to include the Sphaerioidaceae and the Melanconiales.

Coelopidae [INV ZOO] The seaweed flies, a family of myodarian cyclorrhaphous dipteran insects in the subsection Acalypteratae whose larvae breed on decomposing seaweed.

coeloplanula [INV ZOO] A hollow planula having a wall of two layers of cells.

Coelurosauria [PALEON] A group of small, lightly built saurischian dinosaurs in the suborder Theropoda having long necks and narrow, pointed skulls.

Coenagrionidae [INV ZOO] A family of zygopteran insects in the order Odonata.

coenanthium [BOT] An inflorescence characterized by a flat receptacle with up-curved margins.

coenenchyme [INV ZOO] The mesagloea surrounding and uniting the polyps in compound anthozoans. Also known as coenosarc.

Coenobitidae [INV ZOO] A family of terrestrial decapod crustaceans belonging to the Anomura.

coenobium [INV ZOO] A colony of protozoans having a constant size, shape, and cell number, but with undifferentiated cells.

coenocyte [BIOL] A multinucleate mass of protoplasm formed by repeated nucleus divisions without cell fission.

Coenomyidae [INV ZOO] A family of orthorrhaphous dipteran insects in the series Brachycera.

Coenopteridales [PALEOBOT] A heterogeneous group of fernlike fossil plants belonging to the Polypodiophyta.

coenosarc [INV ZOO] 1. The living axial portion of a hydroid colony. 2. *See* coenenchyme.

coenosteum [INV ZOO] The calcareous skeleton of a compound coral or bryozoan colony.

Coenothecalia [INV ZOO] An order of the class Alcyonaria

COELUROSAURIA

Skeleton of *Coelophysis* (about 8 feet, or 2.5 meters long), a Late Triassic theropod from New Mexico. *(After E. H. Colbert, Evolution of the Vertebrates, Wiley, 1955)*

that forms colonies; lacks spicules but has a skeleton composed of fibrocrystalline argonite.

coenotype [BIOL] An organism having the characteristic structure of the group to which it belongs.

coenurosis [VET MED] An infestation by a coenurus, the metacestode of *Taenia* species; most common in sheep, rabbits, and other herbivores.

coenurus [INV ZOO] A metacestode having a large bladder with many daughter cysts arising from its walls; each cyst has one scolex.

coenzyme [BIOCHEM] The nonprotein portion of an enzyme; a prosthetic group which functions as an acceptor of electrons or functional groups.

coenzyme A [BIOCHEM] $C_{21}H_{36}O_{16}N_7P_3S$ A coenzyme in all living cells; required by certain condensing enzymes to act in acetyl or other acyl-group transfer and in fatty-acid metabolism. Abbreviated CoA.

coenzyme I *See* diphosphopyridine nucleotide.

coenzyme II *See* triphosphopyridine nucleotide.

COFFEE

Branch of *Coffea arabica.*

coffee [BOT] Any of various shrubs or small trees of the genus *Coffea* (family Rubiaceae) cultivated for the seeds (coffee beans) of its fruit; most coffee beans are obtained from the Arabian species, *C. arabica.*

cognac [FOOD ENG] Brandy distilled from grapes grown in the Charente and Charente-Maritime departments of France.

cognition [PSYCH] The conscious faculty or process of knowing, of becoming or being aware of thoughts or perceptions, including understanding and reasoning.

cogon [BOT] *Imperate cylindrica.* A grass found in rainforests. Also known as alang-alang.

coherent [BOT] Having similar parts united.

cohesion [BOT] The union of similar plant parts or organs, as of the petals to form a corolla. [SCI TECH] The state or process of sticking together.

coiled tubular gland [ANAT] A structure having a duct interposed between the surface opening and the coiled glandular portion; an example is a sweat gland.

coincidence [GEN] A numerical value equal to the number of double crossovers observed, divided by the number expected.

coir [MATER] A coarse, brown fiber obtained from the husk of the coconut.

coitus [ZOO] The act of copulation. Also known as intercourse.

cola [BOT] *Cola acuminata.* A tree of the sterculia family (Sterculiaceae) cultivated for cola nuts, the seeds of the fruit; extract of cola nuts is used in the manufacture of soft drinks.

colchicine [ORG CHEM] $C_{22}H_{25}O_6N$ An alkaloid extracted from the stem of the autumn crocus; used experimentally to inhibit spindle formation and delay centromere division, and medicinally in the treatment of gout.

COLA

Branch of *Cola acuminata.*

cold agglutination phenomenon [IMMUNOL] Clumping of human blood group O erythrocytes at 0-4°C, but not at body temperature; occurs in primary atypical pneumonia, trypanasomiasis, and other unidentifiable states.

cold agglutinin [IMMUNOL] A nonspecific panagglutinin found in many normal human serums which produce maximum clumping of erythrocytes at 4° and none at 37°C.

cold-blooded [PHYSIO] Having body temperature approximating that of the environment and not internally regulated.

cold hemagglutination [IMMUNOL] A phenomenon caused by the presence of cold agglutinin.

cold torpor [PHYSIO] Condition of reduced body temperature in poikilotherms.

colectomy [MED] Excision of all or a portion of the colon.

Coleochaetaceae [BOT] A family of green algae in the suborder Ulotrichineae; all occur as attached, disklike, or parenchymatous thalli.

Coleodontidae [PALEON] A family of conodonts in the suborder Neurodontiformes.

Coleoidea [INV ZOO] A subclass of cephalopod mollusks including all cephalopods except *Nautilus*, according to certain systems of classification.

Coleophoridae [INV ZOO] The case bearers, narrow-winged moths composing a family of lepidopteran insects in the suborder Heteroneura; named for the silk-and-leaf shell carried by larvae.

Coleoptera [INV ZOO] The beetles, holometabolous insects making up the largest order of the animal kingdom; general features of the Insecta are found in this group.

coleopterous [INV ZOO] In beetles, having hard anterior wings that function as elytra.

coleoptile [BOT] The first leaf of a monocotyledon seedling.

coleorhiza [BOT] The sheath surrounding the radicle in monocotyledons.

Coleorrhyncha [INV ZOO] A monofamilial group of homopteran insects in which the beak is formed at the anteroventral extremity of the face and the propleura form a shield for the base of the beak.

Coleosporaceae [MYCOL] A family of parasitic fungi in the order Uredinales.

colic [MED] **1.** Acute paroxysmal abdominal pain usually caused by smooth muscle spasm, obstruction, or twisting. **2.** In early infancy, paroxysms of pain, crying, and irritability caused by swallowing air, overfeeding, intestinal allergy, and emotional factors.

colic artery [ANAT] Any of the three arteries that supply the colon.

colicin [MICROBIO] A highly specific protein, released by some enteric bacilli, which kills other enteric bacilli.

coliform bacteria [MICROBIO] Colon bacilli, or forms which resemble or are related to them.

Coliidae [VERT ZOO] The colies or mousebirds, composing the single family of the avian order Coliiformes.

Coliiformes [VERT ZOO] A monofamilial order of birds distinguished by long tails, short legs, and long toes, all four of which are directed forward.

colinearity [GEN] The correspondence between the sequence of DNA nucleotide triplets in structural genes and the sequence of amino acids in the polypeptide chains encoded in those genes.

coliphage [VIROL] Any bacteriophage able to infect *Escherichia coli.*

colistin [MICROBIO] $C_{45}H_{85}O_{10}N_{13}$ A basic polypeptide antibiotic produced by *Bacillus colistinus;* consists of an A and B component, active against a broad spectrum of gram-positive microorganisms and some gram-negative microorganisms.

colitis [MED] Inflammation of the large bowel, or colon.

collagen [BIOCHEM] A fibrous protein found in all multicellular animals, especially in connective tissue.

collagenase [BIOCHEM] Any proteinase that decomposes collagen and gelatin.

collagen disease [MED] Any of various clinical syndromes characterized by widespread alterations of connective tissue, including inflammation and fibrinoid degeneration.

collar cell *See* choanocyte.

collard [BOT] *Brassica oleracea* var. *acephala.* A biennial crucifer of the order Capparales grown for its rosette of edible leaves.

COLEODONTIDAE

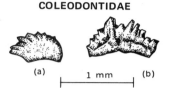

(a)   1 mm   (b)

Representatives of Coleodontidae. *(a) Coleodus,* a typical denticulate blade. *(b) Erismodus. (After illustration in R. R. Shrock and W. H. Twenhofel, Principles of Invertebrate Paleontology, McGraw-Hill, 2d ed., 1953)*

COLLARD

Collard *(Brassica oleracea* var. *acephala).*

COLLATERAL BUD

Collateral bud in a red maple.

COLLEMBOLA

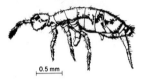

0.5 mm

A collembolan, *Entomobrya cubensis*. (From J. W. Folsom, Proc. U.S. Nat. Mus., 72(6), plate 6, 1927)

COLLENCHYMA

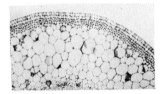

Collenchyma in a transverse section of Jimsonweed stem. (*W. S. Walker*)

COLLOTHECIDAE

Collotheca species.

collarette [ANAT] The line where pupillary and ciliary zones join on the anterior surface of the iris.

collateral bud [BOT] An accessory bud produced beside an axillary bud.

collecting tubule [ANAT] One of the ducts conveying urine from the renal tubules (nephrons) to the minor calyces of the renal pelvis.

collective fruit [BOT] A fruit produced from a complete inflorescence; examples are mulberry and pineapple.

collector [BOT] Any of the pollen-retaining hairs on the stigma or the style of some flowers.

Collembola [INV ZOO] The springtails, an order of primitive insects in the subclass Apterygota having six abdominal segments.

collenchyma [BOT] A primary, or early-differentiated, subepidermal supporting tissue in leaf petioles and vein ribs formed before vascular differentiation.

collenchyme [INV ZOO] A loose mesenchyme that fills the space between ectoderm and endoderm in the body wall of many lower invertebrates, such as sponges.

collencyte [INV ZOO] In Porifera, a cell with reticulopodia.

collet [BOT] A region of demarcation between the hypocotyl and the root.

colleterium [INV ZOO] In insects, mucus-secreting gland of the female reproductive system.

Colletidae [INV ZOO] The colletid bees, a family of hymenopteran insects in the superfamily Apoidea.

colliculus [ANAT] **1.** Any of the four prominences of the corpora quadrigemina. **2.** The elevation of the optic nerve where it enters the retina. **3.** The anterolateral, apical elevation of the arytenoid cartilages.

colligative properties [PHYS CHEM] Those properties that are related by a mathematical function, that is, dependent on the number of particles.

colloblast [INV ZOO] An adhesive cell on the tentacles of ctenophores.

colloid [CHEM] A system of which one phase is made up of particles having dimensions of 10–10,000 angstroms (1–1000 nanometers) and which is dispersed in a different phase.

colloidal osmotic pressure *See* oncotic pressure.

colloid chemistry [PHYS CHEM] The scientific study of matter whose size is approximately 10 to 10,000 angstroms (1 to 1000 nanometers), and which exists as a suspension in a continuous medium, especially a liquid, solid, or gaseous substance.

colloid goiter [MED] A soft, diffuse enlargement of the thyroid gland in which colloid fills the acinar spaces.

Collothecacea [INV ZOO] A monofamilial suborder of mostly sessile rotifers in the order Monogonata; many species are encased in gelatinous tubes.

Collothecidae [INV ZOO] The single family of the Collothecacea.

Collyritidae [PALEON] A family of extinct, small, ovoid, exocyclic Euechinoidea with fascioles or a plastron.

coloboma [MED] A congenital, pathologic, or operative fissure, especially of the eye or eyelid.

colon [ANAT] The lowermost part of the gut of vertebrates, that is, between the large intestine and the anus. Also known as large intestine. [INV ZOO] In insects, the second part of the intestine.

colon bacillus *See* Escherichia coli.

colony [BIOL] A localized population of individuals of the same species which are living either attached or separately. [MICROBIO] A cluster of microorganisms growing on the surface of or within a solid medium; usually cultured from a single cell.

color aberration   *See* chromatic aberration.

Coloradoan   [GEOL] Middle Upper Cretaceous geologic time.

Colorado tick fever   [MED] A nonexanthematous acute viral disease of man occurring in the western United States and transmitted by a bite of the tick *Dermacentor andersoni*; characterized by a short course, intermittent fever, leukopenia, and occasionally meningoencephalitis. Also known as mountain tick fever.

color blindness   [MED] Inability to perceive one or more colors.

color constancy   [PHYSIO] Perception of colors as constant despite changes in sensory input.

colorimeter   [ANALY CHEM] A device for measuring concentration of a known constituent in solution by comparison with colors of a few solutions of known concentration of that constituent. Also known as chromometer.

color index   [PATH] The amount of hemoglobin per erythrocyte relative to normal, equal to the percent normal hemoglobin concentration divided by percent normal erythrocyte count.

color vision   [PHYSIO] The ability to discriminate light on the basis of wavelength composition.

Colossendeidae   [INV ZOO] A family of deep-water marine arthropods in the subphylum Pycnogonida, having long palpi and lacking chelifores, except in polymerous forms.

colostomy   [MED] Surgical formation of an artificial anus by joining the colon to an opening in the anterior abdominal wall.

colostrum   [PHYSIO] The first milk secreted by the mammary gland during the first days following parturition.

colotomy   [MED] Incision of the colon; may be abdominal, lateral, lumbar, or iliac, according to the region of entrance.

colpate   [BOT] Referring to a pollen grain having small and nearly isodiametric apertures.

colposcope   [MED] An instrument for the visual examination of the vagina and cervix; a vaginal speculum.

colpotomy   [MED] Incision of the vagina.

Colubridae   [VERT ZOO] A family of cosmopolitan snakes in the order Squamata.

colulus   [INV ZOO] In spiders, small cone-shaped structure between the anterior spinnerets of the silk gland.

Columbidae   [VERT ZOO] A family of birds in the order Columbiformes composed of the pigeons and doves.

Columbiformes   [VERT ZOO] An order of birds distinguished by a short, pointed bill, imperforate nostrils, and short legs.

columella   [ANAT] *See* stapes. [BIOL] Any part shaped like a column. [BOT] A sterile axial body within the capsules of certain mosses, liverworts, and many fungi.

columnals   [INV ZOO] In Crinoidea, stem ossicles.

columnar epithelium   [HISTOL] Epithelium distinguished by elongated, columnar, or prismatic cells.

columnar stem   [BOT] An unbranched, cylindrical stem bearing a set of large leaves at its summit, as in palms, or no leaves, as in cacti.

column chromatography   [ANALY CHEM] Chromatographic technique of two general types: packed columns usually contain either a granular adsorbent or a granular support material coated with a thin layer of high-boiling solvent (partitioning liquid); open-tubular columns contain a thin film of partitioning liquid on the column walls and have an opening so that gas can pass through the center of the column.

column development chromatography   [ANALY CHEM] Columnar apparatus for separating or concentrating one or more components from a physical mixture by use of adsorbent packing; as the specimen percolates along the length of the adsorbent, its various components are preferentially held at different rates, effecting a separation.

Colydiidae [INV ZOO] The cylindrical bark beetles, a large family of coleopteran insects in the superfamily Cucujoidea.

coma [BOT] **1.** Terminal cluster of bracts, as in pineapple. **2.** Hair tuffs found on some seeds. [MED] Unconsciousness from which the patient cannot be aroused.

Comasteridae [INV ZOO] A family of radially symmetrical Crinozoa in the order Comatulida.

comatose [MED] In a condition of coma; resembling coma.

Comatulida [INV ZOO] The feather stars, an order of free-living echinoderms in the subclass Articulata.

comb [INV ZOO] **1.** A system of hexagonal cells constructed of beeswax by a colony of bees. **2.** A comblike swimming plate in ctenophores. Also known as ctene. [VERT ZOO] A crest of naked tissue on the head of many male fowl.

combined immunological deficiency disease [MED] A severe and usually fatal disease in which the individual lacks not only the T (thymus-derived) cells, which are responsible for graft rejection and for defense against viruses, but also the B (bone-marrow-derived) cells, which are responsible for production of globulins and antibodies. Also known as CID disease.

combine-harvester [AGR] A machine for harvesting legumes and grasses; cuts off the plant tops, then beats and cleans the grain while the machine continues to move across the field.

combining-volumes principle [CHEM] The principle that when gases take part in chemical reactions the volumes of the reacting gases and those of the products (if gaseous) are in the ratio of small whole numbers, provided that all measurements are made at the same temperature and pressure. Also known as Gay-Lussac law.

comedo [MED] A collection of sebaceous material and keratin retained in the hair follicle and excretory duct of the sebaceous gland, whose surface is covered with a black dot caused by oxidation of sebum at the follicular orifice. Also known as blackhead.

comedocarcinoma [MED] A type of adenocarcinoma of the breast in which the ducts are filled with cells which, when expressed from the cut surface, resemble comedos.

Comesomatidae [INV ZOO] A family of free-living nematodes in the superfamily Chromadoroidea found as deposit feeders on soft bottom sediments.

Comleyan [GEOL] Lower Cambrian geologic time.

Commelinaceae [BOT] A family of monocotyledonous plants in the order Commelinales characterized by differentiation of the leaves into a closed sheath and a well-defined, commonly somewhat succulent blade.

Commelinales [BOT] An order of monocotyledonous plants in the subclass Commelinidae marked by having differentiated sepals and petals but lacking nectaries and nectar.

Commelinidae [BOT] A subclass of flowering plants in the class Liliopsida.

commensal [ECOL] An organism living in a state of commensalism.

commensalism [ECOL] An interspecific, symbiotic relationship in which two different species are associated, wherein one is benefited and the other neither benefited nor harmed.

commercial lecithin *See* lecithin.

commissure [BIOL] A joint, seam, or closure line where two structures unite.

commissurotomy [MED] The surgical destruction of a commissure, usually the anterior commissure, particularly in the treatment of certain psychiatric disorders.

COMB

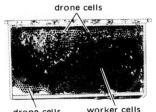

drone cells

drone cells   worker cells

Drone and worker bee cells in comb. *(Walter T. Kelley Co., Clarkson, Ky.)*

common bile duct [ANAT] The duct formed by the union of the hepatic and cystic ducts.

common carotid artery *See* carotid artery.

common cold [MED] A viral disease of humans most frequently caused by the rhinovirus and accompanied by inflammation of the mucous membranes of the nose, throat, and eyes.

common hepatic duct *See* hepatic duct.

common iliac artery *See* iliac artery.

communicable disease [MED] An infectious disease that can be transmitted from one individual to another either directly by contact or indirectly by fomites and vectors.

community [ECOL] Aggregation of organisms characterized by a distinctive combination of two or more ecologically related species; an example is a deciduous forest. Also known as ecological community.

community classification [ECOL] Arrangement of communities into classes with respect to their complexity and extent, their stage of ecological succession, or their primary production.

comose [BOT] Having a tuft of soft hairs.

companion cell [BOT] A specialized parenchyma cell occurring in close developmental and physiologic association with a sieve-tube member.

comparative embryology [EMBRYO] A branch of embryology that deals with the similarities and differences in the development of animals or plants of different orders.

comparative pathology [MED] Investigation and comparison of disease in various animals, including man, to arrive at resemblances and differences which may clarify disease as a phenomenon of nature.

compatibility [IMMUNOL] Ability of two bloods or other tissues to unite and function together. [PHARM] The capacity of two or more ingredients in a medicine to mix without chemical change or loss of therapeutic effectiveness.

compensation [PSYCH] Counterbalancing a weakness or failure in one area by stressing or substituting a strength or success in another area.

compensatory emphysema [MED] Simple, nonobstructive overdistension of lung segments or an entire lung in intrathoracic adaptation to collapse, destruction, or removal of portions of the lung or the opposite lung.

compensatory hypertrophy [MED] An increase in the size of an organ following injury or removal of the opposite paired organ or of part of the same organ.

competence [EMBRYO] The ability of a reacting system to respond to the inductive stimulus during early developmental stages.

competition [ECOL] The inter- or intraspecific interaction resulting when several individuals share an environmental necessity.

competitive enzyme inhibition [BIOCHEM] Prevention of an enzymatic process resulting from the reversible interaction of an inhibitor with a free enzyme.

compital [BOT] **1.** Of the vein of a leaf, intersecting at a wide angle. **2.** Of a fern, bearing sori at the intersection of two veins.

complement [IMMUNOL] A heat-sensitive, complex system in fresh human and other sera which, in combination with antibodies, is important in the host defense mechanism against invading microorganisms.

complemental air [PHYSIO] The amount of air that can still be inhaled after a normal inspiration.

complementary genes [GEN] Nonallelic genes that complement one another's expression in a trait.

COMPOUND LEAF

leaflet

petiole

Odd-pinnately compound leaf.

COMPOUND MICROSCOPE

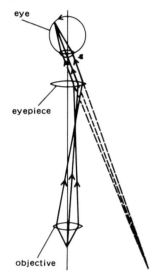

eye

eyepiece

objective

Compound microscope diagram.
(From F. A. Jenkins and H. E.
White, Fundamentals of Optics,
3d ed., McGraw-Hill, 1957)

COMPOUND
TUBULAR-ACINOUS GLAND

One of the distal secreting units
of a compound tubular-acinous
gland.

complementation [GEN] The action of complementary genes.

complement fixation [IMMUNOL] The binding of complement to an antigen-antibody complex so that the complement is unavailable for subsequent reaction.

complement-fixation test [IMMUNOL] A diagnostic test to determine the presence of antigen or antibody in the blood by adding complement to the test system; used especially in diagnosing syphilis.

complete blood count [PATH] Differential and absolute determinations of the numbers of each type of blood cell in a sample and, by extrapolation, in the general circulation.

complete flower [BOT] A flower having all four floral parts, that is, having sepals, petals, stamens, and carpels.

complete leaf [BOT] A dicotyledon leaf consisting of three parts: blade, petiole, and a pair of stipules.

complex [MED] See syndrome. [MINERAL] Composed of many ingredients. [PSYCH] A group of associated ideas with strong emotional tones, which have been transferred from the conscious mind into the unconscious and which influence the personality.

complicate [INV ZOO] Folded lengthwise several times, as applied to insect wings.

Compositae [BOT] A family of dicotyledonous plants in the order Campanulales distinguished by a large calyx.

composite [BOT] **1.** Consisting of diverse, closely packed elements. **2.** A member of the Compositae.

composite nerve [PHYSIO] A nerve containing both sensory and motor fibers.

compost [MATER] A mixture of decaying organic matter used to fertilize and condition the soil.

compound [BIOL] Composed of two or more identical elements, as certain flowers, or the eyes of certain invertebrates. [CHEM] A substance whose molecules consist of unlike atoms and whose constituents cannot be separated by physical means. Also known as chemical compound.

compound acinous gland [ANAT] A structure with spherical secreting units connected to many ducts that empty into a common duct.

compound eye [INV ZOO] An eye typical of crustaceans, insects, centipedes, and horseshoe crabs, constructed of many functionally independent photoreceptor units (ommatidia) separated by pigment cells.

compound gland [ANAT] A secretory structure with many ducts.

compound leaf [BOT] A type of leaf with the blade divided into two or more separate leaflets such as the rose.

compound microscope [OPTICS] A microscope which utilizes two lenses or lens systems; one lens forms an enlarged image of the object, and the second magnifies the image formed by the first.

compound ovary [BOT] An ovary that is composed of more than one carpel.

compound sugar See oligosaccharide.

compound tubular-acinous gland [ANAT] A structure in which the secreting units are simple tubes with acinous side chambers and all are connected to a common duct.

compound tubular gland [ANAT] A structure with branched ducts between the surface opening and the secreting portion.

compressed-air illness See caisson disease.

compression wood [BOT] Dense wood found at the base of some tree trunks and on the undersides of branches.

compulsion [PSYCH] An irresistible, impulsive act performed by an individual against his conscious will and usually arising from an obsession.

compulsive reaction [PSYCH] Behavior disorder in which a person is uncomfortable with conditions of ambiguity and uncertainty.

conarium *See* pineal body.

Concato's disease *See* polyserositis.

concave [SCI TECH] Having a curved form which bulges inward resembling the interior of a sphere or cylinder or a section of these bodies.

concentration [CHEM] In solutions, the mass, volume, or number of moles of solute present in proportion to the amount of solvent or total solution.

concentration-dilution test [PATH] A renal function test to measure the ability of the kidney to concentrate and dilute urine under stress; specific gravity for urine from a normal kidney fluctuates from 1.030 on a restricted fluid intake to 1.003 on a high water intake.

concentration gradient [CHEM] The graded difference in the concentration of a solute throughout the solvent phase.

concentric bundle [BOT] A vascular bundle in which xylem surrounds phloem, or phloem surrounds xylem.

conceptacle [BOT] A cavity which is shaped like a flask with a pore opening to the outside, contains reproductive structures, and is bound in a thallus such as in the brown algae.

conception [BIOL] Fertilization of an ovum by the sperm resulting in the formation of a viable zygote. [PSYCH] The mental process of forming ideas, especially abstract ideas.

conch [INV ZOO] The common name for several species of large, colorful gastropod mollusks of the family Strombidae; the shell is used to make cameos and porcelain.

concha [ANAT] **1.** A shell-like organ, as the hollow portion of the external ear. **2.** Any of the three nasal conchae, designated as superior, middle, or inferior, constituting a medial projection of thin bone from the lateral wall of the nasal cavity. Also known as turbinate bone. [BIOL] A shell.

Conchorhagae [INV ZOO] A suborder of benthonic wormlike animals in the class Kinorhyncha.

Conchostraca [INV ZOO] An order of mussellike crustaceans of moderate size belonging to the subclass Branchiopoda.

concordance [GEN] Similarity in appearance of members of a twin pair with respect to one or more specific traits.

concordance ratio [GEN] The percentage of a person's relatives showing the same trait as the person in question; often computed for identical and fraternal twins.

concrescence [BIOL] Convergence and fusion of parts originally separate, as the lips of the blastopore in embryogenesis.

concrete thinking [PSYCH] Mental processes characterized by literalness and the tendency to be bound to the most immediate and obvious sense impressions, as well as by a lack of generalization and abstraction.

concretionary [PATH] A compact mass of inorganic material formed in a body cavity or in tissue.

concurrent infection [MED] Two or more forms of an infection existing simultaneously.

concussion [MED] A state of shock following traumatic injury, especially cerebral trauma, in which there is temporary functional impairment without physical evidence of damage to impaired tissues.

condenser [OPTICS] A system of lenses or mirrors in an optical projection system, which gathers as much of the light from the source as possible and directs it through the projection lens.

conditional distribution [STAT] If W and Z are random variables with discrete values $w_1, w_2, \ldots,$ and $z_1, z_2, \ldots,$ the conditional distribution of W given $Z = z$ is the distribution

CONCH

Shell of hawk-wing conch.

CONCHOSTRACA

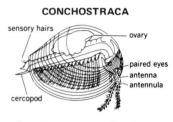

Limnadia lenticularis female, lateral aspect.

CONDENSER

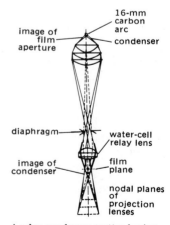

A relay condenser system having a water cell incorporated in second stage.

which assigns to w_i, $i = 1,2, \ldots$, the conditional probability of $W = w_i$ given $Z = z$.

conditional frequency [STAT] If r and s are possible outcomes of an experiment which is performed n times, the conditional frequency of s given that r has occurred is the ratio of the number of times both r and s have occurred to the number of times r has occurred.

conditional probability [STAT] The probability that a second event will be B if the first event is A, expressed as $P(B/A)$.

conditioned reflex [PSYCH] Response of an organism to a stimulus which was inadequate to elicit the response until paired for one or more times with an adequate stimulus.

conditioned stimulus [PSYCH] A stimulus that becomes effective after pairing with an unconditioned stimulus, evoking the conditioned response.

condor [VERT ZOO] *Vultur gryphus*. A large American vulture having a bare head and neck, dull black plumage, and a white neck ruff.

conduction [BOT] Transference of soluble substances from one region of a plant to another. [PHYSIO] Transmission of an impulse along a nerve.

conduction deafness [MED] Deafness due to abnormal conduction of energy to the cochlea.

conductivity [PHYSIO] Ability or capacity to transmit an impulse.

conductor [INV ZOO] In spiders, a projection at the base of the embolus. [PHYSIO] A structure that can transmit an impulse.

conduplicate [BOT] Folded lengthwise and in half with the upper faces together, applied to leaves and petals in the bud.

Condylarthra [PALEON] A mammalian order of extinct, primitive, hoofed herbivores with five-toed plantigrade to semidigitigrade feet.

condyle [ANAT] A rounded bone prominence that functions in articulation. [BOT] The antheridium of certain stoneworts. [INV ZOO] A rounded, articular process on arthropod appendages.

condyloid articulation [ANAT] A joint, such as the wrist, formed by an ovoid surface that fits into an elliptical cavity, permitting all movement except rotation.

condyloma acuminata [MED] A venereal disease characterized by wartlike growths on the genital organs; thought to be of viral origin.

cone [BOT] Either the ovulate or staminate strobilus of a gymnosperm. [HISTOL] A photoceptor of the vertebrate retina that responds differentially to light across the visible spectrum, providing color vision and visual acuity in bright light.

Conewangoan [GEOL] Upper Upper Devonian geologic time.

confabulation [PSYCH] The use of plausible guesses to fill in gaps in memory; characteristic of chronic brain syndromes.

confidence coefficient [STAT] The probability associated with a confidence interval; that is, the probability that the interval contains a given parameter or characteristic. Also known as confidence level.

confidence interval [STAT] An interval which has a specified probability of containing a given parameter or characteristic.

confidence limit [STAT] One of the end points of a confidence interval.

confluence [MED] A uniting, as of neighboring lesions such as vesicles and pustules. [SCI TECH] A flowing together.

conformation [PHYS CHEM] The spatial arrangement of the atoms in a molecule; mainly employed when a given molecule has two or more stable arrangements with the same set of chemical bonds.

congenital [MED] Dating from or existing before birth.

congenital agammaglobulinemia [MED] A congenital deficiency (in serum) of immunoglobulins, characterized clinically by increased susceptibility to bacterial infections; may be a sex-linked recessive tract, affecting male infants, or sporadic, affecting both sexes.

congenital anomaly [MED] A structural or functional abnormality of the human body that develops before birth. Also known as birth defect.

congenital pathology [MED] The study of diseases and defects existing at birth.

congestin [BIOCHEM] A toxin produced by certain sea anemones.

congestion [MED] An abnormal accumulation of fluid, usually blood, but occasionally bile or mucus, within the vessels of an organ or part.

congestive heart failure [MED] A state in which circulatory congestion exists as a result of heart failure.

conglutination [IMMUNOL] The completion of an agglutinating system, or the enhancement of an incomplete one, by the addition of certain substances. [MED] Abnormal union of two contiguous surfaces or bodies.

conglutinin [IMMUNOL] A heat-stable substance in bovine and other serums that aids or causes agglomeration or lysis of certain sensitized cells or particles.

Coniacian [GEOL] Lower Senonian geologic time.

Coniconchia [PALEON] A class name proposed for certain extinct organisms thought to have been mollusks; distinguished by a calcareous univalve shell that is open at one end and by lack of a siphon.

Conidae [INV ZOO] A family of marine gastropod mollusks in the order Neogastropoda containing the poisonous cone shells.

conidiocarp [MYCOL] A group of enclosed conidiophores.

conidiole [MYCOL] A small conidium.

conidiophore [MYCOL] A specialized aerial hypha that produces conidia in certain ascomycetes and imperfect fungi.

conidiospore *See* conidium.

conidium [MYCOL] A unicellular, asexual reproductive spore produced externally upon a conidiophore. Also known as conidiospore.

conifer [BOT] The common name for plants of the order Pinales.

Coniferales [BOT] The equivalent name for Pinales.

Coniferophyta [BOT] The equivalent name for Pinicae.

coniferous [BOT] Cone-bearing.

coniferous forest [ECOL] An area of wooded land predominated by conifers.

conjoint tendon [ANAT] The common tendon of the transverse and internal oblique muscles of the abdomen.

Conjugales [BOT] An order of fresh-water green algae in the class Chlorophyceae distinguished by the lack of flagellated cells, and conjugation being the method of sexual reproduction.

conjugase [BIOCHEM] Any of a group of enzymes which catalyze the breakdown of pteroylglutamic acid.

conjugate acid-base pair [CHEM] An acid and a base related by the ability of the acid to generate the base by loss of a proton.

conjugate division [MYCOL] Division of dikaryotic cells in certain fungi in which the two haploid nuclei divide independently, each daughter cell receiving one product of each nuclear division.

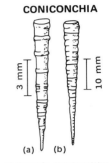

CONICONCHIA

(a) (b)

Early Paleozoic *Tentaculites.* *(a) T. gyracanthus* (Eaton), Upper Silurian. *(b) T. scalariforms* Hall, Middle Devonian. *(Adapted from R. C. Moore, C. G. Lalicker, and A. G. Fischer, Invertebrate Fossils, McGraw-Hill, 1952)*

CONIDIUM

|— 10 μ —|

Two conidia.

conjugated protein [BIOCHEM] A protein combined with a nonprotein group, other than a salt or a simple protein.

conjugation [BOT] Sexual reproduction by fusion of two protoplasts in certain thallophytes to form a zygote. [INV ZOO] Sexual reproduction by temporary union of cells with exchange of nuclear material between two individuals, principally ciliate protozoans. [MICROBIO] Reproduction among colon bacilli by temporary union of cells and transfer of chromosome material.

conjunctiva [ANAT] The mucous membrane covering the eyeball and lining the eyelids.

conjunctivitis [MED] Inflammation of the conjunctiva.

connate [SCI TECH] Born, originated, or produced in a united or fused condition.

connate leaf [BOT] A leaf shaped as though the bases of two opposite leaves had fused around the stem.

CONNATE LEAF

Shape of a connate leaf.

connective [ANAT] A band of nerve tissue connecting two ganglia. [BOT] Tissue between two lobes of an anther. [MYCOL] The structure and the area between conidia.

connective tissue [HISTOL] A primary tissue, distinguished by an abundance of fibrillar and nonfibrillar extracellular components.

connexivum [INV ZOO] In some Hemiptera, the flattened lateral abdominal border.

connivent [BIOL] Converging so as to meet, but not fused into a single part.

Conoclypidae [PALEON] A family of Cretaceous and Eocene exocyclic Euechinoidea in the order Holectypoida having developed aboral petals, internal partitions, and a high test.

Conocyeminae [PALEON] A subfamily of Mesozoan parasites in the family Dicyemidae.

conodont [PALEON] A minute, toothlike microfossil, composed of translucent amber-brown, fibrous or lamellar calcium phosphate; taxonomic identity is controversial.

Conodontiformes [PALEON] A suborder of conodonts from the Ordovician to the Triassic having a lamellar internal structure.

Conodontophoridia [PALEON] The ordinal name for the conodonts.

conoid [SCI TECH] Shaped somewhat like a cone, but not quite conical.

Conopidae [INV ZOO] The wasp flies, a family of dipteran insects in the suborder Cyclorrhapha.

conopodium [BOT] A cone-shaped flower receptacle.

conscience [PSYCH] The moral, self-critical part of oneself wherein have developed, and reside, standards of behavior and performance and value judgments.

consciousness [PSYCH] State of being aware of one's own existence, of one's mental states, and of the impressions made upon one's senses.

conscutum [INV ZOO] In some ticks, the dorsal shield formed by fusion of the scutum and alloscutum.

consensus [SCI TECH] A method of checking or confirming the correctness of an observation or report, based on agreement between different observers.

conservation [ECOL] Those measures concerned with the preservation, restoration, beneficiation, maximization, re-utilization, substitution, allocation, and integration of natural resources.

conservation of energy [PHYS] The principle that energy cannot be created or destroyed, although it can be changed from one form to another; no violation of this principle has been found. Also known as energy conservation.

consortism *See* symbiosis.

constant [SCI TECH] A value that does not change during a particular process.

Constellariidae [PALEON] A family of extinct, marine bryozoans in the order Cystoporata.

constipation [MED] The passage of hard, dry stools.

constitutional anthropology [ANTHRO] A branch of physical anthropology concerned with body composition and constitution of an individual.

constitutive enzyme [BIOCHEM] A type of enzyme always produced by the cell regardless of environmental conditions.

constriction [SCI TECH] Narrowing of a channel or cylindrical member.

constriction disease [PL PATH] A fungus disease of peach trees caused by a species of *Phomopsis* and marked by death of peripheral structures.

constrictive pericarditis [MED] Inflammation and fibrosis of the pericardium resulting in constriction of the heart and restriction of contraction and blood flow. Also known as Pick's disease.

constrictor [PHYSIO] Any muscle that functions to contract or tighten a part of the body.

constructional apraxia *See* optic apraxia.

consumer [ECOL] A nutritional grouping in the food chain of an ecosystem, composed of heterotrophic organisms, chiefly animals, which ingest other organisms or particulate organic matter.

consummatory behavior [PSYCH] Actions that fulfill a motive and cause appetitive behavior to end.

consumption *See* tuberculosis.

contabescence [BOT] Casting off or atrophy of the stamens.

contact dermatitis [MED] An acute or chronic inflammation of the skin resulting from irritation by or sensitization to some substance coming in contact with the skin.

contact inhibition [CYTOL] Cessation of cell division when cultured cells are in physical contact with each other.

contagion [MED] **1.** The process whereby disease spreads from one person to another, by direct or indirect contact. **2.** The bacterium or virus which transmits disease.

contagious abortion [VET MED] Brucellosis in cattle caused by *Brucella abortus* and inducing abortion. Also known as Bang's disease; infectious abortion.

contagious disease [MED] An infectious disease communicable by contact with one suffering from it, with his bodily discharge, or with an object touched by him.

contagious distribution [STAT] A probability distribution which is dependent on a parameter that itself has a probability distribution.

contamination [MICROBIO] Process or act of soiling with bacteria. [PSYCH] The fusion of words, resulting in a new word. [SCI TECH] Something that contaminates.

context [MYCOL] In certain fungi, tissue formed between the hymenium and the true mycelium.

continental drift [GEOL] The concept of continent formation by the fragmentation and movement of land masses on the surface of the earth.

continuous distribution [STAT] Distribution of a continuous population, which is a class of pairs such that the second member of each pair is a value, and the first member of the pair is a proportion density for that value.

contorted [BOT] Twisted; applied to proximate leaves whose margins overlap.

contortuplicate [BOT] Referring to a bud having contorted and plicate leaves.

contour feather [VERT ZOO] Any of the large flight feathers or

CONTOUR FEATHER

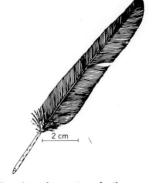

2 cm

Drawing of a contour feather. (From J. C. Welty, *The Life of Birds*, Saunders, 1962)

long tail feathers of a bird. Also known as penna; vane feather.

contour plowing [AGR] Cultivation of land along lines connecting points of equal elevation, to prevent water erosion. Also known as terracing.

contraception [MED] Prevention of impregnation.

contraceptive [MED] Any mechanical device or chemical agent used to prevent conception.

contracted pelvis [MED] A pelvis having one or more major diameters reduced in size, interfering with parturition.

contractile [BIOL] Displaying contraction; having the property of contracting.

contractile vacuole [CYTOL] A tiny, intracellular, membranous bladder that functions in maintaining intra- and extracellular osmotic pressures in equilibrium, as well as excretion of water, such as occurs in protozoans.

contraction [PHYSIO] Shortening of the fibers of muscle tissue.

contracture [ARCH] Narrowing of a section of a column. [MED] **1.** Shortening, as of muscle or scar tissue, producing distortion or deformity or abnormal limitation of movement of a joint. **2.** Retarded relaxation of muscle, as when it is injected with veratrine.

contraindication [MED] A symptom, indication, or condition in which a remedy or a method of treatment is inadvisable or improper.

contralateral [PHYSIO] Opposite; acting in unison with a similar part on the opposite side of the body.

control [STAT] **1.** A test made to determine the extent of error in experimental observations or measurements. **2.** A procedure carried out to give a standard of comparison in an experiment. **3.** Observation made on subjects which have not undergone treatment, to use in comparison with observations made on subjects which have undergone treatment.

contusion [MED] A subcutaneous bruise caused by an injury in which the skin is not broken.

Conularida [PALEON] A small group of extinct invertebrates showing a narrow, four-sided, pyramidal-shaped test.

Conulata [INV ZOO] A subclass of free-living coelenterates in the class Scyphozoa; individuals are described as tetraramous cones to elongate pyramids having tentacles on the oral margin.

Conulidae [PALEON] A family of Cretaceous exocyclic Euechinoidea characterized by a flattened oral surface.

conulus [INV ZOO] A conical projection on the surface of certain Porifera, formed by the principal skeletal elements.

conus [ANAT] Any conical structure, such as the conus arteriosus.

conus arteriosus [EMBRYO] The cone-shaped projection from which the pulmonary artery arises on the right ventricle of the heart in man and mammals.

convalescence [MED] The period and process of recovery after an illness or injury.

convalescent carrier [MED] A person who harbors an infectious agent after recovery from a clinical attack of a disease.

convalescent serum [IMMUNOL] The serum of the blood of one or more patients recovering from an infectious disease; used for prophylaxis of the particular infection.

convergence [ANAT] The coming together of a group of afferent nerves upon a motoneuron of the ventral horn of the spinal cord. [ANTHRO] Independent development of similarities between unrelated cultures. [EVOL] Development of similarities between animals or plants of different groups resulting from adaptation to similar habitats. [PHYSIO]

CONULARIDA

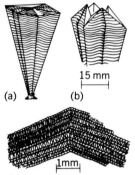

15 mm

(a)   (b)

1mm

(c)

Conularida restorations. *(a)* Attachment disk. *(b)* Distal part of same individual with triangular flaps raised. *(c)* Part of exterior showing the ornamentation. *(Adapted from R. C. Moore, C. G. Lalicker, and A. G. Fischer, Invertebrate Fossils, McGraw-Hill, 1952)*

Turning the eyes inward as an object is brought closer to the eyes.

conversion reaction [PSYCH] The form of hysterical neurosis in which the impulse causing anxiety is converted into functional symptoms of the special senses or voluntary nervous system.

conversion table [SCI TECH] A list of equivalent values for converting from one set of units to another.

conversive heating [MED] The conversion of some form of energy, especially radio waves, into heat for use in thermotherapy.

convex [SCI TECH] Having a curved form which bulges outward, resembling the exterior of a sphere or cylinder or a section of these bodies.

convolute [BIOL] Twisted or rolled together, specifically referring to leaves, mollusk shells, and renal tubules.

convolution [ANAT] A fold, twist, or coil of any organ, especially any one of the prominent convex parts of the brain, separated from each other by depressions or sulci.

Convolvulaceae [BOT] A large family of dicotyledonous plants in the order Polemoniales characterized by internal phloem, the presence of chlorophyll, two ovules per carpel, and plicate cotyledons.

convulsion [MED] An episode of involuntary, generally violent muscular contractions, rhythmically alternated with periods of relaxation; associated with many systematic and neurological diseases.

Coombs serum [IMMUNOL] An immune serum containing antiglobulin that is used in testing for Rh and other sensitizations.

Copeognatha [INV ZOO] An equivalent name for Psocoptera.

Copepoda [INV ZOO] An order of Crustacea commonly included in the Entomostraca; contains free-living, parasitic, and symbiotic forms.

Copodontidae [PALEON] An obscure family of Paleozoic fishes in the order Bradyodonti.

copper [CHEM] A chemical element, symbol Cu, atomic number 29, atomic weight 63.546.

copper blight [PL PATH] A leaf spot disease of tea caused by the fungus *Guignardia camelliae.*

copperhead [VERT ZOO] *Agkistrodon contortrix.* A pit viper of the eastern United States; grows to about 3 feet (90 centimeters) in length and is distinguished by its coppery-brown skin with dark transverse blotches.

copper spot [PL PATH] A fungus disease of lawn grasses caused by *Gloeocercospora sorghi* and marked by coppery-red areas.

coppice [ECOL] A growth of small trees that are repeatedly cut down at short intervals; the new shoots are produced by the old stumps.

coprolite [PALEON] Fossilized feces.

coprophagy [ZOO] Feeding on dung or excrement.

coprophilous [ECOL] Living in dung.

coprophyte [ECOL] A plant that grows on or in dung.

copulation [ZOO] The sexual union of two individuals, resulting in insemination or deposition of the male gametes in close proximity to the female gametes.

copulatory bursa [INV ZOO] **1.** A sac that receives the sperm during copulation in certain insects. **2.** The caudal expansion of certain male nematodes that functions as a clasper during copulation.

copulatory organ [ANAT] An organ employed by certain male animals for insemination.

copulatory spicule *See* spiculum.

copy choice [GEN] An alternative, now disregarded theory

COPEPODA

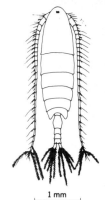

Calanus finmarchicus, a calcanoid free-living copepod.

COPPER

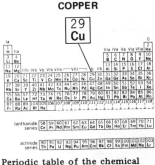

Periodic table of the chemical elements showing the position of copper.

COPULATORY ORGAN

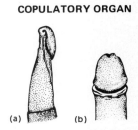

Glans penis of *(a)* bull, *(b)* man.

on breakage and reunion for genetic recombination; it pictures recombination as associated with replication, and assumes that during replication of paired chromosomes, synthesis of a daughter strand on one parental chromosome switches to the corresponding position on the other parental chromosome.

Coraciidae [VERT ZOO] The rollers, a family of Old World birds in the order Coraciiformes.

Coraciiformes [VERT ZOO] An order of predominantly tropical and frequently brightly colored birds.

coracoid [ANAT] One of the paired bones on the posterior-ventral aspect of the pectoral girdle in vertebrates.

coracoid ligament [ANAT] The transverse ligament of the scapula which crosses over the suprascapular notch.

coracoid process [ANAT] The beak-shaped process of the scapula.

coral [INV ZOO] The skeleton of certain solitary and colonial anthozoan coelenterates; composed chiefly of calcium carbonate.

Corallanidae [INV ZOO] A family of sometimes parasitic, but often free-living, isopod crustaceans in the suborder Flabellifera.

Corallidae [INV ZOO] A family of dimorphic coelenterates in the order Gorgonacea.

Corallimorpharia [INV ZOO] An order of solitary sea anemones in the subclass Zoantharia resembling coral in many aspects.

Corallinaceae [BOT] A family of red algae, division Rhodophyta, having compact tissue with lime deposits within and between the cell walls.

coralline [INV ZOO] Any animal that resembles coral, such as a bryozoan or hydroid.

coralline algae [BOT] Red algae belonging to the family Corallinaceae.

corallite [INV ZOO] Skeleton of an individual polyp in a compound coral.

coralloid [BIOL] Resembling coral, or branching like certain coral.

corallum [INV ZOO] Skeleton of a compound coral.

coral reef [GEOL] A ridge or mass of limestone built up of detrital material deposited around a framework of skeletal remains of mollusks, colonial coral, and massive calcareous algae.

Corbiculidae [INV ZOO] A family of fresh-water bivalve mollusks in the subclass Eulamellibranchia; an important food in the Orient.

corbiculum [INV ZOO] **1.** In insects, a fringe of hair on the tibia. **2.** The pollen-collecting apparatus of the leg of a bee.

Cordaitaceae [PALEOBOT] A family of fossil plants belonging to the Cordaitales.

Cordaitales [PALEOBOT] An extensive natural grouping of forest trees of the late Paleozoic.

cordate [BOT] Heart-shaped; generally refers to a leaf base.

cordon [BOT] A plant trained to grow flat against a vertical structure, in a single horizontal shoot or two opposed horizontal shoots.

Cordulegasteridae [INV ZOO] A family of anisopteran insects in the order Odonata.

Coreidae [INV ZOO] The squash bugs and leaf-footed bugs, a family of hemipteran insects belonging to the superfamily Coreoidea.

coremata [INV ZOO] In moths, accessory copulatory organs consisting of paired sacs bearing hairs, and found on the membrane between the seventh and eighth abdominal segments.

coremium [MYCOL] A small bundle of conidiophores in certain imperfect fungi.

coriaceous [BIOL] Leathery, applied to leaves and certain insects.

coriander [BOT] *Coriandrum sativum.* A strong-scented perennial herb in the order Umbellales; the dried fruit is used as a flavoring.

corium [ANAT] *See* dermis. [INV ZOO] Middle portion of the forewing of hemipteran insects.

Corixidae [INV ZOO] The water boatmen, the single family of the hemipteran superfamily Corixoidea.

Corixoidea [INV ZOO] A superfamily of hemipteran insects belonging to the subdivision Hydrocorisae that lack ocelli.

cork [BOT] A protective layer of cells that replaces the epidermis in older plant stems.

corm [BOT] A short, erect, fleshy underground stem, usually broader than high and covered with membranous scales.

cormatose [BOT] Having or producing a corm.

cormidium [INV ZOO] The assemblage of individuals dangling in clusters from the main stem of pelagic siphonophores.

cormous [BOT] Corm-producing.

corn [BOT] *Zea mays.* A grain crop of the grass order Cyperales grown for its edible seeds (technically fruits).

Cornaceae [BOT] A family of dicotyledonous plants in the order Cornales characterized by perfect or unisexual flowers, a single ovule in each locule, as many stamens as petals, and opposite leaves.

Cornales [BOT] An order of dicotyledonous plants in the subclass Rosidae marked by a woody habit, simple leaves, well-developed endosperm, and fleshy fruits.

cornea [ANAT] The transparent anterior portion of the outer coat of the vertebrate eye covering the iris and the pupil. [INV ZOO] The outer transparent portion of each ommatidium of a compound eye.

cornicle [INV ZOO] Either of two protruding horn-shaped dorsal tubes in aphids which secrete a waxy fluid.

corniculate [BIOL] Having small horns or hornlike processes.

corniculate cartilage [ANAT] The cartilaginous nodule on the tip of the arytenoid cartilage.

cornification [PHYSIO] Conversion of stratified squamous epithelial cells into a horny layer and into derivatives such as nails, hair, and feathers.

corn smut [PL PATH] A fungus disease of corn caused by *Ustilago maydis.*

corn sugar *See* dextrose.

cornu [ANAT] A horn or hornlike structure.

corolla [BOT] A collective term for the petals of a flower.

corollate [BOT] Having a corolla.

corolline [BOT] Relating to, resembling, or being borne on a corolla.

corona [BOT] **1.** An appendage or series of fused appendages between the corolla and stamens of some flowers. **2.** The region where stem and root of a seed plant merge. Also known as crown. [INV ZOO] **1.** The anterior ring of cilia in rotifers. **2.** A sea urchin test. **3.** The calyx and arms of a crinoid.

coronal suture [ANAT] The union of the frontal with the parietal bones transversely across the vertex of the skull.

corona radiata [HISTOL] The layer of cells immediately surrounding a mammalian ovum.

coronary artery [ANAT] Either of two arteries arising in the aortic sinuses that supply the heart tissue with blood.

coronary bone [VERT ZOO] The small pastern bone of a horse.

CORM

A corm, a specialized stem.

CORONATAE

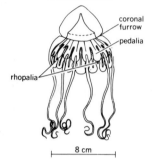

coronal furrow

pedalia

rhopalia

8 cm

Periphylla. (From L. H. Hyman, The Invertebrates, vol. 1, McGraw-Hill, 1940)

COROPHIIDAE

Lateral view of female *Corophium crassicorne* **Bruzelius, a tube-builder.** *(From G. O. Sars, An Account of the Crustacea of Norway, vol. 1, 1895)*

CORPUS CALLOSUM

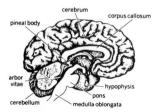

cerebrum

corpus callosum

pineal body

arbor vitae

hypophysis

pons

cerebellum

medulla oblongata

A sagittal section through the human brain showing corpus callosum. *(From H. E. Walter and L. P. Sayles, Biology of the Vertebrates, 3d ed., Macmillan, 1949)*

coronary disease [MED] Any condition that reduces the flow of blood to the heart muscles.

coronary failure [MED] Prolonged precordial pain or discomfort without conventional evidence of myocardial infarction; subendocardial ischemia caused by a disparity between coronary blood flow and myocardial needs; this condition is more commonly referred to as coronary artery insufficiency.

coronary occlusion [MED] Complete blockage of a coronary artery.

coronary sinus [ANAT] A venous sinus opening into the heart's right atrium which drains the cardiac veins.

coronary sulcus [ANAT] A groove in the external surface of the heart separating the atria from the ventricles, containing the trunks of the nutrient vessels of the heart.

coronary thrombosis [MED] Formation of a thrombus in a coronary artery.

coronary valve [ANAT] A semicircular fold of the endocardium of the right atrium at the orifice of the coronary sinus.

coronary vein [ANAT] **1.** Any of the blood vessels that bring blood from the heart and empty into the coronary sinus. **2.** A vein along the lesser curvature of the stomach.

Coronatae [INV ZOO] An order of the class Scyphozoa which includes mainly abyssal species having the exumbrella divided into two parts by a coronal furrow.

coronavirus [VIROL] A major group of animal viruses including avian infectious bronchitis virus and mouse hepatitis virus.

coronoid [BIOL] Shaped like a beak.

coronoid fossa [ANAT] A depression in the humerus into which the apex of the coronoid process of the ulna fits in extreme flexion of the forearm.

coronoid process [ANAT] **1.** A thin, flattened process projecting from the anterior portion of the upper border of the ramus of the mandible, and serving for the insertion of the temporal muscle. **2.** A triangular projection from the upper end of the ulna, forming the lower part of the radial notch.

coronule [INV ZOO] A peripheral ring of spines on some diatom shells.

Corophiidae [INV ZOO] A family of amphipod crustaceans in the suborder Gammaridea.

corpora quadrigemina [ANAT] The inferior and superior colliculi collectively. Also known as quadrigeminal body.

cor pulmonale [MED] Hypertrophy and dilation of the right ventricle secondary to obstruction to the pulmonary blood flow and consequent pulmonary hypertension.

corpus albicans [HISTOL] The white fibrous scar in an ovary; produced by the involution of the corpus luteum.

corpus allatum [INV ZOO] An endocrine structure near the brain of immature arthropods that secretes a juvenile hormone, neotenin.

corpus callosum [ANAT] A band of nerve tissue connecting the cerebral hemispheres in man and higher mammals.

corpus cavernosum [ANAT] The cylinder of erectile tissue forming the clitoris in the female and the penis in the male.

corpus cerebelli [ANAT] The central lobe or zone of the cerebellum; regulates reflex tonus of postural muscles in mammals.

corpus fibrosum [HISTOL] The scarlike structure representing the end result of an involuted ovarian follicle.

corpuscle [ANAT] **1.** A small, rounded body. **2.** An encapsulated sensory-nerve end organ.

corpus luteum [HISTOL] The yellow endocrine body formed in the ovary at the site of a ruptured Graafian follicle.

corpus striatum [ANAT] The caudate and lenticular nuclei, together with the internal capsule which separates them.

corrective therapy [MED] A program, and the techniques, designed to improve or maintain the health of a patient by improving neuromuscular activities and personal health habits and promoting relaxation by adjustment to stresses.

correlation coefficient [STAT] A measurement, which is unchanged by both addition and multiplication of the random variable by positive constants, of the tendency of two random variables X and Y to vary together; it is given by the ratio of the covariance of X and Y to the square root of the product of the variance of X and the variance of Y.

corresponding points [PHYSIO] Any two retinal areas in the respective eyes so that the area in one eye has an identical direction in the opposite retina.

corridor [ECOL] A land bridge that allows free migration of fauna in both directions.

Corrigan's pulse [MED] A pulse characterized by a rapid, forceful ascent (water-hammer quality) and rapid downstroke or descent (collapsing quality); seen with aortic regurgitation and hyperkinetic circulatory states.

Corrodentia [INV ZOO] The equivalent name for Psocoptera.

cortex [ANAT] The outer portion of an organ or structure, such as of the brain and adrenal glands. [BOT] A primary tissue in roots and stems of vascular plants that extends inward from the epidermis to the phloem. [INV ZOO] The peripheral layer of certain protozoans.

corticoid *See* adrenal cortex hormone.

corticolous [ECOL] Living in or growing on bark.

corticosteroid [BIOCHEM] **1.** Any steroid hormone secreted by the adrenal cortex of vertebrates. **2.** Any steroid with properties of an adrenal cortex steroid.

corticosterone [BIOCHEM] $C_{21}H_{30}O_4$ A steroid hormone produced by the adrenal cortex of vertebrates that stimulates carbohydrate synthesis and protein breakdown and is antagonistic to the action of insulin.

corticotrophic [PHYSIO] Having an effect on the adrenal cortex.

corticotropin [BIOCHEM] A hormonal preparation having adrenocorticotropic activity, derived from the adenohypophysis of certain domesticated animals.

cortisol *See* hydrocortisone.

cortisone [BIOCHEM] $C_{21}H_{28}O_5$ A steroid hormone produced by the adrenal cortex of vertebrates that acts principally in carbohydrate metabolism.

Corvidae [VERT ZOO] A family of large birds in the order Passeriformes having stout, long beaks; includes the crows, jays, and magpies.

Corylophidae [INV ZOO] The equivalent name for Orthoperidae.

corymb [BOT] An inflorescence in which the flower stalks arise at different levels but reach the same height, resulting in a flat-topped cluster.

corymbose [BOT] Resembling or pertaining to a corymb.

Corynebacteriaceae [MICROBIO] A family of nonsporeforming, usually nonmotile rod-shaped bacteria in the order Eubacteriales including animal and plant parasites and pathogens.

corynebacteriophage [VIROL] Any bacteriophage able to infect *Corynebacterium diphtheriae*.

Corynebacterium [MICROBIO] A genus of gram-positive, slightly curved rod-shaped bacteria, sometimes containing granules, in the family Corynebacteriaceae.

Corynebacterium diphtheriae [MICROBIO] A facultatively aerobic, nonmotile species of bacteria that causes diphtheria in man. Also known as *Bacillus diphtheriae*; Klebs-Loeffler bacillus.

COSMOID SCALE

enamel
cosmine layer
spongy layer
lamellar bone

Cross section through a cosmoid scale. *(From A. S. Romer, The Vertebrate Body, Saunders, 1962)*

COSSURIDAE

anterior

Cossura, anterior and posterior ends in dorsal view.

Coryphaenidae [VERT ZOO] A family of pelagic fishes in the order Perciformes characterized by a blunt nose and deeply forked tail.

Coryphodontidae [PALEON] The single family of the Coryphodontoidea, an extinct superfamily of mammals.

Coryphodontoidea [PALEON] A superfamily of extinct mammals in the order Pantodonta.

coryza [MED] Inflammation of the mucous membranes of the nose, usually marked by sneezing and discharge of watery mucus.

cosmine [VERT ZOO] The outer layer of cosmoid and ganoid scales; composed of a dentinelike material.

Cosmocercidae [INV ZOO] A group of nematodes assigned to the suborder Oxyurina by some authorities and to the suborder Ascaridina by others.

cosmoid scale [VERT ZOO] A structure in the skin of primitive rhipidistians and dipnoans that is composed of enamel, a dentine layer (cosmine), and laminated bone.

cosmopolitan [ECOL] Having a worldwide distribution wherever the habitat is suitable, with reference to the geographical distribution of a taxon.

Cossidae [INV ZOO] The goat or carpenter moths, a family of heavy-bodied lepidopteran insects in the superfamily Cossoidea having the abdomen extending well beyond the hindwings.

Cossoidea [INV ZOO] A monofamilial superfamily of lepidopteran insects belonging to suborder Heteroneura.

Cossuridae [INV ZOO] A family of fringe worms belonging to the Sedentaria.

Cossyphodidae [INV ZOO] The lively ant guest beetles, a small family of coleopteran insects in the superfamily Tenebrionoidea.

costa [BIOL] A rib or riblike structure. [BOT] The midrib of a leaf. [INV ZOO] The anterior vein of an insect's wing.

Costaceae [BOT] A family of monocotyledonous plants in the order Zingiberales distinguished by having one functional stamen with two pollen sacs and spirally arranged leaves and bracts.

costal cartilage [ANAT] The cartilage occupying the interval between the ribs and the sternum or adjacent cartilages.

costal process [ANAT] An anterior or ventral projection on the lateral part of a cervical vertebra. [EMBRYO] An embryonic rib primordium, the ventrolateral outgrowth of the caudal, denser half of a sclerotome.

costate [BIOL] Having ribs or ridges.

Cotingidae [VERT ZOO] The cotingas, a family of neotropical suboscine birds in the order Passeriformes.

Cottidae [VERT ZOO] The sculpins, a family of perciform fishes in the suborder Cottoidei.

Cottiformes [VERT ZOO] An order set up in some classification schemes to include the Cottoidei.

Cottoidei [VERT ZOO] The mail-cheeked fishes, a suborder of the order Perciformes characterized by the expanded third infraorbital bone.

cotton [BOT] Any plant of the genus *Gossypium* in the order Malvales; cultivated for the fibers obtained from its encapsulated fruits or bolls.

cotton anthracnose [PL PATH] A fungus disease of cotton caused by *Glomerella gossypii* and characterized by reddish-brown to light-colored or necrotic spots.

cotton root rot [PL PATH] A fungus disease of cotton caused by *Phymatotrichum omnivorum* and marked by bronzing of the foliage followed by sudden wilting and death of the plant.

cotton rust [PL PATH] A fungus disease of cotton caused by

Puccinia stakmanii producing low, greenish-yellow or orange elevations on the undersurface of leaves.

cotton wilt [PL PATH] **1.** A fungus disease of cotton caused by *Fusarium vasinfectum* growing in the water-conducting vessels and characterized by wilt, yellowing, blighting, and death. **2.** A fungus blight of cotton caused by *Verticillium albo-atrum* and characterized by yellow mottling of the foliage.

cottonwood [BOT] Any of several poplar trees *(Populus)* having hairy, encapsulated fruit.

cottony rot [PL PATH] A fungus disease of many plants, especially citrus trees, marked by fluffy white growth caused by *Sclerotinia sclerotiorum,* in which there is stem wilt and rot.

cotyledon [BOT] The first leaf of the embryo of seed plants.

cotylocercous cercaria [INV ZOO] A digenetic trematode larva characterized by a sucker or adhesive gland on the tail.

Cotylosauria [PALEON] An order of primitive reptiles in the subclass Anapsida, including the stem reptiles, ancestors of all of the more advanced Reptilia.

cotype *See* syntype.

cougar *See* puma.

cough [MED] A sudden, violent expulsion of air after deep inspiration and closure of the glottis.

Couinae [VERT ZOO] The couas, a subfamily of Madagascan birds in the family Cuculidae.

Coulomb's law [ELEC] The law that the attraction or repulsion between two electric charges acts along the line between them, is proportional to the product of their magnitudes, and is inversely proportional to the square of the distance between them. Also known as law of electrostatic attraction.

coulometer [PHYS CHEM] An electrolytic cell for the precise measurement of electrical quantities or current intensity by quantitative determination of chemical substances produced or consumed.

Coulter counter [MICROBIO] An electronic device for counting the number of cells in a liquid culture.

counterstain [BIOL] A second stain applied to a biological specimen to color elements other than those demonstrated by the principal stain.

countertransference [PSYCH] The conscious or unconscious emotional reaction of the therapist to the patient, which may interfere with psychotherapy.

counting chamber [MICROBIO] An accurately dimensioned chamber in a microslide which can hold a specific volume of fluid and which is usually ruled into units to facilitate the counting under the microscope of cells, bacteria, or other structures in the fluid.

coupling *See* cis arrangement.

Couvinian [GEOL] Lower Middle Devonian geologic time.

covalence [CHEM] The number of covalent bonds which an atom can form.

covalent radius [PHYS CHEM] The effective radius of an atom in a covalent bond.

covariance [STAT] A measurement of the tendency of two random variables, X and Y, to vary together, given by the expected value of the variable $(X - \overline{X})(Y - \overline{Y})$, where $\overline{X}$ and $\overline{Y}$ are the expected values of the variables X and Y respectively.

cover crops [AGR] Crops, especially grasses, grown for the express purpose of preventing and protecting a bare soil surface.

covered smut [PL PATH] A seed-borne smut of certain grain crops caused by *Ustilago hordei* in barley and *U. avenae* in oats.

cover scales [BOT] Small, spirally arranged scales on the axis of a cone of conifers.

COTTONWOOD

Cottonwood poplar *(Populus deltoides)* showing leaf, leaf scar, terminal bud, and twig.

covert [ECOL] A refuge or shelter, such as a coppice, for game animals.

covey [VERT ZOO] **1.** A brood of birds. **2.** A small flock of birds of one kind, used typically of partridge and quail.

cow [AGR] A domestic bovine of any sex or age. [VERT ZOO] A mature female cattle of the genus *Bos*.

cowpea [BOT] *Vigna sinensis*. An annual legume in the order Rosales cultivated for its edible seeds. Also known as black-eye bean.

Cowper's gland *See* bulbourethral gland.

cowpox *See* vaccinia.

cowpox virus [VIROL] The causative agent of cowpox in cattle.

coxa [INV ZOO] The proximal or basal segment of the leg of insects and certain other arthropods which articulates with the body.

coxal cavity [INV ZOO] A cavity in which the coxa of an arthropod limb articulates.

coxal gland [INV ZOO] One of certain paired glands with ducts opening in the coxal region of arthropods.

coxite [INV ZOO] **1.** In insects, one of the paired lateral plates contiguous with the sternum. **2.** The stylus-bearing base of a limb in Thysanura.

coxitis [MED] Inflammation of the hip joint.

coxocerite [INV ZOO] In insects, the basal joint of an antenna.

coxopodite [INV ZOO] The basal joint of a crustacean limb.

coxosternum [INV ZOO] A plate formed by fusion of coxites with the sternum.

coxsackie disease [MED] A variety of syndromes resulting from a coxsackievirus infection.

coxsackievirus [VIROL] A large subgroup of the enteroviruses in the picornavirus group including various human pathogens.

coyote [VERT ZOO] *Canis latrans*. A small wolf native to western North America but found as far eastward as New York State. Also known as prairie wolf.

Cr *See* chromium.

crab [INV ZOO] **1.** The common name for a number of crustaceans in the order Decapoda having five pairs of walking legs, with the first pair modified as chelipeds. **2.** The common name for members of the Merostoma.

crabapple [BOT] Any of several trees of the genus *Malus*, order Rosales, cultivated for their small, edible pomes.

Cracidae [VERT ZOO] A family of New World tropical upland game birds in the order Galliformes; includes the chachalacas, guans, and curassows.

cracked stem [PL PATH] A boron-deficiency disease of celery characterized by brown mottling of leaves and brittleness and cracking of leaf stalks.

cradle cap [MED] Heavy, greasy crusts on the scalp of an infant; seborrheic dermatitis of infants.

Crambiinae [INV ZOO] The snout moths, a subfamily of lepidopteran insects in the family Pyralididae containing small marshland and grassland forms.

cramp [MED] **1.** Painful, involuntary contraction of a muscle, such as a leg or foot cramp that may occur in normal individuals at night or in swimming. **2.** Any cramplike pain, as of the intestine, or that accompanying dysmenorrhea. **3.** Spasm of certain muscles, which may be intermittent or constant, from excessive use.

cranberry [BOT] Any of several plants of the genus *Vaccinium*, especially *V. macrocarpon*, in the order Ericales, cultivated for its small, edible berries.

Cranchiidae [INV ZOO] A family of cephalopod mollusks in the subclass Dibranchia.

CRAB

Male fiddler crab *(Uca pugnax)*, with an enlarged claw.

crane [VERT ZOO] The common name for the long-legged wading birds composing the family Gruidae of the order Gruiformes.

Craniacea [INV ZOO] A family of inarticulate branchiopods in the suborder Craniidina.

cranial capacity [ANAT] The volume of the cranial cavity.

cranial flexure [EMBRYO] A flexure of the embryonic brain.

cranial fossa [ANAT] Any of the three depressions in the floor of the interior of the skull.

cranial index [ANTHRO] The ratio of maximum skull height to maximum skull breadth multiplied by 100.

cranial nerve [ANAT] Any of the paired nerves which arise in the brainstem of vertebrates and pass to peripheral structures through openings in the skull.

Craniata [VERT ZOO] A major subdivision of the phylum Chordata comprising the vertebrates, from cyclostomes to mammals, distinguished by a cranium.

craniectomy [MED] Surgical removal of strips or pieces of the cranial bones.

Craniidina [INV ZOO] A subdivision of inarticulate branchiopods in the order Acrotretida known to possess a pedicle; all forms are attached by cementation.

craniobuccal pouch [EMBRYO] A diverticulum from the buccal cavity in the embryo from which the anterior lobe of the hypophysis is developed. Also known as Rathke's pouch.

cranioclasis [MED] The operation of breaking the fetal head by means of a cranioclast.

cranioclast [MED] A heavy forceps for crushing the fetal head.

craniofacial index [ANTHRO] The ratio of the width of the cranium to the width of the face.

craniology [ANAT] The study of the skull.

craniometer [ANTHRO] An instrument used to measure a skull.

craniometry [ANTHRO] The science of measuring the skull, especially for determining characteristics of a particular race, sex, or somatotype.

cranioplasty [MED] Surgical correction of defects in the cranial bones, usually by implants of metal, plastic material, or bone.

cranioscopy [MED] Examination of the human skull.

cranium [ANAT] That portion of the skull enclosing the brain. Also known as braincase.

craspedon [INV ZOO] A coelenterate medusa stage possessing a velum.

craspedote [INV ZOO] Having a velum, used specifically for velate hydroid medusae.

crassinucellate [BOT] Having the nucellus composed of several layers of cells, at least at the micropylar end of the ovule.

crassula [BOT] A thickened bar on the middle layer between two bordered pits in the tracheids of conifer wood.

Crassulaceae [BOT] A family of dicotyledonous plants in the order Rosales notable for their succulent leaves and resistance to desiccation.

craw [ZOO] **1.** The crop of a bird or insect. **2.** The stomach of a lower animal.

crayfish [INV ZOO] The common name for a number of lobsterlike fresh-water decapod crustaceans in the section Astacura.

creatine [BIOCHEM] $C_4H_9O_2N_3$ α-Methylguanidine-acetic acid; a compound present in vertebrate muscle tissue, principally as phosphocreatine.

creatinine [BIOCHEM] $C_4H_7ON_3$ A compound present in urine, blood, and muscle that is formed from the dehydration of creatine.

CRANIACEA

Pedicle valve of *Crania*, a genus in Craniacea. *(From R. C. Moore, ed., Treatise on Invertebrate Paleontology, pt. H, Geological Society of America, Inc., and University of Kansas Press, 1965)*

creatinuria [MED] The occurrence of creatine in the urine.

Credé procedure [MED] Installation of silver nitrate drops into the eyes of a newborn infant to prevent ophthalmia neonatorum.

creeping disk [INV ZOO] The smooth and adhesive undersurface of the foot or body of a mollusk or of certain other invertebrates, on which the animal creeps.

creeping eruption [MED] A red line of eruption on the skin produced by larva burrowing in the dermis; characterizes the condition of larva migrans.

cremaster [ANAT] A thin muscle that lies along the length of the spermatic cord. [INV ZOO] A spine on the terminal abdominal segment of subterranean insect pupae.

cremocarp [BOT] A dry dehiscent fruit consisting of two indehiscent one-seeded mericarps which separate at maturity and remain pendant from the carpophore.

crena [BIOL] A notch, cleft, deep grove, scallop, or indentation.

crenate [BIOL] Having a scalloped margin; used specifically for foliar structures, shrunken erythrocytes, and shells of certain mollusks.

crenation [PHYSIO] A notched appearance of shrunken erythrocytes; seen when they are exposed to the air or to strong saline solutions.

Crenotrichaceae [MICROBIO] A family of bacteria in the order Chlamydobacteriales having trichomes that are differentiated at the base and tip and attached to a firm substrate.

crenulate [BIOL] Having a minutely crenate margin.

Creodonta [PALEON] A group formerly recognized as a suborder of the order Carnivora.

creosote bush [BOT] *Larrea divaricata*. A bronze-green, xerophytic shrub characteristic of all the American warm deserts.

crepis [INV ZOO] A spicule upon which silica is deposited to form a desma.

crepitation [MED] A noise produced by the rubbing of fractured ends of bones, by cracking joints, and by pressure upon tissues containing abnormal amounts of air, as in cellular emphysema.

crepuscular [ZOO] Active during the hours of twilight or preceding dawn.

cress [BOT] Any of several prostrate crucifers belonging to the order Capparales and grown for their flavorful leaves; includes watercress (*Nasturtium officinale*), garden cress (*Lepidium sativum*), and upland or spring cress (*Barbarea verna*).

crest [BIOL] A ridge of bone, flesh or other tissue. [VERT ZOO] A tuft of feathers on the head of a bird.

Cretaceous [GEOL] The latest system of rocks or period of the Mesozoic Era, between 136 and 65 million years ago.

cretin [MED] An individual afflicted with cretinism.

cretinism [MED] A type of dwarfism caused by hypothyroidism and associated with generalized body changes, including mental deficiency.

crib death [MED] Sudden death of a sleeping infant without apparent cause. Also known as sudden death syndrome.

cribellum [INV ZOO] 1. A small accessory spinning organ located in front of the ordinary spinning organ in certain spiders. 2. A chitinous plate perforated with the openings of certain gland ducts in insects.

Cribrariaceae [BOT] A family of true slime molds in the order Liceales.

cribriform [BIOL] Perforated, like a sieve.

cribriform fascia [ANAT] The sievelike covering of the fossa ovalis of the thigh.

cribriform organ [INV ZOO] A folded membrane bearing

CRETACEOUS

PRECAMBRIAN		
CAMBRIAN		
ORDOVICIAN		
SILURIAN		
DEVONIAN	PALEOZOIC	
Mississippian	CARBONIFEROUS	
Pennsylvanian		
PERMIAN		
TRIASSIC		
JURASSIC	MESOZOIC	
CRETACEOUS		
TERTIARY	CENOZOIC	
QUATERNARY		

Chart showing relationship of Cretaceous to other geologic periods.

papillae and located at interradial angles of certain starfishes.

cribriform plate [ANAT] **1.** The horizontal plate of the ethmoid bone, part of the floor of the anterior cranial fossa. **2.** The bone lining a dental alveolus.

Cricetidae [VERT ZOO] A family of the order Rodentia including hamsters, voles, and some mice.

Cricetinae [VERT ZOO] A subfamily of mice in the family Cricetidae.

cricket [INV ZOO] **1.** The common name for members of the insect family Gryllidae. **2.** The common name for any of several related species of orthopteran insects in the families Tettigoniidae, Gryllotalpidae, and Tridactylidae.

cricoid [ANAT] The signet-ring-shaped cartilage forming the base of the larynx in man and most other mammals.

criminal abortion [MED] Illegal interruption of pregnancy.

crinion-menton [ANTHRO] The measurement from the hairline in the center part of the forehead to the midpoint of the lower edge of the chin.

Crinoidea [INV ZOO] A class of radially symmetrical crinozoans in which the adult body is flower-shaped and is either carried on an anchored stem or is free-living.

Crinozoa [INV ZOO] A subphylum of the Echinodermata comprising radially symmetrical forms that show a partly meridional pattern of growth.

crisis [MED] A turning point in the course of a disease. [PSYCH] The psychological events associated with a specific stage of life, as an identity crisis or developmental crisis.

crissum [VERT ZOO] **1.** The region surrounding the cloacal opening in birds. **2.** The vent feathers covering the circumcloacal region.

crista [BIOL] A ridge or crest.

cristate [BIOL] Having a crista.

critical absorption wavelength [SPECT] The wavelength, characteristic of a given electron energy level in an atom of a specified element, at which an absorption discontinuity occurs.

critical point [PHYS CHEM] **1.** The temperature and pressure at which two phases of a substance in equilibrium with each other become identical, forming one phase. **2.** The temperature and pressure at which two ordinarily partially miscible liquids are consolute.

critical ratio [STAT] The ratio of a particular deviation from the mean value to the standard deviation.

critical solution temperature [PHYS CHEM] The temperature at which a mixture of two liquids, immiscible at ordinary temperatures, ceases to separate into two phases.

critical temperature [PHYS CHEM] The temperature of the liquid-vapor critical point, that is, the temperature above which the substance has no liquid-vapor transition.

critical volume [PHYS] The volume occupied by one mole of a substance at the liquid-vapor critical point, that is, at the critical temperature and pressure.

crochet [INV ZOO] One of the hooks on the proleg of a caterpillar and certain other insect larvae. [VERT ZOO] A projecting protoloph on lophodont molars.

crocodile [VERT ZOO] The common name for about 12 species of aquatic reptiles included in the family Crocodylidae.

Crocodilia [VERT ZOO] An order of the class Reptilia which is composed of large, voracious, aquatic species, including the alligators, caimans, crocodiles, and gavials.

Crocodylidae [VERT ZOO] A family of reptiles in the order Crocodilia including the true crocodiles, false gavial, alligators, and caimans.

Crocodylinae [VERT ZOO] A subfamily of reptiles in the

CRICOID

epiglottis —

arytenoid cartilages

cricoid cartilage

upper ring of trachea

Back view of human laryngeal cartilages and ligaments. *(After Sappey, from J. Symington, ed., Quains' Elements of Anatomy, vol. 2, pt. 2, Longmans, Green, 1914)*

CRINOIDEA

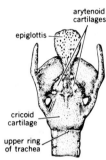

cirri

arm

calyx

node

internode

stem

2 cm

root

Schematic diagram of crinoid.

CRO-MAGNON MAN

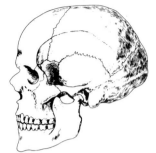

Skull of a Cro-Magnon male.
*(From M. F. Ashley Montagu,
An Introduction to Physical
Anthropology, 2d ed., Charles
C. Thomas, 1951)*

family Crocodylidae containing the crocodiles, *Osteolaemus*, and the false gavial.

Cro-Magnon man [PALEON] **1.** A race of tall, erect Caucasoid men having large skulls; identified from skeletons found in southern France. **2.** A general term to describe all fossils resembling this race that belong to the upper Paleolithic (35,000–8000 B.C.) in Europe.

crop [AGR] A plant or animal grown for its commercial value. [VERT ZOO] A distensible saccular diverticulum near the lower end of the esophagus of birds which serves to hold and soften food before passage into the stomach.

crop rotation [AGR] A method of protecting the soil and replenishing its nutrition by planting a succession of different crops on the same land.

crosier [BOT] Young circinate fern frond. [INV ZOO] A flat spiral shell, as of Spirula. [MYCOL] A hook formed by the terminal cells of ascogenous hyphae.

cross [GEN] The breeding of a male and female of a species; the progeny are known as the F_1 generation.

crossbreed [BIOL] To propagate new individuals by breeding two distinctive varieties of a species.

cross-correlation [STAT] **1.** Correlation between corresponding members of two or more series: if $q_1, \ldots q_n$ and $r_1, \ldots r_n$ are two series, correlation between q_i and r_i, or between q_i and r_{i+j} (for fixed j), is a cross correlation. **2.** Correlation between or expectation of the inner product of two series of random variables, where the difference in indices between the corresponding values of the two series is fixed.

crossed paralysis [MED] Paralysis of the arm and leg on one side, associated with contralateral cranial nerve palsies caused by a brainstem lesion involving cranial nerve nuclei and the ipsilateral pyramidal tract.

cross-eye *See* esotropia.

cross-fertilization [BOT] Fertilization between two separate plants. [ZOO] Fertilization between different kinds of individuals.

crossing over [GEN] The exchange of genetic material between paired homologous chromosomes during meiosis. Also known as crossover.

cross matching [IMMUNOL] Determination of blood compatibility for transfusion by mixing donor cells with recipient serum, and recipient cells with donor serum, and examining for an agglutination reaction.

Crossopterygii [PALEON] A subclass of the class Osteichthyes comprising the extinct lobefins or choanate fishes and represented by one extant species; distinguished by two separate dorsal fins.

Crossosomataceae [BOT] A monogeneric family of xerophytic shrubs in the order Dilleniales characterized by perigynous flowers, seeds with thin endosperm, and small, entire leaves.

crossover *See* crossing over.

crossover experiment [MED] An experiment or clinical investigation in which subjects are divided randomly into at least as many groups as there are kinds of treatment to be given, and then the groups are interchanged until every subject has received each treatment.

cross-pollination [BOT] Transfer of pollen from the anthers of one plant to the stigmata of another plant.

cross-reaction [IMMUNOL] Reaction between an antibody and a closely related, but not complementary, antigen.

cross-tolerance [MED] Tolerance or resistance to the action of a drug brought about by continued use of another drug of similar pharmacologic action.

Crotalidae [VERT ZOO] A family of proglyphodont venomous snakes in the reptilian suborder Serpentes.

crotch [SCI TECH] The angular form made by the parting of two branches, legs, or members.

crotch height [ANTHRO] The measure of the vertical distance from the crotch of a standing subject to the floor.

croup [MED] Any condition of upper-respiratory pathway obstruction in children, especially acute inflammation of the pharynx, larynx, and trachea, characterized by a hoarse, brassy, and stridulent cough and difficulties in breathing.

croup-associated virus [VIROL] A virus belonging to subgroup 2 of the parainfluenza viruses and found in children with croup. Also known as CA virus; laryngotracheobronchitis virus.

crow [VERT ZOO] The common name for a number of predominantly black birds in the genus *Corvus* comprising the most advanced members of the family Corvidae.

crown [ANAT] **1.** The top of the skull. **2.** The portion of a tooth above the gum. [BOT] **1.** The topmost part of a plant or plant part. **2.** *See* corona.

crown fire [FOR] A forest fire burning primarily in the tops of trees and shrubs.

crown gall [PL PATH] A bacterial disease of many plants induced by *Bacterium tumefaciens* and marked by abnormal enlargement of the stem near the root crown.

crown grafting [BOT] A method of vegetative propagation whereby a scion 3-6 inches (8-15 centimeters) long is grafted at the root crown, just below ground level.

crown rot [PL PATH] Any plant disease or disorder marked by deterioration of the stem at or near ground level.

crown rust [PL PATH] A rust disease of oats and certain other grasses caused by varieties of *Puccinia coronata* and marked by light-orange masses of fungi on the leaves.

cruciate [ANAT] Resembling a cross.

Cruciferae [BOT] A large family of dicotyledonous herbs in the order Capparales characterized by parietal placentation; hypogynous, mostly regular flowers; and a two-celled ovary with the ovules attached to the margins of the partition.

cruciform [SCI TECH] Resembling or arranged like a cross.

crude drug [PHARM] **1.** A plant or animal drug containing all principles characteristic of the drug. **2.** The dried leaves, bark, or rhizome of a plant containing therapeutically active principles.

crumena [INV ZOO] In Hemiptera, a sheath into which stylets are retracted.

crura [ANAT] Plural of crus.

crus [ANAT] **1.** The shank of the hindleg, that portion between the femur and the ankle. **2.** Any of various parts of the body resembling a leg or root.

crush kidney *See* lower nephron nephrosis.

crush syndrome [MED] A severe, often fatal condition that follows a severe crushing injury, particularly involving large muscle masses, characterized by fluid and blood loss, shock, hematuria, and renal failure. Also known as compression syndrome.

Crustacea [INV ZOO] A class of arthropod animals in the subphylum Mandibulata having jointed feet and mandibles, two pairs of antennae, and segmented, chitin-encased bodies.

crustecdysone [BIOCHEM] $C_{27}H_{44}O_7$ 20-Hydroxyecdysone, the molting hormone produced by Y organs in crustaceans.

crustose [BOT] Of a lichen, forming a thin crustlike thallus which adheres closely to the substratum of rock, bark, or soil.

crust vegetation [ECOL] Zonal growths of algae, mosses, lichens, or liverworts having variable coverage and a thickness of only a few centimeters.

CROWN FIRE

Characteristic crown fire.
(U.S. Forest Service)

cryobiology [BIOL] The use of low-temperature environments in the study of living plants and animals.

cryogenics [PHYS] The production and maintenance of very low temperatures, and the study of phenomena at these temperatures.

cryoglobulin [PATH] An abnormal protein, usually an immunoglobulin, which precipitates from plasma between 40 and 70°F (4.4 and 21°C).

cryophilous [ECOL] Having a preference for low temperatures. Also known as cryophilic.

cryophyte [ECOL] A plant that forms winter buds below the soil surface.

cryoplankton [BIOL] **1.** Plankton found in glacial and polar regions. **2.** Algae that thrive on snow.

cryoprecipitate [BIOCHEM] The precipitate of a cryoglobulin.

cryosurgery [MED] Selective destruction of tissue by freezing, as the use of a liquid nitrogen probe to the brain in parkinsonism.

cryotherapy [MED] A form of therapy which consists of local or general use of cold.

Cryphaeaceae [BOT] A family of mosses in the order Isobryales distinguished by a rough calyptra.

crypt [ANAT] **1.** A follicle or pitlike depression. **2.** A simple glandular cavity.

cryptic coloration [ZOO] A phenomenon of protective coloration by which an animal blends with the background through color matching or countershading.

Cryptobiidae [INV ZOO] A family of flagellate protozoans in the order Kinetoplastida including organisms with two flagella, one free and one with an undulating membrane.

cryptobiotic [ECOL] Living in concealed or secluded situations.

Cryptobranchidae [VERT ZOO] The giant salamanders and hellbenders, a family of tailed amphibians in the suborder Cryptobranchoidea.

Cryptobranchoidea [VERT ZOO] A primitive suborder of amphibians in the order Urodela distinguished by external fertilization and aquatic larvae.

cryptocarp [BOT] The fruitlike sporophyte stage in red algae.

Cryptocerata [INV ZOO] A division of hemipteran insects in some systems of classification that includes the water bugs (Hydrocorisae).

Cryptochaetidae [INV ZOO] A family of myodarian cyclorrhaphous dipteran insects in the subsection Acalypteratae.

Cryptococcaceae [MYCOL] A family of imperfect fungi in the order Moniliales in some systems of classification; equivalent to the Cryptococcales in other systems.

Cryptococcales [MYCOL] An order of imperfect fungi, in some systems of classification, set up to include the yeasts or yeastlike organisms whose perfect or sexual stage is not known.

cryptococcosis [MED] A yeast infection of man, primarily of the central nervous system, caused by *Cryptococcus neoformans.* Also known as torulosis.

Cryptodira [VERT ZOO] A suborder of the reptilian order Chelonia including all turtles in which the cervical spines are uniformly reduced and the head folds directly back into the shell.

cryptogam [BOT] An old term for nonflowering plants.

cryptomedusa [INV ZOO] The final stage in the reduction of a hydroid medusa to a rudiment having sex cells within the gonophore.

cryptomitosis [INV ZOO] Cell division in certain protozoans in which a modified spindle forms, and chromatin assembles with no apparent chromosome differentiation.

CRYPTOBIIDAE

20 μ

Cryptobia, a genus in Cryptobiidae showing the characteristic two flagella, one free and one with undulating membrane.

Cryptomonadida [BIOL] An order of the class Phytamasti-gophorea considered to be protozoans by biologists and algae by botanists.

Cryptomonadina [BIOL] The equivalent name for Cryptomonadida.

Cryptophagidae [INV ZOO] The silken fungus beetles, a family of coleopteran insects in the superfamily Cucujoidea.

Cryptophyceae [BOT] A class of algae of the Pyrrhophyta in some systems of classification; equivalent to the division Cryptophyta.

Cryptophyta [BOT] A division of the algae in some classification schemes; equivalent to the Cryptophyceae.

cryptophyte [BOT] A plant that produces buds either underwater or underground on corms, bulbs, or rhizomes.

Cryptopidae [INV ZOO] A family of epimorphic centipedes in the order Scolopendromorpha.

cryptorchidism *See* cryptorchism.

cryptorchism [MED] Failure of the testes to descend into the scrotum from the abdomen or inguinal canals. Also known as cryptorchidism.

Cryptostomata [PALEON] An order of extinct bryozoans in the class Gymnolaemata.

cryptotope [IMMUNOL] A determinant (or epitope) of an immunological antigen or immunogen which is initially hidden and becomes functional only when the molecule is broken or degraded.

cryptoxanthin [BIOCHEM] $C_{40}H_{57}O$ A xanthophyll carotenoid pigment found in plants; convertible to vitamin A by many animal livers.

cryptozoite [INV ZOO] The sporozoite stage when living in tissues before entering the blood.

crypts of Lieberkühn [ANAT] Simple, tubular glands which arise as evaginations into the mucosa of the small intestine.

crystal [CRYSTAL] A homogeneous solid made up of an element, chemical compound or isomorphous mixture throughout which the atoms or molecules are arranged in a regularly repeating pattern.

crystal class [CRYSTAL] One of 32 categories of crystals according to the inversions, rotations about an axis, reflections, and combinations of these which leaves the crystal invariant. Also known as symmetry class.

crystal diffraction [SOLID STATE] Diffraction by a crystal of beams of x-rays, neutrons, or electrons whose wavelengths (or de Broglie wavelengths) are comparable with the interatomic spacing of the crystal.

crystal diffraction spectrometer [PHYS] An instrument utilizing diffraction of waves by a crystal lattice to measure wavelengths or crystal properties. Also known as crystal spectrometer. [SOLID STATE] *See* Bragg spectrometer.

crystal face [CRYSTAL] One of the outward planar surfaces which define a crystal and reflect its internal structure. Also known as face.

crystal field theory [PHYS CHEM] The theory which assumes that the ligands of a coordination compound are the sources of negative charge which perturb the energy levels of the central metal ion and thus subject the metal ion to an electric field analogous to that within an ionic crystalline lattice.

crystal growth [CRYSTAL] The growth of a crystal, which involves diffusion of the molecules of the crystallizing substance to the surface of the crystal, diffusion of these molecules over the crystal surface to special sites on the surface, incorporation of molecules into the surface at these sites, and diffusion of heat away from the surface.

crystal habit [CRYSTAL] The size and shape of the crystals in a crystalline solid. Also known as habit.

CRYPTOMONADIDA

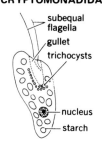

Chilomonas paramecium, an example of a cryptomonad.

CTENOID SCALE

Ctenoid scale from carp. *(From K. F. Lagler et al., Ichthyology, Wiley, 1962)*

CTENOPHORA

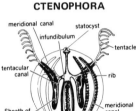

Structure of a cydippid ctenophore.

CTENOTHRISSIFORMES

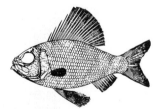

Cretaceous ctenothrissid, *Ctenothrissus radians*; length 10–12 inches (25–30 centimeters). *(After C. Patterson, Phil. Trans. Roy. Soc. London Ser. B. Biol. Sci. no. 739, vol. 247, 1964)*

crystal laser [OPTICS] A laser that uses a pure crystal of ruby or other material for generating a coherent beam of output light.

crystal lattice [CRYSTAL] A lattice from which the structure of a crystal may be obtained by associating with every lattice point an assembly of atoms identical in composition, arrangement, and orientation. Also known as lattice; space lattice.

crystalline lens *See* lens.

crystalline style [INV ZOO] A hyaline rod in the digestive tract of some bivalve mollusks.

crystallization [CRYSTAL] The formation of crystalline substances from solutions or melts.

crystallography [PHYS] The branch of science that deals with the geometric description of crystals and their internal arrangement.

crystalloid [CHEM] Particles in a solution which are able to crystallize under appropriate conditions. [SCI TECH] Resembling a crystal.

crystal spectrometer *See* crystal diffraction spectrometer.

crystal structure [CRYSTAL] The arrangement of atoms or ions in a crystalline solid.

crystal system [CRYSTAL] One of seven categories (cubic, hexagonal, tetragonal, trigonal, orthorhombic, monoclinic, and triclinic) into which a crystal may be classified according to the shape of the unit cell of its Bravais lattice, or according to the dominant symmetry elements of its crystal class.

Cs *See* cesium.

cteinophyte [ECOL] A parasitic plant which destroys its host.

ctene *See* comb.

ctenidium [INV ZOO] **1.** The comb- or featherlike respiratory apparatus of certain mollusks. **2.** A row of spines on the head or thorax of some fleas.

ctenocyst [INV ZOO] A sense organ located at the aboral pole of Ctenophora; functions in balance.

Ctenodrilidae [INV ZOO] A family of fringe worms belonging to the Sedentaria.

ctenoid scale [VERT ZOO] A thin, acellular structure composed of bonelike material and characterized by a serrated margin; found in the skin of advanced teleosts.

Ctenophora [INV ZOO] The comb jellies, a phylum of marine organisms having eight rows of comblike plates as the main locomotory structure.

Ctenostomata [INV ZOO] An order of bryozoans in the class Gymnolaemata recognized as inconspicuous, delicate colonies made up of relatively isolated, short, tubular zooecia with chitinous walls.

Ctenostomatida [INV ZOO] The equivalent name for Odontostomatida.

Ctenothrissidae [PALEON] A family of extinct teleostean fishes in the order Ctenothrissiformes.

Ctenothrissiformes [PALEON] A small order of extinct teleostean fishes; important as a group on the evolutionary line leading from the soft-rayed to the spiny-rayed fishes.

Cu *See* copper.

cubeb [BOT] The dried, nearly ripe fruit (berries) of a climbing vine, *Piper cubeba,* of the pepper family (Piperaceae).

cubic centimeter [MECH] A unit of volume equal to the volume of a cube whose edges are 1 centimeter long. Abbreviated cc; ccm; cm³.

cubic system *See* isometric system.

cubital [ANAT] Pertaining to the elbow or the ulna.

cubitus [ANAT] **1.** The ulna. **2.** The forearm. [INV ZOO] The primary vein of an insect wing.

cuboid [ANAT] The outermost distal tarsal bone in vertebrates. [INV ZOO] Main vein of the wing in many insects, particularly the flies (Diptera). [SCI TECH] Nearly cubic in shape.

cuboidal epithelium [HISTOL] A single-layered epithelium made up of cubelike cells.

Cubomedusae [INV ZOO] An order of coelenterates in the class Scyphozoa distinguished by a cubic umbrella.

cuckoo [VERT ZOO] The common name for about 130 species of primarily arboreal birds in the family Cuculidae; some are social parasites.

Cucujidae [INV ZOO] The flat-back beetles, a family of predatory coleopteran insects in the superfamily Cucujoidea.

Cucujoidea [INV ZOO] A large superfamily of coleopteran insects in the suborder Polyphaga.

Cuculidae [VERT ZOO] A family of perching birds in the order Cuculiformes, including the cuckoos and the roadrunner, characterized by long tails, heavy beaks and conspicuous lashes.

Cuculiformes [VERT ZOO] An order of birds containing the cuckoos and allies, characterized by the zygodactyl arrangement of the toes.

cucullus [BIOL] A hood-shaped structure.

Cucumariidae [INV ZOO] A family of dendrochirotacean holothurian echinoderms in the order Dendrochirotida.

cucumber [BOT] *Cucumis sativus.* An annual cucurbit, in the family Cucurbitaceae grown for its edible, immature fleshy fruit.

cucumber mildew [PL PATH] **1.** A downy mildew of cucumbers and melons caused by *Peronoplasmopara cubensis.* **2.** A powdery mildew of cucumbers and melons caused by *Erysiphe cichoracearum.*

cucumber mosaic [PL PATH] A virus disease of cucumbers and related fruits, producing mottling of terminal leaves and fruits and dwarfing of vines.

cucurbit wilt [PL PATH] A bacterial disease of cucumbers and related plants caused by *Erwinia tracheiphila,* characterized by sudden wilting of the plant.

cuiller [INV ZOO] The spoonlike terminus of a clasper in male insects.

cul-de-sac [ANAT] Blind pouch or diverticulum.

culdoscope [MED] An instrument used to visualize female pelvic organs, introduced through the vagina or a perforation into the retrouterine pouch.

Culicidae [INV ZOO] The mosquitoes, a family of slender, orthorrhaphous dipteran insects in the series Nematocera having long legs and piercing mouthparts.

Culicinae [INV ZOO] A subfamily of the dipteran family Culicidae.

culm [BOT] **1.** A jointed, usually hollow grass stem. **2.** The solid stem of certain monocotyledons, such as the sedges. [MIN ENG] Fine, refuse coal, screened and separated from larger pieces.

culmen [VERT ZOO] The edge of the upper bill in birds.

cultellus [INV ZOO] A knifelike mouthpart in certain bloodsucking flies.

cultigen [BIOL] A cultivated variety or species of organism for which there is no known wild ancestor. Also known as cultivar.

cultivar *See* cultigen.

cultivate [AGR] To prepare soil for the raising of crops.

cultivator [AGR] A farm implement pulled behind a powered machine that is used to break up soil, kill weeds, and create a surface mulch for moisture.

cultural anthropology [ANTHRO] The division of anthropology dealing with the study of all aspects of culture.

culture [ANTHRO] The complex pattern of behavior that

CUCKOO

Black-billed cuckoo, brown above and white underneath with thin red rings around eyes, maximum length 12 inches (30 centimeters).

CUMACEA

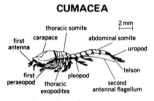

Typical adult male cumacean.

CURCULIONIDAE

Snout beetle. *(From T. I. Storer and R. L. Usinger, General Zoology, 3d ed., McGraw-Hill, 1957)*

CURIUM

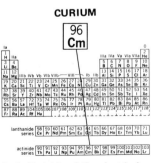

Periodic table of the chemical elements showing the position of curium.

distinguishes a social, ethnic, or religious group. [BIOL] A growth of living cells or microorganisms in a controlled artificial environment.

culture medium [MICROBIO] The nutrients and other organic and inorganic materials used for the growth of microorganisms and plant and animal tissue in culture.

Cumacea [INV ZOO] An order of the class Crustacea characterized by a well-developed carapace which is fused dorsally with at least the first three thoracic somites and overhangs the sides.

cumatophyte [ECOL] A plant that grows under surf conditions.

cumin [BOT] *Cuminum cyminum.* An annual herb in the family Umbelliferae; the fruit is valuable for its edible, aromatic seeds.

cumulus oophorus [HISTOL] The layer of gelatinous, follicle cells surrounding the ovum in a Graafian follicle.

cuneate [BIOL] Wedge-shaped with the acute angle near the base, as in certain insect wings and the leaves of various plants.

cuneiform [ANAT] **1.** Any of three wedge-shaped tarsal bones. **2.** Either of a pair of cartilages lying dorsal to the thyroid cartilage of the larynx. **3.** Wedge-shaped, chiefly referring to skeletal elements.

cuneus [ANAT] A wedge-shaped convolution of the occipital lobe of the brain between the calcarine fissure and the medial part of the parieto-occipital fissure; part of the visual cortex.

cunnus [ANAT] The vulva.

Cupedidae [INV ZOO] The reticulated beetles, the single family of the coleopteran suborder Archostemata.

cupreous [SCI TECH] Containing or resembling copper.

cupula [ANAT] **1.** A cup-shaped structure. **2.** A colorless substance on the elevation of an ampulla of the ear, containing sensory end organs; it coagulates and becomes visible upon applying fixing fluids.

cupule [BOT] **1.** The cup-shaped involucre characteristic of oaks. **2.** A cup-shaped corolla. **3.** The gemmae cup of the Marchantiales. [INV ZOO] A small sucker on the feet of certain male flies.

curare [MATER] A poisonous extract from the plant *Strychnos toxifera* containing alkaloids that produce paralysis of the voluntary muscles by acting on synaptic junctions; used as an adjunct to anesthesia in surgery.

Curculionidae [INV ZOO] The true weevils or snout beetles, a family of coleopteran insects in the superfamily Curculionoidea.

Curculionoidea [INV ZOO] A superfamily of coleopteran insects in the suborder Polyphaga.

Curcurbitaceae [BOT] A family of dicotyledonous herbs or herbaceous vines in the order Violales characterized by an inferior ovary, unisexual flowers, one to five stamens but typically three, and a sympetalous corolla.

Curcurbitales [BOT] The ordinal name assigned to the Curcurbitaceae in some systems of classification.

curd [BOT] The edible flower heads of members of the mustard family such as broccoli. [FOOD ENG] **1.** The clotted portion of soured milk or milk treated with an acid or enzyme; used in making cheese. **2.** Any food resembling milk curd.

curet [MED] An instrument, shaped like a spoon or scoop, for scraping away tissue.

curettage [MED] Scraping of the inside of a body cavity or the hollow of an organ with a curet.

curium [CHEM] An element, symbol Cm, atomic number 96; the isotope of mass 244 is the principal source of this artificially produced element.

curling factor *See* griseofulvin.

Curling's ulcer [MED] An acute gastric ulcer associated with severe skin burns.

curly top [PL PATH] A virus disease of sugarbeets and certain other plants that is transmitted by a leafhopper; affected plants are dwarfed and have curled, upturned leaves.

currant [BOT] A shrubby, deciduous plant of the genus *Ribes* in the order Rosales; the edible fruit, a berry, is borne in clusters on the plant.

currant leaf spot [PL PATH] **1.** An angular leaf spot of currants caused by the fungus *Cercospora angulata.* **2.** An anthracnose of currants caused by *Pseudopeziza ribis* and characterized by brown or black spots.

cursorial [VERT ZOO] Adapted for running.

curve fitting [STAT] The calculation of a curve of some particular character (as a logarithmic curve) that most closely approaches a number of points in a plane.

Cuscutaceae [BOT] A family of parasitic dicotyledonous plants in the order Polemoniales which lack internal phloem and chlorophyll, have capsular fruit, and are not rooted to the ground at maturity.

Cushing's syndrome [MED] A complex of symptoms including facial and truncal obesity, hypertension, edema, and osteoporosis, resulting from oversecretion of adrenocortical hormones.

cushion [BOT] The thickened central area in the prothallus of a fern.

cusp [ANAT] **1.** A pointed or rounded projection on the masticating surface of a tooth. **2.** One of the flaps of a heart valve.

cuspid *See* canine.

cuspidate [BIOL] Having a cusp; terminating in a point.

cutaneous [ANAT] Pertaining to the skin.

cutaneous anaphylaxis [IMMUNOL] Hypersensitivity that is marked by an intense skin reaction following parenteral contact with a sensitizing agent.

cutaneous anthrax *See* malignant pustule.

cutaneous appendage [ANAT] Any of the epidermal derivatives, including the nails, hair, sebaceous glands, mammary glands, and sweat glands.

cutaneous blastomycosis [MED] A form of North American blastomycosis considered by some to be a clinical manifestation of the systemic form.

cutaneous coccidioidomycosis [MED] A primary skin infection by the fungus *Coccidioides immitis;* a skin infection secondary to a pulmonary lesion.

cutaneous leishmaniasis [MED] A parasitic skin infection by *Leishmania tropica* characterized by deep ulcers of the skin and subcutaneous tissue.

cutaneous pain [PHYSIO] A sensation of pain arising from the skin.

cutaneous reaction [MED] **1.** Any change in the outer layers of the skin, as in sunburn or the rash in measles. **2.** Any immediate or delayed immune reaction in the skin resulting from antigen-antibody interaction.

cutaneous sensation [PHYSIO] Any feeling originating in sensory nerve endings of the skin, including pressure, warmth, cold, and pain.

Cuterebridae [INV ZOO] The robust botflies, a family of myodarian cyclorrhaphous dipteran insects in the subsection Calypteratae.

cuticle [ANAT] The horny layer of the nail fold attached to the nail plate at its margin. [BIOL] A noncellular, hardened or membranous secretion of an epithelial sheet, such as the integument of nematodes and annelids, the exoskeleton of

CURRANT

The black currant *(Ribes nigrum).*

arthropods, and the continuous film of cutin on certain plant parts.

cutin [BIOCHEM] A mixture of fatty substances characteristically found in epidermal cell walls and in the cuticle of plant leaves and stems.

cutis *See* dermis.

cutting [BOT] A piece of plant stem with one or more nodes, which, when placed under suitable conditions, will produce roots and shoots resulting in a complete plant.

cuttlefish [INV ZOO] An Old World decapod mollusk of the genus *Sepia;* shells are used to manufacture dentifrices and cosmetics.

Cuvierian organs [INV ZOO] In holothurians, glandular tubes that extend from the cloaca.

Cuvieroninae [PALEON] A subfamily of extinct proboscidean mammals in the family Gomphotheriidae.

Cyamidae [INV ZOO] The whale lice, a family of amphipod crustaceans in the suborder Caprellidea that bear a resemblance to insect lice.

cyanogenetic [SCI TECH] Producing cyanide.

cyanophilous [BIOL] Having an affinity for blue or green dyes.

Cyanophyceae [BOT] The blue-green algae, a class of plants in the division Schizophyta.

cyanophycin [BIOCHEM] A granular protein food reserve in the cells of blue-green algae, especially in the peripheral cytoplasm.

Cyanophyta [BOT] An equivalent name for the Cyanophyceae.

cyanosis [MED] A bluish coloration in the skin and mucous membranes due to deficient levels of oxygen in the blood.

Cyatheaceae [BOT] A family of tropical and pantropical tree ferns distinguished by the location of sori along the veins.

cyathium [BOT] An inflorescence in which the flowers arise from the base of a cuplike involucre.

Cyathoceridae [INV ZOO] The equivalent name for the Lepiceridae.

cyathozooid [INV ZOO] The primary zooid occurring in certain tunicates.

Cycadales [BOT] An ancient order of plants in the class Cycadopsida characterized by tuberous or columnar stems that bear a crown of large, usually pinnate leaves.

Cycadatae *See* Cycadopsida.

Cycadeoidaceae [PALEOBOT] A family of extinct plants in the order Cycadeoidales characterized by sparsely branched trunks and a terminal crown of leaves.

Cycadeoidales [PALEOBOT] An order of extinct plants that were abundant during the Triassic, Jurassic, and Cretaceous periods.

Cycadicae [BOT] A subdivision of large-leaved gymnosperms with stout stems in the plant division Pinophyta; only a few species are extant.

Cycadofilicales [PALEOBOT] The equivalent name for the extinct Pteridospermae.

Cycadophyta [BOT] An equivalent name for Cycadecae elevated to the level of a division.

Cycadophytae [BOT] An equivalent name for Cycadicae.

Cycadopsida [BOT] A class of gymnosperms in the plant subdivision Cycadicae. Formerly known as Cycadatae.

Cyclanthaceae [BOT] The single family of the order Cyclanthales.

Cyclanthales [BOT] An order of monocotyledonous plants composed of herbs; or, seldom, composed of more or less woody plants with leaves that usually have a bifid, expanded blade.

CUTTLEFISH

Drawing of cuttlefish. *(From T. I. Storer and R. L. Usinger, General Zoology, 3d ed., McGraw-Hill, 1957)*

CYCADALES

ovulate cones

Sago palm *(Cycas revoluta)* showing megasporophylls grouped into cones. *(New York Botanical Garden)*

cyclitis [MED] Inflammation of the ciliary body of the eye.

cyclobarbital [PHARM] $C_{12}H_{16}N_2O_3$ A hypnotic and sedative of short duration.

Cyclocystoidea [PALEON] A class of small, disk-shaped, extinct echinozoans in which the lower surface of the body probably consisted of a suction cup.

cyclodialysis [MED] Detaching the ciliary body from the sclera in order to effect reduction of intraocular tension in certain cases of glaucoma, especially in aphakia.

cyclodiathermy [MED] Destruction, by diathermy, of the ciliary body.

cycloid scale [VERT ZOO] A thin, acellular structure which is composed of a bonelike substance and shows annual growth rings; found in the skin of soft-rayed fishes.

cyclomorphosis [ECOL] Cyclic recurrent polymorphism in certain planktonic fauna in response to seasonal temperature or salinity changes.

Cyclophoracea [INV ZOO] A superfamily of gastropod mollusks in the order Prosobranchia.

Cyclophoridae [INV ZOO] A family of land snails in the order Pectinibranchia.

Cyclophyllidea [INV ZOO] An order of platyhelminthic worms comprising most tapeworms of warm-blooded vertebrates.

cyclopia [MED] A congenital anomaly characterized by fusion of the eye sockets with various degrees of fusion of the eyes, to the occurrence of a single median eye.

Cyclopinidae [INV ZOO] A family of copepod crustaceans in the suborder Cyclopoida, section Gnathostoma.

Cyclopoida [INV ZOO] A suborder of small copepod crustaceans.

Cyclopteridae [VERT ZOO] The lumpfishes and snailfishes, a family of deep-sea forms in the suborder Cottoidei of the order Perciformes.

Cyclorhagae [INV ZOO] A suborder of benthonic, microscopic marine animals in the class Kinorhyncha of the phylum Aschelminthes.

Cyclorrhapha [INV ZOO] A suborder of true flies, order Diptera, in which developing adults are always formed in a puparium from which they emerge through a circular opening.

cycloserine [MICROBIO] $C_3H_6O_2N_2$ Broad-spectrum, crystalline antibiotic produced by several species of *Streptomyces*; useful in the treatment of tuberculosis and urinary-tract infections caused by resistant gram-negative bacteria.

cyclosis [CYTOL] Massive rotational streaming of cytoplasm in certain vacuolated cells, such as the stonewort *Nitella* and *Paramecium*.

cyclospermous [BOT] Having a coiled embryo.

cyclospondylous [VERT ZOO] Referring to vertebrae formed of successive concentric cartilaginous layers.

Cyclosporeae [BOT] A class of brown algae, division Phaeophyta, in which there is only a free-living diploid generation.

Cyclosteroidea [PALEON] A class of Middle Ordovician to Middle Devonian echinoderms in the subphylum Echinozoa.

Cyclostomata [INV ZOO] An order of bryozoans in the class Stenolaemata. [VERT ZOO] A subclass comprising the simplest and most primitive of living vertebrates characterized by the absence of jaws and the presence of a single median nostril and an uncalcified cartilaginous skeleton.

cyclothymia [PSYCH] A disposition marked by alterations of mood between elation and depression out of proportion to apparent external events and stimulated, rather, by internal factors.

CYCLOID SCALE

Cycloid scale, a primitive vertebrate scale. *(From K. F. Lagler et al., Ichthyology, Wiley, 1962)*

CYCLORHAGAE

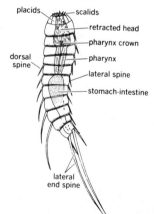

placids — scalids — retracted head — pharynx crown — dorsal spine — pharynx — lateral spine — stomach-intestine — lateral end spine

Side view of *Echinoderella* species, a cyclorhagous kinorhynch.

CYDIPPIDEA

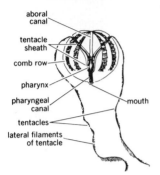

Pleurobranchia, a cydippid ctenophore. (From L. H. Hyman, The Invertebrates, vol. 1, McGraw-Hill, 1940)

CYPRIDACEA

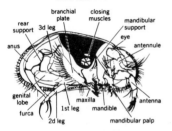

Female Candona suburbana Hoff, an ostracod of the Cypridacea. (After D. J. McGregor and R. V. Kesling, Contributions from the Museum of Paleontology of the University of Michigan, 1969)

CYPRINODONTIDAE

2.5 cm

Striped killifish (Fundulus majalis). (After G. B. Goode, Fishery Industries of the U.S., 1884)

Cydippidea [INV ZOO] An order of the Ctenophora having well-developed tentacles.

Cydnidae [INV ZOO] The ground or burrower bugs, a family of hemipteran insects in the superfamily Pentatomorpha.

cyesis *See* pregnancy.

cylindrarthrosis [ANAT] A joint characterized by rounded articular surfaces.

Cylindrocapsaceae [BOT] A family of green algae in the suborder Ulotrichineae comprising thick-walled, sheathed cells having massive chloroplasts.

cymba [BIOL] A boat-shaped sponge spicule.

cymbium [INV ZOO] In some spiders, the boat-shaped tarsus of the pedipalpus.

cymbocephalic [ANTHRO] Of a head or skull, having an unusually prolonged receding forehead and a protrudent occiput.

cyme [BOT] An inflorescence in which each main axis terminates in a single flower; secondary and tertiary axes may also have flowers, but with shorter flower stalks.

cymose [BOT] Of, pertaining to, or resembling a cyme.

Cymothoidae [INV ZOO] A family of isopod crustaceans in the suborder Flabellifera; members are fish parasites with reduced maxillipeds ending in hooks.

Cynipidae [INV ZOO] A family of hymenopteran insects in the superfamily Cynipoidea.

Cynipoidea [INV ZOO] A superfamily of hymenopteran insects in the suborder Apocrita.

Cynoglossidae [VERT ZOO] The tonguefishes, a family of Asiatic flatfishes in the order Pleuronectiformes.

Cyperaceae [BOT] The sedges, a family of monocotyledonous plants in the order Cyperales characterized by spirally arranged flowers on a spike or spikelet; a usually solid, often triangular stem; and three carpels.

Cyperales [BOT] An order of monocotyledonous plants in the subclass Commelinidae with reduced, mostly wind-pollinated or self-pollinated flowers that have a unilocular, two- or three-carpellate ovary bearing a single ovule.

Cypheliaceae [BOT] A family of typically crustose lichens with sessile apothecia in the order Caliciales.

cyphella [BOT] A small pit on the lower surface of the thallus of certain lichens.

cyphonautes [INV ZOO] The free-swimming bivalve larva of certain bryozoans.

Cyphophthalmi [INV ZOO] A family of small, mitelike arachnids in the order Phalangida.

Cypraecea [INV ZOO] A superfamily of gastropod mollusks in the order Prosobranchia.

Cypraeidae [INV ZOO] A family of colorful marine snails in the order Pectinibranchia.

cypress [BOT] The common name for members of the genus Cupressus and several related species in the order Pinales.

Cypridacea [INV ZOO] A superfamily of mostly fresh-water ostracods in the suborder Podocopa.

Cypridinacea [INV ZOO] A superfamily of ostracods in the suborder Myodocopa characterized by a calcified carapace and having a round back with a downward-curving rostrum.

Cyprinidae [VERT ZOO] The largest family of fishes, including minnows and carps in the order Cypriniformes.

Cypriniformes [VERT ZOO] An order of actinopterygian fishes in the suborder Ostariophysi.

Cyprinodontidae [VERT ZOO] The killifishes, a family of actinopterygian fishes in the order Atheriniformes that inhabit ephemeral tropical ponds.

Cyprinoidei [VERT ZOO] A suborder of primarily fresh-water actinopterygian fishes in the order Cypriniformes having toothless jaws, no adipose fin, and faliciform lower pharyngeal bones.

cypris [INV ZOO] An ostracod-like, free-swimming larval stage in the development of Cirripedia.

Cyrtophorina [INV ZOO] The equivalent name for Gymnostomatida.

cyrtopia [INV ZOO] A type of crustacean larva (Ostracoda) characterized by an elongation of the first pair of antennae and loss of swimming action in the second pair.

cyrtosis [PL PATH] A virus disease of cotton characterized by stunting, distortion, and abnormal branching and coloration.

cyst [MED] A normal or pathologic sac with a distinct wall, containing fluid or other material.

cystacanth [INV ZOO] The infective larva of the Acanthocephala; lies in the hemocele of the intermediate host.

L-cystathionine [BIOCHEM] $C_7H_{14}N_2O_4S$ An amino acid formed by condensation of homocysteine with serine, catalyzed by an enzyme transsulfurase; found in high concentration in the brain of primates.

cystectomy [MED] **1.** Excision of the gallbladder, or part of the urinary bladder. **2.** Removal of a cyst. **3.** Removal of a piece of the anterior capsule of the lens for the extraction of a cataract.

cysteine [BIOCHEM] $C_3H_7O_2NS$ A crystalline amino acid occurring as a constituent of glutathione and cystine.

cystenchyma [INV ZOO] In Porifera, parenchyma characterized by a large vesicular cell structure.

cystencytes [INV ZOO] In Porifera, collencytes which have developed a vesicular structure.

cystic disease [MED] A disorder of women, usually at or near menopause, characterized by the development of large cysts in the breast.

cystic duct [ANAT] The duct of the gallbladder.

cysticercosis [MED] The infestation in man by cysticerci of the genus *Taenia.*

cysticercus [INV ZOO] A larva of tapeworms in the order Cyclophyllidea that has a bladder with a single invaginated scolex.

cystic fibrosis [MED] A hereditary disease of the pancreas transmitted as an autosomal recessive; involves obstructive lesions, atrophy, and fibrosis of the pancreas and lungs, and the production of mucus of high viscosity. Also known as mucoviscidosis.

cysticolous [BIOL] Living within a cyst.

cystidium [MYCOL] A large inflated cell with thick walls in the hymenial layer of some basidiomycetous fungi.

cystine [BIOCHEM] $C_6H_{12}N_2S_2$ A white, crystalline amino acid formed biosynthetically from cysteine.

cystinosis [MED] A congenital metabolic disorder involving sulfur-containing amino acids, usually cystine; characterized by deposits of cystine crystals in the body organs.

cystinuria [MED] The presence in the urine of crystals of cystine together with some lysine, arginine, and ornithine.

cystitis [MED] Inflammation of a fluid-filled organ, especially the urinary bladder.

cystocarp [BOT] A fruiting structure with a special protective envelope, produced after fertilization in red algae.

cystocele [MED] Herniation of the urinary bladder into the vagina.

cystocercous cercariae [INV ZOO] A digenetic trematode larva that can withdraw the body into the tail.

cystography [MED] Radiography of the urinary bladder after the injection of a contrast medium.

Cystoidea [PALEON] A class of extinct crinozoans character-

CYSTEINE

Structural formula of cysteine.

CYSTICERCUS

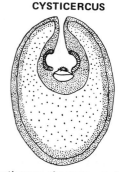

Cysticercus, the metacestode larva of cyclophyllidean tapeworms.

ized by an ovoid body that was either sessile or attached by a short aboral stem.

cystolith [BOT] A concretion of calcium carbonate arising from the cell walls of modified epidermal cells in some flowering plants.

cystoma [MED] A cystic mass, especially in or near the ovary.

cystometer [MED] An instrument used to determine pressure in the urinary bladder under standard conditions.

cyston [INV ZOO] In Siphonophora, a modified dactylozooid that functions in excretion.

Cystoporata [PALEON] An order of extinct, marine bryozoans characterized by cystopores and minutopores.

cystopyelitis [MED] Inflammation of the urinary bladder and the renal pelvis.

cystopyelography [MED] Radiography of the urinary bladder, the ureter, and the renal pelvis after injection of a radiopaque material.

cystopyelonephritis [MED] Inflammation of the urinary bladder, renal pelvis, and renal parenchyma.

cystoscope [MED] An optical instrument for visual examination of the urinary bladder, ureters, and kidneys.

cystospore [BIOL] A zoospore enclosed in a cyst.

cystoureteritis [MED] Inflammation of the urinary bladder and the ureters.

cystozooid [INV ZOO] The body, excluding the head, of a metacestode.

cytase [BIOCHEM] Any of several enzymes in the seeds of cereals and other plants, which hydrolyze the cell-wall material.

Cytheracea [INV ZOO] A superfamily of ostracods in the suborder Podocopa comprising principally crawling and digging marine forms.

Cytherellidae [INV ZOO] The family comprising all living members of the ostracod suborder Platycopa.

cytidine [BIOCHEM] $C_9H_{13}N_3O_5$ Cytosine riboside, a nucleoside composed of one molecule each of cytosine and D-ribose.

cytidylic acid [BIOCHEM] $C_9H_{14}O_8N_3P$ A nucleotide synthesized from the base cytosine and obtained by hydrolysis of nucleic acid. Also known as cytidine monophosphate.

cytocentrum *See* central apparatus.

cytochalasin [BIOCHEM] One of a series of structurally related fungal metabolic products which, among other effects on biological systems, selectively and reversibly block cytokinesis while not affecting karyokinesis; the molecule with minor variations consists of a benzyl-substituted hydroaromatic isoindolone system, which in turn is fused to a small macrolide-like cyclic ring.

cytochemistry [CYTOL] The science concerned with the chemistry of cells and cell components, primarily with the location of chemical constituents and enzymes.

cytochrome [BIOCHEM] Any of the complex protein respiratory pigments occurring within plant and animal cells, usually in mitochondia, that function as electron carriers in biological oxidation.

cytochrome a₃ *See* cytochrome oxidase.

cytochrome oxidase [BIOCHEM] Any of a family of respiratory pigments that react directly with oxygen in the reduced state. Also known as cytochrome a_3.

cytoclesis [EMBRYO] The influence of a group of cells or a placode on differentiation of developing adjacent cells.

cytocrine gland [CYTOL] A cell, especially a melanocyte, that passes its secretion directly to another cell.

cytodiagnosis [PATH] The determination of the nature of an abnormal liquid by the study of cells it contains.

cytogenetics [CYTOL] A branch of molecular pathology combining the methods of cytology and genetics.

cytogenous gland [PHYSIO] A structure that secretes living cells; an example is the testis.

cytokinesis [CYTOL] Division of the cytoplasm following nuclear division.

cytokinin [BIOCHEM] Any of a group of plant hormones which elicit certain plant growth and development responses, especially by promoting cell division.

cytology [BIOL] A branch of the biological sciences which deals with the structure, behavior, growth, and reproduction of cells and the function and chemistry of cell components.

cytolysin [BIOCHEM] A protein or antibody in plasma which causes hemolysis of erythrocytes (hemolysin), or the cytolysis of other types of cells.

cytolysis [PATH] Disintegration or dissolution of cells, usually associated with a pathologic process.

cytolysosome [CYTOL] An enlarged lysosome that contains organelles such as mitochondria.

cytomegalic [MED] Of, pertaining to, or characterizing the greatly enlarged cells with enlarged nuclei and inclusion bodies found in tissues in cytomegalic inclusion disease.

cytomegalic inclusion disease [MED] A virus infection primarily of infants characterized by jaundice, liver enlargement, and circulatory disturbances.

cytomegalovirus [VIROL] An animal virus belonging to subgroup B of the herpesvirus group; causes cytomegalic inclusion disease and pneumonia.

cytomorphosis [CYTOL] All the structural alterations which cells or successive generations of cells undergo from the earliest undifferentiated stage to their final destruction.

cyton [HISTOL] The body of a neuron.

cytopathology [PATH] A branch of pathology concerned with abnormalities within cells.

cytopenia [PATH] A blood cell count below normal.

Cytophagaceae [MICROBIO] A family of slime bacteria, order Myxobacterales, having spherical microcysts surrounded by a thick wall.

cytopharynx [INV ZOO] A channel connecting the surface with the protoplasm in certain protozoans; functions as a gullet in ciliates.

cytophilic antibody [IMMUNOL] A substance capable of combining directly with the receptors of a corresponding antigenic cell. Also known as cytotropic antibody.

cytoplasm [CYTOL] The protoplasm of an animal or plant cell external to the nucleus.

cytoplasmic inheritance [GEN] The control of genetic difference by hereditary units carried in cytoplasmic organelles. Also known as extrachromosomal inheritance.

cytoplasmic streaming [CYTOL] Intracellular movement involving irreversible deformation of the cytoplasm produced by endogenous forces.

cytoproct *See* cytopyge.

cytopyge [INV ZOO] A fixed point for waste discharge in the body of a protozoan, especially a ciliate. Also known as cytoproct.

cytosine [BIOCHEM] $C_4H_5ON_3$ A pyrimidine occurring as a fundamental unit or base of nucleic acids.

cytoskeleton [CYTOL] Protein fibers composing the structural framework of a cell.

cytosome [CYTOL] The cytoplasm of the cell, as distinct from the nucleus.

cytostome [INV ZOO] The mouth-like opening in many unicellular organisms, particularly Ciliophora.

CYTOSINE

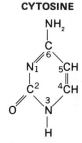

Structural formula of cytosine.

cytotaxis [PHYSIO] Attraction of motile cells by specific diffusible stimuli emitted by other cells.

cytotechnologist [PATH] A person trained to prepare smears of and examine exfoliated cells, referring abnormalities to a physician.

cytotrophoblast [EMBRYO] The inner, cellular layer of a trophoblast, covering the chorion and the chorionic villi during the first half of pregnancy.

cytotropic antibody *See* cytophilic antibody.

cytotropism [BIOL] The tendency of individual cells and groups of cells to move toward or away from each other.

cytozoic [INV ZOO] Of the sporozoan trophozoite, living within a cell.

Czapek's agar [MICROBIO] A nutrient culture medium consisting of salt, sugar, water, and agar; used for certain mold cultures.

Czochralski method [ECOL] Alteration of the original state of a community by human agencies.

D

Dacian [GEOL] Lower upper Pliocene geologic time.

Dacromycetales [MYCOL] An order of jelly fungi in the subclass Heterobasidiomycetidae having branched basidia with the appearance of a tuning fork.

dacryoblennorrhea [MED] Chronic inflammation of the lacrimal sac of the eye accompanied by discharge of mucus.

dacryocyst *See* lacrimal sac.

dacryocystitis [MED] Inflammation of the lacrimal sac.

dacryon [ANAT] The point of the face where the frontomaxillary, the maxillolacrimal, and frontolacrimal sutures meet.

Dactinomycin [MICROBIO] A trade name for actinomycin D.

dactyl [INV ZOO] In scorpions, a terminal ventral projection at the distal end of the praetarsus. [VERT ZOO] A digit, as a finger or toe.

Dactylochirotida [INV ZOO] An order of dendrochirotacean holothurians in which there are 8-30 digitate or digitiform tentacles, which sometimes bifurcate.

dactylognathite [INV ZOO] The distal segment of a maxilliped in crustaceans.

Dactylogyroidea [INV ZOO] A superfamily of trematodes in the subclass Monogenea; all are fish ectoparasites.

dactylopodite [INV ZOO] The distal segment of ambulatory limbs in decapods and of certain limbs in other arthropods.

dactylopore [INV ZOO] Any of the small openings on the surface of Milleporina through which the bodies of the polyps are extended.

Dactylopteridae [VERT ZOO] The flying gurnards, the single family of the perciform suborder Dactylopteroidei.

Dactylopteroidei [VERT ZOO] A suborder of marine shore fishes in the order Perciformes, characterized by tremendously expansive pectoral fins.

dactylopterous [VERT ZOO] Having anterior rays more or less unattached in the pectoral fins.

Dactyloscopidae [VERT ZOO] The sand stargazers, a family of small tropical and subtropical perciform fishes in the suborder Blennioidei.

dactylosternal [VERT ZOO] Of turtles, having marginal fingerlike processes joining the plastron to the carapace.

dactylozooid [INV ZOO] One of the long, defensive polyps of the Milleporina, armed with stinging cells.

dactylus [INV ZOO] The structure of the tarsus of certain insects which follows the first joint; usually consists of one or more joints.

dairy [AGR] A farm concerned with the production of milk. [FOOD ENG] An establishment where milk products are made.

Dakotan [GEOL] Lower Upper Cretaceous geologic time.

Dalatiidae [VERT ZOO] The spineless dogfishes, a family of modern sharks belonging to the squaloid group.

Dallis grass [BOT] The common name for the tall perennial

DACTYLOGYROIDEA

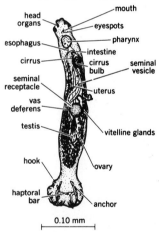

Pseudohaliotrema carbunculus Hargis from the pinfish.

forage grasses composing the genus *Paspalum* in the order Cyperales.

Dalton's law [PHYS] The law that the pressure of a gas mixture is equal to the sum of the partial pressures of the gases composing it. Also known as law of partial pressures.

damping-off [PL PATH] A fungus disease of seedlings and cuttings in which the parasites invade the plant tissues near the ground level, causing wilting and rotting.

Danaidae [INV ZOO] A family of large tropical butterflies, order Lepidoptera, having the first pair of legs degenerate.

dandruff [MED] Scales of dry sebum formed on the scalp in seborrhea.

dandy fever *See* dengue.

Daphoenidae [PALEON] A family of extinct carnivoran mammals in the superfamily Miacoidea.

dark adaptation [PHYSIO] Increase in light sensitivity when the eye remains in the dark.

dark-field illumination [OPTICS] A method of microscope illumination in which the illuminating beam is a hollow cone of light formed by an opaque stop at the center of the condenser large enough to prevent direct light from entering the objective; the specimen is placed at the concentration of the light cone, and is seen with light scattered or diffracted by it.

d'Arsonval current [ELEC] A current consisting of isolated trains of heavily damped high-frequency oscillations of high voltage and relatively low current, used in diathermy.

dart [INV ZOO] A small sclerotized structure ejected from the dart sac of certain snails into the body of another individual as a stimulant before copulation.

dartos [ANAT] A thin layer of smooth muscle attached to skin of the scrotum or of the labia majora.

dart sac [INV ZOO] A dart-forming pouch associated with the reproductive system of certain snails.

darwin [EVOL] A unit of evolutionary rate of change; if some dimension of a part of an animal or plant, or of the whole animal or plant, changes from l_o to l_t over a time of t years according to the formula $l_t = l_o \exp (Et/10^6)$, its evolutionary rate of change is equal to E darwins.

Darwinism [BIOL] The theory of the origin and perpetuation of new species based on natural selection of those offspring best adapted to their environment because of genetic variation and consequent vigor. Also known as Darwin's theory.

Darwin's finch [VERT ZOO] A bird of the subfamily Fringillidae; Darwin studied the variation of these birds and used his data as evidence for his theory of evolution by natural selection.

Darwin's theory *See* Darwinism.

Darwinulacea [INV ZOO] A small superfamily of nonmarine, parthenogenetic ostracods in the suborder Podocopa.

Dasayatidae [VERT ZOO] The stingrays, a family of modern sharks in the batoid group having a narrow tail with a single poisonous spine.

Dascillidae [INV ZOO] The soft-bodied plant beetles, a family of coleopteran insects in the superfamily Dascilloidea.

Dascilloidea [INV ZOO] A superfamily of coleopteran insects in the suborder Polyphaga.

dasheen [BOT] *Colocasia esculenta.* A plant in the order Arales, grown for its edible corm.

Dasycladaceae [BOT] A family of green algae in the order Dasycladales comprising plants formed of a central stem from which whorls of branches develop.

Dasycladales [BOT] An order of lime-encrusted marine algae in the division Chlorophyta, characterized by a thallus composed of nonseptate, highly branched tubes.

DARWINULACEA

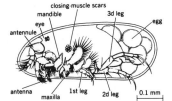

Darwinula stevensoni (Brady and Robertson), a podocopan ostracod of the Darwinulacea. In this parthenogenetic species, eggs and young are protected in a brood space behind the body. *(After R. C. Moore, ed., Treatise on Invertebrate Paleontology, pt. Q, 1961)*

Dasyonygidae [INV ZOO] A family of biting lice, order Mallophaga, that are confined to rodents of the family Procaviidae.

dasypaedes [VERT ZOO] Referring to birds whose young have a covering of down when hatched.

dasyphyllous [BOT] Having leaves with a thick covering of hairs.

Dasypodidae [VERT ZOO] The armadillos, a family of edentate mammals in the infraorder Cingulata.

Dasytidae [INV ZOO] An equivalent name for Melyridae.

Dasyuridae [VERT ZOO] A family of mammals in the order Marsupialia characterized by five toes on each hindfoot.

Dasyuroidea [VERT ZOO] A superfamily of marsupial mammals.

dating [SCI TECH] The use of methods and techniques to fix dates, assign periods of time, and determine age in archeology, biology, and geology.

Daubentoniidae [VERT ZOO] A family of Madagascan prosimian primates containing a single species, the aye-aye.

Daubenton's plane [ANTHRO] A plane passing through the opisthion and the orbital points on the skull.

Davian [GEOL] A subdivision of the Upper Cretaceous in Europe; a limestone formation with abundant hydrocorals, bryozoans, and mollusks in Denmark; marine limestone and nonmarine rocks in southeastern France; and continental formations in the Davian of Spain and Portugal.

Dawsoniales [BOT] An order of mosses comprising rigid plants with erect stems rising from a rhizomelike base.

day neutral [BOT] Reaching maturity regardless of relative length of light and dark periods.

DDT [ORG CHEM] Common name for an insecticide; melting point 108.5°C, insoluble in water, very soluble in ethanol and acetone, colorless, and odorless; especially useful against agricultural pests, flies, lice, and mosquitoes. Also known as dichlorodiphenyltrichloroethane.

dead arm [PL PATH] A fungus disease caused by *Cryptosporella viticola* in which the main lateral shoots are destroyed; common in grapes.

deadly nightshade *See* belladonna.

dead space [ANAT] The space in the trachea, bronchi, and other air passages which contains air that does not reach the alveoli during respiration, the amount of air being about 140 milliliters. Also known as anatomical dead space. [MED] A cavity left after closure of a wound. [PHYSIO] A calculated expression of the anatomical dead space plus whatever degree of overventilation or underperfusion is present; it is alleged to reflect the relationship of ventilation to pulmonary capillary perfusion. Also known as physiological dead space.

deafness [MED] Temporary or permanent impairment or loss of hearing.

dealation [INV ZOO] Casting off of wings, as by female ants after fertilization, or by termites.

deamidase [BIOCHEM] An enzyme that catalyzes the removal of an amido group from a compound.

deaminase [BIOCHEM] An enzyme that catalyzes the hydrolysis of amino compounds, removing the amino group.

death [MED] Cessation of all life functions; can involve the whole organism, an organ, individual cells, or cell parts.

death instinct [PSYCH] In psychoanalytic theory, the unconscious drive which leads the individual toward dissolution and death, and which coexists with the life instinct.

death point [PHYSIO] The limit (as of extremes of temperature) beyond which an organism cannot survive.

death rate *See* mortality rate.

de Beurmann–Gougerot disease *See* sporotrichosis.

de Broglie wave [QUANT MECH] The quantum-mechanical wave associated with a particle of matter.

Debye-Hückel theory [PHYS CHEM] A theory of the behavior of strong electrolytes, according to which each ion is surrounded by an ionic atmosphere of charges of the opposite sign whose behavior retards the movement of ions when a current is passed through the medium.

decafoliate [BOT] Having 10 leaves.

decalcification [CHEM] Loss or removal of calcium or calcium compounds from a calcified material such as bone or soil.

decandrous [BOT] Having 10 stamens.

decanth larva *See* lycophore larva.

Decapoda [INV ZOO] **1.** A diverse order of the class Crustacea including the shrimps, lobsters, hermit crabs, and true crabs; all members have a carapace, well-developed gills, and the first three pairs of thoracic appendages specialized as maxillipeds. **2.** An order of dibranchiate cephalopod mollusks containing the squids and cuttle fishes, characterized by eight arms and two long tentacles.

decarboxylase [BIOCHEM] An enzyme that hydrolyzes the carboxyl radical, COOH.

decerebellate [MED] Lacking the cerebellum, generally by experimental removal.

decerebrate [MED] Lacking the cerebrum either by experimental removal or by disconnection.

decerebrate rigidity [MED] Exaggerated postural tone in the antigravity muscles due to release of vestibular nuclei from cerebral control.

decidua [MED] The endometrium of pregnancy and associated fetal membranes which are cast off at parturition.

deciduous [BIOL] Falling off or being shed at the end of the growing period or season.

deciduous teeth [ANAT] Teeth of a young mammal which are shed and replaced by permanent teeth. Also known as milk teeth.

decile [STAT] Any of the points which divide the total number of items in a frequency distribution into 10 equal parts.

declinate [BIOL] Curved toward one side or downward.

Declomycin [MICROBIO] Trade name for demethylchlortetracycline.

decomposer [ECOL] A heterotrophic organism (including bacteria and fungi) which breaks down the complex compounds of dead protoplasm, absorbs some decomposition products, and releases substances usable by consumers. Also known as microcomposer; microconsumer; reducer.

decompound [BOT] Divided or compounded several times, with each division being compound.

decompression illness *See* aeroembolism.

decorticate [BIOL] Lacking a cortical layer.

decubitus ulcer [MED] An ulcer of the skin and subcutaneous tissues following prolonged lying down, due to pressure on bony protuberances. Also known as bedsore; pressure ulcer.

decumbent [BOT] Lying down on the ground but with an ascending tip, specifically referring to a stem.

decurrent [BOT] Running downward, especially of a leaf base extended past its insertion in the form of a winged expansion.

decussate [BOT] Of the arrangement of leaves, occurring in alternating pairs at right angles. [SCI TECH] To intersect in the form of an X.

dedifferentiation [BIOL] Disintegration of a specialized habit or adaptation. [PHYSIO] Return of a specialized cell or structure to a more general or primitive condition.

deep fascia [ANAT] The fibrous tissue between muscles and forming the sheaths of muscles, or investing other deep, definitive structures, as nerves and blood vessels.

deep hibernation [PHYSIO] Profound decrease in metabolic rate and physiological function during winter, with a body temperature near 0°C, in certain warm-blooded vertebrates. Also known as hibernation.

deep pain [PHYSIO] Pattern of somesthetic sensation of pain, usually indefinitely localized, originating in the viscera, muscles, and other deep tissues.

deep palmar arch [ANAT] The anastomosis between the terminal part of the radial artery and the deep palmar branch of the ulnar artery. Also known as deep volar arch; palmar arch.

deep sleep [PSYCH] The third and fourth stage of the sleep cycle, determined by electroencephalographic recording and characterized by slow brain waves. Also known as slow wave sleep (SWS).

deep volar arch *See* deep palmar arch.

deer [VERT ZOO] The common name for 41 species of even-toed ungulates that compose the family Cervidae in the order Artiodactyla; males have antlers.

defecation [PHYSIO] The process by which fecal wastes that reach the lower colon and rectum are evacuated from the body.

defective virus [VIROL] A virus, such as adeno-associated satellite virus, that can grow and reproduce only in the presence of another virus.

defeminization [PHYSIO] Loss or reduction of feminine attributes, usually due to ovarian dysfunction or removal. [PSYCH] Psychic process involving a deep and permanent change in the character of a woman, resulting in a giving up of feminine feelings, and the assumption of masculine qualities.

defense mechanism [PSYCH] Any psychic device, such as rationalization, denial, or repression, for concealing unacceptable feelings or for protecting oneself against unpleasant feelings, memories, or experiences.

defibrillation [MED] Stopping a local quivering of muscle fibers, especially of the heart.

defibrillator [MED] An electronic instrument used for stopping fibrillation during a heart attack by applying controlled electric pulses to the heart muscles.

deficiency [CYTOL] The absence of a chromosome segment; can be terminal or interstitial.

deficiency disease [MED] Any disease resulting from a dietary deficiency of minerals, vitamins, or essential nutrients.

definite [BOT] **1.** Pertaining to an inflorescence whose primary axis terminates early in flower development. **2.** Having the number of stamens limited to 20.

definite composition law [CHEM] The law that a given chemical compound always contains the same elements in the same fixed proportions by weight. Also known as definite proportions law.

definite proportions law *See* definite composition law.

definitive host [BIOL] The host in which a parasite reproduces sexually.

deflexed [BIOL] Turned sharply downward.

defluvium [PATH] The pathological loss of a part of an animal or plant, as nails or bark.

defoliant [MATER] A chemical sprayed on plants that causes leaves to fall off prematurely.

defoliate [BOT] To remove leaves or cause leaves to fall, especially prematurely.

degenerate code [GEN] A genetic code in which more than

DEER

North American elk *(Cerrus canadensis)*, a deer common in Canada, stands about 5½ feet (1.7 meters) at shoulder.

one triplet sequence of nucleotides (codon) can specify the insertion of the same amino acid into a polypeptide chain.

degeneration [MED] **1.** Deterioration of cellular integrity with no sign of response to injury or disease. **2.** General deterioration of a physical, mental, or moral state.

degenerative arthritis *See* degenerative joint disease.

degenerative disease [MED] General debility and diseases associated with advancing age.

degenerative joint disease [MED] A chronic joint disease characterized pathologically by degeneration of articular cartilage and hypertrophy of bone, clinically by pain on activity which subsides with rest. Also known as degenerative arthritis; hypertrophic arthritis; osteoarthritis; senescent arthritis.

Degeneriaceae [BOT] A family of dicotyledonous plants in the order Magnoliales characterized by laminar stamens; a solitary, pluriovulate, unsealed carpel; and ruminate endosperm.

deglutition [PHYSIO] Act of swallowing.

degree [THERMO] One of the units of temperature or temperature difference in any of various temperature scales, such as the Celsius, Fahrenheit, and Kelvin temperature scales (the Kelvin degree is now known as the kelvin).

degree of freedom [PHYS CHEM] Any one of the variables, including pressure, temperature, composition, and specific volume, which must be specified to define the state of a system. [STAT] A number one less than the number of frequencies being tested with a chi-square test.

dehiscence [BOT] Spontaneous bursting open of a mature plant structure, such as fruit, anther, or sporangium, to discharge its contents. [MED] A defect in the boundary of a bony canal or cavity.

dehydrase [BIOCHEM] An enzyme which catalyzes the removal of water from a substrate.

dehydration [CHEM] Removal of water from any substance.

dehydroascorbic acid [ORG CHEM] $C_6H_6O_6$ A relatively inactive acid resulting from elimination of two hydrogen atoms from ascorbic acid when the latter is oxidized by air or other agents; has potential ascorbic acid activity.

dehydrocholesterol [BIOCHEM] $C_{27}H_{43}OH$ A provitamin of animal origin found in the skin of man, in milk, and elsewhere, which upon irradiation with ultraviolet rays becomes vitamin D.

dehydrogenase [BIOCHEM] An enzyme which removes hydrogen atoms from a substrate and transfers it to an acceptor other than oxygen.

dehydrogenation [CHEM] Removal of hydrogen from a compound.

Deinotheriidae [PALEON] A family of extinct proboscidean mammals in the suborder Deinotherioidea; known only by the genus *Deinotherium.*

Deinotherioidea [PALEON] A monofamilial suborder of extinct mammals in the order Proboscidea.

Deister phase [GEOL] A subdivision of the late Ammerian phase of the Jurassic period between the Kimmeridgian and lower Portlandian.

Deiter's cells [ANAT] **1.** The outer phalangeal cells of the organ of Corti. **2.** Large nerve cells in the lateral vestibular nucleus of the medulla oblongata.

delamination [BIOL] The separation of cells into layers. [EMBRYO] Gastrulation in which the endodermal layer splits off from the inner surface of the blastoderm and the space between this layer and the yolk represents the archenteron.

delayed hypersensitivity [IMMUNOL] Abnormal reactivity in a sensitized individual beginning several hours after contact with the allergen.

delayed speech [MED] A speech disorder characterized by a complete absence of vocalization or vocalization with no communicative value; speech is considered delayed when it fails to develop by the second year, caused by impaired hearing, severe childhood illness, or emotional disturbance.

deletion [GEN] Loss of a chromosome segment of any size, down to a part of a single gene.

deletion mapping [GEN] A procedure for mapping mutants additional to three-factor cross; deletion or multisite mutants have lost a segment of their chromosome; its extent can be found through crosses with other mutants in which the positions of some mutant genes have already been determined.

delirium [MED] Severely disordered mental state associated with fever, intoxication, head trauma, and other encephalopathies.

delirium tremens [MED] Delirium associated with tremors, insomnia, and other physical and neurological symptoms frequently following chronic alcoholism.

Delmontian [GEOL] Upper Miocene or lower Pliocene geologic time.

delomorphous cell *See* parietal cell.

Delphinidae [VERT ZOO] A family of aquatic mammals in the order Cetacea; includes the dolphins.

delphinidin [BIOCHEM] $C_{15}H_{11}O_7Cl$ A purple or brownish-red anthocyanin compound occurring widely in plants.

delta rhythm [PHYSIO] An electric current generated in slow waves with frequencies of 0.5–3 per second from the forward portion of the brain of normal subjects when asleep.

Deltatheridia [PALEON] An order of mammals that includes the dominant carnivores of the early Cenozoic.

delthyrium [INV ZOO] In certain brachiopods, the opening between the hinge and the beak, through which the peduncle is extended.

deltidium [INV ZOO] The plate which covers the delthyrium.

deltoid [ANAT] The large triangular shoulder muscle; originates on the pectoral girdle and inserts on the humerus. [BIOL] Triangular in shape.

deltoid ligament [ANAT] The ligament on the medial side of the ankle joint; the fibers radiate from the medial malleolus to the talus, calcaneus, and navicular bones.

delusion [PSYCH] A conviction based on faulty perceptions, feelings, and thinking.

demarcation potential *See* injury potential.

Dematiaceae [MYCOL] A family of fungi in the order Moniliales; sporophores are not grouped, hyphae are always dark, and the spores are hyaline or dark.

deme [ECOL] A local population in which the individuals freely interbreed among themselves but not with those of other demes.

dementia [PSYCH] Deterioration of intellectual and other mental processes due to organic brain disease.

dementia praecox *See* schizophrenia.

dementia simplex [PSYCH] A subtype of schizophrenia broadly characterized by a slow, progressive deterioration, often combined with mental deficiency.

demersal [BIOL] Living at or near the bottom of the sea.

demethylchlortetracycline [MICROBIO] $C_{21}H_{21}O_8N_2Cl$ A broad-spectrum tetracycline antibiotic produced by a mutant strain of *Streptomyces aureofaciens*.

demilune [BIOL] Crescent-shaped.

demineralization [MED] Removal or loss of minerals and salts from the body, especially by disease.

DELTATHERIDIA

The deltatheridian *Hyaenodon*, original about 4 feet (1.2 meters) long. (*After Scott, from A. S. Romer, Vertebrate Paleontology, 3d ed., University of Chicago Press, 1966*)

DENDROBRANCHIATE GILL

Dendrobranchiate gill of
penaeid shrimp, *Benthesicymus*.
(Smithsonian Institution)

DENDROCERATIDA

Dendritic skeleton of *Dendrilla
cactus. (After Lendenfeld, 1889)*

DENDROIDEA

rhabdosome

Whole colony of a dendroid
graptolite.

demisheath [INV ZOO] One of a pair of covers which protect the ovipositor of an insect.

Demodicidae [INV ZOO] The pore mites, a family of arachnids in the suborder Trombidiformes.

demography [ECOL] The statistical study of populations with reference to natality, mortality, migratory movements, age, and sex, among other social, ethnic, and economic factors.

Demospongiae [INV ZOO] A class of the phylum Porifera, including sponges with a skeleton of one- to four-rayed siliceous spicules, or of spongin fibers, or both.

demyelination [PATH] Destruction of myelin; loss of myelin from nerve sheaths or nerve tracts.

denature [CHEM] **1.** To change a protein by heating it or treating it with alkali or acid so that the original properties such as solubility are changed as a result of the protein's molecular structure being changed in some way. **2.** To add a denaturant, such as methyl alcohol, to grain alcohol to make the grain alcohol poisonous and unfit for human consumption.

dendrite [ANAT] The part of a neuron that carries the unidirectional nerve impulse toward the cell body. Also known as dendron. [CRYSTAL] A crystal having a treelike structure.

Dendrobatinae [VERT ZOO] A subfamily of anuran amphibians in the family Ranidae, including the colorful poisonous frogs of Central and South America.

dendrobranchiate gill [INV ZOO] A respiratory structure of certain decapod crustaceans, characterized by extensive branching of the two primary series.

Dendroceratida [INV ZOO] A small order of sponges of the class Demospongiae; members have a skeleton of spongin fibers or lack a skeleton.

Dendrochirotacea [INV ZOO] A subclass of echinoderms in the class Holothuroidea.

Dendrochirotida [INV ZOO] An order of dendrochirotacean holothurian echinoderms with 10–30 richly branched tentacles.

dendrochronology [GEOL] The science of measuring time intervals and dating events and environmental changes by reading and dating growth layers of trees as demarcated by the annual rings.

Dendrocolaptidae [VERT ZOO] The woodcreepers, a family of passeriform birds belonging to the suboscine group.

dendrogram [BIOL] A genealogical tree; the trunk represents the oldest common ancestor, and the branches indicate successively more recent divisions of a lineage for a group.

dendroid [BIOL] Branched or treelike in form.

Dendroidea [PALEON] An order of extinct sessile, branched colonial animals in the class Graptolithina occurring among typical benthonic fauna.

dendrology [FOR] The division of forestry concerned with the classification, identification, and distribution of trees and other woody plants.

Dendromurinae [VERT ZOO] The African tree mice and related species, a subfamily of rodents in the family Muridae.

dendron *See* dendrite.

dendrophagous [ZOO] Feeding on trees, referring to insects.

dendrophysis [MYCOL] A hyphal thread with arboreal branching in certain fungi.

denervate [MED] To interfere with or cut off the nerve supply to a part of the body, or to remove a nerve; may occur by excision, drugs, or a disease process.

denervation hypersensitivity [PHYSIO] Extreme sensitivity of an organ that has recovered from the removal or interruption of its nerve supply.

dengue [MED] An acute viral disease of man characterized by fever, rash, prostration, and lymphadenopathy; transmitted by the mosquito *Aedes aegypti*. Also known as breakbone fever; dandy fever.

denial [PSYCH] An unconscious defense mechanism in which an individual denies himself recognition of an observation in order to avoid pain or anxiety.

denitrification [MICROBIO] The reduction of nitrate or nitrite to gaseous products such as nitrogen, nitrous oxide, and nitric oxide; brought about by denitrifying bacteria.

denitrifying bacteria [MICROBIO] Bacteria that reduce nitrates to nitrites or nitrogen gas; most are found in soil.

denitrogenate [PHYSIO] To remove nitrogen from the body by breathing nitrogen-free gas.

dense connective tissue [HISTOL] A fibrous connective tissue with an abundance of enlarged collagenous fibers which tend to crowd out the cells and ground substance.

density [MECH] The mass of a given substance per unit volume. [PHYS] The total amount of a quantity, such as energy, per unit of space.

dental [ANAT] Pertaining to the teeth.

dental arch [ANAT] The parabolic curve formed by the cutting edges and masticating surfaces of the teeth.

dental calculi [MED] Calcareous deposits of organic and mineral matter on the teeth. Also known as tartar.

dental caries *See* caries.

dental follicle *See* dental sac.

dental formula [VERT ZOO] An expression of the number and kind of teeth in each half jaw, both upper and lower, of mammals.

Dentaliidae [INV ZOO] A family of mollusks in the class Scaphopoda; members have pointed feet.

dental index [ANTHRO] A ratio of the length of the teeth to the distance from the nasion to the basion multiplied by 100, used to determine the relative size of teeth.

dental pad [VERT ZOO] A firm ridge that replaces incisors in the maxilla of cud-chewing herbivores.

dental papilla [EMBRYO] The mass of connective tissue located inside the enamel organ of a developing tooth, and forming the dentin and dental pulp of the tooth.

dental plate [INV ZOO] A flat plate that replaces teeth in certain invertebrates, such as some worms. [VERT ZOO] A flattened plate that represents fused teeth in parrot fishes and related forms.

dental pulp [HISTOL] The vascular connective tissue of the roots and pulp cavity of a tooth.

dental ridge [EMBRYO] An elevation of the embryonic jaw that forms a cusp or margin of a tooth.

dental sac [EMBRYO] The connective tissue that encloses the developing tooth. Also known as dental follicle.

dentate [BIOL] 1. Having teeth. 2. Having toothlike or conical marginal projections.

dentate fissure *See* hippocampal sulcus.

dentate nucleus [ANAT] An ovoid mass of nerve cells located in the center of each cerebellar hemisphere, which give rise to fibers found in the superior cerebellar peduncle.

denticle [ZOO] A small tooth or toothlike projection, as the type of scale of certain elasmobranchs.

denticulate [ZOO] Having denticles; serrate.

dentigerous [BIOL] Having teeth or toothlike structures.

dentin [HISTOL] A bonelike tissue composing the bulk of a vertebrate tooth; consists of 70% inorganic materials and 30% water and organic matter.

dentinoblast [HISTOL] A mesenchymal cell that forms dentin.

dentinogenesis [PHYSIO] The formation of dentin.

dentinoma [MED] A benign odontogenič tumor made up of dentin.

dentist [MED] One who practices dentistry.

dentistry [MED] A branch of medical science concerned with the prevention, diagnosis, and treatment of diseases of the teeth and adjacent tissues and the restoration of missing dental structures.

dentition [VERT ZOO] The arrangement, type, and number of teeth which are variously located in the oral or in the pharyngeal cavities, or in both, in vertebrates.

denture [MED] A partial or complete prosthetic appliance to replace one or more missing teeth.

deoperculate [BOT] Of mosses and liverworts, to shed the operculum.

deoxidize [CHEM] **1.** To remove oxygen by any of several processes. **2.** To reduce from the state of an oxide.

deoxycholate [BIOCHEM] A salt or ester of deoxycholic acid.

deoxycholic acid [BIOCHEM] $C_{24}H_{40}O_4$ One of the unconjugated bile acids; in bile it is largely conjugated with glycine or taurine.

deoxycorticosterone [BIOCHEM] $C_{21}H_{30}O_3$ A steroid hormone secreted in small amounts by the adrenal cortex.

6-deoxy-L-galactose See L-fucose.

deoxyribonuclease [BIOCHEM] An enzyme that catalyzes the hydrolysis of deoxyribonucleic acid to nucleotides.

deoxyribonucleic acid [BIOCHEM] A linear polymer made up of deoxyribonucleotide repeating units (composed of the sugar 2-deoxyribose, phosphate, and a purine or pyrimidine base) linked by the phosphate group joining the 3′ position of one sugar to the 5′ position of the next; most molecules are double-stranded and antiparallel, resulting in a right-handed helix structure kept together by hydrogen bonds between a purine on one chain and a pyrimidine on another; carrier of genetic information, which is encoded in the sequence of bases; present in chromosomes and chromosomal material of cell organelles such as mitochondria and chloroplasts, and also present in some viruses. Abbreviated DNA.

deoxyribonucleoprotein [BIOCHEM] A protein containing molecules of deoxyribonucleic acid in close association with protein molecules.

deoxyribose [BIOCHEM] $C_5H_{10}O_4$ A pentose sugar in which the hydrogen replaces the hydroxyl groups of ribose; a major constituent of deoxyribonucleic acid.

deoxy sugar [BIOCHEM] A substance which has the characteristics of a sugar, but which shows a deviation from the required hydrogen-to-oxygen ratio.

depauperate [BIOL] Inferiority of natural development or size.

dependence [STAT] The existence of a relationship between frequencies obtained from two parts of an experiment which does not arise from the direct influence of the result of the first part on the chances of the second part but indirectly from the fact that both parts are subject to influences from a common outside factor.

dependency needs [PSYCH] The vital, originally infantile needs for mothering, love, affection, shelter, protection, security, food, and warmth, which are also present in some degree even in adult life.

depersonalization [PSYCH] Loss of the sense of one's identity or of reality concerning the self.

Depertellidae [PALEON] A family of extinct perissodactyl mammals in the superfamily Tapiroidea.

depilation [BIOL] Removal of hair. [BOT] In plants, loss of a hairy covering during maturation.

depolarization [PHYSIO] Reduction in the internal negativity

DEOXYCORTICOSTERONE

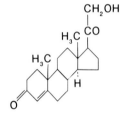

Structural formula of deoxycorticosterone.

of a nerve or muscle cell in the conduction of an impulse.

deposit feeder [INV ZOO] Any animal that feeds on the detritus that collects on the substratum at the bottom of water. Also known as detritus feeder.

deposition potential [PHYS CHEM] The smallest potential which can produce electrolytic deposition when applied to an electrolytic cell.

depressed fracture [MED] A fracture of the skull in which the fractured part is depressed below the normal level.

depression [PSYCH] A mood provoked by conscious awareness of an idea or feeling that was previously pushed into the unconscious.

depressor [ANAT] A muscle that draws a part down.

depressor nerve [PHYSIO] A nerve which, upon stimulation, lowers the blood pressure either in a local part or throughout the body.

depth perception [PHYSIO] Ability to judge spatial relationships.

depurination [BIOCHEM] Detachment of guanine from sugar in a deoxyribonucleic acid molecule.

derealization [PSYCH] Loss of the sense of the reality of people or objects in one's environment.

derepression [MICROBIO] Transfer of microbial cells from an enzyme-repressing medium to a nonrepressing medium.

dermal [ANAT] Pertaining to the dermis.

dermal bone [ANAT] A type of bone that ossifies directly from membrane without a cartilaginous predecessor; occurs only in the skull and shoulder region. Also known as investing bone; membrane bone.

dermal denticle [VERT ZOO] A toothlike scale composed mostly of dentine with a large central pulp cavity, found in the skin of sharks.

dermalia [INV ZOO] Dermal microscleres in sponges.

dermal pore [INV ZOO] One of the minute openings on the surface of poriferans leading to the incurrent canals.

Dermaptera [INV ZOO] An order of small or medium-sized, slender insects having incomplete metamorphosis, chewing mouthparts, short forewings, and cerci.

Dermatemydinae [VERT ZOO] A family of reptiles in the order Chelonia; includes the river turtles.

dermatitis [MED] Inflammation of the skin.

Dermatocarpaceae [BOT] A family of lichens in the order Pyrenulales having an umbilicate or squamulose growth form; most members grow on limestone or calcareous soils.

dermatocranium [ANAT] Bony parts of the skull derived from ossifications in the dermis of the skin.

dermatogen [BOT] The outer layer of primary meristem or the primordial epidermis in embryonic plants. Also known as protoderm.

dermatoglyphics [ANAT] **1.** The integumentary patterns on the surface of the fingertips, palms, and soles. **2.** The study of these patterns.

dermatograph [MED] A crayonlike or similar instrument, used to mark the skin before surgery to outline positions of organs.

dermatologist [MED] A physician who specializes in diseases of the skin.

dermatology [MED] The science of the structure, function, and diseases of the skin.

dermatome [ANAT] An area of skin delimited by the supply of sensory fibers from a single spinal nerve. [EMBRYO] Lateral portion of an embryonic somite from which the dermis will develop. [MED] Instrument for cutting skin for grafting.

dermatomyositis [MED] An inflammatory reaction of unknown cause involving degenerative changes of skin and muscle.

dermatopathic lymphadenitis *See* lipomelanotic reticulosis.

dermatopathology [MED] A branch of pathology concerned with diseases of the skin.

dermatophyte [MYCOL] A fungus parasitic on skin or its derivatives.

dermatophytosis [MED] A fungus infection, such as ringworm, of the skin of humans caused by the organism living in the keratinized tissues, characterized by vesicles, cracking, and scaling.

dermatosclerosis *See* scleroderma.

dermatozoon [ECOL] An animal that is parasitic on skin.

Dermestidae [INV ZOO] The skin beetles, a family of coleopteran insects in the superfamily Dermestoidea, including serious pests of stored agricultural grain products.

Dermestoidea [INV ZOO] A superfamily of coleopteran insects in the suborder Polyphaga.

dermis [ANAT] The deep layer of the skin, a dense connective tissue richly supplied with blood vessels, nerves, and sensory organs. Also known as corium; cutis.

Dermochelidae [VERT ZOO] A family of reptiles in the order Chelonia composed of a single species, the leatherback turtle.

dermoepidermal junction [HISTOL] The area of separation between the stratum basale of the epidermis and the papillary layer of the dermis.

dermographia [MED] A condition in which the skin is peculiarly susceptible to irritation, characterized by elevations or wheals with surrounding erythematous axon reflex flare, caused by tracing a fingernail or a blunt instrument over the skin.

dermoid cyst [MED] A benign cystic teratoma with skin, skin appendages, and their products as the most prominent components, usually involving the ovary or the skin.

dermopharyngeal [VERT ZOO] A plate of membrane bone which supports the pharyngeal teeth in some fishes.

Dermoptera [VERT ZOO] The flying lemurs, an ancient order of primatelike herbivorous and frugivorous gliding mammals confined to southeastern Asia and eastern India.

Derodontidae [INV ZOO] The tooth-necked fungus beetles, a small family of coleopteran insects in the superfamily Dermestoidea.

derris [BOT] Any of certain tropical shrubs in the genus *Derris* in the legume family (Leguminosae), having long climbing branches.

dertrotheca [VERT ZOO] In birds, the horny covering of the maxilla.

Descemet's membrane [HISTOL] A layer of the cornea between the posterior surface of the stroma and the anterior surface of the endothelium which contains collagen arranged on a crystalline lattice.

descending [ANAT] Extending or directed downward or caudally, as the descending aorta. [PHYSIO] In the nervous system, efferent; conducting impulses or progressing down the spinal cord or from central to peripheral.

descending chromatography [ANALY CHEM] A type of paper chromatography in which the sample-carrying solvent mixture is fed to the top of the developing chamber, being separated as it works downward.

descriptive anatomy [ANAT] Study of the separate and individual portions of the body, with regard to form, size, character, and position.

descriptive botany [BOT] The branch of botany that deals with diagnostic characters or systematic description of plants.

desensitization [IMMUNOL] Loss or reduction of sensitivity

DERMESTIDAE

A drawing of a skin beetle. *(From T. I. Storer and R. L. Usinger, General Zoology, 3d ed., McGraw-Hill, 1957)*

to infection or an allergen accomplished by means of frequent, small doses of the antigen. Also known as hyposensitization. [PSYCH] Relief from or removal of a mental complex.

deserticolous [ECOL] Living in a desert.

desexualization [PHYSIO] Depriving an organism of sexual characters or power, as by spaying or castration. [PSYCH] Repression of the sexual drive or rechanneling of sexual energy into areas considered by the individual to be more socially acceptable.

desilker [FOOD ENG] A machine consisting of a series of revolving rolls and brushes for removing silk from ears of corn.

desma [INV ZOO] A branched, knobby spicule in some Demospongiae.

desmacyte [INV ZOO] A bipolar collencyte found in the cortex of certain sponges.

Desmidiaceae [BOT] A family of desmids, mostly unicellular algae in the order Conjugales.

desmochore [ECOL] A plant having sticky or barbed disseminules.

Desmodonta [PALEON] An order of extinct bivalve, burrowing mollusks.

Desmodontidae [VERT ZOO] A small family of chiropteran mammals comprising the true vampire bats.

Desmodoroidea [INV ZOO] A superfamily of marine- and brackish-water-inhabiting nematodes with an annulated, usually smooth cuticle.

desmognathous [VERT ZOO] In some birds, having fused maxillopalatines.

Des Moinesian [GEOL] Lower Middle Pennsylvanian geologic time.

Desmokontae [BOT] The equivalent name for Desmophyceae.

desmolase [BIOCHEM] Any of a group of enzymes which catalyze rupture of atomic linkages that are not cleaved through hydrolysis, such as the bonds in the carbon chain of D-glucose.

desmoneme [INV ZOO] A nematocyst having a long coiled tube which is extruded and wrapped around the prey.

desmopelmous [VERT ZOO] A type of bird foot in which the hindtoe cannot be bent independently because planter tendons are united.

Desmophyceae [BOT] A class of rare, mostly marine algae in the division Pyrrhophyta.

desmoplasia [MED] **1.** The formation and proliferation of connective tissue, frequently in the growth of tumors. **2.** The formation of adhesions.

Desmoscolecida [INV ZOO] An order of the class Nematoda.

Desmoscolecidae [INV ZOO] A family of nematodes in the superfamily Desmoscolecoidea; individuals resemble annelids in having coarse annulation.

Desmoscolecoidea [INV ZOO] A small superfamily of free-living nematodes characterized by a ringed body, an armored head set, and hemispherical amphids.

desmose [INV ZOO] A fibril connecting the centrioles during mitosis in certain protozoans.

desmosome [CYTOL] A surface modification of stratified squamous epithelium formed by the apposition of short cytoplasmic processes of neighboring cells.

Desmostylia [PALEON] An extinct order of large hippopotamuslike, amphibious, gravigrade, shellfish-eating mammals.

Desmostylidae [PALEON] A family of extinct mammals in the order Desmostylia.

Desmothoracida [INV ZOO] An order of sessile and free-

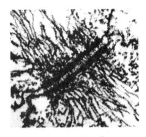

DESMOSOME

Electron micrograph of desmosome in skin of newt. *(Courtesy of D. Kelly)*

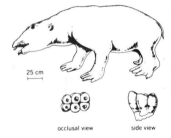

DESMOSTYLIA

25 cm

occlusal view   side view

Restoration of *Desmostylus* from the Miocene of California. *(Redrawn from R. A. Stirton, 1959)*

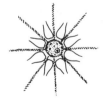

DESMOTHORACIDA

Choanocystis lepidula, a floating species. *(From R. P. Hall, Protozoology, Prentice-Hall, 1953)*

living protozoans in the subclass Heliozoia having a spherical body with a perforate, chitinous test.

desorption [PHYS CHEM] The process of removing a sorbed substance by the reverse of adsorption or absorption.

Desor's larva [INV ZOO] An oval, ciliated larva of certain nemertineans in which the gastrula remains inside the egg membrane.

desquamation [PHYSIO] Shedding; a peeling and casting off, as of the superficial epithelium, mucous membranes, renal tubules, and the skin.

Desulfovibrio [MICROBIO] A genus of strictly anaerobic, motile, rod-shaped bacteria in the family Spirillaceae which reduce sulfates to hydrogen sulfide.

detached meristem [BOT] A meristematic region originating from apical meristem but becoming discontinuous with it because of differentiation of intervening tissue.

determinate [SCI TECH] Bounded by definite limits.

determinate cleavage [EMBRYO] A type of cleavage which separates portions of the zygote with specific and distinct potencies for development as specific parts of the body.

determinate growth [BOT] Growth in which the axis, or central stem, being limited by the development of the floral reproductive structure, does not grow or lengthen indefinitely.

detorsion [INV ZOO] Untwisting of the 180° visceral twist imposed by embryonic torsion on many gastropod mollusks. [MED] Untwisting of an abnormal torsion, as of a ureter or intestine.

detoxification [BIOCHEM] The act or process of removing a poison or the toxic properties of a substance in the body.

detritus feeder *See* deposit feeder.

detrusor [PHYSIO] Any muscle that thrusts down or out.

deuteranomaly [MED] A partial deuteranopia.

deuteranopia [MED] Defective vision consisting of red-green color confusion, with no marked reduction in the brightness of any color.

deuterium [CHEM] The isotope of the element hydrogen with one neutron and one proton in the nucleus; atomic weight 2.0144. Designated D, d, H^2, or ^2H.

deuterocerebrum [INV ZOO] In crustaceans, that portion of the brain which gives rise to the antennular nerves.

deuteroconidium [MYCOL] In dermatophytes, a conidium produced by fission of a hemispore or protoconidium.

deuterogamy [BOT] Secondary pairing of sexual cells or nuclei replacing direct copulation in many fungi, algae, and higher plants.

Deuteromycetes [MYCOL] The equivalent name for Fungi Imperfecti.

Deuterophlebiidae [INV ZOO] A family of dipteran insects in the suborder Cyclorrhapha.

Deuterostomia [ZOO] A division of the animal kingdom which includes the phyla Echinodermata, Chaetognatha, Hemichordata, and Chordata.

deuterozooid [INV ZOO] A zooid formed from a primary zooid by budding.

deutocerebrum [INV ZOO] The median lobes of the insect brain.

deutonymph [INV ZOO] The second nymphal stage in the development of Acaridae.

deutoplasm [EMBRYO] The nutritive yolk granules in egg cells.

deutoscolex [INV ZOO] In some tapeworms, a secondary scolex formed by budding during the bladderworm stage.

deutosternum [INV ZOO] In Acarina, the sternite of the pedipalp-bearing segment of the body.

deutovum [INV ZOO] An undeveloped mite larva released after rupture of the outer eggshell.

devernalization [BOT] Annulment of the vernalization effect.

deviation [EVOL] Evolutionary differentiation involving interpolation of new stages in the ancestral pattern of morphogenesis. [STAT] The difference between any given number in a set and the mean average of those numbers.

Devonian [GEOL] The fourth period of the Paleozoic Era.

dewclaw [VERT ZOO] **1.** A vestigial digit on the foot of a mammal which does not reach the ground. **2.** A claw or hoof terminating such a digit.

dewlap [ANAT] A fleshy or fatty fold of skin on the throat of some humans. [BOT] One of a pair of hinges at the joint of a sugarcane leaf blade. [VERT ZOO] A fold of skin hanging from the neck of some reptiles and bovines.

Dexaminidae [INV ZOO] A family of amphipod crustaceans in the suborder Gammeridea.

dexterotropic [BIOL] Turning toward the right; applied to cleavage, shell formation, and whorl patterns.

dextran [BIOCHEM] Any of the several polysaccharides, $(C_5H_{10}O_5)_n$, that yield glucose units on hydrolysis.

dextrin [BIOCHEM] A polymer of D-glucose which is intermediate in complexity between starch and maltose.

dextro See dextrorotatory.

dextrocardia [MED] The presence of the heart in the right hemithorax, with the cardiac apex directed to the right.

dextrorotatory [OPTICS] Rotating clockwise the plane of polarization of a wave traveling through a medium in a clockwise direction, as seen by an eye observing the light. Abbreviated dextro.

dextrorse [BOT] Twining toward the right.

dextrose [BIOCHEM] $C_6H_{12}O_6 \cdot H_2O$ A dextrorotatory monosaccharide obtained as a white, crystalline, odorless, sweet powder, which is soluble in about one part of water; an important intermediate in carbohydrate metabolism; used for nutritional purposes, for temporary increase of blood volume, and as a diuretic. Also known as corn sugar; grape sugar.

dextrotopic cleavage [EMBRYO] A clockwise spiral cleavage pattern.

diabetes [MED] Any of various abnormal conditions characterized by excessive urinary output, thirst, and hunger; usually refers to diabetes mellitus.

diabetes insipidus [MED] A form of diabetes due to a disfunction of the hypothalamus.

diabetes mellitus [MED] A metabolic disorder arising from a defect in carbohydrate utilization by the body, related to inadequate or abnormal insulin production by the pancreas.

diabetic glomerulosclerosis See intercapillary glomerulosclerosis.

Diacodectidae [PALEON] A family of extinct artiodactyl mammals in the suborder Palaeodonta.

diadelphous stamen [BOT] A stamen that has its filaments united into two sets.

Diadematacea [INV ZOO] A superorder of Euchinoidea having a rigid or flexible test, perforate tubercles, and branchial slits.

Diadematidae [INV ZOO] A family of large euechinoid echinoderms in the order Diadematoida having crenulate tubercles and long spines.

diadematoid [INV ZOO] In Echinoidea, having three primary pore plates, and occasionally a secondary between the aboral and middle primary plates.

Diadematoida [INV ZOO] An order of echinoderms in the superorder Diadematacea with hollow primary radioles and crenulate tubercles.

DEVONIAN

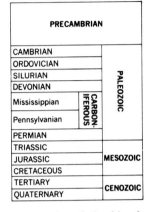

PRECAMBRIAN			
CAMBRIAN		PALEOZOIC	
ORDOVICIAN			
SILURIAN			
DEVONIAN			
Mississippian	CARBONIFEROUS		
Pennsylvanian			
PERMIAN			
TRIASSIC		MESOZOIC	
JURASSIC			
CRETACEOUS			
TERTIARY		CENOZOIC	
QUATERNARY			

Chart showing relationship of Devonian to other periods.

DIADELPHOUS STAMEN

staminode

Cutaway drawing of a flower showing diadelphous stamen.

DIALYSIS

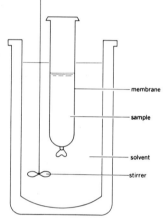

membrane

sample

solvent

stirrer

Equipment for dialysis.

diadromous [BOT] Having venation in the form of fanlike radiations. [VERT ZOO] Of fish, migrating between salt and fresh waters.

Diadumenidae [INV ZOO] A family of anthozoans in the order Actiniaria.

diageotropism [BIOL] Growth orientation of a sessile organism or structure perpendicular to the line of gravity.

diagnosis [MED] Identification of a disease from its signs and symptoms.

diakinesis [CYTOL] The last stage of meiotic prophase, when the chromatids attain maximum contraction and the bivalents move apart and position themselves against the nuclear membrane.

dialysis [PHYS CHEM] A process of selective diffusion through a membrane; usually used to separate low-molecular-weight solutes which diffuse through the membrane from the colloidal and high-molecular-weight solutes which do not.

dialystely [BOT] Having the steles more or less separate in the stem.

dialyzate [CHEM] The material that does not diffuse through the membrane during dialysis; alternatively, it may be considered the material that has diffused.

diamine oxidase [BIOCHEM] A flavoprotein which catalyzes the aerobic oxidation of amines to the corresponding aldehyde and ammonia.

diamond canker [PL PATH] A virus disease that affects the bark of certain stone-fruit trees, resulting in weakening of the trunk and limbs.

diandrous [BOT] Having two stamens.

Dianemaceae [MICROBIO] A family of slime molds in the order Trichales.

Dianulitidae [PALEON] A family of extinct, marine bryozoans in the order Cystoporata.

diapause [PHYSIO] A period of spontaneously suspended growth or development in certain insects, mites, crustaceans, and snails.

diapedesis [MED] Hemorrhage of blood cells, especially erythrocytes, through an intact vessel wall into the tissues.

Diapensiaceae [BOT] The single family of the Diapensiales, an order of flowering plants.

Diapensiales [BOT] A monofamilial order of dicotyledonous plants in the subclass Dilleniidae comprising certain herbs and dwarf shrubs in temperate and arctic regions of the Northern Hemisphere.

Diaphanocephalidae [INV ZOO] A family of parasitic roundworms belonging to the Strongyloidea; snakes are the principal host.

diaphorase [BIOCHEM] Mitochondrial flavoprotein enzymes which catalyze the reduction of dyes, such as methylene blue, by reduced pyridine nucleotides such as reduced diphosphopyridine nucleotide.

diaphragm [ANAT] The dome-shaped partition composed of muscle and connective tissue that separates the abdominal and thoracic cavities in mammals. [OPTICS] Any opening in an optical system which controls the cross section of a beam of light passing through it, to control light intensity, reduce aberration, or increase depth of focus.

diaphragmatic hernia [MED] Protrusion of an abdominal organ through the diaphragm into the thoracic cavity.

diaphragmatic respiration [PHYSIO] Respiration effected primarily by movement of the diaphragm, changing the intrathoracic pressure.

diaphyseal aclasis *See* multiple hereditary exostoses.

diaphysis [ANAT] The shaft of a long bone.

diapophysis [ANAT] The articular portion of a transverse process of a vertebra.

Diapriidae [INV ZOO] A family of hymenopteran insects in the superfamily Proctotrupoidea.

diapsid [VERT ZOO] Pertaining to skulls with distinct supra- and infratemporal fossae.

diarch [BOT] Having two xylem and two phloem bundles.

diarrhea [MED] The passage of loose or watery stools, usually at more frequent than normal intervals.

diarthrosis [ANAT] A freely moving articulation, characterized by a synovial cavity between the bones.

diastase [BIOCHEM] An enzyme that catalyzes the hydrolysis of starch to maltose. Also known as vegetable diastase.

diastasis [MED] Any simple separation of parts normally joined together, as the separation of an epiphysis from the body of a bone without true fracture, or the dislocation of an amphiarthrosis. [PHYSIO] The final phase of diastole, the phase of slow ventricular filling.

diastema [ANAT] A space between two types of teeth, as between an incisor and premolar. [CYTOL] Modified cytoplasm of the equatorial plane prior to cell division.

diastereoisomer [ORG CHEM] One of a pair of optical isomers which are not mirror images of each other. Also known as diastereomer.

diastereomer See diastereoisomer.

diastole [PHYSIO] The rhythmic relaxation and dilation of a heart chamber, especially a ventricle.

diastolic pressure [PHYSIO] The lowest arterial blood pressure during the cardiac cycle; reflects relaxation and dilation of a heart chamber.

diathermy [MED] The therapeutic use of high-frequency electric currents to produce localized heat in body tissues.

diathermy machine [ELECTR] A radio-frequency oscillator, sometimes followed by rf amplifier stages, used to generate high-frequency currents that produce heat within some part of the body for therapeutic purposes.

diathesis [MED] A constitutional or hereditary predisposition to some disease or a structural or mental abnormality.

diatom [INV ZOO] The common name for algae composing the class Bacillariophyceae; noted for the symmetry and sculpturing of the siliceous cell walls.

diatropism [BOT] Growth orientation of certain plant organs that is transverse to the line of action of a stimulus.

Diatrymiformes [PALEON] An order of extinct large, flightless birds having massive legs, tiny wings, and large heads and beaks.

diauxic growth [MICROBIO] The diphasic response of a culture of microorganisms based on a phenotypic adaptation to the addition of a second substrate; characterized by a growth phase followed by a lag after which growth is resumed.

Dibamidae [VERT ZOO] The flap-legged skinks, a small family of lizards in the suborder Sauria comprising three species confined to southeastern Asia.

dibasic [CHEM] **1.** Compounds containing two hydrogens that may be replaced by a monovalent metal or radical. **2.** An alcohol that has two hydroxyl groups, for example, ethylene glycol.

diblastula [INV ZOO] In coelenterates, an embryo that consists of two layers surrounding a central cavity.

Dibranchia [INV ZOO] A subclass of the Cephalopoda containing all living cephalopods except *Nautilus;* members possess two gills and, when present, an internal shell.

dibranchiate [ZOO] Having two gills.

dibucaine [ORG CHEM] $C_{20}H_{29}O_2N_3$ A local anesthetic

DIATOM

Arachnoidiscus ehrenbergii, a concentric diatom with radial symmetry. (*From H. J. Fuller and O. Tippo, College Botany, rev. ed., Holt, 1954*)

DIATRYMIFORMES

Diatryma steini (reconstruction). *(American Museum of Natural History)*

DIBRANCHIA

The luminous deep-sea squid *Lycoteuthis diadema.*

used both as the base and the hydrochloride salt. Also known as cinchocaine.

dicarpellate [BOT] Having two carpels.

dicaryon *See* dikaryon.

dicentric [CYTOL] Having two centromeres.

dicephaly [MED] A severe congenital anomaly in which the infant is born with two distinct heads.

dicerous [INV ZOO] Having two tentacles or two antennae.

dichasium [BOT] A cyme producing two main axes from the primary axis or shoot.

Dichelesthiidae [INV ZOO] A family of parasitic copepods in the suborder Caligoida; individuals attach to the gills of various fishes.

dichlamydeous [BOT] Having both calyx and corolla.

Dichobunidae [PALEON] A family of extinct artiodactyl mammals in the superfamily Dichobunoidea.

Dichobunoidea [PALEON] A superfamily of extinct artiodactyl mammals in the suborder Paleodonta composed of small-to medium-size forms with tri- to quadritubercular bunodont upper teeth.

dichogamy [BIOL] Producing mature male and female reproductive structures at different times.

dichoptous [INV ZOO] Having the margins of the compound eyes separate.

dichotomy [BIOL] **1.** Divided in two parts. **2.** Repeated branching or forking.

dichroism [OPTICS] In certain anisotropic materials, the property of having different absorption coefficients for light polarized in different directions.

dichromatic [BIOL] Having or exhibiting two color phases independently of age or sex.

Dickinsoniidae [PALEON] A family that comprises extinct flat-bodied, mutisegmented coelomates; identified as ediacaran fauna.

Dicksoniaceae [BOT] A family of tree ferns characterized by marginal sori which are terminal on the veins and protected by a bivalved indusium.

Dick test [IMMUNOL] A skin test to determine susceptibility or immunity to scarlet fever.

diclinous [BOT] Having stamens and pistils on different flowers.

dicoccous [BOT] Composed of two adherent one-seeded carpels.

dicotyledon [BOT] Any plant of the class Magnoliopsida, all having two cotyledons.

Dicotyledoneae [BOT] The equivalent name for Magnoliopsida.

Dicranales [BOT] An order of mosses having erect stems, dichotomous branching, and dense foliation.

dicrotic [MED] Pertaining to a secondary pressure wave in an artery on the descending limb of a main wave during diastole of the heart.

dictyoblastospore [MYCOL] A blastospore with both cross and longitudinal septa.

Dictyoceratida [INV ZOO] An order of sponges of the class Demospongiae; includes the bath sponges of commerce.

Dictyonellidina [PALEON] A suborder of extinct articulate brachiopods.

dictyosome [CYTOL] A stack of two or more cisternae; a component of the Golgi apparatus.

Dictyospongiidae [PALEON] A family of extinct sponges in the subclass Amphidiscophora having spicules resembling a one-ended amphidisc (paraclavule).

Dictyosporae [MYCOL] A spore group of the imperfect fungi

DICTYOBLASTOSPORE

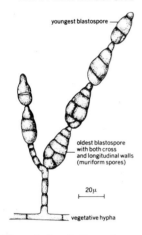

youngest blastospore

oldest blastospore with both cross and longitudinal walls (muriform spores)

20μ

vegetative hypha

Alternaria tenuis. Sporophore with branched chain of dictyoblastospores. *(After G. Goidanich, 1938)*

DICTYOSOME

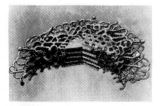

Cutaway view of a model of a portion of a dictyosome composed of six cisternae. Cisternal maturation is depicted from top to bottom.

characterized by multicelled spores with cross and longitudinal septae.

dictyospore [MYCOL] A multicellular spore in certain fungi characterized by longitudinal walls and cross septa.

dictyostele [BOT] A modified siphonostele in which the vascular tissue is dissected into a network of distinct strands; found in certain fern stems.

Dictyosteliaceae [MICROBIO] A family of microorganisms belonging to the Acrasiales and characterized by strongly differentiated fructifications.

Dicyemida [INV ZOO] An order of mesozoans comprising minute, wormlike parasites of the renal organs of cephalopod mollusks.

didactyl [VERT ZOO] Having two dactyls.

didelphic [ANAT] Having a double uterus or genital tract.

Didelphidae [VERT ZOO] The opossums, a family of arboreal mammals in the order Marsupialia.

Didolodontidae [PALEON] A family of extinct medium-sized herbivores in the order Condylarthra.

Didymiaceae [MICROBIO] A family of slime molds in the order Physarales.

Didymosporae [MYCOL] A spore group of the imperfect fungi characterized by two-celled spores.

didymous [BIOL] Occurring in pairs.

didynamous [BOT] Having four stamens occurring in two pairs, one pair long and the other short.

dieback [PL PATH] Of a plant, to die from the top or peripheral parts.

die down [BOT] Normal seasonal death of aboveground parts of herbaceous perennials.

Diego blood group [IMMUNOL] A genetically determined, immunologically distinct group of human erythrocyte antigens recognized by reaction with a specific antibody.

diencephalon [EMBRYO] The posterior division of the embryonic forebrain in vertebrates.

diestrus [PHYSIO] The long, quiescent period following ovulation in the estrous cycle in mammals; the stage in which the uterus prepares for the reception of a fertilized ovum.

diet [BIOL] Food or drink regularly consumed. [MED] Food prescribed, regulated, or restricted as to kind and amount, for therapeutic or other purpose.

Dieterici equation of state [THERMO] An empirical equation of state for gases, $pe^{a/vRT}(v - b) = RT$, where p is the pressure, T is the absolute temperature, v is the molar volume, R is the gas constant, and a and b are constants characteristic of the substance under consideration.

dietetics [MED] The science concerned with applying the principle of nutrition to the feeding of.people under various economic conditions or for therapeutic purposes.

1,1-diethoxyethane *See* acetal.

diethylstilbesterol [BIOCHEM] $C_{18}H_{20}O_2$ A white, crystalline, nonsteroid estrogen that is used therapeutically as a substitute for natural estrogenic hormones. Also known as stilbestrol.

dietician [MED] A person trained in dietetics, or the scientific management of meals for individuals or groups.

Dietl's crisis [MED] Recurrent attacks of radiating pain in the costovertebral angle, accompanied by nausea, vomiting, tachycardia, and hypotension, caused by kinking or twisting of the ureter with intermittent obstructive dilation.

differential blood count *See* differential leukocyte count.

differential calculus [MATH] The study of the manner in which the value of a function changes as one changes the value of the independent variable; includes maximum-minimum problems and expansion of functions into Taylor series.

DICTYOSTELE

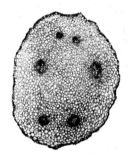

Dictyostele of *Polypodium.*

differential centrifugation [CYTOL] The separation of mixtures such as cellular particles in a medium at various centrifugal forces to separate particles of different density, size, and shape from each other.

differential diagnosis [MED] Distinguishing between diseases of similar character by comparing their signs and symptoms.

differential equation [MATH] An equation expressing a relationship between functions and their derivatives.

differential leukocyte count [PATH] The percentage of each variety of leukocytes in the blood, usually based on counting 100 leukocytes. Also known as differential blood count.

differential reaction rate [PHYS CHEM] The order of a chemical reaction expressed as a differential equation with respect to time; for example, $dx/dt = k(a - x)$ for first order, $dx/dt = k(a - x)(b - x)$ for second order, and so on, where k is the specific rate constant, a is the concentration of reactant A, b is the concentration of reactant B, and dx/dt is the rate of change in concentration for time t.

differential spectrophotometry [SPECT] Spectrophotometric analysis of a sample when a solution of the major component of the sample is placed in the reference cell; the recorded spectrum represents the difference between the sample and the reference cell.

differential threshold [PHYSIO] The smallest difference at which two stimuli can be discriminated.

diffraction [PHYS] Any redistribution in space of the intensity of waves that results from the presence of an object causing variations of either the amplitude or phase of the waves; found in all types of wave phenomena.

diffraction instrument *See* diffractometer.

diffraction ring [OPTICS] Circular light pattern which appears to surround particles in a microscope field.

diffraction symmetry [CRYSTAL] Any symmetry in a crystal lattice which causes the systematic annihilation of certain beams in x-ray diffraction.

diffractometer [PHYS] An instrument used to study the structure of matter by means of the diffraction of x-rays, electrons, neutrons, or other waves. Also known as diffraction instrument.

diffractometry [CRYSTAL] The science of determining crystal structures by studying the diffraction of beams of x-rays or other waves.

diffusate [CHEM] The solute which passes through the semipermeable membrane during dialysis.

diffuse placenta [EMBRYO] A placenta having villi diffusely scattered over most of the surface of the chorion; found in whales, horses, and other mammals.

diffusion [PHYS] The spontaneous movement and scattering of particles (atoms and molecules), of liquids, gases, and solids.

diffusion coefficient [PHYS] The weight of a material, in grams, diffusing across an area of 1 square centimeter in 1 second in a unit concentration gradient. Also known as diffusivity.

digastric [ANAT] Of a muscle, having a fleshy part at each end and a tendinous part in the middle.

Digenea [INV ZOO] A group of parasitic flatworms or flukes constituting a subclass or order of the class Trematoda and having two types of generations in the life cycle.

digenoporous [INV ZOO] Having two genital pores, especially referring to many turbellarians.

Di George's syndrome *See* thymic aplasia.

digestion [PHYSIO] The process of converting food to an

DIGENEA

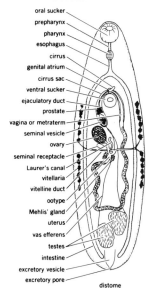

oral sucker
prepharynx
pharynx
esophagus
cirrus
genital atrium
cirrus sac
ventral sucker
ejaculatory duct
prostate
vagina or metraterm
seminal vesicle
ovary
seminal receptacle
Laurer's canal
vitellaria
vitelline duct
ootype
Mehlis' gland
uterus
vas efferens
testes
intestine
excretory vesicle
excretory pore

distome

Diagram of an adult digenetic trematode. *(From R.M. Cable, An Illustrated Laboratory Manual of Parasitology, Burgess, 1940)*

absorbable form by breaking it down to simpler chemical compounds.

digestive enzyme [BIOCHEM] Any enzyme that causes or aids in digestion.

digestive gland [PHYSIO] Any structure that secretes digestive enzymes.

digestive system [ANAT] A system of structures in which food substances are digested.

digestive tract [ANAT] The alimentary canal.

digit [INV ZOO] The distal portion of a chela and a chelicera. [VERT ZOO] Any of the terminal divisions of a limb in vertebrates higher than fishes, as a toe or finger.

digital [VERT ZOO] **1.** Pertaining to a digit. **2.** Resembling a digit.

digitalis [PHARM] The dried leaf of the purple foxglove plant (*Digitalis purpurea*), containing digitoxin and gitoxin; constitutes a powerful cardiac stimulant and diuretic.

digitate [ANAT] Having digits or digitlike processes.

digitigrade [VERT ZOO] Pertaining to animals, such as dogs and cats, which walk on the digits with the posterior part of the foot raised from the ground.

digitinervate [BOT] Having straight veins extending from the petiole like fingers.

digitipinnate [BOT] Having digitate leaves with pinnate leaflets.

digitoxigenin [ORG CHEM] $C_{23}H_{34}O_4$ The steroid aglycone formed by removal of three molecules of the sugar digitoxose from digitoxin.

digitoxin [ORG CHEM] $C_{41}H_{64}O_{13}$ A poisonous steroid glycoside found as the most active principle of digitalis, from the foxglove leaf.

digitus [INV ZOO] In insects, the claw-bearing terminal segment of the tarsus.

diglucoside [BIOCHEM] A compound containing two glucose molecules.

digoxin [ORG CHEM] $C_{41}H_{64}O_{14}$ A crystalline steroid obtained from a foxglove leaf (*Digitalis lanata*); similar to digitalis in pharmacological effects.

digynous [BOT] Having two carpels.

dihybrid [GEN] An organism heterozygous at two loci.

dihydrodiketoanthracene *See* anthraquinone.

dihydroxyacetone phosphate *See* dihydroxyacetonephosphoric acid.

dihydroxyacetonephosphoric acid [BIOCHEM] $C_3H_7O_6P$ A phosphoric acid ester of dehydroxyacetone, produced as an intermediate substance in the conversion of glycogen to lactic acid during muscular contraction. Also known as dihydroxyacetone phosphate.

1,2-dihydroxyanthraquinone *See* alizarin.

dihydroxyphenylalanine [BIOCHEM] $C_9H_{11}NO_4$ An amino acid that can be formed by oxidation of tyrosine; it is converted by a series of biochemical transformations, utilizing the enzyme dopa oxidase, to melanins. Also known as dopa.

2,3-dihydroxypropanal *See* glyceraldehyde.

dikaryon [MYCOL] Also spelled dicaryon. **1.** A pair of distinct, unfused nuclei in the same cell brought together by union of plus and minus hyphae in certain mycelia. **2.** Mycelium containing dikaryotic cells.

diktoma *See* neuroepithelioma.

Dilantin [PHARM] A trademark for diphenlhydantoin, an anticonvulsant in the treatment of epilepsy; the compound is also used as the sodium derivative.

dilation [SCI TECH] The act or process of stretching or expanding.

dilator [PHYSIO] Any muscle, instrument, or drug causing dilation of an organ or part.

dill [BOT] *Anethum graveolens.* A small annual or biennial herb in the family Umbelliferae; the aromatic leaves and seeds are used for food flavoring.

Dilleniaceae [BOT] A family of dicotyledonous trees, woody vines, and shrubs in the order Dilleniales having hypogynous flowers and mostly entire leaves.

Dilleniales [BOT] An order of dicotyledonous plants in the subclass Dilleniidae characterized by separate carpels and numerous stamens.

Dilleniidae [BOT] A subclass of plants in the class Magnoliopsida distinguished by being syncarpous, having centrifugal stamens, and usually having bitegmic ovules and binucleate pollen.

dilution gene [GEN] Any modifier gene that acts to reduce the effect of another gene.

dimercaprol [PHARM] $C_3H_8OS_2$ 2,3-Dimercapto-1-propanol, a colorless, water-soluble oily liquid with a mercaptanlike odor; used as an antidote for arsenic, gold, and mercury poisoning. Also known as British antilewisite (BAL); dithiol.

dimerous [BIOL] Composed of two parts.

diminution [BOT] Increasing simplification of inflorescences on successive branches.

dimorphism [CHEM] Having crystallization in two forms with the same chemical composition. [SCI TECH] Existing in two distinct forms, with reference to two members expected to be identical.

Dimylidae [PALEON] A family of extinct lipotyphlan mammals in the order Insectivora; a side branch in the ancestry of the hedgehogs.

Dinantian [GEOL] Lower Carboniferous geologic time. Also known as Avonian.

dinergate [INV ZOO] An ant of the soldier caste.

Dinidoridae [INV ZOO] A family of hemipteran insects in the superfamily Pentatomoidea.

Dinocerata [PALEON] An extinct order of large, herbivorous mammals having semigraviportal limbs and hoofed, five-toed feet; often called uintatheres.

Dinoflagellata [INV ZOO] The equivalent name for Dinoflagellida.

Dinoflagellida [INV ZOO] An order of flagellate protozoans in the class Phytamastigophorea; most members have fixed shapes determined by thick covering plates.

Dinophilidae [INV ZOO] A family of annelid worms belonging to the Archiannelida.

Dinophyceae [BOT] The dinoflagellates, a class of thallophytes in the division Pyrrhophyta.

Dinornithiformes [PALEON] The moas, an order of extinct birds of New Zealand; all had strong legs with four-toed feet.

dinosaur [PALEON] The name, meaning terrible lizard, applied to the fossil bones of certain large, ancient bipedal and quadripedal reptiles placed in the orders Saurischia and Ornithischia.

diocoel [EMBRYO] The cavity of the diencephalon, which becomes the third brain ventricle.

Dioctophymatida [INV ZOO] An order of parasitic nematode worms in the subclass Enoplia.

Dioctophymoidea [INV ZOO] An order or superfamily of parasitic nematodes characterized by the peculiar structure of the copulatory bursa of the male.

dioecious [BIOL] Having the male and female reproductive organs on different individuals. Also known as dioic.

dioic *See* dioecious.

DINOPHILIDAE

Dinophilus female, dorsal view. *(From Ruebush, Trans. Amer. Micr. Sci., vol. 59, Fig. 1, 1940)*

Diomedeidae [VERT ZOO] The albatrosses, a family of birds in the order Procellariiformes.

dionychous [INV ZOO] Having two claws, as on the tarsus of some spiders.

Diopsidae [INV ZOO] The stalk-eyed flies, a family of myodarian cyclorrhaphous dipteran insects in the subsection Acalypteratae.

diopter [OPTICS] A measure of the power of a lens or a prism, equal to the reciprocal of its focal length in meters. Abbreviated D.

dioptometer [OPTICS] An instrument for determining ocular refraction.

dioptrics [OPTICS] The branch of optics that treats of the refraction of light, especially by the transparent medium of the eye, and by lenses.

diorchism [ANAT] Having two testes.

Dioscoreaceae [BOT] A family of monocotyledonous, leafy-stemmed, mostly twining plants in the order Liliales, having an inferior ovary and septal nectaries and lacking tendrils.

diose [BIOCHEM] A member of a class of monosaccharides that have two carbon atoms; the only compound that belongs to this class is glycolaldehyde.

dipeptidase [BIOCHEM] An enzyme that hydrolyzes a dipeptide.

dipetalous [BOT] Having two petals.

diphosphoglyceric acid [ORG CHEM] $C_3H_8O_9P_2$ An ester of glyceric acid, with two molecules of phosphoric acid, characterized by a high-energy phosphate band.

diphosphopyridine nucleotide [BIOCHEM] $C_{21}H_{27}O_{14}N_7P_2$ An organic coenzyme that functions in enzymatic systems concerned with oxidation-reduction reactions. Abbreviated DPN. Also known as codehydrogenase 1; coenzyme 1; nicotinamide adenine dinucleotide (NAD).

diphtheria [MED] A communicable bacterial disease of man caused by the growth of *Corynebacterium diphtheriae* on any mucous membrane, especially of the throat.

diphycercal [VERT ZOO] Pertaining to a tail fin, having symmetrical upper and lower parts, and with the vertebral column extending to the tip without upturning.

diphyletic [EVOL] Originating from two lines of descent.

Diphyllidea [INV ZOO] A monogeneric order of tapeworms in the subclass Cestoda; all species live in the intestine of elasmobranch fishes.

diphyllous [BOT] Having two leaves.

diphyodont [ANAT] Having two successive sets of teeth, deciduous followed by permanent, as in man.

Diplacanthidae [PALEON] A family of extinct acanthodian fishes in the suborder Diplacanthoidei.

Diplacanthoidei [PALEON] A suborder of extinct acanthodian fishes in the order Climatiiformes.

diplacusis [MED] A difference in the pitch perceptions of the two ears when stimulated by the same sound frequency.

Diplasiocoela [VERT ZOO] A suborder of amphibians in the order Anura typically having the eighth vertebra biconcave.

diplegia [MED] Paralysis of similar parts on the two sides of the body.

dipleurula [INV ZOO] 1. A hypothetical bilaterally symmetrical larva postulated to be an ancestral form of echinoderms and chordates. 2. Any bilaterally symmetrical, ciliated echinoderm larva.

Diplobathrida [PALEON] An order of extinct, camerate crinoids having two circles of plates beneath the radials.

diplobiont [BIOL] An organism characterized by alternating, morphologically dissimilar haploid and diploid generations.

DIPLEURULA

Dipleurula larva.

DIPLOMONADIDA

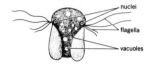

nuclei

flagella

vacuoles

A diplomonad, *Trepomonas rotans. (After McCracken)*

DIPLOPORITA

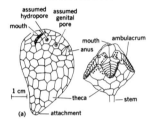

assumed hydropore
assumed genital pore
mouth
mouth
ambulacrum
anus
1 cm
theca
stem
(a)
attachment

Morphological features of representative Diploporita. *(a) Aristocystitis*, from the Ordovician *(after J. Barrande) (b) Asteroblastus*, from the Silurian *(after O. Jaekel).*

DIPLURA

cerci

Campodea folsomi Silvestri (Campodeidae). *(From E. O. Essig, College Entomology, Macmillan, 1942)*

diploblastic [zoo] Having two germ layers, referring to embryos and certain lower invertebrates.

diploblastula [INV ZOO] A two-layered, flagellated larva of certain ceractinomorph sponges. Also known as parenchymella.

diplochlamydeous [BOT] Having a double perianth.

diplococci [MICROBIO] A pair of micrococci.

Diplococcus pneumoniae [MICROBIO] A species of encapulated, nonmotile bacteria that causes pneumonia and other infectious diseases of humans; the type species of the genus. Also known as pneumococci.

diploglossate [VERT ZOO] Pertaining to certain lizards, having the ability to retract the end of the tongue into the basal portion.

diplohaplont [BIOL] An organism characterized by alternating, morphologically similar haploid and diploid generations.

diploidization [GEN] The process of attaining the diploid state.

diploid merogony [EMBRYO] Development of a part of an egg in which the nucleus is the normal diploid fusion product of egg and sperm nuclei.

diploid state [GEN] A condition in which a chromosome set is present in duplicate in a nucleus (2N).

Diplomonadida [INV ZOO] An order of small, colorless protozoans in the class Zoomastigophorea, having a bilaterally symmetrical body with four flagella on each side.

Diplomystidae [VERT ZOO] A family of catfishes in the suborder Siluroidei confined to the waters of Chile and Argentina.

diplont [BIOL] An organism with diploid somatic cells and haploid gametes.

diplophyll [BOT] A leaf with two layers of palisade cells and spongy parechyma between them.

diplopia [MED] A disorder characterized by double vision.

Diplopoda [INV ZOO] The millipeds, a class of terrestrial tracheate, oviparous arthropods; each body segment except the first few bears two pairs of walking legs.

Diploporita [PALEON] An extinct order of echinoderms in the class Cystoidea in which the thecal canals were associated in pairs.

Diplorhina [VERT ZOO] The subclass of the class Agnatha that includes the jawless vertebrates with paired nostrils.

diplosome [CYTOL] A double centriole.

diplospondyly [ANAT] Having two centra in one vertebra.

diplostemonous [BOT] **1.** Having two whorls of stamens, each with the same number of stamens as there are petals, and usually arranged with one stamen opposite each petal and one opposite each sepal. **2.** Having twice as many stamens as petals.

diplotegia [BOT] An inferior fruit having a dry dehiscent pericarp.

diplotene [CYTOL] The stage of meiotic prophase during which pairs of nonsister chromatids of each bivalent repel each other and are kept from falling apart by the chiasmata.

Diplura [INV ZOO] An order of small, primarily wingless insects of worldwide distribution.

Dipneumonomorphae [INV ZOO] A suborder of the order Araneida comprising the spiders common in the United States, including grass spiders, hunting spiders, and black widows.

Dipneusti [VERT ZOO] The equivalent name for Dipnoi.

Dipnoi [VERT ZOO] The lungfishes, a subclass of the Osteichthyes having lungs that arise from a ventral connection in the gut.

Dipodidae [VERT ZOO] The Old World jerboas, a family of mammals in the order Rodentia.

dipolar ion [CHEM] An ion carrying both a positive and a negative charge. Also known as zwitterion.

Diprionidae [INV ZOO] The conifer sawflies, a family of hymenopteran insects in the superfamily Tenthredinoidea.

diprotodont [VERT ZOO] Having two large anterior incisors with the remaining incisors and canines being smaller or absent.

Diprotodonta [VERT ZOO] A proposed order of marsupial mammals to include the phalangers, wombats, koalas, and kangaroos.

Diprotodontidae [PALEON] A family of extinct marsupial mammals.

Dipsacales [BOT] An order of dicotyledonous herbs and shrubs in the subclass Asteridae characterized by an inferior ovary and usually opposite leaves.

Dipsocoridae [INV ZOO] A family of hemipteran insects in the superfamily Dipsocoroidea; members are predators on small insects under bark or in rotten wood.

Dipsocoroidea [INV ZOO] A superfamily of minute, ground-inhabiting hemipteran insects belonging to the subdivision Geocorisae.

Diptera [INV ZOO] The true flies, an order of the class Insecta characterized by possessing only two wings and a pair of balancers.

Dipteriformes [VERT ZOO] The single order of the subclass Dipnoi, the lungfishes.

Dipterocarpaceae [BOT] A family of dicotyledonous plants in the order Theales having mostly stipulate, alternate leaves, a prominently exserted connective, and a calyx that is mostly winged in fruit.

dipterous [BIOL] **1.** Of, related to, or characteristic of Diptera. **2.** Having two wings or winglike structures.

directional selection [GEN] A type of selection within populations which favors a shift in the position of the population mean for a considered character.

directive [INV ZOO] Any of the dorsal and ventral paired mesenteries of certain anthozoan coelenterates.

directive therapy [PSYCH] A method of psychiatric treatment by which the therapist, assuming complete understanding of the patient's needs, endeavors to change the patient's attitudes, behavior, or mode of living.

dirhinic [VERT ZOO] **1.** Having two nostrils. **2.** Pertaining to both nostrils.

disaccharide [BIOCHEM] Any of the class of compound sugars which yield two monosaccharide units upon hydrolysis.

Disasteridae [PALEON] A family of extinct burrowing, exocyclic Euechinoidea in the order Holasteroida comprising mainly small, ovoid forms without fascioles or a plastron.

disc *See* disk.

Discellaceae [MYCOL] A family of fungi of the order Sphaeropsidales, including saprophytes and some plant pathogens.

discifloral [BOT] Having flowers with enlarged, disklike receptacles.

disciform [BIOL] Disk-shaped.

Discinacea [INV ZOO] A family of inarticulate brachiopods in the suborder Acrotretidina.

disclimax [ECOL] A climax community that includes foreign species following a disturbance of the natural climax by man or domestic animals. Also known as disturbance climax.

Discoaster [BOT] A star-shaped coccolith.

discoblastula [EMBRYO] A blastula formed by cleavage of a meroblastic egg; the blastoderm is disk-shaped.

discocarp [MYCOL] A disc-shaped ascocarp.

discocephalous [INV ZOO] Having a sucker on the head.

African lungfish *(Protopterus annectens)*, length to 32 inches (81 centimeters). *(After G. A. Boulenger, Catalogue of the Fresh Water Fishes of Africa in the British Museum, vol. 1, 1909)*

DISCINACEA

A cluster of *Discinisca. (From R. C. Moore, ed., Treatise on Invertebrate Paleontology, pt. H, Geological Society of America, Inc., and University of Kansas Press, 1965)*

DISCOASTER

$3\,\mu$

Discoaster, an example of Coccolithophorida. *(A. McIntyre, Lamont-Doherty Geological Observatory of Columbia University)*

DISCORBACEA

Scanning electron micrograph of a spiral view of coarsely perforate, trochospiral test with terminal aperture and everted lip of *Siphonina* from upper Eocene of Mississippi. *(R. B. MacAdam, Chevron Oil Field Research Co.)*

DISK HARROW

One type of disk harrow. *(Allis-Chalmers)*

discodactylous [VERT ZOO] Having sucking disks on the toes.

discogastrula [EMBRYO] A gastrula formed from a blastoderm.

Discoglossidae [VERT ZOO] A family of anuran amphibians in and typical of the suborder Opisthocoela.

discohexactine [INV ZOO] A sponge spicule with six identical rays that are perpendicular to one another.

discohexaster [INV ZOO] A hexaster having rays with terminal discs.

discoid [BIOL] **1.** Being flat and circular in form. **2.** Any structure shaped like a disc.

discoidal cleavage [EMBRYO] A type of cleavage producing a disc of cells at the animal pole.

Discoidiidae [PALEON] A family of extinct conical or globular, exocyclic Euechinoidea in the order Holectypoida distinguished by the rudiments of internal skeletal partitions.

Discolichenes [BOT] The equivalent name for Lecanorales.

Discolomidae [INV ZOO] The tropical log beetles, a family of coleopteran insects in the superfamily Cucujoidea.

Discomycetes [MYCOL] A group of fungi in the class Ascomycetes in which the surface of the fruiting body is exposed during maturation of the spores.

discopodous [INV ZOO] Having a disk-shaped foot.

Discorbacea [INV ZOO] A superfamily of foraminiferan protozoans in the suborder Rotaliina characterized by a radial, perforate, calcite test and a monolamellar septa.

discorhabd [INV ZOO] A linear sponge spicule with discoid projections or whorls of spines.

discrete [SCI TECH] **1.** Composed of separate and distinct parts. **2.** Having an individually distinct identity.

disease [MED] An alteration of the dynamic interaction between an individual and his environment which is sufficient to be deleterious to the well-being of the individual and produces signs and symptoms.

disinfectant [MATER] A chemical agent that destroys microorganisms but not bacterial spores.

disjunct endemism [PALEON] A type of regionally restricted distribution of a fossil taxon in which two or more component parts are separated by a major physical barrier and hence not readily explicable in terms of present-day geography.

disjunction [CYTOL] Separation of chromatids or homologous chromosomes during anaphase.

disjunctor [MYCOL] A small cellulose body between the conidia of certain fungi, which eventually breaks down and thus frees the conidia.

disk Also spelled disc. [BIOL] Any of various rounded and flattened animal and plant structures. [BOT] An outgrowth of the receptacle that surrounds the base of the ovary.

disk cultivator [AGR] A cultivator consisting of pairs of oppositely inclined disks.

disk flower [BOT] One of the flowers on the disk of a composite plant.

disk furrower [AGR] A furrower in which concave disks, at an angle to the direction of motion, are used to cut the soil.

disk harrow [AGR] A harrow which has two or more opposed gangs of 3–12 disks for cutting clods and trash, destroying weeds, cutting in cover crops, and smoothing and preparing the surface for various farming operations.

disk plow [AGR] A plow consisting of a number of disk blades attached to one axle or gang bolt; used for rapid, shallow plowing.

dislocation [CRYSTAL] A defect occurring along certain lines in the crystal structure and present as a closed ring or a line anchored at its ends to other dislocations, grain boundaries, the surface, or other structural feature. Also known as line defect. [MED] Displacement of one or more bones of a joint.

disomaty [CYTOL] Duplication of chromosomes unaccompanied by nuclear division.

Disomidae [INV ZOO] A family of spioniform annelid worms belonging to the Sedentaria.

disorientation [MED] Mental confusion as to one's normal relationship to his environment, especially time, place, and people; associated with organic brain disorders.

dispermous [BOT] Having two seeds.

dispermy [PHYSIO] Entrance of two spermatozoa into an ovum.

dispersal [BIOL] The scattering or distributing of organisms in the biosphere.

dispersion of a random variable [STAT] The spread of a random variable's distribution about its mean.

displacement [PSYCH] Disguising the goal of a motive by substituting another goal.

displacement chromatography [ANALY CHEM] Variation of column-development or elution chromatography in which the solvent is sorbed more strongly than the sample components; the freed sample migrates down the column, pushed by the solvent.

disruptive selection [GEN] A type of selection within populations in which two or more different genotypes are favored.

dissect [BIOL] To divide, cut, and separate into different parts.

dissecting microscope [OPTICS] Either of two types of optical microscope used to magnify materials undergoing dissection.

disseminated necrotizing periarteritis See polyarteritis nodosa.

disseminule [BIOL] An individual organism or part of an individual adapted for the dispersal of a population of the same organisms.

dissepiment [BOT] A partition which divides a fruit or an ovary into chambers. [PALEON] One of the vertically positioned thin plates situated between the septa in extinct corals of the order Rugosa.

dissociation [MED] Independent, uncoordinated functioning of the atria and ventricles. [MICROBIO] The appearance of a novel colony type on solid media after one or more subcultures of the microorganism in liquid media. [PHYS CHEM] Separation of a molecule into two or more fragments (atoms, ions, radicals) by collision with a second body or by the absorption of electromagnetic radiation. [PSYCH] The segregation of ideas from their affects or feelings, resulting in independent functioning of these components of a person's mental processes.

dissociation constant [PHYS CHEM] A constant whose numerical value depends on the equilibrium between the undissociated and dissociated forms of a molecule; a higher value indicates greater dissociation.

dissociation energy [PHYS CHEM] The energy required for complete separation of the atoms of a molecule.

dissociative reaction [PSYCH] A neurotic disorder leading to gross disorganization of the personality.

dissogeny [ZOO] Having two sexually mature stages, larva and adult, in the life of an individual.

dissolve [CHEM] **1.** To cause to disperse. **2.** To cause to pass into solution.

dissymmetry [SCI TECH] Lack of symmetry.

Distacodidae [PALEON] A family of conodonts in the suborder Conodontiformes characterized as simple curved cones with deeply excavated attachment scars.

DISTACODIDAE

Multioistodus, a typical conodont of the family Distacodidae. *(After illustration in R. R. Shrock and W. H. Twenhofel, Principles of Invertebrate Paleontology, McGraw-Hill, 2d ed., 1953)*

distal [BIOL] Located away from the point of origin or attachment.

distal convoluted tubule [ANAT] The portion of the nephron in the vertebrate kidney lying between the loop of Henle and the collecting tubules.

distemper [VET MED] Any of several contagious virus diseases of mammals, especially the form occurring in dogs, marked by fever, respiratory inflammation, and destruction of myelinated nerve tissue.

distichous [BIOL] Occurring in two vertical rows.

distilled water [CHEM] Water that has been freed of dissolved or suspended solids and organisms by distillation.

distoclusion [MED] Malocclusion of the teeth in which those of the lower jaw are in distal relation to the upper teeth.

distome [INV ZOO] A digenetic trematode characterized by possession of an oral and a ventral sucker.

distribution curve [STAT] The graph of the distribution function of a random variable.

distribution factor [NUCLEO] A term used to express the modification of the effect of radiation in a biological system attributable to the nonuniform distribution of an internally deposited isotope, such as radium's being concentrated in bones.

distribution function *See* distribution of random variable.

distribution of a random variable [STAT] For a discrete random variable, a function (or table) which assigns to each possible value of the random variable the probability that this value will occur; for a continuous random variable x, the monotone nondecreasing function which assigns to each real t the probability that x is less than or equal to t. Also known as distribution function; probability distribution; statistical distribution.

disturbance climax *See* disclimax.

dithiol *See* dimercaprol.

ditokous [VERT ZOO] Producing two eggs or giving birth to two young at one time.

diuresis [MED] Increased excretion of urine.

diuretic [PHARM] Any agent that increases the volume and flow of urine.

diuretic hormone [BIOCHEM] A neurohormone that promotes water loss in insects by increasing the volume of fluid secreted into the Malpighian tubules.

diurnal [BIOL] Active during daylight hours.

divaricate [BIOL] Broadly divergent and spread apart.

divaricator [ZOO] A muscle that causes separation of parts, as of brachiopod shells.

divergent adaptation [EVOL] Adaptation to different kinds of environment that results in divergence from a common ancestral form. Also known as branching adaptation; cladogenic adaptation.

diverticulitis [MED] Inflammation of a diverticulum.

diverticulosis [MED] Presence of many diverticula in the intestine.

diverticulum [MED] An abnormal outpocketing or sac on the wall of a hollow organ.

Divesian *See* Oxfordian.

Dixidae [INV ZOO] A family of orthorrhaphous dipteran insects in the series Nematocera.

dizygotic twins [BIOL] Twins derived from two eggs. Also known as fraternal twins.

Djulfian [GEOL] Upper upper Permian geologic time.

DNA *See* deoxyribonucleic acid.

Docodonta [PALEON] A primitive order of Jurassic mammals of North America and England.

docoglossate [INV ZOO] Having an elongated radula without many marginal teeth, as in limpets.

dodecagynous [BOT] Having 12 pistils.

dodecamerous [BOT] Having the whorls of floral parts in multiples of 12.

dodecandrous [BOT] Having at least 12 stamens.

dodo [VERT ZOO] *Raphus calcullatus.* A large, flightless, extinct bird of the family Raphidae.

doe [VERT ZOO] The adult female deer, antelope, goat, rabbit, or any other mammal of which the male is referred to as buck.

dog [VERT ZOO] Any of various wild and domestic animals identified as *Canis familiaris* in the family Canidae; all are carnivorous and digitigrade, are adapted to running, and have four toes with nonretractable claws on each foot.

dogfish *See* bowfin.

dolabriform [BIOL] Shaped like an ax head.

Dolichopodidae [INV ZOO] The long-legged flies, a family of orthorrhaphous dipteran insects in the series Brachycera.

Dolichothoraci [PALEON] A group of joint-necked fishes assigned to the Arctolepiformes in which the pectoral appendages are represented solely by large fixed spines.

dolioform [BIOL] Barrel-shaped.

doliolaria larva [INV ZOO] A free-swimming larval stage of crinoids and holothurians having an apical tuft and four or five bands of cilia.

Doliolida [INV ZOO] An order of pelagic tunicates in the class Thaliacea; transparent forms, partly or wholly ringed by muscular bands.

dollar spot [PL PATH] A fungus disease of lawn grasses caused by *Sclerotinia homeocarpa* and characterized by small, round, brownish areas which gradually coalesce.

dolphin [VERT ZOO] The common name for about 33 species of cetacean mammals included in the family Delphinidae and characterized by the pronounced beak-shaped mouth.

Domerian [GEOL] Upper Charmouthian geologic time.

domestication [BIOL] The adaptation of an animal or plant through breeding in captivity to a life intimately associated with and advantageous to man.

dominance [GEN] The expression of a heritable trait in the heterozygote such as to make it phenotypically indistinguishable from the homozygote.

dominant allele [GEN] The member of a pair of alleles which is phenotypically indistinguishable in both the homozygous and heterozygous condition.

donkey [VERT ZOO] A domestic ass (*Equus asinus*); a perissodactyl mammal in the family Equidae.

Donnan equilibrium [PHYS CHEM] The particular equilibrium set up when two coexisting phases are subject to the restriction that one or more of the ionic components cannot pass from one phase into the other; commonly, this restriction is caused by a membrane which is permeable to the solvent and small ions but impermeable to colloidal ions or charged particles of colloidal size. Also known as Gibbs-Donnan equilibrium.

Donovan body [MED] The causative microorganism of granuloma inguinale, demonstrated in stained mononuclear cells and characterized by one or two opposite polar chromatin masses.

dopa *See* dihydroxyphenylalanine.

dopamine [BIOCHEM] $C_8H_{11}O_2N$ An intermediate in epinephrine and norepinephrine biosynthesis; the decarboxylation product of dopa.

dopa oxidase [BIOCHEM] An enzyme that catalyzes the oxidation of dihydroxyphenylalanine to melanin; occurs in the skin.

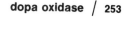

DOLIOLIDA

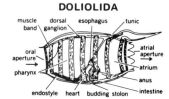

Doliolum, a doliolid, solitary asexual form. *(After Uljanin)*

DOLPHIN

The common dolphin (*Delphinus delphis*), found in many seas throughout the world.

DONKEY

The donkey, sometimes referred to as the burro.

DONOVAN BODY

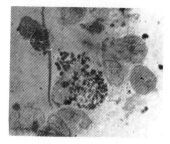

Mononuclear cell, with intramonocytic Donovan body.

Doradidae [VERT ZOO] A family of South American catfishes in the suborder Siluroidei.

Dorilaidae [INV ZOO] The big-headed flies, a family of cyclorrhaphous dipteran insects in the series Aschiza.

Dorippidae [INV ZOO] The mask crabs, a family of brachyuran decapods in the subsection Oxystomata.

dormancy [BOT] A state of quiescence during the development of many plants characterized by their inability to grow, though continuing their morphological and physiological activities.

dormouse [VERT ZOO] The common name applied to members of the family Gliridae; they are Old World arboreal rodents intermediate between squirrels and rats.

dorsal [ANAT] Located near or on the back of an animal or one of its parts.

dorsal aorta [ANAT] The portion of the aorta extending from the left ventricle to the first branch. [INV ZOO] The large, dorsal blood vessel in many invertebrates.

dorsal column [ANAT] A column situated dorsally in each lateral half of the spinal cord which receives the terminals of some afferent fibers from the dorsal roots of the spinal nerves.

dorsal fin [VERT ZOO] A median longitudinal vertical fin on the dorsal aspect of a fish or other aquatic vertebrate.

dorsal lip [EMBRYO] In an amphibian embryo, the margin or lip of the fold of blastula wall marking the dorsal limit of the blastopore during gastrulation and constituting the primary organizer, is necessary to the development of neural tissue, and forms the originating point of chordamesoderm.

dorsiferous [BOT] Of ferns, bearing sori on the back of the frond. [ZOO] Bearing the eggs or young on the back.

dorsifixed [BOT] Having the filament connected to the back of the anther.

dorsiflex [ZOO] To flex or cause to flex in a dorsal direction.

dorsiflexion sign *See* Homan's sign.

dorsigrade [VERT ZOO] Walking on the back of the toes.

dorsocaudad [ANAT] To or toward the dorsal surface and caudal end of the body.

dorsomedial [ANAT] Located on the back, toward the midline.

dorsoposteriad [ANAT] To or toward the dorsal surface and posterior end of the body.

dorsum [ANAT] **1.** The entire dorsal surface of the animal body. **2.** The upper part of the tongue, opposite the velum.

Dorvilleidae [INV ZOO] A family of minute errantian annelids in the superfamily Eunicea.

Dorylaimoidea [INV ZOO] An order or superfamily of nematodes inhabiting soil and fresh water.

Dorylinae [INV ZOO] A subfamily of predacious ants in the family Formicidae, including the army ant (*Eciton hamatum*).

Dorypteridae [PALEON] A family of Permian palaeonisciform fishes sometimes included in the suborder Platysomoidei.

dosage [GEN] The number of genes with a similar action that control a given character. [MED] The prescribed or correct amount of medicine or other therapeutic agent administered to treat a given illness. Also known as dose. [NUCLEO] *See* absorbed dose.

dosage compensation [GEN] A mechanism that equalizes the phenotypic effects of genes located on the X chromosome, which genes therefore exist in two doses in the homogametic sex and one dose in the heterogametic sex.

dose [MED] **1.** The measure, expressed in number of roentgens, of a property of x-rays at a particular place; used in radiology. **2.** *See* dosage. [NUCLEO] *See* absorbed dose.

double blossom [PL PATH] A fungus disease of dewberry and

DORMOUSE

Fat dormouse of Europe, resembling a small squirrel.

blackberry caused by *Fusarium rubi* and characterized by witches'-brooms and enlargement and malformation of the flowers.

double bond [PHYS CHEM] A type of linkage between atoms in which two pair of electrons are shared equally.

double circulation [PHYSIO] A circulatory system in which blood flows through two separate circuits, as pulmonary and systemic.

double fertilization [BOT] In most seed plants, fertilization involving fusion between the egg nucleus and one sperm nucleus, and fusion between the other sperm nucleus and the polar nuclei.

double-loop pattern [ANAT] A whorl type of fingerprint pattern consisting of two separate loop formations and two deltas.

double-work [BOT] In plant propagation, to graft or bud a scion to an intermediate variety that is itself grafted on a stock of still another variety.

Douglas fir [BOT] *Pseudotsuga menziesii.* A large coniferous tree in the order Pinales characterized by bracts extending beyond the cone scales. Also known as red fir.

dove [VERT ZOO] The common name for a number of small birds of the family Columbidae.

Down's syndrome [MED] A syndrome of congenital defects, especially mental retardation, typical facies responsible for the term mongoloid idiocy, or mongolism, and cytogenetic abnormality consisting of trisomy 21 or its equivalent in the form of an unbalanced translocation. Also known as mongolism; trisomy 21 syndrome.

downy mildew [PL PATH] A fungus disease of higher plants caused by members of the family Peronosporaceae and characterized by a white, downy growth on the diseased plant parts.

Dowtonian [GEOL] Uppermost Silurian or lowermost Devonian geologic time.

DPN *See* diphosphopyridine nucleotide.

Dracunculoidea [INV ZOO] An order or superfamily of parasitic nematodes characterized by their habitat in host tissues and by the way larvae leave the host through a skin lesion.

dragonfly [INV ZOO] Any of the insects composing six families of the suborder Anisoptera and having four large, membranous wings and compound eyes that provide keen vision.

dram [MECH] **1.** A unit of mass, used in the apothecaries' system of mass units, equal to ⅛ apothecaries' ounce or 60 grains or 3.8879346 grams. Also known as apothecaries' dram (dram ap); drachm (British). **2.** A unit of mass, formerly used in the United Kingdom, equal to ¹⁄₁₆ ounce (avoirdupois) or approximately 1.77185 grams. Abbreviated dr.

dream [PSYCH] An involuntary series of visual, auditory, or kinesthetic imagery, emotions, and thoughts occurring in the mind during sleep or a sleeplike state, which take the form of a sequence of events or of a story, having a feeling of reality but totally lacking a feeling of free will.

Drepanellacea [PALEON] A monomorphic superfamily of extinct paleocopan ostracods in the suborder Beyrichicopina having a subquadrate carapace, many with a marginal rim.

Drepanellidae [PALEON] A monomorphic family of extinct ostracods in the superfamily Drepanellacea.

Drepanidae [INV ZOO] The hooktips, a small family of lepidopteran insects in the suborder Heteroneura.

Dresbachian [GEOL] Lower Croixan geologic time.

Drilidae [INV ZOO] The false firefly beetles, a family of coleopteran insects in the superfamily Cantharoidea.

Dromadidae [VERT ZOO] A family of the avian order Chara-

DOUBLE-LOOP PATTERN

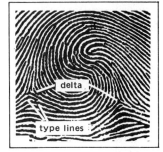

Double-loop pattern in a fingerprint. *(Federal Bureau of Investigation)*

DOUGLAS FIR

Cone on a branch of Douglas fir *(Pseudotsuga menziesii)*.

DRAGONFLY

Winged adult dragonfly.

driiformes containing a single species, the crab plover (*Dromas ardeola*).

dromedary [VERT ZOO] *Camelus dromedarius* The Arabian camel, distinguished by a single hump.

Dromiacea [INV ZOO] The dromiid crabs, a subsection of the Brachyura in the crustacean order Decapoda.

Dromiceidae [VERT ZOO] The emus, a monospecific family of flightless birds in the order Casuariiformes.

drone [INV ZOO] A haploid male bee or ant; one of the three castes in a colony.

droplet infection [MED] Infection by contact with airborne droplets of sputum carrying infectious agents.

dropsy *See* edema.

Droseraceae [BOT] A family of dicotyledonous plants in the order Sarraceniales, distinguished by leaves that do not form pitchers, parietal placentation, and several styles.

Drosophilidae [INV ZOO] The vinegar flies, a family of myodarian cyclorrhaphous dipteran insects in the subsection Acalypteratae, including the fruit fly (*Drosophila melanogaster*).

drug [PHARM] **1.** Any substance used internally or externally as a medicine for the treatment, cure, or prevention of a disease. **2.** A narcotic preparation.

drug idiosyncrasy [MED] A peculiarity of constitution that makes an individual respond differently to a drug or treatment than do most people.

drug resistance [MICROBIO] A decreased reactivity of living organisms to the injurious actions of certain drugs and chemicals.

drug tolerance [MED] Condition that may follow repeated ingestion of a drug in so that the effect produced by the original dose no longer occurs.

drupaceous [BOT] Of, pertaining to, or characteristic of a drupe.

drupe [BOT] A fruit, such as a cherry, having a thin or leathery exocarp, a fleshy mesocarp, and a single seed with a stony endocarp. Also known as stone fruit.

drupelet [BOT] An individual drupe of an aggregate fruit. Also known as grain.

dry gangrene [MED] Local death of a part caused by arterial obstruction without associated venous obstruction or infection.

Dryinidae [INV ZOO] A family of hymenopteran insects in the superfamily Bethyloidea.

Dryomyzidae [INV ZOO] A family of myodarian cyclorrhaphous dipteran insects in the subsection Acalypteratae.

Dryopidae [INV ZOO] The long-toed water beetles, a family of coleopteran insects in the superfamily Dryopoidea.

Dryopoidea [INV ZOO] A superfamily of coleopteran insects in the suborder Polyphaga, including the nonpredatory aquatic beetles.

dry rot [MICROBIO] A rapid decay of seasoned timber caused by certain fungi which cause the wood to be reduced to a dry, friable texture. [PL PATH] Any of various rot diseases of plants characterized by drying of affected tissues.

dry socket [MED] Inflammation of the dental alveolus, especially the inflamed condition following the removal of a tooth. Also known as alveolitis.

D₁ trisomy [MED] A syndrome resulting from the presence in triplicate of chromosomes 13–15; manifested in severe congenital anomalies and usually resulting in death in infancy. Also known as trisomy 13–15.

Duchenne's dystrophy [MED] A sex-linked or autosomal recessive form of muscular dystrophy, which is progressive with pseudohypertrophy.

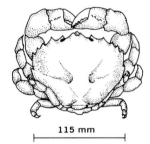

DROMIACEA

115 mm

Dromia erythropus, a dromiid crab. (*Smithsonian Institution*)

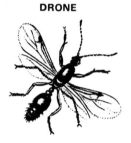

DRONE

Drone, about ³⁄₁₆ inch (0.5 centimeter).

duck [VERT ZOO] The common name for a number of small waterfowl in the family Anatidae, having short legs, a broad, flat bill, and a dorsoventrally flattened body.

duck-billed dinosaur [PALEON] Any of several herbivorous, bipedal ornithopods having the front of the mouth widened to form a ducklike beak.

duckbill platypus *See* platypus.

duck wheat *See* tartary buckwheat.

Ducrey test [IMMUNOL] A skin test to determine past or present infection with *Hemophilus ducreyi.*

duct [ANAT] An enclosed tubular channel for conducting a glandular secretion or other body fluid.

ductless gland *See* endocrine gland.

duct of Cuvier [EMBRYO] Either of the paired common cardinal veins in a vertebrate embryo.

duct of Santorini [ANAT] The dorsal pancreatic duct in a vertebrate embryo; persists in adult life in some species and serves as the pancreatic duct in the adult elasmobranch, pig, and ox.

ductus arteriosus [EMBRYO] Blood shunt between the pulmonary artery and the aorta of the mammalian embryo.

ductus deferens *See* vas deferens.

ductus venosus [EMBRYO] Blood shunt between the left umbilical vein and the right sinus venosus of the heart in the mammalian embryo.

Duffy blood group [IMMUNOL] A genetically determined, immunologically distinct group of human erythrocyte antigens defined by their reaction with anti-Fy^a serum.

Dufour's gland [INV ZOO] In certain hymenopterans, a gland with the duct connected to the sting.

Dugongidae [VERT ZOO] A family of aquatic mammals in the order Sirenia comprising two species, the dugong and the sea cow.

Dugonginae [VERT ZOO] The dugongs, a subfamily of sirenian mammals in the family Dugongidae characterized by enlarged, sharply deflected premaxillae and the absence of nasal bones.

dulse [BOT] Any of several species of red algae of the genus *Rhodymenia* found below the intertidal zone in northern latitudes; an important food plant.

dumping syndrome [MED] An imperfectly understood symptom complex of disagreeable or painful epigastric fullness, nausea, weakness, giddiness, sweating, palpitations, and diarrhea, occurring after meals in patients who have gastric surgery which interferes with the function of the pylorus.

duodenal glands *See* Brunner's glands.

duodenal ulcer [MED] A peptic ulcer occurring in the wall of the duodenum, the first portion of the small intestine.

duodenum [ANAT] The first section of the small intestine of mammals, extending from the pylorus to the jejunum.

duplex uterus [ANAT] A condition in certain primitive mammals, such as rodents and bats, that have two distinct uteri opening separately into the vagina.

duplication [GEN] The presence in a genome of an extra piece of chromosome, usually attached to, or collinearly inserted in, one of the chromosomes of the set.

dura mater [ANAT] The fibrous membrane forming the outermost covering of the brain and spinal cord. Also known as endocranium.

Durham fermentation tube [MICROBIO] A test tube containing lactose or lauryl tryptose and an inverted vial for gas collection; used to test for the presence of coliform bacteria.

Dutch elm disease [PL PATH] A lethal fungus disease of elm trees caused by *Graphium ulmi,* which releases a toxic sub-

DUCK-BILLED DINOSAUR

Restoration of the duck-billed dinosaur *Anatosaurus (Trachodon)* from the Late Cretaceous of North America; this dinosaur was 30–40 feet (9–12 meters) long.

DUPLEX UTERUS

Duplex uterus of the rat. *(From C. K. Weichert, Elements of Chordate Anatomy, 3d ed., McGraw-Hill, 1967)*

stance that destroys vascular tissue; transmitted by a bark beetle.

duvet [MYCOL] A downy coating characteristic of the growth of certain fungi in cultures.

dwarf [BIOL] Being an atypically small form or variety of something. [MED] An abnormally small individual; especially one whose bodily proportions are altered.

dwarf disease [PL PATH] A virus disease marked by the inhibition of fruit production; common in plum trees.

dwarfism [MED] Underdevelopment of the body due to surgical removal of the pituitary gland or hyposecretion of growth hormone.

Dy *See* dysprosium.

dyad [CYTOL] Separation of tetrads at the first meiotic division in gametogenesis in animals and sporogenesis in plants. [SCI TECH] A pair of individuals, structures, or elements.

dynamic ileus *See* spastic ileus.

dysarthria [MED] Impairment of articulation caused by any disorder or lesion affecting the tongue or speech muscles.

dysarthrosis [MED] **1.** Deformity, dislocation, or disease of a joint. **2.** A false joint.

dyschondroplasia *See* enchondromatosis.

dyscrasia [MED] An abnormal state of the body.

dysentery [MED] Inflammation of the intestine characterized by pain, intense diarrhea, and the passage of mucus and blood.

dysgammaglobulinemia [MED] A quantitative or qualitative abnormality of serum globulins.

dysgerminoma [MED] An ovarian tumor composed of large polygonal cells of germ-cell origin, resembling seminoma of the testis, but less malignant. Also known as embryoma of the ovary.

dyshidrosis [MED] Any disturbance in sweat production or excretion.

Dysideidae [INV ZOO] A family of sponges in the order Dictyoceratida.

dyskeratosis [MED] **1.** Imperfect keratinization of individual epidermal cells. **2.** Keratinization of corneal epithelium.

dyskinesia [MED] **1.** Disordered movements of voluntary or involuntary muscles, particularly those seen in disorders of the extrapyramidal system. **2.** Impaired voluntary movements.

dyslexia [MED] Impairment of the ability to read.

dyslogia [MED] **1.** Difficulty in the expression of ideas by speech. **2.** Impairment of reasoning or the faculty to think logically.

dysmenorrhea [MED] Difficult or painful menstruation.

DYSODONTA

2 cm

Dysodont hinge, *Ambonychinia*, Upper Ordovician, Sweden. *(After O. Isberg, 1934)*

Dysodonta [PALEON] An order of extinct bivalve mollusks with a nearly toothless hinge and a ligament in grooves or pits.

Dyson microscope [OPTICS] A type of interference microscope, now obsolete, in which a light ray is split into two parallel beams and then recombined by reflections from surfaces of parallel plates, and one of the beams passes through the object under observation.

dysostosis [MED] Defective formation of bone.

dyspepsia [MED] Disturbed digestion.

dysphagia [MED] Difficulty in swallowing, or inability to swallow, of organic or psychic causation.

dysphasia [MED] Partial aphasia due to a brain lesion.

dysphonia [MED] An impairment of the voice.

dysphoria [MED] **1.** The condition of not feeling well or of being ill at ease. **2.** Morbid impatience and restlessness, anxiety, or fidgetiness.

dysplasia [PATH] Abnormal development or growth, especially of cells.

dyspnea [MED] Difficult or labored breathing.

dysprosium [CHEM] A metallic rare-earth element, symbol Dy, atomic number 66, atomic weight 162.50.

dyssebacia [MED] Plugging of the sebaceous glands, especially around the nose, mouth, and forehead, with a dry, yellowish material.

dysthymia [MED] Any childhood condition caused by malfunction of the thymus. [PSYCH] Any despondent mood or depressive tendency, often associated with hypochondriasis.

dystonia [PHYSIO] Disorder or lack of muscle tonicity.

dystophic [BIOL] Pertaining to an environment that does not supply adequate nutrition.

dystrophy [MED] **1.** Defective nutrition. **2.** Defective or abnormal development or degeneration.

Dytiscidae [INV ZOO] The predacious diving beetles, a family of coleopteran insects in the suborder Adephaga.

DYSPROSIUM

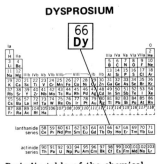

Periodic table of the chemical elements showing the position of dysprosium.

E

e [MATH] The base of the natural logarithms; the number defined by the equation

$$\int_1^e \frac{1}{x}\,dx = 1.$$

eagle [VERT ZOO] Any of several large, strong diurnal birds of prey in the family Accipitridae.

Eagle's medium [MICROBIO] A tissue-culture medium, developed by H. Eagle, containing vitamins, amino acids, inorganic salts and serous enrichments, and dextrose.

ear [ANAT] The receptor organ that sends both auditory information and space orientation information to the brain in vertebrates.

eardrum *See* tympanic membrane.

ear implantation length [ANTHRO] A measure of the distance from the otobasion superior to the otobasion inferior.

ear length [ANTHRO] A measure of the maximum distance along the anterior-posterior axis of the ear.

earlobe [ANAT] The pendulous, fleshy lower portion of the auricle or external ear.

ear rot [PL PATH] Any of several fungus diseases of corn, occurring both in the field and in storage and marked by decay and molding of the ears.

ear shell *See* abalone.

earth pig *See* aardvark.

earthstar [MYCOL] A fungus of the genus *Geastrum* that resembles a puffball with a double peridium, the outer layer of which splits into the shape of a star.

earthworm [INV ZOO] The common name for certain terrestrial members of the class Oligochaeta, especially forms belonging to the family Lumbricidae.

earwax *See* cerumen.

earwig [INV ZOO] The common name for members of the insect order Dermaptera.

eastern equine encephalitis [MED] A mosquito-borne virus infection of horses and mules in the eastern and southern United States caused by a member of arbovirus group A.

Eaton agent [MICROBIO] The name applied to *Mycoplasma pneumoniae* when it was regarded as a virus.

Eaton agent pneumonia [MED] Pneumonitis in man, caused by *Mycoplasma pneumoniae*. Also known as primary atypical pneumonia.

Ebenaceae [BOT] A family of dicotyledonous plants in the order Ebenales, in which a latex system is absent and flowers are mostly unisexual with the styles separate, at least distally.

Ebenales [BOT] An order of woody, sympetalous dicotyledonous plants in the subclass Dilleniidae, having axile placentation and usually twice as many stamens as corolla lobes.

ebony [BOT] Any of several African and Asian trees of the genus *Diospyros*, providing a hard, durable wood.

EAR

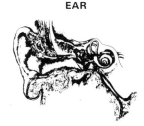

Schematic drawing of the human ear. *(Drawing by M. Brödel, Three Unpublished Drawings of the Anatomy of the Human Ear, Saunders)*

EARTHWORM

External features of earthworm *(Lumbricus terrestris). (From T. I. Storer, General Zoology, 3d ed., McGraw-Hill, 1957)*

EBONY

Twig, leaf, and bud of persimmon *(Diospyros virginiana).*

ECHINISCOIDEA

Echiniscoides sigismundi.

ECHINOIDA

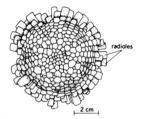

radioles

|__2 cm__|

Colobocentrotus atratus, aboral aspect, a Pacific species adapted for life on wave-exposed coral reefs.

ECHINOSTOME CERCARIA

collar
with spines

A drawing of an echinostome cercaria. *(From R. M. Cable, An Illustrated Laboratory Manual of Parasitology, Burgess, 1958)*

ebracteate [BOT] Without bracts, or much reduced leaves.

ebracteolate [BOT] Without bracteoles.

Ebriida [INV ZOO] An order of flagellate protozoans in the class Phytamastigophorea characterized by a solid siliceous skeleton.

ebulliometer [PHYS CHEM] An instrument used to measure precisely the absolute or differential boiling points of solutions. Also known as ebullioscope.

ebullioscope *See* ebulliometer.

ecblastesis [BOT] Proliferation of the primary axis of an inflorescence.

ecchymosis [MED] A subcutaneous hemorrhage marked by purple discoloration of the skin.

ecdysis [INV ZOO] Molting of the outer cuticular layer of the body, as in insects and crustaceans.

ecdysone [BIOCHEM] The molting hormone of insects.

ecesis [ECOL] Successful naturalization of a plant or animal population in a new environment.

ECG *See* electrocardiogram.

Echeneidae [VERT ZOO] The remoras, a family of perciform fishes in the suborder Percoidei.

echidna [VERT ZOO] A spiny anteater; any member of the family Tachyglossidae.

Echinacea [INV ZOO] A suborder of echinoderms in the order Euechinoidea; individuals have a rigid test, keeled teeth, and branchial slits.

echinate [ZOO] Having a dense covering of spines or bristles.

Echinidae [INV ZOO] A family of echinacean echinoderms in the order Echinoida possessing trigeminate or polyporous plates with the pores in a narrow vertical zone.

Echiniscoidea [INV ZOO] A suborder of tardigrades in the order Heterotardigrada characterized by terminal claws on the legs.

echinococcosis [MED] Infestation by the larva (hydatid) of *Echinococcus granulosis* in man, and in some canines and herbivores. Also known as hydatid disease; hydatidosis.

Echinocystitoida [PALEON] An order of extinct echinoderms in the subclass Perischoechinoidea.

Echinodera [INV ZOO] The equivalent name for Kinorhyncha.

Echinodermata [INV ZOO] A phylum of exclusively marine coelomate animals distinguished from all others by an internal skeleton composed of calcite plates, and a water-vascular system to serve the needs of locomotion, respiration, nutrition, or perception.

Echinoida [INV ZOO] An order of Echinacea with a camarodont lantern, smooth test, and imperforate noncrenulate tubercles.

Echinoidea [INV ZOO] The sea urchins, a class of Echinozoa having a compact body enclosed in a hard shell, or test, formed by regularly arranged plates which bear movable spines.

Echinometridae [INV ZOO] A family of echinoderms in the order Echinoida, including polyporous types with either an oblong or a spherical test.

echinomycin [MICROBIO] $C_{50}H_{60}O_{12}N_{12}S_2$ A toxic polypeptide antibiotic produced by species of *Streptomyces.*

echinopluteus [INV ZOO] The bilaterally symmetrical larva of sea urchins.

Echinosteliaceae [MYCOL] A family of slime molds in the order Echinosteliales.

Echinosteliales [MYCOL] An order of slime molds in the subclass Myxogastromycetidae.

echinostome cercaria [INV ZOO] A digenetic trematode larva characterized by the large anterior acetabulum and a collar with spines.

Echinothuriidae [INV ZOO] A family of deep-water echinoderms in the order Echinothurioida in which the large, flexible test collapses into a disk at atmospheric pressure.

Echinothurioida [INV ZOO] An order of echinoderms in the superorder Diadematacea with solid or hollow primary radioles, diademoid ambulacral plates, noncrenulate tubercles, and the anus within the apical system.

Echinozoa [INV ZOO] A subphylum of free-living echinoderms having the body essentially globoid with meridional symmetry and lacking appendages.

Echiurida [INV ZOO] A small group of wormlike organisms regarded as a separate phylum of the animal kingdom; members have a saclike or sausage-shaped body with an anterior, detachable prostomium.

Echiuridae [INV ZOO] A small family of the order Echiuroinea characterized by a flaplike prostomium.

Echiuroidea [INV ZOO] A phylum of schizocoelous animals.

Echiuroinea [INV ZOO] An order of the Echiurida.

echocardiography [MED] A diagnostic technique for the heart that uses a transducer held against the chest to send high-frequency sound waves which pass harmlessly into the heart; as they strike structures within the heart, they are reflected back to the transducer and recorded on an oscilloscope.

echolalia [MED] The purposeless, often seemingly involuntary repetition of words spoken by another person; a disorder seen in certain psychotic states and in certain organic brain syndromes. Also known as echophrasia.

echolic [MED] Producing abortion or accelerating labor.

echovirus [VIROL] A division of enteroviruses in the picornavirus group; the name is derived from the group designation enteric cytopathogenic human orphan virus.

eclampsia [MED] A disorder occurring during the latter half of pregnancy, characterized by elevated blood pressure, edema, proteinuria, and convulsions or coma.

eclipse period [VIROL] A phase in the proliferation of viral particles during which the virus cannot be detected in the host cell.

ecocline [ECOL] A genetic gradient of adaptability to an environmental gradient; formed by the merger of ecotypes.

ecological association [ECOL] A complex of communities, such as an elm-hackberry association, which develops in accord with variations in physiography, soil, and successional history within the major subdivision of a biotic realm.

ecological community *See* community.

ecological interaction [ECOL] The relation between species that live together in a community; specifically, the effect an individual of one species may exert on an individual of another species.

ecological pyramid [ECOL] A pyramid-shaped diagram representing quantitatively the numbers of organisms, energy relationships, and biomass of an ecosystem; numbers are high for the lowest trophic levels (plants) and low for the highest trophic level (carnivores).

ecological succession [ECOL] A gradual process incurred by the change in the number of individuals of each species of a community and by establishment of new species populations that may gradually replace the original inhabitants.

ecological system *See* ecosystem.

ecology [BIOL] A study of the interrelationships which exist between organisms and their environment. Also known as environmental biology.

ecostate [BIOL] **1.** Lacking costae. **2.** Not costate.

ecosystem [ECOL] A functional system which includes the organisms of a natural community together with their environment. Derived from ecological system.

ecotone [ECOL] A zone of intergradation between ecological communities.

ecotope [ECOL] A specific local habitat.

ecotype [ECOL] A subunit, race, or variety of a plant ecospecies that is restricted to one habitat; equivalent to a taxonomic subspecies.

ecsoma [INV ZOO] The retractile posterior end of certain trematodes.

ectasia [MED] Dilation, especially of a hollow organ.

Ecterocoelia [INV ZOO] The equivalent name for Protostomia.

ectethmoid [ANAT] Either one of the lateral cellular masses of the ethmoid bone.

ecthyma [MED] An inflammatory skin disease characterized by large flat pustules that ulcerate and become crusted, and are surrounded by a distinct inflammatory areola.

ectobronchus [VERT ZOO] A lateral offshoot of the bronchus in birds.

ectocardia [MED] An abnormal position of the heart; it may be outside the thoracic cavity or misplaced within the thorax.

ectochone [INV ZOO] In some Porifera, a funnel-shaped cavity into which the ostia enter.

ectocommensal [ECOL] An organism living on the outer surface of the body of another organism, without affecting its host.

ectocornea [ANAT] The outer layer of the cornea.

ectocyst [INV ZOO] 1. The outer layer of the wall of a zooecium. 2. *See* epicyst.

ectoderm [EMBRYO] The outer germ layer of an animal embryo. Also known as epiblast. [INV ZOO] The outer layer of a diploblastic animal.

ectogenesis [EMBRYO] Development of an embryo or of embryonic tissue outside the body in an artificial environment.

ectogony [BOT] The influence of pollination and fertilization on structures outside the embryo and endosperm; effect may be on color, chemical composition, ripening, or abscission.

ectoloph [VERT ZOO] The ridge extending from the paracone to the metacone on a lophodont molar.

ectomere [EMBRYO] A blastomere that will differentiate into ectoderm.

ectomesoblast [EMBRYO] An undifferentiated layer of embryonic cells from which arises the epiblast and mesoblast.

ectomesoderm [EMBRYO] Mesoderm which is derived from ectoderm and is always mesenchymal; a type of primitive connective tissue.

ectomorph [PSYCH] A somatotype suggested by W. H. Sheldon to describe a person with a thin physique.

ectoparasite [ECOL] A parasite that lives on the exterior of its host.

ectopatagium [VERT ZOO] In bats, that part of the patagium which is attached to the metacarpals and phalanges.

ectophagous [INV ZOO] The larval stage of a parasitic insect which is in the process of development externally on a host.

ectophloic siphonostele [BOT] A type of stele with pith that has the phloem only on the outside of the xylem.

ectophyte [ECOL] A plant which lives externally on another organism.

ectopia [MED] A congenital or acquired positional abnormality of an organ or other part of the body.

ectopic pairing [CYTOL] Pairing between nonhomologous

ECTOPHLOIC SIPHONOSTELE

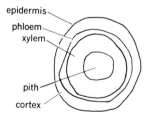

epidermis
phloem
xylem
pith
cortex

Cross section of ectophloic siphonostele.

segments of the salivary gland chromosomes in *Drosophila*, presumably involving mainly heterochromatic regions.

ectoplasm [CYTOL] The outer, gelled zone of the cytoplasmic ground substance in many cells. Also known as ectosarc.

Ectoprocta [INV ZOO] A subphylum of colonial bryozoans having eucoelomate visceral cavities and the anus opening outside the circlet of tentacles.

ectopterygoid [VERT ZOO] A membrane bone located ventrally on the skull, situated behind the palate and extending to the quadrate; found in some fishes and reptiles.

ectosarc *See* ectoplasm.

ectosome [INV ZOO] The outer, cortical layer of a sponge.

ectostosis [PHYSIO] Formation of bone immediately beneath the perichondrium and surrounding and replacing underlying cartilage.

ectostroma [MYCOL] Tissue of a parasitic fungus which penetrates the cortical tissue of the host and bears conidia.

ectosymbiont [ECOL] A symbiont that lives on the surface of or is physically separated from its host.

ectotherm [PHYSIO] A cold-blooded animal.

ectotrophic [BIOL] Obtaining nourishment from outside; applied to certain parasitic fungi that live on and surround the roots of the host plant.

ectozoa [ECOL] Animals which live externally on other organisms.

Ectrephidae [INV ZOO] An equivalent name for Ptinidae.

ectrodactylia [MED] Congenital absence of any of the fingers or toes or parts of them.

eczema [MED] Any skin disorder characterized by redness, thickening, oozing from blisters or papules, and occasional formation of fissures and crusts.

eczematoid reaction [MED] A dermal and epidermal inflammatory response characterized by erythema, edema, vesiculation, and exudation in the acute stage, and by erythema, edema, thickening of the epidermis, and scaling in the chronic stage.

ED₅₀ ED_{50} *See* effective dose 50.

edaphic community [ECOL] A plant community that results from or is influenced by soil factors such as salinity and drainage.

edaphon [BIOL] Flora and fauna in soils.

Edaphosuria [PALEON] A suborder of extinct, lowland, terrestrial, herbivorous reptiles in the order Pelycosauria.

edema [MED] An excessive accumulation of fluid in the cells, tissue spaces, or body cavities due to a disturbance in the fluid exchange mechanism. Also known as dropsy.

Edenian [GEOL] Lower Cincinnatian geologic stage in North America, above the Mohawkian and below Maysvillian.

Edentata [VERT ZOO] An order of mammals characterized by the absence of teeth or the presence of simple prismatic, unspecialized teeth with no enamel.

edentate [VERT ZOO] 1. Lacking teeth. 2. Any member of the Edentata.

edentulous [VERT ZOO] Having no teeth; especially, having lost teeth that were present.

ediacaran fauna [PALEON] The oldest known assemblage of fossil remains of soft-bodied marine animals; first discovered in the Ediacara Hills, Australia.

Edman degradation technique [BIOCHEM] In protein analysis, an approach to amino-end-group determination involving the use of a reagent, phenylisothiocyanate, that can be applied to the liberation of a derivative of the amino-terminal residue without hydrolysis of the remainder of the peptide chain.

Edrioasteroidea [PALEON] A class of extinct Echinozoa having ambulacral radial areas bordered by tube feet, and the

EDENTATA

The giant anteater
(*Myrmecophaga tridactyla*), the
largest of all the edentates and
a native of the tropical grasslands
of Central South America.

EEL

American eel, showing typical
fin structure.

EGG

The egg of a bird. *(After
Schimkewitsch, from L. P. Sayles,
ed., Biology of the Vertebrates,
3d ed., Macmillan, 1949)*

EGGPLANT

Eggplant *(Solanum melongena)*,
cultivar Black Magic. *(Joseph
Harris Co., Rochester, N.Y.)*

mouth and anus located on the upper side of the theca.

educational age [PSYCH] The average achievement of a pupil or student in school subjects based on average performance for a given chronological age as measured by standard
educational tests.

Edwards' syndrome *See* trisomy 18 syndrome.

EEG *See* electroencephalogram.

eel [VERT ZOO] The common name for a number of unrelated
fishes included in the orders Anguilliformes and Cypriniformes; all have an elongate, serpentine body.

effective dose 50 [PHARM] The amount of a drug required to
produce a response in 50% of the subjects to whom the drug
is given. Abbreviated ED$_{50}$. Also known as median effective
dose.

effective half-life [NUCLEO] The half-life of a radioisotope in
a biological organism, resulting from a combination of radioactive decay and biological elimination.

effector [BIOCHEM] Activator of an allosteric enzyme.
[PHYSIO] A structure that is sensitive to a stimulus and causes
an organism or part of an organism to react to the stimulus,
either positively or negatively.

effector organ [PHYSIO] Any muscle or gland that mediates
overt behavior, that is, movement or secretion.

effector system [PHYSIO] A system of effector organs in the
animal body.

efferent [PHYSIO] Carrying or conducting away, as the duct
of an exocrine gland or a nerve.

efflorescence [BOT] The period or process of flowering.
[CHEM] The property of hydrated crystals to lose water of
hydration and crumble when exposed to air.

effuse [BOT] Expanded; spread out in a definite form.

effusion [MED] A pouring out of any fluid into a body cavity
or tissue. [PHYS CHEM] The movement of a gas through an
opening which is small as compared with the average distance
which the gas molecules travel between collisions.
[SCI TECH] **1.** The act or process of leaking or pouring out. **2.**
Any material that is effused.

egest [PHYSIO] **1.** To discharge indigestible matter from the
digestive tract. **2.** To rid the body of waste.

egesta [PHYSIO] All waste substances, solid and liquid, discharged from the body.

egg [CYTOL] **1.** A large, female sex cell enclosed in a porous,
calcareous or leathery shell, produced by birds and reptiles.
2. *See* ovum.

egg apparatus [BOT] A group of three cells, consisting of the
egg and two synergid cells, in the micropylar end of the
embryo sac in seed plants.

egg capsule *See* egg case.

egg case [INV ZOO] **1.** A protective capsule containing the
eggs of certain insects and mollusks. Also known as egg
capsule. **2.** A silk pouch in which certain spiders carry their
eggs. Also known as egg sac. [VERT ZOO] A soft, gelatinous
(amphibians) or strong, horny (skates) envelope containing
the egg of certain vertebrates.

eggplant [BOT] *Solanum melongena.* A plant of the order
Polemoniales grown for its edible egg-shaped, fleshy fruit.

egg raft [ZOO] A floating mass of eggs; produced by a variety
of aquatic organisms.

egg sac [INV ZOO] **1.** The structure containing the eggs of
certain microcrustaceans. **2.** *See* egg case.

egg tooth [VERT ZOO] A toothlike prominence on the tip of
the beak of a bird embryo and the tip of the nose of an
oviparous reptile, which is used to break the eggshell.

eglandular [BIOL] Without glands.

ego [PSYCH] **1.** The self. **2.** The conscious part of the personality that is in contact with reality.

Egyptian henna *See* henna.

Ehrlich's 606 *See* arsphenamine.

eikonometer [OPTICS] A scale used to measure sizes of objects viewed through a microscope, usually attached to the eyepiece so that it is seen superimposed on the image.

Eimeriina [INV ZOO] A suborder of coccidian protozoans in the order Eucoccida in which there is no syzygy and the microgametocytes produce a large number of microgametes.

einsteinium [CHEM] Synthetic radioactive element, symbol Es, atomic number 99; discovered in debris of 1952 hydrogen bomb explosion; now made in cyclotrons.

ejaculation [PHYSIO] The act or process of suddenly discharging a fluid from the body; specifically, the ejection of semen during orgasm.

ejaculatory duct [ANAT] The terminal part of the ductus deferens after junction with the duct of a seminal vesicle, embedded in the prostate gland and opening into the urethra.

ejaculatory sac [INV ZOO] In some insects, an organ that pumps secretions from the vas deferens through the ejaculatory duct to the penis.

ejecta [PHYSIO] Excrement. [SCI TECH] Material which is cast out.

EKG *See* electrocardiogram.

Elaeagnaceae [BOT] A family of dicotyledonous plants in the order Proteales, noted for peltate leaf scales which often give the leaves a silvery-gray appearance.

elaidic acid [ORG CHEM] $CH_3(CH_2)_7CH:CH(CH_2)_7COOH$ A transisomer of an unsaturated fatty acid, oleic acid; crystallizes as colorless leaflets, melts at 44°C, boils at 288°C (100 mm Hg), insoluble in water, soluble in alcohol and ether; used in chromatography as a reference standard. Also known as *trans*-9-octadecenoic acid.

elaioplast [HISTOL] An oil-secreting leucoplast.

Elaphomycetaceae [MYCOL] A family of underground, saprophytic or mycorrhiza-forming fungi in the order Eurotiales characterized by ascocarps with thick, usually woody walls.

Elapidae [VERT ZOO] A family of poisonous reptiles, including cobras, kraits, mambas, and coral snakes; all have a pteroglyph fang arrangement.

Elasipodida [INV ZOO] An order of deep-sea aspidochirotacean holothurians in which there are no respiratory trees and bilateral symmetry is often quite conspicuous.

Elasmidae [INV ZOO] A family of hymenopteran insects in the superfamily Chalcidoidea.

Elasmobranchii [VERT ZOO] The sharks and rays, a subclass of the class Chondrichthyes distinguished by separate gill openings, amphistylic or hyostylic jaw suspension, and ampullae of Lorenzini in the head region.

Elassomatidae [VERT ZOO] The pygmy sunfishes, a family of the order Perciformes.

elastase [BIOCHEM] An enzyme which acts on elastin to change it chemically and render it soluble.

elastic cartilage [HISTOL] A type of cartilage containing elastic fibers in the matrix.

elastic fiber [HISTOL] A homogeneous, fibrillar connective tissue component that is highly refractile and appears yellowish when arranged in bundles.

elasticity [MECH] **1.** The property whereby a solid material changes its shape and size under action of opposing forces, but recovers its original configuration when the forces are removed. **2.** The existence of forces which tend to restore to its original position any part of a medium (solid or fluid) which has been displaced.

EINSTEINIUM

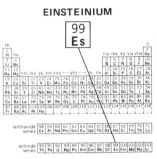

Periodic table of the chemical elements showing the position of einsteinium.

ELASIPODIDA

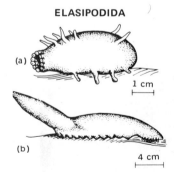

Two examples of bottom-dwelling Elasipodida. *(a) Elpidia.
(b) Psychropotes.*

ELATERIDAE

External features of the click beetle. *(From T. I. Storer and R. L. Usinger, General Zoology, 3d ed., McGraw-Hill, 1957)*

elastic tissue [HISTOL] A type of connective tissue having a predominance of elastic fibers, bands, or lamellae.

elastin [BIOCHEM] An elastic protein composing the principal component of elastic fibers.

elastosis [MED] **1.** Retrogressive change in elastic tissue. **2.** Retrogressive change in cutaneous connective tissue resulting in excessive amounts of material which give the staining reactions for elastin.

elater [BOT] A spiral, filamentous structure that functions in the dispersion of spores in certain plants, such as liverworts and slime molds.

Elateridae [INV ZOO] The click beetles, a large family of coleopteran insects in the superfamily Elateroidea; many have light-producing organs.

Elateroidea [INV ZOO] A superfamily of coleopteran insects in the suborder Polyphaga.

elaterophore [BOT] A tissue bearing elaters, found in some liverworts.

elbow [ANAT] The arm joint formed at the junction of the humerus, radius, and ulna.

elective culture [MICROBIO] A type of microorganism grown selectively from a mixed culture by culturing in a medium and under conditions selective for only one type of organism.

Electra complex [PSYCH] The female analog of the Oedipus complex, that is, the attraction and attachment of a female child to the father.

electrical equivalent [ANALY CHEM] In conductometric analyses of electrolyte solutions, an outside, calibrated current source as compared to (equivalent to) the current passing through the sample under analysis; for example, a Wheatstone-bridge balanced reading.

electric anesthesia [MED] Anesthesia produced by electrical means, as with interrupted direct current.

electric burn [MED] A burn caused by electric current.

electric dipole [ELEC] A localized distribution of positive and negative electricity, without net charge, whose mean positions of positive and negative charges do not coincide.

electric double layer [PHYS CHEM] An electric double layer at a solid-liquid interface; this electric double layer is made up of ions of one charge type which are fixed to the surface of the solid and an equal number of mobile ions of the opposite charge which are distributed through the neighboring region of the liquid; in such a system the movement of liquid causes a displacement of the mobile counterions with respect to the fixed charges on the solid surface. Also known as double layer.

electric eel [VERT ZOO] *Electrophorus electricus.* An eellike cypriniform electric fish of the family Gymnotidae.

electric field [ELEC] **1.** One of the fundamental fields in nature, causing a charged body to be attracted to or repelled by other charged bodies; associated with an electromagnetic wave or a changing magnetic field. **2.** Specifically, the electric force per unit test charge.

electric fish [VERT ZOO] Any of several fishes capable of producing electric discharges from an electric organ.

electric organ [VERT ZOO] An organ consisting of rows of electroplaques which produce an electric discharge.

electric shock [PHYSIO] The sudden pain, convulsion, unconsciousness, or death produced by the passage of electric current through the body.

electrocardiogram [MED] A graphic recording of the electrical manifestations of the heart action as obtained from the body surfaces. Abbreviated ECG; EKG.

electrocardiograph [MED] The instrument used to obtain an electrocardiogram.

electrocardiography [MED] The medical specialty concerned with the production and interpretation of electrocardiograms.

electrocauterization [MED] The application of a direct galvanic current to tissues to cause destruction or coagulation.

electrochemical cell [PHYS CHEM] A combination of two electrodes arranged so that an overall oxidation-reduction reaction produces an electromotive force; includes dry cells, wet cells, standard cells, fuel cells, solid-electrolyte cells, and reserve cells.

electrochemical emf [PHYS CHEM] Electrical force generated by means of chemical action, in man-made cells (such as dry batteries) or by natural means (galvanic reaction).

electrochemical equivalent [PHYS CHEM] The weight in grams of a substance produced or consumed by electrolysis with 100% current efficiency during the flow of a quantity of electricity equal to 1 faraday ($96,487.0 \pm 1.6$ coulombs).

electrochemical potential [PHYS CHEM] The difference in potential that exists when two dissimilar electrodes are connected through an external conducting circuit and the two electrodes are placed in a conducting solution so that electrochemical reactions occur.

electrochemical series [PHYS CHEM] A series in which the metals and other substances are listed in the order of their chemical reactivity or electrode potentials, the most reactive at the top and the less reactive at the bottom. Also known as electromotive series.

electrochemical techniques [PHYS CHEM] The experimental methods developed to study the physical and chemical phenomena associated with electron transfer at the interface of an electrode and solution.

electrochemistry [PHYS CHEM] A branch of chemistry dealing with chemical changes accompanying the passage of an electric current; or with the reverse process, in which a chemical reaction is used to produce an electric current.

electrochromatography [ANALY CHEM] Type of chromatography that utilizes application of an electric potential to produce an electric differential. Also known as electropherography.

electrocoagulation [MED] The coagulation of tissue by means of a high-frequency electric current.

electroconvulsive shock [MED] The technique of eliciting convulsions by applying an electric current through the brain for a brief period.

electrode [ELEC] **1.** An electric conductor through which an electric current enters or leaves a medium, whether it be an electrolytic solution, solid, molten mass, gas, or vacuum. **2.** One of the terminals used in dielectric heating or diathermy for applying the high-frequency electric field to the material being heated.

electrode potential [PHYS CHEM] The voltage existing between an electrode and the solution or electrolyte in which it is immersed; usually, electrode potentials are referred to a standard electrode, such as the hydrogen electrode. Also known as electrode voltage.

electrode voltage *See* electrode potential.

electrodiagnosis [MED] Diagnosis of disease states by recording the spontaneous electrical activity of tissue or organs, or by the response to stimulation of electrically excitable tissue.

electrodialysis [PHYS CHEM] Dialysis that is conducted with the aid of an electromotive force applied to electrodes adjacent to both sides of the membrane.

electroencephalogram [MED] A graphic recording of the electric discharges of the cerebral cortex as detected by electrodes on the surface of the scalp. Abbreviated EEG.

ELECTROLYSIS

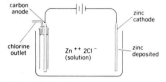

Electrolysis of zinc chloride solution.

electroencephalograph [MED] An instrument used to make electroencephalograms.

electroencephalography [MED] The medical specialty concerned with the production and interpretation of electroencephalograms.

electrolysis [PHYS CHEM] A method by which chemical reactions are carried out by passage of electric current through a solution of an electrolyte or through a molten salt.

electrolyte [PHYS CHEM] A chemical compound which when molten or dissolved in certain solvents, usually water, will conduct an electric current.

electrolytic analysis [ANALY CHEM] Basic electrochemical technique for quantitative analysis of conducting solutions containing oxidizable or reducible material; measurement is based on the weight of material plated out onto the electrode.

electrolytic cell [PHYS CHEM] A cell consisting of electrodes immersed in an electrolyte solution, for carrying out electrolysis.

electrolytic conductivity [PHYS CHEM] The conductivity of a medium in which the transport of electric charges, under electric potential differences, is by particles of atomic or larger size.

electrolytic potential [PHYS CHEM] Difference in potential between an electrode and the immediately adjacent electrolyte, expressed in terms of some standard electrode difference.

electrolytic solution [PHYS CHEM] A solution made up of a solvent and an ionically dissociated solute; it will conduct electricity, and ions can be separated from the solution by deposition on an electrically charged electrode.

electromagnetic energy [ELECTROMAG] The energy associated with electric or magnetic fields.

electromagnetic field [ELECTROMAG] An electric or magnetic field, or a combination of the two, as in an electromagnetic wave.

electromagnetic spectrum [ELECTROMAG] The total range of wavelengths or frequencies of electromagnetic radiation, extending from the longest radio waves to the shortest known cosmic rays.

electromagnetic system of units [ELECTROMAG] A centimeter-gram-second system of electric and magnetic units in which the unit of current is defined as the current which, if maintained in two straight parallel wires having infinite length and being 1 centimeter apart in vacuum, would produce between these conductors a force of 2 dynes per centimeter of length; other units are derived from this definition by assigning unit coefficients in equations relating electric and magnetic quantities. Also known as electromagnetic units (emu).

electromagnetic units *See* electromagnetic system of units.

electromagnetic wave [ELECTROMAG] A disturbance which propagates outward from any electric charge which oscillates or is accelerated; far from the charge it consists of vibrating electric and magnetic fields which move at the speed of light and are at right angles to each other and to the direction of motion.

electromagnetism [PHYS] **1.** Branch of physics relating electricity to magnetism. **2.** Magnetism produced by an electric current rather than by a permanent magnet.

electromotance *See* electromotive force.

electromotive force [PHYS CHEM] **1.** The difference in electric potential that exists between two dissimilar electrodes immersed in the same electrolyte or otherwise connected by ionic conductors. **2.** The resultant of the relative electrode potential of the two dissimilar electrodes at which electrochemical reactions occur. Abbreviated emf. Also known as electromotance.

electromotive series *See* electrochemical series.

electromyogram [MED] **1.** A graphic recording of the electrical response of a muscle to electrical stimulation. **2.** A graphic recording of eye movements during reading.

electromyograph [MED] An instrument used for making electromyograms.

electromyography [MED] A medical specialty concerned with the production and study of electromyograms.

electron [PHYS] **1.** A stable elementary particle which is the negatively charged constituent of ordinary matter, having a mass of about 9.11×10^{-28}g (equivalent to 0.511 MeV), a charge of about -1.602×10^{-19} coulomb, and a spin of $\frac{1}{2}$. Also known as negative electron; negatron. **2.** Collective name for the electron, as in the first definition, and the positron.

electron acceptor [PHYS CHEM] An atom or part of a molecule joined by a covalent bond to an electron donor.

electronarcosis [MED] Profound stupor or unconsciousness produced by passing an electric current through the brain.

electron beam [ELECTR] A narrow stream of electrons moving in the same direction, all having about the same velocity.

electron charge [PHYS] The charge carried by an electron, equal to about -1.602×10^{-19} coulomb, or -4.803×10^{-10} statcoulomb.

electron density [PHYS] **1.** The number of electrons in a unit volume. **2.** When quantum-mechanical effects are significant, the total probability of finding an electron in a unit volume.

electron diffraction [PHYS] The phenomenon associated with the interference processes which occur when electrons are scattered by atoms in crystals to form diffraction patterns.

electron dipole moment *See* electron magnetic moment.

electron distribution [PHYS] A function which gives the number of electrons per unit volume of phase space.

electron donor [PHYS CHEM] An atom or part of a molecule which supplies both electrons of a duplet forming a covalent bond.

electronegative potential [PHYS CHEM] Potential of an electrode expressed as negative with respect to the hydrogen electrode.

electronegativity [PHYS CHEM] The relative ability of an atom or group of atoms to attract electrons to itself.

electronic band spectrum [SPECT] Bands of spectral lines associated with a change of electronic state of a molecule; each band corresponds to certain vibrational energies in the initial and final states and consists of numerous rotational lines.

electronic emission spectrum [SPECT] Spectrum resulting from emission of electromagnetic radiation by atoms, ions, and molecules following excitations of their electrons.

electronic energy curve [PHYS CHEM] A graph of the energy of a diatomic molecule in a given electronic state as a function of the distance between the nuclei of the atoms.

electronic structure [PHYS] The arrangement of electrons in an atom, molecule, or solid, specified by their wave functions, energy levels, or quantum numbers.

electron magnetic moment [ATOM PHYS] The magnetic dipole moment which an electron possesses by virtue of its spin. Also known as electron dipole moment.

electron mass [PHYS] The mass of an electron, equal to about 9.11×10^{-28}g, equivalent to 0.511 MeV. Also known as electron rest mass.

electron microscope [ELECTR] A device for forming greatly magnified images of objects by means of electrons, usually focused by electron lenses.

electron number [ATOM PHYS] The number of electrons in an ion or atom.

electron orbit [PHYS] The path described by an electron.

electron pair [PHYS CHEM] A pair of valence electrons which form a nonpolar bond between two neighboring atoms.

electron paramagnetic resonance [PHYS] Magnetic resonance arising from the magnetic moment of unpaired electrons in a paramagnetic substance or in a paramagnetic center in a diamagnetic substance. Abbreviated EPR. Also known as electron spin resonance (ESR); paramagnetic resonance.

electron rest mass *See* electron mass.

electron shell [ATOM PHYS] **1.** The collection of all the electron states in an atom which have a given principal quantum number. **2.** The collection of all the electron states in an atom which have a given principal quantum number and a given orbital angular momentum quantum number.

electron spectroscopy [SPECT] The study of the energy spectra of photoelectrons or Auger electrons emitted from a substance upon bombardment by electromagnetic radiation, electrons, or ions; used to investigate atomic, molecular, or solid-state structure, and in chemical analysis.

electron spin [QUANT MECH] That property of an electron which gives rise to its angular momentum about an axis within the electron.

electron spin resonance *See* electron paramagnetic resonance.

electron transport system [BIOCHEM] The components of the final sequence of reactions in biological oxidations; composed of a series of oxidizing agents arranged in order of increasing strength and terminating in oxygen.

electron volt [PHYS] A unit of energy equal to the energy acquired by an electron when it passes through a potential difference of 1 volt in a vacuum; it is equal to $(1.602192 \pm 0.000007) \times 10^{-19}$ volt. Abbreviated eV.

electroosmosis [PHYS CHEM] The movement in an electric field of liquid with respect to colloidal particles immobilized in a porous diaphragm or a single capillary tube.

electropherography *See* electrochromatography.

electrophoresis [PHYS CHEM] An electrochemical process in which colloidal particles or macromolecules with a net electric charge migrate in a solution under the influence of an electric current. Also known as cataphoresis.

electrophoretic mobility [BIOCHEM] A characteristic of living cells in suspension and biological compounds (proteins) in solution to travel in an electric field to the positive or negative electrode, because of the charge on these substances.

electrophoretic variants [BIOCHEM] Phenotypically different proteins that are separable into distinct electrophoretic components due to differences in mobilities; an example is erythrocyte acid phosphatase.

electrophrenic respiration [MED] Artificial respiration in which the nerves that control breathing are stimulated electrically through appropriately placed electrodes.

electrophysiology [PHYSIO] The branch of physiology concerned with determining the basic mechanisms by which electric currents are generated within living organisms.

electroplax [VERT ZOO] One of the structural units of an electric organ of some fishes, composed of thin, flattened plates of modified muscle that appear as two large, waferlike, roughly circular or rectangular surfaces.

electropositive [PHYS CHEM] Pertaining to elements, ions, or radicals that tend to give up or lose electrons.

electropositive potential [PHYS CHEM] Potential of an electrode expressed as positive with respect to the hydrogen electrode.

electroshock therapy [MED] Treatment of mental patients by passing an electric current of 85-110 volts through the brain.

electrotaxis [BIOL] Movement of an organism in response to stimulation by electric charges.

electrotonus [PHYSIO] The change of condition in a nerve or a muscle during the passage of a current of electricity.

electrotropism [BIOL] Orientation response of a sessile organism to stimulation by electric charges.

electrovalence [PHYS CHEM] The valence of an atom that has formed an ionic bond.

electrovalent bond *See* ionic bond.

element [CHEM] A substance made up of atoms with the same atomic number; common examples are hydrogen, gold, and iron. Also known as chemical element.

element 104 [CHEM] The first element beyond the actinide series, and the twelfth transuranium element; the atoms of element 104, of mass number 260, were first produced by irradiating plutonium-242 with neon-22 ions in a heavy-ion cyclotron.

element 105 [CHEM] An artificial element whose isotope of mass number 260 was discovered by bombarding californium-249 with nitrogen-15 ions in a heavy-ion linear accelerator.

elementary process [PHYS CHEM] In chemical kinetics, the particular events at the atomic or molecular level which make up an overall reaction.

elephant [VERT ZOO] The common name for two living species of proboscidean mammals in the family Elephantidae; distinguished by the elongation of the nostrils and upper lip into a sensitive, prehensile proboscis.

elephantiasis [MED] A parasitic disease of man caused by the filarial nematode *Wuchereria bancrofti;* characterized by cutaneous and subcutaneous tissue enlargement due to lymphatic obstruction.

Elephantidae [VERT ZOO] A family of mammals in the order Proboscidea containing the modern elephants and extinct mammoths.

elixir [PHARM] A sweetened, aromatic solution, usually hydroalcoholic, sometimes containing soluble medicants; intended for use only as a flavor or vehicle.

elk [VERT ZOO] *Alces alces.* A mammal (family Cervidae) in Europe and Asia that resembles the North American moose but is smaller; it is the largest living deer.

elliptic [SCI TECH] Oval-shaped.

elliptocytosis [MED] A rare hereditary disease of man characterized by the presence of large numbers of oval or elliptic erythrocytes in the circulating blood.

elm [BOT] The common name for hardwood trees composing the genus *Ulmus,* characterized by simple, serrate, deciduous leaves.

Elmidae [INV ZOO] The drive beetles, a small family of coleopteran insects in the superfamily Dryopoidea.

Elopidae [VERT ZOO] A family of fishes in the order Elopiformes, including the tarpon, ladyfish, and machete.

Elopiformes ·[VERT ZOO] A primitive order of actinopterygian fishes characterized by a single dorsal fin composed of soft rays only, cycloid scales, and toothed maxillae.

El Tor vibrio [MICROBIO] Any of the rod-shaped paracholera vibrios; many strains can be agglutinated with anticholera serum.

elusive ulcer *See* Hunner's ulcer.

elytron [INV ZOO] **1.** One of the two sclerotized or leathery anterior wings of beetles which serve to cover and protect the

ELEMENT 104

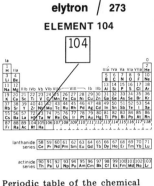

Periodic table of the chemical elements showing the position of element 104.

ELEMENT 105

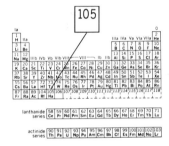

Periodic table of the chemical elements showing the position of element 105.

ELEPHANT

The Indian elephant, recognized by its concave forehead, knobby head, and relatively small ears.

membranous hindwings. **2.** A dorsal scale of certain Polychaeta.

elytrophore [INV ZOO] A structure on the prostomium of certain polychaetes which bears an elytron.

emaciation [MED] A wasted condition of the body; the process of losing flesh so as to become extremely lean.

emarginate [BIOL] Having a margin that is notched or slightly forked.

EMB agar [MICROBIO] A culture medium containing sugar, eosin, and methylene blue, used in the confirming test for coliform bacteria.

Emballonuridae [VERT ZOO] The sheath-tailed bats, a family of mammals in the order Chiroptera.

embalm [MED] To treat a cadaver with antiseptics and preservatives to prevent decay, before burial or dissection.

Embden-Meyerhof pathway *See* glycolytic pathway.

Embiidina [INV ZOO] An equivalent name for Embioptera.

Embioptera [INV ZOO] An order of silk-spinning, orthopteroid insects resembling the grasshoppers; commonly called the embiids or web spinners.

EMBIOPTERA

Body form of a typical embiid
(*Pararhagadochir trachelia* Navas,
female). *(From E. S. Ross, Insects
Close Up; Univeristy of California
Press, 1953)*

Embiotocidae [VERT ZOO] The surfperches, a family of perciform fishes in the suborder Percoidei.

embolectomy [MED] Surgical removal of an embolus.

embolism [MED] The blocking of a blood vessel by an embolus.

embolium [INV ZOO] In certain insects, the outer part of the wing, or the basal part of the hemelytron.

Embolomeri [PALEON] An extinct side branch of slender-bodied, fish-eating aquatic anthracosaurs in which intercentra as well as centra form complete rings.

embolus [MED] A clot or other mass of particulate matter foreign to the bloodstream which lodges in a blood vessel and causes obstruction.

emboly [EMBRYO] Formation of a gastrula by the process of invagination.

EMBRITHOPODA

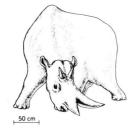

50 cm

Arsinoitherium, the early
Oligocene embrithopod from
Egypt. *(After C. R. Knight)*

Embrithopoda [PALEON] An order established for the unique Oligocene mammal *Arsinoitherium,* a herbivorous animal that resembled the modern rhinoceros.

embryo [BOT] Young sporophyte of a seed plant. [EMBRYO] **1.** An early stage of development in multicellular organisms. **2.** The product of conception up to the third month of human pregnancy.

Embryobionta [BOT] The land plants, a subkingdom of the Plantae characterized by having specialized conducting tissue in the sporophyte (except bryophytes), having multicellular sex organs, and producing an embryo.

embryogenesis [EMBRYO] The formation and development of an embryo from an egg.

embryology [BIOL] The study of the development of the organism from the zygote, or fertilized egg.

embryoma of the ovary *See* dysgerminoma.

embryonal-cell lipoma *See* liposarcoma.

embryonate [EMBRYO] **1.** To differentiate into a zygote. **2.** Containing an embryo.

embryonated egg culture [VIROL] Embryonated hen's eggs inoculated with animal viruses for the purpose of identification, isolation, titration, or for quantity cultivation in the production of viral vaccines.

embryonic differentiation [EMBRYO] The process by which specialized and diversified structures arise during embryogenesis.

embryonic inducer [EMBRYO] The acting system in embryos, which contributes to the formation of specialized tissues by controlling the mode of development of the reacting system.

embryonic induction [EMBRYO] The influence of one cell

group (inducer) over a neighboring cell group (induced) during embryogenesis. Also known as induction.

embryopathy [MED] Any abnormal development of an embryo, either morphological or biochemical.

embryophore [INV ZOO] The cellular covering of the developing onchosphere of certain tapeworms.

Embryophyta [BOT] The equivalent name for Embryobionta.

embryo sac [BOT] The female gametophyte of a seed plant, containing the egg, synergids, and polar and antipodal nuclei; fusion of the antipodals and a pollen generative nucleus forms the endosperm.

embryotomy [MED] Any mutilation of the fetus in the uterus to aid in its removal when natural delivery is impossible.

emesis [MED] The act of vomiting.

emetic [PHARM] Any agent that induces emesis.

emetine [PHARM] $C_{29}H_{40}N_2O_4$ Cephaeline methyl ether, the principal alkaloid of ipecac; a white powder, sparingly soluble in water; it is emetic, diaphoretic, and expectorant, but its chief utility is as an amebicide.

emf *See* electromotive force.

emiocytosis [CYTOL] Fusion of intracellular granules with the cell membrane, followed by discharge of the granules outside of the cell; applied chiefly to the mechanism of insulin secretion. Also known as reverse pinocytosis.

emission electron microscope [ELECTR] An electron microscope in which thermionic, photo, secondary, or field electrons emitted from a metal surface are projected on a fluorescent screen, with or without focusing.

emission flame photometry [ANALY CHEM] A form of flame photometry in which a sample solution to be analyzed is aspirated into a hydrogen-oxygen or acetylene-oxygen flame; the line emission spectrum is formed, and the line or band of interest is isolated with a monochromator and its intensity measured photoelectrically.

emission lines [SPECT] Spectral lines resulting from emission of electromagnetic radiation by atoms, ions, or molecules during changes from excited states to states of lower energy.

emmetropia [MED] Normal vision.

emollient [PHARM] A softening agent, especially for use on skin and mucous membranes; lanolin is widely used as a base.

emotion [PSYCH] A strong mental feeling or affect of the consciousness involving visceral and other physiologic changes.

emphysema [MED] A pulmonary disorder characterized by overdistention and destruction of the air spaces in the lungs.

emphysematous chest [MED] The altered contour of the chest seen in pulmonary emphysema, with increased anteroposterior diameter, flaring at the lower rib margins, low position of the diaphragm, and minimal respiratory motion. Also known as barrel chest.

Empididae [INV ZOO] The dance flies, a family of orthorrhaphous dipteran insects in the series Nematocera.

empirical formula [CHEM] A chemical formula indicating the variety and relative proportions of the atoms in a molecule but not showing the manner in which they are linked together.

empirical probability [STAT] The ratio of the number of times an event has occurred to the total number of trials performed. Also known as a posteriori probability.

empodium [INV ZOO] A small peripheral part located between the claws of the tarsi of many insects and arachnids.

empyema [MED] The presence of pus in the body cavity, hollow organ, or tissue space; when the term is used without qualification, it generally refers to pus in the pleural space.

emu [ELECTROMAG] *See* electromagnetic system of units.

[VERT ZOO] *Dromiceius novae-hollandiae.* An Australian rat-ite bird, the second largest living bird, characterized by rudimentary wings and a feathered head and neck without wattles.

emulsification [CHEM] The process of dispersing one liquid in a second immiscible liquid; the largest group of emulsifying agents are soaps, detergents, and other compounds, whose basic structure is a paraffin chain terminating in a polar group.

emulsion [CHEM] A stable dispersion of one liquid in a second immiscible liquid, such as milk (oil dispersed in water).

Emydidae [VERT ZOO] A family of aquatic and semiaquatic turtles in the suborder Cryptodira.

Enaliornithidae [PALEON] A family of extinct birds assigned to the order Hesperornithiformes, having well-developed teeth found in grooves in the dentary and maxillary bones of the jaws.

enamel [PHYSIO] The hard substance, consisting of over 90% calcium and magnesium salts, which forms a protective cap over the tooth dentin.

enamel organ [EMBRYO] The epithelial ingrowth from the dental lamina which covers the dental papilla, furnishes a mold for the shape of a developing tooth, and forms the dental enamel.

enantiomer *See* enantiomorph.

enantiomorph [CHEM] One of an isomeric pair of crystalline forms or compounds whose molecules are nonsuperimposable mirror images. Also known as enantiomer; optical antipode; optical isomer.

Enantiozoa [INV ZOO] The equivalent name for Parazoa.

enarthrosis [ANAT] A freely movable joint that allows a wide range of motion on all planes. Also known as ball-and-socket joint.

encephalitis [MED] Inflammation of the brain.

encephalitis lethargica [MED] Epidemic encephalitis, probably of viral etiology, characterized by lethargy, ophthalmoplegia, hyperkinesia, and at times residual neurologic disability, particularly parkinsonism with oculogyric crisis. Also known as epidemic encephalitis; sleeping sickness; von Economo's disease.

encephalocele [MED] Hernia of the brain through a congenital or traumatic opening in the cranium.

encephalogram [MED] A roentgenogram of the brain made in encephalography.

encephalography [MED] Roentgenography of the brain following removal of cerebrospinal fluid, by lumbar or cisternal puncture, and its replacement by air or other gas.

encephaloid carcinoma *See* medullary carcinoma.

encephalomalacia [MED] **1.** Infarction of the brain. **2.** Any softening or fragmentation of the brain.

encephalomyelitis [MED] Inflammation of the brain and spinal cord.

encephalomyocarditis [MED] An acute febrile RNA virus disease accompanied by pharyngitis, stiff neck, and hyperactive deep reflexes; certain species of wild rats are the reservoir; human infections range from a mild febrile illness to a severe encephalomyelitis.

encephalopathy [MED] Any disease of the brain.

enchondroma [MED] A benign tumor composed of dysplastic cartilage cells, occurring in the metaphysis of cylindric bones, especially of the hands and feet.

enchondromatosis [MED] A rare disorder principally involving tubular bones, especially those of the feet and hands, characterized by hamartomatous proliferation of cartilage in the metaphysis, indistinguishable in single lesions from enchondromas. Also known as chondrodysplasia; dyschondroplasia; Ollier's disease.

enchymatous [PHYSIO] Of gland cells, distended with secreted material.

Encyrtidae [INV ZOO] A family of hymenopteran insects in the superfamily Chalcidoidea.

encystment [BIOL] The process of forming or becoming enclosed in a cyst or capsule.

Endamoeba coli [INV ZOO] A nonpathogenic ameba that inhabits the human intestinal tract.

endarteritis [MED] Inflammation of the lining (tunica intima) of an artery.

endarteritis obliterans [MED] Endarteritis, particularly of small arteries, accompanied by degeneration of the intima, leading to occlusion of the blood vessel. Also known as obliterating endarteritis.

end bulb *See* bouton.

end bulb of Krause *See* Krause's corpuscle.

Endeidae [INV ZOO] A family of marine arthropods in the subphylum Pycnogonida.

endemic [MED] Peculiar to a certain region, specifically referring to a disease which occurs more or less constantly in any locality.

endemic rural plague *See* sylvatic plague.

endemic typhus *See* murine typhus.

endemism [MED] The state or quality of being endemic.

endergonic [BIOCHEM] Of or pertaining to a biochemical reaction in which the final products possess more free energy than the starting materials; usually associated with anabolism.

endermic [MED] Acting through the skin by absorption, such as medication applied to the skin.

endite [INV ZOO] **1.** One of the appendages on the inner aspect of an arthropod limb. **2.** A ridgelike chewing surface on the inner part of the pedipalpus or maxilla of many arachnids.

endobasion [ANAT] The anteriormost point of the margin of the foramen magnum at the level of its smallest diameter.

endobiotic [ECOL] Referring to an organism living in the cells or tissues of a host.

endocardial fibroelastosis [MED] Fibrous or fibroelastic thickening of the endocardium, of unknown cause.

endocarditis [MED] Inflammation of the endocardium.

endocardium [ANAT] The membrane lining the heart.

endocarp [BOT] The inner layer of the wall of a fruit or pericarp.

endocarpoid [MYCOL] Having discoid ascocarps embedded in the thallus.

endocervicitis [MED] Inflammation of the mucous membrane of the uterine cervix.

endocervix [ANAT] The glandular mucous membrane of the cervix uteri.

endochondral ossification [PHYSIO] The conversion of cartilage into bone. Also known as intracartilaginous ossification.

endochorion [INV ZOO] The inner chorionic layer of insect eggs.

endocommensal [ECOL] A commensal that lives within the body of its host.

endocorpuscular [CYTOL] Located within an erythrocyte.

endocranium [ANAT] **1.** The inner surface of the cranium. **2.** *See* dura mater. [INV ZOO] The processes on the inner surface of the head capsule of certain insects.

endocrine gland [PHYSIO] A ductless structure whose secretion (hormone) is passed into adjacent tissue and then to the bloodstream either directly or by way of the lymphatics. Also known as ductless gland.

endocrine system [PHYSIO] The chemical coordinating system in animals, that is, the endocrine glands that produce hormones.

endocrinology [PHYSIO] The study of the endocrine glands and the hormones that they synthesize and secrete.

endocuticle [INV ZOO] The inner, elastic layer of an insect cuticle.

endocyst [INV ZOO] The soft layer consisting of ectoderm and mesoderm, lining the ectocyst of bryozoans.

endocytic vacuole [CYTOL] A membrane-bound cellular organelle containing extracellular particles engulfed by the mechanisms of endocytosis.

endocytosis [CYTOL] An active process in which extracellular materials are introduced into the cytoplasm of cells by either phagocytosis or pinocytosis.

endoderm [EMBRYO] The inner, primary germ layer of an animal embryo; sometimes referred to as the hypoblast. Also known as entoderm; hypoblast.

endodermis [BOT] The innermost tissue of the cortex of most plant roots and certain stems consisting of a single layer of at least partly suberized or cutinized cells; functions to control the movement of water and other substances into and out of the stele.

endoenzyme [BIOCHEM] An intracellular enzyme, retained and utilized by the secreting cell.

endoergic *See* endothermic.

endogamy [BIOL] Sexual reproduction between organisms which are closely related. [BOT] Pollination of a flower by another flower of the same plant.

endogenous [BIOCHEM] Relating to the metabolism of nitrogenous tissue elements. [GEOL] *See* endogenic. [MED] Pertaining to diseases resulting from internal causes. [PSYCH] Pertaining to mental disorders caused by hereditary or constitutional factors.

endognath [INV ZOO] The inner and main branch of a crustacean's oral appendage.

endolabium [INV ZOO] In insects, a membranous lobe inside the mouth on the middle parts of the anterior of the labium.

endolecithal [INV ZOO] A type of egg found in turbellarians with yolk granules in the cytoplasm of the egg. Also spelled entolecithal.

endolithic [ECOL] Living within rocks, as certain algae and coral.

endolymph [PHYSIO] The lymph fluid found in the membranous labyrinth of the ear.

endolymphatic stromomyosis *See* interstitial endometriosis.

endomeninx [EMBRYO] The internal part of the meninx primitiva that differentiates into the pia mater and arachnoid membrane.

endomere [EMBRYO] A blastomere that forms endoderm.

endometrioma [MED] Endometriosis in which there is a discrete tumor mass.

endometriosis [MED] The presence of endometrial tissue in abnormal locations, including the uterine wall, ovaries, or extragenital sites.

endometritis [MED] Inflammation of the endometrium.

endometrium [ANAT] The mucous membrane lining the uterus.

endomitosis [CYTOL] Division of the chromosomes without dissolution of the nuclear membrane; results in polyploidy or polyteny.

endomixis [INV ZOO] Periodic division and reorganization of the nucleus in certain ciliated protozoans.

endomorph [PSYCH] A somatotype suggested by W. H. Sheldon to describe a person with a rounded physique; associated with viscerotonia.

Endomycetales [MICROBIO] Former designation for Saccharomycetales.

Endomycetoideae [MICROBIO] A subfamily of ascosporogenous yeasts in the family Saccharomycetaceae.

Endomychidae [INV ZOO] The handsome fungus beetles, a family of coleopteran insects in the superfamily Cucujoidea.

endomysium [HISTOL] The connective tissue layer surrounding an individual skeletal muscle fiber.

endoneural fibroma See neurofibroma.

endoneurium [HISTOL] Connective tissue fibers surrounding and joining the individual fibers of a nerve trunk.

endonuclease [BIOCHEM] An enzyme that severs the backbone chain of a strand of deoxyribonucleic acid.

endoparasite [ECOL] A parasite that lives inside its host.

endopeptidase [BIOCHEM] An enzyme that acts upon the centrally located peptide bonds of a protein molecule.

endophagous [INV ZOO] Of an insect larva, living within and feeding upon the host tissues.

endophragm [INV ZOO] In crustaceans, a septum formed by union of cephalic and thoracic apodemes.

endophyte [ECOL] A plant that lives within, but is not necessarily parasitic on, another plant.

endoplasm [CYTOL] The inner, semifluid portion of the cytoplasm.

endoplasmic reticulum [CYTOL] A vacuolar system of the cytoplasm in differentiated cells that functions in protein synthesis and sequestration. Abbreviated ER.

endopleura [BOT] The inner seed coat.

endopleurite [INV ZOO] 1. The portion of a crustacean apodeme which develops from the interepimeral membrane. 2. One of the laterally located parts on the thorax of an insect which fold inward, extending into the body cavity.

endoployploidy [GEN] The presence in a cell of a nucleus with more than the number of diploid chromosomes found in the zygote; occurs, for example, in insects in which most tissues have their own characteristic degree of ploidy.

endopodite [INV ZOO] The inner branch of a biramous crustacean appendage.

Endoprocta [INV ZOO] The equivalent name for Entoprocta.

endoprosthesis [MED] A prosthesis that is used internally.

endopterygoid [VERT ZOO] A paired dermal bone of the roof of the mouth in fishes.

Endopterygota [INV ZOO] A division of the insects in the subclass Pterygota, including those orders which undergo a holometabolous metamorphosis.

endoreduplication [CYTOL] Appearance in mitotic cells of certain chromosomes or chromosome sets in the form of multiples.

end organ [ANAT] The expanded termination of a nerve fiber in muscle, skin, mucous membrane, or other structure.

endosalpingioma See serous cystadenoma.

endosalpinx [ANAT] The mucous membrane that lines the fallopian tube.

endosarc [INV ZOO] The inner, relatively fluid part of the protoplasm of certain unicellular organisms.

endosclerite [INV ZOO] Any of the sclerites forming the endoskeleton of arthropods.

endoscope [MED] An instrument used to visualize the interior of a body cavity or hollow organ.

endosepsis [PL PATH] A fungus disease of figs caused by *Fusarium moniliforme fici*; fruits rot internally.

ENDOTHELIOCHORIAL PLACENTA

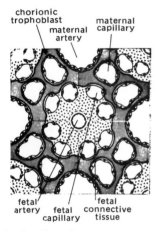

chorionic trophoblast

maternal artery

maternal capillary

fetal artery

fetal capillary

fetal connective tissue

Section through an endotheliochorial placenta.

ENDOTHYRACEA

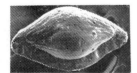

Scanning electron micrograph of *Triticites* from the Pennsylvanian formation of Texas showing the exterior of the fusuline test. *(R. B. MacAdam, Chevron Oil Field Research Co.)*

endosiphuncle [INV ZOO] In some cephalopods, the tube connecting the protoconch with the siphuncle.

endoskeleton [ZOO] An internal skeleton or supporting framework in an animal.

endosmosis [PHYSIO] The passage of a liquid inward through a cell membrane.

endosome [CYTOL] A mass of chromatin near the center of a vesicular nucleus. [INV ZOO] The inner layer of certain sponges.

endosperm [BOT] **1.** The nutritive protein material within the embryo sac of seed plants. **2.** Storage tissue in the seeds of gymnosperms.

endosperm nucleus [BOT] The triploid nucleus formed within the embryo sac of most seed plants by fusion of the polar nuclei with one sperm nucleus.

endospore [BIOL] An asexual spore formed within a cell.

endosternite [INV ZOO] **1.** Internal part of a sternite to which muscles are attached. **2.** The median sternal apodeme.

endosteum [ANAT] The connective tissue lining the medullary cavities of bones.

endostome [BOT] The opening in the inner integument of a bitegmic ovule.

endostracum [INV ZOO] The inner layer of a shell, as of crustaceans and mollusks.

endostyle [INV ZOO] A ciliated groove or pair of grooves in the pharynx of lower chordates.

endosymbiont [ECOL] A symbiont that lives within the body of the host without deleterious effect on the host.

endotergite [INV ZOO] A dorsal plate to which muscles are attached in the insect skeleton.

endothecium [BOT] The middle of three layers that make up an immature anther; becomes the inner layer of a mature anther.

endothelial cell [HISTOL] A type of squamous epithelial cell composing the endothelium.

endotheliochorial placenta [EMBRYO] A type of placenta in which the maternal blood is separated from the chorion by the maternal capillary endothelium; occurs in dogs.

endothelioma [MED] Any tumor arising from, or resembling, endothelium; usually a benign growth, but occasionally a malignant tumor.

endothelium [HISTOL] The epithelial layer of cells lining the heart and vessels of the circulatory system.

Endotheriidae [PALEON] A family of Cretaceous insectivores from China belonging to the Proteutheria.

endothermic [PHYS CHEM] Pertaining to a reaction which absorbs heat. Also known as endoergic.

endothorax [INV ZOO] The system of apodemes in the thorax of a crustacean.

Endothyracea [PALEON] A superfamily of extinct benthic marine foraminiferans in the suborder Fusulinina, having a granular or fibrous wall.

endotoxin [MICROBIO] A toxin that is produced within a microorganism and can be isolated only after the cell is disintegrated.

endotrachea [INV ZOO] The inner chitinous layer of the tracheal tubes in insects.

endotracheal [ANAT] Within the trachea.

endotrophic [BIOL] Obtaining nourishment from within; applied to certain parasitic fungi that live in the root cortex of the host plant.

endozoic [ECOL] Living inside an animal.

end point [ANALY CHEM] That stage in the titration at which an effect, such as a color change, occurs, indicating that a desired point in the titration has been reached.

enema [MED] A rectal injection of liquid for therapeutic, diagnostic, or nutritive purposes.

energy [PHYS] The capacity for doing work.

energy balance [PHYS] The arithmetic balancing of energy inputs versus outputs for an object, reactor, or other processing system; it is positive if energy is released, and negative if it is absorbed. [PHYSIO] The relation of the amount of utilizable energy taken into the body to that which is employed for internal work, external work, and the growth and repair of tissues.

energy metabolism [BIOCHEM] The chemical reactions involved in energy transformations within cells.

energy pyramid [ECOL] An ecological pyramid illustrating the energy flow within an ecosystem.

enervoxe [BOT] Lacking veins, referring specifically to certain leaves.

Engel-Recklinghausen disease *See* osteitis fibrosa cystica.

engram [PHYSIO] A memory imprint; the alteration that has occurred in nervous tissue as a result of an excitation from a stimulus, which hypothetically accounts for retention of that experience. Also known as memory trace.

Engraulidae [VERT ZOO] The anchovies, a family of herring-like fishes in the suborder Clupoidea.

enhancer gene [GEN] Any modifier gene that acts to enhance the action of a nonallelic gene.

Enicocephalidae [INV ZOO] The gnat bugs, a family of hemipteran insects in the superfamily Enicocephaloidea.

Enicocephaloidea [INV ZOO] A superfamily of the Hemiptera in the subdivision Geocorisae containing a single family.

enol [ORG CHEM] An organic compound with a hydroxide group adjacent to a double bond; varies with a ketone form in the effect known as enol-keto tautomerism; an example is the compound $CH_3COH = CHCO_2C_2H_5$.

enophthalmos [MED] Recession of the eyeball into the orbital cavity.

Enopla [INV ZOO] A class or subclass of ribbonlike worms of the phylum Rhynchocoela.

Enoplia [INV ZOO] A subclass of nematodes in the class Adenophorea.

Enoplida [INV ZOO] An order of nematodes in the subclass Enoplia.

Enoplidae [INV ZOO] A family of free-living marine nematodes in the superfamily Enoploidea, characterized by a complex arrangement of teeth and mandibles.

Enoploidea [INV ZOO] A superfamily of small to very large free-living marine nematodes having pocketlike amphids opening to the exterior via slitlike apertures.

Enoploteuthidae [INV ZOO] A molluscan family of deep-sea squids in the class Cephalopoda.

enrichment culture [MICROBIO] A medium of known composition and specific conditions of incubation which favors the growth of a particular type or species of bacterium.

ensiling [AGR] The anaerobic fermentation process used to preserve immature green corn, legumes, grasses, and grain plants; the crop is chopped and packed while at about 70-80% moisture and put into silos or other containers to exclude air.

Entamoeba [INV ZOO] A genus of parasite amebas in the family Endamoebidae, including some species of the genus *Endamoeba* which are parasites of humans and other vertebrates.

Entamoeba histolytica [INV ZOO] A pathogenic ameba, causing amebic dysentery.

Enteletacea [PALEON] A group of extinct articulate brachiopods in the order Orthida.

ENTELETACEA

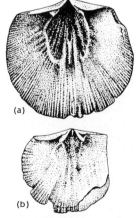

(a)

(b)

Pionodema, (a) pedicle and (b) brachial valve interiors. (From R. C. Moore, ed., Treatise on Invertebrate Paleontology, pt. H, Geological Society of America and University of Kansas Press, 1965)

ENTELODONTIDAE

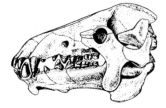

Archaeothesium, Oligocene
entelodont, skull length about
24 inches (61 centimeters).
(After Marsh, 1893)

ENTEROPNEUSTA

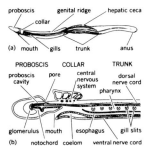

Saccoglossus, the tongue worm
or acorn worm. *(a)* Dorsal view.
(b) Median section of anterior
portion. *(From T. I. Storer and
R. L. Usinger, General Zoology,
4th ed., McGraw-Hill, 1965)*

Entelodontidae [PALEON] A family of extinct palaeodont artiodactyls in the superfamily Entelodontoidea.

Entelodontoidea [PALEON] A superfamily of extinct piglike mammals in the suborder Palaeodonta having huge skulls and enlarged incisors.

entepicondylar [ANAT] Pertaining to the distal or condylar end of the humerus.

enteralgia [MED] Pain in the intestine.

enterectomy [MED] Excision of a part of the intestine.

enteric bacilli [MICROBIO] Microorganisms, especially the gram-negative rods, found in the intestinal tract of man and animals.

enteric cytopathogenic human orphan virus *See* echovirus.

enteritis [MED] Inflammation of the intestinal tract.

Enterobacteriaceae [MICROBIO] A family of bacteria in the order Eubacteriales consisting of straight, rod-shaped, gram-negative, nonsporeforming cells, nonmotile or motile with peritrichous flagella; includes important human and plant pathogens.

enterobiasis [MED] Infestation of the intestinal tract of man with the nematode *Enterobius vermacularis* (pinworm); characterized by mild enteritis.

enterocoel [ZOO] A coelom that arises by mesodermal outpocketing of the archenteron.

Enterocoela [SYST] A section of the animal kingdom that includes the Echinodermata, Chaetognatha, Hemichordata, and Chordata.

enterocolitis [MED] Inflammation of the small intestine and colon.

enterokinase [BIOCHEM] An enzyme which catalyzes the conversion of trypsinogen to trypsin.

enterolith [PATH] A concretion formed in the intestine.

enteron [ANAT] The alimentary canal.

Enteropneusta [INV ZOO] The acorn worms or tongue worms, a class of the Hemichordata; free-living solitary animals with no exoskeleton and with numerous gill slits and a straight gut.

enteroptosis *See* visceroptosis.

enterorrhagia [MED] Intestinal hemorrhage.

enterostome [INV ZOO] In coelenterates, the aboral opening of the actinopharynx which leads to the coelenteron.

enterotoxin [MICROBIO] A toxin produced by *Micrococcus pyogenes* var. *aureus* (*Staphylococcus aureus*) which gives rise to symptoms of food poisoning in man and monkeys.

enterovirus [VIROL] One of the two subgroups of human picornaviruses; includes the polioviruses, the coxsackieviruses, and the echoviruses.

Enterozoa [ZOO] Animals with a digestive tract or cavity; includes all animals except Protozoa, Mesozoa, and Parazoa.

enthalpy [THERMO] The sum of the internal energy of a system plus the product of the system's volume multiplied by the pressure exerted on the system by its surroundings. Also known as heat content; sensible heat; total heat.

entire [BIOL] Having a continuous, unimpaired margin.

Entner-Doudoroff pathway [BIOCHEM] A sequence of reactions for glucose degradation, with the liberation of energy; the distinguishing feature is the formation of 2-keto-3-deoxy-6-phosphogluconate from 6-phosphogluconate and the cleaving of this compound to yield pyruvate and glyceraldehyde-3-phosphate.

entoblast [EMBRYO] A blastomere that differentiates into endoderm.

entobronchus [VERT ZOO] In birds, a dorsal branch of the bronchus.

entoderm *See* endoderm.

Entodiniomorphida [INV ZOO] An order of highly evolved ciliated protozoans in the subclass Spirotrichia, characterized by a smooth, firm pellicle and the lack of external ciliature.

entolecithal *See* endolecithal.

entomochory [ECOL] Dispersal of plant and animal disseminules by insects.

Entomoconchacea [PALEON] A superfamily of extinct marine ostracods in the suborder Myodocopa that are without a rostrum above the permanent aperture.

entomogenous [BIOL] Growing on or in an insect body, as certain fungi.

entomology [INV ZOO] A branch of the biological sciences that deals with the study of insects.

entomophagous [ZOO] Feeding on insects.

entomophilic fungi [MYCOL] Species of fungi that are insect pathogens.

entomophilous [ECOL] Pollinated by insects.

Entomophthoraceae [MYCOL] The single family of the order Entomophthorales.

Entomophthorales [MYCOL] An order of mainly terrestrial fungi in the class Phycomycetes having a hyphal thallus and nonmotile sporangiospores, or conidia.

entomophyte [MYCOL] Any fungus that is parasitic on insects.

Entomostraca [INV ZOO] A group of Crustacea comprising the orders Cephalocarida, Branchiopoda, Ostracoda, Copepoda, Branchiura, and Cirripedia.

Entoniscidae [INV ZOO] A family of isopod crustaceans in the tribe Bopyrina that are parasitic in the visceral cavity of crabs and porcellanids.

entoplastron [VERT ZOO] The anterior median bony plate of the plastron of chelonians.

Entoprocta [INV ZOO] A group of bryozoans, sometimes considered to be a subphylum, having a pseudocoelomate visceral cavity and the anus opening inside the circlet of tentacles.

entrochite [INV ZOO] Stem joint of a crinoid.

entropy [THERMO] Function of the state of a thermodynamic system whose change in any differential reversible process is equal to the heat absorbed by the system from its surroundings divided by the absolute temperature of the system. Also known as thermal charge.

entropy of activation [PHYS CHEM] The difference in entropy between the activated complex in a chemical reaction and the reactants.

entropy of mixing [PHYS CHEM] After mixing substances, the difference between the entropy of the mixture and the sum of the entropies of the components of the mixture.

entropy of transition [PHYS CHEM] The heat absorbed or liberated in a phase change divided by the absolute temperature at which the change occurs.

entypy [EMBRYO] The formation of the amnion in certain mammals by the invagination of the embryonic knob into the yolk sac, without the formation of any amniotic folds.

enucleate [MED] To remove an organ or a tumor in its entirety, as an eye from its socket.

enuresis [MED] Urinary incontinence, especially in the absence of organic cause.

envelope [BIOL] Any enclosing structure, as a membrane, shell, or surrounding leaves. [VERT ZOO] Outer covering of an egg.

environment [ECOL] The sum of all external conditions and influences affecting the development and life of organisms. [PHYS] The aggregate of all the conditions and the influences that determine the behavior of a physical system.

ENTODINIOMORPHIDA

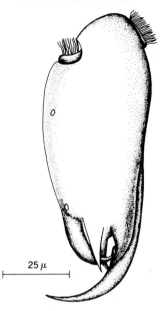

25 μ

Ophryoscolex, an entodiniomorphid.

ENTONISCIDAE

1 mm

Female *Bopyrus squillarum* Latreille.

environmental biology   *See* ecology.

enzootic   [VET MED] **1.** A disease affecting animals in a limited geographic region. **2.** Pertaining to such a disease.

enzyme   [BIOCHEM] Any of a group of catalytic proteins that are produced by living cells and that mediate and promote the chemical processes of life without themselves being altered or destroyed.

enzyme induction   [MICROBIO] The process by which a microbial cell synthesizes an enzyme in response to the presence of a substrate or of a substance closely related to a substrate in the medium.

enzyme inhibition   [BIOCHEM] Prevention of an enzymic process as a result of the interaction of some substance with the enzyme so as to decrease the rate of reaction.

enzyme repression   [BIOCHEM] The process by which the rate of synthesis of an enzyme is reduced in the presence of a metabolite, often the end product of a chain of reactions in which the enzyme in question operates near the beginning.

enzyme unit   [BIOCHEM] The amount of an enzyme that will catalyze the transformation of 10^{-6} mole of substrate per minute or, when more than one bond of each substrate is attacked, 10^{-6} of 1 gram equivalent of the group concerned, under specified conditions of temperature, substrate concentration, and pH number.

enzymology   [BIOCHEM] A branch of science dealing with the chemical nature, biological activity, and biological significance of enzymes.

Eocambrian   [GEOL] Pertaining to the thick sequences of strata conformably underlying Lower Cambrian fossils. Also known as Infracambrian.

Eocanthocephala   [INV ZOO] An order of the Acanthocephala characterized by the presence of a small number of giant subcuticular nuclei.

Eocene   [GEOL] The next to the oldest of the five major epochs of the Tertiary period (in the Cenozoic era).

Eocrinoidea   [PALEON] A class of extinct echinoderms in the subphylum Crinozoa that had biserial brachioles like those of cystoids combined with a theca like that of crinoids.

Eogene   *See* Paleogene.

Eohippus   [PALEON] The earliest, primitive horse, included in the genus *Hyracotherium;* described as a small, four-toed species.

eolotropy   *See* anisotropy.

Eomoropidae   [PALEON] A family of extinct perissodactyl mammals in the superfamily Chalicotherioidea.

Eosentomidae   [INV ZOO] A family of primitive wingless insects in the order Protura that possess spiracles and tracheae.

eosin   [ORG CHEM] $C_{20}H_8O_5Br_4$ **1.** A red fluorescent dye in the form of triclinic crystals that are insoluble in water; used chiefly in cosmetics and as a toner. Also known as bromeosin; bromo acid; eosine; tetrabromofluorescein. **2.** The red to brown crystalline sodium or potassium salt of this dye; used in organic pigments, as a biological stain, and in pharmaceuticals. Also known as eosine; eosine G; eosine Y; eosine yellowish.

eosine G   *See* eosin.

eosine Y   *See* eosin.

eosine yellowish   *See* eosin.

eosinophil   [HISTOL] A granular leukocyte having cytoplasmic granules that stain with acid dyes and a nucleus with two lobes connected by a thin thread of chromatin.

eosinophilia   [MED] A greater than average number of circulating eosinophils. Also known as acidophilia; oxyphilia.

eosinophilic erythroblast   *See* normoblast.

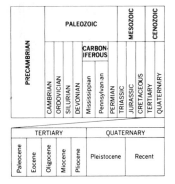

EOCENE

		PALEOZOIC								MESOZOIC				CENOZOIC

A chart showing the position of the Eocene epoch in geologic time.

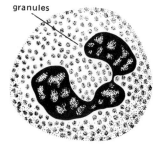

EOSINOPHIL

granules

Diagram of an eosinophil showing the typical two-lobed nucleus.

eosinophilic granuloma [MED] A disease, principally of childhood, characterized by foci of bone inflammation and granulation containing lipids, mononuclear cells, and eosinophils.

eosinophilic pneumonitis *See* Loeffler's syndrome.

Eosuchia [PALEON] The oldest, most primitive, and only extinct order of lepidosaurian reptiles.

Epacridaceae [BOT] A family of dicotyledonous plants in the order Ericales, distinguished by palmately veined leaves, and stamens equal in number with the corolla lobes.

epanthous [MYCOL] Referring to certain fungi which live on flowers.

epaulette [INV ZOO] **1.** Any of the branched or knobbed processes on the oral arms of many Scyphozoa. **2.** The first haired scale at the base of the costal vein in Diptera.

epaxial [BIOL] Above or dorsal to an axis.

epaxial muscle [ANAT] Any of the dorsal trunk muscles of vertebrates.

ependyma [HISTOL] The layer of epithelial cells lining the cavities of the brain and spinal cord. Also known as ependymal layer.

ependymal layer *See* ependyma.

ependymoma [MED] A tumor of the central nervous system whose essential portion consists of cells derived from and resembling ependymal cells. Also known as medulloepithelioma.

ephaptic transmission [PHYSIO] Electrical transfer of activity to a postephaptic unit by the action current of a preephaptic cell.

Ephedra [BOT] A genus of low, leafless, green-stemmed shrubs belonging to the order Ephedrales; source of the drug ephedrine.

Ephedrales [BOT] A monogeneric order of gymnosperms in the subdivision Gneticae.

ephedrine [ORG CHEM] $C_{10}H_{15}NO$ A white, crystalline, water-soluble alkaloid present in several *Ephedra* species and also produced synthetically; a sympathomimetic amine, it is used for its action on the bronchi, blood pressure, blood vessels, and central nervous system.

ephemeral plant [BOT] An annual plant that completes its life cycle in one short moist season; desert plants are examples.

Ephemerida [INV ZOO] An equivalent name for Ephemeroptera.

Ephemeroptera [INV ZOO] The mayflies, an order of exopterygote insects in the subclass Pterygota.

ephidrosis *See* hyperhidrosis.

Ephydridae [INV ZOO] The shore flies, a family of myodarian cyclorrhaphous dipteran insects in the subsection Acalypteratae.

ephyra [INV ZOO] A larval, free-swimming medusoid stage of scyphozoans; arises from the scyphistoma by transverse fission. Also known as ephyrula.

ephyrula *See* ephrya.

epiandrum [INV ZOO] The genital orifice of a male arachnid.

epibasidium [MYCOL] A lengthening of the upper part of each cell of the basidium of various heterobasidiomycetes.

epibenthos [ECOL] Marine life found between the low-water mark and the hundred-fathom line.

epibiotic [ECOL] Living, usually parasitically, on the surface of plants or animals; used especially of fungi.

epiblast *See* ectoderm.

epiblem [BOT] A tissue that replaces the epidermis in most roots and in stems of submerged aquatic plants.

epiboly [EMBRYO] The growing or extending of one part,

EPHEDRA

A species of *Ephedra* growing in Utah. *(Courtesy of Tony Gauba, from National Audubon Society)*

such as the upper hemisphere of a blastula, over and around another part, such as the lower hemisphere, in embryogenesis.

epibranchial [ANAT] Of or pertaining to the segment below the pharyngobranchial region in a branchial arch.

epicalyx [BOT] A ring of fused bracts below the calyx forming a structure that resembles the calyx.

epicanthus [ANAT] A medial fold of skin extending downward from the upper eyelid that hides the inner canthus and carucle; a normal feature in some Asiatic races. [MED] A similar feature occurring as a congenital anomaly, as in Down's syndrome.

epicardium [ANAT] The inner, serous portion of the pericardium that is in contact with the heart. [INV ZOO] A tubular prolongation of the branchial sac in certain ascidians which takes part in the process of budding.

Epicaridea [INV ZOO] A suborder of the Isopoda whose members are parasitic on various marine crustaceans.

epicarp [BOT] The outer layer of the pericarp. Also known as exocarp.

epichilium [BOT] In orchids, the terminal lobe of the lower petal.

epichondrosis [PHYSIO] Development of cartilage on the periosteum, as in the growth of antlers.

epichordal [VERT ZOO] Located upon or above the notochord.

epicnemial [ANAT] Of or pertaining to the anterior portion of the tibia.

epicondyle [ANAT] An eminence on the condyle of a bone.

epicone [INV ZOO] The part anterior to the equatorial groove in a dinoflagellate.

epicotyl [BOT] The embryonic plant stem above the cotyledons.

epicoxite [INV ZOO] In Eurypterida, a small process at the proximal end of the coxa of the second to fifth pairs of appendages.

epicranium [INV ZOO] The dorsal wall of an insect head. [VERT ZOO] The structures covering the cranium in vertebrates.

epicranius [ANAT] The scalp muscle, consisting of a frontal and an occipital portion separated by an aponeurosis.

epictesis [CYTOL] Ability of living cells to concentrate salt solutions.

epicyst [INV ZOO] The outer layer of a cyst wall in encysted protozoans. Also known as ectocyst.

epidemic [MED] A sudden increase in the incidence rate of a disease to a value above normal, affecting large numbers of people and spread over a wide area.

epidemic diarrhea of the newborn [MED] Contagious, fulminating diarrhea with high mortality, seen in newborns; caused by enteropathogenic strains of *Escherichia coli,* strains of *Staphylococcus,* other bacteria, and possibly viruses.

epidemic gastroenteritis [MED] Inflammation of the stomach and intestine, of viral origin; considered to be epidemic when symptoms are manifested by a member of the patient's family within 10 days of the patient's recovery.

epidemic hepatitis *See* infectious hepatitis.

epidemic jaundice *See* infectious hepatitis.

epidemic keratoconjunctivitis [MED] Inflammation of the cornea and conjunctiva, caused by a virus; epidemic by nature.

epidemic neuromyasthenia [MED] A prolonged, debilitating disease of the nervous system of adults; characterized by fatigue, headache, muscle pain, paresis, and emotional and mental disturbances; no etiologic agent has been isolated.

Also known as acute infective encephalomyelitis; Akureyri disease; benign myalgic encephalomyelitis; epidemic vegetative neuritis; Iceland disease.

epidemic pleurodynia [MED] An acute epidemic disease of man, caused by coxsackie B virus; characterized by severe pain in the lower thorax and upper abdomen, and associated with fever and malaise.

epidemic roseola *See* rubella.

epidemic typhus *See* classic epidemic typhus.

epidemic vegetative neuritis *See* epidemic neuromyasthenia.

epidemiology [MED] The study of the mass aspects of disease.

epidermal ridge [ANAT] Any of the minute corrugations of the skin on the palmar and plantar surfaces of man and other primates.

epidermis [BOT] The outermost layer (sometimes several layers) of cells on the primary plant body. [HISTOL] The outer nonsensitive, nonvascular portion of the skin comprising two strata of cells, the stratum corneum and the stratum germinativum.

epidermoid carcinoma *See* squamous-cell carcinoma.

epidermoid cyst [MED] A cyst lined by stratified squamous epithelium without associated cutaneous glands.

epidermolysis [MED] The easy separation of various layers of skin, primarily of the epidermis from the corium, observed in certain pathological conditions.

epidermolysis bullosa [MED] A congenital skin disease characterized by the development of vesicles and bullae upon slight, or even without, trauma.

epididymis [ANAT] The convoluted efferent duct lying posterior to the testis and connected to it by the efferent ductules of the testis.

epididymitis [MED] Inflammation of the epididymis.

epidural [ANAT] Located on or over the dura mater.

epigaster [EMBRYO] The portion of the intestine in vertebrate embryos which gives rise to the colon.

epigastric region [ANAT] The upper and middle part of the abdominal surface between the two hypochondriac regions. Also known as epigastrium.

epigastrium [ANAT] *See* epigastric region. [INV ZOO] The ventral side of mesothorax and metathorax in insects.

epigean [BOT] Pertaining to a plant or plant part that grows above the ground surface. [ZOO] Living near or on the ground surface, applied especially to insects.

epigenesis [EMBRYO] Development in gradual stages of differentiation.

epigenous [BOT] Developing or growing on a surface, especially of a plant or plant part.

epiglottis [ANAT] A flap of elastic cartilage covered by mucous membrane that protects the glottis during swallowing.

epignathous [ZOO] Of the jaw, having the upper component longer than the lower.

epigonium [BOT] An immature sporangial sac in liverworts.

epigynous [BOT] Having the perianth and stamens attached near the top of the ovary; that is, the ovary is inferior.

epigynum [INV ZOO] **1.** The genital pore of female arachnids. **2.** The plate covering this opening.

epihymenium [MYCOL] A thin covering of interwoven hyphae on the hymenium of Basidiomycetes and certain other fungi.

epilabrum [INV ZOO] In Myriapoda, a lateral process of the labrum.

epilemma [HISTOL] The perineurium of very small nerves.

epilepsy [MED] A condition characterized by the paroxysmal

EPIGYNOUS

The arrangement of the sepals, petals, and stamens in an epigynous flower.

recurrence of transient, uncontrollable episodes of abnormal neurological or mental function, or both.

epilimnion [HYD] A fresh-water zone of relatively warm water in which mixing occurs as a result of wind action and convection currents.

epilithic [ECOL] Attached on rocks, as algae or lichens.

epimerase [BIOCHEM] A type of enzyme that catalyzes the rearrangement of hydroxyl groups on a substrate.

epimere [ANAT] The dorsal muscle plate of the lining of a coelomic cavity. [EMBRYO] The dorsal part of a mesodermal segment in the embryo of chordates.

epimeron [INV ZOO] **1.** The posterior plate of the pleuron in insects. **2.** The portion of a somite between the tergum and the insertion of a limb in arthropods.

epimorpha [INV ZOO] Larvae hatched with the adult complement of appendages, all of which are developed.

epimorphosis [PHYSIO] Regeneration in which cell proliferation precedes differentiation.

epimyocardium [EMBRYO] The layer of the embryonic heart from which both the myocardium and epicardium develop.

epimysium [ANAT] The connective-tissue sheath surrounding a skeletal muscle.

epinephrine [BIOCHEM] $C_9H_{13}O_3N$ A hormone secreted by the adrenal medulla that acts to increase blood pressure due to stimulation of heart action and constriction of peripheral blood vessels. Also known as adrenaline.

epineural [ANAT] **1.** Arising from the neural arch. **2.** Any process arising from the neural arch. [INV ZOO] The nervous tissue dorsal to the ventral nerve cord in arthropods.

epineural canal [INV ZOO] A canal that runs between the radial nerve and the epithelium in echinoids and ophiuroids.

epineurium [ANAT] The connective-tissue sheath of a nerve trunk.

epiopticon [INV ZOO] In insects, the middle zone of the optic lobes.

epiostracum [INV ZOO] In Acarina, the thin cuticle covering the exocuticle.

epipelagic [OCEANOGR] Of or pertaining to the portion of oceanic zone into which enough light penetrates to allow photosynthesis.

epipetalous [BOT] Having stamens located on the corolla.

epipetreous [ECOL] Growing on rocks.

epipharynx [INV ZOO] An organ attached beneath the labrium of many insects.

epiphragm [BOT] A membrane covering the aperture of the capsule in certain mosses. [INV ZOO] A membranous or calcareous partition that covers the aperture of certain hibernating land snails.

epiphyll [ECOL] A plant that grows on the surface of leaves.

epiphyseal arch [EMBRYO] The arched structure in the third ventricle of the embryonic brain, which marks the site of development of the pineal body.

epiphyseal plate [ANAT] **1.** The broad, articular surface on each end of a vertebral centrum. **2.** The thin layer of cartilage between the epiphysis and the shaft of a long bone. Also known as metaphysis.

epiphysiolysis [MED] The separation of an epiphysis from the shaft of a bone.

epiphysis [ANAT] **1.** The end portion of a long bone in vertebrates. **2.** *See* pineal body.

epiphyte [ECOL] A plant which grows nonparasitically on another plant or on some nonliving structure, such as a building or telephone pole, deriving moisture and nutrients from the air. Also known as aerophyte.

epiplankton [BIOL] Plankton occurring in the sea from the surface to a depth of about 100 fathoms.

epiplastron [VERT ZOO] In Chelonia, one of the anterior pair of bony plastron plates.

epipleural [ANAT] Arising from a rib. [VERT ZOO] An intramuscular bone arising from and extending between some of the ribs in certain fishes.

epiploic foramen [ANAT] An aperture of the peritoneal cavity, formed by folds of the peritoneum and located between the liver and the stomach. Also known as foramen of Winslow.

epipodite [INV ZOO] A branch of the basal joint of the protopodite of thoracic limbs of many arthropods.

epipodium [BOT] The apical portion of an embryonic phyllopodium. [INV ZOO] **1.** A ridge or fold on the lateral edges of each side of the foot of certain gastropod mollusks. **2.** The elevated ring on an ambulacral plate in Echinoidea.

Epipolasina [INV ZOO] A suborder of sponges in the order Clavaxinellida having radially arranged monactinal or diactinal megascleres.

epiprecoracoid [VERT ZOO] In some Chelonia, a small cartilage at the ventral end of the precoracoid in the pectoral girdle.

epiproct [INV ZOO] A plate above the anus forming the dorsal part of the tenth or eleventh somite of certain insects.

epipterygoid [VERT ZOO] A small bone extending downward from the prootic to the pterygoid bone.

epipubis [VERT ZOO] A single cartilage or bone located in front of the pubis in some vertebrates, particularly in some amphibians.

epirhizous [ECOL] Growing on roots.

episclera [ANAT] The loose connective tissue lying between the conjunctiva and the sclera.

episepalous [BOT] Having stamens growing on or adnate to the sepals.

episiotomy [MED] Medial or lateral incision of the vulva during childbirth, to avoid undue laceration.

episome [GEN] A genetic element in bacteria, presumably a deoxyribonucleic acid fragment, which is not necessary for survival of the organism and which can be integrated in the bacterial chromosome or remain free.

epispadias [MED] A congenital defect of the anterior urethra in which the canal terminates on the dorsum of the penis and posterior to its normal opening.

episperm *See* testa.

epistasis [GEN] The suppression of the effect of one gene by another. [MED] A checking or stoppage of a hemorrhage or other discharge. [PATH] A scum or film of substance floating on the surface of urine.

episternite [INV ZOO] The part of an ovipositor formed from the lateral portion of a somite.

episternum [VERT ZOO] A dermal bone or pair of bones ventral to the sternum of certain fishes and reptiles.

epistome [INV ZOO] **1.** The area between the mouth and the second antennae in crustaceans. **2.** The plate covering this region. **3.** The area between the labrum and the epicranium in many insects. **4.** A flap covering the mouth of certain bryozoans. **5.** The area just above the labrum in certain dipterans.

epithalamus [ANAT] A division of the vertebrate diencephalon including the habenula, the pineal body, and the posterior commissure.

epithallus [BOT] A cortical hyphal layer covering the gonidia in lichens.

epitheca [INV ZOO] **1.** An external, calcareous layer around the basal portion of the theca of many corals. **2.** A protective covering of the epicone. **3.** The outer portion of a diatom frustule.

epitheliochorial placenta [EMBRYO] A type of placenta in which the maternal epithelium and fetal epithelium are in contact. Also known as villous placenta.

epithelioid cell [HISTOL] A macrophage that resembles an epithelial cell. Also known as alveolated cell.

epithelioma [MED] A tumor derived from epithelium; usually a skin cancer, occasionally cancer of a mucous membrane.

epitheliomuscular cell [INV ZOO] An epithelial cell with an elongate base that contains contractile fibrils; common among coelenterates.

epithelium [HISTOL] A primary animal tissue, distinguished by cells being close together with little intercellular substance; covers free surfaces and lines body cavities and ducts.

epithema [VERT ZOO] A horny outgrowth on the beak of certain birds.

epitoke [INV ZOO] The posterior portion of marine polychaetes; contains the gonads.

epitokous [INV ZOO] Pertaining to the heteronereid stage of certain polychaete annelids.

epitope [IMMUNOL] A single determinant of an antigen or immunogen which influences its specificity (immunology).

epitrichium [EMBRYO] The outer layer of the fetal epidermis of many mammals.

epitrochlear [ANAT] Pertaining to a lymph node that lies above the trochlea of the elbow joint.

epitropous [BOT] **1.** Referring to an erect ovule having the raphe in a dorsal position. **2.** Referring to a pendulous ovule having the raphe in a ventral position.

epituberculosis [MED] A massive pulmonary shadow seen in x-ray films in active juvenile tuberculosis, probably caused by bronchial obstruction.

epitympanum [ANAT] The attic of the middle ear, or tympanic cavity.

epivalve [INV ZOO] **1.** The upper or apical shell of certain dinoflagellates. **2.** The upper shell of a diatom.

epixylous [ECOL] Growing on wood; used especially of fungi.

epizoic [BIOL] Living on the body of an animal.

epizoon [ECOL] **1.** An animal that lives on another animal. **2.** An external animal parasite.

epizootic [VET MED] **1.** Affecting many animals of one kind in one region simultaneously; widely diffuse and rapidly spreading. **2.** An extensive outbreak of an epizootic disease.

epizootiology [VET MED] The study of epizootics.

epizygal [INV ZOO] In crinoids, the upper ossicle occurring in a syzygial pair of brachials or columnars.

eponychium [ANAT] The horny layer of the nail fold attached to the nail plate at its margin; represents the remnant of the embryonic condition. [EMBRYO] A horny condition of the epidermis from the second to the eighth month of fetal life, indicating the position of the future nail.

epoophoron [ANAT] A blind longitudinal duct and 10–15 transverse ductules in the mesosalpinx near the ovary which represent remnants of the reproductive part of the mesonephros in the female; homolog of the head of the epididymis in the male. Also known as parovarium; Rosenmueller's organ.

EPR *See* electron paramagnetic resonance.

Epstein-Barr virus [VIROL] Herpeslike virus particles first identified in cultures of cells from Burkett's malignant lymphoma.

epulis [MED] A benign tumorlike lesion of the gingiva.

equal loudness contour [ACOUS] A curve on a graph of

EPITHELIOMUSCULAR CELL

Microscopic view of an epitheliomuscular cell of *Hydra*. *(After L. H. Hyman)*

sound intensity in decibels versus frequency at each point along which sound appears to be equally loud to a listener. Also known as Fletcher-Munson contour.

equally likely cases [STAT] All simple events in a trial have the same probability.

equation [CHEM] A symbolic expression that represents in an abbreviated form the laboratory observations of a chemical change; an equation (such as $2H_2 + O_2 \rightarrow 2H_2O$) indicates what reactants are consumed (H_2 and O_2) and what products are formed (H_2O), the correct formula of each reactant and product, and satisfies the law of conservation of atoms in that the symbols for the number of atoms reacting equals the number of atoms in the products. [MATH] A statement that each of two expressions is equal to the other.

equation of state [PHYS CHEM] A mathematical expression which defines the physical state of a homogeneous substance (gas, liquid, or solid) by relating volume to pressure and absolute temperature for a given mass of the material.

equatorial plate [CYTOL] **1.** The plane across the spindle onto which the centromere of each chromosome comes to lie during the metaphase stage of mitosis. **2.** In the metaphase of meiosis, the plane midway between the two centromeres of each bivalent.

Equidae [VERT ZOO] A family of perissodactyl mammals in the superfamily Equoidea, including the horses, zebras, and donkeys.

equilibrium diagram [PHYS CHEM] A phase diagram of the equilibrium relationship between temperature, pressure, and composition in any system.

equilibrium dialysis [ANALY CHEM] A technique used to determine the degree of ion bonding by protein; the protein solution, placed in a bag impermeable to protein but permeable to small ions, is immersed in a solution containing the diffusible ion whose binding is being studied; after equilibration of the ion across the membrane, the concentration of ion in the protein-free solution is determined; the concentration of ion in the protein solution is determined by subtraction; if binding has occurred, the concentration of ion in the protein solution must be greater.

equine encephalitis [MED] A disease of equines and man caused by one of three viral strains: eastern, western, and Venezuelan equine viruses. Also known as equine encephalomyelitis.

equine encephalomyelitis *See* equine encephalitis.

equipotent [SCI TECH] Equal in capacity or effect.

Equisetales [BOT] The horsetails, a monogeneric order of the class Equisetopsida; the only living genus is *Equisetum*.

Equisetineae [BOT] The equivalent name for Equisetophyta.

Equisetophyta [BOT] A division of the subkingdom Embryobionta represented by a single living genus, *Equisetum.*

Equisetopsida [BOT] A class of the division Equisetophyta whose members made up a major part of the flora, especially in moist or swampy places, during the Carboniferous Period.

equitant [BOT] Of leaves, overlapping transversely at the base.

equivalence law of ordered sampling [STAT] If a random ordered sample of size s is drawn from a population of size N, then on any particular one of the s draws each of the N items has the same probability, $1/N$, of appearing.

equivalent weight [CHEM] The number of parts by weight of an element or compound which will combine with or replace, directly or indirectly, 1.008 parts by weight of hydrogen, 8.00 parts of oxygen, or the equivalent weight of any other element or compound.

EQUISETALES

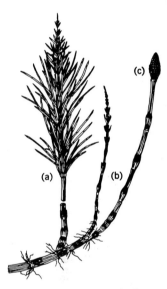

Equisetum arvense, (a) sterile shoot, (b) fertile shoot growing from underground rootstock, (c) cone. (*From E. W. Sinnot and K. S. Wilson, Botany: Principles and Problems, 6th ed., McGraw-Hill, 1963*)

ERBIUM

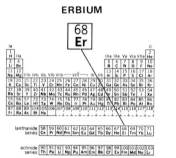

Periodic table of the chemical elements showing the position of erbium.

eqivalve [INV ZOO] In bivalves, having the valves alike in form and size.

Equoidea [VERT ZOO] A superfamily of perissodactyl mammals in the suborder Hippomorpha comprising the living and extinct horses and their relatives.

Er *See* erbium.

ER *See* endoplasmic reticulum.

erbium [CHEM] A trivalent metallic rare-earth element, symbol Er, of the yttrium subgroup, found in euxenite, gadolinite, fergusonite, and xenotine; atomic number 68, atomic weight 167.26, specific gravity 9.051; insoluble in water, soluble in acids; melts at 1400–1500°C.

erectile tissue [ANAT] A spongelike arrangement of vascular spaces; found in the corpus cavernosum of the penis or clitoris.

erection [PHYSIO] The enlarged state of erectile tissue when engorged with blood, as of the penis or clitoris.

erector [PHYSIO] Any muscle that produces erection of a part.

Eremascoideae [BOT] A monogeneric subfamily of ascosporogenous yeasts characterized by mostly septate mycelia, and spherical asci with eight oval to round ascospores.

Erethizontidae [VERT ZOO] The New World porcupines, a family of rodents characterized by sharply pointed, erectile hairs and four functional digits.

Ergasilidae [INV ZOO] A family of copepod crustaceans in the suborder Cyclopoida in which the females are parasitic on aquatic animals, while the males are free-swimming.

ergastic [CYTOL] Pertaining to the nonliving components of protoplasm.

ergastoplasm [CYTOL] A cytoplasm component which shows an affinity for basic dyes; a form of the endoplasmic reticulum.

ergate [INV ZOO] An ant of the worker caste.

ergatogyne [INV ZOO] A female ant that resembles a worker.

ergosterin *See* ergosterol.

ergosterol [BIOCHEM] $C_{28}H_{44}O$ A crystalline, water-insoluble, unsaturated sterol found in ergot, yeast, and other fungi, and which may be converted to vitamin D_2 on irradiation with ultraviolet light or activation with electrons. Also known as ergosterin.

ergot [MYCOL] The dark purple or black sclerotium of the fungus *Claviceps purpurea*. [ORG CHEM] Any of the five optically isomeric pairs of alkaloids obtained from this fungus; only the levorotatory isomers are physiologically active.

ergotism [MED] Acute or chronic intoxication resulting from ingestion of grain infected with ergot fungus, or from chronic use of drugs containing ergot.

Erian [GEOL] Middle Devonian geologic time; a North American provincial series.

Ericaceae [BOT] A large family of dicotyledonous plants in the order Ericales distinguished by having twice as many stamens as corolla lobes.

Ericales [BOT] An order of dicotyledonous plants in the subclass Dilleniidae; plants are generally sympetalous with unitegmic ovules and they have twice as many stamens as petals.

erichthus [INV ZOO] A stomatopod larva.

ericophyte [ECOL] A plant that grows on a heath or moor.

Erinaceidae [VERT ZOO] The hedgehogs, a family of mammals in the order Insectivora characterized by dorsal and lateral body spines.

erineum [PL PATH] An abnormal growth of hairs induced on the epidermis of a leaf by certain mites.

Erinnidae [INV ZOO] A family of orthorrhaphous dipteran insects in the series Brachycera.

Eriocaulaceae [BOT] The single family of the order Eriocaulales.

Eriocaulales [BOT] An order of monocotyledonous plants in the subclass Commelinidae, having a perianth reduced or lacking and having unisexual flowers aggregated on a long peduncle.

Eriococcinae [INV ZOO] A family of homopteran insects in the superfamily Coccoidea; adult females and late instar nymphs have an anal ring.

Eriocraniidae [INV ZOO] A small family of lepidopteran insects in the superfamily Eriocranioidea.

Eriocranioidea [INV ZOO] A superfamily of lepidopteran insects in the suborder Homoneura comprising tiny moths with reduced, untoothed mandibles.

Eriophyidae [INV ZOO] The bud mites or gall mites, a family of economically important plant-feeding mites in the suborder Trombidiformes.

eriophyllous [BOT] Having leaves covered by a cottony pubescence.

erose [BIOL] Having an irregular margin.

erosion [MED] **1.** Surgical removal of tissues by scraping. **2.** Excision of a joint.

erotic [PSYCH] **1.** Pertaining to the libido or sexual passion. **2.** Moved by or arousing sexual desire.

Erotylidae [INV ZOO] The pleasing fungus beetles, a family of coleopteran insects in the superfamily Cucujoidea.

Errantia [INV ZOO] A group of 34 families of polychaete annelids in which the anterior region is exposed and the linear body is often long and is dorsoventrally flattened.

error range [STAT] The difference between the highest and lowest error values; a measure of the uncertainty associated with a number.

Erwinia [MICROBIO] A genus of motile, rod-shaped bacteria in the tribe Erwinieae; these organisms invade living plant tissues and cause dry necroses, galls, wilts, and soft rots.

Erwinieae [MICROBIO] A tribe of phytopathogenic bacteria in the family Enterobacteriaceae, including the single genus *Erwinia.*

erysipelas [MED] An acute, infectious bacterial disease caused by *Streptococcus pyogenes* and characterized by inflammation of the skin and subcutaneous tissues.

Erysipelothrix [MICROBIO] A genus of gram-positive, rod-shaped bacteria of the family Corynebacteriaceae that have a tendency to form long filaments; parasites of mammals, birds, and fish.

Erysiphaceae [MYCOL] The powdery mildews, a family of ascomycetous fungi in the order Erysiphales having light-colored mycelia and conidia.

Erysiphales [MYCOL] An order of ascomycetous fungi which are obligate parasites of seed plants, causing powdery mildew and sooty mold.

erythema [MED] Localized redness of the skin in areas of variable size.

erythema multiforme [MED] An acute inflammatory skin disease characterized by red macules, papules, or tubercles on the extremities, neck, and face.

erythema nodosum [MED] The occurrence of pink to blue, tender nodules on the anterior surfaces of the lower legs; more frequent in women than men.

erythremia *See* erythrocytosis; polycythemia vera.

erythritol [ORG CHEM] $H(CHOH)_4H$ A tetrahydric alcohol; occurs as tetragonal prisms, melting at 121°C, soluble in water; used in medicine as a vasodilator. Also known as antierythrite; erythrite; ethroglucin; erythrol; phycite; tetra-hydroxy butane.

erythroblast [HISTOL] A nucleated cell occurring in bone marrow as the earliest recognizable cell of the erythrocytic series.

erythroblastosis [MED] The abnormal presence of erythroblasts in the blood. [VET MED] A virus disease of birds; considered to be part of the avian leukosis complex in which there is an abnormal number of erythroblasts in the blood.

erythroblastosis fetalis [MED] A form of hemolytic anemia affecting the fetus and newborn infant when a mother is Rh-negative and has developed antibodies against an Rh-positive fetus. Also known as hemolytic disease of newborn.

erythrocruorin [BIOCHEM] Any of the iron-porphyrin protein respiratory pigments found in the blood and tissue fluids of certain invertebrates; corresponds to hemoglobin in vertebrates.

erythrocyte [HISTOL] A type of blood cell that contains a nucleus in all vertebrates but man and that has hemoglobin in the cytoplasm. Also known as red blood cell.

erythrocytopoiesis *See* erythropoiesis.

erythrocytosis [MED] An increase in the number of circulating erythrocytes of more than two standard deviations above the mean normal, usually occurring secondary to hypoxia. Also known as erythremia; polycythemia.

erythromelalgia [MED] A cutaneous vasodilation of the feet or, more rarely, of the hands; characterized by redness, mottling, changes in skin temperature, and neuralgic pains. Also known as acromelalgia; Mitchell's disease.

erythromycin [MICROBIO] A crystalline antibiotic produced by *Streptomyces erythreus* and used in the treatment of gram-positive bacterial infections.

erythrophore [ZOO] A chromatophore containing a red pigment, especially a carotenoid.

erythropia *See* erythropsia.

erythropoiesis [PHYSIO] The process by which erythrocytes are formed. Also known as erythrocytopoiesis.

erythropoietin [BIOCHEM] A hormone, thought to be produced by the kidneys, that regulates erythropoiesis, at least in higher vertebrates.

erythropsia [MED] An abnormality of vision in which all objects appear red; red vision. Also known as erythropia.

erythrosis [MED] **1.** Overproliferation of erythropoietic tissue, as found in polycythemia. **2.** The unusual red skin color of individuals with polycythemia.

Erythroxylaceae [BOT] A homogeneous family of dicotyledonous woody plants in the order Linales characterized by petals that are internally appendiculate, three carpels, and flowers without a disk.

erythroxylon *See* cocaine.

Es *See* einsteinium.

eschar [MED] A dry crust or slough, especially one formed after a thermal or chemical burn.

escharotic [MED] **1.** Caustic. **2.** Producing an eschar.

Escherichia [MICROBIO] A genus of gram-negative, rod-shaped bacteria in the family Enterobacteriaceae that are usually motile by peritrichous flagella.

Escherichia coli [MICRO] The type species of the genus, occurring as part of the normal intestinal flora in vertebrates. Also known as colon bacillus.

Escherichieae [MICROBIO] A tribe of bacteria in the family Enterobacteriaceae defined by the ability to ferment lactose, with the rapid production of acid and visible gas.

Esocidae [VERT ZOO] The pikes, a family of fishes in the order Clupeiformes characterized by an elongated beaklike snout and sharp teeth.

Esocoidei [VERT ZOO] A small suborder of fresh-water fishes in the order Salmoniformes; includes the pikes, mudminnows, and pickerels.

esophageal gland [ANAT] Any of the digestive glands within the submucosa of the esophagus; secretions are chiefly mucus and serve to lubricate the esophagus.

esophageal hiatus [ANAT] The opening in the diaphragm for passage of the esophagus.

esophageal teeth [VERT ZOO] Enamel-tipped hypapophyses of the posterior cervical vertebrae of certain snakes, which penetrate the esophagus and function to break eggshells.

esophagitis [MED] Inflammation of the esophagus.

esophagogastrostomy [MED] Establishment, by surgery, of an anastomosis between the esophagus and the stomach; may be performed by the abdominal route or by transpleural operation.

esophagus [ANAT] The tubular portion of the alimentary canal interposed between the pharynx and the stomach. Also known as gullet.

esotropia [MED] Convergent strabismus, occurring when one eye fixes upon an object and the other deviates inward. Also known as cross-eye.

ESP *See* extrasensory perception.

essential amino acid [BIOCHEM] Any of eight of the 20 naturally occurring amino acids that are indispensable for optimum animal growth but cannot be formed in the body and must be supplied in the diet.

essential hypertension [MED] Elevation of the systemic blood pressure, of unknown origin. Also known as primary hypertension.

ester [ORG CHEM] The compound formed by the elimination of water and the bonding of an alcohol and an organic acid.

esterase [BIOCHEM] Any of a group of enzymes that catalyze the synthesis and hydrolysis of esters.

esthacyte [INV ZOO] A simple sensory cell occurring in certain lower animals, such as sponges. Also spelled aesthacyte.

esthesia [PHYSIO] The capacity for sensation, perception, or feeling. Also spelled aesthesia.

esthesiometer [ENG] An instrument used to measure tactile sensibility by determining the distance by which two points pressed against the skin must be separated in order that they be felt as separate. Also spelled aesthesiometer.

esthesioneuroblastoma *See* neuroepithelioma.

esthesioneuroepithelioma *See* neuroepithelioma.

esthiomene [MED] The chronic ulcerative lesion of the vulva in lymphogranuloma venereum.

estivation [PHYSIO] **1.** The adaptation of certain animals to the conditions of summer, or the taking on of certain modifications, which enables them to survive a hot, dry summer. **2.** The dormant condition of an organism during the summer.

estivoautumnal malaria *See* falciparum malaria.

estradiol [BIOCHEM] $C_{18}H_{24}O_2$ An estrogenic hormone produced by follicle cells of the vertebrate ovary; provokes estrus and proliferation of the human endometrium, and stimulates ICSH (interstitial-cell-stimulating hormone) secretion.

estriol [BIOCHEM] $C_{18}H_{24}O_3$ A crystalline estrogenic hormone obtained from human pregnancy urine.

estrogen [BIOCHEM] Any of various natural or synthetic substances possessing the biologic activity of estrus-producing hormones.

estrogenic hormone [BIOCHEM] A hormone, found principally in ovaries and also in the placenta, which stimulates the accessory sex structures and the secondary sex characteristics in the female.

estrone [BIOCHEM] $C_{18}H_{22}O_2$ An estrogenic hormone produced by follicle cells of the vertebrate ovary; functions the same as estradiol.

estrous cycle [PHYSIO] The physiological changes that take place between periods of estrus in the female mammal.

estrus [PHYSIO] The period in female mammals during which ovulation occurs and the animal is receptive to mating.

ethanal *See* acetaldehyde.

ethanamide *See* acetamide.

ethanoic acid *See* acetic acid.

ethanol [ORG CHEM] C_2H_5OH A colorless liquid, miscible with water, boiling point 78.32°C; used as a reagent and solvent. Also known as alcohol; ethyl alcohol; grain alcohol.

ethanol acid *See* glyoxalic acid.

ether [ELECTROMAG] The medium postulated to carry electromagnetic waves, similar to the way a gas carries sound waves. [ORG CHEM] **1.** One of a class of organic compounds characterized by the structural feature of an oxygen linking two hydrocarbon groups (such as $R-O-R$). **2.** $(C_2H_5)_2O$ A colorless liquid, slightly soluble in water; used as a reagent, intermediate, anesthetic, and solvent. Also known as ethyl ether.

etherize [MED] To produce anesthesia by administration of ether.

ethionine [BIOCHEM] $C_5H_{13}O_2N$ An amino acid that is the ethyl analog of and the biological antagonist of methionine.

Ethiopian zoogeographic region [ECOL] A geographic unit of faunal homogeneity including all of Africa south of the Sahara.

ethmoid bone [ANAT] An irregularly shaped cartilage bone of the skull, forming the medial wall of each orbit and part of the roof and lateral walls of the nasal cavities.

ethmoid notch [ANAT] A space between the two orbital plates of the frontal bone.

ethmoturbinate [ANAT] Of or pertaining to the masses of ethmoid bone which form the lateral and superior portions of the turbinate bones in mammals.

ethnogeography [ANTHRO] The scientific study of the geographic distribution of races, peoples, or cultural groups and their adaptation and relation to the environments in which they live.

ethnology [ANTHRO] The science that deals with the study of the origin, distribution, and relations of races or ethnic groups of mankind.

ethnopsychology [PSYCH] The study of the psychology of races or ethnic groups.

ethology [VERT ZOO] The study of animal behavior in a natural context.

ethyl alcohol *See* ethanol.

etioblast [BOT] An immature chloroplast, containing prolamellar bodies.

etiolation [BOT] The yellowing or whitening of green plant parts grown in darkness.

etiology [MED] Any factors which cause disease. [SCI TECH] A branch of science dealing with the causes of phenomena.

E trisomy *See* trisomy 18 syndrome.

Eu *See* europium.

Eubacteriales [MICROBIO] The true bacteria, an order of the class Schizomycetes; characterized by simple, undifferentiated, rigid cells which are either spherical forms or straight rods.

Eubasidiomycetes [MYCOL] An equivalent name for Homobasidiomycetidae.

Eubrya [BOT] A subclass of the mosses (Bryopsida); the leafy gametophytes arise from buds on the protonemata, which are nearly always filamentous or branched green threads attached to the substratum by rhizoids.

Eubryales [BOT] An order of mosses (Bryatae); plants have the sporophyte at the end of a stem, vary in size from small to robust, and generally grow in tufts.

Eucalyptus [BOT] A large genus of evergreen trees belonging to the myrtle family (Myrtaceae) and occurring in Australia and New Guinea.

eucalyptus gum [PHARM] The dried gummy exudate from *Eucalyptus longirostris* of Australia, composed of kinotannic acid, kino red, glucoside, catechol, and pyrocatechol; used in medicine as an astringent and antidiarrheal agent. Also known as eucalyptus kino; red gum.

eucalyptus kino *See* eucalyptus gum.

Eucarida [INV ZOO] A large superorder of the decapod crustaceans, subclass Malacostraca, including shrimps, lobsters, hermit crabs, and crabs; characterized by having the shell and thoracic segments fused dorsally and the eyes on movable stalks.

Eucaryota [BIOL] Primitive, unicellular organisms having a well-defined nuclear membrane, chromosomes, and mitotic cell division.

eucaryote *See* eukaryote.

Eucestoda [INV ZOO] The true tapeworms, a subclass of the class Cestoda.

Eucharitidae [INV ZOO] A family of hymenopteran insects in the superfamily Chalcidoidea.

euchromatin [CYTOL] The portion of the chromosomes that stains with low intensity, uncoils during interphase, and condenses during cell division.

Eucinetidae [INV ZOO] The plate thigh beetles, a family of coleopteran insects in the superfamily Dascilloidea.

Euclasterida [INV ZOO] An order of asteroid echinoderms in which the arms are sharply distinguished from a small, central disk-shaped body.

Eucleidae [INV ZOO] The slug moths, a family of lepidopteran insects in the suborder Heteroneura.

Euclymeninae [INV ZOO] A subfamily of annelids in the family Maldonidae of the Sedentaria, having well-developed plaques and an anal pore within the plaque.

Eucnemidae [INV ZOO] The false click beetles, a family of coleopteran insects in the superfamily Elateroidea.

Eucoccida [INV ZOO] An order of parasitic protozoans in the subclass Coccidia characterized by alternating sexual and asexual phases; stages of the life cycle occur intracellularly in vertebrates and invertebrates.

Eucoelomata [ZOO] A large sector of the animal kingdom including all forms in which there is a true coelom or body cavity; includes all phyla above Aschelminthes.

Eucommiales [BOT] A monotypic order of dicotyledonous plants in the subclass Hamamelidae; plants have two, unitegmic ovules and lack stipules.

Eudactylinidae [INV ZOO] A family of parasitic copepod crustaceans in the suborder Caligoida; found as ectoparasites on the gills of sharks.

eudoxid *See* eudoxome.

eudoxome [INV ZOO] Cormidium of most calycophoran siphonophores which lead a free existence. Also known as eudoxid.

Euechinoidea [INV ZOO] A subclass of echinoderms in the

EUCALYPTUS

Eucalyptus globulus **showing spray of mature foliage.**

class Echinoidea; distinguished by the relative stability of ambulacra and interambulacra.

eugenics [GEN] The use of practices that influence the hereditary qualities of future generations, with the aim of improving man's genetic future.

Euglenida [INV ZOO] An order of protozoans in the class Phytamastigophorea, including the largest green, noncolonial flagellates.

Euglenidae [INV ZOO] The antlike leaf beetles, a family of coleopteran insects in the superfamily Tenebrionoidea.

Euglenoidina [INV ZOO] The equivalent name for Euglenida.

Euglenophyceae [BOT] The single class of the plant division Euglenophyta.

Euglenophyta [BOT] A division of the plant kingdom including one-celled, chiefly aquatic flagellate organisms having a spindle-shaped or flattened body, naked or with a pellicle.

euglobulin [BIOCHEM] True globulin; a simple protein that is soluble in distilled water and dilute salt solutions.

Eugregarinida [INV ZOO] An order of protozoans in the subclass Gregarinia; parasites of certain invertebrates.

eukaryote [BIOL] A cell with a definitive nucleus. Also spelled eucaryote.

Eulamellibranchia [INV ZOO] The largest subclass of the molluscan class Bivalvia, having a heterodont shell hinge, leaflike gills, and well-developed siphons.

Eulenburg's disease *See* paramyotonia congenita.

eulittoral [OCEANOGR] A subdivision of the benthic division of the littoral zone of the marine environment, extending from high-tide level to about 60 meters, the lower limit for abundant growth of attached plants.

Eulophidae [INV ZOO] A family of hymenopteran insects in the superfamily Chalcidoidea including species that are parasitic on the larvae of other insects.

Eumalacostraca [INV ZOO] A series of the class Crustacea comprising shrimplike crustaceans having eight thoracic segments, six abdominal segments, and a telson.

Eumetazoa [ZOO] A section of the animal kingdom that includes the phyla above the Porifera; contains those animals which have tissues or show some tissue formation and organ systems.

eumitosis [CYTOL] Typical mitosis.

Eumycetes [MYCOL] The true fungi, a large group of microorganisms characterized by cell walls, lack of chlorophyll, and mycelia in most species; includes the unicellular yeasts.

Eumycetozoida [INV ZOO] An order of protozoans in the subclass Mycetozoia; includes slime molds which form a plasmodium.

Eumycophyta [MYCOL] An equivalent name for Eumycetes.

Eumycota [MYCOL] An equivalent name for Eumycetes.

Eunicea [INV ZOO] A superfamily of polychaete annelids belonging to the Errantia.

Eunicidae [INV ZOO] A family of polychaete annelids in the superfamily Eunicea having characteristic pharyngeal armature consisting of maxillae and mandibles.

eunuch [MED] An individual who has undergone complete loss of testicular function.

Euomphalacea [PALEON] A superfamily of extinct gastropod mollusks in the order Aspidobranchia characterized by shells with low spires, some approaching bivalve symmetry.

eupelagic *See* pelagic.

Eupelmidae [INV ZOO] A family of hymenopteran insects in the superfamily Chalcidoidea.

Euphausiacea [INV ZOO] An order of planktonic malacostracans in the class Crustacea possessing photophores which emit a brilliant blue-green light.

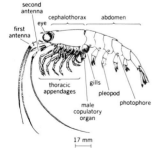

EUPHAUSIACEA

second antenna
cephalothorax   abdomen
first eye
antenna

thoracic appendages   gills
pleopod
male copulatory organ   photophore

17 mm

Diagram of a euphausiid crustacean.

euphenics [GEN] The production of a satisfactory phenotype by means other than eugenics.

Eupheterochlorina [INV ZOO] A suborder of flagellate protozoans in the order Heterochlorida.

Euphorbiaceae [BOT] A family of dicotyledonous plants in the order Euphorbiales characterized by dehiscent fruit having more than one seed and by epitropous ovules.

Euphorbiales [BOT] An order of dicotyledonous plants in the subclass Rosidae having simple leaves and unisexual flowers that are aggregated and reduced.

euphotic [OCEANOGR] Of or constituting the upper levels of the marine environment down to the limits of effective light penetration for photosynthesis.

Euphrosinidae [INV ZOO] A family of amphinomorphan polychaete annelids with short, dorsolaterally flattened bodies.

Euplexoptera [INV ZOO] The equivalent name for Dermaptera.

euploid [GEN] Having a chromosome complement that is an exact multiple of the haploid complement.

eupnea [PHYSIO] Normal or easy respiration rhythm.

Eupodidae [INV ZOO] A family of mites in the suborder Trombidiformes.

Euproopacea [PALEON] A group of Paleozoic horseshoe crabs belonging to the Limulida.

European boreal faunal region [ECOL] A zoogeographic region describing marine littoral faunal regions of the northern Atlantic Ocean between Greenland and the northwestern coast of Europe.

European canker [PL PATH] 1. A fungus disease of apple, pear, and other fruit and shade trees caused by *Nectria galligena* and characterized by cankers with concentric rings of callus on the trunk and branches. 2. A fungus disease of poplars caused by *Dothichiza populea.*

europium [CHEM] A member of the rare-earth elements in the cerium subgroup, symbol Eu, atomic number 63, atomic weight 151.96, steel gray and malleable, melting at 1100–1200°C.

Eurotiaceae [MYCOL] A family of ascomycetous fungi of the order Eurotiales in which the asci are borne in cleistothecia or closed fruiting bodies.

Eurotiales [MYCOL] An order of fungi in the class Ascomycetes bearing ascospores in globose or broadly oval, delicate asci which lack a pore.

Euryalae [INV ZOO] The basket fishes, a family of echinoderms in the subclass Ophiuroidea.

Euryalina [INV ZOO] A suborder of ophiuroid echinoderms in the order Phrynophiurida characterized by a leathery integument.

Euryapsida [PALEON] A subclass of fossil reptiles distinguished by an upper temporal opening on each side of the skull.

eurybathic [ECOL] Living at the bottom of a body of water.

eurycephalic [ANTHRO] Having a cephalic index of 80–84.

Eurychilinidae [PALEON] A family of extinct dimorphic ostracods in the superfamily Hollinacea.

euryene [ANTHRO] Having a short or broad forehead or both with an upper facial index of 45–50.

eurygamous [INV ZOO] Mating in flight, as in many insect species.

Eurylaimi [VERT ZOO] A monofamilial suborder of suboscine birds in the order Passeriformes.

Eurylaimidae [VERT ZOO] The broadbills, the single family of the avian suborder Eurylaimi.

EUPHROSINIDAE

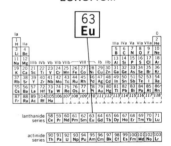

Euphrosine, dorsal view.
(After McIntosh)

EUROPIUM

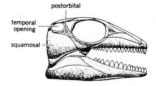

Periodic table of the chemical elements showing the position of europium.

EURYAPSIDA

Lateral view of *Araeoscelis* skull showing temporal opening.
(From A. S. Romer, Vertebrate Paleontology, 3d ed., University of Chicago Press, 1966)

EURYPTERIDA

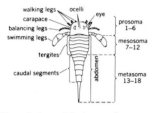

Dorsal view of features of a typical eurypterid. *(Adapted from Clarke and Ruedemann, 1912, from R. R. Shrock and W. H. Twenhofel, Principles of Invertebrate Paleontology, McGraw-Hill, 2d ed., 1953)*

EUSTELE

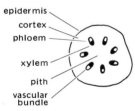

Diagram of a stem in cross section showing tissue arrangements of a eustele.

eurymeric [ANTHRO] Having a broad femur with a platymeric index of 85–100.

Eurymylidae [PALEON] A family of extinct mammals presumed to be the ancestral stock of the order Lagomorpha.

euryon [ANAT] One of the two lateral points functioning as end points to measure the greatest transverse diameter of the skull.

euryphagous [ECOL] Eating a large variety of foods.

Euryphoridae [INV ZOO] A family of copepod crustaceans in the order Caligoida; members are fish ectoparasites.

euryprosopic [ANTHRO] Having a short or broad face or both with a facial index of 80–85.

Eurypterida [PALEON] A group of extinct aquatic arthropods in the subphylum Chelicerata having elongate-lanceolate bodies encased in a chitinous exoskeleton.

Eurypygidae [VERT ZOO] The sun bitterns, a family of tropical and subtropical New World birds belonging to the order Gruiformes.

eurypylous [INV ZOO] Having a wide opening; applied to sponges with wide apopyles opening directly into excurrent canals, and wide prosopyles opening directly from incurrent canals.

eurysome [ANTHRO] Having a broad, thickset body build.

eurytherm [BIOL] An organism that is tolerant of a wide range of temperatures.

Eurytomidae [INV ZOO] The seed and stem chalcids, a family of hymenopteran insects in the superfamily Chalcidoidea.

Eusiridae [INV ZOO] A family of pelagic amphipod crustaceans in the suborder Gammaridea.

eusporangiate [BOT] Having sporogenous tissue derived from a group of epidermal cells.

eustachian tube [ANAT] A tube composed of bone and cartilage that connects the nasopharynx with the middle ear cavity.

eustachian valve *See* canal valve.

eustele [BOT] A modified siphonostele containing collateral or bicollateral vascular bundles; found in most gymnosperm and angiosperm stems.

eusternum [INV ZOO] The anterior sternal plate in insects.

Eusuchia [VERT ZOO] The modern crocodiles, a suborder of the order Crocodilia characterized by a fully developed secondary palate and procoelous vertebrae.

Eusyllinae [INV ZOO] A subfamily of polychaete annelids in the family Syllidae having a thick body and unsegmented cirri.

Eutardigrada [INV ZOO] An order of tardigrades which lack both a sensory cephalic appendage and a club-shaped appendage.

eutely [BIOL] Having the body composed of a constant number of cells, as in certain rotifers.

Euthacanthidae [PALEON] A family of extinct acanthodian fishes in the order Climatiiformes.

euthenics [BIOL] The science that deals with the improvement of the future of man by changing his environment.

Eutheria [VERT ZOO] An infraclass of therian mammals including all living forms except the monotremes and marsupials.

Eutrichosomatidae [INV ZOO] Small family of hymenopteran insects in the superfamily Chalcidoidea.

eutrophication [ECOL] Of bodies of water, the process of becoming better nourished either naturally by processes of maturation or artificially by fertilization.

eutropous [INV ZOO] Referring to certain insects which are adapted to visit only special types of flowers.

eV *See* electron volt.

evagination [BIOL] **1.** The turning inside out of a body part. **2.** The product, or outgrowth, of the turning-out process.

Evaniidae [INV ZOO] The ensign flies, a family of hymenopteran insects in the superfamily Proctotrupoidea.

Eventognathi [VERT ZOO] The equivalent name for Cypriniformes.

evergreen [BOT] Pertaining to a perennially green plant. Also known as aiophyllous.

evolution [BIOL] The processes of biological and organic change in organisms by which descendants come to differ from their ancestors.

ewe [VERT ZOO] A mature female sheep, goat, or related animal, as the smaller antelopes.

Ewing's sarcoma [MED] A primary malignant tumor of bone, usually arising as a central tumor in long bone.

exalate [BOT] Being without winglike appendages.

exanthema [MED] **1.** An eruption on the skin. **2.** Any disease or fever accompanied by a skin eruption.

exanthem subitum [MED] A mild, sometimes epidemic viral disease of young children, with abrupt onset, high fever, and rash. Also known as roseola infantum.

exarch [BOT] A vascular bundle in which the primary wood is centripetal.

exasperate [BIOL] Having a surface roughened by stiff elevations or bristles.

exaspidean [VERT ZOO] Of the tarsal envelope of birds, being continuous around the outer edge of the tarsus.

Excepulaceae [MYCOL] The equivalent name for Discellaceae.

exchange transfusion [MED] The replacement of most or all of the recipient's blood in small amounts at a time by blood from a donor, a technique used particularly in cases of erythroblastosis fetalis, in certain types of poisoning such as salicylism, and occasionally in liver failure. Also known as replacement transfusion.

excipient [PHARM] Any inert substance combined with an active drug for preparing an agreeable or convenient dosage form.

Excipulaceae [MYCOL] The equivalent name for Discellaceae.

excision [MED] The cutting out of a part; removal of a foreign body or growth from a part, organ, or tissue.

excision enzyme [BIOCHEM] A bacterial enzyme that removes damaged dimers from the deoxyribonucleic acid molecule of a bacterial cell following light or ultraviolet radiation or nitrogen mustard damage.

excitation index [SPECT] In emission spectroscopy, the ratio of intensities of a pair of extremely nonhomologous spectra lines; used to provide a sensitive indication of variation in excitation conditions.

excitation purity [ANALY CHEM] The ratio of the departure of the chromaticity of a specified color to that of the reference source, measured on a chromaticity diagram; used as a guide of the wavelength of spectrum color needed to be mixed with a reference color to give the specified color. Also known as purity.

excitation spectrum [SPECT] The graph of luminous efficiency per unit energy of the exciting light absorbed by a photoluminescent body versus the frequency of the exciting light.

excited state [QUANT MECH] A stationary state of higher energy than the lowest stationary state or ground state of a particle or system of particles.

exclusion principle [QUANT MECH] The principle that no two fermions of the same kind may simultaneously occupy the same quantum state. Also known as Pauli exclusion principle.

Excorallanidae [INV ZOO] A family of free-living and parasitic isopod crustaceans in the suborder Flabellifera which have mandibles and first maxillae modified as hooklike piercing organs.

excoriation [MED] Abrasion of a portion of the skin.

excrement [PHYSIO] An excreted substance; the feces.

excrescence [BIOL] **1.** Abnormal or excessive increase in growth. **2.** An abnormal outgrowth.

excretion [PHYSIO] The removal of unusable or excess material from a cell or a living organism.

excretory system [ANAT] Those organs concerned with solid, fluid, or gaseous excretion.

excurrent [BIOL] Flowing out. [BOT] **1.** Having an undivided main stem or trunk. **2.** Having the midrib extending beyond the apex.

exendospermous [BOT] Lacking endosperm. Also known as exalbuminous.

exergonic [BIOCHEM] Of or pertaining to a biochemical reaction in which the end products possess less free energy than the starting materials; usually associated with catabolism.

exfoliation [MED] **1.** The separation of bone or other tissue in thin layers. **2.** A peeling and shedding of the horny layer of the skin.

exfoliative cytology [PATH] The study of cells shed spontaneously from the body surfaces; used principally in the diagnosis of cancer.

exhalation [PHYSIO] The giving off or sending forth in the form of vapor; expiration.

exhaustion delirium [MED] Acute, confusional, delirious reactions brought about by extreme fatigue, long wasting illness, or prolonged insomnia.

exhibitionism [PSYCH] **1.** A sexual perversion in which pleasure is obtained by exposing the genitalia. **2.** In psychoanalysis, gratification of early sexual impulses in young children by physical activity, such as dancing. **3.** Any attracting of attention to oneself.

exine *See* exosporium.

exite [INV. ZOO] A movable appendage or lobe located on the external side of the limb of a generalized arthropod.

exobiology [BIOL] The search for and study of extraterrestrial life.

exocarp *See* epicarp.

exoccipital [ANAT] Lying to the side of the foramen magnum, as the exoccipital bone.

exochorion [INV ZOO] The outer of two layers forming the covering of an insect egg.

exocoel [INV ZOO] The space between pairs of adjacent mesenteries in anthozoan polyps.

Exocoetidae [VERT ZOO] The halfbeaks, a family of actinopterygian fishes in the order Atheriniformes.

exocrine gland [PHYSIO] A structure whose secretion is passed directly or by ducts to its exterior surface, or to another surface which is continuous with the external surface of the gland.

exocuticle [INV ZOO] The middle layer of the cuticle of insects.

exocytosis [CYTOL] Discharge of the contents of old lysosomes or digestive vacuoles into the surrounding environment; characteristic of cells in lower organisms.

exodermis *See* hypodermis.

exoenzyme [BIOCHEM] An enzyme that functions outside the cell in which it was synthesized.

exogamy [GEN] Union of gametes from organisms that are not closely related. Also known as outbreeding.

exogastrula [EMBRYO] An abnormal gastrula that is unable to undergo invagination or further development because of a quantitative increase of presumptive endoderm.

exogenote [GEN] The genetic fragment transferred from the donor to the recipient cell during the process of recombination in bacteria.

exogenous [BIOL] **1.** Due to an external cause; not arising within the organism. **2.** Growing by addition to the outer surfaces. [PHYSIO] Pertaining to those factors in the metabolism of nitrogenous substances obtained from food.

exognathite [INV ZOO] The external branch of an oral appendage of a crustacean.

Exogoninae [INV ZOO] A subfamily of polychaete annelids in the family Syllidae having a short, small body of few segments.

exogynous [BOT] Having the style longer than and exserted beyond the corolla.

exomorphic [BIOL] Pertaining to external appearance of an organism.

exonephric [INV ZOO] Having the excretory organs discharge through the body wall.

exopeptidase [BIOCHEM] An enzyme that acts on the terminal peptide bonds of a protein chain.

exoperidium [MYCOL] The outer layer of the spore case in certain fungi.

exophoria [MED] A type of heterophoria in which the visual lines tend outward.

exophthalmic goiter *See* hyperthyroidism.

exophthalmos [MED] Abnormal protrusion of the eyeball from the orbit.

exopodite [INV ZOO] The outer branch of a biramous crustacean appendage.

exoprosthesis [MED] An externally applied prosthesis.

Exopterygota [INV ZOO] A division of the insect subclass Pterygota including those insects which undergo a hemimetabolous metamorphosis.

exoskeleton [INV ZOO] The external supportive covering of certain invertebrates, such as arthropods. [VERT ZOO] Bony or horny epidermal derivatives, such as nails, hoofs, and scales.

exosmosis [PHYSIO] Passage of a liquid outward through a cell membrane.

exospore [MYCOL] An asexual spore formed by abstriction, as in certain Phycomycetes.

exosporium [BOT] The outer of two layers forming the wall of spores such as pollen and bacterial spores. Also known as exine.

exostome [BOT] The opening through the outer integument of a bitegmic ovule.

exostosis [MED] A benign cartilage-capped protuberance from the surface of long bones but also seen on flat bones, caused by chronic irritation as from infection, trauma, or osteoarthritis.

exotheca [INV ZOO] The tissue external to the theca of corals.

exothecium [BOT] The outer dehiscing cell layer of a sporangium in gymnosperms.

exothermic [PHYS] Indicating liberation of heat. Also known as exoergic.

exotic [ECOL] Not endemic to an area.

exotoxin [MICROBIO] A toxin that is excreted by a microorganism.

expectorant [PHARM] **1.** Tending to promote expectoration. **2.** An agent that promotes expectoration.

experimental psychology [PSYCH] The study of psychological phenomena by experimental methods.

expiration [PHYSIO] Emission of air from the lungs.

exponential growth [MICROBIO] The period of bacterial growth during which cells divide at a constant rate. Also known as logarithmic growth.

exserted [BIOL] Protruding beyond the enclosing structure, such as stamens extending beyond the margin of the corolla.

exsheath [INV ZOO] To escape from the residual membrane of a previous developmental stage, used of the larva of certain nematodes, microfilaria, and so on.

exstipulate [BOT] Lacking stipules.

exstrophy [MED] Eversion; the turning inside out of a part.

extensor [PHYSIO] A muscle that functions to extend a limb or part.

external ankle height [ANTHRO] A measure of the vertical distance taken from the lower end of the fibula to the floor.

external auditory meatus [ANAT] The external passage of the ear, leading to the tympanic membrane in reptiles, birds, and mammals.

external carotid artery [ANAT] An artery which originates at the common carotid and distributes blood to the anterior part of the neck, face, scalp, side of the head, ear, and dura mater.

external ear [ANAT] The portion of the ear that receives sound waves, including the pinna and external auditory meatus.

external fertilization [PHYSIO] Those processes involved in the union of male and female sex cells outside the body of the female.

external gill [ZOO] A gill that is external to the body wall, as in certain larval fishes and amphibians, and in many aquatic insects.

external respiration [PHYSIO] The processes by which oxygen is carried into living cells from the outside environment and by which carbon dioxide is carried in the reverse direction.

exteroceptor [PHYSIO] Any sense receptor at the surface of the body that transmits information about the external environment.

extinction [EVOL] The worldwide death and disappearance of a specific organism or group of organisms. [PHYS CHEM] *See* absorbance. [PSYCH] Decrease in frequency and elimination of a conditioned response if reinforcement of the response is withheld.

extinction coefficient *See* absorptivity.

extirpate [BIOL] To uproot, destroy, make extinct, or exterminate.

extracellular [BIOL] Outside the cell.

extrachromosomal inheritance *See* cytoplasmic inheritance.

extract [CHEM] Material separated from liquid or solid mixture by a solvent. [PHARM] **1.** A pharmaceutical preparation obtained by dissolving the active constituents of a drug with a suitable menstruum, evaporating the solvent, and adjusting to prescribed standards. **2.** A preparation, usually in a concentrated form, obtained by treating plant or animal tissue with a solvent to remove desired odiferous, flavorful, or nutritive components of the tissue.

extraembryonic coelom [EMBRYO] The cavity in the extraembryonic mesoderm; it is continuous with the embryonic coelom in the region of the umbilicus, and is obliterated by growth of the amnion.

extraembryonic membrane *See* fetal membrane.

extrapyramidal system [ANAT] Descending tracts of nerve fibers arising in the cortex and subcortical motor areas of the brain.

extrasensory perception [PSYCH] The alleged phenomenon of perception or awareness of external events in the absence of any sensory stimulation arising from the events. Abbreviated ESP.

extrastaminal [BOT] Located outside the androecium.

extrasystole [MED] Premature beat of the heart.

extrauterine pregnancy [MED] Gestation outside the uterus.

extravasation [MED] The pouring out or eruption of a body fluid from its proper channel or vessel into the surrounding tissue.

extrinsic [ANAT] Functionally connected with but not wholly within the part, referring specifically to certain muscles, as of the eye. [BIOL] Having an effect from the outside.

extrinsic factor *See* vitamin B$_{12}$.

extrophy [MED] Malformation of an organ.

extrorse [BIOL] Directed outward or away from the axis of growth.

extroversion [BIOL] A turning outward. [PSYCH] Turning to things and persons outside oneself rather than to one's own thoughts and feelings.

exudate [MED] **1.** A proteinaceous, cellular material that passes through blood vessel walls into the surrounding tissue in inflammation or a superficial lesion. **2.** Any substance that is exuded.

exumbrella [INV ZOO] The outer, convex surface of the umbrella of jellyfishes.

eye [ZOO] A photoreceptive sense organ that is capable of forming an image in vertebrates and in some invertebrates such as the squids and crayfishes.

eyeball [ANAT] The globe of the eye.

eye-ear plane [ANTHRO] In craniometric study, a position for placing a human skull so that the lower margins of the orbits and the upper margin of the auditory meatus are on the same horizontal plane. Also known as Frankfurt horizontal.

eyeglasses [OPTICS] Optical devices containing corrective lenses for defects of vision or for special purposes.

eyelid [ANAT] A movable, protective section of skin that covers and uncovers the eyeball of many terrestrial animals.

eye socket *See* orbit.

eyespot [BOT] **1.** A small photosensitive pigment body in certain unicellular algae. **2.** A dark area around the hilum of certain seeds, as some beans. [INV ZOO] A simple organ of vision in many invertebrates consisting of pigmented cells overlying a sensory termination. [PL PATH] A fungus disease of sugarcane and certain other grasses caused by *Helminthosporium sacchari* and characterized by yellowish oval lesions on the stems and leaves.

eyestalk [INV ZOO] A movable peduncle bearing a terminal eye in decapod crustaceans.

EYE

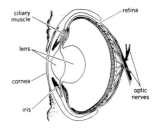

Octopus eye (after J. Wells). *(From R. D. Barnes, Invertebrate Zoology, 2d ed., W. B. Saunders Co., 1968)*

F

F *See* fluorine.

F₁ [GEN] Notation for the first filial generation resulting from a cross.

F₂ [GEN] Notation for the progeny produced by intercrossing members of the F_1. Also known as second generation.

fabella [MED] A small piece of fibrocartilage or a small bone which occasionally develops in the lateral head of the gastrocnemius muscle.

Fabriciinae [INV ZOO] A subfamily of small to minute, colonial, sedentary polychaete annelids in the family Sabellidae.

face [ANAT] The anterior portion of the head, including the forehead and jaws. [CRYSTAL] *See* crystal face.

facet [ANAT] A small plane surface, especially on a bone or a hard body; may be produced by wear, as a worn spot on the surface of a tooth. [INV ZOO] The surface of a simple eye in the compound eye of arthropods and certain other invertebrates.

facial angle [ANTHRO] The angle formed by the union of a line connecting nasion and gnathion with the Frankfort horizontal plane of the head.

facial artery [ANAT] The external branch of the external carotid artery.

facial bone [ANAT] The bone comprising the nose and jaws, formed by the maxilla, zygoma, nasal, lacrimal, palatine, inferior nasal concha, vomer, mandible, and parts of the ethmoid and sphenoid.

facial index [ANTHRO] The ratio of the breadth of the face to its length multiplied by 100.

facial nerve [ANAT] The seventh cranial nerve in vertebrates; a paired composite nerve, with motor elements supplying muscles of facial expression and with sensory fibers from the taste buds of the anterior two-thirds of the tongue and from other sensory endings in the anterior part of the throat.

facies [ANAT] Characteristic appearance of the face in association with a disease or abnormality. [ECOL] The makeup or appearance of a community or species population.

factor I *See* fibrinogen.

factor II *See* prothrombin.

factor III *See* thromboplastin.

factor IV [BIOCHEM] Calcium ions involved in the mechanism of blood coagulation.

factor V *See* proaccelerin.

factor VI [BIOCHEM] An unidentified substance believed to be derived from factor V during blood coagulation.

factor VII [BIOCHEM] A procoagulant, related to prothrombin, that is involved in the formation of a prothrombin-converting principle which transforms prothrombin to thrombin. Also known as stable factor.

factor VIII *See* antihemophilic factor.

factor IX *See* Christmas factor.

factor X *See* Stuart factor.

factor XI [BIOCHEM] A procoagulant present in normal blood but deficient in hemophiliacs. Also known as plasma thromboplastin antecedent (PTA).

factor XII [BIOCHEM] A blood clotting factor effective experimentally only in vitro; deficient in hemophiliacs. Also known as Hageman factor.

facultative aerobe [MICROBIO] An anaerobic microorganism which can grow under aerobic conditions.

facultative anaerobe [MICROBIO] A microorganism that grows equally well under aerobic and anaerobic conditions.

facultative parasite [ECOL] An organism that can exist independently but may be parasitic on certain occasions, such as the flea.

facultative photoheterotroph [MICROBIO] Any bacterium that usually grows anaerobically in light but can also grow aerobically in the dark.

FAD *See* flavin adenine dinucleotide.

Fagaceae [BOT] A family of dicotyledonous plants in the order Fagales characterized by stipulate leaves, seeds without endosperm, female flowers generally not in catkins, and mostly three styles and locules.

Fagales [BOT] An order of dicotyledonous woody plants in the subclass Hamamelidae having simple leaves and much reduced, mostly unisexual flowers.

fagopyrism [VET MED] Photosensitization of the skin and mucous membranes, accompanied by convulsions; produced especially in sheep and swine by feeding on the buckwheat plant, *Fagopyrum sagittatum,* or clovers and grasses containing flavin or carotene and xanthophyll.

Fahrenheit scale [THERMO] A temperature scale; the temperature in degrees Fahrenheit ($°F$) is the sum of 32 plus 9/5 the temperature in degrees Celsius; water at 1 atmosphere pressure freezes very near $32°F$ and boils very near $212°F$.

fairy ring spot [PL PATH] A fungus disease of carnations caused by *Heterosporium echinulatum,* producing bleached spots with concentric dark zones on the leaves.

falciform [BIOL] Sickle-shaped.

falciform ligament [ANAT] The ventral mesentery of the liver; its peripheral attachment extends from the diaphragm to the umbilicus and contains the round ligament of the liver.

falciparum malaria [MED] A severe form of malaria caused by *Plasmodium falciparum* and characterized by sudden attacks of chills, fever, and sweating at irregular intervals; the infecting organism usually localizes in a specific organ, causing capillary blockage. Also known as alged malaria; estivoautumnal malaria; malignant malaria; pernicious malaria.

falcon [VERT ZOO] Any of the highly specialized diurnal birds of prey composing the family Falconidae; these birds have been captured and trained for hunting.

Falconidae [VERT ZOO] The falcons, a family of long-winged predacious birds in the order Falconiformes.

Falconiformes [VERT ZOO] An order of birds containing the diurnal birds of prey, including falcons, hawks, vultures, and eagles.

falculate [ZOO] Curved and with a sharp point.

falling disease [VET MED] A terminal manifestation of copper deficiency in which the animal collapses and dies because of heart failure.

Fallopian tube [ANAT] Either of the paired oviducts that extend from the ovary to the uterus for conduction of the ovum in mammals.

false blossom [PL PATH] **1.** A fungus disease of the cranberry caused by *Exobasidium oxycocci;* erect flower buds are formed which produce malformed flowers that set no fruit. Also known as rosebloom. **2.** A similar virus disease of the cranberry transmitted by the leafhopper, *Scleroracus vaccinii.* Also known as Wisconsin false blossom.

false fruit [BOT] A fruit that develops from the receptacle or other flower part in addition to the ovary, or from the complete inflorescence.

false ligament [ANAT] 'Any peritoneal fold which is not a true supporting ligament.

false rib [ANAT] A rib that is not attached to the sternum directly; any of the five lower ribs on each side in humans.

false ring [BOT] A layer of wood that is less than a full season's growth and often does not form a complete ring.

false smut [PL PATH] **1.** A fungus disease of palm caused by *Graphiola phoenicis* and characterized by small cylindrical protruding pustules, often surrounded by yellowish leaf tissue. **2.** *See* green smut.

falx [ANAT] A sickle-shaped structure.

familial [BIOL] Of, pertaining to, or occurring among the members of a family.

familial aldosterone deficiency [MED] A hereditary metabolic disorder, probably due to a defect in the enzyme involved in dehydrogenation of 18-hydroxycorticosterone to aldosterone, characterized by growth retardation and hypoaldosteronism.

familial dysautonomia [MED] A hereditary disease transmitted as an autosomal recessive and characterized from infancy by evidence of autonomic nervous system dysfunction, including feeding difficulties, absence of overflow tears, indifference to pain, absent corneal reflexes and deep tendon reflexes, and absence of fungiform papillae on the tongue; most common in Jewish children.

familial Mediterranean fever [MED] A hereditary disease of unknown cause characterized by recurrent fever, abdominal and chest pain, arthralgia, and rash, sometimes terminating in renal failure. Abbreviated FMF. Also known as familial recurring polyserositis; periodic disease; periodic peritonitis.

familial osteochondrodystrophy *See* Morquio's syndrome.

familial polyposis [MED] A hereditary condition transmitted as an autosomal dominant and characterized by the appearance of polyps in the small intestine and colon; malignant degeneration is common.

familial recurring polyserositis *See* familial Mediterranean fever.

familial splenic anemia *See* Gaucher's disease.

family [CHEM] A group of elements whose chemical properties, such as valence, solubility of salts, and behavior toward reagents, are similar. [SYST] A taxonomic category based on the grouping of related genera.

famulus [INV ZOO] A sensory seta on the tarsus of certain mites.

fan [AGR] A mechanical device used for winnowing grain. [BIOL] Any structure, such as a leaf or the tail of a bird, resembling an open fan.

Fanconi's anemia [MED] An infantile anemia that resembles pernicious anemia; related to excessive chromosomal breakage and associated with the risk of developing leukemia.

Fanconi's syndrome *See* amino diabetes.

fang [ANAT] The root of a tooth. [VERT ZOO] A long, pointed tooth, especially one of a venomous serpent.

faraday [PHYS] The electric charge required to liberate 1 gram-equivalent of a substance by electrolysis; experimentally equal to $96,487.0 \pm 1.6$ coulombs. Also known as Faraday constant.

Faraday constant *See* faraday.

FARNESOL

Structural formula of farnesol.

farcy *See* glanders.

farinaceous [BIOL] Having a mealy surface covering. [FOOD ENG] **1.** Containing starch or flour. **2.** Having the texture of meal.

Farinales [BOT] An order that includes several groups regarded as orders of the Commelinidae in other systems of classification.

Farinosae [BOT] The equivalent name for Farinales.

farinose [AGR] Yielding farina, a fine meal of vegetable matter. [BIOL] Covered with a white powdery substance.

farm [AGR] A tract of land used for cultivating crops or raising animals.

farmer's lung [MED] An acute pulmonary disorder caused by the inhalation of spores from moldy hay or straw.

farming [AGR] The skills and practices of agriculture.

farnesol [BIOCHEM] $C_{15}H_{25}OH$ A colorless liquid extracted from oils of plants such as citronella, neroli, cyclamen, and tuberose; it has a delicate floral odor, and is an intermediate step in the biological synthesis of cholesterol from mevalonic acid in vertebrates; used in perfumery.

farsightedness *See* hypermetropia.

fascia [HISTOL] Layers of areolar connective tissue under the skin and between muscles, nerves, and blood vessels.

fasciate [BOT] Having bands or stripes.

fasciation [PL PATH] Malformation of plant parts resulting from disorganized tissue growth.

fascicle [BOT] A small bundle, as of fibers or leaves.

fasciculate [BOT] Arranged in tufts or fascicles.

fasciculation potential [PHYSIO] An action potential which is quantitatively comparable to that of a motor unit and which represents spontaneous contraction of a bundle of muscle fibers.

fasciculus [ANAT] A bundle or tract of nerve, muscle, or tendon fibers isolated by a sheath of connective tissues and having common origins, innervation, and functions.

Fasciola hepatica [INV ZOO] A digenetic trematode which parasitizes sheep, cattle, and occasionally humans.

fasciole [INV ZOO] A band of cilia on the test of certain sea urchins.

fascioliasis [MED] The infection of humans with *Fasciola hepatica*.

fasciolopsiasis [MED] The presence of the parasite *Fasciolopsis buski* in a person's small intestine.

fascioscapulohumeral dystrophy [MED] A progressive hereditary form of muscular dystrophy involving atrophy of the muscles of the face, pectoral girdle, and upper arm.

fast chemical reaction [PHYS CHEM] A reaction with a half-life of milliseconds or less; such reactions occur so rapidly that special experimental techniques are required to observe their rate.

fastigiate [BOT] **1.** Having erect branches that are close to the stem. **2.** Becoming narrower at the top. [ZOO] Arranged in a conical bundle.

fat [ANAT] Pertaining to an obese person. [BIOCHEM] Any of the glyceryl esters of fatty acids which form a class of neutral organic compounds. [PHYSIO] The chief component of fat cells and other animal and plant tissues.

fat body [INV ZOO] A nutritional reservoir of fatty tissue surrounding the viscera or forming a layer beneath the integument in the immature larval stages of many insects. [VERT ZOO] A mass of adipose tissue attached to each genital gland in amphibians.

fat cell [HISTOL] The principal component of adipose connective tissue; two types are yellow fat cells and brown fat cells.

fate map [EMBRYO] A graphic scheme indicating the definite spatial arrangement of undifferentiated embryonic cells in accordance with their destination to become specific tissues.

fat embolus [MED] An embolus composed principally of fatty substances.

fatigue [PHYSIO] Exhaustion of strength or reduced capacity to respond to stimulation following a period of activity.

fat-metabolizing hormone *See* ketogenic hormone.

fat necrosis [MED] Pathologic death of adipose tissue often accompanied by soap production from the hydrolyzed fat; associated with pancreatitis.

fatty acid [ORG CHEM] An organic monobasic acid of the general formula $C_nH_{2n+1}COOH$ derived from the saturated series of aliphatic hydrocarbons; examples are palmitic acid, stearic acid, and oleic acid; used as a lubricant in cosmetics and nutrition, and for soaps and detergents.

fatty infiltration [PHYSIO] Infiltration of an organ or tissue with excessive amounts of fats.

fatty metamorphosis [MED] Fatty degeneration, fatty infiltration, or both.

fauces [ANAT] The passage in the throat between the soft palate and the base of the tongue. [BOT] The throat of a calyx, corolla, or similar part.

faucial tonsil *See* palatine tonsil.

fauna [ZOO] 1. Animals. 2. The animal life characteristic of a particular region or environment.

faunal extinction [EVOL] The worldwide death and disappearance of diverse animal groups under circumstances that suggest common and related causes. Also known as mass extinction.

faunal region [ECOL] A division of the zoosphere, defined by geographic and environmental barriers, to which certain animal communities are bound.

favism [MED] An acute hemolytic anemia, usually in persons of Mediterranean area descent, occurring when an individual with glucose-6-phosphate dehydrogenase deficiency of erythrocytes eats the beans or inhales the pollen of *Vicia faba.*

Favositidae [PALEON] A family of extinct Paleozoic corals in the order Tabulata.

favus [MED] A fungal infection of the scalp, usually caused by *Trichophyton schoenleini,* characterized by round, yellow, cup-shaped crusts having a peculiar mousy odor. Also known as tinea favosa.

feather [VERT ZOO] An ectodermal derivative which is a specialized keratinous outgrowth of the epidermis of birds; functions in flight and in providing insulation and protection.

feather rot [PL PATH] A fungus rot of both dead and living tree trunks caused by *Poria subacida* and characterized by the white stringy or spongy nature of the rotted tissue.

feather-veined [BOT] Referring to a leaf in which veins extend at acute angles and in regular series from the midrib.

febrile disease [MED] Any disease associated with or characterized by fever.

feces [PHYSIO] The waste material eliminated by the gastrointestinal tract.

fecundity [BIOL] The innate potential reproductive capacity of the individual organism, as denoted by its ability to form and separate from the body the mature germ cells.

feedback inhibition [BIOCHEM] A cellular control mechanism by which the end product of a series of metabolic reactions inhibits the activity of the first enzyme in the sequence.

Felidae [VERT ZOO] The cats and saber-toothed cats, a family of mammals in the superfamily Feloidea.

FATE MAP

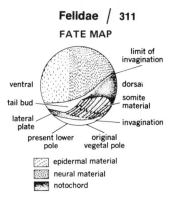

A fate map for the beginning gastrula of a urodele shown in lateral view. *(After W. Vogt, 1926)*

FAVOSITIDAE

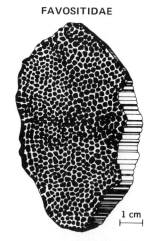

Whole portion of a coral colony of *Favosites gothlandica* Lam from Niagara Limestone (Wenlock) of Owen Sound, Ontario. *(After H. G. Nicholson)*

FEATHER

Detail of the base of a feather. *(From J. Van Tyne and A. J. Berger, Fundamentals of Ornithology, Wiley, 1959)*

FENNEL

Fennel *(Foeniculum vulgare).*
(USDA)

feline [VERT ZOO] 1. Of or relating to the genus *Felis.* 2. Catlike.

Felis [VERT ZOO] The type genus of the Felidae, comprising the true or typical cats, both wild and domestic.

fell-field [ECOL] A culture community of dwarfed, scattered plants or grasses above the timberline.

Feloidea [VERT ZOO] A superfamily of catlike mammals in the order Carnivora.

Felty's syndrome [MED] A complex of symptoms involving rheumatoid arthritis, splenomegaly, lymphadenopathy, and anemia.

female [BOT] A flower lacking stamens. [ZOO] An individual that bears young or produces eggs.

female heterogamety [GEN] The occurrence, in females of a species, of an unequal pair of sex chromosomes.

female homogamety [GEN] The occurrence, in females of a species, of an equal pair of sex chromosomes.

female pseudohermaphroditism *See* gynandry.

feminizing syndrome [MED] Any of a number of symptom complexes in which males tend to take on feminine characteristics due to alterations of adrenocortiocotropin output.

femoral artery [ANAT] The principal artery of the thigh; originates as a continuation of the external iliac artery.

femoral hernia [MED] A hernia that occurs at the passage of the arteries and veins from the abdomen into the legs below the inguinal ligament.

femoral nerve [ANAT] A mixed nerve of the leg; the motor portion innervates muscles of the thigh, and the sensory portion innervates portions of the skin of the thigh, leg, hip, and knee.

femoral ring [ANAT] The abdominal opening of the femoral canal.

femoral vein [ANAT] A vein accompanying the femoral artery.

femorotibial index [ANTHRO] The ratio of the length of the femur to the length of the tibia multiplied by 100.

femur [ANAT] 1. The proximal bone of the hind or lower limb in vertebrates. 2. The thigh bone in humans, articulating with the acetabulum and tibia.

Fenestellidae [PALEON] A family of extinct fenestrated, cryptostomatous bryozoans which abounded during the Silurian.

fenestra [ANAT] An opening in the medial wall of the middle ear. [MED] An opening in a bandage or plaster splint for examination or drainage.

fenestrated membrane [HISTOL] One of the layers of elastic tissue in the tunica media and tunica intima of large arteries.

fenestration [BIOL] 1. A transparent or windowlike break or opening in the surface. 2. The presence of windowlike openings.

fenestrule [INV ZOO] Any of the small openings between branches of a lazy colony of Bryozoa.

fennel [BOT] *Foeniculum vulgare.* A tall perennial herb of the family Umbelliferae; a spice is derived from the fruit.

feral [BIOL] 1. Wild state. 2. Escaped from cultivation or domestication and reverted to the wild state.

ferment [BIOCHEM] An agent that can initiate fermentation and other metabolic processes.

fermentation [MICROBIO] An enzymatic transformation of organic substrates, especially carbohydrates, generally accompanied by the evolution of gas; a physiological counterpart of oxidation, permitting certain organisms to live and grow in the absence of air; used in various industrial processes for the manufacture of products such as alcohols, acids, and cheese by the action of yeasts, molds, and bacteria;

alcoholic fermentation is the best-known example. Also known as zymosis.

fermentation tube [MICROBIO] A culture tube with a vertical closed arm to collect gas formed in a broth culture by microorganisms.

fermium [CHEM] A synthetic radioactive element, symbol Fm, with atomic number 100; discovered in debris of the 1952 hydrogen bomb explosion, and now made in nuclear reactors.

fern [BOT] Any of a large number of vascular plants composing the division Polypodiophyta.

ferredoxins [BIOCHEM] Iron-containing proteins that transfer electrons, usually at a low potential, to flavoproteins; the iron is not present as a heme.

ferric [INORG CHEM] The term for a compound of trivalent iron, for example, ferric bromide, $FeBr_3$.

ferrichrome [MICROBIO] A cyclic hexapeptide that is a microbial hydroxamic acid and is involved in iron transport and metabolism in microorganisms.

ferrihemoglobin [BIOCHEM] Hemoglobin in the oxidized state. Also known as methemoglobin.

ferriporphyrin [BIOCHEM] A red-brown to black complex of iron and porphyrin in which the iron is in the 3+ oxidation state.

Ferrobacillus [MICROBIO] A genus of obligate or facultative chemoautotrophic bacteria which obtain metabolic energy from the oxidation of ferrous iron to the ferric state by oxygen.

ferroporphyrin [BIOCHEM] A red complex of porphyrin and iron in which the iron is in the 2+ oxidation state.

ferrous [CHEM] The term or prefix used to denote compounds of iron in which iron is in the divalent (2+) state.

ferruginous [SCI TECH] **1.** Pertaining to or containing iron. **2.** Having the appearance or color of iron rust (ferric oxide).

fertility [BIOL] The state of or capacity for abundant productivity.

fertility factor [GEN] An episomal bacterial sex factor which determines the role of a bacterium as either a male donor or as a female recipient of genetic material. Also known as F factor; sex factor.

fertilization [PHYSIO] The physicochemical processes involved in the union of the male and female gametes to form the zygote.

fertilization membrane [CYTOL] A membrane that separates from the surface of and surrounds many eggs following activation by the sperm; prevents multiple fertilization.

fertilization tube [BOT] An elongated projection of the antheridium that penetrates the oogonium to allow passage of the male gamete in certain fungi.

fertilizer [MATER] Material that is added to the soil to supply chemical elements needed for plant nutrition.

fertilizin [BIOCHEM] A mucopolysaccharide, derived from the jelly coat of an egg, that plays a role in sperm recognition and the stimulation of sperm motility and metabolic activity.

fescue [BOT] A group of grasses of the genus *Festuca*, used for both hay and pasture.

fetal asphyxia [MED] Deprivation of oxygen to the fetus due to interference with its blood supply.

fetal fat-cell lipoma *See* liposarcoma.

fetal hemoglobin [BIOCHEM] A normal embryonic hemoglobin having alpha chains identical to those of normal adult human hemoglobin, and gamma chains similar to adult beta chains.

fetal membrane [EMBRYO] Any one of the membranous structures which surround the embryo during its development period. Also known as extraembryonic membrane.

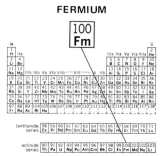

FERMIUM

Periodic table of the chemical elements showing the position of fermium.

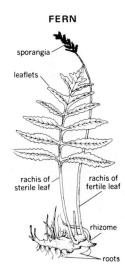

FERN

sporangia

leaflets

rachis of sterile leaf

rachis of fertile leaf

rhizome

roots

The sensitive fern *(Onoclea sensibilis)*, a representative of the Polypodiatae. *(From W. W. Robbins, T. E. Weier, and C. R. Stocking, Botany; An Introduction to Plant Science, 3d ed., Wiley, 1964)*

fetus [EMBRYO] **1.** The unborn offspring of viviparous mammals in the later stages of development. **2.** In human beings, the developing body in utero from the beginning of the ninth week after fertilization through the fortieth week of intrauterine gestation, or until birth.

Feulgen reaction [ANALY CHEM] An aldehyde specific reaction based on the formation of a purple-colored compound when aldehydes react with fuchsin-sulfuric acid; deoxyribonucleic acid gives this reaction after removal of its purine bases by acid hydrolysis; used as a nuclear stain.

fever [MED] An elevation in the central body temperature of warm-blooded animals caused by abnormal functioning of the thermoregulatory mechanisms.

Feyliniidae [VERT ZOO] The limbless skinks, a family of reptiles in the suborder Sauria represented by four species in tropical Africa.

F factor *See* fertility factor.

fiber [BOT] **1.** An elongate, thick-walled, tapering plant cell that lacks protoplasm and has a small lumen. **2.** A very slender root.

fiber crops [AGR] Plants, such as flax, hemp, jute, and sisal, cultivated for their content or yield of fibrous material.

fiber flax [BOT] The flax plant grown in fertile, well-drained, well-prepared soil and cool, humid climate; planted in the early spring and harvested when half the seed pods turn yellow; used in the manufacture of linen.

fibril [BIOL] A small thread or fiber, as a root hair or one of the structural units of a striated muscle.

fibrillation [PHYSIO] An independent, spontaneous, local twitching of muscle fibers.

fibrillose [BIOL] Having fibrils.

fibrin [BIOCHEM] The fibrous, insoluble protein that forms the structure of a blood clot; formed by the action of thrombin.

fibrinase [BIOCHEM] An enzyme that catalyzes the formation of covalent bonds between fibrin molecules. Also known as fibrin-stabilizing factor.

fibrinogen [BIOCHEM] A plasma protein synthesized by the parenchymal cells of the liver; the precursor of fibrin. Also known as factor I.

fibrinoid [BIOCHEM] A homogeneous, refractile, oxyphilic substance occurring in degenerating connective tissue, as in term placentas, rheumatoid nodules, and Aschoff bodies, and in pulmonary alveoli in some prolonged pneumonitides.

fibrinolysin *See* plasmin.

fibrinolysis [PHYSIO] Liquefaction of coagulated blood by the action of plasmin on fibrin.

fibrinous pericarditis [MED] Inflammation of the pericardium involving the deposition of fibrin and leukocytes between the layers of the pericardium; seen in rheumatic carditis and acute infectious diseases.

fibrin-stabilizing factor *See* fibrinase.

fibroadenoma [MED] A benign tumor containing both fibrous and glandular elements.

fibroblast [HISTOL] A stellate connective tissue cell found in fibrous tissue. Also known as a fibrocyte.

fibrocartilage [HISTOL] A form of cartilage rich in dense, closely opposed bundles of collagen fibers; occurs in intervertebral disks, in the symphysis pubis, and in certain tendons.

fibrocyte *See* fibroblast.

fibroid [HISTOL] Composed of fibrous tissue.

fibroid tumor *See* fibroma.

fibroma [MED] A benign tumor composed primarily of fibrous connective tissue. Also known as fibroid tumor.

fibroma molluscum *See* neurofibromatosis.

FIBER FLAX

The type of flax plant grown for fiber, usually taller than the seed flax plant.

fibromatosis [MED] **1.** The occurrence of multiple fibromas. **2.** Localized proliferation of fibroblasts without apparent cause.

fibromyoma [MED] A benign tumor, usually of smooth muscle, with a prominent fibrous stroma; commonly a uterine leiomyoma.

fibromyosis *See* interstitial endometrosis.

fibromyositis *See* myositis.

fibroplasia [MED] The growth of fibrous tissue, as in the second phase of wound healing.

fibrosarcoma [MED] A sarcoma composed of spindle cells that produce collagenous fibrils.

fibrosing adenomatosis *See* sclerosing adenomatosis.

fibrosis [MED] Growth of fibrous connective tissue in an organ or part in excess of that naturally present.

fibrositis [MED] Inflammation of white fibrous connective tissue, usually in a joint region.

fibrous dysplasia [MED] **1.** Extensive formation of fibrous tissue and transformation of bony tissue in one or more bones. **2.** Development of abnormal amounts of fibrous tissue in the mammary glands.

fibrous osteoma *See* ossifying fibroma.

fibrous protein [BIOCHEM] Any of a class of highly insoluble proteins representing the principal structural elements of many animal tissues.

fibula [ANAT] The outer and usually slender bone of the hind or lower limb below the knee in vertebrates; it articulates with the tibia and astragalus in humans, and is ankylosed with the tibia in birds and some mammals.

Fick's law [PHYS] The law that the rate of diffusion of matter across a plane is proportional to the negative of the rate of change of the concentration of the diffusing substance in the direction perpendicular to the plane.

Ficus [BOT] A genus of tropical trees in the family Moraceae including the rubber tree and the fig tree. [INV ZOO] A genus of gastropod mollusks having pear-shaped, spirally ribbed sculptured shells.

Fiedler's myocarditis *See* interstitial myocarditis.

field-emission microscope [ELECTR] A device that uses field emission of electrons or of positive ions (field-ion microscope) to produce a magnified image of the emitter surface on a fluorescent screen.

field-ion microscope [ELECTR] A microscope in which atoms are ionized by an electric field near a sharp tip; the field then forces the ions to a fluorescent screen, which shows an enlarged image of the tip, and individual atoms are made visible; this is the most powerful microscope yet produced. Also known as ion microscope.

fièvre boutonneuse [MED] A mild febrile rickettsial disease of man caused by *Rickettsia conori;* characterized by a rash, tache noire (primary ulcer), and swollen lymph glands. Also known as boutonneuse fever; Marseilles fever.

fifteen-degrees calorie *See* calorie.

fig [BOT] *Ficus carica.* A deciduous tree of the family Moraceae cultivated for its edible fruit, which is a syconium, consisting of a fleshy hollow receptacle lined with pistillate flowers.

Figitidae [INV ZOO] A family of hymenopteran insects in the superfamily Cynipoidea.

Fiji disease [PL PATH] A virus disease of sugarcane; elongated swellings on the underside of leaves precede death of the plant.

filament [BOT] **1.** The stalk of a stamen which supports the anther. **2.** A chain of cells joined end to end, as in certain algae. [INV ZOO] A single silk fiber in the cocoon of a

FIELD-EMISSION MICROSCOPE

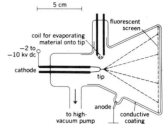

Electron-operated field-emission microscope.

FIÈVRE BOUTONNEUSE

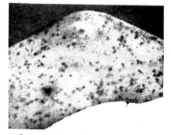

Fièvre boutonneuse, showing rash and tache noire in popliteal area. *(From C. B. Philip, in R. L. Pullen, ed., Communicable Diseases, Lea and Febiger, 1950)*

FIG

Fig *(Ficus carica)*, details of branch and fruit.

silkworm. [SCI TECH] A long, flexible object with a small cross section.

filamentous bacteria [MICROBIO] Bacteria, especially in the order Actinomycetales, whose cells resemble filaments and are often branched.

filaria [INV ZOO] A parasitic filamentous nematode belonging to the order Filaroidea.

filariasis [MED] A disease due to the presence of hairlike nematodes (filariae) in humans, including *Wuchereria bancrofti, W. pacifica,* and *Onchocerca volvulus.*

Filarioidea [INV ZOO] An order of the class Nematoda comprising highly specialized parasites of humans and domestic animals.

filator [INV ZOO] In silkworms, that part of the spinning organ which regulates size of the silk fiber.

filbert [BOT] Either of two European plants belonging to the genus *Corylus* and producing a thick-shelled, edible nut. Also known as hazelnut.

Filicales [BOT] The equivalent name for Polypodiales.

Filicineae [BOT] The equivalent name for Polypodiatae.

Filicornia [INV ZOO] A group of hyperiid amphipod crustaceans in the suborder Genuina having the first antennae inserted anteriorly.

filiform [BIOL] Threadlike or filamentous.

filiform papilla [ANAT] Any one of the papillae occurring on the dorsum and margins of the oral part of the tongue, consisting of an elevation of connective tissue covered by a layer of epithelium.

film [BIOL] A thin, membranous skin, such as a pellicle. [MED] A pathological opacity, as of the cornea.

FILOPLUME

|← 1 cm →|

Drawing of a filoplume showing reduced hairlike vane. *(From J. C. Welty, The Life of Birds, Saunders, 1962)*

filoplume [VERT ZOO] A specialized feather that may be decorative, sensory, or both; it is always associated with papillae of contour feathers.

filopodia [INV ZOO] Filamentous pseudopodia.

filoreticulopodia [INV ZOO] Branched, filamentous pseudopodia.

Filosia [INV ZOO] A subclass of the class Rhizopodea characterized by slender filopodia which rarely anastomose.

filterable virus [VIROL] Virus particles that remain in a fluid after passing through a diatomite or glazed porcelain filter with pores too minute to allow the passage of bacterial cells.

filter bridge [ECOL] A corridor that is limited in that climate or some other factor prevents or restricts migration of certain species.

filter feeder [INV ZOO] A microphagous organism that uses complex filtering mechanisms to trap particles suspended in water.

fimbria *See* pilus.

fimbriate [BIOL] Having a fringe along the edge.

fin [VERT ZOO] A paddle-shaped appendage on fish and other aquatic animals that is used for propulsion, balance, and guidance.

final common pathway *See* lower motor neuron.

finch [VERT ZOO] The common name for birds composing the family Fringillidae.

findspot [ARCHEO] The place where an archeological object has been found.

fin fold [EMBRYO] A median integumentary fold extending along the body of a fish embryo which gives rise to the dorsal, caudal, and anal fins.

finger [ANAT] Any of the four digits on the hand other than the thumb.

fingerprint [ANAT] A pattern of distinctive epidermal ridges on the bulbs of the inside of the end joints of fingers and thumbs.

fingerprinting *See* genetic fingerprinting.

fin rot [VET MED] A bacterial disease of hatchery fishes characterized by necrosis and erosion of the fin tissue.

fin spine [VERT ZOO] A bony process that supports the fins of certain fishes.

fir [BOT] The common name for any tree of the genus *Abies* in the pine family; needles are characteristically flat.

fire blight [PL PATH] A bacterial disease of apple, pear, and related pomaceous fruit trees caused by *Erwinia amylovora;* leaves are blackened, cankers form on the trunk, and flowers and fruits become discolored.

fire disclimax [ECOL] A community that is perpetually maintained at an early stage of succession through recurrent destruction by fire followed by regeneration.

firefly [INV ZOO] Any of various flying insects which produce light by bioluminescence.

first-degree burn [MED] A mild burn characterized by pain and reddening of the skin.

first-filial generation [GEN] The first generation resulting from a cross with all members being heterozygous for characters which differ from those of the parents. Symbolized F_1.

first law of thermodynamics [THERMO] The law that heat is a form of energy, and the total amount of energy of all kinds in an isolated system is constant; it is an application of the principle of conservation of energy.

first-order reaction [PHYS CHEM] A chemical reaction in which the rate of decrease of concentration of component A with time is proportional to the concentration of A.

fish [VERT ZOO] The common name for the cold-blooded aquatic vertebrates belonging to the groups Cyclostomata, Chondrichthyes, and Osteichthyes.

fisher [VERT ZOO] *Martes pennanti.* An arboreal, carnivorous mammal of the family Mustelidae; a relatively large weasel-like animal with dark fur, found in northern North America.

fisheries conservation [ECOL] Those measures concerned with the protection and preservation of fish and other aquatic life, particularly in sea waters.

fishery [ECOL] A place for harvesting fish or other aquatic life, particularly in sea waters.

fish lice [INV ZOO] The common name for all members of the crustacean group Arguloida.

Fissidentales [BOT] An order of the Bryopsida having erect to procumbent, simple or branching stems and two rows of leaves arranged in one plane.

fission [BIOL] A method of asexual reproduction among bacteria, algae, and protozoans by which the organism splits into two or more parts, each part becoming a complete organism. [NUC PHYS] The division of an atomic nucleus into parts of comparable mass; usually restricted to heavier nuclei such as isotopes of uranium, plutonium, and thorium. Also known as atomic fission; nuclear fission.

fissiped [VERT ZOO] **1.** Having the toes separated to the base. **2.** Of or relating to the Fissipeda.

Fissipeda [VERT ZOO] Former designation for a suborder of the Carnivora.

Fissurellidae [INV ZOO] The keyhole limpets, a family of gastropod mollusks in the order Archeogastropoda.

fistula [MED] An abnormal congenital or acquired communication between two surfaces or between a viscus or other hollow structure and the exterior.

Fistuliporidae [PALEON] A diverse family of extinct marine bryozoans in the order Cystoporata.

fistulous withers [VET MED] A chronic inflammation of the withers of a horse accompanied by fluid discharge, which may

FIR

Cone and branches with flat needles of balsam fir (*Abies balsamea*).

FISHER

The fisher *(Martes pennanti)*, an arboreal, nocturnal carnivorous mammal of North America.

FISSIDENTALES

sporophyte

rhizoids

Entire plant of *Fissidens adiantoides.*

FLABELLIFERA

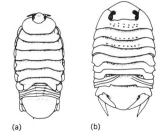

(a)   (b)

Two examples of Flabellifera
showing the caudal fan.
(a) Lironeca epimerias
Richardson. *(b) Tecticeps
renoculis* Richardson.

FLAME EMISSION
SPECTROSCOPY

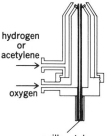

hydrogen
or
acetylene

oxygen

capillary tube
for introduction of
sample solution

Beckman aspirator burner
used to aspirate the solution
directly into oxyhydrogen or
oxyacetylene flame.

FLAME PHOTOMETER

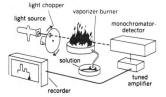

Schematic diagram of atomic
absorption spectrophotometer
for determining metal
concentrations.

be initiated by mechanical injury but depends on bacterial
(*Brucella abortus*) infection for development.

fitness *See* adaptive value.

fix [BIOL] To kill, harden, or preserve a tissue, organ, or
organism by immersion in dilute acids, alcohol, or solutions
of coagulants.

fixative [MATER] A chemical or a mixture of chemicals used
to treat biological specimens before preservation so as to
retain a reasonable facsimile of their appearance when alive.

flabellate [BIOL] Fan-shaped.

Flabellifera [INV ZOO] The largest and morphologically most
generalized suborder of isopod crustaceans; the biramous
uropods are attached to the sides of the abdomen and may
form, with the last abdominal fragment, a caudal fan.

Flabelligeridae [INV ZOO] The cage worms, a family of spioniform worms belonging to the Sedentaria; the anterior part
of the body is often concealed by a cage of setae arising from
the first few segments.

flabellum [INV ZOO] Any structure resembling a fan, as the
epipodite of certain crustacean limbs.

flaccid [BOT] Deficient in turgor. [PHYSIO] Soft, flabby, or
relaxed.

flacherie [INV ZOO] A fatal bacterial disease of caterpillars,
especially silkworms, marked by loss of appetite, dysentery,
and flaccidity of the body; after death the body darkens and
liquefies.

Flacourtiaceae [BOT] A family of dicotyledonous plants in
the order Violales having the characteristics of the more
primitive members of the order.

flagella [BIOL] Relatively long, whiplike, centriole-based locomotor organelles on some motile cells.

Flagellata [INV ZOO] The equivalent name for Mastigophora.

flagellate [BIOL] **1.** Having flagella. **2.** An organism that
propels itself by means of flagella. **3.** Resembling a flagellum.
[INV ZOO] Any member of the protozoan superclass Mastigophora.

flagellated chamber [INV ZOO] An outpouching of the wall
of the central cavity in Porifera that is lined with choanocytes;
connects with incurrent canals through prosophyles.

flagellation [BIOL] The arrangement of flagella on an organism. [PSYCH] Beating or whipping as a means of producing
sexual gratification.

flagellin [MICROBIO] The protein component of bacterial flagella.

flag smut [PL PATH] A smut affecting the leaves and stems of
cereals and other grasses, characterized by formation of sori
within the tissues, which rupture releasing black spore masses
and causing fraying of the infected area.

flame bulb [INV ZOO] The enlarged terminal part of the flame
cell of a protonephridium, consisting of a tuft of cilia.

flame cell [INV ZOO] A hollow cell that contains the terminal
branches of excretory vessels in certain flatworms and rotifers
and some other invertebrates.

flame emission spectroscopy [SPECT] A flame photometry
technique in which the solution containing the sample to be
analyzed is optically excited in an oxyhydrogen or oxyacetylene flame.

flame photometer [SPECT] One of several types of instruments used in flame photometry, such as the emission flame
photometer and the atomic absorption spectrophotometer, in
each of which a solution of the chemical being analyzed is
vaporized; the spectral lines resulting from the light source
going through the vapors enters a monochromator that
selects the band or bands of interest.

FLATFISH

A typical left-eye flounder, with both eyes on left side.

flame photometry [SPECT] A branch of spectrochemical analysis in which samples in solution are excited to produce line emission spectra by introduction into a flame.

flame spectrometry [SPECT] A procedure used to measure the spectra or to determine wavelengths emitted by flame-excited substances.

flame spectrophotometry [SPECT] A method used to determine the intensity of radiations of various wavelengths in a spectrum emitted by a chemical inserted into a flame.

flamingo [VERT ZOO] Any of various long-legged and long-necked aquatic birds of the family Phoenicopteridae characterized by a broad bill resembling that of a duck but abruptly bent downward and rosy-white plumage with scarlet coverts.

flank [VERT ZOO] The part of a quadruped mammal between the ribs and the pelvic girdle.

flash burn [MED] Tissue injury resulting from exposure to high-intensity radiant heat.

flash photolysis [PHYS CHEM] A method of studying fast photochemical reactions in gas molecules; a powerful lamp is discharged in microsecond flashes near a reaction vessel holding the gas, and the products formed by the flash are observed spectroscopically.

flatfish [VERT ZOO] Any of a number of asymmetrical fishes which compose the order Pleuronectiformes; the body is laterally compressed, and both eyes are on the same side of the head.

flatworm [INV ZOO] The common name for members of the phylum Platyhelminthes; individuals are dorsoventrally flattened.

flavan [BIOCHEM] $C_{15}H_{14}O$ 2-Phenylbenzopyran, an aromatic heterocyclic compound from which all flavonoids are derived.

flavanone [BIOCHEM] $C_{15}H_{12}O_2$ A colorless crystalline derivative of flavone.

flavescence [PL PATH] Yellowing or blanching of green plant parts due to diminution of chlorophyll accompanying certain virus disease.

flavin [BIOCHEM] **1.** A yellow dye obtained from the bark of quercitron trees. **2.** Any of several water-soluble yellow pigments occurring as coenzymes of flavoproteins.

flavin adenine dinucleotide [BIOCHEM] $C_{27}H_{33}N_9O_{15}P_2$ A coenzyme that functions as a hydrogen acceptor in aerobic dehydrogenases (flavoproteins). Abbreviated FAD.

flavin mononucleotide *See* riboflavin 5′-phosphate.

flavin phosphate *See* riboflavin 5′-phosphate.

Flavobacterium [MICROBIO] A genus of gram-negative, rod-shaped, motile bacteria in the family Achromobacteraceae that characteristically produce yellow, orange, or red pigmentation.

flavone [BIOCHEM] **1.** Any of a number of ketones composing a class of flavonoid compounds. **2.** $C_{15}H_{10}O_2$ A colorless crystalline compound occurring as dust on the surface of many primrose plants.

flavonoid [BIOCHEM] Any of a series of widely distributed plant constituents related to the aromatic heterocyclic skeleton of flavan.

flavonol [BIOCHEM] **1.** Any of a class of flavonoid compounds that are hydroxy derivatives of flavone. **2.** $C_{16}H_{10}O_2$ A colorless, crystalline compound from which many yellow plant pigments are derived.

flavoprotein [BIOCHEM] Any of a number of conjugated protein dehydrogenases containing flavin that play a role in biological oxidations in both plants and animals; a yellow enzyme.

flax [BOT] *Linum usitatissimum.* An erect annual plant with

FLAVONE

Structural formula of flavone (def. 2).

FLAX

Flowering top of flax plant showing buds and seed capsules. *(USDA)*

FLEA

The sticktight flea, a bird parasite.

FLEXIBILIA

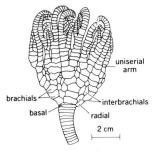

uniserial arm

brachials

interbrachials

basal

radial

2 cm

Talanterocrinus species. *(After F. Bather, 1900)*

FLIGHT FEATHER

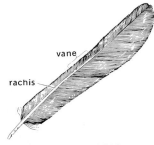

vane

rachis

Full view of a typical flight feather. *(From J. Van Tyne and A. J. Berger, Fundamentals of Ornithology, Wiley, 1959)*

FLOATOBLAST

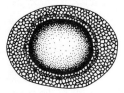

Floatoblast of *Plumatella repens*, in Phylactolaemata, a class of fresh-water Bryozoans.

linear leaves and blue flowers; cultivated as a source of flaxseed and fiber.

flax rust [PL PATH] A disease of flax caused by the rust fungus *Melampsora lini.*

flaxseed [BOT] The seed obtained from the seed flax plant; a source of linseed oil.

flax wilt [PL PATH] A fungus disease of flax caused by *Fusarium oxysporum lini;* diseased plants wilt, yellow, and die.

flea [INV ZOO] Any of the wingless insects composing the order Siphonaptera; most are ectoparasites of mammals and birds.

flea-borne typhus *See* murine typhus.

Fleming's solution [MATER] A tissue fixative made up of a mixture of osmic, chromic and acetic acids.

flesh [ANAT] The soft parts of the body of a vertebrate, especially the skeletal muscle and associated connective tissue and fat.

fleshy fruit [BOT] A fruit having a fleshy pericarp that is usually soft and juicy, but sometimes hard and tough.

Fletcher-Munson contour *See* equal loudness contour.

flex [SCI TECH] To bend.

flexibacteria [MICROBIO] A general term to describe filamentous, gliding, nonphotosynthetic bacteria.

Flexibilia [PALEON] A subclass of extinct stalked or creeping Crinoidea; characteristics include a flexible tegmen with open ambulacral grooves, uniserial arms, a cylindrical stem, and five conspicuous basals and radials.

flexion [BIOL] Act of bending, especially of a joint.

flexion reflex [PHYSIO] An unconditioned, segmental reflex elicited by noxious stimulation and consisting of contraction of the flexor muscles of all joints on the same side. Also known as the nocioceptive reflex.

flexor [PHYSIO] A muscle that bends or flexes a limb or a part.

flexor plate [INV ZOO] In insects, a plate that supports the pretarsus and provides a point of attachment for the claw flexor tendon.

flexuous [BIOL] **1.** Flexible. **2.** Bending in a zigzag manner. **3.** Wavy.

flexure [EMBRYO] A sharp bend of the anterior part of the primary axis of the vertebrate embryo. [VERT ZOO] The last joint of a bird's wing.

flight feather [VERT ZOO] Any of the long contour feathers on the wing of a bird. Also known as remex.

flipper [VERT ZOO] A broad, flat appendage used for locomotion by aquatic mammals and sea turtles.

float [AGR] A device consisting of one or more blades used to level a seedbed. [BIOL] An air-filled sac in many pelagic flora and fauna that serves to buoy up the body of the organism.

floating rib [ANAT] One of the last two ribs in humans which have the anterior end free.

floatoblast [INV ZOO] A free-floating statoblast having a float of air cells.

floccose [BOT] Covered with tufts of woollike hairs.

flocculonodular lobes [ANAT] The pair of lateral cerebellar lobes in vertebrates which function to regulate vestibular reflexes underlying posture; referred to functionally as the vestibulocerebellum.

flocculus [ANAT] A prominent lobe of the cerebellum situated behind and below the middle cerebellar peduncle on each side of the median fissure.

flora [BOT] **1.** Plants. **2.** The plant life characterizing a specific geographic region or environment.

floral axis [BOT] A flower stalk.

floral diagram [BOT] A diagram of a flower in cross section showing the number and arrangement of floral parts.

floret [BOT] A small individual flower that is part of a compact group of flowers, such as the head of a composite plant or inflorescence.

floricome [INV ZOO] A type of branched hexaster spicule.

floriculture [AGR] A segment of horticulture concerned with commercial production, marketing, and retail sale of cut flowers and potted plants, as well as home gardening and flower arrangement.

Florideophyceae [BOT] A class of red algae, division Rhodophyta, having prominent pit connections between cells.

floriferous [BOT] Blooming freely, used principally of ornamental plants.

florigen [BIOCHEM] A plant hormone that stimulates buds to flower.

florivorous [ZOO] Feeding on flowers.

floscelle [INV ZOO] A flowerlike structure around the mouth of some echinoids.

Flosculariacea [INV ZOO] A suborder of rotifers in the order Monogononta having a malleoramate mastax.

Flosculariidae [INV ZOO] A family of sessile rotifers in the suborder Flosculariacea.

flosculous [BOT] 1. Composed of florets. 2. Of a floret, tubular in form.

flosculus [BOT] A floret.

flounder [VERT ZOO] Any of a number of flatfishes in the families Pleuronectidae and Bothidae of the order Pleuronectiformes.

flower [BOT] The characteristic reproductive structure of a seed plant, particularly if some or all of the parts are brightly colored.

flowers of sulfur [PHARM] One of three forms of pharmaceutical sulfur, made by sublimation; the other two forms are precipitated sulfur and washed sulfur. Also known as sublimed sulfur.

fluke [INV ZOO] The common name for more than 40,000 species of parasitic flatworms that form the class Trematoda. [VERT ZOO] A flatfish, especially summer flounder.

fluorescence [ATOM PHYS] 1. Emission of electromagnetic radiation that is caused by the flow of some form of energy into the emitting body and which ceases abruptly when the excitation ceases. 2. Emission of electromagnetic radiation that is caused by the flow of some form of energy into the emitting body and whose decay, when the excitation ceases, is temperature-independent. [NUC PHYS] Gamma radiation scattered by nuclei which are excited to and radiate from an excited state.

fluorescence microscope [OPTICS] A variation of the compound laboratory light microscope which is arranged to admit ultraviolet, violet, and sometimes blue radiations to a specimen, which then fluoresces.

fluorescent antibody [IMMUNOL] An antibody labeled by a fluorescent dye, such as fluorescein.

fluorescent antibody test [IMMUNOL] A clinical laboratory test based on the antigen used in the diagnosis of syphilis and lupus erythematosus and for identification of certain bacteria and fungi, including the tubercle bacillus.

fluorescent staining [CYTOL] The use of fluorescent dyes to mark specific cell structures, such as chromosomes.

fluorine [CHEM] A gaseous or liquid chemical element, symbol F, atomic number 9, atomic weight 18.998; a member of the halide family, it is the most electronegative element and the most chemically energetic of the nonmetallic elements;

FLORAL DIAGRAM

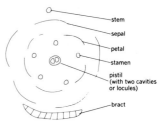

Graphic diagram of a cross section of a flower.

FLOSCULARIACEA

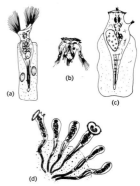

Sessile rotifers of the Flosculariacea. (a) *Floscularia mutabilis.* (b) *Pedalia mira.* (c) *Conchiloides* species. (d) *Lacinularia socialis.*

FLUORINE

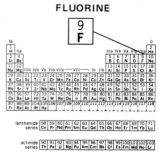

Periodic table of the chemical elements showing the position of fluorine.

FLY

A common black fly *(Simulium)*.

FLYING FISH

The four-winged flying fish
(Parexocoetus mesogaster), with
pectoral fins extended for gliding.

FOLLICLE

Follicles of the milkweed plant.

highly toxic, corrosive, and flammable; used in rocket fuels and as a chemical intermediate.

fluorochromasia [CYTOL] The immediate appearance of fluorescence inside viable cells on exposure to a fluorogenic substrate.

fluorogenic substrate [CHEM] A nonfluorescent material that is acted upon by an enzyme to produce a fluorescent compound.

fluoroscope [ENG] A fluorescent screen designed for use with an x-ray tube to permit direct visual observation of x-ray shadow images of objects interposed between the x-ray tube and the screen.

fly [INV ZOO] The common name for a number of species of the insect order Diptera characterized by a single pair of wings, antennae, compound eyes, and hindwings modified to form knoblike balancing organs, the halters.

flying fish [VERT ZOO] Any of about 65 species of marine fishes which form the family Exocoetidae in the order Atheriniformes; characteristic enlarged pectoral fins are used for gliding.

flyway [VERT ZOO] A geographic migration route for birds, including the breeding and wintering areas that it connects.

Fm *See* fermium.

FMF *See* familial Mediterranean fever.

FMN *See* riboflavin 5'-phosphate.

foal [VERT ZOO] A young horse, especially one under 1 year of age.

focal infection [MED] Infection in a limited area, such as the tonsils, teeth, sinuses, or prostate.

focal seizure [MED] An epileptic manifestation of a restricted nature, usually without loss of consciousness, due to irritation of a localized area of the brain.

foehn sickness [MED] A phenomenon in humans in alpine regions, marked by adverse psychological and physiological effects during prolonged periods of foehn wind.

fog climax [ECOL] A community that deviates from a climatic climax because of the persistent occurrence of a controlling fog blanket.

fold [ANAT] A plication or doubling, as of various parts of the body such as membranes and other flat surfaces.

foliaceous [BOT] Consisting of or having the form or texture of a foliage leaf. [ZOO] Resembling a leaf in growth form or mode.

foliage [BOT] The leaves of a plant.

foliar [BOT] Of, pertaining to, or consisting of leaves.

foliate papilla [VERT ZOO] One of the papillae found on the posterolateral margin of the tongue of many mammals, but vestigial or absent in humans.

foliation [BOT] **1.** The process of developing into a leaf. **2.** The state of being in leaf.

folic acid [BIOCHEM] $C_{19}H_{19}N_7O_6$ A yellow, crystalline vitamin of the B complex; it is slightly soluble in water, usually occurs in conjugates containing glutamic acid residues, and is found especially in plant leaves and vertebrate livers. Also known as pteroylglutamic acid (PGA).

foliferous [BOT] Producing leaves.

foliicolous [BIOL] Growing or parasitic upon leaves, as certain fungi.

foliobranchiate [VERT ZOO] Having leaflike gills.

foliolate [BOT] Having leaflets.

follicetum [BOT] A group of follicles that are lightly coherent to each other.

follicle [BIOL] A deep, narrow sheath or a small cavity. [BOT] A type of dehiscent fruit composed of one carpel opening along a single suture.

follicle-stimulating hormone [BIOCHEM] A protein hormone released by the anterior pituitary of vertebrates which stimulates growth and secretion of the Graafian follicle and also promotes spermatogenesis. Abbreviated FSH.

folliculate [BIOL] Having or composed of follicles.

fomite [MED] An inanimate object which may be contaminated with infectious organisms and thus serve to transmit disease.

fontanelle [ANAT] A membrane-covered space between the bones of a fetal or young skull. [INV ZOO] A depression on the head of termites.

Fontéchevade man [PALEON] A fossil man representing the third interglacial *Homo sapiens* and having browridges and a cranial vault similar to those of modern *Homo sapiens*.

food [BIOL] A material that can be ingested and utilized by the organism as a source of nutrition and energy.

food allergy [IMMUNOL] A hypersensitivity to certain foods.

food-borne disease [MED] Any disease transmitted by contaminated foods.

food chain [ECOL] The scheme of feeding relationships by trophic levels which unites the member species of a biological community.

food microbiology [FOOD ENG] The science that deals with the microorganisms involved in the spoilage, contamination, and preservation of food.

food poisoning [MED] Poisoning due to intake of food contaminated by bacteria or poisonous substances produced by bacteria.

food pyramid [ECOL] An ecological pyramid representing the food relationship among the animals in a community.

food vacuole [CYTOL] A membrane-bound organelle in which digestion occurs in cells capable of phagocytosis. Also known as heterophagic vacuole; phagocytic vacuole.

food web [ECOL] A modified food chain that expresses feeding relationships at various, changing trophic levels.

foot [ANAT] Terminal portion of a vertebrate leg. [INV ZOO] An organ for locomotion or attachment.

foot-and-mouth disease [VET MED] A highly contagious virus disease of cattle, pigs, sheep, and goats that is transmissible to man; characterized by fever, salivation, and formation of vesicles in the mouth and pharynx and on the feet. Also known as hoof-and-mouth disease.

foot breadth [ANTHRO] The measure of the maximum distance across the left foot, when the subject stands with his weight evenly distributed on both feet.

foot gland [INV ZOO] A glandular structure which secretes an adhesive substance in many animals. Also known as pedal gland.

foot length [ANTHRO] A measure of the distance from the heel to the longest toe of the left foot, when the subject stands with the weight evenly distributed on both feet.

foot rot [PL PATH] Any disease that involves rotting of the stem or trunk of a plant. [VET MED] *See* foul foot.

forage [AGR] A vegetable food for domestic animals.

foramen [BIOL] A small opening, orifice, pore, or perforation.

foramen magnum [ANAT] A large oval opening in the occipital bone at the base of the cranium that allows passage of the spinal cord, accessory nerves, and vertebral arteries.

foramen of Magendie [ANAT] The median aperture of the fourth ventricle of the brain.

foramen of Monro *See* interventricular foramen.

foramen of Winslow *See* epiploic foramen.

foramen ovale [ANAT] An opening in the sphenoid for the

FOOD PYRAMID

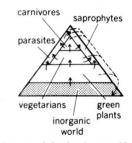

Diagram of the food pyramid showing the general trends in food circulation.

passage of nerves and blood vessels. [EMBRYO] An opening in the fetal heart partition between the two atria.

foramen primum [EMBRYO] A temporary embryonic interatrial opening.

Foraminiferida [INV ZOO] An order of dominantly marine protozoans in the subclass Granuloreticulosia having a secreted or agglutinated shell enclosing the ameboid body.

forb [BOT] A weed or broadleaf herb.

force constant [PHYS CHEM] An expression for the force acting to restrain the relative displacement of the nuclei in a molecule.

forceps [DES ENG] A pincerlike instrument for grasping objects.

forcipate [BIOL] Shaped like forceps; deeply forked.

forcipate trophus [INV ZOO] A type of masticatory apparatus in certain predatory rotifers which resembles forceps and is used for grasping.

Forcipulatida [INV ZOO] An order of echinoderms in the subclass Asteroidea characterized by crossed pedicellariae.

forearm [ANAT] The part of the upper extremity between the wrist and the elbow. Also known as antebrachium.

forearm circumference [ANTHRO] The measure of the circumference taken halfway between the elbow and the wrist.

forearm length [ANTHRO] A measure of the distance from the tip of the elbow to the tip of the middle finger, with the arm flexed at the elbow.

forebrain [EMBRYO] The most anterior expansion of the neural tube of a vertebrate embryo. [VERT ZOO] The part of the adult brain derived from the embryonic forebrain; includes the cerebrum, thalamus, and hypothalamus.

forefinger [ANAT] The index finger; the first finger next to the thumb.

foregut [EMBRYO] The anterior alimentary canal in a vertebrate embryo, including those parts which will develop into the pharynx, esophagus, stomach, and anterior intestine.

forehead [ANAT] The part of the face above the eyes.

forelimb [ANAT] An appendage (as a wing, fin, or arm) of a vertebrate that is, or is homologous to, the foreleg of a quadruped.

forensic medicine [MED] Application of medical evidence or medical opinion for purposes of civil or criminal law.

forest [ECOL] An ecosystem consisting of plants and animals and their environment, with trees as the dominant form of vegetation.

forest conservation [ECOL] Those measures concerned with the protection and preservation of forest lands and resources.

forest ecology [ECOL] The science that deals with the relationship of forest trees to their environment, to one another, and to other plants and to animals in the forest.

forest management [FOR] Measures concerned with the effective organization of a forest to ensure continued production of its goods and services.

forest mapping [FOR] The branch of forestry dealing with the preparation of maps showing the distribution and conformation of individual forest stands.

forest measurement [FOR] The branch of forestry concerned with the measurement of standing trees, cut roundwood, and lumber products.

forest product [FOR] Any material afforded by a forest for commercial use, such as tree products and forage.

forest resources [FOR] Forest land and the trees on it.

forestry [ECOL] The management of forest lands for wood, forages, water, wildlife, and recreation.

forest soil [FOR] The natural medium for growth of tree roots and associated forest vegetation.

FORCIPATE TROPHUS

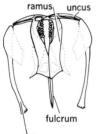

ramus uncus

fulcrum

manubrium

Ventral view of *Dicranophorus* forcipate trophus. *(After Hauer)*

forest stand [FOR] The basic unit of forest mapping; a group of trees that are more or less homogeneous with regard to species composition, density, size, and sometimes habitat.

forest-tundra [ECOL] A temperate and cold savanna which occurs at high altitudes and consists of scattered or clumped trees and a shrub layer of varying coverage.

formaldehyde [ORG CHEM] HCHO The simplest aldehyde; a gas at room temperature, and a poisonous, clear, colorless liquid solution with pungent odor; used to make synthetic resins by reaction with phenols, urea, and melamine, as a chemical intermediate, as an embalming fluid, and as a disinfectant. Also known as formol; methylene oxide.

Formicariidae [VERT ZOO] The antbirds, a family of suboscine birds in the order Passeriformes.

formicarium [BIOL] An artificial ants' nest set up for purposes of study.

formication [MED] An abnormal sensation as of insects crawling in or upon the skin; a common symptom in diseases of the spinal cord and the peripheral nerves; may be a hallucination.

Formicidae [INV ZOO] The ants, social insects composing the single family of the hymenopteran superfamily Formicoidea.

formicivorous [ZOO] Feeding on ants.

Formicoidea [INV ZOO] A monofamilial superfamily of hymenopteran insects in the suborder Apocrita, containing the ants.

formol *See* formaldehyde.

formula [CHEM] **1.** A combination of chemical symbols that expresses a molecule's composition. **2.** A reaction formula showing the interrelationship between reactants and products. [MATH] An equation or rule relating mathematical objects or quantities.

formula weight [CHEM] **1.** The gram-molecular weight of a substance. **2.** In the case of a substance of uncertain molecular weight such as certain proteins, the molecular weight calculated from the composition, assuming that the element present in the smallest proportion is represented by only one atom.

fornix [ANAT] **1.** An arched body or surface. **2.** A concavity or cul-de-sac. [BOT] A small scale, especially in the corolla tube of some plants.

Forssman antibody [IMMUNOL] A heterophile antibody that reacts with Forssman antigen.

Forssman antigen [IMMUNOL] Any of a large group of heterophile antigens occurring in a wide variety of unrelated animals, including horses, dogs, and fowl, and in certain bacteria.

fossa [ANAT] A pit or depression. [VERT ZOO] *Cryptoprocta ferox*. A Madagascan carnivore related to the civets.

fossil [PALEON] The organic remains, traces, or imprint of an organism preserved in the earth's crust since some time in the geologic past.

fossil man [PALEON] Ancient man identified from prehistoric skeletal remains which are archeologically earlier than the Neolithic.

fossorial [VERT ZOO] Adapted for digging.

foulbrood [INV ZOO] The common name for three destructive bacterial diseases of honeybee larvae.

foul foot [VET MED] A feedlot disease of cattle and sheep marked by inflammation and ulceration of the feet; common in wet feedlots. Also called foot rot.

Fourier transform [MATH] For a function $f(t)$, the function $F(x)$ equal to $1/\sqrt{2\pi}$ times the integral over t from $-\infty$ to ∞ of $f(t) \exp (itx)$.

fovea [BIOL] A small depression or pit.

FOX

The gray fox *(Urocynon cineroargenteus),* similar to the red fox but flecked with gray, found from the Great Lakes to South America.

FRANCIUM

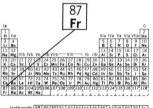

Periodic table of the chemical elements showing the position of francium.

fovea centralis [ANAT] A small, rodless depression of the retina in line with the visual axis, which affords acute vision.

foveal vision [PHYSIO] Vision achieved by looking directly at objects in the daylight so that the image falls on or near the fovea centralis. Also known as photopic vision.

foveola [BIOL] A small pit, especially one in the embryonic gastric mucosa from which gastric glands develop.

foveolate [BIOL] Having small depressions; pitted.

fowl [AGR] A domestic cock or hen, especially an adult hen, as chicken or several gallinaceous birds.

fowl pox [VET MED] A disease of birds caused by a virus and characterized by wartlike nodules on the skin, particularly on the head.

fox [VERT ZOO] The common name for certain members of the dog family (Canidae) having relatively short legs, long bodies, large erect ears, pointed snouts, and long bushy tails.

fp See freezing point.

Fr See francium.

fracture [MED] The breaking of bone, cartilage, or teeth. [SCI TECH] **1.** The act, process, or state of being broken. **2.** The surface appearance of a freshly broken material. **3.** The break produced by fracturing.

fragility test [PATH] A measure of the resistance of red blood cells to osmotic hemolysis in hypotonic salt solutions of graded dilutions.

fragmentation [CYTOL] Amitotic division; a type of asexual reproduction. [PSYCH] Disordered behavior and mental processes.

frambesia See yaws.

francium [CHEM] A radioactive alkali-metal element, symbol Fr, atomic number 87, atomic weight distinguished by nuclear instability; exists in short-lived radioactive forms, the chief isotope being francium-223.

Franconian [GEOL] A North American stage of geologic time; the middle Upper Cambrian.

fraternal twins See dizygotic twins.

freckle [MED] A pigmented macule resulting from focal increase in melanin, usually associated with exposure to sunlight, commonly on the face.

free association [PSYCH] **1.** Spontaneous, consciously unrestricted association of ideas or mental images. **2.** A method used in psychoanalysis to gain an understanding of the organization of the content of the mind.

free-central placenta [BOT] A placenta that consists of a free-standing column attached to the base of a compound, unilocular ovary.

free-central placentation [BOT] Having a placenta which consists of a free-standing projection from the base of a compound, unilocular ovary.

free energy [THERMO] **1.** The internal energy of a system minus the product of its temperature and its entropy. Also known as Helmholtz free energy; Helmholtz function; Helmholtz potential; thermodynamic potential at constant volume; work function. **2.** See Gibbs free energy.

free enthalpy See Gibbs free energy.

freemartin [VERT ZOO] An intersexual, usually sterile female calf twinborn with a male.

free-nuclear [BOT] Having nuclei scattered in the cytoplasm and not separated by cell walls.

free recombination [GEN] Genetic recombination having a frequency of 50%, that is, occurring by independent reassortment.

freestone [BOT] A fruit stone to which the fruit does not cling, as in certain varieties of peach.

freezing microtome [ENG] A microtome used to cut frozen tissue.

freezing point [PHYS CHEM] The temperature at which a liquid and a solid may be in equilibrium. Abbreviated fp.

freezing-point depression [PHYS CHEM] The lowering of the freezing point of a solution compared to the pure solvent; the depression is proportional to the active mass of the solute in a given amount of solvent.

Fregatidae [VERT ZOO] Frigate birds or man-o'-war birds, a family of fish-eating birds in the order Pelecaniformes.

Frenatae [INV ZOO] The equivalent name for Heteroneura.

French measles *See* rubella.

frenulum [ANAT] **1.** A small fold of integument or mucous membrane. **2.** A small ridge on the upper part of the anterior medullary velum. [INV ZOO] A spine on most moths that projects from the hindwings and is held to the forewings by a clasp, thus coupling the wings together.

frenum [ANAT] A fold of tissue that restricts the movements of an organ.

frequency [PHYS] The number of cycles completed by a periodic quantity in a unit time.

frequency polygon [STAT] A graph obtained from a frequency distribution by joining with straight lines points whose abscissae are the midpoints of successive class intervals and whose ordinates are the corresponding class frequencies.

fresh-water ecosystem [ECOL] The living organisms and nonliving materials of an inland aquatic environment.

friction ridge [ANAT] One of the integumentary ridges on the plantar and palmar surfaces of primates.

Friedlander's bacillus *See* Klebsiella pneumoniae.

Friedman test [PATH] A pregnancy test in which a female rabbit is given an intravenous injection of urine from the patient; formation of corpora lutea in the ovaries indicates a positive test.

Friedrich's ataxia [MED] A hereditary sclerosis of the spine with speech impairment, lateral curvature of the spine, and palsy of the lower limbs. Also known as hereditary spinal ataxia.

frigidoreceptor [PHYSIO] A cutaneous sense organ which is sensitive to cold.

Fringillidae [VERT ZOO] The finches, a family of oscine birds in the order Passeriformes.

fritillary [BOT] The common name for plants of the genus *Fritillaria*. [INV ZOO] The common name for butterflies in several genera of the subfamily Nymphalinae.

Froehlich's syndrome *See* adiposogenital dystrophy.

frog [VERT ZOO] The common name for a number of tailless amphibians in the order Anura; most have hindlegs adapted for jumping, scaleless skin, and large eyes.

frogeye [PL PATH] Any of various leaf diseases characterized by formation of concentric rings around the diseased spots.

frond [BOT] **1.** The leaf of a palm or fern. **2.** A foliaceous thallus or thalloid shoot.

frontal angle [ANTHRO] The angle formed by the intersection of lines from the bregma and glabella to the auricular point.

frontal bone [ANAT] Either of a pair of flat membrane bones in vertebrates, and a single bone in humans, forming the upper frontal portion of the cranium; the forehead bone.

frontal crest [ANAT] A median ridge on the internal surface of the frontal bone in humans.

frontal eminence [ANAT] The prominence of the frontal bone above each superciliary ridge in humans.

frontal index [ANTHRO] The ratio of the least to the greatest breadth of the forehead multiplied by 100.

FREQUENCY POLYGON

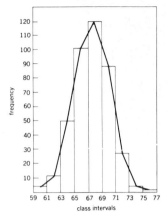

The conversion of a histogram into a frequency polygon by connecting the midpoint value (at the top of each rectangle) with the adjacent midpoint value by straight lines.

FROG

Common frog *(Rana temporaria)*.

frontalis [ANAT] The frontal part of the epicranius muscle.

frontal lobe [ANAT] The anterior portion of a cerebral hemisphere, bounded behind by the central sulcus and below by the lateral cerebral sulcus.

frontal nerve [ANAT] A somatic sensory nerve, attached to the ophthalmic nerve, which innervates the skin of the upper eyelid, the forehead, and the scalp.

frontal plane [ANAT] Any plane parallel with the long axis of the body and perpendicular to the sagittal plane. [MED] In electrocardiography and vectorcardiography, the projection of the vertical axis.

frontal sinus [ANAT] Either of a pair of air spaces within the frontal bone above the nasal bridge.

frostbite [MED] Injury to skin and subcutaneous tissues, and in severe cases to deeper tissues also, from exposure to extreme cold.

frosty mildew [PL PATH] A leaf spot caused by fungi of the genus *Cercosporella* and characterized by pale to white lesions.

frozen section [BIOL] A thin slice of material cut from a frozen sample of tissue or organ.

fructescence [BOT] The period of fruit maturation.

fructification [BOT] 1. The process of producing fruit. 2. A fruit and its appendages. [MYCOL] A sporanginous structure.

fructivorous *See* frugivorous.

D-fructopyranose *See* fructose.

fructose [BIOCHEM] $C_6H_{12}O_5$ The commonest of ketoses and the sweetest of sugars, found in the free state in fruit juices, honey, and nectar of plant glands. Also known as D-fructopyranose.

frugivorous [ZOO] Fruit-eating. Also known as fructivorous.

fruit [BOT] A fully matured plant ovary with or without other floral or shoot parts united with it at maturity.

fruit bud [BOT] A fertilized flower bud that matures into a fruit.

fruit fly [INV ZOO] 1. The common name for those acalypterate insects composing the family Tephritidae. 2. Any insect whose larvae feed on fruit or decaying vegetable matter.

fruiting body [BOT] A specialized, spore-producing organ.

fruiting myxobacteria *See* Myxobacterales.

frustration [PSYCH] The experience of nonfulfillment of some wish or need.

frustration threshold [PSYCH] The point at which an individual feels or shows frustration over inability to achieve an objective.

frustule [INV ZOO] 1. The shell and protoplast of a diatom. 2. A nonciliated planulalike bud in some hydrozoans.

frutescent [BIOL] *See* fruticose. [BOT] Shrublike in habit.

fruticose [BIOL] Resembling a shrub; applied especially to lichens. Also known as frutescent.

FSH *See* follicle-stimulating hormone.

Fucales [BOT] An order of brown algae composing the class Cyclosporeae.

fuchsinophile [BIOL] Having an affinity for the dye fuchsin.

fucoidin [BIOCHEM] A gum composed of L-fucose and sulfate acid ester groups obtained from *Fucus* species and other brown algae.

Fucophyceae [BOT] A class of brown algae.

L-fucopyranose *See* L-fucose.

L-fucose [BIOCHEM] $C_6H_{12}O_5$ A methyl pentose present in some algae and a number of gums and identified in the polysaccharides of blood groups and certain bacteria. Also known as 6-deoxy-L-galactose; L-fucopyranose; L-galacto-methylose; L-rhodeose.

fucoxanthin [BIOCHEM] $C_{40}H_{60}O_6$ A carotenoid pigment; a partial xanthophyll ester found in diatoms and brown algae.

Fucus [BOT] A genus of dichotomously branched brown algae; it is harvested in the kelp industry as a source of algin.

fugacious [BOT] Lasting a short time; used principally to describe plant parts that fall soon after being formed.

fulcrate [BIOL] Having a fulcrum.

fulcrate trophus [INV ZOO] A type of masticatory apparatus in certain rotifers characterized by an elongate fulcrum.

Fulgoroidea [INV ZOO] The lantern flies, a superfamily of homopteran insects in the series Auchenorrhyncha distinguished by the anterior and middle coxae being of equal length and joined to the body at some distance from the median line.

fulmar [VERT ZOO] Any of the oceanic birds composing the family Procellariidae; sometimes referred to as foul gulls because of the foul-smelling substance spat at intruders upon their nests.

fulminate [MED] Of a disease, to come suddenly and follow a severe, intense, and rapid course.

Fulvicin [MICROBIO] A trade name for the antibiotic griseofulvin.

fumagillin [MICROBIO] $C_{26}H_{34}O_7$ An insoluble, crystalline antibiotic produced by a strain of the fungus *Aspergillus fumigatus*.

fumarase [BIOCHEM] An enzyme that catalyzes the hydration of fumaric acid to malic acid, and the reverse dehydration.

Fumariaceae [BOT] A family of dicotyledonous plants in the order Papaverales having four or six stamens, irregular flowers, and no latex system.

Funariales [BOT] An order of mosses; plants are usually annual, are terrestrial, and have stems that are erect, short, simple, or sparingly branched.

functional paralysis *See* hysterical paralysis.

fundic gland [ANAT] Any of the glands of the corpus and fundus of the stomach.

fundus [ANAT] The bottom of a hollow organ.

fungi [MYCOL] Nucleated, usually filamentous, sporebearing organisms devoid of chlorophyll.

fungicide [MATER] An agent that kills or destroys fungi.

fungicolous [ECOL] Living in or on fungi.

fungiform [BIOL] Mushroom-shaped.

fungiform papilla [ANAT] One of the low, broad papillae scattered over the dorsum and margins of the tongue.

Fungi Imperfecti [MYCOL] A class of the subdivision Eumycetes; the name is derived from the lack of a sexual stage.

fungistat [MATER] A compound that inhibits or prevents growth of fungi.

Fungivoridae [INV ZOO] The fungus gnats, a family of orthorrhaphous dipteran insects in the series Nematocera; the larvae feed on fungi.

fungivorous [ZOO] Feeding on or in fungi.

fungus gall [PL PATH] A plant gall resulting from an attack of a parasitic fungus.

funicle *See* funiculus.

funiculus [ANAT] Also known as funicle. **1.** Any structure in the form of a chord. **2.** A column of white matter in the spinal cord. [BOT] The stalk of an ovule. [INV ZOO] A band of tissue extending from the adoral end of the coelom to the adoral body wall in bryozoans.

furca [INV ZOO] A forked process as the last abdominal

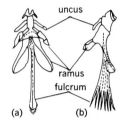

FULCRATE TROPHUS

Fulcrate trophus of *Seison*.
(a) Dorsal view. (b) Lateral view.
(*After de Beauchamp*)

FURCOCERCOUS CERCARIA

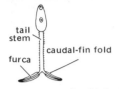

Furcocercous cercaria, obtained either by crushing and examining freshly collected snails or by placing snails in water to permit larvae to escape.

FUSULINIDAE

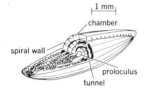

A representative of the Fusulinidae shown in a cutaway diagram.

FUSULININA

Scanning electron micrograph of the surface of the wall of *Earlandia* showing the randomly packed tiny angular calcite grains. (*R. B. MacAdam, Chevron Oil Field Research Co.*)

segment of certain crustaceans, and as part of the spring in collembolans.

furcate [BIOL] Forked.

furcocercous cercaria [INV ZOO] A free-swimming, digenetic trematode larva with a forked tail.

furcula [ZOO] A forked structure, especially the wishbone of fowl.

Furipteridae [VERT ZOO] The smoky bats, a family of mammals in the order Chiroptera having a vestigial thumb and small ears.

Furnariidae [VERT ZOO] The oven birds, a family of perching birds in the superfamily Furnarioidea.

Furnarioidea [VERT ZOO] A superfamily of birds in the order Passeriformes characterized by a predominance of gray, brown, and black plumage.

furuncle [MED] A small cutaneous abscess, usually resulting from infection of a hair follicle by *Staphylococcus aureus*. Also known as boil.

furunculosis [MED] A condition marked by numerous furuncles, or the recurrence of furuncles following healing of a preceding crop.

Fusarium [MYCOL] A genus of fungi in the family Tuberculariaceae having sickle-shaped, multicelled conidia; includes many important plant pathogens.

fusi [INV ZOO] In spiders, organs consisting of two retractile processes which form silk threads.

fusiform [BIOL] Spindle-shaped; tapering toward the ends.

fusiform bacillus [MICROBIO] A bacillus having one blunt and one pointed end, as *Fusobacterium fusiforme*.

fusiform initial cell [BOT] A cell type of the vascular cambium that gives rise to all cells in the vertical system of secondary xylem and phloem.

fusion nucleus [BOT] The triploid, or 3n, nucleus which results from double fertilization and which produces the endosperm in some seed plants.

fusula [INV ZOO] A spindle-shaped, terminal projection of the spinneret of a spider.

Fusulinacea [PALEON] A superfamily of large, marine extinct protozoans in the order Foraminiferida characterized by a chambered calcareous shell.

Fusulinidae [PALEON] A family of extinct protozoans in the superfamily Fusulinacea.

Fusulinina [PALEON] A suborder of extinct rhizopod protozoans in the order Foraminiferida having a monolamellar, microgranular calcite wall.

G

g *See* gram.

Ga *See* gallium.

Gadidae [VERT ZOO] A family of fishes in the order Gadiformes, including cod, haddock, pollock, and hake.

Gadiformes [VERT ZOO] An order of actinopterygian fishes that lack fin spines and a swim bladder duct and have cycloid scales and many-rayed pelvic fins.

gadolinium [CHEM] A rare-earth element, symbol Gd, atomic number 64, atomic weight 157.25; highly magnetic, especially at low temperatures.

gaffkemia [INV ZOO] A septicemic bacterial disease of lobsters caused by *Gaffkya homari.*

galactan [BIOCHEM] Any of a number of polysaccharides composed of galactose units. Also known as galactosan.

galactase [BIOCHEM] A soluble proteolytic enzyme normally present in milk.

galactocele [MED] **1.** A retention cyst caused by obstruction of one or more of the mammary ducts. **2.** A hydrocele with milky contents.

galactogen [BIOCHEM] A polysaccharide, in snails, that yields galactose on hydrolysis.

galactoglucomannan [BIOCHEM] Any of a group of polysaccharides which are prominent components of coniferous woods; they are soluble in alkali and consist of D-glucopyranose and D-mannopyranose units.

galactolipid *See* cerebroside.

galactomannan [BIOCHEM] Any of a group of polysaccharides which are composed of D-galactose and D-mannose units, are soluble in water, and form highly viscous solutions; they are plant mucilages existing as reserve carbohydrates in the endosperm of leguminous seeds.

galactophore [ANAT] A duct that carries milk.

galactorrhea [MED] Excessive flow of milk.

galactosan *See* galactan.

galactose [BIOCHEM] $C_6H_{12}O_6$ A monosaccharide occurring in both levo and dextro forms as a constituent of plant and animal oligosaccharides (lactose and raffinose) and polysaccharides (agar and pectin). Also known as cerebrose.

galactosemia [MED] A congenital metabolic disorder caused by an enzyme deficiency and marked by high blood levels of galactose.

galactosidase [BIOCHEM] An enzyme that hydrolyzes galactosides.

galactoside [BIOCHEM] A glycoside formed by the reaction of galactose with an alcohol; yields galactose on hydrolysis.

galactosuria [MED] Passage of urine containing galactose.

galacturonic acid [BIOCHEM] The monobasic acid resulting from oxidation of the primary alcohol group of D-galactose to carboxyl; it is widely distributed as a constituent of pectins and many plant gums and mucilages.

GADOLINIUM

Periodic table of the chemical elements showing the position of gadolinium.

GALL

Gall formation on tomato caused by *Agrobacterium tumefaciens.* *(Courtesy of Oscar N. Allen)*

GALLIUM

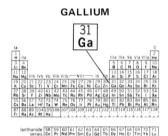

Periodic table of the chemical elements showing the position of gallium.

GAMETANGIAL COPULATION

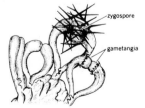

Gametangial copulation, a sexual mechanism in fungi.

galago *See* bushbaby.

Galatheidea [INV ZOO] A group of decapod crustaceans belonging to the Anomura and having a symmetrical abdomen bent upon itself and a well-developed tail fan.

Galaxioidei [VERT ZOO] A suborder of mostly small, freshwater fishes in the order Salmoniformes.

galea [ANAT] The epicranial aponeurosis linking the occipital and frontal muscles. [BIOL] A helmet-shaped structure. [BOT] A helmet-shaped petal near the axis. [INV ZOO] 1. The endopodite of the maxilla of certain insects. 2. A spinning organ on the movable digit of chelicerae of pseudoscorpions.

galeate [BIOL] 1. Shaped like a helmet. 2. Having a galea.

Galen's vein [ANAT] One of the two veins running along the roof the third ventricle that drain the interior of the brain.

Galeritidae [PALEON] A family of extinct exocyclic Euechinoidea in the order Holectypoida, characterized by large ambulacral plates with small, widely separated pore pairs.

gall [MED] A sore on the skin caused by chafing. [PHYSIO] *See* bile. [PL PATH] A large swelling on plant tissues caused by the invasion of parasites, such as fungi or bacteria, following puncture by an insect; insect oviposit and larvae of insects are found in galls.

gallbladder [ANAT] A hollow, muscular organ in humans and most vertebrates which receives dilute bile from the liver, concentrates it, and discharges it into the duodenum.

galleria forest [ECOL] A modified tropical deciduous forest occurring along stream banks.

Galleriinae [INV ZOO] A monotypic subfamily of lepidopteran insects in the family Pyralididae; contains the bee moth or wax worm (*Galleria mellonella*), which lives in beehives and whose larvae feed on beeswax.

gallicolous [BIOL] Producing or inhabiting galls.

Galliformes [VERT ZOO] An order of birds that includes important domestic and game birds, such as turkeys, pheasants, and quails.

gallinaceous [VERT ZOO] Of, pertaining to, or resembling birds of the order Galliformes.

gallium [CHEM] A chemical element, symbol Ga, atomic number 31, atomic weight 69.72. [MET] A silvery-white metal, melting at 29.7°C, boiling at 1983°C.

gallivorous [ZOO] Feeding on the tissues of galls, especially certain insect larvae.

gallnut [PL PATH] A gall resembling a nut.

gallop rhythm [MED] A three-sound sequence resulting from the intensification of the normal third or fourth heart sounds, occurring usually with a rapid ventricular rate.

gallstone [PATH] A nodule formed in the gallbladder or biliary tubes and composed of calcium, cholesterol, or bilirubin, or a combination of these.

Galumnidae [INV ZOO] A family of oribatid mites in the suborder Sarcoptiformes.

galvanic skin response [PHYSIO] The electrical reactions of the skin to any stimulus as detected by a sensitive galvanometer; most often used experimentally to measure the resistance of the skin to the passage of a weak electric current.

galvanotaxis [BIOL] Movement of a free-living organism in response to an electrical stimulus.

galvanotropism [BIOL] Response of an organism to electrical stimulation.

gametangial copulation [MYCOL] Direct fusion of certain fungal gametangia without differentiation of the gametes.

gametangium [BIOL] A structure producing gametes.

gamete [BIOL] A mature germ cell.

gametocyte [HISTOL] An undifferentiated cell from which gametes are produced.

gametogenesis [BIOL] The formation of gametes, or reproductive cells such as ova or sperm.

gametophore [BOT] A branch that bears gametangia.

gametophyll [BOT] A modified leaf that develops sexual organs.

gametophyte [BOT] 1. The haploid generation producing gametes in plants exhibiting metagenesis. 2. An individual plant of this generation.

gamma globulin [IMMUNOL] Any of the serum proteins with antibody activity. Also known as immune globulin; immunoglobulin (Ig).

gamma ray [NUC PHYS] A high-energy photon, especially as emitted by a nucleus in a transition between two energy levels.

Gammaridea [INV ZOO] The scuds or sand hoppers, a suborder of amphipod crustaceans; individuals are usually compressed laterally, are poor walkers, and lack a carapace.

gamma taxonomy [SYST] The third stage in the history of animal systematics, involving renewed interest in the species, with growth of the genetic approach and widespread recognition of the significance of variation.

gamodeme [ECOL] An isolated breeding community.

gamogony [INV ZOO] Spore formation by multiple fission in sporozoans. [ZOO] Sexual reproduction.

gamont [INV ZOO] The gametocyte of sporozoans.

gamopetalous [BOT] Having petals united at their edges. Also known as sympetalous.

gamophyllous [BOT] Having the leaves of the perianth united.

gamosepalous [BOT] Having sepals united at their edges. Also known as synsepalous.

gamostele [BOT] A stele formed by the union of several individual steles.

Gampsonychidae [PALEON] A family of extinct crustaceans in the order Palaeocaridacea.

gangliated cord [ANAT] One of the two main trunks of the sympathetic nervous system, one trunk running along each side of the spinal column.

ganglioma [MED] A form of ganglioneuroma in which neuronal and glial elements appear in about equal proportions.

ganglion [ANAT] A group of nerve cell bodies, usually located outside the brain and spinal cord.

ganglioneuroma [MED] A tumor composed of sympathetic ganglion cells and sheathed nerve fibers.

ganglionitis [MED] Inflammation of a ganglion.

gangosa [MED] Destructive lesions of the nose and hard palate, sometimes more extensive, considered to be the tertiary stage of yaws.

gangplow [AGR] A plow with two or more cutters that turn parallel furrows.

gangrene [MED] A form of tissue death usually occurring in an extremity due to insufficient blood supply.

gangrenous stomatitis *See* noma.

ganoid scale [VERT ZOO] A structure having several layers of enamellike material (ganoin) on the upper surface and laminated bone below.

ganoin [VERT ZOO] The enamellike covering of a ganoid scale.

gape [ANAT] The margin to margin distance between open jaws. [INV ZOO] The space between the margins of a closed mollusk valve.

gar [VERT ZOO] The common name for about seven species of bony fishes in the order Semionotiformes having a slim form, an elongate snout, and close-set ganoid scales.

GANOID SCALE

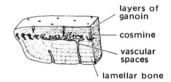

layers of
ganoin

cosmine

vascular
spaces

lamellar bone

Ganoid scale, a type of primitive vertebrate scale. *(From A. S. Romer, The Vertebrate Body, Saunders, 1962)*

GAR

Long-nosed gar, which may grow to a length of 5 feet (1.5 meters).

Garamycin [MICROBIO] Trade name for the antibiotic genta-micin.

Gardner's syndrome [MED] A hereditary disorder transmitted as an autosomal dominant; manifested in childhood by multiple neoplasms, including bony and mesenteric tumors, fatty and fibrous skin, and intestinal polyps.

gargoylism *See* Hurler's syndrome.

garigue [ECOL] A low, open scrubland restricted to limestone sites in the Mediterranean area; characterized by small evergreen shrubs and low trees.

garlic [BOT] *Allium sativum.* A perennial plant of the order Liliales grown for its pungent, edible bulbs.

Gartner's duct [ANAT] The remnant of the embryonic Wolffian duct in the adult female mammal.

gas [PHYS] A phase of matter in which the substance expands readily to fill any containing vessel; characterized by relatively low density.

gas chromatograph [ANALY CHEM] The instrument used in gas chromatography to detect volatile compounds present; also used to determine certain physical properties such as distribution or partition coefficients and adsorption isotherms, and as a preparative technique for isolating pure components or certain fractions from complex mixtures.

gas chromatography [ANALY CHEM] A separation technique involving passage of a gaseous moving phase through a column containing a fixed adsorbent phase; it is used principally as a quantitative analytical technique for volatile compounds.

gas constant [THERMO] The constant of proportionality appearing in the equation of state of an ideal gas, equal to the pressure of the gas times its molar volume divided by its temperature. Also known as gas-law constant.

gas embolus [MED] An embolus composed of a gas resulting from trauma or other causes.

gas gangrene [MED] A localized, but rapidly spreading, necrotizing bacterial wound infection characterized by edema, gas production, and discoloration; caused by several species of *Clostridium.*

gas gland [VERT ZOO] A structure inside the swim bladder of many teleosts which secretes gas into the bladder.

gas-law constant *See* gas constant.

gas-liquid chromatography [ANALY CHEM] A form of gas chromatography in which the fixed phase (column packing) is a liquid solvent distributed on an inert solid support. Abbreviated GLC. Also known as gas-liquid partition chromatography.

gas-liquid partition chromatography *See* gas-liquid chromatography.

Gasserian ganglion [ANAT] A group of nerve cells of the sensory root of the trigeminal nerve. Also known as semilunar ganglion.

gas-solid chromatography [ANALY CHEM] A form of gas chromatography in which the moving phase is a gas and the stationary phase is a surface-active sorbent (charcoal, silica gel, or activated alumina). Abbreviated GSC.

gas sterilization [MICROBIO] Sterilization of heat and liquid-labile materials by means of gaseous agents, such as formaldehyde, ethylene oxide, and β-propiolactone.

Gasteromycetes [MYCOL] A group of basidiomycetous fungi in the subclass Homobasidiomycetidae with enclosed basidia and with basidiospores borne symmetrically on long sterigmata and not forcibly discharged.

Gasterophilidae [INV ZOO] The horse bots, a family of myodarian cyclorrhaphous dipteran insects in the subsection

GAS CHROMATOGRAPH

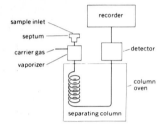

Diagram of a gas chromatograph.

Calypteratae, including individuals that cause myiasis in horses.

gasterospore [BOT] A thick-walled spore produced by a fruiting body.

Gasterosteidae [VERT ZOO] The sticklebacks, a family of actinopterygian fishes in the order Gasterosteiformes.

Gasterosteiformes [VERT ZOO] An order of actinopterygian fishes characterized by a ductless swim bladder, a pelvic fin that is abdominal to subthoracic in position, and an elongate snout.

Gasteruptiidae [INV ZOO] A family of hymenopteran insects in the superfamily Proctotrupoidea.

gastralium [INV ZOO] A microsclere located just beneath the inner cell layer of hexactinellid sponges. [VERT ZOO] One of the riblike structures in the abdomen of certain reptiles.

gastrectomy [MED] Surgical removal of all or part of the stomach.

gastric acid [BIOCHEM] Hydrochloric acid secreted by parietal cells in the fundus of the stomach.

gastric cecum [INV ZOO] One of the elongated pouchlike projections of the upper end of the stomach in insects.

gastric enzyme [BIOCHEM] Any digestive enzyme secreted by cells lining the stomach.

gastric filament [INV ZOO] In scyphozoans, a row of filaments on the surface of the gastric cavity which function to kill or paralyze live prey taken into the stomach. Also known as phacella.

gastric gland [ANAT] Any of the glands in the wall of the stomach that secrete components of the gastric juice.

gastric hypothermia [MED] Cooling of the upper digestive tract; useful in the management of bleeding disorders.

gastric juice [PHYSIO] The digestive fluid secreted by gastric glands; contains gastric acid and enzymes.

gastric mill [INV ZOO] A grinding apparatus consisting of calcareous or chitinous plates in the pharynx or stomach of certain invertebrates.

gastric ostium [INV ZOO] The opening into the gastric pouch in scyphozoans.

gastric pouch [INV ZOO] One of the pouchlike diversions of a scyphozoan stomach.

gastric shield [INV ZOO] A thickening of the stomach wall in some mollusks for mixing the contents.

gastric ulcer [MED] An ulcer of the mucous membrane of the stomach.

gastrin [BIOCHEM] A polypeptide hormone secreted by the pyloric mucosa which stimulates the pancreas to release pancreatic fluid and the stomach to release gastric acid.

gastritis [MED] Inflammation of the stomach.

gastroanastomosis [MED] The surgical formation of a communication between the two pouches of the stomach. Also known as gastrogastrostomy.

gastroblast [INV ZOO] A feeding zooid of a tunicate colony.

gastrocnemius [ANAT] A large muscle of the posterior aspect of the leg, arising by two heads from the posterior surfaces of the lateral and medial condyles of the femur, and inserted with the soleus muscle into the calcaneal tendon, and through this into the back of the calcaneus.

gastrocoele *See* archenteron.

gastrodermis [INV ZOO] The cellular lining of the digestive cavity of certain invertebrates.

gastroduodenitis [MED] Inflammation of the stomach and duodenum.

gastroenteritis [MED] Inflammation of the mucosa of the stomach and intestine.

gastroenterology [MED] The branch of medicine concerned with study of the stomach and intestines.

gastroenterostomy [MED] Surgical formation of a connection between the stomach and small intestine.

gastroepiploic artery [ANAT] Either of two arteries arising from the gastroduodenal and splenic arteries respectively and forming an anastomosis along the greater curvature of the stomach.

gastrogastrostomy *See* gastroanastomosis.

gastrointestinal hormone [BIOCHEM] Any hormone secreted by the gastrointestinal system.

gastrointestinal system [ANAT] The portion of the digestive system including the stomach, intestine, and all accessory organs.

gastrointestinal tract [ANAT] The stomach and intestine.

gastrojejunostomy [MED] Surgical establishment of an anastomosis between the jejunum and the anterior or posterior wall of the stomach.

gastrolith [VERT ZOO] A pebble swallowed by certain animals and retained in the gizzard or stomach, where it serves to grind food.

gastrolysis [MED] The breaking up of adhesions between the stomach and adjacent organs.

Gastromyzontidae [VERT ZOO] A small family of actinopterygian fishes of the suborder Cyprinoidei found in southeastern Asia.

gastropexy [MED] The fixation of a prolapsed stomach in its normal position by suturing it to the abdominal wall or other structure.

gastroplication [MED] An operation for relief of chronic dilation of the stomach by suturing a large horizontal fold in the stomach wall.

gastropod [INV ZOO] A mollusk having a ventral muscular disc adapted for creeping movement.

Gastropoda [INV ZOO] A large, morphologically diverse class of the phylum Mollusca, containing the snails, slugs, limpets, and conchs.

gastropore [INV ZOO] A pore containing a gastrozooid in hydrozoan corals.

gastroptosis [MED] Prolapse or downward displacement of the stomach.

gastroscope [MED] A hollow, tubular instrument used to examine the inside of the stomach by passage through the mouth and esophagus.

gastrosplenic ligament [ANAT] The fold of peritoneum passing from the stomach to the spleen. Also known as gastrosplenic omentum.

gastrosplenic omentum *See* gastrosplenic ligament.

gastrostege [VERT ZOO] One of the large scales on the ventral surface of most snakes.

gastrostome [INV ZOO] The opening of a gastropore.

gastrostomy [MED] The establishment of a fistulous opening into the stomach, with an external opening in the skin; used for artificial feeding.

gastrostyle [INV ZOO] A spiculated projection that extends into the gastrozooid from the base of the gastropore.

Gastrotricha [INV ZOO] A group of microscopic, pseudocoelomate animals considered either to be a class of the Aschelminthes or to constitute a separate phylum.

gastrozooid [INV ZOO] A nutritive polyp of colonial coelenterates, characterized by having tentacles and a mouth.

gastrula [EMBRYO] The stage of development in animals in which the endoderm is formed and invagination of the blastula has occurred.

GASTROTRICHA

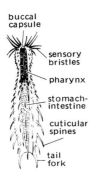

buccal capsule

sensory bristles

pharynx

stomach-intestine

cuticular spines

tail fork

Chaetonotus, a freshwater gastrotrich.

gastrulation [EMBRYO] The process by which the endoderm is formed during development.

Gaucher's cells [PATH] Abnormal macrophages associated with Gaucher's disease.

Gaucher's disease [MED] A rare chronic, probably hereditary disease in which cells loaded with cerebrosides become localized in reticuloendothelial tissue and eventually cause tissue destruction; manifestations include enlargement of the spleen, bronzing of the skin, and anemia. Also known as cerebroside lipoidosis; familial splenic anemia.

Gaussian curve [STAT] The bell-shaped curve corresponding to a population which has a normal distribution. Also known as normal curve.

Gaussian distribution *See* normal distribution.

gavage [MED] The administration of nourishment through a stomach tube.

gavial [VERT ZOO] The name for two species of reptiles composing the family Gavialidae.

Gavialidae [VERT ZOO] The gavials, a family of reptiles in the order Crocodilia distinguished by an extremely long, slender snout with an enlarged tip.

Gaviidae [VERT ZOO] The single, monogeneric family of the order Gaviiformes.

Gaviiformes [VERT ZOO] The loons, a monofamilial order of diving birds characterized by webbed feet, compressed, blade-like tarsi, and a heavy, pointed bill.

Gay-Lussac's law *See* Charles' law; combining-volumes principle.

g-cal *See* calorie.

Gd *See* gadolinium.

Ge *See* germanium.

Gecarcinidae [INV ZOO] The true land crabs, a family of decapod crustaceans belonging to the Brachygnatha.

gecko [VERT ZOO] The common name for more than 300 species of arboreal and nocturnal reptiles composing the family Gekkonidae.

geitonogamy [BOT] Pollination and fertilization of one flower by another on the same plant.

Gekkonidae [VERT ZOO] The geckos, a family of small lizards in the order Squamata distinguished by a flattened body, a long sensitive tongue, and adhesive pads on the toes of many species.

gel [CHEM] A two-phase colloidal system consisting of a solid and a liquid in more solid form than a sol.

Gelastocoridae [INV ZOO] The toad bugs, a family of tropical and subtropical hemipteran insects in the subdivision Hydrocorisae.

gelatin [ORG CHEM] A protein derived from the skin, white connective tissue, and bones of animals; used as a food and in photography, the plastics industry, metallurgy, and pharmaceuticals.

gelatinase [BIOCHEM] An enzyme, found in some yeasts and molds, that liquefies gelatin.

gelatin liquefaction [MICROBIO] Reduction of a gelatin culture medium to the liquid state by enzymes produced by bacteria in a stab culture; used in identifying bacteria.

Gelechiidae [INV ZOO] A large family of minute to small moths in the lepidopteran superfamily Tineoidea, generally having forewings and trapezoidal hindwings.

Gelocidae [PALEON] A family of extinct pecoran ruminants in the superfamily Traguloidea.

gel permeation chromatography [ANALY CHEM] Analysis by chromatography in which the stationary phase consists of beads of porous polymeric material such as a cross-linked

GAVIAL

Indian gavial, fierce looking but not dangerous to man.

GECKO

Banded gecko, which is brown with broad yellow bands, and is economically important as a destroyer of insects.

dextran carbohydrate derivative sold under the trade name Sephadex; the moving phase is a liquid.

geminate [BIOL] Growing in pairs or couples.

geminiflorous [BOT] Having flowers in pairs.

gemma [BOT] A small, multicellular, asexual reproductive body of some liverworts and mosses.

gemmiform [BOT] Resembling a gemma or bud.

gemmiparous [BIOL] Producing a bud or reproducing by a bud.

gemmule [ANAT] A minute dendritic process functioning as a synaptic contact point. [BIOL] Any bud formed by gemmation. [INV ZOO] A cystlike, asexual reproductive structure of many Porifera that germinates when proper environmental conditions exist; it is a protective, overwintering structure which germinates the following spring.

Gempylidae [VERT ZOO] The snake mackerels, a family of the suborder Scombroidei comprising compressed, elongate, or eel-shaped spiny-rayed fishes with caniniform teeth.

Gemuendinoidei [PALEON] A suborder of extinct raylike placoderm fishes in the order Rhenanida.

gena [ANAT] Cheek, or side of the head.

gender identity [PSYCH] The sum of those aspects of personal appearance and behavior culturally attributed to masculinity or femininity.

gene [GEN] The basic unit of inheritance.

gene action [GEN] The functioning of a gene in determining the phenotype of an individual.

gene conversion [GEN] A situation in which gametocytes of an individual that is heterozygous for a pair of alleles undergo meiosis, and the gametes produced are in a 3:1 ratio rather than the expected 2:2 ratio, implying that one allele was converted to the other.

gene flow [GEN] The passage and establishment of genes characteristic of a breeding population into the gene complex of another population through hybridization and backcrossing.

gene frequency [GEN] The number of occurrences of a specific gene within a population.

gene penetrance *See* penetrance.

gene pool [GEN] The total genetic information that a population has.

general adaptation syndrome [PHYSIO] The sum of all nonspecific physiological reactions to prolonged systemic stress; it is divided into three stages: alarm reaction, resistance, and exhaustion.

general anesthesia [MED] Loss of sensation with loss of consciousness, produced by administration of anesthetic drugs.

general paresis [MED] An inflammatory and degenerative disease of the brain caused by infection with *Treponema pallidum*. Also known as syphilitic meningoencephalitis.

generation [BIOL] A group of organisms having a common parent or parents and comprising a single level in line of descent.

generation time [MICROBIO] The time interval required for a bacterial cell to divide or for the population to double.

generator potential [PHYSIO] Depolarization of a receptor cell when acted upon by a physical stimulus.

gene redundancy [GEN] The presence of many copies of one gene within a cell.

Gene's organ [INV ZOO] In ticks, a subscutal or cephalic gland which secretes a viscid substance in which eggs are transferred to the dorsal surface.

gene suppression [GEN] The development of a normal phenotype in a mutant individual or cell due to a second mutation either in the same gene or in a different gene.

genetic analysis [GEN] Resolution of the genetic material into its component elements; in decreasing order of size: the nucleus, the chromosomes, the chromosome arms, the operons, the genes, the codons, and the individual nucleotide pairs.

genetic code [MOL BIO] The genetic information in the nucleotide sequences in deoxyribonucleic acid represented by a four-letter alphabet that makes up a vocabulary of 64 three-nucleotide sequences, or codons; a sequence of such codons (averaging about 100 codons) constructs a message for a polypeptide chain.

genetic drift [GEN] The random fluctuation of gene frequencies from generation to generation that occurs in small populations.

genetic engineering [GEN] The intentional production of new genes and alteration of genomes by the substitution or addition of new genetic material.

genetic fingerprinting [MOL BIO] Identification of chemical entities in animal tissues as indicative of the presence of specific genes. Also known as fingerprinting.

genetic homeostasis [GEN] The tendency of Mendelian populations to maintain a constant genetic composition.

genetic load [GEN] The abnormalities, deformities, and deaths produced in every generation by defective genetic material carried in the gene pool of the human race.

genetic map [GEN] A graphic presentation of the linear arrangement of genes on a chromosome; gene positions are determined by percentages of recombination in linkage experiments. Also known as chromosome map.

genetics [BIOL] The science that is concerned with the study of biological inheritance.

genial *See* mental.

genic hybrid sterility [GEN] Sterility resulting from the interaction of genes in a hybrid to cause disturbances of sex-cell formation or meiosis.

geniculate [SCI TECH] Bent abruptly at an angle, as a bent knee.

geniculate body [ANAT] Any of the four oval, flattened prominences on the posterior inferior aspect of the thalamus; functions as the synaptic center for fibers leading to the cerebral cortex.

geniculate ganglion [ANAT] A mass of sensory and sympathetic nerve cells located along the facial nerve.

geniculum [ANAT] **1.** A small, kneelike, anatomical structure. **2.** A sharp bend in any small organ.

genioglossus [ANAT] An extrinsic muscle of the tongue, arising from the superior mental spine of the mandible.

Geniohyidae [PALEON] A family of extinct ungulate mammals in the order Hyracoidea; all members were medium to large-sized animals with long snouts.

genital atrium [ZOO] A common chamber receiving openings of male, female, and accessory organs.

genital coelom [INV ZOO] In mollusks, the lumina of the gonads.

genital cord [EMBRYO] A mesenchymal shelf bridging the coeloms in mammalian embryos, produced by fusion of the caudal part of the urogenital folds; fuses with the urinary bladder in the male, and is the primordium of the broad ligament and the uterine walls in the female. [INV ZOO] Strands of cells located in the genital canal which are primordial sex cells in crinoids. Also known as genital rachis.

genitalia [ANAT] The organs of reproduction, especially those which are external.

genital orifice *See* genital pore.

genital pore [INV ZOO] A small opening on the side of the head in some gastropods through which the penis is protruded. Also known as genital orifice.

genital rachis *See* genital cord.

genital recess [INV ZOO] A depression between the calyx surface and anal cone in entoprocts which serves as a brood chamber.

genital ridge [EMBRYO] A medial ridge or fold on the ventromedial surface of the mesonephros in the embryo, produced by growth of the peritoneum; the primordium of the gonads and their ligaments.

genital scale [INV ZOO] Any of the small calcareous plates in ophiuroids associated with the buccal shields.

genital segment *See* gonosomite.

genital shield [INV ZOO] In ophiuroids, a support of a bursal slit in the arms located near the arm base.

genital stage [PSYCH] In psychoanalytic theory, the adult personality stage, beginning around 12 years of age and characterized by expression of heterosexual interests.

genital stolon [INV ZOO] Part of the axial complex in ophiuroids.

genital sucker [INV ZOO] In some trematodes, a suckerlike structure surrounding the gonopore.

genital tract [ANAT] The ducts of the reproductive system.

genital tube [INV ZOO] A blood lacuna in crinoids, connected with the subtegminal plexus and suspended in the genital canal.

genitourinary system *See* urogenital system.

genome [GEN] **1.** The genetic endowment of a species. **2.** The haploid set of chromosomes.

genotype [GEN] The genetic constitution of an organism, usually in respect to one gene or a few genes relevant in a particular context. [SYST] The type species of a genus.

gentamicin [MICROBIO] A broad-spectrum antibiotic produced by a species of *Micromonospora.*

Gentianaceae [BOT] A family of dicotyledonous herbaceous plants in the order Gentianales distinguished by lacking stipules and having parietal placentation.

Gentianales [BOT] A family of dicotyledonous plants in the subclass Asteridae having well-developed internal phloem and opposite, simple, mostly entire leaves.

genu *See* knee.

genus [SYST] A taxonomic category that includes groups of closely related species; the principal subdivision of a family.

geobotany [BOT] The study of plants as related to their geologic environment.

Geocorisae [INV ZOO] A subdivision of hemipteran insects containing those land bugs with conspicuous antennae and an ejaculatory bulb in the male.

geocryptophyte [BOT] A plant with underground dormant structures.

geographical botany *See* plant geography.

geographic speciation [EVOL] Evolution of two or more species from a single species following geographic isolation.

geologic age [GEOL] **1.** Any great time period in the earth's history marked by special phases of physical conditions or organic development. **2.** A formal geologic unit of time that corresponds to a stage. **3.** An informal geologic time unit that corresponds to any stratigraphic unit.

geologist [GEOL] An individual who specializes in the geological sciences.

geology [SCI TECH] The study or science of the earth, its history, and its life as recorded in the rocks; includes the study of geologic features of an area, such as the geometry of rock formations, weathering and erosion, and sedimentation.

Geometridae [INV ZOO] A large family of lepidopteran insects in the superfamily Geometroidea that have slender bodies and relatively broad wings; includes measuring worms, loopers, and cankerworms.

Geometroidea [INV ZOO] A superfamily of lepidopteran insects in the suborder Heteroneura comprising small to large moths with reduced maxillary palpi and tympanal organs at the base of the abdomen.

Geomyidae [VERT ZOO] The pocket gophers, a family of rodents characterized by fur-lined cheek pouches which open outward, a stout body with short legs, and a broad, blunt head.

geophagous [ZOO] Feeding on soil, as certain worms.

Geophilomorpha [INV ZOO] An order of centipedes in the class Chilopoda including specialized forms that are blind, epimorphic, and dorsoventrally flattened.

geophilous [ECOL] Living or growing in or on the ground.

geophyte [ECOL] A perennial plant that is deeply embedded in the soil substrata.

Georyssidae [INV ZOO] The minute mud-loving beetles, a family of coleopteran insects belonging to the Polyphaga.

Geosiridaceae [BOT] A monotypic family of monocotyledonous plants in the order Orchidales characterized by regular flowers with three stamens that are opposite the sepals.

geosphere [GEOL] **1.** The solid mass of earth, as distinct from the atmosphere and hydrosphere. **2.** The lithosphere, hydrosphere, and atmosphere combined.

Geospizinae [VERT ZOO] Darwin finches, a subfamily of perching birds in the family Fringillidae.

geotaxis [PHYSIO] Movement of a free-living organism in response to the stimulus of gravity.

geotropism [BOT] Response of a plant to the force of gravity.

Gephyrea [INV ZOO] A class of burrowing worms in the phylum Annelida.

gephyrocercal [VERT ZOO] Having the dorsal and anal fins coming together smoothly at the aborted end of the vertebral column of a fish's tail.

Geraniaceae [BOT] A family of dicotyledonous plants in the order Geraniales in which the fruit is beaked, styles are usually united, and the leaves have stipules.

Geraniales [BOT] An order of dicotyledonous plants in the subclass Rosidae comprising herbs or soft shrubs with a superior ovary and with compound or deeply cleft leaves.

Gerardiidae [INV ZOO] A family of anthozoans in the order Zoanthidea.

gerbil [VERT ZOO] The common name for about 100 species of African and Asian rodents composing the subfamily Gerbillinae.

Gerbillinae [VERT ZOO] The gerbils, a subfamily of rodents in the family Muridae characterized by hindlegs that are longer than the front ones, and a long, slightly haired, usually tufted tail.

geriatrics [MED] The study of the biological and physical changes and the diseases of old age.

germ [BIOL] A primary source, especially one from which growth and development are expected. [MICROBIO] General designation for a microorganism.

germanium [CHEM] A brittle, water-insoluble, silvery-gray metallic element in the carbon family, symbol Ge, atomic number 32, atomic weight 72.59, melting at 959°C.

German measles *See* rubella.

germarium [INV ZOO] The egg-producing portion of an ovary and the sperm-producing portion of a testis in Platyhelminthes and Rotifera.

GERANIALES

A common eastern United States species of geranium *(Geranium maculatum)*, which is characteristic of the order Geraniales. *(Courtesy of A. W. Ambler, from National Audubon Society)*

GERBIL

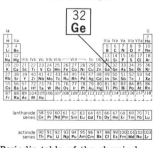

A gerbil, with long tail for balance when it hops.

GERMANIUM

Periodic table of the chemical elements showing the position of germanium.

germ ball [INV ZOO] A group of cells in digenetic trematode miracidial larvae which are embryos.

germ cell [BIOL] An egg or sperm or one of their antecedent cells.

germfree animal [MICROBIO] An animal having no demonstrable, viable microorganisms living in intimate association with it.

germfree isolator [MICROBIO] An apparatus that provides a mechanical barrier surrounding the area in which germfree vertebrates and accessory equipment are housed.

germicide [MATER] An agent that destroys germs.

germiduct [INV ZOO] The oviduct of trematodes.

germigen [INV ZOO] The ovary of trematodes.

germinal epithelium [EMBRYO] The region of the dorsal coelomic epithelium lying between the dorsal mesentery and the mesonephros.

germinal lid [BOT] The operculum of a pollen grain.

germinal vesicle [CYTOL] The enlarged nucleus of the primary oocyte before reduction divisions are complete.

germination [BOT] The beginning or the process of development of a spore or seed.

germ layer [EMBRYO] One of the primitive cell layers which appear in the early animal embryo and from which the embryo body is constructed.

germ-layer theory [EMBRYO] The theory that three primary germ layers, ectoderm, mesoderm, and endoderm, are established in the early embryo and all organs and structures are derived from a specific germ layer.

germovitellarium [INV ZOO] A sex gland which differentiates into a yolk-producing or egg-producing region.

germ theory [MED] The theory that contagious and infectious diseases are caused by microorganisms.

geroderma [MED] The skin of old age, showing atrophy, loss of fat, and loss of elasticity.

gerontology [PHYSIO] The scientific study of aging processes in biological systems, particularly in humans.

Gerrhosauridae [VERT ZOO] A small family of lizards in the suborder Sauria confined to Africa and Madagascar.

Gerridae [INV ZOO] The water striders, a family of hemipteran insects in the subdivision Amphibicorisae having long middle and hind legs and a median scent gland opening on the metasternum.

Gerroidea [INV ZOO] The single superfamily of the hemipteran subdivision Amphibicorisae; all members have conspicuous antennae and hydrofuge hairs covering the body.

Gesneriaceae [BOT] A family of dicotyledonous plants in the order Scrophulariales characterized by parietal placentation, mostly opposite or whorled leaves, and a well-developed embryo.

Gestalt psychology [PSYCH] A school of psychology that views and examines the person as a whole.

gestate [EMBRYO] To carry the young in the uterus from conception to delivery.

gestation period [EMBRYO] The period in mammals from fertilization to birth.

GH *See* growth hormone.

Ghon complex [MED] The combination of a focus of subpleural tuberculosis with associated hilar and mediastinal lymph node tuberculosis.

ghost spot [PL PATH] A disease of tomato characterized by small white rings on the fruit.

giant-cell arteritis [MED] Inflammation of the arteries, particularly the carotid branches, characterized by the appearance of multinucleate giant cells in the exudate. Also known as temporal arteritis.

giant-cell leukemia *See* megakaryocytic leukemia.

giardiasis [MED] Presence of the protozoon *Giardia lamblia* in the human small intestine.

gibberellic acid [BIOCHEM] $C_{18}H_{22}O_6$ A crystalline acid occurring in plants that is similar to the gibberellins in its growth-promoting effects.

gibberellin [BIOCHEM] Any member of a family of naturally derived compounds which have a gibbane skeleton and a broad spectrum of biological activity but are noted as plant growth regulators.

gibbon [VERT ZOO] The common name for seven species of large, tailless primates belonging to the genus *Hylobates*; the face and ears are hairless, and the arms are longer than the legs.

Gibbs free energy [THERMO] The thermodynamic function $G = H - TS$, where H is enthalpy, T absolute temperature, and S entropy; Also known as free energy; free enthalpy; Gibbs function.

Gibbs function *See* Gibbs free energy.

Gibbs-Helmholtz equation [PHYS CHEM] An expression for the influence of temperature upon the equilibrium constant of a chemical reaction, $(d \ln K^0 / dT)_P = \Delta H^0 / RT^2$, where K^0 is the equilibrium constant, ΔH^0 the standard heat of the reaction at the absolute temperature T, and R the gas constant.

gid [VET MED] A chronic brain disease of sheep, less frequently of cattle, characterized by forced movements of circling or rolling, caused by the larval form of the tapeworm *Multiceps multiceps*.

Giemsa stain [CHEM] A stain for hemopoietic tissue and hemoprotozoa consisting of a stock glycerol methanol solution of eosinates of Azure B and methylene blue with some excess of the basic dyes.

gigantism [MED] Abnormal largeness of the body due to hypersecretion of growth hormone.

Giganturoidei [VERT ZOO] A suborder of small, mesopelagic actinopterygian fishes in the order Cetomimiformes having large mouths and strong teeth.

Gila monster [VERT ZOO] The common name for two species of reptiles in the genus *Heloderma* (Helodermatidae) distinguished by a rounded body that is covered with multicolored beaded tubercles, and a bifid protrusible tongue.

gill [VERT ZOO] The respiratory organ of water-breathing animals. Also known as branchia.

gill basket [VERT ZOO] The cartilaginous branchial skeleton of lampreys.

gill cover [VERT ZOO] The fold of skin providing external protection for the gill apparatus of most fishes; it may be stiffened by bony plates and covered with scales.

gill plume [INV ZOO] The gill of most gastropods.

gill raker [VERT ZOO] One of the bony processes on the inside of the branchial arches of fishes which prevents the passage of solid substances through the branchial clefts.

gill rod [VERT ZOO] **1.** An oblique, gelatinous rod that supports the pharynx in cephalochordates. **2.** A branchial ray in some fishes.

gill slit [VERT ZOO] One of a series of external openings of the gill cavity leading from the pharynx to the exterior, when a gill cover is present.

gin [AGR] **1.** A machine used to separate cotton fiber from the seed and waste. **2.** To thus separate cotton fiber.

ginger [BOT] *Zingiber officinale.* An erect perennial herb of the family Zingiberaceae having thick, scaly branched rhizomes; a spice oleoresin is made by an organic solvent extraction of the ground dried rhizome.

GIBBON

Hylobates lar, which is a typical gibbon with extremely long arms and opposable thumb and big toe, and is found in Sumatra and southern Asia.

GILA MONSTER

Gila monster *(Heloderma suspectum),* about 20 inches (50 centimeters) long.

GINSENG

Panax quinquefolius, a ginseng, showing shoot and root.

GIRAFFE

The giraffe of Africa, with greater height in the forepart of the body because of the heavy muscular development at the base of the neck.

gingiva [ANAT] The mucous membrane surrounding the teeth sockets.

gingival crevice [ANAT] The space between the free margin of the gingiva and the surface of a tooth. Also known as gingival sulcus.

gingival sulcus *See* gingival crevice.

gingivectomy [MED] Excision of a portion of the gingiva.

gingivitis [MED] Inflammation of the gingiva.

gingivostomatitis [MED] An inflammation of the gingiva and oral mucosa.

ginglymoarthrodia [ANAT] A composite joint consisting of one hinged and one gliding element.

Ginglymodi [VERT ZOO] An equivalent name for Semionotiformes.

ginglymoid [ANAT] Having the form of a hinge joint.

ginglymus [ANAT] A type of diarthrosis permitting motion only in one plane. Also known as hinge joint.

Ginkgoales [BOT] An order of gymnosperms composing the class Ginkgoopsida with one living species, the dioecious maidenhair tree (*Ginkgo biloba*).

Ginkgoopsida [BOT] A class of the subdivision Pinicae containing the single, monotypic order Ginkgoales.

Ginkgophyta [BOT] The equivalent name for Ginkgoopsida.

ginseng [BOT] The common name for plants of the genus *Panax*, a group of perennial herbs in the family Araliaceae; the aromatic root of the plant has been used medicinally in China.

giraffe [VERT ZOO] *Giraffa camelopardalis.* An artiodactyl mammal in the family Giraffidae characterized by extreme elongation of the neck vertebrae, and two prominent horns on the head.

Giraffidae [VERT ZOO] A family of pecoran ruminants in the superfamily Bovoidea including giraffe, okapi, and relatives.

girdle [ANAT] Either of the ringlike groups of bones supporting the forelimbs (arms) and hindlimbs (legs) in vertebrates. [INV ZOO] **1.** Either of the hooplike bands constituting the sides of the two valves of a diatom. **2.** The peripheral portion of the mantle in chitons.

gitoxin [PHARM] $C_{41}H_{64}O_{14}$ A secondary glycoside derived from *Digitalis purpurea,* crystallizes in stout prisms from chloroform methanol solution, soluble in a mixture of chloroform and alcohol; used in medicine for coronary disease.

gitter cell [PATH] A compound granule cell that is characteristic of certain brain lesions.

gizzard [VERT ZOO] The muscular portion of the stomach of most birds where food is ground with the aid of ingested pebbles.

glabella [ANAT] The bony prominence on the frontal bone joining the supraorbital ridges.

glabello-occipital length [ANTHRO] The distance between the glabella and the opisthocranion.

glabrous [BIOL] Having a smooth surface; specifically, having the epidermis devoid of hair or down.

gladiate [BOT] Sword-shaped.

gladiolus *See* mesosternum.

gland [ANAT] A structure which produces a substance essential and vital to the existence of the organism.

glanders [VET MED] A bacterial disease of equines caused by *Actinobacillus mallei*; involves the respiratory system, skin, and lymphatics. Also known as farcy.

glands of Brunner *See* Brunner's glands.

glands of Leydig [VERT ZOO] Unicellular, epidermal structures of urodele larvae and the adult *Necturus* that secrete a substance which digests the egg capsule and permits hatching.

glandular fever *See* infectious mononucleosis.

glans [ANAT] The conical body forming the distal end of the clitoris or penis. [BOT] **1.** A nut. **2.** A hard, dry, indehiscent one-celled fruit, such as an acorn.

Glareolidae [VERT ZOO] A family of birds in the order Charadriiformes including the ploverlike coursers and the swallowlike pratincoles.

glareous [ECOL] Growing in gravelly soil; refers specifically to plants.

Glasser's disease [VET MED] A generalized bacterial infection of swine caused by *Mycoplasma hyorhinis.*

glass sponge [INV ZOO] A siliceous sponge belonging to the class Hyalospongiae.

glaucoma [MED] A disease of the eye characterized by increased fluid pressure within the eyeball.

glaucous [BOT] Having a white or grayish powdery coating that gives a frosty appearance and rubs off easily.

gleba [MYCOL] The central, sporogenous tissue of the sporophore in certain basidiomycetous fungi.

glebula [BOT] A small prominence on the thallus of a lichen.

gleet [MED] The chronic stage of gonorrheal urethritis, characterized by a slight mucopurulent discharge.

glenoid [ANAT] A smooth, shallow, socketlike depression, particularly of the skeleton.

glenoid cavity [ANAT] The articular surface on the scapula for articulation with the head of the humerus.

gliding bacteria [MICROBIO] The descriptive term for members of the orders Beggiatoales and Myxobacterales; they are motile by means of creeping movements.

gliding joint *See* arthrodia.

gliding motility [MICROBIO] A means of bacterial self-propulsion by slow gliding or creeping movements on the surface of a substrate.

glioma [MED] A malignant tumor derived from the supporting tissue of the central nervous system.

gliosis [MED] Proliferation of neuroglia in the brain or spinal cord, either as a replacement process or in response to a low-grade inflammation.

gliotoxin [MICROBIO] $C_{13}H_{14}O_4N_2S_2$ A heat-labile, bacteriostatic antibiotic produced by species of *Trichoderma* and *Cliocladium* and by *Aspergillus fumigatus.*

Gliridae [VERT ZOO] The dormice, a family of mammals in the order Rodentia.

Glisson's capsule [ANAT] The membranous sheet of collagenous and elastic fibers covering the liver.

Globigerinacea [INV ZOO] A superfamily of foraminiferan protozoans in the suborder Rotaliina characterized by a radial calcite test with bilamellar septa and a large aperture.

globin [BIOCHEM] Any of a class of soluble histone proteins obtained from animal hemoglobins.

globin zinc insulin [PHARM] A preparation of insulin modified by the addition of globin (derived from the hemoglobin of beef blood) and zinc chloride; it has intermediate duration of action.

globoside [BIOCHEM] A glycoside of ceramide containing several sugar residues, but not neuraminic acid; obtained from human, sheep, and hog erythrocytes.

globular protein [BIOCHEM] Any protein that is readily soluble in aqueous solvents.

globulin [BIOCHEM] A heat-labile serum protein precipitated by 50% saturated ammonium sulfate and soluble in dilute salt solutions.

glochid *See* glochidium.

glochidium [BOT] A barbed hair. Also known as glochid. [INV ZOO] The larva of fresh-water mussels in the family Unionidae.

GLOBIGERINACEA

Scanning electron micrograph of *Globigerinoides* from the Holocene of the Caribbean Sea. *(R. B. MacAdam, Chevron Oil Field Research Co.)*

gloea [INV ZOO] An adhesive mucoid substance secreted by certain protozoans and other lower organisms.

glomerule [BOT] A condensed or sessile cyme resembling a composite flower.

glomerulonephritis [MED] Inflammation of the kidney, primarily involving the glomeruli.

glomerulosclerosis [MED] Fibrosis of the renal glomeruli.

glomerulus [ANAT] A tuft of capillary loops projecting into the lumen of a renal corpuscle.

glomus [ANAT] **1.** A fold of the mesothelium arising near the base of the mesentery in the pronephros and containing a ball of blood vessels. **2.** A prominent portion of the choroid plexus of the lateral ventricle of the brain.

glomus aorticum *See* paraaortic body.

glomus caroticum *See* carotid body.

glossa [INV ZOO] A tongue or tonguelike structure in insects, especially the median projection of the labium.

glossalgia [MED] Pain in the tongue.

glossarium [INV ZOO] A thin, pointed glossa, found in certain Diptera.

glossate [INV ZOO] Having a glossa or tonguelike structure.

Glossinidae [INV ZOO] The tsetse flies, a family of cyclorrhaphous dipteran insects in the section Pupipara.

Glossiphoniidae [INV ZOO] A family of small leeches with flattened bodies in the order Rhynchobdellae.

glossitis [MED] Inflammation of the tongue.

glossopalatine nerve [ANAT] The intermediate branch of the facial nerve.

glossopharyngeal nerve [ANAT] The ninth cranial nerve in vertebrates; a paired mixed nerve that supplies autonomic innervation to the parotid gland and contains sensory fibers from the posterior one-third of the tongue and the anterior pharynx.

glossopyrosis [MED] Burning sensation of the tongue.

glossotheca [INV ZOO] That portion of the integument of insect pupae which covers the proboscis.

glottis [ANAT] The opening between the margins of the vocal folds.

glove-and-stocking anesthesia [MED] Loss or diminution of sensation in the hands and feet, corresponding to the areas covered by gloves and stockings.

glove anesthesia [MED] Loss or diminution of sensation in the hands, corresponding to the area covered by gloves.

glucagon [BIOCHEM] The protein hormone secreted by α-cells of the pancreas which plays a role in carbohydrate metabolism. Also known as hyperglycemic factor; hyperglycemic glycogenolytic factor.

glucamine [BIOCHEM] $C_6H_{15}O_4N$ An amine formed by reduction of glucosylamine or of glucose oxime.

glucan [BIOCHEM] A polysaccharide composed of the hexose sugar D-glucose.

glucocerebroside [BIOCHEM] A glycoside of ceramide that contains glucose.

glucocorticoid [BIOCHEM] A corticoid that affects glucose metabolism; secreted principally by the adrenal cortex.

glucogenesis [BIOCHEM] Formation of glucose within the animal body from products of glycolysis.

glucokinase [BIOCHEM] An enzyme that catalyzes the phosphorylation of D-glucose to glucose-6-phosphate.

glucolipid [BIOCHEM] A glycolipid that yields glucose on hydrolysis.

glucomannan [BIOCHEM] A polysaccharide composed of D-glucose and D-mannose; a prominent component of coniferous trees.

gluconeogenesis [BIOCHEM] Formation of glucose within

GLOSSIPHONIIDAE

Glossiphonia complanata, a small leech that sucks blood of aquatic invertebrates, especially snails.

the animal body from substances other than carbohydrates, particularly proteins and fats.

D-glucopyranose *See* glucose.

glucopyranoside [BIOCHEM] Any glucoside that contains a six-membered ring.

glucosamine [BIOCHEM] $C_6H_{13}O_5$ An amino sugar; the most abundant in nature, occurring in glycoproteins and chitin.

glucose [BIOCHEM] $C_6H_{12}O_6$ A monosaccharide; occurs free or combined and is the most common sugar. Also known as cerelose; D-glucopyranose.

glucose phosphate [BIOCHEM] A phosphoric derivative of glucose, as glucose-1-phosphate.

glucose-1-phosphate [BIOCHEM] $C_6H_{12}O_8P$ An ester of glucopyranose in which a phosphate group is attached to carbon atom 1; there are two types: α-D- and β-D-glucose-1-phosphates. Also known as Cori ester.

glucose-6-phosphate [BIOCHEM] $C_6H_{13}O_9P$ An ester of glucose with phosphate attached to carbon atom 6. Also known as Robisonester.

glucose-6-phosphate dehydrogenase [BIOCHEM] The mammalian enzyme that catalyzes the oxidation of glucose-6-phosphate by TPN^+ (triphosphopyridine nucleotide).

glucose tolerance test [PATH] A test to measure the ability of the liver to convert glucose to glycogen.

glucosidase [BIOCHEM] An enzyme that hydrolyzes glucosides.

glucoside [BIOCHEM] One of a group of compounds containing the cyclic forms of glucose, in which the hydrogen of the hemiacetal hydroxyl has been replaced by an alkyl or aryl group.

glucosulfone sodium *See* sodium glucosulfone.

glucosyltransferase [BIOCHEM] An enzyme that catalyzes the glucosylation of hydroxymethyl cytosine; a constituent of bacteriophage deoxyribonucleic acid.

glucuronic acid [BIOCHEM] $C_6H_{10}O_7$ An acid resulting from oxidation of the CH_2OH radical of D-glucose to COOH; a component of many polysaccharides and certain vegetable gums. Also known as glycuronic acid.

glucuronidase [BIOCHEM] An enzyme that catalyzes hydrolysis of glucuronides. Also known as glycuronidase.

glucuronide [BIOCHEM] A compound resulting from the interaction of glucuronic acid with a phenol, an alcohol, or an acid containing a carboxyl group. Also known as glycuronide.

D-glucuronolactone [BIOCHEM] $C_6H_8O_6$ A water-soluble crystalline compound found in plant gums in polymers with other carbohydrates, and an important structural component of almost all fibrous and connective tissues in animals; used in medicine as an antiarthritic.

glue cell *See* adhesive cell.

glume [BOT] One of two bracts at the base of a spikelet of grass.

glumiferous [BOT] Bearing glumes.

Glumiflorae [BOT] An equivalent name for Cyperales.

glumiflorous [BOT] Having flowers with basal glumes or bracts.

glutamate [BIOCHEM] A salt or ester of glutamic acid.

glutamic acid [BIOCHEM] $C_5H_9O_4N$ A dicarboxylic amino acid of the α-ketoglutaric acid family occurring widely in proteins.

glutaminase [BIOCHEM] The enzyme which catalyzes the conversion of glutamine to glutamic acid and ammonia.

glutamine [BIOCHEM] $C_5H_{10}O_3N_2$ An amino acid; the

GLUCOSE-6-PHOSPHATE

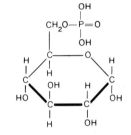

Structural formula for glucose-6-phosphate.

GLUCOSIDE

α-D-Glucoside

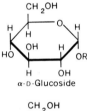

β-D-Glucoside

Structural formulas of two forms of glucoside.

GLUTAMIC ACID

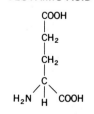

Structural formula of glutamic acid.

monamide of glutamic acid; found in the juice of many plants and essential to the development of certain bacteria.

glutarate [BIOCHEM] The salt or ester of glutaric acid.

glutaric acid [BIOCHEM] $C_5H_5O_4$ A water-soluble, crystalline acid that occurs in green sugarbeets and in water extracts of crude wool.

glutathione [BIOCHEM] $C_{10}H_{17}O_6N_3S$ A widely distributed tripeptide that is important in plant and animal tissue oxidation reactions.

glutelin [BIOCHEM] A class of simple, heat-labile proteins occurring in seeds of cereals; soluble in dilute acids and alkalies.

gluten [BIOCHEM] **1.** A mixture of proteins found in the seeds of cereals; gives dough elasticity and cohesiveness. **2.** An albuminous element of animal tissue.

glutenin [BIOCHEM] A glutelin of wheat.

glutethimide [PHARM] $C_{13}H_{15}NO_2$ A minor or sedative antianxiety tranquilizer that acts as a central nervous system depressant.

gluteus maximus [ANAT] The largest and most superficial muscle of the buttocks.

gluteus medius [ANAT] The muscle of the buttocks lying between the gluteus maximus and gluteus minimus.

gluteus minimus [ANAT] The smallest and deepest muscle of the buttocks.

glutinous [BOT] Having a sticky surface.

glycemia [PHYSIO] The presence of glucose in the blood.

glyceraldehyde [BIOCHEM] $CH_2OHCOHCHO$ A colorless solid, isomeric with dehydroxyacetone; soluble in water and insoluble in organic solvents; an important intermediate in carbohydrate metabolism; used as a chemical intermediate in biochemical research and nutrition. Also known as 2,3-dihydroxypropanal; glyceric aldehyde.

glycerate [BIOCHEM] A salt or ester of glyceric acid.

glyceric acid [BIOCHEM] $C_3H_6O_4$ A hydroxy acid obtained by oxidation of glycerin.

glyceric aldehyde *See* glyceraldehyde.

Glyceridae [INV ZOO] A family of polychaete annelids belonging to the Errantia and characterized by an enormous eversible proboscis.

glyceride [BIOCHEM] An ester of glycerin and an organic acid radical; fats are glycerides of certain long-chain fatty acids.

glycerinated vaccine virus *See* smallpox vaccine.

glycerokinase [BIOCHEM] An enzyme that catalyzes the phosphorylation of glycerol to glycerophosphate during microbial fermentation of propionic acid.

glycerophosphate [BIOCHEM] Any salt of glycerophosphoric acid.

glycerophosphoric acid [BIOCHEM] $C_3H_5(OH)_2OPO_3H_2$ Either of two pale-yellow, water-soluble, isomeric dibasic acids occurring in nature in combined form as cephalin and lecithin.

glycine [BIOCHEM] $C_2H_5O_2N$ A white, crystalline amino acid found as a constituent of many proteins. Also known as aminoacetic acid.

glycocalyx [CYTOL] The outer component of a cell surface, outside the plasmalemma; usually contains strongly acidic sugars, hence it carries a negative electric charge.

glycocholic acid [BIOCHEM] $C_{26}H_{43}NO_6$ A bile obtained by the conjugation of cholic acid with glycine.

glycocyamine [BIOCHEM] $C_3H_7N_3O_2$ A product of interaction of aminocetic acid and arginine, which on transmethylation with methionine is converted to creatine. Also known as guanidine-acetic acid.

glycogen [BIOCHEM] A nonreducing, white, amorphous

GLYCINE

Structural formula of glycine.

polysaccharide found as a reserve carbohydrate stored in muscle and liver cells of all higher animals, as well as in cells of lower animals.

glycogenesis [BIOCHEM] The metabolic formation of glycogen from glucose.

glycogenolysis [BIOCHEM] The metabolic breakdown of glycogen.

glycogenosis [MED] One of several inborn errors in the metabolism of glycogen, classified on the basis of the enzyme deficiency and clinical findings as von Gierke's disease, Pompe's disease, limit dextrinosis, amylopectinosis, McArdle's disease, or Hers' disease.

glycogen storage disease See von Gierke's disease.

glycogen synthetase [BIOCHEM] An enzyme that catalyzes the synthesis of the amylose chain of glycogen.

glycolipid [BIOCHEM] Any of a class of complex lipids which contain carbohydrate residues.

glycolysis [BIOCHEM] The enzymatic breakdown of glucose or other carbohydrate, with the formation of lactic acid or pyruvic acid and the release of energy in the form of adenosinetriphosphate.

glycolytic pathway [BIOCHEM] The principal series of phosphorylative reactions involved in pyruvic acid production in phosphorylative fermentations. Also known as Embden-Meyerhof pathway; hexose diphosphate pathway.

glyconeogenesis [BIOCHEM] The metabolic process of glycogen formation from noncarbohydrate precursors.

glycopeptide See glycoprotein.

glycophyte [BOT] A plant requiring more than 0.5% sodium chloride solution in the substratum.

glycoprotein [BIOCHEM] Any of a class of conjugated proteins containing both carbohydrate and protein units. Also known as glycopeptide.

glycosidase [BIOCHEM] An enzyme that hydrolyzes a glycoside.

glycoside [BIOCHEM] A compound that yields on hydrolysis a sugar (glucose, galactose) and an aglycone; many of the glycosides are therapeutically valuable.

glycosuria [MED] Presence of sugar in the urine.

glycotropic [BIOCHEM] Acting to antagonize the action of insulin.

glycuresis [PHYSIO] Excretion of sugar seen normally in urine.

glycuronic acid See glucuronic acid.

glycuronidase See glucuronidase.

glycuronide See glucuronide.

glyoxalase [BIOCHEM] An enzyme present in various body tissues which catalyzes the conversion of methylglyoxal into lactic acid.

glyoxylic acid [BIOCHEM] $CH(OH)_2COOH$ An aldehyde acid found in many plant and animal tissues, especially unripe fruit.

Glyphocyphidae [PALEON] A family of extinct echinoderms in the order Temnopleuroida comprising small forms with a sculptured test, perforate crenulate tubercles, and diademoid ambulacral plates.

Glyptocrinina [PALEON] A suborder of extinct crinoids in the order Monobathrida.

gm See gram.

Gmelin's test [PATH] A qualitative test for the pigments in bile; test solution is mixed with nitric acid containing nitrous acid; reaction is positive if color appears at the acid-solution junction.

gnat [INV ZOO] The common name for a large variety of biting insects in the order Diptera.

Haemopis grandis, dorsal and ventral view showing the conspicous posterior sucker.

gnathic index [ANTHRO] The ratio of the distance from the nasion to the basion to that from the basion to the alveolar point multiplied by 100.

Gnathiidea [INV ZOO] A suborder of isopod crustaceans characterized by a much reduced second thoracomere, short antennules and antennae, and a straight pleon.

gnathion [ANTHRO] The midpoint of the lower margin of the mandible in humans. [VERT ZOO] The most anterior point of the premaxillae on or near the middle line in certain lower mammals.

gnathite [INV ZOO] A mouth appendage in arthropods.

gnathobase [INV ZOO] **1.** In Crustacea, an inverted masticatory process on the protopodite of appendages near the mouth. **2.** In Arachnoidea, the basal segment of an appendage, having spines turned toward the mouth.

Gnathobdellae [INV ZOO] A suborder of leeches in the order Arhynchobdellae having jaws and a conspicuous posterior sucker; contains most of the important blood-sucking leeches of humans and other warm-blooded animals.

Gnathobelodontinae [PALEON] A subfamily of extinct elephantoid proboscideans containing the shovel-jawed forms of the family Gomphotheriidae.

gnathocephalon [INV ZOO] The part of the insect head lying behind the protocephalon; bears the maxillae and mandibles.

gnathochilarium [INV ZOO] The lower lip of certain arthropods; thought to be fused maxillae.

Gnathodontidae [PALEON] A family of extinct conodonts having platforms with large, cup-shaped attachment scars.

gnathopod [INV ZOO] Any of the crustacean paired thoracic appendages modified for manipulation of food but sometimes functioning in copulatory amplexion.

gnathopodite [INV ZOO] A segmental, modified appendage which serves as a jaw in arthropods.

gnathos [INV ZOO] A mid-ventral plate on the ninth tergum in lepidopterans.

gnathosoma [INV ZOO] In Arachnoidea, the oral region, including appendages.

gnathostegite [INV ZOO] One of a pair of broad plates formed from the outer maxillipeds of some crustaceans, which function to cover other mouthparts.

Gnathostomata [INV ZOO] A suborder of echinoderms in the order Echinoidea characterized by a rigid, exocyclic test and a lantern or jaw apparatus. [VERT ZOO] A group of the subphylum Vertebrata which possess jaws and usually have paired appendages.

gnathostomatous [ZOO] Having jaws.

Gnathostomidae [INV ZOO] A family of parasitic nematodes in the superfamily Spiruroidea; sometimes placed in the superfamily Physalopteroidea.

Gnathostomulida [INV ZOO] Microscopic marine worms of uncertain systematic relationship, mainly characterized by cuticular structures in the pharynx and a monociliated skin epithelium.

gnathotheca [VERT ZOO] In birds, the horny covering of the lower jaw.

gnathothorax [INV ZOO] The thorax and part of the head bearing feeding organs in arthropods, regarded as a primary region of the body.

Gnetales [BOT] A monogeneric order of the subdivision Gneticae; most species are lianas with opposite, oval, entire-margined leaves.

Gnetatae *See* Gnetopsida.

Gneticae [BOT] A subdivision of the division Pinophyta characterized by vessels in the secondary wood, ovules with two integuments, opposite leaves, and an embryo with two cotyledons.

Gnetophyta [BOT] The equivalent name for Gnetopsida.

Gnetopsida [BOT] A class of gymnosperms comprising the subdivision Gneticae.

Gnostidae [INV ZOO] An equivalent name for Ptinidae.

gnotobiology [BIOL] That branch of biology dealing with known living forms; the study of higher organisms in the absence of all demonstrable, viable microorganisms except those known to be present.

gnotobiote [MICROBIO] An individual (host) living in intimate association with another known species (microorganism). **2.** The known microorganism living on a host.

gnu [VERT ZOO] Any of several large African antelopes of the genera *Connochaetes* and *Gorgon* having a large oxlike head with horns that characteristically curve downward and outward and then up, with the bases forming a frontal shield in older individuals.

goal [PSYCH] The object established by a motive.

goat [VERT ZOO] The common name for a number of artiodactyl mammals in the genus *Capra*; closely related to sheep but differing in having a lighter build and hollow, swept-back, sometimes spiral or twisted horns.

Gobiatheriinae [PALEON] A subfamily of extinct herbivorous mammals in the family Uintatheriidae known from one late Eocene genus; characterized by extreme reduction of anterior dentition and by lack of horns.

Gobiesocidae [VERT ZOO] The single family of the order Gobiesociformes.

Gobiesociformes [VERT ZOO] The clingfishes, a monofamilial order of scaleless bony fishes equipped with a thoracic sucking disk which serves for attachment.

Gobiidae [VERT ZOO] A family of perciform fishes in the suborder Gobioidei characterized by pelvic fins united to form a sucking disk on the breast.

Gobioidei [VERT ZOO] The gobies, a suborder of morphologically diverse actinopterygian fishes in the order Perciformes; all lack a lateral line.

goblet cell [HISTOL] A unicellular, mucus-secreting intraepithelial gland that is distended on the free surface. Also known as chalice cell. [INV ZOO] Any of the unicellular choanocytes of the genus *Monosiga*.

goiter [MED] An enlargement of all or part of the thyroid gland; may be accompanied by a hormonal dysfunction.

gold [CHEM] A chemical element, symbol Au, atomic number 79, atomic weight 196.967; soluble in aqua regia; melts at 1065°C.

gold-198 [NUC PHYS] The radioisotope of gold, with atomic number 198 and a half-life of 2.7 days; used in medical treatment of tumors by injecting it in colloidal form directly into tumor tissue.

golden algae [BOT] The common name for members of the class Chrysophyceae.

golden-brown algae [BOT] The common name for members of the division Chrysophyta.

goldfish [VERT ZOO] *Crassius auratus*. An orange cypriniform fish of the family Cyprinidae that can grow to over 18 inches; closely related to the carps.

gold sodium thiosulfate [PHARM] $Na_3Au(S_2O_3)_2 \cdot 2H_2O$ A white crystalline compound that is freely soluble in water; used for treatment of rheumatoid arthritis and lupus erythematosus.

Golgi apparatus [CYTOL] A cellular organelle that is part of the cytoplasmic membrane system; it is composed of regions of stacked cisternae and it functions in secretory processes.

GOLD

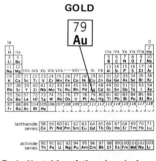

Periodic table of the chemical elements showing the position of gold.

GOLDFISH

Goldfish *(Crassius auratus)*, a common aquarium fish, may grow over 18 inches (46 centimeters) long.

Golgi cell [ANAT] **1.** A nerve cell with long axons. **2.** A nerve cell with short axons that branch repeatedly and terminate near the cell body.

Golgi-Mazzoni's corpuscle [ANAT] A small sensory lamellar corpuscle located in the parietal pleura.

Golgi tendon organ [PHYSIO] Any of the kinesthetic receptors situated near the junction of muscle fibers and a tendon which act as muscle-tension recorders.

Gomphidae [INV ZOO] A family of dragonflies belonging to the Anisoptera.

gomphosis [ANAT] An immovable articulation, as that formed by the insertion of teeth into the bony sockets.

Gomphotheriidae [PALEON] A family of extinct proboscidean mammals in the suborder Elephantoidea consisting of species with shoveling or digging specializations of the lower tusks.

Gomphotheriinae [PALEON] A subfamily of extinct elephantoid proboscideans in the family Gomphotheriidae containing species with long jaws and bunomastodont teeth.

gonad [ANAT] A primary sex gland; an ovary or a testis.

gonadal agenesis [MED] Failure of the gonad to develop, or retrogression of the gonad at very early stages. Also known as gonadal dysgenesis.

gonadal dysgenesis *See* gonadal agenesis.

gonadectomy [MED] Surgical removal of a gonad.

gonadotropic hormone [BIOCHEM] Either of two adenohypophyseal hormones, FSH (follicle-stimulating hormone) or ICSH (interstitial-cell-stimulating hormone), that act to stimulate the gonads.

gonadotropin [BIOCHEM] A substance that acts to stimulate the gonads.

gonapophysis [INV ZOO] A paired, modified appendage of the anal region in insects that functions in copulation, oviposition, or stinging.

Goniadidae [INV ZOO] A family of marine polychaete annelids belonging to the Errantia.

gonidiophore [BIOL] A specialized organ that supports a gonidium.

gonidiophyll [BOT] A specialized leaf that bears gonidia.

gonidium [BIOL] An asexual reproductive cell or group of cells arising in a special organ on or in a gametophyte.

gonimoblast [BOT] A filament arising from the fertilized carpogonium of most red algae.

gonion [ANAT] The tip of the angle of the mandible.

gonioscope [MED] A special optical instrument for studying in detail the angle of the anterior chamber of the eye.

gonitis [MED] Inflammation of the knee joint.

gonocalyx [INV ZOO] The bell of a gonophore in hydroids.

gonococcal arthritis [MED] A blood-borne joint infection by *Neisseria gonorrhoeae* occurring as a manifestation of gonorrhea.

gonococcal epididymitis [MED] Inflammation of the epididymitis due to infection by *Neisseria gonorrhoeae*; a secondary manifestation of gonorrhea.

gonococcus *See* Neisseria gonorrhoeae.

Gonodactylidae [INV ZOO] A family of mantis shrimp in the order Stomatopoda.

gonodendrum [INV ZOO] A branched structure which bears clusters of gonophores.

gonopalpon [INV ZOO] Tentaclelike, sensitive structures associated with cnidarian gonophores.

gonophore [BOT] An elongation of the receptacle extending between the stamens and corolla. [INV ZOO] Reproductive zooid of a hydroid colony.

GONIADIDAE

prostomium

Multiannulate prostomium of *Goniada* in dorsal view.

gonopodium [VERT ZOO] Anal fin modified as a copulatory organ in certain fishes.

Gonorhynchiformes [VERT ZOO] A small order of soft-rayed teleost fishes having fusiform or moderately compressed bodies, single short dorsal and anal fins, a forked caudal fin, and weak toothless jaws.

gonorrhea [MED] A bacterial infection of man caused by the gonococcus (*Neisseria gonorrhoeae*) which invades the mucous membrane of the urogenital tract.

gonorrheal urethritis [MED] Inflammation of the urethra, particularly in males, as the result of gonorrhea.

gonorrheal vulvovaginitis [MED] Inflammation of the vulva and vagina as the result of gonorrhea.

gonosome [INV ZOO] Aggregate of gonophores in a hydroid colony.

gonosomite [INV ZOO] The ninth segment of the abdomen of the male insect. Also known as genital segment.

gonostome [INV ZOO] The part of the genital duct of a coelomate invertebrate known as the coelomic funnel. Also known as coelomostome.

gonostyle [INV ZOO] Gonapophysis of dipteran insects.

gonotheca [INV ZOO] A hyaline protective extension of the perisarc which covers a blastostyle or a gonophore in certain hydroids.

gonotome [EMBRYO] The part of an embryonic somite involved in gonad formation.

gonotype [SYST] The direct offspring of a holotype.

gonozooid [INV ZOO] A zooid of bryozoans and tunicates which produces gametes.

gonys [VERT ZOO] A ridge along the mid-ventral line of the lower mandible of certain birds.

goodness of fit [STAT] The degree to which the observed frequencies of occurrence of events in an experiment correspond to the probabilities in a model of the experiment.

Goodpasture's syndrome [MED] A complex of symptoms associated with acute glomerulonephritis and pulmonary hemorrhage.

goose [VERT ZOO] The common name for a number of waterfowl in the subfamily Anatinae; they are intermediate in size and features between ducks and swans.

gooseberry [BOT] The common name for about six species of thorny, spreading bushes of the genus *Ribes* in the order Rosales, producing small, acidic, edible fruit.

gooseneck barnacle [INV ZOO] Any stalked barnacle, especially of the genus *Lepas*.

gopher [VERT ZOO] The common name for North American rodents composing the family Geomyidae. Also known as pocket gopher.

Gordiidae [INV ZOO] A monogeneric family of worms in the order Gordioidea distinguished by a smooth cuticle.

Gordioidea [INV ZOO] An order of worms belonging to the Nematomorpha in which there is one ventral epidermal cord, a body cavity filled with mesenchymal tissue, and paired gonads.

gordioid larva [INV ZOO] The developmental stage of nematomorphs, free-living for a short time.

Gorgonacea [INV ZOO] The horny corals, an order of the coelenterate subclass Alcyonaria; colonies are fanlike or featherlike with branches spread radially or oppositely in one plane.

gorgonin [BIOCHEM] The protein, frequently containing iodine and bromine, composing the horny skeleton of members of the Gorgonacea; contains iodine and bromine.

Gorgonocephalidae [INV ZOO] A family of ophiuroid echi-

GOOSE

Canada goose *(Branta canadensis)*.

GOOSEBERRY

Gooseberry branch bearing leaves and fruits. *(USDA)*

GOPHER

Pocket gopher of North America.

noderms in the order Phrynophiurida in which the individuals often have branched arms.

gorilla [VERT ZOO] *Gorilla gorilla.* An anthropoid ape, the largest living primate; the two African subspecies are the lowland gorilla and the mountain gorilla.

gossypose *See* raffinose.

Götte's larva [INV ZOO] The larva of polyclads characterized by four ciliated lobes.

gout [MED] A condition of purine metabolism resulting in increased blood levels of uric acid with ultimate deposition as urates in soft tissues around joints.

gr *See* grain.

Graafian follicle [HISTOL] The mature mammalian ovum with its surrounding epithelial cells.

Gracilariidae [INV ZOO] A family of small moths in the superfamily Tineoidea; both pairs of wings are lanceolate and widely fringed.

gracilis [ANAT] A long slender muscle on the medial aspect of the thigh.

graft [BIOL] **1.** To unite to form a graft. **2.** A piece of tissue transplanted from one individual to another or to a different place on the same individual. **3.** An individual resulting from the grafting of parts. [BOT] To unite a scion to an understock in such manner that the two grow together and continue development as a single plant without change in scion or stock.

grain [BOT] **1.** A rounded, granular prominence on the back of a sepal. **2.** *See* cereal. **3.** *See* drupelet. [MECH] A unit of mass in the United States and United Kingdom, common to the avoirdupois, apothecaries', and troy systems, equal to 1/7000 of a pound, or to 6.479891×10^{-5} kilogram. Abbreviated gr.

grain sorghum [BOT] *Sorghum bicolor.* A grass plant cultivated for its grain and to a lesser extent for forage. Also known as nonsaccharine sorghum.

gram [MECH] The unit of mass in the centimeter-gram-second system of units, equal to 0.001 kilogram. Abbreviated g; gm.

gramagrass [BOT] Any grass of the genus *Bouteloua*; pasture grass.

gram-atomic weight [CHEM] The atomic weight of an element expressed in grams, that is, the atomic weight on a scale on which the atomic weight of carbon-12 isotope is taken as 12 exactly.

gram-calorie *See* calorie.

gram-equivalent weight [CHEM] The equivalent weight of an element or compound expressed in grams on a scale in which carbon-12 has an equivalent weight of 3 grams in those compounds in which its formal valence is 4.

gramicidin [MICROBIO] A polypeptide antibacterial antibiotic produced by *Bacillus brevis*; active locally against gram-positive bacteria.

Graminales [BOT] The equivalent name for Cyperales.

Gramineae [BOT] The grasses, a family of monocotyledonous plants in the order Cyperales characterized by distichously arranged flowers on the axis of the spikelet.

graminicolous [ECOL] Living upon grass.

graminifolious [BOT] Having grasslike leaves.

graminivorous [ZOO] Feeding on grasses.

graminoid [BOT] Of or resembling the grasses.

gram-molecular volume [CHEM] The volume occupied by a gram-molecular weight of a chemical in the gaseous state at 0°C and 760 millimeters of pressure.

gram-molecular weight [CHEM] The molecular weight of compound expressed in grams, that is, the molecular weight on a scale on which the atomic weight of carbon-12 isotope is taken as 12 exactly.

gram-negative [MICROBIO] Of bacteria, decolorizing and staining with the counterstain when treated with Gram's stain.

gram-negative diplococci [MICROBIO] Three bacteriologic genera composing the family Neisseriaceae: *Gemella, Veillonella,* and *Neisseria.*

gram-positive [MICROBIO] Of bacteria holding the color of the primary stain when treated with Gram's stain.

Gram's stain [MICROBIO] A differential bacteriological stain; a fixed smear is stained with a slightly alkaline solution of basic dye, treated with a solution of iodine in potassium iodide, and then with a neutral decolorizing agent, and usually counterstained; bacteria stain either blue (gram-positive) or red (gram-negative).

gram-variable [MICROBIO] Pertaining to staining inconsistently with Gram's stain.

grana [CYTOL] A multilayered membrane unit formed by stacks of the lobes or branches of a chloroplast thylakoid.

grand mal [MED] A complete epileptic seizure involving sudden loss of consciousness and tonic convulsion of the skeletal musculature followed by clonic muscular spasms.

granite moss [BOT] The common name for a group of the class Bryatae represented by two Arctic genera and distinguished by longitudinal splitting of the mature capsule into four valves.

Grantiidae [INV ZOO] A family of calcareous sponges in the order Sycettida.

granular [SCI TECH] Having a grainy texture.

granular gland [PHYSIO] A gland that produces and secretes a granular material.

granular leukocyte *See* granulocyte.

granulation [MED] **1.** Tiny red granules made of capillary loops in the base of an ulcer. **2.** Process of granular tissue formation around a focus of inflammation. [PL PATH] Dry, tasteless condition of citrus fruit due to hardening of the juice sacs when fruit is left on trees too late in the season.

granuloblastosis [VET MED] An avian leukosis characterized by the presence of excessive numbers of immature granulocytes in the blood of affected birds.

granulocyte [HISTOL] A leukocyte containing granules in the cytoplasm. Also known as granular leukocyte; polymorph; polymorphonuclear leukocyte.

granulocytic leukemia [MED] A blood disease involving neoplastic transformation of granulocytes, principally the neutrophilic series. Also known as myelogenous leukemia; myeloid leukemia.

granulocytosis [MED] An increase in the number of granulocytes in the circulation.

granuloma [MED] A discrete nodular lesion of inflammatory tissue in which granulation is significant.

granuloma inguinale [MED] An infectious, chronic, destructive granulomatous lesion of humans most frequently localized in the genital and inguinal regions; caused by Donovan bodies (*Donovania granulomatis*).

granuloma pyogenicum [MED] A hemangioma with superimposed inflammation on the skin or other epithelial surfaces.

granulomatosis [MED] Any disease characterized by multiple granulomas.

Granuloreticulosia [INV ZOO] A subclass of the protozoan class Rhizopodea characterized by reticulopodia which often fuse into networks.

Granville wilt [PL PATH] A bacterial wilt of tobacco caused by *Pseudomonas solanacearum.*

GRANA

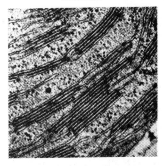

Part of oat leaf chloroplast at high magnification. Grana are seen in side view, revealing profiles of membranes. Thylakoids that interconnect grana are shown, with components of stroma, or ground substance, in which thylakoids lie. The dense granules are chloroplast ribosomes.

GRANITE MOSS

sporophytes

Granite moss (*Andreaea rupestris*), showing gametophyte with sporophytes. (*From G. M. Smith, Cryptogamic Botany, vol. 2, 2d ed., McGraw-Hill, 1955*)

GRAPE

Foliage, tendrils, fruit, and stem
characteristics of the Concord
grape *(Vitis labruscana).*

GRAPEFRUIT

Foliage, flowers, and fruit
of grapefruit.

GRASSHOPPER

0.5 cm

Melanoplus mexicanus, a
grasshopper. *(From Illinois
Natural History Survey)*

grape [BOT] The common name for plants of the genus *Vitis* characterized by climbing stems with cylindrical-tapering tendrils and polygamodioecious flowers; grown for the edible, pulpy berries.

grapefruit [BOT] *Citrus paradisi.* An evergreen tree with a well-rounded top cultivated for its edible fruit, a large, globose citrus fruit characterized by a yellow rind and white, pink, or red pulp.

Graphidaceae [BOT] A family of mosses formerly grouped with lichenized Hysteriales but now included in the order Lecanorales; individuals have true paraphyses.

graphiohexaster [INV ZOO] A type of sponge spicule having long, filamentous processes from four of the branching rays.

Graphiolaceae [MYCOL] A family of parasitic fungi in the order Ustilaginales in which teleutospores are produced in a cuplike fruiting body.

Grapsidae [INV ZOO] The square-backed crabs, a family of decapod crustaceans in the section Brachyura.

graptolite [PALEON] A member of a group of fossil exoskeletons of uncertain affinities.

Graptolithina [PALEON] A class of extinct colonial animals believed to be related to the class Pterobranchia of the Hemichordata.

Graptoloidea [PALEON] An order of extinct animals in the class Graptolithina including branched, planktonic forms described from black shales.

Graptozoa [PALEON] The equivalent name for Graptolithina.

grass [BOT] The common name for all members of the family Gramineae; monocotyledonous plants having leaves that consist of a sheath which fits around the stem like a split tube, and a long, narrow blade.

grasserie [INV ZOO] A polyhedrosis disease of silkworms characterized by spotty yellowing of the skin and internal liquefaction. Also known as jaundice.

grasshopper [INV ZOO] The common name for a number of plant-eating orthopteran insects composing the subfamily Saltatoria; individuals have hindlegs adapted for jumping, and mouthparts adapted for biting and chewing.

grassland [ECOL] Any area of herbaceous terrestrial vegetation dominated by grasses and graminoid species.

grass sickness [VET MED] A disease of horses occurring mainly in Scotland; thought to be caused by a virus similar to the one that causes poliomyelitis in humans.

grass tetany [VET MED] A magnesium-deficiency disease of cows.

gravid [ZOO] **1.** Of the uterus, containing a fetus. **2.** Pertaining to female animals when carrying young or eggs.

Gravigrada [VERT ZOO] The sloths, a group of herbivorous xenarthran mammals in the order Edentata; members are completely hairy and have five upper and four lower prismatic teeth without enamel.

Grawitz's tumor *See* renal-cell carcinoma.

gray blight [PL PATH] A fungus disease of tea caused by *Pestalotia (Pestalozzia) theae,* which invades the tissues and causes the formation of black dots on the leaves.

gray leaf spot [PL PATH] A fungus disease of tomatoes caused by *Stemphylium solani* and characterized by water-soaked brown spots on the leaves that become gray with age.

gray matter [HISTOL] The part of the central nervous system composed of nerve cell bodies, their dendrites, and the proximal and terminal unmyelinated portions of axons.

gray mold [PL PATH] Any fungus disease characterized by a gray surface appearance of the affected part.

gray scab [PL PATH] A fungus disease of willow caused by

Sphaceloma murrayae and characterized by irregular raised leaf spots having grayish-white centers and dark-brown margins.

gray speck [PL PATH] A manganese-deficiency disease of oats characterized by light-green to grayish leaf spots, and later by the buff or light-brown discoloration of the blades.

graywall [PL PATH] A disease of tomatoes thought to be caused by excess sunlight and characterized by translucent grayish-brown streaks or blotches on the fruit and browning of the vascular strands.

grease spot [PL PATH] A fungus disease of turf grasses caused by *Pythium aphanidermatum* and characterized by spots that have a greasy border of blackened leaves and intermingled cottony mycelia. Also known as spot blight.

greater omentum [ANAT] A fold of peritoneum that is attached to the greater curvature of the stomach and hangs down over the intestine and fuses with the mesocolon.

Great Ice Age [GEOL] The Pleistocene epoch.

grebe [VERT ZOO] The common name for members of the family Podicipedidae; these birds have legs set far posteriorly, compressed bladelike tarsi, individually broadened and lobed toes, and a rudimentary tail.

Greeffiellidae [INV ZOO] A superfamily of free-living nematodes in the order Desmoscolecoidea.

green algae [BOT] The common name for members of the plant division Chlorophyta.

green gland *See* antennal gland.

greenhouse [BOT] Glass-enclosed, climate-controlled structure in which young or out-of-season plants are cultivated and protected.

green manure [AGR] Herbaceous plant material plowed into the soil while still green.

green mold [MYCOL] Any fungus, especially *Penicillium* and *Aspergillus* species, that is green or has green spores.

green rosette [PL PATH] A virus disease of the peanut characterized by bunching and yellowing of the leaves with severe stunting of the plant.

green rot [PL PATH] A decay of fallen deciduous trees in which the wood is colored a malachite green by the fungus *Peziza aeruginosa.*

green smut [PL PATH] A fungus disease of rice characterized by enlarged grains covered with a green powder consisting of conidia, and caused by *Ustilaginoidea virens.* Also known as false smut.

greenstick fracture [MED] An incomplete fracture of a long bone, seen in children; the bone is bent but splintered only on the convex side.

green sulfur bacteria [MICROBIO] A physiologic group of green photosynthetic bacteria of the Chloraceae that are capable of using H_2S and other inorganic electron donors.

Gregarinia [INV ZOO] A subclass of the protozoan class Telosporea occurring principally as extracellular parasites of invertebrates.

gressorial [VERT ZOO] Adapted for walking, as certain birds' feet.

Grifulvin [MICROBIO] A trade name for griseofulvin.

Grimmiales [BOT] An order of mosses commonly growing upon rock in dense tufts or cushions and having hygroscopic, costate, usually lanceolate leaves arranged in many rows on the stem.

grindle *See* bowfin.

Grisactin [MICROBIO] A trade name for griseofulvin.

grisein [MICROBIO] $C_{40}H_{61}O_{20}N_{10}SFe$ A red, crystalline, water-soluble antibiotic produced by strains of *Streptomyces griseus.*

GRISEOFULVIN

Structural formula for griseofulvin.

GROUSE

The ruffed grouse *(Bonasa umbellus)*, a popular game bird of North America.

GROWTH CURVE

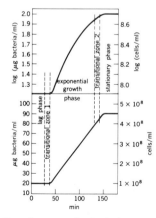

Growth curve of a typical bacterium. Lag phase = 25 minutes, doubling time = 45 minutes, and the specific growth rate = 0.154 min⁻¹.

griseofulvin [MICROBIO] $C_{17}H_{17}O_6Cl$ A colorless, crystalline antifungal antibiotic produced by several species of *Penicillium*. Also known as curling factor.

griseolutein [MICROBIO] Either of two fractions, A or B, of broad-spectrum antibiotics produced by *Streptomyces griseoluteus*; more active against gram-positive than gram-negative microorganisms.

griseomycin [MICROBIO] A white, crystalline antibiotic produced by an actinomycete resembling *Streptomyces griseolus.*

grizzly bear [VERT ZOO] The common name for a number of species of large carnivorous mammals in the genus *Ursus*, family Ursidae.

Groeberiidae [PALEON] A family of extinct rodentlike marsupials.

groin [ANAT] Depression between the abdomen and the thigh.

Gromiida [INV ZOO] An order of protozoans in the subclass Filosia; the test, which is chitinous in some species and thin and somewhat flexible in others, is reinforced with sand grains or siliceous particles.

gross anatomy [ANAT] Anatomy that deals with the naked-eye appearance of tissues.

gross stress reaction [PSYCH] A transient personality disorder in which, under conditions of great or unusual stress, a normal person utilizes neurotic mechanisms to deal with danger.

ground cover [BOT] Prostrate or low plants that cover the ground instead of grass.   [FOR] All forest plants except trees.

ground state [QUANT MECH] The stationary state of lowest energy of a particle or a system of particles.

group psychotherapy [PSYCH] Therapy given to a group of people by a therapist relying on the group effect on the individual and his interactions with the group.

group therapy [PSYCH] A technique of psychotherapy, consisting of a group of persons who discuss personal problems under the guidance of a therapist.

grouse [VERT ZOO] Any of a number of game birds in the family Tetraonidae having a plump body and strong, feathered legs.

growth [MED] Any abnormal, localized increase in cells, such as a tumor.   [PHYSIO] Increase in the quantity of metabolically active protoplasm, accompanied by an increase in cell number or cell size, or both.

growth curve [MICROBIO] A graphic representation of the growth of a bacterial population in which the log of the number of bacteria or the actual number of bacteria is plotted against time.

growth factor [PHYSIO] Any factor, genetic or extrinsic, which affects growth.

growth form [ECOL] The habit of a plant determined by its appearance of branching and periodicity.

growth hormone [BIOCHEM] **1.** A polypeptide hormone secreted by the anterior pituitary which promotes an increase in body size. Abbreviated GH. **2.** Any hormone that regulates growth in plants and animals.

growth rate [MICROBIO] Increase in the number of bacteria in a population per unit time.

growth regulator [BIOCHEM] A synthetic substance that produces the effect of a naturally occurring hormone in stimulating plant growth; an example is dichlorophenoxyacetic acid.

Gruidae [VERT ZOO] The cranes, a family of large, tall, cosmopolitan wading birds in the order Gruiformes.

Gruiformes [VERT ZOO] A heterogeneous order of generally cosmopolitan birds including the rails, coots, limpkins, button quails, sun grebes, and cranes.

Gryllidae [INV ZOO] The true crickets, a family of orthopteran insects in which individuals are dark-colored and chunky with long antennae and long, cylindrical ovipositors.

Grylloblattidae [INV ZOO] A monogeneric family of crickets in the order Orthoptera; members are small, slender, wingless insects with hindlegs not adapted for jumping.

Gryllotalpidae [INV ZOO] A family of North American insects in the order Orthoptera which live in sand or mud; they eat the roots of seedlings growing in moist, light soils.

G₁ stage [CYTOL] The stage between the end of one mitotic cycle and the beginning of DNA synthesis (S stage) for the next cycle.

G₂ stage [CYTOL] The brief stage of the mitotic cycle which follows S and precedes M.

guanidine [BIOCHEM] CH_5N_2 Aminomethanamidine, a product of protein metabolism found in urine. Also known as carbamidine; iminourea.

guanidine-acetic acid *See* glycocyamine.

guanine [BIOCHEM] $C_5H_5ON_5$ A purine base; occurs naturally as a fundamental component of nucleic acids.

guanophore *See* iridocyte.

guanosine [BIOCHEM] $C_{10}H_{13}O_5N_5$ Guanine riboside, a nucleoside composed of guanine and ribose. Also known as vernine.

guanylic acid [BIOCHEM] A nucleotide composed of guanine, a pentose sugar, and phosphoric acid and formed during the hydrolysis of nucleic acid.

guar [BOT] *Cyanopsis psoralioides.* A legume cultivated for forage and for its seed.

guard cell [BOT] Either of two specialized cells surrounding each stoma in the epidermis of plants; functions in regulating stoma size.

Guarnieri body [PATH] Eosinophilic cytoplasmic inclusion bodies found in the epidermal cells of patients with smallpox or chickenpox.

guava [BOT] *Psidium guajava.* A shrub or low tree of tropical America belonging to the family Myrtaceae; produces an edible, aromatic, sweet, juicy berry.

guayule [BOT] *Parthenium argentatum.* A subshrub of the family Compositae that is native to Mexico and the southwestern United States; it has been cultivated as a source of rubber.

gubernaculum [ANAT] A guiding structure, as the fibrous cord extending from the fetal testes to the scrotal swellings. [INV ZOO] **1.** A posterior flagellum of certain protozoans. **2.** A sclerotized structure associated with the copulatory spicules of certain nematodes.

guest insect [ECOL] An insect that lives or breeds in the nest of another.

Guillain-Barré syndrome *See* Landry-Guillain-Barré syndrome.

guinea fowl [VERT ZOO] The common name for plump African game birds composing the family Numididae; individuals have few feathers on the head and neck, but may have a crest of feathers and various fleshy appendages.

guinea pig [VERT ZOO] The common name for several species of wild and domestic hystricomorph rodents in the genus *Cavia*, family Caviidae; individuals are stocky, short-eared, short-legged, and nearly tailless.

guinea worm [INV ZOO] *Dracunculus medinensis.* A parasitic nematode that infects the subcutaneous tissues of humans and other mammals.

guitarfish [VERT ZOO] The common name for fishes composing the family Rhinobatidae.

gula [ANAT] **1.** The upper part of the throat. **2.** The upper

GRYLLIDAE

Grass cricket *(Nemobius fasciatus).* *(From Illinois Natural History Survey)*

GUANINE

Structural formula of guanine.

GUINEA FOWL

The common guinea fowl *(Numida meleagris)*, with red wattles and pearl-green plumage.

GUINEA PIG

The guinea pig, with short legs and large angular claws.

front of the neck. [INV ZOO] A median ventral sclerite of the head of many insects and most beetles which supports the submentum.

gulamentum [INV ZOO] A plate formed by fusion of the gula and the submentum.

gular [ANAT] Of, pertaining to, or situated in the gula or upper throat. [VERT ZOO] A horny shield on the plastron of turtles.

gulfweed [BOT] Brown algae of the genus *Sargassum*.

gull [VERT ZOO] The common name for a number of long-winged swimming birds in the family Laridae having a stout build, a thick, somewhat hooked bill, a short tail, and webbed feet.

gullet [ANAT] *See* esophagus. [INV ZOO] A canal between the cytostome and reservoir that functions in food intake in ciliates.

gum [BOT] A viscid exudate of certain plants and trees.

gumbo *See* okra.

gumme [PATH] A mass of rubberlike necrotic tissue found in any of various organs and tissues in tertiary syphilis.

gummosis [PL PATH] Production of gummy exudates in diseased plants as a result of cell degeneration.

Gunneraceae [BOT] A family of dicotyledonous terrestrial herbs in the order Haloragales, distinguished by two to four styles, a unilocular bitegmic ovule, large inflorescences with no petals, and drupaceous fruit.

gustation [PHYSIO] The act or the sensation of tasting.

gustatoreceptor [ANAT] A taste bud. [PHYSIO] Any sense organ that functions as a receptor for the sense of taste.

gut [ANAT] The intestine. [EMBRYO] The embryonic digestive tube.

Guthrie test [PATH] A screening test for the detection of phenylketonuria in which the inhibition of growth of a strain of *Bacillus subtilis* by a phenylalanine analog is reversed by L-phenylalanine, as found in elevated concentration in the plasma of patients with phenylketonuria.

gutta [BOT] A milky liquid produced by various trees in Malaya. [INV ZOO] A small spot of color on an insect, as on the wing.

guttation [BOT] The discharge of water from a plant surface, especially from a hydathode.

Guttiferae [BOT] A family of dicotyledonous plants in the order Theales characterized by extipulate leaves and conspicuous secretory canals or cavities in all organs.

Guttulinaceae [MICROBIO] A family of microorganisms in the Acrasiales characterized by simple fruiting structures with only slightly differentiated component cells containing little or no cellulose.

Gymnarchidae [VERT ZOO] A monotypic family of electrogenic fishes in the order Osteoglossiformes in which individuals lack pelvic, anal, and caudal fins.

Gymnarthridae [PALEON] A family of extinct lepospondylous amphibians that have a skull with only a single bone representing the tabular and temporal elements of the primitive skull roof.

Gymnoascaceae [MYCOL] A family of ascomycetous fungi in the order Eurotiales including dermatophytes and forms that grow on dung, soil, and feathers.

Gymnoblastea [INV ZOO] A suborder of coelenterates in the order Hydroida comprising hydroids without protective cups around the hydranths and gonozooids.

gymnoblastic [INV ZOO] Having naked medusa buds, referring to anthomedusan hydroids.

gymnocarpic [BOT] Having naked fruits, as lichens with uncovered apothecia, and mosses with expanded hymenia.

gymnocarpous [BOT] Having the hymenium uncovered on the surface of the thallus or fruiting body of lichens or fungi.

gymnocephalous cercaria [INV ZOO] A type of digenetic trematode larva.

Gymnocerata [INV ZOO] An equivalent name for Hydrocorisae.

gymnocidium [BOT] The apophysis or basal swelling of the capsules of certain moss plants.

Gymnocodiaceae [PALEOBOT] A family of fossil red algae.

gymnogynous [BOT] Having a naked ovary.

Gymnolaemata [INV ZOO] A class of ectoproct bryozoans possessing lophophores which are circular in basal outline and zooecia which are short, wide, and vaselike or boxlike.

Gymnonoti [VERT ZOO] An equivalent name for Cypriniformes.

Gymnophiona [VERT ZOO] An equivalent name for Apoda.

Gymnopleura [INV ZOO] A subsection of brachyuran decapod crustaceans including the primitive burrowing crabs with trapezoidal or elongate carapaces, the first pereiopods subchelate, and some or all of the remaining pereiopods flattened and expanded.

gymnopterous [INV ZOO] Referring to insects, having bare wings, that is, without scales.

gymnorhinal [VERT ZOO] Not having feathers covering the nostril region, as in some birds.

gymnosperm [BOT] The common name for members of the division Pinophyta; seed plants having naked ovules at the time of pollination.

Gymnospermae [BOT] The equivalent name for Pinophyta.

Gymnostomatida [INV ZOO] An order of the protozoan subclass Holotrichia containing the most primitive ciliates, distinguished by the lack of ciliature in the oral area.

gymnostomatous [INV ZOO] Without a shell or mantle, as certain mollusks.

Gymnotidae [VERT ZOO] The single family of the suborder Gymnotoidei; eel-shaped fishes having numerous vertebrae, and anus located far forward, and lacking pelvic and developed dorsal fins.

Gymnotoidei [VERT ZOO] A monofamilial suborder of actinopterygian fishes in the order Cypriniformes.

gynaecandrous [BOT] Having staminate and pistillate flowers on the same spike.

gynander [BIOL] A mosaic individual composed of diploid female portions derived from both parents and haploid male portions derived from an extra egg or sperm nucleus.

gynandrophore [BOT] **1.** A prolongation of the main axis that bears a sporophyll. **2.** A gonophore bearing both stamens and a gynoecium.

gynandrous [BOT] Having the stamens and pistils fused, as in certain orchids.

gynandry [PHYSIO] A form of pseudohermaphroditism in which the external sexual characteristics are partly or wholly of the male aspect, but internal female genitalia are present. Also known as female pseudohermaphroditism; virilism.

gynatrium [INV ZOO] In certain insects, the female genital pouch.

gynecology [MED] The branch of the medical science dealing with diseases of women, particularly those affecting the sex organs.

gynetype [SYST] The female holotype.

gynic [BIOL] Female.

gynobase [BOT] A gynoecium-bearing elongation of the receptacle in certain plants.

GYMNOCEPHALOUS CERCARIA

oral sucker — mouth
pharynx — prepharynx
cephalic glands — esophagus
cystogenous gland — ventral sucker
genital primordium — intestine
excretory vesicle — spines
flame cell

Anatomy of gymnocephalous cercaria. *(From R. M. Cable, An Illustrated Laboratory Manual of Parasitology, Burgess, 1958)*

GYMNOPLEURA

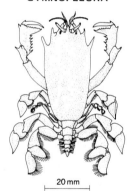

20 mm

Gymnopleuran crab, *Raninoides louisianensis. (Smithsonian Institution)*

GYRODACTYLOIDEA

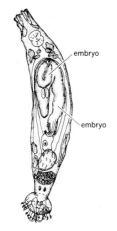

embryo

embryo

Gyrodactylus funduli Hargis from the longnose killfish *(Fundulus similis).*

gynobasic style [BOT] A style attached to the base of an ovary.

gynodioecious [BOT] Dioecious but with some perfect flowers on a plant bearing pistillate flowers.

gynoecium [BOT] The aggregate of carpels in a flower.

gynogenesis [EMBRYO] Development of a fertilized egg through the action of the egg nucleus, without participation of the sperm nucleus.

gynomerogony [EMBRYO] Development of a fragment of a fertilized egg containing the haploid egg nucleus.

gynomonoecious [BOT] Having complete and pistillate flowers on the same plant.

gynophore [BOT] **1.** A stalk that bears the gynoecium. **2.** An elongation of the receptacle between pistil and stamens.

gynostemium [BOT] The column consisting of fused pistils and stamens in orchids.

gypsophilous [ECOL] Flourishing on a gypsum-rich substratum.

gypsy moth [INV ZOO] *Porthetria dispar.* A large lepidopteran insect of the family Lymantriidae that was accidentally imported into New England from Europe in the late 19th century; larvae are economically important as pests of deciduous trees.

Gyracanthididae [PALEON] A family of extinct acanthodian fishes in the suborder Diplacanthoidei.

Gyrinidae [INV ZOO] The whirligig beetles, a family of large coleopteran insects in the suborder Adephaga.

Gyrinocheilidae [VERT ZOO] A monogeneric family of cypriniform fishes in the suborder Cyprinoidei.

Gyrocotylidae [INV ZOO] An order of tapeworms of the subclass Cestodaria; species are intestinal parasites of chimaeroid fishes and are characterized by an anterior eversible proboscis and a posterior ruffled adhesive organ.

Gyrocotyloidea [INV ZOO] A class of trematode worms according to some systems of classification.

Gyrodactyloidea [INV ZOO] A superfamily of ectoparasitic trematodes in the subclass Monogenea; the posterior holdfast is solid and is armed with central anchors and marginal hooks.

gyrogonite [PALEOBOT] A minute, ovoid body that is the residue of the calcareous encrustation about the female sex organs of a fossil stonewort.

gyroma [BOT] **1.** The annulus of a fern. **2.** A discoid or knoblike apothecium in certain lichens.

Gyropidae [INV ZOO] A family of biting lice in the order Mallophaga; members are ectoparasites of South American rodents.

gyrus [ANAT] One of the convolutions (ridges) on the surface of the cerebrum.

H

H *See* hydrogen.

habenula [ANAT] **1.** Stalk of the pineal body. **2.** A ribbonlike structure.

habenular commissure [ANAT] The commissure connecting the habenular ganglia in the roof of the diencephalon.

habenular ganglia [ANAT] Olfactory centers anterior to the pineal body.

habenular nucleus [ANAT] Either of a pair of nerve centers that are located at the base of the pineal body on either side and serve as an olfactory correlation center.

habit [CRYSTAL] *See* crystal habit. [PSYCH] A repetitious behavior pattern.

habitual abortion [MED] Recurring, successive spontaneous abortion.

habituation [MED] **1.** A condition of tolerance to the effects of a drug or a poison, acquired by its continued use; marked by a psychic craving for it when the drug is withdrawn. **2.** Mild drug addiction in which withdrawal symptoms are not severe.

habitus [BIOL] General appearance or constitution of an organism.

hackberry [BOT] **1.** *Celtis occidentalis.* A tree of the eastern United States characterized by corky or warty bark, and by alternate, long-pointed serrate leaves unequal at the base; produces small, sweet, edible drupaceous fruit. **2.** Any of several other trees of the genus *Celtis.*

hackmarack *See* tamarack.

haddock [VERT ZOO] *Melanogrammus aeglefinus.* A fish of the family Gadidae characterized by a black lateral line and a dark spot behind the gills.

hadrocentric [BOT] Having xylem surrounded by phloem.

Hadromerina [INV ZOO] A suborder of sponges in the class Clavaxinellida having monactinal megascleres, usually with a terminal knob at one end.

hadromycosis [PL PATH] Any plant disease resulting from infestation of the xylem by a fungus.

Haeckel's law *See* recapitulation theory.

haematodocha [INV ZOO] A sac in the palpus of male spiders that fills with hemolymph and becomes distended during pairing.

Haemosporina [INV ZOO] A suborder of sporozoan protozoans in the subclass Coccidia; all are parasites of vertebrates, and human malarial parasites are included.

hafnium [CHEM] A metallic element, symbol Hf, atomic number 72, atomic weight 178.49; melting point 2000°C, boiling point above 5400°C.

Hageman factor *See* factor XII.

hagfish [VERT ZOO] The common name for the jawless fishes composing the order Myxinoidea.

H agglutinin [IMMUNOL] An antibody that is type-specific for the flagella of cells or microorganisms.

HACKBERRY

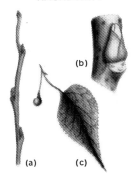

Common hackberry *(Celtis occidentalis). (a)* Twig. *(b)* Lateral bud. *(c)* Leaf.

HAFNIUM

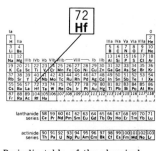

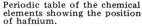

Periodic table of the chemical elements showing the position of hafnium.

HAIR

medulla cortex

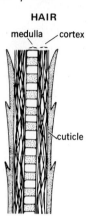

cuticle

Longitudinal section of
a hair shaft.

HAIR CELL

apex

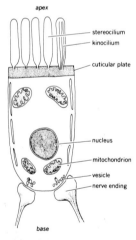

stereocilium
kinocilium

cuticular plate

nucleus

mitochondrion

vesicle
nerve ending

base

Generalized diagram of hair cell
showing the two types of cilia: the
stereocilium and the kinocilium.

hair [ZOO] **1.** A threadlike outgrowth of the epidermis of animals, especially a keratinized structure in mammalian skin. **2.** The hairy coat of a mammal, or of a part of the animal.

hair cell [HISTOL] The basic sensory unit of the inner ear of vertebrates; a columnar, polarized structure with specialized cilia on the free surface.

hair follicle [ANAT] An epithelial ingrowth of the dermis that surrounds a hair.

hair gland [ANAT] Sebaceous gland associated with hair follicles.

hairworm [INV ZOO] The common name for about 80 species of worms composing the class Nematomorpha.

Halacaridae [INV ZOO] A family of marine arachnids in the order Acarina.

Haldane's rule [GEN] The rule that if one sex in a first generation of hybrids is rare, absent, or sterile, then it is the heterogametic sex.

half-cell potential [PHYS CHEM] In electrochemical cells, the electrical potential developed by the overall cell reaction; can be considered, for calculation purposes, as the sum of the potential developed at the anode and the potential developed at the cathode, each being a half-cell.

half-life [CHEM] The time required for one-half of a given material to undergo chemical reactions. [NUCLEO] The average time interval required for one-half of any quantity of identical radioactive atoms to undergo radioactive decay. Also known as half-value period; radioactive half-life.

halibut [VERT ZOO] Either of two large species of flatfishes in the genus *Hippoglossus;* commonly known as a right-eye flounder.

Halichondrida [INV ZOO] A small order of sponges in the class Demospongiae with a skeleton of diactinal or monactinal, siliceous megascleres (or both), a skinlike dermis, and small amounts of spongin.

Halictidae [INV ZOO] The halictid and sweat bees, a family of hymenopteran insects in the superfamily Apoidea.

Haliplidae [INV ZOO] The crawling water beetles, a family of coleopteran insects in the suborder Adephaga.

Halitheriinae [PALEON] A subfamily of extinct sirenian mammals in the family Dugongidae.

Haller's organ [INV ZOO] A chemoreceptor on the tarsus of certain ticks.

hallucination [PSYCH] A perception without an appropriate stimulus.

hallucinogen [PHARM] A substance, such as LSD, that induces hallucinations.

hallucinosis [PSYCH] The condition of being possessed by more or less persistent hallucinations, especially while fully conscious.

hallux [ANAT] The first digit of the hindlimb; the big toe of man.

hallux valgus [MED] A deformity of the great toe, in which the head of the first metatarsal deviates away from the second metatarsal, that is, toward the outside of the foot, and the phalanges are deviated toward the second toe, causing prominence of the metatarsophalangeal joint.

halmophagous [ZOO] Pertaining to organisms which infest and eat stalks or culms of plants.

halobacteria [MICROBIO] Rod-shaped bacteria which display extreme halophilism.

halo blight [PL PATH] A bacterial blight of beans and sometimes other legumes caused by *Pseudomonas phaseolicola* and characterized by water-soaked lesions surrounded by a yellow ring on the leaves, stems, and pods.

halococci [MICROBIO] Coccoid bacteria which display extreme halophilism.

Halocypridacea [INV ZOO] A superfamily of ostracods in the suborder Myodocopa; individuals are straight-backed with a very thin, usually calcified carapace.

halogen [CHEM] Any of the elements of the halogen family, consisting of fluorine, chlorine, bromine, iodine, and astatine.

halonate [MYCOL] Pertaining to a spore surrounded by a colored circle. [PL PATH] A leaf spot surrounded by a halo.

halophile [BIOL] An organism that requires high salt concentrations for growth and maintenance.

halophilism [BIOL] The phenomenon of demand for high salt concentrations for growth and maintenance.

halophyte [ECOL] A plant or microorganism that grows well in soils having a high salt content.

haloplankton [ECOL] Plankton found in the sea. Also known as haliplankton.

Haloragaceae [BOT] A family of dicotyledonous plants in the order Haloragales distinguished by an apical ovary of 2–4 loculi, small inflorescences, and small, alternate or opposite or whorled, exstipulate leaves.

Haloragales [BOT] An order of dicotyledonous plants in the subclass Rosidae containing herbs with perfect or often unisexual, more or less reduced flowers, and a minute or vestigial perianth.

Halosauridae [VERT ZOO] A family of mostly extinct deep-sea teleost fishes in the order Notacanthiformes.

halosere [ECOL] The series of communities succeeding one another, from the pioneer stage to the climax, and commencing in salt water or on saline soil.

haltere [INV ZOO] Either of a pair of capitate filaments representing rudimentary hindwings in Diptera. Also known as balancer.

Halysitidae [PALEON] A family of extinct Paleozoic corals of the order Tabulata.

Hamamelidaceae [BOT] A family of dicotyledonous trees or shrubs in the order Hamamelidales characterized by united carpels, alternate leaves, perfect or unisexual flowers, and free filaments.

Hamamelidae [BOT] A small subclass of plants in the class Magnoliopsida having strongly reduced, often unisexual flowers with poorly developed or no perianth.

Hamamelidales [BOT] A small order of dicotyledonous plants in the subclass Hamamelidae characterized by vessels in the wood and a gynoecium consisting either of separate carpels or of united carpels that open at maturity.

hamartoma [MED] An abnormal condition resulting in the formation of a mass of tissue of disproportionate size and distribution but composed of the normal tissue of the region.

hamate [BIOL] Hook-shaped or hooked.

hammertoe [MED] A condition of the toe, usually the second, in which the proximal phalanx is extremely extended while the two distal phalanges are flexed.

Hamoproteidae [INV ZOO] A family of parasitic protozoans in the suborder Haemosporina; only the gametocytes occur in blood cells.

hamster [VERT ZOO] The common name for any of 14 species of rodents in the family Cricetidae characterized by scent glands in the flanks, large cheek pouches, and a short tail.

hamstring muscles [ANAT] The biceps femoris, semitendinosus, and semimembranosus collectively.

hamulus [VERT ZOO] A hooklike process, especially a small terminal hook on the barbicel of a feather.

hamus [BIOL] A hook or a curved process.

HAMSTER

Hamster, *Cricetus cricetus*, body about 6 inches (15 centimeters) long, used as a pet and a subject in laboratory experiments.

hand [ANAT] The terminal part of the upper extremity modified for grasping.

hand breadth [ANTHRO] The distance between the outside projections of the distal ends of the second and fifth metacarpals of the right hand, with the fingers extended and together.

hand length [ANTHRO] The distance measured from the end of the small wrist bone at the base of the thumb to the tip of the middle finger of the right hand, palm turned up, with the fingers extended and together.

Hand-Schüller-Christian disease [MED] A childhood syndrome characterized by exopthalmos, diabetes insipidus, and softened or punched-out areas in the bones.

hand-shoulder syndrome *See* shoulder-hand syndrome.

hanging-drop preparation [MICROBIO] A technique used in microscopy in which a specimen is placed in a drop of a suitable fluid on a cover slip and the cover slip is inverted over a concavity on a slide.

hangover [MED] Aftereffect following excessive intake of alcohol or certain drugs, such as barbiturates.

Hansen's bacillus *See* Mycobacterium leprae.

Hansen's disease [MED] An infectious disease of humans thought to be caused by *Mycobacterium leprae*; common manifestations are cutaneous and neural lesions. Also known as leprosy.

H antigen [MICROBIO] A general term for microbial flagellar antigens; former designation for species-specific flagellar antigens of *Salmonella*.

hapaxanthous [BOT] Having only a single flowering period.

haplobiont [BOT] A plant that produces only sexual haploid individuals.

haplocaulescent [BOT] Having a simple axis with the reproductive organs on the principal axis.

haplodont [VERT ZOO] Having simple crowns on the molars.

haploid [GEN] Having half of the diploid or full complement of chromosomes, as in mature gametes.

Haplolepidae [PALEON] A family of Carboniferous chondrostean fishes in the suborder Palaeoniscoidei having a reduced number of fin rays and a vertical jaw suspension.

haplomitosis [CYTOL] Type of primitive mitosis in which the nuclear granules form into threadlike masses rather than clearly differentiated chromosomes.

haplont [BOT] A plant with only haploid somatic cells; the zygote is diploid.

haploperistomous [BOT] **1.** Referring to mosses, having only one peristome. **2.** Having a single row of teeth on the peristome.

haplopetalous [BOT] Having a single row of petals.

haplophase [BIOL] Haploid stage in the life cycle of an organism.

haplopore [PALEON] Any randomly distributed pore on the surface of fossil cystoid echinoderms.

Haplosclerida [INV ZOO] An order of sponges in the class Demospongiae including species with a skeleton made up of siliceous megascleres embedded in spongin fibers or spongin cement.

haplosis [CYTOL] Reduction of the chromosome number to half during meiosis.

Haplosporea [INV ZOO] A class of Protozoa in the subphylum Sporozoa distinguished by the production of spores lacking polar filaments.

Haplosporida [INV ZOO] An order of Protozoa in the class Haplosporea distinguished by the production of uninucleate spores that lack both polar capsules and filaments.

haplostele [BOT] A type of protostele with the core of xylem characterized by a smooth outline.

HAPLOSTELE

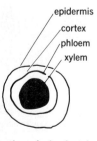

epidermis

cortex

phloem

xylem

Cross section of a haplostele.

haplostemonous [BOT] Having stamens arranged in a single whorl.

haplotype [SYST] The only original species of a genus, thereby being a genotype.

hapten [IMMUNOL] A simple substance that reacts like an antigen in vitro by combining with antibody; may function as an allergen when linked to proteinaceous substances of the tissue.

hapteron [BOT] A disklike holdfast on the stem of certain algae.

haptochlamydeous [BOT] Having the sporophylls protected by rudimentary perianth leaves.

haptoglobin [BIOCHEM] An alpha globulin that constitutes 1–2% of normal blood serum; contains about 5% carbohydrate.

Haptophyceae [BOT] A class of the phylum Chrysophyta that contains the Coccolithophorida.

haptor [INV ZOO] The posterior organ of attachment in certain monogenetic trematodes characterized by multiple suckers and the presence of hooks.

haptospore [BIOL] An adhesive spore.

haptotropism [BIOL] Movement of sessile organisms in response to contact, especially in plants.

Harderian gland [VERT ZOO] An accessory lacrimal gland associated with lower eyelid structures in all vertebrates except land mammals.

hard fiber [BOT] A heavily lignified leaf fiber used in making cordage, twine, and textiles.

hard pad [VET MED] A disease of dogs, probably associated with the canine distemper virus; often characterized by encephalitis and hardening of the foot pads.

hard palate [ANAT] The anterior portion of the roof of the mouth formed by paired palatine processes of the maxillary bones and by the horizontal part of each palate bone.

hard rot [PL PATH] **1.** Any plant disease characterized by lesions with hard surfaces. **2.** A fungus disease of gladiolus caused by *Septoria gladioli* which produces hard-surfaced lesions on the leaves and corms.

Hardy-Weinberg equilibrium [GEN] The state of genotype frequencies described by the Hardy-Weinberg law.

Hardy-Weinberg law [GEN] In an ideal, infinitely large random-mating population in which mutation, selection, and differential migration did not occur, the frequency of alleles at a given locus would remain the same over successive generations; the frequency of the genotypes reaches equilibrium.

hare [VERT ZOO] The common name for a number of lagomorphs in the family Leporidae; they differ from rabbits in being larger with longer ears, legs, and tails.

harelip [MED] A congenital defect, sometimes hereditary, marked by an abnormal cleft between the upper lip and the base of the nose. Also known as cleft lip.

Harpacticoida [INV ZOO] An order of minute copepod crustaceans of variable form, but generally being linear and more or less cylindrical.

harpes [INV ZOO] **1.** A process made of chitin between the claspers of mosquitoes. **2.** In Lepidoptera, claspers of the valves.

Harpidae [INV ZOO] A family of marine gastropod mollusks in the order Neogastropoda.

harrow [AGR] An implement that is pulled over plowed soil to break clods, level the surface, and destroy weeds.

harvester [AGR] A machine used to reap field crops.

Hashimoto's struma *See* struma lymphomatosa.

hastate [BIOL] Shaped like an arrowhead with divergent barbs.

haustellum [INV ZOO] A proboscis modified for sucking.

HARELIP

Drawing of a harelip. *(From L. B. Arey, Developmental Anatomy, 7th ed., Saunders, 1965)*

HARPACTICOIDA

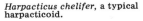

Harpacticus chelifer, a typical harpacticoid.

HARVESTER

Self-propelled forage harvester for corn. *(Sperry New Holland, Division of Sperry Rand Corp.)*

HAWK

Rough-legged hawk *(Buteo lagopus)*.

haustorium [BOT] **1.** An outgrowth of certain parasitic plants which serves to absorb food from the host. **2.** Food-absorbing cell of the embryo sac in nonparasitic plants.

haustrum [ANAT] An outpocketing or pouch of the colon.

Haverhill fever [MED] An acute bacterial infection caused by *Streptobacillus moniliformis*, usually acquired by rat bite, and characterized by acute onset, intermittent fever, erythematous rash, and polyarthritis. Also known as streptobacillary fever.

Haversian canal [HISTOL] The central, longitudinal channel of an osteon containing blood vessels and connective tissue.

Haversian lamella [HISTOL] One of the concentric layers of bone composing a Haversian system.

Haversian system [HISTOL] The term applied to a Haversian canal, with the surrounding concentric lamellae, the lacunae, and the canaliculi.

hawk [VERT ZOO] Any of the various smaller diurnal birds of prey in the family Accipitridae; some species are used for hunting hare and partridge in India and other parts of Asia.

hay [AGR] Forage plants cut and dried for animal feed.

hay fever [MED] An allergic disorder of the nasal membranes and related structures due to sensitization by certain plant pollens. Also known as allergic rhinitis.

haymaker [AGR] A machine for curing hay.

Hay's test [PATH] A test for bile salts; the salts lower the surface tension of water, and therefore a light powder such as flowers of sulfur will not float in a solution containing a high concentration of the salts.

He *See* helium.

head [ANAT] The region of the body consisting of the skull, its contents, and related structures.

headache [MED] A deep form of pain, with a characteristic aching quality, localized in the head.

head breadth [ANTHRO] Greatest horizontal breadth of the head above the ear openings, at whichever level is found, with moderate pressure applied on the caliper points.

head bulb [INV ZOO] A structure armed with spines behind the lips of spiruroid nematodes.

head circumference [ANTHRO] Maximum of three measurements taken above the eyebrows.

head fold [EMBRYO] A ventral fold formed by rapid growth of the head of the embryo over the embryonic disk, resulting in the formation of the foregut accompanied by anteroposterior reversal of the anterior part of the embryonic disk.

head height [ANTHRO] Average perpendicular distance between the tragion point and the mid-longitudinal line on top of the head, measured on both sides, when the head is positioned so that the line between the tragion point and the bottom of the bony orbit is horizontal.

head length [ANTHRO] The distance measured between glabella and opisthocranion, with application of moderate pressure.

head organ [INV ZOO] One of the bulbous structures in the prohaptor of monogenetic trematodes which are openings for adhesive glands.

head process [EMBRYO] The notochord or notochordal plate formed as an axial outgrowth of the primitive node.

head shield [INV ZOO] A conspicuous structure arching over the lips of certain nematodes.

head smut [PL PATH] A fungus disease of corn and sorghum caused by *Sphacelotheca reiliana* which destroys the head of the plant.

health [MED] A state of dynamic equilibrium between an organism and its environment in which all functions of mind and body are normal.

health physics [NUCLEO] The protection of personnel from harmful effects of ionizing radiation by such means as routine radiation surveys, area and personnel monitoring, and protective equipment and procedures.

hearing [PHYSIO] The general perceptual behavior and the specific responses made in relation to sound stimuli.

heart [ANAT] The hollow muscular pumping organ of the cardiovascular system in vertebrates.

heartbeat [PHYSIO] Pulsation of the heart coincident with ventricular systole.

heart block [MED] The cardiac condition resulting from defective transmission of impulses from atrium to ventricle.

heart failure *See* cardiac failure.

heart-lung machine [MED] A machine through which blood is shunted to maintain circulation during heart surgery.

heart rate [PHYSIO] The number of heartbeats per minute.

heartrot [PL PATH] **1.** A rot involving disintegration of the heartwood of a tree. **2.** A fungus disease of beets and rutabagas caused by *Mycosphaerella tabifica* which results in decay of the central tissues of the plant. **3.** A boron-deficiency disease of sugarbeets that causes rot.

heart valve [ANAT] Flaps of tissue that prevent reflux of blood from the ventricles to the atria or from the pulmonary arteries or aorta to the atria.

heartwood [BOT] Xylem of an angiosperm.

heat cramps [MED] Painful voluntary-muscle spasm and cramps following strenuous exercise, usually in persons in good physical condition, due to loss of sodium chloride and water from excessive sweating.

heat exhaustion [MED] A heat-exposure syndrome characterized by weakness, vertigo, headache, nausea, and peripheral vascular collapse, usually precipitated by physical exertion in a hot environment.

heath *See* temperate and cold scrub.

heather [BOT] *Calluna vulgaris.* An evergreen heath of northern and alpine regions distinguished by racemes of small purple-pink flowers.

heat of activation [PHYS CHEM] The increase in enthalpy when a substance is transformed from a less active to a more reactive form at constant pressure.

heat of adsorption [THERMO] The increase in enthalpy when 1 mole of a substance is adsorbed upon another at constant pressure.

heat of association [PHYS CHEM] Increase in enthalpy accompanying the formation of 1 mole of a coordination compound from its constituent molecules or other particles at constant pressure.

heat of combustion [PHYS CHEM] The amount of heat released in the oxidation of 1 mole of a substance at constant pressure, or constant volume. Also known as heat value; heating value.

heat of condensation [THERMO] The increase in enthalpy accompanying the conversion of 1 mole of vapor into liquid at constant pressure and temperature.

heat of cooling [THERMO] Increase in enthalpy during cooling of a system at constant pressure, resulting from an internal change such as an allotropic transformation.

heat of crystallization [THERMO] The increase in enthalpy when 1 mole of a substance is transformed into its crystalline state at constant pressure.

heat of decomposition [PHYS CHEM] The change in enthalpy accompanying the decomposition of 1 mole of a compound into its elements at constant pressure.

heat of dilution [PHYS CHEM] **1.** The increase in enthalpy accompanying the addition of a specified amount of solvent to a solution of constant pressure. Also known as integral heat of dilution; total heat of dilution. **2.** The increase in enthalpy when an infinitesimal amount of solvent is added to a solution at constant pressure. Also known as differential heat of dilution.

heat of dissociation [PHYS CHEM] The increase in enthalpy at constant pressure, when molecules break apart or valence linkages rupture.

heat of formation [PHYS CHEM] The increase in enthalpy resulting from the formation of 1 mole of a substance from its elements at constant pressure.

heat of fusion [THERMO] The increase in enthalpy accompanying the conversion of 1 mole, or a unit mass, of a solid to a liquid at its melting point at constant pressure and temperature. Also known as latent heat of fusion.

heat of hydration [PHYS CHEM] The increase in enthalpy accompanying the formation of 1 mole of a hydrate from the anhydrous form of the compound and from water at constant pressure.

heat of ionization [PHYS CHEM] The increase in enthalpy when 1 mole of a substance is completely ionized at constant pressure.

heat of linkage [PHYS CHEM] The bond energy of a particular type of valence linkage between atoms in a molecule, as determined by the energy required to dissociate all bonds of the type in 1 mole of the compound divided by the number of such bonds in a compound.

heat of reaction [PHYS CHEM] **1.** The negative of the change in enthalpy accompanying a chemical reaction at constant pressure. **2.** The negative of the change in internal energy accompanying a chemical reaction at constant volume.

heat of solidification [THERMO] The increase in enthalpy when 1 mole of a solid is formed from a liquid or, less commonly, a gas at constant pressure and temperature.

heat of solution [PHYS CHEM] The enthalpy of a solution minus the sum of the enthalpies of its components. Also known as integral heat of solution; total heat of solution.

heat of sublimation [THERMO] The increase in enthalpy accompanying the conversion of 1 mole, or unit mass, of a solid to a vapor at constant pressure and temperature. Also known as latent heat of sublimation.

heat of transformation [THERMO] The increase in enthalpy of a substance when it undergoes some phase change at constant pressure and temperature.

heat of vaporization [THERMO] The quantity of energy required to evaporate 1 mole, or a unit mass, of a liquid, at constant pressure and temperature. Also known as enthalpy of vaporization; heat of evaporation; latent heat of vaporization.

heatstroke [MED] A heat-exposure syndrome characterized by hyperpyrexia and prostration due to diminution or cessation of sweating, occurring most commonly in persons with underlying disease.

heaves [VET MED] Chronic emphysema in horses marked by labored breathing due to overdistension of the alveoli. Also known as broken wind.

Heberden-Rosenbach node *See* Heberden's node.

Heberden's arthritis [MED] Degenerative joint disease of the terminal joints of the fingers, producing enlargement (Heberden's nodes) and flexion deformities.

Heberden's node [MED] Nodose deformity of the fingers in degenerative joint disease. Also known as Heberden-Rosenbach node.

Hebrovellidae [INV ZOO] A family of hemipteran insects in the subdivision Amphibicorisae.

hectocotylus [INV ZOO] A specialized appendage of male cephalopods adapted for the transference of sperm.

hedgehog [VERT ZOO] The common name for members of the insectivorous family Erinaceidae characterized by spines on their back and sides.

Hedwigiaceae [BOT] A family of mosses in the order Isobryales.

heel [ANAT] The most posterior part of the human foot.

Heidelberg man [PALEON] An early type of European fossil man known from an isolated lower jaw; considered a variant of *Homo erectus* or an early stock of Neanderthal man.

Heister's valve [ANAT] The spiral valve of the cystic duct.

hekistotherm [ECOL] Plant adapted for conditions of minimal heat; can withstand long dark periods.

HeLa cells [PATH] Human cancer cells maintained in tissue culture since 1953, originally excised from the cervical carcinoma of a patient named Helen Lane.

Helaletidae [PALEON] A family of extinct perissodactyl mammals in the superfamily Tapiroidea.

Helcionellacea [PALEON] A superfamily of extinct gastropod mollusks in the order Aspidobranchia.

Heleidae [INV ZOO] The biting midges, a family of orthorrhaphous dipteran insects in the series Nematocera.

Heliasteridae [INV ZOO] A family of echinoderms in the subclass Asteroidea lacking pentameral symmetry but structurally resembling common asteroids.

Helicinidae [INV ZOO] A family of gastropod mollusks in the order Archeogastropoda containing tropical terrestrial snails.

helicoid [INV ZOO] Of a gastropod shell, shaped like a flat coil or flattened spiral.

helicoid cyme [BOT] A type of determinate inflorescence having a coiled cluster, with flowers on only one side of the axis.

Heliconiaceae [BOT] A family of monocotyledonous plants in the order Zingiberales characterized by perfect flowers with a solitary ovule in each locule, schizocarpic fruit, and capitate stigma.

Helicoplacoidea [PALEON] A class of free-living, spindle- or pear-shaped, plated echinozoans known only from the Lower Cambrian of California.

Helicosporae [MYCOL] A spore group of the Fungi Imperfecti characterized by spirally coiled, septate spores.

helicospore [INV ZOO] Mature spore of the Helicosporida characterized by a peripheral spiral filament.

Helicosporida [INV ZOO] An order of protozoans in the class Myxosporidea characterized by production of spores with a relatively thick, single intrasporal filament and three uninucleate sporoplasms.

helicotrema [ANAT] The opening at the apex of the cochlea through which the scala tympani and the scala vestibuli communicate with each other.

Heligmosomidae [INV ZOO] A family of parasitic roundworms belonging to the Strongyloidea.

Heliolitidae [PALEON] A family of extinct corals in the order Tabulata.

heliophilous [ECOL] Attracted by and adapted for a high intensity of sunlight.

heliophobe [MED] An individual who is extremely sensitive to the sun's rays. [PSYCH] One who has an abnormal fear of the sun's rays.

heliophyte [ECOL] A plant that thrives in full sunlight.

Heliornithidae [VERT ZOO] The lobed-toed sun grebes, a family of pantropical birds in the order Gruiformes.

heliotaxis [BIOL] Orientation movement of an organism in response to the stimulus of sunlight.

HEDGEHOG

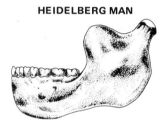

Hedgehog *(Erinaceus europaeus)*, with well-developed ears and eyes, pointed snout, short legs, and spines.

HEIDELBERG MAN

Reconstruction of the Heidelberg jaw. *(From M. F. Ashley Montagu, An Introduction to Physical Anthropology, 2d ed., Charles C. Thomas, 1951)*

HELICOID CYME

Drawing of a helicoid cyme.

HELIUM

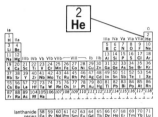

lanthanide series | 58 Ce | 59 Pr | 60 Nd | 61 Pm | 62 Sm | 63 Eu | 64 Gd | 65 Tb | 66 Dy | 67 Ho | 68 Er | 69 Tm | 70 Yb | 71 Lu

actinide series | 90 Th | 91 Pa | 92 U | 93 Np | 94 Pu | 95 Am | 96 Cm | 97 Bk | 98 Cf | 99 Es | 100 Fm | 101 Md | 102 No | 103 Lr

Periodic table of the chemical elements showing the position of helium.

HELLBENDER

The hellbender, with a maximum length of 18 inches (46 centimeters).

heliotrope [BOT] A plant whose flower or stem turns toward the sun.

heliotropism [BIOL] Growth or orientation movement of a sessile organism or part, such as a plant, in response to the stimulus of sunlight.

Heliozoia [INV ZOO] A subclass of the protozoan class Actinopodea; individuals lack a central capsule and have either axopodia or filopodia.

heliozooid [BIOL] Ameboid, but with distinct filamentous pseudopodia.

helium [CHEM] A gaseous chemical element, symbol He, atomic number 2, and atomic weight 4.0026; one of the noble gases in group 0 of the periodic table.

helix [ANAT] The outer margin of the pinna. [BIOL] A spiral structure. [INV ZOO] The coiled spiral form characteristic of certain structures in invertebrates.

hellbender [VERT ZOO] *Cryptobranchus alleganiensis.* A large amphibian of the order Urodela which is the most primitive of the living salamanders, retaining some larval characteristics.

Heller's test [PATH] A test for albumin in urine; presence of albumin is indicated by formation of a white ring at the junction of the solution and a concentrate solution of nitric acid.

Helmholtz free energy *See* free energy.

helminth [INV ZOO] Any parasitic worm.

helminthiasis [MED] Any disease caused by the presence of parasitic worms in the body.

helminthosporin [BIOCHEM] $C_{15}H_{10}O_5$ A maroon, crystalline pigment formed by certain fungi growing on a sugar substrate.

helobial endosperm [BOT] Endosperm in which formation of a partition follows the first division of the triple fusion nucleus, after which two chambers develop, one along the nuclear pattern, the other along the cellular pattern.

Helobiales [BOT] An order embracing most of the Alismatidae in certain systems of classification.

Helodermatidae [VERT ZOO] A family of lizards in the suborder Sauria.

Helodidae [INV ZOO] The marsh beetles, a family of coleopteran insects in the superfamily Dascilloidea.

Helodontidae [PALEON] A family of extinct ratfishes conditionally placed in the order Bradyodonti.

Helomyzidae [INV ZOO] The sun flies, a family of myodarian cyclorrhaphous dipteran insects in the subsection Acalypteratae.

helophyte [ECOL] A marsh plant; buds overwinter underwater.

helophytia [ECOL] Differences in ecological control by fluctuations in water level such as in marshes.

Heloridae [INV ZOO] A family of hymenopteran insects in the superfamily Proctotrupoidea.

Helotiales [MYCOL] An order of fungi in the class Ascomycetes.

Helotidae [INV ZOO] The metallic sap beetles, a family of coleopteran insects in the superfamily Cucujoidea.

helotism [ECOL] Symbiosis in which one organism is a slave to the other, as between certain species of ants.

Helotrephidae [INV ZOO] A family of true aquatic, tropical hemipteran insects in the subdivision Hydrocorisae.

hem-, hema-, hemo-, haem- [HISTOL] Combining form for blood.

hemacytometer *See* hemocytometer.

hemadsorption virus [VIROL] A descriptive term for myxoviruses that agglutinate red blood cells and cause the cells to adsorb to each other. Abbreviated HA virus.

hemagglutination [IMMUNOL] Agglutination of red blood cells.

hemagglutination-inhibition test [IMMUNOL] A test to identify a virus antigen or to quantitate an antibody by adding virus-specific antibody to a mixture of agglutinating virus and red blood cells.

hemagglutinin [IMMUNOL] An erythrocyte-agglutinating antibody.

hemal ring [INV ZOO] A vessel in certain echinoderms, variously located, associated with the coelom and axial gland.

hemal sinus [INV ZOO] The two principal lacunae along the digestive tube in certain echinoderms.

hemal tuft [INV ZOO] Series of fine vessels in echioderms arising from the axial gland.

hemangioendothelioma [MED] A malignant tumor composed of anoplastic endothelial cells. Also known as hemangiosarcoma.

hemangioma [MED] A tumor composed of blood vessels. Also known as capillary angioma.

hemangiopericytoma [MED] A tumor composed of endothelium-lined tubes or cords of cells surrounded by spherical cells with supporting reticulin network.

hemangiosarcoma *See* hemangioendothelioma.

hemapodium [INV ZOO] The dorsal lobe of a parapodium.

hemarthrosis [MED] Passage of blood into a joint.

hematein [BIOCHEM] $C_{16}H_{12}O_6$ A brownish stain and chemical indicator obtained by oxidation of hematoxylin.

hematemesis [MED] Vomiting of blood.

hematidrosis [MED] The appearance of blood or blood products in sweat gland secretions.

hematin [ORG CHEM] $C_{34}H_{33}O_5N_4Fe$ The hydroxide of ferriheme derived from oxidized heme.

hematoblast [HISTOL] An immature erythrocyte.

hematocele [MED] Collection of blood in a body part.

hematochrome [BIOCHEM] A red pigment occurring in green algae, especially when plants are exposed to intense light on subaerial habitats.

hematocrit [PATH] The volume, after centrifugation, occupied by the cellular elements of blood, in relation to the total volume.

hematogenous [PHYSIO] **1.** Pertaining to the production of blood or of its fractions. **2.** Carried by way of the bloodstream. **3.** Originating in blood.

hematoidin crystals [PATH] Yellow to brown crystals in the feces following gastrointestinal hemorrhage.

hematologist [MED] A specialist in the study of blood.

hematology [MED] The science of the blood, its nature, functions, and diseases.

hematoma [MED] A localized mass of blood in tissue; usually it clots and becomes encapsulated by connective tissue.

hematometra [MED] Accumulation of blood in the uterus.

hematomyelia [MED] Hemorrhage into the spinal cord.

hematopathology *See* hemopathology.

hematophagous [ZOO] Feeding on blood.

hematopoiesis [PHYSIO] The process by which the cellular elements of the blood are formed. Also known as hemopoiesis.

hematopoietic tissue [HISTOL] Blood-forming tissue, consisting of reticular fibers and cells. Also known as hemopoietic tissue.

hematopoietin [BIOCHEM] A substance which is produced by the juxtaglomerular apparatus in the kidney and controls the rate of red cell production. Also known as hemopoietin.

hematorrhachis [MED] Hemorrhage into the spinal meninges, producing irritative phenomena.

hematosalpinx [MED] Accumulation of blood in a Fallopian tube. Also known as hemosalpinx.

hematozoon [ECOL] An animal parasitic in blood, such as filarial worms.

hematuria [MED] A pathological condition in which the urine contains blood.

heme [BIOCHEM] $C_{34}H_{32}O_4N_4Fe$ An iron-protoporphyrin complex associated with each polypeptide unit of hemoglobin.

hemelytron [INV ZOO] Proximally hardened anterior wing of certain insects, such as the Hemiptera.

hemeralopia [MED] Day blindness.

hemi- [BIOL] **1.** Prefix for half. **2.** Prefix denoting one side of the body.

hemianesthesia [MED] Loss of sensation on one side of the body.

hemianopsia [MED] Bilateral or unilateral blindness in one-half of the field of vision.

Hemiascomycetes [MYCOL] The equivalent name for Hemiascomycetidae.

Hemiascomycetidae [MYCOL] A subclass of fungi in the class Ascomycetes.

hemiazygous vein [ANAT] A vein on the left side of the vertebral column which drains blood from the left ascending lumbar vein to the azygos vein.

hemiballismus [MED] Sudden, violent, spasmodic movements involving particularly the proximal portions of the extremities of one side of the body; caused by a destructive lesion of the contralateral subthalamic nucleus or its neighboring structures or pathways.

hemicellulose [BIOCHEM] $(C_6H_{10}O_5)_n$ A type of polysaccharide found in plant cell walls in association with cellulose and lignin; it is soluble in and extractable by dilute alkaline solutions. Also known as hexosan.

hemicephaly [MED] Congenital absence of the cerebrum.

Hemichordata [SYST] A group of marine animals categorized as either a phylum of deuterostomes or a subphylum of chordates; includes the Enteropneusta, Pterobranchia, and Graptolithina.

hemichordate [VERT ZOO] Having a rudimentary notochord.

Hemicidaridae [PALEON] A family of extinct Echinacea in the order Hemicidaroida distinguished by a stirodont lantern, and ambulacra abruptly widened at the ambitus.

Hemicidaroida [PALEON] An order of extinct echinoderms in the superorder Echinacea characterized by one very large tubercle on each interambulacral plate.

hemic murmur [MED] Blowing or rasping sound heard in the heart or vessels, usually in association with systole, in abnormal conditions of increased velocity of blood flow.

hemicolectomy [MED] Surgical removal of a portion of the colon.

hemicryptophyte [ECOL] A plant having buds at the soil surface and protected by scales, snow, or litter.

hemicydic [BOT] Of flowers, having the floral leaves arranged partly in whorls and partly in spirals.

hemidiaphragm [ANAT] Lateral half of a diaphragm. [MED] Diaphragm with normal muscle development only on one side.

Hemidiscosa [INV ZOO] An order of sponges in the subclass Amphidiscophora distinguished by birotulates that are hemidiscs with asymmetrical ends.

hemiepiphyte [ECOL] An epiphyte whose seeds germinate on another plant, but whose roots grow into the soil.

hemignathous [VERT ZOO] Having jaws of unequal length, as some fishes and birds.

Hemileucinae [INV ZOO] A subfamily of lepidopteran insects in the family Saturnidae consisting of the buck moths and relatives.

Hemimetabola [INV ZOO] A division of the insect subclass Pterygota; members are characterized by hemimetabolous metamorphosis.

hemimetabolous metamorphosis [INV ZOO] An incomplete metamorphosis; gills are present in aquatic larvae, or naiads.

hemin [BIOCHEM] $C_{34}H_{32}O_4N_4FeCl$ The crystalline salt of ferriheme, containing iron in the ferric state.

hemiparasite [ECOL] A parasite capable of a saprophytic existence, especially certain parasitic plants containing some chlorophyll. Also known as semiparasite.

hemiparesis [MED] Muscle weakness on one side of the body.

hemipelagic sediment [GEOL] Deposits containing terrestrial material and the remains of pelagic organisms, found in the ocean depths.

hemipenis [VERT ZOO] Either of a pair of nonerectile, evertible sacs that lie on the floor of the cloaca in snakes and lizards; used as intromittent organs.

Hemipeplidae [INV ZOO] An equivalent name for Cucujidae.

hemiplegia [MED] Unilateral paralysis of the body.

Hemiprocnidae [VERT ZOO] The crested swifts, a family comprising three species of perching birds found only in southeastern Asia.

Hemiptera [INV ZOO] The true bugs, an order of the class Insecta characterized by forewings differentiated into a basal area and a membranous apical region.

hemipterygoid [VERT ZOO] The part of the pterygoid bone which fuses with the palatine; refers specifically to neognath birds.

Hemisphaeriales [MYCOL] A group of ascomycetous fungi characterized by the wall of the fruit body being a stroma.

hemithorax [ANAT] One side of the chest.

hemitropous [BOT] Having the ovule with the hilum on one side and the micropyle on the opposite side in a plane parallel to the placenta.

Hemizonida [PALEON] A Paleozoic order of echinoderms of the subclass Asteroidea having an ambulacral groove that is well defined by adambulacral ossicles, but with restricted or undeveloped marginal plates.

hemizygous [GEN] Pertaining to the condition or state of having a gene present in a single dose; for instance, in the X chromosome of male mammals.

hemlock [BOT] The common name for members of the genus *Tsuga* in the pine family characterized by two white lines beneath the flattened, needlelike leaves.

hemoblast *See* hemocytoblast.

hemochorial placenta [EMBRYO] A type of placenta having the maternal blood in direct contact with the chorionic trophoblast. Also known as labyrinthine placenta.

hemochromatosis [MED] A disorder of iron metabolism characterized by excessive accumulation of iron in the liver and other tissues and by development of severe cirrhosis.

hemocoel [INV ZOO] An expanded portion of the blood system in arthropods that replaces a portion of the coelom.

hemoconcentration [MED] An increase in the concentration of blood cells resulting from the loss of plasma or water from the bloodstream.

hemoconiosis [MED] Condition of having an abnormal amount of hemoconia in the blood.

hemocyanin [BIOCHEM] A blue respiratory pigment found

HEMLOCK

Branch and cone of Eastern hemlock *(Tsuga canadensis)*.

HEMOCHORIAL PLACENTA

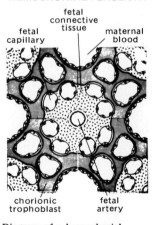

Diagram of a hemochorial placenta found in rodents, many insectivores, and tarsiers.

only in mollusks and in arthropods other than insects.

hemocyte [INV ZOO] A cellular element of blood, especially in invertebrates.

hemocytoblast [HISTOL] A pluripotential blast cell thought to be capable of giving rise to all other blood cells. Also known as hemoblast; stem cell.

hemocytolysis [PHYSIO] The dissolution of blood cells.

hemocytometer [PATH] A specifically designed, ruled and calibrated glass slide used with a microscope to count red and white blood cells. Also spelled hemacytometer.

hemodichorial placenta [EMBRYO] A placenta with a double trophoblastic layer.

hemodynamics [PHYSIO] A branch of physiology concerned with circulatory movements of the blood and the forces involved in circulation.

hemoendothelial placenta [EMBRYO] A placenta having the endothelium of vessels of chorionic villi in direct contact with the maternal blood.

hemoerythrin [BIOCHEM] A red respiratory pigment found in a few annelid and sipunculid worms and in the brachiopod *Lingula.* Also known as hemerythrin.

hemoflagellate [INV ZOO] A parasitic, flagellate protozoan that lives in the blood of the host.

HEMOGLOBIN

Three-dimensional construction representing a hemoglobin molecule bound with two molecules of oxygen (O₂). *(From A. F. Cullis et al., Structures of haemoglobin IX: A three-dimensional Fourier synthesis at 5.5 resolution: Description of the structure, Proc. Roy. Soc. London Ser. A, 265:161, 1962)*

hemoglobin [BIOCHEM] The iron-containing, oxygen-carrying molecule of the red blood cells of vertebrates comprising four polypeptide subunits in a heme group.

hemoglobin A [BIOCHEM] The type of hemoglobin found in normal adults, which moves as a single component in an electrophoretic field, is rapidly denatured by highly alkaline solutions, and contains two titratable sulfhydryl groups per molecule.

hemoglobin C [PATH] A slow-moving abnormal hemoglobin associated with intraerythrocytic crystal formation, target cells, and chronic hemolytic anemia.

hemoglobin E [PATH] An abnormal hemoglobin found in people of Southeast Asia, migrating slightly faster than hemoglobin C; in the homozygous form it causes a mild hemolytic anemia with normochromic target cells.

hemoglobinemia [MED] The presence of hemoglobin in the blood plasma.

hemoglobin H [PATH] An abnormal hemoglobin migrating more rapidly than normal hemoglobin on electrophoresis, and usually associated with thalassemia.

hemoglobin M [PATH] An abnormal hemoglobin associated with hereditary methemoglobinemia, differing from normal hemoglobin in its electrophoretic mobility by the starch-block method.

hemoglobinopathy [MED] Any blood dyscrasia resulting from the genetically determined alteration of the chemical nature of hemoglobin.

hemoglobin S *See* sickle-cell hemoglobin.

hemoglobinuria [MED] A pathological condition in which the urine contains hemoglobin.

hemoglobinuric nephrosis *See* lower nephron nephrosis.

hemogram [PATH] **1.** Erythrocyte and leukocyte count per cubic millimeter of blood plus the differential leukocyte count and hemoglobin level in grams per 100 milliliters of blood. **2.** The differential leukocyte count.

hemohistioblast [HISTOL] The hypothetical reticuloendothelial cell from which all the cells of the blood are eventually differentiated.

hemolymph [INV ZOO] The circulating fluid of the open circulatory systems of many invertebrates.

hemolysin [IMMUNOL] A substance that lyses erythrocytes.

hemolysis [PHYSIO] The lysis, or destruction, of erythrocytes with the release of hemoglobin.

hemolytic anemia [MED] A decrease in the blood concentration of hemoglobin and the number of erythrocytes, due to the inability of the mature erythrocytes to survive in the circulating blood.

hemolytic disease of newborn See erythroblastosis fetalis.

hemolytic jaundice [MED] Accumulation of bile pigments in the plasma as a result of excessive hemolysis.

hemomonochorial placenta [EMBRYO] A placenta with a single trophoblastic layer.

hemoparasite [INV ZOO] A parasitic animal that lives in the blood of a vertebrate.

hemopathology [MED] A branch of medicine dealing with blood diseases. Also known as hematopathology.

hemopathy [MED] Any disease of the blood.

hemopericardium [MED] The presence of blood or bloody effusion in the pericardial sac.

hemoperitoneum [MED] An effusion of blood in the peritoneal cavity.

hemophilia [MED] A rare, hereditary blood disorder marked by a tendency toward bleeding and hemorrhages due to a deficiency of factor VIII.

hemophilic bacteria [MICROBIO] Bacteria of the genera *Hemophilus*, *Bordetella*, and *Moraxella;* all are small, gramnegative, nonmotile, parasitic rods, dependent upon blood factors for growth.

hemopoiesis See hematopoiesis.

hemopoietic tissue See hematopoietic tissue.

hemopoietin See hematopoietin.

hemoporphyrin See hematoporphyrin.

hemorrhage [MED] The escape of blood from the vascular system.

hemorrhagic measles [MED] A grave variety of measles with a hemorrhagic eruption and severe constitutional symptoms. Also known as black measles.

hemorrhoid [MED] A varicosity of the external hemorrhoidal veins, causing painful swelling in the anal region.

hemosalpinx See hematosalpinx.

hemosiderin [BIOCHEM] An iron-containing glycoprotein found in most tissues and especially in liver.

hemostasis [MED] **1.** The arrest of a flow of blood or hemorrhage. **2.** The stopping or slowing of circulation.

hemostat [MED] An instrument to compress a bleeding vessel.

hemostatic [MED] An agent that arrests or checks bleeding, especially by shortening clotting time.

hemothorax [MED] Accumulation of blood in the pleural cavity.

hemotoxin [IMMUNOL] A toxin which causes lysis of blood cells.

hemotrichorial placenta [EMBRYO] A placenta with a triple trophoblastic layer.

hemotrophe [BIOCHEM] The nutritive substance supplied via the placenta to embryos of viviparous animals.

hemp [BOT] *Cannabis sativa.* An Asiatic annual herb cultivated for its tough bast fiber, used for making twines, linenlike fabrics, and canvases; the plant is the source of the drug marijuana.

henbane [BOT] *Hyoscyamus niger.* A poisonous herb containing the toxic alkaloids hyoscyamine and hyoscine; extracts have properties similar to belladonna.

Henderson equation for pH [PHYS CHEM] An equation for the pH of an acid during its neutralization: $pH = pK_a + \log$ [salt]/[acid] where pK_a is the logarithm to base 10 of the

HEMORRHOID

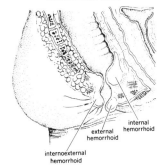

Types of hemorrhoids. *(After Pennington, from W. A. N. Dorland, ed., American Illustrated Medical Dictionary, 19th ed., Saunders, 1942)*

HEMP

Standing crop of hemp at maturity; plants range from 6 to 8 feet (1.8 to 2.4 meters) and have stems about the diameter of a pencil. *(Courtesy of H. L. Dean, University of Iowa)*

HENBANE

Drawing of henbane. *(Adapted from Webster's New International Dictionary, 2d ed., Merriam, 1959)*

reciprocal of the dissociation constant of the acid; the equation is found to be useful for the pH range 4–10, providing the solutions are not too dilute.

Henicocephalidae [INV ZOO] A family of hemopteran insects of uncertain affinities.

henna [BOT] *Lawsonia inermis.* An Old World plant having small opposite leaves and axillary panicles of white flowers; a reddish-brown dye extracted from the leaves is used in hair dyes. Also known as Egyptian henna.

Henry's law [PHYS CHEM] The law that at sufficiently high dilution in a liquid solution, the fugacity of a nondissociating solute becomes proportional to its concentration.

Hensen's node [EMBRYO] Thickening formed by a group of cells at the anterior end of the primitive streak in vertebrate gastrulas.

heparin [BIOCHEM] An acid mucopolysaccharide acting as an antithrombin, antithromboplastin, and antiplatelet factor to prolong the clotting time of whole blood; occurs in a variety of tissues, most abundantly in liver.

hepatectomy [MED] Surgical removal of the liver or a part of it.

hepatic artery [ANAT] A branch of the celiac artery that carries blood to the stomach, pancreas, great omentum, liver, and gallbladder.

hepatic coma [MED] Unconscious state associated with advanced liver disease.

hepatic duct [ANAT] The common duct draining the liver. Also known as common hepatic duct.

hepatic encephalopathy [MED] Behavioral, psychological, and neurological changes associated with advanced liver disease.

hepatic glycogenosis *See* von Gierke's disease.

hepatic plexus [ANAT] Nerve network accompanying the hepatic artery to the liver.

hepatic portal system [ANAT] A system of veins in vertebrates which collect blood from the digestive tract and spleen and pass it through capillaries in the liver.

hepatic vein [ANAT] A blood vessel that drains blood from the liver into the inferior vena cava.

hepatitis [MED] Inflammation of the liver; commonly of viral origin but also occurring in association with syphilis, typhoid fever, malaria, toxemias, and parasitic infestations.

hepatization [PATH] The conversion of tissue into a liverlike substance, as of the lungs during the exudative stage of pneumonia.

hepatoma [MED] A usually malignant neoplasm arising from parenchymal cells of the liver.

hepatomegaly [MED] Enlargement of the liver.

hepatopancreas [INV ZOO] A gland in crustaceans and certain other invertebrates that combines the digestive functions of the liver and pancreas of vertebrates.

hepatorenal syndrome [MED] A complex of syndromes due to hepatic and renal failure, including hyperpyrexia, oliguria, and coma. Also known as Heyd's syndrome.

hepatoscopy [MED] Inspection of the liver, as by laparotomy or peritoneoscopy.

Hepialidae [INV ZOO] A family of lepidopteran insects in the superfamily Hepialoidea.

Hepialoidea [INV ZOO] A superfamily of lepidopteran insects in the suborder Homoneura including medium- to large-sized moths which possess rudimentary mouthparts.

Hepsogastridae [INV ZOO] A family of parasitic insects in the order Mallophaga.

heptagynous [BOT] Having seven pistils.

heptandrous [BOT] Having seven stamens.

heptose [BIOCHEM] Any member of the group of monosaccharides containing seven carbon atoms.

herb [BOT] **1.** A seed plant that lacks a persistent, woody stem aboveground and dies at the end of the season. **2.** An aromatic plant or plant part used medicinally or for food flavoring.

herbaceous [BOT] **1.** Resembling or pertaining to a herb. **2.** Pertaining to a stem with little or no woody tissue.

herbaceous dicotyledon [BOT] A type of dicotyledon in which the primary vascular cylinder forms an ectophloic siphonostele with widely separated vascular strands.

herbaceous monocotyledon [BOT] A type of monocotyledon with a vascular system composed of widely spaced strands arranged in one of four ways.

herbarium [BOT] A collection of plant specimens, pressed and mounted on paper or placed in liquid preservatives, and systematically arranged.

herbicide [MATER] A chemical agent that destroys or inhibits plant growth.

herbicolous [ECOL] Living on herbs.

herbivore [VERT ZOO] An animal that eats only vegetation.

herd instinct [PSYCH] Psychic need for identification with a group.

hereditary deforming chondrodysplasia *See* multiple hereditary exostoses.

hereditary disease [MED] A genetically determined illness transmitted from parent to child.

hereditary hemorrhagic telangiectasia [MED] An inherited disease characterized by dilatation of groups of capillaries and a tendency to hemorrhage. Also known as Osler-Rendu-Weber disease.

hereditary nephritis [MED] A familial disease characterized by recurrent attacks of interstitial inflammation of the kidneys and discharge of blood in the urine.

hereditary spinal ataxia *See* Friedrich's ataxia.

heredity [GEN] The sum of genetic endowment obtained from the parents.

heredofamilial [MED] Referring to a disease or disorder having a familial pattern of occurrence and thought to be hereditary.

heritability [GEN] The proportion of phenotypic variance attributable to genetic effects; it is computed by dividing the genotypic variance by the phenotypic variance.

hermaphrodite [BIOL] An individual (animal, plant, or higher vertebrate) exhibiting hermaphroditism. [PHYSIO] An abnormal condition, especially in humans and other higher vertebrates, in which both male and female reproductive organs are present in the individual.

hermaphroditic *See* monoecious.

hermit crab [INV ZOO] The common name for a number of marine decapod crustaceans of the families Paguridae and Parapaguridae; all lack right-sided appendages and have a large, soft, coiled abdomen.

hernia [MED] Abnormal protrusion of an organ or other body part through its containing wall. Also called rupture.

hernial sac [MED] A pouch of peritoneum containing a herniated organ or other body part.

herniated disk [MED] An intervertebral disk in which the pulpy center has pushed through the fibrocartilage. Also known as slipped disk.

herniotomy [MED] An operation for the relief of irreducible hernia, by cutting through the neck of the sac.

heroin [PHARM] $C_{21}H_{23}O_5N$ A white, crystalline powder made from morphine; the hydrochloride compound is used as a sedative and narcotic.

HERBACEOUS DICOTYLEDON

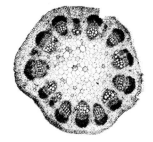

Cross section through the stem of the sunflower showing characteristics of ectophloic siphonostele. *(From J. B. Hill, L. O. Overholts, and H. W. Popp, Botany: A Textbook for Colleges, 2d ed., McGraw-Hill, 1950)*

HERBACEOUS MONOCOTYLEDON

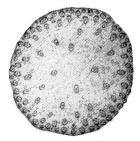

Photomicrograph of transverse section through corn stem. *(From J. B. Hill, L. O. Overholts, and H. W. Popp, Botany: A Textbook for Colleges, 2d ed., McGraw-Hill, 1950)*

HERBARIUM

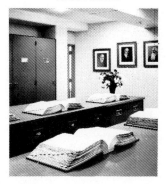

Interior of herbarium. *(Gray Herbarium, Harvard)*

herpes [MED] An acute inflammation of the skin or mucous membranes, characterized by the development of groups of vesicles on an inflammatory base.

herpes simplex [MED] An acute vesicular eruption of the skin or mucous membranes caused by a virus, commonly seen as cold sores or fever blisters.

herpesvirus [VIROL] A major group of deoxyribonucleic acid–containing animal viruses, distinguished by a cubic capsid, enveloped virion, and affinity for the host nucleus as a site of maturation.

herpes zoster [MED] A systemic virus infection affecting spinal nerve roots, characterized by vesicular eruptions distributed along the course of a cutaneous nerve. Also known as shingles; zoster.

herpetic stomatitis [MED] Inflammation of the soft tissues of the mouth characterized by fever blisters.

herpetic tonsillitis [MED] Acute inflammation of the tonsils characterized by fever and vesicles caused by a herpesvirus.

herpetology [VERT ZOO] The study of structure, habits, and classification of reptiles and amphibians.

herring [VERT ZOO] The common name for fishes composing the family Clupeidae; fins are soft-rayed and have no supporting spines, there are usually four gill clefts, and scales are on the body but absent on the head.

Herring body [HISTOL] Any of the distinct colloid masses in the vertebrate pituitary gland, possibly representing greatly dilated endings of nerve fibers.

hertz [PHYS] Unit of frequency; a periodic oscillation has a frequency of n hertz if in 1 second it goes through n cycles. Also known as cycle per second. Abbreviated cps; Hz.

Hesionidae [INV ZOO] A family of small polychaete worms belonging to the Errantia.

hesperidium [BOT] A modified berry, with few seeds, a leathery rind, and membranous extensions of the endocarp dividing the pulp into chambers; an example is the orange.

Hesperiidae [INV ZOO] The single family of the superfamily Hesperioidea comprising butterflies known as skippers because of their rapid, erratic flight.

Hesperioidea [INV ZOO] A monofamilial superfamily of lepidopteran insects in the suborder Heteroneura including heavy-bodied, mostly diurnal insects with clubbed antennae that are bent, curved, or reflexed at the tip.

Hesperornithidae [PALEON] A family of extinct North American birds in the order Hesperornithiformes.

Hesperornithiformes [PALEON] An order of ancient extinct birds; individuals were large, flightless, aquatic diving birds with the shoulder girdle and wings much reduced and the legs specialized for strong swimming.

hesthogenous [VERT ZOO] Referring to birds, covered with down at hatching. Also known as dasypedic.

heteracanthous [VERT ZOO] Having asymmetrical spines in the dorsal fin.

heteractinal [INV ZOO] Pertaining to naillike spicules having a disc of six to eight rays in one plane, and a single stout ray perpendicular to these.

Heteractinida [PALEON] A group of Paleozoic sponges with calcareous spicules; probably related to the Calcarea.

Heterakidae [INV ZOO] A group of nematodes assigned either to the suborder Oxyurina or the suborder Ascaridina.

heterandrous [BOT] Having stamens differing from each other in length or form.

heteroagglutinin [IMMUNOL] An antibody in normal blood serum capable of agglutinating foreign particles and erythrocytes of other species.

HESPERIDIUM

oil gland

rind

A cross section of a hesperidium, an orange.

heteroauxin [BIOCHEM] $C_{10}H_9O_2N$ A plant growth hormone with an indole skeleton.

Heterobasidiomycetidae [MYCOL] A class of fungi in which the basidium either is branched or is divided into cells by cross walls.

heteroblastic [EMBRYO] Arising from different tissues or germ layers, in referring to similar organs in different species.

Heterocapsina [BOT] An order of green algae in the class Xanthophyceae. [INV ZOO] A suborder of yellow-green to green flagellate protozoans in the order Heterochlorida.

heterocarpous [BOT] Producing two distinct types of fruit.

heterocephalous [BOT] Having pistillate and staminate flowers on separate heads, as certain composite plants.

Heterocera [INV ZOO] A formerly recognized suborder of Lepidoptera including all forms without clubbed antennae.

heterocercal [VERT ZOO] Pertaining to the caudal fin of certain fishes and indicating that the upper lobe is larger, with the vertebral column terminating in this lobe.

Heteroceridae [INV ZOO] The variegated mud-loving beetles, a family of coleopteran insects in the superfamily Dryopoidea.

Heterocheilidae [INV ZOO] A family of parasitic roundworms in the superfamily Ascaridoidea.

heterochlamydeous [BOT] Having the perianth differentiated into a distinct calyx and a corolla.

Heterochlorida [INV ZOO] An order of yellow-green to green flagellate oraganisms of the class Phytamastigophorea.

heterochromatin [CYTOL] Specialized chromosome material which remains tightly coiled even in the nondividing nucleus and stains darkly in interphase.

heterochromia [PHYSIO] A condition in which the two irises of an individual have different colors, or in which one iris has two colors.

heterochronism [EMBRYO] Deviation from the normal sequence of organ formation; a factor in evolution.

heterococcolith [BIOL] A coccolith with crystals arranged into boat, trumpet, or basket shapes.

heterocoelous [ANAT] Pertaining to vertebrae with centra having saddle-shaped articulations.

Heterocorallia [PALEON] An extinct small, monofamilial order of fossil corals with elongate skeletons; found in calcareous shales and in limestones.

heterocyst [BOT] Clear, thick-walled cell occurring at intervals along the filament of certain blue-green algae.

heterodactylous [VERT ZOO] Having the first two toes turned backward.

heterodiploid [GEN] A hybrid between two species, usually plants, carrying the full diploid chromosome complements of both species.

heterodont [ANAT] Having teeth that are variable in shape and differentiated into incisors, canines, and molars. [INV ZOO] In bivalves, having two types of teeth on one valve which fit into depressions on the other valve.

Heterodonta [INV ZOO] An order of bivalve mollusks in some systems of classification; hinge teeth are few in number and variable in form.

Heterodontoidea [VERT ZOO] A suborder of sharks in the order Selachii represented by the single living genus *Heterodontus.*

heteroecious [BIOL] Pertaining to forms which pass through different stages of a life cycle in different hosts.

heteroerotism [PSYCH] Sexual desire directed away from one's self or sex.

heterogamete [BIOL] A gamete that differs in size, appear-

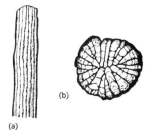

ance, structure, or sex chromosome content from the gamete of the opposite sex. Also known as anisogamete.

heterogametic sex [GEN] That sex of some species in which the two sex chromosomes are different in gene content or size and which therefore produces two or more different kinds of gametes.

heterogamety [GEN] The production of different kinds of gametes by one sex of a species.

heterogamous [BIOL] Of or pertaining to heterogamy.

heterogamy [BIOL] 1. Alternation of a true sexual generation with a parthenogenetic generation. 2. Sexual reproduction by fusion of unlike gametes. Also known as anisogamy. [BOT] Condition of producing two kinds of flowers.

heterogeneity [BIOL] The condition or state of being different in kind or nature.

heterogeneous [CHEM] Pertaining to a mixture of phases such as liquid-vapor, or liquid-vapor-solid. [SCI TECH] Composed of dissimilar or nonuniform constituents.

Heterogeneratae [BOT] A class of brown algae distinguished by a heteromorphic alteration of generations.

heterogenesis [BIOL] Alternation of generations in a complete life cycle, especially the alternation of a dioecious generation with one or more parthenogenetic generations.

heterogenous vaccine [IMMUNOL] A vaccine derived from a source other than the patient.

Heterognathi [VERT ZOO] An equivalent name for Cypriniformes.

heterogony [BIOL] 1. *See* allometry. 2. Alteration of generations in a complete life cycle, especially of a dioecious and hermaphroditic generation. [BOT] Having heteromorphic perfect flowers with respect to lengths of the stamens or styles.

heterograft [IMMUNOL] A tissue or organ obtained from an animal of one species and transplanted to the body of an animal of another species. Also known as heterologous graft.

heterohemolysis [IMMUNOL] Hemolytic amboceptor against the erythrocytes of a species different from that used to obtain the amboceptor.

heterokaryon [MYCOL] A bi- or multinucleate cell having genetically different kinds of nuclei.

heterokaryosis [MYCOL] The condition of a bi- or multinucleate cell having nuclei of genetically different kinds.

heterokont [BIOL] An individual, especially among certain algae, having unequal flagella.

heterolactic fermentation [MICROBIO] A type of lactic acid fermentation by which small yields of lactic acid are produced and much of the sugar is converted to carbon dioxide and other products.

heterolalia [PSYCH] Saying one thing and meaning another as by an unconscious mechanism or in motor aphasia.

heterolateral [ANAT] Of, pertaining to, or located on the opposite side.

heterolecithal [CYTOL] Of an egg, having the yolk distributed unevenly throughout the cytoplasm.

heterologous graft *See* heterograft.

heterologous tumor [MED] A neoplasm composed of tissues that differ from those of the organ at the site of the tumor.

heterolysis [CYTOL] Tissue or cell breakdown by the action of an outside agent or enzyme.

heteromedusoid [INV ZOO] A styloid type of sessile gonophore.

Heteromera [INV ZOO] The equivalent name for Tenebrionoidea.

heteromerous [BOT] Of a flower, having one or more whorls made up of a different number of members than the remaining whorls.

Heteromi [VERT ZOO] An equivalent name for Notacanthiformes.

heteromixis [MYCOL] In Fungi, sexual reproduction which involves the fusion of genetically different nuclei, each from a different thallus.

heteromorphic [CYTOL] Having synoptic or sex chromosomes that differ in size or form. [MED] Differing from the normal in size or morphology. [ZOO] Having a different form at each stage of the life history.

heteromorphosis [BIOL] Regeneration of an organ or part that differs from the original structure at the site. [EMBRYO] Formation of an organ at an abnormal site. Also known as homeosis.

Heteromyidae [VERT ZOO] A family of the mammalian order Rodentia containing the North American kangaroo mice and the pocket mice.

Heteromyinae [VERT ZOO] The spiny pocket mice, a subfamily of the rodent family Heteromyidae.

Heteromyota [INV ZOO] A monospecific order of wormlike animals in the phylum Echiurida.

Heterodontidae [VERT ZOO] The Port Jackson sharks, a family of aberrant modern elasmobranchs in the suborder Heterodontoidea.

Heteronemertini [INV ZOO] An order of the class Anopla; individuals have a middorsal blood vessel and a body wall composed of three muscular layers.

Heteroneura [INV ZOO] A suborder of Lepidoptera; individuals are characterized by fore- and hindwings that differ in shape and venation and by sucking mouthparts.

heteropelmous [VERT ZOO] Having bifid flexor tendons of the toes.

heteropetalous [BOT] Having different types of petals.

heterophagic vacuole   *See* food vacuole.

heterophile agglutination test   [PATH] A test for the presence of heterophile antibodies in the serum produced in infectious mononucleosis; agglutination of sheep red cells is a positive test. Also known as heterophile antibody test; Paul-Bunnell test.

heterophile antibody   [IMMUNOL] Substance that will react with heterophile antigen; found in the serum of patients with infectious mononucleosis.

heterophile antibody test   *See* heterophile agglutination test.

heterophile antigen   [IMMUNOL] A substance that occurs in unrelated species of animals but has similar serologic properties among them. Also known as heterogenetic antigen.

heterophyiasis   [MED] Presence of the minute intestinal fluke *Heterophyes heterophyes* in the small intestine of humans.

Heterophyllidae   [PALEON] The single family of the extinct coelenterate order Heterocorallia.

heterophyllous   [BOT] Having more than one form of foliage leaves on the same plant or stem.

heterophyte   [BOT] A plant that depends upon living or dead plants or their products for food materials.

Heteropiidae   [INV ZOO] A family of calcareous sponges in the order Sycettida.

heteroplasia   [MED] **1.** The presence of a tissue in an abnormal location. **2.** A process whereby tissues are displaced to or developed in locations foreign to their normal habitats.

heteroploidy   [GEN] The condition of a chromosome complement in which one or more chromosomes, or parts of chromosomes, are present in number different from the numbers of the rest.

Heteroporidae   [INV ZOO] A family of trepostomatous-like bryozoans in the order Cyclostomata.

heteropycnosis   [CYTOL] Differential condensation of certain chromosomes, such as sex chromosomes, or chromosome parts.

heterosexuality [PSYCH] Having sexual feeling toward members of the opposite sex.

heterosis [GEN] The increase in size, yield, and performance found in some hybrids, especially of inbred parents. Also known as hybrid vigor.

Heterosoricinae [PALEON] A subfamily of extinct insectivores in the family Soricidae distinguished by a short jaw and hedgehoglike teeth.

Heterospionidae [INV ZOO] A monogeneric family of spioniform worms found in shallow and abyssal depths of the Atlantic and Pacific oceans.

heterospory [BOT] Development of more than one type of spores, especially relating to the microspores and megaspores in ferns and seed plants.

heterostemony [BOT] Presence of two or more different types of stamens in the same flower.

Heterostraci [PALEON] An extinct group of ostracoderms, or armored, jawless vertebrates; armor consisted of bone lacking cavities for bone cells.

heterostyly [BOT] Condition or state of flowers having unequal styles.

Heterotardigrada [INV ZOO] An order of the tardigrades exhibiting wide morphologic variations.

heterothallic [BOT] Pertaining to a mycelium with genetically incompatible hyphae, therefore requiring different hyphae to form a zygospore; refers to fungi and some algae.

heterotopia [ECOL] An abnormal habitat. [MED] Displacement of an organ or other body part from its natural position.

heterotopic [MED] Occurring in an abnormal anatomic location.

Heterotrichida [INV ZOO] A large order of large ciliates in the protozoan subclass Spirotrichia; buccal ciliature is well developed and some species are pigmented.

Heterotrichina [INV ZOO] A suborder of the protozoan order Heterotrichida.

heterotrichous [BOT] Having both prostrate and erect filaments in the thallus, as certain algae. [INV ZOO] Having two types of cilia.

heterotroph [BIOL] An organism that obtains nourishment from the ingestion and breakdown of organic matter.

heterotropia *See* strabismus.

heterotropous [BOT] Pertaining to an ovule that has the hilum and the micropyle at opposite ends in a plane parallel to the placenta.

heteroxenous [BIOL] Requiring more than one host to complete a life cycle.

heterozooid [INV ZOO] Any of the specialized, nonfeeding zooids in a bryozoan colony.

heterozygote [GEN] An individual that has different alleles at one or more loci and therefore produces gametes of two or more different kinds.

heterozygous [GEN] Of or pertaining to a heterozygote.

hexacanth [INV ZOO] Having six hooks; refers specifically to the embryo of certain tapeworms.

Hexacorallia [INV ZOO] The equivalent name for Zoantharia.

hexactin [INV ZOO] A spicule, especially in Porifera, having six equal rays at right angles to each other.

Hexactinellida [INV ZOO] A class of the phylum Porifera which includes sponges with a skeleton made up basically of hexactinal siliceous spicules.

Hexactinosa [INV ZOO] An order of sponges in the subclass Hexasterophora; parenchymal megascleres form a rigid framework and consist of simple hexactins.

HETEROTRICHIDA

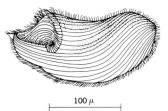

100 μ

Climacostomum, an example of a heterotrich.

hexagynous [BOT] Having six pistils.

Hexanchidae [VERT ZOO] The six- and seven-gill sharks, a group of aberrant modern elasmobranchs in the suborder Notidanoidea.

hexandrous [BOT] Having six stamens.

hexanedioic acid *See* adipic acid.

hexapetalous [BOT] Having or being a perianth comprising six petaloid divisions.

hexapod [INV ZOO] Having six legs.

Hexapoda [INV ZOO] An equivalent name for Insecta.

hexasepalous [BOT] Having six sepals.

hexaster [INV ZOO] A type of hexactin with branching rays that form star-shaped figures.

Hexasterophora [INV ZOO] A subclass of sponges of the class Hexactinellida in which parenchymal microscleres are typically hexasters.

hexenbesen *See* witches'-broom disease.

hexobarbital [PHARM] $C_{12}H_{16}N_2O_3$ A sedative and hypnotic of short duration of action; used also, as the sodium derivative, to induce surgical anesthesia.

hexokinase [BIOCHEM] Any enzyme that catalyzes the phosphorylation of hexoses.

hexosamine [BIOCHEM] A primary amine derived from a hexose by replacing the hydroxyl with an amine group.

hexosan *See* hemicellulose.

hexosan [BIOCHEM] Any monosaccharide that contains six carbon atoms in the molecule.

hexose phosphate [BIOCHEM] Any one of the phosphoric acid esters of a hexose, notably glucose, formed during the metabolism of carbohydrates by living organisms.

Heyd's syndrome *See* hepatorenal syndrome.

Hf *See* hafnium.

Hfr *See* high-frequency recombination.

Hg *See* mercury.

hiascent [BIOL] Gaping.

hiatus hernia [MED] Hernia through the esophageal hiatus, usually of a portion of the stomach.

hibernaculum [BIOL] A winter shelter for plants or dormant animals. [BOT] A winter bud or other winter plant part. [INV ZOO] A winter resting bud produced by a few fresh-water bryozoans which grows into a new colony in the spring.

hibernation [PHYSIO] **1.** Condition of dormancy and torpor found in cold-blooded vertebrates and invertebrates. **2.** *See* deep hibernation.

hickory [BOT] The common name for species of the genus *Carya* in the order Fagales; tall deciduous tree with pinnately compound leaves, solid pith, and terminal, scaly winter buds.

hidradenitis [MED] Inflammation of sweat glands.

hidradenoma papilliferum [MED] Benign tumor of sweat glands, usually on the vulva or perineum.

hidrosis [MED] Abnormally profuse sweat. [PHYSIO] The formation and excretion of sweat.

high-altitude disease *See* mountain sickness.

high-altitude erythremia *See* mountain sickness.

high-energy bond [PHYS CHEM] Any chemical bond yielding a decrease in free energy of at least 5 kilocalories per mole.

high-frequency recombination [MICROBIO] A bacterial cell type, especially *Escherichia coli,* having an integrated F factor and characterized by a high frequency of recombination. Abbreviated Hfr.

hilum [ANAT] *See* hilus. [BOT] Scar on a seed marking the point of detachment from the funiculus.

hilus [ANAT] An opening or recess in an organ, usually for passage of a vessel or duct. Also known as hilum.

Himantandraceae [BOT] A family of dicotyledonous plants

HEXASTEROPHORA

A representative hexasterophoran, *Polylophus.* (*After Shulze, 1887*)

HICKORY

Twig, bud, and leaf of the shagbark hickory, *Carya ovata.*

in the order Magnoliales characterized by several, uniovulate carpels and laminar stamens.

himantioid [MYCOL] Pertaining to a mycelium arranged in spreading fanlike cords.

Himantopterinae [INV ZOO] A subfamily of lepidopteran insects in the family Zygaenidae including small, brightly colored moths with narrow hindwings, ribbonlike tails, and long hairs covering the body and wings.

hindbrain *See* rhombencephalon.

hindgut [EMBRYO] The caudal portion of the embryonic digestive tube in vertebrates.

hinge ligament [INV ZOO] The structure composed of elastic substance connecting the valves of a bivalve.

hinge line [INV ZOO] In a bivalve, the line of articulation between the valves.

hinge plate [INV ZOO] **1.** In bivalve mollusks, the portion of a valve that supports the hinge teeth. **2.** The socket-bearing part of the dorsal valve in brachiopods.

hinge tooth [INV ZOO] A projection on a valve of a bivalve mollusk near the hinge line.

Hiodontidae [VERT ZOO] A family of tropical, fresh-water actinopterygian fishes in the order Osteoglossiformes containing the mooneyes of North America.

hip [ANAT] **1.** The region of the junction of thigh and trunk. **2.** The hip joint, formed by articulation of the femur and hipbone.

hip bone [ANAT] A large broad bone consisting of three parts, the ilium, ischium, and pubis; makes up a lateral half of the pelvis in mammals. Also known as innominate.

Hippidea [INV ZOO] A group of decapod crustaceans belonging to the Anomura and including cylindrical or squarish burrowing crustaceans in which the abdomen is symmetrical and bent under the thorax.

Hippoboscidae [INV ZOO] The louse flies, a family of cyclorrhaphous dipteran insects in the section Pupipara.

hippocampal sulcus [ANAT] A fissure on the brain situated between the para hippocampal gyrus and the fimbria hippocampi. Also known as dentate fissure.

hippocampus [ANAT] A ridge that extends over the floor of the descending horn of each lateral ventricle of the brain.

Hippocrateaceae [BOT] A family of dicotyledonous plants in the order Celastrales distinguished by an extrastaminal disk, mostly opposite leaves, seeds without endosperm, and a well-developed latex system.

hippocrepiform [BIOL] Horseshoe-shaped.

Hippoglossidae [VERT ZOO] A family of actinopterygian fishes in the order Pleuronectiformes composed of the flounders and plaice.

Hippomorpha [VERT ZOO] A suborder of the mammalian order Perissodactyla containing horses, zebras, and related forms.

Hippopotamidae [VERT ZOO] The hippopotamuses, a family of palaeodont mammals in the superfamily Anthracotherioidea.

hippopotamus [VERT ZOO] The common name for two species of artiodactyl ungulates composing the family Hippopotamidae.

Hipposideridae [VERT ZOO] The Old World leaf-nosed bats, a family of mammals in the order Chiroptera.

Hirschsprung's disease [MED] A disease caused by absence of the myenteric ganglion cells in a segment of rectum or distal colon, resulting in spasm of the affected part and dilation of the bowel proximal to the defect.

hirsutism [MED] An abnormal condition characterized by growth of hair in unusual places and in unusual amounts.

HIPPOPOTAMUS

The great African hippopotamus, more at home in water than on land because of its size.

Hirudinea [INV ZOO] A class of parasitic or predatory annelid worms commonly known as leeches; all have 34 body segments and terminal suckers for attachment and locomotion.

Hirudinidae [VERT ZOO] The swallows, a family of passeriform birds in the suborder Oscines.

hispid [BIOL] Having a covering of bristles or minute spines.

histamine [BIOCHEM] $C_5H_9N_3$ An amine derivative of histadine which is widely distributed in human tissues.

Histeridae [INV ZOO] The clown beetles, a large family of coleopteran insects in the superfamily Histeroidea.

Histeroidea [INV ZOO] A superfamily of coleopteran insects in the suborder Polyphaga.

histidase [BIOCHEM] An enzyme found in the liver of higher vertebrates that catalyzes the deamination of histidine to urocanic acid.

histidine [BIOCHEM] $C_6H_9O_2N_3$ A crystalline basic amino acid present in large amounts in hemoglobin and resulting from the hydrolysis of most proteins.

histioblast [ANAT] An immature macrophage.

histiocytoma [MED] **1.** Benign tumor composed of histiocytes. **2.** Dermatofibroma.

histiocytosis [MED] Abnormal proliferation of histiocytes, especially in hematopoietic tissues.

Histioteuthidae [INV ZOO] A family of cephalopod mollusks containing several species of squids.

histochemistry [BIOCHEM] A science that deals with the distribution and activities of chemical components in tissues.

histocompatibility [IMMUNOL] The capacity to accept or reject a tissue graft.

histodifferentiation [EMBRYO] Differentiation of cell groups into tissues.

histogenesis [EMBRYO] The developmental process by which the definite cells and tissues which make up the body of an organism arise from embryonic cells.

histogram [STAT] A graphical representation of a distribution function by means of rectangles whose widths represent intervals into which the range of observed values is divided and whose heights represent the number of observations occurring in each interval.

histoincompatibility [IMMUNOL] The condition in which a recipient rejects a tissue graft.

histologist [ANAT] An individual who specializes in histology.

histology [ANAT] The study of the structure and chemical composition of animal and plant tissues as related to their function.

histolysis [PATH] Disintegration of organic tissue.

histomycosis [MED] Infection of deep tissues by a fungus.

histone [BIOCHEM] Any of the strong, soluble basic proteins of cell nuclei that are precipitated by ammonium hydroxide.

histopathology [PATH] A branch of pathology that deals with tissue changes associated with disease.

histophysiology [PHYSIO] The science of tissue functions.

histoplasmin test [IMMUNOL] Skin test for hypersensitivity reaction to *Histoplasma capsulatum* products in the diagnosis of histoplasmosis.

histoplasmoma [MED] A tumorlike swelling caused by an inflammatory reaction to *Histoplasma capsulatum.*

histoplasmosis [MED] An infectious fungus disease of the lungs of man caused by *Histoplasma capsulatum.*

historadiography [BIOPHYS] A technique for taking x-ray pictures of cells, tissues, or sometimes whole small organisms.

histotome [BIOL] A microtome used to cut tissue sections for microscopic examination.

HIRUDINEA

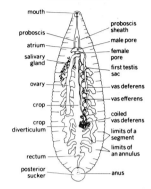

General structure of a leech. Male reproductive system is shown on the right, the female on the left. *(From K. H. Mann, A key to the British freshwater leeches, Freshwater Biol. Assoc. Sci. Publ., 14:3–21, 1954)*

HISTIDINE

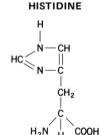

Structural formula of histidine.

HOLASTEROIDA

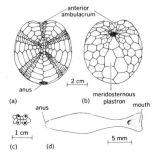

Diagnostic features of
holasteroids. *(a)* Aboral aspect.
(b) Adoral aspect. *(c)* Apical
system. *(d) Echinosigra
paradoxa*, a deep-sea species
of Pourtalesiidae, aboral aspect.
(After T. Mortensen)

HOLLINIDAE

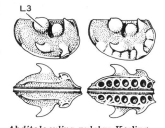

Abditoloculina pulchra Kesling,
male (left) and female (right)
carapaces showing bulbous L3
(lobe 3) typical of the Hollinidae.

HOLMIUM

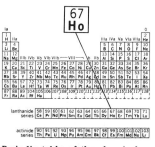

Periodic table of the chemical
elements showing the position
of holmium.

Histriobdellidae [INV ZOO] A small family of errantian polychaete worms that live as ectoparasites on crayfishes.

Ho *See* holmium.

hoarding behavior [VERT ZOO] The carrying of food to the
home nest for storage, in quantities exceeding daily need.

hoarse [MED] Having a harsh, discordant voice, caused by an
abnormal condition of the larynx or throat.

hoary [BOT] Having grayish or whitish color, referring to
leaves.

Hodgkin's disease [MED] A disease characterized by a neoplastic proliferation of atypical histiocytes in one or several
lymph nodes. Also known as lymphogranulomatosis.

Hodotermitidae [INV ZOO] A family of lower (primitive)
termites in the order Isoptera.

Hofbauer cell [HISTOL] A large, possibly phagocytic cell
found in chorionic villi.

hog [AGR] A domestic swine.

hog cholera [VET MED] A fatal infectious virus disease of
swine characterized by fever, diarrhea, and inflammation and
ulceration of the intestine; secondary infection by *Salmonella
cholerae suis* is common. Also known as African swine fever.

hoja blanca [PL PATH] A major virus disease of rice in Cuba
and Venezuela.

Holasteridae [INV ZOO] A family of exocyclic Euechinoidea
in the order Holasteroida; individuals are oval or heart-
shaped, with fully developed pore pairs.

Holasteroida [INV ZOO] An order of exocyclic Euechinoidea
in which the apical system is elongated along the anteroposterior axis and teeth occur only in juvenile stages.

holcodont [VERT ZOO] Having the teeth fixed in a long,
continuous groove.

holdfast [BOT] **1.** A suckerlike base which attaches the thallus of certain algae to the support. **2.** A disklike terminal
structure on the tendrils of various plants used for attachment
to a flat surface. [INV ZOO] An organ by which parasites
such as tapeworms attach themselves to the host.

Holectypidae [PALEON] A family of extinct exocyclic Euechinoidea in the order Holectypoida; individuals are hemispherical.

Holectypoida [INV ZOO] An order of exocyclic Euechinoidea
with keeled, flanged teeth, with distinct genital plates, and
with the ambulacra narrower than the interambulacra on the
adoral side.

holism [BIOL] The view that the whole of a complex system,
such as a cell or organism, is functionally greater than the sum
of its parts. Also known as organicism.

Hollinacea [PALEON] A dimorphic superfamily of extinct
ostracods in the suborder Beyrichicopina including forms
with sulci, lobation, and some form of velar structure.

Hollinidae [PALEON] An extinct family of ostracods in the
superfamily Hollinacea distinguished by having a bulbous
third lobe on the valve.

hollow [SCI TECH] **1.** Having a concave surface. **2.** Having an
interior cavity.

hollow stalk [PL PATH] Any plant disease characterized by
deterioration of the pith in the stalk.

holly [BOT] The common name for the trees and shrubs
composing the genus *Ilex*; distinguished by spiny leaves and
small berries.

hollywood lignumvitae *See* lignumvitae.

holmium [CHEM] A rare-earth element belonging to the yttrium subgroup, symbol Ho, atomic number 67, atomic weight
164.93, melting point 1400–1525°C.

Holobasidiomycetes [MYCOL] An equivalent name for Homobasidiomycetidae.

holobasidium [MYCOL] A spore-producing basidium that is not divided by septa.

holoblastic [EMBRYO] Pertaining to eggs that undergo total cleavage due to the absence of a mass of yolk.

holobranch [VERT ZOO] A gill with a row of filaments on each side of the branchial arch.

holocarpic [BOT] **1.** Having the entire thallus developed into a fruiting body or sporangium. **2.** Lacking rhizoids and haustoria.

holocellulose [BIOCHEM] The total polysaccharide fraction of wood, straw, and so on, that is composed of cellulose and all of the hemicelluloses and that is obtained when the extractives and the lignin are removed from the natural material.

Holocene *See* Recent.

Holocentridae [VERT ZOO] A family of nocturnal beryciform fishes found in shallow tropical and subtropical reefs; contains the squirrelfishes and soldierfishes.

Holocephali [VERT ZOO] The chimaeras, a subclass of the Chondrichthyes, distinguished by four pairs of gills and gill arches, an erectile dorsal fin and spine, and naked skin.

holococcolith [BIOL] A coccolith with simple rhombic or hexagonal crystals arranged like a mosaic.

holocoenosis [ECOL] The nature of the action of the environment on living organisms.

holocrine gland [PHYSIO] A structure whose cells undergo dissolution and are entirely extruded, together with the secretory product.

holoenzyme [BIOCHEM] A complex, fully active enzyme, containing an apoenzyme and a coenzyme.

hologametes [INV ZOO] Fully developed protozoans that take part in sygamy.

hologamy [BIOL] Condition of having gametes similar in size and form to somatic cells. [BOT] Condition of having the whole thallus develop into a gametangium.

holognathous [VERT ZOO] Having the jaw in one piece.

hologony [INV ZOO] Condition of having the germinal area extend the full length of a gonad; refers specifically to certain nematodes.

Holometabola [INV ZOO] A division of the insect subclass Pterygota whose members undergo holometabolous metamorphosis during development.

holometabolous metamorphosis [INV ZOO] Complete metamorphosis, during which there are four stages; the egg, larva, pupa, and imago or adult.

holomorphosis [BIOL] Complete regeneration of a lost body structure.

holomyarian [INV ZOO] Having zones of muscle layers but no muscle cells; refers specifically to certain nematodes.

holonephros [VERT ZOO] Type of kidney having one nephron beside each somite along the entire length of the coelom; seen in larvae of myxinoid cyclostomes.

holophyte [BIOL] An organism that obtains food in the manner of a green plant, that is, by synthesis of organic substances from inorganic substances using the energy of light.

holoplankton [ZOO] Organisms that live their complete life cycle in the floating state.

Holoptychidae [PALEON] A family of extinct lobefin fishes in the order Osteolepiformes.

holorhinal [VERT ZOO] Among birds, having a rounded anterior margin on the nasal bones.

Holostei [VERT ZOO] An infraclass of fishes in the subclass Actinopterygii descended from the Chondrostei and ancestral to the Teleostei.

holostome [INV ZOO] A type of adult digenetic trematode

HOLOCOCCOLITH

A photograph of a holococcolith. *(A. McIntyre, Lamont-Doherty Geological Observatory of Columbia University)*

HOLOSTOME

Diagram of ventral surface of adult holostome showing adhesive organs. *(From R. M. Cable, An Illustrated Laboratory Manual of Parasitology, Burgess, 1940)*

HOLOTHUROIDEA

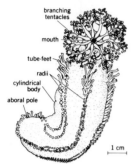

Cucumaria, a representative holothurian.

HOLOTRICHIA

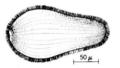

Prorodon, a primitive holotrich.

having a portion of the ventral surface modified as a complex adhesive organ.

Holothuriidae [VERT ZOO] A family of aspidochirotacean echinoderms in the order Aspidochirotida possessing tentacular ampullae and only the left gonad.

Holothuroidea [INV ZOO] The sea cucumbers, a class of the subphylum Echinozoa characterized by a cylindrical body and smooth, leathery skin.

Holothyridae [INV ZOO] The single family of the acarine suborder Holothyrina.

Holothyrina [INV ZOO] A suborder of mites in the order Acarina which are large and hemispherical with a deep-brown, smooth, heavily sclerotized cuticle.

Holotrichia [INV ZOO] A major subclass of the protozoan class Ciliatea; body ciliation is uniform with cilia arranged in longitudinal rows.

holotrichous [BIOL] Having a covering of evenly distributed cilia over the body.

holotype [SYST] A nomenclatural type for the single specimen designated as "the type" by the orginal author at the time of publication of the original description.

holozoic [ZOO] Obtaining food in the manner of most animals, by ingesting complex organic matter.

Holuridae [PALEON] A group of extinct chondrostean fishes in the suborder Palaeoniscoidei distinguished in having lepidotrichia of all fins articulated but not bifurcated, fins without fulcra, and the tail not cleft.

homacanthous [VERT ZOO] Having symmetrical spines in the dorsal fin.

Homacodontidae [PALEON] A family of extinct palaeodont mammals in the superfamily Dichobunoidea.

Homalopteridae [VERT ZOO] A small family of cypriniform fishes in the suborder Cyprinoidei.

Homalorhagae [INV ZOO] A suborder of the class Kinorhyncha having a single dorsal plate covering the neck and three ventral plates on the third zonite.

Homalozoa [INV ZOO] A subphylum of echinoderms characterized by the complete absence of radial symmetry.

Homan's sign [MED] Pain in the calf and popliteal area on passive dorsiflexion of the foot, indicating deep venous thrombosis of the calf. Also known as dorsiflexion sign.

Homaridae [INV ZOO] A family of marine decapod crustaceans containing the lobsters.

homatropine [ORG CHEM] $C_{16}H_{21}O_3N$ An alkaloid that causes pupil dilation and paralysis of accommodation.

homaxial [BIOL] Having all axes equal.

homeosis *See* heteromorphosis.

homeostasis [BIOL] In higher animals, the maintenance of an internal constancy and an independence of the environment.

homeotherm [VERT ZOO] An animal (mammal or bird) which routinely regulates and maintains body temperature.

Hominidae [VERT ZOO] A family of primates in the superfamily Hominoidea containing one living species, *Homo sapiens*.

Hominoidea [VERT ZOO] A superfamily of the order Primates comprising apes and humans.

homiomerous [BOT] Having fungal and algal components of a lichen distributed in about equal proportions in the thallus.

Homo [VERT ZOO] The genus of human beings, including modern humans and many extinct species.

Homobasidiomycetidae [MYCOL] A subclass of basidiomycetous fungi in which the basidium is not divided by cross walls.

homobium [BOT] The mutually beneficial association of fungus and alga in lichens.

homocarpous [BOT] Bearing a single type of fruit.

homocercal [VERT ZOO] Pertaining to the caudal fin of certain fishes which has almost equal lobes, with the vertebral column terminating near the middle of the base.

homochlamydeous [BIOL] Having all members of the perianth similar or not differentiated into calyx or corolla.

homochromy [ZOO] A form of protective coloration whereby the individual blends into the background.

homocysteine [BIOCHEM] $C_4H_9O_2NS$ An amino acid formed in animals by demethylation of methionine.

homocystinuria [MED] A hereditary metabolic disorder in which homocysteine appears in the urine because cystathionine synthetase activity is absent; there is also malpositioning of the lens, and mental retardation.

homodont [VERT ZOO] Having all teeth similar in form; characteristic of nonmammalian vertebrates.

homodynamic [INV ZOO] Developing through continuous successive generations without a diapause; applied to insects.

homoeandrous [BOT] Having uniform stamens.

homoecious [BIOL] Having one host for all stages of the life cycle.

homoeomerous [BOT] Having algae distributed uniformly throughout the thallus of a lichen.

Homo erectus [PALEON] A type of fossil human from the Pleistocene of Java and China representing a specialized side branch in human evolution.

homoerotism [PSYCH] Sexual desire directed toward a member of the same sex; usually sublimated and not expressed.

homogametic sex [GEN] The sex of a species in which the paired sex chromosomes are of equal size and which therefore produces homogametes.

homogamous [BIOL] Of or pertaining to homogamy.

homogamy [BIOL] Inbreeding due to isolation. [BOT] Condition of having all flowers alike.

homogenate [BIOL] A tissue that has been finely divided and mixed.

homogeneous [CHEM] Pertaining to a substance having uniform composition or structure. [SCI TECH] Uniform in structure or composition.

homogentisase [BIOCHEM] The enzyme that catalyzes the conversion of homogentisic acid to fumaryl acetoacetic acid.

homogentisic acid [BIOCHEM] $C_8H_8O_4$ An intermediate product in the metabolism of phenylalanine and tyrosine; found in excess in persons with phenylketonuria and alkaptonuria.

homogony [BOT] Condition of having one type of flower, with stamens and pistil of uniform length.

homograft [BIOL] Graft from a donor transplanted to a genetically dissimilar recipient of the same species. Also known as allograft.

homoiochlamydeous [BOT] Having perianth leaves alike, not differentiated into sepals and petals.

homoiogenetic [EMBRYO] Of a determined part of an embryo, capable of inducing formation of a similar part when grafted into an undetermined field.

Homoistela [PALEON] A class of extinct echinoderms in the subphylum Homalozoa.

homokaryon [MYCOL] A bi- or multinucleate cell having nuclei all of the same kind.

homokaryosis [MYCOL] The condition of a bi- or multinucleate cell having nuclei all of the same kind.

homolecithal [CYTOL] Referring to eggs having small amounts of evenly distributed yolk. Also known as isolecithal.

homologous [BIOL] Pertaining to a structural relation between parts of different organisms due to evolutionary development from the same or a corresponding part, such as the wing of a bird and the pectoral fin of a fish.

homologous serum jaundice [MED] A type of hepatitis caused by a filtrable virus that exists in the blood plasma and may be passed to another person through blood transfusion.

homologous tumor [MED] A neoplasm composed of tissue identical with those of the organ at the site of the tumor.

homomorphism [BOT] Having perfect flowers consisting of only one type.

Homoneura [INV ZOO] A suborder of the Lepidoptera with mandibulate mouthparts, and fore- and hindwings that are similar in shape and venation.

homopetalous [BOT] Having all petals identical.

homophyllous [BOT] Having one kind of leaf.

homoplastic [BIOL] Referring to a graft made from the individual to another of the same species.

homoplasy [BIOL] Correspondence between organs or structures in different organisms acquired as a result of evolutionary convergence or of parallel evolution.

homopolar bond [PHYS CHEM] A covalent bond whose total dipole moment is zero.

Homoptera [INV ZOO] An order of the class Insecta including a large number of sucking insects of diverse forms.

Homo sapiens [VERT ZOO] Modern human species; a large, erect, omnivorous terrestrial biped of the primate family Hominidae.

Homosclerophorida [INV ZOO] An order of primitive sponges of the class Demospongiae with a skeleton consisting of equirayed, tetraxonid, siliceous spicules.

homoserine [BIOCHEM] $C_4H_9O_3N$ An amino acid formed as an intermediate product in animals in the metabolic breakdown of cystathionine to cysteine.

homosexual [BIOL] Of, pertaining to, or being the same sex. [PSYCH] **1.** Of, pertaining to, or exhibiting homosexuality. **2.** One who practices homosexuality.

homosexuality [PSYCH] **1.** State of being sexually attracted to members of the same sex. **2.** A form of homoerotism involving sexual interest without genital expression.

homosexual panic [PSYCH] An acute syndrome that comes as a climax of prolonged tension from unconscious homosexual conflicts or sometimes bisexual tendencies.

homospory [BOT] Production of only one kind of asexual spore.

homothallic [MYCOL] Having genetically compatible hyphae and therefore forming zygospores from two branches of the same mycelium.

homotropous [BOT] Having the radicle directed toward the hilum.

homotype [SYST] A taxonomic type for a specimen which has been compared with the holotype by another than the author of the species and determined by him to be conspecific with it.

homotypy [BIOL] Protective device based on resemblance of shape to the background.

homoxylous [BOT] Pertaining to wood consisting of tracheids and no vessels.

homozygote [GEN] An individual that has identical alleles at one or more loci and therefore produces gametes which are all identical.

homozygous [GEN] Of or pertaining to a homozygote.

honeybee [INV ZOO] *Apis mellifera.* The bee kept for the commercial production of honey; a member of the dipterous family Apidae.

honeycomb [INV ZOO] A mass of wax cells in the form of

hexagonal prisms constructed by honeybees for their brood and honey.

honeycomb coral [PALEON] The common name for members of the extinct order Tabulata; has prismatic sections arranged like the cells of a honeycomb.

honeycomb lung [MED] **1.** Condition of the lung in emphysema. **2.** A lung containing small pus-filled cavities.

honeydew [INV ZOO] The viscous secretion deposited on leaves by many aphids and scale insects; an attractant for ants.

Honey Dew melon [BOT] A variety of muskmelon (*Cucumis melo*) belonging to the Violales; fruit is large, oval, smooth, and creamy yellow to ivory, without surface markings.

honey tube [INV ZOO] Either of a pair of cornicles on the dorsal aspect of one abdominal segment in certain aphids.

hoof [VERT ZOO] **1.** Horny covering for terminal portions of the digits of ungulate mammals. **2.** A hoofed foot, as of a horse.

hoof-and-mouth disease *See* foot-and-mouth disease.

Hookeriales [BOT] An order of the mosses with irregularly branched stems and leaves that appear to be in one plane.

hook gland [INV ZOO] One of the paired longitudinal glands that combine anteriorly to form the head gland in bloodsucking parasitic arthropods.

hookworm [INV ZOO] The common name for parasitic roundworms composing the family Ancylostomidae.

hop [BOT] *Humulus lupulus.* A dioecious liana of the order Urticales distinguished by herbaceous vines produced from a perennial crown; the inflorescence, a catkin, of the female plant is used commercially for beer production.

hophornbeam [BOT] Any tree of the genus *Ostrya* in the birch family recognized by its very scaly bark and the fruit which closely resembles that of the hopvine.

Hoplocarida [INV ZOO] A superorder of the class Crustacea with the single order Stomatopoda.

Hoplonemertini [INV ZOO] An order of unsegmented, ribbonlike worms in the class Enopla; all species have an armed proboscis.

Hoplophoridae [INV ZOO] A family of prawns containing numerous bathypelagic representatives.

hopperburn [PL PATH] A disease of potato and peanut plants caused by a leafhopper which secretes a toxic substance on the leaves, causing browning and shriveling.

hordeolum [MED] A furuncular inflammation of the connective tissue of the eyelids near a hair follicle. Also known as sty.

horizontal plane [ANAT] A transverse plane at right angles to the longitudinal axis of the body.

hormogonium [BOT] Portion of a filament between heterocysts in certain algae; detaches as a reproductive body.

hormone [BIOCHEM] **1.** A chemical messenger produced by endocrine glands and secreted directly into the bloodstream to exert a specific effect on a distant part of the body. **2.** An organic compound that is synthesized in minute quantities in one part of a plant and translocated to another part, where it influences a specific physiological process.

horn [INV ZOO] A tentacle in snails. [VERT ZOO] **1.** The process or structure projecting from the head of many animals; may be made of keratin, bone, or fused hair. **2.** The anterior part of a uterus when the posterior parts are united to form the median corpus uteri. **3.** A tuft of feathers, as in the owl. **4.** A spine in fishes.

hornbeam [BOT] Any tree of the genus *Carpinus* in the birch family distinguished by doubly serrate leaves and by small, pointed, angular winter buds with scales in four rows.

HOOKWORM

A typical hookworm.

HOP

Hop, female inflorescences. (USDA)

HOPHORNBEAM

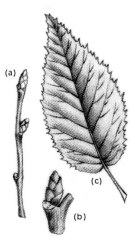

(a)

(c)

(b)

American hophornbeam (*Ostrya virginiana*). (a) Twig. (b) Terminal bud. (c) Leaf.

HORNED TOAD

Horned toad *(Phrynosoma)*, a lizard, uses its tongue to catch insects while on the run.

horned dinosaur [PALEON] Common name for extinct reptiles of the suborder Ceratopsia.

horned liverwort [BOT] The common name for bryophytes of the class Anthocerotae. Also known as hornwort.

horned toad [VERT ZOO] The common name for any of the lizards of the genus *Phrynosoma;* a reptile that resembles a toad but is less bulky.

horned-toad dinosaur [PALEON] The common name for extinct reptiles composing the suborder Ankylosauria.

hornet [INV ZOO] The common name for a number of large wasps in the hymenopteran family Vespidae.

horny coral [INV ZOO] The common name for coelenterate members of the order Gorgonacea.

horse [VERT ZOO] *Equus caballus.* A herbivorous mammal in the family Equidae; the feet are characterized by a single functional digit.

horse chestnut [BOT] *Aesculus hippocastanum.* An ornamental buckeye tree in the order Sapindales, usually with seven leaflets per leaf and resinous buds.

horsehair blight [PL PATH] A fungus disease of tea and certain other tropical plants caused by *Marasmius equicrinis* and characterized by black festoons of mycelia hanging from the branches.

horsepox [VET MED] A disease of horses such as pseudotuberculosis, contagious pustular stomatitis, or a vesicular exanthema.

horseradish [BOT] *Armoracia rusticana.* A perennial crucifer belonging to the order Capparales and grown for its pungent roots, used as a condiment.

horseshoe crab [INV ZOO] The common name for arthropods composing the subclass Xiphosurida, especially the subgroup Limulida.

horseshoe kidney [MED] Congenital fusion of two kidneys at one pole.

horsetail [BOT] The common name for plants of the genus *Equisetum* composing the order Equisetales.

horticulture [BOT] The art and science of growing plants.

host [BIOL] **1.** An organism on or in which a parasite lives. **2.** The dominant partner of a symbiotic or commensal pair.

hostile identification [PSYCH] The assumption by an individual, particularly a child, of socially undesirable characteristics of an older person important to the child so as to gain some special recognition from that person.

hostility [PSYCH] The feeling or display of anger, antagonism, or resistance toward an individual or group.

hot flash [PHYSIO] A sudden transitory sensation of heat, often involving the whole body, due to cessation of ovarian function; a symptom of the climacteric.

hot spot [MOL BIO] A site in a gene at which there is an unusually high frequency of mutation.

housefly [INV ZOO] *Musca domestica.* A dipteran insect with lapping mouthparts commonly found near human habitations; a vector in the transmission of many disease pathogens.

house physician [MED] A physician employed by a hospital.

Houston's valves [ANAT] Transverse folds of the rectal mucosa.

H substance [BIOCHEM] An agent similar to histamine and believed to play a role in local blood vessel response in tissue damage.

Hubbard tank [MED] A large, specially designed tank in which a patient may be immersed for various therapeutic underwater exercises.

huckleberry [BOT] The common name for shrubs of the genus *Gaylussacia* in the family Ericaceae distinguished by an ovary with 10 locules and 10 ovules; the dark-blue berries are edible.

Huhner's test [PATH] An examination of seminal fluid obtained from the vaginal fornix and cervical canal after a specified interval following coitus; used in fertility studies to evaluate spermatozoal survival and activity in the lower female genital tract.

hull [BOT] The outer, usually hard, covering of a fruit or seed. [FOOD ENG] **1.** To remove husks from fruits and seeds, as from ears of corn, nuts, or peas. **2.** To remove the shell of a crustacean or mollusk, as an oyster.

hulless buckwheat *See* tartary buckwheat.

human chorionic gonadotropin [BIOCHEM] A gonadotropic and luteotropic hormone secreted by the chorionic vesicle. Abbreviated HCG. Also known as chorionic gonadotropin.

human ecology [ECOL] The branch of ecology that considers the relations of individual persons and of human communities with their particular environment.

human genetics [GEN] The study of human heredity and variation, including the physical basis, mechanisms of inheritance, inheritance patterns, and population analysis.

humeral [ANAT] Of or pertaining to the humerus or the shoulder region.

humerus [ANAT] The proximal bone of the forelimb in vertebrates; the bone of the upper arm in humans, articulating with the glenoid fossa and the radius and ulna.

humicolous [ECOL] Of or pertaining to plant species inhabiting medium-dry ground.

Humiriaceae [BOT] A family of dicotyledonous plants in the order Linales characterized by exappendiculate petals, usually five petals, flowers with an intrastaminal disk, and leaves lacking stipules.

hummingbird [VERT ZOO] The common name for members of the family Trochilidae; fast-flying, short-legged, weak-footed insectivorous birds with a tubular, pointed bill and a fringed tongue.

humor [PHYSIO] A fluid or semifluid part of the body.

humpback *See* kyphosis.

hunger [PSYCH] The need for food and the physiological and psychological mechanisms regulating food intake.

Hunner's ulcer [MED] A chronic ulcer of the urinary bladder, frequently in association with interstitial cystitis. Also known as elusive ulcer.

Huntington's chorea [MED] A rare hereditary disease of the basal ganglia and cerebral cortex resulting in choreiform (dancelike) movements, intellectual deterioration, and psychosis.

Hurler's syndrome [MED] Mucopolysaccharoidosis I, a hereditary condition transmitted as an autosomal recessive in which there is excessive chondroitin sulfate B and heparin sulfate in the urine and tissues, and which is marked clinically by a complex of symptoms including grotesque skeletal and facial deformities, skin and cardiac changes, clouding of the cornea, and mental deficiency. Also known as gargoylism; lipochondrodystrophy.

Hurthle cells [PATH] Enlarged epithelial cells of the thyroid follicles containing acidophilic cytoplasm, seen most frequently in adenomas.

husk [BOT] The outer coat of certain seeds, particularly if it is a dry, membranous structure.

Hutchinson's freckle *See* melanotic freckle.

Hyaenidae [VERT ZOO] A family of catlike carnivores in the superfamily Feloidea including the hyenas and aardwolf.

Hyaenodontidae [PALEON] A family of extinct carnivorous mammals in the order Deltatheridia.

HUMMINGBIRD

Hummingbird, capable of hovering in flight.

HYAENODONTIDAE

The deltatheridian *Hyaenodon*, original about 4 feet (1.2 meters) long. *(from Scott). (From A. S. Romer, Vertebrate Paleontology, 3d ed., University of Chicago Press, 1966)*

Hyalellidae [INV ZOO] A family of amphipod crustaceans in the suborder Gammaridea.

hyaline [BIOCHEM] A clear, homogeneous, structureless material found in the matrix of cartilage, vitreous body, mucin, and glycogen.

hyaline cartilage [HISTOL] A translucent connective tissue comprising about two-thirds clear, homogeneous matrix with few or no collagen fibrils.

hyaline cast [PATH] A clear, structureless mass of proteinaceous material found in the urine in association with certain kidney diseases.

hyaline membrane [HISTOL] **1.** A basement membrane. **2.** A membrane of a hair follicle between the inner fibrous layer and the outer root sheath.

hyaline membrane disease [MED] A disease occurring during the first few days of neonatal life, characterized by respiratory distress due to formation of a hyalinelike membrane within the alveoli.

hyaline test [INV ZOO] A translucent wall or shell of certain foraminiferans composed of layers of calcite interspersed with separating membranes.

Hyalodictyae [MYCOL] A subdivision of the spore group Dictyosporae characterized by hyaline spores.

Hyalodidymae [MYCOL] A subdivision of the spore group Didymosporae characterized by hyaline spores.

Hyalohelicosporae [MYCOL] A subdivision of the spore group Helicosporae characterized by hyaline spores.

hyaloid artery [EMBRYO] A forward continuation of the central artery of the retina which crosses the vitreous body and ramifies on the posterior surface of the lens capsule.

hyaloid canal [EMBRYO] A canal running postero-anteriorly through the vitreous body through which the hyaloid artery passes in the embryo.

hyaloid fossa [ANAT] A depression for the crystalline lens in the anterior surface of the vitreous body.

hyaloid membrane [ANAT] The limiting membrane surrounding the vitreous body of the eyeball, and forming the suspensory ligament.

Hyalophragmiae [MYCOL] A subdivision of the spore group Phragmosporae characterized by hyaline spores.

hyaloplasm [CYTOL] The optically clear, viscous to gelatinous ground substance of cytoplasm in which formed bodies are suspended.

hyalopterous [INV ZOO] Having transparent wings.

Hyaloscolecosporae [MYCOL] A subdivision of the spore group Scalecosporae characterized by hyaline spores.

Hyalospongia [PALEON] A class of extinct glass sponges, equivalent to the living Hexactinellida, having siliceous spicules made of opaline silica.

Hyalosporae [MYCOL] A subdivision of the spore group Amerosporae characterized by hyaline spores.

Hyalostaurosporae [MYCOL] A subdivision of the spore group Staurosporae characterized by hyaline spores.

hyaluronate [BIOCHEM] A salt or ester of hyaluronic acid.

hyaluronic acid [BIOCHEM] A polysaccharide found as an integral part of the gellike substance of animal connective tissue.

hyaluronidase [BIOCHEM] Any one of a family of enzymes which catalyze the breakdown of hyaluronic acid. Also known as hyaluronate lyase; spreading factor.

Hybodontoidea [PALEON] An ancient suborder of extinct fossil sharks in the order Selachii.

hybrid [GEN] The offspring of genetically dissimilar parents.

hybrid enzyme [BIOCHEM] A form of polymeric enzyme occurring in heterozygous individuals that shows a hybrid

HYALURONIC ACID

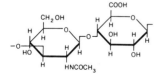

Repeating unit in hyaluronic acid molecule.

molecular form made up of subunits differing in one or more amino acids.

hybridization [GEN] The act or process of producing hybrids.

hybrid merogony [EMBRYO] The fertilization of cytoplasmic fragments of the egg of one species by the sperm of a related species.

hybrid molecule [BIOCHEM] A single molecule, usually protein, peculiar to heterozygotes and containing two structurally different polypeptide chains determined by two different alleles.

hybrid sterility [GEN] Inability to form functional gametes in a hybrid due to disturbances in sex-cell development or in meiosis, caused by incompatible genetic constitution.

hybrid vigor *See* heterosis.

hydathode [BOT] An opening of the epidermis of higher plants specialized for exudation of water.

hydatid [MED] **1.** A cyst formed in tissues due to growth of the larval stage of *Echinococcus granulosus.* **2.** A cystic remnant of an embryonal structure.

hydatiform mole [MED] A benign placental tumor formed as a cystic growth of the chorionic villi. Also known as hydatiform tumor.

hydatiform tumor *See* hydatiform mole.

hydra [INV ZOO] Any species of *Hydra* or related genera, consisting of a simple, tubular body with a mouth at one end surrounded by tentacles, and a foot at the other end for attachment.

Hydrachnellae [INV ZOO] A family of generally fresh-water predacious mites in the suborder Trombidiformes, including some parasitic forms.

hydragogue [MED] Causing the discharge of watery fluid, especially from the bowel.

hydralazine [PHARM] $C_8H_8N_4$ An antihypertensive drug; used as the hydrochloride salt.

hydramnios *See* polyhydramnios.

hydranth [INV ZOO] Nutritive individual in a hydroid colony.

hydrargyrism *See* mercurialism.

hydrarthrosis [MED] An accumulation of fluid in a joint.

hydrase [BIOCHEM] An enzyme that catalyzes removal or addition of water to a substrate without hydrolyzing it.

hydremia [MED] An excessive amount of water in the blood; disproportionate increase in plasma volume as compared with red blood cell volume.

hydric [ECOL] Characterized by or thriving in abundance of moisture.

hydroanemophilous [BOT] Pertaining to or having spores which are ejected and carried away by air currents after the spore-producing structures are moistened.

Hydrobatidae [VERT ZOO] The storm petrels, a family of oceanic birds in the order Procellariiformes.

hydrocarbon [ORG CHEM] One of a very large group of chemical compounds composed only of carbon and hydrogen; the largest source of hydrocarbons is from petroleum crude oil.

hydrocaulus [INV ZOO] Branched, upright stem of a hydroid colony.

hydrocele [MED] Accumulation of fluid in the membranes surrounding the testis.

hydrocephaly [MED] Increased volume of cerebrospinal fluid in the skull.

Hydrocharitaceae [BOT] The single family of the order Hydrocharitales, characterized by an inferior, compound ovary with laminar placentation.

Hydrocharitales [BOT] A monofamilial order of aquatic monocotyledonous plants in the subclass Alismatidae.

HYDATID

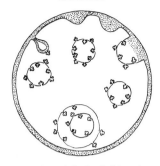

A drawing of the hydatid cyst showing large numbers of attached and free-floating brood capsules, each with one or more scoleces.

HYDRA

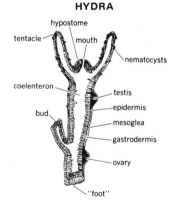

Longitudinal section of *Hydra.* (*From T. I. Storer and R. L. Usinger, General Zoology, 4th ed., McGraw-Hill, 1965*)

hydrocholeresis [MED] Choleresis characterized by an increase of water output, or of a bile relatively low in specific gravity, viscosity, and content of total solids.

hydrochory [BIOL] Dispersal of disseminules by water.

hydrocirrus [INV ZOO] A ring surrounding the mouth in echinoderms.

hydrocladium [INV ZOO] Branchlet of a hydrocaulus.

hydrocoele [INV ZOO] 1. Water vascular system in Echinodermata. 2. Embryonic precursor of the system.

Hydrocorallina [INV ZOO] An order in some systems of classification set up to include the coelenterate groups Milleporina and Stylasterina.

Hydrocorisae [INV ZOO] A subdivision of the Hemiptera containing water bugs with concealed antennae and without a bulbus ejaculatorius in the male.

hydrocortisone [BIOCHEM] $C_{21}H_{30}O_5$ The generic name for 17-hydroxycorticosterone; an adrenocortical steroid occurring naturally and prepared synthetically; its effects are similar to cortisone, but it is more active. Also known as cortisol.

hydrocystoma [MED] A group of clear vesicles, usually located around the eyes, composed of cystic sweat glands.

Hydrodamalinae [VERT ZOO] A monogeneric subfamily of sirenian mammals in the family Dugongidae.

hydroecium [INV ZOO] The closed, funnel-shaped tube at the upper end of coelenterates belonging to the Siphonophora.

hydrogen [CHEM] The first chemical element, symbol H, in the periodic table, atomic number 1, atomic weight 1.00797; under ordinary conditions it is a colorless, odorless, tasteless gas composed of diatomic molecules, H_2; used in manufacture of ammonia and methanol, for hydrofining, for sedulfurization of petroleum products, and to reduce metallic oxide ores.

hydrogenase [BIOCHEM] Enzyme that catalyzes the oxidation of hydrogen.

hydrogen bacteria [MICROBIO] Bacteria capable of obtaining energy from the oxidation of molecular hydrogen.

hydrogen bond [PHYS CHEM] A type of bond formed when a hydrogen atom bonded to atom A in one molecule makes an additional bond to atom B either in the same or another molecule; the strongest hydrogen bonds are formed when A and B are highly electronegative atoms, such as fluorine, oxygen, or nitrogen.

hydrogen equivalent [CHEM] The number of replaceable hydrogen atoms or hydroxyl groups in a molecule of an acid or a base.

hydroid [INV ZOO] 1. The polyp form of a hydrozoan coelenterate. Also known as hydroid polyp; hydropolyp. 2. Any member of the Hydroida.

Hydroida [INV ZOO] An order of coelenterates in the class Hydrozoa including usually colonial forms with a well-developed polyp stage.

hydroid polyp *See* hydroid.

hydrolase [BIOCHEM] Any of a class of enzymes which catalyze the hydrolysis of proteins, nucleic acids, starch, fats, phosphate esters, and other macromolecular substances.

hydrolysis [CHEM] 1. Decomposition or alteration of a chemical substance by water. 2. In aqueous solutions of electrolytes, the reactions of cations with water to produce a weak base or of anions to produce a weak acid.

hydrolytic enzyme [BIOCHEM] A catalyst that acts like a hydrolase.

Hydrometridae [INV ZOO] The marsh treaders, a family of hemipteran insects in the subdivision Amphibicorisae.

hydromorphic [BIOL] Adapted by particular structures to an aquatic environment, for example, organs of water plants.

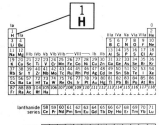

HYDROGEN

Periodic table of the chemical elements showing the position of hydrogen.

hydronasty [BOT] Movement of plants as the result of changes in atmospheric humidity.

hydronephrosis [MED] Accumulation of urine in and distension of the renal pelvis and calyces due to obstructed outflow.

Hydrophilidae [INV ZOO] The water scavenger beetles, a large family of coleopteran insects in the superfamily Hydrophiloidea.

Hydrophiloidea [INV ZOO] A superfamily of coleopteran insects in the suborder Polyphaga.

hydrophilous [ECOL] Inhabiting moist places.

hydrophobia [MED] *See* rabies.

hydrophobic [CHEM] Lacking an affinity for, repelling, or failing to adsorb or absorb water.

Hydrophyllaceae [BOT] A family of dicotyledonous plants in the order Polemoniales distinguished by two carpels, parietal placentation, and generally imbricate corolla lobes in the bud.

hydrophyllium [INV ZOO] A transparent body partly covering the spore sacs of siphonophoran coelenterates.

hydrophyte [BOT] **1.** A plant that grows in a moist habitat. **2.** A plant requiring large amounts of water for growth. Also known as hygrophyte.

hydroplanula [INV ZOO] A coelenterate larval stage between the planula and actinula stages.

hydropolyp *See* hydroid.

hydroponics [BOT] Growing of plants in a nutrient solution with the mechanical support of an inert medium such as sand.

hydropore [INV ZOO] In certain asteroids and echinoids, an opening on the aboral surface of a canal which extends from the ring canal in one of the interradii.

hydropyle [INV ZOO] An area in the cuticular membrane of an embryo structurally adapted for passage of water, as in grasshoppers.

hydrorhiza [INV ZOO] Rootlike structure of a hydroid colony.

hydrosalpinx [MED] A distension of a fallopian tube with fluid.

Hydroscaphidae [INV ZOO] The skiff beetles, a small family of coleopteran insects in the suborder Myxophaga.

hydrosere [ECOL] Community in which the pioneer plants invade open water, eventually forming some kind of soil such as peat or muck.

hydroskeleton [INV ZOO] Water contained within the coelenteron and serving a skeletal function in most coelenterate polyps.

hydrosome [INV ZOO] A hydralike stage in the life history of a coelenterate.

hydrospire [INV ZOO] Either of a pair of flattened tubes composing part of the respiratory system in blastoids.

hydrospore [INV ZOO] Opening into the hydrocoele on the right side in echinoderm larvae.

hydrostome [INV ZOO] The oral chamber of a hydroid polyp.

hydrotaxis [BIOL] Movement of an organism in response to the stimulus of moisture.

hydrotheca [INV ZOO] Cup-shaped portion of the perisarc in some coelenterates that serves to hold and protect a withdrawn hydranth.

hydrotherapy [MED] Treatment of disease by external application of water.

hydrothorax [MED] Collection of serous fluid in the pleural spaces.

hydroureter [MED] Accumulation of urine in and distension of the ureter due to obstructed outflow.

hydroxyamphetamine [PHARM] C_9H_3ON A sympathetic amine used as the hydrobromide salt orally as a drug and locally as a mydriatic and nasal decongestant.

β-hydroxybutyric dehydrogenase [BIOCHEM] The enzyme

HYDROPHILIDAE

A drawing of a water scavenger beetle. *(From T. I. Storer and R. L. Usinger, General Zoology, 3d ed., McGraw-Hill, 1957)*

HYDROXYPROLINE

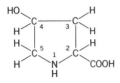

Structural formula of hydroxyproline.

HYLIDAE

North American tree frogs. Green tree frog, *Hyla cinerea* (left), and gray tree frog, *H. versicolor* (right). (*American Museum of Natural History photograph*)

HYLOBATIDAE

A representative of the Hylobatidae.

HYMENOSTOMATIDA

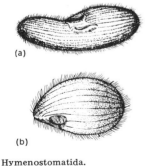

Hymenostomatida.
(*a*) *Paramecium*.
(*b*) *Tetrahymena*.

that catalyzes the conversion of L-β-hydroxybutyric acid to acetoacetic acid by dehydrogenation.

hydroxylase [BIOCHEM] Any of several enzymes that catalyze certain hydroxylation reactions involving atomic oxygen.

hydroxyproline [BIOCHEM] $C_5H_9O_3N$ An amino acid that is essentially limited to structural proteins of the collagen type.

5-hydroxytryptamine *See* serotonin.

8-hydroxyxanthine *See* uric acid.

Hydrozoa [INV ZOO] A class of the phylum Coelenterata which includes the fresh-water hydras, the marine hydroids, many small jellyfish, a few corals, and the Portuguese man-of-war.

hyena [VERT ZOO] An African carnivore represented by three species of the family Hyaenidae that resemble dogs but are more closely related to cats.

Hyeniales [PALEOBOT] An order of Devonian plants characterized by small, dichotomously forked leaves borne in whorls.

Hyeniatae *See* Hyeniopsida.

Hyeniopsida [PALEOBOT] An extinct class of the division Equisetophyta. Formerly known as Hyeniatae.

hygiene [MED] The science that deals with the principles and practices of good health.

Hygrobiidae [INV ZOO] The squeaker beetles, a small family of coleopteran insects in the suborder Adephaga.

hygrochasy [BOT] Moisture-induced dehiscence of seed vessels.

hygroma [MED] A congenital disorder in which a lymph-filled cystic cavity is formed from distended lymphatics.

hygromycin [MICROBIO] $C_{25}H_{33}O_{12}N$ A weakly acidic, soluble antibiotic with a fairly broad spectrum, produced by a strain of *Streptomyces hygroscopicus*.

hygrophilous [ECOL] Living in moist or marshy places.

hygrophyte *See* hydrophyte.

hylaea *See* tropical rainforest.

Hylidae [VERT ZOO] The tree frogs, a large amphibian family in the suborder Procoela; many are adapted to arboreal life, having expanded digital disks.

Hylobatidae [VERT ZOO] A family of anthropoid primates in the superfamily Hominoidea including the gibbon and the siamang of southeastern Asia.

hylophagous [ZOO] Feeding on wood, as termites.

hylophyte [ECOL] A plant inhabiting woods that are usually moist.

hymen [ANAT] A mucous membrane partly closing off the vaginal orifice. Also known as maidenhead.

hymenium [MYCOL] The outer, sporebearing layer of certain fungi or their fruiting bodies.

Hymenomycetes [MYCOL] A group of the Homobasidiomycetidae including forms such as mushrooms and pore fungi in which basidia are formed in an exposed layer (hymenium) and basidiospores are borne asymmetrically on slender stalks.

hymenophore [MYCOL] Portion of a sporophore that bears the hymenium.

hymenopodium [MYCOL] **1.** Tissue beneath the hymenium in certain fungi. **2.** A genus of the Moniliales.

Hymenoptera [INV ZOO] A large order of insects including ants, wasps, bees, sawflies, and related forms; head, thorax and abdomen are clearly differentiated; wings, when present, and legs are attached to the thorax.

hymenopterous [INV ZOO] Having thin, membranous wings, as Hymenoptera.

Hymenostomatida [INV ZOO] An order of ciliated protozoans in the subclass Holotrichia having fairly uniform ciliation and a definite buccal ciliature.

Hynobiidae [VERT ZOO] A family of salamanders in the suborder Cryptobranchoidea.

hyobranchium [VERT ZOO] A Y-shaped bone supporting the tongue and tongue muscles in a snake.

Hyocephalidae [INV ZOO] A monospecific family of hemipteran insects in the superfamily Pentatomorpha.

hyoglossus [ANAT] An extrinsic muscle of the tongue arising from the hyoid bone.

hyoid [ANAT] **1.** A bone or complex of bones at the base of the tongue supporting the tongue and its muscles. **2.** Of or pertaining to structures derived from the hyoid arch.

hyoid arch [EMBRYO] Either of the second pair of pharyngeal segments or gill arches in vertebrate embryos.

hyoid tooth [VERT ZOO] One of a number of teeth on the tongue of fishes.

hyomandibular cleft [EMBRYO] The space between the hyoid arch and the mandibular arch in the vertebrate embryo.

hyomandibular pouch [EMBRYO] A portion of the endodermal lining of the pharyngeal cavity which separates the paired hyoid and mandibular arches in vertebrate embryos.

hyoplastron [VERT ZOO] The second lateral plate in the ventral part of the shell of Chelonia.

Hyopssodontidae [PALEON] A family of extinct mammalian herbivores in the order Condylarthra.

hyostylic [VERT ZOO] Having the jaws and cranium connected by the hyomandibular, as certain fishes.

hypacusia [MED] Impairment of hearing.

hypalgesia [PHYSIO] Diminished sensitivity to pain.

hypandrium [INV ZOO] A plate covering the genitalia on the ninth abdominal segment of certain male insects.

hypanthium [BOT] Expanded receptacle margin to which the sepals, petals, and stamens are attached in some flowers.

hypantrum [VERT ZOO] In reptiles, a notch on the anterior portion of the neural arch that articulates with the hyposphene.

hypaxial musculature [ANAT] The ventral portion of the axial musculature of vertebrates including subvertebral flank and ventral abdominal muscle groups.

hyperacid [PHYSIO] Containing more than the normal concentration of acid in the gastric juice.

hyperactivity [PHYSIO] Excessive or pathologic activity.

hyperadrenalism [MED] Hypersecretion of adrenal hormones marked by increased basal metabolism, decreased sugar tolerance, and glycosuria. Also known as hypercorticism.

hyperadrenocorticism [MED] Hypersecretion of adrenocortical hormones resulting in Cushing's syndrome, or virilism.

hyperaldosteronism [MED] Hypersecretion of aldosterone by the adrenal cortex.

hyperalgesia [PHYSIO] Increased or heightened sensitivity to pain stimulation.

hyperbaric [MED] Pertaining to an anesthetic solution with a specific gravity greater than that of the cerebrospinal fluid.

hyperbaric chamber [ENG] A specially equipped pressure vessel used in medicine and physiological research to administer oxygen at elevated pressures.

hyperbaric medicine [PHARM] Any agent having a specific gravity greater than that of spinal fluid; used for spinal anesthesia.

hyperbilirubinemia [MED] **1.** Excessive amounts of bilirubin in the blood. **2.** A severe, prolonged physiologic jaundice.

hypercalcemia [MED] Excessive amounts of calcium in the blood. Also known as calcemia.

hypercapnia [MED] Excessive amount of carbon dioxide in the blood.

hyperchlorhydria [MED] Excessive secretion of hydrochloric acid in the stomach.

hypercholesteremia [MED] Elevated cholesterol levels in the blood.

hyperchromatic [BIOL] Staining more intensely than normal.

hyperchromatism [PATH] **1.** Excessive pigment formation in the skin. **2.** A condition in which cells or parts of cells stain more intensely than is normal.

hyperchromic [PATH] Pertaining to increased hemoglobin content in erythrocytes due to increased cell thickness, not increased hemoglobin concentration.

hyperchromic anemia [MED] Any of several blood disorders in which erythrocytes show an increase in hemoglobin and a reduction in number.

hypercoagulability [MED] Coagulation of blood more rapidly than normal.

hypercoracoid [VERT ZOO] The upper of two bones at the base of the pectoral fin in teleosts.

hyperdynamic ileus *See* spastic ileus.

hyperemesis [MED] Excessive vomiting.

hyperemesis gravedorium [MED] Pernicious vomiting in pregnancy.

hyperemia [MED] An excess of blood within an organ or tissue caused by blood vessel dilation or impaired drainage, especially of the skin.

hyperergia [IMMUNOL] An altered state of reactivity to antigenic materials, in which the response is more marked than usual; one form of allery or pathergy.

hyperesthesia [PHYSIO] Increased sensitivity or sensation.

hypergamesis [PHYSIO] The process by which the female absorbs excess spermatozoa during fertilization.

hypergammaglobulinemia [MED] Increased blood levels of gamma globulin, usually associated with hepatic disease.

hyperglobulinemia [MED] Increased blood levels of globulin.

hyperglycemia [MED] Excessive amounts of sugar in the blood.

hyperglycinenemia [MED] A hereditary metabolic disorder of males in which blood levels of glycine are excessive, resulting in vomiting, dehydration, osteoporosis, and mental retardation.

hyperhidrosis [MED] Excessive sweating, which may be localized or generalized, chronic or acute, and often accumulating in visible drops on the skin. Also known as ephidrosis; polyhidrosis; sudatoria.

Hyperiidea [INV ZOO] A suborder of amphipod crustaceans distinguished by large eyes which cover nearly the entire head.

hyperimmune antibody [IMMUNOL] An antibody having the characteristics of a blocking antibody.

hyperimmune serum [IMMUNOL] An antiserum that provides a very high degree of immunity due to a high antibody titer.

hyperinsulinism [MED] Condition caused by abnormally high levels of insulin in the blood.

hyperkalemia *See* hyperpotassemia.

hyperkeratosis [MED] **1.** Hypertrophy of the cornea. **2.** Hypertrophy of the horny layer of the skin.

hyperkinesia [MED] Excessive and usually uncontrollable muscle movement.

hyperlipemia [MED] Excessive amounts of fat in the blood.

Hypermastigida [INV ZOO] An order of the multiflagellate protozoans in the class Zoomastigophorea; all inhabit the alimentary canal of termites, cockroaches, and woodroaches.

hypermetabolism [MED] Any state in which there is an increase in basal metabolic rate.

hypermetamorphism [INV ZOO] Type of embryological development in certain insects in which one or more stages have been interpolated between the full-grown larva and the adult.

hypermetropia [MED] A defect of vision resulting from too short an eyeball so that unaccommodated rays focus behind the retina. Also known as farsightedness; hyperopia.

hypermotility [MED] Increased motility, as of the stomach or intestines.

hypernatremia [MED] Excessive amounts of sodium in the blood.

hypernephroma *See* renal-cell carcinoma.

Hyperoartii [VERT ZOO] A superorder in the subclass Monorhina distinguished by the single median dorsal nasal opening leading into a blind hypophyseal sac.

hyperopia *See* hypermetropia.

hyperosmia [MED] An abnormally acute sense of smell.

hyperostosis [MED] Hypertrophy of bony tissue.

Hyperotreti [VERT ZOO] A suborder in the subclass Monorhina distinguished by the nasal opening which is located at the tip of the snout and communicates with the pharynx by a long duct.

hyperoxemia [MED] Extreme acidity of the blood.

hyperparasite [ECOL] An organism that is parasitic on other parasites.

hyperparathyroidism [MED] Condition caused by increased functioning of the parathyroid glands.

hyperpathia [MED] An exaggerated or excessive perception of or response to any stimulus as being disagreeable or painful.

hyperperistalsis [MED] An increase in the rate and depth of the peristaltic waves.

hyperphosphaturia [MED] An excess of phosphates in the urine. Also known as phosphaturia.

hyperpituitarism [MED] Any abnormal condition resulting from overactivity of the anterior pituitary.

hyperplasia [MED] Increase in cell number causing an increase in the size of a tissue or organ.

hyperploid [GEN] Having one or more chromosomes or parts of chromosomes in excess of the haploid number, or of a whole multiple of the haploid number.

hyperploidy [GEN] The condition or state of being hyperploid.

hyperpnea [MED] Increase in depth and rate of respiration.

hyperpotassemia [MED] Excessive amounts of potassium in the blood. Also known as hyperkalemia.

hyperproteinemia [MED] Excessive protein levels in the blood.

hyperpyrexia [MED] Extremely high fever.

hyperreflexia [MED] A condition of abnormally increased reflex action.

hyperresonance [MED] Exaggeration of normal resonance on percussion of the chest; heard chiefly in pulmonary emphysema and pneumothorax.

hypersensitivity [IMMUNOL] The state of being abnormally sensitive, especially to allergens; responsible for allergic reactions.

hypersensitization [IMMUNOL] The process of producing hypersensitivity.

hypersomnia [MED] Excessive sleepiness.

hypersplenism [MED] Condition caused by abnormal spleen activity.

hypertension [MED] Abnormal elevation of blood pressure, generally regarded to be levels of 165 systolic and 95 diastolic.

hyperthecosis [PATH] Abnormal thickening of the inner layer of the Graafian follicle with increased leutein formation.

HYPHOMICROBIACEAE

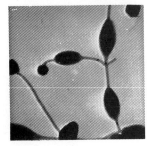

Electron micrograph of
Rhodomicrobium vannielii,
pink to orange in color and
photosynthetic.

HYPNINEAE

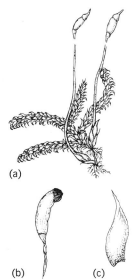

Hypnum reptile. (a) Portion
of plant, with stems shortened.
(b) Urn and peristome.
(c) Faintly bicostate leaf.
*(From W. H. Welch, Mosses of
Indiana, Indiana Department of
Conservation, 1957)*

hyperthermia [PHYSIO] A condition of elevated body temperature.

hyperthyroidism [MED] The constellation of signs and symptoms caused by excessive thyroid hormone in the blood, either from exaggerated functional activity of the thyroid gland or from excessive administration of thyroid hormone, and manifested by thyroid enlargement, emaciation, sweating, tachycardia, exophthalmos, and tremor. Also known as exophthalmic goiter; Grave's disease; thyrotoxicosis; toxic goiter.

hyperthyrotropinism [MED] Excessive thyrotropic hormone secretion by the adenohypophysis.

hypertonia [MED] Abnormal increase in muscle tonicity.

hypertonic [PHYSIO] **1.** Excessive or above normal in tone or tension, as a muscle. **2.** Having an osmotic pressure greater than that of physiologic salt solution or of any other solution taken as a standard.

hypertonic bladder [MED] Hypertonia of the urinary bladder.

hypertonic contracture [MED] Prolonged muscular spasms in spastic paralysis.

Hypertragulidae [PALEON] A family of extinct chevrotainlike pecoran ruminants in the superfamily Traguloidea.

hypertrophic arthritis *See* degenerative joint disease.

hypertrophic gastritis [MED] Chronic inflammation of the stomach with hypertrophy of the mucosa and rugae.

hypertrophy [PATH] Increase in cell size causing an increase in the size of an organ or tissue.

hyperuricemia [MED] Abnormally high level of uric acid in the blood. Also known as lithemia.

hyperventilation [MED] Increase in air intake or of the rate or depth of respiration.

hypervitaminosis [MED] Condition caused by intake of toxic amounts of a vitamin.

hypesthesia [MED] Reduced or subnormal tactile sensibility.

hypha [MYCOL] One of the filaments composing the mycelium of a fungus.

hyphidium [MYCOL] A sterile hymenial structure of hyphal origin.

Hyphochytriales [MYCOL] An order of aquatic fungi in the class Phycomycetes having a saclike to limited hyphal thallus and zoospores with two flagella.

Hyphochytridiomycetes [MYCOL] A class of the true fungi; usually grouped with other classes under the general term Phycomycetes.

hyphoid [MYCOL] Hyphalike.

Hyphomicrobiaceae [MICROBIO] A family of bacteria in the order Hyphomicrobiales; cells occur in free-floating groups with individual cells attached to each other by a slender filament.

Hyphomicrobiales [MICROBIO] An order of bacteria in the class Schizomycetes containing forms that multiply by budding.

Hyphomicrobium [MICROBIO] A genus of nonpigmented, gram-negative, aerobic bacteria in the family Hyphomicrobiaceae; cells usually occur in groups attached by slender filaments.

hyphopodium [MYCOL] Hypha with a haustorium in certain ectoparasitic fungi.

Hypnineae [BOT] A suborder of the Hypnobryales characterized by complanate, glossy plants with ecostate or costate leaves and paraphyllia rarely present.

hypnoanalysis [PSYCH] Technique used in psychotherapy combining hypnosis with psychoanalysis.

Hypnobryales [BOT] An order of mosses composed of procumbent and pleurocumbent plants with usually symmetrical leaves arranged in more than two rows.

hypnosis [PSYCH] An altered state of consciousness in which the individual is more susceptible to suggestion and in which regressive behavior may spontaneously occur.

hypnotherapy [MED] Treatment of disease by means of hypnotism.

hypnotic [PHARM] A drug which induces sleep. Also known as somnificant; soporific.

hypnotism [PSYCH] The practice or study of inducing hypnosis.

hypnotize [PSYCH] To induce a state of hypnosis.

hypo [INORG CHEM] *See* sodium thiosulfate. [PSYCH] Informal term for a hypochondriac or hypochondria.

hypoachene [BOT] A small dry, indehiscent, one-seeded fruit developed from an inferior ovary.

hypoadrenia [MED] Reduced functioning of the adrenal glands. Also known as hypoadrenalism.

hypoadrenocorticism [MED] Lowered or subnormal adrenal cortex activity.

hypoalbuminemia [MED] Abnormally low levels of albumin in the blood.

hypoallergenic [PHARM] Having a low tendency to induce allergic reactions; used particularly for formulated dermatologic preparations.

hypoarion [VERT ZOO] A small lobe in most teleosts below the optic lobes.

hypobaric [MED] Pertaining to an anesthetic solution of specific gravity lower than the cerebrospinal fluid. [PHYS] Having less weight or pressure.

hypobasal [BOT] Located posterior to the basal wall.

hypobranchial musculature [ANAT] The ventral musculature in vertebrates extending from the pectoral girdle forward to the hyoid arch, chin, and tongue.

hypocalcemia [MED] Condition in which there are reduced levels of calcium in the blood.

hypocalcification [MED] Reduction of normal amounts of mineral salts in calcified tissue.

hypocalciuria [MED] Decreased excretion of calcium in the urine.

hypocapnia [MED] Reduced or subnormal blood levels of carbon dioxide.

hypocarp [BOT] A fleshy, modified, sometimes edible fruit, as cashew apple.

hypocentrum [VERT ZOO] The ventral part of the vertebral centrum, arising below the nerve cord by fusion of the lower arcualia of the anterior of the two arches from which the vertebra is formed, in certain fishes and primitive reptiles.

hypocercal [VERT ZOO] Having the notochord ending in the lower lobe of the caudal fin.

hypochil [BOT] Lower portion of the lip in orchids. Also known as hypochillium.

Hypochilidae [INV ZOO] A family of true spiders in the order Araneida.

hypochillium *See* hypochil.

Hypochilomorphae [INV ZOO] A monofamilial suborder of spiders in the order Araneida.

hypochloremia [MED] Reduction in the amount of blood chlorides.

hypochlorhydria [MED] Reduction in the hydrochloric acid content of gastric juice.

hypocholesterolemia [MED] Subnormal levels of serum cholesterol.

hypochondriac [PSYCH] A person affected with hypochondriasis.

hypochondriac region [ANAT] The upper, lateral abdominal region just below the ribs on each side of the body.

hypochondriasis [PSYCH] A chronic condition in which the patient is morbidly concerned with his own health and believes himself suffering from grave bodily diseases.

hypochord [VERT ZOO] A transitory longitudinal rod appearing just below the notochord in amphibian embryos.

hypochromia [PATH] Lack of complete saturation of .the erythrocyte stroma with hemoglobin, as judged by pallor of the unstained or stained erythrocytes when examined microscopically.

hypochromic microcytic anemia [MED] An anemia associated with erythrocytes of reduced size and hemoglobin content.

hypocleidium [ANAT] The median, ventral bone between clavicles. [VERT ZOO] Median process on the wishbone of birds.

hypocone [ANAT] The posterior inner cusp of an upper molar. [INV ZOO] Region of a dinoflagellate posterior to the girdle.

hypoconid [VERT ZOO] The posterobuccal cusp of a lower molar.

hypoconule [VERT ZOO] The fifth or distal cusp of an upper molar.

Hypocopridae [INV ZOO] A small family of coleopteran insects in the superfamily Cucujoidea.

hypocoracoid [VERT ZOO] The lower of two bones at the base of the pectoral fin in teleosts.

hypocotyl [BOT] The portion of the embryonic plant axis below the cotyledon.

hypocrateriform [BIOL] Saucer-shaped.

Hypocreales [MYCOL] An order of fungi belonging to the Ascomycetes and including several entomophilic fungi.

hypodactylum [VERT ZOO] The ventral surface of the toes of a bird.

Hypodermatidae [INV ZOO] The warble flies, a family of myodarian cyclorrhaphous dipteran insects in the subsection Calypteratae.

hypodermic syringe *See* syringe.

hypodermoclysis [MED] Subcutaneous injections of large quantities of fluid for therapeutic purposes.

hypoergia [IMMUNOL] A state of less than normal reactivity to antigenic materials, in which the response is less marked than usual; one form of allergy or pathergy.

hypogammaglobulinemia [MED] Reduced blood levels of gamma globulin.

hypogeous [BIOL] Living or maturing below the surface of the ground.

hypoglossal nerve [ANAT] The twelfth cranial nerve; a paired motor nerve in tetrapod vertebrates innervating tongue muscles; corresponds to the hypobranchial nerve in fishes.

hypoglossal nucleus [ANAT] A long nerve nucleus throughout most of the length of the medulla oblongata; cells give rise to the hypoglossal nerve fibers.

hypoglottis [ANAT] The undersurface of the tongue. [INV ZOO] A division of the labium or lip of beetles.

hypoglycemia [MED] Condition caused by low levels of sugar in the blood.

hypognathous [INV ZOO] Referring to the head of insects whose mouthparts are ventral. [VERT ZOO] Referring to those animals whose lower jaw is slightly longer than the upper.

hypogonadism [MED] Reduced hormonal secretion by the testes or ovaries.

hypogynium [BOT] Structure that supports the ovary in plants such as sedges.

hypogynous [BOT] Having all flower parts attached to the receptacle below the pistil and free from it.

hypoischium [VERT ZOO] A small bony rod passing posteriorly from ischial symphsis and supporting the ventral wall of the cloaca in most lizards.

hypokinesia [MED] Subnormal muscular movements.

hypokinetic syndrome [MED] General decrease in motor functions due to a form of minimal brain dysfunction.

hypomenorrhea [MED] A deficient amount of menstrual flow at the regular period.

hypomere [EMBRYO] The lateral or lower mesodermal plate zone in vertebrate embryos. [INV ZOO] The basal portion of certain sponges that contain no flagellated chambers.

hypomeron [INV ZOO] The lateral inflexed side of the prothorax in Coleoptera.

hypometabolism [MED] Metabolism below the normal rate.

hypomorph [GEN] An allele having an effect similar to the normal allele, but being less active.

hypomotility [MED] Decreased motility, especially of the gastrointestinal tract.

hyponasty [BOT] A nastic movement involving inward and upward bending of a plant part.

hyponatremia [MED] Subnormal or reduced blood sodium levels.

hyponychium [HISTOL] The thickened stratum corneum of the epidermis, which lies under the free edge of the nail.

hypoovarianism [MED] Decrease in ovarian endocrine activity.

hypoparathyroidism [MED] Condition caused by insufficient functioning of the parathyroid gland.

hypopetalous [BOT] Having the petals inserted below, and not adherent to, the pistil.

hypophalangism [MED] Congenital absence of one or more phalanges in a finger or toe.

hypopharynx [ANAT] *See* laryngopharynx. [INV ZOO] A sensory, tonguelike structure on the floor of the mouth of many insects; sometimes modified for piercing.

hypophosphatasia [MED] **1.** Alkaline phosphatase deficiency. **2.** A hereditary metabolic disorder characterized by subnormal amounts of alkaline phosphatase in the tissues.

hypophragm [INV ZOO] The lid or epiphragm that closes the shell opening in some gastropods.

hypophrenia [PSYCH] Mental retardation.

hypophyseal cachexia *See* Simmonds' disease.

hypophysectomy [MED] Surgical removal of the pituitary gland.

hypophysis [ANAT] A small rounded endocrine gland which lies in the sella turcica of the sphenoid bone and is attached to the floor of the third ventricle of the brain in all craniate vertebrates. Also known as pituitary gland.

hypopituitarism [MED] Condition caused by insufficient secretion of pituitary hormones, especially of the adenohypophysis. Also known as panhypopituitarism.

hypopituitary cachexia *See* Simmonds' disease.

hypoplasia [MED] Failure of a tissue or organ to achieve complete development.

hypoplastic dwarf [MED] A normally proportioned individual of subnormal size.

hypoplastron [VERT ZOO] Either of the third pair of lateral bony plates in the plastron of most turtles.

hypopleura [INV ZOO] Sclerite above and in front of the hind coxa in Diptera.

hypoploid [GEN] Having one or more less chromosomes, or

HYPOGYNOUS

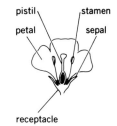

Flower arrangement and parts on receptacle in hypogynous flowers.

parts of chromosomes, than a whole multiple of the haploid number.

hypoploidy [GEN] The condition or state of being hypoploid.

hypoproct [INV ZOO] The extension of the terminal abdominal segment along the medial line beneath the anus, in Diplopoda and some Insecta.

hypoproliferative anemia [MED] Decreased concentration of hemoglobin and number of red blood cells due to subnormal numbers of erythrocyte primordial cells in relation to the stimulus of anemia.

hypoproteinemia [MED] Abnormally low levels of protein in the blood.

hypoprothrombinemia [MED] Deficiency of prothrombin in the blood.

hypopus [INV ZOO] The resting larval stage of certain mites.

hypopygium [INV ZOO] A modified ninth abdominal segment together with the copulatory apparatus in Diptera.

hyporeactive [MED] Characterized by decreased responsiveness to stimuli.

hyporeflexia [MED] A condition in which reflexes are below normal, due to a variety of causes.

hyposensitivity [MED] Condition marked by diminished sensitivity to stimuli.

hyposensitization *See* desensitization.

hypospadias [MED] **1.** Congenital anomaly in which the urethra opens on the ventral surface of the penis or in the perineum. **2.** Congenital anomaly in which the urethra opens into the vagina.

hypospermatogenesis [MED] Decreased sperm production.

hyposphene [VERT ZOO] A process on the neural arch of vertebra of certain reptiles that is shaped like a wedge and fits into the hypantrum.

hypostatic [GEN] Subject to being suppressed, as a gene that can be suppressed by a nonallelic gene.

hyposthenia [MED] Weakness; subnormal strength.

hyposthenuria [MED] The secretion of urine of low specific gravity.

hypostome [INV ZOO] **1.** Projection surrounding the oral aperture in many coelenterate polyps. **2.** Anteroventral part of the head in Diptera. **3.** Median ventral mouthpart in ticks. **4.** Raised area on the posterior oral margin in crustaceans.

hypostracum [INV ZOO] The innermost layer of the cuticle of ticks lying above the hypodermis.

hyposulculus [INV ZOO] A groove of the siphonoglyph below the pharynx in anthozoans.

hyposynergia [MED] Defective coordination.

hypotarsus [VERT ZOO] A process on the tarsometatarsal bone in birds.

hypotension [MED] Abnormally low blood pressure, commonly considered to be levels below 100 diastolic and 40 systolic.

hypothalamic center [ANAT] Any of the neural centers which regulate autonomic functions.

hypothalamoneurohypophyseal tract [ANAT] A bundle of nerve fibers connecting the supraoptic and paraventricular neurons of the hypothalamus with the infundibular stem and neurohypophysis.

hypothalamus [ANAT] The floor of the third brain ventricle; site of production of several substances that act on the adenohypophysis.

hypothallus [BOT] A thin layer of black nonlichenized hyphae which projects beyond the main thallus in lichens.

hypotheca [INV ZOO] **1.** The lower valve of a diatom frustule. **2.** Covering on the hypocone in dinoflagellates.

hypothecium [BOT] The dense layer of hyphae located just below the hymenium in lichens.

hypothenar [ANAT] Of or pertaining to the prominent portion of the palm above the base of the little finger.

hypothermia [PHYSIO] Condition of reduced body temperature in homeotherms.

hypothyroidism [MED] Condition caused by deficient secretion of the thyroid hormone.

hypotonia [MED] Decrease of normal tonicity or tension, especially diminution of intraocular pressure or of muscle tone.

hypotonic [PHYSIO] **1.** Pertaining to subnormal muscle strength or tension. **2.** Referring to a solution with a lower osmotic pressure than physiological saline.

Hypotrichida [INV ZOO] An order of highly specialized protozoans in the subclass Spirotrichia characterized by cirri on the ventral surface and a lack of ciliature on the dorsal surface.

hypotype [SYST] A specimen of a species, which, though not a member of the original type series, is known from a published description or listing.

hypovitaminosis [MED] Condition due to deficiency of an essential vitamin.

hypovolemia [MED] Low blood volume.

hypoxanthine [BIOCHEM] $C_5H_4ON_4$ An intermediate product derived from adenine in the hydrolysis of nucleic acid.

hypoxanthylic acid *See* inosinic acid.

hypoxemia *See* hypoxia.

hypoxia [MED] Oxygen deficiency; any state wherein a physiologically inadequate amount of oxygen is available to or is utilized by tissue, without respect to cause or degree. Also known as hypoxemia.

hypoxic encephalopathy [MED] Brain damage syndrome caused by hypoxia.

hypozygal [INV ZOO] In comatulids, the proximal member of adjacent brachials in an articulation.

hypsicephalic [ANTHRO] Having a high forehead with a length-height index of 62.6 or more.

hypsiconch [ANTHRO] Having high orbits with an orbital index of 89 or more.

hypsicranial [ANTHRO] Having a high skull with a length-height index of 75 or more.

hypsidolichocephalic [ANTHRO] Having a head that is high and narrow, high and long, or high, long, and narrow.

hypsistenocephalic [ANTHRO] Having a very high and narrow head.

hypsodont [VERT ZOO] Of teeth, having crowns that are high or deep and roots that are short.

hypural [VERT ZOO] Of or pertaining to the bony structure formed by fusion of the hemal spines of the last few vertebrae in most teleost fishes.

Hyracodontidae [PALEON] The running rhinoceroses, an extinct family of perissodactyl mammals in the superfamily Rhinoceratoidea.

Hyracoidea [VERT ZOO] An order of ungulate mammals represented only by the conies of Africa, Arabia, and Syria.

hyster- [MED] A combining form that denotes either a relation to or a connection with the uterus, or hysteria.

hysterectomy [MED] Surgical removal of all or part of the uterus.

hysteria [PSYCH] A type of neurosis characterized by extreme emotionalism involving disorders of somatic and psychological functions; the conversion type is associated with neuromuscular and sensory symptoms such as paralysis, tremors, seizures, or blindness, whereas the dissociative

HYPOTRICHIDA

Euplotes, an example of Hypotrichida.

HYPOXANTHINE

Structural formula of hypoxanthine.

displays disorders of consciousness such as amnesia, somnolence, and multiple personalities.

hysteriaceous [MYCOL] Of, belonging to, or resembling the Hysteriales.

Hysteriales [BOT] An order of lichens in the class Ascolichenes including those species with an ascolocular structure.

hysterical anesthesia [MED] Loss of cutaneous pain sensation accompanying hysteria.

hysterical paralysis [MED] Muscle weakness or paralysis without loss of reflex activity, in which no organic nerve lesion can be demonstrated, but which is due to psychogenic factors. Also known as functional paralysis.

hysterics [PSYCH] **1.** Attack of hysteria. **2.** Extreme display of emotions.

hysterogram [MED] A roentgenogram with opacification of the cavity of the uterus by the injection of contrast medium.

hysterography [MED] Roentgenologic examination of the uterus after the introduction of a contrast medium.

hysteromania *See* nymphomania.

hystero-oophorectomy [MED] Surgical removal of the uterus and ovaries.

hysteropexy [MED] Fixation of the uterus by a surgical operation to correct displacement.

hysterorrhaphy [MED] The closure of a uterine incision by suture.

hysterorrhexis [MED] Rupture of the uterus.

hysterosalpingectomy [MED] Surgical removal of the uterus and oviducts.

hysterosalpingography [MED] Roentgenographic examination of the uterus and oviducts after injection of a radiopaque substance.

hysterosalpingo-oophorectomy [MED] The excision of the uterus, oviducts, and ovaries.

hysteroscope [MED] A uterine speculum with a reflector.

hysterosoma [INV ZOO] A body division of an acarid mite composed of the metapodosoma and opisthosoma.

hysterotely [INV ZOO] The appearance or retention of pupal characters in the imago or of larval characters in the pupa or imago.

hysterotomy [MED] **1.** Incision into the uterine wall. **2.** Cesarean section.

Hystricidae [VERT ZOO] The Old World porcupines, a family of Rodentia ranging from southern Europe to Africa and eastern Asia and into the Philippines.

Hystricomorpha [VERT ZOO] A superorder of the class Rodentia.

H zone [HISTOL] The central portion of an A band in a sarcomere; characterized by the presence of myosin filaments.

I

I *See* iodine.

IA *See* international angstrom.

IAA *See* indoleacetic acid.

iatrogenic [MED] An abnormal state or condition produced by a physician in a patient by inadvertent or incorrect treatment.

Iballidae [INV ZOO] A small family of hymenopteran insects in the superfamily Cynipoidea.

I band [HISTOL] The band on either side of a Z line; encompasses portions of two adjacent sarcomeres and is characterized by the presence of actin filaments.

ibis [VERT ZOO] The common name for wading birds making up the family Threskiornithidae and distinguished by a long, slender, downward-curving bill.

I blood group [IMMUNOL] The erythrocyte antigens defined by reactions with anti-I and anti-i antibodies, which occur both in acquired hemolytic anemia and naturally in normal persons of the rare phenotype i.

Icacinaceae [BOT] A family of dicotyledonous plants in the order Celastrales characterized by haplostemonous flowers, pendulous ovules, stipules wanting or vestigial, a polypetalous corolla, valvate petals, and usually one (sometimes three) locules.

Iceland disease *See* epidemic neuromyasthenia.

ice point [PHYS CHEM] The true freezing point of water; the temperature at which a mixture of air-saturated pure water and pure ice may exist in equilibrium at a pressure of 1 standard atmosphere.

ich [VET MED] A dermatitis of fresh-water fishes caused by the invasion of the skin by the ciliated protozoan *Ichthyophthirius multifiliis.*

ichneumon [INV ZOO] The common name for members of the family Ichneumonidae.

Ichneumonidae [INV ZOO] The ichneumon flies, a large family of cosmopolitan, parasitic wasps in the superfamily Ichneumonoidea.

Ichneumonoidea [INV ZOO] A superfamily of hymenopteran insects; members are parasites of other insects.

ichthammol [PHARM] A viscid fluid obtained by the destructive distillation of certain bituminous schists, followed by sulfonation of the distillate and neutralization with ammonia; used as a weak antiseptic and stimulant in skin diseases, and occasionally as an expectorant. Also known as ammonium bithiolicum; ammonium ichthosulfonate; ammonium sulfoichthyolate.

Ichthyobdellidae [INV ZOO] A family of leeches in the order Rhynchobdellae distinguished by cylindrical bodies with conspicuous, powerful suckers.

Ichthyodectidae [PALEON] A family of Cretaceous marine osteoglossiform fishes.

ICHNEUMON

Long-tailed ichneumon
(Megarhyssa lunator), about 6 inches (15 centimeters) long of which more than 4 inches (10 centimeters) is "tail."

ichthyodont [PALEON] A tooth of a fossil fish.

ichthyodorulite [PALEON] A dermal or fin spine of a fossil fish.

Ichthyol [PHARM] A trademark for ichthammol.

ichthyology [VERT ZOO] A branch of vertebrate zoology that deals with the study of fishes.

ichthyophagous [ZOO] Subsisting on a diet of fish.

Ichthyopterygia [PALEON] A subclass of extinct Mesozoic reptiles composed of predatory fish-finned and sea-swimming forms with short necks and a porpoiselike body.

Ichthyornis [PALEON] The type genus of Ichthyornithidae.

Ichthyornithes [PALEON] A superorder of fossil birds of the order Ichthyornithiformes according to some systems of classification.

Ichthyornithidae [PALEON] A family of extinct birds in the order Ichthyornithiformes.

Ichthyornithiformes [PALEON] An order of ancient fossil birds including strong flying species from the Upper Cretaceous that possessed all skeletal characteristics of modern birds.

Ichthyosauria [PALEON] The only order of the reptilian subclass Ichthyopterygia, comprising the extinct predacious fish-lizards; all were adapted to a sea life in having tail flukes, paddles, and dorsal fins.

ichthyosis [MED] A congenital skin disease characterized by dryness and scales, especially on the extensor surfaces of the extremities.

Ichthyostegalia [PALEON] An extinct Devonian order of labyrinthodont amphibians, the oldest known representatives of the class.

Ichthyotomidae [INV ZOO] A monotypic family of errantian annelids in the superfamily Eunicea.

Icosteidae [VERT ZOO] The ragfishes, a family of perciform fishes in the suborder Stromateoidei found in high seas.

icotype [SYST] A typical, accurately identified specimen of a species, but not the basis for a published description.

ICSH *See* luteinizing hormone.

icteric [MED] Pertaining to or characterized by jaundice.

Icteridae [VERT ZOO] The troupials, a family of New World perching birds in the suborder Oscines.

icteroanemia [VET MED] A disease of swine characterized by jaundice, anemia, and erythrocytolysis.

icterogenic [MED] Causing icterus, or jaundice.

icterohematuria [VET MED] A disease of sheep caused by the protozoan *Babesia ovis* and characterized by hemolysis of erythrocytes accompanied by jaundice.

icterohemorrhagic fever *See* Weil's disease.

icterus *See* jaundice.

icterus gravis [MED] Acute yellow atrophy of the liver marked by jaundice and nervous system dysfunctions.

icterus index [PATH] Measure of serum bilirubin levels by comparing the yellow blood serum from a jaundiced patient with the colors of standard potassium dichromate solutions.

Ictidosauria [PALEON] An extinct order of mammallike reptiles in the subclass Synapsida including small carnivorous and herbivorous terrestrial forms.

id [PSYCH] The primitive, psychic energy source of the unconscious.

ID$_{50}$ *See* infective dose 50.

idea [PSYCH] **1.** A mental impression or thought. **2.** An experience or thought not directly due to an external sensory stimulation.

ideal gas [THERMO] Also known as perfect gas. **1.** A gas whose molecules are infinitely small and exert no force on each other. **2.** A gas that obeys Boyle's law (the product of the

ICHTHYOSAURIA

A Jurassic ichthyosaur, much reduced. *(Simplified from E. von Stromer, as used by A. S. Romer, Vertebrate Paleontology, 3d ed., University of Chicago Press, 1966)*

pressure and volume is constant at constant temperature) and Joule's law (the internal energy is a function of the temperature alone).

ideal gas law [THERMO] The equation of state of an ideal gas which is a good approximation to real gases at sufficiently high temperatures and low pressures; that is, $PV = RT$, where P is the pressure, V is the volume per mole of gas, T is the temperature, and R is the gas constant.

ideal solution [CHEM] A solution that conforms to Raoult's law over all ranges of temperature and concentration and shows no internal energy change on mixing and no attractive force between components.

ideas of influence [PSYCH] A clinical manifestation of certain psychotic disorders in which the patients may believe that their thoughts are read, that their limbs move without their consent, or that they are under the control of someone else or some external force or influence.

ideas of reference [PSYCH] A symptom complex in which, through the mechanism of projection, an individual incorrectly believes himself or herself to be the direct object of casual remarks or incidents or of external events.

ideation [PSYCH] Conceptualization of an idea.

ideational apraxia [MED] Inability to perform meaningful motor functions or to use objects properly due to mental confusion caused by diffuse brain disease.

identical twins *See* monozygotic twins.

identification [PSYCH] **1.** The tendency of children to model their behavior after that of one or more selected adults. **2.** A defense mechanism in which a person likens himself or herself to someone else.

identity crisis [PSYCH] The critical period in emotional maturation and personality development, occurring usually during adolescence, which involves the reworking and abandonment of childhood identifications and the integration of new personal and social identifications.

ideomotor [PHYSIO] **1.** Pertaining to involuntary movement resulting from or accompanying some mental activity, as moving the lips while reading. **2.** Pertaining to both ideation and motor activity.

ideotype [SYST] A specimen identified as belonging to a specific taxon but collected from other than the type locality.

idioblast [BOT] A plant cell that differs markedly in shape or function from neighboring cells within the same tissue.

idiochromatin [CYTOL] The portion of the nuclear chromatin thought to function as the physical carrier of genes.

idiocy [PSYCH] The lowest grade of mental deficiency; the individual's mental age is less than 3 years.

idiomuscular [PHYSIO] Pertaining to any phenomenon occurring in a muscle which is independent of outside stimuli.

idiopathic arthritis *See* Reiter's syndrome.

idiopathic colitis [MED] Any form of colitis for which the causative agent is not identified.

idiopathic familial jaundice [MED] A familial form of obstructive jaundice of unknown cause in which there is decreased ability to excrete conjugated bilirubin into the bile ducts.

idiopathic hypercholesterolemia [MED] A genetic derangement of fat metabolism characterized by high blood, cell, and plasma levels of cholesterol.

idiopathic megacolon [MED] Hypertrophy and dilation of the colon.

idiopathic pulmonary hemisiderosis [MED] A disease of unknown etiology characterized by recurrent hemorrhaging from pulmonary capillaries.

idiopathic steatorrhea *See* celiac syndrome.

idiopathic thrombocytopenic purpura [MED] Thrombocytopenic purpura of unknown causes.

idiopathic ulcerative colitis [MED] A form of ulcerative colitis of unknown cause.

idiopathy [MED] **1.** A primary disease; one not a result of any other disease, but of spontaneous origin. **2.** Disease for which no cause is known.

idiosome [CYTOL] **1.** A hypothetical unit of a cell, such as the region of modified cytoplasm surrounding the centriole or centrosome. **2.** A sex chromosome.

Idiostolidae [INV ZOO] A small family of hemipteran insects in the superfamily Pentatomorpha.

idiosyncrasy [MED] A peculiarity of constitution that makes an individual react differently from most persons, to drugs, diet, treatment, or other situations. [PSYCH] Any special or peculiar characteristic or temperament by which a person differs from other persons.

idiot [PSYCH] A person afflicted with idiocy and requiring custodial or protective care.

Idoteidae [INV ZOO] A family of isopod crustaceans in the suborder Valvifera having a flattened body and seven pairs of similar walking legs.

Ig *See* gamma globulin.

iguana [VERT ZOO] The common name for a number of species of herbivorous, arboreal reptiles in the family Iguanidae characterized by a dorsal crest of soft spines and a dewlap; there are only two species of true iguanas.

Iguanidae [VERT ZOO] A family of reptiles in the order Squamata having teeth fixed to the inner edge of the jaws, a nonretractile tongue, a compressed body, five clawed toes, and a long but rarely prehensile tail.

ileitis [MED] Inflammation of the ileum.

ileocecal valve [ANAT] A muscular structure at the junction of the ileum and cecum which prevents reflex of the cecal contents.

ileocolic artery [ANAT] A branch of the superior mesenteric artery that supplies blood to the terminal part of the ileum and the beginning of the colon.

ileocolostomy [MED] Surgical formation of a bypass channel between the ileum and the colon.

ileostomy [MED] Surgical formation of an artificial anus through the abdominal wall into the ileum.

ileum [ANAT] The last portion of the small intestine, extending from the jejunum to the large intestine.

ileus [MED] Acute intestinal obstruction of neurogenic origin.

iliac artery [ANAT] Either of the two large arteries arising by bifurcation of the abdominal aorta and supplying blood to the lower trunk and legs (or hind limbs in quadrupeds). Also known as common iliac artery.

iliac crest height [ANTHRO] A measure of the vertical distance from the top of the iliac crest to the floor while the subject stands.

iliac fascia [ANAT] The fascia covering the pelvic surface of the iliacus muscle.

iliac index [ANTHRO] The ratio multiplied by 100 of the distance between the iliac spines and the distance between the lower margin of the acetabulum and the topmost crest of the ilium.

iliac region *See* inguinal region.

iliacus [ANAT] The portion of the iliopsoas muscle arising from the iliac fossa and sacrum.

iliac vein [ANAT] Any of the three veins on each side of the body which correspond to and accompany the iliac artery.

IDOTEIDAE

5 mm

Idotea neglecta, showing flattened body and seven pairs of legs. *(From G. O. Sars, An Account of the Crustacea of Norway, vol. 2, 1899)*

iliocaudal [VERT ZOO] The muscle that connects the ilium and the tail.

iliococcygeal [ANAT] Pertaining to the ilium and the coccyx.

iliocostalis [ANAT] The lateral portion of the erector spinal muscle that extends the vertebral column and assists in lateral movements of the trunk.

iliofemoral ligament [ANAT] A strong band of dense fibrous tissue extending from the anterior inferior iliac spine to the lesser trochanter and the intertrochanteric line. Also known as Y ligament.

ilioinguinal [ANAT] **1.** Pertaining to the ilium and the groin. **2.** Lying partly within the iliac and partly within the inguinal regions.

iliolumbar [ANAT] Pertaining to both the iliac and lumbar regions.

iliolumbar ligament [ANAT] A fibrous band that radiates laterally from the transverse processes of the fourth and fifth lumbar vertebrae and attaches to the pelvis by two main bands.

iliopsoas [ANAT] The combined iliacus and psoas muscles.

iliotibial tract [ANAT] A thickened portion of the fascia lata extending from the lateral condyle of the tibia to the iliac crest.

ilium [ANAT] Either of a pair of bones forming the superior portion of the pelvis bone in vertebrates.

Illinoian [GEOL] The third glaciation of the Pleistocene in North America, between the Yarmouth and Sangamon interglacial stages.

illness [MED] **1.** The state of being sick. **2.** A sickness, disease, or disorder.

illusion [PSYCH] A false interpretation of a real sensation; a perception that misinterprets the object perceived.

ilotycin [MICROBIO] A trade name for erythromycin.

imaginal disk [INV ZOO] Any of the thickened areas within the sac of the body wall in holometabolous insects which give rise to specific organs in the adult.

imago [INV ZOO] The sexually mature, usually winged stage of insect development. [PSYCH] An unconscious mental picture, usually idealized, of a parent or loved person important in the early development of an individual and carried into adulthood.

imbecile [PSYCH] A person of middle-grade mental deficiency; the individual's mental age is between 3 and 7 years.

imbibition [PHYS CHEM] Absorption of liquid by a solid or a semisolid material.

imbricate [BIOL] Having overlapping edges, such as scales, or the petals of a flower.

iminourea *See* guanidine.

immediate hypersensitivity [IMMUNOL] A type of hypersensitivity in which the response rapidly occurs following exposure of a sensitized individual to the antigen.

immersion electron microscope [ELECTR] An emission electron microscope in which the specimen is a flat conducting surface which may be heated, illuminated, or bombarded by high-velocity electrons or ions so as to emit low-velocity thermionic, photo-, or secondary electrons; these are accelerated to a high velocity in an immersion objective or cathode lens and imaged as in a transmission electron microscope.

immersion foot [MED] A serious and disabling condition of the feet due to prolonged immersion in seawater at 60°F (15.6°C) or lower, but not at freezing temperature.

immersion lens *See* immersion objective.

immersion objective [OPTICS] A high-power microscope objective designed to work with the space between the objective and the cover glass over the object filled with an oil whose index of refraction is nearly the same as that of the objective and the cover glass, in order to reduce reflection losses and increase the index of refraction of the object space. Also known as immersion lens.

immobilize [MED] To render motionless or to fix in place, as by splints or surgery.

immune [IMMUNOL] 1. Safe from attack; protected against a disease by an innate or an acquired immunity. 2. Pertaining to or conferring immunity.

immune body See antibody.

immune globulin See gamma globulin.

immune lysin [IMMUNOL] An antibody that will disrupt a particular type of cell in the presence of complement and cofactors, such as magnesium or calcium ions.

immune precipitation [IMMUNOL] A method of isolating a protein from mixtures by using a specific antibody as the precipitating agent.

immune serum [IMMUNOL] Blood serum obtained from an immunized individual and carrying antibodies.

immunity [IMMUNOL] The condition of a living organism whereby it resists and overcomes infection or disease.

immunization [IMMUNOL] Rendering an organism immune to a specific communicable disease.

immunoblast [IMMUNOL] An immature cell of the plasmacytic series; actively synthesizes antibodies.

immunochemistry [IMMUNOL] A branch of science dealing with the chemical changes associated with immunity factors.

immunodiffusion [IMMUNOL] A serological procedure in which antigen and antibody solutions are permitted to diffuse toward each other through a gel matrix; interaction is manifested by a precipitin line for each system.

immunoelectrophoresis [IMMUNOL] A serological procedure in which the components of an antigen are separated by electrophoretic migration and then made visible by immunodiffusion of specific antibodies.

immunofluorescence [IMMUNOL] Fluorescence as the result of, or identifying, an immune response; a specifically stained antigen fluoresces in ultraviolet light and can thus be easily identified with a homologous antigen.

immunogen [IMMUNOL] A substance which stimulates production of specific antibody or of cellular immunity, and which can react with these products.

immunogenetics [MED] A branch of immunology dealing with the relationships between immunity and genetic factors in disease.

immunogenic [IMMUNOL] Producing immunity.

immunoglobulin See gamma globulin.

immunogranulomatous disease [MED] A condition in which a deviation from the standard immune mechanisms is considered to be associated with widespread granulomatosis.

immunological paralysis See acquired immunological tolerance.

immunologic suppression [IMMUNOL] The use of x-irradiation, chemicals, corticosteroid hormones, or antilymphocyte antisera to suppress antibody production, particularly in graft transplants.

immunologic tolerance [IMMUNOL] 1. A condition in which an animal will accept a homograft without rejection. 2. A state of specific unresponsiveness to an antigen or antigens in adult life as a consequence of exposure to the antigen in utero or in the neonatal period.

immunologist [IMMUNOL] A person who specializes in immunology.

immunology [BIOL] A branch of biological science concerned with the native or acquired resistance of higher animal forms and humans to infection with microorganisms.

Immunopathology [MED] The study of various human and animal diseases in which humoral and cellular immune factors seem important in causing pathological damage to cells, tissues, and the host.

Immunosuppressive [PHARM] Any drug or agent used to suppress antibody production.

Immunotherapy [MED] **1.** *See* serotherapy. **2.** Therapy utilizing immunosuppressives.

Impaction [MED] **1.** The state of being lodged and retained in a body part. **2.** Confinement of a tooth in the jaw so that its eruption is prevented. **3.** A condition in which one fragment of bone is firmly driven into another fragment so that neither can move against the other.

Imparipinnate *See* odd-pinnate.

Impennes [VERT ZOO] A superorder of birds for the order Sphenisciformes in some systems of classification.

Imperfect flower [BOT] A flower lacking either stamens or carpels.

Imperfect gas *See* real gas.

Imperforate [BIOL] Lacking a normal opening.

Imperforate anus [MED] A congenital malformation in which the large intestine ends blindly.

Impermeable [SCI TECH] Not permitting water or other fluid to pass through. Also known as impervious.

Impervious *See* impermeable.

Impetigo [MED] An acute, contagious, inflammatory skin disease caused by streptococcal or staphylococcal infections and characterized by vesicular or pustular lesions.

Implantation [MED] **1.** Placement of a tissue transplant in depth in the body. **2.** Placement in the body of a device for mechanical repair, such as for a ventral hernia or a fracture. **3.** Embedding of an embryo into the endometrium.

Implexed [INV ZOO] In insects, having the integument infolded for muscle attachment.

Impotence [MED] **1.** Inability in the male to perform the sexual act. **2.** Lack of sexual vigor.

Impregnate [MED] To fertilize or cause to become pregnant.

Imprinting [PSYCH] The very rapid development of a response or learning pattern to a stimulus at an early and usually critical period of development; particularly characteristic of some species of birds.

Impulse [PSYCH] A sudden psychogenic urge to act.

Impunctate [BIOL] Lacking pores.

IMVIC test [MICROBIO] A group of four cultural tests used to differentiate genera of bacteria in the family Enterobacteriaceae and to distinguish them from other bacteria; tests are indole, methyl red, Voges-Proskauer, and citrate.

In *See* indium.

Inadequate personality [PSYCH] An individual showing no obvious mental or physical defect, but characterized by inappropriate or inadequate response to intellectual, emotional, and physical demand, and whose behavior pattern shows inadaptability, ineptitude, poor judgment, lack of stamina, and social incompatibility.

Inadunata [PALEON] An extinct subclass of stalked Paleozoic Crinozoa characterized by branched or simple arms that were free and in no way incorporated into the calyx.

Inadunate [BIOL] Not united. [INV ZOO] In crinoids, having the arms free from the calyx.

Inanition [MED] The exhausted, pathologic condition resulting from starvation.

Inaperturate [BIOL] Lacking apertures.

Inarching [BOT] A kind of repair grafting in which two plants

INADUNATA

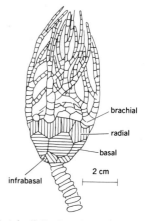

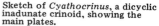

Sketch of *Cyathocrinus*, a dicyclic inadunate crinoid, showing the main plates.

growing on their own roots are grafted together and one plant is severed from its roots after the graft union is established.

Inarticulata [INV ZOO] A class of the phylum Brachiopoda; valves are typically not articulated and are held together only by soft tissue of the living animal.

inarticulate [BIOL] Lacking articulations or joints.

inborn [BIOL] Of or pertaining to a congenital or hereditary characteristic.

inborn errors of metabolism [GEN] Congenital defects in metabolism attributable to variation in genetic constitution.

inbreeding [GEN] Breeding of closely related individuals; self-fertilization, as in some plants, is the most extreme form.

incarcerated hernia [MED] A hernia in which the intestinal loop is permanently trapped in the hernia sac.

incertae sedis [SYST] Placed in an uncertain taxonomic position.

Incirrata [INV ZOO] A suborder of cephalopod mollusks in the order Octopoda.

incised [BIOL] Having a deeply and irregularly notched margin. [MED] Made by cutting, as a wound.

incision [MED] A cut or wound of the body tissue, as an abdominal incision or a vertical or oblique incision.

incisional hernia [MED] Abnormal protrusion of an organ through an operative or accidental incision. Also known as postoperative hernia; posttraumatic hernia.

incisive canal [ANAT] The bifurcated bony passage from the floor of the nasal cavity to the incisive fossa.

incisive foramen [ANAT] One of the two to four openings of the incisive canal on the floor of the incisive fossa.

incisive fossa [ANAT] **1.** A bony pit behind the upper incisors into which the incisive canals open. **2.** A depression on the maxilla at the origin of the depressor muscle of the nose. **3.** A depression of the mandible at the origin of the mentalis muscle.

incisor [ANAT] A tooth specialized for cutting, especially those in front of the canines on the upper jaw of mammals.

included [BOT] Having the stamens and pistils the same height as the petals and thus not extending beyond the corolla.

inclusion [CYTOL] A visible product of cellular metabolism within the protoplasm.

inclusion blennorrhea *See* inclusion conjunctivitis.

inclusion body [VIROL] Any of the abnormal structures appearing within the cell nucleus or the cytoplasm during the course of virus multiplication.

inclusion conjunctivitis [MED] An acute inflammation of the conjunctiva with pus formation caused by a virus and identified from epithelial-cell inclusion bodies in conjunctival scrapings. Also known as inclusion blennorrhea; paratrachoma; swimmer's conjunctivitis; swimming pool conjunctivitis.

inclusion cyst [MED] A cyst formed by the implantation of epithelial tissue into another structure.

inclusion encephalitis [MED] A chronic inflammation of the brain in which large inclusion bodies occur in the nuclei of oligodendria and sometimes in nerve cells.

incompetence [MED] **1.** Insufficiency or inadequacy in performing natural functions. **2.** In forensic medicine, inability to function within the law, as the incompetence of an individual to drive when under the influence of alcohol. **3.** Defective closure of a cardiac valve permitting regurgitation of blood into a cardiac chamber.

incomplete flower [BOT] A flower lacking one or more modified leaves, such as petals, sepals, pistils, or stamens.

incomplete metamorphosis [INV ZOO] Development in the

insect in which a general adult form hatches and develops without a quiescent stage.

Incontinence [MED] Inability to control the natural evacuations, as the feces or the urine; specifically, involuntary evacuation from organic causes.

Incrassate [BIOL] State of being swollen or thickened.

Incretion [PHYSIO] An internal secretion.

Incrustation [PALEON] Fossil formation by encasement in a mineral substance.

Incubation [CHEM] Maintenance of chemical mixtures at specified temperatures for varying time periods to study chemical reactions, such as enzyme activity. [MED] The phase of an infectious disease process between infection by the pathogen and appearance of symptoms. [VERT ZOO] The act or process of brooding or incubating.

Incubation period [MED] The period of time required for the development of symptoms of a disease after infection, or of altered reactivity after exposure to an allergen. [VERT ZOO] The brooding period required to bring an egg to hatching.

Incubator [AGR] A device for the artificial hatching of eggs. [MED] A small chamber with controlled oxygen, temperature, and humidity for newborn infants requiring special care. [MICROBIO] A laboratory cabinet with controlled temperature for the cultivation of bacteria, or for facilitating biologic tests.

Incubous [BOT] The juxtaposition of leaves such that the anterior margins of older leaves overlap the posterior margins of younger leaves.

Incudate [BIOL] Of, pertaining to, or having an incus.

Incumbent [BIOL] Lying on or down.

Incurrent canal [INV ZOO] A canal through which water enters a sponge.

Incurrent siphon *See* inhalant siphon.

Incurvariidae [INV ZOO] A family of lepidopteran insects in the superfamily Incurvarioidea; includes yucca moths and relatives.

Incurvarioidea [INV ZOO] A monofamilial superfamily of lepidopteran insects in the suborder Heteroneura having wings covered with microscopic spines, a single genital opening in the female, and venation that is almost complete.

Incus [ANAT] The middle one of three ossicles in the middle ear. Also known as anvil.

Indeciduate placenta [EMBRYO] A placenta having the maternal and fetal elements associated but not fused.

Indehiscent [BOT] **1.** Remaining closed at maturity, as certain fruits. **2.** Not splitting along regular lines.

Independent assortment [GEN] The random assortment of the alleles at two or more loci on different chromosome pairs or far apart on the same chromosome pair which occurs at meiosis; first discovered by G. Mendel.

Independent events [STAT] Two events in probability such that the occurrence of one of them does not affect the probability of the occurrence of the other.

Independent random variables [STAT] The discrete random variables $X_1, X_2, \ldots, X_n$ are independent if for arbitrary values $x_1, x_2, \ldots, x_n$ of the variables the probability that $X_1 = x_1$ and $X_2 = x_2$, etc., is equal to the product of the probabilities that $X_i = x_i$ for $i = 1, 2, \ldots, n$; random variables which are unrelated.

Indeterminate cleavage [EMBRYO] Cleavage in which all the early cells have the same potencies with respect to development of the entire zygote.

Indeterminate growth [BOT] Growth of a plant in which the axis is not limited by development of a reproductive structure, and therefore growth continues indefinitely.

indeterminate inflorescence [BOT] Floral axis growth by indefinite branching because it is not limited by development of a terminal bud.

index forest [FOR] A forest reaching the highest average in a given locality for density, volume, and increment.

index fossil [PALEON] The ancient remains and traces of an organism that lived during a particular geologic time period and that geologically date the containing rocks.

indicator medium [MICROBIO] A usually solid culture medium containing substances capable of undergoing a color change in the vicinity of a colony which has effected a particular chemical change, such as fermenting a certain sugar.

indicator plant [BOT] A plant used in geobotanical prospecting as an indicator of a certain geological phenomenon.

indigenous [SCI TECH] Existing and having originated naturally in a particular region or environment.

indirect Coombs test *See* Rh blocking test.

indirect developing test *See* Rh blocking test.

indirect hernia [MED] A form of inguinal hernia that passes out of the abdomen through the inguinal canal.

indirect vision *See* peripheral vision.

indium [CHEM] A metallic element, symbol In, atomic number 49, atomic weight 114.82; soluble in acids; melts at 156°C, boils at 1450°C.

individual psychology [PSYCH] A system of psychology in which traits of an individual are compared in terms of striving for superiority and then restated in the form of a composite of this single tendency.

indoleacetic acid [BIOCHEM] $C_{10}H_9O_2N$ A decomposition product of tryptophan produced by bacteria and occurring in urine and feces; used as a hormone to promote plant growth. Abbreviated IAA.

indole test [MICROBIO] A test for the production of indole from tryptophan by microorganisms; a solution of *para*-dimethylaminobenzaldehyde, amyl alcohol, and hydrochloric acid added to the incubated culture of bacteria shows a red color in the alcoholic layer if indole is present.

Indo-Pacific faunal region [ECOL] A marine littoral faunal region extending eastward from the east coast of Africa, passing north of Australia and south of Japan, and ending in the east Pacific south of Alaska.

Indriidae [VERT ZOO] A family of Madagascan prosimians containing wholly arboreal vertical clingers and leapers.

inducer [EMBRYO] The cell group that functions as the acting system in embryonic induction by controlling the mode of development of the reacting system. Also known as inductor.

induction [ELEC] *See* electrostatic induction. [ELECTRO-MAG] *See* electromagnetic induction. [EMBRYO] *See* embryonic induction. [MED] The period from administration of the anesthetic to loss of consciousness by the patient. [SCI TECH] The act or process of causing.

inductor *See* inducer.

inductura [INV ZOO] A layer of lamellar shell material along the inner lip of the aperture in gastropods.

indumentum [BOT] A covering, such as one that is woolly. [MYCOL] A covering of hairs. [VERT ZOO] The plumage covering a bird.

induplicate [BOT] Having the edges turned or rolled inward without twisting or overlapping; applied to the leaves of a bud.

induration [BIOL] The process of hardening, especially by increasing the fibrous elements. [MED] Hardening of a tissue or organ due to an accumulation of blood, inflammation, or neoplastic growth.

INDIUM

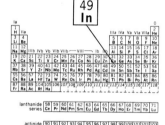

Periodic table of the chemical elements showing the position of indium.

Indusium [ANAT] A covering membrane such as the amnion. [BOT] An epidermal outgrowth covering the sori in many ferns. [MYCOL] The annulus of certain fungi.

Industrial crop [AGR] Any crop that provides materials for industrial processes and products such as soybeans, cotton (lint and seed), flax, and tobacco.

Industrial microbiology [MICROBIO] The study, utilization, and manipulation of those microorganisms capable of economically producing desirable substances or changes in substances, and the control of undesirable microorganisms.

Industrial psychology [PSYCH] Psychology applied to problems in industry, dealing chiefly with the selection, efficiency, and mental health of personnel.

Inequilateral [BIOL] Having the two sides or ends unequal, as the ends of a bivalve mollusk on either side of a line from umbo to gape.

Inert [SCI TECH] Lacking an activity, reactivity, or effect.

Infantile amaurotic familial idiocy *See* Tay-Sachs disease.

Infantile autism [PSYCH] A disorder of children characterized by extreme withdrawal and introspection without regard for reality.

Infantile celiac disease [MED] Celiac syndrome of infants and young children.

Infantile cortical hyperostosis [MED] A condition occurring during the first 3 months of life in which there is fever and painful swelling of the soft tissue of the lower jaw, characterized by periosteal proliferation of the mandible.

Infantile diarrhea [MED] An acute gastrointestinal disease in infants resulting from damage of the intestinal mucosa by an infectious organism.

Infantile eczema [MED] An allergic inflammation of the skin in young children, usually due to common antigens such as food or inhalants.

Infantile genitalia [ANAT] The genital organs of an infant. [MED] Underdevelopment of the adult genitals.

Infantile neuroaxonal dystrophy [MED] A familial disease of the central nervous system occurring early in life and characterized by axonal swellings, arrested development, atrophy of the optic nerves, and eventual blindness.

Infantile paralysis *See* poliomyelitis.

Infantile scurvy [MED] Acute scurvy of infants and young children characterized by periosteal hemorrhage and swelling, especially of long bones. Also known as Cheadle's disease; Moeller-Barlow disease.

Infantile sexuality [PSYCH] An infant's or child's capacity for and enjoyment of activities and experiences that are essentially sexual.

Infantile spasm [MED] A type of seizure seen in infants and young children, characterized by a sudden, brief, massive myoclonic jerk.

Infantilism [MED] Persistence of physical, behavioral, or mental infantile characteristics into childhood, adolescence, or adult life.

Infarct [MED] Localized death of tissue caused by obstructed inflow of arterial blood. Also known as infarction.

Infarction [MED] **1.** Condition or process leading to the development of an infarct. **2.** *See* infarct.

Infect [MED] To cause an infection, as by contamination with or invasion by a pathogen. [MICROBIO] To cause a phage infection of bacteria.

Infection [MED] **1.** Invasion of the body by a pathogenic organism, with or without disease manifestation. **2.** Pathologic condition resulting from invasion of a pathogen.

Infectious abortion *See* contagious abortion.

Infectious anemia [VET MED] A virus disease of horses and mules characterized by intermittent fever, weakness, jaundice, and hemorrhages of mucous membranes; it is often fatal. Also known as swamp fever.

infectious arthritis [MED] An inflammatory joint disease caused by microbial invasion of the articular tissue.

infectious bronchitis [VET MED] A highly contagious respiratory viral disease of chickens.

infectious canine hepatitis [VET MED] An acute inflammatory liver disease of dogs caused by a virus.

infectious chlorosis [PL PATH] A virus disease of plants characterized by yellowing of the green parts.

infectious conjunctivitis [MED] Conjunctivitis due to invasion by a microorganism.

infectious disease [MED] Any disease caused by invasion by a pathogen which subsequently grows and multiplies in the body.

infectious drug resistance [MICROBIO] A type of drug resistance that is transmissible from one bacterium to another by infectivelike agents referred to as resistance factors.

infectious endocarditis [MED] Inflammation of the endocardium due to an infectious microorganism.

infectious hepatitis [MED] An acute infectious virus disease of the liver associated with hepatic inflammation and characterized by fever, liver enlargement, and jaundice. Also known as catarrhal jaundice; epidemic hepatitis; epidemic jaundice; virus hepatitis.

infectious laryngotracheitis [VET MED] A highly contagious respiratory disease of viral etiology affecting chickens.

infectious mononucleosis [MED] A disorder of unknown etiology characterized by irregular fever, pathology of lymph nodes, lymphocytosis, and high serum levels of heterophil antibodies against sheep erythrocytes. Also known as acute benign lymphoblastosis; glandular fever; kissing disease; lymphocytic angina; monocytic angina; Pfeiffer's disease.

infectious myocarditis [MED] Inflammation of the myocardium due to an infectious microorganism.

infectious myxomatosis [VET MED] An infectious virus disease of rabbits characterized by myxomatous lesions.

infectious papillomatosis [VET MED] A virus disease of cattle characterized by the appearance of warts on the body.

infectious rhinitis [MED] Inflammation of the nasal mucous membrane due to an infectious microorganism.

infectious uroarthritis *See* Reiter's syndrome.

infective dose 50 [MICROBIO] The dose of microorganisms required to cause infection in 50% of the experimental animals; a special case of the median effective dose. Abbreviated ID_{50}. Also known as median infective dose.

inferior [BIOL] The lower of two structures.

inferior alveolar artery [ANAT] A branch of the internal maxillary artery supplying the mucous membrane of the mouth and teeth of the lower jaw.

inferior alveolar nerve [ANAT] A branch of the mandibular nerve that innervates the teeth of the lower jaw.

inferior cerebellar peduncle [ANAT] A large bundle of nerve fibers running from the medulla oblongata to the cerebellum. Also known as restiform body.

inferior colliculus [ANAT] One of the posterior pair of rounded eminences arising from the dorsal portion of the mesencephalon.

inferior ganglion [ANAT] **1.** The lower sensory ganglion in the glossopharyngeal nerve. **2.** The lower sensory ganglion on the vagus.

inferiority complex [PSYCH] Repressed unconsious fears and feelings of physical or social inadequacy or both, which may result in excessive anxiety, inability to function, or actual failure.

Inferior mesenteric ganglion [ANAT] A sympathetic ganglion within the inferior mesenteric plexus at the origin of the inferior mesenteric artery.

Inferior ovary [BOT] An ovary that is adnate to and below the other floral parts.

Inferior temporal gyrus [ANAT] A convolution on the temporal lobe of the cerebral hemispheres lying below the middle temporal sulcus and extending to the inferior sulcus.

Inferior vena cava [ANAT] A large vein which drains blood from the iliac veins, lower extremities, and abdomen into the right atrium.

Inferior vena cava syndrome [MED] Edema and venous distention of the abdomen and legs due to obstruction of the inferior vena cava.

Infertility [MED] Involuntary reduction in reproductive ability.

Infestation [MED] The state or condition of having animal parasites on or in the body.

Infiltrating lipoma *See* liposarcoma.

Inflammation [MED] Local tissue response to injury characterized by redness, swelling, pain, and heat.

Inflammatory carcinoma [MED] A carcinoma, usually of the breast, associated with inflammation and characterized by rapid metastasis.

Inflated [BIOL] **1.** Distended, applied to a hollow structure. **2.** Open and enlarged.

Inflected [BOT] Curved or bent sharply inward, downward, or toward the axis. Also known as inflexed.

Inflexed *See* inflected.

Inflorescence [BOT] **1.** A flower cluster. **2.** The arrangement of flowers on a plant.

Influenza [MED] An acute virus disease of the respiratory system characterized by headache, muscle pain, fever, and prostration.

Influenzal meningitis [MED] Inflammation of the meninges caused by *Hemophilus influenzae*.

Influenzal pneumonia [MED] Pneumonia resulting from infection by *Hemophilus influenzae*.

Influenza vaccine [IMMUNOL] A vaccine prepared from formaldehyde-attenuated mixtures of strains of influenza virus.

Influenza virus [VIROL] Any of three immunological types, designated A, B, and C, belonging to the myxovirus group which cause influenza.

Infrabasal [BIOL] Inferior to a basal structure.

Infrabranchial [VERT ZOO] Situated below the gills.

Infracambrian *See* Eocambrian.

Infracentral [ANAT] Located below the centrum.

Infracerebral gland [INV ZOO] A structure lying ventral to the brain in annelids which is thought to produce a hormone that inhibits maturation of the gametes.

Infraciliature [INV ZOO] The neuromotor apparatus, silverline system, or neuroneme system of ciliates.

Infraclass [SYST] A subdivision of a subclass; equivalent to a superorder.

Infraclavicle [VERT ZOO] A bony element of the shoulder girdle located below the cleithrum in some ganoid and crossopterygian fishes.

Infrafoliar [BOT] Located below the leaves.

Infraglenoid tubercle [ANAT] A rough impression below the glenoid cavity, from which the long head of the triceps muscle arises.

Infraorbital [ANAT] Located beneath the orbit.

infrared microscope [OPTICS] A type of reflecting microscope which uses radiation of wavelengths greater than 700 nanometers and is used to reveal detail in materials that are opaque to light, such as molybdenum, wood, corals, and many red-dyed materials.

infrared radiation [ELECTROMAG] Electromagnetic radiation whose wavelengths lie in the range from 0.75 or 0.8 micrometer (the long-wavelength limit of visible red light) to 1000 micrometers (the shortest microwaves).

infrared spectrometer [SPECT] Device used to identify and measure the concentrations of heteroatomic compounds in gases, in many nonaqueous liquids, and in some solids by arc or spark excitation and subsequent measurement of the electromagnetic emissions in the wavelength range of 0.78 to 300 micrometers.

infrared spectrophotometry [SPECT] Spectrophotometry in the infrared region, usually for the purpose of chemical analysis through measurement of absorption spectra associated with rotational and vibrational energy levels of molecules.

infrared spectroscopy [SPECT] The study of the properties of material systems by means of their interaction with infrared radiation; ordinarily the radiation is dispersed into a spectrum after passing through the material.

infraspecies [SYST] A subdivision of a species.

infraspinous [ANAT] Below the spine of the scapula.

infraspinous fossa [ANAT] The recess on the posterior surface of the scapula occupied by the infraspinatus muscle.

infratemporal [ANAT] Situated below the temporal fossa.

infratemporal fossa [ANAT] An irregular space situated below and medial to the zygomatic arch, behind the maxilla and medial to the upper part of the ramus of the mandible.

infructescence [BOT] An inflorescence's fruiting stage.

infundibular canal [INV ZOO] A pathway from the mantle cavity through the funnel for water in cephalopods.

infundibular ganglion [INV ZOO] A branch of pedal ganglion which supplies the funnel in cephalopods.

infundibular process [ANAT] The distal portion of the neural lobe of the pituitary.

infundibulum [ANAT] **1.** A funnel-shaped passage or part. **2.** The stalk of the neurohypophysis.

infusion [MED] The slow injection of a solution into a vein or into subcutaneous or other tissue of the body.

infusoriform larva [INV ZOO] The final larval stage, arising from germ cells within the infusorigen, in the life cycle of dicyemid mesozoans.

infusorigen [INV ZOO] An individual that gives rise to the infusoriform larva in dicyemid mesozoans.

ingesta [BIOL] Food and other substances taken into an animal body.

ingestion [BIOL] The act or process of taking food and other substances into the animal body.

Ingolfiellidea [INV ZOO] A suborder of amphipod crustaceans in which both abdomen and maxilliped are well developed and the head often bears a separate ocular lobe lacking eyes.

ingrown [MED] Of a hair or nail, grown inward so that the normally free end is embedded in or under the skin.

inguinal canal [ANAT] A short, narrow passage between the abdominal ring and the inguinal ring in which lies the spermatic cord in males and the round ligament in females.

inguinal gland [ANAT] Any of the superficial lymphatic glands in the groin.

inguinal hernia [MED] Protrusion of the abdominal viscera through the inguinal canal.

inguinal ligament [ANAT] The thickened lower portion of the aponeurosis of the external oblique muscle extending from the anterior superior spine of the ileum to the tubercle of the pubis and the pectineal line. Also known as Poupart's ligament.

inguinal region [ANAT] The abdominal region occurring on each side of the body as a depression between the abdomen and the thigh. Also known as iliac region.

inhalant canal [INV ZOO] The incurrent canal in sponges and mollusks.

inhalant siphon [INV ZOO] A channel for water intake in the mantle of bivalve mollusks. Also known as incurrent siphon.

inhalation [PHYSIO] The process of breathing in.

inhalator [MED] A device for facilitating the inhalation of a gas or spray, as for providing oxygen or oxygen–carbon dioxide mixtures for respiration in resuscitation.

inheritance [GEN] **1.** The acquisition of characteristics by transmission of germ plasm from ancestor to descendant. **2.** The sum total of characteristics dependent upon the constitution of the fertilized ovum.

inhibiting antibody [IMMUNOL] A substance sometimes produced in the blood of immunized persons which is thought to prevent the expected antigen-reagin reaction.

inhibition [PSYCH] An unconscious mechanism for restraining an impulse by means of an opposing impulse. [SCI TECH] The act of repressing or restraining a physical or chemical action.

inhibition-action balance [PSYCH] Relative balance maintained in every individual between his experiencing of emotional feelings and his outward behavior in response to them.

inhibition index [BIOCHEM] The amount of antimetabolite that will overcome the biological effect of a unit weight of metabolite.

inhibitor [CHEM] A substance which is capable of stopping or retarding a chemical reaction; to be technically useful, it must be effective in low concentration.

iniomi [VERT ZOO] An equivalent name for Salmoniformes.

inion [ANTHRO] The external occipital protuberance of the skull.

injection [MED] **1.** Introduction of a fluid into the skin, vessels, muscle, subcutaneous tissue, or any cavity of the body. **2.** The substance injected.

injury [MED] **1.** A structural or functional stress or trauma that induces a pathologic process. **2.** Damage resulting from the stress.

injury current *See* injury potential.

injury potential [PHYSIO] The potential difference observed between the injured and the noninjured regions of an injured tissue or cell. Also known as demarcation potential; injury current.

ink disease [PL PATH] A fungus disease of the chestnut in Europe caused by *Phytophthora cambivora* which produces black cankers and a black exudate in the trunk. Also known as black canker.

ink sac [INV ZOO] An organ attached to the rectum in many cephalopods which secretes and ejects an inky fluid.

innate [BIOL] Pertaining to a natural or inborn character dependent on genetic constitution. [BOT] Positioned at the apex of a supporting structure. [MYCOL] Embedded in, especially of an organ such as the fruiting body embedded in the thallus of some fungi.

inner cell mass [EMBRYO] The cells at the animal pole of a blastocyst which give rise to the embryo and certain extraembryonic membranes.

inner ear [ANAT] The part of the vertebrate ear concerned with labyrinthine sense and sound reception; consists generally of a bony and a membranous labyrinth, made up of the vestibular apparatus, three semicircular canals, and the cochlea. Also known as internal ear.

innervation [ANAT] The distribution of nerves to a part. [PHYSIO] The amount of nerve stimulation received by a part.

innominate *See* hip bone.

innominate artery [ANAT] The first artery branching from the aortic arch; distributes blood to the head, neck, shoulder, and arm on the right side of the body.

inoculation [BIOL] Introduction of a disease agent into an animal or plant to produce a mild form of disease and render the individual immune. [MICROBIO] Introduction of microorganisms onto or into a culture medium.

inoculum [MICROBIO] A small amount of substance containing bacteria from a pure culture which is used to start a new culture or to infect an experimental animal.

inorganic [INORG CHEM] Pertaining to or composed of chemical compounds that do not contain carbon as the principal element (excepting carbonates, cyanides, and cyanates), that is, matter other than plant or animal.

inorganic chemistry [CHEM] The study of chemical reactions and properties of all the elements and their compounds, with the exception of hydrocarbons, and usually including carbides, oxides of carbon, metallic carbonates, carbon-sulfur compounds, and carbon-nitrogen compounds.

inosculation *See* anastomosis.

inosine [BIOCHEM] $C_{10}H_{12}N_4O_5$ A compound occurring in muscle; a hydrolysis product of inosinic acid.

inosinic acid [BIOCHEM] $C_{10}H_{13}N_4O_8P$ A nucleotide constituent of muscle, formed by deamination of adenylic acid; on hydrolysis it yields hypoxanthine and D-ribose-5-phosphoric acid. Also known as hypoxanthylic acid.

inquiline [BIOL] **1.** An animal living habitually in another animal's home and getting a share of its food. **2.** An insect that develops in a gall produced by an insect of another species. [ZOO] An animal that inhabits the nest of another species.

insanity [PSYCH] **1.** Any mental disorder. **2.** In forensic psychiatry, a mental disorder which prevents one from managing one's affairs, impairs one's ability to distinguish right from wrong, or renders one harmful to oneself or others.

insect [INV ZOO] **1.** A member of the Insecta. **2.** An invertebrate that resembles an insect, such as a spider, mite, or centipede.

Insecta [INV ZOO] A class of the Arthropoda typically having a segmented body with an external, chitinous covering, a pair of compound eyes, a pair of antennae, three pairs of mouthparts, and two pairs of wings.

insect attractant [MATER] A chemical agent, usually associated with an insect's sexual drive, which may be used to attract pests to poisoned bait or for insect surveys.

insect control [ECOL] Regulation of insect populations by biological or chemical means.

insecticide [MATER] A chemical agent that destroys insects.

Insectivora [VERT ZOO] An order of mammals including hedgehogs, shrews, moles, and other forms, most of which have spines.

insectivorous [BIOL] Feeding on a diet of insects.

insectivorous plant [BOT] A plant that captures and digests insects as a source of nutrients by using specialized leaves. Also known as carnivorous plant.

insect pathology [INV ZOO] A biological discipline embracing the general principles of pathology as applied to insects.

insect physiology [INV ZOO] The study of the functional properties of insect tissues and organs.

insemination [BIOL] The planting of seed. [PHYSIO] 1. The introduction of sperm into the vagina. 2. Impregnation.

inserted [BIOL] United or attached to the supporting structure by natural growth.

insertion [BIOL] 1. A place where organs, such as muscles and leaves, are attached. 2. The place where the force of a muscle is applied.

insessorial [VERT ZOO] Adapted for perching, as a bird's foot.

in situ [SCI TECH] In the original location.

insomnia [MED] Sleeplessness; disturbed sleep; prolonged inability to sleep.

insomniac [MED] A person who is susceptible to insomnia.

inspiration [PHYSIO] The drawing in of the breath.

inspirometer [MED] An instrument for measuring the amount of air inspired.

instar [INV ZOO] A stage between molts in the life of arthropods, especially insects.

instep [ANAT] The arch on the medial side of the foot.

instinct [PSYCH] A primary tendency or inborn drive, as toward life, sexual reproduction, and death. [ZOO] A precise form of behavior in which there is an invariable association of a particular series of responses with specific stimuli; an unconditioned compound reflex.

instinctive behavior [ZOO] Any species-typical pattern of responses not clearly acquired through training.

instinctual [PSYCH] Pertaining to an emotional, impulsive, and generally unreasoned behavior or mental process which is a function of the id. [ZOO] Of or pertaining to instincts.

instrumental conditioning See operant conditioning.

insulin [BIOCHEM] A protein hormone produced by the beta cells of the islets of Langerhans which participates in carbohydrate and fat metabolism.

insulinase [BIOCHEM] An enzyme produced by the liver which is able to inactivate insulin.

insulinoma See islet-cell tumor.

insulin shock [MED] Clinical manifestation of hypoglycemia due to excess amounts of insulin in the blood.

insulin shock therapy [MED] Administration of large doses of insulin to induce hypoglycemic comas, followed by administration of glucose, in the treatment of certain psychotic disorders.

insuloma See islet-cell tumor.

integration [MOL BIO] The uptake of a fragment of viral or other DNA and its covalent attachment to, or insertion into, the chromosomal DNA of a recipient cell.

integrifolious [BOT] Having entire leaves.

integument [ANAT] An outer covering, especially the skin, together with its various derivatives.

integumentary musculature [VERT ZOO] Superficial skeletal muscles which are spread out beneath the skin and are inserted into it in some terrestrial vertebrates.

integumentary pattern [ANAT] Any of the features of the skin and its derivatives that are arranged in designs, such as scales, epidermal ridges, feathers coloration, or hair.

integumentary system [ANAT] A system encompassing the integument and its derivatives.

intelligence [PSYCH] 1. The intellect or astuteness of the mind. 2. Ability to recognize and understand qualities and attributes of the physical world and of mankind. 3. Ability to solve problems and engage in abstract thought processes.

intelligence quotient [PSYCH] The numerical designation for intelligence expressed as a ratio of an individual's perform-

INTEGUMENT

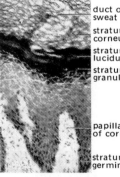

duct of sweat gland
stratum corneum
stratum lucidum
stratum granulosum
papilla of corium
stratum germinativum

The characteristic strata of thick skin of the human finger as seen in cross section at high magnification. *(From J. F. Nonidez and W. F. Windle, Textbook of Histology, McGraw-Hill, 1949)*

ance on a standardized test to the average performance according to age. Abbreviated IQ.

intelligence test [PSYCH] A series of standardized tasks or problems presented to an individual to measure his innate capacity to think, conceive, or reason; examples are the Stanford-Binet test and the Wechsler-Bellevue intelligence test.

intention tremor [MED] A clinical manifestation of certain diseases of the nervous system characterized by involuntary trembling of the limbs brought on by voluntary movements, and ceasing on rest.

interambulacrum [INV ZOO] In echinoderms, an area between two ambulacra.

interarticular [ANAT] Situated between articulating surfaces.

interatrial [ANAT] Located between the atria of the heart.

interatrial septal defect [MED] A congenital malformation of the septum between the atria of the heart.

interatrial septum *See* atrial septum.

interaxillary [BOT] Located within or between the axils of leaves.

intercalary [BOT] Referring to growth occurring between the apex and leaf. [SCI TECH] Inserted between two original components.

intercalary meristem [BOT] A meristem that is forming between regions of permanent or mature meristem.

intercapillary [ANAT] Located between capillaries.

intercapillary glomerulosclerosis [PATH] Nodular eosinophilic hyalin deposits on the periphery of glomeruli in individuals with diabetes. Also known as diabetic glomerulosclerosis; Kimmelstiel-Wilson disease.

intercarpal [ANAT] Located between the carpal bones.

intercavernous sinuses [ANAT] Venous sinuses located on the median line of the dura mater, connecting the cavernous sinuses of each side.

intercellular [HISTOL] Of or pertaining to the region between cells.

intercentrum [VERT ZOO] A type of crescentic intervertebral structure between successive centra in certain reptilian and mammalian tails.

interclavicle [VERT ZOO] A membrane bone in front of the sternum and between the clavicles in monotremes and most reptiles.

intercostal [ANAT] Situated or occurring between adjacent ribs.

intercostal muscles [ANAT] Voluntary muscles between adjacent ribs.

intercostal nerve [ANAT] Any of the branches of the thoracic nerves in the intercostal spaces.

intercourse *See* coitus.

intercrescence [BIOL] A growing together of tissues.

interface [PHYS CHEM] The boundary between any two phases: among the three phases (gas, liquid, and solid), there are five types of interfaces: gas-liquid, gas-solid, liquid-liquid, liquid-solid, and solid-solid.

interference microscope [OPTICS] A microscope used for visualizing and measuring differences in phase or optical paths in transparent or reflecting specimens; it differs from the phase contrast microscope in that the incident and diffracted waves are not separated, but interference is produced between the transmitted wave and another wave which originates from the same source.

interference phenomenon [VIROL] Inhibition by a virus of the simultaneous infection of host cells by some other virus.

interferon [BIOCHEM] A protein produced by intact animal

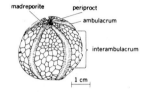

INTERAMBULACRUM

madreporite

periproct

ambulacrum

interambulacrum

1 cm

Morphological features of *Aulechinus grayae*, an echinocystitoid example from the Upper Ordovician of Scotland.

cells when infected with viruses; acts to inhibit viral reproduction and to induce resistance in host cells.

interferonogen [VIROL] A preparation made of inactivated virus particles used as an inoculant to stimulate formation of interferon.

interfoliaceous [BOT] Between a pair of leaves, such as between those which are opposite or verticillate.

intergenic crossing-over [MOL BIO] Recombination between distinct genes or cistrons.

intergenic suppression [GEN] The restoration of a suppressed function or character by a second mutation that is located in a different gene than the original or first mutation.

interkinesis *See* interphase.

interlabium [INV ZOO] A small lobe situated between the lips in certain nematodes.

intermediary metabolism [BIOCHEM] Intermediate steps in the chemical synthesis and breakdown of foodstuffs within body cells.

intermediate ganglion [ANAT] Any of certain small groups of nerve cells found along communicating branches of spinal nerves.

intermediate host [BIOL] The host in which a parasite multiplies asexually.

intermediate lobe [ANAT] The intermediate portion of the adenohypophysis.

intermedin [BIOCHEM] A hormonal substance produced by the intermediate portion of the hypophysis of certain animal species which influences pigmentation; similar to melanocyte-stimulating hormone in humans.

intermembranous ossification [HISTOL] Ossification within connective tissue with no prior formation of cartilage.

intermeningeal [ANAT] Between any two of the three meninges covering the brain and spinal cord.

intermenstrual [PHYSIO] Between periods of menstruation.

intermenstrual flow *See* metrorrhagia.

intermetameric [ANAT] Between adjacent metameres.

intermetatarsal [ANAT] Between adjacent bones of the metatarsus.

intermitotic [CYTOL] Of or pertaining to a stage of the cell cycle between two successive mitoses.

intermolecular force [PHYS CHEM] The force between two molecules; it is that negative gradient of the potential energy between the interacting molecules, if energy is a function of the distance between the centers of the molecules.

intermuscular hernia *See* interstitial hernia.

intermuscular septum [ANAT] A connective-tissue partition between muscles.

internal acoustic meatus [ANAT] An opening in the hard portion of the temporal bone for passage of the facial and acoustic nerves and internal auditory vessels.

internal ankle height [ANTHRO] The measure of the vertical distance from the lower end of the tibia to the floor.

internal capsule [ANAT] A layer of nerve fibers on the outer side of the thalamus and caudate nucleus, which it separates from the lenticular nucleus.

internal carotid [ANAT] A main division of the common carotid artery, distributing blood through three sets of branches to the cerebrum, eye, forehead, nose, internal ear, trigeminal nerve, dura mater, and hypophysis.

internal carotid nerve [ANAT] A sympathetic nerve which forms networks of branches around the internal carotid artery and its branches.

internal ear *See* inner ear.

internal elastic membrane [HISTOL] A sheet of elastin found between the tunica intima and the tunica media in medium- and small-caliber arteries.

internal energy [THERMO] A characteristic property of the state of a thermodynamic system, introduced in the first law of thermodynamics; it includes intrinsic energies of individual molecules, kinetic energies of internal motions, and contributions from interactions between molecules, but excludes the potential or kinetic energy of the system as a whole; it is sometimes erroneously referred to as heat energy.

internal fertilization [PHYSIO] Fertilization of the egg within the body of the female.

internal fistula [ANAT] A fistula which has no opening through the skin.

internal granular layer [HISTOL] The fourth layer of the cerebral cortex.

internal hemorrhage [MED] Bleeding within a body cavity or organ that is concealed from an observer.

internal hernia [MED] A hernia of intraabdominal contents occurring within the abdominal cavity.

internal hydrocephaly See obstructive hydrocephaly.

internal iliac artery [ANAT] The medial terminal division of the common iliac artery.

internalization [PSYCH] A mental mechanism operating outside of and beyond conscious awareness by which certain external attributes, attitudes, or standards are taken within oneself.

internal phloem [BOT] Primary phloem located between the primary xylem and the pith. Also known as intraxylary phloem.

internal secretion [PHYSIO] A secreted substance that is absorbed directly into the blood.

international angstrom [PHYS] A unit of length, equal to $1/6438.4696$ of the wavelength of the red cadmium line in dry air at standard atmospheric pressure, at a temperature of $15°C$ containing 0.03% by volume of carbon dioxide; equal to 1.0000002 angstroms. Abbreviated IA.

international table calorie See calorie

international unit [BIOL] A quantity of a vitamin, hormone, antibiotic, or other biological that produces a specific internationally accepted biological effect.

internode [BIOL] The interval between two nodes, as on a stem or along a nerve fiber.

internuclear [SCI TECH] Located between nuclei.

interoceptor [PHYSIO] A sense receptor located in visceral organs and yielding diffuse sensations.

interocular diameter [ANTHRO] The measure of the distance between the internal canthi.

interorbital [ANAT] Between the orbits of the eyes.

interparietal [ANAT] Between the parietal bones.

interparietal hernia See interstitial hernia.

interphase [CYTOL] Also known as interkinesis. **1.** The period between succeeding mitotic divisions. **2.** The period between the first and second meiotic divisions in those organisms where nuclei are reconstituted at the end of the first division.

interpleurite [INV ZOO] A small sclerite between the pleural sclerites in arthropods.

interproglottid gland [INV ZOO] Any of a number of cell clusters or glands arranged transversely along the posterior margin of the proglottids of certain tapeworms.

interpterygoid [ZOO] A space between palatal plates in certain chordates.

interpulmonary [ANAT] Located between the lungs.

interpupillary diameter [ANTHRO] A measure of the distance between the centers of the pupils, as the subject looks straight ahead.

Interquartile range [STAT] The distance between the top of the lower quartile and the bottom of the upper quartile of a distribution.

Interradial canal [INV ZOO] Any of the radially arranged gastrovascular canals in certain jellyfishes and ctenophores.

Interradius [INV ZOO] The area between two adjacent arms in echinoderms.

Interray [INV ZOO] A division of the radiate body of echinoderms.

Interrenal [ANAT] Located between the kidneys.

Interscapular [ANAT] Between the shoulders or shoulder blades.

Intersegmental [BIOL] Situated between or involving segments. [EMBRYO] Situated between the primordial segments of the embryo.

Intersegmental reflex [PHYSIO] An unconditioned reflex arc connecting input and output by means of afferent pathways in the dorsal spinal roots and efferent pathways in the ventral spinal roots.

Intersex [PHYSIO] An individual who is intermediate in sexual constitution between male and female.

Interspace [ANAT] An interval between the ribs or the fibers or lobules of a tissue or organ.

Interspersion [ECOL] 1. An intermingling of different organisms within a community. 2. The level or degree of intermingling of one kind of organism with others in the community.

Interspinal [ANAT] Situated between or connecting spinous processes: interspinous.

Intersternite [INV ZOO] An intersegmental plate on the ventral surface of the abdomen in insects.

Interstitial [CRYSTAL] A crystal defect in which an atom occupies a position between the regular lattice positions of a crystal. [SCI TECH] Of, pertaining to, or situated in a space between two things.

Interstitial cell [HISTOL] A cell that is not peculiar to or characteristic of a particular organ or tissue but which comprises fibrous tissue binding other cells and tissue elements; examples are neuroglial cells and Leydig cells.

Interstitial cell-stimulating hormone *See* luteinizing hormone.

Interstitial-cell tumor [MED] A benign tumor of the testes composed of Leydig cells. Also known as interstitioma; Leydig-cell tumor.

Interstitial emphysema [MED] Escape of air from the alveoli into the interstices of the lung, commonly due to trauma or violent cough.

Interstitial endometrosis [MED] The presence of endometrial tissue in the form of the stroma thoughout the myometrium. Also known as endolymphatic stromomyosis; fibromyosis; parathelioma; stromal endometriosis; stromal myosis; stromatosis.

Interstitial gland [HISTOL] 1. Groups of Leydig cells which secrete angiogens. 2. Groups of epithelioid cells in the ovarian medulla of some lower animals.

Interstitial hepatitis [MED] Pathologic deterioration and death of parenchymal cells in the liver associated with infiltration of lymphocytes and monocytes in the portal canals. Also known as acute nonsuppurative hepatitis; nonspecific hepatitis.

Interstitial hernia [MED] Protrusion of the intestine between the muscular planes of the abdominal wall. Also known as intermuscular hernia; interparietal hernia.

interstitial implants [MED] Solid or encapsulated radiation sources, made in the form of seeds, wires, or other shapes, to be inserted directly into tissue that is to be irradiated.

interstitial inflammation [MED] Inflammation of the interstitial tissues of an organ.

interstitial keratitis [MED] Inflammation of the cornea in which the iris is almost completely hidden by the diffuse haziness of the corneal tissue.

interstitial lamella [HISTOL] Any of the layers of bone between adjacent Haversian systems.

interstitial myocarditis [MED] Inflammation of the myocardium accompanied by cellular infiltration of interstitial tissues. Also known as Fiedler's myocarditis.

interstitial nephritis *See* pyelonephritis.

interstitial plasma-cell pneumonia *See* Pneumocystis carinii pneumonia.

interstitial pneumonia [MED] Inflammation of the lungs, particularly the stroma, including peribronchial tissue and interalveolar septa.

interstitial radiation [MED] Radiation of tissues by implantation of a radioactive source material.

interstitioma *See* interstitial-cell tumor.

intertemporal [VERT ZOO] Part of the sphenoid complex; a paired membrane bone that fuses with the alisphenoids.

intertergite [INV ZOO] One of the small plates between the tergites of certain insects.

intertubercular sulcus [ANAT] A deep groove on the anterior surface of the upper end of the humerus, separating the greater and lesser tubercles; contains the tendon of the long head of the biceps brachii muscle. Also known as bicipital groove.

intervallum [INV ZOO] The space between the walls of pleosponges.

intervascular [ANAT] Located between or surrounded by blood vessels.

interventricular foramen [ANAT] Either one of the two foramens that connect the third ventricle of the brain with each lateral ventricle. Also known as foramen of Monro.

interventricular septal defect [MED] A congenital malformation of the septum between the ventricles of the heart.

interventricular septum [ANAT] The muscular wall between the heart ventricles. Also known as ventricular septum.

intervertebral [ANAT] Being or located between the vertebrae.

intervillous spaces [HISTOL] Spaces in the placenta which communicate with the maternal blood vessels.

intestinal crura [INV ZOO] The main intestinal branches in certain trematodes.

intestinal digestion [PHYSIO] Conversion of food to an assimilable form by the action of intestinal juices.

intestinal dyspepsia [MED] Disturbed digestion due to diminished secretion of intestinal juices or lack of tonus in the wall of the intestine.

intestinal hormone [BIOCHEM] Either of two hormones, secretin and cholecystokinin, secreted by the intestine.

intestinal juice [PHYSIO] An alkaline fluid composed of the combined secretions of all intestinal glands.

intestinal lipodystrophy *See* Whipple's disease.

intestinal villi [ANAT] Fingerlike projections of the small intestine, composed of a core of vascular tissue covered by epithelium and containing smooth muscle cells and an efferent lacteal end capillary.

intestine [ANAT] The tubular portion of the vertebrate digestive tract, usually between the stomach and the cloaca or anus.

intima [HISTOL] The innermost coat of a blood vessel. Also known as tunica intima.

intine [BOT] The inner membrane of the covering on a pollen grain or spore.

intorsion [BIOL] Inward rotation of a structure about a fixed point or axis.

intoxication [MED] **1.** Poisoning. **2.** The state produced by overindulgence in alcohol.

intraabdominal [ANAT] Occurring or being within the cavity of the abdomen.

intraatrial heart block [MED] A type of heart block which shows a broad, notched P wave of longer than normal duration on the electrocardiographic record.

intracartilaginous ossification *See* endochondral ossification.

intracavernous aneurysm [MED] A dilation of the wall of the internal carotid artery within the cavernous sinus.

intracellular [CYTOL] Within a cell.

intracellular canaliculi [CYTOL] A system of minute canals within certain gland cells which are thought to drain the glandular secretions.

intracellular digestion [PHYSIO] Digestion which takes place within the cytoplasm of the organism, as in many unicellular protozoans.

intracellular enzyme [BIOCHEM] An enzyme that remains active only within the cell in which it is formed. Also known as organized ferment.

intracellular symbiosis [CYTOL] Existence of a self-duplicating unit within the cytoplasm of a cell, such as a kappa particle in *Paramecium*, which seems to be an infectious agent and may influence cell metabolism.

intracervical [ANAT] Located within the cervix of the uterus.

intracistron complementation [GEN] The process whereby two different mutant alleles, each of which determines in homozygotes an inactive enzyme, determine the formation of the active enzyme when present in the same nucleus.

intracortical [ANAT] Occurring or located within the cortex.

intracranial [ANAT] Within the cranium.

intracranial aneurysm [MED] Dilation of a cerebral artery.

intracytoplasmic [CYTOL] Being or occurring within the cytoplasm of a cell.

intradermal [ANAT] Within the skin.

intradermal nevus [MED] A lesion containing melanocytes and located principally or completely within the derma.

intraductal [ANAT] Within a duct.

intradural [ANAT] Within the dura mater.

intraembryonic [EMBRYO] Within the embryo.

intraepidermal [ANAT] Within the epidermis.

intraepidermal epithelioma [MED] Carcinoma in situ, of either the squamous-cell or basal-cell type.

intraesophageal [ANAT] Within the esophagus.

intrafascicular [BOT] Located or occurring within a vascular bundle.

intrafusal fiber [HISTOL] Any of the striated muscle fibers contained in a muscle spindle.

intragenic [MOL BIO] Within a gene, in referring to certain events.

intragenic recombination [MOL BIO] Recombination occurring between the mutons of one cistron.

intragenic suppression [GEN] The restoration of a suppressed function or character as a consequence of a second mutation located within the same gene as the original or first mutation.

intrahepatic [ANAT] Within the liver.

intrajugular process [ANAT] **1.** A small, curved process on some occipital bones which partially or completely divides the jugular foramen into lateral and medial parts. **2.** A small process on the hard portion of the temporal bone which completely or partly separates the jugular foramen into medial and lateral parts.

intralaminar nuclei [ANAT] A diffuse group of nuclei located in the internal medullary lamina of the thalamus.

intraluminal [ANAT] Within the lumen of a structure.

intramarginal [BIOL] Within a margin.

intramedullary [ANAT] **1.** Within the tissues of the spinal cord or medulla oblongata. **2.** Within the bone marrow. **3.** Within the adrenal medulla.

intramembranous [HISTOL] Formed or occurring within a membrane.

intramembranous ossification [HISTOL] Formation of bone tissue directly from connective tissue without a preliminary cartilage stage.

intramural [ANAT] Within the substance of the walls of an organ.

intramuscular [ANAT] Lying within or going into the substance of a muscle.

intraocular [ANAT] Occurring within the globe of the eye.

intraocular pressure [PHYSIO] The hydrostatic pressure within the eyeball.

intraparietal [ANAT] **1.** Within the wall of an organ or cavity. **2.** Within the parietal region of the cerebrum. **3.** Within the body wall.

intrapartum [MED] Occurring during parturition.

intraperitoneal [ANAT] **1.** Within the peritoneum. **2.** Within the peritoneal cavity.

intrapetiolar [BOT] **1.** Enclosed by the base of the petiole. **2.** Between the petiole and the stem.

intrapulmonic [ANAT] Being or occurring within the lungs.

intrastaminal [BOT] Located inside the androecium.

intrathecal [ANAT] Within the subarachnoid space.

intratracheal [ANAT] Being or occurring within the trachea.

intrauterine [ANAT] Being or occurring within the uterus.

intravaginal [ANAT] **1.** Being or occurring within the vagina. **2.** Located within a tendon sheath. [BOT] Located within a sheath, referring to branches of grass.

intravenous [ANAT] Located within, or going into, the veins.

intraventricular heart block [MED] Prolongation of the process of ventricular excitation, measured by the QRS complex of the electrocardiogram. Also known as arborization block; parietal block; peri-infarction block.

intravital [BIOL] Occurring while the cell or organism is alive.

intravital stain [BIOL] A nontoxic dye injected into the body to selectively stain certain cells or tissues.

intrinsic factor [BIOCHEM] A substance, produced by the stomach, which combines with the extrinsic factor (vitamin B_{12}) in food to yield an antianemic principle; lack of the intrinsic factor is believed to be a cause of pernicious anemia. Also known as Castle's intrinsic factor.

intrinsic viscosity [PHYS CHEM] The ratio of a solution's specific viscosity to the concentration of the solute, extrapolated to zero concentration. Also known as limiting viscosity number.

introgressive hybridization [GEN] The spreading of genes of a species into the gene complex of another due to hybridization between numerically dissimilar populations, the extensive backcrossing preventing formation of a single stable population.

introitus [ANAT] An opening or entryway, especially the opening into the vagina.

introjection [PSYCH] The symbolic absorption into and toward oneself of concepts and feelings generated toward another person or object; motivates irrational behavior toward oneself.

intromission [ZOO] The act or process of inserting one body into another, specifically, of the penis into the vagina.

introrse [BIOL] Turned inward or toward the axis.

introspection [PSYCH] The act of looking into one's own mind and feelings.

introspective diplopia *See* physiologic diplopia.

introversion [MED] The act or process of turning in upon itself, as a hollow organ. [PSYCH] Preoccupation with the self associated with diminished interest in external events.

introvert [PSYCH] An individual whose interests are self-directed, and not directed toward the outside world. [ZOO] **1.** A structure capable of introversion. **2.** To turn inward.

intubation [MED] The introduction of a tube into a hollow organ to keep it open, especially into the larynx to ensure the passage of air.

inulase [BIOCHEM] An enzyme produced by certain molds that catalyzes the conversion of inulin to levulose.

inulin [BIOCHEM] A polysaccharide made up of polymerized fructofuranose units.

inunction [MED] Act of applying an oil or fatty material, especially rubbing an ointment into the skin as a therapeutic measure.

in utero [EMBRYO] Within the uterus, referring to the fetus.

invagination [EMBRYO] The enfolding of a part of the wall of the blastula to form a gastrula. [PHYSIO] **1.** The act of ensheathing or becoming ensheathed. **2.** The process of burrowing or enfolding to form a hollow space within a previously solid structure, as the invagination of the nasal mucosa within a bone of the skull to form a paranasal sinus.

invasion [MED] **1.** The phase of an infectious disease during which the pathogen multiplies and is distributed; precedes signs and symptoms. **2.** The process by which microorganisms enter the body.

inversion [CHEM] Change of a compound into an isomeric form. [GEN] A type of chromosomal rearrangement in which two breaks take place in a chromosome and the fragment between breaks rotates 180° before rejoining. [MED] The act or process of turning inward or upside down.

invertase *See* saccharase.

Invertebrata [INV ZOO] A division of the animal kingdom including all animals which lack a spinal column; has no taxonomic status.

invertebrate [INV ZOO] An animal lacking a backbone and internal skeleton.

invertebrate zoology [ZOO] A branch of biology that deals with the study of Invertebrata.

inverted microscope [OPTICS] A microscope in which the body of the microscope, including the objective and the ocular, are below the stage, the illumination for transmitted light is above the stage, and with opaque materials, the vertical illuminator is used under the stage near the objective.

invertin *See* saccharase.

investing bone *See* dermal bone.

inviscid fluid [FL MECH] A fluid which has no viscosity; it therefore can support no shearing stress, and flows without energy dissipation. Also known as ideal fluid; nonviscous fluid; perfect fluid.

in vitro [BIOL] Pertaining to a biological reaction taking place in an artificial apparatus.

in vivo [BIOL] Pertaining to a biological reaction taking place in a living cell or organism.

involucel [BOT] A small, secondary involucre.

involucrate [BOT] Having an involucre.

involucre [BOT] Bracts forming one or more whorls at the base of an inflorescence or fruit in certain plants.

involucrum [ANAT] **1.** The covering of a part. **2.** New bone laid down by periosteum around a sequestrum in osteomyelitis.

involuntary fibers *See* smooth muscle fibers.

involuntary muscle [PHYSIO] Muscle not under the control of the will; usually consists of smooth muscle tissue and lies within a vescus, or is associated with skin.

involute [BIOL] Being coiled, curled, or rolled in at the edge.

involution [BIOL] A turning or rolling in. [EMBRYO] Gastrulation by ingrowth of blastomeres around the dorsal lip. [MED] **1.** The retrogressive change to their normal condition that organs undergo after fulfilling their functional purposes, as the uterus after pregnancy. **2.** The period of regression or the processes of decline or decay which occur in the human constitution after middle life.

involutional melancholia *See* involutional psychosis.

involutional psychosis [PSYCH] A prolonged psychotic reaction occurring in late middle life, characterized by depression and paranoid ideas. Also known as involutional melancholia.

involvucel [BOT] A secondary involucre.

involvucellate [BOT] Possessing an involvucel.

iodine [CHEM] A nonmetallic halogen element, symbol I, atomic number 53, atomic weight 126.9044; melts at 114°C, boils at 184°C; the poisonous, corrosive, dark plates or granules are readily sublimed; insoluble in water, soluble in common solvents; used as germicide and antiseptic, in dyes, tinctures, and pharmaceuticals, in engraving lithography, and as a catalyst and analytical reagent.

iodine-131 [NUC PHYS] A radioactive, artificial isotope of iodine, mass number 131; its half-life is 8 days with beta and gamma radiation; used in medical and industrial radioactive tracer work; moderately radiotoxic.

iodine solution *See* tincture of iodine.

iodine test [ANALY CHEM] Placing a few drops of potassium iodide solution on a sample to detect the presence of starch; test is positive if sample turns blue.

iodine tincture *See* tincture of iodine.

iodized oil [MATER] A thick, viscous, oily liquid with a garliclike odor that is an iodine addition product of vegetable oil, containing about 40% organically combined iodine; used in medicine as a radiopaque medium for radiography.

iodoacetic acid [ORG CHEM] CH_2ICOOH White or colorless crystals that are soluble in water and alcohol, and melt at 82-83°C; used in biological research for its inhibitive effect on enzymes.

iodophilic [CYTOL] A characteristic of certain cytoplasmic inclusions and vacuoles which stain darkly in iodine solution.

iodopsin [BIOCHEM] The visual pigment found in the retinal cones, consisting of retinene, combined with photopsin.

ion [CHEM] An isolated electron or positron or an atom or molecule which by loss or gain of one or more electrons has acquired a net electric charge.

ion exchange [PHYS CHEM] A chemical reaction in which mobile hydrated ions of a solid are exchanged, equivalent for equivalent, for ions of like charge in solution; the solid has an open, fishnetlike structure, and the mobile ions neutralize the charged, or potentially charged, groups attached to the solid matrix; the solid matrix is termed the ion exchanger.

ion-exchange chromatography [ANALY CHEM] A chromato-

IODINE

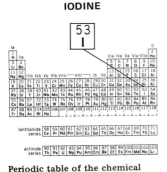

Periodic table of the chemical elements showing the position of iodine.

graphic procedure in which the stationary phase consists of ion-exchange resins which may be acidic or basic.

ion exclusion [CHEM] Ion-exchange resin system in which the mobile ions in the resin-gel phase electrically neutralize the immobilized charged functional groups attached to the resin, thus preventing penetration of solvent electrolyte into the resin-gel phase; used in separations where electrolyte is to be excluded from the resin, but not nonpolar materials, as the separation of salt from nonpolar glycerin.

ion-exclusion chromatography [ANALY CHEM] Chromatography in which the adsorbent material is saturated with the same mobile ions (cationic or anionic) as are present in the sample-carrying eluent (solvent), thus repelling the similar sample ions.

ionic bond [PHYS CHEM] A type of chemical bonding in which one or more electrons are transferred completely from one atom to another, thus converting the neutral atoms into electrically charged ions; these ions are approximately spherical and attract one another because of their opposite charge. Also known as electrovalent bond.

ionic membrane [CHEM ENG] Semipermeable membrane that conducts electricity; the application of an electric field to the membrane achieves an electrophoretic movement of ions through the membrane; used in electrodialysis.

ionic radii [PHYS CHEM] Radii which can be assigned to ions because the rapid variation of their repulsive interaction with distance makes them repel like hard spheres; these radii determine the dimensions of ionic crystals.

ionic strength [PHYS CHEM] A measure of the average electrostatic interactions among ions in an electrolyte; it is equal to one-half the sum of the terms obtained by multiplying the molality of each ion by its valence squared.

ionium [NUC PHYS] A naturally occurring radioisotope, symbol Io, of thorium, atomic weight 230.

ionization [CHEM] A process by which a neutral atom or molecule loses or gains electrons, thereby acquiring a net charge and becoming an ion; occurs as the result of the dissociation of the atoms of a molecule in solution ($NaCl \rightarrow Na^+ + Cl^-$) or of a gas in an electric field ($H_2 \rightarrow 2H^+$).

ionization constant [PHYS CHEM] Analog of the dissociation constant, where $k = [H^+][A^-]/[HA]$; used for the application of the law of mass action to ionization; in the equation HA represents the acid, such as acetic acid.

ionization potential [ATOM PHYS] The energy per unit charge needed to remove an electron from a given kind of atom or molecule to an infinite distance; usually expressed in volts. Also known as ion potential.

ionization radiation *See* ionizing radiation.

ionizing radiation [NUCLEO] **1.** Particles or photons that have sufficient energy to produce ionization directly in their passage through a substance. Also known as ionization radiation. **2.** Particles that are capable of nuclear interactions in which sufficient energy is released to produce ionization.

ion microscope *See* field-ion microscope.

ionophore [BIOCHEM] Any of a class of compounds, generally cyclic, having the ability to carry ions across lipid barriers due to the property of cation selectivity; examples are valinomycin and nonactin.

ion-permeable membrane [MATER] A film or sheet of a substance which is preferentially permeable to some species or types of ions.

ion potential *See* ionization potential.

Iospilidae [INV ZOO] A small family of pelagic polychaetes assigned to the Errantia.

ipecac [BOT] Any of several low, perennial, tropical South

IPECAC

Ipecac (*Cephaelis ipecacuanha*).

IRIDIUM

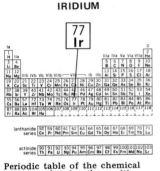

Periodic table of the chemical elements showing the position of iridium.

IRIS

The iris, an ornamental flower.

IRON

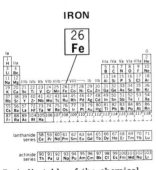

Periodic table of the chemical elements showing the position of iron.

American shrubs or half shrubs in the genus *Cephaelis* of the family Rubiaceae; the dried rhizome and root, containing emetine, cephaeline, and other alkaloids, is used as an emetic and expectorant.

Ipidae [INV ZOO] The equivalent name for Scolytidae.

IQ *See* intelligence quotient.

Ir *See* iridium.

Iridaceae [BOT] A family of monocotyledonous herbs in the order Liliales distinguished by three stamens and an inferior ovary.

iridectomy [MED] Surgical removal of part of the iris.

iridium [CHEM] A metallic element, symbol Ir, atomic number 77, atomic weight 192.2, in the platinum group; insoluble in acids, melting at 2454°C.

iridocyclitis [MED] Inflammation of the iris and the ciliary body.

iridocyclochoroiditis [MED] Inflammation of the iris, the ciliary body, and the choroid. Also known as uveitis.

iridocyte [HISTOL] A specialized cell in the integument of certain animal species which is filled with iridescent crystals of guanine and a variety of lipophores. Also known as guanophore; iridophore.

iridophore *See* iridocyte.

iris [ANAT] A pigmented diaphragm perforated centrally by an adjustable pupil which regulates the amount of light reaching the retina in vertebrate eyes. [BOT] Any plant of the genus *Iris*, the type genus of the family Iridaceae, characterized by linear or sword-shaped leaves, erect stalks, and bright-colored flowers with the three inner perianth segments erect and the outer perianth segments drooping.

Irish potato *See* potato.

iron [CHEM] A silvery-white metallic element, symbol Fe, atomic number 26, atomic weight 55.847, melting at 1530°C.

iron-binding protein [BIOCHEM] A serum protein, such as hemoglobin, for the transport of iron ions.

iron-deficiency anemia [MED] Hypochromic microcytic anemia due to excessive loss, deficient intake, or poor absorption of iron. Also known as nutritional hypochromic anemia.

iron metabolism [BIOCHEM] The chemical and physiological processes involved in absorption of iron from the intestine and in its role in erythrocytes.

iron-porphyrin protein [BIOCHEM] Any protein containing iron and porphyrin; examples are hemoglobin, the cytochromes, and catalase.

ironwood [BOT] Any of a number of hardwood trees in the United States, including the American hornbeam, the buckwheat, and the eastern hophornbeam.

irradiation [BIOPHYS] Subjection of a biological system to sound waves of sufficient intensity to modify their structure or function. [ENG] The exposure of a material, object, or patient to x-rays, gamma rays, ultraviolet rays, or other ionizing radiation.

irradiation cataract [MED] A cataract that develops slowly following prolonged or intense irradiation, as by radium or roentgen rays. Also known as cyclotron cataract; radiation cataract.

irradiation cystitis [MED] Inflammation of the urinary bladder following radiation therapy of pelvic organs.

irregular [BOT] Lacking symmetry, as of a flower having petals unlike in size or shape.

irregular cleavage [EMBRYO] Division of a zygote into random masses of cells, as in certain coelenterates.

irregular connective tissue [HISTOL] A loose or dense connective tissue with fibers irregularly woven and irregularly distributed; collagen is the dominant fiber type.

Irregularia [INV ZOO] An artificial assemblage of echinoderms in which the anus and periproct lie outside the apical system, the ambulacral plates remain simple, the primary radioles are hollow, and the rigid test shows some degree of bilateral symmetry.

Irreversible energy loss [THERMO] Energy transformation process in which the resultant condition lacks the driving potential needed to reverse the process; the measure of this loss is expressed by the entropy increase of the system.

Irritability [PHYSIO] **1.** A condition or quality of being excitable; the power of responding to a stimulus. **2.** A condition of abnormal excitability of an organism, organ, or part, when it reacts excessively to a slight stimulation.

Irritable colon [MED] Any of several disturbed colonic functions associated with anxiety or emotional stress. Also known as adaptive colitis; mucous colitis; spastic colon; unstable colon.

Irritant [BOT] An external stimulus or change in environment which provokes a response.

Isandrous [BOT] Having stamens that are similar and equal in number to the petals.

Isanthous [BOT] Having regular flowers.

Ischemia [MED] Localized tissue anemia as a result of obstruction of the blood supply or to vasoconstriction.

Ischemic neuropathy [MED] Nerve lesions characterized by numbness, tingling, and pain with loss of sensory and motor functions of the parts involved, due to obstruction of the blood supply to the nerves.

Ischemic paralysis [MED] Impaired motor function due to obstructed circulation to the area.

Ischemic tubulorrhexis *See* lower nephron nephrosis.

Ischiocavernosus [ANAT] Pertaining to the ischium and one or both of the corpora cavernosa of the penis or clitoris.

Ischioflexorius [VERT ZOO] The posterior muscle in the thigh of a salamander.

Ischiopodite [INV ZOO] The segment nearest the basipodite of walking legs in certain crustaceans. Also known as ischium.

Ischiorectal region [ANAT] The region between the ischium and the rectum.

Ischium [ANAT] Either of a pair of bones forming the dorsoposterior portion of the vertebrate pelvis; the inferior part of the pelvis in humans upon which the body rests in sitting. [INV ZOO] *See* ischiopodite.

Ischium-pubis index [ANAT] The ratio (length of pubis × 100/length of ischium) by which the sex of an adult pelvis may usually be determined; the index is greater than 90 in females, and less than 90 in males.

Ischnacanthidae [PALEON] The single family of the acanthodian order Ischnacanthiformes.

Ischnacanthiformes [PALEON] A monofamilial order of extinct fishes of the order Acanthodii; members were slender, lightly armored predators with sharp teeth, deeply inserted fin spines, and two dorsal fins.

Isectolophidae [PALEO] A family of extinct ceratomorph mammals in the superfamily Tapiroidea.

Island of Langerhans *See* islet of Langerhans.

Island of Reil [ANAT] The insula of the cerebral hemisphere.

Islet-cell carcinoma [MED] A metastatic tumor of pancreatic cells of the islet of Langerhans in the pancreas.

Islet-cell tumor [MED] A benign tumor of the pancreatic islet cells. Also known as insulinoma; insuloma; Langerhansian adenoma.

Islet of Langerhans [HISTOL] A mass of cell cords in the pancreas that is of an endocrine nature, secreting insulin and

ISOCITRIC ACID

CH$_2$COOH
|
CHCOOH
|
CHOHCOOH

Structural formula of isocitric acid.

ISOELECTRIC POINT

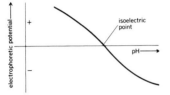

Graph showing the isoelectric point where particles are electrophoretically inert.

ISOETALES

Isoetes, the entire plant. *(From A. J. Eames, in E. W. Sinnott and K. S. Wilson, Botany: Principles and Problems, 6th ed., McGraw-Hill, 1963)*

a minor hormone like lipocaic. Also known as island of Langerhans; islet of the pancreas.

islet of the pancreas   *See* islet of Langerhans.

isoagglutinin   [IMMUNOL] An agglutinin which acts upon the red blood cells of members of the same species. Also known as isohemagglutinin.

isoagglutinogen   [IMMUNOL] An antigen which is specific to the erythrocytes of some individuals in a species.

isoalloxazine mononucleotide   *See* riboflavin 5'-phosphate.

isoantibody   [IMMUNOL] An antibody, found only in some members of a species, which acts upon cells or cell components of other members of the same species.

isoantigen   [IMMUNOL] An antigen in an individual capable of stimulating production of a specific antibody in another member of the same species.

isobiochore   [ECOL] A boundary line on a map connecting world environments that have similar floral and faunal constituents.

isobryales   [BOT] An order of mosses in which the plants are slender to robust and up to 90 centimeters in length.

isocarpic   [BOT] Having the same number of carpels and perianth divisions.

isocercal   [VERT ZOO] Of the tail fin of a fish, having the upper and lower lobes symmetrical and the vertebral column gradually tapering.

isochela   [INV ZOO] 1. A chela having two equally developed parts. 2. A chelate spicule with both ends identical.

isochromosome   [CYTOL] An abnormal chromosome with a medial centromere and identical arms formed as a result of transverse, rather than longitudinal, splitting of the centromere.

isocitric acid   [BIOCHEM] $HOOCCH_2CH(COOH)CH(OH)$-$COOH$ An isomer of citric acid that is involved in the Krebs tricarboxylic acid cycle in bacteria and plants. Also known as 2-hydroxy-1,2,3-propane-tricarboxylic acid.

isocrinida   [INV ZOO] An order of stalked articulate echinoderms with nodal rings of cirri.

isodiametric   [BIOL] Having equal diameters or dimensions.

isodont   [VERT ZOO] 1. Having all teeth alike. 2. Of a snake, having the maxillary teeth of equal length.

isodulcitol   *See* rhamnose.

isoelectric point   [PHYS CHEM] The pH value of the dispersion medium of a colloidal suspension at which the colloidal particles do not move in an electric field.

isoenzyme   [BIOCHEM] Any of the electrophoretically distinct forms of an enzyme, representing different polymeric states but having the same function. Also known as isozyme.

isoetaceae   [BOT] The single family assigned to the order Isoetales in some systems of classification.

isoetales   [BOT] A monotypic order of the class Isoetopsida containing the single genus *Isoetes*, characterized by long, narrow leaves with a spoonlike base, spirally arranged on an underground cormlike structure.

isoetatae   *See* Isoetopsida.

isoetopsida   [BOT] A class of the division Lycopodiophyta; members are heterosporous and have a distinctive appendage, the ligule, on the upper side of the leaf near the base. Formerly known as Isoetatae.

isoflavone   [BIOCHEM] $C_{15}H_{10}O_2$ A colorless, crystalline ketone, occurring in many plants, generally in the form of a hydroxy derivative.

isogamete   [BIOL] A reproductive cell that is morphologically similar in both male and female and cannot be distinguished on form alone.

isogamy [BIOL] Sexual reproduction by union of gametes or individuals of similar form or structure.

isogeneratae [BOT] A class of brown algae distinguished by having an isomorphic alternation of generations.

isogony [BIOL] Growth of parts at such a rate as to maintain relative size differences.

isohemagglutinin See isoagglutinin.

isohemolysin [IMMUNOL] A hemolysin produced by an individual injected with erythrocytes from another individual of the same species.

isohemolysis [IMMUNOL] Hemolysis induced by the action of an isohemolysin.

isoimmunization [IMMUNOL] Immunization of an individual by the introduction of antigens from another individual of the same species.

isokont [INV ZOO] Having flagella or cilia of equal length.

isolating mechanism [ECOL] Any endogenous or environmental change which tends to separate two populations and to permit their evolutionary divergence.

isolation [CHEM] Separation of a pure chemical substance from a compound or mixture; as in distillation, precipitation, or absorption. [MED] Separation of an individual with a communicable disease from other, healthy individuals. [MICROBIO] Separation of an individual or strain from a natural, mixed population. [PHYSIO] Separation of a tissue, organ, system, or other part of the body for purposes of study. [PSYCH] Dissociation of a memory or thought from the emotions or feelings associated with it.

isolecithal See homolecithal.

isoleucine [BIOCHEM] $C_6H_{13}O_2$ An essential monocarboxylic amino acid occurring in most dietary proteins.

isomerase [BIOCHEM] An enzyme that catalyzes isomerization reactions.

isomerism [BIOL] The condition of having two or more comparable parts made up of identical numbers of similar segments. [CHEM] The phenomenon whereby certain chemical compounds have structures that are different although the compounds possess the same elemental composition. [NUC PHYS] The occurrence of nuclear isomers.

isomerous [BIOL] Characterized by isomerism.

isometopidae [INV ZOO] A family of hemipteran insects in the superfamily Cimicimorpha.

isometric particle [VIROL] A plant virus particle that appears at first sight to be spherical when viewed in the electron microscope, but which is actually an icosahedron, possessing 20 sides.

isometric system [CRYSTAL] The crystal system in which the forms are referred to three equal, mutually perpendicular axes. Also known as cubic system.

isoniazid [PHARM] $C_6H_7N_3O$ A drug used as a tuberculostatic. Also known as isonicotinic acid hydrazide.

isonicotinic acid hydrazide See isoniazid.

isopathic principle [PSYCH] The rule according to which the cause cures the effect, as when a feeling of guilt is relieved by an exhibition of guilt, namely hate.

isopedin [VERT ZOO] The inner-layer bony material in cosmoid and ganoid scales.

isophyllous [BOT] Having foliage leaves of similar form on a plant or stem.

isopoda [INV ZOO] An order of malacostracan crustaceans characterized by a cephalon bearing one pair of maxillipeds in addition to the antennae, mandibles, and maxillae.

isoprecipitin [IMMUNOL] A precipitin effective only against the serum of individuals of the same species from which it is derived.

ISOGENERATAE

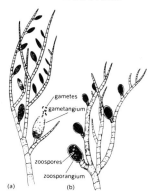

gametes

gametangium

zoospores

zoosporangium

(a)   (b)

Ectocarpus, an example of Isogeneratae showing the isomorphic alternation of generations. *(a)* Branched filament with multicellular gametangia. A gametangium is releasing biflagellate gametes. *(b)* Branched filament with unicellular zoosporangia with zoospores. *(From H. J. Fuller and O. Tippo, College Botany, rev. ed., Holt, 1954)*

isopropylarterenol *See* isoproterenol.

isoproterenol [PHARM] $C_{11}H_{17}NO_3$ A sympathomimetic amine used as a bronchodilator. Also known as isopropylarterenol.

Isoptera [INV ZOO] An order of Insecta containing morphologically primitive forms characterized by gradual metamorphosis, lack of true larval and pupal stages, biting and prognathous mouthparts, two pairs of subequal wings, and the abdomen joined broadly to the thorax.

isopygous [INV ZOO] Having a pygidium and a cephalon of equal size, as in certain trilobites.

Isospondyli [VERT ZOO] A former equivalent name for Clupeiformes.

isospore [BIOL] A spore that does not display sexual dimorphism.

isostemonous [BOT] Having the number of stamens of a flower equal to the number of perianth divisions.

isotonic [PHYSIO] **1.** Having uniform tension, as the fibers of a contracted muscle. **2.** Of a solution, having the same osmotic pressure as the fluid phase of a cell or tissue.

isotonic sodium chloride solution *See* normal saline.

isotope [NUC PHYS] One of two or more atoms having the same atomic number but different mass number.

isotope-dilution analysis [ANALY CHEM] Variation on paperchromatography analysis; a labeled radioisotope of the same type as the one being quantitated is added to the solution, then quantitatively analyzed afterward via radioactivity measurement.

isotope effect [PHYS CHEM] The effect of difference of mass between isotopes of the same element on nonnuclear physical and chemical properties, such as the rate of reaction or position of equilibrium, of chemical reactions involving the isotopes. [SOLID STATE] Variation of the transition temperatures of the isotopes of a superconducting element in inverse proportion to the square root of the atomic mass.

isotope exchange reaction [CHEM] A chemical reaction in which interchange of the atoms of a given element between two or more chemical forms of the element occurs, the atoms in one form being isotopically labeled so as to distinguish them from atoms in the other form.

isotope farm [BOT] A carbon-14 growth chamber, or greenhouse, arranged as a closed system in which plants can be grown in a carbon-14 dioxide (CO_2^{14}) atmosphere and thus become labeled with C^{14}; isotope farms also can be used with other materials, such as heavy water (D_2O), phosphorus-35 (P^{35}), and so forth, to produce biochemically labeled compounds.

isotopic indicator *See* isotopic tracer.

isotopic label *See* isotopic tracer.

isotopic tracer [CHEM] An isotope of an element, either radioactive or stable, a small amount of which may be incorporated into a sample material (the carrier) in order to follow the course of that element through a chemical, biological, or physical process, and also follow the larger sample. Also known as isotopic indicator; isotopic label; label; tag.

isotropic [BIOL] Having a tendency for equal growth in all directions. [CYTOL] An ovum lacking any predetermined axis.

isotypes [GEN] A series of antigens, for example, blood types, common to all members of a species but differentiating classes and subclasses within the species.

isozyme *See* isoenzyme.

isthmus [BIOL] A passage or constricted part connecting two parts of an organ.

Istiophoridae [VERT ZOO] The billfishes, a family of oceanic perciform fishes in the suborder Scombroidei.

Isuridae [VERT ZOO] The mackerel sharks, a family of pelagic, predacious galeoids distinguished by a heavy body, nearly symmetrical tail, and sharp, awllike teeth.

Itaconic acid [ORG CHEM] CH_2:$C(COOH)CH_2COOH$ A colorless crystalline compound that decomposes at 165°C, prepared by fermentation with *Aspergillus terreus;* used as an intermediate in organic synthesis and in resins and plasticizers. Also known as methylene succinic acid.

Itch [PHYSIO] An irritating cutaneous sensation allied to pain.

Ithomiinae [INV ZOO] The glossy-wings, a subfamily of weak-flying lepidopteran insects having on the wings broad, transparent areas in which the scales are reduced to short hairs.

Itonididae [INV ZOO] The gall midges, a family of orthorrhaphous dipteran insects in the series Nematocera; most are plant pests.

Ixoderm [MYCOL] A viscous layer covering the pileus of certain fungi and consisting of interlaced ends of hyphae which become gelatinous.

Ixodides [INV ZOO] The ticks, a suborder of the Acarina distinguished by spiracles behind the third or fourth pair of legs.

J

jaagsiekte [VET MED] A contagious disease of sheep, sometimes of goats and guinea pigs, resembling the more benign and diffuse forms of bronchiolar carcinoma in humans.

Jacanidae [VERT ZOO] The jacanas or lily-trotters, the single family of the avian superfamily Jacanoidea.

Jacanoidea [VERT ZOO] A monofamilial superfamily of colorful marsh birds distinguished by greatly elongated toes and claws, long legs, a short tail, and a straight bill.

jackal [VERT ZOO] **1.** *Canis aureus.* A wild dog found in southeastern Europe, southern Asia, and northern Africa. **2.** Any of various similar Old World wild dogs; they resemble wolves but are smaller and more yellowish.

jacket crown [MED] An artificial crown of a tooth consisting of a covering of porcelain or resin.

Jacksonian epilepsy [MED] Recurrent Jacksonian seizures.

Jacksonian seizure [MED] A focal seizure originating in one part of the motor or sensorimotor cortex and manifested usually by spasmodic contractions of or crawling or burning sensations of the skin; may become generalized, leading to loss of consciousness.

Jacobsoniidae [INV ZOO] The false snout beetles, a small family of coleopteran insects in the superfamily Dermestoidea.

Jacobson's cartilage *See* vomeronasal cartilage.

Jacobson's organ [VERT ZOO] An olfactory canal in the nasal mucosa which ends in a blind pouch; it is highly developed in reptiles and vestigial in humans.

jaculator [BOT] A hooklike placental process of certain fruits which helps in expelling the seeds.

Jadassohn's nevus *See* nevus sebaceus.

jaguar [VERT ZOO] *Felis onca.* A large, wild cat indigenous to Central and South America; it is distinguished by a buff-colored coat with black spots, and has a relatively large head and short legs.

Jamaica bayberry *See* bayberry.

Japanese encephalitis [MED] A human viral infection epidemic in Japan, transmitted by the common house mosquito (*Culex pipiens*) and characterized by severe inflammation of the brain.

Japygidae [INV ZOO] A family of wingless insects in the order Diplura with forcepslike anal appendages; members attack and devour small soil arthropods.

jaundice [INV ZOO] *See* grasserie. [MED] Yellow coloration of the skin, mucous membranes, and secretions resulting from hyperbile-rubinemia. Also known as icterus.

jaundice of newborn [MED] Jaundice in infants during the first few days after birth, due to various causes.

Java black rot [PL PATH] A fungus disease of stored sweet potatoes caused by *Diplodia tubericola*; the inside of the root becomes black and brittle.

JAVA MAN

Lateral view of the cranium of *Homo erectus* II, one of the first specimens of Java man. *(Carnegie Institution of Washington, as used in M. F. Ashley Montagu, An Introduction to Physical Anthropology, 2d ed., Charles C. Thomas, 1951)*

JERBOA

Jerboa, with body 3–6 inches (7–15 centimeters) long and tail up to 8 inches (20 centimeters) long.

Java man [PALEON] An overspecialized, apelike form of *Homo sapiens* from the middle Pleistocene having a small brain capacity, low cranial vault, and massive browridges.

jaw [ANAT] Either of two bones forming the skeleton of the mouth of vertebrates: the upper jaw or maxilla, and the lower jaw or mandible.

jawless vertebrate [VERT ZOO] The common name for members of the Agnatha.

jejunitis [MED] Inflammation of the jejunum.

jejunostomy [MED] The making of an artificial opening through the abdominal wall into the jejunum.

jejunum [ANAT] The middle portion of the small intestine, extending between the duodenum and the ileum.

jellyfish [INV ZOO] Any of various free-swimming marine coelenterates belonging to the Hydrozoa or Scyphozoa and having a bell- or bowl-shaped body. Also known as medusa.

jelly fungus [MYCOL] The common name for many members of the Heterobasidiomycetidae, especially the orders Tremallales and Dacromycetales, distinguished by a jellylike appearance or consistency.

Jennerian vaccine *See* smallpox vaccine.

Jenner's stain *See* May-Grünwald stain.

jenny [VERT ZOO] **1.** A female animal, as a jenny wren. **2.** A female donkey.

Jensen's sarcoma [VET MED] A transmissible malignant tumor originally observed in a rat inoculated with acid-fast bacteria from a cow with pseudotuberculous enteritis.

jerboa [VERT ZOO] The common name for 25 species of rodents composing the family Dipodidae; all are adapted for jumping, having extremely long hindlegs and feet.

jimsonweed [BOT] *Datura stramonium.* A tall, poisonous annual weed having large white or violet trumpet-shaped flowers and globose prickly fruits. Also known as apple of Peru.

JND *See* just-noticeable difference.

Johne's disease [VET MED] A chronic inflammation of the intestinal tract of sheep, cattle, and deer caused by *Mycobacterium paratuberculosis.*

Johnston's organ [INV ZOO] In Diptera, an organ in the second segment of the antenna that is concerned with the sense of balance and position in space.

joint [ANAT] A contact surface between two individual bones. Also known as articulation.

joint capsule [ANAT] A sheet of fibrous connective tissue enclosing a synovial joint.

joint marginal distribution [STAT] The distribution obtained by summing the joint distribution of three random variables over all possible values of one of these variables.

joint mouse [PATH] A small, loose body within a synovial joint, frequently calcified, derived from synovial membrane, organized fibrin fragments of articular cartilage, or arthritic osteophytes. Also known as joint body.

Jonathan freckle [PL PATH] A storage disease of apples characterized by small circular discolorations in the skin.

Jonathan spot [PL PATH] A disease of apples characterized by circular depressed necrotic areas around the lenticels.

Joppeicidae [INV ZOO] A monospecific family of hemipteran insects included in the Pentatomorpha; found in the Mediterranean regions.

jordanon *See* microspecies.

Jordan's rule [EVOL] The rule that organisms which are closely related tend to occupy adjacent rather than identical or distant ranges. [VERT ZOO] The rule that fishes in areas of low temperatures tend to have more vertebrae than those in warmer waters.

jugal [ANAT] Pertaining to the zygomatic bone. [VERT ZOO] In lower vertebrates, a bone lying below the orbit of the eye.

Jugatae [INV ZOO] The equivalent name for Homoneura.

jugate [BIOL] Structures which are joined together.

Juglandaceae [BOT] A family of dicotyledonous plants in the order Juglandales having unisexual flowers, a solitary basal ovule in a unilocular inferior ovary, and pinnately compound, exstipulate leaves.

Juglandales [BOT] An order of dicotyledonous plants in the subclass Hamamelidae distinguished by compound leaves; includes hickory, walnut, and butternut.

jugular [ANAT] Of or pertaining to the region of the neck above the clavicle.

jugular compression [PATH] A test for a spinal subarachnoid block by noting the rate of rise and fall of the spinal fluid pressure following compression and release of the jugular veins.

jugular foramen [ANAT] An opening in the cranium formed by the jugular notches of the occipital and temporal bones for passage of an internal jugular vein, the ninth, tenth, and eleventh cranial nerves, and the inferior petrosal sinus.

jugular process [ANAT] A rough process external to the condyle of the occipital bone.

jugular vein [ANAT] The vein in the neck which drains the brain, face, and neck into the innominate.

jugulum [INV ZOO] In insects, the small process on the forewing and hindwing that holds them together during flight. [VERT ZOO] The part of a bird's neck where it enters the breast.

jugum [BOT] One pair of opposite leaflets of a pinnate leaf. [INV ZOO] **1.** The most posterior and basal portion of the wing of an insect. **2.** A crossbar connecting the two arms of the brachidium in some brachiopods.

Juncaceae [BOT] A family of monocotyledonous plants in the order Juncales characterized by an inflorescence of diverse sorts, vascular bundles with abaxial phloem, and cells without silica bodies.

Juncales [BOT] An order of monocotyledonous plants in the subclass Commelinidae marked by reduced flowers and capsular fruits with one too many anatropous ovules per carpel.

junctional nevus [MED] A skin lesion containing nevus cells at the junction of the epidermis and dermis.

Jungermanniales [BOT] The leafy liverworts, an order of bryophytes in the class Marchantiatae characterized by chlorophyll-containing, ribbonlike or leaflike bodies and an undifferentiated thallus.

Jungian psychology *See* analytic psychology.

jungle [ECOL] An impenetrable thicket of second-growth vegetation replacing tropical rain forest that has been disturbed; lower growth layers are dense.

jungle yellow fever [MED] A form of yellow fever endemic in forested areas of Brazil.

Jura *See* Jurassic.

Jurassic [GEOL] Also known as Jura. **1.** The second period of the Mesozoic era of geologic time. **2.** The corresponding system of rocks.

just-noticeable difference [PSYCH] A subjective scale used in psychophysical tests, defined by C. Fechner as the subjective unit, and the absolute threshold as the zero point of the subjective scale; for example, the subjective intensity of a particular brightness of light would be specified when it was given as 100 just-noticeable differences above the absolute threshold. Abbreviated JND. Also known as difference limen; difference threshold.

JURASSIC

PRECAMBRIAN		
CAMBRIAN		
ORDOVICIAN		PALEOZOIC
SILURIAN		
DEVONIAN		
Mississippian	CARBONIFEROUS	
Pennsylvanian		
PERMIAN		
TRIASSIC		
JURASSIC		MESOZOIC
CRETACEOUS		
TERTIARY		CENOZOIC
QUATERNARY		

Chart showing the geologic periods and eras, with the Jurassic in the Mesozoic.

jute [BOT] Either of two Asiatic species of tall, slender, half-shrubby annual plants, *Corchorus capsularis* and *C. olitorius*, in the family Malvaceae, useful for their fiber.

juvenile cell *See* metamyelocyte.

juvenile hormone *See* neotenin.

juvenile melanoma [MED] A benign compound nevus in young people; resembles a malignant melanoma histologically.

Jynginae [VERT ZOO] The wrynecks, a family of Old World birds in the order Piciformes; a subfamily of the Picidae in some systems of classification.

K

K *See* potassium.

kaempferol [BIOCHEM] A flavonoid with a structure similar to that of quercetin but with only one hydroxyl in the B ring; acts as an enzyme cofactor and causes growth inhibition in plants.

Kahler's disease *See* multiple myeloma.

Kahn flocculation test [PATH] A macroscopic precipitin test for identification of the antibody resulting from syphilitic infection, by using an antigen prepared from normal beef heart.

kaki [BOT] *Diospyros kaki.* The Japanese persimmon; it provides a type of ebony wood that is black with gray, yellow, and brown streaks, has a close, even grain, and is very hard.

kakidrosis [MED] Secretion of sweat having a disagreeable odor.

kala azar [MED] Visceral leishmaniasis due to the protozoan *Leishmania donovani,* transmitted by certain sandflies (*Phlebotomus*); characterized by chronic, irregular fever, enlargement of the spleen and liver, emaciation, anemia, and leukopenia.

kale [BOT] Either of two biennial crucifers, *Brassica oleracea* var. *acephala* and *B. fimbriata,* in the order Capparales, grown for the nutritious curled green leaves.

kallidin I *See* bradykinin.

Kalotermitidae [INV ZOO] A family of relatively primitive, lower termites in the order Isoptera.

kalymma [INV ZOO] The part of the outer layer of certain radiolarians that has vacuoles.

Kamptozoa [INV ZOO] An equivalent name for Entoprocta.

kanamycin [MICROBIO] $C_{18}H_{36}O_{11}N_4$ A water-soluble, basic antibiotic produced by strains of *Streptomyces kanamyceticus;* the sulfate salt is effective in infections caused by gram-negative bacteria.

kangaroo [VERT ZOO] Any of various Australian marsupials in the family Macropodidae generally characterized by a long, thick tail that is used as a balancing organ, short forelimbs, and enlarged hindlegs adapted for jumping.

Kansan glaciation [GEOL] The second glaciation of the Pleistocene epoch in North America; began about 400,000 years ago, after the Aftonian and before the Yarmouth interglacials.

Kansasii disease [MED] A mycobacterial tuberculosislike infection caused by *Mycobacterium kansasii,* an orange-yellow acid-fast bacterium.

kapok [BOT] Silky fibers that surround the seeds of the kapok or ceiba tree. Also known as ceiba; Java cotton; silk cotton.

kapok tree [BOT] *Ceiba pentandra.* A tree of the family Bombacaceae which produces pods containing seeds covered with silk cotton. Also known as silk cotton tree.

kappa particle [CYTOL] A self-duplicating nucleoprotein particle found in various strains of *Paramecium* and thought

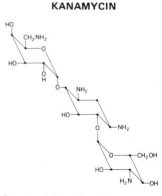

to function as an infectious agent; classed as an intracellular symbiont, occupying a position between the viruses and the bacteria and organelles.

Karumiidae [INV ZOO] The termitelike beetles, a small family of coleopteran insects in the superfamily Cantharoidea distinguished by having a tenth tergum.

karyocyte *See* normoblast.

karyogamy [CYTOL] Fusion of gametic nuclei, as in fertilization.

karyokinesis [CYTOL] Nuclear division characteristic of mitosis.

karyolymph [CYTOL] The clear material composing the ground substance of a cell nucleus.

karyolysis [CYTOL] Dissolution of a cell nucleus.

karyomastigont [INV ZOO] Pertaining to members of the protozoan order Oxymonadida; individuals can be uni- or multinucleate, and unattached forms give rise to two pairs of flagella.

karyoplasm *See* nucleoplasm.

karyoplasmic ratio *See* nucleocytoplasmic ratio.

karyorrhexis [CYTOL] Fragmentation of a nucleus with scattering of the pieces in the cytoplasm.

karyotype [CYTOL] The normal diploid or haploid complement of chromosomes, with respect to size, form, and number, characteristic of an individual, species, genus, or other grouping.

Kathlaniidae [INV ZOO] A family of nematodes assigned to the Ascaridina by some authorities and to the Oxyurina by others.

Kayser-Fleischer ring [PATH] A ring of golden-brown or brownish-green pigment behind the limbic border of the cornea, due to the deposition of copper.

Kazanian [GEOL] A European stage of geologic time: Upper Permian (above Kungurian, below Tatarian).

kcal *See* kilocalorie.

keel [VERT ZOO] The median ridge on the breastbone in certain birds. Also known as carina.

Keflin [MICROBIO] The trade name for cephalothin.

Keilor skull [PALEON] An Australian fossil type specimen of *Homo sapiens* from the Pleistocene.

Kell blood group system [IMMUNOL] A family of antigens found in erythrocytes and designated K, k, Kp^a, Kp^b, and Ku; antibodies to the K antigen, which occurs in about 10% of the population of England, have been associated with hemolytic transfusion reactions and with hemolytic disease.

keloid [MED] A firm, elevated fibrous formation of tissue at the site of a scar.

keloid acne [MED] Acnelike infection of hair follicles, especially on the nape of the neck, resulting in hard, white or reddish keloids and scarring. Also known as folliculitis keloidalis.

kelp [BOT] The common name for brown seaweed belonging to the Laminariales and Fucales.

kelvin [THERMO] A unit of absolute temperature equal to 1/273.16 of the absolute temperature of the triple point of water. Symbolized K. Formerly known as degree Kelvin.

Kelvin absolute temperature scale [THERMO] A temperature scale in which the ratio of the temperatures of two reservoirs is equal to the ratio of the amount of heat absorbed from one of them by a heat engine operating in a Carnot cycle to the amount of heat rejected by this engine to the other reservoir; the temperature of the triple point of water is defined as 273.16K. Also known as Kelvin temperature scale.

Kelvin temperature scale [THERMO] **1.** An International Practical Temperature Scale which agrees with the Kelvin absolute temperature scale within the limits of experimental determination. **2.** *See* Kelvin absolute temperature scale.

kenozooecium [INV ZOO] The outer, nonliving, hardened portion of a kenozooid.

kenozooid [INV ZOO] A type of bryozoan heterozooid possessing a slender tubular or boxlike chamber, completely enclosed and lacking an aperture.

Kentucky coffee tree [BOT] *Gymnocladus dioica.* An extremely tall, dioecious tree of the order Rosales readily recognized when in fruit by its leguminous pods containing heavy seeds, once used as a coffee substitute.

Kenyapithecus [PALEON] An early member of Hominidae from the Miocene.

keratectomy [MED] Surgical removal of a portion of the cornea.

keratin [BIOCHEM] Any of various albuminoids characteristic of epidermal derivatives, such as nails and feathers, which are insoluble in protein solvents, have a high sulfur content, and generally contain cystine and arginine as the predominating amino acids.

keratinized tissue [HISTOL] Any tissue with a high keratin content, such as the epidermis or its derivatives.

keratinocyte [HISTOL] A specialized epidermal cell that synthesizes keratin.

keratinous degeneration [CYTOL] The occurrence of keratin granules in the cytoplasm of a cell, other than a keratinocyte.

keratitis [MED] Inflammation of the cornea.

keratitis rosacea [MED] The occurrence of small, sterile infiltrates at the periphery of the cornea.

keratoconjunctivitis [MED] Concurrent inflammation of the cornea and the conjunctiva. Also known as shipyard eye.

keratohyalin [HISTOL] Granules in the stratum granulosum of keratinized stratified squamous epithelium which become keratin.

keratomalacia [MED] Degeneration of the cornea characterized by infiltration and keratinization of the epithelium, eventually leading to thinning and perforation of the cornea; generally occurs in vitamin-A deficiency.

keratoplasty [MED] A plastic operation on the cornea, especially the transplantation of a portion of the cornea.

keratosis [MED] Any disease of the skin characterized by an overgrowth of the cornified epithelium.

Kerguelen faunal region [ECOL] A marine littoral faunal region comprising a large area surrounding Kerguelen Island in the southern Indian Ocean.

kernel [BOT] **1.** The inner portion of a seed. **2.** A whole grain or seed of a cereal plant, such as corn or barley.

kernel blight [PL PATH] Any of several fungus diseases of barley caused chiefly by *Gibberella zeae, Helminthosporium sativum,* and *Alternaria* species shriveling and discoloring the grain.

kernel spot [PL PATH] A fungus disease of the pecan kernel caused by *Coniothyrium caryogenum* and characterized by dull-brown roundish spots.

kernicterus [PATH] Deposition of bilirubin in the gray matter of the brain and spinal cord, especially the basal ganglia, accompanied by nerve cell degeneration.

Kernig's sign [MED] In meningeal irritation, with the patient lying face up and the thigh flexed at the hip, the pain and spasm of the hamstring muscles when an attempt is made to completely extend the leg at the knee.

keto- [ORG CHEM] Organic chemical prefix for the keto or carbonyl group, C:O, as in a ketone.

ketoacidosis *See* ketosis.

ketogenesis [BIOCHEM] Production of ketone bodies.

KENTUCKY COFFEE TREE

Branch of Kentucky coffee tree *(Gymnocladus dioica).*

ketogenic hormone [BIOCHEM] A factor originally derived from crude anterior hypophysis extract which stimulated fatty-acid metabolism; now known as a combination of adrenocorticotropin and the growth hormone. Also known as fat-metabolizing hormone.

ketogenic substance [BIOCHEM] Any foodstuff which provides a source of ketone bodies.

ketoglutaric acid [BIOCHEM] $C_5H_6O_5$ A dibasic keto acid occurring as an intermediate product in carbohydrate and protein metabolism.

ketohexose [BIOCHEM] Any monsaccharide composed of a six-carbon chain and containing one ketone group.

ketolase [BIOCHEM] A type of enzyme that catalyzes cleavage of carbohydrates at the carbonyl carbon position.

ketolysis [BIOCHEM] Dissolution of ketone bodies.

ketone [ORG CHEM] One of a class of chemical compounds of the general formula $RR'CO$, where R and R′ are alkyl, aryl, or heterocyclic radicals; the groups R and R′ may be the same or different, or incorporated into a ring; the ketones, acetone, and methyl ethyl ketone are used as solvents, and ketones in general are important intermediates in the synthesis of organic compounds.

ketone body [BIOCHEM] Any of various ketones which increase in blood and urine in certain conditions, such as diabetic acidosis, starvation, and pregnancy. Also known as acetone body.

ketonemia *See* acetonemia.

ketonuria [MED] Presence of ketone bodies in the urine.

ketose [BIOCHEM] A carbohydrate that has a ketone group.

ketosis [MED] Excess amounts of ketones in the body, especially associated with diabetes mellitus. Also known as ketoacidosis.

ketosteroid [BIOCHEM] One of a group of neutral steroids possessing keto substitution, which produces a characteristic red color with *m*-dinitrobenzene in an alkaline solution; these compounds are principally metabolites of adrenal cortical and gonadal steroids.

Keuper [GEOL] A European stage of geologic time, especially in Germany; Upper Triassic.

Keweenawan [GEOL] The younger of two Precambrian time systems that constitute the Proterozoic period in Michigan and Wisconsin.

key [SYST] An arrangement of the distinguishing features of a taxonomic group to serve as a guide for establishing relationships and names of unidentified members of the group.

kg-cal *See* kilocalorie.

khellin [PHARM] $C_{14}H_{12}O_5$ A synthetic compound that crystallizes from methanol solution, has a bitter taste, melts at 154–155°C, and is more soluble in water than in organic solvents; used in medicine as an antispasmodic, a coronary vasodilator, and a bronchodilator. Also spelled chellin; khellin. Also known as visammin.

Kidd blood group system [IMMUNOL] The erythrocyte antigens defined by reactions to anti-Jk[a] antibodies, originally found in the mother (Mrs. Kidd) of the erythroblastotic infant, and to anti-Jk[b] antibodies.

kidney [ANAT] Either of a pair of organs involved with the elimination of water and waste products from the body of vertebrates; in humans they are bean-shaped, about 5 inches (12.7 centimeters) long, and are located in the posterior part of the abdomen behind the peritoneum.

Kienböck's disease *See* osteochondrosis.

killed vaccine [IMMUNOL] A suspension of killed microorganisms used as antigens to produce immunity.

KETONE

Structural formula of a ketone.

killer whale [VERT ZOO] *Orcinus orca.* A predatory, cosmopolitan cetacean mammal, about 30 feet (9 meters) long, found only in cold waters.

kilo- [MATH] A prefix meaning thousand.

kilocalorie [THERMO] A unit of heat energy equal to 1000 calories. Abbreviated kcal. Also known as kilogram-calorie (kg-cal); large calorie (Cal).

kilogram-calorie *See* kilocalorie.

Kimmelstiel-Wilson disease *See* intercapillary glomerulosclerosis.

Kimmeridgian [GEOL] A European stage of geologic time; middle Upper Jurassic, above Oxfordian, below Portlandian.

kinase [BIOCHEM] Any enzyme that catalyzes phosphorylation reactions.

Kinderhookian [GEOL] Lower Mississippian geologic time, above the Chautauquan of Devonian, below Osagian.

kineplasty [MED] An amputation of a limb in which tendons are arranged in the stump to permit their use in moving parts of the prosthetic appliance. Also spelled cineplasty.

kinesiatrics [MED] The treatment of disease by systematic active or passive movements. Also known as kinesitherapy; kinetotherapy.

kinesis [PHYSIO] The general term for physical movement, including that induced by stimulation, for example, light.

kinesitherapy *See* kinesiatrics.

kinesthesis [PHYSIO] The system of sensitivity present in the muscles and their attachments.

kinesthetic apraxia *See* motor apraxia.

kinetic [SCI TECH] Pertaining to or producing motion.

kinetic energy [MECH] The energy which a body possesses because of its motion; in classical mechanics, equal to one-half of the body's mass times the square of its speed.

kinetin [BIOCHEM] $C_{10}H_9ON_5$ A cytokinin formed in many plants which has a stimulating effect on cell division.

kinetoblast [INV ZOO] The outer covering of aquatic larvae that has cilia and functions in locomotion.

kinetochore *See* centromere.

kinetoplast [CYTOL] A genetically autonomous, membrane-bound organelle associated with the basal body at the base of flagella in certain flagellates, such as trypanosomes. Also known as parabasal body.

Kinetoplastida [INV ZOO] An order of colorless protozoans in the class Zoomastigophorea having pliable bodies and possessing one or two flagella in some stage of their life.

kinetosome *See* basal body.

kinetotherapy *See* kinesiatrics.

kingdom [SYST] One of the primary divisions that include all living organisms: most authorities recognize two, the animal kingdom and the plant kingdom, while others recognize three or more, such as Protista, Plantae, Animalia, and Mychota.

kinin [PHARM] Any of several pharmacologically active polypeptides that act as hypotensives, contracting isolated smooth muscles and increasing capillary permeability; an example is bradykinin.

kinocilium [CYTOL] A type of cilium containing one central pair of microfibrils and nine peripheral pairs; they extend from the apex of hair cells in all vertebrate ears except mammals.

kinomere *See* centromere.

kinoplasm [CYTOL] The substance of the protoplasm that is thought to form astral rays and spindle fibers.

Kinorhyncha [INV ZOO] A class of the phylum Aschelminthes consisting of superficially segmented microscopic marine animals lacking external ciliation.

Kinosternidae [VERT ZOO] The mud and musk turtles, a

KINETIN

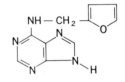

Structural formula for kinetin.

family of chelonian reptiles in the suborder Cryptodira found in North, Central, and South America.

Kirkbyacea [PALEON] A monomorphic superfamily of extinct ostracods in the suborder Beyrichicopina, all of which are reticulate.

Kirkbyidae [PALEON] A family of extinct ostracods in the superfamily Kirkbyacea in which the pit is reduced and lies below the middle of the valve.

kissing disease *See* infectious mononucleosis.

kitol [ORG CHEM] $C_{40}H_{60}O_2$ One of the provitamins of vitamin A derived from whale liver oil; crystallizes from methanol solution.

kiwi [VERT ZOO] The common name for three species of nocturnal ratites of New Zealand composing the family Apterygidae; all have small eyes, vestigial wings, a long slender bill, and short powerful legs.

Kjeldahl method [ANALY CHEM] Quantitative analysis of organic compounds to determine nitrogen content by interaction with concentrated sulfuric acid; ammonia is distilled from the NH_4SO_4 formed.

Klebsiella [MICROBIO] A genus of nonmotile, gram-negative, rod-shaped bacteria in the family Enterobacteriaceae; species are human pathogens.

Klebsiella pneumoniae [MICROBIO] An encapsulated pathogenic bacterium that causes severe pneumonitis in humans. Formerly known as Friedlander's bacillus; pneumobacillus.

Klebs-Loeffler bacillus *See* Corynebacterium diphtheriae.

klendusity [BOT] The tendency of a plant to resist disease due to a protective covering, such as a thick cuticle, that prevents inoculation.

kleptomania [PSYCH] An obsessive desire to steal; stolen objects are usually petty and useless, being of symbolic value only.

kleptophobia [PSYCH] **1.** An abnormal fear of thieves or of being robbed. **2.** An abnormal fear of becoming a kleptomaniac.

Klinefelter's syndrome [MED] A complex of symptoms associated with hypogonadism in males as an accompaniment of an anomaly of the sex chromosomes; somatic cells are found to have a Y chromosome and more than one X chromosome.

Kline flocculation test [PATH] A microscopic precipitin test for identification of the antibody resulting from syphilitic infection.

klinotaxis [BIOL] Positive orientation movement of a motile organism induced by a stimulus.

Kloedenellacea [PALEON] A dimorphic superfamily of extinct ostracods in the suborder Kloedenellocopina having the posterior part of one dimorph longer and more inflated than the other dimorph.

Kloedenellocopina [PALEON] A suborder of extinct ostracods in the order Paleocopa characterized by a relatively straight dorsal border with a gently curved or nearly straight ventral border.

knee [ANAT] **1.** The articulation between the femur and the tibia in humans. Also known as genu. **2.** The corresponding articulation in the hindlimb of a quadrupedal vertebrate.

kneecap *See* patella.

kneecap height [ANTHRO] A measure of the vertical distance taken from the floor at the base of the heel to the top of the muscle mass near the end of the thigh bone as the subject sits with both feet on the floor.

Kneriidae [VERT ZOO] A small family of tropical African fresh-water fishes in the order Gonorynchiformes.

koala [VERT ZOO] *Phascolarctos cinereus.* An arboreal marsupial mammal of the family Phalangeridae having large hairy

KOALA

The koala, a slow-moving arboreal animal with opposable digits and a rudimentary tail.

ears, gray fir, and two clawed toes opposing three others on each limb.

Koch's postulates [MICROBIO] A set of laws elucidated by Robert Koch: the microorganism identified as the etiologic agent must be present in every case of the disease; the etiologic agent must be isolated and cultivated in pure culture; the organism must produce the disease when inoculated in pure culture into susceptible animals; a microorganism must be observed in and recovered from the experimentally diseased animal.

Kohler's disease *See* osteochondrosis.

kohlrabi [BOT] A biennial crucifer, designated *Brassica caulorapa* and *B. oleracea* var. *caulo-rapa*, of the order Capparales grown for its edible turniplike, enlarged stem.

koilonychia [MED] Spoon-shaped deformity of the fingernails, which may be familial or associated with a disease, such as iron-deficiency anemia. Also known as spoon nail.

koilorachic [MED] Having the lumbar spinal region concave ventrally.

Kölliker's canal [INV ZOO] A canal extending from the otocyst toward the exterior, as in certain cephalopods.

Kolmer test [PATH] A complement-fixation test for syphilis and other diseases.

Komodo dragon [VERT ZOO] *Varanus komodoensis.* A predatory reptile of the family Varanidae found only on the island of Komodo; it is the largest living lizard and may grow to 10 feet.

Koonungidae [INV ZOO] A family of Australian crustaceans in the order Anaspidacea with sessile eyes and the first thoracic limb modified for digging.

Koplik's sign [MED] Small red spots surrounded by white areas seen in the mucous membrane of the mouth in the prodromal stage of measles. Also known as Koplik's spots.

Koplik's spots *See* Koplik's sign.

Kr *See* krypton.

kraurosis [MED] A progressive, sclerosing, shriveling process of the skin, due to glandular atrophy.

Krause's corpuscle [ANAT] One of the spheroid nerve-end organs resembling lamellar corpuscles, but having a more delicate capsule; found especially in the conjunctiva, the mucosa of the tongue, and the external genitalia; they are believed to be cold receptors. Also known as end bulb of Krause.

Krause's gland [ANAT] Any of the accessory lacrimal glands found in the upper eyelid.

Krebs cycle [BIOCHEM] A sequence of enzymatic reactions involving oxidation of a two-carbon acetyl unit to carbon dioxide and water to provide energy for storage in the form of high-energy phosphate bonds. Also known as citric acid cycle; tricarboxylic acid cycle.

Krebs-Henseleit cycle [BIOCHEM] A cyclic reaction pathway involving the breakdown of arginine to urea in the presence of arginase.

Krebspest [INV ZOO] A fatal fungus disease of crayfish caused by *Aphanomyces mystaci.*

krill [INV ZOO] A name applied to planktonic crustaceans that constitute the diet of many whales, particularly whalebone whales.

Krukenberg's tumor [MED] Bilateral carcinoma of the ovaries; originally described as primary, but now denoting a metastatic form usually of gastric origin.

krummholz [ECOL] Stunted alpine forest vegetation. Also known as elfinwood.

krypton [CHEM] A colorless, inert gaseous element, symbol

KOHLRABI

Kohlrabi *(Brassica caulorapa),* cultivar Early White Vienna. *(Joseph Harris Co., Rochester, N.Y.)*

KRYPTON

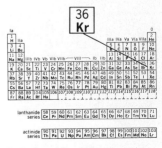

Periodic table of the chemical elements showing the position of krypton.

Kr, atomic number 36, atomic weight 83.80; it is odorless and tasteless; used to fill luminescent electric tubes.

kudzu [BOT] Any of various perennial vine legumes of the genus *Pueraria* in the order Rosales cultivated principally as a forage crop.

Kuehneosauridae [PALEON] The gliding lizards, a family of Upper Triassic reptiles in the order Squamata including the earliest known aerial vertebrates.

KUMQUAT

Fruit cluster and foliage of Nagami kumquat *(Fortunella margarita)*. *(J. Horace McFarland Co.)*

kumquat [BOT] A citrus shrub or tree of the genus *Fortunella* in the order Sapindales grown for its small, flame- to orange-colored edible fruit having three to five locules filled with an acid pulp, and a sweet, pulpy rind.

Kungurian [GEOL] A European stage of geologic time; Middle Permian, above Artinskian, below Kazanian.

Kupffer cell [HISTOL] One of the fixed macrophages lining the hepatic sinusoids.

Kurtoidei [VERT ZOO] A monogeneric suborder of perciform fishes having a unique ossification that encloses the upper part of the swim bladder, and an occipital hook in the male for holding eggs during brooding.

kurtorachic [MED] Having the lumbar spinal region concave dorsally.

kurtosis [STAT] The extent to which a frequency distribution is concentrated about the mean or peaked; it is sometimes defined as the ratio of the fourth moment of the distribution to the square of the second moment.

Kutorginida [PALEON] An order of extinct brachiopod mollusks that is unplaced taxonomically.

K virus [VIROL] A group 2 papovavirus affecting rats and mice.

kwashiorkor [MED] A nutritional deficiency disease in infants and young children, mainly in the tropics, caused primarily by a diet low in proteins and rich in carbohydrates. Also known as nutritional dystrophy.

kymograph [MED] A device for recording internal body movements by making tracings with a stylus on a revolving smoked drum.

kymography [MED] A recording the movements of internal organs by a kymograph.

kynurenic acid [BIOCHEM] $C_{10}H_7O_3N$ A product of tryptophan metabolism found in the urine of mammals.

kynurenine [BIOCHEM] $C_{10}H_{12}O_3N_2$ An intermediate product of tryptophan metabolism occurring in the urine of mammals.

kyphos [MED] The anteroposterior hump in the spine occurring in kyphoscoliosis.

kyphoscoliosis [MED] Lateral curvature of the spine accompanied by rotation of the vertebrae.

kyphosis [MED] Angular curvature of the spine, usually in the thoracic region. Also known as humpback; hunchback.

L

La *See* lanthanum.

labellate [BIOL] Having a labellum.

labellum [BOT] The median membrane of the corolla of an orchid often differing in size and morphology from the other two petals. [INV ZOO] **1.** A prolongation of the labrum in certain beetles and true bugs. **2.** In Diptera, either of a pair of sensitive fleshy lobes consisting of the expanded end of the labium.

labial gland [ANAT] Any of the small, tubular mucous and serous glands underneath the mucous membrane of mammalian lips. [INV ZOO] A salivary gland, or modification thereof, opening at the base of the labium in certain insects.

labial palp [INV ZOO] **1.** Either of a pair of fleshy appendages on either side of the mouth of certain bivalve mollusks. **2.** A jointed appendage attached to the labium of certain insects.

labial papilla [INV ZOO] Any of the sensory bristles around the mouth of many nematodes; they are jointed projections of the cuticle.

Labiatae [BOT] A large family of dicotyledonous plants in the order Lamiales; members are typically aromatic and usually herbaceous or merely shrubby.

labiate [ANAT] Having liplike margins that are thick and fleshy. [BIOL] Having lips. [BOT] Having the limb of a tubular calyx or corolla divided into two unequal overlapping parts.

labiatiflorous [BOT] Having the corolla divided into two liplike parts, one longer than the other.

labile [PSYCH] Unstable in mood. [SCI TECH] Also known as metastable. **1.** Readily changed, as by heat, oxidation, or other processes. **2.** Moving from place to place.

labile factor *See* proaccelerin.

lability [PSYCH] Very rapid fluctuations in intensity and modality of emotions; seen in the affective reaction or in certain organic brain disorders.

labiosternite [INV ZOO] A median area between the palpigers on the head of an insect.

labiostipes [INV ZOO] A segment of the basal part of the labium in insects.

labium [BIOL] **1.** A liplike structure. **2.** The lower lip, as of a labiate corolla or of an insect.

labium majus [ANAT] Either of the two outer folds surrounding the vulva in the female.

labium minus [ANAT] Either of the two inner folds, at the inner surfaces of the labia majora, surrounding the vulva in the female.

laboratory [SCI TECH] A place for experimental study.

Laboulbeniales [MYCOL] An order of ascomycetous fungi made up of species that live primarily on the external surfaces of insects.

LABIATAE

Monarda fistulosa, an eastern North American species of wild bergamot, showing the characteristic opposite leaves and irregular corolla of the family Labiatae. *(Courtesy of A. W. Ambler, National Audubon Society)*

Labridae [VERT ZOO] The wrasses, a family of perciform fishes in the suborder Percoidei.

labrum [INV ZOO] **1.** The upper lip of certain arthropods, lying in front of or above the mandibles. **2.** The outer edge of a gastropod shell.

labyrinth [ANAT] **1.** Any body structure full of intricate cavities and canals. **2.** The inner ear.

labyrinthine placenta *See* hemochorial placenta.

labyrinthine reflex [PHYSIO] The involuntary response to stimulation of the vestibular apparatus in the inner ear.

labyrinthine syndrome *See* Ménière's syndrome.

labyrinthitis [MED] Inflammation of the labyrinth of the inner ear.

labyrinthodont [PALEON] Having teeth with greatly convoluted dentin.

Labyrinthodontia [PALEON] A subclass of fossil amphibians descended from crossopterygian fishes, ancestral to reptiles, and antecedent to at least part of other amphibian types.

Labyrinthulia [INV ZOO] A subclass of the protozoan class Rhizopoda containing mostly marine, ovoid to spindle-shaped, uninucleate organisms that secrete a network of filaments (slime tubes) along which they glide.

Labyrinthulida [INV ZOO] The single order of the protozoan subclass Labyrinthulia.

laccase [BIOCHEM] Any of a class of plant oxidases which catalyze the oxidation of phenols.

laccate [BIOL] Having a lacquered appearance.

Lacciferinae [INV ZOO] A subfamily of scale insects in the superfamily Coccoidea in which the male lacks compound eyes, the abdomen is without spiracles in all stages, and the apical abdominal segments of nymphs and females do not form a pygidium.

lacerate [MED] To inflict a wound by tearing.

lacerated [BIOL] Having a deeply and irregularly incised margin or apex.

laceration [MED] A wound made by tearing.

Lacertidae [VERT ZOO] A family of reptiles in the suborder Sauria, including all typical lizards, characterized by movable eyelids, a fused lower jaw, homodont dentition, and epidermal scales.

lacertus [ANAT] A small bundle of fibers.

lacrimal [ANAT] Pertaining to tears, tear ducts, or tear-secreting organs.

lacrimal apparatus [ANAT] The functional and structural mechanisms for secreting and draining tears; includes the lacrimal gland, lake, puncta, canaliculi, sac, and nasolacrimal duct.

lacrimal bone [ANAT] A small bone located in the anterior medial wall of the orbit, articulating with the frontal, ethmoid, maxilla, and inferior nasal concha.

lacrimal canal *See* nasolacrimal canal.

lacrimal canaliculus [ANAT] A small tube lined with stratified squamous epithelium which runs vertically a short distance from the punctum of each eyelid and then turns horizontally in the lacrimal part of the lid margin to the lacrimal sac. Also known as lacrimal duct.

lacrimal duct *See* lacrimal canaliculus.

lacrimal gland [ANAT] A compound tubuloalveolar gland that secretes tears. Also known as tear gland.

lacrimal lake [ANAT] The space where tears collect in the inner canthus of the eye.

lacrimal sac [ANAT] The dilation at the upper end of the nasolacrimal duct within the medial canthus of the eye. Also known as dacryocyst.

LABYRINTHULIA

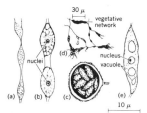

Labyrinthulia. *Labyrinthula zopfi: (a)* portion of living network, *(b)* two organisms stained, *(c)* encysted stage *(after Valkoanov). L. macrocystis: (d)* vegetative network, *(e)* single organism stained *(after Cienkowski). (From R. P. Hall, Protozoology, Prentice-Hall, 1953)*

lacrimation [PHYSIO] 1. Normal secretion of tears. 2. Excessive secretion of tears, as in weeping.

lacrimose [MYCOL] Having tear-shaped appendages, as the gills of certain fungi.

lactalbumin [BIOCHEM] A simple protein contained in milk which resembles serum albumin and is of high nutritional quality.

lactase [BIOCHEM] An enzyme that catalyzes the hydrolysis of lactose to dextrose and galactose.

lactase deficiency syndrome [MED] Diarrhea induced by ingestion of a lactose-containing food such as milk, secondary to a congenital or acquired deficiency of lactase.

lactate [ORG CHEM] A salt or ester of lactic acid in which the acidic hydrogen of the carboxyl group has been replaced by a metal or an organic radical. [PHYSIO] To secrete milk.

lactate dehydrogenase [BIOCHEM] A zinc-containing enzyme which catalyzes the oxidation of several α-hydroxy acids to corresponding α-keto acids.

lactation [PHYSIO] Secretion of milk by the mammary glands.

lacteal [ANAT] One of the intestinal lymphatics that absorb chyle. [PHYSIO] Pertaining to or resembling milk.

lactescent [BIOL] Having a milky appearance. [PHYSIO] Secreting milk or a milklike substance.

lactic acid [BIOCHEM] $C_3H_6O_3$ A hygroscopic α-hydroxy acid, occurring in three optically isomeric forms: L form, in blood and muscle tissue as a product of glucose and glycogen metabolism; D form, obtained by fermentation of sucrose; and DL form, a racemic mixture present in foods prepared by bacterial fermentation, and also made synthetically. Also known as 2-hydroxypropanoic acid; α-hydroxypropanoic acid.

lactic dehydrogenase [BIOCHEM] An enzyme that catalyzes the dehydrogenation of L-lactic acid to pyruvic acid. Abbreviated LDH.

lactic dehydrogenase virus [VIROL] A virus of the rubella group which infects mice.

lactiferous [PHYSIO] Forming or carrying milk.

lactin *See* lactose.

lactivorous [ZOO] Feeding on milk.

Lactobacillaceae [MICROBIO] A family of sugar-fermenting bacteria in the order Eubacteriales including both spherical and rod-shaped forms.

Lactobacilleae [MICROBIO] A tribe of rod-shaped bacteria in the family Lactobacillaceae.

lactoflavin *See* riboflavin.

lactogenic hormone *See* prolactin.

lactoglobulin [BIOCHEM] A crystalline protein fraction of milk, which is soluble in half-saturated ammonium sulfate solution and insoluble in pure water.

lactonase [BIOCHEM] The enzyme that catalyzes the hydrolysis of 6-phosphoglucono-Δ-lactone to 6-phosphogluconic acid in the pentose phosphate pathway.

lactose [BIOCHEM] $C_{12}H_{22}O_{11}$ A disaccharide composed of D-glucose and D-galactose which occurs in milk. Also known as lactin; milk sugar.

lactosuria [MED] Presence of lactose in the urine.

lacuna [BIOL] A small space or depression. [HISTOL] A cavity in the matrix of bone or cartilage which is occupied by the cell body.

lacunar system [INV ZOO] A series of intercommunicating spaces branching from two longitudinal vessels in the hypodermis of many acanthocephalins.

Lacydonidae [INV ZOO] A benthic family of pelagic errantian polychaetes.

LACTIC ACID

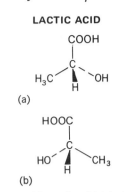

(a)

(b)

Structural formulas of *(a)* dextro form and *(b)* levo form of lactic acid.

LACTOBACILLEAE

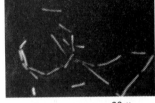

├─ 20 μ ─┤

Photomicrograph showing morphology of *Lactobacillus brevis,* tribe Lactobacilleae.

LACTOSE

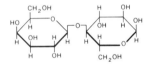

The α-D anomer of lactose, the usual form in which lactose is obtained.

LACUNAR SYSTEM

Lacunar system of *Moniliformis,* dorsal view, showing regular circular branches. *(After Meyer in L. H. Hyman, The Invertebrates, vol. 3, McGraw-Hill, 1951)*

Ladinian [GEOL] A European stage of geologic time: upper Middle Triassic (above Anisian, below Carnian).

Laemobothridae [INV ZOO] A family of lice in the order Mallophaga including parasites of aquatic birds, especially geese and coots.

Laennec's cirrhosis *See* portal cirrhosis.

lagena [VERT ZOO] The curved, flasklike organ of hearing in lower vertebrates; corresponds to the cochlea.

Lagenidiales [MYCOL] An order of aquatic fungi belonging to the class Phycomycetales characterized by a saclike to limited hyphal thallus and zoospores having two flagella.

lageniform [BIOL] Flask-shaped.

lager beer [FOOD ENG] A variety of beer produced by bottom fermentation and allowed to age from 6 weeks to 6 months; it is usually dry, light in color, and well carbonated.

Lagomorpha [VERT ZOO] The order of mammals including rabbits, hares, and pikas; differentiated from rodents by two pairs of upper incisors covered by enamel, vertical or transverse jaw motion, three upper and two lower premolars, fused tibia and fibula, and a spiral valve in the cecum.

lag phase [MICROBIO] The period of physiological activity and diminished cell division following the addition of inoculum of bacteria to a new culture medium.

Lagriidae [INV ZOO] The long-jointed bark beetles, a family of coleopteran insects in the superfamily Tenebrionoidea.

Lagynacea [INV ZOO] A superfamily of foraminiferan protozoans in the suborder Allogromiina having a free or attached test that has a membranous to tectinous wall and a single, ovoid, tubular, or irregular chamber.

lalopathy [MED] Any disorder of speech or disturbance of language.

laloplegia [MED] Inability to speak, caused by paralysis of the muscles concerned in speech, except those of the tongue.

Lamarckism [EVOL] The theory that organic evolution takes place through the inheritance of modifications caused by the environment, and by the effects of use and disuse of organs.

lamb [VERT ZOO] A young sheep.

lamb dysentery [VET MED] A bacterial infection and inflammation of the intestinal tract of lambs caused by *Clostridium perfringens*, chiefly along the English-Scottish border.

lamella [ANAT] A thin scale or plate.

lamellar bone [HISTOL] Any bone with a microscopic structure consisting of thin layers or plates.

lamellar chloroplast [CYTOL] A type of chloroplast in which the layered structure extends more or less uniformly through the whole chloroplast body.

Lamellibranchiata [INV ZOO] An equivalent name for Bivalvia.

lamellibranchiate [INV ZOO] **1.** Having bilateral platelike gills. **2.** Having the body bilaterally compressed and symmetrical, like a bivalve.

lamellicorn [INV ZOO] Having the antennal joints laterally expanded into flat plates, as in beetles.

Lamellisabellidae [INV ZOO] A family of marine animals in the order Thecanephria.

Lamiaceae [BOT] An equivalent name for Labiatae.

Lamiales [BOT] An order of dicotyledonous plants in the subclass Asteridae marked by its characteristic gynoecium, consisting of usually two biovulate carpels, with each carpel divided between the ovules by a false partition, or with the two halves of the carpel seemingly wholly separate.

lamina [ANAT] A thin sheet or layer of tissue; a scalelike structure.

lamina basalis [EMBRYO] The basal plate of the neural tube.

lamina cribrosa [ANAT] **1.** The portion of the sclera which is perforated for the passage of the optic nerve. **2.** The fascia covering the saphenous opening in the thigh. **3.** The anterior or posterior perforated space of the brain. **4.** The perforated plates of bone through which pass branches of the cochlear part of the vestibulocochlear nerve.

lamina orbitalis [ANAT] The thin plate of bone which encloses the ethmoid cells and forms a major portion of the median wall of the orbit.

laminal placentation [BOT] Condition in which the ovules occur on the inner surface of the carpels.

laminar [SCI TECH] **1.** Arranged in thin layers. **2.** Pertaining to viscous streamline flow without turbulence.

Laminariophyceae [BOT] A class of algae belonging to the division Phaeophyta.

lamina suprachoroidea [ANAT] The thin connective tissue membrane that unites the choroid and sclera of the eye.

lamina terminalis [ANAT] The layer of gray matter in the brain connecting the optic chiasma and the anterior commissure where the latter becomes continuous with the rostral lamina.

lamination [MED] An operation in embryotomy in which the skull is cut in slices. [SCI TECH] Arrangement in layers.

laminectomy [MED] Surgical removal of the lateral portion of the neural arch from one or more vertebrae.

laminography *See* sectional radiography.

lampbrush chromosome [CYTOL] An exceptionally large chromosome characterized by fine lateral projections which are associated with active ribonucleic acid and protein synthesis.

lamprey [VERT ZOO] The common name for all members of the order Petromyzonida.

Lampridiformes [VERT ZOO] An order of teleost fishes characterized by a compressed, often ribbonlike body, fins composed of soft rays, a ductless swim bladder, and protractile maxillae among other distinguishing features.

Lampyridae [INV ZOO] The firefly beetles, a large cosmopolitan family of coleopteran insects in the superfamily Cantharoidea.

lanatoside [BIOCHEM] Any of three natural glycosides from the leaves of *Digitalis lanata*; on hydrolysis with acid, it yields one molecule of D-glucose, three molecules of digitoxose, and one molecule of acetic acid; all three glycosides are cardioactive.

Lancefield groups [MICROBIO] Antigenically determined categories for classification of β-hemolytic streptococci.

lanceolate [BIOL] Shaped like the head of a lance.

Lanceolidae [INV ZOO] A family of bathypelagic amphipod crustaceans in the suborder Hyperiidea.

lancet [MED] A sharp-pointed, double-edged cutting instrument used to make small incisions.

Landenian [GEOL] A European stage of geologic time: upper Paleocene (above Montian, below Ypresian of Eocene).

Landry-Guillain-Barré syndrome [MED] A diffuse motorneuron paresis, rapid in onset, and usually ascending and symmetrical in distribution, with proximal involvement greater than distal, and motor deficits greater than sensory. Also known as Guillain-Barré syndrome; Landry's paralysis.

Landry's paralysis *See* Landry-Guillain-Barré syndrome.

Langerhans cell [HISTOL] **1.** A type of cytotrophoblast in the human chorionic vesicle which is thought to secrete chorionic gonadotropin. **2.** A highly branched dendritic cell of the mammalian epidermis showing a lobulated nucleus and a diagnostic organelle resembling a tennis racket.

Langerhansian adenoma *See* islet-cell tumor.

Lange's nerve [INV ZOO] One of the paired cords of nervous

LAMPRIDIFORMES

Oarfish *(Regalecus glesne)*, an example of Lampridiformes; length to over 20 feet (6 meters). *(After D. S. Jordan and B. W. Evermann, The Fishes of North and Middle America, U.S. Nat. Mus. Bull. no. 47, 1900)*

LAMPYRIDAE

Drawing of a firefly beetle; there are 11109 species. *(From T. I. Storer and R. L. Usinger, General Zoology, 3d ed., McGraw-Hill 1957)*

LANTERN FISH

2.5 cm

Drawing of the lantern fish
(*Myctophym punctatum*);
luminous glands are shown
as small circles.

LANTHANUM

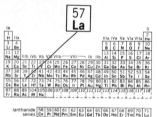

Periodic table of the chemical
elements showing the position
of lanthanum.

LARCH

Cone and needles of Western
larch (*Larix occidentalis*).

tissue lying in the wall of the radial perihemal canal of asteroids.

Lang's vesicle [INV ZOO] A seminal bursa in many polyclad flatworms.

Languriidae [INV ZOO] The lizard beetles, a cosmopolitan family of coleopteran insects in the superfamily Cucujoidea.

Laniatores [INV ZOO] A suborder of arachnids in the order Phalangida having flattened, often colorful bodies and found chiefly in tropical areas.

lanosterol [BIOCHEM] $C_{30}H_{50}O$ An unsaturated sterol occurring in wool fat and yeast.

lantern fish [VERT ZOO] The common name for the deep-sea teleost fishes composing the family Myctophidae and distinguished by luminous glands that are widely distributed upon the body surface.

lanthanide series [CHEM] Rare-earth elements of atomic numbers 57 through 71; their chemical properties are similar to those of lanthanum, atomic number 57.

lanthanum [CHEM] A chemical element, symbol La, atomic number 57, atomic weight 138.91; it is the second most abundant element in the rare-earth group.

Lanthonotidae [VERT ZOO] A family of lizards (Sauria) belonging to the Anguimorpha line; restricted to North Borneo.

lanugo [ANAT] A downy covering of hair, especially that seen on the fetus or persisting on the adult body.

laparotomy [MED] A surgical incision through the abdominal wall into the abdominal cavity.

lapidicolous [ECOL] Living under a stone.

lapillus [VERT ZOO] A small otolith in the saclike portion of the labyrinthine inner ear in teleosts.

lappet [ZOO] A lobe or flaplike projection, such as on the margin of a jellyfish or the wattle of a bird.

larch [BOT] The common name for members of the pine genus *Larix*, with deciduous needles and short, spurlike branches which annually bear a crown of needles.

larch canker [PL PATH] A destructive fungus disease of larch and sometimes pine and fir caused by *Dasyscypha willkommii* and characterized by flat, depressed cankers on the twigs and branches.

large calorie *See* kilocalorie.

large intestine *See* colon.

Largidae [INV ZOO] A family of hemipteran insects in the superfamily Pentatomorpha.

Laridae [VERT ZOO] A family of birds in the order Charadriiformes composed of the gulls and terns.

Larinae [VERT ZOO] A subfamily of birds in the family Laridae containing the gulls and characterized by a thick, slightly hooked beak, a square tail, and a stout white body, with shades of gray on the back and the upper wing surface.

larva [INV ZOO] An independent, immature, often vermiform stage that develops from the fertilized egg and must usually undergo a series of form and size changes before assuming characteristic features of the parent.

Larvacea [INV ZOO] A class of the subphylum Tunicata consisting of minute planktonic animals in which the tail, with dorsal nerve cord and notochord, persists throughout life.

Larvaevoridae [INV ZOO] The tachina flies, a large family of dipteran insects in the suborder Cyclorrhapha distinguished by a thick covering of bristles on the body; most are parasites of arthropods.

larva migrans [INV ZOO] Fly larva, *Hypoderma* or *Gastrophilus*, that produces a creeping eruption in the dermis. [MED] Infestation of the dermis by various burrowing nematode larvae, producing a creeping eruption that may become contaminated with bacteria.

larvicide [MATER] A pesticide used to kill larva.

larviporous [INV ZOO] Feeding on larva, referring especially to insects.

laryngeal pouch [VERT ZOO] A lateral saclike expansion of the cavity of the larynx that is greatly developed in certain monkeys.

laryngectomy [MED] Surgical removal of all or part of the larynx.

laryngismus stridulus [MED] **1.** Spasmodic croup. **2.** The laryngeal spasm sometimes seen in hypocalcemic states.

laryngitis [MED] Inflammation of the larynx.

laryngology [MED] The science of anatomy, physiology, and diseases of the larynx.

laryngopharynx [ANAT] The lower portion of the pharynx, lying adjacent to the larynx. Also known as hypopharynx.

laryngoscope [MED] A tubular instrument, combining a light system and a telescopic system, used in the visualization of the interior larynx and adaptable for diagnostic, therapeutic, and surgical procedures.

laryngotracheal groove [EMBRYO] A channel in the floor of the pharynx serving as the anlage of the respiratory system.

laryngotracheitis [MED] Inflammation of the larynx and trachea.

laryngotracheobronchitis [MED] Acute inflammation of the mucosa of the larynx, trachea, and bronchi.

laryngotracheobronchitis virus *See* croup-associated virus.

larynx [ANAT] The complex of cartilages and related structures at the opening of the trachea into the pharynx in vertebrates; functions in protecting the entrance of the trachea, and in phonation in higher forms.

laser [OPTICS] A device that uses the maser principle of amplification of electromagnetic waves by stimulated emission of radiation and operates in the optical or infrared regions. Derived from light amplification by stimulated emission of radiation. Also known as optical maser.

laser photocoagulator [MED] A laser combined with an ophthalmoscope for directing bursts of coherent light through a human eye to burn selected points on a detached retina; subsequent healing of the burns causes scars that weld the retina back into position.

Lasiocampidae [INV ZOO] The tent caterpillars and lappet moths, a family of cosmopolitan (except New Zealand) lepidopteran insects in the suborder Heteroneura.

lassa fever [MED] A sometimes fatal disease of humans, caused by Lassa virus and characterized by high fever, muscle aches, mouth ulcers, petechial rash, pneumonia, and cardiac and renal damage.

lasso cell *See* adhesive cell.

late blight [PL PATH] A fungus blight disease in which symptoms do not appear until late in the growing season and vary for different species.

latency [MED] The stage of an infectious disease, other than the incubation period, in which there are no clinical signs or symptoms. [PHYSIO] The period between the introduction of and the response to a stimulus. [PSYCH] The phase between the Oedipal period and adolescence, characterized by an apparent cessation of psychosexual development.

latent bud [BOT] An axillary bud' whose development is inhibited, sometimes for many years, due to the influence of apical and other buds. Also known as dormant bud.

latent heat [THERMO] The amount of heat absorbed or evolved by 1 mole, or a unit mass, of a substance during a

LARYNX

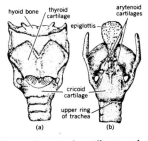

Human laryngeal cartilages and ligaments. *(a)* Front view. *(b)* Back view. *(After Sappey, from J. Symington, ed., Quain's Elements of Anatomy, vol. 2, pt. 2, Longmans, Green, 1914)*

change of state (such as fusion, sublimation or vaporization) at constant temperature and pressure.

latent heat of fusion *See* heat of fusion.

latent heat of sublimation *See* heat of sublimation.

latent heat of vaporization *See* heat of vaporization.

latent homosexuality [PSYCH] Homosexual tendencies present in the unconscious but not felt or expressed overtly.

latent period [MED] Any stage of an infectious disease in which there are no clinical signs of symptoms of the infection. [PHYSIO] The period between the introduction of a stimulus and the response to it. [VIROL] The initial period of phage growth after infection during which time virus nucleic acid is manufactured by the host cell.

latent-virus infection [MED] A chronic, inapparent virus infection in which a virus-host equilibrium is established.

laterad [ANAT] Toward the lateral aspect.

lateral [ANAT] At, pertaining to, or in the direction of the side; on either side of the medial vertical plane. [MIN ENG] **1.** In horizon mining, a hard heading branching off a horizon along the strike of the seams. **2.** A horizontal mine working.

lateral bud [BOT] Any bud that develops on the side of a stem.

lateral fissure [ANAT] A deep cleft dividing the temporal lobe of the cerebral cortex from the frontal and parietal lobes.

lateral hermaphroditism [MED] A form of human hermaphroditism in which there is an ovary on one side and a testis on the other.

lateral lemniscus [ANAT] The secondary auditory pathway arising in the cochlear nuclei and terminating in the inferior colliculus and medial geniculate body.

lateral line [INV ZOO] A longitudinal lateral line along the sides of certain oligochaetes consisting of cell bodies of the layer of circular muscle. [VERT ZOO] A line along the sides of the body of most fishes, often distinguished by differently colored scales, which marks the lateral line organ.

lateral-line organ [VERT ZOO] A small, pear-shaped sense organ in the skin of many fishes and amphibians that is sensitive to pressure changes in the surrounding water.

lateral-line system [VERT ZOO] The complex of lateral-line end organs and nerves in skin on the sides of many fishes and amphibians.

lateral meristem [BOT] Strips or cylinders of dividing cells located parallel to the long axis of the organ in which they occur; the lateral meristem functions to increase the diameter of the organ.

lateral root [BOT] A root branch arising from the main axis.

lateral sclerosis *See* amyotrophic lateral sclerosis; primary lateral sclerosis.

lateral thrombus *See* mural thrombus.

lateral ventricle [ANAT] The cavity of a cerebral hemisphere; communicates with the third ventricle by way of the interventricular foramen.

laterigrade [INV ZOO] Having a sideways walk, such as a crab.

laterosternite [INV ZOO] Any of the small bony plates at the side of the eusternum, as in Dermaptera and Isoptera.

laterotergite [INV ZOO] Any of the small bony plates adjoining the tergum of the abdominal segments in some crustaceans and insects.

latex [BOT] A colloid produced by some plants in which terpenes are dispersed in water.

Lathridiidae [INV ZOO] The minute brown scavenger beetles, a large cosmopolitan family of coleopteran insects in the superfamily Cucujoidea.

lathyrism [MED] Poisoning produced by ingestion of vetch

LATERAL-LINE ORGAN

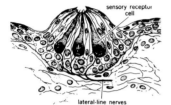

sensory receptor cell

lateral-line nerves

Schematic drawing of a lateral-line sense organ in skin of adult salamander, the common aquatic vermilion spotted newt. Sensory receptor cells, surrounded by supporting cells and skin epithelium, terminate in hairs projecting into skin pores. Lateral-line nerves terminate around bases of the sensory cells.

(*Lathyrus*) and characterized by spastic paraplegia and decreased connective-tissue tensile strength. Also known as neurolathyrism.

laticifer [BOT] A latex duct found in the mid-cortex of certain plants.

latiferous [BOT] Containing or secreting latex.

Latimeridae [VERT ZOO] A family of deep-sea lobefin fishes (Coelacanthiformes) known from a single living species, *Latimeria chalumnae*.

Latin square [MATH] An $n \times n$ square array of n different symbols, each symbol appearing once in each row and once in each column; these symbols prove useful in ordering the observations of an experiment.

lattice [CRYSTAL] A regular periodic arrangement of points in three-dimensional space; it consists of all those points P for which the vector from a given fixed point to P has the form $n_1 \mathbf{a} + n_2 \mathbf{b} + n_3 \mathbf{c}$, where n_1, n_2, and n_3 are integers, and $\mathbf{a}$, $\mathbf{b}$, and $\mathbf{c}$ are fixed, linearly independent vectors. Also known as periodic lattice; space lattice.

lattissimus dorsi [ANAT] The widest muscle of the back; a broad, flat muscle of the lower back that adducts and extends the humerus, is used to pull the body upward in climbing, and is an accessory muscle of respiration.

Lattorfian *See* Tongrian.

Laugiidae [PALEON] A family of Mesozoic fishes in the order Coelacanthiformes.

Lauraceae [BOT] The laurel family of the order Magnoliales distinguished by definite stamens in series of three, a single pistil, and the lack of petals.

Lauratonematidae [INV ZOO] A family of marine nematodes of the superfamily Enoploidea; many females possess a cloaca.

laurel forest *See* temperate rainforest.

Laurence-Moon-Biedl syndrome [MED] A hereditary endocrine disorder of the pituitary or other hypothalamic structures transmitted as a dominant mutant gene and characterized principally by girdle-type obesity, mental retardation, and hypogenitalism.

laurisilva *See* temperate rainforest.

Lauxaniidae [INV ZOO] A family of myodarian cyclorrhaphous dipteran insects in the subsection Acalypteratae; larvae are leaf miners.

law of definite composition *See* law of definite proportion.

law of definite proportion [CHEM] The law that a given chemical compound always contains the same elements in the same fixed proportion by weight. Also known as law of definite composition.

law of large numbers [STAT] The law that if, in a collection of independent identical experiments, $N(B)$ represents the number of occurrences of an event B in n trials, and p is the probability that B occurs at any given trial, then for large enough n it is unlikely that $N(B)/n$ differs from p by very much. Also known as Bernoulli theorem.

law of mass action [CHEM] The law stating that the rate at which a chemical reaction proceeds is directly proportional to the molecular concentrations of the reacting compounds.

law of minimum [BIOL] The law that those essential elements for which the ratio of supply to demand (A/N) reaches a minimum will be the first to be removed from the environment by life processes; it was proposed by J. Von Liebig, who recognized phosphorus, nitrogen, and potassium as minimum in the soil; in the ocean the corresponding elements are phosphorus, nitrogen, and silicon. Also known as Liebig's law of minimum.

law of partial pressures *See* Dalton's law.

LATIMERIDAE

Living coelacanth, *Latimeria chalumnae*; 5 feet (1.5 meters). (*After P. P. Grasse, ed., Traite de Zoologie, tome 13, fasc. 3, 1958*)

LAWRENCIUM

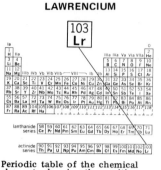

Periodic table of the chemical elements showing the position of lawrencium.

LEAD

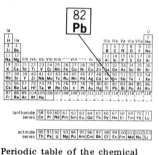

Periodic table of the chemical elements showing the position of lead.

law of specificity of bacteria *See* Koch's postulate.

lawrencium [CHEM] A chemical element, symbol Lr, atomic number 103; two isotopes have been discovered, mass number 257 or 258 and mass number 256.

laxative [PHARM] An agent that stimulates bowel movement and relieves constipation.

layering [BOT] A propagation method by which root formation is induced on a branch or a shoot attached to the parent stem by covering the part with soil. [ECOL] A stratum of plant forms in a community, such as mosses, shrubs, or trees in a bog area.

LD$_{50}$ *See* lethal dose 50.

LDH *See* lactic dehydrogenase.

lead [CHEM] A chemical element, symbol Pb, atomic number 82, atomic weight 207.19.

lead encephalopathy [MED] Degeneration of the neurons of the brain accompanied by cerebral edema, due to lead poisoning.

lead palsy *See* lead polyneuropathy.

lead poisoning [MED] Poisoning due to ingestion or absorption of lead over a prolonged period of time; characterized by colic, brain disease, anemia, and inflammation of peripheral nerves.

lead polyneuropathy [MED] A distal polyneuropathy, affecting mainly the wrist and hand, seen principally in adults with chronic lead poisoning; characterized by weakness, paresthesias, pain, and glove-and-stocking anesthesia. Also known as lead palsy.

leaf [BOT] A modified aerial appendage which develops from a plant stem at a node, usually contains chlorophyll, and is the principal organ in which photosynthesis and transpiration occur.

leaf blight [PL PATH] Any of various blight diseases which cause browning, death, and falling of the leaves.

leaf blotch [PL PATH] A plant disease characterized by discolored areas in the leaves with indistinct or diffuse margins.

leaf bud [BOT] A bud that produces a leafy shoot.

leaf cast [PL PATH] Any of several diseases of conifers characterized by falling of the needles.

leaf curl [PL PATH] A fungus or viral disease of plants marked by the curling of leaves.

leaf cushion [BOT] The small part of the thickened leaf base that remains after abscission in various conifers, and also in some extinct plants.

leaf drop [PL PATH] Premature falling of leaves, associated with disease.

leafhopper [INV ZOO] The common name for members of the homopteran family Cicadellidae.

leaflet [BOT] 1. A division of a compound leaf. 2. A small or young foliage leaf.

leaf miner [INV ZOO] Any of the larvae of various insects which burrow into and eat the parenchyma of leaves.

leaf mottle [PL PATH] A fungus disease characterized by chlorotic mottling of the leaves; for example, caused by *Verticillium dahliae* in sun-flower.

leaf-nosed [VERT ZOO] Having a leaflike membrane on the nose, as certain bats.

leaf roll [PL PATH] Any of several virus diseases characterized by upward or inward rolling of the leaf margins.

leaf rot [PL PATH] Any plant disease characterized by breakdown of leaf tissues; for example, caused by *Pellicularia koleroga* in coffee.

leaf rust [PL PATH] Any rust disease primarily affecting leaves; common in coffee, alfalfa, and wheat, barley, and other cereals.

leaf scald [PL PATH] A bacterial disease of sugarcane caused by *Bacterium albilineans* which invades the vascular tissues, causing creamy or grayish streaking and withering of the leaves.

leaf scar [BOT] A mark on a stem, formed by secretion of suberin and a gumlike substance, showing where a leaf has abscised.

leaf scorch [BOT] Any of several disorders and fungus diseases marked by a burned appearance of the leaves; for example, caused by the fungus *Diplocarpon earliana* in strawberry.

leaf spot [PL PATH] Any of various diseases or disorders characterized by the appearance of well-defined discolored spots on the leaves.

leaf stripe [PL PATH] Any of various plant diseases characterized by striped discolorations on the foliage.

leaf trace [BOT] The remnant of a vascular bundle where it passes from the stem to the petiole.

leak [PL PATH] A watery rot of fruits and vegetables caused by various fungi, such as *Rhizopus nigricans* in strawberry.

leaky mutant gene [GEN] An allele with reduced activity relative to that of the normal allele.

learning [PSYCH] The gathering, processing, storage, and recall of information received through the senses.

least squares method [STAT] A technique of fitting a curve close to some given points which minimizes the sum of the squares of the deviations of the given points from the curve.

leather rot [PL PATH] A hard rot of strawberry caused by the fungus *Phytophthora cactorum*.

leberidocytes [INV ZOO] Glycogen-containing cells that develop from and regress to leukocytes in the blood of Arachnida at molting.

Lecanicephaloidea [INV ZOO] An order of tapeworms of the subclass Cestoda distinguished by having the scolex divided into two portions; all species are intestinal parasites of elasmobranch fishes.

Lecanoraceae [BOT] A temperate and boreal family of lichens in the order Lecanorales characterized by a crustose thallus and a distinct thalloid rim on the apothecia.

Lecanorales [BOT] An order of the Ascolichenes having open, discoid apothecia with a typical hymenium and hypothecium.

lechriodont [VERT ZOO] Having vomerine and pterygoid teeth arranged in a nearly transverse row.

Lecideaceae [BOT] A temperate and boreal family of lichens in the order Lecanorales; members lack a thalloid rim around the apothecia.

lecithin [BIOCHEM] Any of a group of phospholipids having the general composition $CH_2OR_1 \cdot CHOR_2 \cdot CH_2OPO_2 OHR_3$, in which R_1 and R_2 are fatty acids and R_3 is choline, and with emulsifying, wetting, and antioxidant properties. [MATER] **1.** A mixture of phosphatides and oil obtained by drying the separate gums from the degumming of soybean oil; consists of the phosphatides (lecithin), cephalin, other fatlike phosphorus-containing compounds, and 30–35% entrained soybean oil; may be treated to produce more refined grades; used in foods, cosmetics, and paints. Also known as commercial lecithin; crude lecithin; soybean lecithin; soy lecithin. **2.** A waxy mixture of phosphatides obtained by refining commercial lecithin to remove the soybean oil and other materials; used in pharmaceuticals. Also known as refined lecithin.

lecithinase [BIOCHEM] An enzyme that catalyzes the breakdown of a lecithin into its constituents.

lecithinase A [BIOCHEM] An enzyme that catalyzes the re-

LECANICEPHALOIDEA

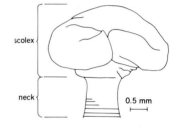

scolex

neck

0.5 mm

Anterior end of a lecanicephaloid tapeworm showing scolex and neck.

LEECH

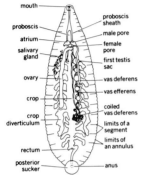

General structure of a leech.
Male reproductive system is
shown on the right, the female
on the left. *(From K. H. Mann,
A key to the British freshwater
leeches, Freshwater Biol. Assoc.
Sci. Publ., 14:3–21, 1954)*

moval of only one fatty acid from lecithin, yielding lipoleci-
thin.

lecithinase C [BIOCHEM] An enzyme that catalyzes the removal of the nitrogenous base of lecithin to produce the base
and a phosphatidic acid.

lecithinase D [BIOCHEM] An enzyme that catalyzes the removal of the phosphorylated base from lecithins, producing
α-β-diglyceride.

le cri du chat syndrome [MED] A complex of congenital
malformations resulting from a deletion in chromosome 4 or
5 and characterized by mental retardation and the production
of a catlike cry.

lectoallotype [SYST] A specimen differing in sex from the
lectotype and subsequently chosen from the original material.

lectotype [SYST] A specimen selected as the type of a species
or subspecies if the type was not designated by the author of
the classification.

Lecythidaceae [BOT] The single family of the order Lecythidales.

Lecythidales [BOT] A monofamilial order of dicotyledonous
tropical woody plants in the subclass Dilleniidae; distinguished by entire leaves, valvate sepals, separate petals,
numerous centrifugal stamens, and a syncarpous, inferior
ovary with axile placentation.

Lederberg technique [MICROBIO] A method for rapid isolation of individual bacterial cells for demonstrating the spontaneous origin of bacterial mutants.

Ledian [GEOL] Lower upper Eocene geologic time. Also
known as Auversian.

Leeaceae [BOT] A family of dicotyledonous plants in the
order Rhamnales distinguished by solitary ovules in each
locule, simple to compound leaves, a small embryo, and
hypogynous flowers.

leech [INV ZOO] The common name for members of the
annelid class Hirudinea.

leek [BOT] *Allium porrum.* A biennial herb known only by
cultivation; grown for its mildly pungent succulent leaves and
thick cylindrical stalk.

LE factor [PATH] A substance in body fluids, especially the
blood, of patients with systemic lupus erythematosus, and
sometimes in other diseases. Derived from lupus erythematosus factor.

left ventricular thrust *See* apex impulse.

leg [ANAT] The lower extremity of a human limb, between
the knee and the ankle. [ZOO] An appendage or limb used
for support and locomotion.

legena [VERT ZOO] An appendage of the sacculus containing
sensory areas in the inner ear of tetrapods; termed the cochlea
in humans.

legume [BOT] A dry, dehiscent fruit derived from a single
simple pistil; common examples are alfalfa, beans, peanuts,
and vetch.

legume forage [AGR] Any plant of the legume family (Leguminosae) used for livestock feed, grazing, hay, or silage.

Leguminosae [BOT] The legume family of the plant order
Rosales characterized by stipulate, compound leaves, 10 or
more stamens, and a single carpel; many members harbor
symbiotic nitrogen-fixing bacteria in their roots.

Leiodidae [INV ZOO] The round carrion beetles, a cosmopolitan family of coleopteran insects in the superfamily
Staphylinoidea; commonly found under decaying bark.

leiomyofibroma [MED] A benign tumor composed of smooth
muscle cells and fibrocytes.

leiomyoma [MED] A benign tumor composed of smooth muscle cells.

leiomyosarcoma [MED] A malignant tumor composed of anaplastic smooth muscle cells.

leiosporous [MYCOL] Having smooth spores.

Leishman-Donovan bodies [PATH] Small, oval protozoans lacking flagella and undulating membranes, found within macrophages of the skin, liver, and spleen in leishmanial infections such as kala-azar and mucocutaneous leishmaniasis.

Leishmania [INV ZOO] A genus of flagellated protozoan parasites that are the etiologic agents of several diseases of humans, such as leishmaniasis.

leishmaniasis [MED] Any of several infections caused by *Leishmania* species.

Leitneriales [BOT] A monofamilial order of flowering plants in the subclass Hamamelidae; members are simple-leaved, dioecious shrubs with flowers in catkins, and have a superior, pseudomonomerous ovary with a single ovule.

Lelapiidae [INV ZOO] A family of calcaronean sponges in the order Sycettida characterized by a rigid skeleton composed of tracts or bundles of modified triradiates.

lemma [BOT] Either of the upper pair of bracts enclosing the flower of a grass spikelet.

lemming [VERT ZOO] The common name for the small burrowing rodents composing the subfamily Microtinae.

Lemnaceae [BOT] The duckweeds, a family of monocotyledonous plants in the order Arales; members are small, free-floating, thalloid aquatics with much reduced flowers that form a miniature spadix.

lemniscus [ANAT] A secondary sensory pathway of the central nervous system, usually terminating in the thalamus.

lemon [BOT] *Citrus limon.* A small evergreen tree belonging to the order Sapindales cultivated for its acid citrus fruit which is a modified berry called a hesperidium.

lemur [VERT ZOO] The common name for members of the primate family Lemuridae; characterized by long tails, foxlike faces, and scent glands on the shoulder region and wrists.

Lemuridae [VERT ZOO] A family of prosimian primates of Madagascar belonging to the Lemuroidea; all members are arboreal forest dwellers.

Lemuroidea [VERT ZOO] A suborder or superfamily of Primates including the lemurs, tarsiers, and lorises, or sometimes simply the lemurs.

lengthening reaction [PHYSIO] Sudden inhibition of the stretch reflex when extensor muscles are subjected to an excessive degree of stretching by forceful flexion of a limb.

length-height index [ANTHRO] The ratio of the basion-bregma height of the skull to its greatest length multiplied by 100.

lens [ANAT] A transparent, encapsulated, nearly spherical structure located behind the pupil of vertebrate eyes, and in the complex eyes of many invertebrates, that focuses light rays on the retina. Also known as crystalline lens.

lens placode [EMBRYO] The ectodermal anlage of the lens of the eye; its formation is induced by the presence of the underlying optic vesicle.

lentic [ECOL] Of or pertaining to still waters such as lakes, reservoirs, ponds, and bogs. Also spelled lenitic.

lenticel [BOT] A loose-structured opening in the periderm beneath the stomata in the stem of many woody plants that facilitates gas transport.

lenticular [SCI TECH] Having the shape of a lentil or double convex lens.

lentiginose [ANAT] Of or pertaining to pigment spots in the skin; freckled.

lentil [BOT] *Lens esculenta.* A seminvy annual legume having

LEMMING

A lemming, a small burrowing rodent with short legs and stout claws adapted for digging.

LEMON

Fruit and branch of lemon tree (*Citrus limon*).

LENTICEL

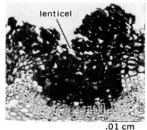

A lenticel as seen in a cross section of the stem periderm of *Sambucus nigra*.

pinnately compound, vetchlike leaves; cultivated for its thin, lens-shaped, edible seed.

Leonardian [GEOL] A North American provincial series: Lower Permian (above Wolfcampian, below Guadalupian).

leopard [VERT ZOO] *Felis pardus.* A species of wildcat in the family Felidae found in Africa and Asia; the coat is characteristically buff-colored with black spots.

Leotichidae [INV ZOO] A small Oriental family of hemipteran insects in the superfamily Leptopodoidea.

Lepadomorpha [INV ZOO] A suborder of barnacles in the order Thoracica having a peduncle and a capitulum which is usually protected by calcareous plates.

leper [MED] A person afflicted with Hansen's disease.

Leperditicopida [PALEON] An order of extinct ostracods characterized by very thick, straight-backed valves which show unique muscle scars and other markings.

Leperditillacea [PALEON] A superfamily of extinct paleocopan ostracods in the suborder Kloedenellocopina including the unisulcate, nondimorphic forms.

Lepiceridae [INV ZOO] Horn's beetle, a family of Central American coleopteran insects composed of two species.

Lepidocentroida [INV ZOO] The name applied to a polyphyletic assemblage of echinoids now regarded as members of the Echinocystitoida and Echinothurioida.

Lepidodendrales [PALEOBOT] The giant club mosses, an order of extinct lycopods (Lycopodiopsida) consisting primarily of arborescent forms characterized by dichotomous branching, small amounts of secondary vascular tissue, and heterospory.

lepidomorium [VERT ZOO] A small scale, or a unit of a composite scale, having a bony base, a conical crown of dentin, and a pulp cavity, and sometimes covered with enamel.

lepidophyllous [BOT] Having scaly leaves.

Lepidoptera [INV ZOO] A large order of scaly-winged insects, including the butterflies, skippers, and moths; adults are characterized by two pairs of membranous wings and sucking mouthparts, featuring a prominent, coiled proboscis.

Lepidorsirenidae [VERT ZOO] A family of slender, obligate air-breathing, eellike fishes in the order Dipteriformes having small thin scales, slender ribbonlike paired fins, and paired ventral lungs.

Lepidosaphinae [INV ZOO] A family of homopteran insects in the superfamily Coccoidea having dark-colored, noncircular scales.

Lepidosauria [VERT ZOO] A subclass of reptiles in which the skull structure is characterized by two temporal openings on each side which have reduced bony arcades, and by the lack of an antorbital opening in front of the orbit.

Lepidotrichidae [INV ZOO] A family of wingless insects in the order Thysanura.

Lepismatidae [INV ZOO] A family of silverfish in the order Thysanura characterized by small or missing compound eyes.

Lepisostei [VERT ZOO] An equivalent name for Semionotiformes.

Lepisosteidae [VERT ZOO] A family of fishes in the order Semionotiformes.

Lepisosteiformes [VERT ZOO] An equivalent name for Semionotiformes.

Leporidae [VERT ZOO] A family of mammals in the order Lagomorpha including the rabbits and hares.

Lepospondyli [PALEON] A subclass of extinct amphibians including all forms in which the vertebral centra are formed by ossification directly around the notochord.

lepospondylous [VERT ZOO] Having the notochord enclosed

by cylindrical vertebrae shaped like an hourglass in longitudinal section.

leproma [MED] The cutaneous nodular lesion of leprosy.

lepromatous leprosy [MED] A severely debilitating form of Hansen's disease characterized by the presence of multiple nodular lesions (lepromata) on the skin.

lepromin [IMMUNOL] An emulsion of ground lepromata containing the leprosy bacillus; used for intradermal skin tests in Hansen's disease.

leprosarium [MED] An institution for the treatment of lepers.

Leptaleinae [INV ZOO] A subfamily of the Formicidae including largely arboreal ant forms which inhabit plants in tropical and subtropical regions.

Leptictidae [PALEON] A family of extinct North American insectivoran mammals belonging to the Proteutheria which ranged from the Cretaceous to middle Oligocene.

leptine [ANTHRO] Having a high, a narrow, or a high and narrow forehead with an upper facial index of 55-60 on the skull and of 53-57 on the living person.

Leptinidae [INV ZOO] The mammal nest beetles, a small European and North American family of coleopteran insects in the superfamily Staphylinoidea.

Leptocardii [ZOO] The equivalent name for Cephalochordata.

leptocephalous larva [VERT ZOO] The marine larva of the fresh-water European eel *Anguilla vulgaris.*

leptocephaly [ANTHRO] Abnormal narrowness and tallness of the head or skull.

leptocercal [VERT ZOO] Of the tail of a fish, tapering to a long, slender point.

Leptochoeridae [PALEON] An extinct family of palaeodont artiodactyl mammals in the superfamily Dichobunoidea.

Leptodactylidae [VERT ZOO] A large family of frogs in the suborder Procoela found principally in the American tropics and Australia.

leptodactylous [VERT ZOO] Having slender toes, as certain birds.

Leptodiridae [INV ZOO] The small carrion beetles, a cosmopolitan family of coleopteran insects in the superfamily Staphylinoidea.

Leptolepidae [PALEON] An extinct family of fishes in the order Leptolepiformes representing the first teleosts as defined on the basis of the advanced structure of the caudal skeleton.

Leptolepiformes [PALEON] An extinct order of small, ray-finned teleost fishes characterized by a relatively strong, ossified axial skeleton, thin cycloid scales, and a preopercle with an elongated dorsal portion.

Leptomedusae [INV ZOO] A suborder of hydrozoan coelenterates in the order Hydroida characterized by the presence of a hydrotheca.

leptomeninges [ANAT] The pia mater and arachnoid considered together.

leptomeningitis [MED] Inflammation of the leptomeninges of the brain or the spinal cord.

Leptomitales [MYCOL] An order of aquatic Phycomycetes characterized by a hyphal thallus, or basal rhizoids and terminal hyphae, and zoospores with two flagella.

Leptopodidae [INV ZOO] A tropical and subtropical family of hemipteran insects in the superfamily Leptopodoidea distinguished by the spiny body and appendages.

Leptopodoidea [INV ZOO] A superfamily of hemipteran insects in the subdivision Geocorisae.

leptorrhine [ANTHRO] Having a long, narrow nose with a

LEPTOLEPIDAE

Leptolepis dubia (Blainville) restoration, scales omitted. (From A. S. Woodward, Catalogue of the Fossil Fishes in the British Museum (Natural History), pt. 3, 1895)

LEPTOSTROMATACEAE

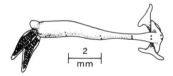

shield-shaped
pycnidium
pore conidium

300 μ

8 μ

Pycnidia and conidia of
Leptothyrium vulgare.

LERNAEIDAE

2
mm

Lernaea cyprinacea L., the
female, showing the branched
processes of the head that
anchor the parasite deep in the
flesh of the host.

nasal index of less than 47 on the skull or of less than 70 on the living person.

Leptosomatidae [VERT ZOO] The cuckoo rollers, a family of Madagascan birds in the order Coraciiformes composed of a single species distinguished by the downy covering on the newly hatched young.

Leptospira [MICROBIO] A genus of aerobic spirochetes in the family Treponemataceae characterized by twisted filaments with hooked or recurved extremities.

leptospirosis [MED] Infection with spirochetes of the genus *Leptospira.*

leptospirosis icterohemorrhagia *See* Weil's disease.

leptostaphyline [ANTHRO] Having a high, narrow palate with a palatal index of less than 80 on the skull.

Leptostraca [INV ZOO] A primitive group of crustaceans considered as one of a series of Malacostraca distinguished by an additional abdominal somite that lacks appendages, and a telson bearing two movable articulated prongs.

Leptostromataceae [MYCOL] A family of fungi of the order Sphaeropsidales; pycnidia are black and shield-shaped, circular or oblong, and slightly asymmetrical; included are some fruit-tree pathogens.

leptotene [CYTOL] The first stage of meiotic prophase, when the chromosomes appear as thin threads having well-defined chromomeres.

Leptotyphlopidae [VERT ZOO] A family of small, harmless, burrowing circumtropical snakes (Serpentes) in the order Squamata; teeth are present only on the lower jaw and are few in number.

leptus [INV ZOO] The six-legged mite larva.

Lepus [VERT ZOO] The type genus of the family Leporidae, comprising the typical hares.

Lernaeidae [INV ZOO] A family of copepod crustaceans in the suborder Caligoida; all are fixed ectoparasites, that is, they penetrate the skin of fresh-water fish.

Lernaeopodidae [INV ZOO] A family of ectoparasitic crustaceans belonging to the Lernaeopodoida; individuals are attached to the walls of the fishes' gill chambers by modified second maxillae.

Lernaeopodoida [INV ZOO] The fish maggots, a group of ectoparasitic crustaceans characterized by a modified postembryonic development reduced to two or three stages, a free-swimming larva, and the lack of external signs of physical maturity in adults.

lesbian [PSYCH] **1.** Pertaining to female homosexuality. **2.** A female homosexual.

Lesch-Nyhan syndrome [MED] A hereditary disease of male children, transmitted as an X-linked recessive, characterized by hyperuricemia, deficiency of hypoxanthine-guanine phosphoribosyl transferase, mental retardation, spastic cerebral palsy, choreathetosis, and self-mutilating biting.

lesion [BIOL] A structural or functional alteration due to injury or disease.

Leskeineae [BOT] A suborder of mosses in the order Hypnobryales; plants are not complanate, paraphyllia are frequently present, leaves are costate, and alar cells are not generally differentiated.

lespedeza [BOT] Any of various legumes of the genus *Lespedeza* having trifoliate leaves, small purple pea-shaped blossoms, and one seed per pod.

lesser circulation *See* pulmonary circulation.

lesser omentum [ANAT] A fold of the peritoneum extending from the lesser curvature of the stomach to the transverse hepatic fissure.

Lestidae [INV ZOO] A family of odonatan insects belonging to the Zygoptera; distinguished by the wings being held in a V position while at rest.

Lestoniidae [INV ZOO] A monospecific family of hemipteran insects in the superfamily Pentatomorpha found only in Australia.

lethal [BIOL] Causing death.

lethal dose 50 [PHARM] The dose of a substance which is fatal to 50% of the test animals. Abbreviated LD$_{50}$. Also known as median lethal dose.

lethal gene [GEN] A gene mutation that causes premature death in heterozygotes if dominant, and in homozygotes if recessive. Also known as lethal mutation.

lethal mutation *See* lethal gene.

lethargy [MED] A morbid condition of drowsiness or stupor; mental torpor.

Letterer-Siwe disease [MED] A fatal disease of infants and young children, of unknown etiology, characterized by hyperplasia of the reticuloendothelial system without lipid storage.

lettuce [BOT] *Lactuca sativa.* An annual plant of the order Asterales cultivated for its succulent leaves; common varieties are head lettuce, leaf or curled lettuce, romaine lettuce, and iceberg lettuce.

Leucaltidae [INV ZOO] A family of calcinean sponges in the order Leucettida having numerous small, interstitial, flagellated chambers.

Leucascidae [INV ZOO] A family of calcinean sponges in the order Leucettida having a radiate arrangement of flagellated chambers.

Leucettida [INV ZOO] An order of calcareous sponges in the subclass Calcinea having a leuconoid structure and a distinct dermal membrane or cortex.

leucine [BIOCHEM] $C_6H_{13}O_2N$ A monocarboxylic essential amino acid obtained by hydrolysis of protein-containing substances such as milk.

leucine amino peptidase [BIOCHEM] An enzyme that acts on peptides to catalyze the release of terminal amino acids, especially leucine residues, having free α-amino groups.

Leucodontineae [BOT] A family of mosses in the order Isobryales with foliated branches, often bearing catkins.

leucon [INV ZOO] A type of sponge having the choanocytes restricted to flagellated chambers inserted between the incurrent and excurrent canals, and a reduced or absent paragastric cavity.

leucophore [HISTOL] A white reflecting chromatophore.

leucoplast [BOT] A nonpigmented plastid; capable of developing into a chromoplast.

Leucosiidae [INV ZOO] The purse crabs, a family of true crabs belonging to the Oxystomata.

leucosin [BIOCHEM] A simple protein of the albumin type found in wheat and other cereals.

Leucosoleniida [INV ZOO] An order of calcareous sponges in the subclass Calcaronea characterized by an asconoid structure and the lack of a true dermal membrane or cortex.

Leucospidae [INV ZOO] A small family of hymenopteran insects in the superfamily Chalcidoidea distinguished by a longitudinal fold in the forewings.

leucosporous [MYCOL] Having white spores.

Leucothoidae [INV ZOO] A family of amphipod crustaceans in the suborder Gammaridea including semiparasitic and commensal species.

Leucotrichaceae [MICROBIO] A family of short, cylindrical bacteria in the order Beggiatoales occurring as long, unbranched, nonmotile trichomes attached to solid substrates.

leucovorin [PHARM] Folinic acid used as a calcium salt to

LEUCALTIDAE

Leucaltis clathria showing the small chambers. *(After Polejaeff, 1883)*

LEUCINE

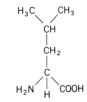

Structural formula of leucine.

LEUCOSIIDAE

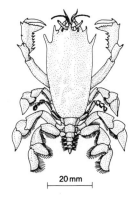

20 mm

Drawing of a purse crab, *Persephona punctata.* *(Smithsonian Institution)*

counteract the toxic effects of folic acid antagonists and for treatment of megaloblastic anemias.

leukemia [MED] Any of several diseases of the hemopoietic system characterized by uncontrolled leukocyte proliferation. Also known as leukocythemia.

leukemia virus *See* leukovirus.

leukemoid [MED] Similar to leukemia, that is, characterized by the presence of immature leukocytes in the blood, but of a different etiology.

leukoblast [HISTOL] An immature cell from which leukocytes are formed.

leukocarpous [BOT] Having white fruit.

leukocidin [BIOCHEM] A toxic substance released by certain bacteria which destroys leukocytes.

leukocyte [HISTOL] A colorless, ameboid blood cell having a nucleus and granular or nongranular cytoplasm. Also known as white blood cell; white corpuscle.

leukocythemia *See* leukemia.

leukocytolysin [BIOCHEM] A leukocyte-disintegrating lysin.

leukocytopenia *See* leukopenia.

leukocytopoiesis [PHYSIO] Formation of leukocytes.

leukocytosis [MED] Elevation of the leukocyte count to values above the normal limit.

leukoderma [MED] Defective skin pigmentation, especially the congenital absence of pigment in patches or bands.

leukodystrophy [MED] A condition thought to result from an inborn error of metabolism and characterized by progressive degeneration of the white matter of the cerebrum, or by defective buildup of myelin.

leukoencephalitis [MED] Inflammation of the white matter of the cerebrum.

leukoerythroblastic anemia *See* myelophthisic anemia.

leukoerythroblastosis *See* myelophthisic anemia.

leukolymphosarcoma *See* leukosarcoma.

leukoma [MED] A large and dense opacity of the cornea as a result of an ulcer, wound, or inflammation, which presents an appearance of ground glass.

leukonychia [MED] Whitish discoloration or spotting of the fingernails.

leukopenia [MED] A reduction in the leukocyte count to values below the normal limit. Also known as leukocytopenia.

leukoplakia [MED] Formation of thickened white patches on mucous membranes, particularly of the mouth and vulva.

leukopoiesis [HISTOL] The formation of leukocytes. Also known as leukocytopoiesis.

leukorrhea [MED] A whitish, mucopurulent discharge from the female genital canal.

leukosarcoma [MED] Lymphosarcoma accompanied by a small number of anaplastic lymphoid cells in the peripheral blood. Also known as leukolymphosarcoma; sarcoleukemia.

leukosis [MED] An excess of white blood cells.

leukotomy *See* lobotomy.

leukovirus [VIROL] A major group of animal viruses including those causing leukemia in birds, mice, and rats. Also known as leukemia virus.

leurocristine *See* vincristine.

levator [MED] An instrument used for raising a depressed portion of the skull. [PHYSIO] Any muscle that raises or elevates a part.

level above threshold [PHYSIO] Also known as sensation level. **1.** The pressure level of a sound in decibels above its threshold of audibility for the individual observer. **2.** In general, the level of any psychophysical stimulus, such as light, above its threshold of perception.

level of significance of a test [STAT] The probability of false rejection of the null hypothesis. Also known as significance level.

leveret [VERT ZOO] A young hare.

levigate *See* glabrous.

levo form [PHYS CHEM] An optical isomer which induces levorotation in a beam of plane polarized light.

levorotation [OPTICS] Rotation of the plane of polarization of plane polarized light in a counterclockwise direction, as seen by an observer facing in the direction of light propagation. Also known as levulorotation.

levotropic cleavage [EMBRYO] Spiral cleavage with the cells displaced counterclockwise.

levulorotation *See* levorotation.

levulose [BIOCHEM] Levorotatory D-fructose.

levulose tolerance test [PATH] A liver function test based on the observation that blood sugar increases in cases of hepatic disease following oral administration of levulose.

Lewis acid [CHEM] A substance that can accept an electron pair from a base; thus, $AlCl_3$, BF_3, and SO_3 are acids.

Lewis base [CHEM] A substance that can donate an electron pair; examples are the hydroxide ion, OH^-, and ammonia, NH_3.

Lewis blood group system [IMMUNOL] An antigen, designated by Le^a, first recognized in a Mrs. Lewis, occurring in about 22% of the population, detected by anti-Le^a antibodies; primarily composed of soluble antigens of serum and body fluids like saliva, with secondary absorption by erythrocytes.

Leydig cell [HISTOL] One of the interstitial cells of the testes; thought to produce androgen.

Leydig-cell tumor *See* interstitial-cell tumor.

LGV *See* lymphogranuloma venereum.

Li *See* lithium.

liana [BOT] A woody or herbaceous climbing plant with roots in the ground.

Lias *See* Liassic.

Liassic [GEOL] The Lower Jurassic period of geologic time. Also known as Lias.

Libellulidae [INV ZOO] A large family of odonatan insects belonging to the Anisoptera and distinguished by a notch on the posterior margin of the eyes and a foot-shaped anal loop in the hindwing.

libido [PSYCH] 1. Sexual desire. 2. The sum total of all instinctual forces; psychic energy or drive usually associated with the sexual instinct.

Libman-Sacks endocarditis [MED] Inflammation of, accompanied by the presence of watery vegetations on, the endocardium; complicates systemic lupus erythematosus.

Libytheidae [INV ZOO] The snout butterflies, a family of cosmopolitan lepidopteran insects in the suborder Heteroneura distinguished by long labial palps; represented in North America by a single species.

Liceaceae [MYCOL] A family of plasmodial slime molds in the order Liceales.

Liceales [MYCOL] An order of plasmodial slime molds in the subclass Myxogastromycetidae.

lichen [BOT] The common name for members of the Lichenes.

Lichenes [BOT] A group of organisms consisting of fungi and algae growing together symbiotically.

Lichenes Imperfecti [BOT] A class of the Lichenes containing species with no known method of sexual reproduction.

lichenification [MED] The process whereby the skin becomes leathery and hardened; often the result of chronic pruritis and

LICHENES

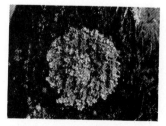

A picture of the foliose (leaflike) type of plant body of the *Parmelia* species. *(From H. J. Fuller and O. Tippo, College Botany, rev. ed., Holt, 1954)*

the irritation produced by scratching or rubbing eruptions.

lichenology [BOT] The study of lichens.

lichenophagous [ZOO] Animals which feed on lichens.

Lichnophorina [INV ZOO] A suborder of ciliophoran protozoans belonging to the Heterotrichida.

licorice [BOT] *Glycyrrhiza glabra.* A perennial herb of the legume family (Leguminosae) cultivated for its roots, which when dried provide a product used as a flavoring in medicine, candy, and tobacco and in the manufacture of shoe polish.

lidocaine [ORG CHEM] $C_{14}H_{22}N_2O$ A crystalline compound, used as a local anesthetic. Also known as lignocaine.

Liebig's law of the minimum *See* law of minimum.

life cycle [BIOL] The functional and morphological stages through which an organism passes between two successive primary stages.

life form [ECOL] The form characteristically taken by a plant at maturity.

life zone [ECOL] A portion of the earth's land area having a generally uniform climate and soil, and a biota showing a high degree of uniformity in species composition and adaptation.

ligament [HISTOL] A flexible, dense white fibrous connective tissue joining, and sometimes encapsulating, the articular surfaces of bones.

ligamentum nuchae *See* nuchal ligament.

ligand [CHEM] The molecule, ion, or group bound to the central atom in a chelate or a coordination compound; an example is the ammonia molecules in $[Co(NH_3)_6]^{3+}$.

ligase [BIOCHEM] An enzyme that catalyzes the union of two molecules, involving the participation of a nucleoside triphosphate which is converted to a nucleoside diphosphate or monophosphate. Also known as synthetase.

ligation [MED] Surgical tying of vessels or ducts with a ligature.

ligature [MED] A cord or thread used for tying vessels and ducts.

light [OPTICS] Electromagnetic radiation with wavelengths capable of causing the sensation of vision, ranging approximately from 4000 (extreme violet) to 7700 angstroms (extreme red). Also known as light radiation; visible radiation. **2.** More generally, electromagnetic radiation of any wavelength; thus, the term is sometimes applied to infrared and ultraviolet radiation.

light adaptation [PHYSIO] The disappearance of dark adaptation; the chemical processes by which the eyes, after exposure to a dim environment, become accustomed to bright illumination, which initially is perceived as quite intense and uncomfortable.

light microscope *See* optical microscope.

light radiation *See* light.

light reflex [PHYSIO] The postural orientation response of certain aquatic forms stimulated by the source of light; receptors may be on the ventral or dorsal surface.

Ligiidae [INV ZOO] A family of primitive terrestrial isopods in the suborder Oniscoidea.

ligneous [BIOL] Of, pertaining to, or resembling wood.

lignicolous [BIOL] Living in or on wood.

lignify [BOT] To convert cell wall constituents into wood or woody tissue by chemical and physical changes.

lignin [BIOCHEM] A substance that together with cellulose forms the woody cell walls of plants and cements them together.

lignivorous [ZOO] Feeding on wood, as various insects.

lignocaine *See* lidocaine.

lignocellulose [BIOCHEM] Any of a group of substances in woody plant cells consisting of cellulose and lignin.

lignosa [BOT] Woody vegetation.

lignumvitae [BOT] *Guaiacum sanctum*. A medium-sized evergreen tree in the order Sapindales that yields a resin or gum known as gum guaiac or resin of guaiac. Also known as hollywood lignumvitae.

ligulate [BOT] **1.** Strap-shaped. **2.** Having ligules.

ligule [BOT] **1.** A small outgrowth in the axis of the leaves in Selaginellales. **2.** A thin outgrowth of a foliage leaf or leaf sheath. [INV ZOO] A small lobe on the parapodium of certain polychaetes.

likelihood ratio [STAT] The probability of a random drawing of a specified sample from a population, assuming a given hypothesis about the parameters of the population, divided by the probability of a random drawing of the same sample, assuming that the parameters of the population are such that this probability is maximized.

Liliaceae [BOT] A family of monocotyledonous plants in the order Liliales distinguished by six stamens and an inferior ovary.

Liliales [BOT] An order of monocotyledonous plants in the subclass Liliidae having the typical characteristics of the subclass.

Liliatae *See* Liliopsida.

Liliidae [BOT] A subclass of the Liliopsida; all plants are syncarpous and have a six-membered perianth, with all members petaloid.

Liliopsida [BOT] The monocotyledons, making up a class of the Magnoliophyta; characterized generally by a single cotyledon, parallel-veined leaves, and stems and roots lacking a well-defined pith and cortex.

lily [BOT] **1.** Any of the perennial bulbous herbs with showy unscented flowers constituting the genus *Lilium*. **2.** Any of various other plants having similar flowers.

Limacidae [INV ZOO] A family of gastropod mollusks containing the slugs.

limb [ANAT] An extremity or appendage used for locomotion or prehension, such as an arm or a leg. [BOT] A large primary tree branch.

limbate [BIOL] Having a part of one color bordered with a different color.

limb bud [EMBRYO] A mound-shaped lateral proliferation of the embryonic trunk; the anlage of a limb.

limbic system [ANAT] The inner edge of the cerebral cortex in the medial and temporal regions of the cerebral hemispheres.

limb-kinetic apraxia *See* motor apraxia.

limbus [BIOL] A border clearly defined by its color or structure, as the margin of a bivalve shell or of the cornea of the eye.

lime [BOT] *Citrus aurantifolia*. A tropical tree with elliptic oblong leaves cultivated for its acid citrus fruit which is a hesperidium.

limicolous [ECOL] Living in mud.

limivorous [ZOO] Feeding on mud, as certain annelids, for the organic matter it contains.

Limnebiidae [INV ZOO] The minute moss beetles, a family of coleopteran insects in the superfamily Hydrophiloidea.

limnetic [ECOL] Of, pertaining to, or inhabiting the pelagic region of a body of fresh water.

Limnichidae [INV ZOO] The minute false water beetles, a cosmopolitan family of coleopteran insects in the superfamily Dryopoidea.

Limnocharitaceae [BOT] A family of monocotyledonous

LILIACEAE

Carolina lily *(Lilium michauxii)*, a characteristic member of the family Liliaceae in the order Liliales. The six-membered perianth, with all members petaloid, is typical of the family, order, and subclass (Liliidae). *(Courtesy of John H. Gerard, from National Audubon Society)*

LIME

Tahiti (Persian) lime *(Citrus aurantifolia)*.

plants in the order Alismatales characterized by schizogenous secretory canals, multiaperturate pollen, several or many ovules, and a horseshoe-shaped embryo.

limnology [ECOL] The science of the life and conditions for life in lakes, ponds, and streams.

Limnomedusae [INV ZOO] A suborder of hydrozoan coelenterates in the order Hydroida characterized by naked hydroids.

limnophyte [ECOL] A plant that lives in ponds.

limnoplankton [BIOL] Plankton found in fresh water, especially in lakes.

Limnoriidae [INV ZOO] The gribbles, a family of isopod crustaceans in the suborder Flabellifera that burrow into submerged marine timbers.

limoniform [BOT] Lemon-shaped.

limpet [INV ZOO] Any of several species of marine gastropod mollusks composing the families Patellidae and Acmaeidae which have a conical and tentlike shell with ridges extending from the apex to the border.

Limulacea [INV ZOO] A group of horseshoe crabs belonging to the Limulida.

Limulida [INV ZOO] A subgroup of Xiphosurida including all living members of the subclass.

Limulodidae [INV ZOO] The horseshoe crab beetles, a family of coleopteran insects in the superfamily Staphylinoidea.

Limulus [INV ZOO] The horseshoe crab; the type genus of the Limulacea.

Linaceae [BOT] A family of herbaceous or shrubby dicotyledonous plants in the order Linales characterized by mostly capsular fruit, stipulate leaves, and exappendiculate petals.

Linales [BOT] An order of dicotyledonous plants in the subclass Orsidae containing simple-leaved herbs or woody plants with hypogynous, regular, syncarpous flowers having five to many stamens which are connate at the base.

lincomycin [MICROBIO] $C_{18}H_{34}O_6N_2S \cdot HCl$ Monobasic crystalline antibiotic, produced by *Streptomyces lincolnensis*, that is active as lincomycin hydrochloride mainly toward gram-positive microorganisms.

linden *See* basswood.

linea alba [ANAT] A tendinous ridge extending in the median line of the abdomen from the pubis to the tiphoid process and formed by the blending of aponeuroses of the oblique and transverse muscles of the abdomen.

lineage [GEN] Descent from a common progenitor.

linear molecule [PHYS CHEM] A molecule whose atoms are arranged so that the bond angle between each is 180°; an example is carbon dioxide, CO_2.

Lineidae [INV ZOO] A family of the Heteronemertini.

line of Gennari [ANAT] A band of white fibers in the cortex of the visual projection area of the occipital lobe. Also known as stripe of Gennari.

Lineolaceae [MICROBIO] A family of bacteria in some systems of classification that includes coenocytic members (*Lineola*) of the Caryophanales.

lineolate [BIOL] Marked with fine lines.

lingua [INV ZOO] **1.** A tonguelike process in insects. **2.** The floor of the mouth of mites.

lingual artery [ANAT] An artery originating in the external carotid and supplying the tongue.

lingual gland [ANAT] A serous, mucous, or mucoserous gland lying deep in the mucous membrane of the mammalian tongue.

lingual nerve [ANAT] A branch of the mandibular nerve having somatic sensory components and innervating the

LIMPET

Two views of the rough keyhole limpet *(Diodora aspera)* of the Pacific Coast, maximum length 2 inches (5 centimeters).

mucosa of the floor of the mouth and the anterior two-thirds of the tongue.

lingual tonsil [ANAT] An aggregation of lymphoid tissue composed of 35–100 separate tonsillar units occupying the posterior part of the tongue surface.

Linguatulida [INV ZOO] The equivalent name for Pentastomida.

Linguatuloidea [INV ZOO] A suborder of pentastomid arthropods in the order Porocephalida; characterized by an elongate, ventrally flattened, annulate, posteriorly attenuated body, simple hooks on the adult, and binate hooks in the larvae.

lingula [ANAT] A tongue-shaped organ, structure, or part thereof.

Lingulacea [INV ZOO] A superfamily of inarticulate brachiopods in the order Lingulida characterized by an elongate, biconvex calcium phosphate shell, with the majority having a pedicle.

lingulate [BIOL] Tongue- or strap-shaped.

Lingulida [INV ZOO] An order of inarticulate brachiopods represented by two living genera, *Lingula* and *Glottidia*.

liniment [PHARM] A heat-generating liquid that is thinner than ointment and is applied to the skin with friction.

linin net [CYTOL] The reticulum composed of chromatinic or oxyphilic substances in a cell nucleus.

linkage [GEN] Failure of nonallelic genes to recombine at random as a result of their being located within the same chromosome or chromosome fragment.

linkage group [GEN] The genes located on a single chromosome.

linoleic acid [BIOCHEM] $C_{17}H_{31}COOH$ A yellow unsaturated fatty acid, boiling at 229°C (14 mm Hg), occurring as a glyceride in drying oils; obtained from linseed, safflower, and tall oils; a principal fatty acid in plants, and considered essential in animal nutrition; used in medicine, feeds, paints, and margarine. Also known as linolic acid; 9,12-octadecadienoic acid.

linolenic acid [BIOCHEM] $C_{17}H_{29}COOH$ One of the principal unsaturated fatty acids in plants and an essential fatty acid in animal nutrition; a colorless liquid that boils at 230°C (17 mm Hg), soluble in many organic solvents; used in medicine and drying oils. Also known as 9,12,15-octadecatrienoic acid.

linolic acid *See* linoleic acid.

lion [VERT ZOO] *Felis leo.* A large carnivorous mammal of the family Felidae distinguished by a tawny coat and blackish tufted tail, with a heavy blackish or dark-brown mane in the male.

Liopteridae [INV ZOO] A small family of hymenopteran insects in the superfamily Cynipoidea.

lip [ANAT] A fleshy fold above and below the entrance to the mouth of mammals. [MED] The margin of an open wound. [SCI TECH] The edge of a hollow cavity or container.

Liparidae [INV ZOO] The equivalent name for Lymantriidae.

lipase [BIOCHEM] An enzyme that catalyzes the hydrolysis of fats or the breakdown of lipoproteins.

lipemia [MED] The presence of a fine emulsion of fatty substance in the blood. Also known as lipidemia; lipoidemia.

Liphistiidae [INV ZOO] A family of spiders in the suborder Liphistiomorphae in which the abdomen shows evidence of true segmentation by the presence of tergal and sternal plates.

Liphistiomorphae [INV ZOO] A suborder of arachnids in the order Araneida containing families with a primitively segmented abdomen.

lipid [BIOCHEM] One of a class of compounds which contain long-chain aliphatic hydrocarbons and their derivatives, such

LINGULACEA

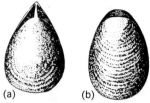

(a) (b)

Internal molds of Cambrian lingulacean *Lingulella*. *(a)* Pedicle valve. *(b)* Brachial valve. *(From C. D. Walcott, Cambrian Brachiopoda, USGS Monogr. no. 51, 1912)*

as fatty acids, alcohols, amines, amino alcohols, and alde-hydes; includes waxes, fats, and derived compounds. Also known as lipin; lipoid.

lipidemia *See* lipemia.

lipid histiocytosis [MED] **1.** Any collection of histiocytes containing lipids. **2.** *See* Niemann-Pick disease.

lipid metabolism [BIOCHEM] The physiologic and metabolic processes involved in the assimilation of dietary lipids and the synthesis and degradation of lipids.

lipid nephrosis [MED] A chronic kidney disease of children associated with thickening of the basement membranes of glomeruli and characterized by edema, presence of protein in the urine, and abnormally high blood levels of albumin and cholesterol.

lipidosis [MED] The generalized deposition of fat or fatty substances in reticuloendothelial cells. Also known as lipoi-dosis.

lipid pneumonia [MED] **1.** Pneumonia resulting from aspira-tion of oily substances, such as nose drops. **2.** Deposition of lipids in tissues of chronically inflamed lungs. Also known as lipoid pneumonia.

lipid proteinosis [MED] A hereditary disorder characterized by extracellular deposits of phospholipid-protein conjugate involving various areas of the body, including the skin and air passages.

lipid storage disease [MED] Any of various rare diseases characterized by the accumulation of large histiocytes con-taining lipids throughout reticuloendothelial tissues; exam-ples are Goucher's disease, Niemann-Pick disease, and amau-rotic familial idiocy.

lipin [BIOCHEM] **1.** A compound lipid, such as a cerebroside. **2.** *See* lipid.

lipoblastoma *See* liposarcoma.

lipochrome [BIOCHEM] Any of various fat-soluble pigments, such as carotenoid, occurring in natural fats. Also known as chromolipid.

lipodystrophy [MED] A disturbance of fat metabolism in which the subcutaneous fat disappears over some regions of the body, but is unaffected in others.

lipofuscin [BIOCHEM] Any of a group of lipid pigments found in cardiac and smooth muscle cells, in macrophages, and in parenchyma and interstitial cells; differential reactions in-clude sudanophilia, Nile blue staining, fatty acid, glycol, and ethylene.

lipoic acid [BIOCHEM] $C_8H_{14}O_2S_2$ A compound which par-ticipates in the enzymatic oxidative decarboxylation of α-keto acids in a stage between thiamine pyrophosphate and coen-zyme A.

lipoid [BIOCHEM] **1.** A fatlike substance. **2.** *See* lipid.

lipoidemia *See* lipemia.

lipoidosis *See* lipidosis.

lipoid pneumonia *See* lipid pneumonia.

lipolysis [BIOCHEM] The process of breaking fat up.

lipoma [MED] A benign tumor composed of fat cells.

lipomatosis [MED] **1.** Multiple lipomas. **2.** Obesity.

lipomelanotic reticulosis [MED] A form of lymph node hy-perplasia characterized by preservation of the architectural structure, inflammatory exudate, and hyperplasia of the reticulum cells which show phagocytosis of hemosiderin, melanin, and occasionally fat. Also known as dermatopathic lymphadenitis.

Lipomycetoideae [MICROBIO] A subfamily of oxidative yeasts in the family Saccharomycetaceae characterized by budding cells and a saclike appendage which develops into an ascus.

lipomyxoma *See* liposarcoma.

lipophore [HISTOL] A chromatophore which contains lipochrome.

lipopolysaccharide [BIOCHEM] Any of a class of conjugated polysaccharides consisting of a polysaccharide combined with a lipid.

lipoprotein [BIOCHEM] Any of a class of conjugated proteins consisting of a protein combined with a lipid.

liposarcoma [MED] A sarcoma originating in adipose tissue. Also known as embryonal-cell lipoma; fetal fat-cell lipoma; infiltrating lipoma; lipoblastoma; lipomyxoma; myxolipoma; myxoma lipomatodes.

Lipostraca [PALEON] An order of the subclass Branchiopoda erected to include the single fossil species *Lepidocaris rhyniensis*.

lipotropic [BIOCHEM] Having an affinity for lipid compounds. [PHARM] Having a preventive or curative effect on the deposition of excessive fat in abnormal sites.

lipotropic hormone [BIOCHEM] Any hormone having lipolytic activity on adipose tissue.

Lipotyphla [VERT ZOO] A group of insectivoran mammals composed of insectivores which lack an intestinal cecum and in which the stapedial artery is the major blood supply to the brain.

lipoxidase [BIOCHEM] An enzyme catalyzing the oxidation of the double bonds of an unsaturated fatty acid.

liquefaction [PHYS] A change in the phase of a substance to the liquid state; usually, a change from the gaseous to the liquid state, especially of a substance which is a gas at normal pressure and temperature.

liquid [PHYS] A state of matter intermediate between that of crystalline substances and gases in which a substance has the capacity to flow under extremely small shear stresses and conforms to the shape of a confining vessel, but is relatively incompressible, lacks the capacity to expand without limit, and can possess a free surface.

liquid chromatography [ANALY CHEM] A form of chromatography employing a liquid as the moving phase and a solid or a liquid on a solid support as the stationary phase; techniques include column chromatography, gel permeation chromatography, and partition chromatography.

liquid-vapor equilibrium [PHYS CHEM] The equilibrium relationship between the liquid and its vapor phase for a partially vaporized compound or mixture at specified conditions of pressure and temperature; for mixtures, it is expressed by $K = x/y$, where K is the equilibrium constant, x the mole fraction of a key component in the vapor, and y the mole fraction of the same key component in the liquid. Also known as vapor-liquid equilibrium.

lirella [BOT] A long, narrow apothecium with a medial longitudinal furrow, occurring in certain lichens.

Liriopeidae [INV ZOO] The phantom craneflies, a family of dipteran insects in the suborder Orthorrhapha distinguished by black and white banded legs.

Lissamphibia [VERT ZOO] A subclass of Amphibia including all living amphibians; distinguished by pedicellate teeth and an operculum-plectrum complex of the middle ear.

lissencephalous [VERT ZOO] Having an almost smooth cerebrum with few or no convolutions.

listeriosis [MED] A bacterial disease of humans and some animals caused by *Listeria monocytogenes* (Corynebacteriaceae); occurs primarily as meningitis or granulomatosis infantiseptica in humans, and takes many forms, such as meningoencephalitis, distemperlike disease, or generalized infection, in animals.

LITHIUM

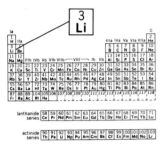

Periodic table of the chemical elements showing the position of lithium.

litchi *See* lychee.

liter [MECH] A unit of volume or capacity, equal to 1 decimeter cubed, or 0.001 cubic meter, or 1000 cubic centimeters. Abbreviated l.

lithemia *See* hyperuricemia.

lithiasis [MED] The formation of calculi in the body.

Lithistida [PALEON] An order of fossil sponges in the class Demospongia having a reticulate skeleton composed of irregular and knobby siliceous spicules.

lithite [INV ZOO] A calcareous secretion found in association with organs of hearing or balance such as otocysts, lithocysts, or tentaculocysts of many invertebrates.

lithium [CHEM] A chemical element, symbol Li, atomic number 3, atomic weight 6.939; an alkali metal.

Lithobiomorpha [INV ZOO] An order of chilopods in the subclass Pleurostigmophora; members are anamorphic and have 15 leg-bearing trunk segments, and when eyes are present, they are ocellar.

lithocyst [BOT] Epidermal plant cell in which cytoliths are formed. [INV ZOO] One of the minute sacs containing lithites in many invertebrates; thought to function in audition and orientation.

lithocyte [INV ZOO] A special cell in anthomedusae containing a statolith.

lithodesma [INV ZOO] A small shell-like plate found in certain bivalves and connected with the resilium.

Lithodidae [INV ZOO] The king crabs, a family of anomuran decapods in the superfamily Paguridea distinguished by reduced last pereiopods and by the asymmetrical arrangement of the abdominal plates in the female.

lithodomous [ZOO] Burrowing in rock.

lithogenesis [PATH] The process of formation of calculi or stones.

lithopedion [MED] A retained fetus that has become calcified.

lithophagous [ZOO] Feeding on stone, as certain mollusks.

lithophilous [ECOL] Growing or living on stones or rocks.

lithophyll [PALEOBOT] A fossil leaf, or the fossil impression of a leaf.

lithophyte [ECOL] A plant that grows on rock.

lithosere [ECOL] A succession of plant communities that originate on rock.

lithostyle [INV ZOO] A static organ in Narcomedusae. Also known as tentaculocyst.

lithotomous [INV ZOO] Boring into stone, as certain mollusks.

lithotomy [MED] Surgical removal of a calculus.

lithuria [MED] A condition marked by excess of uric (lithic) acid or its salts in the urine.

Litopterna [PALEON] An order of hoofed, herbivorous mammals confined to the Cenozoic of South America; characterized by a skull without expansion of the temporal or squamosal sinuses, a postorbital bar, primitive dentition, and feet that were three-toed or reduced to a single digit.

little cherry disease [PL PATH] A virus disease of sweet cherries characterized by small, angular pointed fruits which retain the bright red color of immaturity and never reach mature size.

little leaf [PL PATH] Any of various plant diseases and disorders characterized by chlorotic, underdeveloped, and sometimes distorted leaves.

little peach disease [PL PATH] A virus disease of the peach tree in which the fruit is dwarfed and delayed in ripening, the leaves yellow, and the tree dies.

Little's disease [MED] Spastic diplegia of infants which is characterized by spasticity of the lower extremities; involves degenerative and atrophic cerebral changes as well as congenital malformation.

littoral zone [ECOL] Of or pertaining to the biogeographic zone between the high- and low-water marks.

Littorinacea [PALEON] An extinct superfamily of gastropod mollusks in the order Prosobranchia.

Littorinidae [INV ZOO] The periwinkles, a family of marine gastropod mollusks in the order Pectinibranchia distinguished by their spiral, globular shells.

lituate [BOT] Having a forked member or part with the ends turned slightly outward, as in certain fungi.

Lituolacea [INV ZOO] A superfamily of benthic marine foraminiferans in the suborder Textulariina having a multilocular, rectilinear, enrolled or uncoiled test with a simple to labyrinthic wall.

liver [ANAT] A large vascular gland in the body of vertebrates, consisting of a continuous parenchymal mass covered by a capsule; secretes bile, manufactures certain blood proteins and enzymes, and removes toxins from the systemic circulation.

liver failure [MED] Severe functional disability of the liver marked clinically by a variety of signs and symptoms, including jaundice, coma, and abnormal blood levels of such things as ammonia, bilirubin, and alkaline phosphatase.

liver fluke [INV ZOO] Any trematode, especially *Clonorchis sinensis*, that lodges in the biliary passages within the liver.

liver phosphorylase [BIOCHEM] An enzyme that catalyzes the breakdown of liver glycogen to glucose-1-phosphate.

liverwort [BOT] The common name for members of the Marchantiatae.

live-virus vaccine [IMMUNOL] A suspension of attenuated live viruses injected to produce immunity.

living fossil [BIOL] A living species belonging to an ancient stock otherwise known only as fossils.

livor mortis [PATH] The reddish-blue discoloration of the cadaver that occurs in the dependent portions of the body due to gradual gravitational flow of unclotted blood.

lizard [VERT ZOO] Any reptile of the suborder Sauria.

lizard-hipped dinosaur [PALEON] The name applied to members of the Saurichia because of the comparatively unspecialized three-pronged pelvis.

LK virus [VIROL] A type of equine herpesvirus.

llama [VERT ZOO] Any of three species of South American artiodactyl mammals of the genus *Lama* in the camel family; differs from the camel in being smaller and lacking a hump.

Llandellian [GEOL] Upper Middle Ordovician geologic time.

Llandoverian [GEOL] Lower Silurian geologic time.

llano [ECOL] A savannah of Spanish America and the southwestern United States generally having few trees.

Llanvirnian [GEOL] Lower Middle Ordovician geologic time.

loach [VERT ZOO] The common name for fishes composing the family Cobitidae; most are small and many are eel-shaped.

lobar pneumonia [MED] An acute febrile disease involving one or more lobes of the lung, usually following pneumococcal infection.

lobar sclerosis [MED] Neuroglial proliferation accompanied by atrophy of a cerebral lobe leading to mental and neurological deficits; most common in infants and children who have suffered prolonged hypoxia.

Lobata [INV ZOO] An order of the Ctenophora in which the body is helmet-shaped.

lobate [BIOL] Having lobes. [VERT ZOO] Of a fish, having the skin of the fin extend onto the bases of the fin rays.

lobe [BIOL] A rounded projection on an organ or body part.

LIVERWORT

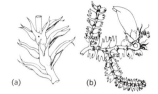

Leafy liverworts of the Jungermanniales. *(a) Herberta*, showing three ranks of equal bifid leaves *(after Muller)*. *(b) Lepidozia*, ventral aspect, showing reduced ventral leaves *(after A. Lorenz)*. *(From E. W. Sinnot and K. S. Wilson, Botany: Principles and Problems, 5th ed., McGraw-Hill, 1955)*

LIZARD

Fence lizard *(Sceloporus undulatus)* often runs along fences and trees in the eastern United States.

LLAMA

A South American llama.

LOBATA

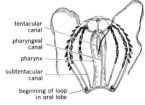

tentacular canal
pharyngeal canal
pharynx
subtentacular canal
beginning of loop in oral lobe

Bolinopsis mikado.

LOBSTER

The lobster, showing the stout pincers on the front legs which are used to crush prey.

LOCUST

Leaf scar, branch, and leaves of the black locust *(Robinia pseudoacacia).*

LOCUSTIDAE

Drawing of the grasshopper *(Melanoplus mexicanus).*

lobectomy [MED] Surgical removal of a lobe of an organ, particularly of a lung.

lobefin fish [VERT ZOO] The common name for members composing the subclass Crossopterygii.

loblolly pine [BOT] *Pinus taeda.* A hard yellow pine of the central and southeastern United States having a reddish-brown fissured bark, needles in groups of three, and a full bushy top.

lobopodia [INV ZOO] Broad, thick pseudopodia.

Lobosia [INV ZOO] A subclass of the protozoan class Rhizopodea generally characterized by lobopodia.

lobotomy [MED] An operative section of the fibers between the frontal lobes of the brain. Also known as leukotomy; prefrontal lobotomy.

lobster [INV ZOO] The common name for several bottom-dwelling decapod crustaceans making up the family Homaridae which are commercially important as a food item.

lobular pneumonia *See* bronchopneumonia.

lobule [BIOL] **1.** A small lobe. **2.** A division of a lobe.

local anesthetic [PHARM] A drug which induces loss of sensation only in the region to which it is applied.

local immunity [IMMUNOL] Immunity localized in a specific tissue or region of the body.

locellus [BOT] **1.** In some legumes, a secondary compartment of a unilocular ovary that is formed by a false partition. **2.** One of the two cavities of a pollen sac.

lochia [MED] The discharge from the uterus and vagina during the first few weeks after labor.

lociation [ECOL] One of the subunits of a faciation, distinguished by the relative abundance of a dominant species.

lockjaw *See* tetanus.

loco disease [VET MED] Poisoning in livestock resulting from ingestion of selenium-containing plants (loco weed); characterized by atrophy, delirium, convulsions, and stupor, often terminating in death.

locomotion [SCI TECH] Progressive movement, as of an animal or a vehicle.

locomotor ataxia *See* tabes dorsalis.

locomotor rods [INV ZOO] Knobbed or hooked rods occurring on the lower surface of certain Nematoda and used for crawling.

locomotor system [ZOO] Appendages and associated parts, such as muscles, joints, and bones, concerned with motor activities and locomotion of the animal body.

loco weed [BOT] Any species of *Astragalus* containing selenium taken up from the soil.

loculate [BIOL] Having, or divided into, loculi.

locule [BOT] A cavity that contains a seed in an ovary or fruit.

loculicidal [BOT] Dehiscing along the midrib or outer median line of the locules.

loculus [BIOL] A small cavity or chamber.

locus [GEN] The fixed position of a gene in a chromosome.

locust [BOT] Either of two species of commercially important trees, black locust *(Robinia pseudoacacia)* and honey locust *(Gladitsia triacanthos),* in the family Leguminosae. [INV ZOO] The common name for various migratory grasshoppers of the family Locustidae.

Locustidae [INV ZOO] A family of insects in the order Orthoptera; antennae are usually less than half the body length, hindlegs are adapted for jumping, and the ovipositor is multipartite.

lodicule [BOT] A small scale at the base of the ovary in grasses.

Iodix [INV ZOO] A ventral sclerite of the seventh abdominal segment in Lepidoptera; covers the genital plate.

Loeffler's syndrome [MED] Extensive infiltration of the lung by eosinophils, and eosinophilia of the peripheral circulation. Also known as eosinophilic pneumonitis.

Loganiaceae [BOT] A family of mostly woody dicotyledonous plants in the order Gentianales; members lack a latex system and have fully united carpels and axile placentation.

logarithm [MATH] **1.** The real-valued function $\log u$ defined by $\log u = v$ if $e^v = u$, e^v denoting the exponential function. Also known as hyperbolic logarithm; Naperian logarithm; natural logarithm. **2.** An analog in complex variables relative to the function e^z.

logarithmic distribution [STAT] A frequency distribution whose value at any integer $n = 1, 2, \ldots$ is $\lambda^n / (-n) \log (1 - \lambda)$, where λ is fixed.

logarithmic growth *See* exponential growth.

logarithmic transformation [STAT] The replacement of a variate y with a new variate $z = \log y$ or $z = \log (y + c)$, where c is a constant; this operation is often performed when the resulting distribution is normal, or if the resulting relationship with another variable is linear.

logistic curve [STAT] **1.** A type of growth curve, representing the size of a population y as a function of time t: $y = k / (1 + e^{-kbt})$, where k and b are positive constants. Also known as Pearl-Reed curve. **2.** More generally, a curve representing a function of the form $y = k / (1 + e^{cf(t)})$, where c is a constant and $f(t)$ is some function of time.

logomania [MED] Logorrhea so excessive as to be a form of a manic state; new words may be invented to keep up the garrulity.

logoplegia [MED] Loss of ability to articulate, usually due to paralysis of the speech organs.

logorrhea [MED] Excessive, usually rapid, incoherent, and uncontrollable talkativeness.

logotype [SYST] The selection or designation of a genotype after the generic name was published.

loiasis [MED] A filariasis of tropical Africa, caused by the filaria *Loa*, and characterized by diurnal periodicity of microfilariae in the blood, and transient cutaneous swelling caused by migrating adult worms.

loment [BOT] A pod or legume that is constricted between the seeds.

Lonchaeidae [INV ZOO] A family of minute myodarian cyclorrhaphous dipteran insects in the subsection Acalypteratae.

long bone [ANAT] A bone in which the length markedly exceeds the width, as the femur or the humerus.

longicollous [BIOL] Having a long beak or neck.

longicorn [INV ZOO] Having long feelers or antennae, as certain beetles.

Longidorinae [INV ZOO] A subfamily of nematodes belonging to the Dorylaimoidea including economically important plant parasites.

longipennate [VERT ZOO] **1.** Having long wings. **2.** Having long feathers.

Longipennes [VERT ZOO] An equivalent name for Charadriiformes.

longitudinal fissure [ANAT] The midline cleft dividing the cerebrum into two identical halves.

longitudinal section [SCI TECH] A section taken through the lengthwise dimension of a structure, organism, or other object.

Long's coefficient [PATH] The number 2.6, multiplied by the last two figures of the urine specific gravity, determined at

25°C, to derive the number of grams of solids per liter of urine.

loon [VERT ZOO] The common name for birds composing the family Gaviidae, all of which are fish-eating diving birds.

loop of Henle [ANAT] The U-shaped portion of a renal tubule formed by a descending and an ascending tubule. Also known as Henle's loop.

loop pattern [ANAT] A type of fingerprint pattern in which one or more of the ridges enter on either side of the impression, recurve, touch, or pass an imaginary line from the delta to the core and terminate on the entering side.

loose connective tissue [HISTOL] A type of irregularly arranged connective tissue having a relatively large amount of ground substance.

loose kernel smut [PL PATH] A type of kernel smut disease distinguished by the ruptured spore-containing gall.

Looser-Milkman syndrome *See* Milkman's syndrome.

Lopadorrhynchidae [INV ZOO] A small family of pelagic polychaete annelids belonging to the Errantia.

Lophialetidae [PALEON] A family of extinct perissodactyl mammals in the superfamily Tapiroidea.

Lophiiformes [VERT ZOO] A modified order of actinopterygian fishes distinguished by the reduction of the first dorsal fin to a few flexible rays, the first of which is on the head and bears a terminal bulb; includes anglerfish and allies.

Lophiodontidae [PALEON] An extinct family of perissodactyl mammals in the superfamily Tapiroidea.

lophocaltrops [INV ZOO] A sponge spicule with crested or branched rays.

lophocercous [VERT ZOO] Having a ridgelike caudal fin that lacks rays.

lophocyte [INV ZOO] A specialized cell of uncertain function beneath the dermal membrane of certain Demospongiae which bears a process terminating in a tuft of fibrils.

lophodont [VERT ZOO] Having molar teeth whose grinding surfaces have transverse ridges.

Lophogastrida [INV ZOO] A suborder of free-swimming marine crustaceans in the order Mysidacea characterized by imperfect fusion of the sixth and seventh abdominal somites, seven pairs of gills and brood lamellae, and natatory, biramous pleopods.

lophophore [INV ZOO] A food-gathering organ consisting of a fleshy basal ridge or lobe, from which numerous ciliated tentacles arise; found in Bryozoa, Phoronida, and Brachiopoda.

lophoselenodont [VERT ZOO] Having cheek teeth with crescentic ridges on the grinding surface.

lophosteon [VERT ZOO] The ridge of the keel on the sternum of birds.

lophotrichous [CYTOL] Having a polar tuft of flagella.

Loranthaceae [BOT] A family of dicotyledonous plants in the order Santalales in which the ovules have no integument and are embedded in a large, central placenta.

lorate [BIOL] Strap-shaped.

lordosis [MED] Exaggerated forward curvature of the lumbar spine.

L organisms [MICROBIO] Pleomorphic forms of bacteria occurring spontaneously, or favored by agents such as penicillin, which lack cell walls and grow in minute colonies; transition may be reversible under certain conditions.

lorica [INV ZOO] A hard shell or case in certain invertebrates, as in many rotifers and protozoans; functions as an exoskeleton.

Loricaridae [VERT ZOO] A family of catfishes in the suborder Siluroidei found in the Andes.

LOOP PATTERN

Loop fingerprint patterns.
(a) Twelve-count ulna loop.
(b) Radial loop. *(Federal Bureau of Investigation)*

LOPADORRHYNCHIDAE

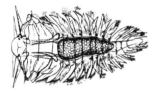

Lopadorrhynchus, dorsal view. *(After McIntosh)*

LOPHOGASTRIDA

Lophogaster typicus, adult male.

Loricata [INV ZOO] An equivalent name for Polyplacophora.

loricate [INV ZOO] Of, pertaining to, or having a lorica.

loris [VERT ZOO] Either of two slow-moving, nocturnal, arboreal primates included in the family Lorisidae: the slender loris (*Loris tardigradus*) and the slow loris (*Nycticebus coucang*).

Lorisidae [VERT ZOO] A family of prosimian primates comprising the lorises of Asia and the galagos of Africa.

lorum [INV ZOO] **1.** The piece of the lower jaw bearing the submentum in certain insects. **2.** The dorsal plate protecting the pedicle in spiders.

lotic [ECOL] Of or pertaining to swiftly moving waters.

louping ill [VET MED] A virus disease of sheep, similar to encephalomyelitis, transmitted by the tick *Ixodes racinus.* Also known as ovine encephalomyelitis; trembling ill.

louping-ill virus [VIROL] A group B arbovirus that is infectious in sheep, monkeys, mice, horses, and cattle.

louse [INV ZOO] The common name for the apterous ectoparasites composing the orders Anoplura and Mallophaga.

Lower Austral life zone [ECOL] A term used by C. H. Merriam to describe the southern portion of the Austral life zone, characterized by accumulated temperatures of 18,000°F (10,000°C).

Lower Cambrian [GEOL] The earliest epoch of the Cambrian period of geologic time, ending about 540,000,000 years ago.

Lower Cretaceous [GEOL] The earliest epoch of the Cretaceous period of geologic time, extending from about 140- to 120,000,000 years ago.

Lower Devonian [GEOL] The earliest epoch of the Devonian period of geologic time, extending from about 400- to 385,000,000 years ago.

Lower Jurassic [GEOL] The earliest epoch of the Jurassic period of geologic time, extending from about 185- to 170,000,000 years ago.

Lower Mississippian [GEOL] The earliest epoch of the Mississippian period of geologic time, beginning about 350,000,-000 years ago.

lower motor neuron [ANAT] An efferent neuron which has its body located in the anterior gray column of the spinal cord or in the brainstem nuclei, and its axon passing by way of a peripheral nerve to skeletal muscle. Also known as final common pathway.

lower motor neuron disease [MED] An injury to any part of a lower motor neuron, characterized by flaccid paralysis of the muscle, diminished or absent reflexes, and progressive atrophy of the muscle.

lower nephron nephrosis [MED] Retrogressive kidney changes following traumatic injury and other conditions producing shock; sometimes accompanied by distal and collecting tubule necrosis. Also known as acute tubular necrosis; crush kidney; hemoglobinuric nephrosis; ischemic tubulorrhexis.

Lower Ordovician [GEOL] The earliest epoch of the Ordovician period of geologic time, extending from about 490- to 460,000,000 years ago.

Lower Pennsylvanian [GEOL] The earliest epoch of the Pennsylvanian period of geologic time, beginning about 310,000,-000 years ago.

Lower Permian [GEOL] The earliest epoch of the Permian period of geologic time, extending from about 275- to 260,000,000 years ago.

lower rib height [ANTHRO] The measure of the vertical distance taken from the lower edge of the last front-attached rib to the floor as the subject stands.

Lower Silurian [GEOL] The earliest epoch of the Silurian

LOUSE

The head louse of humans, a pale-gray-brown insect with black margins and a maximum length of $\frac{1}{10}$ inch (0.25 centimeter).

LUCIFERIN

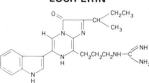

Structural formula of *Cypridina* luciferin.

LUMBAR VERTEBRAE

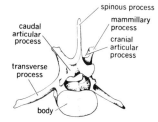

Fifth lumbar vertebra of the dog, the caudal lateral aspect showing the body or centrum. *(From M. E. Miller, G. C. Christensen, and H. E. Evans, Anatomy of the Dog, Saunders, 1964)*

LUMBRICUS

Cross section of the earthworm. *(Lumbricus terrestris). (From T. I. Storer, General Zoology, 3d ed., McGraw-Hill, 1957)*

period of geologic time, beginning about 420,000,000 years ago.

Lower Sonoran life zone [ECOL] A term used by C. H. Merriam to describe the climate and biotic communities of subtropical deserts and thorn savannas in the southwestern United States.

Lower Triassic [GEOL] The earliest epoch of the Triassic period of geologic time, extending from about 230- to 215,000,000 years ago.

loxodont [VERT ZOO] Having molar teeth with shallow hollows between the ridges.

loxolophodont [VERT ZOO] Having crests on the molar teeth that connect three of the tubercles and with the fourth or posterior inner tubercle being rudimentary or absent.

Loxonematacea [PALEON] An extinct superfamily of gastropod mollusks in the order Prosobranchia.

LSD *See* lysergic acid diethylamide.

LSD-25 *See* lysergic acid diethylamide.

Lu *See* lutetium.

Lucanidae [INV ZOO] The stag beetles, a cosmopolitan family of coleopteran insects in the superfamily Scarabaeoidea.

luciferase [BIOCHEM] An enzyme that catalyzes the oxidation of luciferin.

luciferin [BIOCHEM] A species-specific pigment in many luminous organisms that emits heatless light when combined with oxygen.

Luciocephalidae [VERT ZOO] A family of fresh-water fishes in the suborder Anabantoidei.

Ludlovian [GEOL] A European stage of geologic time; Upper Silurian, below Gedinnian of Devonian, above Wenlockian.

Ludwig's angina [MED] Acute streptococcal cellulitis of the floor of the mouth.

Luidiidae [INV ZOO] A family of echinoderms in the suborder Paxillosina.

lumbago [MED] Backache in the lumbar or lumbosacral region. [VET MED] *See* azoturia.

lumbar artery [ANAT] Any of the four or five pairs of branches of the abdominal aorta opposite the lumbar region of the spine; supplies blood to loin muscles, skin on the sides of the abdomen, and the spinal cord.

lumbar nerve [ANAT] Any of five pairs of nerves arising from lumbar segments of the spinal cord; characterized by motor, visceral sensory, somatic sensory, and sympathetic components; they innervate the skin and deep muscles of the lower back and the lumbar plexus.

lumbar vertebrae [ANAT] Those vertebrae located between the lowest ribs and the pelvic girdle in all vertebrates.

lumbodorsal fascia [ANAT] The sheath of the erector spinae muscle alone, or the sheaths of the erector spinae and the quadratus lumborum muscles.

lumbosacral plexus [ANAT] A network formed by the anterior branches of lumbar, sacral, and coccygeal nerves which for descriptive purposes are divided into the lumbar, sacral, and pudendal plexuses.

Lumbricidae [INV ZOO] A family of annelid worms in the order Oligochaeta; includes the earthworm.

Lumbriclymeninae [INV ZOO] A subfamily of mud-swallowing sedentary worms in the family Maldanidae.

Lumbriculidae [INV ZOO] A family of aquatic annelids in the order Oligochaeta.

Lumbricus [INV ZOO] A genus of earthworms recognized as the type genus of the family Lumbricidae.

Lumbrineridae [INV ZOO] A family of errantian polychaetes in the superfamily Eunicea.

lumen [SCI TECH] The space within a tube.

lumpy jaw *See* actinomycosis.

lunate [BIOL] Crescent-shaped

lunette [VERT ZOO] The thin membranous moon-shaped lower eyelid of a snake.

lung [ANAT] Either of the paired air-filled sacs, usually in the anterior or anteroventral part of the trunk of most tetrapods, which function as organs of respiration.

lung bud [EMBRYO] A primary outgrowth of the embryonic trachea; the anlage of a primary bronchus and all its branches.

lungfish [VERT ZOO] The common name for members of the Dipnoi; all have lungs that arise from a ventral connection with the gut.

lunule [BIOL] A crescent-shaped organ, structure, or mark.

lupine [BOT] A leguminous plant of the genus *Lupinus* with an upright stem, leaves divided into several digitate leaflets, and terminal racemes of pea-shaped blossoms.

lupulin [BIOCHEM] Resinous compound produced in the glandular scales of hops.

lupus erythematosus [MED] An acute or subacute febrile collagen disease characterized by a butterfly-shaped rash over the cheeks and perilingual erythema.

lupus erythematosus factor *See* LE factor.

lupus vulgaris [MED] True tuberculosis of the skin; a slow-developing, scarring, and deforming disease, often asymptomatic, frequently involving the face, and occurring in a wide variety of appearances.

Lusitanian [GEOL] Lower Jurassic geologic time.

lutein [BIOCHEM] **1.** A dried, powdered preparation of corpus luteum. **2.** *See* xanthophyll.

luteinizing hormone [BIOCHEM] A glycoprotein hormone secreted by the adenohypophysis of vertebrates which stimulates hormone production by interstitial cells of gonads. Also known as interstitial-cell-stimulating hormone (ICSH).

luteoma [MED] A tumor of the ovary composed of cells resembling those of the corpus luteum.

luteotropic hormone *See* prolactin.

lutetium [CHEM] A chemical element, symbol Lu, atomic number 71, atomic weight 174.97; a very rare metal and the heaviest member of the rare-earth group.

Lutheran blood group [IMMUNOL] The erythrocyte antigens defined by reactions with an antibody designated anti-Lu^a, initially detected in the serum of a multiply transfused patient with lupus erythematosus, who developed antibodies against erythrocytes of a donor named Lutheran, and by anti-Lu^b.

Lutjanidae [VERT ZOO] The snappers, a family of perciform fishes in the suborder Percoidei.

lyase [BIOCHEM] An enzyme that catalyzes the nonhydrolytic cleavage of its substrate with the formation of a double bond; examples are decarboxylases.

Lycaenidae [INV ZOO] A family of heteroneuran lepidopteran insects in the superfamily Papilionoidea including blue, gossamer, hairstreak, copper, and metalmark butterflies.

Lycaeninae [INV ZOO] A subfamily of the Lycaenidae distinguished by functional prothoracic legs in the male.

lychee [BOT] A tree of the genus *Litchi* in the family Sapindaceae, especially *L. chinensis* which is cultivated for its edible fruit, a one-seeded berry distinguished by the thin, leathery, rough pericarp that is red in most varieties. Also spelled litchi.

lychnisc [INV ZOO] A hexactin in which the central part of the spicule is surrounded by a system of 12 struts.

Lychniscosa [INV ZOO] An order of sponges in the subclass Hexasterophora in which parenchymal megascleres form a rigid framework and are all or in part lychniscs.

lycopene [BIOCHEM] $C_{40}H_{50}$ A red, crystalline hydrocar-

LUPINE

A field crop of lupine near maturity.

LUPUS ERYTHEMATOSUS

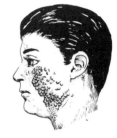

Manifestation of lupus erythematosus. *(From W. A. N. Dorland, American Illustrated Medical Dictionary, 24th ed., Saunders, 1965)*

LUTETIUM

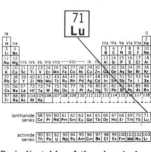

Periodic table of the chemical elements showing the position of lutetium.

bon that is the coloring matter of certain fruits, as tomatoes; it is isomeric with carotene.

lycoperdonosis [MED] A respiratory disease caused by inhalation of spores from the puffball mushroom, *Lycoperdon*.

lycophore larva [INV ZOO] A larva of certain cestodes characterized by cilia, large frontal glands, and 10 hooks. Also known as decanth larva.

Lycopodiales [BOT] The type order of Lycopodiopsida.

Lycopodiatae *See* Lycopodiopsida.

Lycopodineae [BOT] The equivalent name for Lycopodiopsida.

Lycopodiophyta [BOT] A division of the subkingdom Embryobionta characterized by a dominant independent sporophyte, dichotomously branching roots and stems, a single vascular bundle, and small, simple, spirally arranged leaves.

Lycopodiopsida [BOT] The lycopods, the type class of Lycophodiophyta. Formerly known as Lycopodiatae.

lycopodium [MATER] A yellow powder prepared from the spores of *Lycopodium clavatum*; used as a desiccant and absorbent.

Lycopsida [BOT] Former subphylum of the Embryophyta now designated as the division Lycopodiophyta.

Lycoriidae [INV ZOO] A family of small, dark-winged dipteran insects in the suborder Orthorrhapha.

Lycosidae [INV ZOO] A family of hunting spiders in the suborder Dipneumonomorphae that actively pursue their prey.

Lycoteuthidae [INV ZOO] A family of squids.

Lyctidae [INV ZOO] The large-winged beetles, a large cosmopolitan family of coleopteran insects in the superfamily Bostrichoidea.

Lygaeidae [INV ZOO] The lygaeid bugs, a large family of phytophagous hemipteran insects in the superfamily Lygaeoidea.

Lygaeoidea [INV ZOO] A superfamily of pentatomorphan insects having four-segmented antennae and ocelli.

Lyginopteridaceae [PALEOBOT] An extinct family of the Lyginopteridales including monostelic pteridosperms having one or two vascular traces entering the base of the petiole.

Lyginopteridales [PALEOBOT] An order of the Pteridospermae.

Lyginopteridatae [PALEOBOT] The equivalent name for Pteridospermae.

Lymantriidae [INV ZOO] The tussock moths, a family of heteroneuran lepidopteran insects in the superfamily Noctuoidea; the antennae of males is broadly pectinate and there is a tuft of hairs on the end of the female abdomen.

Lymexylonidae [INV ZOO] The ship timber beetles composing the single family of the coleopteran superfamily Lymexylonoidea.

Lymexylonoidea [INV ZOO] A monofamilial superfamily of wood-boring coleopteran insects in the suborder Polyphaga characterized by a short neck and serrate antennae.

lymnocryptophyte [ECOL] A helophyte, or plant that lives in bogs.

lymph [HISTOL] The colorless fluid which circulates through the vessels of the lymphatic system.

lymphadenitis [MED] Inflammation of lymph nodes.

lymphadenoid goiter *See* struma lymphomatosa.

lymphadenoma [MED] A tumorlike enlargement of a lymph node.

lymphadenomatosis [MED] A malignant neoplasm of lymphoid tissue.

lymphadenopathy [MED] Enlargement or disease of lymph nodes.

lymphadenosis [MED] Neoplasia or hyperplasia of lymph nodes.

lymphagogue [PHARM] An agent that stimulates lymph flow.

lymphangiectasis [MED] Dilation in the wall of a lymphatic vessel.

lymphangiectomy [MED] Surgical removal of a pathologic lymphatic channel, as for cancer.

lymphangioendothelial sarcoma *See* lymphangiosarcoma.

lymphangioendothelioma [MED] A tumor composed of aggregations of lymphatic vessels, between which are large mononuclear cells presumed to be endothelial cells.

lymphangiofibroma [MED] A benign tumor composed of lymphangiomatous and fibromatous elements.

lymphangioma [MED] An abnormal mass of lymphatic vessels.

lymphangiosarcoma [MED] A sarcoma whose parenchymal cells form vascular channels resembling lymphatics. Also known as lymphangioendothelial sarcoma.

lymphangitis [MED] Inflammation of lymphatic vessels.

lymphatic system [ANAT] A system of vessels and nodes conveying lymph in the vertebrate body, beginning with capillaries in tissue spaces and eventually forming the thoracic ducts which empty into the subclavian veins.

lymphatic tissue [HISTOL] Tissue consisting of networks of lymphocytes and reticular and collagenous fibers. Also known as lymphoid tissue.

lymphedema [MED] Edema resulting from lymph vessel obstruction.

lymph gland *See* lymph node.

lymph heart [VERT ZOO] A muscular expansion of a lymphatic vessel which contracts, driving lymph to the veins, as in amphibians.

lymph node [ANAT] An aggregation of lymphoid tissue surrounded by a fibrous capsule; found along the course of lymphatic vessels. Also known as lymph gland.

lymphoblast [HISTOL] Precursor of a lymphocyte.

lymphoblastic leukemia *See* acute lymphocytic leukemia.

lymphoblastic reticulosarcoma [MED] A sarcoma of reticulum cells.

lymphoblastoma [MED] A malignant neoplasm of lymphoid tissue composed of lymphoblasts.

lymphoblastosis [MED] An excessive number of lymphoblasts in peripheral blood; occasionally found in tissues.

lymphocyte [HISTOL] An agranular leukocyte formed primarily in lymphoid tissue; occurs as the principal cell type of lymph and composes 20–30% of the blood leukocytes.

lymphocytic angina *See* infectious mononucleosis.

lymphocytic choriomeningitis [MED] An acute viral meningitis caused by a specific virus endemic in mice; characterized clinically by rapid onset of symptoms of meningeal irritation, pleocytosis and often a rise in protein in the cerebrospinal fluid, and a short, benign course with recovery.

lymphocytic leukemia [MED] A type of leukemia in which lymphocytic cells predominate.

lymphocytic lymphoma [MED] A malignant neoplasm of lymphoid tissue composed predominantly of lymphocytic cells.

lymphocytic sarcoma *See* lymphosarcoma.

lymphocytoma [MED] A malignant neoplasm of lymphoid tissue composed principally of cells which resemble mature lymphocytes.

lymphocytopenia [MED] Reduction of the absolute number of lymphocytes per unit volume of peripheral blood. Also known as lymphopenia.

LYMPH NODE

Diagram of a lymph node. *(After C. Toldt, An Atlas of Human Anatomy for Students and Physicians, vol. 1, 2d ed., Macmillan, 1941)*

lymphocytosis [MED] An abnormally high lymphocyte count in the blood.

lymphoepithelioma [MED] A squamous-cell carcinoma of the nasopharynx whose parenchymal cells resemble elements of the reticuloendothelial system.

lymphogranuloma inguinale *See* lymphogranuloma venereum.

lymphogranulomatosis *See* Hodgkin's disease.

lymphogranuloma venereum [MED] A systemic infectious venereal disease caused by a microorganism belonging to the PLT-Bedsonia group, characterized by enlargement of inguinal lymph nodes and genital ulceration. Abbreviated LGV. Also known as lymphogranuloma inguinale; lymphopathia venereum; venereal bubo.

lymphoid cell [HISTOL] A mononucleocyte that resembles a leukocyte.

lymphoid hemoblast of Pappenheim *See* pronormoblast.

lymphoid tissue *See* lymphatic tissue.

lymphoma [MED] Any neoplasm, usually malignant, of the lymphoid tissues.

lymphomatosis [MED] Multiple malignant lymphomas.

lymphopathia venereum *See* lymphogranuloma venereum.

lymphopenia *See* lymphocytopenia.

lymphopoiesis [PHYSIO] The production of lymph.

lymphosarcoma [MED] A malignant lymphoma composed of anaplastic lymphoid cells resembling lymphocytes or lymphoblasts, according to the degree of differentiation. Also known as lymphocytic sarcoma.

lymph sinus [ANAT] One of the tracts of diffuse lymphatic tissue between the cords and nodules, and between the septa and capsule of a lymph node.

lymph vessel [ANAT] A tubular passage for conveying lymph.

LYNX

Lynx canadensis, with tufted ears, large broad feet, long legs, and a short tail.

lynx [VERT ZOO] Any of several wildcats of the genus *Lynx* having long legs, short tails, and usually tufted ears; differs from other felids in having 28 instead of 30 teeth.

Lyomeri [VERT ZOO] The equivalent name for Saccopharyngiformes.

Lyon hypothesis [GEN] The assumption that mammalian females are X-chromosome mosaics as a result of the inactivation of one X chromosome in some embryonic cells and the other in the rest.

Lyonnet's glands [INV ZOO] A pair of accessory silk glands in the larvae of Lepidoptera.

Lyon phenomenon [GEN] In the normal human female, the rendering of two X chromosomes inactive, or at least largely so, at an early stage in embryogenesis.

lyophilization [BIOPHYS] Rapid freezing of a material, especially biological specimens for preservation, at a very low temperature followed by rapid dehydration by sublimation in a high vacuum.

Lyopomi [VERT ZOO] An equivalent name for Notacanthiformes.

lyra [ANAT] A triangular area on the ventral surface of the corpus callosum joining the lateral pillars of the fornix. [INV ZOO] A series of chitinous rods constituting part of the stridulating organ in certain spiders.

LYSARETIDAE

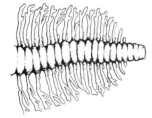

Anterior end of *Iphitime* in dorsal view.

Lysaretidae [INV ZOO] A family of errantian polychaete worms in the superfamily Eunicea.

lyse [CYTOL] To undergo lysis.

lysergic acid diethylamide [ORG CHEM] $C_{15}H_{15}N_2CON$-$(C_2H_5)_2$ A psychotomimetic drug synthesized from compounds derived from ergot. Abbreviated LSD; LSD-25.

Lysianassidae [INV ZOO] A family of pelagic amphipod crustaceans in the suborder Gammaridea.

lysigenous [BIOL] Of or pertaining to the space formed following lysis of cells.

lysin [IMMUNOL] A substance, particularly antibodies, capable of lysing a cell.

lysine [BIOCHEM] $C_6H_{14}O_2N_2$ An essential, basic amino acid obtained from many proteins by hydrolysis.

Lysiosquillidae [INV ZOO] A family of crustaceans in the order Stomatopoda.

lysis [CYTOL] Dissolution of a cell or tissue by the action of a lysin. [MED] **1.** Gradual decline in the manifestations of a disease, especially an infectious disease. Also known as defervescence. **2.** Gradual fall of fever.

lysogeny [MICROBIO] Lysis of bacteria, with the liberation of bacteriophage particles.

lysosome [CYTOL] A specialized cell organelle surrounded by a single membrane and containing a mixture of hydrolytic (digestive) enzymes.

lysozyme [BIOCHEM] An enzyme present in certain body secretions, principally tears, which acts to hydrolyze certain bacterial cell walls.

lyssacine [INV ZOO] An early stage of the skeletal network in hexactinellid sponges.

Lyssacinosa [INV ZOO] An order of sponges in the subclass Hexasterophora in which parenchymal megascleres are typically free and unconnected but are sometimes secondarily united.

lytic infection [MICROBIO] Penetration of a host cell by lytic phage.

lytic phage [VIROL] Any phage that cause host cells to lyse.

lytic reaction [CYTOL] A reaction that leads to lysis of a cell.

lytta [VERT ZOO] A rodlike structure of muscle, fatty and connective tissue, or cartilage, lying within the longitudinal axis of the tongue of carnivorous mammals.

LYSINE

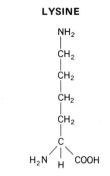

Structural formula of lysine.

M *See* molarity.

macadamia nut [BOT] The hard-shelled seed obtained from the fruit of a tropical evergreen tree, *Macadamia ternifolia*.

macaque [VERT ZOO] The common name for 12 species of Old World monkeys composing the genus *Macaca*, including the Barbary ape and the rhesus monkey.

macaw [VERT ZOO] The common name for large South and Central American parrots of the genus *Ara* and related genera; individuals are brilliantly colored with a long tail, a hooked bill, and a naked area around the eyes.

Machaeridea [INV ZOO] A class of homolozoan echinoderms in older systems of classification.

Machilidae [INV ZOO] A family of insects belonging to the Thysanura having large compound eyes and ocelli and a monocondylous mandible of the scraping type.

machopolyp *See* machozooid.

machozooid [INV ZOO] A defensive polyp equipped with stinging organs in certain hydroid colonies. Also known as machopolyp.

mackerel [VERT ZOO] The common name for perciform fishes composing the subfamily Scombroidei of the family Scombridae, characterized by a long slender body, pointed head, and large mouth.

mackerel shark [VERT ZOO] The common name for isurid galeoid elasmobranchs making up the family Isuridae; heavy-bodied fish with sharp-edged, awllike teeth and a nearly symmetrical tail.

macrandrous [BOT] Having both antheridia and oogonia on the same plant; used especially for certain green algae.

Macraucheniidae [PALEON] A family of extinct herbivorous mammals in the order Litopterna; members were proportioned much as camels are, and eventually lost the vertebral arterial canal of the cervical vertebrae.

macrencephaly [MED] The condition of having an abnormally large brain.

Macristiidae [VERT ZOO] A family of oceanic teleostean fishes assigned by some zoologists to the order Ctenothrissiformes.

macro- [SCI TECH] Prefix meaning large.

macroblast of Naegeli *See* pronormoblast.

macroblepharia [MED] The condition of having abnormally large eyelids.

macrobrachia [MED] The condition of having excessive arm development.

macrocephalus [MED] An individual with an abnormally large head.

macrocephaly [MED] The condition of having an abnormally large head.

macrocheiria [MED] The condition of having abnormal hand enlargement.

MACADAMIA NUT

(a)

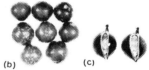

(b)   (c)

Macadamia integrifolia. (a) Mature nuts. *(b)* Nuts without husks. *(c)* Nuts in husk showing method of dehiscence. *(From R. A. Jaynes, ed., Handbook of North American Nut Trees, Humphrey Press, 1969)*

MACKEREL

Common mackerel *(Scomber scombrus),* an economically important food fish.

MACRAUCHENIIDAE

Skull and jaw of *Theosodon garrettorum,* an early Miocene macraucheniid litoptern, Santa Cruz Formation, Patagonia, Argentina. *(After W. Scott, 1910)*

macrocnemic [INV ZOO] Having the sixth protocneme or primary pair of mesenteries perfect, referring specifically to Zoanthidae.

macroconsumer [ECOL] A large consumer which ingests other organisms or particulate organic matter. Also known as biophage.

macrocrania [MED] The condition of having abnormally large skull size compared with face size.

macrocyclic [MYCOL] Of a rust fungus, having binuclear spores as well as teliospores and sporidia, or having a life cycle that is long or complex. [ORG CHEM] An organic molecule with a large ring structure usually containing over 15 atoms.

Macrocypracea [INV ZOO] A superfamily of marine ostracods in the suborder Podocopa having all thoracic legs different from each other, greatly reduced furcae, and long, thin Zenker's organs.

macrocyst [MED] A large cyst, visible to the naked eye.

macrocyte [HISTOL] An erythrocyte whose diameter or mean corpuscular volume exceed that of the mean normal by more than two standard deviations. Also known as macronormocyte.

macrocytic anemia [MED] A form of anemia characterized by the presence of macrocytes in the blood.

macrocytosis [MED] Presence of macrocytes in the blood.

macrodactyly [MED] The condition of having abnormally large fingers or toes.

Macrodasyoidea [INV ZOO] An order of wormlike invertebrates of the class Gastrotricha characterized by distinctive, cylindrical adhesive tubes in the cuticle which are moved by delicate muscle strands.

macrodontia [MED] The condition of having abnormally large teeth.

macroevolution [EVOL] The larger course of evolution by which the categories of animal and plant classification above the species level have been evolved from each other and have differentiated into the forms within each.

macrofauna [ECOL] **1.** Widely distributed fauna. **2.** Fauna of a macrohabitat. [ZOO] Animals visible to the naked eye.

macroflora [BOT] Plants visible to the naked eye. [ECOL] **1.** Widely distributed flora. **2.** Flora of a macrohabitat.

macrofollicular adenoma [MED] **1.** A benign tumor of the thyroid with enlarged follicles. **2.** A type of malignant lymphoma.

macrofossil [PALEON] A fossil large enough to be observed with the naked eye.

macrogamete [BIOL] The larger, usually female gamete produced by a heterogamous organism.

macroglia [HISTOL] That portion of the neuroglia composed of astrocytes.

macroglobulin [BIOCHEM] Any gamma globulin with a sedimentation constant of 195.

macroglobulinemia [MED] **1.** Abnormal increase in macroglobulins in the blood. **2.** A disease characterized by proliferation of lymphocytes and plasmocytes and abnormally high macroglobulin blood levels.

macroglossia [MED] Enlargement of the tongue.

macrogyria [MED] The condition of having congenitally enlarged brain convolutions.

macrohabitat [ECOL] An extensive habitat presenting considerable variation of the environment, containing a variety of ecological niches, and supporting a large number and variety of complex flora and fauna.

Macrolepidoptera [INV ZOO] A former division of Lepidoptera that included the larger moths and butterflies.

macrolide antibiotic [MICROBIO] A basic antibiotic characterized by a macrocyclic ring structure.

macromastia [MED] The condition of having abnormally enlarged breasts.

macromelia [MED] The condition of having abnormally large arms or legs.

macromere [EMBRYO] Any of the large blastomeres composing the vegetative hemisphere of telolecithal morulas and blastulas.

macromesentery [INV ZOO] Any of the larger, complete mesenteries of anthozoans.

macromolecular [ORG CHEM] Composed of or characterized by large molecules.

macromolecule [ORG CHEM] A large molecule in which there is a large number of one or several relatively simple structural units, each consisting of several atoms bonded together.

macronormocyte *See* macrocyte.

macronucleus [INV ZOO] A large, densely staining nucleus of most ciliated protozoans, believed to influence nutritional activities of the cell.

macronutrient [BIOCHEM] An element, such as potassium and nitrogen, essential in large quantities for plant growth.

macrophage [HISTOL] A large phagocyte of the reticuloendothelial system. Also known as a histiocyte.

macrophagy [BIOL] Feeding on large particulate matter.

macrophreate [INV ZOO] A comatulid with a large, deep cavity in the calyx.

macrophyllous [BOT] Having large or long leaves.

macrophyte [ECOL] A macroscopic plant, especially one in an aquatic habitat.

macroplankton [ECOL] Large organisms, such as jellyfish, which drift with the water.

Macropodidae [VERT ZOO] The kangaroos, a family of Australian herbivorous mammals in the order Marsupialia.

macropodous [BOT] **1.** Having a large or long hypocotyl. **2.** Having a long stem or stalk.

macroprosopus [MED] An individual with an abnormally large face.

macropsia [MED] A disturbance of vision in which objects seem larger than they are. Also known as megalopia.

macropterous [ZOO] Having large or long wings or fins.

Macroscelidea [VERT ZOO] A monofamilial order of mammals containing the elephant shrews and their allies.

Macroscelididae [VERT ZOO] The single, African family of the mammalian order Macroscelidea.

macrosclereids [BOT] Columnar sclereids of relatively large size such as in the coat of certain seeds.

macroscopic [SCI TECH] Large enough to be observed by the naked eye.

macroseptum [INV ZOO] A primary septum in certain anthozoans.

macrosiphon [INV ZOO] A large internal siphon in certain cephalopods.

macrosporangiophore [BOT] A structure that bears a macrosporangium.

macrosporangium [BOT] A spore case in which macrospores are produced. Also known as megasporangium.

macrospore [BOT] The larger of two spore types produced by heterosporous plants; the female gamete. Also known as megaspore. [INV ZOO] The larger gamete produced by certain radiolarians; the female gamete.

macrosporozoite [INV ZOO] In sporozoans, a larger endogamous sporozoite.

Macrostomida [INV ZOO] An order of rhabdocoels having a

simple pharynx, paired protonephridia, and a single pair of longitudinal nerves.

Macrostomidae [INV ZOO] A family of rhabdocoels in the order Macrostomida; members are broad and flattened in shape and have paired sex organs.

macrostylous [BOT] **1.** Having long styles. **2.** Having long styles and short stamens.

Macrotermitinae [INV ZOO] A subfamily of termites in the family Termitidae.

macrothermophyte *See* megathermophyte.

macrotrichia [INV ZOO] Large setae on the body or wings of insects.

macrotype [INV ZOO] An arrangement of mesenteries composed of mostly macromesenteries, in Anthozoa.

Macrouridae [VERT ZOO] The grenadiers, a family of actinopterygian fishes in the order Gadiformes in which the body tapers to a point, and the dorsal, caudal, and anal fins are continuous.

Macroveliidae [INV ZOO] A family of hemipteran insects in the subdivision Amphibicorisae.

Macrura [INV ZOO] A group of decapod crustaceans in the suborder Reptantia including eryonids, spiny lobsters, and true lobsters; the abdomen is extended and bears a well-developed tail fan.

macrurous [ZOO] Having a long tail.

macula [ANAT] Any anatomical structure having the form of a spot or strain.

macula densa [MED] A thickening of the epithelium of the ascending limb of the loop of Henle.

macula germinitiva [CYTOL] The nucleolus of the ovum.

macula lutea [ANAT] A yellow spot on the retina; the area of maximum visual acuity, being made up almost entirely of retinal cones.

maculate [BOT] Marked with speckles or spots.

maculopapule [MED] A small, circumscribed, discolored elevation of the skin; a macule and papule combined.

madarosis [MED] Loss of the eyelashes or eyebrows.

Madreporaria [INV ZOO] The equivalent name for Scleractinia.

madreporite [INV ZOO] A delicately perforated sieve plate at the distal end of the stone canal in echinoderms.

madura foot *See* mycetoma.

maduromycosis *See* mycetoma.

Maestrichtian [GEOL] A European stage of geologic time: Upper Cretaceous (above Menevian, below Fastiniogian).

Magelonidae [INV ZOO] A monogeneric family of spioniform annelid worms belonging to Sedentaria.

Magendie's foramen [ANAT] The median opening of the fourth ventricle.

magenstrasse [ANAT] Gastric canal.

maggot [INV ZOO] Larva of a dipterous insect.

Magnamycin [MICROBIO] A trade name for the antibiotic carbomycin.

magnesium [CHEM] A metallic element, symbol Mg, atomic number 12, atomic weight 24.312.

magnetic field [ELECTROMAG] One of the elementary fields in nature; it is found in the vicinity of a magnetic body or current-carrying medium and, along with electric field, in a light wave; charges moving through a magnetic field experience the Lorentz force.

magnetic nuclear resonance *See* nuclear magnetic resonance.

magnetochemistry [PHYS CHEM] A branch of chemistry which studies the interrelationship between the bulk mag-

MAGELONIDAE

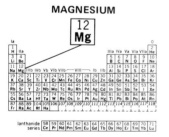

Anterior end of a species of *Magelona* in dorsal view.

MAGNESIUM

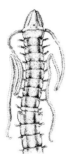

Periodic table of the chemical elements showing the position of the element magnesium.

netic properties of a substance and its atomic and molecular structure.

magnification [OPTICS] A measure of the effectiveness of an optical system in enlarging or reducing an image; the magnification may be lateral, longitudinal, or angular.

magnifying power [OPTICS] The ratio of the tangent of the angle subtended at the eye by an image formed by an optical system, to the tangent of the angle subtended at the eye by the corresponding object at a distance for convenient viewing.

magnocellular [CYTOL] Having large cell bodies; said of various nuclei of the central nervous system.

Magnolia [BOT] A genus of trees, the type genus of the Magnoliaceae, with large, chiefly white, yellow, or pinkish flowers, and simple, entire, usually large evergreen or deciduous alternate leaves.

Magnoliaceae [BOT] A family of dicotyledonous plants of the order Magnoliales characterized by hypogynous flowers with few to numerous stamens, stipulate leaves, and uniaperturate pollen.

Magnoliales [BOT] The type order of the subclass Magnoliidae; members are woody plants distinguished by the presence of spherical ethereal oil cells and by a well-developed perianth of separate tepals.

Magnoliatae *See* Magnoliopsida.

Magnoliidae [BOT] A primitive subclass of flowering plants in the class Magnoliopsida generally having a well-developed perianth, numerous centripetal stamens, and bitegmic, crassinucellate ovules.

Magnoliophyta [BOT] The angiosperms, a division of vascular seed plants having the ovules enclosed in an ovary and well-developed vessels in the xylem.

Magnoliopsida [BOT] The dicotyledons, a class of flowering plants in the division Magnoliophyta generally characterized by having two cotyledons and net-veined leaves, with vascular bundles borne in a ring enclosing a pith. Formerly known as Magnoliatae.

magnum [ANAT] Large, as in foramen magnum.

mahogany [BOT] Any of several tropical trees in the family Meliaceae of the Geraniales. [MATER] The hard wood of these trees, especially the red or yellow-brown wood of the West Indies mahogany tree (*Swietenia mahagoni*).

maidenhead *See* hymen.

Maindroniidae [INV ZOO] A family of wingless insects belonging to the Thysanura proper.

maize [BOT] *Zea mays.* Indian corn, a tall cereal grass characterized by large ears.

Majidae [INV ZOO] The spider, or decorator, crabs, a family of decapod crustaceans included in the Brachyura; members are slow-moving animals that often conceal themselves by attaching seaweed and sessile animals to their carapace.

major hysteria [PSYCH] A conversion reaction manifested in movements that suggest a generalized convulsion.

major operation [MED] An extensive, difficult, and potentially dangerous surgical procedure, usually requiring general anesthesia.

makroskelic [ANTHRO] Being long-legged relative to the trunk length, with a skelic index of 95–100.

mal- [SCI TECH] A combining form meaning bad, wrong, irregular, abnormal, inadequate.

mala [INV ZOO] **1.** The lobe of the maxilla in some insects. **2.** The grinding part of the mandible in some insects. **3.** The third mandibular segment of certain myriapods.

Malachiidae [INV ZOO] An equivalent name for Malyridae.

malacia [MED] Abnormal softening of tissues of an organ or other body structure.

MAGNOLIA

Leaf and twig of the cucumber tree, *Magnolia acuminata.*

MAGNOLIACEAE

Flower of the tulip tree (*Liriodendron tulipifera*), a characteristic eastern American species of the family of Magnoliaceae, order Magnoliales. (*Photograph by F. W. Westlake, from National Audubon Society*)

Malacobothridia [INV ZOO] A subclass of worms in the class Trematoda; members typically have one or two soft, flexible suckers and are endoparasitic in vertebrates and invertebrates.

Malacocotylea [INV ZOO] The equivalent name for Digenea.

malacology [INV ZOO] The study of mollusks.

malacoplakia [MED] The accumulation of modified histiocytes (malacoplakia cells) to produce soft, pale, elevated plaques, usually in the urinary bladder of middle-aged women.

malacoplakia cell [PATH] A large histiocyte, occasionally multinucleate and containing Michaelis-Gutmann calcospherules in the cytoplasm; seen in malacoplakia.

Malacopoda [INV ZOO] A subphylum of invertebrates in the phylum Oncopoda.

malacopterous [VERT ZOO] Having soft fins.

Malacopterygii [VERT ZOO] An equivalent name for Clupeiformes in older classifications.

Malacostraca [INV ZOO] A large, diversified subclass of Crustacea including shrimps, lobsters, crabs, sow bugs, and their allies; generally characterized by having a maximum of 19 pairs of appendages and trunk limbs which are sharply differentiated into thoracic and abdominal series.

malacostracous [INV ZOO] Having a soft shell.

maladie du sommeil *See* African sleeping sickness.

maladjustment [PSYCH] Failure to conform or inadequate conformity due to the inability or a lack of motivation to change one's feelings or attitudes to adjust to the demands of the environment.

malady [MED] A disorder, disease, or illness.

malaise [MED] A general state of ill-being or the feeling of poor health.

Malapteruridae [VERT ZOO] A family of African catfishes in the suborder Siluroidei.

malar [ANAT] Of or pertaining to the cheek or to the zygomatic bone.

malar bone *See* zygomatic bone.

malaria [MED] A group of human febrile diseases with a chronic relapsing course caused by hemosporidian blood parasites of the genus *Plasmodium*, transmitted by the bite of the *Anopheles* mosquito.

malaria pigment [PATH] Dark-brown, amorphous, microcrystalline and birefringent pigment found in parasitized erythrocytes, especially with malarial parasites, and in littoral phagocytes of spleen, liver, and bone marrow.

malar stripe [VERT ZOO] **1.** The area extending from the corner of the mouth backward and down in birds. **2.** Area on side of throat below the base of the lower mandible.

malassimilation [MED] Faulty or inadequate assimilation.

Malayan filariasis [MED] Filariasis of man caused by *Brugia malayi*, transmitted by *Mansonia* and *Anopheles* mosquitoes.

Malcidae [INV ZOO] A small family of Ethiopian and Oriental hemipternan insects in the superfamily Pentatomorpha.

Malcodermata [INV ZOO] The equivalent name for Cantharoidea.

Maldanidae [INV ZOO] The bamboo worms, a family of mud-swallowing annelids belonging to the Sedentaria.

Maldaninae [INV ZOO] A subfamily of the Maldanidae distinguished by cephalic and anal plaques with the anal aperture located dorsally.

mal de pinto *See* pinta.

male [BOT] A flower lacking pistils. [ZOO] **1.** Of or pertaining to the sex that produces spermatozoa. **2.** An individual of this sex.

male climacteric [PHYSIO] A condition presumably due to

loss of testicular function, associated with an elevated urinary excretion of gonadotropins and symptoms of loss of sexual desire and potency, hot flashes, and vasomotor instability.

male heterogamety [GEN] Having paired sex chromosomes of unequal sizes in the male.

male homogamety [GEN] Having the paired sex chromosomes of equal size in the male.

malella [INV ZOO] In myriapods, the distal toothed process of the outer stripes of the deutomala.

male pseudohermaphroditism *See* androgyny.

male sterility [PHYSIO] The condition in which male gametes are absent, deficient in number, or nonfunctional.

male Turner's syndrome *See* Ullrich-Turner syndrome.

malformation [MED] A deformity of a part of the body resulting from abnormal development.

malfunction [SCI TECH] Failure to function normally.

malic acid [BIOCHEM] $COOH \cdot CH_2 \cdot CHOH \cdot COOH$ Hydroxysuccinic acid: a dibasic hydroxy acid existing in two optically active isomers and a racemic form; found in apples and many other fruits.

malic dehydrogenase [BIOCHEM] An enzyme in the Krebs cycle that catalyzes the conversion of L-malic acid to oxaloacetic acid.

malignant [MED] **1.** Endangering the life or health of an individual. **2.** Of or pertaining to the growth and proliferation of certain neoplasms which terminate in death if not checked by treatment.

malignant adenoma [MED] A neoplasm which resembles an adenoma cytologically and gives rise to metastases.

malignant catarrh [VET MED] A catarrhal fever of cattle caused by a virus and characterized by acute inflammation and edema of the respiratory and digestive systems.

malignant disease [MED] Any disease that endangers the life of an individual over a short period of time, especially cancer.

malignant edema [VET MED] Inflammatory edema associated with certain infections, especially an acute wound infection in wild and domestic animals.

malignant embolus [MED] A blood-borne mass of malignant cells which have become dislodged from the parent neoplasm.

malignant glaucoma [MED] A form of glaucoma associated with severe pain and rapidly leading to blindness.

malignant hypertension [MED] A severe form of hypertension with a rapid course leading to progressive cardiac and renal vascular disease. Also known as accelerated hypertension.

malignant malaria *See* falciparum malaria.

malignant pustule [MED] The commonest form of anthrax in man, resulting from contamination of the skin; characterized by a necrotic pustule surrounded by an area of edema and vesicles containing yellow fluid. Also known as cutaneous anthrax.

malignant rhabdomyoma *See* rhabdomyosarcoma.

malinger [MED] To pretend or exaggerate illness in order to avoid responsibilities.

malleate trophus [INV ZOO] A type of crushing masticatory apparatus in rotifers that are incidentally predatory, such as brachionids.

mallee *See* tropical scrub.

malleolus [ANAT] A projection on the distal end of the tibia and fibula at the ankle.

malleoramate trophus [INV ZOO] An intermediate type of rotiferan masticatory apparatus having a looped manubrium and teeth on the incus (comprising the fulcrum and rami); developed for grinding.

malleus [ANAT] The outermost, hammer-shaped ossicle of

MALLEORAMATE TROPHUS

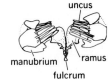

Malleoramate trophus of a bdelloid rotifier, ventral view.

MALLOPHAGA

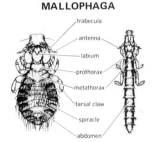

Morphology of two typical mallophagans. *(After V. L. Kellogg, Smithsonian Institution)*

MALTOSE

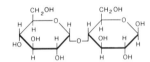

Formula for maltose.

MALVACEAE

The swamp rose mallow *(Hibiscus moscheutos)*, a common eastern American species, family Malvaceae, order Malvales. *(Courtesy of A. W. Ambler, from National Audubon Society)*

the middle ear; attaches to the tympanic membrane and articulates with the incus.

Mallophaga [INV ZOO] Biting lice, a comparatively small order of wingless insects characterized by five-segmented antennae, distinctly developed mandibles, one or two terminal claws on each leg, and a prothorax developed as a distinct segment.

Mallory bodies [PATH] Oval acidophilic hyalin inclusion bodies seen in the cytoplasm of hepatic cells in Laennec's cirrhosis.

Mallory-Weiss syndrome [MED] Painless vomiting of blood secondary to lacerations of the distal esophagus and esophagogastric junction, usually a result of prolonged violent vomiting, coughing, or hiccuping.

Malm [GEOL] The Upper Jurassic geologic series, above Dogger and below Cretaceous.

malnutrition [MED] Defective nutrition due to inadequate intake of nutrients or to their faulty digestion, assimilation, or metabolism.

malocclusion [MED] Faulty occlusion of the teeth.

malonyl urea *See* barbituric acid.

Malpighiaceae [BOT] A family of dicotyledonous plants in the order Polygalales distinguished by having three carpels, several fertile stamens, five petals that are commonly fringed or toothed, and indehiscent fruit.

Malpighian corpuscle [ANAT] **1.** A lymph nodule of the spleen. **2.** *See* renal corpuscle.

Malpighian layer [HISTOL] The germinative layer of the epidermis.

Malpighian pyramid *See* renal pyramid.

Malpighian tubule [INV ZOO] Any of the blind tubes that open into the posterior portion of the gut in most insects and certain other arthropods and excrete matter or secrete substances such as silk.

malposition [MED] Abnormal position of an organ or other body structure, or of the fetus.

malpractice [MED] Improper or injurious medical or surgical treatment, through carelessness, ignorance, or intent.

malpresentation [MED] Abnormal position of the child at birth, making normal delivery difficult or impossible.

Malta fever *See* brucellosis.

maltase [BIOCHEM] An enzyme that catalyzes the conversion of maltose to dextrose.

Malthusianism [BIOL] The theory that population increases more rapidly than the food supply unless held in check by epidemics, wars, or similar phenomena.

maltobiose *See* maltose.

maltose [BIOCHEM] $C_{12}H_{22}O_{11}$ A crystalline disaccharide that is a product of the enzymatic hydrolysis of starch, dextrin, and glycogen; does not appear to exist free in nature. Also known as maltobiose; malt sugar.

maltosuria [MED] Presence of maltose in the urine.

malt sugar *See* maltose.

malunion [MED] Faulty union of the pieces of a fractured bone.

Malvaceae [BOT] A family of herbaceous dicotyledons in the order Malvales characterized by imbricate or contorted petals, mostly unilocular anthers, and minutely spiny, multiporate pollen.

Malvales [BOT] An order of flowering plants in the subclass Dilleniidae having hypogynous flowers with valvate calyx, mostly separate petals, numerous centrifugal stamens, and a syncarpous pistil.

mamm-, mammo- [ANAT] A combining form meaning breast.

mamma [ANAT] A milk-secreting organ characterizing all mammals.

mamma aberrans [MED] A supernumerary breast.

mammal [VERT ZOO] A member of Mammalia.

Mammalia [VERT ZOO] A large class of warm-blooded vertebrates containing animals characterized by mammary glands, a body covering of hair, three ossicles in the middle ear, a muscular diaphragm separating the thoracic and abdominal cavities, red blood cells without nuclei, and embryonic development in the allantois and amnion.

mammary [ANAT] Of or pertaining to the mamma, or breast.

mammary gland [PHYSIO] A highly modified sebaceous gland that secretes milk; a unique anatomical feature of mammals.

mammary lymphatic plexus [ANAT] A network of anastomosing lymphatic vessels in the walls of the ducts and between the lobules of the mamma; also functions to drain skin, areola, and nipple.

mammary region [ANAT] The space on the anterior surface of the chest between a line drawn through the lower border of the third rib and one drawn through the upper border of the xiphoid cartilage.

mammary ridge [EMBRYO] An ectodermal thickening forming a longitudinal elevation on the chest between the limbs from which the mammary glands develop.

mammary-stimulating hormone [BIOCHEM] **1.** Estrogen and progesterone considered together as the hormones which induce proliferation of the mammary ductile and acinous elements respectively. **2.** *See* prolactin.

mammectomy *See* mastectomy.

mammillary [ANAT] **1.** Of or pertaining to the nipple. **2.** Breast- or nipple-shaped.

mammillary body [ANAT] Either of two small, spherical masses of gray matter at the base of the brain in the space between the hypophysis and oculomotor nerve, which receive and relay olfactory impulses.

mammillary line [ANAT] A vertical line passing through the center of the nipple.

mammillary process [ANAT] One of the tubercles on the posterior part of the superior articular processes of the lumbar vertebrae.

mammillitis [MED] Inflammation of the nipple.

mammogen *See* prolactin.

mammogenic hormone [BIOCHEM] **1.** Any hormone that stimulates or induces development of the mammary gland. **2.** *See* prolactin.

mammography [MED] Radiographic examination of the breast, performed with or without injection of the glandular ducts with a contrast medium.

mammoplasia [PHYSIO] Development of breast tissue.

mammoplasty [MED] Plastic surgery performed to alter the shape of the breast.

mammoth [PALEON] Any of various large Pleistocene elephants having long, upcurved tusks and a heavy coat of hair.

mammotropin *See* prolactin.

Mammutinae [PALEON] A subfamily of extinct proboscidean mammals in the family Mastodontidae.

man [ANTHRO] **1.** *Homo sapiens.* A member of the human race. **2.** An adult human male.

mandarin [BOT] A large and variable group of citrus fruits in the species *Citrus reticulata* and some of its hybrids; many varieties of the trees are compact with willowy twigs and small, narrow, pointed leaves; includes tangerines, King oranges, Temple oranges, and tangelos.

mandible [ANAT] **1.** The bone of the lower jaw. **2.** The lower

MANDRILL

The mandrill *(Mandrillus sphinx).*

MANGANESE

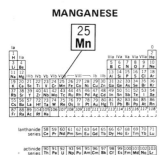

Periodic table of the chemical elements showing the position of manganese.

MANGO

Fruit on a branch of the mango tree.

jaw. [INV ZOO] Any of various mouthparts in many invertebrates designed to hold or bite into food.

mandibular arch [EMBRYO] The first visceral arch in vertebrates.

mandibular cartilage [EMBRYO] The bar of cartilage supporting the mandibular arch.

mandibular fossa [ANAT] The depression in the temporal bone that articulates with the mandibular condyle.

mandibular gland *See* submandibular gland.

mandibular nerve [ANAT] A mixed nerve branch of the trigeminal nerve; innervates various structures of the lower jaw and face.

Mandibulata [INV ZOO] A subphylum of Arthropoda; members possess a pair of mandibles which characterize the group.

mandrill [VERT ZOO] *Mandrillus sphinx.* An Old World cercopithecoid monkey found in West-central Africa and characterized by large red callosities near the ischium and by blue ridges on each side of the nose in males.

manganese [CHEM] A metallic element, symbol Mn, atomic weight 54.938, atomic number 25; a transition element whose properties fall between those of chromium and iron.

mange [VET MED] Infestation of the skin of mammals by certain mites (Sarcoptoidea) which burrow into the epidermis; characterized by multiple lesions accompanied by severe itching.

mango [BOT] *Mangifera indica.* A large evergreen tree of the sumac family (Anacardiaceae), native to southeastern Asia, but now cultivated in Africa, tropical America, Florida, and California for its edible fruit, a thick-skinned, yellowish-red, fleshy drupe.

mangrove [BOT] A tropical tree or shrub of the genus *Rhizophora* characterized by an extensive, impenetrable system of prop roots which contribute to land building. [MATER] Liquid derived from the mangrove tree *Rhizophora mucronata* and used in the leather tanning industry.

mangrove swamp [ECOL] A tropical or subtropical marine swamp distinguished by the abundance of low to tall trees, especially mangrove trees.

mania [PSYCH] Excessive enthusiasm or excitement; a violent desire or passion; manifestation of a psychotic disorder.

-mania [PSYCH] A combining form denoting obsession, abnormal preoccupation, or compulsion.

manic-depressive psychosis [PSYCH] A severe disturbance of affect characterized by extreme and pathological elation alternating with severe dejection, both of which may last for months or years.

Manidae [VERT ZOO] The pangolins, a family of mammals comprising the order Pholidota.

manifest content [PSYCH] Any idea, feeling, or action considered to be the conscious expression of repressed motives or desires, particularly the remembered content of a dream or fantasy which conceals and distorts the unconscious meaning.

manifest stimulus [PSYCH] The obvious or external basis for anxiety, fear, or dread, such as the immediate or apparent cause for a phobic reaction.

manihot *See* cassava.

manikin [MED] **1.** A model of a term fetus; used in the teaching of obstetrics. **2.** A model of an adult human used to teach first aid and basic nursing skills.

Manila hemp *See* abaca.

manioc *See* cassava.

manipulation [MED] Skillful use of the hands in moving body parts, as reducing a dislocation, or changing the position of a fetus.

manna [MATER] The concrete, yellowish, saccharine exudation of the flowering ash (*Fraxinus ornus*); contains mannitol,
sugar, mucilage, and resin and has been used as a mild
laxative.

mannan [BIOCHEM] Any of a group of polysaccharides composed chiefly or entirely of D-mannose units.

manometer [ENG] A double-leg liquid-column gage used to
measure the difference between two fluid pressures.

manometry [ENG] The use of manometers to measure gas
and vapor pressures.

mansonelliasis [MED] A parasitic infection of humans by the
filarioid nematode *Mansonella ozzardi*.

Mantidae [INV ZOO] A family of predacious orthopteran
insects characterized by a long, slender prothorax bearing a
pair of large, grasping legs, and a freely moving head with
large eyes.

mantis [INV ZOO] The common name for insects comprising
the family Mantidae.

mantle [ANAT] Collectively, the convolutions, corpus callosum, and fornix of the brain. [BIOL] An enveloping layer, as
the external body wall lining the shell of many invertebrates,
or the external meristematic layers in a stem apex. [VERT
ZOO] The back and wing plumage of a bird if distinguished
from the rest of the plumage by a uniform color.

mantle cavity [INV ZOO] The space between mantle and body
proper in bivalve mollusks.

mantle lobe [INV ZOO] Either of the flaps on the dorsal and
ventral sides of the mantle in bivalve mollusks.

Mantodea [INV ZOO] An order equivalent to the family Mantidae in some systems of classification.

Mantoux test [IMMUNOL] An intradermal test for tuberculin
sensitivity, that is, for past or present infection with tubercle
bacilli.

manubrium [ANAT] **1.** The triangular cephalic portion of the
sternum in humans and certain other mammals. **2.** The
median anterior portion of the sternum in birds. **3.** The
process of the malleus. [BOT] A cylindrical cell that projects
inward from the middle of each shield composing the antheridium in stoneworts. [INV ZOO] The elevation bearing the
mouth in hydrozoan polyps.

manure [MATER] Animal excreta collected from stables and
barnyards with or without litter; used to enrich the soil.

manus [ANAT] The hand of a human or the forefoot of a
quadruped. [INV ZOO] The proximal enlargement of the
propodus of the chela of arthropods.

manus valga [MED] Clubhand with deviation of the ulna.

manyplies *See* omasum.

maple [BOT] Any of various broad-leaved, deciduous trees of
the genus *Acer* in the order Sapindales characterized by
simple, opposite, usually palmately lobed leaves and a fruit
consisting of two long-winged samaras. [MATER] The hard,
light-colored, close-grained wood, especially from sugar
maple (*A. saccharum*).

maple syrup urine disease [MED] A hereditary metabolic
disorder caused by deficiency of branched-chain keto acid
decarboxylase; characterized by the maple-syrup-like odor of
urine. Also known as branched-chain ketoaciduria.

mapping [GEN] The determination of the arrangement of
genes among and along chromosomes.

maquis [ECOL] A type of vegetation composed of shrubs, or
scrub, usually not exceeding 3 meters in height, the majority
having small, hard, leathery, often spiny or needlelike
drought-resistant leaves and occurring in areas with a Mediterranean climate.

Marantaceae [BOT] A family of monocotyledonous plants in

MANTIS

The praying mantis
(*Stagomantis carolina*).(*Illinois
Natural History Survey*)

MAPLE

Twig, leaf, and terminal and
axillary buds of the sugar maple
(*Acer saccharum*).

the order Zingiberales characterized by one functional stamen with a single functional pollen sac, solitary ovules in each locule, and mostly arillate seeds.

marantic [MED] **1.** Of or pertaining to marasmus. **2.** Of or pertaining to slowed circulation.

marantic endocarditis [MED] Nonbacterial thrombic endocarditis, usually associated with neoplasm or other debilitating disease.

marasmus [MED] Chronic severe wasting of the tissues of the body, particularly in children, due to malnutrition.

Marattiaceae [BOT] A family of ferns coextensive with the order Marattiales.

Marattiales [BOT] An ancient order of ferns having massive eusporangiate sporangia in sori on the lower side of the circinate leaves.

marble bone disease *See* osteopetrosis.

Marburg virus [VIROL] A large virus transmitted to humans by the grivet monkey (*Cercopithecus aethiops*).

marcescent [BOT] Withering without falling off.

Marcgraviaceae [BOT] A family of dicotyledonous shrubs or vines in the order Theales having exstipulate leaves with scanty or no endosperm, two integuments, and highly modified bracts.

Marchantiales [BOT] The thallose liverworts, an order of the class Marchantiopsida having a flat body composed of several distinct tissue layers, smooth-walled and tuberculate-walled rhizoids, and male and female sex organs borne on stalks on separate plants.

Marchantiopsida [BOT] The liverworts, a class of lower green plants; the plant body is usually a thin, prostrate thallus with rhizoids on the lower surface.

mare [VERT ZOO] A mature female horse or other equine.

Marfan's syndrome [MED] A hereditary connective-tissue disorder transmitted as an autosomal dominant; manifested by skeletal and ocular changes and by congenital heart disease.

Margaritiferidae [INV ZOO] A family of gastropod mollusks with nacreous shells that provide an important source of commercial pearls.

Margarodinae [INV ZOO] A subfamily of homopteran insects in the superfamily Coccoidea in which abdominal spiracles are present in all stages of development.

marginal blight [PL PATH] A bacterial disease of lettuce caused by *Pseudomonas marginalis*, characterized by brownish marginal discoloration of the foliage.

marginal chlorosis [PL PATH] A virus disease characterized by yellowing or blanching of leaf margins; common disease of peanut plants.

marginal placentation [BOT] Arrangement of ovules near the margins of carpels.

marginal sinus [ANAT] **1.** One of the small, bilateral sinuses of the dura mater which skirt the edge of the foramen magnum, usually uniting posteriorly to form the occipital sinus. **2.** *See* terminal sinus. [EMBRYO] An enlarged venous sinus incompletely encircling the margin of the placenta.

marginal ulcer [MED] A peptic ulcer of the jejunum on the efferent margin of a gastrojejunostomy.

marginate [BOT] Having a distinct margin or border.

maricolous [ECOL] Living in the sea.

mariculture [AGR] The cultivation of marine organisms, plant and animal, for purposes of human consumption.

Marie's ataxia [MED] A hereditary ataxia combining features of cerebellar, posterior column, and pyramidal tract lesions, with onset after age 20, normal or exaggerated deep tendon

MARCHANTIALES

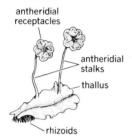

antheridial receptacles

antheridial stalks

thallus

rhizoids

Marchantia, a genus in Marchantiales; example of male gametophyte (antheridial plant). *(From H. J. Fuller and O. Tippo, College Botany, Holt, 1949)*

MARGINAL PLACENTATION

Types of marginal placentation; ovules in black.

reflexes, and frequently optic atrophy and oculomotor palsies, but no clubfeet or scoliosis.

Marie's disease [MED] Rheumatic spondylitis involving the spine only, or invading the shoulders and hips.

Marie-Strümpell disease *See* rheumatoid spondylitis.

marihuana *See* marijuana.

marijuana [BOT] The Spanish name for the dried leaves and flowering tops of the hemp plant (*Cannabis sativa*), which have narcotic ingredients and are smoked in cigarettes. Also spelled marihuana.

marine biocycle [ECOL] A major division of the biosphere composed of all biochores of the sea.

marine biology [BIOL] A branch of biology that deals with those living organisms which inhabit the sea.

marine littoral faunal region [ECOL] A geographically determined division of that portion of the zoosphere composed of marine animals.

marine microbiology [MICROBIO] The study of the microorganisms living in the sea.

Marinesco-Sjögren-Garland syndrome [MED] A hereditary, congenital form of cerebral ataxia transmitted as an autosomal recessive; characterized by mental retardation, cataracts, minor skeletal anomalies, and hypertension.

marjoram [BOT] Any of several perennial plants of the genera *Origanum* and *Majorana* in the mint family, Labiatae; the leaves are used as a food seasoning.

marked anatomy *See* pathologic anatomy.

marmoset [VERT ZOO] Any of 10 species of South American primates belonging to the family Callithricidae; individuals are primitive in that they have claws rather than nails and a nonprehensile tail.

marmot [VERT ZOO] Any of several species of stout-bodied, short-legged burrowing rodents of the genus *Marmota* in the squirrel family Sciuridae.

marrow [ANAT] The soft tissues contained in the medullary canals of long bones and in the interstices of cancellous bone.

Marseilles fever *See* fièvre boutonneuse.

marsh [ECOL] A transitional land-water area, covered at least part of the time by estuarine or coastal waters, and characterized by aquatic and grasslike vegetation, especially without peatlike accumulation.

Marsileales [BOT] A small monofamilial order of heterosporous, leptosporangiate ferns (Polypodiophyta); leaves arise on long stalks from the rhizome, and sporangia are enclosed in modified folded leaves or leaf segments called sporocarps.

Marsipobranchii [VERT ZOO] An equivalent name for Cyclostomata.

marsupial [VERT ZOO] **1.** A member of the Marsupialia. **2.** Having a marsupium. **3.** Of, pertaining to, or constituting a marsupium.

Marsupialia [VERT ZOO] The single order of the mammalian infraclass Metatheria, characterized by the presence of a marsupium in the female.

marsupialization [MED] Surgical evacuation of pancreatic, hydatid, and other cysts when complete removal and closure are not possible.

Marsupicarnivora [VERT ZOO] An order proposed to include the polydactylous and polyprotodont carnivorous superfamilies of Marsupialia.

marsupium [VERT ZOO] A fold of skin that forms a pouch enclosing the mammary glands on the abdomen of most marsupials.

marten [VERT ZOO] Any of seven species of carnivores of the genus *Martes* in the family Mustelidae which resemble the weasel but are larger and of a semiarboreal habit.

MARIJUANA

Branch of the marijuana
(*Cannabis sativa*) plant.

MARMOSET

The common marmoset, about
9 inches (23 centimeters) long
with a 12-inch (30-centimeter)
tail, an inhabitant of the mouth
of the Amazon River.

MARMOT

Though short-legged and heavy,
the marmot runs, jumps, and
climbs with speed and agility.

masculine [BIOL] Having an appearance or qualities distinctive for a male.

masculine pelvis [ANAT] A female pelvis similar to the normal male pelvis in having a deeper cavity and more conical shape. Also known as android pelvis.

masculine protest [PSYCH] The struggle to dominate, exhibited primarily by women but to some extent also by men, with the desire to escape identification with the feminine role.

masculinize [PHYSIO] To cause a female or a sexually immature animal to take on male secondary sex characteristics.

masculinoma [MED] Adrenocorticoid adenoma of the ovary.

mask face [MED] An expressionless face seen in certain degenerative and inflammatory diseases of the basal ganglia and the extrapyramidal system; voluntary movements are near normal while involuntary movements are infrequent.

masochism [PSYCH] Pleasure derived from experiencing physical or psychological pain.

mass action law [PHYS CHEM] The law that the rate of a chemical reaction for a uniform system at constant temperature is proportional to the concentrations of the substances reacting. Also known as Guldberg and Waage law.

massage [MED] The act of rubbing, kneading, or stroking the superficial parts of the body with the hand or with an instrument, for therapeutic purposes.

masseter [ANAT] The masticatory muscle, arising from the zygomatic arch and inserted into the lower jaw.

mass extinction *See* faunal extinction.

mass reflex [PHYSIO] A spread of reflexes suggesting lack of control by higher cortical centers; seen in normal newborns, in persons under the influence of drugs or in severe emotional states, and in encephalopathy or high spinal cord transections.

mass spectrograph [ENG] A mass spectroscope in which the ions fall on a photographic plate which after development shows the distribution of particle masses.

mass spectrometer [ENG] A mass spectroscope in which a slit moves across the paths of particles with various masses, and an electrical detector behind it records the intensity distribution of masses.

mass spectrometry [ANALY CHEM] An analytical technique for identification of chemical structures, determination of mixtures, and quantitative elemental analysis, based on application of the mass spectrometer.

mass spectroscope [ENG] An instrument used for determining the masses of atoms or molecules, in which a beam of ions is sent through a combination of electric and magnetic fields so arranged that the ions are deflected according to their masses.

mast-, masto- [ANAT] A combining form denoting breast; denoting mastoid.

Mastacembeloidei [VERT ZOO] The spiny eels, a suborder of perciform fishes that are eellike in shape and have the pectoral girdle suspended from the vertebral column.

mastalgia [MED] Pain in the breast.

mastatrophy [MED] Atrophy of the breast.

mastax [INV ZOO] The muscular pharynx in rotifers.

mast cell [HISTOL] A connective-tissue cell with numerous large, basophilic, metachromatic granules in the cytoplasm.

mast-cell disease *See* mastocytosis.

mastectomy [MED] Surgical removal of the breast. Also known as mammectomy.

masticate [PHYSIO] To chew.

masticatory [PHARM] A medicine to be chewed but not swallowed.

Mastigamoebidae [INV ZOO] A family of ameboid protozo-

MASS SPECTROMETER

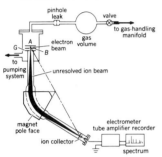

Schematic diagram of mass spectrometer tube. Electric field caused by potential difference of several volts between plates *A* and *B* draws ions through slit in *B*. Ions are further accelerated by potential difference of hundreds or thousands of volts between *B* and *G*.

MASTACEMBELOIDEI

Spiny eel *(Mastacembelus circumcinctus)*; length to 7 inches (18 centimeters). *(After H. M. Smith, The Fresh-Water Fishes of Siam or Thailand, U.S. Nt. Mus. Bull. no. 188, 1945)*

ans possessing one or two flagella, belonging to the order Rhizomastigida.

Mastigophora [INV ZOO] A superclass of the Protozoa characterized by possession of flagella.

mastitis [MED] Inflammation of the breast.

mastocytoma [VET MED] A local proliferation of mast cells forming a tumorous nodule; seen most frequently in dogs, but occasionally noted in humans.

mastocytosis [MED] Excessive mast cell proliferation. Also known as mast-cell disease.

mastodon [PALEON] A member of the Mastodontidae, especially the genus *Mammut.*

Mastodontidae [PALEON] An extinct family of elephantoid proboscideans that had low-crowned teeth with simple ridges and without cement.

mastodynia [MED] A type of cystic hyperplasia of the breast marked by an increase of connective tissue in the breast, without a proportionate increase in glandular epithelium.

mastoid [ANAT] **1.** Breast-shaped. **2.** The portion of the temporal bone where the mastoid process is located.

mastoid air cell *See* mastoid cell.

mastoid antrum [ANAT] An air-filled space between the upper portion of the middle ear and the mastoid cells.

mastoid canaliculus [ANAT] A small canal opening just above the stylomastoid foramen; gives passage to the auricular branch of the vagus nerve.

mastoid cell [ANAT] One of the compartments in the mastoid portion of the temporal bone, connected with the mastoid antrum and lined with a mucous membrane. Also known as mastoid air cell; mastoid sinus.

mastoid foramen [ANAT] A small opening behind the mastoid process.

mastoid fossa [ANAT] The depression behind the suprameatal spine on the lateral surface of the temporal bone.

mastoiditis [MED] Inflammation of the mastoid cells.

mastoid process [ANAT] A nipple-shaped, inferior projection of the mastoid portion of the temporal bone.

mastoid sinus *See* mastoid cell.

mastopathy [MED] Any disease or pain of the mammary gland. Also known as mazopathy.

mastopexy [MED] Surgical fixation of a pendulous breast. Also known as mazopexy.

mastoplasty [MED] Plastic surgery on the breast.

mastorrhagia [MED] Hemorrhage from the breast.

mastoscirrhus [MED] Hardening of breast tissue, usually indicative of cancer.

Mastotermitidae [INV ZOO] A family of lower termites in the order Isoptera with a single living species, in Australia.

mastotomy [MED] Incision of the breast.

match [IMMUNOL] To select blood donors whose erythrocytes are compatible with those of the recipient.

matching distribution [STAT] The distribution of number of matches obtained if N tickets labeled 1 to N are drawn at random one at a time and laid in a row, and a match is counted when a ticket's label matches its position.

mate [BIOL] **1.** To pair for breeding. **2.** To copulate.

materia medica [PHARM] **1.** The science that treats of the sources, properties, and preparation of medicinal substances. **2.** The materials from which medicinal substances are prepared. **3.** A treatise on the subject.

maternal [BIOL] Of, pertaining to, or related to a mother.

maternal effect [GEN] Determination of characters of the progeny by the maternal parent; mediated by the genetic constitution of the mother.

maternal impressions [PSYCH] The congenital developmen-

MASTIGOPHORA

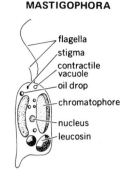

Ochromonas ludibunda, a holophytic and holozoic mastigophoran.

tal effects formerly thought to be produced upon the fetus in the uterus by mental impressions of a vivid character received by the mother during pregnancy.

maternal inheritance [GEN] The acquisition of characters transmitted through the cytoplasm of the egg.

maternal mortality rate [MED] The number of deaths reported as due to puerperal causes in a calendar year per 100 live births reported in the same year and place.

maternal placenta [EMBRYO] The outer placental layer, developed from the decidua basalis.

maternity [BIOL] **1.** Motherhood. **2.** The state of being pregnant.

mathematical biology [BIOL] A discipline that encompasses all applications of mathematics, computer technology, and quantitative theorizing to biological systems, and the underlying processes within the systems.

mathematical model [MATH] **1.** A mathematical representation of a process, device, or concept by means of a number of variables which are defined to represent the inputs, outputs, and internal states of the device or process, and a set of equations and inequalities describing the interaction of these variables. **2.** A mathematical theory or system together with its axioms.

mating [BIOL] The meeting of individuals for sexual reproduction.

matrix [HISTOL] **1.** The intercellular substance of a tissue. Also known as ground substance. **2.** The epithelial tissue from which a toenail or fingernail develops. [MYCOL] The substrate on or in which fungus grows.

matroclinous [BIOL] Deriving from the maternal parent.

maturation [BIOL] **1.** The process of coming to full development. **2.** The final series of changes in the growth and formation of germ cells.

mature [BIOL] **1.** Fully grown and developed. **2.** Ripe. [FOOD ENG] Having attained the final state of processing, as certain wines. [PSYCH] Having the emotional qualities of a well-adjusted adult.

Maunoir's hydrocele [MED] A congenital lymphatic cyst of the neck.

Mauriac syndrome [MED] A complex of symptoms associated with diabetes mellitus in children including retarded growth, obesity, and enlargement of the liver, probably related to inadequate control of the condition.

Mauritius hemp [BOT] A hard fiber obtained from the leaves of the cabuya, grown on the island of Mauritius; not a true hemp.

maxilla [ANAT] **1.** The upper jawbone. **2.** The upper jaw. [INV ZOO] Either of the first two pairs of mouthparts posterior to the mandibles in certain arthropods.

maxillary air sinus *See* maxillary sinus.

maxillary antrum *See* maxillary sinus.

maxillary arch *See* palatomaxillary arch.

maxillary artery [ANAT] A branch of the external carotid artery which supplies the deep structures of the face (internal maxillary) and the side of the face and nose (external maxillary).

maxillary nerve [ANAT] A somatic sensory branch of the trigeminal nerve; innervates the meninges, the skin of the upper portion of the face, upper teeth, and mucosa of the nose, palate, and cheeks.

maxillary process of the embryo [EMBRYO] An outgrowth of the dorsal part of the mandibular arch that forms the lateral part of the upper lip, the upper cheek region, and the upper jaw except the premaxilla.

maxillary sinus [ANAT] A paranasal air cavity in the body of the maxilla. Also known as maxillary air sinus; maxillary antrum.

maxilliped [INV ZOO] One of the three pairs of crustacean appendages immediately posterior to the maxillae.

maxillo-alveolar index [ANTHRO] Ratio of the breadth to the length of the alveolar arch, multiplied by 100.

maximum breathing capacity [PHYSIO] The greatest volume of air an individual can breathe voluntarily in 10–30 seconds; expressed as liters per minute. Abbreviated MBC.

maximum likelihood method [STAT] A technique in statistics where the likelihood distribution is so maximized as to produce an estimate to the random variables involved.

maxoplasia [MED] Degenerative disease of the breast.

mayfly [INV ZOO] The common name for insects composing the order Ephemeroptera.

May-Grünwald stain [MATER] A saturated solution of methylene blue eosinate in methyl alcohol; used to stain blood. Also known as Jenner's stain.

maz-, mazo- [EMBRYO] A combining form denoting placenta.

mazaedium [BOT] The fruiting body of certain lichens, with the spores lying in a powdery mass in the capitulum. [MYCOL] A slimy layer on the hymenial surface of some ascomycetous fungi.

maze [PSYCH] A network of paths, blind alleys, and compartments; used in intelligence tests and in experimental psychology for developing learning curves.

mazopathy [MED] **1.** Disease of the placenta. **2.** *See* mastopathy.

mazopexy *See* mastopexy.

MBC *See* maximum breathing capacity.

McArdle's syndrome [MED] A hereditary metabolic disorder caused by deficiency of muscle phosphorylase, with abnormal glycogen deposition in skeletal muscle leading to muscle fiber destruction. Also known as myophosphorylase deficiency glycogenosis.

McBurney's incision [MED] A short diagonal incision in the lower right quadrant in which the muscle fibers are separated rather than cut; used for appendectomy.

McBurney's point [ANAT] A point halfway between the umbilicus and the anterior superior iliac spine; a point of extreme tenderness in appendicitis.

M-component hypergammaglobulinemia [MED] A form of hypergammaglobulinemia characterized by a single, prominent, more or less narrow band occurring from the slow gamma to the fast alpha-1 region of the electrophoretic strip.

Md *See* mendelevium.

mean [MATH] The average or expected value.

mean deviation *See* average deviation.

mean rank method [STAT] A method of handling data which has the same observed frequency occurring at two or more consecutive ranks; it consists of assigning the average of the ranks as the rank for the common frequency.

mean square deviation [STAT] A measure of the extent to which a collection $v_1, v_2, \ldots, v_n$ of numbers is unequal; it is given by the expression $(1/n)[(v_1 - \bar{v})^2 + \ldots + (v_n - \bar{v})^2]$, where $\bar{v}$ is the mean of the numbers.

Meantes [VERT ZOO] The mud eels, a small suborder of the Urodela including three species of aquatic eellike salamanders with only anterior limbs.

measles [MED] An acute, highly infectious viral disease with cough, fever, and maculopapular rash; the appearance of Koplik spots on the oral mucous membranes marks the onset. Also known as rubeola.

measles encephalitis [MED] Acute disseminated encephalitis following measles.

MAYFLY

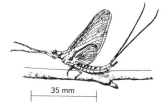

Mayfly, adult male, *Hexagenia* species, in usual resting position. *(From A. H. Morgan, Field Book of Ponds and Streams, Putnam, 1930)*

measles immune globulin [IMMUNOL] Sterile human globulin used to provide passive immunization against measles.

measles virus vaccine [IMMUNOL] A suspension of live attenuated or inactivated measles virus used for active immunization against measles.

measly [FOOD ENG] Of meat, containing larval tapeworms. [MED] Infected with measles.

meatal plate [EMBRYO] A mass of ectodermal cells on the bottom of the branchial groove in a 2-month embryo.

meatotomy [MED] Incision into and enlargement of a meatus.

meatus [ANAT] A natural opening or passage in the body.

mechanical dysmenorrhea [MED] Painful menstruation due to mechanical obstruction of the discharge of menstrual fluids. Also known as obstructive dysmenorrhea.

mechanical stage [ENG] A stage on a microscope provided with a mechanical device for positioning or changing the position of a slide.

mechanoreceptor [PHYSIO] A receptor that provides the organism with information about such mechanical changes in the environment as movement, tension, and pressure.

Meckel's cartilage [EMBRYO] The cartilaginous axis of the mandibular arch in the embryo and fetus.

Meckel's diverticulum [EMBRYO] The persistent blind end of the yolk stalk forming a tube connected with the lower ileum.

meconium [EMBRYO] A greenish mass of mucous, desquamated epithelial cells, lanugo, and vernix caseosa that collects in the fetal intestine, becoming the first fecal discharge of the newborn.

meconium ileus [MED] Intestinal obstruction in the newborn with cystic fibrosis due to trypsin deficiency.

Mecoptera [INV ZOO] The scorpion flies, a small order of insects; adults are distinguished by the peculiar prolongation of the head into a beak, which bears chewing mouthparts.

mecystasis [PHYSIO] Increase in muscle length with maintenance of the original degree of tension.

media [HISTOL] The middle, muscular layer in the wall of a vein, artery, or lymph vessel.

mediad [ANAT] Toward the median line or plane of the body or of a part of the body.

medial [ANAT] **1.** Being internal as opposed to external (lateral). **2.** Toward the midline of the body. [SCI TECH] Located in the middle.

medial arteriosclerosis [MED] Calcification of the tunica media of small and medium-sized muscular arteries. Also known as medial calcinosis; Mönckeberg's arteriosclerosis.

medial calcinosis *See* medial arteriosclerosis.

medial lemniscus [ANAT] A lemniscus arising in the nucleus gracilis and nucleus cuneatus of the brain, crossing immediately as internal arcuate fibers, and terminating in the posterolateral ventral nucleus of the thalamus.

medial necrosis [MED] Death of cells in the tunica media of arteries. Also known as medionecrosis.

median [SCI TECH] Located in the middle. [STAT] An average of a series of quantities or values; specifically, the quantity or value of that item which is so positioned in the series, when arranged in order of numerical quantity or value, that there are an equal number of items of greater magnitude and lesser magnitude.

median effective dose *See* effective dose 50.

median infective dose *See* infective dose 50.

median lethal dose *See* lethal dose 50.

median maxillary cyst [MED] Cystic dilation of embryonal inclusions in the incisive fossa or between the roots of the central incisors. Also known as nasopalatine cyst.

median nasal process [EMBRYO] The region below the fron-

MECOPTERA

|―2 mm―|

Male scorpion fly *(Panorpa)*.

tonasal sulcus between the olfactory sacs; forms the bridge and mobile septum of the nose and various parts of the upper jaw and lip.

median nerve test [MED] A test for loss of function of the median nerve by having the patient abduct the thumb at right angles to the palm with fingertips in contact and forming a pyramid.

mediastinitis [MED] Inflammation of the mediastinum.

mediastinum [ANAT] **1.** A partition separating adjacent parts. **2.** The space in the middle of the chest between the two pleurae.

medical bacteriology [MED] A branch of medical microbiology that deals with the study of bacteria which affect human health, especially those which produce disease.

medical climatology [MED] The study of the relation between climate and disease.

medical entomology [MED] The study of insects that are vectors for diseases and parasitic infestations in humans and domestic animals.

medical ethics [MED] Principles and moral values of proper medical conduct.

medical examiner [MED] A professionally qualified physician duly authorized and charged by a governmental unit to determine facts concerning causes of death, particularly deaths not occurring under natural circumstances, and to testify thereto in courts of law.

medical geography [MED] The study of the relation between geographic factors and disease.

medical history [MED] An account of a patient's past and present state of health obtained from the patient or relatives.

medical microbiology [MED] The study of microorganisms which affect human health.

medical mycology [MED] A branch of medical microbiology that deals with fungi that are pathogenic to humans.

medical parasitology [MED] A branch of medical microbiology which deals with the relationship between humans and those animals which live in or on them.

medical protozoology [MED] A branch of medical microbiology that deals with the study of Protozoa which are parasites of humans.

medical radiography [MED] The use of x-rays to produce photographic images for visualizing internal anatomy as an aid in diagnosis.

medication [MED] **1.** A medicinal substance. **2.** Treatment by or administration of a medicine.

medicinal [MED] Of, pertaining to, or having the nature of medicine.

medicine [MED] **1.** Any agent administered for the treatment of disease. **2.** The science and art of treating and healing.

medionecrosis [MED] Necrosis occurring in the tunica media of an artery.

Mediterranean anemia *See* thalassemia.

Mediterranean faunal region [ECOL] A marine littoral faunal region including that offshore portion of the Atlantic Ocean from northern France to near the Equator.

Mediterranean fever *See* brucellosis.

medius [ANAT] The middle finger.

medulla [ANAT] **1.** The central part of certain organs and structures such as the adrenal glands and hair. **2.** Marrow, as of bone or the spinal cord. **3.** *See* medulla oblongata. [BOT] **1.** Pith. **2.** The central spongy portion of some fungi.

medulla oblongata [ANAT] The somewhat pyramidal, caudal portion of the vertebrate brain extending from the pons to the spinal cord. Also known as medulla.

medullary canal [ANAT] The central cavity of a long bone.

medullary carcinoma [MED] A form of poorly differentiated adenocarcinoma, usually of the breast, grossly well circumscribed, gray-pink, and firm. Also known as encephaloid carcinoma.

medullary cord [ANAT] Dense lymphatic tissue separated by sinuses in the medulla of a lymph node. [EMBRYO] A primary invagination of the germinal epithelium of the embryonic gonad that differentiates into rete testis and seminiferous tubules or into rete ovarii.

medullary layer [BOT] A thick subcortical layer of interwoven hyphae in the thallus of some lichens.

medullary plate [EMBRYO] The platelike dorsal zone of the epiblast in the early vertebrate embryo; the primordium of the nervous system.

medullary rays [BOT] Strands of connective tissue in the stele of certain cryptogams and dicotyledons extending outward from the medulla, and sometimes separating vascular bundles.

medullary sheath [ANAT] A layer of white myelin surrounding the axis cylinder of a medullated nerve fiber. [BOT] A sheath of protoxylem surrounding the pith of certain stems.

medulloblastoma [MED] A malignant neoplasm of the brain with a tendency to metastasize in the meninges.

Medullosaceae [PALEOBOT] A family of seed ferns; these extinct plants all have large spirally arranged petioles with numerous vascular bundles.

medusa *See* jellyfish.

megacanthopore [INV ZOO] A large prominent tube, commonly projecting as a spine in a mature region of a bryozoan colony.

Megachilidae [INV ZOO] The leaf-cutting bees, a family of hymenopteran insects in the superfamily Apoidea.

Megachiroptera [VERT ZOO] The fruit bats, a group of Chiroptera restricted to the Old World; most species lack a tail, but when present it is free of the interfemoral membrane.

megacolon [MED] Hypertrophy and dilation of the colon associated with prolonged constipation.

Megadermatidae [VERT ZOO] The false vampires, a family of tailless bats with large ears and a nose leaf; found in Africa, Australia, and the Malay Archipelago.

megagametophyte [BOT] The female gametophyte; develops from a megaspore.

megakaryocyte [HISTOL] A giant bone-marrow cell characterized by a large, irregularly lobulated nucleus; precursor to blood platelets.

megakaryocytopenia *See* megakaryophthisis.

megakaryophthisis [MED] A scarcity of megakaryocytes in the bone marrow. Also known as megakaryocytopenia.

megaloblast [PATH] A large nucleated erythroblast appearing in bone marrow in vitamin B_{12} or folic acid deficiency.

megaloblastic anemia [MED] Anemia characterized by the occurrence of megaloblasts in the bone marrow and blood.

megaloblast of Sabin *See* pronormoblast.

megalocardia [MED] Abnormal enlargement of the heart.

megalocephaly [MED] The condition of having a head whose maximum fronto-occipital circumference is greater than two standard deviations above the mean for age and sex.

Megalodontoidea [INV ZOO] A superfamily of hymenopteran insects in the suborder Symphyta.

megalomania [PSYCH] The delusion of greatness and omnipotence characterizing certain psychotic reactions.

Megalomycteroidei [VERT ZOO] The mosaic-scaled fishes, a monofamilial suborder of the Cetomimiformes; members are rare species of small, elongate deep-sea fishes with degenerate eyes and irregularly disposed scales.

MEGACHILIDAE

Drawing of a leaf-cutting beetle. *(From T. I. Storer and R. L. Usinger, General Zoology, 3d ed., McGraw-Hill, 1957)*

megalopia *See* macropsia.

megalopore [INV ZOO] One of the large pores in the dorsal plates of some chitons, leading to photosensitive organs.

megalops larva [INV ZOO] A preimago stage of certain crabs having prominent eyes and chelae.

Megaloptera [INV ZOO] A suborder included in the order Neuroptera by some authorities.

Megalopygidae [INV ZOO] The flannel moths, a small family of lepidopteran insects in the suborder Heteroneura.

megalosphere [INV ZOO] The initial, large-chambered shell of sexual individuals of certain dimorphic species of Foraminifera.

megaloureter [MED] Abnormal enlargement of a ureter.

-megaly [MED] A combining form denoting abnormal enlargement.

Megamerinidae [INV ZOO] A family of myodarian cyclorrhaphous dipteran insects in the subsection Acalypteratae.

megaphyll [BOT] A relatively large leaf, according to an arbitrary classification of leaf sizes based on the blade area in millimeters.

megaphyllous [BOT] Having large leaves or leaflike extensions.

megaphyte [ECOL] A soft shrub with a thick stem; it is morphologically similar to an herb.

Megapodiidae [VERT ZOO] The mound birds and brush turkeys, a family of birds in the order Galliformes; distinguished by their method of incubating eggs in mounds of dirt or in decomposing vegetation.

megarectum [MED] Abnormal enlargement of the rectum.

megasclere [INV ZOO] A large sclerite.

megasporangium *See* macrosporangium.

megaspore *See* macrospore.

megasporophyll [BOT] A leaf bearing megasporangia.

megathermophyte [ECOL] A plant that requires great heat and abundant moisture for normal growth. Also known as macrothermophyte.

Megathymiinae [INV ZOO] The giant skippers, a subfamily of lepidopteran insects in the family Hesperiidae.

megazoospore [INV ZOO] A large zoospore.

megistotherm [ECOL] A plant that lives best at a more or less uniformly high temperature.

Mehlis' gland [INV ZOO] One of the large unicellular glands around the ootype of flatworms.

Meibomian cyst *See* chalazion.

Meibomian gland *See* tarsal gland.

meibomianitis [MED] Inflammation of the tarsal glands.

Meig's syndrome [MED] A complex of symptoms associated with ovarian fibroma including abnormal accumulation of serous fluid in the pleural and peritoneal cavities.

Meinertellidae [INV ZOO] A family of wingless insects belonging to the Microcoryphia.

meiocyte [CYTOL] A cell undergoing meiotic division.

meiosis [CYTOL] A type of cell division occurring in diploid or polyploid tissues that results in a reduction in chromosome number, usually by half.

meiotherm [ECOL] A plant that lives best in cool-temperate surroundings.

Meissner's corpuscle [ANAT] An ovoid, encapsulated cutaneous sense organ presumed to function in touch sensation in hairless portions of the skin.

Meissner's plexus *See* submucous plexus.

mel-, melo- [SCI TECH] A combining form denoting dark or black; denoting or pertaining to melanin.

Melamphaidae [VERT ZOO] A family of bathypelagic fishes in the order Beryciformes.

MEISSNER'S CORPUSCLE

Drawing of Meissner's corpuscle. *(From F. A. Geldard, Human Senses, Wiley, 1953)*

Melampsoraceae [MYCOL] A family of parasitic fungi in the order Uredinales in which the teleutospores are laterally united to form crusts or columns.

melancholia [PSYCH] A disordered mental condition of psychotic proportion characterized by severe depression.

Melanconiaceae [MYCOL] The single family of the order Melanconiales.

Melanconiales [MYCOL] An order of the class Fungi Imperfecti including many plant pathogens commonly causing anthracnose; characterized by densely aggregated cnidophores on an acervulus.

Melandryidae [INV ZOO] The false darkling beetles, a family of coleopteran insects in the superfamily Tenebrionoidea.

melanin [BIOCHEM] Any of a group of brown or black pigments occurring in plants and animals.

melaniridosome [VERT ZOO] A chromatophore having melanophore and iridophore components; common in the corium of teleost fishes.

melanoblast [HISTOL] **1.** Precursor cell of melanocytes and melanophores. **2.** An immature pigment cell in certain vertebrates. **3.** A mature cell that elaborates melanin.

melanoblastoma [MED] A malignant tumor composed principally of melanoblasts.

melanocarcinoma [MED] A malignant melanoma derived from epithelial tissue.

melanocyte [HISTOL] A cell containing dark pigments.

melanocyte-stimulating hormone [BIOCHEM] A protein substance secreted by the intermediate lobe of the pituitary of man which causes dispersion of pigment granules in the skin; similar to intermedins in other vertebrates. Abbreviated MSH. Also known as melanophore-dilating principle; melanophore hormone.

melanocytoma [MED] A benign tumor composed principally of melanocytes.

melanocytosis [MED] An excessive number of melanocytes.

melanoderma [MED] Abnormal darkening of the skin.

melanogen [BIOCHEM] A colorless precursor of melanin.

melanogenesis [BIOCHEM] The formation of melanin.

melanoglosia [MED] Any blackening of the tongue associated with certain disorders and diseases in animal and man.

melanoid [MED] Dark-colored; resembling melanin.

melanoma [MED] **1.** A malignant tumor composed of anaplastic melanocytes. **2.** A benign or malignant tumor composed of melanocytes.

melanomatosis [MED] **1.** Widespread distribution of melanoma. **2.** Diffuse melanotic pigmentation of the meninges.

melanophage [HISTOL] A phagocytic cell which engulfs and contains melanin.

melanophore [HISTOL] A type of chromatophore containing melanin.

melanophore-dilating principle *See* melanocyte-stimulating hormone.

melanophore hormone *See* melanocyte-stimulating hormone.

melanoprotein [BIOCHEM] A conjugated protein in which melanin is the associated chromagen.

melanosarcoma [MED] A malignant melanoma.

melanose [PL PATH] **1.** A fungus disease of grapevine caused by *Septoria ampelina*; leaves are infected and fall off. **2.** A fungus disease of citrus trees and fruits caused by *Diaporthe citri*, characterized by hard, brown, usually gummy elevations on the rind, twigs, and leaves.

melanosis coli [PATH] Melanotic pigmentation of the mucosa of the colon in large numbers of minute foci.

melanosis iridis [PATH] Abnormal melanotic pigmentation of the iris.

melanosome [CYTOL] An organelle which contains melanin and in which tyrosinase activity is not demonstrable.

melanotic cancer [MED] A neoplasm producing excessive quantities of melanin.

melanotic freckle [MED] An unevenly pigmented macule that sometimes develops on the skin, usually on the face, of an individual beyond middle life and enlarges progressively. Also known as Hutchinson's freckle.

melanuria [MED] The presence of black pigment in the urine.

Melasidae [INV ZOO] The equivalent name for Eucnemidae.

Melastomataceae [BOT] A large family of dicotyledonous plants in the order Myrtales characterized by an inferior ovary, axile placentation, up to twice as many stamens as petals (or sepals), anthers opening by terminal pores, and leaves with prominent, subparallel longitudinal ribs.

Meleagrididae [VERT ZOO] The turkeys, a family of birds in the order Galliformes characterized by a bare head and neck.

melena [MED] The discharge of stools colored black by altered blood.

Meliaceae [BOT] A family of dicotyledonous plants in the order Sapindales characterized by mostly exstipulate, alternate leaves, stamens mostly connate by their filaments, and syncarpous flowers.

Melinae [VERT ZOO] The badgers, a subfamily of carnivorous mammals in the family Mustelidae.

Melinninae [INV ZOO] A subfamily of sedentary annelids belonging to the Ampharetidae which have a conspicuous dorsal membrane, with or without dorsal spines.

melioidosis [VET MED] An endemic bacterial disease, primarily of rodents but occasionally communicable to humans, caused by *Pseudomonas pseudomallei* and characterized by infectious granulomas.

Meliolaceae [MYCOL] The sooty molds, a family of ascomycetous fungi in the order Erysiphales, with dark mycelia and conidia.

meliphagous [BIOL] Feeding on honey.

melitose *See* raffinose.

melitriose *See* raffinose.

Melittidae [INV ZOO] A family of hymenopteran insects in the superfamily Apoidea.

melituria [MED] The presence of sugar in the urine.

Meloidae [INV ZOO] The blister beetles, a large cosmopolitan family of coleopteran insects in the superfamily Meloidea; characterized by soft, flexible elytra and the strongly vesicant properties of the body fluids.

Meloidea [INV ZOO] A superfamily of coleopteran insects in the suborder Polyphaga.

melon [BOT] Either of two soft-fleshed edible fruits, muskmelon or watermelon, or varieties of these.

melting point [THERMO] **1.** The temperature at which a solid of a pure substance changes to a liquid. Abbreviated mp. **2.** For a solution of two or more components, the temperature at which the first trace of liquid appears as the solution is heated.

Melusinidae [INV ZOO] A family of orthorrhaphous dipteran insects in the series Nematocera.

Melyridae [INV ZOO] The soft-winged flower beetles, a large family of cosmopolitan coleopteran insects in the superfamily Cleroidea.

Membracidae [INV ZOO] The treehoppers, a family of homopteran insects included in the series Auchenorrhyncha having a pronotum that extends backward over the abdomen, and a vertical upper portion of the head.

MELINAE

The American badger *(Taxidea taxus).*

MELOIDAE

Blister beetle. *(From T. I. Storer and R. L. Usinger, General Zoology, 3d ed., McGraw-Hill, 1957)*

MEMBRACIDAE

(a) (b)

Membracids on stems. *(a)* Adult. *(b)* Nymphs. *(Courtesy of C. H. Hanson)*

membrane [CHEM ENG] **1.** The medium through which the fluid stream is passed for purposes of filtration. **2.** The ion-exchange medium used in dialysis, diffusion, osmosis and reverse osmosis, and electrophoresis. [HISTOL] A thin layer of tissue surrounding a part of the body, separating adjacent cavities, lining cavities, or connecting adjacent structures.

membrane bone *See* dermal bone.

membranelle [INV ZOO] **1.** An undulating membrane composed of a row of fused cilia in some ciliate Protozoa. **2.** The ciliated band in the tornaria larva.

membrane potential [PHYSIO] A potential difference across a living cell membrane.

membranous glomerulonephritis [MED] A type of glomerulonephritis characterized by thickening of the basement membrane due to deposition of electron-dense material.

membranous labyrinth [ANAT] The membranous portion of the inner ear of vertebrates.

membranous pregnancy [MED] Gestation in which there has been a rupture of the amniotic sac and the fetus is in direct contact with the wall of the uterus.

membranous urethra [ANAT] The part of the urethra between the two facial layers of the urogenital diaphragm.

membranule [INV ZOO] A small, opaque space in the anal area of the wing of some dragonflies.

memory [PSYCH] The recollection of past events or sensations, or the performance of previously learned skills without practice.

memory trace [PHYSIO] *See* engram. [PSYCH] An experience intentionally forgotten but not fully repressed, which may result in the development of a neurotic conflict.

men-, meno- [PHYSIO] A combining form denoting menses.

menacme [PHYSIO] The period of a woman's life during which menstruation persists.

menarche [PHYSIO] The onset of menstruation.

mendelevium [CHEM] Synthetic radioactive element, symbol Md, with atomic number 101; made by bombarding lighter elements with light nuclei accelerated in cyclotrons.

Mendelian genetics [GEN] Scientific study of heredity as related to or in accordance with Mendel's laws.

Mendelian population [GEN] A group of individuals who interbreed according to a certain system of mating; the total genic content of the individuals of the group is called gene pool.

Mendelian ratio [GEN] The ratio of occurrence of various phenotypes in any cross involving Mendelian characters.

Mendelism [GEN] The basic laws of inheritance as formulated by Mendel.

Mendel's laws [GEN] Two basic principles of genetics formulated by Mendel: the law of segregation and law of independent assortment.

Menetrier's disease [MED] Benign, diffuse hypertrophic gastritis; symptoms include vomiting, diarrhea, weight loss, and excessive secretion of mucus.

Ménière's syndrome [MED] A disease of the inner ear characterized by deafness, vertigo, and tinnitus; possibly an allergic process. Also known as labyrinthine syndrome.

meninges [ANAT] The membranes that cover the brain and spinal cord; there are three in mammals and one or two in submammalian forms.

meninginococcemia [MED] **1.** The presence of meningococci in the blood. **2.** A clinical disorder consisting of fever, skin hemorrhages, varying degrees of shock, and meningococci in the blood.

meningioma [MED] A localized tumor composed of menin-

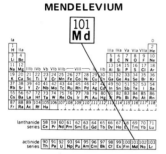

MENDELEVIUM

Periodic table of the chemical elements showing the position of mendelevium.

geal cells, involving the meninges and other central nervous system structures. Also known as meningothelioma.

meningism [MED] A condition in which signs and symptoms suggest meningitis, but clinical evidence for the disease is absent. Also known as meningismus.

meningismus *See* meningism.

meningitis [MED] Inflammation of the meninges of the brain and spinal cord, caused by viral, bacterial, and protozoan agents.

meningocele [MED] Hernia of the meninges through a defect in the skull or vertebral column, forming a cyst filled with cerebrospinal fluid.

meningococcal meningitis [MED] Inflammation of the meninges caused by the bacterium *Neisseria meningitidis* (meningococcus).

meningococcus [MICROBIO] Common name for *Neisseria meningitidis.*

meningocyte [HISTOL] One of the epithelioid cells lining the subarachnoid space; may become phagocytic.

meningoencephalitis [MED] Inflammation of the brain and its meninges.

meningoencephalocele [MED] A protrusion of the brain and its membranes through a defect in the skull.

meningoencephalomyelitis [MED] Combined inflammation of the meninges, brain, and spinal cord.

meningoencephalopathy [MED] Disease of the brain and its meninges.

meningomyelitis [MED] Inflammation of the spinal cord and its membranes.

meningomyocele [MED] Hernia of the spinal cord and its meninges through a defect in the vertebral column.

meningothelioma *See* meningioma.

meningothelium [HISTOL] Epithelium of the arachnoid which envelops the brain.

meningovascular [MED] Involving both the meninges and the cerebral blood vessels.

meningovascular syphilis [MED] Syphilis of the central nervous system involving the formation of gummas of the leptomeninges and endarteritis of cerebral vessels.

meninx [ANAT] Any one of the three membranes covering the brain and spinal cord.

meninx primaria [INV ZOO] A thick sheath of connective tissue surrounding the central nervous system of some lower vertebrates, such as Anura.

meninx primitiva [VERT ZOO] A membrane enclosing the central nervous system, as in cyclostomes and elasmobranchs.

meninx secundaria [VERT ZOO] A pigmented membrane surrounding the central nervous system of some lower vertebrates, such as Anura.

menisc-, menisco- [SCI TECH] A combining form denoting crescentic, sickle-shaped, semilunar; denoting meniscus, semilunar cartilage.

meniscectomy [MED] Surgical removal of a meniscus or semilunar cartilage.

meniscitis [MED] Inflammation of the semilunar cartilages.

Meniscotheriidae [PALEON] A family of extinct mammals of the order Condylarthra possessing selenodont teeth and molarized premolars.

meniscus [ANAT] A crescent-shaped body, especially an interarticular cartilage. [FL MECH] The free surface of a liquid which is near the walls of a vessel and which is curved because of surface tension.

Menispermaceae [BOT] A family of dicotyledonous woody vines in the order Ranunculales distinguished by mostly alternate, simple leaves, unisexual flowers, and a dioecious habit.

menometrorrhagia [MED] Excessive uterine bleeding during menstruation, plus irregular uterine bleeding at other times.

menopause [PHYSIO] The natural physiologic cessation of menstruation, usually occurring in the last half of the fourth decade. Also known as climacteric.

menoplania [MED] Bleeding during menstruation from a part of the body other than the uterus.

Menoponidae [INV ZOO] A family of biting lice (Mallophaga) adapted to life only upon domestic and sea birds.

menorrhagia [MED] Excessive bleeding during menstruation. Also known as hypermenorrhea.

menorrhalgia [MED] Pelvic pain occurring at the menstrual period.

menorrhea [PHYSIO] The normal flow of the menses. [MED] Excessive menstruation.

menostasis [MED] Suppression of the menstrual flow.

menostaxis [MED] Prolonged menstruation.

mensa [VERT ZOO] Grinding surface of a tooth.

menses *See* menstruation.

menstrual age [EMBRYO] The age of an embryo or fetus calculated from the first day of the mother's last normal menstruation preceding pregnancy.

menstrual cycle [PHYSIO] The periodic series of changes associated with menstruation and the intermenstrual cycle; menstrual bleeding indicates onset of the cycle.

menstrual period [PHYSIO] The time of menstruation.

menstruate [PHYSIO] To discharge the products of menstruation.

menstruation [PHYSIO] The periodic discharge of sanguineous fluid and sloughing of the uterine lining in women from puberty to the menopause. Also known as menses.

menstruum [MATER] A solvent, commonly one that extracts certain principles from entire plant or animal tissues.

mental [ANAT] Pertaining to the chin. Also known as genial. [PSYCH] **1.** Pertaining to the mind, psyche, or inner self. **2.** Pertaining to the intellectual or cognitive functions. **3.** Imaginary or unreal, as when a pain is said to be purely mental.

mental aberration [PSYCH] A departure from normal mental function.

mental adjustment [PSYCH] The act or process by which an individual adapts his attitudes, traits, or feelings to the social environment.

mental age [PSYCH] The degree of mental development of an individual in terms of the chronological age of the average individual of equivalent mental ability; specifically, a score derived from intelligence tests.

mental deficiency [PSYCH] A condition characterized by intellectual retardation, social inadequacy, and persistent dependency.

mental health [PSYCH] A relatively enduring state of being in which an individual has effected an integration of his instinctual drives in a way that is reasonably satisfying to himself as reflected in his zest for living and his feeling of self-realization.

mental hygiene [PSYCH] That branch of hygiene dealing with the preservation of mental and emotional health.

mental illness [PSYCH] Any form of mental aberration; usually refers to a chronic or prolonged disorder in which there are wide deviations from the normal.

mental retardation [PSYCH] An abnormal slowness of mental function and behavior patterns relative to age and development.

mental telepathy [PSYCH] A form of extrasensory perception in which one person is aware of an external event through direct sensory perception, and another person, not in the same place, also becomes aware of the event but not through direct sensory perception.

mentation [PSYCH] Mental activity.

Menthaceae [BOT] An equivalent name for Labiatae.

menton-philtrum [ANTHRO] A measure of the distance from the midpoint of the lower edge of the chin to the midpoint of the philtrum or vertical groove of the upper lip.

menton-supramentale [ANTHRO] The measurement of the distance taken from the midpoint of the lower edge of the chin to the supramentale, or the angle between the chin and the lower lip.

mentum [ANAT] The chin. [BOT] A projection formed by union of the sepals at the base of the column in some orchids. [INV ZOO] **1.** A projection between the mouth and foot in certain gastropods. **2.** The median or basal portion of the labium in insects.

Menurae [VERT ZOO] A small suborder of suboscine perching birds restricted to Australia, including the lyrebirds and scrubbirds.

Menuridae [VERT ZOO] The lyrebirds, a family of birds in the suborder Menurae notable for their vocal mimicry.

meprobromate [PHARM] $C_9H_{18}N_2O_4$ The compound 2-methyl-2-*n*-propyl-1,3-propanediol dicarbamate, a tranquilizer with anticonvulsant, muscle relaxant, and sedative actions.

mer-, mero- [SCI TECH] A combining form meaning part or partial.

meralluride [PHARM] $C_9H_{16}HgN_2O_6$ A diuretic consisting of succinamic acid and theophylline, in approximately molecular proportions, administered as the sodium derivative.

Meramecian [GEOL] A North American provincial series of geologic time: Upper Mississippian (above Osagian, below Chesterian).

meraspis [PALEON] Advanced larva of a trilobite; stage in which the pygidium begins to form.

merbromin [ORG CHEM] $C_{20}H_8O_6Na_2Br_2Hg$ A green crystalline powder that gives a deep-red solution in water; used as an antiseptic. [PHARM] The disodium salt of 2′,7′,-dibromo-4′-(hydroxymercuri)-fluorescein, an organomercurial antibacterial agent applied topically.

mercapt-, mercapto- [CHEM] A combining form denoting the presence of the thiol (SH) group.

mercaptoacetic acid *See* thioglycollic acid.

mercurialism [MED] Chronic type of mercury poisoning. Also known as hydrargyrism.

mercurial nephrosis [MED] Nephrosis caused by poisoning with mercury bichloride.

mercurial tremor [MED] A fine muscular tremor observed in persons with mercurialism or poisoning by other heavy metals.

Mercurochrome [PHARM] A trademark for merbromin.

mercury [CHEM] A metallic element, symbol Hg, atomic number 80, atomic weight 200.59, existing at room temperature as a silvery, heavy liquid. Also known as quicksilver.

merganser [VERT ZOO] Any of several species of diving water fowl composing a distinct subfamily of Anatidae and characterized by a serrate bill adapted for catching fish.

mericarp [BOT] An individual, one-seeded carpel of a schizocarp.

meridional canal [INV ZOO] A canal into which adradial canals open in Ctenophora.

meridional furrow [EMBRYO] A longitudinal groove extending between the poles of a segmenting egg.

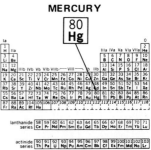

Periodic table of the chemical elements showing the position of mercury.

Meridosternata [INV ZOO] A suborder of echinoderms including various deep-sea forms of sea urchins.

meristele [BOT] A unit of vascular tissue in a polystele; supplies the leaf.

meristem [BOT] Formative plant tissue composed of undifferentiated cells capable of dividing and giving rise to other meristematic cells as well as to specialized cell types; found in growth areas.

meristematic ring [BOT] The layer of meristematic tissue between the cortex and the pith which gives rise to vascular tissues.

meristic [BIOL] Pertaining to a change in number or in geometric relation of parts of an organism. [ZOO] Of, pertaining to, or divided into segments.

Merkel's corpuscles [ANAT] Touch receptors consisting of flattened platelets at the tips of certain cutaneous nerves.

mermaid's purse [VERT ZOO] The horny or leathery egg envelope of elasmobranchs.

Mermithidae [INV ZOO] A family of filiform nematodes in the superfamily Mermithoidea; only juveniles are parasitic.

Mermithoidea [INV ZOO] A superfamily of nematodes composed of two families, both of which are invertebrate parasites.

meroblast [INV ZOO] The stage between the schizont and the merozoite in some Sporozoa.

meroblastic [EMBRYO] Of or pertaining to an ovum that undergoes incomplete cleavage due to large amounts of yolk.

merocerite [INV ZOO] In crustaceans, the fourth antennal segment.

merocrine [PHYSIO] Pertaining to glands in which the secretory cells undergo cytological changes without loss of cytoplasm during secretion.

merognathite [INV ZOO] In crustaceans, the fourth segment of the gnathite.

merogony [EMBRYO] The normal or abnormal development of a part of an egg following cutting, shaking, or centrifugation of the egg before or after fertilization.

meromorphosis [BIOL] Regeneration of a part that is smaller than the one lost.

meromyarian [INV ZOO] Having few muscle cells in each quadrant as seen in cross section; applied especially to nematodes.

meromyosin [BIOCHEM] Protein fragments of a myosin molecule, produced by enzymatic digestion.

meron [INV ZOO] **1.** In insects, the posterior part of the coxa. **2.** In dipterans, the sclerite between the middle and hind coxae, or proximal to the hind coxa.

meront [INV ZOO] The product of schizogony.

Meropidae [VERT ZOO] The bee-eaters, a family of brightly colored Old World birds in the order Coraciiformes.

meroplankton [BIOL] Plankton composed of floating developmental stages (that is, eggs and larvae) of the benthos and nekton organisms. Also known as temporary plankton.

meropodite [INV ZOO] **1.** In Crustacea, the fourth segment of a thoracic appendage. **2.** The femur of spiders.

merosporangium [MYCOL] A process on the apex of a sporangiophore which produces a row of spores, as in certain Mucorales.

Merostomata [INV ZOO] A class of primitive arthropods of the subphylum Chelicerata distinguished by their aquatic mode of life and the possession of abdominal appendages which bear respiratory organs; only three living species are known.

Merozoa [INV ZOO] The equivalent name for Cestoda.

merozoite [INV ZOO] An ameboid trophozoite in some sporozoans produced from a schizont by schizogony.

merozoon [INV ZOO] A piece of a unicellular animal that contains part of the macronucleus; formed by artificial division, that is, fragmentation.

merozygote [GEN] A zygote which received only part of the genetic material from one of the parents (typically in bacteria).

merrythought [VERT ZOO] In birds, the furcula, formed by fusion of the clavicles; the "wishbone."

merthiolate [PHARM] A trademark for thimerosal.

merycism *See* rumination.

Merycoidodontidae [PALEON] A family of extinct tylopod ruminants in the superfamily Merycoidodontoidea.

Merycoidodontoidea [PALEON] A superfamily of extinct ruminant mammals in the infraorder Tylopoda which were exceptionally successful in North America.

Mesacanthidae [PALEON] An extinct family of primitive acanthodian fishes in the order Acanthodiformes distinguished by a pair of small intermediate spines, large scales, superficially placed fin spines, and a short branchial region.

mesarch [BOT] Having metaxylem on both sides of the protoxylem in a siphonostele. [ECOL] Originating in a mesic environment.

mesarteritis [MED] Inflammation of the tunica media of an artery.

mescal buttons [BOT] The dried tops from the cactus *Lophophora williamsii*; capable of producing inebriation and hallucinations.

mescaline [ORG CHEM] $C_{11}H_{17}NO_3$ The alkaloid 3,4,5-trimethoxyphenethylamine, found in mescal buttons; produces unusual psychic effects and visual hallucinations.

mesectoderm [EMBRYO] The portion of the mesenchyme arising from ectoderm.

mesencephalon [EMBRYO] The middle portion of the embryonic vertebrate brain; gives rise to the cerebral peduncles and the tectum. Also known as midbrain.

mesenchymal cell [HISTOL] An undifferentiated cell found in mesenchyme and capable of differentiating into various specialized connective tissues.

mesenchymal epithelium [HISTOL] A layer of squamous epithelial cells lining subdural, subarachnoid, and perilymphatic spaces, and the chambers of the eyeball.

mesenchymal hyalin [PATH] A form of hyalin which results from degeneration or necrosis of nonepithelial tissue, usually of muscle, as in Zenker's hyaline necrosis, or of blood vessels.

mesenchymal tissue [EMBRYO] Undifferentiated tissue composed of branching cells embedded in a coagulable fluid matrix.

mesenchyme [EMBRYO] That part of the mesoderm from which all connective tissues, blood vessels, blood, lymphatic system proper, and the heart are derived.

mesenchymoma [MED] A tumor composed of cells resembling those of embryonic mesenchyme, or of mesenchyme with its derivatives.

mesendoderm [EMBRYO] Embryonic tissue which differentiates into mesoderm and endoderm.

mesenteric [ANAT] Of or pertaining to the mesentery.

mesenteric artery [ANAT] Either of two main arterial branches arising from the abdominal aorta: the inferior, supplying the descending colon and the rectum, and the superior, supplying the small intestine, the cecum, and the ascending and transverse colon.

mesenteric lymphadenitis [MED] Inflammation of the lymph nodes in the mesentery.

mesenteron [EMBRYO] *See* midgut. [INV ZOO] Central gastric cavity in an actinozoan.

mesentery [ANAT] A fold of the peritoneum that connects the intestine with the posterior abdominal wall.

mesentoderm [EMBRYO] **1.** The entodermal portion of the mesoderm. **2.** Undifferentiated tissue from which entoderm and mesoderm are derived. **3.** That part of the mesoderm which gives rise to certain structures of the digestive tract.

mesepimeron [INV ZOO] In insects, the epimeron of the mesothorax.

mesepisternum [INV ZOO] The sclerite below the anterior spiracle in Diptera.

mesethmoid [ANAT] A bone or cartilage in the center of the ethmoid region of the vertebrate skull; usually constitutes the greater portion of the nasal septum.

mesic [ECOL] **1.** Of or pertaining to a habitat characterized by a moderate amount of water. **2.** Of or pertaining to a mesophyte.

mesmerism [PSYCH] Hypnotism induced by animal magnetism, a supposed force passing from operator to man.

mesoappendix [ANAT] The mesentery of the vermiform appendix.

mesobenthos [OCEANOGR] Of or pertaining to the sea bottom at depths of 180–900 meters (100–500 fathoms).

mesobilirubin [BIOCHEM] $C_{33}H_{40}O_6N_4$ Yellow, crystalline by-product of bilirubin reduction.

mesobilirubinogen [BIOCHEM] $C_{33}H_{44}O_6N_4$ Colorless, crystalline by-product of bilirubin reduction; may be converted to urobilin, stercobilinogen, or stercobilin.

mesobiliverdin [BIOCHEM] $C_{28}H_{38}O_6N_4$ A structural isomer of phycoerythrin and phycocyanobilin released by certain biliprotein by treatment with alkali.

mesoblast [EMBRYO] Undifferentiated mesoderm of the middle layer of the embryo.

mesoblastema *See* mesoderm.

mesobronchus [VERT ZOO] In birds, the primary trunk of the bronchus.

mesocardium [ANAT] Epicardium covering the blood vessels which enter and leave the heart. [EMBRYO] The mesentery supporting the embryonic heart.

mesocarp [BOT] The middle layer of the pericarp.

mesocecum [ANAT] The mesentery sometimes connecting the cecum with the right iliac fossa.

mesocercaria [INV ZOO] The developmental stage in the second intermediate host of *Alaria*, a digenetic trematode.

mesochilium [BOT] In orchids, the middle part of the labellum.

mesocoel [ANAT] The ventricle of the mesencephalon.

mesocolon [ANAT] The part of the mesentery that is attached to the colon.

mesoconch [ANTHRO] Having moderately rounded orbits with an orbital index of 83 to 89.

mesocotyl [BOT] The portion of the axis between the cotyledon and the coleoptile in grass seedlings.

mesocranial [ANTHRO] Having a medium-sized skull with a cranial index of 75–80.

mesocycle [BOT] The layer of tissue between the xylem and the phloem of a monostelic stem.

mesoderm [EMBRYO] The third germ layer, lying between the ectoderm and endoderm; gives rise to the connective tissues, muscles, urogenital system, vascular system, and the epithelial lining of the coelom. Also known as mesoblastema.

mesodermal tumor [MED] A tumor composed of cells normally derived from the mesoderm.

mesodont [ZOO] Having medium-sized teeth.

MESOBILIVERDIN

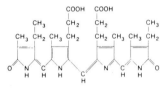

Structural formula for mesobiliverdin.

mesogaster [ANAT] The mesentery of the stomach.

mesogastrium [ANAT] The middle abdominal region. [EMBRYO] The embryonic mesentery by which the stomach is attached to the dorsal abdominal wall.

Mesogastropoda [INV ZOO] The equivalent name for Pectinibranchia.

mesoglea [INV ZOO] The gelatinous layer between the ectoderm and endoderm in coelenterates and certain sponges.

mesognathic [ANTHRO] Designating a condition of the upper jaw in which it has a mild degree of anterior projection with respect to the profile of the facial skeleton, when the skull is oriented on the Frankfort horizontal plane; having a gnathic index of 98.0 to 102.9.

mesoinositol *See* myoinositol.

mesolamella [INV ZOO] A thin gelatinous membrane between the epidermis and gastrodermis in hydrozoans.

mesomere [EMBRYO] The muscle-plate region between the epimere and hypomere in vertebrates.

mesometrium [ANAT] The part of the broad ligament attached directly to the uterus.

mesomorph [PSYCH] A somatotype characterized by an athletic physique.

mesonephric duct [EMBRYO] The efferent duct of the mesonephros. Also known as Wolffian duct.

mesonephric fold *See* mesonephric ridge.

mesonephric ridge [EMBRYO] A fold of the dorsal wall of the coelom lateral to the mesentery formed by development of the mesonephros. Also known as mesonephric fold.

mesonephroma [MED] Any of several benign or malignant tumors of the genital tract thought to be derived from mesonephros. Also known as teratoid adenocystoma.

mesonephroma ovarii [MED] A malignant ovarian mesonephroma.

mesonephros [EMBRYO] One of the middle of three pairs of embryonic renal structures in vertebrates; persists in adult fish and is replaced by the metanephros in higher forms.

mesonotum [INV ZOO] In insects, the dorsal portion of the mesothorax.

Mesonychidae [PALEON] A family of extinct mammals of the order Condylarthra.

mesopetalum [BOT] The labellum of an orchid.

mesophile [BIOL] An organism, as certain bacteria, that grows at moderate temperature.

mesophily [ECOL] Physiological response of organisms living in environments with moderate temperatures and a fairly high, constant amount of moisture.

mesophragma [INV ZOO] **1.** A phragma of the mesothorax, descending from the postscutellum, in certain insects. **2.** In some crustaceans, a process of the endosternum that forms an arch over the sternal canal.

mesophyll [BOT] Parenchymatous tissue between the upper and lower epidermal layers in foliage leaves.

mesophyte [ECOL] A plant requiring moderate amounts of moisture for optimum growth.

mesoplankton [ECOL] Plankton found at depths below the zone of penetration of photosynthetically effective light.

mesoplastron [VERT ZOO] One of a pair of plates, one on each side, between the hyoplastron and the hypoplastron of certain turtles.

mesopleurite [INV ZOO] In Diptera, a lateral sclerite of the mesothorax.

mesopleuron [INV ZOO] A pleuron of the mesothorax in insects.

mesopodium [BOT] The petiole region of a leaf. [INV ZOO] The middle part of the foot in mollusks.

mesopore [PALEON] A tube paralleling the autopore or chamber in fossil bryozoans.

mesoprescutum [INV ZOO] The prescutum of the mesothorax in insects.

mesopterygium [VERT ZOO] The middle one of three basal cartilages in the pectoral fin of sharks and rays.

mesopterygoid [VERT ZOO] **1.** Pertaining to a process of the pterygoid bone in birds. **2.** Pertaining to a part in teleosts which articulates anteriorly with the palatine bone, posteriorly with the metapterygoid bone, and laterally with the pterygoid bone. **3.** The space between the pterygoids in mammals.

mesoptic vision [PHYSIO] Vision in which the human eye's spectral sensitivity is changing from the photoptic state to the scotoptic state.

mesoptile [VERT ZOO] A down feather of the second set in birds having two sets; it follows the protoptile and precedes the metoptile or teleoptile.

mesorchium [EMBRYO] The mesentery that supports the embryonic testis in vertebrates.

mesorrhine [ANTHRO] Having a nose of moderate size: nasal index is 47–51 on the skull and 70–85 on the living person.

mesosalpinx [ANAT] The portion of the broad ligament forming the mesentery of the uterine tube.

Mesosauria [PALEON] An order of extinct aquatic reptiles known from a single genus, *Mesosaurus,* characterized by a long snout, numerous slender teeth, small forelimbs, and webbed hindfeet.

mesoscutellum [INV ZOO] The scutellum of the mesothorax in insects.

mesoscutum [INV ZOO] The scutum of the mesothorax in insects.

mesosere [ECOL] A sere originating in a mesic habitat and characterized by mesophytes.

mesosoma [INV ZOO] **1.** The anterior portion of the abdomen in certain arthropods. **2.** The middle of the body of some invertebrates, especially when the phylogenetic segmentation pattern cannot be determined.

mesosome [MICROBIO] An extension of the cell membrane within a bacterial cell; possibly involved in cross-wall formation, cell division, and the attachment of daughter chromosomes following deoxyribonucleic acid replication.

mesosperm [BOT] The integument that encloses the nucellus of the ovule.

mesospore [MYCOL] Also known as mesosporium. **1.** A one-celled teleutospore in certain rust fungi. **2.** The middle coat of a spore having three coats.

mesosporium *See* mesospore.

mesosternum [ANAT] The middle portion of the sternum in vertebrates. Also known as gladiolus. [INV ZOO] The ventral portion of the mesothorax in insects.

Mesostigmata [INV ZOO] The mites, a suborder of the Acarina characterized by a single pair of breathing pores (stigmata) located laterally in the middle of the idiosoma between the second and third, or third and fourth, legs.

Mesosuchia [PALEON] A suborder of extinct crocodiles of the Late Jurassic and Early Cretaceous.

Mesotaeniaceae [BOT] The saccoderm desmids, a family of fresh-water algae in the order Conjugales; cells are oval, cylindrical, or rectangular and have simple, undecorated walls in one piece.

Mesotardigrada [INV ZOO] An order of tardigrades which combines certain echiniscoidean features with eutardigradan characters.

mesotarsus [INV ZOO] The tarsus of a middle limb of insects.

MESOSAURIA

Restoration of *Mesosaurus.*
(*After McGregor*)

mesotergum [INV ZOO] Median arched portion or axis of the body of a trilobite.

mesotheca [INV ZOO] The middle lamina of bifoliate bryozoan colonies.

mesothecium [BOT] The middle layer of the investment of an anther sac.

mesothelioma [MED] A primary benign or malignant tumor composed of cells resembling the mesothelium.

mesothelium [ANAT] The simple squamous-cell epithelium lining the pleural, pericardial, peritoneal, and scrotal cavities. [EMBRYO] The lining of the wall of the primitive body cavity situated between the somatopleure and splanchnopleure.

mesotherm [ECOL] A plant that grows successfully at moderate temperatures.

mesothorax [INV ZOO] The middle of three somites composing the thorax in insects.

mesovarium [ANAT] A fold of the peritoneum that connects the ovary with the broad ligament.

Mesoveliidae [INV ZOO] The water treaders, a small family of hemipteran insects in the subdivision Amphibicorisae having well-developed ocelli.

mesoxyalyurea *See* alloxan.

Mesozoa [INV ZOO] A division of the animal kingdom sometimes ranked intermediate between the Protozoa and the Metazoa; composed of two orders of small parasitic, wormlike organisms.

Mesozoic [GEOL] A geologic era from the end of the Paleozoic to the beginning of the Cenozoic; commonly referred to as the Age of Reptiles.

mesozooid [INV ZOO] A type of bryozoan heterozooid that produces slender tubes (mesozooecia or mesopores), internally subdivided by many closely spaced diaphragms, that open as tiny polygonal apertures.

mesquite [BOT] Any plant of the genus *Prosopis*, especially *P. juliflora*, a spiny tree or shrub bearing sugar-rich pods; an important livestock feed.

messenger ribonucleic acid [BIOCHEM] A linear sequence of nucleotides which is transcribed from and complementary to a single strand of deoxyribonucleic acid and which carries the information for protein synthesis to the ribosomes. Abbreviated m-RNA.

metabiosis [ECOL] An ecological association in which one organism precedes and prepares a suitable environment for a second organism.

metabolic [PHYSIO] Of or pertaining to metabolism.

metabolic disorder [MED] Any disorder that involves an alteration in the normal metabolism of carbohydrates, lipids, proteins, water, and nucleic acids; evidenced by various syndromes and diseases.

metabolism [PHYSIO] The physical and chemical processes by which foodstuffs are synthesized into complex elements (assimilation, anabolism), complex substances are transformed into simple ones (disassimilation, catabolism), and energy is made available for use by an organism.

metabolite [BIOCHEM] A product of intermediary metabolism.

metabolize [PHYSIO] To transform by metabolism; to subject to metabolism.

metacarpus [ANAT] The portion of a hand or forefoot between the carpus and the phalanges.

metacentric [CYTOL] Having the centromere near the middle of the chromosome.

metacercaria [INV ZOO] Encysted cercaria of digenetic trematodes; the infective form.

MESOZOA

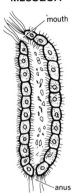

Salinella, in longitudinal section. *(After Frenzel, 1892)*

MESOZOIC

PRECAMBRIAN		
CAMBRIAN		
ORDOVICIAN		
SILURIAN		PALEOZOIC
DEVONIAN		
Mississippian	CARBONIFEROUS	
Pennsylvanian		
PERMIAN		
TRIASSIC		
JURASSIC		MESOZOIC
CRETACEOUS		
TERTIARY		CENOZOIC
QUATERNARY		

Chart showing the position of the Mesozoic era in relation to the other eras and to the periods of geologic time.

metacestode [INV ZOO] Encysted larva of a tapeworm; occurs in the intermediate host.

Metachlamydeae [BOT] An artificial group of flowering plants, division Magnoliophyta, recognized in the Englerian system of classification; consists of families of dicotyledons in which petals are characteristically fused, forming a sympetalous corolla.

metachromasia [CHEM] **1.** The property exhibited by certain pure dyestuffs, chiefly basic dyes, of coloring certain tissue elements in a different color, usually of a shorter wavelength absorption maximum, than most other tissue elements. **2.** The assumption of different colors or shades by different substances when stained by the same dye. Also known as metachromatism.

metachromatic granules [CYTOL] Granules which assume a color different from that of the dye used to stain them.

metachromatic leukodystrophy [MED] A hereditary degenerative disease transmitted as an autosomal recessive, due to sulfatase A deficiency, with excess accumulation of sulfated lipids responsible for metachromasia in various tissues. Abbreviated MLD. Also known as sulfatide lipidosis.

metachromatic stain [MATER] A stain which changes apparent color when absorbed by certain cell constituents.

metachromatism *See* metachromasia.

metachronous [CYTOL] Pertaining to the consecutive beating of adjacent cilia, resulting in a wavelike motion.

metachrosis [VERT ZOO] The ability of some animals to change color by the expansion and contraction of chromatophores.

metacneme [INV ZOO] A secondary mesentery in many zoantharians.

metacoele [EMBRYO] The coelom proper, developing in the lateral plate of mesoderm.

metacone [VERT ZOO] **1.** The posterior of three cusps of primitive upper molars. **2.** The posteroexternal cusp of an upper molar in higher vertebrates, especially mammals.

metaconid [VERT ZOO] The posteroexternal cusp of a lower molar in mammals; corresponds with the metacone.

metaconule [VERT ZOO] A posterior secondary cusp on an upper molar.

Metacopina [PALEON] An extinct suborder of ostracods in the order Podocopida.

metacoracoid [VERT ZOO] The posterior portion of the coracoid.

metacromion [ANAT] The posterior process of the acromion process.

metadiscoidal [EMBRYO] Referring to a placenta in which the villi are scattered at first but later become restricted to a disc, as in humans and monkeys.

metagenesis [BIOL] The phenomenon in which one generation of certain plants and animals reproduces asexually, followed by a sexually reproducing generation. Also known as alternation of generations.

metagnathous [INV ZOO] Having mouthparts specialized for biting in larvae and for sucking in the adults, as certain insects. [VERT ZOO] Having the points of the beak crossed, as crossbills.

metagranulocyte *See* metamyelocyte.

metakaryocyte *See* normoblast.

metal fume fever [MED] A febrile influenzalike occupational disorder following the inhalation of finely divided particles and fumes of metallic oxides. Also known as brass chills; brass founder's ague; galvo; Monday fever; metal ague; polymer fume fever; spelter shakes; teflon shakes; zinc chills.

metallic bond [PHYS CHEM] The type of chemical bond that is present in all metals, and may be thought of as resulting from a sea of valence electrons which are free to move throughout the metal lattice.

metalloporphyrin [BIOCHEM] A compound, such as heme, consisting of a porphyrin combined with a metal such as iron, copper, silver, zinc, or magnesium.

metalloprotein [BIOCHEM] A protein enzyme containing a metallic atom as an inherent portion of its molecule.

metaloph [VERT ZOO] Posterior crest of a molar, joining the metacone, the metaconule, and the hypocone.

metamer [ORG CHEM] One of two or more chemical compounds that exhibits isomerism with the others.

metamere [ZOO] One of the linearly arranged similar segments of the body of metameric animals. Also known as somite.

metamerism [ZOO] The condition of an animal body characterized by the repetition of similar segments (metameres), exhibited especially by arthropods, annelids, and vertebrates in early embryonic stages and in certain specialized adult structures. Also known as segmentation.

metamorphosis [BIOL] **1.** A structural transformation. **2.** A marked structural change in an animal during postembryonic development. [MED] A degenerative change in tissue or organ structure.

metamyelocyte [HISTOL] A granulocytic cell intermediate in development between the myelocyte and granular leukocyte; characterized by a full complement of cytoplasmic granules and a bean-shaped nucleus. Also known as juvenile cell; metagranulocyte.

metanauplius [INV ZOO] A primitive larval stage of certain decapod crustaceans characterized by seven pairs of appendages; follows the nauplius stage.

metanephridium [INV ZOO] A type of nephridium consisting of a tubular structure lined with cilia which opens into the coelomic cavity.

metanephrine [BIOCHEM] An inactive metabolite of epinephrine (3-*O*-methylepinephrine) that is excreted in the urine; it is recovered and measured as a test for pheochromocytoma.

metanephros [EMBRYO] One of the posterior of three pairs of vertebrate renal structures; persists as the definitive or permanent kidney in adult reptiles, birds, and mammals.

metanitricyte *See* normoblast.

metanotum [INV ZOO] The notum or tergum of the metathorax of an insect.

metaphase [CYTOL] **1.** The phase of mitosis during which centromeres are arranged on the equator of the spindle. **2.** The phase of the first meiotic division when centromeric regions of homologous chromosomes come to lie equidistant on either side of the equator.

metaphragma [INV ZOO] A phragma of the metathorax in insects.

metaphyseal aclasis *See* multiple hereditary exostoses.

metaphysis *See* epiphyseal plate.

Metaphyta [BIOL] A kingdom set up to include mosses, ferns, and other plants in some systems of classification.

metaplasia [PATH] Transformation of one form of tissue to another.

metaplasm [CYTOL] The ergastic substance of protoplasm.

metapleure [VERT ZOO] A fold of the abdominal integument in certain primitive Chordata.

metapleuron [INV ZOO] The pleuron of the metathorax in insects.

metapodeon [INV ZOO] That portion of the abdomen posterior to the podeon in insects.

metapodium [ANAT] **1.** The metatarsus in bipeds. **2.** The

METANAUPLIUS

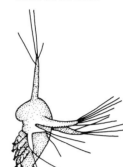

0.2 mm

Lateral view of metanauplius larva of penaeid shrimp. *(Smithsonian Institution)*

metatarsus and metacarpus in quadrupeds. [INV ZOO] Posterior portion of the foot of a mollusk.

metapodosoma [INV ZOO] Portion of the body bearing the third and fourth pairs of legs in Acarina.

metapophysis [VERT ZOO] A projection of the anterior articular process of a vertebra, especially in the lumbar region, in certain vertebrates.

metapostscutellum [INV ZOO] The postscutellum of the metathorax in insects.

metaprescutum [INV ZOO] The prescutum of the metathorax in insects.

metapterygoid [VERT ZOO] An element, as a bone, located behind the pterygoid in certain lower vertebrates.

metarteriole [ANAT] A blood vessel intermediate between an arteriole and a true capillary. Also known as precapillary.

metarubricyte *See* normoblast.

metascutellum [INV ZOO] The scutellum of the metathorax in insects.

metascutum [INV ZOO] The scutum of the metathorax in insects.

metaseptum [BIOL] A secondary septum.

metasicula [INV ZOO] The succeeding part of the sicula or colonial tube of graptolites.

metasoma [INV ZOO] The posterior region of the body of certain invertebrates, a term used especially when the phylogenetic segmentation pattern cannot be identified.

metastable ion [ANALY CHEM] In mass spectroscopy, an ion formed by a secondary dissociation process in the analyzer tube (formed after the parent or initial ion has passed through the accelerating field).

metastable phase [PHYS CHEM] Existence of a substance as either a liquid, solid, or vapor under conditions in which it is normally unstable in that state.

metastasis [MED] Transfer of the causal agent (cell or microorganism) of a disease from a primary focus to a distant one through the blood or lymphatic vessels.

metastasize [MED] To be transferred by metastasis.

metastatic anemia *See* myelophthisic anemia.

metasternum [INV ZOO] The ventral portion of the metathorax in insects.

metastoma [INV ZOO] Median plate posterior to the mouth in certain crustaceans and related arthropods.

Metastrongylidae [INV ZOO] A family of roundworms belonging to the Strongyloidea; species are parasitic in sheep, cattle, horses, dogs, and other domestic animals.

metatarsal [ANAT] Of or pertaining to the metatarsus.

metatarsalgia [MED] Tenderness and burning pain in the metatarsal region.

metatarsus [ANAT] The part of a foot or hindfoot between the tarsus and the phalanges.

Metatheria [VERT ZOO] An infraclass of therian mammals including a single order, the Marsupialia; distinguished by a small braincase, a total of 50 teeth, the inflected angular process of the mandible, and a pair of marsupial bones articulating with the pelvis.

metathorax [INV ZOO] Posterior segment of the thorax in insects.

metatroch [INV ZOO] A segmented larval form following the trochophore in annelids.

metatype [SYST] A topotype of the same species as the holotype.

metaxylem [BOT] Primary xylem differentiated after and distinguished from protoxylem by thicker tracheids and vessels with pitted or reticulated walls.

Metazoa [ZOO] The multicellular animals that make up the

METATROCH

Metatroch of planktonic sabellarian larva, showing early segmentation and long swimming setae. *(From O. Hartman, Polychaetous annelids, Paraonidae, Magelonidae, Longosomidae, Ctenodrilidae and Sabellariidae, Allan Hancock Pacific Expedition, 10(3):311–389, 1944)*

major portion of the animal kingdom; cells are organized in layers or groups as specialized tissues or organ systems.

metazoea [INV ZOO] The last zoea of certain decapod crustaceans; metamorphoses into a megalopa.

metencephalon [EMBRYO] The cephalic portion of the rhombencephalon; gives rise to the cerebellum and pons.

metenteron [INV ZOO] Any of the radial digestive chambers of a sea anemone or other coelenterate.

Meteoriaceae [BOT] A family of mosses in the order Isobryales in which the calyptra is frequently hairy.

metepimeron [INV ZOO] The epimeron of the metathorax in insects.

metepisternum [INV ZOO] The episternum of the metathorax in insects.

meter-kilogram-second system [MECH] A metric system of units in which length, mass, and time are fundamental quantities, and the units of these quantities are the meter, the kilogram, and the second respectively. Abbreviated mks system.

metestrus [PHYSIO] The beginning of the luteal phase following estrus.

methadone [PHARM] $C_{21}H_{27}NO$ The compound 6-(dimethylamino)-4,4-diphenyl-3-heptanone, a narcotic analgesic, administered in the hydrochloride form for maintenance treatment of heroin addiction.

methanogenesis [BIOCHEM] The biosynthesis of the hydrocarbon methane; common in certain bacteria. Also known as bacterial methanogenesis.

Methanomonadaceae [MICROBIO] A family of bacteria in the suborder Pseudomonadineae; members are gram-negative rods able to use carbon monoxide (*Carboxydomonas*), methane (*Methanomonas*), and hydrogen (*Hydrogenomonas*) as their sole source of energy for growth.

methemoglobin *See* ferrihemoglobin.

methemoglobinemia [MED] The presence of methemoglobin in the blood.

methemoglobinuria [MED] The presence of methemoglobin in the urine.

methionine [BIOCHEM] $C_5H_{11}O_2NS$ An essential amino acid; furnishes both labile methyl groups and sulfur necessary for normal metabolism.

metochy [ECOL] The relationship between a neutral guest insect and its host.

metopion [ANAT] The point of crossing of the line connecting the highest points of the frontal eminences with the sagittal plane.

metosteon [VERT ZOO] The ossified posterior portion of the sternum in birds.

metoxenous [ECOL] Being parasitic on different hosts at different stages in the life history.

metraterm [INV ZOO] The distal portion of the uterus in trematodes.

metria [MED] **1.** Any pathologic condition of the uterus. **2.** Any uterine inflammation occurring between childbirth and the 6 weeks following.

metric system [MECH] A system of units used in scientific work throughout the world and employed in general commercial transactions and engineering applications; its units of length, time, and mass are the meter, second, and kilogram respectively, or decimal multiples and submultiples thereof.

Metridiidae [INV ZOO] A family of zoantharian coelenterates in the order Actiniaria.

metriocranic [ANTHRO] Having a skull that is moderately high compared with its width, with a breadth-height index of 92 to 98.

METAZOEA

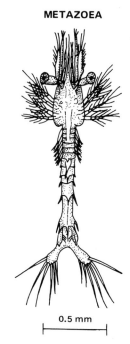

0.5 mm

Metazoea larva of a penaeid shrimp.

METHIONINE

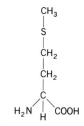

Structural formula of methionine.

MIACOIDEA

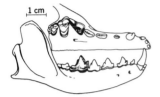

| 1 cm |

Right side of *Miacis* jaw shown in ventral (upper) and lateral (lower) views.

MICROAUTORADIOGRAPH

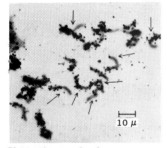

10 μ

Photomicrograph of an autoradiograph of *Vicia faba* chromosomes in metaphase labeled with tritiated thymidine. In the chromosome pairs indicated by the arrows, one chromosome is labeled but the other is not. *(Courtesy of W. L. Hughes)*

MICROCERCOUS CERCARIA

Drawing of a microcercous cercaria showing small tail. *(From R. M. Cable, An Illustrated Laboratory Manual of Parasitology, Burgess, 1958)*

metritis [MED] Inflammation of the uterus, usually involving both the endometrium and myometrium.

metrorrhagia [MED] Uterine bleeding during the intermenstrual cycle. Also known as intermenstrual flow; polymenorrhea.

metrorrhea [MED] Any pathologic discharge from the uterus.

metrorrhexis [MED] Rupture of the uterine wall.

metrosalpingitis [MED] Inflammation of the uterus and oviducts.

metrostaxis [MED] Slight, chronic bleeding from the uterus.

metula [MYCOL] A spore-bearing branch of a conidiophore having flask-shaped outgrowths.

Meyliidae [INV ZOO] A family of free-living nematodes in the superfamily Desmoscolecoidea.

Meziridae [INV ZOO] A family of hemipteran insects in the superfamily Aradoidea.

Mg *See* magnesium.

Miacidae [PALEON] The single, extinct family of the carnivoran superfamily Miacoidea.

Miacoidea [PALEON] A monofamilial superfamily of extinct carnivoran mammals; a stem group thought to represent the progenitors of the earliest member of modern carnivoran families.

micelle [MOL BIO] A submicroscopic structural unit of protoplasm built up from polymeric molecules.

Michaelis constant [BIOCHEM] A constant K_m such that the initial rate of reaction V, produced by an enzyme when the substrate concentration is high enough to saturate the enzyme, is related to the rate of reaction v at a lower substrate concentration c by the formula $V = v(1 + K_m/c)$.

micracanthopore [INV ZOO] Small, minute tubes projecting from the surface of bryozoan colonies.

micrencephaly [MED] The condition of having an abnormally small brain. Also spelled microencephaly.

microabscess [MED] A small abscess.

microaerophilic [MICROBIO] Pertaining to those microorganisms requiring free oxygen but in very low concentration for optimum growth.

microanatomy [ANAT] Anatomical study of microscopic tissue structures.

microaneurysm [MED] Dilation of the wall of a capillary, characteristic of certain disease entities.

microautoradiograph [GRAPHICS] An image which is produced by placing a specimen containing radioactive material (usually a radioactive tracer) in close contact with a photographic film and optically enlarging the developed image.

microbalance [ENG] A small, light type of analytical balance that can weigh loads of up to 0.1 gram to the nearest microgram.

microbe [MICROBIO] A microorganism, especially a bacterium of a pathogenic nature.

microbicide [MATER] An agent that kills microbes.

microbiology [BIOL] The science and study of microorganisms, including protozoans, algae, fungi, bacteria, viruses, and rickettsiae.

microbiota [BIOL] Microscopic flora and fauna.

microcentrum [CYTOL] The centrosome, or a group of centrosomes, functioning as the dynamic center of a cell.

microcephalus [MED] An individual with microcephaly.

microcephaly [MED] The condition of having an abnormally small head, with a circumference less than two standard deviations below the mean.

microceratous [INV ZOO] Having short antennae.

microcercous cercaria [INV ZOO] A cercaria with a very short broad tail.

microchaeta [INV ZOO] A small bristle on the body of some insects.

microchemistry [BIOCHEM] The chemistry of individual cells and minute organisms. [CHEM] The study of chemical reactions, using small quantities of materials, frequently less than 1 milligram or 1 milliliter, and often requiring special small apparatus and microscopical observation.

Microchiroptera [VERT ZOO] A suborder of the mammalian order Chiroptera composed of the insectivorous bats.

microcirculation [PHYSIO] The flow of blood or lymph in the vessels of the microcirculatory system.

microcneme [INV ZOO] Microsepta in certain anemones.

Micrococcaceae [MICROBIO] A family of spherical, gram-positive bacteria in the order Eubacteriales characterized by chemoorganotrophic energy metabolism.

microconsumer *See* decomposer.

Microcotyloidea [INV ZOO] A superfamily of ectoparasitic trematodes in the subclass Monogenea.

Microcyprini [VERT ZOO] The equivalent name for Cyprinodontiformes.

microcyst [MED] A very small cyst.

microcyte [MED] A red blood cell whose diameter or mean corpuscular volume or both are more than two standard deviations below the normal mean. Also known as micro-erythrocyte.

microcythemia [MED] Blood characterized by the presence of small red blood cells.

microcytic anemia [MED] Any form of anemia in which small erythrocytes occur in the blood.

microcytosis [MED] A blood disorder characterized by a preponderance of microcytes.

microdactyly [MED] A condition of abnormal smallness of fingers or toes.

microdissection [BIOL] Dissection under a microscope.

Microdomatacea [PALEON] An extinct superfamily of gastropod mollusks in the order Aspidobranchia.

microelectrode [ENG] An electrode small enough to record electrical activity in a single neuron or sensory cell.

microencephaly *See* micrencephaly.

microenvironment [ECOL] The specific environmental factors in a microhabitat.

microerythrocyte *See* microcyte.

microevolution [EVOL] **1.** Evolutionary processes resulting from the accumulation of minor changes over a relatively short period of time; evolutionary changes due to gene mutation. **2.** Evolution of species.

microfibril [MOL BIOL] The submicroscopic unit of a microscopic cellular fiber.

microfilaria [INV ZOO] Slender, motile prelarval forms of filarial nematodes measuring 150–300 micrometers in length; adult filaria are mammalian parasites.

microflora [BOT] Microscopic plants. [ECOL] The flora of a microhabitat.

microfluoroscope [ENG] A fluoroscope in which a very fine-grained fluorescent screen is optically enlarged.

microforge [ENG] In micromanipulation techniques, an optical-mechanical device for controlling the position of needles or pipets in the field of a low-power microscope by a simple micromanipulator.

microfossil [PALEON] A small fossil which is studied and identified by means of the microscope.

microgamete [BIOL] The smaller, or male gamete produced by heterogametic species.

microgametoblast [INV ZOO] The stage following the mi-

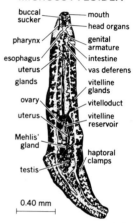

MICROCOTYLOIDEA

buccal sucker — mouth
— head organs
pharynx — genital armature
esophagus — intestine
uterus — vas deferens
glands — vitelline glands
ovary — vitelloduct
uterus — vitelline reservoir
Mehlis' gland
— haptoral clamps
testis

0.40 mm

Ventral view of *Heteraxinoides xanthophilis* (Hargis), an ectoparasite of the spot fish (*Leiostomus xanthurus*).

crogametocyte and preceding the microgamete in certain Sporozoa.

microgametocyte [BIOL] A cell that gives rise to microgametes.

microgametophyte [BOT] The male gametophyte produced from a microspore.

microgamy [BIOL] Sexual reproduction by fusion of the small male and female gametes in certain protozoans and algae.

microgastria [MED] A condition of abnormal smallness of the stomach.

microgenesis [BIOL] Abnormally small development of a part.

microgenitalism [MED] Having abnormally small genitalia.

microglia [HISTOL] Small neuroglia cells of the central nervous system having long processes and exhibiting ameboid and phagocytic activity under certain pathologic conditions.

microglossia [MED] A condition of abnormal smallness of the tongue.

micrognathia [MED] A condition of abnormal smallness of the jaws, particularly the mandible.

microgyria [MED] A condition of abnormal smallness of the gyri of the brain.

microhabitat [ECOL] A small, specialized, and effectively isolated location.

Microhylidae [VERT ZOO] A family of anuran amphibians in the suborder Diplasiocoela including many heavy-bodied forms with a pointed head and tiny mouth.

microincineration [CHEM] Reduction of small quantities of organic substances to ash by application of heat.

microinfarct [MED] A very small infarct.

microinjection [CYTOL] Injection of cells with solutions by using a micropipet.

microinvasion [MED] Invasion by tumor, especially a squamous-cell carcinoma of the uterine cervix, a very short distance into the tissues beneath the point of origin.

Microlepidoptera [INV ZOO] A former division of Lepidoptera.

microlith [MED] A calculus of microscopic size.

microlithiasis [MED] The presence of numerous microliths.

microlithiasis alveolaris pulmonum [MED] A rare form of pulmonary calcification of unidentified etiology in which microliths, and larger osseous nodules, are found.

Micromalthidae [INV ZOO] A family of coleopteran insects in the superfamily Cantharoidea; the single species is the telephone pole beetle.

micromania [PSYCH] A delusional state in which the patient believes himself diminutive in size and mentally inferior.

micromanipulation [BIOL] The techniques and practice of microdissection, microvivisection, microisolation, and microinjection.

micromanipulator [ENG] A device for holding and moving fine instruments for the manipulation of microscopic specimens under a microscope.

micromere [EMBRYO] A small blastomere of the upper or animal hemisphere in eggs that undergo uneven cleavage.

micromesentery [INV ZOO] A secondary, incomplete mesentery occurring in zoantharians.

micrometer [ENG] **1.** An instrument attached to a telescope or microscope for measuring small distances or angles. **2.** A caliper for making precise measurements; a spindle is moved by a screw thread so that it touches the object to be measured; the dimension can then be read on a scale. Also known as micrometer caliper. [MECH] A unit of length equal to one-millionth of a meter. Abbreviated μm. Also known as micron (μ).

micron   *See* micrometer.

micronekton   [ECOL] Active pelagic crustaceans and other forms intermediate between thrusting nekton and feebler-swimming plankton.

micront   [INV ZOO] A small cell produced by schizogony; gives rise to microgametes.

micronucleus   [INV ZOO] The smaller, reproductive nucleus in multinucleate protozoans.

micronutrient   [BIOCHEM] Trace elements and compounds required by living systems only in minute amounts.

microorganism   [MICROBIO] A microscopic organism, including bacteria, protozoans, yeast, viruses, and algae.

micropaleontology   [PALEON] A branch of paleontology that deals with the study of microfossils.

microparasite   [BIOL] A microscopic parasite.

Micropezidae   [INV ZOO] A family of myodarian cyclorrhaphous dipteran insects in the subsection Acalypteratae.

microphage   [HISTOL] A small phagocyte, especially a neutrophil.

microphagy   [BIOL] Feeding on minute organisms or particles.

microphakia   [MED] Congenital condition of abnormal smallness of the crystalline lens.

microphthalmus   [MED] A condition characterized by an abnormally small eyeball. Also known as nanophthalmus.

microphyll   [BOT] A relatively small leaf, according to an arbitrary classification of leaf sizes based on the blade area in millimeters.

microphyllous   [BOT] **1.** Having small leaves. **2.** Having leaves with a single, unbranched vein.

Microphysidae   [INV ZOO] A palearctic family of hemipteran insects in the subfamily Cimicimorpha.

microphyte   [ECOL] **1.** A microscopic plant. **2.** A dwarfed plant due to unfavorable environmental conditions.

microplankton   [BIOL] Microscopic plankton, somewhat larger than nanoplankton.

micropore   [INV ZOO] A small pore in the shell of a chiton which contains a sense organ.

microprobe   [SPECT] An instrument for chemical microanalysis of a sample, in which a beam of electrons is focused on an area less than a micrometer in diameter, and the characteristic x-rays emitted as a result are dispersed and analyzed in a crystal spectrometer to provide a qualitative and quantitative evaluation of chemical composition.

micropsia   [MED] A visual disturbance in which objects appear undersized.

micropterous   [INV ZOO] In some insects, having small hindwings that are hidden by the tegmina.   [VERT ZOO] Having small or rudimentary fins.

Micropterygidae   [INV ZOO] The single family of the lepidopteran superfamily Micropterygoidea; members are minute moths possessing toothed, functional mandibles and lacking a proboscis.

Micropterygoidea   [INV ZOO] A monofamilial superfamily of lepidopteran insects in the suborder Homoneura.

Micropygidae   [INV ZOO] A family of echinoderms in the order Diadematoida that includes only one genus, *Micropyga*, which has noncrenulate tubercles and umbrellalike outer tube feet.

micropyle   [BOT] A minute opening in the integument at the tip of an ovule through which the pollen tube commonly enters; persists in the seed as an opening or a scar between the hilum and point of radicle.

MICROSAURIA

Microbrachis, a microsaur from the Late Pennsylvanian epoch. *(After Steen)*

MICROSCLERE

Shapes of various microscleres.

microradiogram [PHYS] A two-dimensional x-ray image of a sample, produced by one type of x-ray microscope used in microradiography; all levels of the sample object are imaged into essentially a single focal plane for subsequent micropho-tographic enlargement.

microradiograph [GRAPHICS] An enlarged radiographic image on photographic film produced either by increasing the distance from specimen to photographic plate to secure inherent enlargement of divergent x-ray beams, or by optical enlargement of a developed image.

microrefractometry [OPTICS] The measurement of refractive indices of microscopic objects; this is often done by immers-ing an object in a series of mediums of graded refractive index until one is found that makes the object invisible in a phase-contrast microscope.

Microsauria [PALEON] An order of Carboniferous and early Permian lepospondylous amphibians.

microschizont [INV ZOO] A male schizont.

microsclere [INV ZOO] A minute sclerite in Porifera.

microscope [OPTICS] An instrument through which minute objects are enlarged by means of a lens or lens system; principal types include optical, electron, and x-ray.

microscope stage [OPTICS] The platform on which speci-mens are placed for microscopic examination.

microscopic [OPTICS] *See* microscopical. [SCI TECH] Of ex-tremely small size.

microscopical [OPTICS] Also known as microscopic. **1.** Of or pertaining to the microscope. **2.** Visible only under a microscope.

microscopical diagnosis [PATH] Identification of a disease by microscopic examination of specimens taken from the patient.

microscopist [SCI TECH] An individual skilled in the use of the microscope.

microscopy [OPTICS] The interpretive application of micro-scope magnification to the study of materials that cannot be properly seen by the unaided eye.

microseptum [INV ZOO] An incomplete or imperfect mesen-tery in zoantharians.

microsere [ECOL] A sere in a microhabitat.

microsome [CYTOL] **1.** A fragment of the endoplasmic re-ticulum. **2.** A minute granule of protoplasm.

microspecies [ECOL] A small, localized species population that is clearly differentiated from related forms. Also known as jordanon.

microspectrograph [SPECT] A microspectroscope provided with a photographic camera or other device for recording the spectrum.

microspectrophotometer [SPECT] A split-beam or double-beam spectrophotometer including a microscope for the localization of the object under study, and capable of carrying out spectral analyses within the dimensions of a single cell.

microspectroscope [SPECT] An instrument for analyzing the spectra of microscopic objects, such as living cells, in which light passing through the sample is focused by a compound microscope system, and both this light and the light which has passed through a reference sample are dispersed by a prism spectroscope, so that the spectra of both can be viewed simultaneously.

microsphere [INV ZOO] The initial chamber of a foraminifer-an.

Microsporaceae [BOT] A monogeneric family of green algae in the suborder Ulotrichineae; the chloroplast is a parietal network.

microsporangium [BOT] A sporangium bearing microspores.

microspore [BOT] The smaller spore of heterosporous plants; gives rise to the male gametophyte.

Microsporida [INV ZOO] The single order of the class Microsporidea.

Microsporidae [INV ZOO] The equivalent name for Sphaeriidae.

Microsporidea [INV ZOO] A class of Cnidospora characterized by the production of minute spores with a single intrasporal filament or one or two intracapsular filaments and a single sporoplasm; mainly intracellular parasites of arthropods and fishes.

microsporidiosis [VET MED] Infection with microsporidians.

microsporocyte [BOT] A microspore mother cell.

microsporophyll [BOT] A sporophyll bearing microsporangia.

microstylous [BOT] Having short styles, referring specifically to heterostylous flowers.

microsurgery [BIOL] Surgery on single cells by micromanipulation.

microteliospore [MYCOL] A spore produced by a microtelium.

microtelium [MYCOL] The sorus of a microcyclic rust fungi.

microtherm [ECOL] A plant requiring a mean annual temperature range of 0-14°C for optimum growth.

Microtinae [VERT ZOO] A subfamily of rodents in the family Muridae that includes lemmings and muskrats.

microtome [ENG] An instrument for cutting thin sections of tissues or other materials for microscopical examination.

microtomy [BIOL] Cutting of thin sections of specimens with a microtome.

Microtragulidae [PALEON] A group of saltatorial caenolistoid marsupials that appeared late in the Cenozoic and paralleled the small kangaroos of Australia.

microtrichia [INV ZOO] Small hairs on the integument of various insects, especially on the wings.

microtubule [CYTOL] One of the thin rods or tubes contained by certain cytoplasmic components, such as in cilia or the mitotic spindle; they may occur free or in bundles in the cytoplasm, where they help to maintain cell shape.

microvillus [CYTOL] One of the filiform processes that form a brush border on the surfaces of certain specialized cells, such as intestinal epithelium.

microwave [ELECTROMAG] An electromagnetic wave which has a wavelength between about 0.3 and 30 centimeters, corresponding to frequencies of 1-100 gigahertz; however, there are no sharp boundaries distinguishing microwaves from infrared and radio waves.

microwave spectroscope [SPECT] An instrument used to observe the intensity of microwave radiation emitted or absorbed by a substance as a function of frequency, wavelength, or some related variable.

microwave spectroscopy [SPECT] The methods and techniques of observing and the theory for interpreting the selective absorption and emission of microwaves at various frequencies by solids, liquids, and gases.

microwave spectrum [ELECTROMAG] The range of wavelengths or frequencies of electromagnetic radiation that are designated microwaves. [SPECT] A display, photograph, or plot of the intensity of microwave radiation emitted or absorbed by a substance as a function of frequency, wavelength, or some related variable.

microzooid [INV ZOO] A free-swimming individual produced by budding in Vorticella and other ciliated Protozoa.

microzoospermia [MED] A condition of abnormal smallness of sperm in the semen.

MICROTUBULE

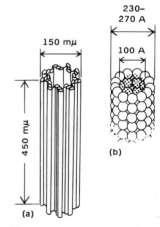

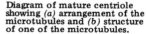

Diagram of mature centriole showing *(a)* arrangement of the microtubules and *(b)* structure of one of the microtubules.

micrurgy [SCI TECH] The art and science of using minute tools in a magnified field.

mictic [BIOL] **1.** Requiring or produced by sexual reproduction. **2.** Of or pertaining to eggs which without fertilization develop into males and with fertilization develop into amictic females, as occurs in rotifers.

micturition *See* urination.

midaxillary line [ANAT] A perpendicular line drawn downward from the apex of the axilla.

midbrain [ANAT] Those portions of the adult brain derived from the embryonic midbrain. [EMBRYO] The middle portion of the embryonic vertebrate brain. Also known as mesencephalon.

midclavicular line [ANAT] A vertical line parallel to and midway between the midsternal line and a vertical line drawn downward through the outer end of the clavicle.

Middle Cambrian [GEOL] The geologic epoch between Upper and Lower Cambrian, beginning approximately 540,000,-000 years ago.

Middle Cretaceous [GEOL] The geologic epoch between the Upper and Lower Cretaceous, beginning approximately 120,000,000 years ago.

Middle Devonian [GEOL] The geologic epoch between the Upper and Lower Devonian, beginning approximately 385,000,000 years ago.

middle ear [ANAT] The middle portion of the ear in higher vertebrates; in mammals it contains three ossicles and is separated from the external ear by the tympanic membrane and from the inner ear by the oval and round windows.

Middle Jurassic [GEOL] The geologic epoch between the Upper and Lower Jurassic, beginning approximately 170,000,000 years ago.

middle lobe syndrome [MED] A complex of symptoms due to enlarged lymph nodes compressing the bronchus of the right middle lobe, causing atelectasis, bronchiectasis, or chronic pneumonitis of the lobe.

Middle Mississippian [GEOL] The geologic epoch between the Upper and Lower Mississippian.

Middle Ordovician [GEOL] The geologic epoch between the Upper and Lower Ordovician, beginning approximately 460,000,000 years ago.

Middle Pennsylvania [GEOL] The geologic epoch between the Upper and Lower Pennsylvanian.

Middle Permian [GEOL] The geologic epoch between the Upper and Lower Permian, beginning approximately 260,-000,000 years ago.

Middle Silurian [GEOL] The geologic epoch between the Upper and Lower Silurian.

Middle Triassic [GEOL] The geologic epoch between the Upper and Lower Triassic, beginning approximately 215,-000,000 years ago.

midget [MED] An individual who is abnormally small, but otherwise normal.

midgut [EMBRYO] The middle portion of the digestive tube in vertebrate embryos. Also known as mesenteron. [INV ZOO] The mesodermal intermediate part of an invertebrate intestine.

midrib [BOT] The large central vein of a leaf.

migraine [MED] Recurrent paroxysmal vascular headache, commonly having unilateral onset and often associated with nausea and vomiting.

migrant [ZOO] An animal that moves from one habitat to another.

migration [VERT ZOO] Periodic movement of animals to new areas or habitats.

Mikulicz's disease [MED] Enlargement of salivary and lacrimal glands from any of various causes.

mildew [MYCOL] **1.** A whitish growth on plants, organic matter, and other materials caused by a parasitic fungus. **2.** Any fungus producing such growth.

mild mental retardation [PSYCH] Subnormal general intellectual functioning in which the intelligence quotient is approximately 52–67.

miliaria [MED] An acute inflammatory skin disease, the lesions consisting of vesicles and papules, which may be accompanied by a prickling or tingling sensation. Also known as heat rash; prickly heat.

Milichiidae [INV ZOO] A family of myodarian cyclorrhaphous dipteran insects in the subsection Acalypteratae.

milieu interieur [PHYSIO] The fundamental concept that the living organism exists in an aqueous internal environment which bathes all tissues and provides a medium for the elementary exchange of nutrients and waste.

milieu therapy [PSYCH] The treatment of mental disorder or maladjustment by making substantial changes in a patient's immediate life circumstances and environment in a way that will enhance the effectiveness of other forms of therapy. Also known as situation therapy.

Miliolacea [INV ZOO] A superfamily of marine or brackish foraminiferans in the suborder Miliolina characterized by an imperforate test wall of tiny, disordered calcite rhombs.

Miliolidae [INV ZOO] A family of foraminiferans in the superfamily Miliolacea.

Miliolina [INV ZOO] A suborder of the Foraminiferida characterized by a porcelaneous, imperforate calcite wall.

milk [PHYSIO] **1.** The whitish fluid secreted by the mammary gland for the nourishment of the young; composed of carbohydrates, proteins, fats, mineral salts, vitamins, and antibodies. **2.** Any whitish fluid in nature resembling milk, as coconut milk.

milk-alkali syndrome [MED] A complex of symptoms associated with prolonged excessive intake of milk and soluble alkali, including hypercalcemia, renal insufficiency, milk alkalosis, conjunctivitis, and calcinosis. Also known as Burnett's syndrome; milk-drinker's syndrome.

milk-drinker's syndrome *See* milk-alkali syndrome.

milk factor [BIOCHEM] A filtrable, noncellular agent in the milk and tissues of certain strains of inbred mice; transmitted from the mother to the offspring by nursing. Also known as Bittner milk factor.

milk fever [MED] A fever occurring during the first six weeks after childbirth, believed to be caused by puerperal infection.

milk intolerance [MED] Extreme sensitivity to milk due to allergy to milk protein or lactose deficiency; characterized by diarrhea, abdominal cramps, and vomiting.

milk leg *See* phlegmasia alba dolens.

milk line *See* mammary ridge.

Milkman's syndrome [MED] Decreased tubular reabsorption of phosphate, resulting in osteomalacia which gives a peculiar striped appearance (multiple pseudofractures) to the bones in roentgenograms. Also known as Looser-Milkman syndrome.

milk sugar *See* lactose.

milk teeth *See* deciduous teeth.

milkweed [BOT] Any of several latex-secreting plants of the genus *Asclepias* in the family Asclepiadaceae.

milky disease [INV ZOO] A bacterial disease of Japanese beetle larvae or related grubs caused by *Bacillus papilliae* and *B. lentimorbus* that penetrate the intestine and sporulate in the body cavity; blood of the grub eventually turns milky white.

Milleporina [INV ZOO] An order of the class Hydrozoa

MILIOLIDAE

Test representative of Miliolidae. *(From L. H. Hyman, The Invertebrates, vol. 1, McGraw-Hill, 1940)*

MILLEPORINA

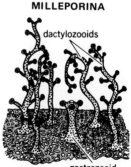

Polyps of Milleporina. *(From L. H. Hyman, The Invertebrates, vol. 1, McGraw-Hill, 1940)*

known as the stinging corals; they resemble true corals because of a calcareous exoskeleton.

Miller-Abbott tube [MED] A double-lumen rubber tube having a balloon at the end, inserted through the nasal passage and passed through the pylorus to locate and treat intestinal obstructions.

Miller indices [CRYSTAL] Three integers identifying a type of crystal plane; the intercepts of a plane on the three crystallographic axes are expressed as fractions of the crystal parameters; the reciprocals of these fractions, reduced to integral proportions, are the Miller indices. Also known as crystal indices.

Miller law [AGR] A law administered by the Food and Drug Administration that regulates the production and use of agricultural fungicides in the United States, and will not allow materials to leave poisonous residues on edible crops.

millet [BOT] A common name applied to at least five related members of the grass family grown for their edible seeds.

milli-micro- *See* nano-.

millimicron [MECH] A unit of length equal to one-thousandth of a micron (or micrometer) or one-billionth of a meter. Abbreviated mμ.

millipede [INV ZOO] The common name for members of the arthropod class Diplopoda.

Milroy's disease [MED] Familial chronic lymphedema of the lower extremities.

milt [PHYSIO] The secretion of the testis of fishes. [VERT ZOO] The testis of fishes in breeding condition when it is filled with secretion.

mimetic [ZOO] Pertaining to or exhibiting mimicry.

mimicry [ZOO] Assumption of color, form, or behavior patterns by one species of another species, for camouflage and protection.

Mimidae [VERT ZOO] The mockingbirds, a family of the Oscines in the order Passeriformes.

Mimosoideae [BOT] A subfamily of the legume family, Leguminosae; members are largely woody and tropical or subtropical with regular flowers and usually numerous stamens.

min *See* minim.

mind [PSYCH] **1.** The sum total of the neural processes which receive, code, and interpret sensations, recall and correlate stored information, and act on it. **2.** The state of consciousness. **3.** The understanding, reasoning, and intellectual faculties and processes considered as a whole. **4.** The psyche, or the conscious, subconscious, and unconscious considered together.

mineral dressing *See* beneficiation.

mineralocorticoid [BIOCHEM] A steroid hormone secreted by the adrenal cortex that regulates mineral metabolism and, secondarily, fluid balance.

minim [MECH] A unit of volume in the apothecaries' measure; equals $\frac{1}{60}$ fluidram (approximately 0.061612 cubic centimeter) or about 1 drop (of water). Abbreviated min.

minimal brain dysfunction syndrome [MED] A complex of learning and behavioral disabilities seen primarily in children of near-average or above-average intelligence exhibiting also deviations of function of the central nervous system.

minimum frontal diameter [ANTHRO] The smallest distance measured between the temporal crests, with only moderate pressure.

minimum lethal dose [PHARM] The amount of an injurious agent which is the average of the smallest dose that kills and the largest dose that fails to kill.

mink [VERT ZOO] Any of three species of slender-bodied

MINK

The American mink *(Mustela vison)*, the largest mink. It is valued for its dark, soft, thick fur.

aquatic carnivorous mammals in the genus *Mustela* of the family Mustelidae.

Minnesota multiphasic personality inventory [PSYCH] An empirical scale of an individual's personality based mainly on his own yes-or-no responses to a questionnaire of 550 items; included are special validating scales which measure the individual's test-taking attitude and degree of frankness. Abbreviated MMPI. Also known as multiphasic personality inventory (MPI).

Minnesota preschool scale [PSYCH] A verbal and nonverbal test designed to measure the learning ability of children from 18 months to 6 years of age.

minnow [VERT ZOO] The common name for any fresh-water fish composing the family Cyprinidae, order Cypriniformes.

minor surgery [MED] Any superficial surgical procedure involving little hazard to the life of the patient and not requiring general anesthesia.

Miocene [GEOL] A geologic epoch of the Tertiary period, extending from the end of the Oligocene to the beginning of the Pliocene.

Miosireninae [PALEON] A subfamily of extinct sirenian mammals in the family Dugongidae.

miosis [PHYSIO] Contraction of the pupil of the eye.

miotic [PHARM] **1.** Causing miosis. **2.** Any agent that causes miosis. [PHYSIO] Of or pertaining to miosis.

miracidium [INV ZOO] The ciliated first larva of a digenetic trematode; forms a sporocyst after penetrating intermediate host tissues.

Mirapinnatoidei [VERT ZOO] A suborder of tiny oceanic fishes in the order Cetomimiformes.

Miridae [INV ZOO] The largest family of the Hemiptera; included in the Cimicomorpha, it contains herbivorous and predacious plant bugs which lack ocelli and have a cuneus and four-segmented antennae.

Miripinnati [VERT ZOO] The equivalent name for Marapinnatoidei.

miscarriage [MED] Expulsion of the fetus before it is viable.

miscegenation [ANTHRO] Intermarriage between different races.

misogamy [PSYCH] A feeling of revulsion and repugnance toward marriage.

misogyny [PSYCH] Hatred of women.

misology [PSYCH] Unreasoning aversion to intellectual or literary matters, or to argument or speaking.

misoneism [PSYCH] Hatred of new circumstances or things, or change.

misopedia [PSYCH] Morbid hatred of all children, but especially of one's own.

missed abortion [MED] A condition in which a fetus weighing less than 500 grams dies and remains in the uterus for an extended period of time.

missense codon [GEN] A mutant codon that directs the incorporation of a different amino acid and results in the synthesis of a protein with a sequence in which one amino acid has been replaced by a different one; in some cases the mutant protein may be unstable or less active.

missense mutation [MOL BIO] A mutation that converts a codon coding for one amino acid to a codon coding for another amino acid.

Mississippian [GEOL] A large division of late Paleozoic geologic time, after the Devonian and before the Pennsylvanian, named for a succession of highly fossiliferous marine strata consisting largely of limestones found along the Mississippi River between southeastern Iowa and southern Illinois;

MINNOW

The common shiner *(Notropis cornutus)* which ranges eastward from the Rocky Mountains.

MIOCENE

Position of the Miocene epoch in relation to other epochs and to the periods and eras of geologic time.

MISSISSIPPIAN

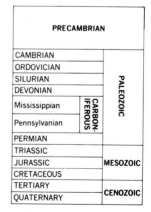

Chart showing position of the Mississippian in geologic time.

approximately equivalent to the European Lower Carboniferous.

Missourian [GEOL] A North American provincial series of geologic time: lower Upper Pennsylvanian (above Desmoinesian, below Virgilian).

mistletoe [BOT] **1.** *Viscum album.* The true, Old World mistletoe having dichotomously branching stems, thick leathery leaves, and waxy-white berries. **2.** Any of several species of green hemiparasitic plants of the family Loranthaceae.

Mitchell's disease *See* erythromelalgia.

mite [INV ZOO] The common name for the acarine arthropods composing the diverse suborders Onychopalpida, Mesostigmata, Trombidiformes, and Sarcoptiformes.

miticide [MATER] An agent that kills mites. Also known as acaricide.

mitochondria [CYTOL] Minute cytoplasmic organelles in the form of spherical granules, short rods, or long filaments found in almost all living cells; submicroscopic structure consists of an external membrane system.

mitogenesis [CYTOL] **1.** Induction of mitosis. **2.** Formation as a result of mitosis.

mitogenetic [CYTOL] Inducing or stimulating mitosis.

mitomycin [MICROBIO] A complex of three antibiotics (mitomycin A, B, and C) produced by *Streptomyces caespitosus.*

mitosis [CYTOL] Nuclear division involving exact duplication and separation of the chromosome threads so that each of the two daughter nuclei carries a chromosome complement identical to that of the parent nucleus.

mitosome [CYTOL] A threadlike structure in the cytoplasm of secondary spermatocytes, said to arise from the spindle fibers and to form the middle piece and tail envelope of a spermatozoon.

mitotic cycle [CYTOL] The five phases—interphase or resting phase, prophase, metaphase, anaphase, and telophase—a cell goes through when cell division occurs.

mitotic index [CYTOL] The number of cells undergoing mitosis per thousand cells.

mitotic inhibitor [CYTOL] A compound that inhibits mitosis.

mitotic poison [CYTOL] A compound that prevents or affects the completion of mitosis.

mitotic recombination [GEN] Recombination occurring at mitosis, usually as a rare accident; it takes place by means of two rare processes so far identified: somatic crossing-over and chromosome misdistribution.

mitra [BOT] A helmet-shaped part of the calyx or the corolla. [MYCOL] A thick, round pileus of certain fungi, as mushrooms.

mitral cell [ANAT] A large nerve cell in the olfactory bulb.

mitral commissurotomy [MED] Any of several surgical procedures performed to relieve mitral stenosis.

mitral stenosis [MED] Obstruction of the mitral valve, usually due to narrowing of the orifice.

mitral valve [ANAT] The atrioventricular valve on the left side of the heart.

mitriform [BIOL] Shaped like a miter.

mittelschmerz [MED] Pain or discomfort in the lower abdomen in women occurring midway in the intermenstrual interval, thought to be secondary to the irritation of the pelvic peritoneum by fluid or blood escaping from the point of ovulation in the ovary.

mixed aphasia [PSYCH] A combination of two or more forms of aphasia with impairment or loss of language function.

mixed arthritis [MED] A combination of features of rheumatoid arthritis and degenerative joint disease seen in the same patient or the same joint.

MISTLETOE

Mistletoe *(Phoradendron flavescens).*

MITOCHONDRIA

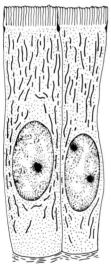

Drawing of intestinal epithelial cells showing numerous filamentous mitochondria concentrated in the apical cytoplasm.

mixed cryoglobulin [BIOCHEM] A cryoglobulin with a monoclonal component made of immunoglobulin belonging to two different classes, one of which is monoclonal.

mixed forest [FOR] A forest consisting of two or more types of trees, with no more than 80% of the most common tree.

mixed gland [PHYSIO] A gland that secretes more than one substance, especially a gland containing both mucous and serous components.

mixed laterality [PSYCH] The tendency, when there is a choice, to prefer to use parts of one side of the body for certain tasks and parts of the opposite side for others.

mixed nerve [PHYSIO] A nerve containing both sensory and motor components.

mixipterygium [VERT ZOO] The clasper of elasmobranchs.

Mixodectidae [PALEON] A family of extinct insectivores assigned to the Proteutheria; a superficially rodentlike group confined to the Paleocene of North America.

mixolimnion [HYD] The upper layer of a meromictic lake, characterized by low density and free circulation; this layer is mixed by the wind.

mixotrophic [BIOL] Obtaining nutrition by combining autotrophic and heterotrophic mechanisms.

mixture [PHARM] A liquid medicine prepared by adding insoluble substances to a liquid medium, usually with a suspending agent. [SCI TECH] The product of mixing; components are not in a fixed proportion to each other.

mks system *See* meter-kilogram-second system.

MLD *See* metachromatic leukodystrophy.

MMPI *See* Minnesota multiphasic personality inventory.

Mn *See* manganese.

mnemonic [PSYCH] **1.** Aiding or pertaining to memory. **2.** A device, such as combinations of letters, pictures, or words, to stimulate recall of the facts they represent.

mnemonics [PSYCH] The science of the cultivation of memory functions using systematic methods.

Mnesarchaeidae [INV ZOO] A family of lepidopteran insects in the suborder Homoneura; members are confined to New Zealand.

Mo *See* molybdenum.

Mobilina [INV ZOO] A suborder of ciliophoran protozoans in the order Peritrichida.

Mobulidae [VERT ZOO] The devil rays, a family of batoids that are surface feeders and live mostly on plankton.

modifier gene [GEN] A gene that alters the phenotypic expression of a nonallelic gene.

modiolus [ANAT] The central axis of the cochlea.

Moeller-Barlow disease *See* infantile scurvy.

Moeritheriidae [PALEON] The single family of the extinct order Moeritherioidea.

Moeritherioidea [PALEON] A suborder of extinct sirenian mammals considered as primitive proboscideans by some authorities and as a sirenian offshoot by others.

moiety [CHEM] A part or portion of a molecule, generally complex, having a characteristic chemical or pharmacological property.

moist-heat sterilization [ENG] Sterilization with steam under pressure, as in an autoclave, pressure cooker, or retort; most bacteriological media are sterilized by autoclaving at 121°C, with 15 pounds pressure, for 20 minutes or more.

molality [CHEM] Concentration given as moles per 1000 grams of solvent.

molal solution [CHEM] Concentration of a solution expressed in moles of solute divided by 1000 grams of solvent.

molar [ANAT] **1.** A tooth adapted for grinding. **2.** Any of the three pairs of cheek teeth behind the premolars on each side of the jaws in man.

molarity [CHEM] Measure of the number of gram-molecular weights of a compound present (dissolved) in 1 liter of solution; it is indicated by M, preceded by a number to show solute concentration.

molar solution [CHEM] Aqueous solution that contains 1 mole (gram-molecular weight) of solute in 1 liter of the solution.

molasses [FOOD ENG] A brown viscid syrup prepared from raw sugar during sugar manufacturing processes.

mold [MYCOL] Any of various woolly fungus growths. [PALEON] An impression made in rock or earth material by an inner or outer surface of a fossil shell or other organic structure; a complete mold would be the hollow space.

moldboard plow [AGR] A plow equipped with a curved iron plate (moldboard) that lifts and turns the soil. Also known as turnplow.

MOLE

The European common mole, with thick black fur, a reduced tail, and spadelike forelimbs.

mole [CHEM] An amount of substance of a system which contains as many elementary units as there are atoms of carbon in 0.012 kilogram of the pure nuclide carbon-12; the elementary unit must be specified and may be an atom, molecule, ion, electron, photon, or even a specified group of such units. [MED] **1.** A mass formed in the uterus by the maldevelopment of all or part of the embryo or of the placenta and membranes. **2.** A fleshy, pigmented nevus. [VERT ZOO] Any of 19 species of insectivorous mammals composing the family Talpidae; the body is stout and cylindrical, with a short neck, small or vestigial eyes and ears, a long naked muzzle, and forelimbs adapted for digging.

molecular adhesion [PHYS CHEM] A particular manifestation of intermolecular forces which causes solids or liquids to adhere to each other; usually used with reference to adhesion of two different materials, in contrast to cohesion.

molecular biology [BIOL] That part of biology which attempts to interpret biological events in terms of the physico-chemical properties of molecules in a cell.

molecular dipole [PHYS CHEM] A molecule having an electric dipole moment, whether it is permanent or produced by an external field.

molecular genetics [MOL BIO] The approach which deals with the physics and chemistry of the processes of inheritance.

molecular orbital [PHYS CHEM] A wave function describing an electron in a molecule.

molecular pathology [PATH] The study of the bases and mechanisms of disease on a molecular or chemical level.

molecular structure [PHYS CHEM] The manner in which electrons and nuclei interact to form a molecule, as elucidated by quantum mechanics and a study of molecular spectra.

molecular vibration [PHYS CHEM] The theory that all atoms within a molecule are in continuous motion, vibrating at definite frequencies specific to the molecular structure as a whole as well as to groups of atoms within the molecule; the basis of spectroscopic analysis.

molecular weight [CHEM] The sum of the atomic weights of all the atoms in a molecule.

molecule [CHEM] A group of atoms held together by chemical forces; the atoms in the molecule may be identical as in H_2, S_2, and S_8, or different as in H_2O and CO_2; a molecule is the smallest unit of matter which can exist by itself and retain all its chemical properties.

mole fraction [CHEM] The ratio of the number of moles of a substance in a mixture or solution to the total number of moles of all the components in the mixture or solution.

Molidae [VERT ZOO] A family of marine fishes, including some species of sunfishes, in the order Perciformes.

Mollicutes [MICROBIO] A class containing the single order Mycoplasmatales; name refers to the absence of a true cell wall and the plasticity of the outer membrane.

Moll's glands [ANAT] Modified sweat glands found in the eyelids. Also known as ciliary glands.

Mollusca [INV ZOO] One of the divisions of phyla of the animal kingdom containing snails, slugs, octopuses, squids, clams, mussels, and oysters; characterized by a shell-secreting organ, the mantle, and a radula, a food-rasping organ located in the forward area of the mouth.

molluscicide [MATER] An agent that kills mollusks.

molluscum contagiosum [MED] A viral disease of the skin, characterized by one or more discrete, waxy, dome-shaped nodules with frequent umbilication.

mollusk [INV ZOO] Any member of the Mollusca.

Molossidae [VERT ZOO] The free-tailed bats, a family of tropical and subtropical insectivorous mammals in the order Chiroptera.

Molpadida [INV ZOO] An order of sea cucumbers belonging to the Apodacea and characterized by a short, plump body bearing a taillike prolongation.

Molpadidae [INV ZOO] The single family of the echinoderm order Molpadida.

molt [PHYSIO] To shed an outer covering as part of a periodic process of growth.

molting hormone [BIOCHEM] Any of several hormones which activate molting in arthropods.

molybdenum [CHEM] A chemical element, symbol Mo, atomic number 42, and atomic weight 95.95.

moment of inertia [MECH] The sum of the products formed by multiplying the mass (or sometimes, the area) of each element of a figure by the square of its distance from a specified line. Also known as rotational inertia.

Momotidae [VERT ZOO] The motmots, a family of colorful New World birds in the order Coraciiformes.

monacanthid [INV ZOO] Having a single row of ambulacral spines, as in certain starfishes.

monactine [INV ZOO] A single-rayed spicule in the sponges.

monad [BIOL] One of a group of free-living individuals. [BOT] A single pollen grain.

monadelphous [BOT] Having the filaments of the stamens united into one set.

Monadidae [INV ZOO] A family of flagellated protozoans in the order Kinetoplastida having two flagella of uneven length.

monandrous [BOT] Having one stamen.

Mönckeberg's arteriosclerosis *See* medial arteriosclerosis.

Monday fever *See* metal fume fever.

Monera [BIOL] A kingdom that includes the bacteria and blue-green algae in some classification schemes.

monestrous [PHYSIO] Having a single estrous cycle per year.

Monge's disease *See* mountain sickness.

Mongolian spot [MED] A focal bluish-gray discoloration of the skin of the lower back, also aberrantly on the face, present at birth and fading gradually.

mongolism *See* Down's syndrome.

mongoloid [ANTHRO] Characteristic of a Far Eastern racial stock and often American Indians; features yellow complexion, a broad flat face with small nose and prominent cheekbones, and eyes that have an epicanthal fold. [MED] Having physical characteristics associated with Down's syndrome.

mongoose [VERT ZOO] The common name for 39 species of carnivorous mammals which are members of the family Viveridae; they are plantigrade animals and have a long

MOLYBDENUM

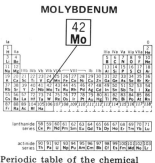

Periodic table of the chemical elements showing the position of molybdenum.

MONADELPHOUS

Drawing of a monadelphous stamen.

MONGOOSE

The golden-brown mongoose, a stout, weasellike animal which grows to a length of 15–18 inches (34–45 centimeters), excluding the tail.

slender body, short legs, nonretractile claws, and scent glands.

Monhysterida [INV ZOO] An order of aquatic nematodes in the subclass Chromadoria.

Monhysteroidea [INV ZOO] A superfamily of free-living nematodes in the order Monhysterida characterized by single or paired outstretched ovaries, circular to cryptospiral amphids, and a stoma which is usually shallow and unarmed.

Moniliaceae [MYCOL] A family of fungi in the order Moniliales; sporophores are usually lacking, but when present they are aggregated into fascicles, and hyphae and spores are hyaline or brightly colored.

Moniliales [MYCOL] An order of fungi of the Fungi Imperfecti containing many plant pathogens; asexual spores are always formed free on the surface of the material on which the organism is living, and never occur in either pycnidia or acervuli.

moniliasis *See* candidiasis.

monilicorn [INV ZOO] Having antennae which look like a chain of beads.

moniliform [BIOL] Constructed with contractions and expansions at regular alternating intervals, giving the appearance of a string of beads.

monimolimnion [HYD] The dense bottom stratum of a meromictic lake; it is stagnant and does not mix with the water above.

monimostylic [VERT ZOO] Having the quadrate united with the squamosal, and sometimes with other bones, as in certain reptiles.

monitor [VERT ZOO] Any of 27 carnivorous, voracious species of the reptilian family Varanidae characterized by a long, slender forked tongue and a dorsal covering of small, rounded scales containing pointed granules.

monkey [VERT ZOO] Any of several species of frugivorous and carnivorous primates which compose the families Cercopithecidae and Cebidae in the suborder Anthropoidea; the face is typically flattened and hairless, all species are pentadactyl, and the mammary glands are always in the pectoral region.

monoacid [CHEM] Compound with a single acid group, such as hydrochloric acid, HCl, or 2-naphthol-7-sulfonic acid, $C_{10}H_6(OH)(SO_3H)$.

monoamine [ORG CHEM] An amine compound that has only one amino group.

monoamine oxidase [BIOCHEM] An enzyme that catalyzes oxidative deamination of monoamines.

monoamine oxidase inhibitor [PHARM] Any drug, such as isocarboxazid and tranylcypromine, that inhibits monoamine oxidase and therby leads to an accumulation of the amines on which the enzyme normally acts.

monobasic [CHEM] Pertaining to an acid with one displaceable hydrogen atom, such as hydrochloric acid, HCl.

monoblast [HISTOL] A motile cell of the spleen and bone marrow from which monocytes are derived.

monoblastic leukemia *See* acute monocytic leukemia.

Monoblepharidales [MYCOL] An order of aquatic fungi in the class Phycomycetes; distinguished by a mostly hyphal thallus and zoospores with one posterior flagellum.

Monobothrida [PALEON] An extinct order of monocyclic camerate crinoids.

monocardiogram *See* vectorcardiogram.

monocarpic [BOT] Referring to a plant that blooms only once and then dies. Also known as hypaxanthic.

monocarpous [BOT] Having a single ovary.

monochasium [BOT] A cymose inflorescence producing one branch on each main axis.

MONKEY

The spider monkey *(Ateles)*, with slender body, long prehensile tail, long muscular limbs, and no thumbs.

monochlamydous [BOT] Referring to flowers having only one set of floral envelopes, that is, either a calyx or a corolla.

monochromat [MED] An individual who suffers from total color blindness even at high light levels; such persons are typically deficient or lacking in cone receptors, so that their form vision is also poor.

monoclimax [ECOL] A climax community controlled primarily by one factor, as climate.

monoclinic [BOT] Having both stamens and pistils in the same flower.

monoclonal cryoglobulin [BIOCHEM] A cryoglobulin composed of immunoglobin with only one class or subclass of heavy and light chain.

monocondylar [VERT ZOO] Having only one occipital condyle, as the skull of reptiles and birds.

monocotyledon [BOT] Any plant of the class Liliopsida; all have a single cotyledon.

Monocotyledoneae [BOT] The equivalent name for Liliopsida.

monocrepid [INV ZOO] A desma formed by secondary deposits of silica on a monaxon.

monocular vision [MED] Sight with one eye.

Monocyathea [PALEON] A class of extinct parazoans in the phylum Archaeocyatha containing single-walled forms.

monocyte [HISTOL] A large (about 12 micrometers), agranular leukocyte with a relatively small, eccentric, oval or kidney-shaped nucleus.

monocytic angina *See* infectious mononucleosis.

monocytic leukemia [MED] A form of leukemia in which monocytic cells are predominant in the blood. Also known as myelomonocytic leukemia.

monocytoma [MED] A neoplasm composed principally of monocytes, usually anaplastic.

monocytopenia [MED] Reduction in the number of circulating monocytes per unit volume of blood to below the minimum normal levels.

monocytosis [MED] Increase in the number of circulating monocytes per unit volume of blood to above the maximum normal levels.

monodactylous [ZOO] Having a single digit or claw.

Monodellidae [INV ZOO] A monogeneric family of crustaceans in the order Thermosbaenacea distinguished by seven pairs of biramous pereiopods on thoracomeres 2-8, and by not having the telson united to the last pleonite.

monodelphic [VERT ZOO] **1.** Having a single genital tract, in the female. **2.** Having a single uterus.

monodont [VERT ZOO] Having only one, persistent tooth, as the male narwhal.

monoecious [BOT] **1.** Having both staminate and pistillate flowers on the same plant. **2.** Having archegonia and antheridia on different branches. [ZOO] Having male and female reproductive organs in the same individual. Also known as hermaphroditic.

Monoedidae [INV ZOO] An equivalent name for Colydiidae.

Monogenea [INV ZOO] A diverse subclass of the Trematoda which are principally ectoparasites of fishes; individuals have enlarged anterior and posterior holdfasts with paired suckers anteriorly and opisthaptors posteriorly.

Monogenoidea [INV ZOO] A class of the Trematoda in some systems of classification; equivalent to the Monogenea of other systems.

Monogonota [INV ZOO] An order of the class Rotifera, characterized by the presence of a single gonad in both males and females.

monogony [BIOL] Asexual reproduction.

MONOCYTE

nucleus

Diagrammatic representation of a monocyte.

MONODELLIDAE

Drawing of a male *Monodella halophila* Karaman showing the seven pairs of appendages and the characteristic telson. *(After S. Karaman)*

monogynous [BOT] Having only one pistil. [VERT ZOO] **1.** Having only one female in a colony. **2.** Consorting with only one female.

monohybrid [GEN] A hybrid individual heterozygous for one gene or a single character.

monoideism [PSYCH] A mental condition marked by the domination of a single idea; persistent and thorough preoccupation with one idea, but seldom an idea that is complete.

monolophous [INV ZOO] Referring to spicules characterized by one ray that is forked or branched like a crest.

monomerosomatous [INV ZOO] Having fused body segments, as in certain insects.

Monommidae [INV ZOO] A family of coleopteran insects in the superfamily Tenebrionoidea.

monomorphic [BIOL] Having or exhibiting only a single form.

mononuclear [CYTOL] Having only one nucleus.

mononucleosis [MED] Any of various conditions marked by an abnormal increase in monocytes in the peripheral blood.

monophagous [ZOO] Subsisting on a single kind of food. Also known as monotrophic.

Monophisthocotylea [INV ZOO] An order of the Monogenea in which the posthaptor is without discrete multiple suckers or clamps.

Monophlebinae [INV ZOO] A subfamily of the homopteran superfamily Coccoidea distinguished by a dorsal anus.

monophyletic [EVOL] Developed from a single common ancestral form.

monophyllous [BOT] **1.** Having a single leaf. **2.** Having a one-piece calyx.

monophyodont [VERT ZOO] Having only one set of teeth throughout life.

Monopisthocotylea [INV ZOO] An order of trematode worms in the subclass Pectobothridia.

Monoplacophora [INV ZOO] A group of shell-bearing mollusks represented by few living forms; considered to be a sixth class of mollusks.

monoplegia [MED] Paralysis involving a single limb, muscle, or group of muscles.

monoploid [GEN] **1.** Having only one set of chromosomes. **2.** Having the haploid number of chromosomes.

monopodial [BOT] Stem branching in which there are lateral shoots on a primary axis.

monopodium [BOT] A primary axis that continues to grow while giving off successive lateral branches.

Monopylina [INV ZOO] A suborder of radiolarian protozoans in the order Oculosida in which pores lie at one pole of a single-layered capsule.

monopyrenous [BOT] Having a single stone, as a fruit.

monorchid [ANAT] **1.** Having one testis. **2.** Having one testis descended into the scrotum.

Monorhina [VERT ZOO] The subclass of Agnatha that includes the jawless vertebrates with a single median nostril.

monosaccharide [BIOCHEM] A carbohydrate which cannot be hydrolyzed to a simpler carbohydrate; a polyhedric alcohol having reducing properties associated with an actual or potential aldehyde or ketone group; classified on the basis of the number of carbon atoms, as triose (3C), tetrose (4C), pentose (5C), and so on.

monosepalous [BOT] **1.** Having only one sepal. **2.** Having all the sepals united.

Monosigales [BOT] A botanical order equivalent to the Choanoflagellida in some systems of classification.

monosiphonous [BIOL] Having a single central tube, as in

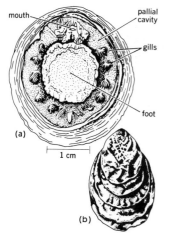

MONOPLACOPHORA

mouth · pallial cavity

gills

foot

(a)

1 cm

(b)

Living and fossil Monoplacophora. *(a) Neopilina galatheae* Lemche. *(b) Tryblidium reticulatum* Lindström. *(Adapted from R. C. Moore, ed., Treatise on Invertebrate Paleontology, pt. 1, 1957)*

the thallus of certain filamentous algae or the hydrocaulus of some hydrozoans.

monospermous [BOT] Having or producing one seed.

monosporangium [BOT] A sporangium producing monospores.

monospore [BOT] A simple or undivided nonmotile asexual spore; produced by the diploid generation of some algae.

monostelic [BOT] Having only one stele running through the entire axis.

monostome [INV ZOO] A cercaria having only one mouth or sucker.

Monostylifera [INV ZOO] A suborder of the Hoplonemertini characterized by a single stylet.

Monotomidae [INV ZOO] The equivalent name for Rhizophagidae.

Monotremata [VERT ZOO] The single order of the mammalian subclass Prototheria containing unusual mammallike reptiles, or quasi-mammals.

Monotropaceae [BOT] A family of dicotyledonous herbs or half shrubs in the order Ericales distinguished by a small, scarcely differentiated embryo without cotyledons, lack of chlorophyll, leaves reduced to scales, and anthers opening by longitudinal slits.

monotrophic *See* monophagous.

monotype [BIOL] A single type of organism that constitutes a species or genus.

monozygotic twins [BIOL] Twins which develop from a single fertilized ovum. Also known as identical twins.

mons [ANAT] An eminence.

mons pubis [ANAT] The eminence of the lower anterior abdominal wall above the superior rami of the pubic bones.

monster [MED] A congenitally malformed fetus which is incapable of properly performing the vital functions, or which exhibits marked structural differences from the normal.

Monstrilloida [INV ZOO] A suborder or order of microscopic crustaceans in the subclass Copepoda; adults lack a second antenna and mouthparts, and the digestive tract is vestigial.

mons veneris [ANAT] The mons pubis of the female.

montane [ECOL] Of, pertaining to, or being the biogeographic zone composed of moist, cool slopes below the timberline and having evergreen trees as the dominant lifeform.

Montgomery's tubercles [ANAT] Elevations in the areola of the nipple due to apocrine sweat glands; most prominent during pregnancy and lactation.

Montian [GEOL] A European stage of geologic time: Paleocene (above Danian, below Thanetian).

Monticellidae [INV ZOO] A family of tapeworms in the order Proteocephaloidea, in which some or all of the organs are in the cortical mesenchyme; catfish parasites.

monticulus [ANAT] The median dorsal portion of the cerebellum.

moor *See* bog.

moose [VERT ZOO] An even-toed ungulate of the genus *Alces* in the family Cervidae; characterized by spatulate antlers, long legs, a short tail, and a large head with prominent overhanging snout.

mor *See* ectohumus.

Moraceae [BOT] A family of dicotyledonous woody plants in the order Urticales characterized by two styles or style branches, anthers inflexed in the bud, and secretion of a milky juice.

moral idiocy [PSYCH] Inability to understand moral principles and values and to act in accordance with them, appar-

MONOSTOME

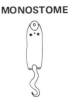

Drawing of monostome type of cercaria showing the one sucker. *(From R. M. Cable, An Illustrated Laboratory Manual of Parasitology, Burgess, 1958)*

MONOTROPACEAE

The Indian pipe *(Monotropa uniflora)*, a widespread North American and Eurasian species of the family Monotropaceae in the order Ericales. *(Photograph by Arthur Cronquist)*

MONSTRILLOIDA

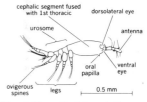

Monstrilla reticulata, adult female, shown in lateral view. *(From C. C. Davis, 1949)*

ently without impairment of the reasoning and intellectual faculties.

moral masochism [PSYCH] Masochism in which there is a need for punishment arising from unconscious sexual desires and reactivation of the Oedipus complex, characterized by self-destructive acts or the provocation of punishment from authority figures.

morbidity [MED] **1.** The quantity or state of being diseased. **2.** The conditions inducing disease. **3.** The ratio of the number of sick individuals to the total population of a community.

mordant [CHEM] An agent, such as alum, phenol, or aniline, that fixes dyes to tissues, cells, textiles, and other materials by combining with the dye to form an insoluble compound. Also known as dye mordant.

Mordellidae [INV ZOO] The tumbling flower beetles, a family of coleopteran insects in the superfamily Meloidea.

mores [ECOL] Groups of organisms preferring the same physical environment and having the same reproductive season.

morgue [MED] A place where dead bodies are held pending identification and disposition.

moribund [BIOL] **1.** In a dying or deathlike state. **2.** In a state of suspended life functions; dormant.

Moridae [VERT ZOO] A family of actinopterygian fishes in the order Gadiformes.

Morinae [VERT ZOO] The deep-sea cods, a subfamily of the Moridae.

Mormyridae [VERT ZOO] A large family of electrogenic fishes belonging to the Osteoglossiformes; African river and lake fishes characterized by small eyes, a slim caudal peduncle, and approximately equal dorsal and anal fins in most.

Mormyriformes [VERT ZOO] Formerly an order of fishes which are now assigned to the Osteoglossiformes.

morning sickness [MED] Morning nausea associated with early pregnancy.

moron [PSYCH] A mentally defective individual with a mental age between 7 and 12 years; a child with an IQ between 50 and 69.

Moro reflex [PHYSIO] The startle reflex observed in normal infants from birth through the first few months, consisting in abduction and extension of all extremities, followed by flexion and abduction of the extremities.

morphallaxis [PHYSIO] Regeneration whereby one part is transformed into another by reorganization of tissue fragments rather than by cell proliferation.

Morphinae [INV ZOO] A subfamily of large tropical butterflies in the family Nymphalidae.

morphine [PHARM] $C_{17}H_{19}NO_3 \cdot H_2O$ A white crystalline narcotic powder, melting point 254°C, an alkaloid obtained from opium; used in medicine in the form of a hydrochloride or sulfate salt.

morphinism [MED] **1.** The condition caused by the habitual use of morphine. **2.** The morphine habit.

morphogenesis [EMBRYO] The transformation involved in the growth and differentiation of cells and tissue. Also known as topogenesis.

morphology [BIOL] A branch of biology that deals with structure and form of an organism at any stage of its life history.

Morquio's syndrome [MED] A hereditary disease transmitted as an autosomal recessive and characterized by large quantities of keratosulfate in urine, dwarfism, and a typical facies with broad mouth, prominent maxilla, short nose, and widely spaced teeth. Also known as Brailsford-Morquio syndrome; familial osteochondrodystrophy; mucopolysaccharidosis IV.

mortality rate [MED] For a given period of time, the ratio of the number of deaths occurring per 1000 population. Also known as death rate.

morula [EMBRYO] A solid mass of blastomeres formed by cleavage of the eggs of many animals; precedes the blastula or gastrula, depending on the type of egg. [INV ZOO] A cluster of immature male gametes in which differentiation occurs outside the gonad; common in certain annelids.

Moruloidea [INV ZOO] The only class of the phylum Mesozoa; embryonic development in the organisms proceeds as far as the morula or stereoblastula stage.

mosaic [BIOL] An organism or part made up of tissues or cells exhibiting mosaicism. [EMBRYO] An egg in which the cytoplasm of early cleavage cells is of the type which determines its later fate. [SCI TECH] A surface pattern made by the assembly and arrangement of many small pieces.

mosaicism [GEN] The coexistence in an individual of somatic cells of genetically different types; it is caused by gene or chromosome mutations, especially nondisjunction, after fertilization, by double fertilization, or by fusion of embryos.

mosasaur [PALEON] Any reptile of the genus *Mosasaurus*; large, aquatic, fish-eating lizards from the Cretaceous which are related to the monitors but had paddle-shaped limbs.

Moschcowitz's disease *See* thrombotic thrombocytopenic purpura.

mosquito [INV ZOO] Any member of the dipterous subfamily Culicinae; a slender fragile insect, with long legs, a long slender abdomen, and narrow wings.

moss [BOT] Any plant of the class Bryatae, occurring in nearly all damp habitats except the ocean.

moss forest *See* temperate rainforest.

Motacillidae [VERT ZOO] The pipits, a family of passeriform birds in the suborder Oscines.

moth [INV ZOO] Any of various nocturnal or crepuscular insects belonging to the lepidopteran suborder Heteroneura; typically they differ from butterflies in having the antennae feathery and rarely clubbed, a stouter body, less brilliant coloration, and proportionally smaller wings.

motile [BIOL] Capable of spontaneous movement. [PSYCH] An individual characterized by motor-type mental imagery which takes the form of inner feelings of action.

motion sickness [MED] A complex of symptoms, including nausea, vertigo, and vomiting, occurring as the result of random multidirectional accelerations of a vehicle.

motivation [PSYCH] The comparatively spontaneous drive, force, or incentive, which partly determines the direction and strength of the response of a higher organism to a given situation; it arises out of the internal state of the organism.

motoneuron *See* motor neuron.

motor [PHYSIO] **1.** That which causes action or movement. **2.** Pertaining to efferent nerves which innervate muscles and glands.

motor alexia [MED] Inability to read aloud, while comprehension of the written word is preserved.

motor aphasia [MED] A form of aphasia in which the patient knows what he wishes to say but is unable to get the words out, and is able to perceive and comprehend both spoken and written language but is unable to repeat what he sees or hears; due principally to a brain lesion.

motor apraxia [MED] Inability to carry out, on command, a complex or skilled movement, though the purpose thereof is clear to the patient. Also known as kinesthetic apraxia; limb-kinetic apraxia.

motor area [ANAT] The ascending frontal gyrus containing nerve centers for voluntary movement; characterized by the

MOTH

The luna moth *(Tropaea luna).*

presence of Betz cells. Also known as Broadman's area 4; motor cortex; pyramidal area.

motor ataxia [MED] Inability to coordinate the muscles, which becomes apparent only on body movement.

motor cell [BOT] *See* bulliform cell. [PHYSIO] An efferent nerve cell in the anterior horn of the spinal cord.

motor cortex *See* motor area.

motor end plate [ANAT] A specialized area beneath the sarcolemma where functional contact is made between motor nerve fibers and muscle fibers.

motor nerve [PHYSIO] A nerve composed wholly or principally of motor fibers.

motor neuron [PHYSIO] An efferent nerve cell. Also known as motoneuron.

motor speech area [ANAT] The cortical area located in the triangular and opercular portions of the inferior frontal gyrus; in right-handed people it is more developed on the left side.

motor system [PHYSIO] Any portion of the nervous system that regulates and controls the contractile activity of muscle and the secretory activity of glands.

motor unit [ANAT] The axon of an anterior horn cell, or the motor fiber of a cranial nerve, together with the striated muscle fibers innervated by its terminal branches.

mountain lion *See* puma.

mountain sickness [MED] A disease occurring in persons living at high altitudes when homeostatic adjustments to the lowered atmospheric oxygen tension fail or develop disproportionately. Also known as high-altitude disease; high-altitude erythremia; Monge's disease; seroche.

mountain tick fever *See* Colorado tick fever.

mouse [VERT ZOO] Any of various rodents which are members of the families Muridae, Heteromyidae, Cricetidae, and Zapodidae; characterized by a pointed snout, short ears, and an elongated body with a long, slender, sparsely haired tail.

mouse deer *See* chevrotain.

mouth [ANAT] The oral or buccal cavity and its related structures.

mouth breadth [ANTHRO] The measure of the distance from one to the other corner of the mouth in a natural relaxed position.

mouth-to-mouth resuscitation [MED] A method of artificial respiration in which the rescuer's mouth is placed over the victim's mouth and air is blown forcefully into the victim's lungs every few seconds to inflate them.

moving-boundary electrophoresis [ANALY CHEM] A U-tube variation of electrophoresis analysis that uses buffered solution so that all ions of a given species move at the same rate to maintain a sharp, moving front (boundary).

mp *See* melting point.

MPI *See* Minnesota multiphasic personality inventory.

m-RNA *See* messenger ribonucleic acid.

MSH *See* melanocyte-stimulating hormone.

M stage [CYTOL] The stage of the mitotic cycle following G_2 and preceding G_1 in which mitosis actually takes place.

muc-, muci-, muco- [ZOO] A combining form denoting pertaining to mucus, mucin, mucosa.

Mucedinaceae [MYCOL] The equivalent name for Moniliaceae.

mucigen [BIOCHEM] A substance from which mucin is derived; contained in mucus-secreting epithelial cells.

mucin [BIOCHEM] A glycoprotein constituent of mucus and various other secretions of man and lower animals.

mucinosis [MED] Accumulations of materials containing mucin or mucinous substances in the skin; sometimes accompanied by papule and nodule formation.

MOUSE

The house mouse *(Mus musculus)*, usually gray with a long, sparsely haired tail, maximum overall length 7 inches (18 centimeters).

mucinous cyst *See* mucous cyst.

mucocutaneous [ANAT] Pertaining to a mucous membrane and the skin, and to the line where these join.

mucoid [BIOCHEM] Any of various glycoproteins, similar to mucins but differing in solubilities and precipitation properties and found in cartilage, in the crystalline lens, and in white of egg. **2.** Resembling mucus. [MICROBIO] Pertaining to large colonies of bacteria characterized by being moist and sticky.

mucolytic [BIOCHEM] Effecting the dissolution, liquefaction, or dispersion of mucus and mucopolysaccharides.

mucopolysaccharide [BIOCHEM] Any of a group of polysaccharides containing an amino sugar and uronic acid; a constituent of mucoproteins, glycoproteins, and blood-group substances.

mucopolysaccharidosis [MED] Any of several inborn metabolic disorders involving mucopolysaccharides; the six types are MPS I, Hurler's syndrome; MPS II, Hunter's syndrome; MPS III, Sanfillipo's syndrome; MPS IV, Morquio's syndrome; MPS V, Scheil's syndrome; and MPS VI, Maroteaux-Lamy's syndrome.

mucoprotein [BIOCHEM] Any of a group of glycoproteins containing a sugar, usually chondroitinsulfuric or mucoitinsulfuric acid, combined with amino acids or polypeptides.

mucopurulent [MED] Containing mucus and pus.

Mucorales [MYCOL] An order of terrestrial fungi in the class Phycomycetes, characterized by a hyphal thallus and nonmotile sporangiospores, or conidiospores.

mucormycosis [MED] An acute, usually fulminating fungus infection of man caused by several genera of Mucorales, including *Absidia, Rhizopus,* and *Mucor.*

mucosa [HISTOL] A mucous membrane.

mucosanguineous [MED] Containing mucus and blood.

mucoserous [PHYSIO] Containing or producing both mucus and serum.

mucous [PHYSIO] Of or pertaining to mucus; secreting mucus.

mucous cell [PHYSIO] A mucus-secreting cell.

mucous colitis *See* irritable colon.

mucous connective tissue [HISTOL] A type of loose connective tissue in which the ground substance is especially prominent and soft; occurs in the umbilical cord.

mucous cyst [MED] A retention cyst of a gland, containing a secretion rich in mucin. Also known as mucinous cyst.

mucous degeneration [MED] Any retrogressive change associated with abnormal production of mucus.

mucous gland [PHYSIO] A gland that secretes mucus.

mucous membrane [HISTOL] The type of membrane lining cavities and canals which have communication with air; it is kept moist by glandular secretions. Also known as tunica mucosa.

mucoviscidosis *See* cystic fibrosis.

mucro [BIOL] An abrupt, sharp terminal tip or process.

mucronate [BIOL] Terminated abruptly by a sharp terminal tip or process.

mucus [PHYSIO] A viscid fluid secreted by mucous glands, consisting of mucin, water, inorganic salts, epithelial cells, and leukocytes, held in suspension.

mudfish *See* bowfin.

mud puppy [VERT ZOO] Any of several American salamanders of the genera *Necturus* and *Proteus* making up the family Proteidae; distinguished by having both lungs and gills as an adult.

MUD PUPPY

Necturus maculosus, the most common species of mud puppy found in the eastern United States and Canada.

Mugilidae [VERT ZOO] The mullets, a family of perciform fishes in the suborder Mugiloidei.

Mugiloidei [VERT ZOO] A suborder of fishes in the order Perciformes; individuals are rather elongate, terete fishes with a short spinous dorsal fin that is well separated from the soft dorsal fin.

mulberry [BOT] Any of various trees of the genus *Morus* (family Moraceae), characterized by milky sap and simple, often lobed alternate leaves.

mule [VERT ZOO] The sterile hybrid offspring of the male ass and the mare, or female horse.

Müllerian duct *See* paramesonephric duct.

Müllerian duct cyst [MED] A congenital cyst arising from vestiges of the Müllerian ducts.

Müllerian mimicry [ZOO] Mimicry between two aposometic species.

Müller's larva [INV ZOO] The ciliated larva characteristic of various members of the Polycladida; resembles a modified ctenophore.

multi- [SCI TECH] A prefix meaning many.

multicellular [BIOL] Consisting of many cells.

multicipital [BIOL] Having many heads or branches arising from one point.

multifid [BIOL] Divided into many lobes.

multiglandular [ANAT] Of or pertaining to several glands.

Multillidae [INV ZOO] An economically important family of Hymenoptera; includes the cow killer, a parasite of bumblebee pupae.

multilocular [BIOL] Having many small chambers or vesicles.

multiphasic personality inventory *See* Minnesota multiphasic personality inventory.

multiple-anomaly syndrome [MED] Any syndrome associated with numerous congenital abnormalities.

multiple cancellous exostoses *See* multiple hereditary exostoses.

multiple cartilaginous exostoses *See* multiple hereditary exostoses.

multiple colloid goiter *See* adenomatous goiter.

multiple hereditary exostoses [MED] An inherited form of exostosis, revealing itself at several sites in childhood or adolescence. Also known as diaphyseal aclasis; hereditary deforming chondrodysplasia; metaphyseal aclasis; multiple cancellous exostoses; multiple cartilaginous exostoses.

multiple myeloma [MED] A primary bone malignancy characterized by diffuse osteoporosis, anemia, hyperglobulinemia, and other clinical features. Also known as Kahler's disease.

multiple neuritis *See* polyneuritis.

multiple neurofibroma *See* neurofibromatosis.

multiple neurofibromatosis *See* neurofibromatosis.

multiple personality [PSYCH] A personality capable of dissociation into several or many other personalities at the same time, whereby the delusion is entertained that the one person is many separate persons; a symptom in schizophrenic patients.

multiple pregnancy [MED] Being pregnant with more than one fetus.

multiple sclerosis [MED] A degenerative disease of the nervous system of unknown cause in which there is demyelination followed by gliosis.

multiple serositis *See* polyserositis.

Multituberculata [PALEON] The single order of the nominally mammalian suborder Allotheria; multituberculates had enlarged incisors, the coracoid bones were fused to the scapula, and the lower jaw consisted of the dentary bone alone.

MULTITUBERCULATA

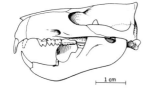

1 cm

Skull and jaw of *Ptilodus*, an early Tertiary multituberculate. *(After G. G. Simpson, 1937)*

multituberculate [VERT ZOO] Of teeth, having several or many simple conical cusps.

mummification [MED] **1.** Drying of a part of the body into a hard mass. **2.** Dry gangrene.

mumps [MED] An acute contagious viral disease characterized chiefly by painful enlargement of a parotid gland.

mumps orchitis [MED] Inflammation of the testis due to the mumps virus.

Munchausen syndrome [PSYCH] A personality disorder in which the patient describes dramatic but false symptoms or simulates acute illness, happily undergoing examinations, hospitalization, and diagnostic and therapeutic manipulations, and upon discovery of the real nature of his case often leaves without notice and moves on to another hospital.

mural thrombus [MED] A thrombus attached to the wall of a blood vessel or mural endocardium. Also known as lateral thrombus.

muramic acid [BIOCHEM] An organic acid found in the mucopeptide (murein) in the cell walls of bacteria and blue-green algae.

muramidase [BIOCHEM] Lysozyme when acting as an enzyme on the hydrolysis of the muramic acid-containing mucopeptide in the cell walls of some bacteria.

Murchisoniacea [PALEON] An extinct superfamily of gastropod mollusks in the order Prosobranchia.

murein [BIOCHEM] The peptidoglycan of bacterial cell walls.

Muricacea [INV ZOO] A superfamily of gastropod mollusks in the order Prosobranchia.

muricate [ZOO] Covered with sharp, hard points.

Muricidae [INV ZOO] A family of predatory gastropod mollusks in the order Neogastropoda; contains the rock snails.

Muridae [VERT ZOO] A large diverse family of relatively small cosmopolitan rodents; distinguished from closely related forms by the absence of cheek pouches.

muriform [BIOL] **1.** Resembling the arrangement of courses in a brick wall, especially having both horizontal and vertical septa. **2.** Pertaining to or resembling a rat or mouse.

Murinae [VERT ZOO] A subfamily of the Muridae which contains such forms as the striped mouse, house mouse, harvest mouse, and field mouse.

murine plague [VET MED] Infection of the rat by the bacterium *Pasteurella pestis;* transmitted from rat to rat and from rat to man by a flea.

murine typhus [MED] A relatively mild, acute, febrile illness of worldwide distribution caused by *Rickettsia mooseri,* transmitted from rats to man by the flea and characterized by headache, macular rash, and myalgia. Also known as endemic typhus; flea-borne typhus; rat typhus; shop typhus; urban typhus.

murmur [MED] A blowing or roaring heart sound heard through the wall of the chest.

Murray Valley encephalitis [MED] An acute inflammation of the brain and spinal cord caused by a virus; confined to Australia and New Guinea. Also known as Australian X disease.

Musaceae [BOT] A family of monocotyledonous plants in the order Zingiberales characterized by five functional stamens, unisexual flowers, spirally arranged leaves and bracts, and fleshy, indehiscent fruit.

muscarinism [MED] Poisoning due to ingestion of certain mushrooms.

Musci *See* Bryopsida.

Muscicapidae [VERT ZOO] A family of passeriform birds assigned to the Oscines; includes the Old World flycatchers or fantails.

MULTITUBERCULATE

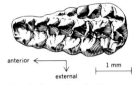

anterior ← external 1 mm

Occlusal view of a first upper molar of a multituberculate mammal.

MUSHROOM

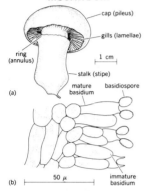

cap (pileus)

gills (lamellae)

ring (annulus)

1 cm

stalk (stipe)

(a)

mature basidiospore
basidium

50 μ

(b)

immature basidium

Agaricus bisporus, the common cultivated mushroom. *(a)* Basidiocarp. *(b)* Cross section of gill, showing basidia in various stages of development.

MUSK-OX

The musk-ox *(Ovibos moschatus)*; flattened horns, implanted low in the skull, are used defensively.

MUSKRAT

The muskrat *(Ondatra zibethica).*

Muscidae [INV ZOO] A family of myodarian cyclorrhaphous dipteran insects in the subsection Calypteratae; includes the houseflies, stable flies, and allies.

muscle [ANAT] A contractile organ composed of muscle tissue that changes in length and effects movement when stimulated. [HISTOL] A tissue composed of cells containing contractile fibers; three types are smooth, cardiac, and skeletal.

muscle banners [INV ZOO] Folds of mesogloea on the mesenteries of anthozoans; they support retractor muscles.

muscle fiber [HISTOL] The contractile cell or unit of which muscle is composed.

muscle hemoglobin *See* myoglobin.

muscular atrophy [MED] Degenerative reduction of muscle size, especially skeletal muscles, due to a lesion involving either the cell body or axon of the lower motor neuron.

muscular dystrophy [MED] A group of diseases characterized by degeneration of or injury to individual muscle cells, not primarily involving the nerve supply; the most common form is Duchenne-Greisinger disease.

muscularis externa [HISTOL] The layer of the digestive tube consisting of smooth muscles.

muscularis mucosae [HISTOL] Thin, deep layer of smooth muscle in some mucous membranes, as in the digestive tract.

muscular system [ANAT] The muscle cells, tissues, and organs that effect movement in all vertebrates.

musculoaponeurotic [HISTOL] Composed of muscle and of fibrous connective tissue in the form of a membrane.

musculocutaneous [ANAT] Of or pertaining to muscles and skin.

musculocutaneous nerve [ANAT] A branch of the brachial plexus with both motor and somatic sensory components; innervates flexor muscles of the upper arm, and skin of the lateral aspect of the forearm.

mushroom [MYCOL] 1. A fungus belonging to the basidiomycetous order Agaricales. 2. The fruiting body (basidiocarp) of such a fungus.

Musidoridae [INV ZOO] A family of orthorrhaphous dipteran insects in the series Brachycera distinguished by spear-shaped wings.

musk [PHYSIO] Any of various strong-smelling substances obtained from the musk glands of musk deer or similar animals; used in the form of a tincture as a fixative for perfume.

musk bag *See* musk gland.

muskeg [ECOL] A peat bog or tussock meadow, with variably woody vegetation.

musk gland [VERT ZOO] A large preputial scent gland of the musk deer and various other animals, including skunk and musk-ox. Also known as musk bag.

muskmelon [BOT] *Cucumis melo.* The edible, fleshy, globular to long-tapered fruit of a trailing annual plant of the order Violales; surface is uniform to broadly sutured to wrinkled, and smooth to heavily netted, and flesh is pale green to orange; varieties include cantaloupe, Honey Dew, Casaba, and Persian melons.

musk-ox [VERT ZOO] *Ovibos moschatus.* An even-toed ungulate which is a member of the mammalian family Bovidae; a heavy-set animal with a shag pilage, splayed feet, and flattened horns set low on the head.

muskrat [VERT ZOO] *Ondatra zibethica.* The largest member of the rodent subfamily Microtinae; essentially a water rat with a laterally flattened, long, naked tail, a broad blunt head with short ears, and short limbs.

Musophagidae [VERT ZOO] The turacos, an African family of birds of uncertain affinities usually included in the order Cuculiformes; resemble the cuckoos anatomically but have two unique pigments, turacin and turacoverdin.

mustard [BOT] Any of several annual crucifers belonging to the genus *Brassica* of the order Capparales; leaves are lyrately lobed, flowers are yellow, and pods have linear beaks; the mustards are cultivated for their pungent seed and edible foliage, and the seeds of *B. niger* are used as a condiment, prepared as a powder, paste, or oil.

Mustilidae [VERT ZOO] A large, diverse family of low-slung, long-bodied carnivorous mammals including minks, weasels, and badgers; distinguished by having only one molar in each upper jaw, and two at the most in the lower jaw.

mutagen [GEN] An agent that raises the frequency of mutation above the spontaneous rate.

mutant [GEN] An individual bearing an allele that has undergone mutation and is expressed in the phenotype.

mutarotation [CHEM] A change in the optical rotation of light that takes place in the solutions of freshly prepared sugars.

mutase [BIOCHEM] An enzyme able to catalyze a dismutation or a molecular rearrangement.

mutation [GEN] An abrupt change in the genotype of an organism, not resulting from recombination; genetic material may undergo qualitative or quantitative alteration, or rearrangement.

mutation hotspot [GEN] Any of the regions of the DNA sequence of a gene that mutate much more frequently than the rest.

mutation rate [GEN] The average frequency per gamete with which a mutation takes place.

mutator gene [GEN] A gene in which some mutant alleles increase the mutation rates of some or all other genes.

Mutillidae [INV ZOO] The velvet ants, a family of hymenopteran insects in the superfamily Scolioidea.

mutualism [ECOL] Mutual interactions between two species that are beneficial to both species.

myalgia [MED] Pain in the muscles.

myasthenia [MED] Muscular weakness.

myasthenia gravis [MED] A muscle disorder of unknown etiology characterized by varying degrees of weakness and excessive fatigability of voluntary muscle.

myasthenia reaction [MED] The electromyographic reaction observed in myasthenia gravis in which there is a gradual loss of intensity and duration for the tetanic contraction, and a gradual diminution in amplitude and frequency of motor unit discharges until the muscle is fatigued.

myasthenic crisis [MED] Profound myasthenia and respiratory paralysis associated with myasthenia gravis.

myatonia [MED] Lack of muscle tone.

Mycelia Sterilia [MYCOL] An order of fungi of the class Fungi Imperfecti distinguished by the lack of spores; certain members are plant pathogens.

mycelium [MYCOL] A mass of fungal filaments (hyphae) which may be septate and compose the vegetative body of a fungus.

Mycetaeidae [INV ZOO] The equivalent name for Endomychidae.

mycetocyte [INV ZOO] **1.** One of the cells clustered together to form a mycetome. **2.** An individual cell functioning like a mycetome.

mycetoma [MED] A chronic fungus or bacterial infection, usually of the feet, resulting in swelling. Also known as madura foot; maduromycosis.

mycetome [INV ZOO] One of the specialized structures in the body of certain insects for holding endosymbionts.

MUSTARD

Pod or silique (fruit) of a mustard plant.

MUTILLIDAE

Drawing of a velvet ant. *(From T. I. Storer and R. L. Usinger, General Zoology, 3d ed., McGraw-Hill, 1957)*

MYCORRHIZA

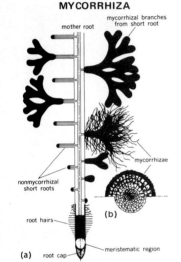

mycorrhizal branches from short root

mother root

mycorrhizae

nonmycorrhizal short roots

(b)

root hairs

(a)   root cap

meristematic region

Development of mycorrhizae on a pine root. Solid areas indicate absorbing surfaces. *(a)* The main axis, a mother root. *(b)* Cross section, representing a mycorrhizal root. *(From P. J. Kramer, Plant and Soil Water Relationships, McGraw-Hill, 1949)*

Mycetophagidae [INV ZOO] The hairy fungus beetles, a cosmopolitan family of coleopteran insects in the superfamily Cucujoidea.

mycetophagous [BIOL] Feeding on fungi.

Mycetozoa [BIOL] A zoological designation for organisms that exhibit both plant and animal characters during their life history (Myxomycetes); equivalent to the botanical Myxomycophyta.

Mycetozoia [INV ZOO] A subclass of the protozoan class Rhizopodea.

Mycobacteriaceae [MICROBIO] A family of nonsporeforming, nonmotile, aerobic gram-positive bacteria belonging to the order Actinomycetales.

mycobacterial disease [MED] Any disease caused by species of *Mycobacterium*.

Mycobacterium [MICROBIO] A genus of acid-fast, rod-shaped bacteria in the family Mycobacteriaceae.

Mycobacterium leprae [MICROBIO] Bacterium that causes leprosy. Also known as Hansen's bacillus.

Mycobacterium tuberculosis [MICROBIO] Bacterium that causes tuberculosis; a slow-growing, drug-sensitive species. Also known as tubercle bacillus.

mycobactin [BIOCHEM] Any compound produced by some strains, and required for growth by other strains, of *Mycobacteria*.

mycocecidium [PL PATH] A gall caused by fungi.

mycology [BOT] The branch of botany that deals with the study of fungi.

mycomycin [MICROBIO] $C_{13}H_{10}O_2$ An antibiotic produced by *Nocardia acidophilus* and a species of *Actinomyces*; characterized as a highly unsaturated aliphatic acid that shows strong activity against *Mycobacterium tuberculosis*.

mycophagous [ZOO] Feeding on fungi.

Mycophiformes [VERT ZOO] An equivalent name for Salmoniformes.

Mycoplasmataceae [MICROBIO] The single family of the Mycoplasmatales.

Mycoplasmatales [MICROBIO] An order of the class Mollicutes; organisms are gram-negative, generally nonmotile, nonsporing bacteria which lack a true cell wall.

mycopremna [BOT] A rhizome that contains symbiotic fungi, as in some orchids.

mycorrhiza [BOT] A mutual association in which the mycelium of a fungus invades the roots of a seed plant.

mycostatin [MICROBIO] Trade name for the antibiotic mystatin.

Mycota [MYCOL] An equivalent name for Eumycetes.

mycotic stomatitis *See* thrush.

Myctophidae [VERT ZOO] The lantern fishes, a family of deep-sea forms of the suborder Myctophoidei.

Myctophoidei [VERT ZOO] A large suborder of marine salmoniform fishes characterized by having the upper jaw bordered only by premaxillae, and lacking a mesocoracoid arch in the pectoral girdle.

Mydaidae [INV ZOO] The mydas flies, a family of orthorrhaphous dipteran insects in the series Brachycera.

mydriasis [MED] Prolonged dilation of the pupil of the eye.

mydriatic [PHARM] An agent which produces dilation of the pupil, such as eucatropine hydrochloride.

myel-, myelo- [ANAT] A combining form for bone marrow, spinal cord.

myelencephalon [EMBRYO] The caudal portion of the hindbrain; gives rise to the medulla oblongata.

myelin [BIOCHEM] A soft, white fatty substance that forms a sheath around certain nerve fibers.

myelin sheath [HISTOL] An investing cover of myelin around the axis cylinder of certain nerve fibers.

myelitis [MED] **1.** Inflammation of the spinal cord. **2.** Inflammation of the bone marrow.

myeloblast [HISTOL] The youngest precursor cell for blood granulocytes, having a nucleus with finely granular chromatin and nucleoli and intensely basophilic cytoplasm.

myeloblastic leukemia *See* acute granulocytic leukemia.

myeloblastoma [MED] A malignant tumor composed of myeloblasts.

myeloblastosis [MED] Diffuse proliferation of myeloblasts, with involvement of blood, bone marrow, and other tissues and organs.

myelocele [ANAT] The canal of the spinal cord. [MED] Spina bifida, with protrusion of the spinal cord.

myelocyte [HISTOL] A motile precursor cell of blood granulocytes found in bone marrow.

myelocytoma [MED] A malignant plasmacytoma.

myelocytosis [MED] The presence of myelocytes in the blood.

myelodysplasia [MED] Abnormal spinal cord development, especially the lumbosacral portion.

myeloencephalitis [MED] Inflammation of the brain and spinal cord.

myelofibrosis [PATH] Growth of white, fibrous connective tissue in the bone marrow.

myelogenous leukemia *See* granulocytic leukemia.

myelogram [MED] Roentgenogram of the spinal cord, made by myelography. [PATH] Differential cell study of material extracted from bone marrow.

myeloid [ANAT] **1.** Of or pertaining to bone marrow. **2.** Of or pertaining to the spinal cord.

myeloid leukemia *See* granulocytic leukemia.

myeloid metaplasia [MED] The occurrence of hemopoietic tissue in abnormal places in the body.

myeloid myeloma [MED] A malignant plasmacytoma.

myeloid reaction [MED] Increased numbers of granulocytes in the bone marrow and peripheral circulation, often with the appearance of immature granulocytes in the blood.

myeloid tissue [HISTOL] Red bone marrow attached to argyrophile fibers which form wide meshes containing scattered fat cells, erythroblasts, myelocytes, and mature myeloid elements.

myeloma [MED] A primary tumor of the bone marrow composed of any of the bone marrow cell types.

myelomalacia [MED] Softening of the spinal cord.

myelomeningitis [MED] Inflammation of the spinal cord and its meninges.

myelomonocyte [HISTOL] **1.** A monocyte developing in bone marrow. **2.** A blood cell intermediate between monocytes and granulocytes.

myelomonocytic leukemia *See* monocytic leukemia.

myelopathic anemia *See* myelophthisic anemia.

myelophthisic anemia [MED] An anemia associated with space-occupying disorders of the bone marrow. Also known as leukoerythroblastic anemia; leukoerythroblastosis; metastatic anemia; myelopathic anemia; myelosclerotic anemia; osteosclerotic anemia.

myelophthisis [MED] **1.** Loss of bone marrow. **2.** Atrophy of the spinal cord.

myeloplegia [MED] Spinal paralysis.

myelopoiesis [PHYSIO] The process by which blood cells form in the bone marrow.

myelosclerosis [MED] **1.** Multiple sclerosis of the spinal cord. **2.** Hardening of the bone marrow.

myelosclerotic anemia *See* myelophthisic anemia.

myenteric [HISTOL] Of or pertaining to the muscular coat of the intestine.

myenteric plexus [ANAT] A network of nerves between the circular and longitudinal layers of the muscular coat of the digestive tract. Also known as Auerbach's plexus.

myenteron [HISTOL] The muscular coat of the intestine.

Mygalomorphae [INV ZOO] A suborder of spiders (Araneida) including American tarantulas, trap-door spiders, and purse-web spiders; the tarantulas may attain a leg span of 10 inches (25 centimeters).

myiasis [MED] Infestation of vertebrates by the larvae, or maggots, of flies.

Mylabridae [INV ZOO] The equivalent name for Bruchidae.

Myliogatidae [VERT ZOO] The eagle rays, a family of batoids which may reach a length of 15 feet (4.6 meters).

Mymaridae [INV ZOO] The fairy flies, a family of hymenopteran insects in the superfamily Chalcidoidea.

myoblast [EMBRYO] A precursor cell of a muscle fiber.

myocardial infarct [MED] An infarct in heart muscle.

myocardiopathy [MED] Disease of the myocardium. Also known as cardiomyopathy.

myocarditis [MED] Inflammation of the myocardium.

myocardium [HISTOL] The muscular tissue of the heart wall.

myocardosis [MED] Any noninflammatory disease of the myocardium.

myoclonic epilepsy [MED] Recurrent irregular, arrhythmic clonic muscle spasms, usually occurring more frequently in the morning or on going to sleep and often associated with other types of seizures.

myoclonic status [MED] Continual clonic spasms lasting an hour or more.

myoclonus [MED] **1.** Clonic muscle spasm. **2.** Any disorder characterized by scattered, irregular, arrhythmic muscle spasms.

myocoel [EMBRYO] Portion of the coelom enclosed in a myotome.

myocomma [HISTOL] A ligamentous connection between successive myomeres. Also known as myoseptum.

myocyte [HISTOL] **1.** A contractile cell. **2.** A muscle cell.

Myodaria [INV ZOO] A section of the Schizophora series of cyclorrhaphous dipterans; in this group adult antennae consist of three segments, and all families except the Conopidae have the second cubitus and the second anal veins united for almost their entire length.

Myodocopa [INV ZOO] A suborder of the order Myodocopida; includes exclusively marine ostracods distinguished by possession of a heart.

Myodocopida [INV ZOO] An order of the subclass Ostracoda.

Myodopina [INV ZOO] The equivalent name for Myodocopa.

myodystrophy [MED] Muscle degeneration.

myoelastic fiber [HISTOL] An elastic fiber associated with the smooth muscles in bronchi and bronchioles.

myoelectric potential [PHYSIO] The electrical potential created by muscle action.

myofascitis [MED] Muscular pain of obscure nature and origin in the lower back.

myofibril [CYTOL] A contractile fibril in a muscle cell. [INV ZOO] *See* myoneme.

myofilament [CYTOL] The structural unit of muscle proteins in a muscle cell.

myofrisk [INV ZOO] A contractile structure surrounding the spines of certain radiolarians.

myoglobin [BIOCHEM] A hemoglobinlike iron-containing

protein pigment occurring in muscle fibers. Also known as muscle hemoglobin; myohemoglobin.

myoglobinuria [MED] The presence of myoglobin in the urine.

myohematin [BIOCHEM] A cytochrome respiratory enzyme allied to hematin.

myohemoglobin *See* myoglobin.

myoinositol [BIOCHEM] The commonest isomer of inositol. Also known as mesionositol.

myokinase [BIOCHEM] An enzyme that catalyzes the reversible transfer of phosphate groups in adenosinediphosphate; occurs in muscle and other tissues.

myolipoma [MED] A benign tumor composed of adipose and smooth muscle cells.

myology [MED] The study of muscles in both the normal and diseased states.

myoma [MED] **1.** A benign uterine tumor composed principally of smooth muscle cells. **2.** Any neoplasm originating in muscle.

myomalacia [MED] Degeneration, with softening, of muscle tissue.

myomere [EMBRYO] A muscle segment differentiated from the myotome, which divides to form the epimere and hypomere.

myometritis [MED] Inflammation of the myometrium.

myometrium [HISTOL] The muscular tissue of the uterus.

Myomorpha [VERT ZOO] A suborder of rodents recognized in some systems of classification.

myoneme [INV ZOO] A contractile fibril in a protozoan. Also known as myofibril.

myoneural junction [ANAT] The point of junction of a motor nerve with the muscle which it innervates. Also known as neuromuscular junction.

myopathia *See* myopathy.

myopathic facies [MED] An expressionless face with sunken cheeks and a drooping lower lip characteristic of patients with myopathies, especially myotonic dystrophy.

myopathy [MED] Any disease of the muscles. Also known as myopathia.

myopericarditis [MED] A combination of myocarditis and pericarditis.

myophagia [PATH] The invasion of degenerated muscle sarcoplasm by histiocytes.

myophosphorylase deficiency glycogenosis *See* McArdle's syndrome.

myophrisk [INV ZOO] A myoneme or contractile element in Protozoa.

myopia [MED] A condition in which the focal image is formed in front of the retina of the eye. Also known as nearsightedness.

myoplasty [MED] Plastic surgery performed on muscle tissue.

Myopsida [INV ZOO] A natural assemblage of cephalopod mollusks considered as a suborder in the order Teuthoida according to some systems of classification, and a group of the Decapoda according to other systems; the eye is covered by the skin of the head in all species.

myopsychopathy [MED] Any disease of the muscles associated with mental retardation or loss of intellect.

myorhythmia [MED] Muscle tremor with a rate of 2-4 per second, and irregular intervals between cycles.

myosarcoma [MED] A sarcoma derived from muscle.

myoseptum *See* myocomma.

myosin [BIOCHEM] A muscle protein, comprising up to 50% of the total muscle proteins; combines with actin to form actomycin.

myositis [MED] Inflammation of muscle. Also known as fibromyositis.

myositis ossificans [MED] Muscle inflammation with bone formation in muscle, tendons, or ligaments.

myostatic reflex *See* stretch reflex.

myosynovitis [MED] Inflammation of synovial membranes and surrounding musculature.

myotasis [PHYSIO] Stretching of a muscle.

myotome [ANAT] A group of muscles innervated by a single spinal nerve. [EMBRYO] The muscle plate that differentiates into myomeres.

myotonia [MED] Tonic muscular spasm occurring after injury or infection.

myotonic [MED] Of, pertaining to, or characterized by myotonia.

Myriangiales [MYCOL] An order of parasitic fungi of the class Ascomycetes which produce asci at various levels in uniascal locules within stromata.

Myriapoda [INV ZOO] Informal designation for those mandibulate arthropods having two body tagmata, one pair of antennae, and more than three pairs of adult ambulatory appendages.

Myricaceae [BOT] The single family of the plant order Myricales.

Myricales [BOT] An order of dicotyledonous plants in the subclass Hamamelidae, marked by its simple, resinous-dotted, aromatic leaves, and a unilocular ovary with two styles and a single ovule.

Myrientomata [INV ZOO] The equivalent name for the Protura.

myringitis [MED] Inflammation of the tympanic membrane.

Myriotrochidae [INV ZOO] A family of holothurian echinoderms in the order Apodida, distinguished by eight or more spokes in each wheel-shaped spicule.

myrmecology [BIOL] The study of ants.

Myrmecophagidae [VERT ZOO] A small family of arboreal anteaters in the order Edentata.

myrmecophile [ECOL] An organism, usually a beetle, that habitually inhabits the nest of ants.

myrmecophyte [ECOL] A plant that houses and benefits from the habitation of ants.

Myrmeleontidae [INV ZOO] The ant lions, a family of insects in the order Neuroptera; larvae are commonly known as doodlebugs.

Myrmicinae [INV ZOO] A large diverse subfamily of ants (Formicidae); some members are inquilines and have no worker caste.

myrosin [BIOCHEM] A plant enzyme that hydrolyzes the glucoside sinigrin and is thus involved in the formation of mustard oil in brassicaceous plants.

myrrh [MATER] A gum resin of species of myrrh (*Commiphora*); partially soluble in water, alcohol, and ether; used in dentifrices, perfumery, and pharmaceuticals.

Myrsinaceae [BOT] A family of mostly woody dicotyledonous plants in the order Primulales characterized by flowers without staminodes, a schizogenous secretory system, and gland-dotted leaves.

Myrtaceae [BOT] A family of dicotyledonous plants in the order Myrtales characterized by an inferior ovary, numerous stamens, anthers usually opening by slits, and fruit in the form of a berry, drupe, or capsule.

Myrtales [BOT] An order of dicotyledonous plants in the subclass Rosidae characterized by opposite, simple, entire leaves and perigynous to epigynous flowers with a compound pistil.

MYRIOTROCHIDAE

50 μ

An eight-spoked spicule from the skin of an echinoderm in the Myriotrochidae.

Mysida [INV ZOO] A suborder of the crustacean order Mysidacea characterized by fusion of the sixth and seventh abdominal somites in the adult, lack of gills, and other specializations.

Mysidacea [INV ZOO] An order of free-swimming Crustacea included in the division Pericarida; adult consists of 19 somites, each bearing one pair of functionally modified, biramous appendages, and the carapace envelops most of the thorax and is fused dorsally with up to four of the anterior thoracic segments.

mysis [INV ZOO] A larva of certain higher crustaceans, characterized by biramous thoracic appendages.

Mystacinidae [VERT ZOO] A monospecific family of insectivorous bats (Chiroptera) containing the New Zealand short-tailed bat; hindlegs and body are stout, and fur is thick.

Mystacocarida [INV ZOO] An order of primitive Crustacea; the body is wormlike and the cephalothorax bears first and second antennae, mandibles, and first and second maxillae.

mystax [INV ZOO] A cluster or row of hairs above the mouth of certain insects.

Mysticeti [VERT ZOO] The whalebone whales, a suborder of the mammalian order Cetacea, distinguished by horny filter plates of suspended from the upper jaws.

Mytilacea [INV ZOO] A suborder of bivalve mollusks in the order Filibranchia.

Mytilidae [INV ZOO] A family of mussels in the bivalve order Anisomyaria.

myx-, myxo- [ZOO] A combining form denoting mucus, mucous, mucin, mucinous.

myxadenitis [MED] Inflammation of mucous glands.

myxameba [BIOL] An independent ameboid cell of the vegetative phase of Acrasiales.

myxedema [MED] A condition caused by hypothyroidism characterized by a subnormal basal metabolic rate, dry coarse hair, loss of hair, mental dullness, anemia, and slowed reflexes.

Myxicolinae [INV ZOO] A subfamily of sedentary polychaete annelids in the family Sabellidae.

Myxiniformes [VERT ZOO] The equivalent name for the Myxinoidea.

Myxinoidea [VERT ZOO] The hagfishes, an order of eellike, jawless vertebrates (Agnatha) distinguished by having the nasal opening at the tip of the snout and leading to the pharynx, with barbels around the mouth and 6–15 pairs of gill pouches.

myxoadenoma [MED] An adenoma of a mucous gland.

Myxobacterales [MICROBIO] The slime bacteria, an order of bacteria of the class Schizomycetes; vegetative cells are flexible, slender motile rods that tend to swarm and arrange themselves in parallel rows prior to aggregating to form a fruiting body. Also known as fruiting myxobacteria.

myxochondrofibrosarcoma [MED] A sarcoma composed of anaplastic myxoid, chondroid, and fibrous cells.

Myxococcaceae [MICROBIO] A family of slime bacteria (Myxobacterales) which have spherical microcysts surrounded by a thick wall.

myxofibroma of nerve sheath *See* neurofibroma.

myxoflagellate [INV ZOO] A flagellate stage following the myxamoeba stage in development of Myxomycetes or Mycetozoa.

Myxogastromycetidae [MYCOL] A large subclass of plasmodial slime molds (Myxomycetes).

myxolipoma *See* liposarcoma.

myxoma [MED] A benign tumor composed of mucinous connective tissue.

MYSIDA

Neomysis integer, adult male mysid.

MYSTACOCARIDA

Derocheilocaris typicus, lateral view.

MYXOBACTERALES

Myxococcus fulvus, vegetative cells. (After N. A. Woods, from A. T. Henrici and E. J. Ordal, The Biology of Bacteria, 3d ed., Heath, 1948)

MYXOSPORIDEA

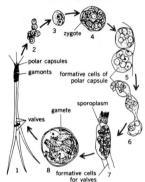

The life cycle of *Triactinomyxon
legeri*, a representative member
of the Actinomyxida showing:
1, mature spore; 2, liberated
gamonts; 3, gamonts pairing;
4, zygote formation; 5, spore
formation from zygote; 6–7,
later stages in spore formation.
*(After MacKinnon and Adam,
1924)*

MYZOSTOMARIA

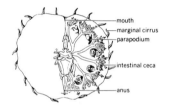

Myzostomum (Myzostomidae)
in ventral view, with organ
systems indicated. *(From
R. R. von Stummer-Traunfels,
Myzostomida, Kukenthal's
Handbuch der Zoologie,
Bd. 3, Lief. 2, Bogen 9–17 pp.
132–210, 1927)*

myxoma lipomatodes *See* liposarcoma.

myxomatosis [VET MED] A virus disease of rabbits producing
fever, skin lesions resembling myxomas, and mucoid swelling
of mucous membranes.

Myxomycetes [BIOL] Plasmodial (acellular or true) slime
molds, a class of microorganisms of the division Mycota; they
are on the borderline of the plant and animal kingdoms and
have a noncellular, multinucleate, jellylike, creeping, assimilative stage (the plasmodium) which alternates with a myxameba stage.

Myxomycophyta [BOT] An order of microorganisms, equivalent to the Mycetozoia of zoological classification.

Myxophaga [INV ZOO] A suborder of the Coleoptera.

Myxophyceae [BOT] An equivalent name for the Cyanophyceae.

myxosarcoma [MED] A sarcoma whose parenchyma is composed of anaplastic myxoid cells.

myxosporangium [MYCOL] The fruit body or sporangium of
Myxomycetes.

myxospore [MYCOL] 1. A spore embedded in a slimy disintegration of the hypha. 2. A spore of the Myxomycetes.

Myxosporida [INV ZOO] An order of the protozoan class
Myxosporidea characterized by the production of spores with
one or more valves and polar capsules, and by possession of
a single sporoplasm.

Myxosporidea [INV ZOO] A class of the protozoan subphylum Cnidospora; members are parasitic in some fish, a few
amphibians, and certain invertebrates.

myxovirus [VIROL] A group of ribonucleic-acid animal viruses characterized by hemagglutination and hemadsorption;
includes influenza and fowl plague viruses and the paramyxo-
viruses.

Myzopodidae [VERT ZOO] A monospecific order of insectivorous bats (Chiroptera) containing the Old World disk-
winged bat of Madagascar; characterized by long ears and by
a vestigial thumb with a monostalked sucking disk.

myzorhynchus [INV ZOO] An apical sucker on the scolex of
certain tapeworms.

Myzostomaria [INV ZOO] An aberrant group of Polychaeta;
most are greatly depressed, broad, and very small, and true
segmentation is delayed or absent in the adult; all are
parasites of echinoderms.

Myzostomidae [INV ZOO] A monogeneric family of the Myzostomaria.

N

N _See_ normality.

Na _See_ sodium.

N.A. _See_ numerical aperture.

Nabidae [INV ZOO] The damsel bugs, a family of hemipteran insects in the superfamily Cimicimorpha.

Nabothian cyst [MED] Cystic distention of the Nabothian glands of the uterine cervix.

Nabothian glands [ANAT] Mucous glands of the uterine cervix.

nacre [INV ZOO] An iridescent inner layer of many mollusk shells.

NAD _See_ diphosphopyridine nucleotide.

Naegeli-type leukemia [MED] A type of monocytic leukemia in which the leukoyctes resemble cells of the granulocytic series.

naiad [INV ZOO] The nymph stage of Hemimetabola.

nail [ANAT] The horny epidermal derivative covering the dorsal aspect of the terminal phalanx of each finger and toe. [MED] A metallic rod with one blunt end and one sharp end, used surgically to anchor bone fragments.

nailhead spot [PL PATH] A fungus rot of tomato caused by _Alternaria tomato_ and marked by small brown to black sunken spots on the fruit.

nail-patella syndrome [GEN] A genetic disorder inherited as an autosomal dominant that is part of a linkage group with ABO blood group genes; characterized by defects in the nails and abnormalities of the elbows and other bones, skin, eyes, and kidneys.

Najadaceae [BOT] A family of monocotyledonous, submerged aquatic plants in the order Najadales distinguished by branching stems and opposite or whorled leaves.

Najadales [BOT] An order of aquatic and semiaquatic flowering plants in the subclass Alismatidae; the perianth, when present, is not differentiated into sepals and petals, and the flowers are usually not individually subtended by bracts.

naked bud [BOT] A bud covered only by rudimentary foliage leaves.

Namanereinae [INV ZOO] A subfamily of largely fresh-water errantian annelids in the family Nereidae.

Namurian [GEOL] A European stage of geologic time; divided into a lower stage (Lower Carboniferous or Upper Mississippian) and an upper stage (Upper Carboniferous or Lower Pennsylvanian).

nanism [MED] Dwarfed stature due to arrested development.

nannoplankton [BIOL] Minute plankton; the smallest plankton, including algae, bacteria, and protozoans.

nano- [BIOL] A prefix meaning dwarfed. [MATH] A prefix representing 10^{-9}, which is 0.000000001 or one-billionth of the unit adjoined. Also known as milli-micro- (deprecated usage).

NAKED BUD

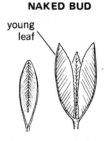

young leaf

Closed and open naked buds of the hobblebush.

nanocephalus [MED] A fetus with an undersized head.

nanometer [MECH] A unit of length equal to one-billionth of a meter, or 10^{-9} meter. Also known as nanon.

nanon *See* nanometer.

nanophanerophyte [ECOL] A shrub not exceeding 2 meters in height.

nanophthalmus *See* microphthalmus.

nanozooid [INV ZOO] Dwarf zooid; bryozoan heterozooid possessing only a single tentacle.

nape [ANAT] The back of the neck.

Naperian logarithm *See* logarithm.

napex [ANAT] That portion of the scalp just below the occipital protuberance.

napiform [BOT] Turnip-shaped, referring to roots.

narcissism [PSYCH] Excessive self-love.

narco- [MED] Combining form meaning numbness, narcosis, or stupor.

narcoanalysis [PSYCH] Induction of a reversible sleep by intravenous injections of drugs such as amobarbital or thiopental sodium in order to elicit memories and feelings not expressed by the person in a wakeful state because of resistance.

narcolepsy [MED] A disorder of sleep mechanism characterized by two or more of four distinct symptoms: uncontrollable periods of daytime drowsiness, cataleptic attacks of muscular weakness, sleep paralysis, and vivid nocturnal or hypnogogic hallucinations.

narcomania [MED] Morbid physiologic or psychologic craving for narcotics to avoid painful stimuli.

Narcomedusae [INV ZOO] A suborder of hydrozoan coelenterates in the order Trachylina; the hydroid generation is represented by an actinula larva.

narcosis [MED] Drug-produced state of profound stupor, unconsciousness, or arrested activity.

narcosis therapy [MED] Prolonged, drug-induced sleep as treatment for certain mental disorders. Also known as sleep therapy.

narcospasm [MED] Spasm accompanied by stupor.

narcosynthesis [MED] Psychotherapeutic treatment under partial anesthesia, in which abreaction is a significant factor in obtaining positive results.

narcotic [PHARM] A drug which in therapeutic doses diminishes awareness of sensory impulses, especially pain, by the brain; in large doses, it causes stupor, coma, or convulsions.

narcotine *See* noscapine.

naris *See* nostril.

narrow-angle glaucoma [MED] Increased intraocular tension due to a block of the angle of the anterior chamber from contact of the iris by the trabecula. Also known as obstructive glaucoma.

narrow-spectrum antibiotic [MICROBIO] An antibiotic effective against a limited number of microorganisms.

narwhal [VERT ZOO] *Monodon monoceros.* An arctic whale characterized by lack of a dorsal fin, and by possession in the male of a long, twisted, pointed tusk (or rarely, two tusks) which is a source of ivory.

nasal [ANAT] Of or pertaining to the nose.

nasal base breadth [ANTHRO] The distance measured across the alae (wings) of the nose, when they are in rest position.

nasal bone [ANAT] Either of two rectangular bone plates forming the bridge of the nose; they articulate with the frontal, ethmoid, and maxilla bones.

nasal bridge breadth [ANTHRO] The width measured between the junctures of the cheekbone and nasal bone, just inside the internal canthi.

nasal bridge salient [ANTHRO] The distance measured from the tip of the bony bridge in the midline of the nose to the juncture of the bony sidewall with the cheek.

nasal cavity [ANAT] Either of a pair of cavities separated by a septum and located between the nasopharynx and anterior nares.

nasal crest [ANAT] **1.** The linear prominence on the medial border of the palatal process of the maxilla. **2.** The linear prominence on the medial border of the palatine bone. **3.** The linear prominence on the internal border of the nasal bone and forming part of the nasal septum.

nasal height [ANTHRO] The height of the nose measured from the nasion to the middle of the lower margin of the anterior nares.

nasal index [ANTHRO] The ratio, × 100, of the greatest width of the anterior nasal openings of the skull to the height of the nasal skeleton.

nasal pit *See* olfactory pit.

nasal process of the frontal bone [ANAT] The downward projection of the nasal part of the frontal bone which terminates as the nasal spine.

nasal process of the maxilla [ANAT] Frontal process of the maxilla.

nasal root breadth [ANTHRO] The distance measured between junctures of the cheekbone and the nasal bone, just inside the internal canthi.

nasal root salient [ANTHRO] The distance measured between the internal canthus and the nasion.

nasal septum [ANAT] The partition separating the two nasal cavities.

nasal tip height [ANTHRO] The distance measured between the subnasale and the pronasale.

nasal tip salient [ANTHRO] The distance measured from the nasal wing to the pronasale.

nascent [CHEM] Pertaining to an atom or simple compound at the moment of its liberation from chemical combination, when it may have greater activity than in its usual state.

Nasellina [INV ZOO] The equivalent name for Monopylina.

nasion [ANTHRO] Midpoint of the nasofrontal suture.

nasion-menton [ANTHRO] The distance measured between the nasion and the midpoint of the lower edge of the chin.

nasolacrimal canal [ANAT] The bony canal that lodges the nasolacrimal duct. Also known as lacrimal canal.

nasolacrimal duct [ANAT] The membranous duct lodged within the nasolacrimal canal; it gives passage to the tears from the lacrimal sac to the inferior meatus of the nose.

nasolacrimal groove [EMBRYO] The furrow, the maxillary, and the lateral nasal processes of the embryo.

nasopalatine cyst *See* median maxillary cyst.

nasopalatine duct [EMBRYO] A canal between the oral and nasal cavities of the embryo at the point of fusion of the maxillary and palatine processes.

nasopharynx [ANAT] The space behind the posterior nasal orifices, above a horizontal plane through the lower margin of the palate.

nastic movement [BOT] Movement of a flat plant part, oriented relative to the plant body and produced by diffuse stimuli causing disproportionate growth or increased turgor pressure in the tissues of one surface.

Nasutitermitinae [INV ZOO] A subfamily of termites in the family Termitidae, characterized by having the cephalic glands open at the tip of an elongated tube which projects anteriorly.

Natalidae [VERT ZOO] The funnel-eared bats, a monogeneric family of small, tropical American insectivorous bats (Chiroptera) with large, funnellike ears.

natant [BIOL] Floating or swimming on the surface of water.

Natantia [INV ZOO] A suborder of decapod crustaceans comprising shrimp and related forms characterized by a long rostrum and a ventrally flexed abdomen.

Naticacea [INV ZOO] A superfamily of gastropod mollusks in the order Prosobranchia.

Naticidae [INV ZOO] A family of gastropod mollusks in the order Pectinibranchia comprising the moon-shell snails.

native [BIOL] Grown, produced or originating in a specific region or country. [GEOCHEM] Pertaining to an element found in nature in a nongaseous state.

natremia [MED] Excessive amounts of sodium in the blood.

natriuretic [PHARM] A medicinal agent which inhibits reabsorption of cations, particularly sodium, from urine.

natural immunity [IMMUNOL] Native immunity possessed by the individuals of a race, strain, or species.

natural resource [MATER] A deposit of minerals, water, or other materials furnished by nature.

natural science [SCI TECH] Collectively, the branches of science dealing with objectively measurable phenomena pertaining to the transformations and relationships of energy and matter; includes biology, physics, and chemistry.

natural selection [EVOL] Darwin's theory of evolution, according to which organisms tend to produce progeny far above the means of subsistence; in the struggle for existence that ensues, only those progeny with favorable variations survive; the favorable variations accumulate through subsequent generations, and descendants diverge from their ancestors.

Naucoridae [INV ZOO] A family of hemipteran insects in the superfamily Naucoroidea.

Naucoroidea [INV ZOO] The creeping water bugs, a superfamily of hemipteran insects in the subdivision Hydrocorisae; they are suboval in form, with chelate front legs.

nauplius [INV ZOO] A larval stage characteristic of many groups of Crustacea; the oval, unsegmented body has three pairs of appendages: uniramous antennules, biramous antennae, and mandibles.

nausea [MED] Feeling of discomfort in the stomach region, accompanied by aversion to food and a tendency to vomit.

Nautilidae [INV ZOO] A monogeneric family of cephalopod mollusks in the order Nautiloidea; *Nautilus pompilius* is the only well-known living species.

Nautiloidea [INV ZOO] A primitive order of tetrabranchiate cephalopods; shells are external and smooth, being straight or coiled and chambered with curved transverse septa.

navel [ANAT] The umbilicus.

navel height [ANTHRO] The vertical distance, of a standing subject, measured from the center of the navel to the floor.

navicular [ANAT] A boat-shaped bone, especially the lateral bone on the radial side of the proximal row of the carpus. [BIOL] Resembling or having the shape of a boat.

navicular cells [PATH] Boat-shaped squamous epithelial cells filled with glycogen and prominent in the exfoliated cells of the uterine cervix of pregnant women.

Nb *See* niobium.

Nd *See* neodymium.

Ne *See* neon.

neallotype [SYST] A type specimen that, compared with the holotype, is of the opposite sex, and was collected and described later.

Neanderthal man [PALEON] A type of fossil human that is a subspecies of *Homo sapiens* and is distinguished by a low

NAUPLIUS

Nauplius of the shrimp *Penaeus*; in addition to their sensory and feeding functions, the three pairs of appendages are organs of locomotion for this free-swimming larva. *(From T. I. Storer and R. L. Usinger, General Zoology, 4th ed., McGraw-Hill, 1965)*

NAUTILIDAE

Shell of *Nautilus pompilius*, which may be up to 10 inches (25 centimeters) in diameter.

broad braincase, continuous arched browridges, projecting occipital region, short limbs, and large joints.

Nearctic fauna [ECOL] The indigenous animal communities of the Nearctic zoogeographic region.

Nearctic zoogeographic region [ECOL] The zoogeographic region that includes all of North America to the edge of the Mexican Plateau.

near-infrared radiation [ELECTROMAG] Infrared radiation having a relatively short wavelength, between 0.75 and about 2.5 micrometers (some scientists place the upper limit from 1.5 to 3 micrometers), at which radiation can be detected by photoelectric cells, and which corresponds in frequency range to the lower electronic energy levels of molecules and semiconductors. Also known as photoelectric infrared radiation.

near-infrared spectrophotometry [ANALY CHEM] Spectrophotometry at wavelengths in the near-infrared region, generally using instruments with quartz prisms in the monochromators and lead sulfide photoconductor cells as detectors to observe absorption bands which are harmonics of bands at longer wavelengths.

nearsightedness *See* myopia.

near-ultraviolet radiation [ELECTROMAG] Ultraviolet radiation having relatively long wavelength, in the approximate range from 300 to 400 nanometers.

near wilt [PL PATH] A fungus disease of peas caused by *Fusarium oxysporum pisi;* affects scattered plants and develops more slowly than true wilt.

Nebaliacea [INV ZOO] A small, marine order of Crustacea in the subclass Leptostraca distinguished by a large bivalve shell, without a definite hinge line, an anterior articulated rostrum, eight thoracic and seven abdominal somites, a pair of articulated furcal rami, and the telson.

neck [ANAT] The usually constricted communicating column between the head and trunk of the vertebrate body.

neck breadth [ANTHRO] The diameter of the neck measured halfway between the otobasion inferior and the shoulder.

neck depth [ANTHRO] The diameter of the neck between the tip of the thyroid cartilage and the back of the neck, measured perpendicular to the axis of the neck, with contact only.

Neckeraceae [BOT] A family of mosses in the order Isobryales distinguished by undulate leaves.

neck rot [PL PATH] A fungus disease of onions caused by species of *Botrytis* and characterized by rotting of the leaves just above the bulb.

necr-, necro- [MED] Combining form denoting death.

necrobiosis [MED] Death of a cell or group of cells under either normal or pathologic conditions.

necrocytosis [PATH] Cellular death.

necrogenous [ECOL] Living in dead bodies.

Necrolestidae [PALEON] An extinct family of insectivorous marsupials.

necrophagous [ZOO] Feeding on dead bodies.

necrophile [PSYCH] A person affected with necrophilia.

necrophilia [PSYCH] 1. Longing for death. 2. *See* necrophilism.

necrophilism [PSYCH] Also known as necrophilia. 1. Unnatural obsession with and usually erotic attraction for dead bodies. 2. Sexual violation of a corpse.

necrophobia [PSYCH] Abnormal dread of death and of dead bodies.

necropsy [MED] To perform an autopsy.

necrosis [MED] Death of a cell or group of cells as a result of injury, disease, or other pathologic state.

necrotic [MED] Pertaining to, causing, or undergoing necrosis.

necrotic enteritis [VET MED] A bacterial infection of young swine caused by *Salmonella suipestifer* or *S. choleraesuis* and characterized by fever and necrotic and ulcerative inflammation of the intestine.

necrotic ring spot [PL PATH] A virus leaf spot of cherries marked by small, dark water-soaked rings which may drop out, giving the leaf a tattered appearance.

necrotize [MED] To undergo necrosis; to become necrotic.

necrozoospermia [MED] A condition in which spermatozoa are immobile.

nectar [BOT] A sugar-containing liquid secretion of the nectaries of many flowers.

nectarine [BOT] A smooth-skinned, fuzzless fruit originating as a spontaneous somatic mutation of the peach, *Prunus persica* and *P. persica* var. *nectarina*.

nectary [BOT] A secretory organ or surface modification of a floral organ in many flowers, occurring on the receptacle, in and around ovaries, on stamens, or on the perianth; secretes nectar.

nectocalyx [INV ZOO] A swimming bell of a siphonophore. Also known as nectophore.

nectocyst [INV ZOO] The cavity within a nectocalyx.

Nectonematoidea [INV ZOO] A monogeneric order of worms belonging to the class Nematomorpha, characterized by dorsal and ventral epidermal chords, a pseudocoele, and dorsal and ventral rows of bristles; adults are parasites of true crabs and hermit crabs.

nectophore *See* nectocalyx.

nectopod [INV ZOO] An appendage or limb specialized for swimming, as in certain mollusks.

nectosome [INV ZOO] The part of a complex siphonophore that bears swimming bells.

Nectridea [PALEON] An order of extinct lepospondylous amphibians characterized by vertebrae in which large fan-shaped hemal arches grow directly downward from the middle of each caudal centrum.

Nectrioidaceae [MYCOL] The equivalent name for Zythiaceae.

Necturus [VERT ZOO] A genus of mud puppies in the family Proteidae.

Needham's sac [INV ZOO] A dilation of male genital duct containing spermatophores in certain cephalopods.

needle [BOT] A slender-pointed leaf, as of the firs and other evergreens.

negative afterimage [PHYSIO] An afterimage that is seen on a bright background and is complementary in color to the initial stimulus.

negative assortative mating [GEN] The mating of unlike individuals with a frequency greater than under random mating.

negative electron *See* electron.

negative empathy [PSYCH] Empathy which takes place against a certain resistance or unwillingness.

negative feedback [CONT SYS] Feedback in which a portion of the output of a circuit, device, or machine is fed back 180° out of phase with the input signal, resulting in a decrease of amplification so as to stabilize the amplification with respect to time or frequency, and a reduction in distortion and noise. Also known as inverse feedback; reverse feedback; stabilized feedback. [SCI TECH] Feedback which tends to reduce the output in a system.

negative interference [GEN] A crossover exchange between

NECTRIDEA

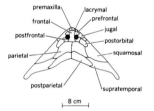

Skull of a lower Permian nectridian, *Diplocaulus. (From E. H. Colbert, Evolution of the Vertebrates, Wiley, 1955)*

homologous chromosomes which increases the likelihood of another in the same vicinity.

negative ion [CHEM] An atom or group of atoms which by gain of one or more electrons has acquired a negative electric charge.

negative phase [IMMUNOL] The temporary quantitative reduction of serum antibodies immediately following a second inoculation of antigen.

negative pressure [PHYS] A way of expressing vacuum; a pressure less than atmospheric or the standard 760 mm Hg.

negative reinforcement [PSYCH] A stimulus or event that strengthens an avoidance response to an unpleasant stimulus when it follows the response.

negative staining [BIOL] A method in microscopy for demonstrating the form of cells, bacteria, and other small objects by staining the ground rather than the objects.

negative tropism [BIOL] Tropism in which the organism moves away from the stimulus.

negativism [PSYCH] Indifference, opposition, or resistance to suggestions, or persistent refusal to do as asked, without apparent or objective reasons.

negatron *See* electron.

Negri bodies [PATH] Acidophil cytoplasmic inclusion bodies in neurons, considered diagnostic of rabies.

Neididae [INV ZOO] A small family of thread-legged hemipteran insects in the superfamily Lygaeoidea.

Neisseriaceae [MICROBIO] A family of parasitic, gram-negative cocci in the order Eubacteriales; several species are human pathogens.

Neisseria gonorrhoeae [MICROBIO] The bacterial organism that causes gonorrhea; the type species of the genus. Also known as *Diplococcus gonorrhoeae;* gonococcus.

nekton [INV ZOO] Free-swimming aquatic animals, essentially independent of water movements.

Nelumbonaceae [BOT] A family of flowering aquatic herbs in the order Nymphaeales characterized by having roots, perfect flowers, alternate leaves, and triaperturate pollen.

Nemataceae [BOT] A family of mosses in the order Hookeriales distinguished by having perichaetial leaves only.

nemathecium [BOT] A wartlike protuberance on the thallus of red algae; contains tetraspores, antheridia, or cystocarps.

Nemathelminthes [INV ZOO] A subdivision of the Amera which comprised the classes Rotatoria, Gastrotrichia, Kinorhyncha, Nematoda, Nematomorpha, and Acanthocephala.

nematicide [MATER] A chemical used to kill plant-parasitic nematodes.

Nematocera [INV ZOO] A series of dipteran insects in the suborder Orthorrhapha; adults have antennae that are usually longer than the head, and the flagellum consists of 10–65 similar segments.

nematocyst [INV ZOO] An intracellular effector organelle in the form of a coiled tube which may be rapidly everted in food gathering or defense by coelenterates.

Nematoda [INV ZOO] A group of unsegmented worms which have been variously recognized as an order, class, and phylum.

nematode [INV ZOO] **1.** Any member of the Nematoda. **2.** Of or pertaining to the Nematoda.

Nematodonteae [BOT] A group of mosses included in the subclass Eubrya in which there may be faint transverse bars on the peristome teeth.

nematogen [INV ZOO] A reproductive phase of the Dicyemida during which vermiform larvae are formed asexually from the germ cells in the axial cells.

NEMATOGEN

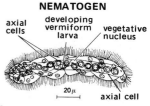

Young stem nematogen, with three axial cells, of *Dicyema schulzianum. (After H. Nouvel)*

Nematognathi [VERT ZOO] The equivalent name for Siluriformes.

nematogone [BOT] Any of the thin-walled propagative cells in the gemmae of certain mosses.

Nematoidea [INV ZOO] An equivalent name for Nematoda.

nematology [INV ZOO] The study of nematodes.

Nematomorpha [INV ZOO] A group of the Aschelminthes or a separate phylum that includes the horsehair worms.

Nematophytales [PALEOBOT] A group of fossil plants from the Silurian and Devonian periods that bear some resemblance to the brown seaweeds (Phaeophyta).

nematosphere [INV ZOO] The enlarged end of a tentacle in some sea anemones.

Nematosporoideae [BOT] A subfamily of the Saccharomycetaceae containing parasitic yeasts; two genera have been studied in culture: *Nematospora* with asci that contain eight spindle-shaped ascospores, and *Metschnikowia* whose asci contain one or two needle-shaped ascospores.

nematozooid [INV ZOO] A zooid bearing organs of defense, in hydroids and siphonophores.

Nemertea [INV ZOO] An equivalent name for Rhynchocoela.

Nemertina [INV ZOO] An equivalent name for Rhynchocoela.

Nemertinea [INV ZOO] An equivalent name for Rhynchocoela.

Nemestrinidae [INV ZOO] The hairy flies, a family of dipteran insects in the series Brachycera of the suborder Orthorrhapha.

Nemichthyidae [VERT ZOO] A family of bathypelagic, eellike amphibians in the order Apoda.

Nemognathinae [INV ZOO] A subfamily of the coleopteran family Meloidae; members have greatly elongate maxillae that form a poorly fitted tube.

nemoral [ECOL] Pertaining to or inhabiting a grove or wooded area.

Neoanthropinae [PALEON] A subfamily of the Hominidae in some systems of classification, set up to include *Homo sapiens* and direct ancestors of *H. sapiens.*

neoblast [INV ZOO] Any of various undifferentiated cells in annelids which migrate to and proliferate at sites of repair and regeneration.

neocarpy [BOT] Fruit production by an immature plant.

Neocathartidae [PALEON] An extinct family of vulturelike diurnal birds of prey (Falconiformes) from the Upper Eocene.

neocerebellum [ANAT] Phylogenetically, the most recent part of the cerebellum; receives cerebral cortex impulses via the corticopontocerebellar tract.

Neocomian [GEOL] A European stage of Lower Cretaceous geologic time; includes Berriasian, Valanginian, Hauterivian, and Barremian.

neocortex [ANAT] Phylogenetically the most recent part of the cerebral cortex; includes all but the olfactory, hippocampal, and piriform regions of the cortex.

neodymium [CHEM] A metallic element, symbol Nd, with atomic weight 144.24, atomic number 60; a member of the rare-earth group of elements.

Neogastropoda [INV ZOO] An order of gastropods which contains the most highly developed snails; respiration is by means of ctenidia, the nervous system is concentrated, an operculum is present, and the sexes are separate.

Neogene [GEOL] An interval of geologic time incorporating the Miocene and Pliocene of the Tertiary period; the Upper Tertiary.

Neognathae [VERT ZOO] A superorder of the avian order

NEODYMIUM

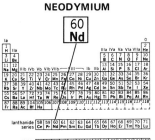

Periodic table of the chemical elements showing the position of neodymium.

Neornithes, characterized as flying birds with fully developed wings and sternum with a keel, fused caudal vertebrae, and absence of teeth.

Neogregarinida [INV ZOO] An order of sporozoan protozoans in the subclass Gregarinia which are insect parasites.

neomycin [MICROBIO] The collective name for several colorless antibiotics produced by a strain of *Streptomyces fradiae;* the commercial fraction ($C_{23}H_{46}N_6O_{13}$) has a broad spectrum of activity.

neon [CHEM] A gaseous element, symbol Ne, atomic number 10, atomic weight 20.183; a member of the family of noble gases in the zero group of the periodic table.

neonatal [MED] Pertaining to a newborn infant.

neonatal impetigo [MED] A type of impetigo occurring in the newborn, characterized by bullae and caused by staphylococci or sometimes streptococci.

neonatal line [ANAT] A prominent incremental line formed in the neonatal period in the enamel and dentin of a deciduous tooth or of a first permanent molar.

neonatal mortality rate [MED] The number of deaths reported among infants under 1 month of age in a calendar year per 1000 live births reported in the same year and place.

neonatal myasthenia [MED] Muscle weakness and ineffective motor activities in infants born of myasthenic mothers.

neonate [MED] A newborn infant.

neonatology [MED] The study of the newborn up to 2 months of age.

neonychium [VERT ZOO] **1.** One of the protective pads enclosing fetal claws in unguliculate vertebrates and of some other mammals. **2.** A horny pad covering a claw in birds before hatching.

neopallium [ANAT] Phylogenetically, the new part of the cerebral cortex; formed from the region between the pyriform lobe and the hippocampus, it comprises the nonolfactory region.

neopalynology [BOT] A field of palynology concerned with extant microorganisms and disassociated microscopic parts of megaorganisms.

neoplasia [MED] **1.** Formation of a neoplasm or tumor. **2.** Formation of new tissue.

neoplasm [MED] An aberrant new growth of abnormal cells or tissues; a tumor.

Neopseustidae [INV ZOO] A family of Lepidoptera in the superfamily Eriocranioidea.

Neoptera [INV ZOO] A section of the insect subclass Pterygota; members have a muscular and articular mechanism allowing the wings to be flexed over the abdomen when at rest.

Neopterygii [VERT ZOO] An equivalent name for Actinopterygii.

Neorhabdocoela [INV ZOO] A group of the Rhabdocoela comprising fresh-water, marine, or terrestrial forms, with a bulbous pharynx, paired protonephridia, sexual reproduction, and ventral gonopores.

Neornithes [VERT ZOO] A subclass of the class Aves containing all known birds except the fossil *Archaeopteryx.*

neossoptile [VERT ZOO] A downy feather on most newly hatched birds.

neostigmine [PHARM] A quaternary ammonium cation used as the bromide ($C_{12}H_{19}BrN_{20}O_2$) and methylsulfate ($C_{13}H_{22}N_2O_6S$) salts; has anticholinesterase activity.

neotenin [BIOCHEM] A hormone secreted by cells of the corpus allatum in arthropod larvae and nymphs; inhibits the development of adult characters. Also known as juvenile hormone.

NEON

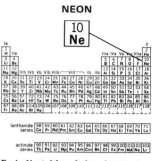

Periodic table of the chemical elements showing the position of neon.

neoteny [BOT] A condition in which a plant attains sexual maturity while some vegetative features remain permanently juvenile. [VERT ZOO] A phenomenon peculiar to some salamanders, in which large larvae become sexually mature while still retaining gills and other larval features.

Neotropical zoogeographic region [ECOL] A zoogeographic region that includes Mexico south of the Mexican Plateau, the West Indies, Central America, and South America.

neotype [SYST] A specimen selected as type subsequent to the original description when the primary types are known to be destroyed; a nomenclatural type.

neounitarian theory of hematopoiesis [HISTOL] A theory that under certain conditions, such as in tissue culture or in pathologic states, lymphocytes or cells resembling lymphocytes can become multipotent.

Nepenthaceae [BOT] A family of dicotyledonous plants in the order Sarraceniales; includes many of the pitcher plants.

nephr-, nephro- [ANAT] Combining form denoting kidney.

nephrectomy [MED] Surgical removal of a kidney.

nephric tubule *See* uriniferous tubule.

nephridioblast [INV ZOO] An ecodermal precursor cell of a nephridium in certain animals.

nephridioduct [INV ZOO] The duct of a nephridium, sometimes serving as a common excretory and genital outlet.

nephridiopore [INV ZOO] The external opening of a nephridium.

nephridiostome [INV ZOO] The ciliated opening of a nephridium into the coelomic cavity.

nephridium [INV ZOO] Any of various paired excretory structures present in the Platyhelminthes, Rotifera, Rhynchocoela, Acanthocephala, Priapuloidea, Entoprocta, Gastrotricha, Kinorhyncha, Cephalochorda, and some Archiannelida and Polychaeta.

nephritic [MED] **1.** Pertaining to or affected with nephritis. **2.** Pertaining to or affecting the kidney.

nephritis [MED] Inflammation of the kidney.

nephroabdominal [ANAT] Of or pertaining to the kidneys and abdomen.

nephroblastoma *See* Wilms' tumor.

nephrocalcinosis [PATH] Deposition of calcium salts in the kidney tubules.

nephrocoel [ANAT] The cavity of a nephrotome.

nephrocyte [INV ZOO] Any of various cells which store excretions and then move to the body surface to discharge these wastes.

nephrodystrophy *See* nephrosis.

nephrogenic [EMBRYO] **1.** Having the potential to develop into kidney tissue. **2.** Of renal origin.

nephrogenic cord [EMBRYO] The longitudinal cordlike mass of mesenchyme derived from the mesomere or nephrostomal plate of the mesoderm, from which develop the functional parts of the pronephros, mesonephros, and metanephros.

nephrogenic tissue [EMBRYO] The tissue of the nephrogenic cord derived from the nephrotome plate that forms the blastema or primordium from which the embryonic and definitive kidneys develop.

nephrolithiasis [PATH] Formation of renal calculi.

nephrolithotomy [MED] Excision of renal calculi from the kidney.

nephrology [MED] The study of the kidney, including diseases.

nephrolysis [MED] **1.** Dissolution of kidney tissue by the action of a nephrolysin. **2.** Surgical detachment of a kidney from surrounding adhesions.

nephroma [MED] A tumor of the kidney.

nephromegaly [MED] Enlargement of the kidney.

nephromixium [INV ZOO] A compound nephridium composed of flame cells and the coelomic funnel; functions as both an excretory organ and a genital duct.

nephron [ANAT] The functional unit of a kidney, consisting of the glomerulus with its capsule and attached uriniferous tubule.

nephropathy [MED] **1.** Any disease of the kidney. **2.** *See* nephrosis.

nephropexy [MED] Fixation of a floating kidney by means of surgery.

Nephropidae [INV ZOO] The true lobsters, a family of decapod crustaceans in the superfamily Nephropidea.

Nephropidea [INV ZOO] A superfamily of the decapod section Macrura including the true lobsters and crayfishes, characterized by a rostrum and by chelae on the first three pairs of pereiopods, with the first pair being noticeably larger.

nephroptosis [MED] Prolapse of the kidney.

nephrorrhaphy [MED] **1.** The stitching of a floating kidney to the posterior wall of the abdomen or to the loin. **2.** Suturing a wound in the kidney.

nephros [ANAT] The kidney.

nephrosclerosis [MED] Sclerosis of the renal arteries and arterioles.

nephrosis [PATH] Degenerative or retrogressive renal lesions, distinct from inflammation (nephritis) or vascular involvement (nephrosclerosis), especially as applied to tubular lesions (tubular nephritis). Also known as nephrodystrophy; nephropathy.

nephrostome [INV ZOO] The funnel-shaped opening of a nephridium into the coelom.

nephrotic [MED] Pertaining to or affected by nephroses.

nephrotic edema [MED] A type of edema occurring in persons with chronic lepoid nephrosis or the nephrotic stage of glomerular nephritis.

nephrotic syndrome [MED] A complex of symptoms, including proteinuria, hyperalbuminemia, and hyperlipemia, resulting from damage to the basement membrane of glomeruli.

nephrotome [EMBRYO] The narrow mass of embryonic mesoderm connecting somites and lateral mesoderm, from which the pronephros, mesonephros, metanephros, and their ducts develop.

nephrotomy [MED] Incision of the kidney.

Nephtyidae [INV ZOO] A family of errantian annelids of highly opalescent colors, distinguished by an eversible pharynx.

Nepidae [INV ZOO] The water scorpions, a family of hemipteran insects in the superfamily Nepoidea, characterized by a long breathing tube at the tip of the abdomen, chelate front legs, and a short stout beak.

Nepoidea [INV ZOO] A superfamily of hemipteran insects in the subdivision Hydrocorisae.

Nepticulidae [INV ZOO] The single family of the lepidopteran superfamily Nepticuloidea.

Nepticuloidea [INV ZOO] A monofamilial superfamily of heteroneuran Lepidoptera; members are tiny moths with wing spines, and the females have a single genital opening.

neptunium [CHEM] A chemical element, symbol Np, atomic number 93, atomic weight 237.0482; a member of the actinide series of elements.

Nereidae [INV ZOO] A large family of mostly marine errantian annelids that have a well-defined head, elongated body with many segments, and large complex parapodia on most segments.

NEPTUNIUM

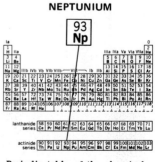

Periodic table of the chemical elements showing the position of neptunium.

Nerillidae [INV ZOO] A family of archiannelids characterized by well-developed parapodia and setae.

Neritacea [INV ZOO] A superfamily of gastropod mollusks in the order Aspidobranchia.

neritic [OCEANOGR] Of or pertaining to the region of shallow water adjoining the seacoast and extending from low-tide mark to a depth of about 200 meters.

Neritidae [INV ZOO] A family of primitive marine, freshwater, and terrestrial snails in the order Archaeogastropoda.

Nernst equation [PHYS CHEM] The relationship showing that the electromotive force developed by a dry cell is determined by the activities of the reacting species, the temperature of the reaction, and the standard free-energy change of the overall reaction.

nerve [ANAT] A bundle of nerve fibers or processes held together by connective tissue.

nerve block [PHYSIO] Interruption of impulse transmission through a nerve.

nerve cell *See* neuron.

nerve cord [INV ZOO] Paired, ventral cords of nervous tissue in certain invertebrates, such as insects or the earthworm. [ZOO] Dorsal, hollow tubular cord of nervous tissue in chordates.

nerve deafness [MED] Deafness due to an abnormality of the sense organs or of the nerves involved in hearing.

nerve eminence [VERT ZOO] In some fishes, a sense organ comprising a group of cells and connected with the lateral line system.

nerve ending [ANAT] **1.** The structure on the distal end of an axon. **2.** The termination of a nerve.

nerve fiber [CYTOL] The long process of a neuron, usually the axon.

nerve impulse [PHYSIO] The transient physicochemical change in the membrane of a nerve fiber which sweeps rapidly along the fiber to its termination, where it causes excitation of other nerves, muscle, or gland cells, depending on the connections and functions of the nerve.

nerve net [INV ZOO] A network of continuous nerve cells characterized by diffuse spread of excitation, local and equipotential autonomy, spatial attenuation of conduction, and facilitation; occurs in coelenterates and certain other invertebrates.

nerve pentagon [INV ZOO] A nerve ring composed of five segments around the mouth of echinoderms.

nerve tracing [MED] A method used by chiropractors by which nerves are located and their pathologies are studied.

nerve tract [ANAT] A bundle of nerve fibers having the same general origin and destination.

nervicolous [ECOL] Living on leaf veins.

nervous [BIOL] **1.** Of or pertaining to nerves. **2.** Originating in or affected by nerves. **3.** Affecting or involving nerves. [PSYCH] A state or condition of nervousness.

nervousness [PSYCH] Hyperexcitability of the nervous system; characterized generally by restless or impulsive behavior, shaken mental poise, and an uncomfortable awareness of self.

nervous system [ANAT] A coordinating and integrating system which functions in the adaptation of an organism to its environment; in vertebrates, the system consists of the brain, brainstem, spinal cord, cranial and peripheral nerves, and ganglia.

nervous tissue [HISTOL] The nerve cells and neuroglia of the nervous system.

nervure [INV ZOO] One of the riblike structures supporting the membranous wings of insects.

nervus lateralis [VERT ZOO] In fishes, a branch of the vagus nerve which connects the lateral line organ with the brain.

Nesiotinidae [INV ZOO] A family of bird-infesting biting lice (Mallophaga) that are restricted to penguins.

Nesophontidae [PALEON] An extinct family of large, shrewlike lipotyphlans from the Cenozoic found in the West Indies.

net blotch [PL PATH] A fungus disease of barley caused by *Helminthosporium teres* and marked by spotting of the foliage.

nettle [BOT] A prickly or stinging plant of the family Urticaceae, especially in the genus *Urtica.*

nettle cell *See* cnidoblast.

net-veined [BIOL] Having a network of veins, as a leaf or an insect wing.

neural arc [PHYSIO] A nerve circuit consisting of effector and receptor with intercalated neurons between them.

neural arch *See* vertebral arch.

neural canal [EMBRYO] The embryonic vertebral canal.

neural crest [EMBRYO] Ectoderm composing the primordium of the cranial, spinal, and autonomic ganglia and adrenal medulla, located on either side of the neural tube.

neural ectoderm [EMBRYO] Embryonic ectoderm which will form the neural tube and neural crest.

neural fold [EMBRYO] Either of a pair of dorsal longitudinal folds of the neural plate which unite along the midline, forming the neural tube.

neuralgia [MED] Severe, stabbing, paroxysmal pain along the pathway of a nerve. Also known as neurodynia.

neural gland [INV ZOO] A mass which is near the nerve ganglion in ascidians and which may be a homolog of the hypohysis of vertebrates.

neural groove [EMBRYO] A longitudinal groove between the neural folds of the vertebrate embryo before the neural tube is completed.

neural plate [EMBRYO] The thickened dorsal plate of ectoderm that differentiates into the neural tube.

neural spine [ANAT] The spinous process of a vertebra.

neural tube [EMBRYO] The embryonic tube that differentiates into brain and spinal cord.

neuraminic acid [BIOCHEM] $C_9H_{17}NO_8$ An amino acid, the aldol condensation product of pyruvic acid and N-acetyl-D-mannosamine, regarded as the parent acid of a family of widely distributed acyl derivatives known as sialic acids.

neuraminidase [MICROBIO] A bacterial enzyme that acts to split salic acid from neuraminic acid glycosides.

neurapophysis [EMBRYO] Either of two projections on each embryonic vertebra which unite to form the neural arch.

neurapraxia [MED] Injury to a nerve in which there is localized degeneration of the myelin sheath with transient nerve block.

neurasthenia [MED] A group of symptoms, now generally subsumed in the neurasthenic neurosis, formerly ascribed to debility or exhaustion of the nerve centers.

neurasthenic neurosis [PSYCH] A neurotic disorder characterized by chronic complaints of easy fatigability, lack of energy, weakness, various aches and pains, and sometimes exhaustion. Also known as psychophysiologic nervous system reaction.

neurectomy [MED] Surgical removal of a portion of a nerve.

neurenteric canal [EMBRYO] A temporary duct connecting the neural tube and primitive gut in certain vertebrate and tunicate embryos.

neurilemma [HISTOL] A thin tissue covering the axon directly, or covering the myelin sheath when present, of peripheral nerve fibers.

neurilemmoma [MED] A solitary, encapsulated benign tumor originating in the neurilemma of peripheral, cranial, and sympathetic nerves. Also known as schwannoma.

neurine [BIOCHEM] $CH_2=CHN(CH_3)_3OH$ A very poisonous, syrupy liquid with fishy aroma; soluble in water and alcohol; a product of putrefaction of choline in brain tissue and bile, and in cadavers. Also known as trimethylvinylammonium hydroxide.

neurinomatosis *See* neurofibromatosis.

neurite *See* axon.

neuritis [MED] Degenerative or inflammatory nerve lesions associated with pain, hypersensitivity, anesthesia or paresthesia, paralysis, muscular atrophy, and loss of reflexes in the innervated part of the body.

neuroanatomy [ANAT] The study of the anatomy of the nervous system and nerve tissue.

neuroarthropathy [MED] Joint disease associated with disease of the nervous system.

neuroastrocytoma [MED] Ganglioneuroma, especially when on the floor of the third brain ventricle and in the temporal lobes and exhibiting neuronal elements within predominant astrocytic elements.

neurobiotaxis [EVOL] Hypothetical migration of nerve cells and ganglia toward regions of maximum stimulation during phylogenetic development.

neuroblast [EMBRYO] Embryonic, undifferentiated neuron, derived from neural plate ectoderm.

neuroblastoma [MED] A malignant neoplasm composed of anaplastic sympathicoblasts; occurs usually in the adrenal medulla of children.

neuroblastomatosis *See* neurofibromatosis.

neurobrucellosis [MED] Brucellosis with neurologic involvement, manifested by signs and symptoms of meningitis, encephalitis, radiculitis, or neuritis.

neurochemistry [BIOCHEM] Chemistry of the nervous system.

neurochorioretinitis [MED] Chorioretinitis combined with optic neuritis.

neurochoroiditis [MED] Choroiditis combined with optic neuritis.

neurocirculatory [ANAT] Pertaining to both the nervous and the vascular systems.

neurocirculatory asthenia [MED] A syndrome characterized by dyspnea, palpitation, chest pain, fatigue, and faintness.

neurocoele [ANAT] The system of cavities and ventricles in the brain and spinal cord.

neurocranium [ANAT] The portion of the cranium which forms the braincase.

neurocutaneous [ANAT] 1. Concerned with both the nerves and skin. 2. Pertaining to innervation of the skin.

neurocyte [CYTOL] The body of a nerve cell.

neurodermatitis [MED] A skin disorder characterized by localized, often symmetrical, patches of pruritic dermatitis with lichenification, occurring in persons of nervous temperament.

neurodermatosis [MED] A skin disease which is presumed to have a psychogenic component or basis.

Neurodontiformes [PALEON] A suborder of Conodontophoridia having a lamellar internal structure.

neurodynia *See* neuralgia.

neuroendocrine [BIOL] Pertaining to both the nervous and endocrine systems, structurally and functionally.

neuroendocrinology [BIOL] The study of the structural and functional interrelationships between the nervous and endocrine systems.

neuroepidermal [BIOL] Pertaining to both the nerves and epidermis, structurally and functionally.

neuroepithelioma [MED] A tumor resembling primitive medullary epithelium, containing cells of small cuboidal or columnar form with a tendency to form true rosettes, occurring in the retina, central nervous system, and occasionally in peripheral nerves. Also known as diktoma; esthesioneuroblastoma; esthesioneuroepithelioma.

neurofibril [CYTOL] A fibril of a neuron, usually extending from the processes and traversing the cell body.

neurofibroma [MED] A tumor characterized by the diffuse proliferation of peripheral nerve elements. Also known as endoneural fibroma; myxofibroma of nerve sheath; neurofibromyxoma; perineural fibroblastoma; perineural fibroma.

neurofibromatosis [MED] A hereditary disease characterized by the presence of neurofibromas in the skin or along the pathway of peripheral nerves. Also known as fibroma molluscum; multiple neurofibroma; multiple neurofibromatosis; neurinomatosis; neuroblastomatosis; Smith-Recklinghausen's disease.

neurofibromyxoma *See* neurofibroma.

neurofibrosarcoma [MED] A malignant tumor composed of interlacing bundles of anaplastic spindle-shaped cells which resemble those of nerve sheaths.

neurogenesis [EMBRYO] The formation of nerves.

neurogenic [BIOL] **1.** Originating in nervous tissue. **2.** Innervated by nerves. [MED] Caused or affected by a trauma, dysfunction, or disease of the nervous system.

neurogenic bladder [MED] A urinary bladder disorder due to lesions of the central or peripheral nervous system.

neurogenic shock [MED] Shock caused by vasodilation leading to low blood pressure and serious reduction in venous return and in cardiac output; due to such causes as injury to the central nervous system, spinal anesthesia, or reflex.

neuroglia [HISTOL] The nonnervous, supporting elements of the nervous system.

neurohemal organ [ZOO] Any of various structures in vertebrates and some invertebrates that consist of clusters of bulbous, secretion-filled axon terminals of neurosecretory cells which function as storage-and-release centers for neurohormones.

neurohormone [BIOCHEM] A hormone produced by nervous tissue.

neurohumor [BIOCHEM] A hormonal transmitter substance, such as acetylcholine, released by nerve endings in the transmission of impulses.

neurohypophysis [ANAT] The neural portion or posterior lobe of the hypophysis.

neurolathyrism *See* lathyrism.

neuroleptic [PHARM] **1.** A drug that is useful in the treatment of mental disorders, especially psychoses. **2.** Pertaining to the actions of such a drug.

neuroleptoanalgesia [MED] A state of analgesic consciousness produced by the administration of neuroleptic drugs, allowing painless surgery to be performed on a wakeful subject.

neurologist [MED] A person versed in neurology, usually a physician who specializes in the diagnosis and treatment of disorders of the nervous system and the study of its functioning.

neurology [MED] The study of the anatomy, physiology, and disorders of the nervous system.

neuroma [MED] A tumor of the nervous system.

neuromast [VERT ZOO] A lateral-line sensory organ in fishes and other lower vertebrates consisting of a cluster of receptor cells connected with nerve fibers.

neuromere [EMBRYO] An embryonic segment of the central nervous system in vertebrates.

neuromuscular [BIOL] Pertaining to both nerves and muscles, functionally and structurally.

neuromuscular junction *See* myoneural junction.

neuromyasthenia [MED] Fatigue, headache, intense muscle pain, slight or transient muscle weakness, mental disturbances, objective signs in neurologic examination but usually normal cerebrospinal fluid findings, occurring in epidemics and thought to be viral in origin. Also known as benign myalgic encephalomyelitis.

neuromyelitis [MED] Inflammation of the spinal cord and of nerves.

neuron [HISTOL] A nerve cell, including the cell body, axon, and dendrites.

neuron doctrine [BIOL] A doctrine that the neuron is the basic structural and functional unit of the nervous system, and that it acts upon another neuron through the synapse.

neuroneme [INV ZOO] A nerve fibril lying parallel to a myoneme in infusorians.

neuronitis [MED] Inflammation of a neuron; particularly, neuritis involving the cells and roots of spinal nerves.

neuropathy [MED] Any disease affecting neurons.

neuropharmacology [MED] The science dealing with the action of drugs on the nervous system.

neurophysiology [PHYSIO] The study of the functions of the nervous system.

neuropil [HISTOL] Nervous tissue consisting of a fibrous network of nonmyelinated nerve fibers; gray matter with few nerve cell bodies; usually a region of synapses between axons and dendrites.

neuroplasm [CYTOL] Protoplasm of nerve cells.

neuropodium [CYTOL] A terminal branch of an axon.

neuropore [EMBRYO] A terminal aperture of the neural tube before complete closure at the 20–25 somite stage.

neuropsychology [PSYCH] A system of psychology based on neurology.

neuropsychopathy [MED] A mental disease based upon or manifesting itself in disorders or symptoms of the nervous system.

Neuroptera [INV ZOO] An order of delicate insects having endopterygote development, chewing mouthparts, and soft bodies.

neuropterous [INV ZOO] Having a network of nervures in the wings.

neuroradiology [MED] The roentgenology of neurologic disease.

neurorrhexis [MED] Surgical tearing away of a nerve from its origin, as in the treatment of neuralgia.

neurosarcoma [MED] A sarcoma composed of elements resembling those of the nervous system, or thought to be neurogenic.

neurosclerosis [MED] Hardening of nervous tissue.

neurosecretion [PHYSIO] The synthesis and release of hormones by nerve cells.

neurosis [PSYCH] A category of emotional maladjustments characterized by some impairment of thinking and judgment, with anxiety as the chief symptom.

neurosurgery [MED] Surgery of the nervous system.

neurosyphilis [MED] Syphilitic infection of the nervous system.

neuroticism [PSYCH] A neurotic condition, character, or trait.

neurotic personality [PSYCH] An individual who exhibits

symptoms or manifestations intermediate between normal character traits and true neurotic features.

neurotoxin [BIOCHEM] A poisonous substance in snake venom that acts as a nervous system depressant.

neurotrophic ulcer [MED] A decubitus ulcer due to trophic disturbances following interruption or disease of afferent nerve fibers plus the factor of external trauma.

neurotropic [BIOL] Having an affinity for nerve tissue.

neurovaricosis [MED] The formation or the presence of a varicosity on a nerve fiber.

neurovascular [BIOL] Pertaining structurally and functionally to both the nervous and vascular structures.

neurula [EMBRYO] An early embryonic stage of Chordata in which nervous tissue begins to differentiate and the neural tube is formed.

neurulation [EMBRYO] Differentiation of nerve tissue and formation of the neural tube.

neuston [BIOL] Minute organisms that float or swim on surface water or on a surface film of water.

neutral atom [ATOM PHYS] An atom in which the number of electrons that surround the nucleus is equal to the number of protons in the nucleus, so that there is no net electric charge.

neutralism [ECOL] A neutral interaction between two species, that is, one having no evident effect on either species.

neutralize [CHEM] To make a solution neutral (neither acidic nor basic, pH of 7) by adding a base to an acidic solution, or an acid to a basic solution.

neutralizing antibody [IMMUNOL] An antibody that reduces or abolishes some biological activity of a soluble antigen or of a living microorganism.

neutral molecule [PHYS CHEM] A molecule in which the number of electrons surrounding the nuclei is the same as the total number of protons in the nuclei, so that there is no net electric charge.

neutron [PHYS] An elementary particle which has approximately the same mass as the proton but lacks electric charge, and is a constituent of all nuclei having mass number greater than 1.

neutropenia [MED] Abnormally low number of neutrophils in the peripheral circulation.

neutrophil [HISTOL] A large granular leukocyte with a highly variable nucleus, consisting of three to five lobes, and cytoplasmic granules which stain with neutral dyes and eosin.

neutrophilia [BIOL] Affinity for neutral dyes. [MED] An abnormal increase in leukocytes in the tissues or peripheral circulation.

neutrophilic leukemia [MED] Granulocytic leukemia in which the leukocytes resemble cells of the neutrophilic series.

neutrophilous [BIOL] Preferring an environment free of excess acid or base.

nevocarcinoma [MED] A malignant melanoma.

nevus [MED] A lesion containing melanocytes.

nevus sebaceus [MED] A nevus formed by an aggregate of sebaceous glands. Also known as Jadassohn's nevus.

newborn [MED] Born recently; said of human infants less than a month old, especially of those only a few days old.

Newcastle disease [VET MED] An acute viral disease of fowls, with respiratory, gastrointestinal, and central nervous system involvement; may be transmitted to human beings as a mild conjunctivitis. Also known as avian pneumoencephalitis; avian pseudoplague; Philippine fowl disease.

Newcastle virus [VIROL] A ribonucleic acid hemagglutinating myxovirus responsible for Newcastle disease.

newt [VERT ZOO] Any of the small, semiaquatic salamanders

NEUTROPHIL

nucleus

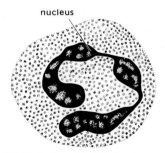

Diagram of neutrophil showing multilobed nucleus.

NICKEL

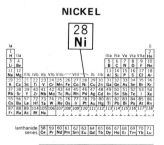

Periodic table of the chemical elements showing the position of nickel.

NICOTINIC ACID

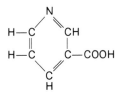

Structural formula of nicotinic acid.

of the genus *Triturus* in the family Salamandridae; all have an aquatic larval stage.

Newtonian fluid [FL MECH] A simple fluid in which the state of stress at any point is proportional to the time rate of strain at that point; the proportionality factor is the viscosity coefficient.

Newton's laws of motion [MECH] Three fundamental principles (called Newton's first, second, and third laws) which form the basis of classical, or Newtonian, mechanics, and have proved valid for all mechanical problems not involving speeds comparable with the speed of light and not involving atomic or subatomic particles.

NGU *See* nongonococcal urethritis.

Ni *See* nickel.

niacin *See* nicotinic acid.

niacinamide *See* nicotinamide.

Niagaran [GEOL] A North American provincial geologic series, in the Middle Silurian.

niche [ECOL] The status of an organism in its biotic and abiotic environment.

nickel [CHEM] A chemical element, symbol Ni, atomic number 28, atomic weight 58.71.

Nicoletiidae [INV ZOO] A family of the insect order Thysanura proper.

Nicomachinae [INV ZOO] A subfamily of the limnivorous sedentary annelids in the family Maldanidae.

nicotinamide [BIOCHEM] $C_6H_6ON_2$ Crystalline basic amide of the vitamin B complex that is interconvertible with nicotinic acid in the living organism; the amide of nicotinic acid. Also known as niacinamide.

nicotinamide adenine dinucleotide *See* diphosphopyridine nucleotide.

nicotinic acid [BIOCHEM] $C_6H_5NO_2$ A component of the vitamin B complex; a white, water-soluble powder stable to heat, acid, and alkali; used for the treatment of pellagra. Also known as niacin.

nictitating membrane [VERT ZOO] A membrane of the inner angle of the eye or below the eyelid in many vertebrates, and capable of extending over the eyeball.

nidamental gland [ZOO] Any of various structures that secrete covering material for eggs or egg masses.

nidation [PHYSIO] 1. Renewal of the uterine lining between menstrual periods. 2. Embedding of the fertilized ovum in the mucous membrane of the uterus.

nidicolous [ZOO] 1. Spending a short time in the nest after hatching. 2. Sharing the nest of another species.

nidus [MED] A focus of infection. [ZOO] A nest or breeding place.

Niemann-Pick disease [MED] A hereditary sphingolipidosis due to an enzyme deficiency resulting in abnormal accumulation of sphingomyelin; symptoms include anemia, enlargement of the liver, spleen, and lymph nodes, gastrointestinal disturbances, and various neurologic deficits. Also known as lipid hystiocytosis.

night ape *See* bushbaby.

night blindness [MED] Reduced dark adaptation resulting from vitamin A deficiency or from retinitis pigmentosa or other peripheral retinal disease. Also known as nyctalopia.

night sweat [MED] Drenching perspiration occurring at night or during sleep in the course of certain febrile diseases.

nigrescent [BIOL] Blackish.

nihilism [MED] Pessimism in regard to the efficacy of treatment, particularly the use of drugs. [PSYCH] The content of delusions encountered in depressed or melancholic states; the patient insists that his inner organs no longer exist, and that his relatives have passed away.

Nilionidae [INV ZOO] The false ladybird beetles, a family of coleopteran insects in the superfamily Tenebrionoidea.

niobium [CHEM] A chemical element, symbol Nb, atomic number 41, atomic weight 92.906.

nipple [ANAT] The conical projection in the center of the mamma, containing the outlets of the milk ducts.

Nippotaeniidea [INV ZOO] An order of tapeworms of the subclass Cestoda including some internal parasites of certain fresh-water fishes; the head bears a single terminal sucker.

Nissl bodies [CYTOL] Chromophil granules of nerve cells which ultrastructurally are composed of large ribosomes.

Nitelleae [BOT] A tribe of stoneworts, order Charales, characterized by 10 cells in two tiers of five each composing the apical crown.

Nitidulidae [INV ZOO] The sap-feeding beetles, a large family of coleopteran insects in the superfamily Cucujoidea; individuals have five-jointed tarsi and antennae with a terminal three-jointed clavate expansion.

nitrate [CHEM] **1.** A salt or ester of nitric acid. **2.** Any compound containing the NO_3^- radical.

nitride [INORG CHEM] Compound of nitrogen and a metal, such as Mg_3N_2.

nitrification [MICROBIO] Formation of nitrous and nitric acids or salts by oxidation of the nitrogen in ammonia; specifically, oxidation of ammonium salts to nitrites and oxidation of nitrites to nitrates by certain bacteria.

Nitrobacteraceae [MICROBIO] The nitrifying bacteria, a family of the order Pseudomonadales which live autotrophically, derive energy from nitrification, and obtain carbon for growth from carbon dioxide.

nitrogen [CHEM] A chemical element, symbol N, atomic number 7, atomic weight 14.0067; it is a gas, diatomic (N_2) under normal conditions; about 78% of the atmosphere is N_2; in the combined form the element is a constituent of all proteins.

nitrogenase [BIOCHEM] An enzyme that catalyzes a six-electron reduction of N_2 in the process of nitrogen fixation.

nitrogen balance [PHYSIO] The difference between nitrogen intake (as protein) and total nitrogen excretion for an individual.

nitrogen cycle *See* carbon-nitrogen cycle.

nitrogen fixation [CHEM ENG] Conversion of atmospheric nitrogen into compounds such as ammonia, calcium cyanamide, or nitrogen oxides by chemical or electric-arc processes. [MICROBIO] Assimilation of atmospheric nitrogen by heterotrophic bacteria. Also known as dinitrogen fixation.

nitrogen narcosis [MED] Narcosis caused by gaseous nitrogen at high pressure in the blood; produced in divers breathing air at depths of 100 feet (30 meters) or more. Also known as rapture of the deep.

nitrogenous fertilizer [MATER] Fertilizer materials, natural or synthesized, containing nitrogen available for fixation by vegetation, such as potassium nitrate, KNO_3, or ammonium nitrate, NH_4NO_3.

nitrophilous [ECOL] Living in nitrogenous soils.

nitrophyte [ECOL] A nitrophilous plant.

NMR *See* nuclear magnetic resonance.

nobelium [CHEM] A chemical element, symbol No, atomic number 102, atomic weight 254 when the element is produced in the laboratory; a synthetic element, in the actinium series.

nocardiosis [MED] Infection by species of the fungus *Nocardia* characterized by spreading granulomatous lesions.

nocioceptive reflex *See* flexion reflex.

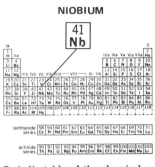

NIOBIUM

Periodic table of the chemical elements showing the position of niobium.

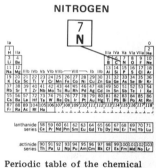

NITROGEN

Periodic table of the chemical elements showing the position of nitrogen.

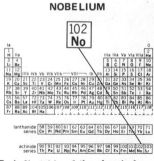

NOBELIUM

Periodic table of the chemical elements showing the position of nobelium.

noct-, nocti-, nocto-, noctu- [SCI TECH] Combining form meaning night.

noctalbuminuria [MED] Excretion of protein in night urine only.

Noctilionidae [VERT ZOO] The fish-eating bats, a tropical American monogeneric family of the Chiroptera having small eyes and long, narrow wings.

Noctuidae [INV ZOO] A large family of dull-colored, medium-sized moths in the superfamily Noctuoidea; larva are mostly exposed foliage feeders, representing an important group of agricultural pests.

Noctuoidea [INV ZOO] A large superfamily of lepidopteran insects in the suborder Heteroneura; most are moderately large moths with reduced maxillary palpi.

nocturnal [BIOL] Being active or occurring only at night.

nocturnal emission [PHYSIO] Normal, involuntary seminal discharge occurring during sleep in males after puberty.

nocturnal enuresis [MED] Involuntary nocturnal urination during sleep.

nodal rhythm [PHYSIO] A cardiac rhythm characterized by pacemaker function originating in the atrioventricular node, with a heart rate of 40-70 per minute.

nodal tachycardia [MED] A cardiac arrhythmia characterized by a heart rate of 140-220 per minute.

nodal tissue [HISTOL] **1.** Tissue from the sinoatrial node, and the atrioventricular node and bundle and its branches, composed of a dense network of Purkinje fibers. **2.** Tissue from a lymph node.

node [ANAT] **1.** A knob or protuberance. **2.** A small, rounded mass of tissue, such as a lymph node. **3.** A point of constriction along a nerve. [BOT] A point of attachment for a leaf on a stem.

node of Ranvier [ANAT] The region of a local constriction in a myelinated nerve; formed at the junction of two Schwann cells.

Nodosariacea [INV ZOO] A superfamily of Foraminiferida in the suborder Rotaliina characterized by a radial calcite test wall with monolamellar septa, and a test that is coiled, uncoiled, or spiral about the long axis.

nodose [BIOL] Having many or noticeable protuberances; knobby.

nodular [SCI TECH] Occurring in the form of small, rounded lumps.

nodular goiter *See* adenomatous goiter.

nodule [ANAT] **1.** A small node. **2.** A small aggregation of cells. [MED] A primary skin lesion, seen as a circumscribed solid elevation.

nodules of the semilunar valves [ANAT] Small nodes in the midregion of the pulmonary and aortic semilunar valves.

nodulose [BIOL] Having minute nodules or fine knobs.

Noeggerathiales [PALEOBOT] A poorly defined group of fossil plants whose geologic range extends from Upper Carboniferous to Triassic.

nogalamycin [MICROBIO] An antineoplastic antibiotic produced by *Streptomyces nogalaster*.

noma [MED] Spreading gangrene beginning in a mucous membrane; considered to be a malignant form of infection by fusospirochetal organisms. Also known as gangrenous stomatitis.

nomenclature [SCI TECH] A systematic arrangement of the distinctive names employed in any science.

nomen dubium [SYST] A proposed taxonomic name invalid because it is not accompanied by a definition or description of the taxon to which it applies.

nomen nudum [SYST] A proposed taxonomic name invalid

because the accompanying definition or description of the taxon cannot be interpreted satisfactorily.

nonaperaturate [BIOL] Lacking apertures.

noncommunicating hydrocephaly *See* obstructive hydrocephaly.

nondisjunction mosaic [GEN] A population of cells with different chromosome numbers produced when one chromosome is lost during mitosis or when both members of a pair of chromosomes are included in the same daughter nucleus; can occur during embryogenesis or adulthood.

nongonococcal urethritis [MED] Human urethral inflammation not associated with common bacterial pathogens; thought to be caused by bacteria of the Bedsonia group. Abbreviated NGU.

nongranular leukocyte [HISTOL] A white blood cell, such as a lymphocyte or monocyte, with clear homogeneous cytoplasm.

nonideal gas [STAT MECH] A gas whose molecules have significant interaction, more than that needed to bring about the equilibrium.

Nonionacea [INV ZOO] A superfamily of Foraminiferida in the suborder Orbitoidacea, characterized by a granular calcite test wall with monolamellar septa, and a planispiral to trochospiral test.

non-Newtonian fluid [FL MECH] A fluid whose flow behavior departs from that of a Newtonian fluid, so that the rate of shear is not proportional to the corresponding stress. Also known as non-Newtonian system.

non-Newtonian system *See* non-Newtonian fluid.

nonodontogenic cyst [MED] Any oral cyst that develops from epithelium which has been sequestered in bony or soft-tissue suture lines during embryonic development.

nonparalytic poliomyelitis [MED] Infection by poliomyelitis virus accompanied by upper respiratory or gastrointestinal symptoms, muscular pain and stiffness, and mild fever.

nonprotein nitrogen [BIOCHEM] The nitrogen fraction in the body tissues, excretions, and secretions, not precipitated by protein precipitants.

nonpsychotic organic brain syndrome [MED] Organic brain syndrome in which there is no apparent psychosis.

nonsaccharine sorghum *See* grain sorghum.

nonsense mutation [MOL BIO] A mutation that changes a codon that codes for one amino acid into a codon that does not specify any amino acid (a nonsense codon).

nonspecific [MED] Not attributable to any one definite cause, as a disease not caused by one particular microorganism, or an immunity not conferred by a specific antibody. [PHARM] Of medicines or therapy, not counteracting any one causative agent.

nonspecific hepatitis *See* interstitial hepatitis.

nonspecific immunity [IMMUNOL] Resistance attributable to factors other than specific antibodies, including genetic, age, or hormonal factors.

nonstriated muscle fiber *See* smooth muscle fiber.

nonvenereal syphilis [MED] Syphilis not acquired during sexual intercourse.

noosphere *See* anthroposphere.

noradrenaline *See* norepinephrine.

norepinephrine [BIOCHEM] $C_8H_{11}O_3N$ A hormone produced by chromaffin cells of the adrenal medulla; acts as a vasoconstrictor and mediates transmission of sympathetic nerve impulses. Also known as noradrenaline.

normal curve *See* Gaussian curve.

normal distribution [STAT] The most commonly occurring probability distributions have the form

$$(1/\sigma\sqrt{2\pi}) \int_{-\infty}^{u} \exp(-u^2/2)du, \; u = (x-e)/\sigma$$

where e is the mean and σ is the variance. Also known as Gauss' error curve; Gaussian distribution.

normality [CHEM] Measure of the number of gram-equivalent weights of a compound per liter of solution. Abbreviated N.

normalize [STAT] To carry out a normal transformation on a variate.

normal saline [PHYSIO] U.S. Pharmacopoeia title for a sterile solution of sodium chloride in purified water, containing 0.9 gram of sodium chloride in 100 milliliters; isotonic with body fluids. Also known as isotonic sodium chloride solution; normal salt solution; physiological saline; physiological salt solution; physiological sodium chloride solution; sodium chloride solution.

normal salt solution *See* normal saline.

normal solution [CHEM] An aqueous solution containing one equivalent of the active reagent in grams in 1 liter of the solution.

normal transformation [STAT] A transformation on a variate that converts it into a variate which has a normal distribution.

normoblast [HISTOL] The smallest of the nucleated precursors of the erythrocyte; slightly larger than a mature adult erythrocyte. Also known as acidophilic erythroblast; arthochromatic erythroblast; eosinophilic erythroblast; karyocyte; metakaryocyte; metanitricyte; metarubricyte.

normochromatic [CYTOL] Pertaining to cells of the erythrocytic series which have a normal staining color; attributed to the presence of a full complement of hemoglobin.

normochromic [CYTOL] Pertaining to erythrocytes which have a mean corpuscular hemoglobin (MCH), or color, index and a mean corpuscular hemoglobin concentration (MCHC), or saturation, index within two standard deviations above or below the mean normal.

normocyte [HISTOL] An erythrocyte having both a diameter and a mean corpuscular volume (MCV) within two standard deviations above or below the mean normal.

normothermia [PHYSIO] A state of normal body temperature.

North American blastomycosis [MED] A type of blastomycosis caused by the diphasic fungus *Blastomyces dermatitidis*; two recognized forms are cutaneous and systemic.

northern anthracnose [PL PATH] A fungus disease of red and crimson clovers in North America, Asia, and Europe caused by *Kabatulla caulivora*; depressed, linear brown lesions form on the stems and petioles.

noscapine [PHARM] $C_{22}H_{23}NO_7$ A white, colorless, tasteless alkaloid obtained from opium; used as a nonaddicting antitussive. Formerly known as narcotine.

nose [ANAT] The nasal cavities and the structures surrounding and associated with them in all vertebrates.

nose height [ANTHRO] The distance measured between the nasion and the subnasale.

nose leaf [VERT ZOO] A leaflike expansion of skin on the nose of certain bats; believed to have a tactile function.

nose length [ANTHRO] The distance measured between the nasion and the pronasale.

nosocomial [MED] **1.** Pertaining to a hospital. **2.** Of disease, caused or aggravated by hospital life.

Nosodendridae [INV ZOO] The wounded-tree beetles, a small family of coleopteran insects in the superfamily Dermestoidea.

nostopathy [PSYCH] Pathogenic homecoming, that is, stress-precipitated illness as observed in individuals who have spent a considerable length of time in institutions such as hospitals or prisons.

nostril [ANAT] One of the external orifices of the nose. Also known as naris.

nostrum [PHARM] A quack medicine.

Notacanthidae [VERT ZOO] A family of benthic, deep-sea teleosts in the order Notacanthiformes, including the spiny eel.

Notacanthiformes [VERT ZOO] An order of actinopterygian fishes whose body is elongated, tapers posteriorly, and has no caudal fin.

notch graft [BOT] A plant graft in which the scion is inserted in a narrow slit in the stock.

Noteridae [INV ZOO] The burrowing water beetles, a small family of coleopteran insects in the suborder Adephaga.

Nothosauria [PALEON] A suborder of chiefly marine Triassic reptiles in the order Sauropterygia.

Notidanoidea [VERT ZOO] A suborder of rare sharks in the order Selachii; all retain the primitive jaw suspension of the order.

Notiomastodontinae [PALEON] A subfamily of extinct elephantoid proboscidean mammals in the family Gomphotheriidae.

Notioprogonia [PALEON] A suborder of extinct mammals comprising a diversified archaic stock of Notoungulata.

notochord [VERT ZOO] An elongated dorsal cord of cells which is the primitive axial skeleton in all chordates; persists in adults in the lowest forms (*Branchiostoma* and lampreys) and as the nuclei pulposi of the intervertebral disks in adult vertebrates.

notochordal canal [EMBRYO] A canal formed by a continuation of the primary pit into the head process of mammalian embryos; provides a temporary connection between the yolk sac and amnion.

notochordal plate [EMBRYO] A plate of cells representing the root of the head process of the embryo after the embryo becomes vesiculated.

Notodelphyidiformes [INV ZOO] A tribe of the Gnathostoma in some systems of classification.

Notodelphyoida [INV ZOO] A small group of crustaceans bearing a superficial resemblance to many insect larvae as a result of uniform segmentation, comparatively small trunk appendages, and crowding of inconspicuous oral appendages into the anterior portion of the head.

Notodontidae [INV ZOO] The puss moths, a family of lepidopteran insects in the superfamily Noctuoidea, distinguished by the apparently three-branched cubitus.

Notogaean [ECOL] Pertaining to or being a biogeographic region including Australia, New Zealand, and the southwestern Pacific islands.

Notommatidae [INV ZOO] A family of rotifers in the order Monogonota including forms with a cylindrical body covered by a nonchitinous cuticle and with a slender posterior foot.

Notomyotina [INV ZOO] A suborder of echinoderms in the order Phanerozonida in which the upper marginals alternate in position with the lower marginals, and each tube foot has a terminal sucking disk.

Notonectidae [INV ZOO] The backswimmers, a family of aquatic, carnivorous hemipteran insects in the superfamily Notonectoidea; individuals swim ventral side up, aided in breathing by an air bubble.

Notonectoidea [INV ZOO] A superfamily of Hemiptera in the subdivision Hydrocorisae.

notopodium [INV ZOO] The dorsal branch of a parapodium in certain annelids.

NOTACANTHIFORMES

Spiny eel (*Notacanthus nasus*). (*After D. S. Jordan and B. W. Evermann, The Fishes of North and Middle America, U.S. Nat. Mus. Bull. no. 47, 1900*)

NOTHOSAURIA

25 cm

Ceresiosaurus calcagnii, ventral view; skeleton of a nothosaur with elongated neck; Triassic, Switzerland. (*From B. Peyer, 1944*)

NOTODELPHYOIDA

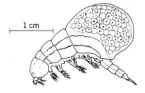

1 cm

Doropygus psyllus Thorell, female; eggs are incubated in swollen part of thorax.

NOTOSTRACA

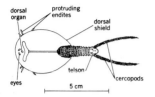

Lepidurus arcticus, female, dorsal aspect.

NOTOUNGULATA

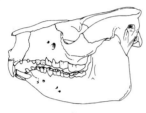

Skull and jaw of *Adinotherium ovinum*, an early Miocene toxodontid notoungulate from the Santa Cruz formation of Patagonia, Argentina. *(After W. Scott)*

Notopteridae [VERT ZOO] The featherbacks, a family of actinopterygian fishes in the order Osteoglossiformes; bodies are tapered and compressed, with long anal fins that are continuous with the caudal fin.

Notoryctidae [PALEON] An extinct family of Australian insectivorous mammals in the order Marsupialia.

Notostigmata [INV ZOO] The single suborder of the Opilioacriformes, an order of mites.

Notostigmophora [INV ZOO] A subclass or suborder of the Chilopoda, including those centipedes embodying primitive as well as highly advanced characters, distinguished by dorsal respiratory openings.

Notostraca [INV ZOO] The tadpole shrimps, an order of crustaceans generally referred to the Branchiopoda, having a cylindrical trunk that consists of 25-44 segments, a dorsoventrally flattened dorsal shield, and two narrow, cylindrical cercopods on the telson.

Nototheniidae [VERT ZOO] A family of perciform fishes in the suborder Blennioidei, including most of the fishes of the permanently frigid waters surrounding Antarctica.

Notoungulata [PALEON] An extinct order of hoofed herbivorous mammals, characterized by a skull with an expanded temporal region, primitive dentition, and primitive feet with five toes, the weight borne mainly by the third digit.

notum [INV ZOO] The dorsal portion of a thoracic segment in insects.

novobiocin [MICROBIO] $C_{30}H_{36}O_{11}N_2$ A moderately broadspectrum antibiotic produced by strains of *Streptomyces niveus* and *S. spheroides*; it is a dibasic acid and is converted either to the monosodium salt or to the calcium acid salt for pharmaceutical use.

Novocain [PHARM] A trade name for procaine hydrochloride.

Np *See* neptunium.

nucellus [BOT] The oval central mass of tissue in the ovule; contains the embryo sac.

nucha [ANAT] The nape of the neck.

nuchal ligament [ANAT] An elastic ligament extending from the external occipital protuberance and middle nuchal line to the spinous process of the seventh cervical vertebra. Also known as ligamentum nuchae.

nuchal organ [INV ZOO] Any of various sense organs on the prostomium of many annelids, which are sensitive to changes in the immediate environment of the individual.

nuchal rigidity [MED] Stiffness in the nape of the neck, often accompanied by pain and spasm on attempts to move the head; the most common sign of meningitis.

nuchal tentacle [INV ZOO] Any of various filiform or thick, fleshy tactoreceptors on anterior segments of many annelids.

nuciferous [BOT] Bearing nuts.

nucivorous [BIOL] Feeding on nuts.

nuclear endosperm [BOT] Endosperm in which the formation of partitions is delayed until after several or many mitotic divisions so that there is a free-nuclear stage in early ontogeny.

nuclear magnetic moment [NUC PHYS] The magnetic dipole moment of an atomic nucleus; a vector whose scalar product with the magnetic flux density gives the negative of the energy of interaction of a nucleus with a magnetic field.

nuclear magnetic resonance [PHYS] A phenomenon exhibited by a large number of atomic nuclei, in which nuclei in a static magnetic field absorb energy from a radio-frequency field at certain characteristic frequencies. Abbreviated NMR. Also known as magnetic nuclear resonance.

nuclear membrane [CYTOL] The envelope surrounding the

cell nucleus, separating the nucleoplasm from the cytoplasm; composed of two membranes and contains numerous pores.

nuclear physics [PHYS] The study of the characteristics, behavior, and internal structures of the atomic nucleus.

nuclear sclerosis [MED] Hardening of the ocular lens nucleus.

nuclease [BIOCHEM] An enzyme that catalyzes the splitting of nucleic acids to nucleotides, nucleosides, or the components of the latter.

nucleic acid [BIOCHEM] A large, acidic, chainlike molecule containing phosphoric acid, sugar, and purine and pyrimidine bases; two types are ribonucleic acid and deoxyribonucleic acid.

nuclein [BIOCHEM] Any of a poorly defined group of nucleic acid protein complexes occurring in cell nuclei.

nucleocytoplasmic ratio [CYTOL] The ratio between the measured cross-sectional area or estimated volume of the nucleus of a cell to the volume of its cytoplasm. Also known as karyoplasmic ratio.

nucleolar [CYTOL] Of or pertaining to the nucleolus.

nucleolus [CYTOL] A small, spherical body composed principally of protein and located in the metabolic nucleus. Also known as plasmosome.

nucleoplasm [CYTOL] The protoplasm of a nucleus. Also known as karyoplasm.

nucleoprotein [BIOCHEM] Any member of a class of conjugated proteins in which molecules of nucleic acid are closely associated with molecules of protein.

nucleoreticulum [CYTOL] Any type of network found within a nucleus.

nucleosidase [BIOCHEM] An enzyme that catalyzes the hydrolysis of a nucleoside to its component pentose and its purine or pyrimidine base.

nucleoside [BIOCHEM] The glycoside resulting from removal of the phosphate group from a nucleotide; consists of a pentose sugar linked to a purine or pyrimidine base.

nucleosome [CYTOL] A morphologically repeating unit of DNA containing 190 base pairs of DNA folded together with eight histone molecules. Also known as v-body.

nucleospindle [CYTOL] A mitotic spindle derived from nuclear material.

nucleotidase [BIOCHEM] Any of a group of enzymes which split phosphoric acid from nucleotides, leaving nucleosides.

nucleotide [BIOCHEM] An ester of a nucleoside and phosphoric acid; the structural unit of a nucleic acid.

nucleus [CYTOL] A small mass of differentiated protoplasm rich in nucleoproteins and surrounded by a membrane; found in most animal and plant cells, contains chromosomes, and functions in metabolism, growth, and reproduction. [HISTOL] A mass of nerve cells in the central nervous system. [NUC PHYS] The central, positively charged, dense part of an atom. Also known as atomic nucleus. [SCI TECH] A central mass about which accretion takes place.

nuculanium [BOT] An indehiscent fleshy fruit, as a grape, differing from a berry in being superior.

Nuda [INV ZOO] A class of the phylum Ctenophora distinguished by the lack of tentacles.

Nudechiniscidae [INV ZOO] A family of heterotardigrades in the suborder Echiniscoidea characterized by a uniform cuticle.

Nudibranchia [INV ZOO] A suborder of the Opisthobranchia containing the sea slugs; these mollusks lack a shell and a mantle cavity, and the gills are variable in size and shape.

nudibranchiate [INV ZOO] Not having a protective covering on the gills.

nudicaudate [VERT ZOO] Having a naked tail.

nudicaulous [BOT] Pertaining to or having stems without leaves.

nudiflorous [BOT] Not having glands or hairs on flowers.

nudism [PSYCH] Intolerance for wearing clothing, with a morbid tendency for the individual to remove his clothing.

null hypothesis [STAT] The hypothesis that there is no validity to the specific claim that two variations (treatments) of the same thing can be distinguished by a specific procedure.

numerical taxonomy [SYST] The numerical evaluation of the affinity or similarity between taxonomic units and the ordering of these units into taxa on the basis of their affinities.

Numididae [VERT ZOO] A family of birds in the order Galliformes commonly known as guinea fowl; there are few if any feathers on the neck or head, but there may be a crest of feathers and various fleshy appendages.

nurse cell [HISTOL] A cell type of the ovary of many animals which nourishes the developing egg cell.

nurse graft [BOT] A plant graft in which the scion remains united with the stock only until roots develop on the scion.

nursing [MED] The application of the principles of physical, biological, and social sciences in the physical and mental care of people.

nut [BOT] **1.** A fruit which has at maturity a hard, dry shell enclosing a kernel consisting of an embryo and nutritive tissue. **2.** An indehiscent, one-celled, one-seeded, hard fruit derived from a single, simple, or compound ovary.

nutation [BOT] Rhythmic change in the position of growing plant organs caused by variation in the growth rates on different sides of the growing apex.

nutgall [PL PATH] A nutlike gall.

nutlet [BOT] The stone of a drupaceous fruit.

nutmeg [BOT] *Myristica fragrans.* A dark-leafed evergreen tree of the family Myristicaceae cultivated for the golden-yellow fruits which resemble apricots; a delicately flavored spice is obtained from the kernels inside the seeds.

nutmeg liver [MED] Chronic passive hyperemia of the liver; the cut surface of the diseased organ resembles the cut surface of a nutmeg.

nutrient [BIOL] Providing nourishment.

nutrient foramen [ANAT] The opening into the canal which gives passage to the blood vessels of the medullary cavity of a bone.

nutrition [BIOL] The science of nourishment, including the study of nutrients that each organism must obtain from its environment in order to maintain life and reproduce.

nutritional anemia [MED] Anemia resulting from certain nutritional deficiencies.

nutritional dystrophy *See* kwashiorkor.

nutritional hypochromic anemia *See* iron-deficiency anemia.

Nuttalliellidae [INV ZOO] A family of ticks (Ixodides) containing one rare African species, *Nuttalliella namaqua,* morphologically intermediate between the families Argasidae and Ixodidae.

nux vomica [BOT] The seed of *Strychnos nux-vomica,* an Indian tree of the family Loganiaceae; contains the alkaloid strychnine, and was formerly used in medicine.

Nyctaginaceae [BOT] A family of dicotyledonous plants in the order Caryophyllales characterized by an apocarpous, monocarpous, or syncarpous gynoecium, sepals joined to a tube, a single carpel, and a cymose inflorescence.

nyctalopia *See* night blindness.

nyctanthous [BOT] Flowering nocturnally.

Nycteridae [VERT ZOO] The slit-faced bats, a monogeneric family of insectivorous chiropterans having a simple, well-

NUTMEG

Nutmeg *(Myristica fragrans),* mature fruits. *(USDA)*

developed nose leaf, and large ears joined together across the forehead.

Nyctibiidae [VERT ZOO] A family of birds in the order Caprimulgiformes including the neotropical potoos.

nyctinasty [BOT] A nastic movement in higher plants associated with diurnal light and temperature changes.

nyctipelagic [ECOL] Rising to the surface of the sea during the night.

Nyctribiidae [INV ZOO] The bat tick flies, a family of myodarian cyclorrhaphous dipteran insects in the subsection Acalypteratae.

nymph [INV ZOO] Any immature larval stage of various hemimetabolic insects.

nympha [INV ZOO] **1.** The edge of the valve to which the hinge ligaments are attached in bivalve mollusks. **2.** A pair of sclerites below the epigynal plate in mites.

Nymphaeaceae [BOT] A family of dicotyledonous plants in the order Nymphaeales distinguished by the presence of roots, perfect flowers, alternate leaves, and uniaperturate pollen.

Nymphaeales [BOT] An order of flowering aquatic herbs in the subclass Magnoliidae; all lack cambium and vessels and have laminar placentation.

Nymphalidae [INV ZOO] The four-footed butterflies, a family of lepidopteran insects in the superfamily Papilionoidea; prothoracic legs are atrophied, and the well-developed patagia are heavily sclerotized.

Nymphalinae [INV ZOO] A subfamily of the lepidopteran family Nymphalidae.

nymphochrysalis [INV ZOO] A pupalike resting stage between the larva and the nymph in certain mites.

nymphomania [PSYCH] Excessive sexual desire on the part of a woman. Also known as hysteromania.

Nymphonidae [INV ZOO] A family of marine arthropods in the subphylum Pycnogonida; members have chelifores, five-jointed palpi, and ten-jointed ovigers.

nymphosis [INV ZOO] The process of developing into a nymph or a pupa.

Nymphulinae [INV ZOO] A subfamily of the lepidopteran family Pyralididae which is notable because some species are aquatic.

Nysmyth's membrane [ANAT] The primary enamel cuticle which is the transitory remnants of the enamel organ and oral epithelium covering the enamel of a tooth after eruption.

Nyssaceae [BOT] A family of dicotyledonous plants in the order Cornales characterized by perfect or unisexual flowers with imbricate petals, a solitary ovule in each locule, a unilocular ovary, and more stamens than petals.

nystagmus [MED] Involuntary oscillatory movement of the eyeballs.

nystatin [MICROBIO] $C_{46}H_{77}NO_{19}$ An antifungal antibiotic produced by *Streptomyces noursei*; used for the treatment of infections caused by *Candida (Monilia) albicans.*

NYMPHAEACEAE

A common eastern American species of water lily *(Nymphaea odorata)* in the family Nymphaeaceae of the order Nymphaeales. The large and indefinite number of tepals, stamens, and carpels is characteristic of the family and a large part of the order. *(Photograph by Hugh Spencer, National Audubon Society)*

O

O *See* oxygen.

oak [BOT] Any tree of the genus *Quercus* in the order Fagales, characterized by simple, usually lobed leaves, scaly winter buds, a star-shaped pith, and its fruit, the acorn, which is a nut; the wood is tough, hard, and durable, generally having a distinct pattern.

oak wilt [PL PATH] A fungus disease of oak trees caused by *Chalara quercina*, characterized by wilting and yellow and red discoloration of the leaves progressing from the top downward and inward.

O antigen [MICROBIO] A somatic antigen of certain flagellated microorganisms.

oat [BOT] Any plant of the genus *Avena* in the family Graminae, cultivated as an agricultural crop for its seed, a cereal grain, and for straw.

obclavate [BIOL] Inversely clavate.

obcordate [BOT] Referring to a leaf, heart-shaped with the notch apical.

obdiplastemonous [BOT] Having the stamens arranged in two whorls, with members of the outer whorl positioned opposite the petals.

obduction [MED] The act or instance of performing a postmortem examination.

obelion [ANTHRO] The point where the line which joins the parietal foramens crosses the sagittal suture.

Obermayer's reagent [CHEM] A 0.4% solution of ferric chloride in concentrated hydrochloric acid; used to test for indican in urine, with a pale-blue or deep-violet color indicating positive.

obese [ANAT] Extremely fat.

obfuscation [PSYCH] Mental confusion.

objective [OPTICS] The first lens, lens system, or mirror through which light passes or from which it is reflected in an optical system; many scientists exclude mirrors from the definition.

objective sign [MED] A sign which can be detectable by someone other than the patient.

object relationship [PSYCH] The attitudes and responses of one person toward another; the capacity of an individual to react appropriately to and to accept and love other people.

oblanceolate [SCI TECH] Inversely lanceolate.

obligate [BIOL] Restricted to a specified condition of life, as an obligate parasite.

oblique [ANAT] Referring to a muscle, positioned obliquely and having one end that is not attached to bone. [BOT] Referring to a leaf, having the two sides of a blade unequal. [SCI TECH] Having a slanted direction or position.

obliterating endarteritis *See* endarteritis obliterans.

obliteration [MED] **1.** Complete removal of an organ or other body part by disease or surgical excision. **2.** Closure of a

OAK

Terminal bud, leaf, and twig of white oak *(Quercus alba)*.

lumen. **3.** Loss of memory or consciousness of specific events.

obliterative appendicitis [MED] Obliteration of the lumen of the appendix by fibrofatty tissue.

Obolellida [PALEON] A small order of Early and Middle Cambrian inarticulate brachiopods, distinguished by a shell of calcium carbonate.

obovate [BIOL] Inversely ovate.

obsession [PSYCH] Persistence of or anxious preoccupation with an idea or emotion recognized as unreasonable by the individual.

obsessive-compulsive neurosis [PSYCH] A neurotic disorder in which anxiety relates to obsessions against which the individual fights but which he cannot control and by which he is dominated.

obsessive-compulsive personality [PSYCH] A behavioral disorder in which a person is generally characterized by chronic, excessive concern with conformity or adherence to standards, resulting in inhibited, inflexible behavior, inability to relax, and the performance of an inordinate amount of work.

obsolete [BIOL] A part of an organism that is imperfect or indistinct, compared with a corresponding part of similar organisms.

obstetric [MED] **1.** Of or pertaining to obstetrics. **2.** Of or pertaining to pregnancy and childbirth.

obstetrical analgesia [MED] Analgesia induced to diminish or obliterate the pain of childbirth.

obstetrician [MED] One who practices obstetrics.

obstetrics [MED] The branch of medicine that deals with pregnancy, labor, and the puerperium.

obstipation [MED] Constipation that is difficult to relieve.

obstruction [MED] Occlusion or stenosis of hollow viscera, ducts, and vessels.

obstructive atelectasis [MED] Collapse of all or a portion of the lung due to bronchial obstruction or occlusion. Also known as absorption atelectasis.

obstructive dysmenorrhea *See* mechanical dysmenorrhea.

obstructive emphysema [MED] Overdistension of the lung due to partial obstruction of the air passages, which permits air to enter the alveoli but which resists expiration of the air.

obstructive glaucoma *See* narrow-angle glaucoma.

obstructive hydrocephaly [MED] Accumulation of cerebrospinal fluid in the brain ventricles caused by obstruction of the passage of the fluid from the ventricles to the subarachnoid space. Also known as internal hydrocephaly; noncommunicating hydrocephaly.

obstructive jaundice [MED] Jaundice caused by mechanical obstruction of the biliary passages, preventing the outflow of bile.

obtund [MED] To make dull or reduce, as to obtund sensibility.

obturation [MED] **1.** The closing of an opening or passage. **2.** A form of intestinal obstruction in which the lumen is occupied by its normal contents or by foreign bodies.

obturator [ANAT] **1.** Pertaining to that which closes or stops up, as an obturator membrane. **2.** Either of two muscles, originating at the pubis and ischium, which rotate the femur laterally. [MED] A solid wire or rod contained within a hollow needle or cannula.

obturator artery [ANAT] A branch of the internal iliac; it branches into the pubic and acetabular arteries.

obturator foramen [ANAT] A large opening in the pelvis, between the ischium and the pubis, that gives passage to vessels and nerves; it is partly closed by a fibrous obturator membrane.

obturator membrane [ANAT] **1.** A fibrous membrane closing the obturator foramen of the pelvis. **2.** A thin membrane between the crura and foot plates of the stapes.

obturator nerve [ANAT] A mixed nerve arising in the lumbar plexus; innervates the adductor, gracilis, and obturator externus muscles, and the skin of the medial aspect of the thigh, hip, and knee joints.

obtuse [BOT] Of a leaf, having a blunt or rounded free end.

obvallate [BIOL] Surrounded by or as if by a wall.

obverse [BIOL] Having the apex wider than the base, as certain leaves.

obvolute [BIOL] Overlapping.

occipital arch [INV ZOO] A part of an insect cranium lying between the occipital suture and postoccipital suture.

occipital artery [ANAT] A branch of the external carotid which branches into the mastoid, auricular, sternocleidomastoid, and meningeal arteries.

occipital bone [ANAT] The bone which forms the posterior portion of the skull, surrounding the foramen magnum.

occipital condyle [ANAT] An articular surface on the occipital bone which articulates with the atlas. [INV ZOO] A projection on the posterior border of an insect head which articulates with the lateral neck plates.

occipital crest [ANAT] Either of two transverse ridges connecting the occipital protuberances with the foramen magnum.

occipital ganglion [INV ZOO] One of a pair of ganglia located just posterior to the brain in insects.

occipital lobe [ANAT] The posterior lobe of the cerebrum having the form of a three-sided pyramid.

occipital pole [ANAT] The tip of the occipital lobe of the brain.

occipital protuberance [ANAT] A prominence on the surface of the occipital bone to which the ligamentum nuchae is attached.

occipitofrontalis [ANAT] A muscle in two parts, the frontal (inserting in the skin of the forehead) and the occipital (inserting in the galea sponeurotica).

occiput [ZOO] The back of the head of an insert or vertebrate.

occlusal disharmony [MED] Increased or maldirected occlusal force on individual teeth or groups of teeth causing a malposition or functional aberration.

occlusion [ANAT] The relationship of the masticatory surfaces of the maxillary teeth to the masticatory surfaces of the mandibular teeth when the jaws are closed. [MED] A closing or shutting up. [PHYSIO] The deficit in muscular tension when two afferent nerves that share certain motoneurons in the central nervous system are stimulated simultaneously, as compared to the sum of tensions when the two nerves are stimulated separately.

occult blood [PATH] Blood in body products such as feces, not detectable on gross examination.

occult hydrocephaly [MED] A syndrome in which the brain ventricular system is enlarged while cerebrospinal fluid pressure remains normal, causing dementia, disturbances of equilibrium, and disorders of sphincter control.

occult virus [VIROL] A virus whose presence is assumed but which cannot be recovered.

occupational acne [MED] Acne acquired from regular exposure to acnegenic materials in certain industries; disappears when the cause is removed.

occupational disease [MED] A functional or organic disease caused by factors arising from the operations or materials of an individual's industry, trade, or occupation.

OCELLUS

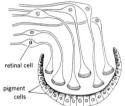

retinal cell

pigment
cells

Simple pigment cup ocellus of
flatworm. *(After L. Hyman;
from R. D. Barnes, Invertebrate
Zoology, 2d ed., W. B. Saunders
Co., 1968)*

OCTOPUS

Octopus bairdi with a body length
of 3 inches (7.6 centimeters)
and arms up to 40 inches (1
meter) long, webbed for one-
third of their length.

occupational medicine [MED] The branch of medicine which deals with the relationship of humans to their occupations, for the purpose of the prevention of disease and injury and the promotion of optimal health, productivity, and social adjustment.

occupational neurosis [PSYCH] Any neurotic disorder manifested by the individual's inability to use those parts of his body commonly employed in his occupation, such as a writer's inability to write due to a painful feeling of fatigue in the hand.

occupational therapy [MED] The teaching of skills or the use of selected occupations for therapeutic or rehabilitation purposes.

Oceanian [ECOL] Of or pertaining to the zoogeographic region that includes the archipelagos and islands of the central and south Pacific.

oceanic zone [OCEANOGR] The biogeographic area of the open sea.

oceanodromous [VERT ZOO] Of a fish, migratory in salt water.

ocellus [INV ZOO] A small, simple invertebrate eye composed of photoreceptor cells and pigment cells.

ocelot [VERT ZOO] *Felis pardalis.* A small arboreal wild cat, of the family Felidae, characterized by a golden head and back, silvery flanks, and rows of somewhat metallic spots on the body.

Ochnaceae [BOT] A family of dicotyledonous plants in the order Theales, characterized by simple, stipulate leaves, a mostly gynobasic style, and anthers that generally open by terminal pores.

Ochoan [GEOL] A North American provincial series: uppermost Permian (above Guadalupian, below Lower Triassic).

Ochotonidae [VERT ZOO] A family of the mammalian order Lagomorpha; members are relatively small, and all four legs are about equally long.

ochroleucous [BIOL] Pale ocher or buff colored.

ochronosis [MED] A blue or brownish-blue discoloration of cartilage and connective tissue, especially around joints, caused by melanotic pigment.

Ochteridae [INV ZOO] The velvety shorebugs, the single family of the hemipteran superfamily Ochteroidea.

Ochteroidea [INV ZOO] A monofamilial tropical and subtropical superfamily of hemipteran insects in the subdivision Hydrocorisae; individuals are black with a silky sheen, and the antennae are visible from above.

ocrea [BOT] A tubular stipule or pair of coherent stipules.

octactine [INV ZOO] A modified hexactin having eight equal rays.

octagynous [BOT] Having eight pistils.

octandrous [BOT] Having eight stamens.

Octocorallia [INV ZOO] The equivalent name for Alcyonaria.

octopetalous [BOT] Having eight petals.

octophore [MYCOL] A modified ascus having eight radially arranged spores.

Octopoda [INV ZOO] An order of the dibranchiate cephalopods, characterized by having eight arms equipped with one to three rows of suckers.

Octopodidae [INV ZOO] The octopuses, in family of cephalopod mollusks in the order Octopoda.

octopus [INV ZOO] Any member of the genus *Octopus* in the family Octopodidae; the body is round with a large head and eight partially webbed arms, each bearing two rows of suckers, and there is no shell.

octosepalous [BOT] Having eight sepals.

ocular [BIOL] Of or pertaining to the eye.

ocular lobe [INV ZOO] A projecting thoracic lobe occurring in some beetles.

ocular plate [INV ZOO] One of the plates at the end of the ambulacra in sea urchins.

ocular skeleton [VERT ZOO] A rigid structure in most sub-mammalian vertebrates consisting of a cup of hyaline carti-lage enclosing the posterior part of the eye, and a thin-walled ring of intramembranous bones in the edge of the sclera at its junction with the cornea.

oculist *See* ophthalmologist.

oculoglandular tularemia [MED] Infection by *Pasteurella tu-larensis* which, in addition to the usual symptoms of tulare-mia, causes swollen eyelids, conjunctivitis, swollen lymph nodes, and ulcers on the conjunctivae.

oculomotor [PHYSIO] Pertaining to eye movement. **2.** Per-taining to the oculomotor nerve.

oculomotor nerve [ANAT] The third cranial nerve; a paired somatic motor nerve arising in the floor of the midbrain, which innervates all extrinsic eye muscles except the lateral rectus and superior oblique, and furnishes autonomic fibers to the ciliary and pupillary sphincter muscles within the eye.

oculomotor nucleus [ANAT] A nucleus in the floor of the midbrain that gives rise to motor fibers of the oculomotor nerve.

oculomotor paralysis [MED] Paralysis of the oculomotor nerve.

Oculosida [INV ZOO] An order of the protozoan subclass Radiolaria; pores are restricted to certain areas in the central capsule, and an olive-colored material is always present near the astropyle.

odd-pinnate [BOT] Of a compound leaf, having a single leaflet at the tip of the petiole with leaflets on both sides of the petiole. Also known as imparipinnate.

odds ratio [STAT] The ratio of the probability of occurrence of an event to the probability of the event not occurring.

Odiniidae [INV ZOO] A family of cyclorrhaphous myodarian dipteran insects in the subsection Acalypteratae.

Odobenidae [VERT ZOO] A family of carnivorous mammals in the suborder Pinnipedia; contains a single species, the walrus (*Odobenus rosmarus*).

Odonata [INV ZOO] The dragonflies, an order of the class Insecta, characterized by a head with large compound eyes, and wings with clear or transparent membranes traversed by networks of veins.

odont-, odonto- [VERT ZOO] A combining form meaning tooth.

odontectomy [MED] Surgical excision of a tooth.

odontexesis [MED] Removal of deposits from the surface of teeth.

odontoblast [HISTOL] One of the elongated, dentin-forming cells covering the dental papilla.

odontoblastoma *See* ameloblastic odontoma.

Odontoceti [VERT ZOO] The toothed whales, a suborder of cetacean mammals distinguished by a single blowhole.

odontoclast [HISTOL] A multinuclear cell concerned with resorption of the roots of milk teeth.

odontogenesis [EMBRYO] Formation of teeth.

odontogenic [HISTOL] **1.** Pertaining to the origin and devel-opment of teeth. **2.** Originating in tissues associated with teeth.

odontogenic cyst [MED] A cyst originating in tissues associ-ated with teeth.

odontogenic fibroma [MED] A benign tumor originating in the mesenchymal derivatives of the tooth germ.

Odontognathae [PALEON] An extinct superorder of the avian

ODONATA

Adult dragonfly showing the four membranous wings.

subclass Neornithes, including all large, flightless aquatic forms and other members of the single order Hesperornithiformes.

odontoid [BIOL] Toothlike.

odontoid process [ANAT] A toothlike projection on the anterior surface of the axis vertebra with which the atlas articulates.

odontology [VERT ZOO] A branch of science that deals with the formation, development, and abnormalities of teeth.

odontoma [MED] A benign tumor representing a developmental excess, composed of mesodermal or octodermal tooth-forming tissue, alone or in association with the calcified derivatives of these structures.

odontophore [INV ZOO] A structure in the mouth of most mollusks, except bivalves, which supports the radula.

Odontostomatida [INV ZOO] An order of the protozoan subclass Spirotrichia; individuals are compressed laterally and possess very little ciliature.

Oecophoridae [INV ZOO] A family of small to moderately small moths in the lepidopteran superfamily Tineoidea, characterized by a comb of bristles, the pecten, on the scape of the antennae.

Oedemeridae [INV ZOO] The false blister beetles, a large family of coleopteran insects in the superfamily Tenebrionoidea.

oedipal [PSYCH] Pertaining to the Oedipus complex.

Oedipus complex [PSYCH] In psychoanalytic theory, the attraction and attachment of the child to the parent of the opposite sex, accompanied by feelings of envy and hostility toward the parent of the child's sex, whose displeasure and punishment the child so fears that the child represses his feelings toward the parent of opposite sex.

Oedogoniales [BOT] An order of fresh-water algae in the division Chlorophyta; characterized as branched or unbranched microscopic filaments with a basal holdfast cell.

Oegophiurida [INV ZOO] An order of echinoderms in the subclass Ophiuroidea, represented by a single living genus; members have few external skeletal plates and lack genital bursae, dorsal and ventral arm plates, and certain jaw plates.

Oegopsida [INV ZOO] A suborder of cephalopod mollusks in the order Decapoda of one classification system, and in the order Teuthoidea of another system.

Oepikellacea [PALEON] A dimorphic superfamily of extinct ostracods in the order Paleocopa, distinguished by convex valves and the absence of any trace of a major sulcus in the external configuration.

Oestridae [INV ZOO] A family of cyclorrhaphous myodarian dipteran insects in the subsection Calypteratae.

offshoot [BIOL] A shoot branching laterally from the main stem.

oidiophore [MYCOL] A hypha that produces oidia.

oidium [MYCOL] One of the small, thin-walled spores with flat ends produced by autofragmentation of the vegetative hyphae in certain Eumycetes.

Oikomonadidae [INV ZOO] A family of protozoans in the order Kinetoplastida containing organisms that have a single flagellum.

oil gland *See* uropygial gland.

Oiluvium *See* Pleistocene.

ointment [PHARM] A semisolid preparation used for a protective and emollient effect or as a vehicle for the local or endermic administration of medicaments; ointment bases are composed of various mixtures of fats, waxes, animal and vegetable oils, and solid and liquid hydrocarbons.

okapi [VERT ZOO] *Okapia johnstoni.* An artiodactylous mam-

OEDOGONIALES

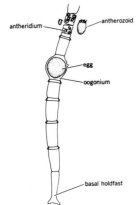

antheridium — antherozoid

— egg

— oogonium

— basal holdfast

Oedogonium, an attached, unbranched, microscopic filament arising from a basal holdfast cell.

OIDIUM

|— 10 μ —|

Two oidia showing the flat ends of this thin-walled spore.

OIKOMONADIDAE

20 μ

Oikomonas, a genus in the Oikomonadidae, showing the single flagellum characteristic of the family.

mal in the family Giraffidae; has a hazel coat with striped hindquarters, and the head shape, lips, and tongue are the same as those of the giraffe, but the neck is not elongate.

okra [BOT] *Hibiscus esculentus.* A tall annual plant grown for its edible immature pods. Also known as gumbo.

Olacaceae [BOT] A family of dicotyledonous plants in the order Santalales characterized by dry or fleshy indehiscent fruit, the presence of petals, stamens, and chlorophyll, and a 2-5-celled ovary.

Oldhaminidina [PALEON] A suborder of extinct articulate brachiopods in the order Strophomenida distinguished by a highly lobate brachial valve seated within an irregular convex pedicle valve.

Oleaceae [BOT] A family of dicotyledonous plants in the order Scrophulariales characterized generally by perfect flowers, two stamens, axile to parietal or apical placentation, a four-lobed corolla, and two ovules in each locule.

oleandomycin [MICROBIO] $C_{35}H_{61}O_{12}N$ A macrolide antibiotic produced by *Streptomyces antibioticus*; active mainly against gram-positive microorganisms. Also known as matromycin.

olecranon [ANAT] The large process at the distal end of the ulna that forms the bony prominence of the elbow and receives the insertion of the triceps muscle.

Olenellidae [PALEON] A family of extinct arthropods in the class Trilobita.

Olethreutidae [INV ZOO] A family of moths in the superfamily Tortricoidea whose hindwings usually have a fringe of long hairs along the basal part of the cubitus.

olfaction [PHYSIO] **1.** The function of smelling. **2.** The sense of smell.

olfactoreceptor [PHYSIO] A structure which is a receptor for the sense of smell.

olfactory aura [MED] Prodromal disagreeable olfactory sensation preceding or characterizing an epileptic attack.

olfactory bulb [VERT ZOO] The bulbous distal end of the olfactory tract located beneath each anterior lobe of the cerebrum; well developed in lower vertebrates.

olfactory cell [PHYSIO] One of the sensory nerve cells in the olfactory epithelium.

olfactory foramen [ANAT] Any of the openings in the cribriform plate of the ethmoid bone through which pass the fila olfactoria of the olfactory nerves.

olfactory gland [PHYSIO] A type of serous gland in the nasal mucous membrane.

olfactory lobe [VERT ZOO] A lobe projecting forward from the inferior surface of the frontal lobe of each cerebral hemisphere, including the olfactory bulb, tracts, and trigone; well developed in most vertebrates, but reduced in man.

olfactory nerve [ANAT] The first cranial nerve; a paired sensory nerve with its origin in the olfactory lobe and formed by processes of the olfactory cells which lie in the nasal mucosa; greatly reduced in man.

olfactory organ [PHYSIO] Any of the small chemoreceptors in the mucous membrane lining the upper part of the nasal gravity which receive stimuli interpreted as odors.

olfactory pit [EMBRYO] A depression near the olfactory placode in the embryo that develops into part of the nasal cavity. Also known as nasal pit.

olfactory region [ANAT] The area on and above the superior conchae and on the adjoining nasal septum where the mucous membrane has olfactory epithelium and olfactory glands.

olfactory stalk [ANAT] The structure that connects the olfactory bulb to the cerebrum of the vertebrate brain.

olfactory tract [ANAT] A narrow tract of white nerve fibers

OKRA

Okra *(Hibiscus esculentus),* branch with pods.

OLDHAMINIDINA

Dorsal view of the shell of *Leptodus.*

OLIGOCENE

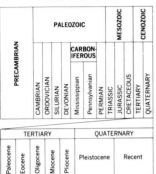

		PALEOZOIC									MESOZOIC	CENOZOIC
PRECAMBRIAN	CAMBRIAN	ORDOVICIAN	SILURIAN	DEVONIAN	Mississippian	Pennsylvanian	PERMIAN	TRIASSIC	JURASSIC	CRETACEOUS	TERTIARY	QUATERNARY
					CARBON-IFEROUS							

TERTIARY					QUATERNARY	
Paleocene	Eocene	Oligocene	Miocene	Pliocene	Pleistocene	Recent

Chart showing position of Oligocene epoch in relation to the eras and periods of geologic time.

OLIGOTRICHIDA

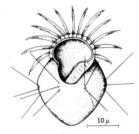

Halteria, an example of an oligotrichid. The long bristles are used in a kind of jumping motion.

OLIVE

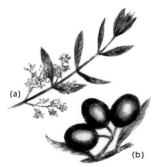

Olive *(Olea europeae)* branches. *(a)* Bearing small white flowers. *(b)* Bearing drupes, or fruit.

originating in the olfactory bulb and extending posteriorly to the anterior perforated substance, where it enlarges to form a lateral root (olfactory trigone).

olig-, oligo- [SCI TECH] A combining form denoting few, scant, or deficiency.

oligemia [MED] A state in which the total blood volume is reduced.

Oligobrachiidae [INV ZOO] A monotypic family of the order Athecanephria.

Oligocene [GEOL] The third of the five major worldwide divisions (epochs) of the Tertiary period (Cenozoic era), extending from the end of the Eocene to the beginning of the Miocene.

Oligochaeta [INV ZOO] A class of the phylum Annelida including worms that exhibit both external and internal segmentation, and setae which are not borne on parapodia.

oligochromemia *See* anemia.

oligocythemia [MED] A reduction in the total number of red blood cells in the body.

oligodendroglia [HISTOL] Small neuroglial cells with spheroidal or ovoid nuclei and fine cytoplasmic processes with secondary divisions.

oligodendroglioma [MED] A slowly growing, large, well-defined cerebral glioma, composed of small cells with richly chromatic nuclei and scanty, poorly staining cytoplasm.

oligolecithal [CYTOL] Containing very little yolk.

oligomenorrhea [MED] Abnormally infrequent menstruation.

Oligomera [INV ZOO] A subphylum of the phylum Vermes comprising groups with two or three coelomic divisions.

oligomerous [BOT] Having one or more whorls with fewer members than other whorls of the flower.

oligomycin [MICROBIO] $C_{25}H_{40-42}O_7$ An antifungal antibiotic produced by an actinomycete resembling *Streptomyces diastachromogenes;* the colorless, hexagonal crystals are soluble in many organic solvents.

oligophagous [ZOO] Eating only a limited variety of foods.

Oligopygidae [PALEON] An extinct family of exocyclic Euechinoidia in the order Holectypoida which were small ovoid forms of the Early Tertiary.

oligosaccharide [BIOCHEM] A sugar composed of two or more monosaccharide units joined by glycosidic bonds. Also known as compound sugar.

oligospermia [MED] Scarcity of spermatozoa in the semen.

oligothermic [BIOL] Tolerant of relatively low temperatures.

Oligotrichida [INV ZOO] A minor order of the Spirotrichia; the body is round in cross section, and the adoral zone of membranelles is often highly developed at the oral end of the organism.

oligotrophic [HYD] Of a lake, lacking plant nutrients and usually containing plentiful amounts of dissolved oxygen without marked stratification.

oliguria [MED] Diminished excretion of urine.

olivaceous [BIOL] 1. Resembling an olive. 2. Olive colored.

olivary nucleus [ANAT] A prominent, convoluted gray band that opens medially and occupies the upper two-thirds of the medulla oblongata.

olive [BOT] Any plant of the genus *Olea* in the order Schrophulariales, especially the evergreen olive tree (*O. europeae*) cultivated for its drupaceous fruit, which is eaten ripe (black olives) and unripe (green), and is of high oil content.

olive knot [PL PATH] A bacterial disease of the olive caused by *Pseudomonas sevastonoi* and characterized by excrescences on the foliage and branches, and sometimes on the trunk. Also known as olive tubercle.

olive oil [MATER] Pale- or greenish-yellow edible oil; main components are olein and palmitin; soluble in ether, chloroform, and carbon disulfide; derived from the pulp of olive tree fruit; used in foods, ointments, linaments, and soaps, as a lubricant, and for tanning. Also known as Florence oil; lucca oil; sweet oil.

olive tubercle *See* olive knot.

Olividae [INV ZOO] A family of snails in the gastropod order Neogastropoda.

Ollier's disease *See* enchondromatosis.

omasum [VERT ZOO] The third chamber of the ruminant stomach where the contents are mixed to a more or less homogeneous state. Also known as manyplies; psalterium.

ombrophilous [ECOL] Able to thrive in areas of abundant rainfall.

ombrophobous [ECOL] Unable to live in the presence of long, continuous rain.

omentum [ANAT] A fold of the peritoneum connecting or supporting abdominal viscera.

ommatidium [INV ZOO] The structural unit of a compound eye, composed of a cornea, a crystalline cone, and a receptor element connected to the optic nerve.

ommatophore [INV ZOO] A movable peduncle that bears an eye, as in snails.

omnicolous [BOT] Being able to grow on different substrata, referring specifically to lichens.

omnivore [ZOO] An organism that eats both animal and vegetable matter.

omohyoid [ANAT] **1.** Pertaining conjointly to the scapula and the hyoid bone. **2.** A muscle attached to the scapula and the hyoid bone.

Omophronidae [INV ZOO] The savage beetles, a small family of coleopteran insects in the suborder Adephaga.

omosternum [ANAT] A bony or cartilaginous interarticular structure between the sternum and each clavicle in certain mammals. [VERT ZOO] A median bony element of the amphibian sternum, extending anteriorly from the ventral ends of the precoracoids and having the episternum at its anterior end.

omphalodisc [BOT] An apothecium having a small protuberance in the center, as in certain lichens.

omphalogenesis [EMBRYO] Development of the yolk sac.

omphaloidium [BOT] The scar at the hilum of a seed.

omphalomesenteric artery *See* vitelline artery.

omphalomesenteric duct *See* vitelline duct.

omphalomesenteric vein *See* vitelline vein.

omphaloproptosis [MED] Abnormal protrusion of the navel.

Omphralidae [INV ZOO] A family of orthorrhaphous dipteran insects in the series Nematocera.

Onagraceae [BOT] A family of dicotyledonous plants in the order Myrtales characterized generally by an inferior ovary, axile placentation, twice as many stamens as petals, a four-nucleate embryo sac, and many ovules.

Onchidiidae [INV ZOO] An intertidal family of sluglike pulmonate mollusks of the order Systellommatophora in which the body is oval or lengthened, with the convex dorsal integument lacking a mantle cavity or shell.

onchocerciasis [MED] Infection with the filaria *Onchocerca volvulus*; results in skin tumors, papular dermatitis, and ocular complications.

onchogryposis [MED] A thickened, ridged, and curved condition of a nail.

oncholysis [MED] A slow process of loosening of a nail from its bed, beginning at the free edge and progressing gradually toward the root.

OMMATIDIUM

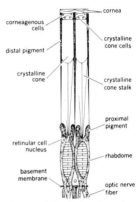

Two ommatidia from the compound eye of the crayfish *Astacus. (After H. Bernhards, from R. D. Barnes, Invertebrate Zoology, 2d ed., W. B. Saunders Co., 1968)*

ONION

Onion flower and an oblate type of bulb.

ONISCOIDEA

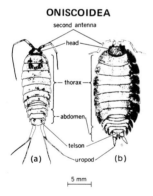

Genera of oniscoideans. *(a) Ligia*, the shore slater found above tidewater mark. *(b) Porcellio*, the sow bug or wood louse.

onchomycosis [MED] Any fungus disease of the nail.

onchosphere [INV ZOO] The hexacanth embryo identified as the earliest differentiated stage of cyclophyllidean tapeworms.

oncocyte [HISTOL] A columnar-shaped cell with finely granular eosinophilic cytoplasm, found in salivary and certain endocrine glands, nasal mucosa, and other locations.

oncocytoma [MED] A benign tumor composed principally of oncocytes; usually occurs in salivary glands.

oncogenesis [MED] Processes of tumor formation.

oncology [MED] The study of the causes, development, characteristics, and treatment of tumors.

Oncopoda [INV ZOO] A phylum of the superphylum Articulata.

oncotic pressure [PHYSIO] Also known as colloidal osmotic pressure. **1.** The osmotic pressure exerted by colloids in a solution. **2.** The pressure exerted by plasma proteins.

oncotomy [MED] Surgical incision of a tumor, abscess, or other swelling.

onion [BOT] **1.** *Allium cepa.* A biennial plant in the order Liliales cultivated for its edible bulb. **2.** Any plant of the genus *Allium.*

onion scab *See* onion smudge.

onion smudge [PL PATH] A fungus disease of the onion caused by *Colletotrichum circinans* and characterized by black concentric integral rings or smutty spots on the bulb scales. Also known as onion scab.

onion smut [PL PATH] A fungus disease of onion, especially seedlings, caused by *Urocystis cepulae* and characterized by elongate black blisters on the scales and foliage.

Oniscoidea [INV ZOO] A terrestrial suborder of the Isopoda; the body is either dorsoventrally flattened or highly vaulted, and the head, thorax, and abdomen are broadly joined.

ontogeny [EMBRYO] The origin and development of an organism from conception to adulthood.

Onuphidae [INV ZOO] A family of tubicolous, herbivorous, scavenging errantian annelids in the superfamily Eunicea.

onych-, onycho- [ZOO] A combining form denoting claw or nail.

onychia [MED] Inflammation of the nail matrix.

onychium [ANAT] The layer of tissue under the nail. [INV ZOO] A false articulation that bears claws at the end of the tarsus in some spiders.

Onychodontidae [PALEON] A family of Lower Devonian lobefin fishes in the order Osteolepiformes.

Onychopalpida [INV ZOO] A suborder of mites in the order Acarina.

Onychophora [INV ZOO] A phylum of wormlike animals that combine features of both the annelids and the arthropods.

onychorrhexis [MED] Longitudinal striation of the nail plate, with or without the formation of fissures.

Onygenaceae [MYCOL] A family of ascomycetous fungi in the order Eurotiales comprising forms that inhabit various animal substrata, such as horns and hoofs.

oocyst [INV ZOO] The encysted zygote of some Sporozoa.

oocyte [HISTOL] An egg before the completion of maturation.

oogamete [BIOL] A large, nonmotile female gamete containing reserve materials.

oogamous [BIOL] Of sexual reproduction, characterized by fusion of a motile sperm with an oogamete.

oogenesis [PHYSIO] Processes involved in the growth and maturation of the ovum in preparation for fertilization.

oogonium [BOT] The unisexual female sex organ in oogamous algae and fungi. [HISTOL] A descendant of a primary germ cell which develops into an oocyte.

ookinete [INV ZOO] The elongated, mobile zygote of certain Sporozoa, as that of the malaria parasite.

oolemma *See* zona pellucida.

oology [VERT ZOO] A branch of zoology concerned with the study of eggs, especially bird eggs.

Oomycetes [MYCOL] A class of the Phycomycetes comprising the biflagellate water molds and downy mildews.

oophagous [ZOO] Feeding or living on eggs.

oophoritis [MED] Inflammation of the ovaries.

ooplasm [CYTOL] Cytoplasm of an egg.

oospore [BOT] A spore which is produced by heterogamous fertilization and from which the sporophytic generation develops.

oostegite [INV ZOO] In many crustaceans, a platelike expansion of the basal segment of a thoracic appendage that aids in forming an egg receptacle.

oostegopod [INV ZOO] A thoracic leg bearing an oostegite in crustaceans.

ootheca [BOT] A sporangium. [INV ZOO] An egg case, as in insects.

ootid [CYTOL] One of four cells into which an oocyte divides at maturation; develops into an ovum.

Opalinata [INV ZOO] A superclass of the subphylum Sarcomastigophora containing highly specialized forms which resemble ciliates.

Opegraphaceae [BOT] A family of the Hysteriales characterized by elongated ascocarps; members are crustose on bark and rocks.

open-angle glaucoma [MED] Bilateral, increased intraocular tension due to reduced aqueous outflow but with the angle open and the aqueous in free contact with the trabecula.

open tuberculosis [MED] Tuberculosis in which tubercle bacilli are being discharged from the body; tuberculosis capable of transmission to other persons.

operant conditioning [PSYCH] A form of learning in which the subject, in a given situation, tends to respond in a way that produces rewarding effects, reinforcing previous pleasurable experiences. Also known as instrumental conditioning; reinforcement conditioning.

operative ankylosis *See* arthrodesis.

operator [GEN] A sequence at one end of an operon on which a repressor acts, thus regulating the transcription of the operon.

operculum [ANAT] **1.** The soft tissue partially covering the crown of an erupting tooth. **2.** That part of the cerebrum which borders the lateral fissure. [BIOL] **1.** A lid, flap, or valve. **2.** A lidlike body process.

operon [GEN] A functional unit composed of a number of adjacent cistrons on the chromosome; its transcription is regulated by a receptor sequence, the operator, and a repressor.

Opheliidae [INV ZOO] A family of limivorous worms belonging to the annelid group Sedentaria.

Ophiacodonta [VERT ZOO] A suborder of extinct reptiles in the order Pelycosauria, including primitive, partially aquatic carnivores.

Ophidiidae [VERT ZOO] A family of small actinopterygian fishes in the order Gadiformes, comprising the cusk eels and brotulas.

Ophiocanopidae [INV ZOO] A family of asterozoan echinoderms in the subclass Ophiuroidea.

Ophiocistioidea [PALEON] A small class of extinct Echino-

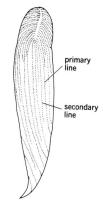

OPALINATA

primary line

secondary line

An opalinid *(Protoopalina axonucleata)* showing primary and secondary lines of blepharoplasts.

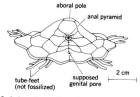

OPHIOCISTIOIDEA

aboral pole

anal pyramid

tube-feet (not fossilized)

supposed genital pore

2 cm

Volchovia volborthi (Ordovician, Soviet Union).

OPHIOGLOSSIDAE

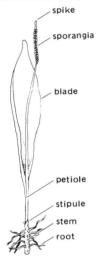

Adder's-tongue fern
(Ophioglossum) with a single leaf.

OPHIOPLUTEUS

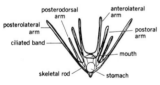

Bilateral symmetrical larva of the
brittle stars.

OPHIUROIDEA

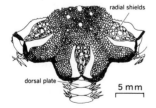

Pectinura cylindrica,
ophiuroid showing central disk
and a part of one of the arms
with the dorsal plate showing.

zoa in which the domed aboral surface of the test was roofed by polygonal plates and carried an anal pyramid.

Ophioglossales [BOT] An order of ferns in the subclass Ophioglossidae.

Ophioglossidae [BOT] The adder's-tongue ferns, a small subclass of the class Polypodiopsida; the plants are homosporous and eusporangiate and are distinguished by the arrangement of the sporogenous tissue in the characteristic fertile spike of the sporophyte.

Ophiomyxidae [INV ZOO] The single family of the echinoderm suborder Ophiomyxina distinguished by a soft, unprotected integument.

Ophiomyxina [INV ZOO] A monofamilial suborder of ophiuroid echinoderms in the order Phrynophiurida.

ophiopluteus [INV ZOO] The pluteus larva of brittle stars.

Ophiurida [INV ZOO] An order of echinoderms in the subclass Ophiuroidea in which the vertebrae articulate by means of ball-and-socket joints, and the arms, which do not branch, move mainly from side to side.

Ophiuroidea [INV ZOO] The brittle stars, a subclass of the Asterozoa in which the arms are usually clearly demarcated from the central disk and perform whiplike locomotor movements.

ophryon [ANTHRO] The point where the sagittal plane intersects an arc drawn horizontally from the frontotemporalis across the frontal bone.

ophthalmectomy [MED] Excision, or enucleation, of the eye.

ophthalmia [MED] Inflammation of the eye, especially involving the conjunctiva.

ophthalmia neonatorum [MED] Inflammation of the eyes in the newborn contracted during passage through the birth canal; may be gonorrheal or purulent.

ophthalmia nodosa [MED] Inflammation of the eye due to lodging of caterpillar hairs in the conjunctiva, cornea, or iris.

ophthalmic [ANAT] Of or pertaining to the eye.

ophthalmic nerve [ANAT] A sensory branch of the trigeminal nerve which supplies the lacrimal glands, upper eyelids, skin of the forehead, and anterior portion of the scalp, meninges, nasal mucosa, and frontal, ethmoid, and sphenoid air sinuses.

ophthalmodynamometer [MED] An instrument which measures the pressure necessary to collapse the retinal arteries.

ophthalmological [PHARM] Any drugs used in the treatment of eye disease.

ophthalmologist [MED] A physician who specializes in ophthalmology. Also known as oculist.

ophthalmology [MED] The study of the anatomy, physiology, and diseases of the eye.

ophthalmomalacia [MED] Abnormal softness or subnormal tension of the eye.

ophthalmometer [OPTICS] **1.** An instrument for measuring refractive errors, especially astigmatism. **2.** An instrument for measuring the capacity of the chamber of the eye. **3.** An instrument for measuring the eye as a whole.

ophthalmopod *See* eyestalk.

ophthalmorrhexis [MED] Rupture of the eyeball.

ophthalmoscope [OPTICS] An instrument, consisting essentially of a concave mirror with a hole in it and fitted with lenses of different powers, for examining the interior of the eye through the pupil.

opiate [PHARM] **1.** A sleep-inducing drug. **2.** Any narcotic. **3.** An opium preparation. **4.** Any tranquilizing agent.

Opilioacaridae [INV ZOO] The single family of moderately large mites of the suborder Notostigmata which comprises the Opiliocariformes.

Opilioacariformes [INV ZOO] A small monofamilial order of the Acari comprising large mites characterized by long legs and by the possession of a pretarsus on the pedipalp, with prominent claws.

opisthaptor [INV ZOO] A posterior adhesive organ in monogenetic trematodes.

opisthion [ANAT] The median point on the posterior margin of the foramen magnum.

Opisthobranchia [INV ZOO] A subclass of the class Gastropoda containing the sea hares, sea butterflies, and sea slugs; generally characterized by having gills, a small external or internal shell, and two pairs of tentacles.

Opisthocoela [VERT ZOO] A suborder of the order Anura; members have opisthocoelous trunk vertebrae, and the adults typically have free ribs.

opisthocoelous [ANAT] Of, related to, or being a vertebra with the centrum convex anteriorly and concave posteriorly.

Opisthocomidae [VERT ZOO] A family of birds in the order Galliformes, including the hoatzins.

opisthocranion [ANTHRO] The point, wherever it may lie in the sagittal plane on the occipital bone, which marks the posterior extremity of the longest diameter of the skull, measured from the glabella.

opisthogenesis [ZOO] Development which proceeds forward from the posterior end of the body.

opisthoglossal [VERT ZOO] Having the tongue fixed anteriorly and free posteriorly.

opisthognathous [INV ZOO] Having the mouthparts ventral and posterior to the cranium. [VERT ZOO] Having retreating jaws.

opisthonephros [VERT ZOO] The fundamental adult kidney in amphibians and fishes.

Opisthopora [INV ZOO] An order of the class Oligochaeta distinguished by meganephridiostomal, male pores opening posteriorly to the last testicular segment.

opisthotic [ANAT] Of, relating to, or being the posterior and inferior portions of the bony elements in the inner ear capsule.

opisthotonus [MED] A condition, caused by a tetanic spasm of the back muscles, in which the trunk is arched forward while the head and lower limbs are bent backward.

opium [PHARM] A narcotic obtained from the unripe capsules of the opium poppy (*Papaver somniferum*); crude extract contains alkaloids such as morphine (5–15%), narcotine (2–8%), and codeine (0.1–2.5%).

Opomyzidae [INV ZOO] A family of cyclorrhaphous myodarian dipteran insects in the subsection Acalypteratae.

opossum [VERT ZOO] Any member of the family Didelphidae in the order Marsupialia; these mammals are arboreal and mainly omnivorous, and have many incisors, with all teeth pointed and sharp.

Oppenheim's disease *See* amyotonia congenita.

opposite [BOT] **1.** Located side by side. **2.** Of leaves, being in pairs on an axis with each member separated from the other of the pair by half the circumference of the axis.

opsonic action [IMMUNOL] The effect produced upon susceptible microorganisms and other cells by opsonins, which renders them vulnerable to phagocytes.

opsonic index [IMMUNOL] A numerical measure of the opsonic activity of sera, expressed as the ratio of the average number of bacteria engulfed per phagocytic cell in immune serum compared with the corresponding value for normal serum.

opsonin [IMMUNOL] A substance in blood serum that renders bacteria more susceptible to phagocytosis by leukocytes.

OPOSSUM

Common opossum, with long prehensile tail used to climb and to gather twigs and leaves for nest building.

opsonize [IMMUNOL] To render microorganisms susceptible to phagocytosis.

-opsy [MED] A combining form denoting examination, or denoting a condition of vision.

optical axis [ANAT] An imaginary straight line passing through the midpoint of the cornea (anterior pole) and the midpoint of the retina (posterior pole).

optical isomerism [PHYS CHEM] Existence of two forms of a molecule such that one is a mirror image of the other; the two molecules differ in that rotation of light is equal but in opposite directions.

optical microscope [OPTICS] An instrument used to obtain an enlarged image of a small object, utilizing visible light; in general it consists of a light source, a condenser, an objective lens, and an ocular or eyepiece, which can be replaced by a recording device. Also known as light microscope; photon microscope.

optical-righting reflex *See* visual-righting reflex.

optical rotation [OPTICS] Rotation of the plane of polarization of plane-polarized light, or of the major axis of the polarization ellipse of elliptically polarized light by transmission through a substance or medium.

optical spectra [SPECT] Electromagnetic spectra for wavelengths in the ultraviolet, visible and infrared regions, ranging from about 10 nanometers to 1 millimeter, associated with excitations of valence electrons of atoms and molecules, and vibrations and rotations of molecules.

optical spectrograph [SPECT] An optical spectroscope provided with a photographic camera or other device for recording the spectrum made by the spectroscope.

optical spectrometer [SPECT] An optical spectroscope that is provided with a calibrated scale either for measurement of wavelength or for measurement of refractive indices of transparent prism materials.

optical spectroscope [SPECT] An optical instrument, consisting of a slit, collimator lens, prism or grating, and a telescope or objective lens, which produces an optical spectrum arising from emission or absorption of radiant energy by a substance, for visual observation.

optical spectroscopy [SPECT] The production, measurement, and interpretation of optical spectra arising from either emission or absorption of radiant energy by various substances.

optical staining *See* Rheinberg illumination.

optic apraxia [MED] A form of apraxia in which the individual fails to represent spatial relations correctly in drawing or construction by other means. Also known as constructional apraxia.

optic bulb [VERT ZOO] The peripheral expansion of the optic vesicle in embryos which later invaginates to form the optic cup.

optic canal [ANAT] The channel at the apex of the orbit, the anterior termination of the optic groove, just beneath the lesser wing of the sphenoid bone; it gives passage to the optic nerve and ophthalmic artery.

optic capsule [VERT ZOO] A cartilaginous capsule that develops around the eye in elasmobranchs and higher vertebrate embryos.

optic chiasma [ANAT] The partial decussation of the optic nerves on the undersurface of the hypothalamus.

optic cup [EMBRYO] A two-layered depression formed by invagination of the optic vesicle from which the pigmented and sensory layers of the retina will develop.

optic disk [ANAT] The circular area in the retina that is the site of the convergence of fibers from the ganglion cells of the retina to form the optic nerve.

optic gland [INV ZOO] Either of a pair of endocrine glands in the octopus and squid which are found near the brain and produce a substance which causes gonadal maturation.

optician [ENG] A maker of optical instruments or lenses.

optic lobe [ANAT] One of the anterior pair of colliculi of the mammalian corpora quadrigemina. [INV ZOO] A lateral lobe of the forebrain in certain arthropods. [VERT ZOO] Either of the corpora bigemina of lower vertebrates.

optic nerve [ANAT] The second cranial nerve; a paired sensory nerve technically consisting of three layers of special nerve cells in the retina of the eye; fibers converge to form the optic tracts.

optics [PHYS] **1.** Narrowly, the science of light and vision. **2.** Broadly, the study of the phenomena associated with the generation, transmission, and detection of electromagnetic radiation in the spectral range extending from the long-wave edge of the x-ray region to the short-wave edge of the radio region, or in wavelength from about 1 nanometer to about 1 millimeter.

optic stalk [EMBRYO] The constriction of the optic vesicle which connects the embryonic eye and forebrain in vertebrates.

optic tectum [VERT ZOO] The roof of the mesencephalon constituting a major visual center and association area of the brain of premature vertebrates.

optic tract [ANAT] The band of optic nerve fibers running from the optic chiasma to the lateral geniculate body and midbrain.

optic vesicle [EMBRYO] An evagination of the lateral wall of the forebrain in vertebrate embryos which precedes formation of the optic cup.

optocoel [ANAT] The cavity in the optic lobe.

optometrist [MED] One who measures the degrees of visual powers, without the aid of cycloplegic or mydriatic agents.

optometry [MED] Measurement of visual powers.

oral [ANAT] Of or pertaining to the mouth.

oral arm [INV ZOO] In a jellyfish, any of the prolongations of the distal end of the manubrium.

oral cavity [ANAT] The cavity of the mouth.

oral character [PSYCH] A Freudian term applied to persons who have undergone an unusual degree of oral stimulation during the developmental period and are characterized by an attitude of carefree indifference and by dependence on a mother figure.

oral contraceptive [PHARM] Any medication taken by mouth that renders a woman nonfertile as long as the medication is continued.

oral disc [INV ZOO] The flattened upper or free end of the body of a polyp that has the mouth in the center and tentacles around the margin.

oral erotic stage [PSYCH] In psychoanalysis, the first, or receptive, part of the oral phase of psychosexual development, dominated by sucking and lasting for the first 6 to 9 months of life.

oral erotism [PSYCH] The primordial pleasurable experience of nursing, reappearing in usually disguised and sublimated form in later life.

oral groove [INV ZOO] A depressed, groovelike peristome.

oral personality [PSYCH] An individual who is mouth-centered far beyond the age when the oral phase should have been passed, and who exhibits oral erotism and sadism in disguised and sublimated form.

oral phase [PSYCH] In psychoanalysis, the initial stage in the

ORANGUTAN

The orangutan *(Pongo pygmaeus)*, a long-armed primate that frequently moves about on all fours.

ORBINIIDAE

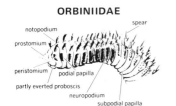

Anterior parts of the body of *Phylo*, a genus of Orbiniidae.

ORCHIDACEAE

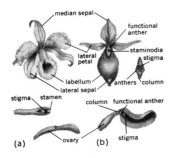

The two common floral types found in orchids. *(a) Cattleya. (b) Cypripedium.*

psychosexual development of a child, extending from birth to about 18 months, in which the mouth is the focus of sensations, interests, and activities and hence the source of gratification and security.

orange [BOT] Any of various evergreen trees of the genus *Citrus*, cultivated for the edible fruit, which is a berry with an aromatic, leathery rind containing numerous oil glands.

orangeophile [HISTOL] A type of acidophile cell of the anterior lobe of the adenohypophysis, presumed to elaborate growth hormone.

orangutan [VERT ZOO] *Pongo pygmaeus*. The largest of the great apes, a long-armed primate distinguished by long sparse reddish-brown hair, naked face and hands and feet, and a large laryngeal cavity which appears as a pouch below the chin.

orbicular [SCI TECH] Having the form of a sphere or orb.

Orbiniidae [INV ZOO] A family of polychaete annelids belonging to the Sedentaria; the prostomium is exposed, and the thorax and abdomen are weakly separated.

Orbiniinae [INV ZOO] A subfamily of sedentary polychaete annelids in the family Orbiniidae.

orbit [ANAT] The bony cavity in the lateral front of the skull beneath the frontal bone which contains the eyeball. Also known as eye socket.

orbital fossa [INV ZOO] A depression from which the eyestalk arises on the front of the carapace of crustaceans.

orbital index [ANTHRO] The ratio of the orbital height, taken at right angles to the orbital width between the upper and lower orbital margins, times 100, to the orbital width, taken between the maxillofrontale and lateral margin of the orbit in such a manner that the line of the orbital width bisects the plane of the orbital entrance.

Orbitoidacea [INV ZOO] A superfamily of foraminiferan protozoans in the suborder Rotaliina characterized by a low trochospire or a planispiral, uncoiled or branching test composed of radial calcite with bilamellar septa.

orch-, orchi-, orchid-, orchido-, orchio- [ZOO] A combining form denoting testis.

orchard [AGR] A group of fruit-bearing, nut-bearing, or sugar maple trees under cultivation.

orchid [BOT] Any member of the family Orchidaceae; plants have complex, specialized irregular flowers usually with only one or two stamens.

Orchidaceae [BOT] A family of monocotyledonous plants in the order Orchidales characterized by irregular flowers and a great number of microscopic seeds having no endosperm contained within a capsule.

Orchidales [BOT] An order of monocotyledonous plants in the subclass Liliidae; plants are mycotropic and sometimes nongreen with numerous tiny seeds that have an undifferentiated embryo and little or no endosperm.

orchiectomy [MED] Surgical removal of one or both testes.

orchil [MATER] Dark-brownish-red coloring matter derived from lichens as paste or aqueous extract; main components are orcin and orcein; used as carpet-yarn dye. Also known as orseille.

orchiopexy [MED] Surgical fixation of a testis.

order [CHEM] A classification of chemical reactions, in which the order is described as first, second, third, or higher, according to the number of molecules (one, two, three, or more) which appear to enter into the reaction; decomposition of H_2O_2 to form water and oxygen is a first-order reaction. [SYST] A taxonomic category ranked below the class and above the family, made up either of families, subfamilies, or suborders.

Ordovician [GEOL] The second period of the Paleozoic era, above the Cambrian and below the Silurian, from approximately 500 million to 440 million years ago.

Orectolobidae [VERT ZOO] An ancient isurid family of galeoid sharks, including the carpet and nurse sharks, which are primarily bottom feeders with small teeth and a blunt rostrum with barbels near the mouth.

oreodont [PALEON] Any member of the family Merycoidodontidae.

organ [ANAT] A differentiated structure of an organism composed of various cells or tissues and adapted for a specific function.

organelle [CYTOL] A specialized subcellular structure, such as a mitochondrion, having a special function; a condensed system showing a high degree of internal order and definite limits of size and shape.

organic [ORG CHEM] Of chemical compounds, based on carbon chains or rings and also containing hydrogen with or without oxygen, nitrogen, or other elements.

organic acid [ORG CHEM] A chemical compound with one or more carboxyl radicals (COOH) in its structure; examples are butyric acid, $CH_3(CH_2)_2COOH$, maleic acid, $HOOCCH$-$CHCOOH$, and benzoic acid, C_6H_5COOH.

organic brain syndrome [MED] A mental condition of multiple etiologies, resulting in diffuse impairment of brain tissue function and manifested by a complex of symptoms including impaired judgment and intellectual function, and often somatic and motor dysfunctions.

organic chemistry [CHEM] The study of the composition, reactions, and properties of carbon-chain or carbon-ring compounds or mixtures thereof.

organic evolution [EVOL] The processes of change in organisms by which descendants come to differ from their ancestors, and a history of the sequence of such changes.

organicism *See* holism.

organism [BIOL] An individual constituted to carry out all life functions.

organismic psychology [PSYCH] A movement in psychology based on the theory that the individual is made up of elements composing a single organized system, and an element in the system cannot be evaluated independently of its position within the system.

organized ferment *See* intracellular enzyme.

organizer [EMBRYO] Any part of the embryo which exerts a morphogenetic stimulus on an adjacent part or parts, as in the induction of the medullary plate by the dorsal lip of the blastopore.

organizing pneumonia [MED] Pneumonia in which the healing process is characterized by organization and cicatrization of the exudate rather than by resolution and resorption. Also known as unresolved pneumonia.

organ of Corti [ANAT] A specialized structure located on the basilar membrane of the mammalian cochlea, which contains rods of Corti and hair cells connected to ganglia of the cochlear nerve. Also known as spiral organ.

organ of Leydig [VERT ZOO] Two large accumulations of lymphoid tissues which run longitudinally the length of the esophagus in selachian fishes.

organogenesis [EMBRYO] The formation of an organ.

organometallic compound [ORG CHEM] Molecules containing carbon-metal linkage; a compound containing an alkyl or aryl radical bonded to a metal, such as tetraethyllead, $Pb(C_2H_5)_4$.

organotropic [MICROBIO] Of microorganisms, localizing in

ORDOVICIAN

PRECAMBRIAN		
CAMBRIAN		PALEOZOIC
ORDOVICIAN		
SILURIAN		
DEVONIAN		
Mississippian	CARBONIFEROUS	
Pennsylvanian		
PERMIAN		
TRIASSIC		MESOZOIC
JURASSIC		
CRETACEOUS		
TERTIARY		CENOZOIC
QUATERNARY		

Chart showing the position of the Ordovician period in relation to the other periods and to the eras of geologic time.

or entering the body by way of the viscera or, occasionally, somatic tissue.

organs of Zuckerkandl *See* aortic paraganglion.

orgasm [PHYSIO] The intense, diffuse, and subjectively pleasurable sensation experienced during sexual intercourse or genital manipulation, culminating in the male with seminal ejaculation and in the female with uterine contractions, warm suffusion, and pelvic throbbing sensations.

orgasmolepsy [MED] Sudden loss of muscle tone during orgasm, accompanied by a transitory loss of consciousness.

Oribatei [INV ZOO] A heavily sclerotized group of free-living mites in the suborder Sarcoptiformes which serve as intermediate hosts of tapeworms.

Oribatulidae [INV ZOO] A family of oribatid mites in the suborder Sarcoptiformes.

Oriental zoogeographic region [ECOL] A zoogeographic region which encompasses tropical Asia from the Iranian Peninsula eastward through the East Indies to, and including, Borneo and the Philippines.

orientation [PHYS CHEM] The arrangement of radicals in an organic compound in relation to each other and to the parent compound. [PSYCH] Determination of one's relation to the environment.

orifice [SCI TECH] An aperture or hole.

ormer *See* abalone.

Ormyridae [INV ZOO] A small family of hemipteran insects in the superfamily Chalcidoidea.

Orneodidae [INV ZOO] A small family of lepidopteran insects in the superfamily Tineoidea; adults have each wing divided into six featherlike plumes.

ornithine [BIOCHEM] $C_5H_{12}O_2N_2$ An amino acid occurring in the urine of some birds, but not found in native proteins.

Ornithischia [PALEON] An order of extinct terrestrial reptiles, popularly known as dinosaurs; distinguished by a four-pronged pelvis, and a median, toothless predentary bone at the front of the lower jaw.

ornithology [VERT ZOO] The study of birds.

ornithophilous [ECOL] Referring to flowers that are pollinated by birds.

Ornithopoda [PALEON] A suborder of extinct reptiles in the order Ornithischia including all bipedal forms in the order.

Ornithorhynchidae [VERT ZOO] A monospecific order of monotremes containing the semiaquatic platypus; characterized by a duck-billed snout, horny plates instead of teeth in the adult, and a flattened, well-developed tail.

ornithosis [MED] Any form of psittacosis originating in birds other than psittacines.

Oromericidae [VERT ZOO] An extinct family of camellike tylopod ruminants in the superfamily Cameloidea.

oropharynx [ANAT] The oral pharynx, located between the lower border of the soft palate and the larynx.

orophyte [ECOL] Any plant that grows in the subalpine region.

Oroya fever [MED] The severe form of Carrion's disease, characterized by a sudden, severe, and rapid course, often fatal anemia, and remittent fever.

orseille *See* orchil.

Orthacea [PALEON] An extinct group of articulate brachiopods in the suborder Orthidina in which the delthyrium is open.

Ortheziinae [INV ZOO] A subfamily of homopteran insects in the superfamily Coccoidea having abdominal spiracles present in all stages and a flat anal ring bearing pores and setae in immature forms and adult females.

Orthida [PALEON] An order of extinct articulate brachiopods

ORTHACEA

delthyrium

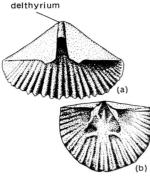

Shell of *Hesperothis.* *(a)* Pedicle valve posterior. *(b)* Brachial valve interior.

which includes the oldest known representatives of the class.

Orthidina [PALEON] The principal suborder of the extinct Orthida, including those articulate brachiopods characterized by biconvex, finely ribbed shells with a straight hinge line and well-developed interareas on both valves.

orthocephaly [ANTHRO] The condition of having the skull with a vertical index of 70.1-75.

orthochromatic [BIOL] Having normal staining characteristics.

orthodentin [ANAT] Dentin peculiar to mammalian teeth and containing processes, but not cell bodies, of odontoblasts.

orthodontics [MED] A branch of dentistry that deals with the prevention and treatment of malocclusion.

orthognathic [ANTHRO] Pertaining to a condition of the upper jaw in which it is in an approximately vertical relationship to the profile of the facial skeleton, when the skull is oriented on the Frankfort horizontal plane; having a gnathic index of 97.9 or less.

orthokinesis [BIOL] Random movement of a motile cell or organism in response to a stimulus.

Orthonectida [INV ZOO] An order of Mesozoa; orthonectids parasitize various marine invertebrates as multinucleate plasmodia, and sexually mature forms are ciliated organisms.

orthopedics [MED] The branch of surgery concerned with corrective treatment of musculoskeletal deformities, diseases, and ailments by manual and instrumental measures.

Orthoperidae [INV ZOO] The minute fungus beetles, a family of coleopteran insects in the superfamily Cucujoidea.

orthopnea [MED] A condition in which there is difficulty in breathing except when sitting or standing upright.

Orthopsida [INV ZOO] An order of echinoderms in the subclass Euechinoidea.

Orthopsidae [PALEON] A family of extinct echinoderms in the order Hemicidaroida distinguished by a camarodont lantern.

Orthoptera [INV ZOO] A heterogeneous order of generalized insects with gradual metamorphosis, chewing mouthparts, and four wings.

orthopterous [INV ZOO] Having straight, folded hindwings, as in grasshoppers.

Orthorrhapha [INV ZOO] A suborder of the Diptera; in this group of flies, the adult escapes from the puparium through a T-shaped opening.

orthoscope [MED] **1.** An instrument for examination of the eye through a layer of water, whereby the curvature and hence the refraction of the cornea is neutralized. **2.** An instrument used in drawing the projections of skulls.

orthostatic [MED] Pertaining to or caused by standing upright.

orthotonus [MED] Tetanic muscle spasm in which the body assumes a posture of rigid straightness.

orthotropism [BOT] The tendency of a plant to grow with the longer axis oriented vertically.

orthotropous [BOT] Having a straight ovule with the micropyle at the end opposite the stalk.

Orussidae [INV ZOO] A small family of hymenopteran insects in the superfamily Siricoidea.

Os *See* osmium.

Osagean [GEOL] A provincial series of geologic time in North America; Lower Mississippian (above Kinderhookian, below Meramecian).

osage orange [BOT] *Maclura pomifera.* A tree in the mulberry family of the Urticales characterized by yellowish bark, milky sap, simple entire leaves, strong axillary thorns, and aggregate green fruit about the size and shape of an orange.

ORTHONECTIDA

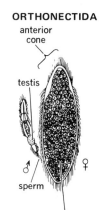

anterior cone

testis

sperm

ovocytes

The orthonectid *Rhopalura ophiocomae*, male discharging sperm near genital pore of female.

OSAGE ORANGE

Leaf and branch of *Maclura pomifera.*

Oscillatoriales [BOT] An order of blue-green algae (Cyano-phyceae) which are filamentous and truly multicellular.

Oscillospiraceae [MICROBIO] A family of large, gram-nega-tive, motile bacteria of the order Caryophanales; motility is lost on exposure to oxygen.

Oscines [VERT ZOO] The songbirds, a suborder of the order Passeriformes.

osculum [INV ZOO] An excurrent orifice in Porifera.

Osgood-Schlatter disease *See* osteochondrosis.

Osler-Rendu-Weber disease *See* hereditary hemorrhagic te-langiectasia.

osmeterium [INV ZOO] A bifurcated protrusible structure on the first segment of the thorax of some butterfly larvae.

osmiophil [BIOL] Staining with osmic acid.

osmium [CHEM] A chemical element, symbol Os, atomic number 76, atomic weight 190.2. [MET] A hard white metal of rare natural occurrence.

osmolality [CHEM] The molality of an ideal solution of a nondissociating substance that exerts the same osmotic pres-sure as the solution being considered.

osmolarity [CHEM] The molarity of an ideal solution of a nondissociating substance that exerts the same osmotic pres-sure as the solution being considered.

osmole [CHEM] **1.** The unit of osmolarity equal to the osmo-larity of a solution that exerts an osmotic pressure equal to that of an ideal solution of a nondissociating substance that has a concentration of 1 mole of solute per liter of solution. **2.** The unit of osmolality equal to the osmolality of a solution that exerts an osmotic pressure equal to that of an ideal solution of a nondissociating substance that has a concentra-tion of 1 mole of solute per kilogram of solvent.

osmometer [ANALY CHEM] A device for measuring molecu-lar weights by measuring the osmotic pressure exerted by solvent molecules diffusing through a semipermeable mem-brane.

osmoreceptor [PHYSIO] One of a group of structures in the hypothalamus which respond to changes in osmotic pressure of the blood by regulating the secretion of the neurohypophy-seal antidiuretic hormone.

osmoregulatory mechanism [PHYSIO] Any physiological mechanism for the maintenance of an optimal and constant level of osmotic activity of the fluid in and around the cells.

osmosis [PHYS CHEM] The transport of a solvent through a semipermeable membrane separating two solutions of differ-ent solute concentration, from the solution that is dilute in solute to the solution that is concentrated.

osmotic fragility [PHYSIO] Susceptibility of red blood cells to lyses when placed in dilute (hypotonic) salt solutions.

osmotic gradient *See* osmotic pressure.

osmotic pressure [PHYS CHEM] **1.** The applied pressure re-quired to prevent the flow of a solvent across a membrane which offers no obstruction to passage of the solvent, but does not allow passage of the solute, and which separates a solution from the pure solvent. **2.** The applied pressure required to prevent passage of a solvent across a membrane which separates solutions of different concentration, and which allows passage of the solute, but may also allow limited passage of the solvent. Also known as osmotic gradient.

osmotic shock [PHYSIO] The bursting of cells suspended in a dilute salt solution.

osseous [ANAT] Bony; composed of or resembling bone.

osseous system [ANAT] The skeletal system of the body.

osseous tissue [HISTOL] Bone tissue.

ossicle [ANAT] Any of certain small bones, as those of the

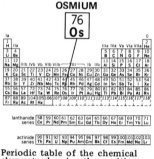

OSMIUM

76
Os

Periodic table of the chemical elements showing the position of osmium.

middle ear. [INV ZOO] Any of various calcareous bodies.

ossify [PHYSIO] To form or turn into bone.

ossifying fibroma [MED] A benign bone tumor derived from ossiferous connective tissue. Also known as fibrous osteoma; osteogenic fibroma.

ost-, oste-, osteo- [ANAT] A combining form meaning bone.

Ostariophysi [VERT ZOO] A superorder of actinopterygian fishes distinguished by the structure of the anterior four or five vertebrae which are modified as an encasement for the bony ossicles connecting the inner ear and swim bladder.

Osteichthyes [VERT ZOO] The bony fishes, a class of fishlike vertebrates distinguished by having a bony skeleton, a swim bladder, a true gill cover, and mesodermal ganoid, cycloid, or ctenoid scales.

osteitis [MED] Inflammation of bone.

osteitis fibrosa cystica [MED] Generalized skeletal demineralization due to an increased rate of bone destruction resulting from hyperparathyroidism. Also known as Engel-Recklinghausen disease; osteitis fibrosa generalisata.

osteoarthritis *See* degenerative joint disease.

osteoarthropathy [MED] Any disease of bony articulations.

osteoblast [HISTOL] A bone-forming cell of mesenchymal origin.

osteochondritis [MED] Inflammation of both bone and cartilage.

osteochondroma [MED] A benign hamartomatous tumor originating in bone or cartilage.

osteochondrosis [MED] A disease characterized by avascular necrosis of ossification centers followed by regeneration. Also known as Calvé's disease; Kienböck's disease; Kohler's disease; Osgood-Schlatter disease; Scheuermann's disease.

osteoclasis [MED] Forcible fracture of a long bone without open operation, to correct a deformity. [PHYSIO] **1.** Destruction of bony tissue. **2.** Bone resorption.

osteoclast [HISTOL] A large multinuclear cell associated with bone resorption. [MED] A large surgical apparatus through which leverage can be exerted to effect osteoclasis.

osteocranium [ANAT] The bony skull, as distinguished from the chondrocranium.

osteocyte [HISTOL] A bone cell.

osteodentin [HISTOL] A type of dentin which closely resembles bone in structure.

osteodermia [MED] A condition characterized by ossification within the skin.

osteodystrophy [MED] Any defective bone formation, as in rickets or dwarfism.

osteofibrosis [MED] Fibrosis of bone.

osteogenesis [PHYSIO] Formation or histogenesis of bone.

osteogenesis imperfecta [MED] A disease inherited as an autosomal dominant and characterized by hypoplasia of osteoid tissue and collagen, resulting in bone fractures.

osteogenesis imperfecta congenita [MED] A form of osteogenesis imperfecta in which fractures occur at or before birth.

osteogenic fibroma *See* ossifying fibroma.

osteogenic sarcoma *See* osteosarcoma.

Osteoglossidae [VERT ZOO] The bony tongues, a family of actinopterygian fishes in the order Osteoglossiformes.

Osteoglossiformes [VERT ZOO] An order of soft-rayed, actinopterygian fishes distinguished by paired, usually bony rods at the base of the second gill arch, a single dorsal fin, no adipose fin, and a usually abdominal pelvic fin.

osteoid [HISTOL] The young hyaline matrix of true bone in which the calcium salts are deposited.

osteolathyrism [MED] Degeneration of bone collagen result-

OSTEOLEPIFORMES

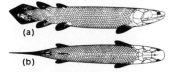

(a)

(b)

Devonian osteolepid
Gyroptychius agassizi, with
a length of 18 inches (46
centimeters), shown in
(a) lateral and *(b)* dorsal aspects.
*(Restorations, after E. Jarvik, On
the Morphology and Taxonomy
of the Middle Devonian
Osteolepid Fishes of Scotland,
Kgl. Svenska Vetenskapsakad,
Handl., Tredje Serien, Band 25,
no. 1, Stockholm, 1948)*

OSTRACODERM

|— 5 cm —|

The ostracoderm *Hemicyclaspis*,
a cephalaspid, a Lower Devonian
jawless vertebrate. *(From E. H.
Colbert, Evolution of the
Vertebrates, Wiley, 1955)*

ing from experimental administration of β-aminoproprioni-trile.

Osteolepidae [PALEON] A family of extinct fishes in the order Osteolepiformes.

Osteolepiformes [PALEON] A primitive order of fusiform lobefin fishes, subclass Crossopterygii, generally character-ized by rhombic bony scales, two dorsal fins placed well back on the body, and a well-ossified head covered with large dermal plating bones.

osteology [ANAT] The study of anatomy and structure of bone.

osteolysis [MED] Degeneration of bone tissue. [PHYSIO] Resorption of bone.

osteoma [MED] A benign bone tumor, especially in mem-brane bones of the skull.

osteomalacia [MED] Failure of bone to ossify due to a re-duced amount of available calcium. Also known as adult rickets.

osteometry [ANAT] The study of the size and proportions of the osseous system.

osteomyelitis [MED] Inflammation of bone tissue and bone marrow.

osteon [HISTOL] A Haversian system with its lamellae and Haversian canal.

osteonephropathy [MED] Any syndrome involving bone changes accompanying kidney disease.

osteopath [MED] A physician who specializes in osteopathy.

osteopathy [MED] **1.** A school of healing which teaches that the body is a vital mechanical organism whose structural and functional integrity are coordinate and interdependent, the abnormality of either constituting disease. **2.** Any disease of bone.

osteopetrosis [MED] A rare developmental error of un-known cause but of familial tendency, characterized chiefly by excessive radiographic density of most or all of the bones. Also known as marble bone disease.

osteopoikilosis [MED] A bone affection of unknown cause and no symptoms, characterized by ellipsoidal, dense foci in all bones.

osteoporosis [MED] Deossification with absolute decrease in bone tissue, resulting in enlargement of marrow and Haver-sian spaces, decreased thickness of cortex and trabeculae, and structural weakness.

osteosarcoma [MED] A malignant tumor principally com-posed of anaplastic cells of mesenchymal derivation. Also known as osteogenic sarcoma.

osteosclereid [INV ZOO] A sclereid with knobs on both ends.

osteosclerotic anemia *See* myelophthisic anemia.

osteoscute [VERT ZOO] A bony external scale or plate on the body surface, as in labyrinthodonts and armadillos.

Osteostraci [PALEON] An order of extinct jawless verte-brates; they were mostly small, with the head and part of the body encased in a solid armor of bone, and the posterior part of the body and the tail covered with thick scales.

osteotomy [MED] **1.** Surgical division of a bone. **2.** Making a section of a bone for the purpose of correcting a deformity.

ostiole [BIOL] A small orifice or pore.

ostium [BIOL] A mouth, entrance, or aperture.

Ostomidae [INV ZOO] The bark-gnawing beetles, a family of coleopteran insects in the superfamily Cleroidea.

Ostracoda [INV ZOO] A subclass of the class Crustacea con-taining small, bivalved aquatic forms; the body is unsegment-ed and there is no true abdominal region.

ostracoderm [PALEON] Any of various extinct jawless verte-brates covered with an external skeleton of bone which together with the Cyclostomata make up the class Agnatha.

Ostreidae [INV ZOO] A family of bivalve mollusks in the order Anisomyaria containing the oysters.

ostrich [VERT ZOO] *Struthio camelus.* A large running bird with soft plumage, naked head, neck and legs, small wings, and thick powerful legs with two toes on each leg; the only living species of the Struthioniformes.

Ostwald viscometer [ENG] A viscometer in which liquid is drawn into the higher of two glass bulbs joined by a length of capillary tubing, and the time for its meniscus to fall between calibration marks above and below the upper bulb is compared with that for a liquid of known viscosity.

ot-, oto- [ANAT] A combining form meaning ear.

otalgia [MED] Pain in the ear.

Otariidae [VERT ZOO] The sea lions, a family of carnivorous mammals in the superfamily Canoidea.

Othniidae [INV ZOO] The false tiger beetles, a small family of coleopteran insects in the superfamily Tenebrionoidea.

otic [ANAT] Of or pertaining to the ear or a part thereof.

-otic [SCI TECH] A suffix meaning of, pertaining to, characterized by, or causing the process.

otic capsule [EMBRYO] A cartilaginous capsule surrounding the auditory vesicle during development, later fusing with the spheroid and occipital cartilages.

otic ganglion [ANAT] The nerve ganglion located immediately below the foramen ovale of the sphenoid bone.

otidium [INV ZOO] The otocyst in mollusks.

Otitidae [INV ZOO] A family of cyclorrhaphous myodarian dipteran insects in the subsection Acalypteratae.

otitis [MED] Inflammation of the ear.

otitis externa [MED] Inflammation of the external ear.

otitis media [MED] Inflammation of the middle ear.

otocrypt [INV ZOO] An open invagination of the integument on the foot of some mollusks.

otocyst [EMBRYO] The auditory vesicle of vertebrate embryos. [INV ZOO] An auditory vesicle, otocell, or otidium in some invertebrates.

otolaryngology [MED] A branch of medicine that deals with the ear, nose, and throat. Also known as otorhinolaryngology.

otolith [ANAT] A calcareous concretion on the end of a sensory hair cell in the vertebrate ear and in some invertebrates.

otology [MED] A branch of medicine that deals with the ear and its diseases.

otomycosis [MED] Fungus infection of the external ear, usually caused by *Aspergillus niger* and *A. fumigatus.*

Otopheidomenidae [INV ZOO] A family of parasitic mites in the suborder Mesostigmata.

otoporpae [INV ZOO] Lines of cnidoblasts on the exumbrella of hydromedusae.

otorhinolaryngology *See* otolaryngology.

otosclerosis [MED] Sclerosis of the inner ear, causing a progressive increase in deafness.

otoscope [MED] An apparatus designed for examination of the ear and for rendering the tympanic membrane visible.

otter [VERT ZOO] Any of various members of the family Mustelidae, having a long thin body, short legs, a somewhat flattened head, webbed toes, and a broad flattened tail; all are adapted to aquatic life.

outcrossing *See* heterosis.

outpatient [MED] A patient who comes to the hospital or clinic for diagnosis and treatment but who does not occupy a bed in the institution.

OSTRICH

The ostrich *(Struthio camelus),* a large African running bird, grows to 8 feet (2 meters).

OTTER

North American otter *(Lutra canadensis),* with supple body, webbed feet, and thick water-resistant fur.

ovalbumin [BIOCHEM] The major, conjugated protein of egg-white.

oval window [ANAT] The membrane-covered opening into the inner ear of tetrapods, to which the ossicles of the middle ear are connected.

ovarian [ANAT] Of or pertaining to the ovaries.

ovarian agenesis [MED] Failure of the ovaries to develop.

ovarian dysmenorrhea [MED] Dysmenorrhea caused by ovarian disease.

ovarian follicle [HISTOL] An ovum and its surrounding follicular cells, found in the ovarian cortex.

ovarian insufficiency [MED] Deficient functioning of the ovaries, leading to amenorrhea, oligomenorrhea, or abnormal dysfunctional uterine bleeding.

ovariectomy [MED] Excision of an ovary.

ovariole [INV ZOO] The tubular structural unit of an insect ovary.

ovary [ANAT] A glandular organ that produces hormones and gives rise to ova in female vertebrates. [BOT] The enlarged basal portion of a pistil that bears the ovules in angiosperms.

ovate [BIOL] Egg-shaped. [BOT] Referring to leaves, attached by the broad end.

ovejector [INV ZOO] The muscular terminal portion of the female genital tract in nematodes.

overdominance [GEN] Monohybrid heterosis, that is, the phenomenon of the phenotype being more pronounced in the heterozygote than in either homozygote with respect to a single specified pair of alleles.

overdose [MED] An excessive dose of medicine.

ovicell [INV ZOO] A broad chamber in certain bryozoans.

ovicyst [INV ZOO] The pouch of a tunicate in which the eggs develop.

oviduct [ANAT] A tube that serves to conduct ova from the ovary to the exterior or to an intermediate organ such as the uterus. Also known in mammals as Fallopian tube; uterine tube.

oviger [INV ZOO] A modified leg used for carrying eggs in some pycnogonids.

ovine encephalomyelitis *See* louping ill.

oviparous [VERT ZOO] Producing eggs that develop and hatch externally.

oviposit [INV ZOO] To lay eggs, referring specifically to insects.

ovipositor [INV ZOO] A specialized structure in many insects for depositing eggs. [VERT ZOO] A tubular extension of the genital orifice in most fishes.

ovisac [INV ZOO] A capsule or receptacle for eggs.

ovotesticular hermaphroditism [MED] A rare form of hermaphroditism in which an ovotestis is present on one or both sides.

ovoviviparous [VERT ZOO] Producing eggs that develop internally and hatch before or soon after extrusion.

ovulation [PHYSIO] Discharge of an ovum or ovule from the ovary.

ovule [BOT] A structure in the ovary of a seed plant that develops into a seed following fertilization.

ovum [CYTOL] A female gamete. Also known as egg.

Oweniidae [INV ZOO] A family of limivorous polychaete annelids of the Sedentaria.

owl [VERT ZOO] Any of a number of diurnal and nocturnal birds of prey composing the order Strigiformes; characterized by a large head, more or less forward-directed large eyes, a short hooked bill, and strong talons.

Oxalidaceae [BOT] A family of dicotyledonous plants in the

OWENIIDAE

Myriochele. (a) Entire ovigerous individual. *(b)* The tapering tube in which the mud-swallowing worm is securely contained.

OWL

Great horned owl *(Bubo virginianus)*, a nocturnal bird of prey.

order Geraniales, generally characterized by regular flowers, two or three times as many stamens as sepals or petals, a style which is not gynobasic, and the fruit which is a beakless, loculicidal capsule.

oxalosis [MED] A rare hereditary metabolic disorder, inherited as an autosomal recessive, in which glyoxylic acid metabolism is impaired, resulting in overproduction of oxalic acid and deposition of calcium oxalate in body tissues.

oxaluria [MED] The presence of oxalic acid or oxalates in the urine.

Oxamycin [MICROBIO] A trade name for the antibiotic cycloserine.

oxea [INV ZOO] A rod-shaped sponge spicule with sharp points at both ends.

Oxfordian [GEOL] A European stage of geologic time, in the Upper Jurassic (above Callovian, below Kimmeridgean). Also known as Divesian.

Oxford unit [MICROBIO] The minimum quantity of penicillin which, when dissolved in 50 milliliters of a meat broth, is sufficient to inhibit completely the growth of a test strain of *Micrococcus aureus*; equivalent to the specific activity of 0.6 microgram of the master standard penicillin.

oxidase [BIOCHEM] An enzyme that catalyzes oxidation reactions by the utilization of molecular oxygen as an electron acceptor.

oxidation [CHEM] **1.** A chemical reaction that increases the oxygen content of a compound. **2.** A chemical reaction in which a compound or radical loses electrons, that is in which the positive valence is increased.

oxidation potential [PHYS CHEM] The difference in potential between an atom or ion and the state in which an electron has been removed to an infinite distance from this atom or ion.

oxidation-reduction potential *See* redox potential.

oxidation-reduction reaction [CHEM] An oxidizing chemical change, where an element's positive valence is increased (electron loss), accompanied by a simultaneous reduction of an associated element (electron gain).

oxidation state [CHEM] The number of electrons to be added (or subtracted) from an atom in a combined state to convert it to elemental form. Also known as oxidation number.

oxidizing agent [CHEM] Compound that gives up oxygen easily, removes hydrogen from another compound, or attracts negative electrons.

oxidoreductase [BIOCHEM] An enzyme catalyzing a reaction in which two molecules of a compound interact so that one molecule is oxidized and the other reduced, with a molecule of water entering the reaction.

oximeter [MED] A photoelectric photometer used to measure the oxygenated fraction of the hemoglobin in blood which is either circulating in a particular tissue of an intact animal or human being, or during, or shortly after, its withdrawal from the vascular system, by observation of the absorption of light transmitted through or reflected from the blood.

oximetry [PHYSIO] Measurement of the degree of oxygen saturation of the blood.

Oxyaenidae [PALEON] An extinct family of mammals in the order Deltatheridea; members were short-faced carnivores with powerful jaws.

oxyaster [INV ZOO] A stellate sponge spicule having pointed rays.

oxycephaly [MED] A condition in which the head assumes a roughly conical shape due to premature closure of the coronal or lambdoid sutures, or to artificial pressure on the frontal and occipital regions of the infant's head. Also known as acrocephaly.

OXYGEN

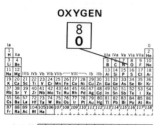

Periodic table of the chemical elements showing the position of oxygen.

OXYMONADIDA

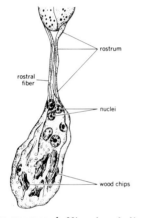

An oxymonad, *Microrhopalodina inflata.* The pliable necklike rostrum attaches the organism to the host's intestinal wall. The ingested wood chips taken in by the host are shown. *(After McCracken)*

OZAWAINELLIDAE

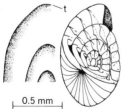

A representative of the Ozawainellidae. The initial wall, the tectum (t), is shown on the left, and the calcareous shell on the right.

oxygen [CHEM] A gaseous chemical element, symbol 0, atomic number 8, and atomic weight 15.9994; an essential element in cellular respiration and in combustion processes; the most abundant element in the earth's crust, and about 20% of the air by volume.

oxygenase [BIOCHEM] An oxidoreductase that catalyzes the direct incorporation of oxygen into its substrate.

oxygen mask [ENG] A mask that covers the nose and mouth and is used to administer oxygen.

oxyhemoglobin [BIOCHEM] The red crystalline pigment formed in blood by the combination of oxygen and hemoglobin, without the oxidation of iron.

oxylophyte [ECOL] A plant that thrives in or is restricted to acid soil.

Oxymonadida [INV ZOO] An order of xylophagous protozoans in the class Zoomastigophorea; colorless flagellate symbionts in the digestive tract of the roach *Cryptocercus* and of certain termites.

oxyntic cell *See* parietal cell.

oxypetalous [BOT] Having sharp-pointed petals.

oxyphilia *See* eosinophilia.

oxyphyte [ECOL] A plant that thrives on acid soil.

oxyphytia [ECOL] Discordant habitat control due to an excessively acidic substratum.

Oxystomata [INV ZOO] A subsection of the Brachyura, including those true crabs in which the first pair of pereiopods is chelate, and the mouth frame is triangular and forward.

Oxystomatidae [INV ZOO] A family of free-living marine nematodes in the superfamily Enoploidea, distinguished by amphids that are elongated longitudinally.

oxytetracycline [MICROBIO] $C_{22}H_{24}O_9N_2$ A crystalline, amphoteric, broad-spectrum antibiotic produced by *Streptomyces rimosus*; produced commercially by fermentation.

oxytocic [MED] Hastening parturition. [PHARM] A drug that hastens parturition.

oxytocin [BIOCHEM] $C_{43}H_{66}O_{12}N_{12}S_2$ A polypeptide hormone secreted by the neurohypophysis that stimulates contraction of the uterine muscles.

oxytylote [INV ZOO] A straight thin sponge spicule having a point at one end and a knob at the other.

Oxyurata [INV ZOO] The equivalent name for Oxyurina.

Oxyuridae [INV ZOO] A family of the nematode superfamily Oxyuroidea.

Oxyurina [INV ZOO] A suborder of nematodes in the order Ascaridida.

Oxyuroidea [INV ZOO] A superfamily of the class Nematoda.

oyster [INV ZOO] Any of various bivalve mollusks of the family Ostreidae; the irregular shell is closed by a single adductor muscle, the foot is small or absent, and there is no siphon.

Ozawainellidae [PALEON] A family of extinct protozoans in the superfamily Fusulinacea.

ozonium [MYCOL] **1.** A barren mycelium. **2.** A dense mycelium.

P

P *See* phosphorus; poise.

P₁ [GEN] The parental generation; parents of the F_1 generation.

Pa *See* protactinium.

PABA *See para*-aminobenzoic acid.

paca [VERT ZOO] Any of several rodents of the genus *Cuniculus,* especially *C. paca,* with a white-spotted brown coat, found in South and Central America.

Pacchionian bodies *See* arachnoidal granulations.

pacemaker [MED] An electric device that functions to regulate the pace of the heartbeat. [PHYSIO] Any body structure, such as the sinoatrial node, that functions in the establishment and maintenance of a rhythmic activity.

pachycephalosaur [PALEON] A bone-headed dinosaur, composing the family Pachycephalosauridae.

Pachycephalosauridae [PALEON] A family of ornithischian dinosaurs characterized by a skull with a solid rounded mass of bone 10 centimeters thick above the minute brain cavity.

pachyderm [VERT ZOO] Any of various nonruminant hooved mammals characterized by thick skin, including the elephants, hippopotamuses, rhinoceroses, and others.

pachydermatous [MED] Abnormally thick-skinned.

pachydermia [MED] Abnormal thickening of the skin.

pachyglossal [VERT ZOO] Of lizards, having a thick tongue.

pachyglossia [MED] Abnormal thickness of the tongue.

pachygyria [MED] A malformation of the brain characterized by its being too broad in form.

pachymeningitis [MED] Inflammation of the dura mater.

pachynema *See* pachytene.

pachytene [CYTOL] The third stage of meiotic prophase during which paired chromosomes thicken, each chromosome splits into chromatids, and breakage and crossing over between nonsister chromatids occur. Also known as pachynema.

Pacific faunal region [ECOL] A marine littoral faunal region including offshore waters west of Central America, running from the coast of South America at about 5° south latitude to the southern tip of California.

Pacific temperate faunal region [ECOL] A marine littoral faunal region including a narrow zone in the North Pacific Ocean, from Indochina to Alaska and along the west coast of the United States to about 40° north latitude.

Pacinian corpuscle [ANAT] An encapsulated lamellar sensory nerve ending that functions as a kinesthetic receptor.

packet [BIOL] A cluster of organisms in the form of a cube resulting from cell division in three planes.

pactamycin [MICROBIO] An antitumor and antibacterial antibiotic produced by *Streptomyces pactum* var. *pactum.*

pad [ANAT] A small circumscribed mass of fatty tissue, as in

PACHYCEPHALOSAURIDAE

Head of *Stegocerus,* a member of the Pachycephalosauridae, Late Cretaceous, Canada. Dome above level of eyes is solid bone.

terminal phalanges of the fingers or the underside of the toes of an animal, such as a dog.

paedogamy [INV ZOO] A type of autogamy in certain protozoans whereby there is mutual fertilization of gametes derived from a single cell.

paedogenesis [ZOO] Reproduction in immature or larval stages, as the axolotyl or certain Diptera.

paedomorphosis [EVOL] Phylogenetic change in which adults retain juvenile characters, accompanied by an increased capacity for further change; indicates potential for further evolution.

Paenungulata [VERT ZOO] A superorder of mammals, including proboscideans, xenungulates, and others.

Paeoniaceae [BOT] A monogeneric family of dicotyledonous plants in the order Dilleniales; members are mesophyllic shrubs characterized by cleft leaves, flowers with an intrastaminal disk, and seeds having copious endosperm.

Paget's cells [PATH] Large, epithelial cells with clear cytoplasm found in certain breast and skin cancers.

Paget's disease [MED] **1.** A type of carcinoma of the breast that involves the nipple or areola and the larger ducts, characterized by the presence of Paget's cells. **2.** Osseous hyperplasia simultaneous with accelerated deossification. **3.** An apocrine gland skin cancer, composed principally of Paget's cells.

Paguridae [INV ZOO] The hermit crabs, a family of decapod crustaceans belonging to the Paguridea.

Paguridea [INV ZOO] A group of anomuran decapod crustaceans in which the abdomen is nearly always asymmetrical, being either soft and twisted or bent under the thorax.

pain [PHYSIO] Patterns of somesthetic sensation, generally unpleasant, or causing suffering or distress.

pain spot [PHYSIO] Any of the small areas of skin overlying the endings of either very small myelinated (delta) or unmyelinated (C) nerve fibers whose stimulation, depending on the intensity and duration, results in the sensation of either pain or itching.

pain threshold [PHYSIO] The lowest limit for the perception of pain sensations.

pair [SCI TECH] A set of two things that are identical or nearly so, or are designed to function as a unit.

paired-associate learning [PSYCH] Learning established by responding with a specific word or syllable to another word or syllable.

Palaeacanthaspidoidei [PALEON] A suborder of extinct, placoderm fishes in the order Rhenanida; members were primitive, arthrodire-like species.

Palaeacanthocephala [INV ZOO] An order of the Acanthocephala including parasitic worms characterized by fragmented nuclei in the hypodermis, lateral placement of the chief lacunar vessels, and proboscis hooks arranged in long rows.

Palaechinoida [PALEON] An extinct order of echinoderms in the subclass Perischoechinoidea with a rigid test in which the ambulacra bevel over the adjoining interambulacra.

Palaemonidae [INV ZOO] A family of decapod crustaceans in the group Caridea.

Palaeocaridacea [INV ZOO] An order of crustaceans in the superorder Syncarida.

Palaeocaridae [INV ZOO] A family of the crustacean order Palaeocaridacea.

palaeocerebellum [VERT ZOO] Brain region phylogenetically preceding the cerebellum.

Palaeoconcha [PALEON] An extinct order of simple, smooth-hinged bivalve mollusks.

PAEONIACEAE

The garden peony *(Paeonia lactiflora)*, member of the family Paeoniaceae in the order Dilleniales. *(Courtesy of F. E. Westlake, from National Audubon Society)*

PALAEACANTHOCEPHALA

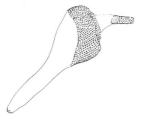

Corynosoma reductum, a palaeacanthocephalan. The anterior end of the body is provided with spines of various forms which aid in attachment to the host animal. *(From H. J. Van Cleave, Acanthocephala of North American Mammals, University of Illinois Press, 1953)*

Palaeocopida [PALEON] An extinct order of crustaceans in the subclass Ostracoda characterized by a straight hinge and by the anterior location for greatest height of the valve.

Palaeognathae [VERT ZOO] The ratites, making up a super-order of birds in the subclass Neornithes; merged with the Neognathae in some systems of classification.

Palaeoisopus [PALEON] A singular, monospecific, extinct arthropod genus related to the pycnogonida, but distinguished by flattened anterior appendages.

Palaeomastodontinae [PALEON] An extinct subfamily of elaphantoid proboscidean mammals in the family Mastodontidae.

Palaeomerycidae [PALEON] An extinct family of pecoran ruminants in the superfamily Cervoidea.

Palaeonemertini [INV ZOO] A family of the class Anopla distinguished by the two- or three-layered nature of the body-wall musculature.

Palaeonisciformes [PALEON] A large extinct order of chondrostean fishes including the earliest known and most primitive ray-finned forms.

Palaeoniscoidei [PALEON] A suborder of extinct fusiform fishes in the order Palaeonisciformes with a heavily ossified exoskeleton and thick rhombic scales on the body surface.

Palaeopantopoda [PALEON] A monogeneric order of extinct marine arthropods in the subphylum Pycnogonida.

Palaeopneustidae [INV ZOO] A family of deep-sea echinoderms in the order Spatangoida characterized by an oval test, long spines, and weakly developed fascioles and petals.

Palaeopterygii [VERT ZOO] An equivalent name for the Actinopterygii.

Palaeoryctidae [PALEON] A family of extinct insectivorous mammals in the order Deltatheridia.

Palaeospondyloidea [PALEON] An ordinal name assigned to the single, tiny fish *Palaeospondylus,* known only from Middle Devonian shales in Carthness, Scotland.

Palaeotheriidae [PALEON] An extinct family of perissodactylous mammals in the superfamily Equoidea.

palaeotheriodont [VERT ZOO] Being or having lophodont teeth with longitudinal external tubercles that are connected with inner tubercles by transverse oblique crests.

palaeotropical *See* paleotropical.

palama [VERT ZOO] The membranous web on the feet of aquatic birds.

palatal index [ANTHRO] The ratio, multiplied by 100, of the length to the breadth of the hard palate.

palate [ANAT] The roof of the mouth.

palatine bone [ANAT] Either of a pair of irregularly L-shaped bones forming portions of the hard palate, orbits, and nasal cavities.

palatine canal [ANAT] One of the canals in the palatine bone, giving passage to branches of the descending palatine nerve and artery.

palatine gland [ANAT] Any of numerous small oral glands on the palate of mammals.

palatine process [ANAT] A thick process that projects horizontally mediad from the medial aspect of the maxilla. [EMBRYO] An outgrowth on the ventromedial aspect of the maxillary process that develops into the definite palate.

palatine suture [ANAT] The median suture joining the bones of the palate.

palatine tonsil [ANAT] Either of a pair of almond-shaped aggregations of lymphoid tissue embedded between folds of tissue connecting the pharynx and posterior part of the tongue with the soft palate. Also known as faucial tonsil; tonsil.

PALAEOISOPUS

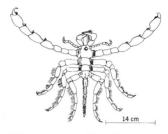

Palaeoisopus problematicus.
(After Broili)

PALAEONISCOIDEI

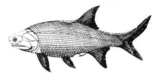

Elonichthys robisoni Hibbert (Palaeoniscoidei), a Lower Carboniferous palaeoniscoid from Scotland, which attained a length of 12 inches (30 centimeters). (*After R. H. Traquair, The ganoid fishes of the British Carboniferous formations, Palaeontographical Society, vol. 55, 1901*)

PALAEOSPONDYLOIDEA

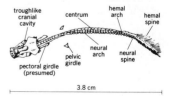

Skeleton of Middle Devonian *Palaeospondylus,* shown dorsally. *(After J. A. Moy-Thomas)*

PALEOCENE

													MESOZOIC	CENOZOIC
PRECAMBRIAN	PALEOZOIC													
						CARBON-IFEROUS								
	CAMBRIAN	ORDOVICIAN	SILURIAN	DEVONIAN	Mississippian	Pennsylvanian	PERMIAN	TRIASSIC	JURASSIC	CRETACEOUS	TERTIARY	QUATERNARY		

TERTIARY					QUATERNARY	
Paleocene	Eocene	Oligocene	Miocene	Pliocene	Pleistocene	Recent

Chart showing the relationship of the Paleocene to the eras and periods of geologic time.

PALEOCOPA

1 mm

Exterior and interior views of right valve of *Eurychilina subradiata* Ulrich, showing wide complete frill; from the Ordovician of Minnesota.

PALEOZOIC

PRECAMBRIAN		
CAMBRIAN		
ORDOVICIAN		
SILURIAN		PALEOZOIC
DEVONIAN		
Mississippian	CARBON-IFEROUS	
Pennsylvanian		
PERMIAN		
TRIASSIC		
JURASSIC		MESOZOIC
CRETACEOUS		
TERTIARY		CENOZOIC
QUATERNARY		

Chart showing the relationship of the Paleozoic to the other eras and to the periods of geologic time.

palatomaxillary arch [ANAT] An arch formed by the palatine, maxillary, and premaxillary bones. Also known as maxillary arch.

palatomaxillary index [ANTHRO] An index denoting the form of the dental arch and palate, expressed by the formula: palatomaxillary width multiplied by 100, divided by the palatomaxillary length.

palatoquadrate [VERT ZOO] A series of bones or a cartilaginous rod constituting part of the roof of the mouth or upper jaw of most nonmammalian vertebrates.

palea [BOT] **1.** The upper, enclosing bract of a grass flower. **2.** A chaffy scale found on the receptacle of the disk flowers of some composite plants. [INV ZOO] One of the enlarged flattened setae forming the operculum of the tube of certain polychaete worms.

Paleaeodonta [VERT ZOO] A suborder of artiodactylous mammals including piglike forms such as the extinct "giant pigs" and the hippopotami.

Palearctic [ECOL] Pertaining to a biogeographic region including Europe, northern Asia and Arabia, and Africa north of the Sahara.

paleate [BOT] Having a covering of chaffy scales, as some rhizomes.

paleoanthropology [ANTHRO] A branch of anthropology concerned with the study of fossil man.

paleobiochemistry [PALEON] The study of chemical processes used by organisms that lived in the geologic past.

paleobioclimatology [PALEON] The study of climatological events affecting living organisms for millennia or longer.

paleobotany [PALEON] The branch of paleontology concerned with the study of ancient and fossil plants and vegetation of the geologic past.

Paleocene [GEOL] A major worldwide division (epoch) of geologic time of the Tertiary period; extends from the end of the Cretaceous period to the Eocene epoch.

Paleocharaceae [PALEOBOT] An extinct group of fossil plants belonging to the Charophyta distinguished by sinistrally spiraled gyrogonites.

Paleocopa [PALEON] An order of extinct ostracods distinguished by a long, straight hinge.

paleoecology [PALEON] The ecology of the geologic past.

Paleogene [GEOL] A geologic time interval comprising the Oligocene, Eocene, and Paleocene of the lower Tertiary period. Also known as Eogene.

paleomagnetics [GEOPHYS] The study of the direction and intensity of the earth's magnetic field throughout geologic time.

Paleonthropinae [PALEON] A former subfamily of fossil man in the family Hominidae; set up to include the Neanderthalers together with Rhodesian man.

paleontology [BIOL] The study of life of the past as recorded by fossil remains.

paleopalynology [PALEON] A field of palynology concerned with fossils of microorganisms and of dissociated microscopic parts of megaorganisms.

Paleoparadoxidae [PALEON] A family of extinct hippopotamuslike animals in the order Desmostylia.

Paleoptera [INV ZOO] A section of the insect subclass Pterygota including primitive forms that are unable to flex their wings over the abdomen when at rest.

paleotropical [ECOL] Of or pertaining to a biogeographic region that includes the Oriental and Ethiopian regions. Also spelled palaeotropical.

Paleozoic [GEOL] The era of geologic time from the end of the Precambrian (600 million years before present) until the beginning of the Mesozoic era (225 million years before present).

paleozoology [PALEON] The branch of paleontology concerned with the study of ancient animals as recorded by fossil remains.

palette [INV ZOO] In male beetles, the modified cupule-bearing tarsus of the anterior leg.

pali [INV ZOO] In madrepore corals, a series of small columns projecting vertically from the theca base toward the stomodaeum.

palilalia [MED] Pathologic repetition of words or phrases.

Palinuridae [INV ZOO] The spiny lobsters or langoustes, a family of macruran decapod crustaceans belonging to the Scyllaridea.

palisade cell [BOT] One of the columnar cells of the palisade mesophyll which contain numerous chloroplasts.

palisade mesophyll [BOT] A tissue system of the chlorenchyma in well-differentiated broad leaves composed of closely spaced palisade cells oriented parallel to one another, but with their long axes perpendicular to the surface of the blade.

palladium [CHEM] A chemical element, symbol Pd, atomic number 46, atomic weight 106.4.

pallanesthesia [MED] Absence of pallesthesia, or vibration sense.

pallial artery [INV ZOO] The artery that supplies blood to the mantle of a mollusk.

pallial chamber [INV ZOO] The mantle cavity in mollusks.

pallial line [INV ZOO] A mark on the inner surface of a bivalve shell caused by attachment of the mantle.

pallial nerve [INV ZOO] One of the pair of dorsal nerves that innervate the mantle in mollusks.

pallial sinus [INV ZOO] An inward bend in the posterior portion of the pallial line in bivalve mollusks.

palliative [PHARM] **1.** Having a soothing or relieving quality. **2.** A drug that soothes or relieves symptoms of a disease.

pallium [ANAT] The cerebral cortex. [INV ZOO] The mantle of a mollusk or brachiopod.

Pallopteridae [INV ZOO] A family of myodarian cyclorrhaphous dipteran insects in the subsection Acalypteratae.

pallor [MED] Paleness, especially of the skin and mucous membranes.

palm [ANAT] The flexor or volar surface of the hand. [BOT] Any member of the monocotyledonous family Arecaceae; most are trees with a slender, unbranched trunk and a terminal crown of large leaves that are folded between the veins.

Palmales [BOT] An equivalent name for Arecales.

palmar [ANAT] Of or pertaining to the palm of the hand.

palmar aponeurosis [ANAT] Bundles of fibrous connective tissue which radiate from the tendons of the deep fascia of the forearm toward the proximal ends of the fingers.

palmar arch *See* deep palmar arch; superficial palmar arch.

palmaria [INV ZOO] The third brachials occurring in Crinoidea.

palmar reflex [PHYSIO] Flexion of the fingers when the palm of the hand is irritated.

palmate [BOT] Having lobes, such as on leaves, that radiate from a common point. [VERT ZOO] Having webbed toes. [ZOO] Having the distal portion broad and lobed, resembling a hand with the fingers spread.

palmatilobate [BOT] Pertaining to leaves that are palmate and have rounded lobes with divisions midway to the base.

palmatipartite [BOT] Pertaining to leaves that are palmate and have divisions more than midway to the base.

palmella [BOT] A sedentary stage of certain algae in which

PALLADIUM

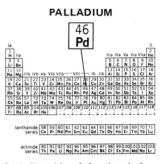

Periodic table of the chemical elements showing the position of palladium.

PALM

The common coconut palm (*Cocos nucifera*). The coconut fruit can be seen on the tree hanging below the branches.

the cells undergo division within a jellylike mass and produce motile gametes.

palm nut [BOT] The edible seed of the African oil palm (*Elaeis guineensis*).

palmula [INV ZOO] The terminal lobe or process between the paired claws on the feet of insects.

Palmyridae [INV ZOO] A mongeneric family of errantian polychaete annelids.

palp [INV ZOO] Any of various sensory, usually fleshy appendages near the oral aperture of certain invertebrates.

palpable [MED] **1.** Capable of being felt or touched. **2.** Evident.

palpacle [INV ZOO] Any of the tentacles of a dactylozooid.

palpal organ [INV ZOO] An organ on the terminal joint of each pedipalp of a male spider which functions to convey sperm to the female genital orifice.

palpation [MED] Diagnostic examination by touch.

Palpatores [INV ZOO] A suborder of long-legged arachnids in the order Phalangida.

palpebra [ANAT] The eyelid.

palpebral disk [VERT ZOO] A scale, often transparent, covering the eyelid of certain lizards.

palpebral fissure [ANAT] The opening between the eyelids.

palpebral fold [ANAT] A fold formed by the reflection of the conjunctiva from the eyelid onto the eye.

Palpicornia [INV ZOO] The equivalent name for Hydrophiloidea.

palpiger [INV ZOO] The palpi-bearing portion of an insect labium.

Palpigradida [INV ZOO] An order of rare tropical and warm-temperate arachnids; all are minute, whitish, eyeless animals with an elongate body that terminates in a slender, multisegmented flagellum set with setae.

palpimacula [INV ZOO] Any of the sensory areas on the labial palps of some insects.

palpitate [MED] To flutter, or beat abnormally fast; applied especially to the rate of the heartbeat.

palpocil [INV ZOO] A fine, filamentous tactile hair.

palpus [INV ZOO] **1.** A process on a mouthpart of an arthropod that has a tactile or gustatory function. **2.** Any similar process on other invertebrates.

palsy [MED] Any of various special types of paralysis, such as cerebral palsy.

paludicole [ECOL] Living in marshes.

palustrine [ECOL] Being, living, or thriving in a marsh.

palynology [PALEON] The study of spores, pollen, microorganisms, and microscopic fragments of megaorganisms that occur in sediments.

pamabrom [PHARM] $C_{11}H_{18}BrN_5O_3$ A water-soluble, fine white powder, decomposing at 300°C; used in medicine as a diuretic. Also known as 2-amino-2-methyl-1-propanol-8-bromotheophyllinate.

pamaquine [PHARM] $C_{19}H_{29}N_3O$ An antimalarial drug, used as the salt.

pampa [ECOL] An extensive plain in South America, usually covered with grass.

Pamphiliidae [INV ZOO] The web-spinning sawflies, a family of hymenopteran insects in the superfamily Megalodontoidea.

pamprodactylous [VERT ZOO] Having the toes turned forward, as of certain birds.

panacinar emphysema [MED] Emphysema characterized by diffuse destruction of one lung.

panagglutinin [IMMUNOL] An agglutinin lacking specificity, which agglutinates erythrocytes of various types.

Panama disease [PL PATH] A fungus disease of banana caused by invasion of the vascular system by *Fusarium oxysporum cubense*, resulting in yellowing and wilting of the foliage and ultimate death of the shoots.

panarteritis [MED] **1.** Arteritis involving all the coats of an artery. **2.** *See* polyarteritis.

panarthritis [MED] Inflammation of several joints.

pancarditis [MED] Carditis involving the endocardium, myocardium, and pericardium.

Pancarida [INV ZOO] A superorder of the subclass Malacostraca; the cylindrical, cruciform body lacks an external division between the thorax and pleon and has the cephalon united with the first thoracomere.

pancreas [ANAT] A composite gland in most vertebrates that produces and secretes digestive enzymes, as well as at least two hormones, insulin and glucagon.

pancreatectomy [MED] Surgical removal of the pancreas.

pancreatic diarrhea [MED] Diarrhea due to deficiency of pancreatic digestive enzymes, characterized by the passage of large, greasy stools having a high fat and nitrogen content.

pancreatic diverticulum [EMBRYO] One of two diverticula (dorsal and ventral) from the embryonic duodenum or hepatic diverticulum that form the pancreas or its ducts.

pancreatic duct [ANAT] The main duct of the pancreas formed from the dorsal and ventral pancreatic ducts of the embryo.

pancreatic juice [PHYSIO] The thick, transparent, colorless secretion of the pancreas.

pancreatic lipase *See* steapsin.

pancreatin [BIOCHEM] A cream-colored, amorphous powder obtained from the fresh pancreas of a hog; contains amylopsin, trypsin, steapsin, and other enzymes.

pancreatitis [MED] Inflammation of the pancreas.

pancreozymin [BIOCHEM] A crude extract of the intestinal mucosa that stimulates secretion of pancreatic juice.

pancytopenia [MED] Abnormally low numbers of all formed elements in the blood.

panda [VERT ZOO] Either of two Asian species of carnivores in the family Procyonidae; the red panda (*Ailurus fulgens*) has long, thick, red fur, with black legs; the giant panda (*Ailuropoda melanoleuca*) is white, with black legs and black patches around the eyes.

Pandanaceae [BOT] The single, pantropical family of the plant order Pandanales.

Pandanales [BOT] A monofamilial order of monocotyledonous plants; members are more or less arborescent and sparingly branched, with numerous long, firm, narrow, parallel-veined leaves that usually have spiny margins.

Pandaridae [INV ZOO] A family of dimorphic crustaceans in the suborder Caligoida; members are external parasites of sharks.

pandemic [MED] Epidemic occurring over a widespread geographic area.

Pandionidae [VERT ZOO] A monospecific family of birds in the order Falconiformes; includes the osprey (*Pandion haliaetus*), characterized by a reversible hindtoe, well-developed claws, and spicules on the scales of the feet.

pandurate [BOT] Of a leaf, having the outline of a fiddle.

panendoscope [MED] A modification of the cystoscope, utilizing a Foroblique lens system, permitting adequate visualization of both the urinary bladder and the urethra.

Paneth cells *See* cells of Paneth.

Pangea [GEOL] Postulated former supercontinent supposedly composed of all the continental crust of the earth, and later fragmented by drift into Laurasia and Gondwana.

PANCARIDA

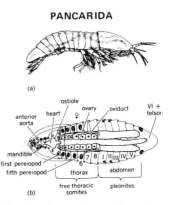

Female *Thermosbaena mirabilis* Monod. *(a)* Lateral view *(after A. Bzuun). (b)* Ventral section *(after R. Siewing).*

PANDA

The giant panda, noted for its black and white patches of fur.

PANDARIDAE

Pandarus satyrus Dana, male.

PANGOLIN

Giant pangolin *(Manis gigantea)*, a native to tropical Africa, maximum length 5 feet (1.5 meters).

PANICLE

Drawing of a panicle showing the branching of this type of inflorescence.

pangene [CYTOL] A hypothetical heredity-controlling protoplasmic particle proposed by Darwin.

pangenesis [BIOL] Darwin's comprehensive theory of heredity and development, according to which all parts of the body give off gemmules which aggregate in the germ cells; during development, they are sorted out from one another and give rise to parts similar to those of their origin.

pangolin [VERT ZOO] Any of seven species composing the mammalian family Manidae; the entire dorsal surface of the body is covered with broad, horny scales, the small head is elongate, and the mouth is terminal in the snout.

panhypopituitarism *See* hypopituitarism.

panicle [BOT] A branched or compound raceme in which the secondary branches are often racemose as well.

panmixis [BIOL] Random mating within a breeding population; in a closed population this results in a high degree of uniformity.

Panmycin [MICROBIO] A trade name for tetracycline.

pannose [BIOL] Having a felty or woolly texture.

pannus [MED] **1.** Vascularization accompanied by deposition of connective tissue beneath the cornal epithelium. **2.** Overgrowth of connective tissue on the articular surface of a diarthrodial joint.

panophthalmitis [MED] Inflammation of all the tissues of the eyeball.

pansporoblast [INV ZOO] A sporont of cnidosporan protozoans that contains two sporoblasts.

panthalassic [ECOL] Living in both coastal and offshore waters.

Pantodonta [PALEON] An extinct order of mammals which included the first large land animals of the Tertiary.

Pantodontidae [VERT ZOO] A family of fishes in the order Osteoglossiformes; the single, small species is known as African butterflyfish because of its expansive pectoral fins.

Pantolambdidae [PALEON] A family of middle to late Paleocene mammals of North America in the superfamily Pantolambdoidea.

Pantolambdodontidae [PALEON] A family of late Eocene mammals of Asia in the superfamily Pantolambdoidea.

Pantolambdoidea [PALEON] A superfamily of extinct mammals in the order Pantodonta.

Pantolestidae [PALEON] An extinct family of large aquatic insectivores referred to the Proteutheria.

pantophagous [ZOO] Feeding on a variety of foods.

Pantophthalmidae [INV ZOO] The wood-boring flies, a family of orthorrhaphous dipteran insects in the series Brachycera.

Pantopoda [INV ZOO] The equivalent name for Pycnogonida.

pantothenate [BIOCHEM] A salt or ester of pantothenic acid.

pantothenic acid [BIOCHEM] $C_9H_{17}O_5N$ A member of the vitamin B complex that is essential for nutrition of some animal species. Also known as vitamin B_3.

Pantotheria [PALEON] An infraclass of carnivorous and insectivorous Jurassic mammals; early members retained many reptilian features of the jaws.

panuveitis [MED] Inflammation of the entire uveal tract.

panzootic [VET MED] Affecting many animals of different species.

papain [BIOCHEM] An enzyme preparation obtained from the juice of the fruit and leaves of the papaya *(Carica papaya)*; contains proteolytic enzymes.

Papanicolaou's stains [CHEM] A group of stains used on exfoliated cells, particularly those from the vagina, for examination and diagnosis.

Papanicolaou test [PATH] A technique for the detection of precancerous and early noninvasive cancer by the staining and examination of exfoliated cells; used especially in the diagnosis of uterine cervical and endometrial cancer. Also known as Pap test.

Papaveraceae [BOT] A family of dicotyledonous plants in the order Papaverales, with regular flowers, numerous stamens, and a well-developed latex system.

Papaverales [BOT] An order of dicotyledonous plants in the subclass Magnoliidae, marked by a syncarpous gynoecium, parietal placentation, and only two sepals.

papaverine [ORG CHEM] $C_{20}H_{21}O_4N$ A white, crystalline alkaloid, melting at 147°C; soluble in acetone and chloroform, insoluble in water; used as a smooth muscle relaxant and weak analgesic, usually as the water-soluble hydrochloride salt. Also known as 6,7-dimethoxy-1-veratrylisoquinoline.

paper chromatography [ANALY CHEM] Procedure for analysis of complex chemical mixtures by the progressive absorption of the components of the unknown sample (in a solvent) on a special grade of paper.

paper electrochromatography [ANALY CHEM] Variation of paper electrophoresis in which the electrolyte-impregnated absorbent paper is suspended vertically and the electrodes are connected to the sides of the paper, producing a current at right angles to the downward movement of the unknown sample.

paper electrophoresis [ANALY CHEM] A variation of paper chromatography in which an electric current is applied to the ends of the electrolyte-impregnated absorbent paper, thus moving chargeable molecules of the unknown sample toward the appropriate electrode.

Papilionidae [INV ZOO] A family of lepidopteran insects in the superfamily Papilionoidea; members are the only butterflies with fully developed forelegs bearing an epiphysis.

Papilionoidea [INV ZOO] A superfamily of diurnal butterflies (Lepidoptera) with clubbed antennae, which are rounded at the tip, and forewings that always have two or more veins.

Papilionoideae [BOT] A subfamily of the family Leguminosae with characteristic irregular flowers that have a banner, two wing petals, and two lower petals united to form a boat-shaped keel.

papilla [BIOL] A small, nipplelike eminence.

papilla of Vater *See* ampulla of Vater.

papillary adenoma of ovary *See* serous cystadenoma.

papillary carcinoma [MED] A carcinoma characterized by fingerlike outgrowths.

papillary muscle [ANAT] Any of the muscular eminences in the ventricles of the heart from which the chordae tendineae arise.

papillate [BIOL] **1.** Having or covered with papillae. **2.** Resembling a papilla.

papilledema [MED] Edema of the optic disk. Also known as choked disk.

papillocystoma *See* serous cystadenoma.

papilloma [MED] A growth pattern of epithelial tumors in which the proliferating epithelial cells grow outward from a surface, accompanied by vascularized cores of connective tissue, to form a branching structure.

papillomatosis [MED] Widespread formation of papillomas.

papillomatous [MED] Characterized by or pertaining to a papilloma.

papovavirus [VIROL] A deoxyribonucleic acid–containing group of animal viruses, including papilloma and vacuolating viruses.

pappataci fever *See* phlebotomus fever.

pappose [BOT] Having a pappus.

PAPAVERACEAE

Oriental poppy *(Papaver orientale)*, of the family Papaveraceae in the order Papaverales. *(Photograph by John H. Gerard, from National Audubon Society)*

PAPER CHROMATOGRAPHY

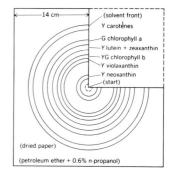

Paper chromatogram of chlorophyll showing ring separation. The circular type shown is one kind of paper chromatogram. *G* signifies green; *Y*, yellow.

Pappotheriidae [PALEON] A family of primitive, tenreclike Cretaceous insectivores assigned to the Proteutheria.

pappus [BOT] An appendage or group of appendages consisting of a modified perianth on the ovary or fruit of various seed plants; adapted to dispersal by wind and other means.

paprika [BOT] *Capsicum annuum.* A type of pepper with nonpungent flesh, grown for its long red fruit from which a dried, ground condiment is prepared.

Pap test *See* Papanicolaou test.

papula [BIOL] A small papilla.

papule [MED] A solid circumscribed elevation of the skin varying from less than 0.1 to 1 centimeter in diameter.

papulonecrotic [MED] Papule formation with a tendency to central necrosis; applied especially to a variety of skin tuberculosis.

paraaortic body [ANAT] One of the small masses of chromaffin tissue lying along the abdominal aorta. Also known as glomus aorticum.

parabasal body *See* kinetoplast.

parabiosis [BIOL] Experimental joining of two individuals to study the effects of one partner upon the other.

parabronchi [VERT ZOO] In birds, the tertiary tubes of the lungs; their terminations are embedded in the lung mesenchyme.

Paracanthopterygii [VERT ZOO] A superorder of teleost fishes, including the codfishes and allied groups.

paracele [ANAT] The lateral ventricle of the brain.

paracentesis [MED] Puncture of the wall of a fluid-filled cavity by means of a hollow needle to draw off the contents.

paracentric inversion [GEN] A type of chromosomal alteration that occurs within one arm of a chromosome and does not span the centromere.

paracme [EVOL] The decline of a species or race following the peak of development.

paracoccidioidomycosis *See* South American blastomycosis.

paracondyloid [VERT ZOO] A process on the outer side of each condyle of the occipital bone in the skull of certain mammals.

paracone [VERT ZOO] **1.** The anterior cusp of a primitive tricuspid upper molar. **2.** The principal anterior, external cusp of an upper molar in higher forms.

paraconid [VERT ZOO] **1.** The cusp of a primitive lower molar corresponding to the paracone. **2.** The anterior, internal cusp of a lower molar in higher forms.

Paracrinoidea [PALEON] A class of extinct Crinozoa characterized by the numerous, irregularly arranged plates, uniserial armlike appendages, and no clear distinction between adoral and aboral surfaces.

Paracucumidae [INV ZOO] A family of holothurian echinoderms in the order Dendrochirotida; the body is invested with plates and has a simplified calcareous ring.

paracystitis [MED] Inflammation of the connective tissue surrounding the urinary bladder.

parademe [INV ZOO] A secondary apodeme that develops on the edge of a sclerite.

paradidymis [ANAT] Atrophic remains of the paragenital tubules of the mesonephros, occurring near the convolutions of the epididymal duct.

paradoxical embolus [MED] An embolus which is transported to the circulation in peripheral arteries through septal defect in the heart, usually a patent foramen ovale.

paraesophageal cyst [MED] A bronchogenic cyst intimately connected with the esophageal wall, containing cartilage, and

PARACENTRIC INVERSION

a b c d e f g → a bd ce f g

Schematic drawing of the change taking place in paracentric inversion.

usually filled with a mucoid material and desquamated epithelial cells.

parafacialia [INV ZOO] Narrow parts of the head capsule between the frontal suture and the eyes, as in certain Diptera.

parafrons [INV ZOO] The area between the eyes and the frontal suture in certain insects.

parafrontals [INV ZOO] The continuation of genae between the eyes and the frontal suture in insects.

paraganglion [ANAT] Any of various isolated chromaffin bodies associated with structures such as the abdominal aorta, heart, kidney, and gonads. Also known as chromaffin body.

paragaster [INV ZOO] A central cavity into which the gastric ostia open in Porifera.

paragastric [ANAT] Located near the stomach. [INV ZOO] A cavity in Porifera into which radial canals open, and which opens to the outside through the cloaca.

paragastrula [INV ZOO] A stage of the amphiblastula in Porifera when flagellated cells invaginate into a sphere of rounded cells.

paragglutination [IMMUNOL] Agglutination of colon bacteria with the serum of a patient infected, or recovering from an infection, with dysentery bacilli.

paraglesia [MED] Sensation which is abnormal or disordered.

paraglossa [INV ZOO] A process on each side of the ligula in insects.

paragnath [INV ZOO] **1.** One of the paired leaflike lobes of the metastoma situated behind the mandibles in most crustaceans. **2.** One of the paired lobes of the hypopharynx in certain insects. **3.** One of the small, sharp and hard jaws of certain annelids.

paragranuloma [MED] The least aggressive form of Hodgkin's disease.

paraheliotropism [BOT] Tropism in which a plant tends to turn the leaf edges toward intense light in order to protect the surfaces.

parainfluenza [MED] A viral condition similar to or resulting from influenza. [MICROBIO] An organism exhibiting growth characteristics of *Hemophilus influenzae.*

parakeet [VERT ZOO] Any of various small, slender species of parrots with long tails in the family Psittacidae.

parakeratosis [PATH] Incomplete keratinization of epidermal cells characterized by retention of nuclei of cells attaining the level of the stratum corneum.

paralalia [MED] Disturbance of the faculty of speech, characterized by distortion of sounds, or the habitual substitution of one sound for another.

paralexia [MED] Transposition or substitution of words or syllables in reading.

paralgesia [MED] **1.** Paresthesia characterized by pain. **2.** Any perverted and disagreeable cutaneous sensation, as of formication, cold, or burning.

paralimnic [ECOL] Pertaining to or living on lake shores.

parallel evolution [EVOL] Evolution of similar characteristics in different groups of organisms.

parallel-veined [BOT] Of a leaf, having the veins parallel, or nearly parallel, to each other.

paralutein cells [HISTOL] Epithelioid cells of the corpus luteum.

paralysis [MED] Complete or partial loss of motor or sensory function.

paralysis agitans *See* parkinsonism.

paralytic secretion [PHYSIO] Glandular secretion occurring in a denervated gland.

PARAMECIUM

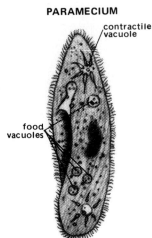

contractile vacuole

food vacuoles

A paramecium. Two contractile vacuoles and numerous food vacuoles are shown.

paralytic spinal poliomyelitis [MED] An acute inflammatory virus disease chiefly involving the anterior horns of the gray matter of the spinal cord.

paramastigote [INV ZOO] Having one long primary flagellum and a short accessory flagellum, as some mastigophorans.

Parameciidae [INV ZOO] A family of ciliated protozoans in the order Holotrichia; the body has differentiated anterior and posterior ends and is bounded by a hard but elastic pellicle.

paramecium [INV ZOO] A single-celled protozoan belonging to the family Parameciidae.

paramedical [MED] Having a supplementary or secondary relation to medicine.

Paramelina [VERT ZOO] An order of marsupials that includes the bandicoots in some systems of classification.

paramere [BIOL] One half of a bilaterally symmetrical animal or somite.   [INV ZOO] Any of several paired structures of an insect, especially those on the ninth abdominal segment.

paramesonephric duct [EMBRYO] An embryonic genital duct; in the female, it is the anlage of the oviducts, uterus, and vagina; in the male, it degenerates, leaving the appendix testes. Also known as Müllerian duct.

paramethadione [PHARM] $C_7H_{11}NO_3$ An anticonvulsant primarily useful in the treatment of petit mal epilepsy.

paramo [ECOL] A biological community, essentially a grassland, covering extensive high areas in equatorial mountains of the Western Hemisphere.

paramutation [GEN] A mutation in which one member of a heterozygous pair of alleles permanently changes its partner allele.

paramylum [BIOCHEM] A reserve, starchlike carbohydrate of various protozoans and algae.

paramyotonia congenita [MED] A heredofamilial condition characterized by recurrent muscular stiffness and weakness (myotonia) on exposure to cold, as well as on mechanical irritation; transmitted as an autosomal dominant and considered to be a variety of the hyperkalemic form of periodic paralysis. Also known as Eulenburg's disease; myotonia congenita intermittens.

paramyxovirus [VIROL] A subgroup of myxoviruses, including the viruses of mumps, measles, parainfluenza, and Newcastle disease; all are ribonucleic acid–containing viruses and possess an ether-sensitive lipoprotein envelope.

paranasal sinus [ANAT] Any of the paired sinus cavities of the human face; includes the frontal, ethmoid, and sphenoid sinuses.

paranephritis [MED] 1. Inflammation of the adrenal gland. 2. Inflammation of the connective tissue adjacent to the kidney.

paranesthesia [MED] Anesthesia of the body below the waist.

paranoia [PSYCH] A rare form of paranoid psychosis characterized by the slow development of a complex, internally logical system of persecutory or grandiose delusions.

paranoid personality [PSYCH] An individual characterized by the tendency to be hypersensitive, rigid, extremely self-important, and jealous, to project hostile feelings so that he easily becomes suspicious of others and is quick to blame them or attribute evil motives to them.

paranoid schizophrenia [PSYCH] A form of schizophrenia in which delusions of persecution or grandeur (or both), hallucinations, and ideas of reference predominate and sometimes are systematized.

paranosmia [MED] A deviation in odor sensitivity involving change in odor quality.

paranthropophytia [ECOL] Discrepant control of regions or areas due to immediate and continuous or periodic interference, as by certain cultivation practices.

Paranyrocidae [PALEON] An extinct family of birds in the order Anseriformes, restricted to the Miocene of South Dakota.

Paraonidae [INV ZOO] A family of small, slender polychaete annelids belonging to the Sedenteria.

Paraparchitacea [PALEON] A superfamily of extinct ostracods in the suborder Kloedenellocopina including nonsulcate, nondimorphic forms.

paraparesis [MED] Partial paralysis of the lower extremities.

parapertussis [MED] An acute bacterial respiratory infection similar to mild pertussis and caused by *Bordetella pertussis*.

paraphimosis [MED] **1.** Retraction and constriction, especially of the prepuce behind the glans penis. Also known as Spanish collar. **2.** Retraction of the eyelid behind the eyeball.

paraphyllium [BOT] A branching outgrowth that arises between leaves or from their bases, in mosses.

paraphysis [BOT] A sterile filament borne among the sporogenous or gametogenous organs in many cryptogams. [VERT ZOO] A median evagination of the roof of the telencephalon of some lower vertebrates.

paraplasm *See* hyaloplasm.

paraplegia [MED] Paralysis of the lower limbs.

parapleuron [INV ZOO] The episternum of the metathorax of an insect.

parapodium [INV ZOO] **1.** One of the short, paired processes on the sides of the body segments in certain annelids. **2.** A lateral expansion of the foot in gastropod mollusks.

parapolar cell [INV ZOO] Either of the first two trunk cells in the development of certain Mesozoa.

paraproct [INV ZOO] A plate on each side of the anus in diplopods and some insects.

parapsychology [PSYCH] The study of the phenomena of extrasensory perception and psychokinesis.

Parasaleniidae [INV ZOO] A family of echinacean echinoderms in the order Echinoida composed of oblong forms with trigeminate ambulacral plates.

Paraselloidea [INV ZOO] A group of the Asellota that contains forms in which the first pleopods of the male are coupled along the midline, and are lacking in the female.

Paraseminotidae [PALEON] A family of Lower Triassic fishes in the order Palaeonisciformes.

parasexual cycle [GEN] A series of events leading to genetic recombination in vegetative or somatic cells; it was first described in filamentous fungi; there are three essential steps: heterokaryosis; fusion of unlike haploid nuclei in the heterokaryon to yield heterozygous diploid nuclei; and recombination and segregation at mitosis by two independent processes, mitotic crossing-over and loss of chromosomes.

parasite [BIOL] An organism that lives in or on another organism of different species from which it derives nutrients and shelter.

parasitemia [MED] The presence of parasites in the blood.

Parasitica [INV ZOO] A group of hymenopteran insects that includes four superfamilies of the Apocrita: Ichneumonoidea, Chalcidoidea, Cynipoidea, and Proctotrupoidea; some are pytophagous, while others are parasites of other insects.

parasitic castration [BIOL] Destruction of the reproductive organs by parasites.

parasitic stomatitis *See* thrush.

parasitoidism [BIOL] Systematic feeding by an insect larva on living host tissues so that the host will live until completion of larval development.

parasitology [BIOL] A branch of biology which deals with those organisms, plant or animal, which have become dependent on other living creatures.

parasphenoid [VERT ZOO] A bone in the base of the skull of many vertebrates.

parastacidae [INV ZOO] A family of crayfishes assigned to the Nephropoidea.

parastemon [BOT] A sterile stamen.

parasternum [VERT ZOO] The abdominal ribs, considered collectively, in certain reptiles.

parastyle [VERT ZOO] A small cusp anterior to the paracone of an upper molar.

Parasuchia [PALEON] The equivalent name for Phytosauria.

parasympathetic [ANAT] Of or pertaining to the craniosacral division of the autonomic nervous system.

parasympathetic nervous system [ANAT] The craniosacral portion of the autonomic nervous system, consisting of preganglionic nerve fibers in certain sacral and cranial nerves, outlying ganglia, and postganglionic fibers.

parasympathomimetic [PHARM] Of drugs, having an effect similar to that produced when the parasympathetic nerves are stimulated.

parataxic distortion [PSYCH] A perceptual or judgmental distortion of interpersonal relations resulting from the observer's need to pattern his responses on previous experiences and thus defend himself against anxiety.

parathecium [BOT] The peripheral layer of the hypha in lichens. [MYCOL] The peripheral layer of the apothecium, as in cup fungi.

parathelioma *See* interstitial endometrosis.

parathormone [BIOCHEM] A polypeptide hormone that functions in regulating calcium and phosphate metabolism. Also known as parathyroid hormone.

Parathuramminacea [PALEON] An extinct superfamily of foraminiferans in the suborder Fusulinina, with a test having a globular or tubular chamber and a simple, undifferentiated wall.

parathyroidectomy [MED] Excision of a parathyroid gland.

parathyroid gland [ANAT] A paired endocrine organ located within, on, or near the thyroid gland in the neck region of all vertebrates except fishes.

parathyroid hormone *See* parathormone.

paratrachoma *See* inclusion conjunctivitis.

paratrichosis [MED] A condition in which the hair is either imperfect in growth or develops in abnormal places.

paratroch [INV ZOO] The ciliated band encircling the anus in certain trichophore larvae.

paratuberculosis *See* Johne's disease.

paratype [SYST] A specimen other than the holotype which is before the author at the time of original description and which is designated as such or is clearly indicated as being one of the specimens upon which the original description was based.

paratyphoid fever [MED] A bacterial disease of man resembling typhoid fever and caused by *Salmonella paratyphi*.

paraxial [SCI TECH] Lying near the axis.

paraxonic [VERT ZOO] Pertaining to a state or condition wherein the axis of the foot lies between the third and fourth digits.

Parazoa [INV ZOO] A name proposed for a subkingdom of animals which includes the sponges (Porifera).

paregoric [PHARM] Camphorated opium tincture, a preparation of opium, camphor, benzoic acid, anise oil, glycerin, and diluted alcohol; used mainly as an antiperistaltic.

Pareiasauridae [PALEON] A family of large, heavy-boned

terrestrial reptiles of the late Permian, assigned to the order Cotylosauria.

parenchyma [BOT] A tissue of higher plants consisting of living cells with thin walls that are agents of photosynthesis and storage; abundant in leaves, roots, and the pulp of fruit, and found also in leaves and stems. [HISTOL] The specialized epithelial portion of an organ, as contrasted with the supporting connective tissue and nutritive framework.

parenchymalia [INV ZOO] Parenchymal spicules of Hexactinelida.

parenchymal jaundice [MED] Any of various forms of jaundice in which the disease is due in part to damaged liver cells.

parenchymella *See* diploblastula.

parenchymula [INV ZOO] The flagellate larva of calcinean sponges in which there is a cavity filled with gelatinous connective tissue.

parental generation [GEN] Parents of the F_1 generation.

parenteral [MED] Outside the intestine; not via the alimentary tract.

parent figure [PSYCH] A person who represents essential but not necessarily ideal attributes of a father or mother and who is the object of the attitudes and responses of an individual in a parent-child relationship.

paresis [MED] **1.** A slight paralysis. **2.** Incomplete loss of muscular power. **3.** Weakness of a limb.

paresthesia [MED] Tingling, crawling, or burning sensation of the skin.

Pareulepidae [INV ZOO] A monogeneric family of errantian polychaete annelids.

parietal [ANAT] Of or situated on the wall of an organ or other body structure. [BOT] Of a plant part, having a peripheral location or orientation; in particular, attached to the main wall of an ovary.

parietal block *See* intraventricular heart block.

parietal bone [ANAT] The bone that forms the side and roof of the cranium.

parietal cell [HISTOL] One of the peripheral, hydrochloric acid-secreting cells in the gastric fundic glands. Also known as acid cell; delomorphous cell; oxyntic cell.

Parietales [BOT] An order of plants in the Englerian system; families are placed in the order Violales in other systems.

parietal lobe [ANAT] The cerebral lobe of the brain above the lateral cerebral sulcus and behind the central sulcus.

parietal peritoneum [ANAT] The portion of the peritoneum lining the interior of the body wall.

parietal placenta [BOT] A placenta located on the walls of or on the partial partitions of a compound, unilocular ovary.

parietal pleura [ANAT] The pleura lining the inner surface of the thoracic cavity.

paripinnate [BOT] Of leaves, pinnate without a terminal leaflet.

parkinsonism [MED] A clinical state characterized by tremor at a rate of three to eight tremors per second, with "pill-rolling" movements of the thumb common, muscular rigidity, dyskinesia, hypokinesia, and reduction in number of spontaneous and autonomic movements; produces a masked facies, disturbances of posture, gait, balance, speech, swallowing, and muscular strength. Also known as paralysis agitans; Parkinson's disease.

Parkinson's disease *See* parkinsonism.

parkland *See* temperate woodland; tropical woodland.

Parmeliaceae [BOT] The foliose shield lichens, a family of the order Lecanorales.

Parnidae [INV ZOO] The equivalent name for Dryopidae.

parocciput [INV ZOO] In insects, a thickening of the occiput which articulates with the neck sclerites.

paromomycin [MICROBIO] A broad-spectrum antibiotic produced by *Streptomyces rimosus forma paromomycinus;* it is effective in the treatment of intestinal amebiasis in humans.

paronychia [MED] A suppurative inflammation about the margin of a nail.

paronychium *See* perionychium.

parotic process [VERT ZOO] **1.** A process formed by fusion of exoccipital, opisthotic, and prootic elements in adult lizards. **2.** A process formed by fusion of opisthotic and pterotic elements in the skull of some fishes.

parotid duct [ANAT] The duct of the parotid gland. Also known as Stensen's duct.

parotidectomy [MED] Excision of a parotid gland.

parotid gland [ANAT] The salivary gland in front of and below the external ear; the largest salivary gland in humans; a compound racemose serous gland that communicates with the mouth by Steno's duct.

parotitis [MED] Inflammation of the parotid glands.

parovarian cyst [MED] A cyst of mesonephric origin arising between the layers of the mesosalpinx, adjacent to the ovary.

parovarium *See* epoophoron.

paroxysm [MED] **1.** A sudden attack, or the periodic crisis in the progress of a disease. **2.** A spasm, convulsion, or seizure. [PSYCH] A sudden, uncontrollable emotional outburst.

PARROT

A parrot, with short hooked beak.

parrot [VERT ZOO] Any member of the avian family Psittacidae, distinguished by the short, stout, strongly hooked beak.

pars [ANAT] A part.

pars anterior [ANAT] The major secretory portion of the anterior lobe of the adenohypophysis. Also known as pars distalis.

pars distalis *See* pars anterior.

pars intermedia [ANAT] The intermediate lobe of the adenohypophysis.

parsley [BOT] *Petroselinum crispum.* A biennial herb of European origin belonging to the order Umbellales; grown for its edible foliage.

pars nervosa [ANAT] The inferior subdivision of the neurohypophysis. Also known as pars neuralis.

pars neuralis *See* pars nervosa.

parsnip [BOT] *Pastinaca sativa.* A biennial herb of Mediterranean origin belonging to the order Umbellales; grown for its edible thickened taproot.

pars tuberalis [ANAT] A pair of processes that grow forward or upward along the stalk of the adenohypophysis.

parthenocarpy [BOT] Production of fruit without fertilization.

parthenogenesis [INV ZOO] A special type of sexual reproduction in which an egg develops without entrance of a sperm; common among rotifers, aphids, thrips, ants, bees, and wasps.

parthenogonidium [INV ZOO] A zooid or a gonidium of a protozoan colony, such as *Volvox*, with functions of asexual reproduction.

parthenokaryogamy [BIOL] Fusion of two female haploid nuclei.

parthenomerogony [EMBRYO] Development of a nucleated fragment of an unfertilized egg following parthenogenetic stimulation.

partial cleavage [EMBRYO] Cleavage in which only part of the egg divides into blastomeres.

partial pressure [PHYS] The pressure that would be exerted by one component of a mixture of gases if it were present alone in a container.

partial reinforcement [PSYCH] **1.** In classical conditioning, the reinforcement of some portion of unconditioned responses. **2.** In operant learning, the reinforcement of some portion of instrumental responses.

partial veil [MYCOL] A membrane of the young sporophore of certain mushrooms that extends from the stem to the edge of the pileus and then is ruptured during growth to become the annulus around the stem or the cortina on the margin of the pileus in the mature sporophore.

partition chromatography [ANALY CHEM] Chromatographic procedure in which the stationary phase is a high-boiling liquid spread as a thin film on an inert support, and the mobile phase is a vaporous mixture of the components to be separated in an inert carrier gas.

partition coefficient [ANALY CHEM] In the equilibrium distribution of a solute between two liquid phases, the constant ratio of the solute's concentration in the upper phase to its concentration in the lower phase. Symbolized K.

partition function [STAT MECH] **1.** The integral, over the phase space of a system, of the exponential of $(-E/kT)$, where E is the energy of the system, k is Boltzmann's constant, and T is the temperature; from this function all the thermodynamic properties of the system can be derived. **2.** In quantum statistical mechanics, the sum over allowed states of the exponential of $(-E/kT)$. Also known as sum of states; sum over states.

part learning [PSYCH] Learning a task by dividing it into smaller units and memorizing each unit separately.

partridge [VERT ZOO] Any of the game birds comprising the genera *Alectoris* and *Perdix* in the family Phasianidae.

patroclinous [BIOL] Deriving from the paternal parent.

parturient [MED] **1.** In labor; giving birth. **2.** Of or pertaining to parturition.

parturifacient [MED] **1.** Inducing labor. **2.** An agent that induces labor.

parturiometer [MED] An instrument to determine the progress of labor by measuring the expulsive force of the uterus.

parturition [MED] The process of giving birth.

parulis [MED] A subperiosteal abscess arising from dental structures.

parvovirus [VIROL] The equivalent name for picodnavirus.

Passalidae [INV ZOO] The peg beetles, a family of tropical coleopteran insects in the superfamily Scarabaeoidea.

Passeres [VERT ZOO] The equivalent name for Oscines.

Passeriformes [VERT ZOO] A large order of perching birds comprising two major divisions: Suboscines and Oscines.

Passifloraceae [BOT] A family of dicotyledonous, often climbing plants in the order Violales; flowers are polypetalous and hypogynous with a corona, and seeds are arillate with an oily endosperm.

passive-aggressive personality [PSYCH] A personality disorder characterized by the passive expression of hostility and aggressiveness, as by stubbornness, pouting, or inefficiency.

passive anaphylaxis [IMMUNOL] Anaphylaxis elicited by temporary sensitization with antibodies followed by injection of the corresponding sensitizing antigen.

passive congestion [MED] An increased content of blood in an organ or other body part due to impaired return of venous blood.

passive cutaneous anaphylaxis [IMMUNOL] The vascular reaction at the site of intradermally injected antibody when, 3 hours later, the specific antigen, usually mixed with Evans blue dye, is injected intravenously.

passive-dependent personality [PSYCH] A character disorder marked by a behavioral pattern characterized by a lack of self-confidence, indecisiveness, and a tendency to cling to and seek support from others.

passive immunity [IMMUNOL] **1.** Immunity acquired by injection of antibodies in another individual or in an animal. **2.** Immunity acquired by the fetus by the transfer of maternal antibodies through the placenta.

Pasteur effect [MICROBIO] Inhibition of fermentation by supplying an abundance of oxygen to replace anerobic conditions.

pasteurellosis *See* hemorrhagic septicemia.

Pasteuriaceae [MICROBIO] A family of stalked bacteria in the order Hyphomicrobiales.

pasteurizer [ENG] An apparatus used for pasteurization of fluids.

patagium [VERT ZOO] **1.** A membrane or fold of skin extending between the forelimbs and hindlimbs of flying squirrels, flying lizards, and other arboreal animals. **2.** A membrane or fold of skin on a bird's wing anterior to the humeral and radioulnar bones.

patch budding [BOT] Budding in which a small rectangular patch of bark bearing the scion (a bud) is fitted into a corresponding opening in the bark of the stock.

patch test [IMMUNOL] A test in which material is applied and left in contact with intact skin surfaces for 48 hours in order to demonstrate tissue sensitivity.

patella [ANAT] A sesamoid bone in front of the knee, developed in the tendon of the quadriceps femoris muscle. Also known as kneecap.

Patellacea [PALEON] An extinct superfamily of gastropod mollusks in the order Aspidobranchia which developed a cap-shaped shell and were specialized for clinging to rock.

Patellidae [INV ZOO] The true limpets, a family of gastropod mollusks in the order Archeogastropoda.

patent [MED] Open; exposed.

patent medicine [PHARM] A medicine, generally trademarked, whose composition is incompletely disclosed.

Paterinida [PALEON] A small extinct order of inarticulated brachiopods, characterized by a thin shell of calcium phosphate and convex valves.

paternity test [IMMUNOL] Identification of the blood groups of a mother, her child, and a putative father in order to establish the probability of paternity or nonpaternity; actually, only nonpaternity can be established.

Paterson-Kelly syndrome *See* Plummer-Vinson syndrome.

pathergy [IMMUNOL] Either a subnormal response to an allergen or an unusually intense one in which the individual becomes sensitive not only to the specific substance but to others.

pathogen [MED] A disease-producing agent; usually refers to living organisms.

pathogenesis [MED] The origin and course of development of disease.

pathogenic [MED] **1.** Producing or capable of producing disease. **2.** Pertaining to pathogenesis.

pathognomonic [MED] Characteristic of a given disease, enabling it to be distinguished from other diseases.

pathologic anatomy [PATH] The study of structural changes caused by disease. Also known as morbid anatomy.

pathologist [MED] A person who specializes in the study and practice of pathology.

pathology [MED] The study of the causes, nature, and effects of diseases and other abnormalities.

pathophobia [PSYCH] Exaggerated dread of death.

pathopsychology [PSYCH] The branch of science dealing with mental processes, particularly as manifested by abnormal cognitive, perceptual, and intellectual functioning, during the course of mental disorders.

paturon [INV ZOO] The basal segment of a chelicera in arachnids.

Paucituberculata [VERT ZOO] An order of marsupial mammals in some systems of classification, including the opossum, rats, and polydolopids.

paunch [ANAT] In colloquial usage, the cavity of the abdomen and its contents. [VERT ZOO] *See* rumen.

paurometabolous metamorphosis [INV ZOO] A simple, gradual, direct metamorphosis in which immature forms resemble the adult except in size and are referred to as nymphs.

Pauropoda [INV ZOO] A class of the Myriapoda distinguished by bifurcate antennae, 12 trunk segments with 9 pairs of functional legs, and the lack of eyes, spiracles, tracheae, and a circulatory system.

Paussidae [INV ZOO] The flat-horned beetles, a family of coleopteran insects in the suborder Adephaga.

Pavlov's pouch [PHYSIO] A small portion of stomach, completely separated from the main stomach, but retaining its vagal nerve branches, which communicates with the exterior; used in the long-term investigation of gastric secretion, and particularly in the study of conditioned reflexes.

paw [VERT ZOO] The foot of an animal, especially a quadruped having claws.

paxilla [INV ZOO] A pillarlike spine in certain starfishes that sometimes has a flattened summit covered with spinules.

Paxillosida [INV ZOO] An order of the Asteroidea in some systems of classification, equivalent to the Paxillosina.

Paxillosina [INV ZOO] A suborder of the Phanerozonida with pointed tube feet which lack suckers, and with paxillae covering the upper body surface.

Pb *See* lead.

PBI *See* protein-bound iodine.

PBI test *See* protein-bound iodine test.

P blood group [IMMUNOL] A system of immunologically distinct, genetically determined erythrocyte antigens first defined by their reaction with anti-P, and immune rabbit antiserum, and later broadened to include related antigens.

PC *See* phosphocreatine.

Pd *See* palladium.

pea [BOT] **1.** *Pisum sativum.* The garden pea, an annual leafy leguminous vine cultivated for its smooth or wrinkled, round edible seeds which are borne in dehiscent pods. **2.** Any of several related or similar cultivated plants.

peach [BOT] *Prunus persica.* A low, spreading, freely branching tree of the order Rosales, cultivated in less rigorous parts of the temperate zone for its edible fruit, a juicy drupe with a single large seed, a pulpy yellow or white mesocarp, and a thin firm epicarp.

peanut [BOT] *Arachis hypogaea.* A low, branching, self-pollinated annual legume cultivated for its edible seed, which is a one-loculed legume formed beneath the soil in a pod.

pear [BOT] Any of several tree species of the genus *Pyrus* in the order Rosales, cultivated for their fruit, a pome that is wider at the apical end and has stone cells throughout the flesh.

pearl [PATH] **1.** Rounded masses of concentrically arranged squamous epithelial cells, seen in some carcinomas. **2.** Mucous casts of the bronchi or bronchioles found in the sputum of asthmatic persons.

pearly tumor *See* cholesteatoma.

pébrine [INV ZOO] A contagious protozoan disease of silkworms and other caterpillars caused by *Nosema bombycis.*

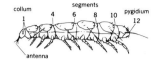

PAUROPODA

Pauropus silvaticus Tiegs, fully extended adult, enlarged, showing the 9 pairs of legs and the 12 trunk segments indicated by number.

PAXILLA

Arm of a starfish, *Persephonaster neozelanicus*, showing modified spines covering the surface.

PEAR

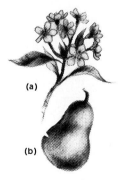

Pear *(Pyrus communis).* *(a)* Flower cluster. *(b)* Fruit.

PECAN

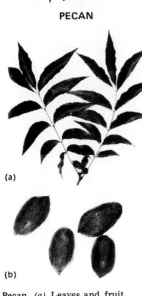

(a)

(b)

Pecan. *(a)* Leaves and fruit.
(b) Hulled nuts.

PECTINIBRANCHIA

2 mm

Shell of the snail *Orthonema*,
from the Pennsylvanian-Permian.
*(From R. R. Shrock and W. H.
Twenhofel, Principles of
Invertebrate Paleontology, 2d
ed., McGraw-Hill, 1953)*

PEDICELLARIA

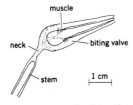

muscle

neck — biting valve

stem   1 cm

Tridentate type of echinoid
pedicellaria showing the beak
made of three movable jaws.

pecan [BOT] *Carya illinoensis.* A large deciduous hickory tree in the order Fagales which produces an edible, oblong, thin-shelled nut.

peccary [VERT ZOO] Either of two species of small piglike mammals in the genus *Tayassu,* composing the family Tayassuidae.

pecking order [PSYCH] A social hierarchy of prestige, dominance, or authority. [VERT ZOO] A hierarchy of social dominance within a flock of poultry where each bird is allowed to peck another lower in the scale and must submit to pecking by one of higher rank.

Pecora [VERT ZOO] An infraorder of the Artiodactyla; includes those ruminants with a reduced ulna and usually with antlers, horns, or deciduous horns.

pecten [ZOO] Any of various comblike structures possessed by animals.

Pectenidae [INV ZOO] A family of bivalve mollusks in the order Anisomyaria; contains the scallops.

pectic acid [BIOCHEM] A complex acid, partially demethylated, obtained from the pectin of fruits.

pectin [BIOCHEM] A purified carbohydrate obtained from the inner portion of the rind of citrus fruits, or from apple pomace; consists chiefly of partially methoxylated polygalacturonic acids.

Pectinariidae [INV ZOO] The cone worms, a family of polychaete annelids belonging to the Sedentaria.

pectinase [BIOCHEM] An enzyme that catalyzes the transformation of pectin into sugars and galacturonic acid.

pectinesterase [BIOCHEM] An enzyme that catalyzes the hydrolytic breakdown of pectins to pectic acids.

pectineus [ANAT] A muscle arising from the pubis and inserted on the femur.

Pectinibranchia [INV ZOO] An order of gastropod mollusks which contains many families of snails; respiration is by means of ctenidia, the nervous system is not concentrated, and sexes are separate.

Pectobothridia [INV ZOO] A subclass of parasitic worms in the class Trematoda, characterized by caudal hooks or hard posterior suckers or both.

pectoral fin [VERT ZOO] One of the pair of fins of fishes corresponding to forelimbs of a quadruped.

pectoral girdle [ANAT] The system of bones supporting the upper or anterior limbs in vertebrates. Also known as shoulder girdle.

pectoralis major [ANAT] The large muscle connecting the anterior aspect of the chest with the shoulder and upper arm.

pectoralis minor [ANAT] The small, deep muscle connecting the third to fifth ribs with the scapula.

pedal [BIOL] Of or pertaining to the foot.

pedal disk [INV ZOO] The broad, flat base of many sea anemones, used for attachment to a substrate.

pedal ganglion [INV ZOO] One of the paired ganglia supplying nerves to the foot muscles in most mollusks.

pedal gland *See* foot gland.

pediatrician [MED] A physician who specializes in pediatrics.

pediatrics [MED] The branch of medicine that deals with the growth and development of the child through adolescence, and with the care, treatment, and prevention of diseases, injuries, and defects of children.

pedicel [BOT] **1.** The stem of a fruiting or sporebearing organ. **2.** The stem of a single flower. [ZOO] A short stalk in an animal body.

pedicellaria [INV ZOO] In echinoids and starfishes, any of various small grasping organs in the form of a beak carried on a stalk.

pedicellate [BIOL] Having a pedicel.

Pedicellinea [INV ZOO] The single order of the class Calyssozoa, including all entoproct bryozoans.

pedicle [ANAT] A slender process acting as a foot or stalk (as the base of a tumor), or the basal portion of an organ that is continuous with other structures.

pediculosis [MED] Infestation with lice, especially of the genus *Pediculus.*

pedigree [GEN] The ancestral line of an individual.

Pedilidae [INV ZOO] The false ant-loving flower beetles, a family of coleopteran insects in the superfamily Tenebrionoidea.

Pedinidae [INV ZOO] The single family of the order Pedinoida.

Pedinoida [INV ZOO] An order of Diadematacea making up those forms of echinoderms which possess solid spines and a rigid test.

Pedionomidae [VERT ZOO] A family of quaillike birds in the order Gruiformes.

Pedipalpida [INV ZOO] Former order of the Arachnida; these animals are now placed in the orders Uropygi and Amblypygi.

pedipalpus [INV ZOO] One of the second pair of appendages of an arachnid.

pedodontics [MED] The branch of dentistry concerned with the care of children's teeth.

pedology [MED] The science of the study of the physiological as well as the psychological aspects of childhood.

pedometer [ENG] An instrument for measuring and weighing a newborn child.

pedophilia [PSYCH] Love of children by adults for sexual purposes.

peduncle [ANAT] A band of white fibers joining different portions of the brain. [BOT] **1.** A flower-bearing stalk. **2.** A stalk supporting the fruiting body of certain thallophytes. [INV ZOO] The stalk supporting the whole or a large part of the body of certain crinoids, brachiopods, and barnacles.

pedunculate [BIOL] **1.** Having or growing on a peduncle. **2.** Being attached to a peduncle.

Pegasidae [VERT ZOO] The single family of the order Pegasiformes.

Pegasiformes [VERT ZOO] The sea moths or sea dragons, a small order of actinopterygian fishes; the anterior of the body is encased in bone, and the nasal bones are enlarged to form a rostrum that projects well forward of the mouth.

peg graft [BOT] A graft made by driving a scion of leafless dormant wood with wedge-shaped base into an opening in the stock and sealing with wax or other material.

Peisidicidae [INV ZOO] A monogeneric family of polychaete annelids belonging to the Errantia.

Peking man [PALEON] *Sinanthropus pekinensis.* An extinct human type; the braincase was thick, with a massive basal and occipital torus structure and heavy browridges.

pelage [VERT ZOO] The coat of mammals.

pelagic [OCEANOGR] Pertaining to water of the open portion of an ocean, above the abyssal zone and beyond the outer limits of the littoral zone.

pelargonidin [BIOCHEM] An anthocyanidin pigment obtained by hydrolysis of pelargonin in the form of its red-brown crystalline chloride, $C_{15}H_{11}ClO_5$.

pelargonin [BIOCHEM] An anthocyanin obtained from the dried petals of red pelargoniums or blue cornflowers in the form of its red crystalline chloride, $C_{27}H_{31}ClO_{15}$.

PEGASIFORMES

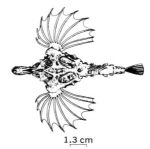

1.3 cm

Sea moth *(Pegasus draconis)*, of the order Pegasiformes. *(After D. S. Jordan and J. O. Snyder, vol. 24, Leland Stanford University Contributions to Biology, 1901)*

PEKING MAN

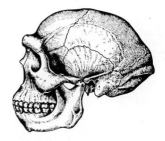

Reconstruction of female *Homo erectus pekinensis* skull, showing the heavy browridges. *(After F. Weidenreich, from M. F. Ashley Montagu, An Introduction to Physcial Anthropology, 2d ed., Charles C. Thomas, 1951)*

PELICAN

Great white pelican (*Pelicanus erythrorhynchos*).

Pelecanidae [VERT ZOO] The pelicans, a family of aquatic birds in the order Pelecaniformes.

Pelecaniformes [VERT ZOO] An order of aquatic, fish-eating birds characterized by having all four toes joined by webs.

Pelecanoididae [VERT ZOO] The diving petrels, a family of oceanic birds in the order Procellariiformes.

Pelecinidae [INV ZOO] The pelecinid wasps, a monospecific family of hymenopteran insects in the superfamily Proctotrupoidea.

Pelecypoda [INV ZOO] The equivalent name for Bivalvia.

pelican [VERT ZOO] Any of several species of birds composing the family Pelecanidae, distinguished by the extremely large bill which has a distensible pouch under the lower mandible.

pellagra [MED] A disease caused by nicotinic acid deficiency characterized by skin lesions, inflammation of the soft tissues of the mouth, diarrhea, and central nervous system disorders.

pellet [AGR] A small, cylindrical, compressed mass of livestock feed. [PHARM] A small pill. [SCI TECH] A small spherical or cylindrical body. [VERT ZOO] A mass of undigestible material regurgitated by a carnivorous bird.

pellicle [CYTOL] A plasma membrane. [INV ZOO] A thin protective membrane, as 6n certain protozoans.

pellicularia disease [PL PATH] A fungus disease of coffee and other tropical plants caused by *Pellicularia koleroga* and characterized by leaf spots.

pellion [INV ZOO] A ring of plates which support suckers of echinoids.

Pelmatozoa [INV ZOO] A division of the Echinodermata made up of those forms which are anchored to the substrate during at least part of their life history.

Pelobatidae [VERT ZOO] A family of frogs in the suborder Anomocoela, including the spadefoot toads.

Pelodytidae [VERT ZOO] A family of frogs in the suborder Anomocoela.

Pelogonidae [INV ZOO] The equivalent name for Ochteridae.

Pelomedusidae [VERT ZOO] The side-necked or hidden-necked turtles, a family of the order Chelonia.

pelophilous [ECOL] Growing on clay.

Pelopidae [INV ZOO] A family of oribatid mites, order Sarcoptiformes.

Peloridiidae [INV ZOO] The single family of the homopteran series Coleorrhyncha.

pelotherapy [MED] Therapeutic treatment with earth or mud.

pelta [BOT] The shieldlike apothecium occurring in certain lichens. [INV ZOO] A membrane near the blepharoplast in some flagellates.

peltate [BOT] Of leaves, having the petiole attached to the lower surface instead of the base.

pelvic cavity *See* pelvis.

pelvic fin [VERT ZOO] One of the pair of fins of fishes corresponding to the hindlimbs of a quadruped.

pelvic girdle [ANAT] The system of bones supporting the lower limbs, or the hindlimbs, of vertebrates.

pelvic index [ANAT] The ratio of the anteroposterior diameter to the transverse diameter of the pelvis.

pelvis [ANAT] **1.** The main, basin-shaped cavity of the kidney into which urine is discharged by nephrons. **2.** The basin-shaped structure formed by the hipbones together with the sacrum and coccyx, or caudal vertebrae. **3.** The cavity of the bony pelvis. Also known as pelvic cavity.

pelviscope [MED] An endoscope for examination of the pelvic organs of the female.

Pelycosauria [PALEON] An extinct order of primitive, mammallike reptiles of the subclass Synapsida, characterized by a temporal fossa that lies low on the side of the skull.

pemphigus [MED] An acute or chronic disease of the skin characterized by the appearance of bullae, which develop in crops or in continuous succession.

pemphigus contageosus [MED] A vesicular dermatitis endemic in tropical areas, chiefly affecting the armpits and groin.

Penaeidea [INV ZOO] A primitive section of the Decapoda in the suborder Natantia; in these forms, the pleurae of the first abdominal somite overlap those of the second, the third legs are chelate, and the gills are dendrobranchiate.

penetrance [GEN] The proportion of individuals carrying a dominant gene in the heterozygous condition or a recessive gene in the homozygous condition in which the specific phenotypic effect is apparent. Also known as gene penetrance.

penetrant [INV ZOO] A large barbed nematocyst that pierces the body of the prey and injects a paralyzing agent.

penetration gland [INV ZOO] A gland at the anterior end of certain cercariae that secretes a histolytic substance.

penguin [VERT ZOO] Any member of the avian order Sphenisciformes; structurally modified wings do not fold and they function like flippers, the tail is short, feet are short and webbed, and the legs are set far back on the body.

penial seta [INV ZOO] **1.** One of a pair of chitinoid setae located near the anus in nematodes. **2.** Any of the setae near the aperture of the vas deferens in earthworms.

penicillate [BIOL] Having a tuft of fine hairs.

penicillin [MICROBIO] **1.** The collective name for salts of a series of antibiotic organic acids produced by a number of *Penicillium* and *Aspergillus* species; active against most grampositive bacteria and some gram-negative cocci. **2.** *See* benzyl penicillin sodium.

penicillinase [BIOCHEM] A bacterial enzyme that hydrolyzes and inactivates penicillin.

penicillin G₁ potassium *See* benzyl penicillin potassium.

Penicillium [MYCOL] A genus of fungi in the family Moniliaceae characterized by erect branching conidiophores having terminal tufts of club-shaped cells from which conidia are formed.

Peniculina [INV ZOO] A suborder of the Hymenostomatida.

penis [ANAT] The male organ of copulation in man and certain other vertebrates. Also known as phallus.

penis envy [PSYCH] The envy of the young female child for the penis which she does not possess, or which she thinks she has lost.

penna *See* contour feather.

Pennales [BOT] An order of diatoms (Bacillariophyceae) in which the form is often circular, and the markings on the valves are radial.

pennate [BIOL] **1.** Wing-shaped. **2.** Having wings. **3.** Having feathers.

Pennatulacea [INV ZOO] The sea pens, an order of the subclass Alcyonaria; individuals lack stolons and live with their bases embedded in the soft substratum of the sea.

Pennellidae [INV ZOO] A family of copepod crustaceans in the suborder Caligoida; skin-penetrating external parasites of various marine fishes and whales.

penniculus [ANAT] A tuft of arterioles in the spleen. [BIOL] A brush-shaped structure.

Pennsylvanian [GEOL] A division of late Paleozoic geologic time, extending from 320 to 280 million years ago, varyingly considered to rank as an independent period or as an epoch

PENGUIN

Emperor penguin with its young. The largest species of penguins, it may reach a height of 4 feet (1.2 meters) and weigh up to 100 pounds (45 kilograms).

PENICILLIN

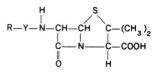

Basic structural formula for penicillin showing the β-lactamthiazolide ring system common to all penicillins; proper combination of substituents at R and Y results in any one of nine penicillins.

PENNSYLVANIAN

PRECAMBRIAN		
CAMBRIAN		
ORDOVICIAN		
SILURIAN		PALEOZOIC
DEVONIAN		
Mississippian	CARBONIFEROUS	
Pennsylvanian		
PERMIAN		
TRIASSIC		
JURASSIC		MESOZOIC
CRETACEOUS		
TERTIARY		CENOZOIC
QUATERNARY		

Chart showing position of the Pennsylvanian in relation to the periods and eras of geologic time.

of the Carboniferous period; named for outcrops of coal-bearing rock formations in Pennsylvania.

pentacrinoid [INV ZOO] The larva of a feather star.

pentactinal [ZOO] Having five rays or branches.

pentacula [INV ZOO] The five-tentacled stage in the life history of echinoderms.

pentadactyl [VERT ZOO] Having five digits on the hand or foot.

pentadelphous [BOT] Having the stamens in five sets with the filaments more or less united within each set.

pentagynous [BOT] Having five styles.

pentalacunar node [BOT] A node characterized by five leaf gaps.

pentalogy [MED] Five symptoms or defects which together characterize a disease or syndrome.

Pentamerida [PALEON] An extinct order of articulate brachiopods.

Pentameridina [PALEON] A suborder of extinct brachiopods in the order Pentamerida; dental plates associated with the brachiophores were well developed, and their bases enclosed the dorsal adductor muscle field.

pentamerous [BOT] Having each whorl of the flower consisting of five members, or a multiple of five.

pentandrous [BOT] Having five stamens.

pentapetalous [BOT] Having five petals.

pentaquine [PHARM] $C_{18}H_{27}N_3O$ An antimalarial drug generally given with quinine; used as the phosphate salt.

pentarch [BOT] Having five xylem groups and five phloem groups arranged in an alternating pattern.

pentasepalous [BOT] Having five sepals.

Pentastomida [INV ZOO] A class of bloodsucking parasitic arthropods; the adult is vermiform, and there are two pairs of hooklike, retractile claws on the cephalothorax.

Pentatomidae [INV ZOO] The true stink bugs, a family of hemipteran insects in the superfamily Pentatomoidea.

Pentatomoidea [INV ZOO] A subfamily of the hemipteran group Pentatomorpha distinguished by marginal trichobothria and by antennae which are usually five-segmented.

Pentatomorpha [INV ZOO] A large group of hemipteran insects in the subdivision Geocorisae in which the eggs are nonoperculate, a median spermatheca is present, accessory salivary glands are tubular, and the abdomen has trichobothria.

pentobarbital sodium [PHARM] $C_{11}H_{17}N_2NaO_3$ A short- to intermediate-acting barbiturate; used as a hypnotic and sedative drug. Also known as sodium pentobarbitone.

pentose [BIOCHEM] Any one of a class of carbohydrates containing five atoms of carbon.

pentose phosphate pathway [BIOCHEM] A pathway by which glucose is metabolized or transformed in plants and microorganisms; glucose-6-phosphate is oxidized to 6-phosphogluconic acid, which then undergoes oxidative decarboxylation to form ribulose-5-phosphate, which is ultimately transformed to fructose-6-phosphate.

pentosuria [MED] The presence of pentose in the urine.

Pentothal [PHARM] Trademark for the ultra-short-acting barbiturate thiopental sodium. Also known as Pentothal sodium; sodium Pentothal.

Pentothal sodium *See* Pentothal.

pentyl *See* amyl.

pentylenetetrazol [PHARM] $C_6H_{10}N_4$ A central nervous system stimulant used as an analeptic.

Penutian [GEOL] A North American stage of geologic time: lower Eocene (above Bulitian, below Ulatasian).

pepo [BOT] A fleshy indehiscent berry with many seeds and a hard rind; characteristic of the Cucurbitaceae.

pepper [BOT] Any of several warm-season perennials of the genus *Capsicum* in the order Polemoniales, especially *C. annum* which is cultivated for its fruit, a many-seeded berry with a thickened integument. [FOOD ENG] Any of various spices and condiments obtained from the fruits of plants of the genus *Piper*.

peppermint [BOT] Any of various aromatic herbs of the genus *Mentha* in the family Labiatae, especially *M. piperita*.

pepsin [BIOCHEM] A proteolytic enzyme found in the gastric juice of mammals, birds, reptiles, and fishes.

pepsinogen [BIOCHEM] The precursor of pepsin, found in the stomach mucosa.

peptic [PHYSIO] **1.** Of or pertaining to peptin. **2.** Of or pertaining to digestion.

peptic ulcer [MED] An ulcer involving the mucosa, submucosa, and muscular layer on the lower esophagus, stomach, or duodenum, due in part at least to the action of acid-pepsin gastric juice.

peptidase [BIOCHEM] An enzyme that catalyzes the hydrolysis of peptides to amino acids.

peptide [BIOCHEM] A compound of two or more amino acids joined by peptide bonds.

peptide bond [ORG CHEM] A bond in which the carboxyl group of one amino acid is condensed with the amino group of another to form a $-CO \cdot NH-$ linkage. Also known as peptide linkage.

peptide linkage *See* peptide bond.

peptone [BIOCHEM] A water-soluble mixture of proteoses and amino acids derived from albumin, meat, or milk; used as a nutrient and to prepare nutrient media for bacteriology.

Peracarida [INV ZOO] A superorder of the Eumalacostraca; these crustaceans have the first thoracic segment united with the head, the cephalothorax usually larger than the abdomen, and some thoracic segments free from the carapace.

Peramelidae [VERT ZOO] The bandicoots, a family of insectivorous mammals in the order Marsupialia.

percentile [STAT] A value in the range of a set of data which separates the range into two groups so that a given percentage of the measures lies below this value.

perception [PHYSIO] Recognition in response to sensory stimuli; the act or process by which the memory of certain qualities of an object is associated with other qualities impressing the senses, thereby making possible recognition of the object.

perch [VERT ZOO] **1.** Any member of the family Percidae. **2.** The common name for a number of unrelated species of fish belonging to the Centrarchidae, Anabantoidei, and Percopsiformes.

Percidae [VERT ZOO] A family of fresh-water actinopterygian fishes in the suborder Percoidei; comprises the true perches.

Perciformes [VERT ZOO] The typical spiny-rayed fishes, comprising the largest order of vertebrates; characterized by fin spines, a swim bladder without a duct, usually ctenoid scales, and 17 or fewer caudal fin rays.

Percoidei [VERT ZOO] A large, typical suborder of the order Perciformes; includes over 50% of the species in this order.

Percomorphi [VERT ZOO] An equivalent, ordinal name for the Perciformes.

Percopsidae [VERT ZOO] A family of fishes in the order Percopsiformes.

Percopsiformes [VERT ZOO] A small order of actinopterygian fishes characterized by single, ray-supported dorsal and

PEPPERMINT

Flowering peppermint *(Mentha piperita)* branch and two separate leaves showing the characteristic shape.

PERCH

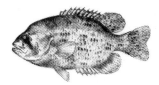

Rock bass *(Ambloplites rupestris)*, a perch of the family Centrarchidae.

PERCOPSIFORMES

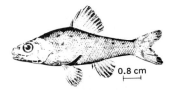

0.8 cm

Sand roller *(Percopsis transmontana)*. *(After D. S. Jordan and B. W. Evermann, The Fishes of North and Middle America, U.S. Nat. Mus. Bull. no. 47, 1900)*

anal fins and a subabdominal pelvic fin with three to eight soft rays.

percussion [MED] The act of striking or firmly tapping the surface of the body with a finger or a small hammer to elicit sounds, or vibratory sensations, of diagnostic value.

pereiopod [INV ZOO] One of the locomotory limbs on the thorax of Malacostraca.

perennial [BOT] A plant that lives for an indefinite period, dying back seasonally and producing new growth from a perennating part.

perennibranchiate [VERT ZOO] Having gills that persist throughout life, as certain Amphibia.

perfect flower [BOT] A flower having both stamens and pistils.

perfect fluid *See* inviscid fluid.

perfect gas *See* ideal gas.

perfoliate [BOT] Pertaining to the form of a leaf having its base united around the stem. [INV ZOO] Pertaining to the form of certain insect antennae having the terminal joints expanded and flattened to form plates which encircle the stalk.

perforatorium *See* acrosome.

perfusion [PHYSIO] The pumping of a fluid through a tissue or organ by way of an artery.

Pergidae [INV ZOO] A small family of hymenopteran insects in the superfamily Tenthredinoidea.

perianal [ANAT] Situated or occurring around the anus.

perianth [BOT] The calyx and corolla considered together.

periapical cyst *See* radicular cyst.

periappendicitis [MED] Inflammation of the tissue around the vermiform process, or of the serosal region of the vermiform appendix.

periarteritis [MED] Inflammation of the outer coat of an artery and of the periarterial tissues.

periarteritis nodosa *See* polyarteritis nodosa.

periblast [EMBRYO] The nucleated layer of cytoplasm that surrounds the blastodisk of an egg undergoing discoidal cleavage.

periblastula [EMBRYO] The blastula of a centrolecithal egg, formed by superficial segmentation.

periblem [BOT] A layer of primary meristem which produces the cortical cells.

pericardial [ANAT] **1.** Of or pertaining to the pericardium. **2.** Located around the heart.

pericardial cavity [ANAT] A potential space between the inner layer of the pericardium and the epicardium of the heart.

pericardial fluid [PHYSIO] The fluid in the pericardial cavity.

pericarditis [MED] Inflammation of the pericardium.

pericardium [ANAT] The membranous sac that envelops the heart; it contains 5-20 grams of clear serous fluid.

pericarp [BOT] The wall of a fruit, developed by ripening and modification of the ovarian wall.

pericentric inversion [GEN] A type of chromosome aberration in which chromosome material involving both arms of the chromosome is inverted, thus spanning the centromere.

perichaetium [BOT] One of the cluster of membranes or leaves enveloping the sex organs of bryophytes.

pericholangitis [MED] Inflammation of the tissues around the bile ducts or the interlobular bile capillaries.

perichondrium [ANAT] The fibrous connective tissue covering cartilage, except at joints.

pericladium [BOT] The clasping part of a sheathing petiole.

periclinal chimera [GEN] A plant carrying a mixture of cells of two distinct species.

PERFOLIATE

Shape of a perfoliate leaf.

PERICENTRIC INVERSION

a b cd e f g → a b cf e d g

Schematic drawing of a pericentric inversion in a chromosome; the white dot is the chromomere.

periclinium [BOT] The involucre forming part of a composite flower.

pericoronitis [MED] Inflammation of the tissue surrounding the coronal portion of the tooth, usually a partially erupted third molar.

pericranium [ANAT] The periosteum on the outer surface of the cranial bones.

pericycle [BOT] The outer boundary of the stele of plants; may not be present as a distinct layer of cells.

pericystium [ANAT] The tissues surrounding a bladder. [MED] The vascular wall of a cyst.

pericyte [HISTOL] A mesenchymal cell found around a capillary; it may or may not be contractile.

periderm [BOT] A group of secondary tissues forming a protective layer which replaces the epidermis of many plant stems, roots, and other parts; composed of cork cambium, phelloderm, and cork. [EMBRYO] The superficial transient layer of epithelial cells of the embryonic epidermis.

peridium [BOT] The outer investment of the sporophore of many fungi.

perifollicular [HISTOL] Surrounding a follicle.

perigonium [BOT] The perianth of a liverwort. [INV ZOO] The sac containing the generative bodies in the gonophore of a hydroid.

perigynium [BOT] A fleshy cup- or tubelike structure surrounding the archegonium of various bryophytes.

perigynous [BOT] Bearing the floral organs on the rim of an expanded saucer- or cup-shaped receptacle or hypanthium.

perihepatitis [MED] Inflammation of the peritoneum and tissues surrounding the liver.

peri-infarction block *See* intraventricular heart block.

perikaryon [CYTOL] **1.** The body of a nerve cell, containing the nucleus. **2.** A cytoplasmic mass surrounding a nucleus.

Perilampidae [INV ZOO] A family of hymenopteran insects in the superfamily Chalcidoidea.

perilymph [PHYSIO] The fluid separating the membranous from the osseous labyrinth of the internal ear.

perimetrium [ANAT] The serous covering of the uterus.

perimysium [ANAT] The connective tissue sheath eveloping a muscle or a bundle of muscle fibers.

perineum [ANAT] **1.** The portion of the body included in the outlet of the pelvis, bounded in front by the pubic arch, behind by the coccyx and sacrotuberous ligaments, and at the sides by the tuberosities of the ischium. **2.** The region between the anus and the scrotum in the male, between the anus and the posterior commissure of the vulva in the female.

perineural [ANAT] Situated around nervous tissue or a nerve.

perineural fibroblastoma *See* neurofibroma.

perineural fibroma *See* neurofibroma.

perineurium [ANAT] The connective tissue sheath covering a primary bundle of nerve fibers (fasciculus).

periocular [ANAT] Surrounding the eye.

period [CHEM] A family of elements with consecutive atomic numbers in the periodic table and with closely related properties; for example, chromium through copper. [GEOL] A unit of geologic time constituting a subdivision of an era; the fundamental unit of the standard geologic time scale. [PHYS] The duration of a single repetition of a cyclic phenomenon.

periodic disease *See* familial Mediterranean fever.

periodic peritonitis *See* familial Mediterranean fever.

periodic table [CHEM] A table of the elements, written in sequence in the order of atomic number or atomic weight and arranged in horizontal rows (periods) and vertical columns (groups) to illustrate the occurrence of similarities in the

PERICYCLE

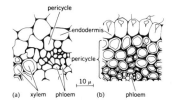

Pericycles. *(a)* Part of transection of root of *Actaea alba* Mill, including xylem and phloem. Pericycle thin-walled and one cell in radial dimension. *(b)* Part of transection of root of *Smilax herbacea* L., including phloem. Pericycle thick-walled and four or five cells in radial dimension (brace).

PERIDERM

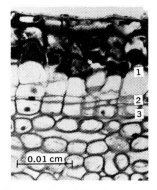

Cross section of the stem periderm of *Sambucus nigra:* (1) phellem; (2) phellogen; (3) phelloderm.

PERIGYNOUS

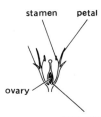

Perigynous arrangement of flower parts on the receptacle.

properties of the elements as a periodic function of the sequence.

periodontitis [MED] Inflammation of the periodontium.

periodontium [ANAT] The tissues surrounding a tooth.

periodontoclasia [MED] Any periodontal disease that results in the destruction of the periodontium.

periodontosis [MED] A degenerative disturbance of the periodontium, characterized by degeneration of connective-tissue elements of the periodontal ligament and by bone resorption.

perionychium [ANAT] The border of epidermis surrounding an entire nail. Also known as paronychium.

periople [VERT ZOO] The thin waxy outer layer of the hoof of equines.

periosteum [ANAT] The fibrous membrane enveloping bones, except at joints and the points of tendonous and ligamentous attachment.

periostitis [MED] Inflammation of the periosteum.

periostracum [INV ZOO] A protective layer of chitin covering the outer portion of the shell in many mollusks, especially fresh-water forms.

periotic [ANAT] **1.** Situated about the ear. **2.** Of or pertaining to the parts immediately about the internal ear.

peripheral [ANAT] Pertaining to or located at or near the surface of a body or an organ. [SCI TECH] Remote from the center; marginal; on the periphery.

peripheral nervous system [ANAT] The autonomic nervous system, the cranial nerves, and the spinal nerves including their associated sensory receptors.

peripheral vision [PHYSIO] The act of seeing images that fall upon parts of the retina outside the macula lutea. Also known as indirect vision.

periphlebitis [MED] Inflammation of the tissues around a vein or of the adventitia of a vein.

periphyton [ECOL] Sessile biotal components of a fresh-water ecosystem.

periplasm [CYTOL] **1.** Cytoplasm around the aster. **2.** Cytoplasm surrounding the yolk of a centrolecithal ovum. [MYCOL] In fungi, the region of an oogonium outside the oosphere.

periplast [CYTOL] **1.** A cell membrane. **2.** A pellicle covering ectoplasm. [HISTOL] The stroma of an animal organ. [INV ZOO] The ectoplasm of a flagellate.

peripneustic [INV ZOO] Having stigmata along the sides of the body, as in insect larvae.

periproct [INV ZOO] The area surrounding the anus of echinoids.

Periptychidae [PALEON] A family of extinct herbivorous mammals in the order Condylarthra distinguished by specialized, fluted teeth.

Peripylina [INV ZOO] An equivalent name for Porulosida.

perisarc [INV ZOO] The outer integument of a hydroid.

Periscelidae [INV ZOO] A family of myodarian cyclorrhaphous dipteran insects in the subsection Acalypteratae.

Perischoechinoidea [INV ZOO] A subclass of principally extinct echinoderms belonging to the Echinoidea and lacking stability in the number of columns of plates that make up the ambulacra and interambulacra.

perisperm [BOT] In a seed, the nutritive tissue that is derived from the nucellus and deposited on the outside of the embryo sac.

perisporangium [BOT] The membrane covering a sorus.

Perissodactyla [VERT ZOO] An order of exclusively herbivorous mammals distinguished by an odd number of toes and mesaxonic feet, that is, with the axis going through the third toe.

peristalsis [PHYSIO] The rhythmic progressive wave of muscular contraction in tubes, such as the intestine, provided with both longitudinal and transverse muscular fibers.

peristethium [INV ZOO] The mesosternum of an insect.

peristome [BOT] The fringe around the opening of a moss capsule. [INV ZOO] The area surrounding the mouth of various invertebrates.

perithecium [MYCOL] A spherical, cylindrical, or oval ascocarp which usually opens by a terminal slit or pore.

peritoneal cavity [ANAT] The potential space between the visceral and parietal layers of the peritoneum.

peritoneoscope [MED] A long, slender endoscope equipped with sheath, obturator, biopsy forceps, a sphygmomanometer bulb and tubing, scissors, and a syringe; introduced into the peritoneal cavity through a small incision in the abdominal wall permitting visualization of the gas-inflated peritoneal cavity.

peritoneoscopy *See* laparoscopy.

peritoneum [ANAT] The serous membrane enveloping the abdominal viscera and lining the abdominal cavity.

peritonitis [MED] Inflammation of the peritoneum.

peritonsillar abscess [MED] An abscess forming in acute tonsillitis around one or both tonsils.

Peritrichia [INV ZOO] A specialized subclass of the class Ciliatea comprising both sessile and mobile forms.

Peritrichida [INV ZOO] The single order of the protozoan subclass Peritrichia.

peritrichous [INV ZOO] Of certain protozoans, having spirally arranged cilia around the oral disk. [MICROBIO] Of bacteria, having a uniform distribution of flagella on the body surface.

perityphlitis [MED] Inflammation of the peritoneum surrounding the cecum and vermiform appendix.

perivitelline space [CYTOL] In mammalian ova, the space formed between the ovum and the zona pellucida at the time of maturation, into which the polar bodies are given off.

perizonium [BOT] The membrane of an auxospore in diatoms.

perlèche [MED] An inflammatory condition occurring at the angles of the mouth with resultant fissuring.

permanent teeth [ANAT] The second set of teeth of a mammal, following the milk teeth; in humans, the set of 32 teeth consists of 8 incisors, 4 canines, 8 premolars, and 12 molars.

permeable membrane [CHEM] A thin sheet or membrane of material through which selected liquid or gas molecules or ions will pass, either through capillary pores in the membrane or by ion exchange; used in dialysis, electrodialysis, and reverse osmosis.

permease [BIOCHEM] Any of a group of enzymes which mediate the phenomenon of active transport.

Permian [GEOL] The last period of geologic time in the Paleozoic era, from 280 to 225 million years ago.

pernicious anemia [MED] A megaloblastic macrocytic anemia resulting from lack of vitamin B_{12}, secondary to gastric atrophy and loss of intrinsic factor necessary for vitamin B_{12} absorption, and accompanied by degeneration of the posterior and lateral columns of the spinal cord.

pernicious malaria *See* falciparum malaria.

perniosis [MED] Any dermatitis resulting from chilblain.

Perognathinae [VERT ZOO] A subfamily of the rodent family Heteromyidae, including the pocket and kangaroo mice.

Peronosporales [MYCOL] An order of aquatic and terrestrial

PERITRICHIA

Trichodina, a mobile peritrich.

PERMIAN

PRECAMBRIAN		
CAMBRIAN		PALEOZOIC
ORDOVICIAN		PALEOZOIC
SILURIAN		PALEOZOIC
DEVONIAN		PALEOZOIC
Mississippian	CARBONIFEROUS	PALEOZOIC
Pennsylvanian	CARBONIFEROUS	PALEOZOIC
PERMIAN		PALEOZOIC
TRIASSIC		MESOZOIC
JURASSIC		MESOZOIC
CRETACEOUS		MESOZOIC
TERTIARY		CENOZOIC
QUATERNARY		CENOZOIC

Chart showing position of Permian period in relation to other periods and to the eras of geologic time.

PETALICHTHYIDA

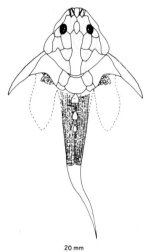

20 mm

Lunaspis broilii from the Lower Devonian of Germany; restoration in dorsal aspect. *(After Walter Gross)*

phycomycetous fungi with a hyphal thallus and zoospores with two flagella.

Perothopidae [INV ZOO] A small family of coleopteran insects in the superfamily Elateroidea found only in the United States.

peroxidase [BIOCHEM] An enzyme that catlyzes reactions in which hydrogen peroxide is an electron acceptor.

Persian melon [BOT] A variety of muskmelon (*Cucumis melo*) in the order Violales; the fruit is globular and without sutures, and has dark-green skin, thin abundant netting, and firm, thick, orange flesh.

persistent [BOT] Of a leaf, withering but remaining attached to the plant during the winter.

personality [PSYCH] The totality of traits and modes of behavior that characterize the individual and his relation to others.

personality disorder [PSYCH] Any of various disorders characterized by abnormal behavior rather than by neurotic, psychotic, or mental disturbances.

perspiration [PHYSIO] **1.** The secretion of sweat. **2.** *See* sweat.

pertussis [MED] An infectious inflammatory bacterial disease of the air passages, caused by *Hemophilus pertussis* and characterized by explosive coughing ending in a whooping inspiration. Also known as whooping cough.

pessary [MED] An appliance of varied form placed in the vagina for uterine support or contraception. [PHARM] Any suppository or other form of medication placed in the vagina for therapeutic purposes.

pessulus [VERT ZOO] A bar composed of cartilage or bone that crosses the windpipe of a bird at its division into bronchi.

pesticide [MATER] A chemical agent that destroys pests. Also known as biocide.

pestilence [MED] **1.** Any epidemic contagious disease. **2.** Infection with the plague organism *Pasteurella pestis*.

petal [BOT] One of the sterile, leaf-shaped flower parts that make up the corolla.

Petalichthyida [PALEON] A small order of extinct dorsoventrally flattened fishes belonging to the class Placodermi; the external armor is in two shields of large plates.

Petalodontidae [PALEON] A family of extinct cartilaginous fishes in the order Bradyodonti distinguished by teeth with deep roots and flattened diamond-shaped crowns.

Petaluridae [INV ZOO] A family of dragonflies in the suborder Anisoptera.

petasma [INV ZOO] A modified endopodite of the first abdominal appendage in a male decapod crustacean.

petechiae [MED] Hemorrhages the size of the head of a pin.

petiole [BOT] The stem which supports the blade of a leaf.

petiolule [BOT] The stalk of a leaflet of a compound leaf.

petit mal [MED] A generalized epileptic seizure of the absence type, that is, characterized by different degrees of impaired consciousness.

petrel [VERT ZOO] A sea bird of the families Procellariidae and Hydrobatidae, generally small to medium-sized with long wings and dark plumage with white areas near the rump.

petri dish [MICROBIO] A shallow glass or plastic dish with a loosely fitting overlapping cover used for bacterial plate cultures and plant and animal tissue cultures.

Petriidae [INV ZOO] A small family of coleopteran insects in the superfamily Tenebrionoidea.

petroleum microbiology [MICROBIO] Those aspects of microbiological science and engineering of interest to the petroleum industry, including the role of microbes in petroleum formation, and the exploration, production, manufacturing, storage, and food synthesis from petroleum.

Petromyzonida [VERT ZOO] The lampreys, an order of eel-like, jawless vertebrates (Agnatha) distinguished by a single, dorsal nasal opening, and the mouth surrounded by an oral disk and provided with a rasping tongue.

Petromyzontiformes [VERT ZOO] The equivalent name for Petromyzonida.

petrosal nerve [ANAT] Any of several small nerves passing through the petrous part of the temporal bone and usually attached to the geniculate ganglion.

petrosal process [ANAT] A sharp process of the sphenoid bone located below the notch for the passage of the abducens nerve, which articulates with the apex of the petrous portion of the temporal bone and forms the medial boundary of the foramen lacerum.

Petrosaviaceae [BOT] A small family of monocotyledonous plants in the order Triuridales characterized by perfect flowers, three carpels, and numerous seeds.

Peyer's patches [HISTOL] Aggregates of lymph nodules beneath the epithelium of the ileum.

Pfeiffer's disease *See* infectious mononucleosis.

pH [CHEM] A term used to describe the hydrogen-ion activity of a system; it is equal to $-\log a_H{}^+$; here $a_H{}^+$ is the activity of the hydrogen ion; in dilute solution, activity is essentially equal to concentration and pH is defined as $-\log_{10}[H^+]$, where $[H^+]$ is hydrogen-ion concentration in moles per liter; a solution of pH 0 to 7 is acid, pH of 7 is neutral, pH over 7 to 14 is alkaline.

phacella *See* gastric filament.

Phaenocephalidae [INV ZOO] A monospecific family of coleopteran insects in the superfamily Cucujoidea, found only in Japan.

Phaenothontidae [VERT ZOO] The tropic birds, a family of fish-eating aquatic forms in the order Pelecaniformes.

Phaeocoleosporae [MYCOL] A spore group of the Fungi Imperfecti with dark filiform spores.

Phaeodictyosporae [MYCOL] A spore group of the Fungi Imperfecti with dark muriform spores.

Phaeodidymae [MYCOL] A spore group of the Fungi Imperfecti with dark two-celled spores.

Phaeodorina [INV ZOO] The equivalent name for Tripylina.

Phaeohelicosporae [MYCOL] A spore group of the Fungi Imperfecti with dark, spirally coiled, septate spores.

Phaeophragmiae [MYCOL] A spore group of the Fungi Imperfecti with dark three- to many-celled spores.

Phaeophyta [BOT] The brown algae, constituting a division of plants; the plant body is multicellular, varying from a simple filamentous form to a complex, sometimes branched body having a basal attachment.

Phaeosporae [MYCOL] A spore group of Fungi Imperfecti characterized by dark one-celled, nonfiliform spores.

Phaeostaurosporae [MYCOL] A spore group of the Fungi Imperfecti with dark star-shaped or forked spores.

phage *See* bacteriophage.

phage cross [VIROL] Multiple infection of a single bacterium by phages that differ at one or more genetic sites, leading to the production of recombinant progeny phage.

phagocyte [CYTOL] An ameboid cell that engulfs foreign material.

phagocytic vacuole *See* food vacuole.

phagocytin [BIOCHEM] A type of bactericidal agent present within phagocytic cells.

phagocytosis [CYTOL] A mechanism by which certain ani-

PHAEODICTYOSPORAE

Characteristic dark muriform spores of *Stemphylium*, a member of the Phaeodictyosporae.

PHAEODIDYMAE

Characteristic spores of *Diplodia* in the Phaeodidymae.

mal cells, such as protozoans and leukocytes, engulf and carry particles into the cytoplasm.

phagosome [CYTOL] A closed intracellular vesicle containing material captured by phagocytosis.

phagotroph [INV ZOO] An organism that ingests nutrients by phagocytosis.

Phalacridae [INV ZOO] The shining flower beetles, a family of coleopteran insects in the superfamily Cucujoidea.

Phalacrocoracidae [VERT ZOO] The cormorants, a family of aquatic birds in the order Pelecaniformes.

Phalaenidae [INV ZOO] The equivalent name for Noctuidae.

Phalangeridae [VERT ZOO] A family of marsupial mammals in which the marsupium is well developed and opens anteriorly, the hindfeet are syndactylous, and the hallux is opposable and lacks a claw.

Phalangida [INV ZOO] An order of the class Arachnida characterized by an unsegmented cephalothorax broadly joined to a segmented abdomen, paired chelate chelicerae, and paired palpi.

phalanx [ANAT] One of the bones of the fingers or toes.

Phalaropodidae [VERT ZOO] The phalaropes, a family of migratory shore birds characterized by lobate toes and by reversal of the sex roles with respect to dimorphism and care of the young.

phallic stage [PSYCH] In psychoanalytic theory, the stage in development during which the child becomes interested in his or her sexual organs and forms an attachment to the parent of the opposite sex.

Phallostethidae [VERT ZOO] A family of actinopterygian fishes in the order Atheriniformes.

Phallostethiformes [VERT ZOO] An equivalent name for Atheriniformes.

phallus [ANAT] *See* penis. [EMBRYO] An undifferentiated embryonic structure derived from the genital tubercle that differentiates into the penis in males and the clitoris in females.

phanerogam [BOT] A seed plant or flowering plant.

phanerophyte [ECOL] A perennial tree or shrub with dormant buds borne on aerial shoots.

Phanerorhynchidae [PALEON] A family of extinct chondrostean fishes in the order Palaeonisciformes having vertical jaw suspension.

Phanerozonida [INV ZOO] An order of the Asteroidea in which the body margins are defined by two conspicuous series of plates and in which pentamerous symmetry is generally constant.

Phanodermatidae [INV ZOO] A family of free-living nematodes in the superfamily Enoploidea.

phaosome [INV ZOO] A light-sensitive organelle in certain epidermal cells of annelids.

Pharetronida [INV ZOO] An order of calcareous sponges in the subclass Calcinea characterized by a leuconoid structure.

pharmaceutical chemistry [CHEM ENG] The chemistry of drugs and of medicinal and pharmaceutical products.

pharmacodynamics [PHARM] The science that deals with the actions of drugs.

pharmacogenetics [GEN] The science of genetically determined variations in drug responses.

pharmacognosy [PHARM] The science of crude drugs.

pharmacologic pyrogen [PHARM] A naturally occurring pharmacologic agent, such as serotonin or a catecholamine that controls body temperature; it can cause fever when injected under experimental conditions.

pharmacology [CHEM] The science dealing with the nature and properties of drugs, particularly their actions.

PHALANGERIDAE

Hindfoot of *Trichosurus*, representing the five-toed, syndactylous condition (digits II and III are bound in one web of skin).

pharmacopoeia [PHARM] A book containing a selected list of medicinal substances and their dosage forms, providing also a description and the standards for purity and strength for each.

pharmacotherapy [MED] The treatment of disease by means of drugs.

pharmacy [MED] **1.** The art and science of the preparation and dispensation of drugs. **2.** A place where drugs are dispensed.

pharyngeal aponeurosis [ANAT] The fibrous submucous layer of the pharynx.

pharyngeal bursa [EMBRYO] A small pit caudal to the pharyngeal tonsil, resulting from the ingrowth of epithelium along the course of the degenerating tip of the notochord of the vertebrate embryo.

pharyngeal cleft [EMBRYO] One of the paired open clefts on the sides of the embryonic pharynx between successive visceral arches in vertebrates.

pharyngeal plexus [ANAT] **1.** A nerve plexus innervating the pharynx. **2.** A plexus of veins situated at the side of the pharynx.

pharyngeal pouch [EMBRYO] One of the five paired sacculations in the lateral aspect of the pharynx in vertebrate embryos. Also known as visceral pouch.

pharyngeal tonsil *See* adenoid.

pharyngeal tooth [VERT ZOO] A tooth developed on the pharyngeal bone in many fishes.

pharyngitis [MED] Inflammation of the pharynx.

Pharyngobdellae [INV ZOO] A family of leeches in the order Arhynchobdellae distinguished by the lack of jaws.

pharyngology [MED] The science of the pharyngeal mechanism, functions, and diseases.

pharyngoscope [MED] An instrument for examining the pharynx.

pharynx [ANAT] A chamber at the oral end of the vertebrate alimentary canal, leading to the esophagus.

phase [CHEM] Portion of a physical system (liquid, gas, solid) that is homogeneous throughout, has definable boundaries, and can be separated physically from other phases. [THERMO] The type of state of a system, such as solid, liquid, or gas.

phase diagram [PHYS CHEM] A graphical representation of the equilibrium relationships between phases (such as vapor-liquid, liquid-solid) of a chemical compound, mixture of compounds, or solution. [THERMO] **1.** A graph showing the pressures at which phase transitions between different states of a pure compound occur, as a function of temperature. **2.** A graph showing the temperatures at which transitions between different phases of a binary system occur, as a function of the relative concentrations of its components.

phase equilibria [PHYS CHEM] The equilibrium relationships between phases (such as vapor, liquid, solid) of a chemical compound or mixture under various conditions of temperature, pressure, and composition.

Phasianidae [VERT ZOO] A family of game birds in the order Galliformes; typically, members are ground feeders, have bare tarsi and copious plumage, and lack feathers around the nostrils.

phasmid [INV ZOO] One of a pair of lateral caudal pores which function as chemoreceptors in certain nematodes.

Phasmidae [INV ZOO] A family of the insect order Orthoptera including the walking sticks and leaf insects.

Phasmidea [INV ZOO] An equivalent name for Secernentea.

Phasmidia [INV ZOO] An equivalent name for Secernentea.

pheasant [VERT ZOO] Any of various large sedentary game

PHARYNGOBDELLAE

Dorsal and ventral view of *Erpobdella punctato*, a jawless leech common in lakes and streams in the Northern Hemisphere.

PHASMIDAE

Walking stick (*Diapheromera femorata*). (*From Illinois Natural History Survey*)

birds with long tails in the family Phasianidae; sexual dimorphism is typical of the group.

phellem [BOT] Cork; the outer tissue layer of the periderm.

phelloderm [BOT] Layers of parenchymatous cells formed as inward derivatives of the phellogen.

phellogen [BOT] The meristematic portion of the periderm, consisting of one layer of cells that initiate formation of the cork and secondary cortex tissue.

Phenacodontidae [PALEON] An extinct family of large herbivorous mammals in the order Condylarthra.

phenetic [BIOL] Pertaining to the physical characteristics of an individual, without consideration for its genetic makeup.

Phengodidae [INV ZOO] The fire beetles, a New World family of coleopteran insects in the superfamily Cantharoidea.

phenobarbital [PHARM] $C_{12}H_{12}N_2O_3$ A crystalline compound, 5-ethyl-5-phenylbarbituric acid, with a slightly bitter taste, melting at 174–178°C; soluble in water, alcohol, chloroform, and ether; used in medicine as a long-acting sedative, anticonvulsant, and hypnotic. Also known as phenobarbitone; phenylethylmalonylurea.

phenobarbitone *See* phenobarbital.

phenocopy [GEN] The nonhereditary alteration of a phenotype to a form imitating a mutant trait; caused by external conditions during development.

phenogenetics [GEN] The study of the phenotypic effects of the genetic material. Also known as physiological genetics.

phenol [ORG CHEM] **1.** C_6H_5OH White, poisonous, corrosive crystals with sharp, burning taste; melts at 43°C, boils at 182°C; soluble in alcohol, water, ether, carbon disulfide, and other solvents; used to make resins and weed killers, and as a solvent and chemical intermediate. Also known as carbolic acid; phenylic acid. **2.** A chemical compound based on the substitution product of phenol, for example, ethylphenol $(C_2H_4C_4H_5OH)$, the ethyl substitute of phenol.

phenol-coefficient method [CHEM] A method for evaluating water-miscible disinfectants in which a test organism is added to a series of dilutions of the disinfectant; the phenol coefficient is the number obtained by dividing the greatest dilution of the disinfectant killing the test organism by the greatest dilution of phenol showing the same result.

phenolphthalein [ORG CHEM] $(C_6H_4OH)_2COC_6H_4CO$ Pale-yellow crystals; soluble in alcohol, ether, and alkalies, insoluble in water; used as an acid-base indicator (carmine-colored to alkalies, colorless to acids) for titrations, as a laxative and dye, and in medicine.

phenotype [GEN] The observable characters of an organism.

phenotypic lag [GEN] Delay in the expression of a newly acquired character.

phenylalanine [BIOCHEM] $C_9H_{11}O_2N$ An essential amino acid, obtained in the levo form by hydrolysis of proteins (as lactalbumin); converted to tyrosine in the normal body. Also known as α-aminohydrocinnamic acid; α-amino-β-phenylpropionic acid; β-phenylalanine.

β-phenylalanine *See* phenylalanine.

phenylbutazone [ORG CHEM] $C_{19}H_{20}O_2N_2$ White or light-yellow powder with aromatic aroma and bitter taste; melts at 107°C; slightly soluble in water, soluble in acetone; used in medicine as an analgesic and antipyretic. Also known as butazolidine; 4-butyl-1,2-diphenyl-3,5-pyrazolindinedione.

phenylephrine [PHARM] $C_9H_{13}NO_2$ A sympathomimetic amine, used in its hydrochloride salt form as a vasoconstrictor.

phenylethylmalonylurea *See* phenobarbital.

phenylhydrazine [ORG CHEM] $C_6H_5NHNH_2$ Poisonous, oily liquid, boiling at 244°C; soluble in alcohol, ether,

---

20 cm

Skeleton of *Ectoconus*, an early Paleocene phenacodont condylarth. *(After G. G. Simpson)*

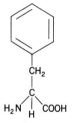

CH_2

C

H_2N H $COOH$

Structural formula of phenylalanine.

chloroform, and benzene, slightly soluble in water; used in analytical chemistry to detect sugars and aldehydes, and as a chemical intermediate. Also known as hydrazinobenzene.

phenylketonuria [MED] A hereditary disorder of metabolism, transmitted as an autosomal recessive, in which there is a lack of the enzyme phenylalanine hydroxylase, resulting in excess amounts of phenylalanine in the blood and of excess phenyl-pyruvic and other acids in the urine. Abbreviated PKU. Also known as phenylpyruvic oligophrenia.

phenylpyruvic acid [BIOCHEM] $C_6H_5CH_2 \cdot CO \cdot COOH$ A keto acid, occurring as a metabolic product of phenylalanine.

phenylpyruvic oligophrenia *See* phenylketonuria.

phenylthiocarbamide *See* phenylthiourea.

phenylthiourea [ORG CHEM] $C_6H_5NHCSNH_2$ A crystal-line compound that has either a bitter taste or is tasteless, depending on the heredity of the taster; used in human genetics studies. Also known as phenylthiocarbamide.

pheochromoblast [HISTOL] A precursor of a pheochromo-cyte.

pheochromocytoma [MED] A tumor of the sympathetic nervous system composed principally of chromaffin cells; found most often in the adrenal medulla.

pheromone [PHYSIO] Any substance secreted by an animal which influences the behavior of other individuals of the same species.

phialospore [MYCOL] One of a chain of spores produced successively on phialides.

Philadelphia chromosome [PATH] An abnormally small G-group chromosome found in the hematopoietic cells of most patients with chronic granulocytic leukemia.

Philippine fowl disease *See* Newcastle disease.

Philomycidae [INV ZOO] A family of pulmonate gastropods composed of slugs.

-philous [SCI TECH] Suffix meaning having an affinity for.

philtrum [ANAT] The depression on the upper lip immediately below the nasal septum.

philtrum-otobasion inferior [ANTHRO] A measure of the distance from the midpoint of the philtrum to the otobasion inferior.

phimosis [MED] Elongation of the prepuce and constriction of the orifice, so that the foreskin cannot be retracted to uncover the glans penis.

phlebitis [MED] Inflammation of a vein.

phlebography [MED] **1.** X-ray photography of a vein or veins following intravenous injection of a radiopaque substance. **2.** Recording of venous pulsations.

phlebolith [MED] A calculus in a vein.

phlebosclerosis [MED] **1.** Sclerosis of a vein. **2.** Chronic phlebitis.

phlebotaxis [BIOL] Movement of a simple motile organism in response to the presence of blood.

phlebothrombosis [MED] A venous thrombus not associated with inflammation of the vein.

phlebotomus fever [MED] An acute viral infection, transmitted by the fly *Phlebotomus papatosii* and characterized by fever, pains in the head and eyes, inflammation of the conjunctiva, leukopenia, and general malaise. Also known as Chitral fever; pappataci fever; sandfly fever; three-day fever.

phlegm [PHYSIO] A viscid, stringy mucus, secreted by the mucosa of the air passages.

phlegmasia alba dolens [MED] A painful swelling of the leg usually seen postpartum, due to femoral vein thrombophlebitis or lymphatic obstruction. Also known as milk leg.

phlegmon [MED] Pyogenic inflammation with infiltration

and spread in the tissues, seen with invasive organisms which produce hyaluronidase and fibrinolysins.

phleomycin [MICROBIO] An antibacterial antibiotic produced by *Streptomyces verticillatus;* antitumor activity has also been demonstrated.

Phloeidae [INV ZOO] The bark bugs, a small neotropical family of hemipteran insects in the superfamily Pentatomoidea.

phloem [BOT] A complex, food-conducting vascular tissue in higher plants; principal conducting cells are sieve elements. Also known as bast; sieve tissue.

phloem necrosis [PL PATH] A pathological state in a plant in which the phloem undergoes brown discoloration and disintegration.

phobia [PSYCH] A disproportionate, obsessive, persistent, and unrealistic fear of an external situation or object, symbolically taking the place of an internal unconscious conflict.

phobic neurosis [PSYCH] A neurotic disorder characterized by a persistent phobia which frequently interferes with the individual's activities and creates tension and anxiety, sometimes accompanied by physical manifestation.

Phocaenidae [VERT ZOO] The porpoises, a family of marine mammals in the order Cetacea.

Phocidae [VERT ZOO] The seals, a pinniped family of carnivoran mammals in the superfamily Canoidea.

Phodilidae [VERT ZOO] A family of birds in the order Strigiformes; the bay owl (*Pholidus bodius*) is the single species.

Phoenicopteridae [VERT ZOO] The flamingos, a family of long-legged, long-necked birds in the order Ciconiiformes.

Phoenicopteriformes [VERT ZOO] An order comprising the flamingos in some systems of classification.

Phoeniculidae [VERT ZOO] The African wood hoopoes, a family of birds in the order Coraciiformes.

Pholadidae [INV ZOO] A family of bivalve mollusks in the subclass Eulamellibranchia; individuals may have one or more dorsal accessory plates, and the visceral mass is attached to the valves in the dorsal portion of the body.

Pholidophoridae [PALEON] A generalized family of extinct fishes belonging to the Pholidophoriformes.

Pholidophoriformes [PALEON] An extinct actinopterygian group composed of mostly small fusiform marine and freshwater fishes of an advanced holostean level.

pholidosis [VERT ZOO] The pattern or character of scales on a scaled animal.

Pholidota [VERT ZOO] An order of mammals comprising the living pangolins and their fossil predecessors; characterized by an elongate tubular skull with no teeth, a long protrusive tongue, strong legs, and five-toed feet with large claws.

Phomaceae [MYCOL] The equivalent name for Sphaerioidaceae.

Phomales [MYCOL] The equivalent name for Sphaeropsidales.

phonation [PHYSIO] Production of speech sounds.

phonocardiography [MED] A graphic recording of heart sounds.

phonoreception [PHYSIO] The perception of sound through specialized sense organs.

phony disease [PL PATH] A virus disease of the peach characterized by dwarfing, abnormal darkening of leaves, and a light crop of small, highly colored fruit; tree stops bearing fruit after a few years.

phoresy [ECOL] A type of commensalism which involves the transporation of one organism (the guest) by a larger organism (the host) of a different species.

PHOLIDOPHORIDAE

Pholidophorus bechei Agassiz, Lower Jurassic of England: length to 8 inches (20 centimeters). *(From D. Rayner, The structure of certain Jurassic holostean fishes with special reference to their neurocrania, Phil. Trans. Roy. Soc. London, Ser. B, no. 601, Cambridge University Press, 1948)*

Phoridae [INV ZOO] The hump-backed flies, a family of cyclorrhaphous dipteran insects in the series Aschiza.

phosphatase [BIOCHEM] An enzyme that catalyzes the hydrolysis and synthesis of phosphoric acid esters and the transfer of phosphate groups from phosphoric acid to other compounds.

phosphate [CHEM] **1.** Generic term for any compound containing a phosphate group (PO_4^{3-}), such as potassium phosphate, K_3PO_4. **2.** Generic term for a phosphate-containing fertilizer material.

phosphatide *See* phospholipid.

phosphaturia *See* hyperphosphaturia.

phosphocreatine [BIOCHEM] $C_4H_{10}N_3O_5P$ Creatine phosphate, a phosphoric acid derivative of creatine which contains an energy-rich phosphate bond; it is present in muscle and other tissues, and during the anaerobic phase of muscular contraction it hydrolyzes to creatine and phosphate and makes energy available. Abbreviated PC.

phosphoenolpyruvic acid [BIOCHEM] $CH_2=O(OPO_3H_2)-COOH$ A high-energy phosphate formed by dehydration of 2-phosphoglyceric acid; it reacts with adenosinediphosphate to form adenosinetriphosphate and enolpyruvic acid.

phosphoglucoisomerase [BIOCHEM] An enzyme that catalyzes the conversion of galactose-1-phosphate to glucose-1-phosphate.

phosphoglucomutase [BIOCHEM] An enzyme that catalyzes the conversion of glucose-1-phosphate to glucose-6-phosphate.

phospholipase [BIOCHEM] An enzyme that catalyzes a hydrolysis of a phospholipid, especially a lecithinase that acts in this manner on a lecithin.

phospholipid [BIOCHEM] Any of a class of esters of phosphoric acid containing one or two molecules of fatty acid, an alcohol, and a nitrogenous base. Also known as phosphatide.

phosphomonoesterase [BIOCHEM] An enzyme catalyzing hydrolysis of phosphoric acid esters containing one ester linkage.

phosphoprotein [BIOCHEM] Any of a class of proteins that are linked with phosphoric acid.

phosphorescence [ATOM PHYS] **1.** Luminescence that persists after removal of the exciting source. Also known as afterglow. **2.** Luminescence whose decay, upon removal of the exciting source, is temperature-dependent.

phosphoric acid [INORG CHEM] H_3PO_4 Water-soluble, transparent crystals, melting at 42°C; used as a fertilizer, in soft drinks and flavor syrups, pharmaceuticals, water treatment, and animal feeds and to pickle and rust-proof metals. Also known as orthophosphoric acid.

phosphorolysis [BIOCHEM] A reaction by which elements of phosphoric acid are incorporated into the molecule of a compound.

phosphorus [CHEM] A nonmetallic element, symbol P, atomic number 15, atomic weight 30.98; used to manufacture phosphoric acid, in phosphor bronzes, incendiaries, pyrotechnics, matches, and rat poisons.

phosphorylase [BIOCHEM] An enzyme that catalyzes the formation of glucose-1-phosphate (Cori ester) from glycogen and inorganic phosphate; it is widely distributed in animals, plants, and microorganisms.

phosphorylation [ORG CHEM] The esterification of compounds with phosphoric acid.

phosphotransacetylase [BIOCHEM] An enzyme that catalyzes the reversible transfer of an acetyl group from acetyl coenzyme A to a phosphate, with formation of acetyl phosphate.

PHOSPHORUS

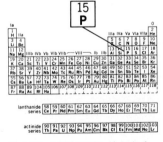

Periodic table of the chemical elements showing the position of phosphorus.

photic region [ECOL] The uppermost layer of a body of water that receives enough sunlight to permit the occurrence of photosynthesis.

Photidae [INV ZOO] A family of amphipod crustaceans in the suborder Gammaridea.

photoautotrophic [BIOL] Pertaining to organisms which derive energy from light and manufacture their own food.

photochemical reaction [PHYS CHEM] A chemical reaction influenced or initiated by light, particularly ultraviolet light, as in the chlorination of benzene to produce benzene hexachloride.

photochemistry [PHYS CHEM] The study of the effects of light on chemical reactions.

photocoagulator [MED] An instrument that uses a xenon flash lamp and an associated train of optics to focus an intense beam of light on a detached retina for the purpose of inducing coagulation and a lesion that welds the retina back into position.

photodinesis [CYTOL] Protoplasmic streaming which is induced by light.

photoelectric colorimetry [ANALY CHEM] Measurement of the colorant concentration in a solution by means of the tristimulus values of three primary light filter-photocell combinations.

photoionization [PHYS CHEM] The removal of one or more electrons from an atom or molecule by absorption of a photon of visible or ultraviolet light. Also known as atomic photoelectric effect.

photoluminescence [ATOM PHYS] Luminescence stimulated by visible, infrared, or ultraviolet radiation.

photometric titration [ANALY CHEM] A titration in which the titrant and solution cause the formation of a metal complex accompanied by an observable change in light absorbance by the titrated solution.

photomicrograph [GRAPHICS] A micrograph produced by photography.

photon [QUANT MECH] A massless particle, the quantum of the electromagnetic field, carrying energy, momentum, and angular momentum. Also known as light quantum.

photonasty [BOT] Nastic movement in response to light.

photoperiodism [PHYSIO] The physiological responses of an organism to the length of night or day or both.

photophilic [BIOL] Thriving in full light.

photophobia [PSYCH] An abnormal fear of light.

photophobic [BIOL] 1. Avoiding light. 2. Exhibiting negative phototropism.

photophore gland [VERT ZOO] A highly modified integumentary gland which develops into a luminous organ composed of a lens and a light-emitting gland; occurs in deep-sea teleosts and elasmobranchs.

photophosphorylase [BIOCHEM] An enzyme that is associated with the surface of a thylakoid membrane and is involved in the final stages of adenosinetriphosphate production by photosynthetic phosphorylation.

photophosphorylation [BIOCHEM] Phosphorylation induced by light energy in photosynthesis.

photophygous [BIOL] Thriving in shade.

photopic vision *See* foveal vision.

photopigment [BIOCHEM] A pigment that is unstable in the presence of light of appropriate wavelengths, such as the chromophore pigment which combines with opsins to form rhodopsin in the rods and cones of the vertebrate eye.

photoreactivation [GEN] The ability of cells irradiated with ultraviolet to repair defects in DNA (caused by irradiation) if the cells were exposed to visible light after uv irradiation.

photoreception [PHYSIO] The process of absorption of light energy by plants and animals and its utilization for biological functions, such as photosynthesis and vision.

photoreceptor [PHYSIO] A highly specialized, light-sensitive cell or group of cells containing photopigments.

photorespiration [BIOCHEM] Respiratory activity taking place in plants during the light period; CO_2 is released and O_2 is taken up, but no useful form of energy, such as adenosine-triphosphate, is derived.

photosynthesis [BIOCHEM] Synthesis of chemical compounds in light, especially the manufacture of organic compounds (primarily carbohydrates) from carbon dioxide and a hydrogen source (such as water), with simultaneous liberation of oxygen, by chlorophyll-containing plant cells.

phototaxis [BIOL] Movement of a motile organism or free plant part in response to light stimulation.

phototroph [BIOL] An organism that utilizes light as a source of metabolic energy.

phototropism [BOT] A growth-mediated response of a plant to stimulation by visible light.

phototropy *See* phototropism.

Phoxichilidiidae [INV ZOO] A family of marine arthropods in the subphylum Pycnogonida; typically, chelifores are present, palpi are lacking, and ovigers have five to nine joints in males only.

Phoxocephalidae [INV ZOO] A family of amphipod crustaceans in the suborder Gammaridea.

Phractolaemidae [VERT ZOO] A family of tropical African fresh-water fishes in the order Gonorynchiformes.

phragma [INV ZOO] In insects and arthropods, an infolded process or septum of the wall of the thorax.

Phragmobasidiomycetes [MYCOL] An equivalent name for Heterobasidiomycetidae.

phragmocone [INV ZOO] In belemnites and other mollusks, the conical internal shell divided internally into chambers by a series of septa perforated by a siphuncle.

phragmoid [BOT] Having septae perpendicular to the long axis, as the conidia of certain fungi.

phragmoplast [CYTOL] A thin barrier which is formed across the spindle equator in late cytokinesis in plant cells and within which the cell wall is laid down.

phragmosome [CYTOL] A differentiated cytoplasmic partition in which the phragmoplast and cell plate develop during cell division in plant cells.

Phragmosporae [MYCOL] A spore group of the Fungi Imperfecti with three- to many-celled spores.

phragmospore [BOT] A spore having two or more septa.

Phreatoicidae [INV ZOO] A family of isopod crustaceans in the suborder Phreatoicoidea in which only the left mandible retains a lacinia mobilis.

Phreatoicoidea [INV ZOO] A suborder of the Isopoda having a subcylindrical body that appears laterally compressed, antennules shorter than the antennae, and the first thoracic segment fused with the head.

phreatophyte [ECOL] A plant with a deep root system which obtains water from the groundwater or the capillary fringe above the water table.

phrenectomy [MED] Resection of a section of a phrenic nerve or removal of an entire phrenic nerve.

phrenic nerve [ANAT] A nerve, arising from the third, fourth, and fifth cervical (cervical plexus) segments of the spinal cord; innervates the diaphragm.

phrynoderma [MED] Dryness of the skin with follicular hyperkeratosis, caused by vitamin A deficiency.

Phrynophiurida [INV ZOO] An order of the Ophiuroidea in

PHREATOICOIDEA

Onchotelson brevicaudatus (Smith), adult male.

PHYCOCYANOBILIN

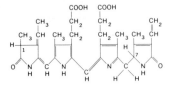

Structural formula of phycocyanobilin.

PHYLLOBRANCHIATE GILL

The phyllobranchiate gill of the mud shrimp (*Thalassina*). (*Smithsonian Institution*)

PHYLLOLEPIDA

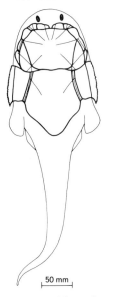

50 mm

Phyllolepis orvini from the Upper Devonian of Greenland. The restoration of the dermal armor is shown in dorsal aspect. The outline of the front of the head and the body is hypothetical, but shows how the fish might have appeared in life. (*After Erik Stensiö*)

which the vertebrae usually articulate by means of hourglass-shaped surfaces, and the arms are able to coil upward or downward in the vertical plane.

phthisis [MED] **1.** Any disease characterized by emaciation and loss of strength. **2.** *See* tuberculosis.

Phycitinae [INV ZOO] A large subfamily of moths in the family Pyralididae in which the frenulum of the female is a simple spine rather than a bundle of bristles.

phycobilin [BIOCHEM] Any of various protein-bound pigments which are open-chain tetrapyrroles and occur in some groups of algae.

phycocyanin [BIOCHEM] A blue phycobilin.

phycocyanobilin [BIOCHEM] $C_{31}H_{38}O_2N_4$ Phycobilin with an ethylidene side chain ($=CH-CH_3$) and only one asymmetric carbon atom (C_1).

phycoerythrin [BIOCHEM] A red phycobilin.

phycoerythrobilin [BIOCHEM] $C_{31}H_{38}O_2N_4$ Phycobilin with seven conjugated double bonds, an ethylidine side chain ($=CH-CH_3$), and two asymmetric carbon atoms (C_1 and C_7).

Phycomycetes [MYCOL] A primitive class of true fungi belonging to the Eumycetes; they lack regularly spaced speta in the actively growing portions of the plant body, and have the sporangiospore, produced in the sporangium by cleavage, as the fundamental, asexual reproductive unit.

Phycosecidae [INV ZOO] A small family of coleopteran insects of the superfamily Cucujoidea, including five species found in New Zealand, Australia, and Egypt.

phylactocarp [INV ZOO] A modified branch of the hydrocladium in Hydromedusae, for protection of the gonotheca.

Phylactolaemata [INV ZOO] A class of fresh-water ectoproct bryozoans; individuals have lophophores which are U-shaped in basal outline, and relatively short, wide zooecia.

phylad [EVOL] A group of closely related species.

phyletic evolution [EVOL] The gradual evolution of population without separation into isolated parts.

phyllary [BOT] A bract of the involucre of a composite plant.

phyllidium [INV ZOO] A leaf-shaped or cuplike outgrowth from the side of the scolex of some tapeworms.

Phyllobothrioidea [INV ZOO] The equivalent name for Tetraphyllidea.

phyllobranchiate gill [INV ZOO] A type of decapod crustacean gill with flattened branches, or lamellae usually arranged in two opposite series.

phylloclade [BOT] A flattened stem that fulfills the same functions as a leaf.

phyllocyst [INV ZOO] The rudimentary cavity of a hydrophyllium.

phyllode [BOT] A broad, flat petiole that replaces the blade of a foliage leaf. [INV ZOO] A petal-shaped group of ambulacra near the mouth of certain echinoderms.

Phyllodocidae [INV ZOO] A leaf-bearing family of errantian annelids in which the species are often brilliantly iridescent and are highly motile.

phyllody [BOT] Metamorphosis of the petiole into a foliage leaf.

Phyllogoniaceae [BOT] A family of mosses in the order Isobryales in which the leaves are equitant.

Phyllolepida [PALEON] A monogeneric order of placoderms from the late Upper Devonian in which the armor is broad and low with a characteristic ornament of concentric and transverse ridges on the component plates.

phyllophagous [BIOL] Feeding on leaves.

Phyllophoridae [INV ZOO] A family of dendrochirotacean

holothurians in the order Dendrochirotida having a rather naked skin and a complex calcareous ring.

phyllopodium [BOT] The axis of a leaf.

phyllopodous [INV ZOO] Having leaflike swimming appendages, as Branchiopoda.

phyllorhiza [BOT] A young leaf bearing a root.

phyllosoma [INV ZOO] A flat, transparent, long-legged larval stage of various spiny lobsters.

phyllosperm [BOT] Having seed born on leaves, as in pteridophytes and cycads.

phyllospondylous [VERT ZOO] Of vertebrae, having a hypocentrum but no pleurocentra; the neural arch extends ventrad to enclose the notochord and form transverse processes which articulate with the ribs.

phyllosporous [BOT] Having the ovules located on specialized or modified leaves.

Phyllosticales [MYCOL] An equivalent name for Sphaeropsidales.

Phyllostomatidae [VERT ZOO] The New World leaf-nosed bats (Chiroptera), a large tropical and subtropical family of insect- and fruit-eating forms with narrow, pointed ears.

phyllotaxy [BOT] The arrangement of leaves on a stem.

Phylloxerinae [INV ZOO] A subfamily of homopteran insects in the family Chermidae in which the sexual forms lack mouthparts, and the parthenogenetic females have a beak but the digestive system is closed, and no honeydew is produced.

phylogeny [EVOL] The evolutionary or ancestral history of organisms.

phylum [SYST] A major taxonomic category in classifying animals (and plants in some systems), composed of groups of related classes.

phyma [MED] **1.** A tumor of new growth of varying size, composed of any of the structures of the skin or subcutaneous tissue. **2.** A localized plastic exudate larger than a tubercle; a circumscribed swelling of the skin.

Phymatidae [INV ZOO] A family of carnivorous hemipteran insects characterized by strong, thick forelegs.

phymatosis [MED] Any disease characterized by the formation of phymas or nodules.

Phymosomatidae [INV ZOO] A family of echinacean echinoderms in the order Phymosomatoida with imperforate crenulate tubercles; one surviving genus is known.

Phymosomatoida [INV ZOO] An order of Echinacea with a stirodont lantern and diademoid ambulacral plates.

physa [INV ZOO] The rounded basal portion of the body of certain sea anemones.

Physalopteridae [INV ZOO] A family of parasitic nematodes in the superfamily Spiruroidea.

Physaraceae [MYCOL] A family of slime molds in the order Physarales.

Physarales [MYCOL] An order of Myxomycetes in the subclass Myxogastromycetidae.

physical anthropology [ANTHRO] The science that deals with the biological aspects of man and their relation to his historical or cultural aspects.

physical medicine [MED] A consultative, diagnostic, and therapeutic medical specialty, coordinating and integrating the use of physical and occupational therapy and physical reconditioning in the professional management of the diseased and injured.

physical therapy [MED] The treatment of disease and injury by physical means.

physician [MED] An individual authorized to practice medicine.

physics [SCI TECH] The study of those aspects of nature which can be understood in a fundamental way in terms of elementary principles and laws.

physiognomy [PSYCH] The prediction of personality functioning from facial appearances and expression.

physiological biophysics [BIOPHYS] An area of biophysics concerned with the use of physical mechanisms to explain the behavior and the functioning of living organisms or parts thereof, and with the response of living organisms to physical forces.

physiological dead space *See* dead space.

physiological ecology [ECOL] The study of biological processes and growth under natural or simulated environments.

physiological genetics *See* phenogenetics.

physiological psychology [PSYCH] The study of the physiological mechanisms or correlates of behavior.

physiological saline *See* normal saline.

physiological salt solution *See* normal saline.

physiological sodium chloride solution *See* normal saline.

physiologic diplopia [PHYSIO] A normal phenomenon in which there is formation of images in noncorresponding retinal points, giving a perception of depth. Also known as introspective diplopia.

physiologic tremor [PHYSIO] A tremor in normal individuals, caused by fatigue, apprehension, or overexposure to cold.

physiology [BIOL] The study of the basic activities that occur in cells and tissues of living organisms by using physical and chemical methods.

physoclistous [VERT ZOO] Not having a channel connecting the swim bladder with the digestive tract, as in most teleosts.

Physopoda [INV ZOO] The equivalent name for Thysanoptera.

Physosomata [INV ZOO] A superfamily of amphipod crustaceans in the suborder Hyperiidea; the eyes are small or rarely absent, and the inner plates of the maxillipeds are free at the apex.

physostomous [VERT ZOO] Having a pneumatic duct connecting the swim bladder with the digestive tract, as in ganoids.

Phytalmiidae [INV ZOO] A family of myodarian cyclorrhaphous dipteran insects in the subsection Acalypteratae.

Phytamastigophorea [INV ZOO] A class of the subphylum Sarcomastigophora, including green and colorless phytoflagellates.

phytase [BIOCHEM] An enzyme occurring in plants, especially cereals, which catalyzes hydrolysis of phytic acid to inositol and phosphoric acid.

phytocenose [ECOL] The aggregation of plants characterized by a distinctive combination of species in a particular habitat.

phytochorology *See* plant geography.

phytochrome [BIOCHEM] A protein plant pigment which serves to direct the course of plant growth and development in response variously to the presence or absence of light, to photoperiod, and to light quality.

phytogeography *See* plant geography.

phytohemagglutinin [BIOCHEM] A mucoprotein derivative of the stringbean (*Phaseolus vulgaris*); stimulates mammalian lymphocytes to mitosis.

phytohormone *See* plant hormone.

Phytomastigina [INV ZOO] The equivalent name for Phytamastigophorea.

Phytomonadida [INV ZOO] The equivalent name for Volvocida.

phytoparasite [ECOL] A parasitic plant.

phytopathogen [ECOL] An organism that causes a disease in a plant.

phytophagous [ZOO] Feeding on plants.

phytoplankton [ECOL] Planktonic plant life.

Phytosauria [PALEON] A suborder of Late Triassic long-snouted aquatic thecodonts resembling crocodiles but with posteriorly located external nostrils, absence of a secondary palate, and a different structure of the pelvic and pectoral girdles.

Phytoseiidae [INV ZOO] A family of the suborder Mesostigmata.

phytosis [PL PATH] **1.** Infection with a phytoparasite. **2.** Any disease caused by a phytoparasite.

phytosterol [BIOCHEM] Any of various sterols obtained from plants, including ergosterol and stigmasterol.

phytotoxin [BIOCHEM] **1.** A substance toxic to plants. **2.** A toxin produced by plants.

pia arachnoid [VERT ZOO] The outer meninx of certain submammalian forms having two membranes covering the brain and spinal cord.

Piacention *See* Plaisancian.

pia mater [ANAT] The vascular membrane covering the surface of the brain and spinal cord.

piastrenemia *See* thrombocytosis.

pi bonding [PHYS CHEM] Covalent bonding in which the greatest overlap between atomic orbitals is along a plane perpendicular to the line joining the nuclei of the two atoms.

Picidae [VERT ZOO] The woodpeckers, a large family of birds in the order Piciformes; adaptive modifications include a long tongue and hyoid mechanism, and stiffened tail feathers.

Piciformes [VERT ZOO] An order of birds characterized by the peculiar arrangement of the tendons of the toes.

Picinae [VERT ZOO] The true woodpeckers, a subfamily of the Picidae.

pico- [MATH] A prefix meaning 10^{-12}; used with metric units. Also known as micromicro-.

picodnavirus [VIROL] A group of deoxyribonucleic acid-containing animal viruses including the adeno-satellite viruses.

picornavirus [VIROL] A viral group made up of small (18–30 nanometers), ether-sensitive viruses that lack an envelope and have a ribonucleic acid genome; among subgroups included are enteroviruses and rhinoviruses, both of human origin.

Picrodendraceae [BOT] A small family of dicotyledonous plants in the order Juglandales characterized by unisexual flowers borne in catkins, four apical ovules in a superior ovary, and trifoliate leaves.

picrotoxin [BIOCHEM] $C_{30}H_{34}O_{13}$ A poisonous, crystalline plant alkaloid found primarily in *Cocculus indicus*; used as a stimulant and convulsant drug.

Picumninae [VERT ZOO] The piculets, a subfamily of the avian family Picidae.

piece-root grafting [BOT] Grafting in which each piece of a cut seedling root is used as a stock.

Pierce's disease [PL PATH] A virus disease of grapes in which there is mottling between the veins of leaves, early defoliation, and early ripening and withering of the fruit.

Pieridae [INV ZOO] A family of lepidopteran insects in the superfamily Papilionoidea including white, sulfur, and orange-tip butterflies; characterized by the lack of a prespiracular bar at the base of the abdomen.

Piesmatidae [INV ZOO] The ash-gray leaf bugs, a family of hemipteran insects belonging to the Pentatomorpha.

PHYTOSAURIA

Nicrosaurus, a quadrupedal thecodont. *(After E. H. Colbert, Evolution of Vertebrates, Wiley, 1955)*

PIG

The Yorkshire pig, a domestic breed with a high-quality carcass, raised for its excellent bacon.

PIKA

A pika, a rock-dwelling diurnal mammal which grows to the size of a guinea pig.

PIKE

Esox lucius. One of two species of fish community called a pike.

Piesmidae [INV ZOO] A small family of hemipteran insects in the superfamily Lygaeoidea.

piezoelectric effect [SOLID STATE] **1.** The generation of electric polarization in certain dielectric crystals as a result of the application of mechanical stress. **2.** The reverse effect, in which application of a voltage between certain faces of the crystal produces a mechanical distortion of the material.

piezoelectricity [SOLID STATE] Electricity or electric polarization resulting from the piezoelectric effect.

pig [VERT ZOO] Any wild or domestic mammal of the superfamily Suoidea in the order Artiodactyla; toes terminate in nails which are modified into hooves, the tail is short, and the body is covered sparsely with hair which is frequently bristlelike.

pigeon [VERT ZOO] Any of various stout-bodied birds in the family Columbidae having short legs, a bill with a horny tip, and a soft cere.

pigeon milk [PHYSIO] A milky glandular secretion of the crop of pigeons that is regurgitated to feed newly hatched young.

pigment [BIOCHEM] Any coloring matter in plant or animal cells.

pigmentation [PHYSIO] The normal color of the body and its organs, resulting from a summation of the natural color of the tissue, the pigments deposited therein, and the pigments carried through the blood bathing the tissue.

pigment cell [CYTOL] Any cell containing deposits of pigment.

pika [VERT ZOO] Any member of the family Ochotonidae, which includes 14 species of lagomorphs resembling rabbits but having a vestigial tail and short, rounded ears.

pike [VERT ZOO] Any of about five species of predatory fish which compose the family Esocidae in the order Clupeiformes; the body is cylindrical and compressed, with cycloid scales that have deeply scalloped edges.

Pilacraceae [MYCOL] A family of Basidiomycetes.

Pilargidae [INV ZOO] A family of small, short, depressed errantian polychaete annelids.

pileus [BIOL] The umbrella-shaped upper cap of mushrooms and other basidiomycetous fungi.

Pilidae [INV ZOO] A family of fresh-water snails in the order Pectinibranchia.

pilidium [BOT] A hemispherical apothecium of certain lichens. [INV ZOO] The helmet-shaped larva of Nemertea.

Pilifera [VERT ZOO] Collective designation for animals with hair, that is, mammals.

pill [PHARM] A small, solid dosage form of a globular, ovoid, or lenticular shape, containing one or more medicinal substances.

pilomotor nerve [ANAT] A nerve causing contraction of one of the arrectoris pilorum muscles.

pilomotor reflex [PHYSIO] Erection of the hairs of the skin (gooseflesh) in response to chilling or irritation of the skin or to an emotional stimulus.

pilosebaceous [ANAT] Pertaining to the hair follicles and sebaceous glands, as the pilosebaceous apparatus, comprising the hair follicle and its attached gland.

pilosis [MED] The abnormal or excessive development of hair.

Piltdown man [PALEON] An alleged fossil man based on fragments of a skull and mandible that were eventually discovered to constitute a skillful hoax.

pilus [ANAT] A hair. [BIOL] A fine, slender, hairlike body. [MICROBIO] Any filamentous appendage other than flagella on certain gram-negative bacteria. Also known as fimbria.

pimento [BOT] *Capsicum annuum.* A type of pepper in the order Polemoniales grown for its thick, sweet-fleshed red fruit.

pimple [MATER] A small, conical elevation on the surface of a plastic. [MED] A small pustule or papule.

piña [BOT] A fiber obtained from the large leaves of the pineapple plant. Also known as pineapple fiber.

pinacocyte [INV ZOO] A flattened polygonal cell occurring in the dermal epithelium of sponges, and lining the exhalant canals.

Pinales [BOT] An order of gymnospermous woody trees and shrubs in the class Pinopsida, including pine, spruce, fir, cypress, yew, and redwood; the largest plants are the conifers.

Pinatae *See* Pinopsida.

pincer [INV ZOO] A grasping apparatus, as on the anterior legs of a lobster, consisting of two hinged jaws.

pinch graft [MED] A small, full-thickness graft lifted from the donor area by a needle, and cut free with a razor.

pine [BOT] Any of the cone-bearing trees composing the genus *Pinus;* characterized by evergreen leaves (needles), usually in tight clusters of two to five.

pineal body [ANAT] An unpaired, elongated, club-shaped, knoblike or threadlike organ attached by a stalk to the roof of the vertebrate forebrain. Also known as conarium; epiphysis.

pineapple [BOT] *Ananas sativus.* A perennial plant of the order Bromeliales with long, swordlike, usually rough-edged leaves and a dense head of small abortive flowers; the fruit is a sorosis that develops from the fleshy inflorescence and ripens into a solid mass, covered by the persistent bracts and crowned by a tuft of leaves.

pineapple fiber *See* piña.

pine nut [BOT] The edible seed borne in the cone of various species of pine (*Pinus*), such as stone pine (*P. pinea*) and piñon pine (*P. cembroides* var. *edulis*).

pinfeather [VERT ZOO] A young, underdeveloped feather, especially one still enclosed in a cylindrical horny sheath which is afterward cast off.

pinguecula [ANAT] A small patch of yellowish-white connective tissue located on the conjunctiva, between the cornea and the canthus of the eye.

Pinicae [BOT] A large subdivision of the Pinophyta, comprising woody plants with a simple trunk and excurrent branches, simple, usually alternative, needlelike or scalelike leaves, and wood that lacks vessels and usually has resin canals.

pink disease [PL PATH] A fungus disease of the bark of rubber, cacao, citrus, coffee, and other trees caused by *Corticium salmonicolor* and characterized by a pink covering of hyphae on the stems and branches.

pinkeye [MED] **1.** A contagious, mucopurulent conjunctivitis. **2.** *See* catarrhal conjunctivitis.

pink root [PL PATH] A fungus disease of onion and garlic caused by various organisms, especially species of *Phoma* and *Fusarium*; marked by red discoloration of the roots.

pink rot [PL PATH] **1.** A fungus disease of potato tubers caused by *Phytophtora erythroseptica* and characterized by wet rot and pink color of the cut surfaces of the tuber upon exposure to air. **2.** A rot disease of apples caused by the fungus *Tricothecium roseum.* **3.** A watery soft rot of celery caused by the fungus *Sclerotinia sclerotiorum.*

pinna [ANAT] The cartilaginous, projecting flap of the external ear of vertebrates. Also known as auricle.

pinnate [BOT] Having parts arranged like a feather, branching from a central axis.

pinnate muscle [ANAT] A muscle having a central tendon

PINALES

Branch of pine with leaves and ovulate cones. *(From J. B. Hill et al., Botany, 3d ed., McGraw-Hill, 1960)*

PINEAPPLE

Pineapple *(Ananas sativus)* fruit with tuft of leaves. *(USDA)*

onto which many short, diagonal muscle fibers attach at rather acute angles.

pinnatilobate [BOT] Having pinnately lobed leaves.

pinnatipartite [BOT] Having the leaves lobed three-quarters of the way to the midrib.

pinnatiped [VERT ZOO] Having lobed toes, as occurs in certain birds.

pinnatisect [BOT] Having leaves lobed almost to the base or midrib.

Pinnipedia [VERT ZOO] A suborder of aquatic mammals in the order Carnivora, including walruses and seals.

Pinnotheridae [INV ZOO] The pea crabs, a family of decapod crustaceans belonging to the Brachygnatha.

pinnulate [BIOL] Having pinnules.

pinnule [BIOL] The secondary branch of a plumelike or pinnate organ.

pinocytosis [CYTOL] A form of active transport of water and dissolved substances across a cell membrane involving the internalization of a fluid-filled vacuole.

Pinophyta [BOT] The gymnosperms, a division of seed plants characterized as vascular plants with roots, stems, and leaves, and with seeds that are not enclosed in an ovary but are borne on cone scales or exposed at the end of a stalk.

Pinopsida [BOT] A class of gymnospermous plants in the subdivision Pinicae characterized by entire-margined or slightly toothed, narrow leaves. Formerly known as Pinatae.

pinosome [CYTOL] A closed intracellular vesicle containing material captured by pinocytosis.

pinta [MED] A disease of the skin seen most frequently in tropical America, characterized by dyschromic changes and hyperkeratosis in patches of the skin; caused by the spirochete *Treponema carateum*. Also known as carate; mal de pinto; piquite; purupuru; quitiqua.

pinulus [INV ZOO] A sponge spicule, usually with five rays, one of which develops numerous small spines.

Pinus [BOT] The type genus of the family Pinaceae; the true pines, coniferous trees of north temperate regions having early deciduous primary leaves, needlelike secondary leaves, and cones with woody scales.

pinworm [INV ZOO] *Enterobius vermicularis.* A phasmid nematode of the superfamily Oxyuroidea; causes enterobiasis. Also known as human threadworm; seatworm.

pioneer [ECOL] An organism that is able to establish itself in a barren area and begin an ecological cycle.

Piophilidae [INV ZOO] The skipper flies, a family of myodarian cyclorrhaphous dipteran insects in the subsection Acalypteratae.

Piperaceae [BOT] A family of dicotyledonous plants in the order Piperales characterized by alternate leaves, a solitary ovule, copious perisperm, and scanty endosperm.

Piperales [BOT] An order of dicotyledonous herbaceous plants marked by ethereal oil cells, uniaperturate pollen, and reduced crowded flowers with orthotropous ovules.

piperoxan [PHARM] $C_{14}H_{19}NO_2$ An adrenergic blocking agent that has been used, as the hydrochloride salt, for diagnosis of pheochromocytoma.

pipet [CHEM] Graduated or calibrated tube which may have a center reservoir (bulb); used to transfer known volumes of liquids from one vessel to another; types are volumetric or transfer, graduated, and micro.

Pipidea [VERT ZOO] A family of frogs sometimes included in the suborder Opisthocoela, but more commonly placed in its own suborder, Aglossa; a definitive tongue is lacking, and free ribs are present in the tadpole but they fuse to the vertebrae in the adult.

Pipridae [VERT ZOO] The manakins, a family of colorful, neotropical suboscine birds in the order Passeriformes.

piptoblast [INV ZOO] A statoblast that is free but has no float.

piquite *See* pinta.

piriformis [ANAT] A muscle arising from the front of the sacrum and inserted into the greater trochanter of the femur.

Piroplasmea [INV ZOO] A class of parasitic protozoans in the superclass Sarcodina; includes the single genus *Babesia*.

Pisces [VERT ZOO] The fish and fishlike vertebrates, including the classes Agnatha, Placodermi, Chondrichthyes, and Osteichthyes.

piscivorous [ZOO] Feeding on fishes.

Pisionidae [INV ZOO] A small family of errantian polychaete annelids; allies of the scale bearers.

pistachio [BOT] *Pistacia vera.* A small, spreading dioecious evergreen tree with leaves that have three to five broad leaflets, and with large drupaceous fruit; the edible seed consists of a single green kernel covered by a brown coat and enclosed in a tough shell.

pistil [BOT] The ovule-bearing organ of angiosperms; consists of an ovary, a style, and a stigma.

pistillate [BOT] **1.** Having a pistil. **2.** Having pistils but no stamens.

pistillidium [BOT] The female sex organ in bryophytes, pteridophytes, and gymnosperms.

pistillody [BOT] Metamorphosis of a flower organ into pistils.

pistillum [INV ZOO] A muscle mass within a chitinous tube in the aurophore of a medusoid colony.

pit [BOT] **1.** A cavity in the secondary wall of a plant cell, formed where secondary deposition has failed to occur, and the primary wall remains uncovered; two main types are simple pits and bordered pits. **2.** The stone of a drupaceous fruit.

pit chamber [BOT] The cavity of a bordered pit.

pitcher plant [BOT] Any of various insectivorous plants of the families Sarraceniaceae and Nepenthaceae; the leaves form deep pitchers in which water collects and insects are drowned and digested.

pith [BOT] A central zone of parenchymatous tissue that occurs in most vascular plants and is surrounded by vascular tissue.

pit membrane [BOT] The middle layer of a plant cell wall; forms the floor of pits of adjacent cells.

Pittidae [VERT ZOO] The pittas, a homogeneous family of brightly colored suboscine birds with an erectile crown of feathers, in the suborder Tyranni.

pitting [MED] **1.** The formation of pits; in the fingernails, a consequence and sign of psoriasis. **2.** The preservation for a short time of indentations on the skin made by pressing with the finger; seen in pitting edema.

pitting edema [MED] Edema of such degree that the skin can be temporarily indented by pressure with the fingers.

pituicyte [HISTOL] The characteristic cell of the neurohypophysis; these cells are pigmented and fusiform and are probably derived from neuroglial cells.

pituitary [ANAT] Of or pertaining to the hypophysis. [PHYSIO] Secreting phlegm or mucus (archaic usage).

pituitary dwarfism [MED] Stunted growth due to a deficiency of the primary growth hormone; characterized clinically by growth failure in early life, and in older persons by deficient subcutaneous fat with loose, wrinkled skin and precocious senility.

pituitary gland *See* hypophysis.

Pityaceae [PALEOBOT] A family of fossil plants in the order

PIPTOBLAST

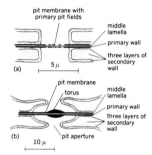

Piptoblast of *Fredericella sultana.*

PIT

Diagrams of pit pairs in the secondary walls of plants. *(a)* Simple. *(b)* Bordered.

Cordaitales known only as petrifactions of branches and wood.

pityriasis [MED] A fine, branny desquamation of the skin.

pivot joint [ANAT] A diarthrosis that permits a rotation of one bone around another; an example is the articulation of the atlas with the axis. Also known as trochoid.

PK *See* psychokinesis.

PKU *See* phenylketonuria.

placebo [MED] A preparation, devoid of pharmacologic effect, given to patients for psychologic effect, or as a control in evaluating a medicinal believed to have a pharmacologic action.

placenta [BOT] A plant surface bearing a sporangium. [EMBRYO] A vascular organ that unites the fetus to the wall of the uterus in all mammals except marsupials and monotremes.

placenta accreta [MED] A placenta that has partially grown into the myometrium of the uterus.

placental barrier [EMBRYO] The tissues intervening between the maternal and the fetal blood of the placenta, which prevent or hinder certain substances or organisms from passing from mother to fetus.

placentation [BOT] The arrangement of ovules and placenta in the ovary of a plant. [EMBRYO] The formation and fusion of the placenta to the uterine wall.

placode [EMBRYO] A platelike epithelial thickening, frequently marking, in the embryo, the anlage of an organ or part.

Placodermi [PALEON] A large and varied class of Paleozoic fishes characterized by a complex bony armor covering the head and the front portion of the trunk.

Placodontia [PALEON] A small order of Triassic marine reptiles of the subclass Euryapsida characterized by flat-crowned teeth in both the upper and lower jaws and on the palate.

Placothuriidae [INV ZOO] A family of holothurian echinoderms in the order Dendrochirotida; individuals are invested in plates and have a complex calcareous ring mechanism.

placula [INV ZOO] An embryonic stage of an ascidian. [VERT ZOO] A flattened blastula having a small segmentation cavity.

Plagiaulacida [PALEON] A primitive, monofamilial suborder of multituberculate mammals distinguished by their dentition (dental formula I 3/0 C 0/0 Pm 5/4 M 2/2), having cutting premolars and two rows of cusps on the upper molars.

Plagiaulacidae [PALEON] The single family of the extinct mammalian suborder Plagiaulacida.

plagiocephaly [MED] A type of strongly asymmetric cranial deformation, in which the anterior portion of one side and the posterior portion of the opposite side of the skull are developed more than their counterparts so that the maximum length of the skull is not in the midline but on a diagonal.

plagiodont [VERT ZOO] Of a snake, having obliquely set, or two converging series of, palatal teeth.

Plagiosauria [PALEON] An aberrant Triassic group of labyrinthodont amphibians.

plagiosere [ECOL] An atypical sere that has deviated from its normal course due to external intervention, as by human activity.

plague [MED] **1.** An infectious bacterial disease of rodents and humans caused by *Pasteurella pestis*, transmitted to man by the bite of an infected flea (*Xenopsylla cheopis*) or by inhalation. Also known as black death; bubonic plague. **2.** Any contagious, malignant, epidemic disease.

plagula [INV ZOO] The ventral plate that protects the pedicle in spiders.

Plaisancian [GEOL] A European stage of geologic time: low-

PLACODONTIA

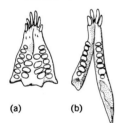

(a)   (b)

Dentition of *Paraplacodus broilii*, Triassic, Switzerland. *(a)* Upper jaw. *(b)* Lower jaw. *(After B. Peyer, 1935)*

er Pliocene (above Pontian of Miocene, below Astian). Also known as Piacention; Plaisanzian.

Plaisanzian *See* Plaisancian.

planaria [INV ZOO] Any flatworm of the turbellarian order Tricladida; the body is broad and dorsoventrally flattened, with anterior lateral projections, the auricles, and a pair of eyespots on the anterior dorsal surface.

Planck's constant [QUANT MECH] A fundamental physical constant, the elementary quantum of action; the ratio of the energy of a photon to its frequency, it is equal to $6.62620 \pm 0.00005 \times 10^{-34}$ joule-second. Symbolized h.

planidium [INV ZOO] A first-stage legless larva of various insects in the orders Diptera and Hymenoptera.

planigraphy *See* sectional radiography.

Planipennia [INV ZOO] A suborder of insects in the order Neuroptera in which the larval mandibles are modified for piercing and for sucking.

plankton [ECOL] Passively floating or weakly motile aquatic plants and animals.

plankton net [ENG] A net for collecting plankton.

planoblast [INV ZOO] The medusa form of a hydrozoan.

planospiral [INV ZOO] Having the shell coiled in one plane, used particularly of foraminiferans and mollusks.

planospore [BIOL] A motile spore.

planozygote [BIOL] A motile zygote.

plant [BOT] Any organism belonging to the kingdom Plantae, generally distinguished by the presence of chlorophyll, a rigid cell wall, and abundant, persistent, active embryonic tissue, and by the absence of the power of locomotion.

Plantae [BOT] The plant kingdom.

Plantaginaceae [BOT] The single family of the plant order Plantaginales.

Plantaginales [BOT] An order of dicotyledonous herbaceous plants in the subclass Asteridae, marked by small hypogynous flowers with a persistent regular corolla and four petals.

plantar [ANAT] Of or relating to the sole of the foot.

plantaris [ANAT] A small muscle of the calf of the leg; origin is the lateral condyle of the femur, and insertion is the calcaneus; flexes the knee joint.

plantar reflex [PHYSIO] Flexion of the toes in response to stroking of the outer surface of the sole, from heel to little toe.

plant fermentation [BIOCHEM] A form of plant metabolism in which carbohydrates are partially degraded without the consumption of molecular oxygen.

plant geography [BOT] A major division of botany, concerned with all aspects of the spatial distribution of plants. Also known as geographical botany; phytochorology; phytogeography.

plant hormone [BIOCHEM] An organic compound that is synthesized in minute quantities by one part of a plant and translocated to another part, where it influences physiological processes. Also known as phytohormone.

plantigrade [VERT ZOO] Pertaining to walking with the whole sole of the foot touching the ground.

plant key [BOT] An analytical guide to the identification of plants, based on the use of contrasting characters to subdivide a group under study into branches.

plant kingdom [BOT] The worldwide array of plant life constituting a major division of living organisms.

plant pathology [BOT] The branch of botany concerned with diseases of plants.

plant physiology [BOT] The branch of botany concerned with the processes which occur in plants.

plant societies [ECOL] Assemblages of plants which constitute structural parts of plant communities.

PLANARIA

Planaria, a bilaterally symmetrical, dorsoventrally flattened, triploblastic organism.

plantula [INV ZOO] A small, cushionlike structure on the ventral surface of the segments of insect tarsi.

plant virus [VIROL] A virus that replicates only within plant cells.

planula [INV ZOO] The ciliated, free-swimming larva of coelenterates.

Planuloidea [INV ZOO] The equivalent name for Moruloidea.

plaque [MED] **1.** A patch, or an abnormal flat area on any internal or external body surface. **2.** A localized area of atherosclerosis. [VIROL] A clear area representing a colony of viruses on a plate culture formed by lysis of the host cell.

plasma [HISTOL] The fluid portion of blood or lymph.

plasma cell *See* plasmacyte.

plasmacyte [HISTOL] A fairly large, generally ovoid cell with a small, eccentrically placed nucleus; the chromatin material is adherent to the nuclear membrane and the cytoplasm is agranular and deeply basophilic everywhere except for a clear area adjacent to the nucleus in the area of the cytocentrum. Also known as plasma cell.

plasmagel [CYTOL] The outer, gelated zone of protoplasm in a pseudopodium.

plasmalemma *See* cell membrane.

plasma membrane *See* cell membrane.

plasmapheresis [MED] The withdrawal of blood from a donor to obtain plasma, its components, or the nonerythrocytic formed elements of blood, followed by the return of the erythrocytes to the donor.

plasmasol [CYTOL] The inner, solated zone of protoplasm in a pseudopodium.

plasmaspore [BOT] An adhesive spore occurring in a sporangium.

plasma thromboplastin antecedent *See* factor XI.

plasma thromboplastin component *See* Christmas factor.

plasmid [GEN] An extrachromosomal genetic element found among various strains of *Escherichia coli* and other bacteria.

plasmin [BIOCHEM] A proteolytic enzyme in plasma which can digest many proteins through the process of hydrolysis. Also known as fibrinolysin.

plasminogen [BIOCHEM] The inert precursor, or zymogen, of plasmin. Also known as profibrinolysin.

plasmodesma [CYTOL] An intercellular bridge, thought to be strands of cytoplasm connecting two cells.

Plasmodiidae [INV ZOO] A family of parasitic protozoans in the suborder Haemosporina inhabiting the erythrocytes of the vertebrate host.

Plasmodiophorida [INV ZOO] An order of the protozoan subclass Mycetozoia occurring as endoparasites of plants.

Plasmodiophoromycetes [MYCOL] A class of the Fungi.

plasmoditrophoblast *See* syncytiotrophoblast.

plasmodium [MICROBIO] The noncellular, multinucleate, jellylike, ameboid, assimilative stage of the Myxomycetes.

Plasmodium [INV ZOO] A genus of protozoans in the family Plasmodiidae in which all the true malarial parasites are placed.

Plasmodroma [INV ZOO] A subphylum of the Protozoa, including Mastigophora, Sarcodina, and Sporozoa, in some taxonomic systems.

plasmogamy [INV ZOO] Fusion of protoplasts, without nuclear fusion, to form a multinucleate mass; occurs in certain protozoans.

plasmolysis [PHYSIO] Shrinking of the cytoplasm away from the cell wall due to exosmosis by immersion of a plant cell in a solution of higher osmotic activity.

plasmoptysis [CYTOL] Bursting forth of cytoplasm from a cell due to rupture of the cell wall.

plasmosome *See* nucleolus.

plasmotomy [INV ZOO] Subdivision of a plasmodium into two or more parts.

plastic surgery [MED] Surgical repair, replacement, or alteration of lost, injured, or deformed parts of the body by transfer of tissue.

plastid [CYTOL] One of the specialized cell organelles containing pigments or protein materials, often serving as centers of special metabolic activities; examples are chloroplasts and leukoplasts.

plastron [INV ZOO] The ventral plate of the cephalothorax of spiders. [VERT ZOO] The ventral portion of the shell of tortoises and turtles.

Platanaceae [BOT] A small family of monoecious dicotyledonous plants in which flowers have several carpels which are separate, three or four stamens, and more or less orthotropous ovules, and leaves are stipulate.

Plataspidae [INV ZOO] A family of shining, oval hemipteran insects in the superfamily Pentatomoidea.

plate budding [BOT] Plant budding by inserting a rectangular scion with a bud under a flap of bark on the stock in such a manner that the exposed wood on the stock is covered.

platelet *See* thrombocyte.

platinum [CHEM] A chemical element, symbol Pt, atomic number 78, atomic weight 195.09.

platmeric [ANTHRO] Of a thighbone, being flattened laterally, with a platymeric index of 75 to 85.

Platyasterida [INV ZOO] An order of Asteroidea in which traces of metapinnules persist, the ossicles of the arm skeleton being arranged in two growth gradient systems.

Platybelondoninae [PALEON] A subfamily of extinct elephantoid mammals in the family Gomphotheriidae consisting of species with digging specializations of the lower tusks.

platycelous [VERT ZOO] Of a vertebra, having a flat or concave ventral surface and a convex dorsal surface.

Platycephalidae [VERT ZOO] The flatheads, a family of perciform fishes in the suborder Cottoidei.

Platyceratacea [PALEON] A specialized superfamily of extinct gastropod mollusks which adapted to a coprophagous life on crinoid calices.

platycnemic index [ANTHRO] The ratio, multiplied by 100, of the anteroposterior diameter to the lateral diameter of the shinbone.

Platycopa [INV ZOO] A suborder of ostracod crustaceans in the order Podocopida including marine forms with two pairs of thoracic legs.

Platycopina [INV ZOO] The equivalent name for Platycopa.

Platyctenea [INV ZOO] An order of the ctenophores whose members are sedentary or parasitic; adults often lack ribs and are flattened due to shortening of the main axis.

platydactyl [VERT ZOO] Having flat digits, as certain tailless amphibians.

Platygasteridae [INV ZOO] A family of hymenopteran insects in the superfamily Proctotrupoidea.

Platyhelminthes [INV ZOO] A phylum of invertebrates composed of bilaterally symmetrical, nonsegmented, dorsoventrally flattened worms characterized by lack of coelom, anus, circulatory and respiratory systems, and skeleton.

platymeric index [ANTHRO] The index, multiplied by 100, of the anteroposterior diameter to the lateral diameter of the femur.

platypellic [ANTHRO] Referring to or descriptive of a broad pelvis having a pelvic index of less than 90.

Platypodidae [INV ZOO] The ambrosia beetles, a family of coleopteran insects in the superfamily Curculionoidea.

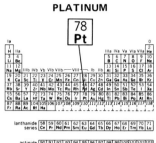

PLATINUM

Periodic table of the chemical elements showing the position of platinum.

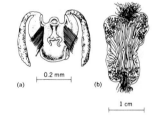

PLATYCTENEA

Coeloplana bocki. (*a*) Larva. (*b*) Adult.

PLATYPUS

The platypus, a primitive aquatic mammal, showing the webbed digits which aid in swimming.

PLATYSOMIDAE

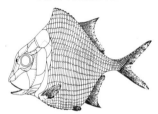

Platysomus parvulus Agassiz, an Upper Carboniferous platysomid from Britain which attained a length of 5 inches (13 centimeters). *(Modified from R. H. Traquair, On the structure and affinities of the Platysomidae, Trans. Roy. Soc. Edinburgh, vol. 29, 1879)*

PLEISTOCENE

PRECAMBRIAN	PALEOZOIC								MESOZOIC			CENOZOIC	
						CARBON-IFEROUS							
	CAMBRIAN	ORDOVICIAN	SILURIAN	DEVONIAN	Mississippian	Pennsylvanian	PERMIAN	TRIASSIC	JURASSIC	CRETACEOUS	TERTIARY	QUATERNARY	

TERTIARY					QUATERNARY	
Paleocene	Eocene	Oligocene	Miocene	Pliocene	Pleistocene	Recent

Chart showing the relationship of the Pleistocene epoch to the periods and eras of geologic time.

Platypsyllidae [INV ZOO] The equivalent name for Leptinidae.

platypus [VERT ZOO] *Ornithorhynchus anatinus.* A monotreme, making up the family Ornithorhynchidae, which lays and incubates eggs in a manner similar to birds, and retains some reptilian characters; the female lacks a marsupium. Also known as duckbill platypus.

platysma [ANAT] A subcutaneous muscle of the neck, extending from the face to the clavicle; muscle of facial expression.

Platysomidae [PALEON] A family of extinct palaeonisciform fishes in the suborder Platysomoidei; typically, the body is laterally compressed and rhombic-shaped, with long dorsal and anal fins.

Platysomoidei [PALEON] A suborder of extinct deep-bodied marine and fresh-water fishes in the order Palaeonisciformes.

platyspondylia [MED] A rare congenital skeletal defect marked by abnormally shaped vertebrae.

Platysternidae [VERT ZOO] The big-headed turtles, a family of Asiatic fresh-water Chelonia with a single species (*Platysternon megacephalum*), characterized by a large head, hooked mandibles, and a long tail.

play therapy [PSYCH] A form of treatment, used particularly with children, in which a child's play, as with dolls in the presence of a therapist, is used as a medium for expression and communication.

pleasure principle [PSYCH] The instinctive attempt to avoid pain, discomfort, or unpleasant situations; the desire to obtain maximum gratification with minimum effort.

Plecoptera [INV ZOO] The stoneflies, an order of primitive insects in which adults differ only slightly from immature stages, except for wings and tracheal gills.

Plectascales [MYCOL] An equivalent name for Eurotiales.

plectoderm [MYCOL] Tissue composed of densely interwoven branched hyphae and covering a fruit body.

Plectognathi [VERT ZOO] The equivalent name for Tetraodontiformes.

Plectoidea [INV ZOO] A superfamily of small, free-living nematodes characterized by simple spiral amphids or variants thereof, elongate cylindroconoid stoma, and reflexed ovaries.

plectonephridia [INV ZOO] Nephridia consisting of networks of fine excretory tubules lying on the body wall and septa of certain oligochaetes.

plectostele [BOT] A protostele that has the xylem divided into plates.

Pleidae [INV ZOO] A family of hemipteran insects in the superfamily Pleoidea.

pleiochasium [BOT] The axis of a cymose inflorescence bearing more than two lateral branches.

pleiomerous [BOT] Having more than the usual number of parts, as of petals.

pleiotropy [GEN] The quality of a gene having more than one phenotypic effect.

Pleistocene [GEOL] An epoch of geologic time of the Quaternary period, following the Tertiary and before the Holocene. Also known as Ice Age; Oiluvium.

pleocytosis [MED] Increase of cells in the cerebrospinal fluid.

pleodont [VERT ZOO] Having solid teeth.

Pleoidea [INV ZOO] A superfamily of suboval hemipteran insects belonging to the subdivision Hydrocoriseae.

pleomorphism [BIOL] The occurrence of more than one distinct form of an organism in a single life cycle. [CRYSTAL] *See* polymorphism.

pleon [INV ZOO] The abdominal region of crustaceans.

pleopod [INV ZOO] An abdominal appendage in certain crustaceans that is modified for swimming.

Pleosporales [BOT] The equivalent name for the lichenized Pseudophaeriales.

plerocercoid [INV ZOO] The infective metacestode of certain cyclophyllidean tapeworms; distinguished by a solid body.

plerome [BOT] Central core of primary meristem which gives rise to all cells of the stele from the pericycle inward.

Plesiocidaroida [PALEON] An extinct order of echinoderms assigned to the Euechinoidea.

plesiomorph [EVOL] The original character of a branching phyletic lineage, found in the ancestral forms.

Plesiosauria [PALEON] A group of extinct reptiles in the order Sauropterygia constituting a highly specialized offshoot of the nothosaurs.

plesiotype [SYST] A specimen or specimens on which subsequent descriptions are based.

Plethodontidae [VERT ZOO] A large family of salamanders in the suborder Salamandroidea characterized by the absence of lungs and the presence of a fine groove from nostril to upper lip.

pleura [ANAT] The serous membrane covering the lung and lining the thoracic cavity.

Pleuracanthodii [PALEON] An order of Paleozoic sharklike fishes distinguished by two-pronged teeth, a long spine projecting from the posterior braincase, and direct backward extension of the tail.

pleural cavity [ANAT] The potential space included between the parietal and visceral layers of the pleura.

pleural rib *See* ventral rib.

pleuranthous [BOT] Having the inflorescences on lateral axes, rather than on main axis.

pleurapophysis [ANAT] One of the lateral processes of a vertebra, corresponding morphologically to a rib.

pleurisy [MED] Inflammation of the pleura. Also known as pleuritis.

pleurite [INV ZOO] A sclerite of the pleuron.

pleuritis *See* pleurisy.

pleurobranchia [INV ZOO] A gill that arises from the lateral wall of the thorax in certain arthropods.

pleurocarpous [BOT] Having the sporophyte in leaf axils along the side of the stem or on lateral branches; refers specifically to mosses.

pleurocentrum [VERT ZOO] One of the paired dorsal and lateral elements of the vertebral centrum of many fishes and fossil amphibians, formed from the dorsal arcualia.

Pleuroceridae [INV ZOO] A family of fresh-water snails in the order Pectinibranchia.

Pleurocoelea [INV ZOO] An extinct superfamily of gastropod mollusks of the order Opisthobranchia in which the shell, mantle cavity, and gills were present.

Pleurodira [VERT ZOO] A suborder of turtles (Chelonia) distinguished by spines on the posterior cervical vertebrae so that the head is retractile laterally.

pleurodontia [VERT ZOO] Attachment of the teeth to the inner surface of the jawbone.

pleurodynia [MED] Severe paroxysmal pain and tenderness of the intercostal muscles.

pleurolophocercous cercaria [INV ZOO] A larval digenetic trematode distinguished by a long, powerful tail with a pair of fin folds, a protrusible oral sucker, and pigmented dorsal eyespots.

Pleuromeiaceae [PALEOBOT] A family of plants in the order Pleuromiales, but often included in the Isoetales due to a phylogenetic link.

PLEURACANTHODII

├─ 30 cm ─┤

Xenacanthus (Pleuracanthus), Carboniferous and Permian sharklike form; note the spine projecting from the posterior braincase. *(After Fritsch)*

PLEURONECTIFORMES

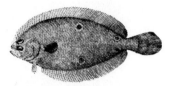

Fourspot flounder *(Paralichthys oblongus)*, of the Pleuronectiformes. *(After G. B. Goode, Fishery Industries of the United States, sect. 1, 1884)*

PLEUROPNEUMONIALIKE ORGANISM

Electron micrograph of a goat strain PPLO. *(From E. Klieneberger-Nobel and F. W. Cuckow, J. Gen. Microbiol., 12:95–99, 1955)*

PLIOCENE

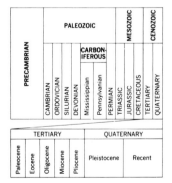

PRECAMBRIAN	PALEOZOIC								MESOZOIC			CENOZOIC
	CAMBRIAN	ORDOVICIAN	SILURIAN	DEVONIAN	CARBON-IFEROUS Mississippian	Pennsylvanian	PERMIAN	TRIASSIC	JURASSIC	CRETACEOUS	TERTIARY	QUATERNARY

TERTIARY					QUATERNARY	
Paleocene	Eocene	Oligocene	Miocene	Pliocene	Pleistocene	Recent

Chart showing the relationship of the Pliocene to the eras and periods of geologic time.

Pleuromeiales [PALEOBOT] An order of Early Triassic lycopods consisting of the genus *Pleuromeia*; the upright branched stem had grasslike leaves and a single terminal strobilus.

pleuron [INV ZOO] The lateral portion of a single thoracic segment in arthropods.

Pleuronectiformes [VERT ZOO] The flatfishes, an order of actinopterygian fishes distinguished by the loss of bilateral symmetry.

Pleuronematina [INV ZOO] A suborder of the Hymenostomatida.

pleuroperitoneal cavity [VERT ZOO] The body cavity containing both the lungs and the abdominal viscera in all pulmonate vertebrates except mammals.

pleuropneumonia [MED] Combined pleurisy and pneumonia. [VET MED] An infectious disease of cattle producing pleural and lung inflammation, caused by *Mycoplasma* species.

pleuropneumonialike organism [MICROBIO] Any of a poorly defined group of microorganisms classified in the order Mycoplasmatales, including the smallest organisms capable of independent life, and comparable in size to the large filterable viruses. Abbreviated PPLO.

pleuropodium [INV ZOO] Either of a pair of lateral glandular processes on the first segment of the abdomen of some insect embryos.

pleurospore [MYCOL] A spore formed laterally on a basidium.

pleurosteon [VERT ZOO] A lateral process of the sternum in young birds; develops into the costal process.

Pleurostigmophora [INV ZOO] A subclass of the centipedes, in some taxonomic systems, distinguished by lateral spiracles.

Pleurotomariacea [PALEON] An extinct superfamily of gastropod mollusks in the order Aspidobranchia.

plexus [ANAT] A network of interlacing nerves or anastomosing vessels.

plica [BIOL] A fold, as of skin or a leaf.

Pliensbachian [GEOL] A European stage of geologic time: Lower Jurassic (above Sinemurian, below Toarcian).

Pliocene [GEOL] A worldwide epoch of geologic time of the Tertiary period, extending from the end of the Miocene to the beginning of the Pleistocene.

Pliohyracinae [PALEON] An extinct subfamily of ungulate mammals in the family Procaviidae.

ploidy [GEN] Number of complete chromosome sets in a nucleus.

Plokiophilidae [INV ZOO] A small family of predacious hemipteran insects in the superfamily Cimicoidea; individuals live in the webs of spiders and embiids.

Plotosidae [VERT ZOO] A family of Indo-Pacific salt-water catfishes (Siluriformes).

plow [AGR] An implement consisting of a share, moldboard, and landside attached to a frame; used to cut, lift, turn, and pulverize soil in preparation for a seedbed.

plum [BOT] Any of various shrubs or small trees of the genus *Prunus* that bear smooth-skinned, globular to oval, drupaceous stone fruit.

plumage [VERT ZOO] The entire covering of feathers of a bird.

Plumatellina [INV ZOO] The single order of the ectoproct bryozoan class Phylactolaemata.

Plumbaginaceae [BOT] The leadworts, the single family of the order Plumbaginales.

Plumbaginales [BOT] An order of dicotyledonous plants in the subclass Caryophyllidae; flowers are pentamerous with fused petals, trinucleate pollen, and a compound ovary containing a single basal ovule.

plumbism [MED] Lead poisoning.

plum blotch [PL PATH] A fungus disease of plums caused by *Phyllosticta congesta* and characterized by minute brown or gray angular leaf spots and brown or gray blotches on the fruit.

plumicome [BIOL] A spicule with plumelike tufts.

plumicorn [VERT ZOO] A hornlike tuft of feathers on the head of certain birds.

plumiped [VERT ZOO] A bird having feathered feet.

Plummer-Vinson syndrome [MED] Dysphagia koilonychia, gastric achlorhydria, glossitis, and hypochromic microcytic anemia caused by iron deficiency. Also known as Paterson-Kelly syndrome; sideropenic dysphagia.

plumose [VERT ZOO] Having feathers or plumes.

plum pocket [PL PATH] A mild fungus disease of plums, caused either by *Taphrina pruni* or *T. communia*, in which the stone of the fruit is aborted.

plumule [BOT] The primary bud of a plant embryo. [VERT ZOO] A down feather.

pluteus [INV ZOO] The free-swimming, bilaterally symmetrical, easel-shaped larva of ophiuroids and echinoids.

plutonism [MED] A disease caused by exposure to plutonium, manifested in experimental animals by graying of the hair, liver degeneration, and tumor formation.

plutonium [CHEM] A reactive metallic element, symbol Pu, atomic number 94, in the transuranium series of elements; the first isotope to be identified was plutonium-239; used as a nuclear fuel, to produce radioactive isotopes for research, and as the fissile agent in nuclear weapons.

pluviilignosa [ECOL] A tropical rain forest.

Pm *See* promethium.

pneumatic [VERT ZOO] **1.** In birds, referring to bones that are penetrated by canals which are connected with the respiratory system. **2.** In certain fishes, the duct between the swim bladder and the alimentary tract.

pneumatocele [MED] **1.** Herniation of the lung. **2.** A sac or tumor containing gas; especially the scrotum filled with gas.

pneumatophore [BOT] **1.** An air bladder in marsh plants. **2.** A submerged or exposed erect root that functions in the respiration of certain marsh plants. [INV ZOO] The air sac of a siphonophore.

pneumatosis [MED] The presence of air or gas in abnormal situations in the body.

pneumaturia [MED] The voiding of urine containing free gas.

pneumoangiography [MED] The outline of the vessels of the lung by means of radiopaque material, for roentgenographic visualization.

pneumobacillus *See* Klebsiella pneumoniae.

pneumococci *See* Diplococcus pneumoniae.

pneumoconiosis [MED] Any lung disease caused by dust inhalation.

pneumocystis carinii pneumonia [MED] A lung infection in man caused by the protozoans *Pneumocystis carinii.* Also known as interstitial plasma-cell pneumonia.

pneumoencephalography [MED] A method of visualizing the ventricular system and subarachnoid pathways of the brain by roentgenography after removal of spinal fluid followed by the injection of air or gas into the subarachnoid space.

pneumoenteritis [MED] Inflammation of the lungs and of the intestine.

pneumography [MED] **1.** Roentgenography of the lung. **2.** The recording of the respiratory excursions.

PLUTEUS

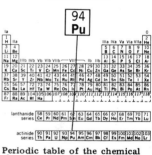

ciliated band

Drawing of a pluteus larva.

PLUTONIUM

Periodic table of the chemical elements showing the position of plutonium.

pneumohemothorax [MED] The presence of air or gas and blood in the thoracic cavity.

pneumolithiasis [MED] The occurrence of calculi or concretions in a lung.

pneumomycosis [MED] Any disease of the lungs caused by a fungus.

pneumonectomy [MED] Surgical removal of an entire lung.

pneumonia [MED] An acute or chronic inflammation of the lungs caused by numerous microbial, immunological, physical, or chemical agents, and associated with exudate in the alveolar lumens.

pneumonic plague [MED] A virulent type of plague in man, with lung involvement.

pneumonitis [MED] Inflammation of the lung.

pneumonolysis [MED] The loosening of any portion of lung adherent to the chest wall; a form of collapse therapy used in the treatment of pulmonary tuberculosis.

pneumopericardium [MED] The presence of air in the pericardial cavity.

pneumoperitoneum [MED] **1.** The presence of air or gas in the peritoneal cavity. **2.** Injection of a gas into the peritoneal cavity as a diagnostic or therapeutic measure.

pneumotaxis [BIOL] Movement of an organism in response to stimulation by carbon dioxide.

pneumothorax [MED] The presence of air or gas in the pleural cavity.

Po *See* polonium.

Poaceae [BOT] The equivalent name for Gramineae.

Poales [BOT] The equivalent name for Cyperales.

pock [MED] A pustule of an eruptive fever, especially of smallpox.

pocket gopher *See* gopher.

pod [BOT] A dry dehiscent fruit; a legume.

Podargidae [VERT ZOO] The heavy-billed frogmouths, a family of Asian and Australian birds in the order Caprimulgiformes.

pod blight [PL PATH] A fungus disease of legumes caused by *Diaporthe* species.

podeon [INV ZOO] The slender middle portion of the abdomen of Hymenoptera; unites the propodeon and the metapodeon.

podetium [BOT] A stalklike outgrowth of thallus which bears the apothecium in certain lichens.

podiatrist *See* chiropodist.

Podicipedide [VERT ZOO] The single family of the avian order Podicipediformes.

Podicipediformes [VERT ZOO] The grebes, an order of swimming and diving birds distinguished by dense, silky plumage, a rudimentary tail, and toes that are individually broadened and lobed.

Podicipitiformes [VERT ZOO] The equivalent name for Podicipediformes.

podite [INV ZOO] A segment of a limb of an arthropod.

podobranch [INV ZOO] A gill of a crustacean attached to the basal segment of a thoracic limb.

Podocopa [INV ZOO] A suborder of fresh-water ostracod crustaceans in the order Podocopida in which the inner lamella has a calcified rim joining the outer lamella along a chitinous zone of concrescence, and the two valves fit together firmly.

Podocopida [INV ZOO] An order of the Ostracoda; contains all fresh-water ostracods and is divided into the suborders Podocopa, Metacopina, and Platycopina.

Podocopina [INV ZOO] The equivalent name for Podocopa.

podocyst [INV ZOO] A sinus in the foot of certain gastropod mollusks.

pododerm [VERT ZOO] The dermal layer of a hoof, underneath the horny layer, in equines.

Podogona [INV ZOO] The equivalent name for Ricinuleida.

podogynium [BOT] A stalk that supports the gynoecium.

podomere [INV ZOO] A segment of an arthropod limb.

podosoma [INV ZOO] That portion of the body in Arachnoidea which bears the four pairs of walking legs.

Podostemaceae [BOT] The single family of the order Podostemales.

Podostemales [BOT] An order of dicotyledonous plants in the subclass Rosidae; plants are submerged aquatics with modified, branching shoots and small, perfect flowers having a superior ovary and united carpels.

pod rot [PL PATH] A fungus disease of cacao caused by *Monilia roreri* and characterized by lesions on the pods.

Poeciliidae [VERT ZOO] A family of fishes in the order Atheriniformes including the live-bearers, such as guppies, swordtails, and mollies.

Poecilosclerida [INV ZOO] An order of sponges of the class Demospongiae in which the skeleton includes two or more types of megascleres.

Poeobiidae [INV ZOO] A monotypic family of spioniform worms (*Poeobius meseres*) belonging to the Sedentaria and found in the North Pacific Ocean.

pogonion [ANAT] The most anterior point of the chin on the symphysis of the mandible.

pogonochore [BOT] A type of plant that produces plumed disseminules.

Pogonophora [INV ZOO] The single class of the phylum Brachiata; the elongate body consists of three segments, each with a separate coelom; there is no mouth, anus, or digestive canal, and sexes are separate.

poikilocyte [HISTOL] An irregularly shaped erythrocyte.

poikilocytosis [MED] A condition in which erythrocytes are distorted in shape.

poikilotherm [ZOO] An animal, such as reptiles, fishes, and invertebrates, whose body temperature varies with and is usually higher than the temperature of the environment; a cold-blooded animal.

point mutation [GEN] Mutation of a single gene due to addition, loss, replacement, or change of sequence in one or more base pairs of the deoxyribonucleic acid of that gene.

poise [FL MECH] A unit of dynamic viscosity equal to the dynamic viscosity of a fluid in which there is a tangential force 1 dyne per square centimeter resisting the flow of two parallel fluid layers past each other when their differential velocity is 1 centimeter per second per centimeter of separation. Abbreviated P.

poison [CHEM] A substance that exerts inhibitive effects on catalysts, even when present only in small amounts; for example, traces of sulfur or lead will poison platinum-based catalysts. [MATER] A substance that in relatively small doses has an action that either destroys life or impairs seriously the functions of organs or tissues.

poison canal [INV ZOO] The duct conveying secretion of poison glands outward from the poison sac in Hymenoptera.

poison gland [VERT ZOO] Any of various specialized glands in certain fishes and amphibians which secrete poisonous mucuslike substances.

poison hemlock [BOT] *Conium maculatum.* A branching biennial poisonous herb that contains a volatile alkaloid, coniine, in its fruits and leaves.

poison ivy [BOT] Any of several climbing, shrubby, or arbo-

POEOBIIDAE

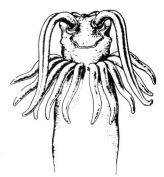

Anterior end of *Poeobius* (Poeobiidae) in dorsal view.

POISON HEMLOCK

Flowering branch of the poison hemlock. (*Adapted from Webster's New International Dictionary, 2d ed., Merriam, 1959*)

POISON IVY

Three-foliolate leaf of poison ivy.

rescent plants of the genus *Rhus* in the sumac family (Anacardiaceae); characterized by ternate leaves, greenish flowers, and white berries that produce an irritating oil.

poison oak [BOT] Any of several bushy poison ivy plants or shrubby poison sumacs.

poisonous plant [BOT] Any of about 400 species of vascular plants containing principles which initiate pathological conditions in man and animals.

poison sumac [BOT] *Rhus vernix*. A tall bush of the sumac family (Anacardiaceae) bearing pinnately compound leaves with 7–13 entire leaflets, and drooping, axillary clusters of white fruits that produce an irritating oil.

polar bear [VERT ZOO] *Thalarctos maritimus*. A large aquatic carnivore found in the polar regions of the Northern Hemisphere.

polar body [CYTOL] One of the small bodies cast off by the oocyte during maturation.

polar covalent bond [PHYS CHEM] A bond in which a pair of electrons is shared in common between two atoms, but the pair is held more closely by one of the atoms.

polarimetric analysis [ANALY CHEM] A method of chemical analysis based on the optical activity of the substance being determined; the measurement of the extent of the optical rotation of the substance is used to identify the substance or determine its quantity.

polarized electromagnetic radiation [ELECTROMAG] Electromagnetic radiation in which the direction of the electric field vector is not random.

polarized light [OPTICS] Polarized electromagnetic radiation whose frequency is in the optical region.

polar nucleus [BOT] One of the two nuclei in the center of the embryo sac of a seed plant which fuse to form the endosperm nucleus.

polarographic analysis [ANALY CHEM] An electroanalytical technique in which the current through an electrolysis cell is measured as a function of the applied potential; the apparatus consists of a potentiometer for adjusting the potential, a galvanometer for measuring current, and a cell which contains two electrodes, a reference electrode whose potential is constant and an indicator electrode which is commonly the dropping mercury electrode. Also known as polarography.

polarography *See* polarographic analysis.

pole blight [PL PATH] A destructive disease of white pines characterized by shortening of the needle-bearing stems, yellowing and shortening of needles, and copious flow of resin.

Polemoniaceae [BOT] A family of autotrophic dicotyledonous plants in the order Polemoniales distinguished by lack of internal phloem, corolla lobes that are convolute in the bud, three carpels, and axile placentation.

Polemoniales [BOT] An order of dicotyledonous plants in the subclass Asteridae, characterized by sympetalous flowers, a regular, usually five-lobed corolla, and stamens equal in number and alternate with the petals.

polian vesicle [INV ZOO] Interradial reservoirs connecting with the ring vessel in most asteroids and holothuroids.

poliomyelitis [MED] An acute infectious viral disease which in its most serious form involves the central nervous system and, by destruction of motor neurons in the spinal cord, produces flaccid paralysis. Also known as Heine-Medin disease; infantile paralysis.

poliovirus vaccine [IMMUNOL] A vaccine prepared from one or all three types of polioviruses in a live or attenuated state.

pollen [BOT] The small male reproductive bodies produced in pollen sacs of the seed plants.

POISON SUMAC

Poison sumac fruits and compound leaf *(Rhus vernix).*

pollen basket [INV ZOO] Hairs on the back of the tibia of worker bees which function in transporting pollen.

pollen count [BOT] The number of grains of pollen that collect on a specified area (often taken as 1 square centimeter) in a specified time.

pollen sac [BOT] **1.** The pollen-containing chamber of the anther. **2.** The microsporangium of a seed plant.

pollen tube [BOT] The tube produced by the wall of a pollen grain which enters the embryo sac and provides a passage through which the male nuclei reach the female nuclei.

pollex [ANAT] The thumb.

pollination [BOT] The transfer of pollen from a stamen to a pistil; fertilization in flowering plants.

pollinium [BOT] A cluster of coherent pollen grains that are transported as a unit during pollination.

pollution [ECOL] Destruction or impairment of the purity of the environment. [PHYSIO] Emission of semen at times other than during coitus.

polonium [CHEM] A chemical element, symbol Po, atomic number 84; all polonium isotopes are radioactive; polonium-210 is the naturally occurring isotope found in pitchblende.

polyadelphous [BOT] Having the filaments united into more than two bundles.

polyandrous [BOT] Describing the condition where a female consorts with several males.

Polyangiaceae [MICROBIO] A family of bacteria in the order Myxobacterales in which microcysts are not spherical or surrounded by a wall but are contained in round or oval cysts; usually, complex fruiting bodies are formed.

polyarch [BOT] Having many bundles of protoxylem.

polyarteritis [MED] Inflammation of several arteries simultaneously. Also known as panarteritis.

polyarteritis nodosa [MED] A systemic disease characterized by widespread inflammation of small- and medium-sized arteries in which some of the foci are nodular. Also known as disseminated necrotizing periarteritis; periarteritis nodosa.

polyarthritis [MED] Inflammation of several joints simultaneously.

polyatomic molecule [CHEM] A chemical molecule with three or more atoms.

polyaxon [INV ZOO] A spicule that is laid down along several axes.

Polybrachiidae [INV ZOO] A family of sedentary marine animals in the order Thecanephria.

polycarp [INV ZOO] A gonad on the inner surface of the mantle in some ascidians.

Polychaeta [INV ZOO] The largest class of the phylum Annelida, distinguished by paired, lateral, fleshy appendages (parapodia) provided with setae, on most segments.

polychasium [BOT] A cymose branch system having more than two branches arising at about the same point.

Polycirrinae [INV ZOO] A subfamily of polychaete annelids in the family Terebellidae.

Polycladida [INV ZOO] A class of marine Turbellaria whose leaflike bodies have a central intestine with radiating branches, many eyes, and tentacles in most species.

polyclimax [ECOL] A climax community under the controlling influence of many environmental factors, including soils, topography, fire, and animal interactions.

polyclonal mixed cryoglobulin [BIOCHEM] A cryoglobulin made of heterogeneous immunoglobin molecules belonging to two or more different classes, and sometimes additional serum proteins.

Polycopidae [INV ZOO] The single family of the suborder Cladocopa.

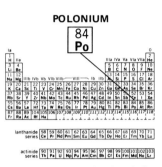

POLONIUM

Periodic table of the chemical elements showing the position of polonium.

Polyctenidae [INV ZOO] A family of hemipteran insects in the superfamily Cimicoidea; the individuals are bat ectoparasites which resemble bedbugs but lack eyes and have ctenidia and strong claws.

polycyesis [MED] Multiple pregnancy.

polycystic kidney [MED] A usually hereditary, congenital, and bilateral disease in which a large number of cysts are present on the kidney.

polycythemia [MED] A condition characterized by an increased number of erythrocytes in the circulation.

polycythemia vera [MED] An absolute increase in all blood cells derived from bone marrow, especially erythrocytes.

polydactyly [MED] The condition of having supernumerary fingers or toes.

polydipsia [MED] Excessive thirst.

Polydolopidae [PALEON] A Cenozoic family of rodentlike marsupial mammals.

polyembryony [ZOO] A form of sexual reproduction in which two or more offspring are derived from a single egg.

polyestrous [PHYSIO] Having several periods of estrus in a year.

Polygalaceae [BOT] A family of dicotyledonous plants in the order Polygalales distinguished by having a bicarpellate pistil and monadelphous stamens.

polygalacturonase [BIOCHEM] An enzyme that catalyzes the hydrolysis of glycosidic linkage of polymerized galacturonic acids.

Polygalales [BOT] An order of dicotyledonous plants in the subclass Rosidae characterized by its simple leaves and usually irregular, hypogynous flowers.

polygamous [BOT] Having both perfect and imperfect flowers on the same plant. [VERT ZOO] Having more than one mate at one time.

polygene [GEN] One of a group of nonallelic genes that collectively control a quantitative character.

Polygnathidae [PALEON] A family of Middle Silurian to Cretaceous conodonts in the suborder Conodontiformes, having platforms with small pitlike attachment scars.

Polygonaceae [BOT] The single family of the order Polygonales.

Polygonales [BOT] An order of dicotyledonous plants in the subclass Caryophyllidae characterized by well-developed endosperm, a unilocular ovary, and often trimerous flowers.

polygoneutic [ZOO] Having more than one brood in a season.

polyhedral disease *See* polyhedrosis.

polyhedrosis [INV ZOO] Any of several virus diseases of insect larvae characterized by the breakdown of tissues and presence of polyhedral granules. Also known as polyhedral disease.

polyhidrosis *See* hyperhidrosis.

polyhydramnios [MED] An excessive volume of amniotic fluid. Also known as hydramnios.

polymastigote [INV ZOO] Having flagella grouped in a tuft.

polymenorrhea *See* metrorrhagia.

polymer [ORG CHEM] **1.** Substance made of giant molecules formed by the union of simple molecules (monomers); for example polymerization of ethylene forms a polyethylene chain, or condensation of phenol and formaldehyde (with production of water) forms phenol-formaldehyde resins.

Polymera [INV ZOO] Formerly a subphylum of the Vermes; equivalent to the phylum Annelida.

polymerase [BIOCHEM] An enzyme that links nucleotides together to form polynucleotide chains.

polymerization [CHEM] **1.** The bonding of two or more

POLYGNATHIDAE

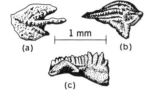

1 mm

(a)   (b)

(c)

Toothlike shapes of polygnaths show their platformlike appearance. *(a) Ancyrodella. (b) Palmatolepis. (c) Polygnathus.*

POLYGONALES

Polygonum hydropiper, eastern American smartweed, characteristic of order Polygonales. Sheathing stipules and many tiny flowers are typical of group. *(Photograph by A. W. Ambler, from National Audubon Society)*

monomers to produce a polymer. **2.** Any chemical reaction that produces such a bonding.

polymorphism [BIOL] **1.** Occurrence of different forms of individual in a single species. **2.** Occurrence of different structural forms in a single individual at different periods in the life cycle. [CRYSTAL] The property of a chemical substance crystallizing into two or more forms having different structures, such as diamond and graphite. Also known as pleomorphism. [GEN] The coexistence of genetically determined distinct forms in the same population, even the rarest of them being too common to be maintained solely by mutation; human blood groups are an example.

polymorphonuclear leukocyte *See* granulocyte.

polymyarian [INV ZOO] Referring to the cross-sectional appearance of muscle cells in a nematode, having many cells in each quadrant.

polymyositis [MED] Inflammation of many muscles simultaneously.

polymyxin [MICROBIO] Any of the basic polypeptide antibiotics produced by certain strains of *Bacillus polymyxa*.

Polynemidae [VERT ZOO] A family of perciform shore fishes in the suborder Mugiloidei.

polyneuritis [MED] Degenerative or inflammatory lesions of several nerves simultaneously, usually symmetrical. Also known as multiple neuritis.

Polynoidae [INV ZOO] The largest family of polychaetes, included in the Errantia and having a body of varying size and shape that is covered with elytra.

polynucleotide [BIOCHEM] A linear sequence of nucleotides.

Polyodontidae [INV ZOO] A family of tubicolous, often large-bodied errantian polychaetes with characteristic cephalic and parapodial structures.

polyoma virus [VIROL] A small deoxyribonucleic acid virus normally causing inapparent infection in mice, but experimentally capable of producing parotid tumors and a wide variety of other tumors.

polyopia [MED] A condition in which more than one image of an object is formed upon the retina.

Polyopisthocotylea [INV ZOO] An order of the trematode subclass Monogenea having a solid posterior holdfast bearing suckers or clamps.

polyorrhymenitis *See* polyserositis.

polyp [INV ZOO] A sessile coelenterate individual having a hollow, somewhat cylindrical body, attached at one end, with a mouth surrounded by tentacles at the free end; may be solitary (hydra) or colonial (coral). [MED] A smooth, rounded or oval mass projecting from a membrane-covered surface.

polyparium [INV ZOO] The connecting tissue forming the common base of a colony of polyps.

polypectomy [MED] Surgical excision of a polyp.

polypeptide [BIOCHEM] A chain of amino acids linked together by peptide bonds but with a lower molecular weight than a protein; obtained by synthesis, or by partial hydrolysis of protein.

polypetalous [BOT] Having distinct petals, in reference to a flower or a corolla. Also known as choripetalous.

Polyphaga [INV ZOO] A suborder of the order Coleoptera; members are distinguished by not having the hind coxae fused to the metasternum and by lacking notopleural sutures.

polyphagous [ZOO] Feeding on many different kinds of plants or animals.

polyphenol oxidase [BIOCHEM] A copper-containing enzyme that catalyzes the oxidation of phenol derivatives to quinones.

polyphyletic [EVOL] Having more than one ancestral stock.

POLYNOIDAE

Harmothoe of the Polynoidae, dorsal view. *(After McIntosh)*

POLYSOME

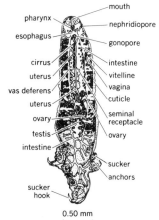

A helical free polysome is bound together by a single messenger-RNA molecule, which follows the helix. Successive ribosomes along the messenger have proceeded farther in production of polypeptide chains.

POLYSTOMATOIDEA

Ventral view of *Heteronchocotyle leucas* Hargis from the bull shark (*Carcharhinus leucas*).

polyphyodont [VERT ZOO] Having teeth which may be constantly replaced.

Polyplacophora [INV ZOO] The chitons, an order of mollusks in the class Amphineura distinguished by an elliptical body with a dorsal shell that comprises eight calcareous plates overlapping posteriorly.

polyploidy [GEN] The occurrence of related forms possessing chromosome numbers which are three or more times the haploid number.

Polypodiales [BOT] The true ferns; the largest order of modern ferns, distinguished by being leptosporangiate and by having small sporangia with a definite number of spores.

Polypodiatae *See* Polypodiopsida.

Polypodiophyta [BOT] The ferns, a division of the plant kingdom having well-developed roots, stems, and leaves that contain xylem and phloem and show well-developed alternation of generations.

Polypodiopsida [BOT] A class of the division Polypodiophyta; stems of these ferns bear several large, spirally arranged, compound leaves with sporangia grouped in sori on their undermargins. Formerly known as Polypodiatae.

polyprotodont [VERT ZOO] Having four or five incisors on each side of the upper jaw, and one or two less on the lower jaw.

Polypteridae [VERT ZOO] The single family of the order Polypteriformes.

Polypteriformes [VERT ZOO] An ancient order of actinopterygian fishes distinguished by thick, rhombic, ganoid scales with an enamellike covering, a slitlike spiracle behind the eye, a symmetrical caudal fin, and a dorsal series of free, spinelike finlets.

polyribosome *See* polysome.

polysaccharide [BIOCHEM] A carbohydrate composed of many monosaccharides.

polysepalous [BOT] Having separate sepals. Also known as chorisepalous.

polyserositis [MED] Widespread, chronic fibrosing inflammation of serous membranes, especially in the upper abdomen. Also known as chronic hyperplastic perihepatitis; Concato's disease; multiple serositis; Pick's disease; polyorrhymenitis.

polysome [CYTOL] A complex of ribosomes bound together by a single messenger ribonucleic acid molecule. Also known as polyribosome.

polysomy [GEN] The occurrence in a nucleus of one or more individual chromosomes in a number higher than that of the remainder.

polyspermy [PHYSIO] Penetration of the egg by more than one sperm.

polyspondyly [VERT ZOO] Condition of having vertebral parts multiple where myotome has been lost.

polystele [BOT] A stele consisting of vascular units in the parenchyma.

polystemonous [BOT] Characterized by many stamens.

Polystomatoidea [INV ZOO] A superfamily of monogeneid trematodes characterized by strong suckers and hooks on the posterior end.

Polystylifera [INV ZOO] A suborder of the Hoplonemertini distinguished by many stylets.

polytene chromosome [GEN] A giant, multistranded, cable-like chromosome composed of many identical chromosomes having their chromomeres in register and produced by polyteny.

Polytrichales [BOT] An order of ascocarpous perennial

mosses; rigid, simple stems are highly developed and arise from a prostrate subterranean rhizome.

polytrichous [INV ZOO] Having an even coat of cilia covering the body, as certain infusorians.

polyuria [MED] The passage of copious amounts of urine.

polyvalent [CHEM] An ion or radical with more than one valency, such as the sulfate ion, SO_4^{--}. Also known as multivalent; polygen. [IMMUNOL] **1.** Of antigens, having many combining sites or determinants. **2.** Pertaining to vaccines composed of mixtures of different organisms, and to the resulting mixed antiserum.

polyxylic [BOT] Having many strands of xylem and several concentric vascular rings.

Polyzoa [INV ZOO] The equivalent name for Bryozoa.

polyzoarium [INV ZOO] **1.** The skeletal system occurring in a polyzoan colony. **2.** A polyzoan colony.

polyzooid [INV ZOO] One of the individuals forming a polyzoan colony.

Pomacentridae [VERT ZOO] The damselfishes, a family of perciform fishes in the suborder Percoidei.

Pomadasyidae [VERT ZOO] The grunts and sweetlips, a family of perciform fishes in the suborder Percoidei.

Pomatiasidae [INV ZOO] A family of land snails in the order Pectinibranchia.

Pomatomidae [VERT ZOO] A monotypic family of the Perciformes containing the bluefish (*Pomatomus saltatrix*).

pome [BOT] An inferior, indehiscent fleshy fruit having two or more cells.

pomegranate [BOT] *Punica granatum.* A small, deciduous ornamental tree of the order Myrtales cultivated for its fruit, which is a reddish, pomelike berry containing numerous seeds embedded in crimson pulp.

Pompilidae [INV ZOO] The spider wasps, the single family of the superfamily Pompiloidea.

Pompiloidea [INV ZOO] A monofamilial superfamily of hymenopteran insects in the suborder Apocrita with oval abdomen and strong spinose legs.

ponderosa pine [BOT] *Pinus ponderosa.* A hard pine tree of western North America; attains a height of 150-225 feet (46-69 meters) and has long, dark-green leaves in bundles of two to five and tawny, yellowish bark.

Ponerinae [INV ZOO] A subfamily of tropical carnivorous ants (Formicidae) in which pupae characteristically form in cocoons.

Pongidae [VERT ZOO] A family of anthropoid primates in the superfamily Hominoidea; includes the chimpanzee, gorilla, and orangutan.

pons [ANAT] **1.** A process or bridge of tissue connecting two parts of an organ. **2.** A convex white eminence located at the base of the brain; consists of fibers receiving impulses from the cerebral cortex and sending fibers to the contralateral side of the cerebellum.

pontine flexure [EMBRYO] A flexure in the embryonic brain concave dorsally, occurring in the region of the myelencephalon.

Pontodoridae [INV ZOO] A monotypic family of pelagic polychaetes assigned to the Errantia.

poplar [BOT] Any tree of the genus *Populus*, family Salicaceae, marked by simple, alternate leaves, scaly buds, bitter bark, and flowers and fruits in catkins.

popliteal artery [ANAT] A continuation of the femoral artery in the posterior portion of the thigh above the popliteal space and below the buttock.

popliteal nerve [ANAT] Either of two branches of the sciatic nerve in the lower part of the thigh; larger branch continues

PONGIDAE

Typical anthropoid ape of the family Pongidae.

POPPY

Flower and leaves of the opium poppy *(Papaver somniferum).*

PORCUPINE

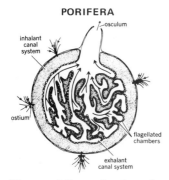

The Canadian porcupine *(Erethizon dorsatum),* about 3½ feet (1 meter) long, with long white hairs scattered through the brownish-black fur.

PORIFERA

inhalant canal system

osculum

ostium

flagellated chambers

exhalant canal system

Diagram of the canal system of a young fresh-water sponge. *(After Ankel, 1950)*

as the tibial nerve, and the smaller branch continues as the peroneal nerve.

popliteal space [ANAT] A diamond-shaped area behind the knee joint.

popliteal vein [ANAT] A vein passing through the popliteal space, formed by merging of the tibial veins and continuing to become the femoral vein.

popliteus [ANAT] **1.** The ham or hinder part of the knee joint. **2.** A muscle on the back of the knee joint.

poppy [BOT] Any of various ornamental herbs of the genus *Papaver*, family Papaveraceae, with large, showy flowers; opium is obtained from the fruits of the opium poppy (*P. somniferum*).

population [BIOL] A group of organisms occupying a specific geographic area or biome.

population dispersal [BIOL] The process by which groups of living organisms expand the space or range within which they live.

population dispersion [BIOL] The spatial distribution at any particular moment of the individuals of a species of plant or animal.

population dynamics [BIOL] The aggregate of processes that determine the size and composition of any population.

population genetics [GEN] The study of both experimental and theoretical consequences of Mendelian heredity on the population level; includes studies of gene frequencies, genotypes, phenotypes, and mating systems.

Porcellanasteridae [INV ZOO] A family of essentially deepwater forms in the suborder Paxillosina.

Porcellanidae [INV ZOO] The rock sliders, a family of decapod crustaceans of the group Anomura which resemble true crabs but are distinguished by the reduced, chelate fifth pereiopods and the well-developed tail fan.

porcupine [VERT ZOO] Any of about 26 species of rodents in two families (Hystricidae and Erethizontidae) which have spines or quills in addition to regular hair.

pore [BIOL] Any minute opening by which matter passes through a wall or membrane.

pore canal [INV ZOO] One of the minute spiral tubules that pass through the cuticle, but not the epicuticle, of insects.

pore fungus [MYCOL] The common name for members of the families Boletaceae and Polyporaceae in the group Hymenomycetes; sporebearing surfaces are characteristically within tubes or pores.

porencephaly [MED] A condition in which the cavity of a lateral ventricle extends to the surface of the cerebral hemisphere; may result from brain tissue destruction or maldevelopment.

pore rhombs [INV ZOO] Canals grouped in half rhombs on each of two adjoining plates of calyx in Cystidea.

poricidal [BIOL] Referring to a structure that opens by many pores.

Porifera [INV ZOO] The sponges, a phylum of the animal kingdom characterized by the presence of canal systems and chambers through which water is drawn in and released; tissues and organs are absent.

porocyte [INV ZOO] One of the perforated, tubular cells which constitute the wall of the incurrent canals in certain Porifera.

porogamy [BOT] Passage of the pollen tube through the micropyle of an ovule in a seed plant.

poroids [BOT] Minute pores in the theca of dinoflagellates and diatoms.

porosis [MED] Condition characterized by increased porosity, as of bone.

Poroxylaceae [PALEOBOT] A monogeneric family of extinct plants included in the Cordaitales.

porphin [BIOCHEM] A heterocyclic ring consisting of four pyrrole rings linked by methine $(-CH=)$ bridges; the basic structure of chlorophyll, hemoglobin, the cytochromes, and certain other related substances.

porphobilinogen [BIOCHEM] $C_{10}H_{14}O_4N_2$ Dicarboxylic acid derived from pyrrole; a product of hemoglobin breakdown that gives the urine a Burgundy-red color.

porphyria [MED] A usually hereditary, pathologic disorder of porphyrin metabolism characterized by porphyrinuria and photosensitivity.

porphyrin [BIOCHEM] A class of red-pigmented compounds with a cyclic tetrapyrrolic structure in which the four pyrrole rings are joined through their α-carbon atoms by four methene bridges $(=C-)$; the porphyrins form the active nucleus of chlorophylls and hemoglobin.

porpoise [VERT ZOO] Any of several species of marine mammals of the family Phocaenidae which have small flippers, a highly developed sonar system, and smooth, thick, hairless skin.

porta hepatis [ANAT] The transverse fissure of the liver through which the portal vein and hepatic artery enter the liver and the hepatic ducts leave.

portal [ANAT] **1.** Of or pertaining to the porta hepatis. **2.** Pertaining to the portal vein or system.

portal circulation [PHYSIO] The passage of venous blood through a portal system.

portal cirrhosis [MED] Replacement of normal liver structure by abnormal lobules of liver cells, often hyperplastic, delimited by bands of fibrous tissue, giving the gross appearance of a finely nodular surface. Also known as Laennec's cirrhosis.

portal hypertension [MED] Portal venous pressure in excess of 20 mm Hg, resulting from intrahepatic or extrahepatic portal venous compression or occlusion.

portal system [ANAT] A system of veins that break into a capillary network before returning the blood to the heart.

portal vein [ANAT] Any vein that terminates in a network of capillaries.

Portlandian [GEOL] A European geologic stage of the Upper Jurassic, above Kimmeridgian, below Berriasian of Cretaceous.

Portuguese man-of-war [INV ZOO] Any of several brilliantly colored tropical siphonophores in the genus *Physalia* which possess a large float and extremely long tentacles.

Portulacaceae [BOT] A family of dicotyledonous plants in the order Caryophyllales distinguished by a syncarpous gynoecium, few, cyclic tepals and stamens, two sepals, and two to many ovules.

Portunidae [INV ZOO] The swimming crabs, a family of the Brachyura having the last pereiopods modified as swimming paddles.

port-wine stain [MED] A congenital hemangioma characterized by one or more red to purplish patches, usually on the face.

Porulosida [INV ZOO] An order of the protozoan subclass Radiolaria in which the central capsule shows many pores.

position effect [GEN] **1.** Change in expressivity of a gene associated with chromosome aberrations. **2.** Inherent gene expression as influenced by neighboring genes.

positive afterimage [PHYSIO] An afterimage persisting after the eyes are closed or turned toward a dark background, and of the same color as the stimulating light.

positive assortative mating [GEN] The mating of individuals

PORPOISE

The common porpoise, a cetacean found in North American and European coastal waters.

PORTUGUESE MAN-OF-WAR

Float and tentacles of a Portuguese man-of-war. *(From T. I. Storer and R. L. Usinger, General Zoology, 3d ed., McGraw-Hill, 1957)*

alike in phenotype or genotype or both at a rate greater than that expected by chance.

positive feedback [CONT SYS] Feedback in which a portion of the output of a circuit or device is fed back in phase with the input so as to increase the total amplification. Also known as reaction (British usage); regeneration; regenerative feedback; retroaction (British usage).

positive interference [GEN] The reduction, by one crossover exchange, of the likelihood of another crossover in its vicinity.

positive ion [CHEM] An atom or group of atoms which by loss of one or more electrons has acquired a positive electric charge; occurs on ionization of chemical compounds as H^+ from ionization of hydrochloric acid, HCl.

positive reinforcement [PSYCH] A stimulus or event that strengthens a positive response to a pleasant stimulus when it follows the response.

positive tropism [BIOL] Orientation movement toward a source of a stimulus.

postabdomen [INV ZOO] In scorpions, the posterior five abdominal segments. **2.** The anal tubercle in spiders.

postcava [ANAT] The inferior or ascending vena cava.

postcentral gyrus [ANAT] The cerebral convolution that lies immediately posterior to the central sulcus and extends from the longitudinal fissure above the posterior ramus of the lateral sulcus.

postcentral sulcus [ANAT] The first sulcus of the parietal lobe of the cerebrum, lying behind and roughly parallel to the central sulcus.

postclavicle [VERT ZOO] A membrane bone in the shoulder girdle of some higher ganoids and teleosts.

postencephalitic parkinsonism [MED] The parkinsonian syndrome occurring as a sequel to lethargic encephalitis within a variable period, from days to many years, after the acute process.

posterior [ZOO] **1.** The hind end of an organism. **2.** Toward the back, or hinder end, of the body.

posterior chamber [ANAT] The space in the eye between the posterior surface of the iris and the ciliary body, and the lens.

postfrons [INV ZOO] That portion of the frons lying posterior to the base line of the antennae in insects.

posthitis [MED] Inflammation of the prepuce.

postmentum [INV ZOO] The base of labium that is attached to the cranium in insects.

postmortem [ADP] Any action taken after an operation is completed to help analyze that operation. [MED] Occurring after death.

postnatal [MED] Subsequent to birth.

postnecrotic cirrhosis [MED] Cirrhosis, usually due to toxic agents or viral hepatitis, characterized by necrosis of liver cells, regenerating nodules of hepatic tissue, and the presence of large bands of connective tissue.

postoperative hernia *See* incisional hernia.

postpartum [MED] Following childbirth.

postpatagium [VERT ZOO] In birds, a small fold of skin extending between the upper arm and the trunk.

postprandial [MED] After a meal.

poststernellum [INV ZOO] The most posterior portion of an insect sternite.

poststernite [INV ZOO] The posterior sclerite of the sternum of insects.

postsynaptic membrane [HISTOL] The excitable membrane bounding the dendritic portion of a synapse.

posttraumatic hernia *See* incisional hernia.

Potamogalinae [VERT ZOO] An aberrant subfamily of West African tenrecs (Tenrecidae).

Potamogetonaceae [BOT] A large family of monocotyledonous plants in the order Najadales characterized by a solitary, apical or lateral ovule, usually two or more carpels, flowers in spikes or racemes, and four each of tepals and stamens.

Potamogetonales [BOT] The equivalent name for Najadales.

Potamonidae [INV ZOO] A family of fresh-water crabs included in the Brachyura.

potamoplankton [BIOL] Plankton found in rivers.

potassium [CHEM] A chemical element, symbol K, atomic number 19, atomic weight 39.102; an alkali metal. Also known as kalium.

potassium penicillin G₁ *See* benzyl penicillin potassium.

potato [BOT] *Solanum tuberosum.* An erect herbaceous annual that has a round or angular aerial stem, underground lateral stems, pinnately compound leaves, and white, pink, yellow, or purple flowers occurring in cymose inflorescences; produces an edible tuber which is a shortened, thickened underground stem having nodes (eyes) and internodes. Also known as Irish potato; white potato.

potential energy [MECH] The capacity to do work that a body or system has by virtue of its position or configuration.

potentiometer [ENG] A device for the measurement of an electromotive force by comparison with a known potential difference.

potentiometric titration [ANALY CHEM] Solution titration in which the end point is read from the electrode-potential variations with the concentrations of potential-determining ions, following the Nernst concept. Also known as constant-current titration.

potentiometry [ELEC] Use of a potentiometer to measure electromotive forces, and the applications of such measurements.

Pottiales [BOT] An order of mosses distinguished by erect stems, lanceolate to broadly ovate or obovate leaves, a strong, mostly percurrent or excurrent costa, and a cucullate calyptra.

poultry [AGR] Domesticated fowl grown for their meat and eggs.

Poupart's ligament *See* inguinal ligament.

pour-plate culture [MICROBIO] A technique for pure-culture isolation of bacteria; liquid, cooled agar in a test tube is inoculated with one loopful of bacterial suspension and mixed by rolling the tube between the hands; subsequent transfers are made from this to a second test tube, and from the second to a third; contents of each tube are poured into separate petri dishes; pure cultures can be isolated from isolated colonies appearing on the plates after incubation.

Pourtalesiidae [INV ZOO] A family of exocyclic Euechinoidea in the order Holasteroida, including those forms with a bottle-shaped test.

powdery mildew [MYCOL] A fungus characterized by production of abundant powdery conidia on the host; a member of the family Erysiphaceae or the genus *Oidium*. [PL PATH] A plant disease caused by a powdery mildew fungus.

powdery scab [PL PATH] A fungus disease of potato tubers caused by *Spongospora subterranea* and characterized by nodular discolored lesions, which burst and expose masses of powdery fungus spores.

pox [MED] A vesicular or pustular exanthematic disease that may leave pit scars.

poxvirus [VIROL] A deoxyribonucleic acid–containing animal virus group including the viruses of smallpox, molluscum contagiosum, and various animal pox and fibromas.

POTASSIUM

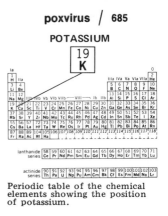

Periodic table of the chemical elements showing the position of potassium.

POTATO

Irish potato. *(a)* Flowering stem. *(b)* Tuber.

POTTIALES

The entire plant of the moss *Tortula ruralis.*

PRAIRIE DOG

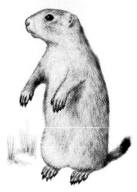

A prairie dog *(Cynomys ludovicianus)* has the habit of sitting erect at the entrance to its burrow.

PRASEODYMIUM

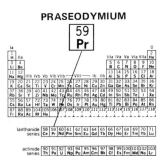

Periodic table of the chemical elements showing the position of praseodymium.

PRECAMBRIAN

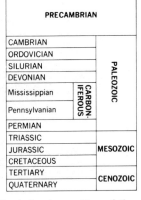

Chart showing position of the Precambrian in relation to the eras and periods of geologic time.

PPLO   *See* pleuropneumonialike organism.

Pr   *See* praseodymium.

prairie dog   [VERT ZOO] The common name for three species of stout, fossorial rodents belonging to the genus *Cynomys* in the family Sciuridae; all have a short, flat tail, small ears, and short limbs terminating in long claws.

prairie wolf   *See* coyote.

praseodymium   [CHEM] A chemical element, symbol Pr, atomic number 59, atomic weight 140.91; a metallic element of the rare-earth group.

Prasinovolvocales   [BOT] An order of green algae in which there are lateral appendages in the flagellum.

pratal   [ECOL] Pertaining to meadows, referring specifically to the flora of rich humid grasslands.

pratincolous   [ECOL] **1.** Living in meadows. **2.** Living in low grass.

preadaptation   [EVOL] Possession by an organism or group of organisms, specialized to one mode of life, of characters which favor easy adaptation to a new environment.

preantenna   [INV ZOO] One of the pair of antennae on the first segment in Onychophora.

prebasilare   [INV ZOO] In certain Diplopoda, a transverse sclerite between the mentum of the gnathochilarium and the first body sternite.

Precambrian   [GEOL] All geologic time prior to the beginning of the Paleozoic era (before 600 million years ago); equivalent to about 90% of all geologic time.

precentral gyrus   [ANAT] The cerebral convolution that lies between the precentral sulcus and the central sulcus and extends from the superomedial border of the hemisphere to the posterior ramus of the lateral sulcus.

precipitate   [CHEM] **1.** A substance separating, in solid particles, from a liquid as the result of a chemical or physical change; **2.** To form a precipitate.

precipitation   [CHEM] The process of producing a separable solid phase within a liquid medium; represents the formation of a new condensed phase, such as a vapor or gas condensing to liquid droplets; a new solid phase gradually precipitates within a solid alloy as a result of slow, inner chemical reaction; in analytical chemistry, precipitation is used to separate a solid phase in an aqueous solution.

precipitin test   [IMMUNOL] An immunologic test in which a specific reaction between antigen and antibody results in a visible precipitate.

preclimax   [ECOL] The plant community followed immediately by the climax community.

precognition   [PSYCH] A form of extrasensory perception involving foreknowledge of a future event.

precollagenous fiber   *See* reticular fiber.

precoracoid   [VERT ZOO] An anterior and ventral element of the pectoral girdle in front of the coracoid in many amphibians and reptiles.

predation   [BIOL] The killing and eating of an individual of one species by an individual of another species.

predator   [ZOO] An animal that preys on other animals as a source of food.

prednisolone   [BIOCHEM] $C_{21}H_{28}O_5$ A glucocorticoid that is a dehydrogenated analog of hydrocortisone.

prednisone   [PHARM] $C_{21}H_{26}O_5$ An adrenocortical steroid drug, obtained in crystalline form, that is an analog of cortisone.

preeclampsia   [MED] A toxemia occurring in the latter half of pregnancy, characterized by an acute elevation of blood pressure and usually by edema and proteinuria, but without convulsions or coma. Also known as toxemia of pregnancy.

preen gland *See* uropygial gland.

prefloration [BOT] The form and arrangement of the leaves within the flower bud.

prefoliation *See* vernation.

prefrontal [ANAT] Situated in the anterior part of the frontal lobe of the brain. [VERT ZOO] **1.** Of or pertaining to a bone of some vertebrate skulls, located anterior and lateral to the frontal bone. **2.** Of, pertaining to, or being a scale or plate in front of the frontal scale on the head of some reptiles and fishes.

prefrontal lobotomy *See* lobotomy.

pregnancy [MED] The state of being pregnant, from conception to childbirth.

pregnancy test [PATH] Any biologic or chemical procedure used to diagnose pregnancy.

pregnanediol [BIOCHEM] $C_{21}H_{36}O_2$ A metabolite of progesterone, present in urine during the progestational phase of the menstrual cycle and also during pregnancy.

pregnenolone [BIOCHEM] $C_{21}H_{32}O_2$ A steroid ketone that is formed as an oxidation product of cholesterol, stigmasterol, and certain other steroids.

prehallux [VERT ZOO] A rudimentary addition digit on the hindlimb.

prehensile [VERT ZOO] Adapted for seizing, grasping, or plucking, especially by wrapping around some object.

prelogging [FOR] Cutting down and removing small trees before large trees are logged.

premaxilla [ANAT] Either of two bones of the upper jaw of vertebrates located in front of and between the maxillae.

prementum [INV ZOO] The part of the labium lying in front of the mentum and bearing the ligula and labial palps in insects.

premolar [ANAT] In each quadrant of the permanent human dentition, one of the two teeth between the canine and the first molar; a bicuspid.

prenatal [MED] Existing or occurring before birth.

preoperculum [VERT ZOO] The flat membrane bone of the operculum or gill cover of most fishes.

prepatagium [VERT ZOO] A fold of skin that extends between the upper arm and the forearm of birds.

prepenna [VERT ZOO] A nestling down feather which precedes an adult contour feather.

prepharynx [INV ZOO] In trematodes, a narrow thin-walled tube connecting the oral sucker and the pharynx.

preplumula [VERT ZOO] A nestling down feather which precedes an adult down feather.

prepollex [VERT ZOO] A rudimentary or additional digit on the preaxial side of the thumb of certain amphibians and mammals.

prepsychotic [PSYCH] Of or pertaining to the mental state that precedes or is potentially capable of precipitating a psychotic disorder.

prepuce [ANAT] **1.** The foreskin of the penis, a fold of skin covering the glans penis. **2.** A similar fold over the glans clitoridis.

preputial gland *See* Tyson's gland.

presbycusis [MED] A condition of diminished auditory acuity associated with old age.

presbyophrenia [PSYCH] A variety of senile dementia in which apparent mental alertness is combined with disorientation of place and loss of memory.

presbyopia [MED] Diminished ability to focus the eye on near objects due to gradual loss of elasticity of the crystalline lens with age.

prescutum [INV ZOO] The anterior part of the tergum of a segment of the thorax in insects.

presoma [INV ZOO] The anterior portion of an invertebrate that lacks a definitive head structure.

presphenoid [VERT ZOO] In many vertebrates and humans, a cranial bone united with the basisphenoid and forming the anterior part of the sphenoid.

pressure [MECH] A type of stress which is exerted uniformly in all directions; its measure is the force exerted per unit area.

pressure point [PHYSIO] A point of marked sensibility to pressure or weight, arranged like the temperature spots, and showing a specific end apparatus arranged in a punctate manner and connected with the pressure sense.

pressure ulcer *See* decubitus ulcer.

presternum [INV ZOO] The first division of the sternum of a segment of the thorax in insects. [VERT ZOO] The manubrium or anterior part of sternum in mammals.

presynaptic [ANAT] Located near or occurring before a synapse.

presynaptic membrane [HISTOL] The excitable membrane bounding the axon terminal adjacent to the dendrite at a synapse.

pretarsus [INV ZOO] The terminal part of an insect leg.

prevomer [VERT ZOO] **1.** The vomer of nonmammalian vertebrates. **2.** In monotremes, a membrane bone of the floor of the nasal cavities.

priapism [MED] Persistent erection of the penis, usually unaccompanied by sexual desire, as seen in certain pathologic conditions.

Priapulida [INV ZOO] A minor phylum of wormlike marine animals; the body is made up of three distinct portions (proboscis, trunk, and caudal appendage) and is often covered with spines and tubercles, and the mouth is surrounded by concentric rows of teeth.

Priapuloidea [INV ZOO] An equivalent name for Priapulida.

prickle cell [CYTOL] A cell characterized by intercellular bridges; especially, an epidermal cell between the basal and granular layers.

prickly heat *See* miliaria.

primary [CHEM] A term used to distinguish basic compounds from similar or isomeric forms; in organic compounds, for example, RCH_2OH is a primary alcohol, R_1R_2CHOH is a secondary alcohol, and $R_1R_2R_3COH$ is a tertiary alcohol; in inorganic compounds, for example, NaH_2PO_4 is primary sodium phosphate, Na_2HPO_4 is the secondary form, and Na_3PO_4 is the tertiary form. [VERT ZOO] Of or pertaining to quills on the distal joint of a bird wing.

primary atypical pneumonia *See* Eaton agent pneumonia.

primary hypertension *See* essential hypertension.

primary lateral sclerosis [MED] A sclerotic disease of the crossed pyramidal tracts of the spinal cord, characterized by paralysis of the limbs, with rigidity, increased tendon reflexes, and absence of sensory and nutritive disorders. Also known as lateral sclerosis.

primary lesion [MED] **1.** In syphilis, tuberculosis, and cowpox, a chancre. **2.** In dermatology, the earliest clinically recognizable manifestation of cutaneous disease, such as a macule, papule, vesicle, pustule, or wheal.

primary meristem [BOT] Meristem which is derived directly from embryonic tissue and which gives rise to epidermis, vascular tissue, and the cortex.

primary phloem [BOT] Phloem derived from apical meristem.

primary reinforcement [PSYCH] **1.** In classical conditioning, presenting an unconditioned stimulus immediately following

PRIAPULIDA

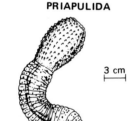

3 cm

Adult form of *Priapulus*.

a conditioned stimulus. **2.** In operant conditioning, presenting an incentive immediately following the operant response.

primary root [BOT] The first plant root to develop; derived from the radicle.

primary succession *See* prisere.

primary syphilis [MED] The first stage of the venereal disease, characterized clinically by a painless ulcer, or chancre, at the point of infection and painless, discrete regional adenopathy.

primary tissue [BOT] Plant tissue formed during primary growth. [HISTOL] Any of the four fundamental tissues composing the vertebrate body.

primary xylem [BOT] Xylem derived from apical meristem.

Primates [VERT ZOO] The order of mammals to which man belongs; characterized in terms of evolutionary trends by retention of a generalized limb structure and dentition, increasing digital mobility, replacement of claws by flat nails, development of stereoscopic vision, and progressive development of the cerebral cortex.

primine [BOT] The outermost integument of an ovule.

Primitiopsacea [PALEON] A small dimorphic superfamily of extinct ostracods in the suborder Beyrichicopina; the velum of the male was narrow and uniform, but that of the female was greatly expanded posteriorly.

primitive gut [EMBRYO] The tubular structure in embryos which differentiates into the alimentary canal.

primitive streak [EMBRYO] A dense, opaque band of ectoderm in the bilaminar blastoderm associated with the morphogenetic movements and proliferation of the mesoderm and notochord; indicates the first trace of the vertebrate embryo.

primordial gut *See* archenteron.

primordium *See* anlage.

Primulaceae [BOT] A family of dicotyledonous plants in the order Primulales characterized by a herbaceous habit and capsular fruit with two to many seeds.

Primulales [BOT] An order of dicotyledonous plants in the subclass Dilleniidae distinguished by sympetalous flowers, stamens located opposite the corolla lobes, and a compound ovary with a single style.

Prioniodidae [PALEON] A family of conodonts in the suborder Conodontiformes having denticulated bars with a large denticle at one end.

Prioniodinidae [PALEON] A family of conodonts in the suborder Conodontiformes characterized by denticulated bars or blades with a large denticle in the middle third of the specimen.

prionodont [VERT ZOO] Having many simple, similar teeth set in a row like sawteeth.

prisere [ECOL] The ecological succession of vegetation that occurs in passing from barren earth or water to a climax community. Also known as primary succession.

Pristidae [VERT ZOO] The sawfishes, a family of modern sharks belonging to the batoid group.

Pristiophoridae [VERT ZOO] The saw sharks, a family of modern sharks often grouped with the squaloids which have a greatly extended rostrum with enlarged denticles along the margins.

proaccelerin [BIOCHEM] A labile procoagulant in normal plasma but deficient in the blood of patients with parahemophilia; essential for rapid conversion of prothrombin to thrombin. Also known as factor V; labile factor.

proamnion [EMBRYO] The part of the embryonic area at the sides and in front of the head of the developing amniote embryo, which remains without mesoderm for a considerable period.

PRIMATES

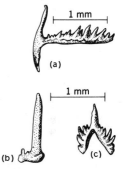

Representative Primates.
(a) Member of the family Tupaiidae. *(b)* Man, of the family Hominidae.

PRIONIODIDAE

Three examples of Prioniodidae:
(a) Ligonodina. (b) Hibbardella.
(c) Hindeodella.

PRIONIODINIDAE

Two examples of the
Prioniodinidae: *(a) Bryantodus;*
(b) Lewistownella.

Proanura [PALEON] Triassic forerunners of the Anura.

proatlas [VERT ZOO] A bone between the atlas and the skull in certain reptiles.

probability [STAT] The probability of an event is the ratio of the number of times it occurs to the large number of trials that take place; the mathematical model of probability is a positive measure which gives the measure of the space the value 1.

probability density function [STAT] A real-valued function whose integral over any set gives the probability that a random variable has values in this set. Also known as density function; frequency function.

probability deviation *See* probable error.

probability ratio test [STAT] Testing a simple hypothesis against a simple alternative by using the ratio of the probability of each simple event under the alternative to the probability of the event under the hypothesis.

probable error [STAT] The error that is exceeded by a variable with a probability of $\frac{1}{2}$. Also known as probability deviation.

probend [GEN] The clinically affected individual who is being studied with respect to a pedigree of interest to human genetics.

Proboscidea [VERT ZOO] An order of herbivorous placental mammals characterized by having a proboscis, incisors enlarged to become tusks, and pillarlike legs with five toes bound together on a broad pad.

proboscis [INV ZOO] A tubular organ of varying form and function on a large number of invertebrates, such as insects, annelids, and tapeworms. [VERT ZOO] The flexible, elongated snout of certain mammals.

procaine *See* procaine base.

procaine base [ORG CHEM] $C_6H_4NH_2COOCH_2CH_2$ $N(C_2H_5)_2$ Water-insoluble, light-sensitive, odorless, white powder, melting at 60°C; soluble in alcohol, ether, chloroform, and benzene; used in medicine as a local anesthetic. Also known as *para*-aminobenzoyldiethylaminoethanol base; planocaine base; procaine.

procambium [BOT] The tissue which gives rise to vascular bundles.

Procampodeidae [INV ZOO] A family of the insect order Diplura.

procarp [BOT] The female organ of red seaweeds, consisting of the carpogonium, trichogyne, and auxiliary cells.

Procaviidae [VERT ZOO] A family of mammals in the order Hyracoidea including the hyraxes.

Procaviinae [VERT ZOO] A subfamily of ungulate mammals in the family Procaviidae.

Procellariidae [VERT ZOO] A family of birds in the order Procellariiformes comprising the petrels, fulmars, and shearwaters.

Procellariiformes [VERT ZOO] An order of oceanic birds characterized by tubelike nostril openings, webbed feet, dense plumage, compound horny sheath of the bill, and, often, a peculiar musky odor.

procephalon [INV ZOO] The part of an insect's head that lies anteriorly to the segment in which the mandibles are located.

procercoid [INV ZOO] The solid parasitic larva of certain eucestodes, such as pseudophyllideans, that develops in the body of the intermediate host.

procoagulant [BIOCHEM] Any of blood clotting factors V to VIII; accelerates the conversion of prothrombin to thrombin in the presence of thromboplastin and calcium.

Procoela [VERT ZOO] A suborder of the Anura characterized by a procoelous vertebral column and a free coccyx articulating with a double condyle.

procoelous [VERT ZOO] The form of a vertebra that is concave anteriorly and convex posteriorly.

Procolophonia [PALEON] A subclass of extinct cotylosaurian reptiles.

proctiger [INV ZOO] The cone-shaped, reduced terminal segment of the abdomen of an insect which contains the anus.

proctitis [MED] Inflammation of the anus or rectum.

proctodaeum [EMBRYO] The posterior part of the alimentary canal in embryo, formed by invagination of the anus.

proctology [MED] A branch of medicine concerned with the structure and disease of the anus, rectum, and sigmoid colon.

proctoscope [MED] An instrument for inspecting the anal canal and rectum.

proctosigmoidectomy [MED] The abdominoperineal excision of the anus and rectosigmoid, usually with the formation of an abdominal colostomy.

Proctotrupidae [INV ZOO] A family of hymenopteran insects in the superfamily Proctotrupoidea.

Proctotrupoidea [INV ZOO] A superfamily of parasitic Hymenoptera in the suborder Apocrita.

procumbent [BOT] Having stems that lie flat on the ground but do not root at the nodes. [SCI TECH] **1.** Lying stretched out. **2.** Slanting forward. **3.** Lying face down.

Procyonidae [VERT ZOO] A family of carnivoran mammals in the superfamily Canoidea, including raccoons and their allies.

prodentin [EMBRYO] A cap of uncalcified matrix covering the tooth cusps before dentin formation.

Prodinoceratinae [PALEON] A subfamily of extinct herbivorous mammals in the family Untatheriidae; animals possessed a carnivorelike body of moderate size.

prodrome [MED] **1.** An early or premonitory manifestation of impending disease before the specific symptoms begin. **2.** An aura.

producer [ECOL] An autotrophic organism of the ecosystem; any of the green plants.

Productinida [PALEON] A suborder of extinct articulate brachiopods in the order Strophomenida characterized by the development of spines.

proenzyme *See* zymogen.

proepimeron [INV ZOO] The epimeron of an insect prothorax.

proerythroblast of Ferrata *See* pronormoblast.

proestrus [PHYSIO] The beginning of the follicular phase of estrus.

proeusternum [INV ZOO] In Diptera, the sclerite between propleura, forming the ventral part of the prothorax.

profibrinolysin *See* plasminogen.

profunda [ANAT] Deep-seated; applied to certain arteries.

profundal zone [ECOL] The region occurring below the limnetic zone and extending to the bottom in lakes deep enough to develop temperature stratification.

Proganosauria [PALEON] The equivalent name for Mesosauria.

progeny [BIOL] Offspring; descendants.

progestational hormone [BIOCHEM] **1.** The natural hormone progesterone, which induces progestational changes of the uterine mucosa. **2.** Any derivative or modification or progesterone having similar actions.

progesterone [BIOCHEM] $C_{21}H_{30}O_2$ A steroid hormone produced in the corpus luteum, placenta, testes, and adrenals; plays an important physiological role in the luteal phase of the menstrual cycle and in the maintenance of pregnancy; it is an intermediate in the biosynthesis of androgens, estrogens, and the corticoids.

PROCUMBENT

The procumbent stem of purslane.

PROGESTERONE

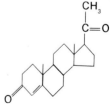

Structural formula of progesterone.

PROJAPYGIDAE

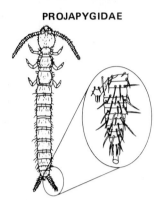

Anajapyx vesiculosus Silvestri, showing the cerci which contain a silk gland.

PROLACERTIFORMES

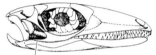

incomplete lower
temporal arcade

Lateral view of *Prolacerta* skull, Lower Triassic. *(From A. S. Romer, Vertebrate Paleontology, 3d ed., University of Chicago Press, 1966)*

PROMETHIUM

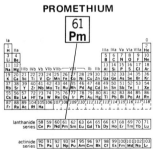

Periodic table of the chemical elements showing the position of promethium.

proglottid [INV ZOO] One of the segments of a tapeworm.

prognathic [ANTHRO] A condition of the upper jaw in which it projects anteriorly with respect to the profile of the facial skeleton, when the skull is oriented on the Frankfort horizontal plane; having a gnathic index of 103.0 or more.

prognosis [MED] A prediction as to the course and outcome of a disease, injury, or developmental abnormality.

Progymnospermopsida [PALEON] A class of plants intermediate between ferns and gymnosperms; comprises the Denovian genus *Archaeopteris*.

prohaptor [INV ZOO] The anterior attachment organ of a typical monogenetic trematode.

Projapygidae [INV ZOO] A family of wingless insects in the order Diplura.

projection area [ANAT] An area of the cortex connected with lower centers of the brain by projection fibers.

projection fibers [ANAT] Fibers joining the cerebral cortex to lower centers of the brain, and vice versa.

projective technique [PSYCH] A procedure used to identify and evaluate an individual's characteristic modes of thought and behavior, his personality traits, attitudes, and motivation, by means of an objective test.

projective test [PSYCH] Observation of a subject's responses to various test materials presented in a relatively unstructured, yet standard situation.

prokaryote [CYTOL] **1.** A primitive nucleus, where the deoxyribonucleic acid–containing region lacks a limiting membrane. **2.** Any cell containing such a nucleus, such as the bacteria and the blue-green algae.

Prolacertiformes [PALEON] A suborder of extinct terrestrial reptiles in the order Eosuchia distinguished by reduction of the lower temporal arcade.

prolactin [BIOCHEM] A protein hormone produced by the adenohypophysis; stimulates lactation and promotes functional activity of the corpus luteum. Also known as lactogenic hormone; luteotropic hormone; mammary-stimulating hormone; mammogen; mammogenic hormone; mammotropin.

prolamin [BIOCHEM] Any of the simple proteins, such as zein, found in plants; soluble in strong alcohol, insoluble in absolute alcohol and water.

prolapse [MED] The falling or sinking down of a part or organ.

proleg [INV ZOO] An unjointed appendage on the abdomen of arthropod larvae.

proliferate [BIOL] To increase by frequent reproduction.

proliferative arthritis *See* rheumatoid arthritis.

proline [BIOCHEM] $C_5H_9O_2$ A heterocyclic amino acid occurring in essentially all proteins, and as a major constituent in collagen proteins.

prolinemia [MED] A rare hereditary disease caused by absence of the degradative enzymes that convert proline to glutamic acid, and characterized by a high content of proline in blood and urine with consequent mental retardation and renal malfunction.

proloculus [INV ZOO] The first chamber of the test in foraminiferans.

prometaphase [CYTOL] A stage between prophase and metaphase in mitosis in which the nuclear membrane disappears and the spindle forms.

promethium [CHEM] A chemical element, symbol Pm, atomic number 61; atomic weight of the most abundant isotope is 147; a member of the rare-earth group of metals.

promoter region [GEN] The sites along DNA where RNA

polymerase molecules specifically bind and start transcription.

promuscis [INV ZOO] The proboscis of Hemiptera.

promycelium [MYCOL] A mycelium that develops from a zygospore and gives rise to a sporangium or to a sporidium.

promyelocyte [HISTOL] The earliest myelocyte stage derived from the myeloblast.

pronate [ANAT] **1.** To turn the forearm so that the palm of the hand is down or toward the back. **2.** To turn the sole of the foot outward with the lateral margin of the foot elevated; to evert.

pronator [PHYSIO] A muscle which pronates, as the muscles of the forearm attached to the ulna and radius.

pronephros [EMBRYO] One of the anterior pair of renal organs in higher vertebrate embryos; the pair initiates formation of the archinephric duct.

pronghorn [VERT ZOO] *Antilocapra americana.* An antelope-like artiodactyl composing the family Antilocapridae; the only hollow-horned ungulate with branched horns present in both sexes.

pronormoblast [HISTOL] A nucleated erythrocyte precursor with scanty basophilic cytoplasm without hemoglobin. Also known as lymphoid hemoblast of Pappenheim; macroblast of Naegeli; megaloblast of Sabin; proerythroblast of Ferrata; prorubricyte; rubriblast; rubricyte.

pronotum [INV ZOO] The dorsal part of the prothorax of insects.

pronucleus [CYTOL] One of the two nuclear bodies of a newly fertilized ovum, the male pronucleus and the female pronucleus, the fusion of which results in the formation of the germinal (cleavage) nucleus.

prootic [VERT ZOO] The anterior bone of the otic capsule in the skull of vertebrates.

propagule [BOT] **1.** A reproductive structure of brown algae. **2.** A propagable shoot.

Propalticidae [INV ZOO] A family of coleopteran insects of the superfamily Cucujoidea found in Old World tropics and Pacific islands.

2-propanone *See* acetone.

properdin [IMMUNOL] A macroglobin of normal plasma capable of killing various bacteria and viruses in the presence of complement and magnesium ions.

prophage [VIROL] Integrated unit formed by union of the provirus into the bacterial genome.

prophase [CYTOL] The initial stage of mitotic or meiotic cell division in which chromosomes are condensed from the nuclear material and split logitudinally to form pairs.

prophylaxis [MED] The prevention of disease.

Propionibacteriaceae [MICROBIO] A family of nonmotile, nonsporulating, gram-positive anaerobic rod-shaped bacteria in the order Eubacteriales that ferment glucose and other substrates; propionic acid is a characteristic end product of the fermentation.

proplastid [BOT] Precursor body of a cell plastid.

propleuron [INV ZOO] A pleuron of the prothorax in insects.

propodeon [INV ZOO] A segment of the abdomen anterior to the podeon in Hymenoptera.

propodite [INV ZOO] The sixth leg joint of certain crustaceans. Also known as propodus.

propodium [INV ZOO] The small anterior part of the foot of a mollusk.

propodosoma [INV ZOO] In Acarina, that portion of the body bearing the first and second pairs of legs.

propodus *See* propodite.

proprioception [PHYSIO] The reception of internal stimuli.

PROPLASTID

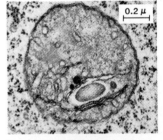

Proplastid in cell of broad bean root tip. Double membranous envelope is shown, and rudimentary nature of internal membrane system is obvious.

proprioceptor [PHYSIO] A sense receptor that signals spatial position and movements of the body and its members, as well as muscular tension.

prop root [BOT] A root that serves to support or brace the plant. Also known as brace root.

propterygium [VERT ZOO] The anterior of the three principal basal cartilages forming a support of one of the paired fins of sharks, rays, and certain other fishes.

proptosis [MED] A falling downward or forward, especially of an eyeball.

propygidium [INV ZOO] The dorsal plate located anterior to the pygidium in coleopterans.

Prorastominae [PALEON] A subfamily of extinct dugongs (Dugongidae) which occur in the Eocene of Jamaica.

prorennin *See* renninogen.

Prorhynchidae [INV ZOO] A family of turbellarians in the order Alloeocoela.

prorubricyte *See* pronormoblast.

Prosauropoda [PALEON] A division of the extinct reptilian suborder Sauropodomorpha; they possessed blunt teeth, long forelimbs, and extremely large claws on the first finger of the forefoot.

proscolex [INV ZOO] A stage in the development of a tapeworm comprising a rounded, fluid-filled cyst.

prosethmoid [VERT ZOO] A median anterior cranial bone of teleosts.

Prosobranchia [INV ZOO] The largest subclass of the Gastropoda; generally, respiration is by means of ctenidia, an operculum is present, there is one pair of tentacles, and the sexes are separate.

prosocoel [VERT ZOO] The cavity of the forebrain in a primitive vertebrate embryo.

prosodus [INV ZOO] A canal leading from an incurrent canal to a flagellated chamber in Porifera.

prosoma [INV ZOO] The anterior part of the body of mollusks and other invertebrates; primitive segmentation is not apparent.

Prosopora [INV ZOO] An order of the class Oligochaeta comprising mesonephridiostomal forms in which there are male pores in the segment of the posterior testes.

prosopyle [INV ZOO] The opening into a flagellated chamber from an inhalant canal in sponges.

prostaglandin [BIOCHEM] Any of various physiologically active compounds containing 20 carbon atoms and formed from essential fatty acids; found in highest concentrations in normal human semen; activities affect the nervous system, circulation, female reproductive organs, and metabolism.

prostate [ANAT] A glandular organ that surrounds the urethra at the neck of the urinary bladder in the male.

prostatectomy [MED] Surgical removal of all or part of the prostate.

prostatitis [MED] Inflammation of the prostate.

prosternum [INV ZOO] The ventral part of the prothorax of insects.

prostheca [INV ZOO] Inner movable lobe of the mandibles in the larvae of certain beetles. [MICROBIO] One of the appendages that are part of the wall in bacteria in the genus *Caulobacter*.

prosthecate bacteria [MICROBIO] Single-celled microorganisms that differ from typical unicellular bacteria in having one or more appendages which extend from the cell surface; the best-known genus is *Caulobacter*.

prosthesis [MED] An artificial substitute for a missing part of the body, such as a substitute hand, leg, eye, or denture.

PROSTHECATE BACTERIA

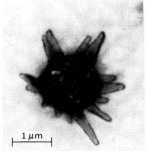

Photograph of a bacterial cell taken with the electron microscope. Note the 14 prosthecae extending from the cell and the transparent gas vesicles inside the cell.

prosthodontics [MED] The science and practice of replacement of missing dental and oral structures.

Prostigmata [INV ZOO] The equivalent name for Trombidiformes.

prostomium [INV ZOO] The portion of the head anterior to the mouth in annelids and mollusks.

protactinium [CHEM] A chemical element, symbol Pa, atomic number 91; the third member of the actinide group of elements; all the isotopes are radioactive; the longest-lived isotope is protactinium-231.

protagonist *See* agonist.

protamine [BIOCHEM] Any of the simple proteins that are combined with nucleic acid in the sperm of certain fish, and that upon hydrolysis yield basic amino acids; used in medicine to control hemorrhage, and in the preparation of an insulin form to control diabetes.

protandry [PHYSIO] That condition in which an animal is first a male and then becomes a female.

protanomaly *See* protanopia.

protanopia [MED] Partial color blindness in which there is defective red vision; green sightedness. Also known as protanomaly.

Proteaceae [BOT] A large family of dicotyledonous plants in the order Proteales, notable for often having a large cluster of small or reduced flowers.

Proteales [BOT] An order of dicotyledonous plants in the subclass Rosidae marked by its strongly perigynous flowers, a four-lobed, often corollalike calyx, and reduced or absent true petals.

protease [BIOCHEM] An enzyme that digests proteins.

protective coloration [ZOO] A color pattern that blends with the environment and increases the animal's probability of survival.

Proteeae [MICROBIO] A tribe of the Enterobactereaceae comprising the genus *Proteus*; characteristically, representatives are motile, ferment dextrose with gas production, and produce urease.

protegulum [INV ZOO] The smooth biconcave embryonic shell of a brachiopod.

Proteida [VERT ZOO] A suborder coextensive with Proteidae in some classification systems.

Proteidae [VERT ZOO] A family of the amphibian suborder Salamandroidea; includes the neotenic, aquatic *Necturus* and *Proteus* species.

protein [BIOCHEM] Any of a class of high-molecular-weight polymer compounds composed of a variety of α-amino acids joined by peptide linkages.

proteinase [BIOCHEM] A type of protease which acts directly on native proteins in the first step of their conversion to simpler substances.

protein-bound iodine [BIOCHEM] Iodine bound to blood protein. Abbreviated PBI.

protein-bound iodine test [PATH] A test of thyroid function that reflects the level of circulating thyroid hormone by determination of the level of protein-bound iodine in the blood. Abbreviated PBI test.

proteinuria [MED] The presence of protein in the urine.

Proteocephalidae [INV ZOO] A family of tapeworms in the order Proteocephaloidea in which the reproductive organs are within the central mesenchyme of the segment.

Proteocephaloidea [INV ZOO] An order of tapeworms of the subclass Cestoda in which the holdfast organ bears four suckers and, frequently, a suckerlike apical organ.

proteolysin [BIOCHEM] A lysin that produces proteolysis.

PROTACTINIUM

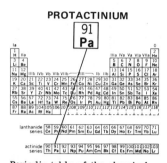

Periodic table of the chemical elements showing the position of protactinium.

proteolysis [BIOCHEM] Fragmentation of a protein molecule by addition of water to the peptide bonds.

proteolytic enzyme [BIOCHEM] Any enzyme that catalyzes the breakdown of protein.

Proteomyxida [INV ZOO] The single order of the Proteomyxidia.

Proteomyxidia [INV ZOO] A subclass of Actinopodea including protozoan organisms which lack protective coverings or skeletal elements and have reticulopodia, or filopodia.

proteoplast [CYTOL] A type of cell plastid containing crystalline, fibrillar, or amorphous masses of protein.

proteose [BIOCHEM] One of a group of derived proteins intermediate between native proteins and peptones; soluble in water, not coagulable by heat, but precipitated by saturation with ammonium or zinc sulfate.

proteranthous [BOT] Flowering before the appearance of foliage leaves.

proteroglyph [VERT ZOO] Having specialized fang teeth in the anterior part of the upper jaw.

proterosoma [INV ZOO] In Acarina, the body region comprising the gnathosoma and propodosoma.

Proterostomia [ZOO] That part of the animal kingdom in which cleavage of the egg is of the determinate type; includes all bilateral phyla except Echinodermata, Chaetognatha, Pogonophora, Hemichordata, and Chordata.

Proterosuchia [PALEON] A suborder of moderate-sized thecodont reptiles with lightly built triangular skulls, downturned snouts, and palatal teeth.

Proterotheriidae [PALEON] A group of extinct herbivorous mammals in the order Litopterna which displayed an evolutionary convergence with the horses in their dentition and in reduction of the lateral digits of their feet.

Proterozoic *See* Algonkian.

Proteutheria [VERT ZOO] A group of primatelike insectivores that contains the living tree shrews.

prothallium [BOT] The gametophyte of a pteridophyte in the form of a flat green thallus with thizoids.

prothetely [INV ZOO] The development of pupal or of imaginal characters in an insect larva.

prothoracic gland [INV ZOO] One of the paired glands in the prothorax of insects which produce ecdysone.

prothorax [INV ZOO] The first thoracic segment of an insect; bears the first pair of legs.

prothrombin [BIOCHEM] An inactive plasma protein precursor of thrombin. Also known as factor II; thrombinogen.

prothrombin factor *See* vitamin K.

prothrombin time [PATH] A one-stage clotting test based on the time required for clotting to occur after the addition of tissue thromboplastin and calcium to decalcified plasma.

Protista [BIOL] A proposed kingdom to include all unicellular organisms lacking a definite cellular arrangement, such as bacteria, algae, diatoms, and fungi.

Protoariciinae [INV ZOO] A subfamily of polychaete annelids in the family Orbiniidae.

protobasidium [MYCOL] A basidium that produces a four-celled mycelium; a sporidium is developed by abstriction from each cell.

protoblast [CYTOL] A cell lacking a membrane. [EMBRYO] **1.** The single-cell stage of an embryo. **2.** A blastomere from which a definite organ or part is developed.

protoblema [MYCOL] A layer of tissue covering the teleoblema and constituting the primary veil of certain fungi.

Protobranchia [INV ZOO] A small and primitive order in the class Bivalvia; the hinge is taxodont in all but one family, there is a central ligament pit, and the anterior and posterior adductor muscles are nearly equal in size.

protocephalon [INV ZOO] **1.** That part of the cephalothorax comprising the head in Malacostraca. **2.** The first of six segments that make up the head in insects.

Protoceratidae [PALEON] An extinct family of pecoran ruminants in the superfamily Traguloidea.

protocercal [VERT ZOO] Having the caudal fin divided into two equal lobes.

protocerebrum [INV ZOO] **1.** The anterior pair of the ganglionic centers in Crustacea. **2.** The anterior part of the brain in insects.

Protochordata [INV ZOO] The equivalent name for Hemichordata.

protocneme [INV ZOO] Any of the six pairs of primary mesenteries of Zoantharia.

Protococcaceae [BOT] A monogeneric family of green algae in the suborder Ulotrichineae in which reproduction is entirely vegetative.

Protococcida [INV ZOO] A small order of the protozoan subclass Coccidia; all are invertebrate parasites, and only sexual reproduction is known.

protoconch [INV ZOO] The shell of larval mollusks.

protocone [VERT ZOO] The inner cusp of an upper molar.

protoconid [VERT ZOO] The external cusp of a lower molar.

protoconule [VERT ZOO] The anterior intermediate cusp of an upper molar.

protocorm [BOT] A swelling of the rhizophore which precedes root formation, as in certain club mosses.

protocranium [INV ZOO] The posterior part of the epicranium in insects.

Protocucujidae [INV ZOO] A small family of coleopteran insects in the superfamily Cucujoidea found in Chile and Australia.

protoderm *See* dermatogen.

Protodonata [PALEON] An extinct order of huge dragonfly-like insects found in Permian rocks.

Protodrilidae [INV ZOO] A family of annelids belonging to the Archiannelida.

protoepiphyte [ECOL] A plant that grows on, and gets all its nourishment from, another plant.

Protoeumalacostraca [PALEON] The stem group of the crustacean series Eumalacostraca.

protogyny [PHYSIO] A condition in hermaphroditic or dioecious organisms in which the female reproductive structures mature before the male structures.

protoloph [VERT ZOO] The anterior transverse crest of an upper molar.

protomala [INV ZOO] The mandible of a myriapod.

Protomastigida [INV ZOO] The equivalent name for Kinetoplastida.

Protomonadina [INV ZOO] An order of flagellates, subclass Mastigophora, with one or two flagella, including many species showing protoplasmic collars ringing the base of the flagellum.

Protomonida [INV ZOO] The equivalent name for Protomonadina.

Protomyzostomidae [INV ZOO] A family of parasitic polychaetes belonging to the Myzostomaria and known for three species from Japan and the Murman Sea.

proton [PHYS] An elementary particle that is the positively charged constituent of ordinary matter and, together with the neutron, is a building stone of all atomic nuclei; its mass is approximately 938 MeV (million electron volts) and spin $\frac{1}{2}$.

protonema [BOT] A green, filamentous structure that originates from an asexual spore of mosses and some liverworts and that gives rise by budding to a mature plant.

protonephridium [INV ZOO] **1.** A primitive excretory tube in many invertebrates. **2.** The duct of a flame cell.

Protophyta [BOT] A division of the plant kingdom, according to one system of classification, set up to include the bacteria, the blue-green algae, and the viruses.

protoplasm [CYTOL] The colloidal complex of protein that composes the living material of a cell.

protoplast [CYTOL] The living portion of a cell considered as a unit; includes the cytoplasm, the nucleus, and the plasma membrane.

protopod [INV ZOO] Having feet or legs on the anterior segments.

protopodite [INV ZOO] The basal segment of a crustacean limb bearing an endopodite or exopodite, or both, at its distal extremity.

Protopteridales [PALEOBOT] An extinct order of ferns, class Polypodiatae.

Protosireninae [PALEON] An extinct superfamily of sirenian mammals in the family Dugongidae found in the middle Eocene of Egypt.

Protospondyli [VERT ZOO] An equivalent name for Semionotiformes.

protostele [BOT] A stele consisting of a solid rod of xylem surrounded by phloem.

protosternum [INV ZOO] The sternite of the cheliceral segment of the prosoma in Acarina.

Protostomia [INV ZOO] A major division of bilateral animals; includes most worms, arthropods, and mollusks.

Protosuchia [PALEON] A suborder of extinct crocodilians from the Late Triassic and Early Jurassic.

prototheca [INV ZOO] A cup-shaped skeletal plate at the aboral end of a coral embryo; it is the first skeletal structure formed.

Prototheria [VERT ZOO] A small subclass of Mammalia represented by a single order, the Monotremata.

prototroch [INV ZOO] The band of cilia characteristic of a trochophore larva.

prototroph [BIOL] An organism which has no additional nutritional requirements other than those of the wild type.

Prototrupoidea [INV ZOO] A superfamily of the Hymenoptera.

protoxylem [BOT] The part of the primary xylem that differentiates from the procambium and is formed during elongation of an embryonic plant organ.

Protozoa [INV ZOO] A diverse phylum of eukaryotic microorganisms; the structure varies from a simple uninucleate protoplast to colonial forms, the body is either naked or covered by a test, locomotion is by means of pseudopodia or cilia or flagella, there is a tendency toward universal symmetry in floating species and radial symmetry in sessile types, and nutrition may be phagotrophic or autotrophic or saprozoic.

protozoology [INV ZOO] That branch of biology which deals with the Protozoa.

Protrachaeta [INV ZOO] The equivalent name for Onychophora.

protractor [ANAT] An extensor muscle. [MED] A surgical instrument formerly used to remove foreign material and devitalized tissue from a wound.

protrypsin *See* trypsinogen.

Protura [INV ZOO] An order of primitive wingless insects belonging to the subclass Apterygota; individuals are elongate and eyeless, lack antennae, and are from pale amber to

PROTOSTELE

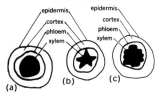

Cross sections of three types of protostele. *(a)* Haplostele. *(b)* Actinostele. *(c)* Plectostele.

PROTURA

0.25 cm

Acerentulus barberi, a proturan insect. *(From H. S. Ewing, Ann. Entomol. Soc. Amer., 33(3):497, 1940)*

white in color; anamorphosis is characteristic of the group.

proventriculus [INV ZOO] **1.** A sac anterior to the gizzard in earthworms. **2.** A dilation of the foregut anterior to the midgut of Mandibulata. [VERT ZOO] The true stomach of a bird, usually separated from the gizzard by a constriction.

provirus [VIROL] The phage genome.

provitamin [BIOCHEM] A vitamin precursor; assumes vitamin activity upon activation or chemical change.

proximal [ANAT] Near the body or the median line of the body.

proximal convoluted tubule [ANAT] The convoluted portion of the vertebrate nephron lying between Bowman's capsule and the loop of Henle; functions in the resorption of sugar, sodium and chloride ions, and water.

proximoceptor [PHYSIO] An exteroceptor involved in taste or cutaneous sensations.

pruinose [BOT] Describing a surface covered with whitish particles or globules.

pruritus [MED] Localized or generalized itch due to irritation of sensory nerve endings.

psalterium *See* omasum.

Psammodontidae [PALEON] A family of extinct cartilaginous fishes in the order Bradyodonti in which the upper and lower dentitions consisted of a few large quadrilateral plates arranged in two rows meeting in the midline.

Psammodrilidae [INV ZOO] A small family of spioniform worms belonging to the Sedentaria.

psammoma [MED] A tumor, usually a meningioma, which contains psammoma bodies.

psammomatus papilloma *See* serous cystadenoma.

psammophore [INV ZOO] One of the rows of hairs below the mandibles and sides of the head in desert ants; functions in removal of sand grains.

psammophyte [ECOL] Thriving (as a plant) on sandy soil.

psammosere [ECOL] A plant succession that originates in sand, as on dunes.

Pselaphidae [INV ZOO] The ant-loving beetles, a large family of coleopteran insects in the superfamily Staphylinoidea.

Psephenidae [INV ZOO] The water penny beetles, a small family of coleopteran insects in the superfamily Dryopoidea.

Pseudaliidae [INV ZOO] A family of roundworms belonging to the Strongyloidea which occur as parasites of whales and porpoises.

pseudanthium [BOT] A cluster of small or reduced flowers which collectively simulate a single flower.

pseudoalleles [GEN] Closely linked genes that behave as alleles and can be separated by crossing over.

pseudoarticulation [INV ZOO] The incomplete subdivision of a segment, or a groove that has the appearance of a joint, as in the limbs of arthropods.

pseudobasidium [MYCOL] A large thick-walled basidium, constituting a resting spore.

Pseudoborniales [PALEOBOT] An order of fossil plants found in Middle and Upper Devonian rocks.

pseudobulb [BOT] In orchids, a thickened internode for storage of water and food reserves.

pseudocarp [BOT] A false fruit, that is, one formed from other parts in addition to the ovary.

pseudocentrum [VERT ZOO] A centrum formed by fusion of the dorsal or dorsal and ventral arcualia, as in tailed amphibians.

pseudocoele [INV ZOO] A space between the body wall and internal organs that is not formed by gastrulation and lacks a cellular lining.

Pseudocoelomata [INV ZOO] A group comprising the animal

PSEUDOBORNIALES

Pseudobornia node with a whorl of three leaves; has rhizomes and stems up to 6 centimeters wide and over a meter long. *(Modified from A. G. Nathorst)*

phyla Entoprocta, Aschelminthes, and Acanthocephala; characterized by a pseudocoelom.

pseudoconch [VERT ZOO] A structure formed above and posterior to the true concha in crocodiles.

pseudoconidium [MYCOL] A spore formed on the lateral projection of a pseudomycelium of certain yeasts.

pseudoconjugation [INV ZOO] Conjugation of Sporozoa in which two individuals join temporarily and without true fusion.

pseudocostate [BOT] Having a marginal vein uniting all others.

pseudoculus [INV ZOO] An oval area on each side of the head of Pauropoda; may be a receptor for mechanical vibrations.

Pseudocycnidae [INV ZOO] A family of the Caligoida which comprises external parasites on the gills of various fishes.

pseudocyesis [MED] A condition characterized by amenorrhea, enlargement of the abdomen, and other symptoms simulating gestation, due to an emotional disorder.

pseudocyst [INV ZOO] In Sporozoa, a residual mass of protoplasm which swells and ruptures to liberate spores.

Pseudodiadematidae [INV ZOO] A family of Jurassic and Cretaceous echinoderms in the order Phymosomatoida which had perforate crenulate tubercles.

pseudodont [VERT ZOO] Having false teeth, that is, teeth made of horny material, as monotremes.

pseudogamy [BIOL] The activation of an ovum by a spermatozoon which does not fertilize the ovum. [MYCOL] The union of hyphae from different thalli.

pseudogaster [INV ZOO] An apparent gastral cavity of certain sponges, which opens to the exterior by way of a pseudosculum and has true oscula opening into it.

pseudoglanders [VET MED] An infectious bacterial lymphangitis of horses and other equines caused by *Corynebacterium pseudotuberculosis* and characterized by ulcerating nodules of the lymph nodes in the legs.

pseudoheart [INV ZOO] One of the contractile vessels which pump blood from the dorsal to the ventral vessel in annelids.

pseudohermaphroditism [PHYSIO] A condition in humans which simulates hermaphroditism, with gynandry in females and androgyny in males.

Pseudomonadaceae [MICROBIO] A family of rod-shaped to ovoid, gram-negative bacteria in the suborder Pseudomonadineae in which pigments, other than photosynthetic pigments, are common.

Pseudomonadales [MICROBIO] An order of ovoid, rodshaped, comma-shaped, or spiral bacteria in the class Schizomycetes; cells are characteristically rigid and motile by means of polar flagella.

Pseudomonadineae [MICROBIO] A suborder of bacteria of the order Pseudomonadales including those families whose cells do not contain photosynthetic pigments.

pseudomonocarpous [BOT] Having the seeds retained in leafbases until they are liberated, as in cycads.

pseudomonocotyledonous [BOT] Having two cotyledons united to give the appearance of one.

pseudomonomerous ovary [BOT] An ovary which appears to be composed of a single carpel, but which has developed along evolutionary lines from a compound ovary.

pseudomucinous cystadenoma [MED] A benign ovarian tumor composed of columnar mucin-producing cells lining multilocular cysts filled with mucinous material.

pseudomycelium [MYCOL] A structure consisting of chains or groups of adherent cells of yeasts.

pseudomycorrhiza [BOT] An association of conifer roots

with parastitic fungi in the absence of mycorrhizal fungi.

pseudonavicella [INV ZOO] In Sporozoa, a small boat-shaped spore containing sporozoites.

pseudoosculum [INV ZOO] The exterior opening from the pseudogaster.

pseudoparalysis [MED] An apparent motor paralysis that is caused by voluntary inhibition of motor impulses because of pain or other organic or psychic causes.

pseudoparasite [MED] Something in the blood that is mistaken for a parasite.

pseudopenis [INV ZOO] The protruded evagination of the deferent duct, in certain male oligochaetes.

pseudoperianth [BOT] 1. An envelope investing the archegonium of certain liverworts. 2. A false perianth.

Pseudophoracea [INV ZOO] An extinct superfamily of gastropod mollusks in the order Aspidobranchia.

Pseudophyllidea [INV ZOO] An order of tapeworms of the subclass Cestoda, parasitic principally in the intestine of cold-blooded vertebrates.

pseudoplasmodium [INV ZOO] An aggregate of amebas resembling a plasmodium.

pseudopodium [BOT] A slender, leafless branch of the gametophyte in certain Bryatae. [CYTOL] Temporary projection of the protoplast of ameboid cells in which cytoplasm streams actively during extension and withdrawal. [INV ZOO] Foot of a rotifer.

pseudopore [INV ZOO] A small opening between the outermost tube and the intercanal system of certain Porifera.

Pseudoscorpionida [INV ZOO] An order of terrestrial Arachnida having the general appearance of miniature scorpions without the postabdomen and sting.

Pseudosphaeriales [BOT] An order of the class Ascolichenes, shared by the class Ascomycetes; the ascocarp is flask-shaped and lined with a layer of interwoven, branched pseudoparaphyses.

Pseudosporidae [INV ZOO] A family of the protozoan subclass Proteomyxidia; flagellated stages invade Volvocidae and filamentous algae and become amebas.

pseudostratified epithelium [HISTOL] A type of epithelium in which all cells reach to the basement membrane but some extend toward the surface only part way, while others reach the surface.

Pseudosuchia [PALEON] A suborder of extinct reptiles of the order Thecodontia comprising bipedal, unarmored or feebly armored forms which resemble dinosaurs in many skull features but retain a primitive pelvis.

Pseudothelphusidae [INV ZOO] A family of fresh-water crabs belonging to the Brachyura.

Pseudotriakidae [VERT ZOO] The false catsharks, a family of galeoids in the carcharinid line.

pseudotuberculosis [MED] A bacterial infection in man and many animals caused by *Pasteurella pseudotuberculosis;* may be severe in man with septicemia and symptoms resembling typhoid fever.

pseudovelum [INV ZOO] A velum lacking muscular and nerve cells, in Scyphozoa. [MYCOL] In fungi, a pseudoveil, formed by the union of contemporaneous outgrowths from the pileus and the stipe, and protecting the immature hymenium.

pseudovum [BIOL] An ovum that can develop without being fertilized.

psilate [BOT] Lacking ornamentation; generally applied to pollen.

Psilidae [INV ZOO] The rust flies, a family of myodarian

PSEUDOSCORPIONIDA

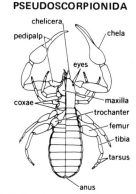

Adult Pseudoscorpionida: ventral view, left side; dorsal view, right side. *(From H. S. Pratt, A Manual of the Common Invertebrate Animals, rev. ed., McGraw-Hill, 1951)*

PSEUDOSTRATIFIED EPITHELIUM

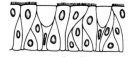

Arrangement of cells in pseudostratified epithelium.

cyclorrhaphous dipteran insects in the subsection Acalypteratae.

Psilophytales [PALEOBOT] A group formerly recognized as an order of fossil plants.

Psilophytineae [PALEON] The equivalent name for Rhyniopsida.

Psilopsida [BOT] A subdivision of the Tracheophyta.

Psilorhynchidae [VERT ZOO] A small family of actinopterygian fishes belonging to the Cyprinoidei.

psilosis [MED] Falling out of the hair.

Psilotales [BOT] The equivalent name for Psilotophyta.

Psilotatae [BOT] A class of the Psilotophyta.

Psilotophyta [BOT] A division of the plant kingdom represented by three living species; the life cycle is typical of the vascular cryptogams.

Psittacidae [VERT ZOO] The single family of the Psittaciformes.

Psittaciformes [VERT] The parrots, a monofamilial order of birds that exhibit zygodactylism and have a strong hooked bill.

psittacosis [MED] Pneumonia and generalized infection of man and of birds caused by agents of the PLT-Bedsonia group; transmitted to man by psittacine birds.

psoas [ANAT] Either of two muscles: psoas major which arises from the bodies and transverse processes of the lumbar vertebrae and is inserted into the lesser trochanter of the femur, and psoas minor which arises from the bodies and transverse processes of the lumbar vertebrae and is inserted on the pubis.

Psocoptera [INV ZOO] An order of small insects in which wings may be present or absent, tarsi are two- or three-segmented, cerci are absent, and metamorphosis is gradual.

Psolidae [INV ZOO] A family of echinoderms in the order Dendrochirotida characterized by a ventral adhesive sucker and a U-shaped gut, with the mouth and anus opening upward on the adoral surface.

Psophiidae [VERT ZOO] The trumpeters, a family of birds in the order Gruiformes.

psoriasis [MED] A usually chronic, often acute inflammatory skin disease of unknown cause; characterized by dull red, well-defined lesions covered by silvery scales which when removed disclose tiny capillary bleeding points.

psorosis [PL PATH] A virus disease of tangerine, grapefruit, and sweet orange trees characterized by scaly bark, a gummy exudate, retarded growth, small yellow leaves, and dieback of twigs. Also known as scaly bark.

psorosperm [INV ZOO] **1.** The resistant spore of myxosporidians. **2.** A minute parasitic organism, especially sporozoans.

psyche [PSYCH] The mind or self as a functional entity.

psychedelic [PSYCH] Of, pertaining to, or producing a psychic state.

psychiatric social worker [PSYCH] A specialist who utilizes the techniques of both social work and psychiatry to serve the community.

psychiatrist [MED] A person who specializes in psychiatry; a licensed physician trained in psychiatry.

psychiatry [MED] The medical science that deals with the origins, diagnosis, and treatment of mental and emotional disorders.

Psychidae [INV ZOO] The bagworms, a family of lepidopteran insects in the superfamily Tineoidea; males are large, hairy moths, but females are degenerate, wingless, and legless and live in bag-shaped cases.

psychoanalysis [PSYCH] A technique used in the treatment of neuroses and other emotional disorders which relies upon

PSOLIDAE

anus

2 cm

tube-feet

sole

Psolus phantapus. (After T. Mortensen, 1927)

PSYCHIDAE

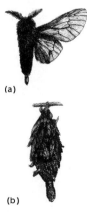

(a)

(b)

Thyridopteryx ephemeraeformis Haworth. *(a)* Male. *(b)* Male case.

free associations of the patient to bring ideas and experiences from the unconscious to the conscious divisions of the psyche.

psychobiology [PSYCH] The school of psychiatry and psychology in which the individual is considered as the sum of his environment as well as being considered a physical organism.

Psychodidae [INV ZOO] The moth flies, a family of orthorrhaphous dipteran insects in the series Nematocera.

psychodynamics [PSYCH] The study of human behavior from the point of view of motivation and drives, depending largely on the functional significance of emotion, and based on the assumption that an individual's total personality and reactions at any given time are the product of the interaction between his genetic constitution and his environment.

psychogalvanic reflex [PHYSIO] A variation in the electric conductivity of the skin in response to emotional stimuli, which cause changes in blood circulation, secretion of sweat, and skin temperature.

psychogenic [MED] Of psychic origin.

psychokinesis [PSYCH] The alleged ability of an individual to exert a mental influence on physical events in advance of their occurrence. Abbreviated PK.

psychologist [PSYCH] **1.** An individual who has made a professional study of, and usually thereafter professionally engages in, psychology. **2.** Specifically, an individual with the minimum professional qualifications set forth by an intraprofessionally recognized psychological association.

psychology [BIOL] **1.** The science that deals with the functions of the mind and the behavior of an organism in relation to its environment. **2.** The mental activity characteristic of a person or a situation.

psychometrics [PSYCH] The mathematical and statistical treatment of psychological data.

psychomotor [PSYCH] Pertaining to both mental and motor activity.

psychoneurotic reaction [PSYCH] A behavior disorder in which a person is unusually anxious, or incapacitated in his work and his relations with other people.

psychopathic personality [PSYCH] An emotionally immature individual characterized by pronounced defects in judgment and prone to impulsive, generally amoral or antisocial behavior.

psychopathology [PSYCH] The systematic study of mental diseases.

psychopharmacology [PSYCH] The science that deals with the action of drugs on mental function.

psychophysics [PSYCH] **1.** The study of mental processes by physical methods. **2.** The study of the relations of stimuli to the sensations they produce.

psychophysiologic nervous system reaction *See* neurasthenic neurosis.

psychosis [PSYCH] An impairment of mental functioning to the extent that it interferes grossly with an individual's ability to meet the ordinary demands of life, characterized generally by severe affective disturbance, profound introspection, and withdrawal from reality, formation of delusions or hallucinations, and regression presenting the appearance of personality disintegration.

psychosomatic [MED] Of or pertaining to the interrelationship between mental processes and somatic functions.

psychosurgery [MED] The branch of medicine that deals with the treatment of various psychoses, severe neuroses, and chronic painful conditions by means of operative procedures on the brain.

psychotaxis [PSYCH] Involuntary adjustment of a person's

PSYLLIDAE

0.5 mm

Anterior dorsum of a psyllid.

PTERIDOSPERMAE

1 m

Reconstruction of a seed fern. Note appearance of adventitious roots on the lower portion of the stem.

PTEROBRANCHIA

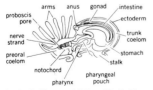

Cephalodiscus. (After W. Patten, from H. V. Neal and H. W. Rand, Comparative Anatomy, Blakiston-McGraw-Hill, 1936)

thoughts and behavior in order to retain the agreeable and avoid the disagreeable; an ego defense mechanism.

psychotherapy [PSYCH] The use of psychological means in the treatment of emotional and mental disorders.

psychotomimetic [PSYCH] **1.** Mimicking a psychotic disorder. **2.** Pertaining to any drug or compound, such as lysergic acid diethylamide or mescaline, which can induce a psychoticlike state.

psychotropic [PSYCH] Pertaining to any drug or agent having a particular affinity for or effect on the psyche.

psychrophile [BIOL] An organism that thrives at low temperatures.

Psyllidae [INV ZOO] The jumping plant lice, a family of the Homoptera in the series Sternorrhyncha in which adults have a transverse head with protuberant eyes and three ocelli, 6- to 10-segmented antennae, and wings with reduced but conspicuous venation.

Pt *See* platinum.

PTA *See* factor XI.

ptarmigan [VERT ZOO] Any of various birds of the genus *Lagopus* in the family Tetraonidae; during the winter, plumage is white and hairlike feathers cover the feet.

PTC *See* Christmas factor.

pteralia [INV ZOO] Axillary sclerites which form the articulation of insect wings.

Pteraspidomorphi [VERT ZOO] The equivalent name for Diplorhina.

Pterasteridae [INV ZOO] A family of deep-water echinoderms in the order Spinulosida distinguished by having webbed spine fins.

Pteridophyta [BOT] The equivalent name for Polypodiophyta.

Pteridospermae [PALEOBOT] Seed ferns, a class of the Cycadicae comprising extinct plants characterized by naked seeds borne on large fernlike fronds.

Pteridospermophyta [PALEOBOT] The equivalent name for Pteridospermae.

Pteriidae [INV ZOO] Pearl oysters, a family of bivalve mollusks which have nacreous shells.

pterinophore [CYTOL] A yellow to orange chromatophore that contains pterine pigment.

pterion [ANTHRO] The region surrounding the sphenoparietal suture where the frontal bone, parietal bone, squama temporalis, and greater wing of the sphenoid bone come together most closely.

Pterobranchia [INV ZOO] A group of small or microscopic marine animals regarded as a class of the Hemichordata; all are sessile, tubicolous organisms with a U-shaped gut and three body segments.

pterocarpous [BOT] Having winged fruit.

pterochore [BOT] A type of plant that produces winged disseminules.

Pteroclidae [VERT ZOO] The sandgrouse, a family of graminivorous birds in the order Columbiformes; mainly an Afro-Asian group resembling pigeons and characterized by cryptic coloration, usually corresponding with the soil color of the habitat.

pterodactyl [PALEON] The common name for members of the extinct reptilian order Pterosauria.

Pterodactyloidea [PALEON] A suborder of Late Jurassic and Cretaceous reptiles in the order Pterosauria distinguished by lacking tails and having increased functional wing length due to elongation of the metacarpels.

pteroic acid [BIOCHEM] $C_{14}H_{12}N_6O_3$ A crystalline amino acid formed by hydrolysis of folic acid or other pteroylglutamic acids.

Pteromalidae [INV ZOO] A family of hymenopteran insects in the superfamily Chalcidoidea.

Pteromedusae [INV ZOO] A suborder of hydrozoan coelenterates in the order Trachylina characterized by a modified, bipyramidal medusae.

pteropegum [INV ZOO] The socket of an insect wing.

Pterophoridae [INV ZOO] The plume moths, a family of the lepidopteran superfamily Pyralidoidea in which the wings are divided into featherlike plumes, maxillary palpi are lacking, and the legs are long.

Pteropidae [INV ZOO] The fruit bats, a large family of the Chiroptera found in Asia, Australia, and Africa.

pteropleurite [INV ZOO] A sclerite of the thorax between the wing insertion and mesopleurite, in Diptera.

Pteropoda [INV ZOO] The sea butterflies, an order of pelagic gastropod mollusks in the subclass Opisthobranchia in which the foot is modified into a pair of large fins and the shell, when present, is thin and glasslike.

Pteropodidae [VERT ZOO] A family of fruit-eating bats in the suborder Megachiroptera, characterized by primitive ears and by shoulder joints.

pteropod ooze [GEOL] A pelagic sediment containing at least 45% calcium carbonate in the form of tests of marine animals, particularly pteropods.

Pteropsida [BOT] A large group of vascular plants characterized by having parenchymatous leaf gaps in the stele and by having leaves which are thought to have originated in the distant past as branched stem systems.

Pterosauria [PALEON] An extinct order of flying reptiles of the Mesozoic era belonging to the subclass Archosauria; the wing resembled that of a bat, and a large heeled sternum supported strong wing muscles.

pterospermous [BOT] Having winged seeds.

pteroylglutamic acid *See* folic acid.

pterygium [MED] **1.** A triangular mass of mucous membrane growing on the conjunctiva, usually near the inner canthus. **2.** Overgrowth of the cuticle forward on the nail. [VERT ZOO] A generalized vertebrate limb.

pterygobranchiate [INV ZOO] Having feathery gills, as certain crustaceans.

pterygoid bone [VERT ZOO] A rodlike bone or group of bones forming a portion of the palatoquadrate arch in lower vertebrates.

pterygomandibular [ANAT] Pertaining to the pterygoid process and the mandible.

pterygomaxillary [ANAT] Pertaining to the pterygoid process and the maxilla.

pterygopalatine [ANAT] Situated between the pterygoid process of the sphenoid bone and the palatine bone.

pterygopalatine fossa [ANAT] The gap between the pterygoid process of the sphenoid bone and the maxilla and palatine bone.

pterygoquadrate [EMBRYO] Of, pertaining to, or being the first branchial arch in lower vertebrate embryos; gives rise to most of the upper jaw. [VERT ZOO] A cartilage comprising the dorsal half of the mandibular arch of certain fishes.

Ptiliidae [INV ZOO] The feather-winged beetles, a family of coleopteran insects in the superfamily Staphylinoidea.

ptilinum [INV ZOO] A vesicle on, or a bladderlike expansion of, the head of a fly emerging from the pupa.

Ptilodactylidae [INV ZOO] The toed-winged beetles, a family of the Coleoptera in the superfamily Dryopoidea.

PTEROPHORIDAE

Platypilia carduidactyla Riley, showing plumelike wings and long legs.

Ptilodontoidea [PALEON] A suborder of extinct mammals in the order Multituberculata.

Ptinidae [INV ZOO] The spider beetles, a family of coleopteran insects in the superfamily Bostrichoidea.

ptosis [MED] Prolapse, abnormal depression, or falling down of an organ or part; applied especially to drooping of the upper eyelid, from paralysis of the third cranial nerve.

ptyalase *See* ptyalin.

ptyalin [BIOCHEM] A diastatic enzyme found in saliva which catalyzes the hydrolysis of starch to dextrin, maltose, and glucose, and the hydrolysis of sucrose to glucose and fructose. Also known as ptyalase; salivary amylase; salivary diastase.

Ptychodactiaria [INV ZOO] An order of the zoantharian anthozoans of the phylum Coelenterata known only from two genera, *Ptychodactis* and *Dactylanthus*.

Ptychomniaceae [BOT] A family of mosses in the order Isobryales distinguished by an eight-ribbed capsule.

Ptyctodontida [PALEON] An order of Middle and Upper Devonian fishes of the class Placodermi in which both the head and trunk shields are present, and the joint between them is a well-differentiated and variable structure.

ptyxis [BOT] The form resulting when developing leaves are folded or rolled up in the bud.

Pu *See* plutonium.

puberty [PHYSIO] The period at which the generative organs become capable of exercising the function of reproduction; signalized in the boy by a change of voice and discharge of semen, in the girl by the appearance of the menses.

puberulic acid [BIOCHEM] $(HO)_3(C_7H_2O)COOH$ A keto acid formed as a metabolic product of certain species of *Penicillium;* has some germicidal activity against gram-positive bacteria.

pubescence [BIOL] A downy or hairy covering on some plants and certain insects.

pubic arch [ANAT] The arch formed by the conjoined rami of the pubis and ischium.

pubic crest [ANAT] The crest extending from the pubic tubercle to the medial extremity of the pubis.

pubic height [ANTHRO] A measure of the vertical distance from the pubis to the floor taken when the subject is in a standing position.

pubic symphysis [ANAT] The fibrocartilaginous union of the pubic bones. Also known as symphysis pubis.

pubis [ANAT] The pubic bone, the portion of the hipbone forming the front of the pelvis.

public health [MED] **1.** The state of health of a community or of a population. **2.** The art and science dealing with the protection and improvement of community health.

puerperal sepsis [MED] A toxic condition caused by infection in the birth canal, occurring as a complication or sequel of pregnancy.

puerperium [MED] **1.** The state of a woman during labor or immediately after delivery. **2.** The period from delivery to the time when the uterus returns to normal size, about 6 weeks.

pullorum disease [VET MED] A highly fatal disease of chickens and other birds caused by *Salmonella pullorum,* characterized by weakness, lassitude, lack of appetite, and whitish or yellowish diarrhea. Also known as bacillary white diarrhea; white diarrhea.

pulmonary anthrax [MED] A form of anthrax in man caused by the inhalation of dust containing *Bacillus anthracis* spores.

pulmonary artery [ANAT] A large artery that conducts venous blood from the heart to the lungs of tetrapods.

pulmonary aspergillosis [MED] A form of aspergillosis

PTYCTODONTIDA

Rhamphodopsis threiplandi **from the Middle Devonian of Scotland.** Restoration of skeleton with outline of body in lateral aspect. *(After Roger S. Miles)*

caused by *Aspergillus fumigatus*, with symptoms resembling chronic tuberculosis.

pulmonary circulation [PHYSIO] The circulation of blood through the lungs for the purpose of oxygenation and the release of carbon dioxide. Also known as lesser circulation.

pulmonary edema [MED] An effusion of fluid into the alveoli and interstitial spaces of the lungs.

pulmonary plexus [ANAT] A nerve plexus composed chiefly of vagal fibers situated on the anterior and posterior aspects of the bronchi and accompanying them into the substance of the lung.

pulmonary stenosis [MED] Narrowing of the orifice of the pulmonary artery.

pulmonary valve [ANAT] A valve consisting of three semilunar cusps situated between the right ventricle and the pulmonary trunk.

pulmonary vein [ANAT] A large vein that conducts oxygenated blood from the lungs to the heart in tetrapods.

Pulmonata [INV ZOO] A subclass of the gastropod mollusks containing the "lung"-bearing snails; gills have been lost and in their place the mantle cavity has become a pulmonary sac.

pulp [ANAT] A mass of soft spongy tissue in the interior of an organ. [BOT] The soft succulent portion of a fruit.

pulp cavity [ANAT] The space within the central part of a tooth containing the dermal pulp and made up of the pulp chamber and a root canal.

pulp chamber [ANAT] The coronal portion of the central cavity of a tooth.

pulpotomy [MED] Surgical removal of the pulp of a tooth.

pulsation [PHYSIO] A beating or throbbing, usually rhythmic, as of the heart or an artery.

pulse [PHYSIO] **1.** The regular, recurrent, palpable wave of arterial distention due to the pressure of the blood ejected with each contraction of the heart. **2.** A single wave.

pulsellum [INV ZOO] A flagellum located at the posterior end of a protozoan.

pulse pressure [PHYSIO] The difference between the systolic and diastolic blood pressure.

pulse rate [PHYSIO] The number of pulsations of an artery per minute.

pulvillus [INV ZOO] A small cushion or cushionlike pad, often covered with short hairs, on an insect's foot between the claws of the last segment.

pulvinus [BOT] A cushionlike enlargement of the base of a petiole which functions in turgor movements of leaves.

puma [VERT ZOO] *Felis concolor.* A large, tawny brown wild cat (family Felidae) once widespread over most of the Americas. Also known as American lion; catamount; cougar; mountain lion.

pumpkin [BOT] Any of several prickly vines with large lobed leaves and yellow flowers in the genus Cucurbita of the order Violales; the fruit is orange-colored and large, with a firm rind.

puna [ECOL] An alpine biological community in the central portion of the Andes Mountains of South America characterized by low-growing, widely spaced plants that lack much green color most of the year.

punctate [BIOL] Dotted; full of minute points.

punctate sensitivity [PHYSIO] Greater sensitivity in some spots of the skin than in others.

pupa [INV ZOO] The quiescent, intermediate form assumed by an insect that undergoes complete metamorphosis; it follows the larva and precedes the adult stages and is enclosed in a hardened cuticle or a cocoon.

puparium [INV ZOO] The pupal casing.

PUNA

Landscape of the puna in the Lake Titicaca area, Bolivia.

pupate [INV ZOO] **1.** To develop into a pupa. **2.** To pass through a pupal stage.

pupil [ANAT] The contractile opening in the iris of the vertebrate eye.

pupillary reflex [PHYSIO] **1.** Contraction of the pupil in response to stimulation of the retina by light. Also known as Whytt's reflex. **2.** Contraction of the pupil on accommodation for close vision, and dilation of the pupil on accommodation for distant vision. **3.** Contraction of the pupil on attempted closure of the eye. Also known as Westphal-Pilcz reflex; Westphal's pupillary reflex.

Pupipara [INV ZOO] A section of cyclorrhaphous dipteran insects in the Schizophora series in which the young are born as mature maggots ready to become pupae.

pupiparous [INV ZOO] Bringing forth young that are developed to the pupa stage, as certain parasitic insects.

Purbeckian [GEOL] A stage of geologic time in Great Britain: uppermost Jurassic (above Bononian, below Cretaceous).

pure culture [MICROBIO] A culture that contains cells of one kind, all progeny of a single cell.

pure forest [FOR] A forest in which one species makes up 80% or more of the total number of trees.

purine [BIOCHEM] A heterocyclic compound containing fused pyrimidine and imidazole rings; adenine and guanine are the purine components of nucleic acids and coenzymes.

Purkinje cell [HISTOL] Any of the cells of the cerebral cortex with large, flask-shaped bodies forming a single cell layer between the molecular and granular layers.

Purkinje effect [PHYSIO] When illumination is reduced to a low level, slowly enough to allow adaptation by the eye, the sensation produced by the longer-wave stimuli (red, orange) decreases more rapidly than that produced by shorter-wave stimuli (violet, blue). Also spelled Parkinje effect.

Purkinje fibers [HISTOL] Modified cardiac muscle fibers composing the terminal portion of the conducting system of the heart.

puromycin [MICROBIO] $C_{22}H_{29}O_5N_7$ A colorless, crystalline broad-spectrum antibiotic produced by a strain of *Streptomyces*.

purple blotch [PL PATH] A fungus disease of onions, garlic, and shallots caused by *Alternaria porri* and characterized by small white spots which become large purplish blotches.

purple-top [PL PATH] A virus disease of potato plants characterized by purplish or chlorotic discoloration of the top shoots, swelling of axillary branches, and severe wilting.

purpura [MED] Spontaneous hemorrhages into tissues such as joints, skin, and mucosal surfaces.

purulent [MED] Consisting of, containing, or forming pus.

purupuru *See* pinta.

pus [MED] A viscous, creamy, pale-yellow or yellow-green fluid produced by liquefaction necrosis in a neutrophil-rich exudate.

pustule [MED] A small, circumscribed, pus-filled elevation on the skin. [PL PATH] A blisterlike mark on a leaf due to rupture of surface tissues overlying spore masses of a parasitic fungus.

Pustulosa [PALEON] An extinct suborder of echinoderms in the order Phanerozonida found in the Paleozoic.

pusule [INV ZOO] A noncontractile fluid-filled vacuole emptied by means of a duct; found in dinoflagellates.

putamen [VERT ZOO] The membrane lining the shell of a bird's egg.

putrefaction [BIOCHEM] Decomposition of organic matter, particularly the anaerobic breakdown of proteins by bacteria, with the production of foul-smelling compounds.

PURINE

Structural formula of purine.

pyarthrosis [MED] Suppuration involving a joint.

pycnidiospore [MYCOL] A conidium produced by a pycnidium.

pycnidium [MYCOL] A cavity that bears pycnidiospores in certain fungi.

pycniospore [MYCOL] A haploid spore of a rust fungus that fuses with a haploid hypha of opposite sex to produce dikaryotic aeciospores.

pycnium [MYCOL] A flask-shaped fruit body of a rust fungus formed in clusters just beneath the surface of a host tissue.

Pycnodontiformes [PALEON] An extinct order of specialized fishes characterized by a laterally compressed, disk-shaped body, long dorsal and anal fins, and an externally symmetrical tail.

Pycnogonida [INV ZOO] The sea spiders, a subphylum of marine arthropods in which the body is reduced to a series of cylindrical trunk somites supporting the appendage.

Pycnogonidae [INV ZOO] A family of the Pycnogonida lacking both chelifores and palpi and having six to nine jointed ovigers in the male only.

pycnosis [CYTOL] Nuclear degeneration, characterized by nuclear condensation, and clumping of chromosomes.

pyelitis [MED] Inflammation of the renal pelvis.

pyelonephritis [MED] The disease process resulting from the effects of infections of the parenchyma and the pelvis of the kidney. Also known as interstitial nephritis.

pyemia [MED] A disease state due to the presence of pyogenic microorganisms in the blood and the formation, wherever these organisms lodge, of embolic or metastatic abscesses.

Pygasteridae [PALEON] The single family of the extinct order Pygasteroida.

Pygasteroida [PALEON] An order of extinct echinoderms in the superorder Diadematacea having four genital pores, noncrenulate tubercles, and simple ambulacral plates.

pygidium [INV ZOO] **1.** A caudal shield on the abdomen of some Arthropoda. **2.** The terminal body segment of many invertebrates.

pygochord [INV ZOO] A ventral median ridge of the intestinal epithelium in certain Enteropneusta.

Pygopodidae [VERT ZOO] The flap-footed lizards, a family of the suborder Sauria.

pygostyle [VERT ZOO] A compressed upturned bone at the posterior end of the vertebral column of birds, formed by the fusion of several vertebrae.

pyknosis [PATH] The polymerization and contraction of the nuclear chromosomal components.

pylephlebitis [MED] Inflammation of the portal vein.

pylome [INV ZOO] An aperture for emission of pseudopodia and intake of food in some Sarcodina.

pyloric caecum [INV ZOO] **1.** One of the tubular pouches that open into the ventriculus of an insect. **2.** One of the paired tubes having lateral glandular diverticula in each ray of a starfish. [VERT ZOO] One of the tubular pouches that open from the pyloric end of the stomach into the alimentary canal of most fishes.

pyloric sphincter [ANAT] The thickened ring of circular smooth muscle at the lower end of the pyloric canal of the stomach.

pyloric stenosis [MED] Obstruction of the pyloric opening of the stomach due to hypertrophy of the pyloric sphincter.

pylorospasm [MED] Spasm of the pylorus.

pylorus [ANAT] The orifice of the stomach communicating with the small intestine.

pyobacillosis [VET MED] A bacterial infection of sheep, swine, or rarely cattle caused by *Corynebacterium pyogenes;*

PYCNODONTIFORMES

Coelodus costae Heckel, a pycnodont from Lower Cretaceous of Italy; length to 4 inches (10 centimeters). *(After A. S. Woodward)*

PYCNOGONIDA

8 mm

Pycnogonum stearnsi, a common littoral pycnogonid of the Pacific Coast.

usually marked by abscess formation, but in sheep takes the form of chronic purulent pneumonia.

pyocyanin [MICROBIO] $C_{13}H_{10}N_{20}$ An antibiotic substance forming blue crystals, produced by *Pseudomonas aeruginosa;* active against many bacteria and fungi.

pyoderma [MED] Any pus-producing skin lesion or lesions, used in reference to groups of furuncles, pustules, or even carbuncles.

pyonephritis [MED] Suppurative inflammation of a kidney.

pyonephrosis [MED] Replacement of renal tissue by abscesses.

pyorrhea [MED] A purulent discharge.

Pyralidae [INV ZOO] The equivalent name for Pyralididae.

Pyralididae [INV ZOO] A large family of moths in the lepidopteran superfamily Pyralidoidea; the labial palpi are well developed, and the legs are usually long and slender.

Pyralidinae [INV ZOO] A subfamily of the Pyralididae.

Pyralidoidea [INV ZOO] A superfamily of the Lepidoptera belonging to the Heteroneura and including long-legged, slender-bodied moths with well-developed maxillary palpi.

pyramid [ANAT] **1.** Any conical eminence of an organ. **2.** A body of the longitudinal nerve fibers of the corticospinal tract on each side of the anterior median fissure of the medulla oblongata.

pyramidal area *See* motor area.

pyramidal system [ANAT] The corticospinal and corticobulbar tracts.

Pyramidellidae [INV ZOO] A family of gastropod mollusks in the order Tectibranchia; the operculum is present in this group.

pyramid of numbers [ECOL] The concept that an organism making up the base of a food chain is numerically abundant while each succeeding member of the chain is represented by successively fewer individuals; uses feeding relationship as a basis for the quantitative analysis of an ecological system.

Pyraustinae [INV ZOO] A large subfamily of the Pyralididae containing relatively large, economically important moths.

pyrene [BOT] The seed of a fleshy fruit surrounded by the inner fruit wall or endocarp.

pyrenoid [BOT] A colorless body found within the chromatophore of certain algae; a center for starch formation and storage.

Pyrenolichenes [BOT] The equivalent name for Pyrenulales.

Pyrenulaceae [BOT] A family of the Pyrenulales; all species are crustose and most common on tree bark in the tropics.

Pyrenulales [BOT] An order of the class Ascolichenes including only those lichens with perithecia that contain true paraphyses and unitunicate asci.

pyrexia [MED] Elevation of temperature above the normal; fever.

Pyrgotidae [INV ZOO] A family of myodarian cyclorrhaphous dipteran insects in the subsection Acalypteratae.

pyridoxine hydrochloride [BIOCHEM] $C_8H_{11}NO_3 \cdot HCl$ A crystalline compound, decomposing at about 208°C; used in medicine in vitamin therapy. Also known as pyridoxal hydrochloride; vitamin B_6 hydrochloride.

pyriform [BIOL] Pear-shaped.

pyrimidine [BIOCHEM] $C_4H_4N_2$ A heterocyclic organic compound containing nitrogen atoms at positions 1 and 3; naturally occurring derivatives are components of nucleic acids and coenzymes.

pyrogen [BIOCHEM] A group of substances thought to be polysaccharides of microbial origin that produce an increase in body temperature when injected in man and some animals.

PYRIMIDINE

Structural formula of pyrimidine.

pyromania [PSYCH] A monomania for setting or watching fires.

pyrophosphatase [BIOCHEM] An enzyme catalyzing hydrolysis of esters containing two or more molecules of phosphoric acid to form a simpler phosphate ester.

Pyrosomida [INV ZOO] An order of pelagic tunicates in the class Thaliacea in which species form tubular swimming colonies and are often highly luminescent.

Pyrotheria [PALEON] An extinct monofamilial order of primitive, mastodonlike, herbivorous, hoofed mammals restricted to the Eocene and Oligocene deposits of South America.

Pyrotheriidae [PALEON] The single family of the Pyrotheria.

Pyrrhocoridae [INV ZOO] A family of hemipteran insects belonging to the superfamily Pyrrhocoroidea.

Pyrrhocoroidea [INV ZOO] A superfamily of the Pentatomorpha.

Pyrrhophyta [BOT] A small division of motile, generally unicellular flagellate algae characterized by the presence of yellowish-green to golden-brown plastids and by the general absence of cell walls.

pyruvate [BIOCHEM] Salt of pyruvic acid, such as sodium pyruvate, $NaOOCCOCH_3$.

pyruvic acid [BIOCHEM] Important intermediate in protein and carbohydrate metabolism; liquid with acetic-acid aroma; melts at 11.8°C; miscible with alcohol, ether, and water; used in biochemical research. Also known as acetyl formic acid; α-ketopropionic acid; pyroracemic acid.

Pythidae [INV ZOO] An equivalent name for Salpingidae.

python [VERT ZOO] The common name for members of the reptilian subfamily Pythoninae.

Pythoninae [VERT ZOO] A subfamily of the reptilian family Boidae distinguished anatomically by the skull structure and the presence of a pair of vestigial hindlegs in the form of stout, movable spurs.

pyuria [MED] The presence of pus in the urine.

pyxidium [BOT] A capsular fruiting body dehiscing around its circumference, thus causing the upper part to fall off.

PYROSOMIDA

Colony of *Pyrosoma atlanticum.*
(From Metcalf and Hopkins, after Ritter)

PYROTHERIA

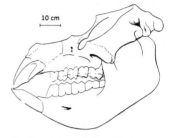

10 cm

Skull and jaw of *Pyrotherium sorondi*, an early Oligocene pyrothere from South America. *(After F. Loomis, 1914)*

PYTHON

A python, which kills its prey by constriction.

Q

Q fever [MED] An acute, febrile infectious disease of man, characterized by sudden onset and patchy pneumonitis, and caused by a bacterialike organism, *Coxiella burneti.*

QRS complex [MED] The electrocardiographic deflection representing ventricular depolarization; the initial downward deflection is termed a *Q* wave; the initial upward deflection, an *R* wave; and the downward deflection following the *R* wave, an *S* wave. Also known as ventricular depolarization complex.

quadrant [ANAT] One of the four regions into which the abdomen may be divided for purposes of physical diagnois. [PHYSIO] A sector of one-fourth of the field of vision of one or both eyes.

quadrate bone [VERT ZOO] A small element forming part of the upper jaw joint on each side of the head in vertebrates below mammals.

quadratojugal [VERT ZOO] A small bone connecting the quadrate and jugal bones on each side of the skull in many lower vertebrates.

quadriceps [ANAT] Four-headed, as a muscle.

quadriceps femoris [ANAT] The large extensor muscle of the thigh, combining the rectus femoris and vastus muscles.

quadrigeminal body *See* corpora quadrigemina.

Quadrijugatoridae [PALEON] A monomorphic family of extinct ostracods in the superfamily Hollinacea.

quadriplegia [MED] Paralysis affecting the four extremities of the body; may be spastic or flaccid.

quadrivalent [CYTOL] An association in polyploids of four homologous chromosomes at diplotene, diakinesis, or metaphase I of meiosis.

quadrumanous [VERT ZOO] Having both hindfeet and front feet constructed like hands, as most Primates except humans.

quadruped [VERT ZOO] An animal that has four legs.

quadruple point [PHYS CHEM] Temperature at which four phases are in equilibrium, such as a saturated solution containing an excess of solute.

quagmire *See* bog.

quail [VERT ZOO] Any of several migratory game birds in the family Phasianidae.

qualitative analysis [ANALY CHEM] The analysis of a gas, liquid, or solid sample or mixture to identify the elements, radicals, or compounds composing the sample.

quantasome [CYTOL] One of the highly ordered array of units that has a "cobblestone" appearance in electron micrographs of the lamella of chloroplasts, and thought to be the most probable site of the light reaction in photosynthesis.

quantitative analysis [ANALY CHEM] The analysis of a gas, liquid, or solid sample or mixture to determine the precise percentage composition of the sample in terms of elements, radicals, or compounds.

QUANTASOME

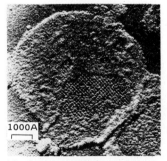

Membranes containing chlorophyll taken from a spinach chloroplast. This chromium-shadowed preparation shows that the membrane is composed of a highly ordered array of units, or quantasomes. *(After R. B. Park, courtesy of Science, 144(3621), 1964)*

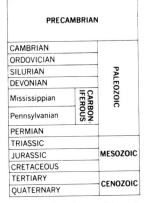

QUATERNARY

PRECAMBRIAN		
CAMBRIAN		
ORDOVICIAN		
SILURIAN		PALEOZOIC
DEVONIAN		
Mississippian	CARBON-IFEROUS	
Pennsylvanian		
PERMIAN		
TRIASSIC		
JURASSIC	MESOZOIC	
CRETACEOUS		
TERTIARY	CENOZOIC	
QUATERNARY		

Chart showing the position of the Quaternary in relation to other periods and to the eras of geologic time.

QUILLWORT

A quillwort *(Isoetes nuttallii).* *(From A. W. Haupt, Plant Morphology, McGraw-Hill, 1953)*

quantitative genetics [GEN] The study of continuously varying traits, such as those of the intellect and personality, which cannot be categorized as dichotomies.

quantitative inheritance [GEN] The acquisition of characteristics, such as height, weight, and intelligence, which show a quantitative and continuous type of variation.

quantum evolution [EVOL] A special but extreme case of phyletic evolution; the rapid evolution that takes place when relatively sudden and drastic change occurs in the environment or when organisms spread into new habitats where conditions differ from those to which they are adapted; the organisms must then adapt quickly to the new conditions if they are to survive.

quantum mechanics [PHYS] The modern theory of matter, of electromagnetic radiation, and of the interaction between matter and radiation; it differs from classical physics, which it generalizes and supersedes, mainly in the realm of atomic and subatomic phenomena. Also known as quantum theory.

quantum theory *See* quantum mechanics.

quantum theory of valence [PHYS CHEM] The theory of valence based on quantum mechanics; it accounts for many experimental facts, explains the stability of a chemical bond, and allows the correlation and prediction of many different properties of molecules not possible in earlier theories.

quantum yield [PHYS CHEM] For a photochemical reaction, the number of moles of a stated reactant disappearing, or the number of moles of a stated product produced, per einstein of light of the stated wavelength absorbed.

quarantine [MED] Limitation of freedom of movement of susceptible individuals who have been exposed to communicable disease, for a period of time equal to the incubation period of the disease.

Quaternary [GEOL] The second period of the Cenozoic geologic era, following the Tertiary, and including the last 2-3 million years.

quaternary system [PHYS CHEM] An equilibrium relationship between a mixture of four (four phases, four components, and so on).

quebracho [BOT] Any of a number of South American trees belonging to different genera in the order Sapindales, but all being a valuable source of wood, bark, and tannin.

queen [INV ZOO] A mature, fertile female in a colony of ants, bees, or termites, whose function is to lay eggs.

Queensland tick typhus [MED] A benign infectious disease of humans found in rural northeastern Australia, caused by the bacterialike microorganism *Rickettsia australis* and presumed to be carried by the tick *Ixodes holocyclus.*

quellung reaction [MICROBIO] Swelling of the capsule of a bacterial cell, caused by contact with serum containing antibodies capable of reacting with polysaccharide material in the capsule; applicable to *Pneumococcus, Klebsiella,* and *Hemophilus.*

quill [VERT ZOO] The hollow, horny shaft of a large stiff wing or tail feather.

quill knob [VERT ZOO] A tubercle on the ulna of birds, for attachment of the fibrous ligaments connecting with a quill follicle.

quillwort [BOT] The common name for plants of the genus *Isoetes.*

quinacrine [PHARM] $C_{23}H_{30}ClN_3O$ Formerly an important antimalarial drug but now used in the treatment of giardiasis, tapeworm infections, amebiasis, and a variety of other conditions.

quinalbarbitone *See* secobarbital.

quince [BOT] *Cydonia oblonga.* A deciduous tree of the order

Rosales characterized by crooked branching, leaves that are densely hairy on the underside and solitary white or pale-pink flowers; fruit is an edible pear- or apple-shaped tomentose pome.

quinine [PHARM] $C_{20}H_{24}N_2O_2$ An alkaloid of cinchona, used principally as an antimalarial drug.

quinoa [BOT] *Chenopodium quinoa.* An annual herb of the family Chenopodiaceae grown at high altitudes in South America for the highly nutritious seeds.

quinquefoliate [BOT] Of a leaf, having five leaflets.

quintuplet [BIOL] One of five children who have been born at one birth.

quitiqua *See* pinta.

R

r *See* roentgen.

R *See* roentgen.

rabbit [VERT ZOO] Any of a large number of burrowing mammals in the family Leporidae.

rabies [VET MED] An acute, encephalitic viral infection transmitted to humans by the bite of a rabid animal. Also known as hydrophobia.

raccoon [VERT ZOO] Any of 16 species of carnivorous nocturnal mammals belonging to the family Procyonidae; all are arboreal or semiarboreal and have a bushy, long ringed tail.

race [ANTHRO] **1.** A distinctive human type possessing characteristic traits that are transmissible by descent. **2.** Descendants of a common ancestor. [BIOL] **1.** An infraspecific taxonomic group of organisms, such as subspecies or microspecies. **2.** A fixed variety or breed.

racemase [BIOCHEM] Any of a group of enzymes that catalyze racemization reactions.

raceme [BOT] An inflorescence on which flowers are borne on stalks of equal length on an unbranched main stalk that continues to grow during flowering.

racemic mixture [CHEM] A compound which is a mixture of equal quantities of dextrorotatory and levorotatory isomers of the same compound, and therefore is optically inactive.

racemic 1-phenyl-2-aminopropane *See* amphetamine.

racemose [ANAT] Of a gland, compound and shaped like a bunch of grapes, with freely branching ducts that terminate in acini. [BOT] Bearing, or occurring in the form of, a raceme.

rachiglossate radula [INV ZOO] A radula of certain gastropod mollusks which has one or three longitudinal series of teeth, each of which may bear many cusps.

rachilla [BOT] The axis of a grass spikelet.

rachiodont [VERT ZOO] Pertaining to egg-eating snakes having well-developed hypophyses on the anterior thoracic vertebrae, which function as teeth.

rachis [ANAT] The vertebral column. [BIOL] An axial structure such as the axis of an inflorescence, the central petiole of a compound leaf, or the central cord of an ovary in Nematoda.

rad [NUCLEO] The standard unit of absorbed dose, equal to energy absorption of 100 ergs per gram (0.01 joule per kilogram); supersedes the roentgen as the unit of dosage.

radial [SCI TECH] **1.** Directed or diverging from a center. **2.** Raylike.

radial artery [ANAT] A branch of the brachial artery in the forearm; principal branches are the radial recurrent and the main artery of the thumb.

radial canal [INV ZOO] **1.** One of the numerous canals that radiate from the spongocoel in certain Porifera. **2.** Any of the canals extending from the coelenteron to the circular canal in the margin of the umbrella in jellyfishes. **3.** A canal radiating

RACCOON

A semiarboreal raccoon which uses tree hollows for breeding and nesting.

RACHIGLOSSATE RADULA

Rachiglossate radula, from *Murex* (marine gastropod). *(From R. R. Shrock and W. H. Twenhofel, Principles of Invertebrate Paleontology, 2d ed., McGraw-Hill, 1953)*

from the circumoral canal along each ambulacral area in many echinoderms.

radial chromatography [ANALY CHEM] A circular disk of absorbent paper which has a strip (wick) cut from edge to center to dip into a solvent; the solvent climbs the wick, touches the sample, and resolves it into concentric rings (the chromatogram). Also known as circular chromatography; radial-paper chromatography.

radial cleavage [EMBRYO] A cleavage pattern characterized by formation of a mass of cells that show radial symmetry.

radial nerve [ANAT] A large nerve that arises in the brachial plexus and branches to enervate the extensor muscles and skin of the posterior aspect of the arm, forearm, and hand.

radial notch [ANAT] A depression on the lateral surface of the coronoid process of the ulna where the head of the radius articulates.

radial-paper chromatography *See* radial chromatography.

radial symmetry [SCI TECH] An arrangement of usually similar parts in a regular pattern around a central axis.

Radiata [INV ZOO] Members of the Eumetazoa which have a primary radial symmetry; includes the Coelenterata and Ctenophora.

radiation [PHYS] **1.** The emission and propagation of waves transmitting energy through space or through some medium; for example, the emission and propagation of electromagnetic, sound, or elastic waves. **2.** The energy transmitted by waves through space or some medium; when unqualified, usually refers to electromagnetic radiation. Also known as radiant energy. **3.** A stream of particles, such as electrons, neutrons, protons, α-particles, or high-energy photons, or a mixture of these.

radiation biochemistry [BIOCHEM] The study of the response of the constituents of living matter to radiation.

radiation biology *See* radiobiology.

radiation biophysics [BIOPHYS] The study of the response of organisms to ionizing radiations and to ultraviolet light.

radiation cataract *See* irradiation cataract.

radiation cytology [CYTOL] An aspect of biology that deals with the effects of radiations on living cells.

radiation dermatitis *See* radiodermatitis.

radiation detection instrument [NUCLEO] Any device that detects and records the characteristics of ionizing radiation.

radiation genetics [GEN] The study of the genetic effects of radiation. Also known as radiogenetics.

radiation hazard [MED] Health hazard arising from exposure to ionizing radiation.

radiation sickness [MED] **1.** Illness, usually manifested by nausea and vomiting, resulting from the effects of therapeutic doses of radiation. **2.** Radiation injury following exposure to excessive doses of radiation, such as the explosion of an atomic bomb.

radiation source [NUCLEO] Usually a man-made, sealed source of radioactivity used in teletherapy, radiography, as a power source for batteries, or in various types of industrial gages; machines such as accelerators, and radioisotopic generators and natural radionuclides may also be considered as sources.

radiation therapy [MED] The use of ionizing radiation or radioactive substances to treat disease. Also known as actinotherapy; radiotherapy.

radical [BOT] **1.** Of, pertaining to, or proceeding from the root. **2.** Arising from the base of a stem or from an underground stem. [CHEM] A stable group of atoms found as part of the molecules of a number of compounds, organic or

RADIAL CLEAVAGE

Radial cleavage pattern found in invertebrates.

inorganic; examples are CH_3^- (methyl), $C_2H_5^-$ (ethyl), SO_4^{--} (sulfate), and ClO_3^- (chlorate).

radiciflorous [BOT] Having flowers that arise at the extreme base of the stem.

radicle [BOT] The embryonic root of a flowering plant.

radicolous [BIOL] Living in or on plant roots.

radicotomy *See* rhizotomy.

radiculitis [MED] Inflammation of a nerve root.

radioactive [NUC PHYS] Exhibiting radioactivity or pertaining to radioactivity.

radioactive age determination *See* radiometric dating.

radioactive carbon dating *See* radiocarbon dating.

radioactive isotope *See* radioisotope.

radioactive source [NUCLEO] Any quantity of radioactive material intended for use as a source of ionizing radiation.

radioactive tracer [NUCLEO] A radioactive isotope which, when attached to a chemically similar substance or injected into a biological or physical system, can be traced by radiation detection devices, permitting determination of the distribution or location of the substance to which it is attached. Also known as radiotracer.

radioautography *See* autoradiography.

radiobiology [BIOL] Study of the scientific principles, mechanisms, and effects of the interaction of ionizing radiation with living matter. Also known as radiation biology.

radiocarbon *See* carbon-14.

radiocarbon dating [NUCLEO] A method of estimating the age of carbon-bearing materials which have formed within the last 70,000 years and which have utilized carbon dioxide from the atmosphere or some other portion of the earth's dynamic reservoir, based on the decay of carbon-14. Also known as carbon-14 dating; radioactive carbon dating.

radiodermatitis [MED] Degenerative changes in the skin following excessive exposure to ionizing radiation. Also known as radiation dermatitis.

radioecology [ECOL] The interdisciplinary study of organisms, radionuclides, ionizing radiation, and the environment.

radio-frequency spectroscopy [SPECT] The branch of spectroscopy concerned with the measurement of the intervals between atomic or molecular energy levels that are separated by frequencies from about 10^5 to 10^9 hertz, as compared to the frequencies that separate optical energy levels of about 6×10^{14} hertz.

radiogenetics *See* radiation genetics.

radiogenic age determination *See* radiometric dating.

radiogenic dating *See* radiometric dating.

radiography [GRAPHICS] The technique of producing a photographic image of an opaque specimen by transmitting a beam of x-rays or γ-rays through it onto an adjacent photographic film; the image results from variations in thickness, density, and chemical composition of the specimen; used in medicine and industry.

radiohumeral index [ANTHRO] The ratio, multiplied by 100, of the length of the radius to the length of the humerus of the human arm.

radioiodine [NUC PHYS] Any radioactive isotope of iodine, especially iodine-131; used as a tracer to determine the activity and size of the thyroid gland, and experimentally, to destroy the thyroid glands of animals.

radioisotope [NUC PHYS] An isotope which exhibits radioactivity. Also known as radioactive isotope; unstable isotope.

Radiolaria [INV ZOO] A subclass of the protozoan class Actinopodea whose members are noted for their siliceous skeletons and characterized by a membranous capsule which separates the outer from the inner cytoplasm.

RADIOGRAPHY

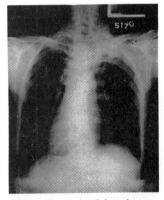

Chest radiograph of foundry worker made with intense beam from rotating-target x-ray tube, showing nodules in lungs which are caused by silicosis, and shadows of skeleton, heart, and stomach.

RADIOLARIA

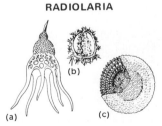

Radiolaria. Skeletons representing (a, b) certain Monopylina (or Nasellina) and (c) certain Periphylina (or Spumellina). (After Haeckel)

RADIOLE

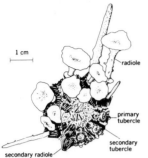

1 cm

radiole

primary tubercle

secondary tubercle

secondary radiole

Morphological features of *Goniocidaris parasol* showing radioles.

RADISH

Radish *(Raphanus sativus)*, cultivar Red Boy. *(Joseph Harris Co., Rochester, N.Y.)*

RADIUM

| 88 |
| Ra |

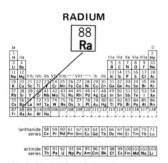

Periodic table of the chemical elements showing the position of radium.

radiole [INV ZOO] A spine on a sea urchin.

radiologist [MED] A physician who specializes in the use of radiant energy in the diagnosis and treatment of disease.

radiology [MED] The medical science concerned with radioactive substances, x-rays, and other ionizing radiations, and the application of the principles of this science to diagnosis and treatment of disease.

radiomedial [INV ZOO] A cross vein between the radius and the medius of an insect wing.

radiometric analysis [ANALY CHEM] Quantitative chemical analysis that is based on measurement of the absolute disintegration rate of a radioactive component having a known specific activity.

radiometric dating [NUCLEO] A technique for measuring the age of an object or sample of material by determining the ratio of the concentration of a radioisotope to that of a stable isotope in it; for example, the ratio of carbon-14 to carbon-12 reveals the approximate age of bones, pieces of wood, and other archeological specimens. Also known as isotopic age determination; nuclear age determination; radioactive age determination; radioactive dating; radiogenic age determination; radiogenic dating.

radiomimetic activity [BIOL] The radiationlike effects of certain chemicals, such as nitrogen mustard, urethane, and fluorinated pyrimidines.

radiomimetic substances [CHEM] Chemical substances which cause biological effects similar to those caused by ionizing radiation.

radiophosphorus [NUC PHYS] The radioactive isotope of phosphorus, with mass number 32 and a half-life of 14.3 days; it is used in the form of sodium phosphate in the treatment of polycythemia vera, as an antineoplastic agent, in various ways as a diagnostic agent, and also as a tracer.

radio pill [ELECTR] A device used in biotelemetry for monitoring the physiologic activity of an animal, such as pH values of stomach acid; an example is the Heidelberg capsule.

radioresistance [BIOL] The resistance of organisms or tissues to the harmful effects of various radiations.

radiotherapy *See* radiation therapy.

radiotracer *See* radioactive tracer.

radish [BOT] *Raphanus sativus.* **1.** An annual or biennial crucifer belonging to the order Capparales. **2.** The edible, thickened hypocotyl of the plant.

radium [CHEM] A radioactive member of group IIA, symbol Ra, atomic number 88; the most abundant naturally occurring isotope has mass number 226 and a half-life of 1620 years. A highly toxic solid that forms water-soluble compounds; decays by emission of α, β, and γ-radiation; melts at 700°C, boils at 1140°C; turns black in air; used in medicine, in industrial radiography, and as a source of neutrons and radon.

radium needle [NUCLEO] A radium cell in the form of a needle, usually of platinum-iridium or gold alloy, designed primarily for insertion in tissue.

radius [ANAT] The outer of the two bones of the human forearm or of the corresponding part in vertebrates other than fish.

radon breath analysis [MED] Examination of exhaled air for the presence of radon to determine the presence and quantity of radium in the human body.

radula [INV ZOO] A filelike ribbon studded with horny or chitinous toothlike structures, found in the mouth of all classes of mollusks except Bivalvia.

raffinase [BIOCHEM] An enzyme that hydrolyzes raffinose, yielding fructose in the reaction.

raffinose [BIOCHEM] $C_{18}H_{32}O_{16} \cdot 5H_2O$ A white, crystalline trisaccharide found in sugarbeets, cottonseed meal, and molasses; yields glucose, fructose, and galactose on complete hydrolysis. Also known as gossypose; melitose; melitriose.

Rafflesiales [BOT] A small order of dicotyledonous plants; members are highly specialized, nongreen, rootless parasites which grow from the roots of the host.

Raillietiellidae [INV ZOO] A small family of parasitic arthropods in the order Cephalobaenida.

Rainey's corpuscle [INV ZOO] The sickle-shaped spore of an encysted sarcosporidian.

rainforest [ECOL] A forest of broad-leaved, mainly evergreen, trees found in continually moist climates in the tropics, subtropics, and some parts of the temperate zones.

rain rot [VET MED] Weeping dermatitis accompanied by swelling of the skin and loss of hair in sheep exposed to prolonged periods of rain.

Rajidae [VERT ZOO] The skates, a family of elasmobranchs included in the batoid group.

Rajiformes [VERT ZOO] The equivalent name for Batoidea.

râle [MED] An abnormal sound accompanying the normal sounds of respiration within the air passages and heard on auscultation of the chest.

Rallidae [VERT ZOO] A large family of birds in the order Gruiformes comprising rails, gallinules, and coots.

ram [VERT ZOO] A male sheep or goat.

Ramapithecinae [PALEON] A subfamily of Hominidae including the protohominids of the Miocene and Pliocene.

Ramapithecus [PALEON] The genus name given to a fossilized upper jaw fragment found in the Siwalik hills, India; closely related to the family of man.

ramate [BIOL] Having branches.

ramentum [BOT] A thin brownish scale consisting of a single layer of cells and occurring on the leaves and young shoots of many ferns.

ramicolous [BOT] Living on twigs.

ramicorn [INV ZOO] Having the antennae branched, as some insects.

ramie [BOT] *Boehmeria nivea.* A shrub or half-shrub of the nettle family (Urticaceae) cultivated as a source of a tough, strong, durable, lustrous natural woody fiber resembling flax, obtained from the phloem of the plant; used for high-quality papers and fabrics. Also known as China grass; rhea.

ramiflorous [BOT] Having the flowers on branches.

ramoconidium [MYCOL] A fungal spore produced from part of a conidiophore.

Ramon flocculation test [IMMUNOL] A method of standardizing antitoxins; a toxin-antitoxin flocculation that is a precipitin reaction in which the end point is the zone of optimal proportion; that is, the zone in which there is no uncombined antigen or antibody.

ramose [BIOL] Having lateral divisions or branches.

Ramphastidae [VERT ZOO] The toucans, a family of birds with large, often colorful bills in the order Piciformes.

ramus [ANAT] A slender bone process branching from a large bone. [VERT ZOO] The barb of a feather. [ZOO] The branch of a structure such as a blood vessel, nerve, arthropod appendage, and so on.

random error [STAT] An error that can be predicted only on a statistical basis.

random experiments [STAT] Experiments which do not always yield the same result when repeated under the same conditions.

randomization [STAT] Assigning subjects to treatment groups by use of tables of random numbers.

RAMIE

Branch of the ramie shrub
(Boehmeria nivea); grows to a
height of 4–7 feet (1.2–2 meters).

RANUNCULACEAE

Colorado columbine
(Aquilegia coerulea) of family
Ranunculaceae, order
Ranunculales. Nectar is borne
in long, hollow spur that
projects from base of each petal.
*(U.S. Forest Service photograph
by C. A. Kutzleb)*

RAPE

Rape *(Brassica napus)*, the Dwarf
Essex variety. *(From L. H. Bailey,
ed., The Standard Cyclopedia of
Horticulture, vol. 3, Macmillan,
1937)*

randomized blocks [STAT] An experimental design in which the various treatments are reproduced in each of the blocks and are randomly assigned to the units within the blocks, permitting unbiased estimates of error to be made.

randomized test [STAT] Acceptance or rejection of the null hypothesis by use of a random variable to decide whether an observation causes rejection or acceptance.

random mating [GEN] The mating of individuals in a population at the rate expected by chance.

random ordered sample [STAT] An ordered sample of size s drawn from a population of size N such that the probability of any particular ordered sample is the reciprocal of the number of permutations of N things taken s at a time.

random sampling [STAT] A sampling from some population where each entry has an equal chance of being drawn.

Ranidae [VERT ZOO] A family of frogs in the suborder Diplasiocoela including the large, widespread genus *Rana*.

Rankine temperature scale [THERMO] A scale of absolute temperature; the temperature in degrees Rankine (°R) is equal to $\frac{9}{5}$ of the temperature in kelvins and to the temperature in degrees Fahrenheit plus 459.67.

rank of an observation [STAT] The number assigned to an observation if a collection of observations is ordered from smallest to largest and each observation is given the number corresponding to its place in the order.

rank tests [STAT] Tests which use the ranks of observations with respect to one another rather than the observations themselves.

ranula [MED] A retention cyst of a salivary gland.

Ranunculaceae [BOT] A family of dicotyledonous herbs in the order Ranunculales distinguished by alternate leaves with net venation, two or more distinct carpels, and numerous stamens.

Ranunculales [BOT] An order of dicotyledons in the subclass Magnoliidae characterized by its mostly separate carpels, triaperturate pollen, herbaceous or only secondarily woody habit, and frequently numerous stamens.

rape [BOT] *Brassica napus.* A plant of the cabbage family in the order Capparales; the plant does not form a compact head, the leaves are bluish-green, deeply lobed, and curled, and the small flowers produce black seeds; grown for forage.

raphania [MED] A disease thought to be due to chronic ingestion of the poison in seeds of the wild radish.

raphe [ANAT] A broad seamlike junction between two lateral halves of an organ or other body part. [BOT] **1.** The part of the funiculus attached along its full length to the integument of an anatropous ovule, between the chalaza and the attachment to the placenta. **2.** The longitudinal median line or slit on a diatom valve.

Raphidae [VERT ZOO] A family of birds in the order Columbiformes that included the dodo (*Raphus calcullatus*); completely extirpated during the 17th and early 18th centuries.

raphide [BOT] One of the long, needle-shaped crystals, usually consisting of calcium oxalate, occurring as a metabolic by-product in certain plant cells.

rapid-eye-movement sleep [PSYCH] That part of the sleep cycle during which the eyes move rapidly, accompanied by a loss of muscle tone and a low-amplitude encephalogram recording; most dreaming occurs during this stage of sleep. Abbreviated REM sleep.

raptorial [ZOO] **1.** Living on prey. **2.** Adapted for snatching or seizing prey, as birds of prey.

rapture of the deep *See* nitrogen narcosis.

rash [MED] A lay term for nearly any skin eruption, but more commonly for acute inflammatory dermatoses.

rasorial [ZOO] Adapted for scratching for food; applied to birds.

raspberry [BOT] Any of several species of upright shrubs of the genus *Rubus*, with perennial roots and prickly biennial stems, in the order Rosales; the edible black or red juicy berries are aggregate fruits, and when ripe they are easily separated from the fleshy receptacle.

rastellus [INV ZOO] A group of teeth in the chelicera of Arachnida.

rat [VERT ZOO] The name applied to over 650 species of mammals in several families of the order Rodentia; they differ from mice in being larger and in having teeth modified for gnawing.

rataria larva [INV ZOO] The second, hourglass-shaped, free-swimming larva of the siphonophore *Velella*.

rat-bite fever [MED] Either of two diseases transmitted by the bite of a rat: spirillary rat-bite fiver and streptobacillary fever.

rate constant [PHYS CHEM] Numerical constant in a rate-of-reaction equation; for example, $r_A = kC_A{}^aC_B{}^bC_C{}^c$, where C_A, C_B, and C_C are reactant concentrations, k is the rate constant (specific reaction rate constant), and a, b, and c are empirical constants.

ratfish [VERT ZOO] The common name for members of the chondrichthyan order Chimaeriformes.

Rathke's pouch *See* craniobuccal pouch.

rationalization [PSYCH] A defense mechanism against difficult and unpleasant situations in which the individual attempts to use plausible means to justify or defend the unacceptable situations.

ratite [BOT] Having a sternum without a keel.

ratites [VERT ZOO] A group of flightless, mostly large, running birds comprising several orders and including the emus, cassowaries, kiwis, and ostriches.

rattan [BOT] Any of several long-stemmed, climbing palms, especially of the genera *Calanius* and *Daemonothops*; stem material is used to make walking sticks, wickerwork, and cordage.

rattlesnake [VERT ZOO] Any of a number of species of the genera *Sistrurus* or *Crotalus* distinguished by the characteristic rattle on the end of the tail.

rat typhus *See* murine typhus.

Raunkiaer system [BOT] A classification system for plant life-forms based on the position of perennating buds in relation to the soil surface.

Rauwolfia [BOT] A genus of mostly poisonous, tropical trees and shrubs of the dogbane family (Apocynaceae); certain species yield substances used as emetics and cathartics, while *R. serpentina* is a source of alkaloids used as tranquilizers.

raw humus *See* ectohumus.

ray [PHYS] A moving particle or photon of ionizing radiation. [VERT ZOO] Any of about 350 species of the elasmobranch order Batoidea having flattened bodies with large pectoral fins attached to the side of the head, ventral gill slits, and long, spikelike tails.

rayfin fish [VERT ZOO] The common name for members of the Actinopterygii.

ray flower [BOT] One of the small flowers with a strap-shaped corolla radiating from the margin of the head of a capitulum.

ray initial [BOT] One of the cells of the cambium which divide to produce new phloem and xylem ray cells.

Raynaud's disease [MED] A usually bilateral disease of blood vessels, especially of the extremities; excited by cold or emotion, characterized by intermittent pallor, cyanosis, and redness, and generally accompanied by pain.

Rb *See* rubidium.

RASPBERRY

A red raspberry branch, Loudon variety. *(USDA)*

RAT

The brown Norway or common rat *(Rattus norvegicus)*. One of the worst pests in the United States, it destroys billions of dollars worth of farm products annually.

RAY

Devil ray, which is named for the anterior cephalic fins, extensions of the pectoral fins.

RECENT

Chart showing the position of the Recent epoch in relation to the various periods of geologic time.

RBE   *See* relative biological effectiveness.

RDS   *See* respiratory distress syndrome of newborn.

Re   *See* rhenium.

reactants   [CHEM] The molecules that act upon one another to produce a new set of molecules (products); for example, in the reaction $HCl + NaOH \rightarrow NaCl + H_2O$, the HCl and NaOH are the reactants.

reaction   [CONT SYS] *See* positive feedback.   [MECH] The equal and opposite force which results when a force is exerted on a body, according to Newton's third law of motion.   [NUC PHYS] *See* nuclear reaction.

reaction formation   [PSYCH] A defense mechanism, characterized by the development of conscious, socially acceptable activity which is the antithesis of repressed or rejected unconscious desires.

reaction time   [PHYSIO] The interval between application of a stimulus and the beginning of the response.

reaction wood   [BOT] An abnormal development of a tree and therefore its wood as the result of unusual forces acting on it, such as an atypical gravitational pull.

reagent   [ANALY CHEM] A substance, chemical, or solution used in the laboratory to detect, measure, or otherwise examine other substances, chemicals, or solutions; grades include ACS (American Chemical Society standards), reagent (for analytical reagents), CP (chemically pure), USP (U.S. Pharmacopeia standards), NF (National Formulary standards), and purified, technical (for industrial use).

reagin   [IMMUNOL] **1.** An antibody which occurs in human atopy, such as hay fever and asthma, and which readily sensitizes the skin. **2.** An antibody which reacts in various serologic tests for syphilis.

reality principle   [PSYCH] The concept that the pleasure principle is normally modified by the demands of the external environment and that the individual adjusts to these inescapable requirements so that he ultimately secures satisfaction of his instinctual wishes.

recapitulation theory   [BIOL] The biological theory that an organism passes through developmental stages resembling various stages in the phylogeny of its group; ontogeny recapitulates phylogeny. Also known as biogenetic law; Haeckel's law.

Recent   [GEOL] An epoch of geologic time (late Quaternary) following the Pleistocene; referred to as Holocene in several European countries.

receptacle   [BIOL] An organ that functions as a repository. [BOT] **1.** The peduncle of a racemose inflorescence. **2.** The torus of a flower. **3.** The modified end of an algal thallus branch that contains conceptacles.

receptive aphasia   *See* sensory aphasia.

receptor   [BIOCHEM] A site or structure in a cell which combines with a drug or other biological to produce a specific alteration of cell function.   [PHYSIO] A sense organ.

recessive   [GEN] **1.** An allele that is not expressed phenotypically when present in the heterozygous condition. **2.** An organism homozygous for a recessive gene.

reciprocal translocation   [CYTOL] The special case of translocation in which two segments exchange positions.

reclinate   [BOT] Curved downward from the apex to the base, referring specifically to an ovule suspended from a funiculus.

recombination   [GEN] **1.** The occurrence of gene combinations in the progeny that differ from those of the parents as a result of independent assortment, linkage, and crossing-over. **2.** The production of genetic information in which there are elements of one line of descent replaced by those of another line, or additional elements.

recombination mosaic [GEN] A mosaic produced as the result of somatic crossing-over.

recon [GEN] The smallest deoxyribonucleic acid unit capable of recombination.

recovery room [MED] A hospital room in which surgical patients are kept during the period immediately following an operation for care and recovery from anesthesia.

recruitment [PHYSIO] A serial discharge from neurons innervating groups of muscle fibers.

Recticornia [INV ZOO] A family of amphipod crustaceans in the superfamily Genuina containing forms in which the first antennae are straight, arise from the anterior margin of the head, and have few flagellar segments.

rectocele [MED] Bulging or herniation of the rectum into the vagina. Also known as vaginal protocele.

rectrix [VERT ZOO] One of the stiff tail feathers used by birds to control direction of flight.

rectum [ANAT] The portion of the large intestine between the sigmoid flexure and the anus.

rectus [ANAT] Having a straight course, as certain muscles.

rectus abdominis [ANAT] The long flat muscle of the anterior abdominal wall which, as vertical fibers, arises from the pubic crest and symphysis, and is inserted into the cartilages of the fifth, sixth, and seventh ribs.

rectus femoris [ANAT] A division of the quadriceps femoris inserting in the patella and ultimately into the tubercle of the tibia.

rectus oculi [ANAT] Any of four muscles (superior, inferior, lateral, and medial) of the eyeball, running forward from the optic foramen and inserted into the sclerotic coat.

recumbent [BOT] Of or pertaining to a plant or plant part that tends to rest on the surface of the soil.

recurved [SCI TECH] Curving inward to backward.

red algae [BOT] The common name for members of the phylum Rhodophyta.

red blood cell *See* erythrocyte.

redia [INV ZOO] A larva produced within the miracidial sporocyst of certain digenetic trematodes which may give rise to daughter rediae or to cercariae.

redifferentiation [PHYSIO] The return to a position of greater specialization in actual and potential functions, or the developing of new characteristics.

red leaf [PL PATH] Any of various nonparasitic plant diseases marked by red discoloration of the foliage.

red nucleus [HISTOL] A mass of reticular fibers in the gray matter of the tegmentum of the mesencephalon of higher vertebrates; it receives fibers from the cerebellum of the opposite side and gives rise to rubrospinal tract fibers of the opposite side.

redox potential [PHYS CHEM] Voltage difference at an inert electrode immersed in a reversible oxidation-reduction system; measurement of the state of oxidation of the system. Also known as oxidation-reduction potential.

redox potentiometry [ANALY CHEM] Use of neutral electrode probes to measure the solution potential developed as the result of an oxidation or reduction reaction.

redox system [CHEM] A chemical system in which reduction and oxidation (redox) reactions occur.

red ring [PL PATH] A nematode disease of the coconut palm caused by *Aphelenchoides cocophilus* and marked by the appearance of red rings in the stem cross section.

red rot [PL PATH] Any of several fungus diseases of plants characterized by red patches on stems or leaves; common in sugarcane, sisal, and various evergreen and deciduous trees.

red rust [PL PATH] An algal disease of certain subtropical plants, such as tea and citrus, caused by the green alga *Cephaleuros virescens* and characterized by a rusty appearance of the leaves or twigs.

red stele [PL PATH] A fatal fungus disease of strawberries caused by *Phytophthora fragariae* invading the roots, producing redness and rotting with consequent dwarfing and wilting of the plant.

red stripe [PL PATH] **1.** A fungus decay of timber caused by *Polyporus vaporarius* and characterized by reddish streaks. **2.** A bacterial disease of sugarcane caused by *Xanthomonas rubrilineans* and characterized by red streaks in the young leaves, followed by invasion of the vascular system.

red thread [PL PATH] A fungus disease of turf grasses caused by *Cortecium fuciforme* and characterized by the appearance of red stromata on the pinkish hyphal threads.

red tide [BIOL] A reddish discoloration of coastal surface waters due to concentrations of certain toxin-producing dinoflagellates. Also known as red water.

redtop grass [BOT] One of the bent grasses, *Agrostis alba* and its relatives, which grow on a wide variety of soils; it is a perennial, spreads slowly by rootstocks, and has top growth 2-3 feet (60-90 centimeters) tall.

reducing sugar [ORG CHEM] Any of the sugars that because of their free or potentially free aldehyde or ketone groups, possess the property of readily reducing alkaline solutions of many metallic salts such as copper, silver, or bismuth; examples are the monosaccharides and most of the disaccharides, including maltose and lactose.

reduction [CHEM] **1.** Reaction of hydrogen with another substance. **2.** Chemical reaction in which an element gains an electron (has a decrease in positive valence).

reduction potential [PHYS CHEM] The potential drop involved in the reduction of a positively charged ion to a neutral form or to a less highly charged ion, or of a neutral atom to a negatively charged ion.

redundancy [GEN] **1.** Repetition of a specified deoxyribonucleic acid sequence in a nucleus. **2.** Multiplicity of codons for individual amino acids.

Reduviidae [INV ZOO] The single family of the hemipteran group Reduvioidea; nearly all have a stridulatory furrow on the prosternum, ocelli are generally present, and the beak is three-segmented.

Reduvioidea [INV ZOO] The assassin bugs or conenose bugs, a monofamilial group of hemipteran insects in the subdivision Geocorisae.

red water [BIOL] *See* red tide. [VET MED] **1.** A babesiasis of cattle characterized by hematuria following release of hemoglobin by destruction of erythrocytes. **2.** A chronic disease of cattle attributed to oxalic acid in the forage; hematuria results from escape of blood from lesions in the bladder. **3.** An acute febrile septicemia of cattle, and sometimes horses, sheep, and swine, caused by the bacterium *Clostridium hemolyticum* and characterized by hemoglobinuria and sometimes intestinal hemorrhages.

redwater fever [VET MED] Any of several diseases of cattle, such as babesiasis or blackwater, caused by species of *Babesia* and characterized by hematuria.

redwood [BOT] *Sequoia sempervirens.* An evergreen tree of the pine family; it is the tallest tree in the Americas, attaining 350 feet (107 meters); its soft heartwood is a valuable building material.

reed [BOT] Any tall grass with a slender jointed stem.

reef [GEOL] **1.** A ridge- or moundlike layered sedimentary rock structure built almost exclusively by organisms. **2.** An

REDWOOD

Leaves and cone of redwood
(Sequoia sempervirens).

offshore chain or range of rock or sand at or near the surface of the water.

reepithelialization [MED] **1.** Regrowth of epithelial tissue over a denuded surface. **2.** Surgical placement of a graft of epithelium over a denuded surface.

reference electrode [PHYS CHEM] A nonpolarizable electrode that generates highly reproducible potentials; used for pH measurements and polarographic analyses; examples are the calomel electrode, silver-silver chloride electrode, and mercury pool.

referred pain [MED] Pain felt in one area but originating in another area.

reflecting microscope [OPTICS] A microscope whose objective is composed of two mirrors, one convex and the other concave; its imaging properties are independent of the wavelength of light, allowing it to be used even for infrared and ultraviolet radiation.

reflex [PHYSIO] An automatic response mediated by the nervous system.

reflex arc [ANAT] A chain of neurons composing the anatomical substrate or pathway of the unconditioned reflex.

reflex bladder [MED] A urinary bladder controlled only by the simple reflex arc through the sacral part of the spinal cord.

reflexed [BOT] Turned abruptly backward.

reflex epilepsy [MED] Seizure brought on by specific sensory stimuli.

reforestation [FOR] Establishment of a new forest by seeding or planting seedlings on forest land that fails to restock naturally.

refractory [MED] Resisting treatment. [PHYSIO] Resisting stimulation, as a muscle or nerve immediately after responding to a stimulus.

Refsum's disease [MED] A familial disorder characterized by visual disturbances, ataxia, neuritic changes, and cardiac damage, associated with high blood level of phytinic acid.

regeneration [BIOL] The replacement by an organism of tissues or organs which have been lost or severely injured. [CONT SYS] *See* positive feedback.

regional anatomy [ANAT] The detailed study of the anatomy of a part or region of the body of an animal.

regional anesthesia *See* regional block anesthesia.

regional block anesthesia [MED] Anesthesia limited to a part or region of the body by injecting an anesthetic around the nerves that supply the area to block conduction from the area. Also known as regional anesthesia.

regional enteritis [MED] Inflammation of isolated segments of the small intestine, with involved parts becoming thick-walled and rigid. Also known as regional enterocolitis; regional ileitis.

regional enterocolitis *See* regional enteritis.

regional ileitis *See* regional enteritis.

regression [PSYCH] A mental state and a mode of adjustment to difficult and unpleasant situations, characterized by behavior of a type that had been satisfying and appropriate at an earlier stage of development but which no longer befits the age and social status of the individual. [STAT] Given two stochastically dependent random variables, regression functions measure the mean expectation of one relative to the other.

regular [BOT] Having radial symmetry, referring to a flower.

regular connective tissue [HISTOL] Connective tissue in which the fibers are arranged in definite patterns.

Regularia [INV ZOO] An assemblage of echinoids in which the anus and periproct lie within the apical system; not considered a valid taxon.

REINDEER

Reindeer do not display sexual dimorphism; males are larger than females, and reach a length of about 6 feet (1.8 meters) and a shoulder height of about 4 feet (1.2 meters).

regulative egg [EMBRYO] An egg in which unfertilized fragments develop as complete, normal individuals.

regulator gene [GEN] A gene that controls the rate of transcription of one or more other genes.

regurgitation [MED] Reverse circulation of blood in the heart due to defective functioning of the valves. [PHYSIO] Bringing back into the mouth undigested food from the stomach.

rehabilitation [MED] The restoration to a disabled individual of maximum independence commensurate with his limitations by developing his residual capacity.

Rehfuss tube [MED] A stomach tube designed for the removal of specimens of gastric contents for analysis after administration of a test meal.

Reighardiidae [INV ZOO] A monotypic family of arthropods in the order Cephalobaenida; the posterior end of the organism is rounded, without lobes, and the cuticula is covered with minute spines.

reindeer [VERT ZOO] *Rangifer tarandus.* A migratory ruminant of the deer family (Cervidae) which inhabits the Arctic region and has a circumpolar distribution; characteristically, both sexes have antlers and are brown with yellow-white areas on the neck and chest.

reinfection [MED] A second infection after recovery from an earlier infection with the same kind of organism.

Reissner's membrane [ANAT] The anterior wall of the cochlear duct, which separates the cochlear duct from the scala vestibuli. Also known as vestibular membrane of Reissner.

Reiter's syndrome [MED] The triad of idiopathic nongonococcal urethritis, conjunctivitis, and subacute or chronic polyarthritis. Also known as arthritis urethritica; idiopathic arthritis; infectious uroarthritis.

relapsing fever [MED] An acute infectious disease caused by various species of the spirochete *Borrelia,* characterized by episodes of fever which subside spontaneously and recur over a period of weeks.

relative biological effectiveness [NUCLEO] A factor used to compare the biological effectiveness of different types of ionizing radiation; it is the inverse ratio of the amount of absorbed radiation required to produce a given effect to a standard (or reference) radiation required to produce the same effect. Abbreviated RBE.

relative erythrocytosis [MED] A form of erythrocytosis that occurs when the concentration of erythrocytes in the circulating blood increases through loss of plasma.

relative frequency [STAT] The ratio of the number of occurrences of a given type of event or the number of members of a population in a given class to the total number of events or the total number of members of the population.

relaxin [BIOCHEM] A hormone found in the serum of humans and certain other animals during pregnancy; probably acting with progesterone and estrogen, it causes relaxation of pelvic ligaments in the guinea pig.

reliability [STAT] **1.** The amount of credence placed in a result. **2.** The precision of a measurement, as measured by the variance of repeated measurements of the same object.

relict [BIOL] A persistent, isolated remnant of a once-abundant species.

Relizean stage [GEOL] A subdivision of the Miocene in the California-Oregon-Washington area.

relogging [FOR] An operation in which small trees are salvaged, often for pulpwood, after the large trees are logged.

rem [NUCLEO] A unit of ionizing radiation, equal to the amount that produces the same damage to humans as 1 roentgen of high-voltage x-rays. Derived from roentgen equivalent man.

remex *See* flight feather.

REM sleep *See* rapid-eye-movement sleep.

renal artery [ANAT] A branch of the abdominal or ventral aorta supplying the kidneys in vertebrates.

renal calculus [MED] A concretion in the kidney.

renal-cell carcinoma [MED] A malignant renal tumor composed principally of large, often hyalin, polygonal cells. Also known as clear-cell carcinoma; Grawitz's tumor; hypernephroma.

renal corpuscle [ANAT] The glomerulus together with its Bowman's capsule in the renal cortex. Also known as Malpighian corpuscle.

renal dwarfism [MED] Dwarfism due to any of various chronic diseases of the kidney in children.

renal failure [MED] Severe malfunction of the kidneys, producing uremia and the resulting constitutional symptoms.

renal papilla [ANAT] A fingerlike projection into the renal pelvis through which the collecting tubules discharge.

renal pyramid [ANAT] Any of the conical masses composing the medullary substance of the kidney. Also known as Malpighian pyramid.

renal rickets [MED] The metabolic bone disease due to increased bone resorption resulting from the acidosis and secondary hyperparathyroidism of renal insufficiency.

renal tubular acidosis [MED] Defective hydrogen-ion excretion in the renal tubules, resulting in hyperchloremic acidosis and inadequate acidification of the urine.

renal tubule [ANAT] One of the glandular tubules which elaborate urine in the kidneys.

renal vein [ANAT] A vein which returns blood from the kidney to the vena cava.

renette [INV ZOO] An excretory cell found in certain nematodes.

renin [BIOCHEM] A proteolytic enzyme produced in the afferent glomerular arteriole which reacts with the plasma component hypertensinogen to produce angiotensin II.

rennet [VERT ZOO] The lining of the stomach of certain animals, especially the fourth stomach in ruminants.

rennin [BIOCHEM] An enzyme found in the gastric juice of the fourth stomach of calfs; used for coagulating milk casein in cheesemaking. Also known as chymosin.

renninogen [BIOCHEM] The zymogen of rennin. Also known as prorennin.

renopericardial [INV ZOO] Pertaining to a ciliated canal that connects the kidney and the pericardium in higher mollusks.

reovirus [VIROL] A group of ribonucleic acid-containing animal viruses, including agents of encephalitis and phlebotomus fever.

rep [NUCLEO] A unit of ionizing radiation, equal to the amount that causes absorption of 93 ergs per gram of soft tissue. Derived from roentgen equivalent physical. Also known as parker; tissue roentgen.

repair synthesis [MOL BIO] Enzymatic excision and replacement of regions of damaged deoxyribonucleic acid, as in repair of thymine dimers by ultraviolet irradiation.

repand [BOT] Having a margin that undulates slightly, referring to a leaf.

repent [BOT] Of a stem, creeping along the ground and rooting at the nodes.

repetitive DNA [GEN] Short segments of DNA occurring in more than one copy per genome; some segments, less than 20 nucleotides long, may recur, with more than 10 copies in a genome, for instance in mammals; other segments may recur in only a few copies, and others have an intermediate recurrence.

RENAL CORPUSCLE

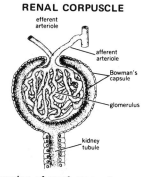

Drawing of renal corpuscle.
(*From C. K. Weichert, Anatomy of the Chordates, 4th ed., McGraw-Hill, 1970*)

REPENT

Ground ivy showing repent type of stem.

replacement transfusion   *See* exchange transfusion.

replica plating   [MICROBIO] A method for the isolation of nutritional mutants in microorganisms; colonies are grown from a microorganism suspension previously exposed to a mutagenic agent, on a complete medium in a petri dish; a velour surface is used to transfer the impression of all these colonies to a petri dish containing a minimal medium; colonies that do not grow on the minimal medium are the mutants.

replicating fork   [MOL BIO] The Y-shaped region of a chromosome that is a growing point in the replication of deoxyribonucleic acid.

replication   [ANALY CHEM] The formation of a faithful mold or replica of a solid that is thin enough for penetration by an electron microscope beam; can use plastic (such as collodion) or vacuum deposition (such as of carbon or metals) to make the mold.   [MOL BIO] Duplication, as of a nucleic acid, by copying from a molecular template.   [VIROL] Multiplication of phage in a bacterial cell.

replicon   [GEN] A genetic element characterized by possession of the structural gene that controls a specific initiator and receptor locus (replicator) for its action.

replum   [BOT] A thin wall separating the two valves or chambers of certain fruits.

repression   [PSYCH] A defense mechanism whereby ideas, feelings, or desires, in conflict with the individual's conscious self-image or motives, are unconsciously dismissed from consciousness.

repressor   [BIOCHEM] An end product of metabolism which represses the synthesis of enzymes in the metabolic pathway.   [GEN] The product of a regulator gene that acts to repress the transcription of another gene.

reproduction   [BIOL] The mechanisms by which organisms give rise to other organisms of the same kind.

reproductive behavior   [ZOO] The behavior patterns in different types of animals by means of which the sperm is brought to the egg and the parental care of the resulting young insured.

reproductive system   [ANAT] The structures concerned with the production of sex cells and perpetuation of the species.

Reptantia   [INV ZOO] A suborder of the crustacean order Decapoda including all decapods other than shrimp.

reptile   [VERT ZOO] Any member of the class Reptilia.

Reptilia   [VERT ZOO] A class of terrestrial vertebrates composed of turtles, tuatara, lizards, snakes, and crocodileans; characteristically they lack hair, feathers, and mammary glands, the skin is covered with scales, they have a three-chambered heart, and the pleural and peritoneal cavities are continuous.

repulsion arrangement   *See* repulsion linkage.

repulsion linkage   [GEN] The occurrence in a heterozygote of two or more linked mutants of both dominant and recessive alleles on the same member of a homologous chromosome pair. Also known as repulsion arrangement; repulsion phase; trans arrangement; trans phase.

repulsion phase   *See* repulsion linkage.

RES   *See* reticuloendothelial system.

Resedaceae   [BOT] A family of dicotyledonous herbs in the order Capparales having irregular, hypogynous flowers.

reserpine   [PHARM] $C_{33}H_{40}N_2O_9$ An alkaloid extracted from certain species of *Rauwolfia* and used as a sedative and antihypertensive.

reserve cell   [HISTOL] **1.** One of the small, undifferentiated epithelial cells at the base of the stratified columnar lining of the bronchial tree. **2.** A chromophobe cell.

residual air   *See* residual volume.

residual volume   [PHYSIO] Air remaining in the lungs after the most complete expiration possible; it is elevated in diffuse obstructive emphysema and during an attack of asthma. Also known as residual air.

resilium   [INV ZOO] The horny flexible hinge occurring in a bivalve.

resin duct   [BOT] A canal (intercellular space) lined with secretory cells that release resins into the canal; common in gymnosperms.

resistance factor   *See* R factor.

resistance transfer factor   [GEN] A carrier of genetic information in bacteria which is considered to control the ability of self-replication and conjugal transfer of R factors. Abbreviated RTF.

resolution   *See* resolving power.

resolving power   [OPTICS] A quantitative measure of the ability of an optical instrument to produce separable images of different points on an object; usually, the smallest angular or linear separation of two object points for which they may be resolved according to the Rayleigh criterion. Also known as resolution. [PHYS] A measure of the ability of a mass spectroscope to separate particles of different masses, equal to the ratio of the average mass of two particles whose mass spectrum lines can just be completely separated, to the difference in their masses. [SPECT] A measure of the ability of a spectroscope or interferometer to separate spectral lines of nearly equal wavelength, equal to the average wavelength of two equally strong spectral lines whose images can barely be separated, divided by the difference in wavelengths; for spectroscopes, the lines must be resolved according to the Rayleigh criterion; for interferometers, the wavelengths at which the lines have half of maximum intensity must be equal. Also known as resolution.

respiration   [PHYSIO] **1.** The processes by which tissues and organisms exchange gases with their environment. **2.** The act of breathing with the lungs, consisting of inspiration and expiration.

respiratory arrest   [MED] Sudden cessation of spontaneous respiration due to failure of the respiratory center.

respiratory center   [PHYSIO] A large area of the brain involved in regulation of respiration.

respiratory distress syndrome of newborn   [MED] A disease occurring during the first days of life, characterized by respiratory distress and cyanosis; a hyaline membrane lines the alveoli when the disease persists for more than several hours. Abbreviated RDS.

respiratory epithelium   [HISTOL] The ciliated pseudostratified epithelium lining the respiratory tract.

respiratory pigment   [BIOCHEM] Any of various conjugated proteins that function in living organisms to transfer oxygen in cellular respiration.

respiratory quotient   [PHYSIO] The ratio of volumes of carbon dioxide evolved and oxygen consumed during a given period of respiration. Abbreviated RQ.

respiratory system   [ANAT] The structures and passages involved with the intake, expulsion, and exchange of oxygen and carbon dioxide in the vertebrate body.

respiratory tree   [ANAT] The trachea, bronchi, and bronchioles. [INV ZOO] Either of a pair of branched tubular appendages of the cloaca in certain holothurians that is thought to have a respiratory function.

restiform body   *See* inferior cerebellar peduncle.

resting cell   [CYTOL] An interphase cell.

resting potential   [PHYSIO] The potential difference between the interior cytoplasm and the external aqueous medium of the living cell.

resting spore [BIOL] A spore that remains dormant for long periods before germination, withstanding adverse conditions; usually invested in a thickened cell wall.

Restionaceae [BOT] A large family of monocotyledonous plants in the order Restionales characterized by unisexual flowers, wholly cauline leaves, unilocular anthers, and a more or less open inflorescence.

Restionales [BOT] An order of monocotyledonous plants in the subclass Commelinidae having reduced flowers and a single, pendulous, orthotropous ovule in each of the one to three locules of the ovary.

restoration [ECOL] A conservation measure involving the correction of past abuses that have impaired the productivity of the resources base.

restriction enzyme [MOL BIO] One of an increasingly inden- tified class of enzymes, each of which attaches to double- stranded DNA at a specific sequence of nucleotides and cleaves it at or near that site.

restriction of ego [PSYCH] A defense mechanism for escap- ing anxiety by avoiding situations consciously perceived as dangerous or uncomfortable.

resupinate [BOT] Inverted, usually through 180°, so as to appear upside down or reversed.

resuscitation [MED] Restoration of consciousness or life functions after apparent death.

retardation [MED] Slow mental or physical functioning. [OPTICS] In interference microscopy, the difference in optical path between the light passing through the specimen and the light bypassing the specimen. Also known as optical-path difference.

rete [BIOL] **1.** A network. **2.** A plexus.

rete cord [EMBRYO] One of the deep, anastomosing strands of cells of the medullary cords of the vertebrate embryo, that form the rete testis or the rete ovarii.

rete mirabile [VERT ZOO] A network of small blood vessels that are formed by the branching of a large vessel and that usually reunite into a single trunk; believed to have an oxygen-storing function in certain aquatic fauna.

retentate [PHYS CHEM] The substance retained by a semiper- meable membrane during dialysis.

retention cyst [MED] A cyst caused by obstructed outflow of secretion from a gland.

retention index [ANALY CHEM] In gas chromatography, the relationship of retention volume with arbitrarily assigned numbers to the compound being analyzed; used to indicate the volume retention behavior during analysis.

retention time [ANALY CHEM] In gas chromatography, the time at which the center, or maximum, of a symmetrical peak occurs on a gas chromatogram.

retention volume [ANALY CHEM] In gas chromatography, the product of retention time and flow rate.

rete ovarii [ANAT] Vestigial tubules or cords of cells near the hilus of the ovary, corresponding with the rete testis, but not connected with the mesonephric duct.

rete testis [ANAT] The network of anastomosing tubules in the mediastinum testis.

reticular cell *See* reticulocyte.

reticular degeneration [PATH] Rupture of epidermal cells with formation of multilocular bullae due to intracellular edema.

reticular fiber [HISTOL] Any of the delicate, branching argen- tophile fibers conspicuous in the connective tissue of lym- phatic tissue, myeloid tissue, the red pulp of the spleen, and most basement membranes. Also known as argentaffin fiber; argyrophil lattice fiber; precollagenous fiber.

reticular formation [ANAT] The portion of the central nervous system which consists of small islands of gray matter separated by fine bundles of nerve fibers running in every direction.

Reticulariaceae [MYCOL] A family of plasmodial slime molds in the order Liceales.

reticular system *See* reticuloendothelial system.

reticular tissue [HISTOL] Connective tissue having reticular fibers as the principal element.

reticulate [BIOL] Having or resembling a network of fibers, veins, or lines. [GEN] Of or relating to evolutionary change resulting from genetic recombination between strains in an interbreeding population.

reticulin [BIOCHEM] A protein isolated from reticular fibers.

reticulocyte [HISTOL] Also known as reticular cell. **1.** A large, immature red blood cell, having a reticular appearance when stained due to retention of portions of the nucleus. **2.** A cell of reticular tissue.

reticuloendothelial granulomatosis [MED] A group of rare diseases characterized by generalized reticuloendothelial hyperplasia with or without intracellular lipid deposition.

reticuloendothelial system [ANAT] The macrophage system, including all phagocytic cells such as histiocytes, macrophages, reticular cells, monocytes, and microglia, except the granular white blood cells. Abbreviated RES. Also known as reticular system; hematopoietic system.

reticulopodia [INV ZOO] Pseudopodia in the form of a branching network.

Reticulosa [PALEON] An order of Paleozoic hexactinellid sponges with a branching form in the subclass Hexasterophora.

reticulospinal tract [ANAT] Nerve fibers descending from large cells of the reticular formation of the pons and medulla into the spinal cord.

reticulum [BIOL] A fine network. [VERT ZOO] The second stomach in ruminants.

retina [ANAT] The photoreceptive layer and terminal expansion of the optic nerve in the dorsal aspect of the vertebrate eye.

retinaculum [BOT] A small glandular mass at the base of the stalk of a pollinium in orchids. [INV ZOO] **1.** A small hook that holds the egg sac in position in cirripedes. **2.** A structure on the forewings of certain butterflies and moths that hooks to the frenulum. **3.** An organ that articulates with the furcula on the abdomen of a springtail.

retinal astigmatism [MED] Astigmatism due to changes in the localization of the fixation point.

retinal pigment *See* rhodopsin.

retinal retinitis *See* vascular retinopathy.

retinene [BIOCHEM] A pigment extracted from the retina, which turns yellow by the action of light; the chief carotenoid of the retina.

retinitis [MED] Inflammation of the retina.

retinitis pigmentosa [MED] A hereditary affection inherited as a sex-linked recessive and characterized by slowly progressing atrophy of the retinal nerve layers, and clumping of retinal pigment, followed by attenuation of the retinal arterioles and waxy atrophy of the optic disks.

retinoblastoma [MED] A malignant tumor of the sensory layer of the retina.

retinochoroiditis [MED] Inflammation of the retina and choroid.

retinol *See* vitamin A.

RETORTAMONADIDA

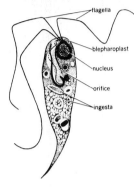

Chilomastix aulastomi.

retinopathy [MED] Any pathologic condition involving the retina.

retinoschisis [MED] **1.** Separation with hole formation of the layers composing the retina. **2.** A congenital anomaly characterized by cleavage of the retina.

retinula [INV ZOO] The receptor element at the inner end of the ommatidium in a compound eye.

Retortamonadida [INV ZOO] An order of parasitic flagellate protozoans belonging to the class Zoomastigophorea, having two or four flagella and a complex blepharoplast-centrosome-axostyle apparatus.

retractile [BIOL] Pertaining to a part or organ that can be drawn inward, such as feelers or claws, or the proboscis of many invertebrates.

retractor [ANAT] A muscle that draws a limb or other body part toward the body. [MED] A clawlike instrument for holding tissues away from the surgical field.

retroaction *See* positive feedback.

retrocerebral gland [INV ZOO] Any of various endocrine glands located behind the brain in insects which function in postembryonic development and metamorphosis.

retroflexion [ANAT] The state of being bent backward. [MED] A condition in which the uterus is bent backward on itself, producing a sharp angle in its longitudinal axis at the junction of the cervix and the fundus.

retrograde amnesia [MED] Loss of memory for events occurring prior to, but not after, the onset of a current disease or trauma.

retrogression [MED] Going backward, as in degeneration or atrophy of tissues. [PSYCH] Return to infantile behavior.

retrolental fibroplasia [MED] An oxygen-induced disease of the retina in premature infants characterized by formation of an opaque membrane behind the lens of the eye.

retrorse [BIOL] Bent downward or backward.

retrostalsis [PHYSIO] Reverse peristalsis.

retroversion [ANAT] A turning back. [MED] A condition in which the uterus is tilted backward without any change in the angle of its longitudinal axis.

retuse [BOT] Having a rounded apex with a slight, central notch.

reversed-phase partition chromatography [ANALY CHEM] Paper chromatography in which the low-polarity phase (such as paraffin, paraffin jelly, or grease) is put onto the support (paper) and the high-polarity phase (such as water, acids, or organic solvents) is allowed to flow over it.

reverse feedback *See* negative feedback.

reverse graft [BOT] A plant graft made by inserting the scion in an inverted position.

reverse mutation [GEN] A mutation in a mutant allele which makes it capable of producing the nonmutant phenotype; may actually restore the original deoxyribonucleic-acid sequence of the gene or produce a new one which has a similar effect. Also known as back mutation.

reverse passive anaphylaxis [IMMUNOL] Hypersensitivity produced when the antigen is injected first, then followed in several hours by the specific antibody, causing shock.

reverse pinocytosis *See* emiocytosis.

reversible electrode [PHYS CHEM] An electrode that owes its potential to unit charges of a reversible nature, in contrast to electrodes used in electroplating and destroyed during their use.

revolute [BOT] Rolled backward and downward.

R factor [GEN] A self-replicating, infectionlike agent that carries genetic information and transmits drug resistance from bacterium to bacterium by conjugation of cell. Also known as resistance factor.

Rh *See* rhodium.

Rhabdiasoidea [INV ZOO] An order or superfamily of parasitic nematodes.

rhabdion [INV ZOO] One of the sclerotized segments lining the buccal cavity of nematodes.

rhabdite [INV ZOO] A small rodlike or fusiform body secreted by epidermal or parenchymal cells of certain turbellarians and trematodes.

Rhabditia [INV ZOO] A subclass of nematodes in the class Secernentea.

Rhabditidia [INV ZOO] An order of nematodes in the subclass Rhabditia including parasites of man and domestic animals.

Rhabditoidea [INV ZOO] A superfamily of small to moderate-sized nematodes in the order Rhabditidia with small, pore-like, anteriorly located amphids, and esophagus with corpus, isthmus, and valvulated basal bulb.

Rhabdocoela [INV ZOO] Formerly an order of the Turbellaria, and now divided into three orders, Catenulida, Macrostomida, and Neorhabdocoela.

rhabdoid [BIOL] Having the shape of a rod.

rhabdolith [BOT] A minute coccolith having a shield surmounted by a long stem and found at all depths in the ocean, from the surface to the bottom. [INV ZOO] A calcareous supporting rod found in some Protozoa.

rhabdome [INV ZOO] The central translucent cylinder in the retinula of a compound eye.

rhabdomere [INV ZOO] The refracting element in a rhabdome.

rhabdomyoblastoma *See* rhabdomyosarcoma.

rhabdomyoma [MED] A benign tumor of skeletal muscle.

rhabdomyosarcoma [MED] A malignant tumor of skeletal muscle in the extremities composed of anaplastic muscle cells. Also known as malignant rhabdomyoma; rhabdomyoblastoma.

Rhabdophorina [INV ZOO] A suborder of ciliates in the order Gymnostomatida.

rhabdosome [INV ZOO] A colonial graptolite that develops from a single individual.

rhabdovirus [VIROL] A group of ribonucleic acid-containing animal viruses, including rabies virus and certain infective agents of fish and insects.

rhabdus [INV ZOO] A uniaxial sponge spicule.

Rhachitomi [PALEON] A group of extinct amphibians in the order Temnospondyli in which pleurocentra were retained.

rhachitomous [VERT ZOO] Being, having, or pertaining to vertebrae with centra whose parts do not fuse.

Rhacophoridae [VERT ZOO] A family of arboreal frogs in the suborder Diplasiocoela.

Rhacopilaceae [BOT] A family of mosses in the order Isobryales generally having dimorphous leaves with smaller dorsal leaves and a capsule that is plicate when dry.

Rhaetian [GEOL] A European stage of geologic time; the uppermost Triassic (above Norian, below Hettangian of Jurassic). Also known as Rhaetic.

Rhaetic *See* Rhaetian.

Rhagionidae [INV ZOO] The snipe flies, a family of predatory orthorrhaphous dipteran insects in the series Brachycera that are brownish or gray with spotted wings.

rhagon [INV ZOO] A pyramid-shaped, colonial sponge having an osculum at the apex and flagellated chambers in the upper wall only.

Rhamnaceae [BOT] A family of dicotyledonous plants in the order Rhamnales characterized by a solitary ovule in each

RHAMPHORHYNCHOIDEA

Restoration of Jurassic pterosaur skeleton of *Rhamphorhynchus* by S. W. Williston; length about 30 centimeters. *(After O. C. Marsh)*

RHENIUM

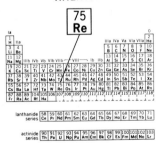

Periodic table of the chemical elements showing the position of rhenium.

locule, free stamens, simple leaves, and flowers that are hypogynous to perigynous or epigynous.

Rhamnales [BOT] An order of dicotyledonous plants in the subclass Rosidae having a single set of stamens, opposite the petals, usually a well-developed intrastamenal disk, and two or more locules in the ovary.

rhamnose [BIOCHEM] $C_6H_{12}O_5$ A deoxysugar occurring free in poison sumac, and in glycoside combination in many plants. Also known as isodulcitol.

rhamphoid [BIOL] Beak-shaped.

Rhamphorhynchoidea [PALEON] A Jurassic suborder of the Pterosauria characterized by long, slender tails with an expanded tip.

rhampotheca [VERT ZOO] The horny sheath covering a bird's beak.

Rh antigen *See* Rh factor.

Rh blocking serum [IMMUNOL] A serum that reacts with Rh-positive blood without causing agglutination, but which blocks the action of anti-Rh serums that are subsequently introduced.

Rh blocking test [IMMUNOL] A test for the detection of Rh antibody in plasma wherein erythrocytes having the Rh antigen are incubated in the patient's serum so that the antibodies may be adsorbed on these cells, which are then employed in the antiglobulin test. Also known as indirect Coombs test; indirect developing test.

Rh blood group [IMMUNOL] The extensive, genetically determined system of red blood cell antigens defined by the immune serum of rabbits injected with rhesus monkey erythrocytes, or by human antiserums. Also known as rhesus blood group.

rhea [BOT] *See* ramie. [VERT ZOO] The common name for members of the avian order Rheiformes.

Rheidae [VERT ZOO] The single family of the avian order Rheiformes.

Rheiformes [VERT ZOO] The rheas, an order of South American running birds; called American ostriches, they differ from the true ostrich in their smaller size, feathered head and neck, three-toed feet, and other features.

Rheinberg illumination [OPTICS] An illumination technique used in optical microscopes that is a modification of the dark-field method; the central disk is transparent and colored; an annulus of a complementary color fills the remaining condenser aperture; the specimen is seen in the color of the annulus against the background of the central disk. Also known as optical staining.

Rhenanida [PALEON] An order of extinct marine fishes in the class Placodermi distinguished by mosaics of small bones between the large plates in the head shield.

rhenium [CHEM] A metallic element, symbol Re, atomic number 75, atomic weight 186.2; a transition element.

rheophile [ECOL] Living or thriving in running water.

rheophilous bog [ECOL] A bog which draws its source of water from drainage.

rheoplankton [ECOL] Plankton found in flowing water.

rheoreceptors [VERT ZOO] Any of various sense organs in the integument of fishes and certain amphibians which are sensitive to the stimulation of water current.

rheotaxis [BIOL] Movement of a motile cell or organism in response to the direction of water currents.

rheotropism [BIOL] Orientation response of an organism to the stimulus of a flowing fluid, as water.

rhesus blood group *See* Rh blood group.

rhesus factor *See* Rh factor.

rhesus macaque *See* rhesus monkey.

rhesus monkey [VERT ZOO] *Macaque mulatta.* An agile, gregarious primate found in southern Asia and having a short tail, short limbs of almost equal length, and a stocky build. Also known as rhesus macaque.

rheumatic arteritis [MED] A type of allergic arteritis associated with acute rheumatic fever.

rheumatic carditis [MED] Inflammation of the heart resulting from rheumatic fever.

rheumatic encephalopathy [MED] An inflammatory reaction of the brain and the smaller arteries of the cerebral cortex associated with rheumatic fever.

rheumatic endocarditis [MED] Inflammation of the endocardium in acute rheumatic fever, usually involving heart valves.

rheumatic fever [MED] A febrile disease occurring in childhood as a delayed sequel of infection by *Streptococcus hemolyticus,* group A; characterized by arthritis, carditis, nosebleeds, and chorea.

rheumatic pneumonia [MED] Pneumonia associated with acute rheumatic fever.

rheumatism [MED] Any combination of muscle or joint pain, stiffness, or discomfort arising from nonspecific disorders.

rheumatoid arthritis [MED] A chronic systemic inflammatory disease of connective tissue in which symptoms and changes predominate in articular and related structures. Also known as atrophic arthritis; chronic infectious arthritis; proliferative arthritis.

rheumatoid nodules [MED] Subcutaneous lateral foci of fibrinoid degeneration or necrosis surrounded by mononuclear cells in a regular palisade arrangement, occurring usually in association with rheumatoid arthritis or rheumatic fever.

rheumatoid spondylitis [MED] A chronic progressive arthritis of young men, affecting mainly the spine and sacroiliac joints, leading to fusion and deformity. Also known as Marie-Strümpell disease.

Rh factor [IMMUNOL] Any of several red blood cell antigens originally identified in the blood of rhesus monkeys. Also known as Rh antigen; rhesus factor.

Rhigonematidae [INV ZOO] A family of nematodes in the superfamily Oxyuroidea.

Rhincodontidae [VERT ZOO] The whale sharks, a family of essentially tropical galeoid elasmobranchs in the isurid line.

rhinencephalon [ANAT] The anterior olfactory portion of the vertebrate brain.

rhinitis [MED] Inflammation of the mucous membranes in the nose.

Rhinobatidae [VERT ZOO] The guitarfishes, a family of elasmobranchs in the batoid group.

Rhinoceratidae [VERT ZOO] A family of perissodactyl mammals in the superfamily Rhinoceratoidea, comprising the living rhinoceroses.

Rhinoceratoidea [VERT ZOO] A superfamily of perissodactyl mammals in the suborder Ceratomorpha including living and extinct rhinoceroses.

rhinoceros [VERT ZOO] The common name for the odd-toed ungulates composing the family Rhinoceratidae, characterized by massive, thick-skinned limbs and bodies, and one or two horns which are composed of a solid mass of hairs attached to the bony prominence of the skull.

Rhinochimaeridae [VERT ZOO] A family of ratfishes, order Chimaeriformes, distinguished by an extremely elongate rostrum.

Rhinocryptidae [VERT ZOO] The tapaculos, a family of ground-inhabiting suboscine birds in the suborder Tyranni characterized by a large, movable flap which covers the nostrils.

RHINOCEROS

Black rhinoceros *(Diceros bicornis)* with thick gray skin and a mobile upper lip shaped like a parrot beak.

rhinogenous [MED] Originating in the nose.

rhinolaryngology [MED] The science of the anatomy, physiology, and pathology of the nose and larynx.

rhinolaryngoscope [MED] A scope containing a mirror and a light, used to examine the nose and larynx.

rhinology [MED] The science of the anatomy, functions, and diseases of the nose.

Rhinolophidae [VERT ZOO] The horseshoe bats, a family of insect-eating chiropterans widely distributed in the Eastern Hemisphere and distinguished by extremely complex, horseshoe-shaped nose leaves.

rhinopharyngitis [MED] Inflammation of the nose and pharynx or of the nasopharynx.

rhinophore [INV ZOO] An olfactoreceptor of certain land mollusks, usually borne on a tentacle.

rhinoplasty [MED] A plastic operation on the nose.

Rhinopomatidae [VERT ZOO] The mouse-tailed bats, a small family of insectivorous chiropterans found chiefly in arid regions of northern Africa and southern Asia and characterized by long, wirelike tails and rudimentary nose leaves.

Rhinopteridae [VERT ZOO] The cow-nosed rays, a family of batoid sharks having a fleshy pad at the front end of the head and a well-developed poison spine.

rhinorrhea [MED] **1.** A mucous discharge from the nose. **2.** Escape of cerebrospinal fluid through the nose.

rhinoscleroma [MED] A chronic infectious bacterial disease caused by *Klebsiella rhinoscleromatis* and characterized by hard nodules and plaques of inflamed tissue in the nose and adjacent areas.

rhinoscope [MED] An instrument for examining the nasal cavities.

Rhinotermitidae [INV ZOO] A family of lower termites of the order Isoptera.

rhinotheca [VERT ZOO] The horny sheath on the upper part of a bird's bill.

rhinovirus [VIROL] A subgroup of the picornavirus group including small, ribonucleic acid–containing forms which are not inactivated by ether.

Rhipiceridae [INV ZOO] The cedar beetles, a family of coleopteran insects in the superfamily Elateroidea.

Rhipidistia [VERT ZOO] The equivalent name for Osteolepiformes.

rhipidium [BOT] A fan-shaped inflorescence with cymose branching in which branches lie in the same plane and are suppressed alternately on each side.

rhipidoglossate [INV ZOO] Having a radula with many teeth arranged in a fanlike pattern, as in ear shells.

Rhipiphoridae [INV ZOO] The wedge-shaped beetles, a family of coleopteran insects in the superfamily Meloidea.

rhizanthous [BOT] Producing flowers directly from the root.

rhizautoicous [BOT] Of mosses, having the antheridial branch and the archegonial branch connected by rhizoids.

Rhizobiaceae [MICRO] A family of rod-shaped, gram-negative, soil-inhabiting bacteria consisting of two genera, *Rhizobium* and *Agrobacterium*, in the order Eubacteriales.

rhizocarpous [BOT] Pertaining to perennial herbs having perennating underground parts from which stems and foliage arise annually.

Rhizocephala [INV ZOO] An order of crustaceans which parasitize other crustaceans; adults have a thin-walled sac enclosing the visceral mass and show no trace of segmentation, appendages, or sense organs.

Rhizochloridina [INV ZOO] A suborder of flagellate protozoans in the order Heterochlorida.

rhizocorm [BOT] An underground stem which is a single-jointed rhizome, popularly a bulb.

rhizodermis [BOT] The outermost tissue layer of a root.

Rhizodontidae [PALEON] An extinct family of lobefin fishes in the order Osteolepiformes.

rhizoid [BOT] A rootlike structure which helps to hold the plant to a substrate; found on fungi, liverworts, lichens, mosses, and ferns.

Rhizomastigida [INV ZOO] An order of the protozoan class Zoomastigophorea; all species are microscopic and ameboid, and have one or two flagella.

Rhizomastigina [INV ZOO] The equivalent name for Rhizomastigida.

rhizome [BOT] An underground horizontal stem, often thickened and tuber-shaped, and possessing buds, nodes, and scalelike leaves.

rhizomorph [BOT] A rootlike structure, characteristic of many basidiomycetes, consisting of a mass of densely packed and intertwined hyphae.

rhizomycelium [MYCOL] A rhizoid mycelium that connects the reproductive bodies in certain Phycomycetes.

Rhizophagidae [INV ZOO] The root-eating beetles, a family of minute coleopteran insects in the superfamily Cucujoidea.

rhizophagous [BIOL] Feeding on roots.

Rhizophoraceae [BOT] A family of dicolyledonous plants in the order Cornales distinguished by opposite, stipulate leaves, two ovules per locule, folded or convolute bud petals, and a berry fruit.

rhizophore [BOT] A leafless, downward-growing dichotomous *Selaginella* shoot that has tufts of adventitious roots at the apex.

rhizoplast [CYTOL] A delicate fiber or thread running between the nucleus and the blepharoplast in cells bearing flagella.

rhizopod [INV ZOO] An anastomosing rootlike pseudopodium.

Rhizopodea [INV ZOO] A class of the protozoan superclass Sarcodina in which pseudopodia may be filopodia, lobopodia, or reticulopodia, or may be absent.

rhizosphere [GEOL] The soil region subject to the influence of plant roots and characterized by a zone of increased microbiological activity.

Rhizostomeae [INV ZOO] An order of the class Scyphozoa having the umbrella generally higher than it is wide with the margin divided into many lappets but not provided with tentacles.

rhizotomy [MED] Surgical division of any root, as of a nerve. Also known as radicotomy.

Rhodesian man [PALEON] A type of fossil man inhabiting southern and central Africa during the late Pleistocene; the skull was large and low, marked by massive browridges, with a cranial capacity of 1300 cubic centimeters or less.

Rhodesian trypanosomiasis [MED] A fulminating form of African sleeping sickness caused by *Trypanosoma rhodesiense*, transmitted by the tsetse fly, and characterized by parasitemia, edema, lymphadenitis, and myocarditis. Also known as East African sleeping sickness.

Rhodininae [INV ZOO] A subfamily of limivorous worms in the family Maldanidae.

rhodium [CHEM] A chemical element, symbol Rh, atomic number 45, atomic weight 102.905.

Rhodobacteriineae [MICROBIO] A suborder of the order Pseudomonadales comprising all of the photosynthetic, or phototrophic, bacteria except those of the genus *Rhodomicrobium*.

RHIZOSTOMEAE

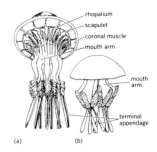

Rhizostomeae. *(a) Rhizostoma;* the bell may reach 2 feet **(61 centimeters) in diameter.** *(b) Mastigias. (From L. Hyman, The Invertebrates, vol. 1, McGraw-Hill, 1940)*

RHODESIAN MAN

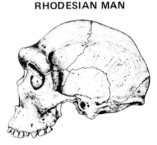

Skull of Rhodesian man *(Homo erectus rhodesiensis),* late Pleistocene. *(From M. F. Ashley Montagu, An Introduction to Physical Anthropology, 2d ed., Charles C. Thomas, 1951)*

RHODIUM

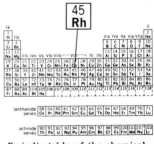

Periodic table of the chemical elements showing the position of rhodium.

RHYNCHOBDELLAE

(a)   (b)

Examples of Rhynchobdellae.
(a) Glossiphonia complanata.
(b) Placobdella parasitica.

RHYNCHOCEPHALIA

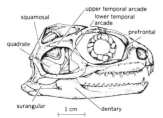

Skull of a living rhynchocephalian,
Sphenodon, a typical diapsid.
*(From A. S. Romer, Vertebrate
Paleontology, 3d ed., University
of Chicago Press, 1966)*

RHYNCHOCOELA

Tubulanus capistratus from the
Pacific coast; size ranges from
5 to 30 millimeters. *(From
L. H. Hyman, The Invertebrates,
vol. 2, McGraw-Hill, 1951)*

Rhodophyceae [BOT] A class of algae belonging to the division or subphylum Rhodophyta.

Rhodophyta [BOT] The red algae, a large diverse phylum or division of plants distinguished by having an abundance of the pigment phycoerythrin.

rhodoplast [BOT] A reddish chromatophore occurring in red algae.

rhodopsin [BIOCHEM] A deep-red photosensitive pigment contained in the rods of the retina of marine fishes and most higher vertebrates. Also known as retinal pigment; visual purple.

rhodoxanthin [BIOCHEM] $C_{40}H_{50}O_2$ A xanthophyll carotenoid pigment.

Rhoipteleaceae [BOT] A monotypic family of dicotyledonous plants in the order Juglandales having pinnately compound leaves, and flowers in triplets with four sepals and six stamens, and the lateral flowers female but sterile.

rhombencephalon [EMBRYO] The most caudal of the primary brain vesicles in the vertebrate embryo. Also known as hindbrain.

Rhombifera [PALEON] An extinct order of Cystoidea in which the thecal canals crossed the sutures at the edges of the plates, so that one-half of any canal lay in one plate and the other half on an adjoining plate.

rhombocoele [EMBRYO] The cavity of the rhombencephalon.

rhombogen [INV ZOO] A form of reproductive individual of the mesozoan order Dicyemida found in the sexually mature host which arises from nematogens and gives rise to free-swimming infusorigens.

rhomboporoid cryptostome [PALEON] Any of a group of extinct bryozoans in the order Cryptostomata that built twiglike colonies with zooecia opening out in all directions from the central axis of each branch.

Rhopalidae [INV ZOO] A family of pentatomorphan hemipteran insects in the superfamily Coreoidea.

rhopalium [INV ZOO] A sense organ found on the margin of a discomedusan.

Rhopalocera [INV ZOO] Formerly a suborder of Lepidoptera comprising those forms with clubbed antennae.

rhopalocercous cercaria [INV ZOO] A free-swimming digenetic trematode larva distinguished by a very wide tail.

Rhopalodinidae [INV ZOO] A family of holothurian echinoderms in the order Dactylochirotida in which the body is flask-shaped, the mouth and anus lying together.

Rhopalosomatidae [INV ZOO] A family of hymenopteran insects in the superfamily Scolioidea.

rhubarb [BOT] *Rheum rhaponticum.* A herbaceous perennial of the order Polygoniales grown for its thick, edible petioles.

Rhynchobdellae [INV ZOO] An order of the class Hirudinea comprising leeches that possess an eversible proboscis and lack hemoglobin in the blood.

Rhynchocephalia [VERT ZOO] An order of lepidosaurian reptiles represented by a single living species, *Sphenodon punctatus*, and characterized by a diapsid skull, teeth fused to the edges of the jaws, and an overhanging beak formed by the upper jaw.

rhynchocoel [INV ZOO] A cavity that holds the inverted proboscis in nemertinean worms.

Rhynchocoela [INV ZOO] A phylum of bilaterally symmetrical, unsegmented, ribbonlike worms having an eversible proboscis and a complete digestive tract with an anus.

rhynchodaeum [INV ZOO] The part of the proboscis lying anterior to the brain in nemertinean worms.

Rhynchodina [INV ZOO] A suborder of ciliate protozoans in the order Thigmotrichida.

Rhynchonellida [INV ZOO] An order of articulate brachiopods; typical forms are dorsibiconvex, the posterior margin is curved, the dorsal interarea is absent, and the ventral one greatly reduced.

rhynchophorous [ZOO] Having a beak.

Rhynchosauridae [PALEON] An extinct family of generally large, stout, herbivorous lepidosaurian reptiles in the order Rhynchocephalea.

rhynchostome [INV ZOO] The anterior terminal opening through which the proboscis is everted, in Nemertea.

Rhynchotheriinae [PALEON] A subfamily of extinct elaphantoid mammals in the family Gomphotheriidae comprising the beak-jawed mastodonts.

Rhyniatae *See* Rhyniopsida

Rhyniophyta [PALEOBOT] A subkingdom of the Embryobionta including the relatively simple, uppermost Silurian-Devonian vascular plants.

Rhyniopsida [PALEOBOT] A class of extinct plants in the subkingdom Rhyniophyta characterized by leafless, usually dichotomously branched stems that bore terminal sporangia. Formerly known as Rhyniatae.

Rhynochetidae [VERT ZOO] A monotypic family of gruiform birds containing only the kagu of New Caledonia.

Rhysodidae [INV ZOO] The wrinkled bark beetles, a family of coleopteran insects in the suborder Adephaga.

rib [ANAT] One of the long curved bones forming the wall of the thorax in vertebrates. [BOT] A primary vein in a leaf.

riboflavin [BIOCHEM] $C_{17}H_{20}N_4O_6$ A water-soluble, yellow-orange fluorescent pigment that is essential to human nutrition as a component of the coenzymes flavin mononucleotide and flavin adenine dinucleotide. Also known as lactoflavin; vitamin B_2; vitamin G.

riboflavin 5′-phosphate [BIOCHEM] $C_{17}H_{21}N_4O_9P$ The phosphoric acid ester of riboflavin. Also known as flavin phosphate; flavin mononucleotide; FMN; isoalloxazine mononucleotide; vitamin B_2 phosphate.

D-ribo-2-ketohexose *See* allulose.

D-riboketose *See* ribulose.

ribonuclease [BIOCHEM] $C_{587}H_{909}N_{171}O_{197}S_{12}$ An enzyme that catalyzes the depolymerization of ribonucleic acid.

ribonucleic acid [BIOCHEM] A long-chain, usually single-stranded nucleic acid consisting of repeating nucleotide units containing four kinds of heterocyclic, organic bases: adenine, cytosine, quanine, and uracil; they are conjugated to the pentose sugar ribose and held in sequence by phosphodiester bonds; involved intracellularly in protein synthesis. Abbreviated RNA.

ribonucleoprotein [BIOCHEM] Any of a large group of conjugated proteins in which molecules of ribonucleic acid are closely associated with molecules of protein.

ribose [BIOCHEM] $C_5H_{10}O_5$ A pentose sugar occurring as a component of various nucleotides, including ribonucleic acid.

riboside [BIOCHEM] Any glycoside containing ribose as the sugar component.

ribosomal ribonucleic acid [BIOCHEM] Any of three large types of ribonucleic acid found in ribosomes: 5S RNA, with molecular weight 40,000; 14-16S RNA, with molecular weight 600,000; and 18-22S RNA with molecular weight 1,200,000. Abbreviated r-RNA.

ribosome [CYTOL] One of the small, complex particles composed of various proteins and three molecules of ribonucleic acid which synthesize proteins within the living cell.

ribulose [BIOCHEM] $C_5H_{10}O_5$ A pentose sugar that exists only as a syrup; synthesized from arabinose by isomerization

RHYNCHONELLIDA

Hypothyridina; posterior view of shell showing posterior curved margin of shell. *(From R. C. Moore, ed., Treatise on Invertebrate Paleontology, pt. H, Geological Society of America, Inc., and University of Kansas Press, 1965)*

RHYNCHOTHERIINAE

Arrangement of tusks in the gomphotheriid *Rhynchotherium.* *(After Osborn)*

RIBULOSE

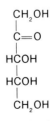

Structural formula of ribulose.

with pyridine; important in carbohydrate metabolism. Also known as D-erythropentose; D-riboketose.

rice [BOT] *Oryza sativa.* An annual cereal grass plant of the order Cyperales, cultivated as a source of human food for its carbohydrate-rich grain.

Richmondian [GEOL] A North American stage of geologic time: Upper Ordovician (above Maysvillian, below Lower Silurian).

Ricinidae [INV ZOO] A family of bird lice, order Mallophaga, which occur on numerous land and water birds.

Ricinuleida [INV ZOO] An order of rare, ticklike arachnids in which the two anterior pairs of appendages are chelate, and the terminal segments of the third legs of the male are modified as copulatory structures.

rickets [MED] A disorder of calcium and phosphorus metabolism affecting bony structures, due to vitamin D deficiency.

Rickettsiaceae [MICROBIO] A family of typhuslike and related agents in the order Rickettsiales.

Rickettsiales [MICROBIO] An order of very small bacteria which are mostly obligate intracellular parasites of animals; placed in the class Schizomycetes.

rickettsialpox [MED] An acute febrile disease caused by the rickettsial organism *Rickettsia akari* and transmitted from the mouse to humans by the mite *Allodermanyssus sanguineus;* characterized by rash, a primary ulcer, and often swelling of glands.

rickettsiosis [MED] Any disease caused by rickettsiae.

rictus [VERT ZOO] The mouth aperture in birds.

Riedel's disease [MED] A form of chronic thyroiditis with irregular localized areas of stony, hard fibrosis.

rifampicin [MICROBIO] An antibacterial and antiviral antibiotic; action depends upon its preferential inhibition of bacterial ribonucleic acid polymerase over animal-cell RNA polymerase.

Rift Valley fever [MED] A toxic generalized febrile virus disease of man and animals in South and East Africa, transmitted by a mosquito, and characterized by headache, photophobia, myalgia, and anorexia.

rigor [MED] **1.** Stiffness. **2.** A chill associated with muscular contraction and tremor.

rigor mortis [PATH] Stiffening and rigidity of the musculature occurring after death, beginning within 5-10 hours, and disappearing after 3-4 days.

rim blight [PL PATH] A fungus disease of tea caused by members of the genus *Cladosporium* and characterized by yellow discoloration of the leaf margins followed by browning.

rind [BOT] **1.** The bark of a tree. **2.** The thick outer covering of certain fruits.

rinderpest [VET MED] An acute, contagious, and often fatal virus disease of cattle, sheep, and goats which is characterized by fever and the appearance of ulcers on the mucous membranes of the intestinal tract.

ring bark [BOT] Tree bark having cylindrical formations of phellogen.

ring canal [INV ZOO] In echinoderms, the circular tube of the water-vascular system that surrounds the esophagous.

ring cell [BOT] A thick-walled cell of the annulus of a fern sporangium.

ring centriole [CYTOL] A perforated disc at the end or in the middle of a spermatozoon through which the axial filament passes.

ring deoxyribonucleic acid *See* circular deoxyribonucleic acid.

ringent [BOT] Having widely separated, gaping lips. [ZOO] Gaping irregularly.

Ringer's solution [CHEM] A solution of 0.86 gram sodium chloride, 0.03 gram potassium chloride, and 0.033 gram calcium chloride in boiled, purified water, used topically as a physiological salt solution.

ring rot [PL PATH] **1.** A fungus disease of the sweet potato root caused by *Rhizopus stolonifer* and marked by rings of dry rot. **2.** A bacterial disease of potatoes caused by *Corynebacterium sepedonicum* and characterized by brown discoloration of the annular vascular tissue.

ring spot [PL PATH] Any of various virus and fungus diseases of plants characterized by the appearance of a discolored, annular lesion.

ring system [ORG CHEM] Arbitrary designation of certain compounds as closed, circular structures, as in the six-carbon benzene ring; common rings have four, five, and six members, either carbon or some combination of carbon, nitrogen, oxygen, sulfur, or other elements.

ring test [IMMUNOL] The simplest of the precipitin tests for antigen-antibody reaction; the solution containing antigen is layered on a solution containing antibody; a white disk or precipitate forms at the point where the two solutions diffuse until optimum concentration for precipitation is reached.

ring vessel [INV ZOO] A part of the water-vascular system in echinoderms; it is the circular canal around the mouth into which the stone canal empties, and from which a radial water vessel traverses to each of five radii.

ringworm [MED] A fungus infection of skin, hair, or nails producing annular lesions with elevated margins. Also known as tinea.

Riodininae [INV ZOO] A subfamily of the lepidopteran family Lycaenidae in which prothoracic legs are nonfunctional in the male.

riparian [BIOL] Living or located on a riverbank.

ripe [BOT] Of fruit, fully developed, having mature seed and so usable as food. [FOR] Of timber or a forest, ready to be cut.

risorius [ANAT] A muscle of the cheek that has its insertion at the angle of the mouth.

Riss [GEOL] **1.** A European stage of geologic time: Pleistocene (above Mindel, below Würm). **2.** The third stage of glaciation of the Pleistocene in the Alps.

Rissoacea [PALEON] An extinct superfamily of gastropod mollusks.

ristocetin [MICROBIO] The generic name for an antibiotic produced by *Nocardia lurida* that is effective in test tube experiments against such microorganisms as *Diplococcus pneumoniae, Streptococcus hemolyticus,* and *Clostridium tetani.*

rivulose [BOT] Marked by irregular, narrow lines.

RNA *See* ribonucleic acid.

roach [INV ZOO] An insect of the family Blattidae; the body is wide and flat, the anterior part of the thorax projects over the head, and antennae are long and filiform, with many segments. Also known as cockroach.

Robertinacea [INV ZOO] A superfamily of marine, benthic foraminiferans in the suborder Rotaliina characterized by a trochospiral or modified test with a wall of radial aragonite, and having bilamellar septa.

Roccilaceae [BOT] A family of fruticose species of Hysterales that grow profusely on trees and rocks along the coastlines of Portugal, California, and western South America.

rock shell [INV ZOO] The common name for a large number of gastropod mollusks composing the family Muridae and

RING VESSEL

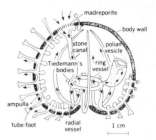

Schematic diagram of water-vascular system of an echinoid showing ring vessel in the lower center. Arrows show direction of water flow.

ROACH

American roach *(Periplantea americana)* has long well-developed wings, and often flies.

ROCK SHELL

The festive rock shell *(Murex festivus).* It has an elaborately ornamented shell and grows over 3 inches (8 centimeters) long.

characterized by having conical shells with various sculpturing.

Rocky Mountain spotted fever [MED] An acute, infectious, typhuslike disease of man caused by the rickettsial organism *Rickettsia rickettsi* and transmitted by species of hard-shelled ticks; characterized by sudden onset of chills, headache, fever, and an exanthem on the extremities. Also known as American spotted fever; tick fever; tick typhus.

rod [HISTOL] One of the rod-shaped sensory bodies in the retina which are sensitive to dim light.

rodent [VERT ZOO] The common name for members of the order Rodentia.

Rodentia [VERT ZOO] An order of mammals characterized by a single pair of ever-growing upper and lower incisors, a maximum of five upper and four lower cheek teeth on each side, and free movement of the lower jaw in an anteroposterior direction.

rodenticide [MATER] A chemical agent used to kill rodents.

rod weeder [AGR] A type of equipment used to prepare the soil during harrowing; it is a power-driven rod, usually square in cross section, which also operates below the surface of loose soil, killing weeds and maintaining the soil in loose mulched condition; adapted to large operations and used in dry areas in the northwestern United States.

roentgen [NUCLEO] An exposure dose of x- or γ-radiation such that the electrons and positrons liberated by this radiation produce, in air, when stopped completely, ions carrying positive and negative charges of 2.58×10^{-4} coulomb per kilogram of air. Abbreviated R (formerly r). Also spelled röntgen.

roentgen equivalent man *See* rem.

roentgen equivalent physical *See* rep.

roentgenography [PHYS] Radiography by means of x-rays.

roentgenotherapy *See* x-ray therapy.

roentgen rays *See* x-rays.

roentgen spectrometry *See* x-ray spectrometry.

roll-tube technique [MICROBIO] A pure-culture technique, employed chiefly in tissue culture, in which, during incubation, the test tubes are held in a wheellike instrument at an angle of about 15° from the horizontal and the wheel is rotated vertically about once every 2 minutes.

Romberg's sign [MED] **1.** A sign for obturator hernia in which there is pain radiating to the knee. **2.** A sign for loss of position sense in which the patient cannot maintain equilibrium when standing with feet together and eyes closed.

röntgen *See* roentgen.

rooster [VERT ZOO] An adult male of certain birds and fowl, such as pheasants and ptarmigans.

root [BOT] The absorbing and anchoring organ of a vascular plant; it bears neither leaves nor flowers and is usually subterranean.

root canal [ANAT] The cavity within the root of a tooth, occupied by pulp, nerves, and vessels.

root cap [BOT] A thick, protective mass of parenchymal cells covering the meristematic tip of the root.

root hair [BOT] One of the hairlike outgrowths of the root epidermis that function in absorption.

root knot [PL PATH] Any of various plant diseases caused by root-knot nematodes which produce gall-like enlargements on the roots.

root-mean-square error [STAT] The square root of the second moment corresponding to the frequency function of a random variable.

root pressure [BOT] The force which makes water rise in the axial stele of a plant.

root rot [PL PATH] Any of various plant diseases characterized by decay of the roots.

rootstock [BOT] A root or part of a root used as the stock for grafting.

rootworm [INV ZOO] **1.** An insect larva that feeds on plant roots. **2.** A nematode that infests the roots of plants.

Roproniidae [INV ZOO] A small family of hymenopteran insects in the superfamily Proctotrupoidea.

Rorschach test [PSYCH] A projective psychologic test in which the subject describes what he imagines seeing in a series of 10 standard inkblots of varying designs and colors.

rosacea [MED] A chronic skin disorder of middle age characterized by redness, papules, and oiliness.

Rosaceae [BOT] A family of dicotyledonous plants in the order Rosales typically having stipulate leaves and perigynous flowers, numerous stamens, and several or many separate carpels.

rosaceous [BOT] **1.** Having five petals arranged in a circle. **2.** Resembling a rose.

Rosales [BOT] A morphologically diffuse order of dicotyledonous plants in the subclass Rosidae.

rose [BOT] A member of the genus *Rosa* in the rose family (Rosaceae); plants are erect, climbing, or trailing shrubs, generally prickly stemmed, and bear alternate, odd-pinnate single leaves.

rosebloom *See* false blossom.

rose hip [BOT] The ripened fruit of a rose plant.

rosemary [BOT] *Rosmarinus officinalis.* A fragrant evergreen of the mint family from France, Spain, and Portugal; leaves have a pungent bitter taste and are used as an herb and in perfumes.

Rosenmueller's organ *See* epoophoron.

roseola infantum *See* exanthem subitum.

roseola typhosa [MED] The rose-colored eruption characteristic of typhus or typhoid fever.

rosette [BIOL] Any structure or marking resembling a rose. [PL PATH] Any of various plant diseases in which the leaves become clustered in the form of a rosette.

rosette organ [INV ZOO] **1.** In certain ascidians, a ventral stolon from which buds are constricted off. **2.** In ctenophores, the ciliated gastrodermal cells surrounding openings between the gastrovascular canals and mesoglea.

Rosidae [BOT] A large subclass of the class Magnoliatae; most have a well-developed corolla with petals separate from each other, binucleate pollen, and ovules usually with two integuments.

rostellum [BIOL] The anterior, flattened region of the scolex of armed tapeworms.

rostral gland [INV ZOO] The labial gland of spiders. [VERT ZOO] The premaxillary portion of the labial gland, as in snakes.

Rostratulidae [VERT ZOO] A small family of birds in the order Charadriiformes containing the painted snipe; females are more brightly colored than males.

rostrum [BIOL] A beak or beaklike process.

rot [PL PATH] Any plant disease characterized by breakdown and decay of plant tissue.

Rotaliacea [INV ZOO] A superfamily of foraminiferans in the suborder Rotaliina characterized by a planispiral or trochospiral test having apertural pores and composed of radial calcite, with secondarily bilamellar septa.

rotate [BOT] Of a sympetalous corolla, having a short tube and petals radiating like the spokes of a wheel.

Rotatoria [INV ZOO] The equivalent name for Rotifera.

röteln *See* rubella.

ROSACEAE

A common eastern American species of wild rose *(Rosa carolina),* in the family Rosaceae of the order Rosales. *(Photograph by A. W. Ambler, from National Audubon Society)*

ROTALIACEA

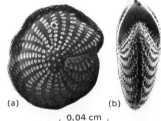

(a)   (b)

0.04 cm

Scanning electron micrograph of *Elphidium,* suborder Rotaliina, superfamily Rotaliacea, from Recent of Adriatic Sea. *(a)* Side view showing large sutural pores that open into canal system. *(b)* Edge view with series of apertural pores at lower margin of final chamber face and imperforate peripheral band. *(R. B. MacAdam, Chevron Oil Field Research Co.)*

Rotifera [INV ZOO] A class of the phylum Aschelminthes distinguished by the corona, a retractile trochal disk provided with several groups of cilia and located on the head.

Rotliegende [GEOL] A European series of geologic time: Lower and Middle Permian.

rough bark [PL PATH] **1.** Any of various virus diseases of woody plants characterized by roughening and often splitting of the bark. **2.** A fungus disease of apples caused by *Phomopsis mali* and characterized by rough cankers on the bark.

rough colony [MICROBIO] A flattened, irregular, and wrinkled colony of bacteria indicative of decreased capsule formation and virulence.

rouleau [PATH] A roll of erythrocytes resembling a stack of coins.

round ligament [ANAT] **1.** A flattened band extending from the fovea on the head of the femur to attach on either side of the acetabular notch between which it blends with the transverse ligament. **2.** A fibrous cord running from the umbilicus to the notch in the anterior border of the liver; represents the remains of the obliterated umbilical vein.

round window [ANAT] A membrane-covered opening between the middle and inner ears in amphibians and mammals through which energy is dissipated after traveling in the membranous labyrinth.

roundworm [INV ZOO] The name applied to nematodes.

Rous sarcoma [VET MED] A fibrosarcoma that can be produced in chickens, pheasants, and ducklings inoculated with the filterable, ribonucleic acid Rous virus.

Rovamycin [MICROBIO] A trade name for spiramycin.

RQ *See* respiratory quotient.

r-RNA *See* ribosomal ribonucleic acid.

RTF *See* resistance transfer factor.

Ru *See* ruthenium.

rubber tree [BOT] *Hevea brasiliensis.* A tall tree of the spurge family (Euphorbiaceae) from which latex is collected and coagulated to produce rubber.

rubella [MED] An infectious viral disease of humans characterized by coldlike symptoms, fever, and transient, generalized pale-pink rash. Its occurrence in early pregnancy is associated with congenital abnormalities.

rubeola *See* measles.

Rubiaceae [BOT] The single family of the plant order Rubiales.

Rubiales [BOT] An order of dicotyledonous plants marked by their inferior ovary, regular or nearly regular corolla, and opposite leaves with interpetiolar stipules or whorled leaves without stipules.

rubidium [CHEM] A chemical element, symbol Rb, atomic number 37, atomic weight 85.47; a reactive alkali metal; salts of the metal may be used in glass and ceramic manufacture.

rubriblast *See* pronormoblast.

rubricyte *See* pronormoblast.

ruby laser [OPTICS] An optically pumped solid-state laser that uses a ruby crystal to produce an intense and extremely narrow beam of coherent red light.

ruddy turnstone [VERT ZOO] *Arenaria interpes.* A member of the avian order Charadriiformes that perform transpacific flights during their migration.

ruderal [ECOL] **1.** Growing on rubbish, or waste or disturbed places. **2.** A plant that thrives in such a habitat.

rudistids [PALEON] Fossil sessile bivalves that formed reefs during the Cretaceous in the southern Mediterranean or the Tethyan belt.

ruff [VERT ZOO] A fringe of hair or feathers on the neck.

RUBBER TREE

Branch of the rubber tree.

RUBIDIUM

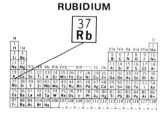

Periodic table of the chemical elements showing the position of rubidium.

Ruffini cylinder [ANAT] A cutaneous nerve ending suspected as the mediator of warmth.

rufous [BOT] Having a reddish-brown color.

ruga [ANAT] A wrinkle, fold, elevation, or ridge of tissue as in the mucosa of the stomach, vagina, and palate.

Rugosa [PALEON] An order of extinct corals having either simple or compound skeletons with internal skeletal structures consisting mainly of three elements, the septa, tabulae, and dissepiments.

rugose [BIOL] Having a wrinkled surface.

rugose mosaic [PL PATH] A virus disease of potatoes marked by dwarfed, wrinkled, and mottled leaves and resulting in premature death.

rumen [VERT ZOO] The first chamber of the ruminant stomach. Also known as paunch.

ruminant [PHYSIO] Characterized by the act of regurgitation and rechewing of food. [VERT ZOO] A mammal belonging to the Ruminantia.

Ruminantia [VERT ZOO] A suborder of the Artiodactyla including sheep, goats, camels, and other forms which have a complex stomach and ruminate their food.

ruminate [BOT] Of endosperm that is mottled, or irregularly ridged and sulcate, looking as if chewed.

rumination [MED] Voluntary regurgitation of food from the stomach, followed by remastication and swallowing in emotionally or mentally disturbed persons. Also known as merycism. [PHYSIO] Regurgitation and remastication of food in preparation for true digestion in ruminants. [PSYCH] An obsessional preoccupation with a single idea or system of ideas.

runcinate [BOT] Pinnately cut with downward-pointing lobes.

runner [BOT] A horizontally growing, sympodial stem system; adventitious roots form near the apex, and a new runner emerges from the axil of a reduced leaf. Also known as stolon.

running bird [VERT ZOO] Any of the large, flightless, heavy birds usually categorized as ratites.

rupestrine [ECOL] Growing or living on rocks.

rupicolous [ECOL] Living among or growing on rocks.

rupture *See* hernia.

Russell bodies [PATH] Hyaline eosinophilic globules 4–5 micrometers in diameter, thought to be particles of antibody globulin, occurring in the cytoplasm of plasma cells in chronic inflammatory exudates.

Russell's viper *See* tic polonga.

rust [PL PATH] Any plant disease caused by rust fungi (Uredinales) and characterized by reddish-brown lesions on the plant parts.

rusty blotch [PL PATH] A fungus disease of barley caused by *Helminthosporium californicum* and characterized by brown blotches on the foliage.

rusty mottle [PL PATH] A virus disease of cherry characterized by retarded development of blossoms and leaves in the spring, followed by necrotic spotting and shot-holing of the foliage with considerable defoliation.

rut [PHYSIO] The period during which the male animal has a heightened mating drive.

rutabaga [BOT] *Brassica napobrassica.* A biennial crucifer of the order Capparales probably resulting from the natural crossing of cabbage and turnip and characterized by a large, edible, yellowish fleshy root.

Rutaceae [BOT] A family of dicotyledonous plants in the order Sapindales distinguished by mostly free stamens and glandular-punctate leaves.

RUFFINI CYLINDER

Longitudinal section through a Ruffini cylinder. *(From F. A. Geldard, Human Senses, Wiley, 1953)*

RUNNING BIRD

A kiwi with typical furlike plumage, vestigial wings, and a long, slender slightly curved bill.

RUTHENIUM

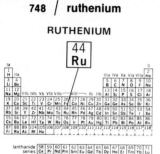

Periodic table of the chemical
elements showing the position
of ruthenium.

ruthenium [CHEM] A chemical element, symbol Ru, atomic number 44, atomic weight 101.07.

rye [BOT] *Secale cereale.* A cereal plant of the order Cyperales cultivated for its grain, which contains the most desirable gluten, next to wheat.

rye buckwheat *See* tartary buckwheat.

Rynchopidae [VERT ZOO] The skimmers, a family of birds in the order Charadriiformes distinguished by a knifelike lower beak that is longer and narrower than the upper one.

Rytiodinae [VERT ZOO] A subfamily of trichechiform sirenians in the family Dugongidae.

S

S   *See* sulfur.

Sabellariidae   [INV ZOO] The sand-cementing worms, a family of polychaete annelids belonging to the Sedentaria and characterized by a compact operculum formed of setae of the first several segments.

Sabellidae   [INV ZOO] A family of sedentary polychaete annelids often occurring in intertidal depths but descending to great abyssal depths; one of two families that make up the feather-duster worms.

Sabellinae   [INV ZOO] A subfamily of the Sabellidae including the most numerous and largest members.

Sabin vaccine   [IMMUNOL] A live-poliovirus vaccine that is administered orally.

sable   [VERT ZOO] *Martes zibellina.* A carnivore of the family Mustelidae; a valuable fur-bearing animal, quite similar to the American marten.

Sabouraud's agar   [MICROBIO] A peptone-maltose agar used as a culture medium for pathogenic fungi, especially the dermatophytes.

sac   [BIOL] A soft-walled cavity within a plant or animal, often containing a special fluid and usually having a narrow opening or none at all.

saccate   [BOT] Having a saclike or pouchlike form.

saccharase   [BIOCHEM] An enzyme that catalyzes the hydrolysis of disaccharide to monosaccharides, specifically of sucrose to dextrose and levulose. Also known as invertase; invertin; sucrase.

Saccharomyces   [MYCOL] A genus of unicellular yeasts in the subfamily Saccharomycetoideae which have simple, budding cells, a pseudomycelium, or both, and one to four round to oval spores per ascus.

Saccharomycetaceae   [MYCOL] The single family of the order Saccharomycetales.

Saccharomycetales   [MYCOL] An order of the subclass Hemiascomycetidae comprising typical yeasts, characterized by the presence of naked asci in which spores are formed by free cells.

Saccharomycetoideae   [MYCOL] A subfamily of Saccharomycetacae in which spores may be hat-, sickle-, or kidney-shaped, or round or oval.

saccharopinuria   [MED] An inborn error of amino acid metabolism characterized by abnormally high levels of saccharopine in the urine.

saccharose   *See* sucrose.

Saccoglossa   [INV ZOO] An order of gastropod mollusks belonging to the Opisthobranchia.

Saccopharyngiformes   [VERT ZOO] Formerly an order of actinopterygian fishes, the gulpers, now included in the Anguilliformes.

Saccopharyngoidei   [VERT ZOO] The gulpers, a suborder of

SABELLARIIDAE

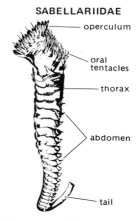

- operculum
- oral tentacles
- thorax
- abdomen
- tail

Sabellaria **in right lateral view.**

SABLE

The sable *(Martes zibellina)*, which has an elongated body reaching 25 inches (60 centimeters) in length, excluding the tail.

actinopterygian fishes in the order Anguilliformes having degenerative adaptations, including loss of swim bladder, opercle, branchiostegal ray, caudal fin, scales, and ribs.

saccular aneurysm [MED] A saclike arterial dilation communicating with the artery by a relatively small opening.

sacculus [ANAT] The smaller, lower saclike chamber of the membranous labyrinth of the vertebrate ear.

saccus *See* vesicle.

sac fungus [MYCOL] The common name for members of the class Ascomycetes.

sacral block [MED] Anesthesia induced by injection of an anesthetic through the caudal hiatus.

sacral nerve [ANAT] Any of five pairs of spinal nerves in the sacral region which innervate muscles and skin of the lower back, lower extremities, and perineum, and branches to the hypogastric and pelvic plexuses.

sacral vertebrae [ANAT] Three to five fused vertebrae that form the sacrum in most mammals; amphibians have one sacral vertebra, reptiles usually have two, and birds have 10–23 fused in the synsacrum.

sacrococcygeus [ANAT] One of two inconstant thin muscles extending from the lower sacral vertebrae to the coccyx.

sacroiliac [ANAT] Pertaining to the sacrum and the ilium.

sacrospinous [ANAT] Pertaining to the sacrum and the spine of the ischium.

sacrum [ANAT] A triangular bone, consisting in man of five fused vertebrae, located below the last lumbar vertebra, above the coccyx, and between the hipbones.

sadism [PSYCH] Sexual perversion in which pleasure is derived from inflicting physical or mental cruelty upon another.

safflower [BOT] *Carthamnus tinctorius.* An annual thistlelike herb belonging to the composite family (Compositae); the leaves are edible, flowers yield dye, and seeds yield a cooking oil.

saffron [BOT] *Crocus sativus.* A crocus of the iris family (Iridaceae); the source of a yellow dye used for coloring food and medicine.

Sagartiidae [INV ZOO] A family of zoantharians in the order Actiniaria.

sage [BOT] *Salvia officinalis.* A half-shrub of the mint family (Labiatae); the leaves are used as a spice.

sagebrush [BOT] Any of various hoary undershrubs of the genus *Artemisia* found on the alkaline plains of the western United States.

Sagenocrinida [PALEON] A large order of extinct, flexible crinoids that occurred from the Silurian to the Permian.

Saghathiinae [PALEON] An extinct subfamily of hyracoids in the family Procaviidae.

sagitta [MATH] The distance between the midpoint of an arc and the midpoint of its chord. [VERT ZOO] The larger of two otoliths in the ear of most fishes.

sagittal [ZOO] In the median longitudinal plane of the body, or parallel to it.

Sagittariidae [VERT ZOO] A family of birds in the order Falconiformes comprising a single species, the secretary bird, noted for its nuchal plumes resembling quill pens stuck behind an ear.

sagittate [BOT] Shaped like an arrowhead, especially referring to leaves.

sagittocyst [INV ZOO] A cyst in the epidermis of certain turbellarians containing a single spindle-shaped needle.

sailfish [VERT ZOO] Any of several large fishes of the genus *Istiophorus* characterized by a very large dorsal fin that is highest behind its middle.

Saint Louis encephalitis [MED] A mosquito-borne arbovirus

SAFFLOWER

Safflower *(Carthamnus tinctorius). (USDA)*

SAGE

Sage *(Salvia officinalis). (USDA)*

infection of the central nervous system, occurring in the central and western United States and in Florida.

Saint Vitus dance [MED] Chorea associated with rheumatic fever. Also known as Sydenham's chorea.

Sakmarian [GEOL] A European stage of geologic time; the lowermost Permian, above Stephanian of Carboniferous and below Artinskian.

salamander [VERT ZOO] The common name for members of the order Urodela.

Salamandridae [VERT ZOO] A family of urodele amphibians in the suborder Salamandroidea characterized by a long row of prevomerine teeth.

Salamandroidea [VERT ZOO] The largest suborder of the Urodela characterized by teeth on the roof of the mouth posterior to the openings of the nostrils.

Salangidae [VERT ZOO] A family of soft-rayed fishes, in the suborder Galaxioidei, which live in estuaries of eastern Asia.

Saldidae [INV ZOO] The shore bugs, a family of predacious hemipteran insects in the superfamily Saldoidea.

Saldoidea [INV ZOO] A superfamily of the hemipteran group Leptopodoidea.

Saleniidae [INV ZOO] A family of echinoderms in the order Salenioida distinguished by imperforate tubercles.

Salenioida [INV ZOO] An order of the Echinacea in which the apical system includes one or several large angular plates covering the periproct.

Salicaceae [BOT] The single family of the order Salicales.

Salicales [BOT] A monofamilial order of dicotyledonous plants in the subclass Dilleniidae; members are dioecious, woody plants, with alternate, simple, stipulate leaves and plumose-hairy mature seeds.

salicylamide [ORG CHEM] $C_6H_4(OH)CONH_2$ Pinkish or white crystals; soluble in alcohol, ether, chloroform, and hot water; melts at 193°C; used in medicine as an analgesic, antipyretic, and antirheumatic drug. Also known as *ortho*-hydroxybenzamide.

salicylate [ORG CHEM] A salt of salicylic acid with the formula $C_6H_4(OH)COOM$, where M is a monovalent metal; for example, $NaC_7H_5O_3$, sodium salicylate.

salicylic acid [ORG CHEM] $C_6H_4(OH)(COOH)$ White crystals with sweetish taste; soluble in alcohol, acetone, ether, benzene, and turpentine, slightly soluble in water; discolored by light; melts at 158°C; used as a chemical intermediate and in medicine, dyes, perfumes, and preservatives. Also known as *ortho*-hydroxybenzoic acid.

salicylism [MED] A syndrome produced by excessive doses of salicylates; characterized by dizziness, headache, and nausea.

Salientia [VERT ZOO] The equivalent name for Anura.

saliva [PHYSIO] The opalescent, tasteless secretions of the oral glands.

salivarium [INV ZOO] In insects, the recess of the preoral cavity into which the salivary duct opens.

salivary amylase *See* ptyalin.

salivary diastase *See* ptyalin.

salivary gland [PHYSIO] A gland that secretes saliva, such as the sublingual or parotid.

salivary gland chromosomes [CYTOL] Polytene chromosomes found in the interphase nuclei of salivary glands in the larvae of Diptera; chromosomes in the larva undergo complete somatic pairing to form two homologous polytene chromosomes fused side by side.

salivation [MED] Mild mercury poisoning suffered by workmen in amalgamation plants. [PHYSIO] Excessive secretion of saliva.

SALICALES

Eastern American cottonwood (*Populus deltoides*), showing female catkins with ripening capsules and cottony seeds. Note characteristic three carpels in opened capsule at lower left; more common number in other species of the order is two. (*Photograph by John H. Gerard, from National Audubon Society*)

SALMON

Coho salmon *(Oncorhynchus kisutch)*.

SALPIDA

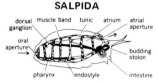

Salp, *Thalia democratica* (Salpida), solitary asexual form. *(After Metcalf)*

Salk vaccine [IMMUNOL] A killed-virus vaccine administered for active immunization against poliomyelitis.

salmon [VERT ZOO] The common name for a number of fish in the family Salmonidae which live in coastal waters of the North Atlantic and North Pacific and breed in rivers tributary to the oceans.

Salmonella [MICROBIO] A genus of gram-negative, rod-shaped pathogenic bacteria of the family Enterobacteriaceae that are usually motile by peritrichous flagella and lack proteolytic enzymes. Also known as *Eberthella.*

Salmonelleae [MICROBIO] A tribe of the Enterobacteriaceae comprising the pathogenic genera *Salmonella* and *Shigella.*

salmonellosis [MED] Infection with any species of *Salmonella.*

Salmonidae [VERT ZOO] A family of soft-rayed fishes in the suborder Salmonoidei including the trouts, salmons, white-fishes, and graylings.

Salmoniformes [VERT ZOO] An order of soft-rayed fishes comprising salmon and their allies; the stem group from which most higher teleostean fishes evolved.

Salmonoidei [VERT ZOO] A suborder of the Salmoniformes comprising forms having an adipose fin.

Salmopercae [VERT ZOO] An equivalent name for Percopsiformes.

Salpida [INV ZOO] An order of tunicates in the class Thaliacea including transparent forms ringed by muscular bands.

Salpingidae [INV ZOO] The narrow-waisted bark beetles, a family of coleopteran insects in the superfamily Tenebrionoidea.

salpingitis [MED] 1. Inflammation of the fallopian tube. 2. Inflammation of the eustachian tube.

salpingo-oophoritis [MED] Inflammation of the fallopian tubes and ovaries.

salsoline [PHARM] $C_{11}H_{15}NO_2$ A compound that crystallizes from alcohol solution, melts at 221°C, soluble in hot alcohol and chloroform; used in medicine as an antihypertensive agent.

salt [CHEM] The reaction product when a metal displaces the hydrogen of an acid; for example, $H_2SO_4 + 2NaOH \rightarrow Na_2SO_4$ (a salt) + $2H_2O$.

saltatorial [ZOO] Adapted for leaping.

salt bridge [PHYS CHEM] A bridge of a salt solution, usually potassium chloride, placed between the two half-cells of a galvanic cell, either to reduce to a minimum the potential of the liquid junction between the solutions of the two half-cells or to isolate a solution under study from a reference half-cell and prevent chemical precipitations.

salt gland [VERT ZOO] A compound tubular gland, located around the eyes and nasal passages in certain marine turtles, snakes, and birds, which copiously secretes a watery fluid containing a high percentage of salt.

Salticidea [INV ZOO] The jumping spiders, a family of predacious arachnids in the suborder Dipneumonomorphae having keen vision and rapid movements.

saltmarsh [ECOL] A maritime habitat found in temperate regions, but typically associated with tropical and subtropical mangrove swamps, in which excess sodium chloride is the predominant environmental feature.

salt-spray climax [ECOL] A climax community along exposed Atlantic and Gulf seacoasts composed of plants able to tolerate the harmful effects of salt picked up and carried by onshore winds from seawater.

Salvarsan [PHARM] A trademark for arsephenamine, an early antisyphilitic drug.

Salviniales [BOT] A small order of heterosporous, leptosporangiate ferns (division Polypodiophyta) which float on the surface of the water.

samara [BOT] A dry, indehiscent, winged fruit usually containing a single seed, such as sugar maple (*Acer saccharum*).

samarium [CHEM] Group III rare-earth metal, atomic number 62, symbol Sm; melts at 1350°C, tarnishes in air, ignites at 200–400°C.

Sambonidae [INV ZOO] A family of pentastomid arthropods in the suborder Porocephaloidea of the order Porocephalida.

sample [SCI TECH] Representative fraction of material tested or analyzed in order to determine the nature, composition, and percentage of specified constituents, and possibly their reactivity. [STAT] A selection of a certain collection from a larger collection.

sample splitter [ENG] An instrument, generally constructed of acrylic resin, designed to subdivide a total sample of marine plankton while maintaining a quantitatively correct relationship between the various phyla in the sample.

Samythinae [INV ZOO] A subfamily of sedentary polychaete annelids in the family Ampharetidae having a conspicuous dorsal membrane.

Sandalidae [INV ZOO] The equivalent name for Rhipiceridae.

sandalwood [BOT] 1. Any species of the genus *Santalum* of the sandalwood family (Santalaceae) characterized by a fragrant wood. 2. *S. album.* A parasitic tree with hard, close-grained, aromatic heartwood used in ornamental carving and cabinetwork.

sand dollar [INV ZOO] The common name for the flat, disk-shaped echinoderms belonging to the order Clypeasteroida.

sandfly [INV ZOO] Any of various small biting Diptera, especially of the genus *Phlebotomus*, which are vectors for phlebotomus (sandfly) fever.

sandfly fever *See* phlebotomus fever.

sand hopper [INV ZOO] The common name for gammaridean crustaceans found on beaches.

sandpiper [VERT ZOO] Any of various small birds that are related to plovers and that frequent sandy and muddy shores in temperate latitudes; bill is moderately long with a soft, sensitive tip, legs and neck are moderately long, and plumage is streaked brown, gray, or black above and is white below.

sand shark [VERT ZOO] Any of various shallow-water predatory elasmobranchs of the family Carchariidae. Also known as tiger shark.

sane [PSYCH] Of sound mind.

Sanfilippo's syndrome [MED] A hereditary metabolic disorder, transmitted as an autosomal recessive, characterized by excessive amounts of heparitin sulfate in the urine, and manifested by minor skeletal changes and slight hepatomegaly.

sanguivorous [ZOO] Feeding on blood.

sanidaster [INV ZOO] A rod-shaped spicule having spines at intervals along its length.

San Joaquin Valley fever *See* coccidioidomycosis.

SA node *See* sinoauricular node.

Santalaceae [BOT] A family of parasitic dicotyledonous plants in the order Santalales characterized by dry or fleshy indehiscent fruit, plants with chlorophyll, petals absent, and ovules without integument.

Santalales [BOT] An order of dicotyledonous plants in the subclass Rosidae characterized by progressive adaptation to parasitism, accompanied by progressive simplification of the ovules.

sap [BOT] The fluid part of a plant which circulates through

SAMARIUM

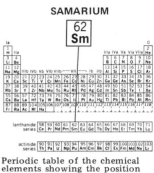

Periodic table of the chemical elements showing the position of samarium.

SANDALWOOD

Sandalwood branch with foliage and fruit.

SAND DOLLAR

Sand dollar. The nearly circular body may reach 3 inches (7.6 centimeters) in diameter. The characteristic pattern of a five-part flower is seen on the aboral surface.

the vascular system and is composed of water, gases, salts, and organic products of metabolism.

saphenous nerve [ANAT] A somatic sensory nerve arising from the femoral nerve and innervating the skin of the medial aspect of the leg, foot, and knee joint.

Sapindaceae [BOT] A family of dicotyledonous plants in the order Sapindales distinguished by mostly alternate leaves, usually one and less often two ovules per locule, and seeds lacking endosperm.

Sapindales [BOT] An order of mostly woody dicotyledonous plants in the subclass Rosidae with compound or lobed leaves and polypetalous, hypogynous to perigynous flowers with one or two sets of stamens.

sapling [BOT] A young tree with a trunk less than 4 inches (10 centimeters) in diameter at a point approximately 4 feet (1.2 meters) above the ground.

saponin [ORG CHEM] Any of numerous plant glycosides characterized by foaming in water and by producing hemolysis when water solutions are injected into the bloodstream; used as beverage foam producer, textile detergent and sizing, soap substitute, and emulsifier.

Sapotaceae [BOT] A family of dicotyledonous plants in the order Ebenales characterized by a well-developed latex system.

saprobic [BOT] Living on decaying organic matter; applied to plants and microorganisms.

saprogen [BIOL] An organism that lives on nonliving organic matter.

Saprolegniales [MYCOL] An order of aquatic fungi belonging to the class Phycomycetes, having a mostly hyphal thallus and zoospores with two flagella.

saprophage [BIOL] An organism that lives on decaying organic matter.

saprophyte [BOT] A plant that lives on decaying organic matter.

saprovore [ZOO] A detritus-eating animal.

saprozoic [ZOO] Feeding on decaying organic matter; applied to animals.

sapwood [BOT] The younger, softer, outer layers of a woody stem, between the cambium and heartwood. Also known as alburnum.

Sapygidae [INV ZOO] A family of hymenopteran insects in the superfamily Scolioidea.

sarcenchyma [BOT] Parenchyma with sparse, granular ground substance.

sarcocarp [BOT] The fleshy part of a fruit.

sarcochore [BOT] A plant dispersing minute, light disseminules.

sarcoderm [BOT] The layer of fleshy tissue between a seed and its external covering.

Sarcodina [INV ZOO] A superclass of Protozoa in the subphylum Sarcomastigophora in which movement involves protoplasmic flow, often with recognizable pseudopodia.

sarcoglia [CYTOL] The protoplasm occurring at a myoneural junction.

sarcoidosis [MED] A disease of unknown etiology characterized by granulomatous lesions, somewhat resembling true tubercles, but showing little or no necrosis, affecting the lymph nodes, skin, liver, spleen, heart, skeletal muscles, lungs, bones in distal parts of the extremities (osteitis cystica of Jüngling), and other structures, and sometimes by hyperglobulinemia, cutaneous anergy, and hypercalcinuria.

sarcolemma [HISTOL] The thin connective tissue sheath enveloping a muscle fiber.

sarcoleukemia *See* leukosarcoma.

sarcoma [MED] A malignant tumor arising in connective tissue and composed principally of anaplastic cells that resemble those of supportive tissues.

sarcoma botryoides [MED] A malignant mesenchymoma that forms grapelike structures; most common in the vagina of infants.

Sarcomastigophora [INV ZOO] A subphylum of Protozoa comprising forms that possess flagella or pseudopodia or both.

sarcomatosis [MED] Multiple growths resembling a sarcoma and found in various parts of the body.

sarcomatrix [INV ZOO] The protoplasmic zone of digestion and assimilation in a radiolarian.

sarcomere [HISTOL] One of the segments defined by Z disks in a skeletal muscle fibril.

Sarcophagidae [INV ZOO] A family of the myodarian orthorrhaphous dipteran insects in the subsection Calypteratae comprising flesh flies, blowflies, and scavenger flies.

sarcophagous [BIOL] Feeding on flesh.

sarcoplasm [HISTOL] Hyaline, semifluid interfibrillar substance of striated muscle tissue.

sarcoplasmic reticulum [CYTOL] Collectively, the cysternae of a single muscle fiber.

Sarcopterygii [VERT ZOO] A subclass of Osteichthyes, including Crossopterygii and Dipnoi in some systems of classification.

Sarcoptiformes [INV ZOO] A suborder of the Acarina including minute globular mites without stigmata.

sarcosoma [INV ZOO] The fleshy portion of an anthozoan.

Sarcosporida [INV ZOO] An order of Protozoa of the class Haplosporea which comprises parasites in skeletal and cardiac muscle of vertebrates.

sarcosporidiosis [VET MED] A disease of mammals other than man caused by muscle infestation by sporozoans of the order Sarcosporida.

sarcostyle [INV ZOO] A fibril or column of muscular tissue.

sarcotubule [CYTOL] A tubular invagination of a muscle fiber.

sardine [VERT ZOO] **1.** *Sardina pilchardus.* The young of the pilchard, a herringlike fish in the family Clupeidae found in the Atlantic along the European coasts. **2.** The young of any of various similar and related forms which are processed and eaten as sardines.

sarkomycin [MICROBIO] $C_7H_8O_3$ An antibiotic produced by an actinomycete which acts as a carcinolytic agent.

Sarmatian [GEOL] A European stage of geologic time: the upper Miocene, above Tortonian, below Pontian.

sarmentocymarin [BIOCHEM] A cardioactive, steroid glycoside from the seeds of *Strophanthus sarmentosus;* on hydrolysis it yields sarmentogenin and sarmentose.

sarmentogenin [BIOCHEM] $C_{23}H_{34}O_5$ The steroid aglycon of sarmentocymarin; isometric with digitoxigenin, and characterized by a hydroxyl group at carbon number 11.

sarmentose [BOT] Producing slender, prostrate stems or runners.

sarmentum [BOT] The thin stem of a climber or runner.

Sarothriidae [INV ZOO] The equivalent name for Jacobsoniidae.

Sarraceniaceae [BOT] A small family of dicotyledonous plants in the order Sarraceniales in which leaves are modified to form pitchers, placentation is axile, and flowers are perfect with distinct filaments.

Sarraceniales [BOT] An order of dicotyledonous herbs or shrubs in the subclass Dilleniidae; plants characteristically

SARRACENIACEAE

An eastern American species of the pitcher plant *(Sarracenia purpurea)*, in the family Sarraceniaceae. *(Photograph of Henry M. Mayer, from National Audubon Society)*

SARSAPARILLA

Smilax aristolochiaefolia, which yields a flavoring material known as Mexican sarsaparilla.

SASSAFRAS

Sassafras albidum, twig, terminal bud, and leaf.

have alternate, simple leaves that are modified for catching insects, and grow in waterlogged soils.

sarsaparilla [BOT] Any of various tropical American vines of the genus *Smilax* (family Liliaceae) found in dense, moist jungles; a flavoring material used in medicine and soft drinks is obtained from the dried roots of at least four species.

sartorius [ANAT] A large muscle originating in the anterior superior iliac spine and inserting in the tibia; flexes the hip and knee joints, and rotates the femur laterally.

sassafras [BOT] *Sassafras albidum.* A medium-sized tree of the order Magnoliales recognized by the bright-green color and aromatic odor of the leaves and twigs.

satellite [CYTOL] A chromosome segment distant from but attached to the rest of the chromosome by an achromatic filament.

satellite band [MOL BIO] A fraction of the deoxyribonucleic acid (DNA) of an organism which has a different density from the rest and is therefore separable as a band in density gradient centrifugation; these bands are usually made up of highly repetitive sequences of DNA.

satellite cell [HISTOL] One of the neurilemmal cells surrounding nerve cells in the peripheral nervous system.

satellitosis [MED] A condition, associated with inflammatory and degenerative diseases of the central nervous system, in which satellite cells increase around the nerve cells.

saturated hydrocarbon [ORG CHEM] A saturated carbon-hydrogen compound with all carbon bonds filled; that is, there are no double or triple bonds as in olefins and acetylenics.

Saturniidae [INV ZOO] A family of medium- to large-sized moths in the superfamily Saturnioidea including the giant silkworms, characterized by reduced, often vestigial, mouthparts and strongly bipectinate antennae.

Saturnioidea [INV ZOO] A superfamily of medium- to very-large-sized moths in the suborder Heteroneura having the frenulum reduced or absent, reduced mouthparts, no tympanum, and pectinate antennae.

Satyrinae [INV ZOO] A large, cosmopolitan subfamily of lepidopterans in the family Nymphalidae, containing the wood nymphs, meadow browns, graylings, and arctics, characterized by bladderlike swellings at the bases of the forewing veins.

Sauria [VERT ZOO] The lizards, a suborder of the Squamata, characterized generally by two or four limbs but sometimes none, movable eyelids, external ear openings, and a pectoral girdle.

Saurichthyidae [PALEON] A family of extinct chondrostean fishes bearing a superficial resemblance to the Aspidorhynchiformes.

Saurischia [PALEON] The lizard-hipped dinosaurs, an order of extinct reptiles in the subclass Archosauria characterized by an unspecialized, three-pronged pelvis.

Sauropoda [PALEON] A group of fully quadrupedal, seemingly herbivorous dinosaurs from the Jurassic and Cretaceous periods in the suborder Sauropodomorpha; members had small heads, spoon-shaped teeth, long necks and tails, and columnar legs.

Sauropodomorpha [PALEON] A suborder of extinct reptiles in the order Saurischia, including large, solid-limbed forms.

Sauropterygia [PALEON] An order of Mesozoic marine reptiles in the subclass Euryapsida.

Saururaceae [BOT] A family of dicotyledonous plants in the order Piperales distinguished by mostly alternate leaves, two to ten ovules per carpel, and carpels distinct or united into a compound ovary.

savanna [ECOL] Any of a variety of physiognomically or

environmentally similar vegetation types in tropical and extratropical regions; all contain grasses and one or more species of trees of the families Leguminosae, Bombacaceae, Bignoniaceae, or Dilleniaceae.

savanna-woodland *See* tropical woodland.

sawfish [VERT ZOO] Any of several elongate viviparous fishes of the family Pristidae distinguished by a dorsoventrally flattened elongated snout with stout toothlike projections along each edge.

saxicolous [ECOL] Living or growing among rocks.

Saxifragaceae [BOT] A family of dicotyledonous plants in the order Rosales which are scarcely or not at all succulent and have two to five carpels usually more or less united, and leaves not modified into pitchers.

Sb *See* antimony.

Sc *See* scandium.

scab [MED] Crusty exudate covering a wound or ulcer during the healing process.

scabies [MED] A contagious skin disorder caused by the mite *Sarcoptes scabiei* burrowing beneath the skin, causing the formation of multiform lesions with intense itching.

scabrous [BIOL] Having a rough surface covered with stiff hairs or scales.

scala [ANAT] A subdivision of the cavity of the cochlea; especially, one of the perilymphatic spaces.

scala media [ANAT] The middle channel of the cochlea, filled with endolymph and bounded above by Reissner's membrane and below by the basilar membrane. Also known as cochlear duct.

scalariform [BIOL] Resembling a ladder; having transverse markings or bars.

scala tympani [ANAT] The lowest channel in the cochlea of the ear; filled with perilymph.

scala vestibuli [ANAT] The uppermost channel of the cochlea; filled with perilymph.

scale [BOT] The bract of a catkin. [VERT ZOO] A flat calcified or cornified platelike structure on the skin of most fishes and of some tetrapods.

scale insect [INV ZOO] Any of various small, structurally degenerate homopteran insects in the superfamily Coccoidea which resemble scales on the surface of a host plant.

scalenus [ANAT] One of three muscles in the neck, arising from the transverse processes of the cervical vertebrae, and inserted on the first two ribs.

scale scar [BOT] A mark left on a stem after bud scales have fallen off.

Scalibregmidae [INV ZOO] A family of mud-swallowing worms belonging to the Sedentaria and found chiefly in sublittoral and great depths.

scallion *See* shallot.

scallop [INV ZOO] Any of various bivalve mollusks in the family Pectinidae distinguished by radially ribbed valves with undulated margins.

scalp [ANAT] The integument that covers the cranium.

scalpel [DES ENG] A small, straight, very sharp knife (or detachable blade for a knife), used for dissecting.

Scalpellidae [INV ZOO] A primitive family of barnacles in the suborder Lepadomorpha having more than five plates.

scaly bark *See* psorosis.

scaly leg [VET MED] A highly contagious disease of poultry caused by the sarcoptid mite *Knemidokoptes mutans.*

scandent [BOT] Climbing by stem-roots or tendrils.

scandium [CHEM] A metallic group III element, symbol Sc, atomic number 21; melts at 1200°C; found associated with rare-earth elements.

SCALE SCAR

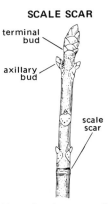

Position of scale scar on the twig of a buckeye tree.

SCALLOP

Typical ribbed valve of a scallop.

SCANDIUM

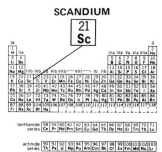

Periodic table of the chemical elements showing the position of scandium.

scanning electron microscope [ELECTR] A type of electron microscope in which a beam of electrons, a few hundred angstroms in diameter, systematically sweeps over the specimen; the intensity of secondary electrons generated at the point of impact of the beam on the specimen is measured, and the resulting signal is fed into a cathode-ray-tube display which is scanned in synchronism with the scanning of the specimen.

scansorial [BOT] Adapted for climbing.

Scapanorhychidae [VERT ZOO] The goblin sharks, a family of deep-sea galeoids in the isurid line having long, sharp teeth and a long, pointed rostrum.

scape [BOT] A peduncle arising at ground level or underground in acaulescent plants. [INV ZOO] **1.** The peduncle of the bananeer of Diptera. **2.** The basal joint of an antenna, when longer than the other joints, in insects. [VERT ZOO] The shaft of a feather.

scapha [ANAT] The furrow of the auricle between the helix and the antihelix.

Scaphidiidae [INV ZOO] The shining fungus beetles, a family of coleopteran insects in the superfamily Staphylinoidea.

scaphocephaly [MED] A condition of the skull characterized by elongation and narrowing, and a projecting, keellike sagittal suture, caused by its premature closure.

scaphocerite [INV ZOO] The scalelike exopodite of the second antenna in decapods.

scaphognathite [INV ZOO] The epipodite of the second maxilla of decapods; regulates the flow of water through the respiratory chamber.

scaphoid [ANAT] A boat-shaped bone of the carpus.

Scaphopoda [INV ZOO] A class of the phylum Mollusca in which the soft body fits the external, curved and tapering, nonchambered, aragonitic shell which is open at both ends.

scapula [ANAT] The large, flat, triangular bone forming the back of the shoulder. Also known as shoulder blade.

scapulus [INV ZOO] A modified submarginal region in some sea anemones.

scapus [BIOL] The stem, shaft, or column of a structure.

scar [MED] A permanent mark on the skin or other tissue, formed from connective-tissue replacement of tissue destroyed by a wound or disease process.

SCARABAEIDAE

A drawing of a lamellicorn beetle. *(From T. I. Storer and R. L. Usinger, General Zoology, 3d ed., McGraw-Hill, 1957)*

Scarabaeidae [INV ZOO] The lamellicorn beetles, a large cosmopolitan family of coleopteran insects in the superfamily Scarabaeoidea including the Japanese beetle and other agricultural pests.

Scarabaeoidea [INV ZOO] A superfamily of Coleoptera belonging to the suborder Polyphaga.

scarabiasis [MED] Invasion of the intestine by the dung beetle, characterized by anorexia, emaciation, and disturbance of the gastrointestinal tract.

Scaridae [VERT ZOO] The parrotfishes, a family of perciform fishes in the suborder Percoidei which have the teeth of the jaw generally coalescent.

scarification [MED] The operation of making numerous small, superficial incisions in skin or other tissue.

scarious [BOT] Having a thin, membranous texture.

scarlet fever [MED] An acute, contagious bacterial disease caused by *Streptococcus hemolyticus;* characterized by a papular, or rough, bright-red rash over the body, with fever, sore throat, headache, and vomiting occurring 2-3 days after contact with a carrier.

scarlet fever streptococcus antitoxin [IMMUNOL] A sterile aqueous solution of antitoxins obtained from the blood of animals immunized against group A beta hemolytic streptococci toxin; formerly used in the treatment of, and to produce immunity against, scarlet fever.

scarlet fever streptococcus toxin [IMMUNOL] Toxic filtrate of cultures of *Streptococcus pyogenes* responsible for the characteristic rash of scarlet fever; the toxin is used in the Dick test.

Scarpa's fascia [ANAT] The deep, membranous layer of the superficial fascia of the lower abdomen.

scar tissue [MED] Contracted, dense connective tissue that is formed by the healing process of a wound or diseased tissue.

Scatopsidae [INV ZOO] The minute black scavenger flies, a family of orthorrhaphous dipteran insects in the series Nematocera.

scavenger [ECOL] An organism that feeds on carrion, refuse, and similar matter.

Scelionidae [INV ZOO] A family of small, shining wasps in the superfamily Proctotrupoidea, characterized by elbowed, 11- or 12-segmented antennae.

scent gland [VERT ZOO] A specialized skin gland of the tubuloalveolar or acinous variety which produces substances having peculiar odors; found in many mammals.

Schardinger dextrin [BIOCHEM] Any of several water-soluble, nonreducing, dextrorotatory polysaccharides that have a low molecular weight and that are obtained by the action of *Bacillus macerans* on a starch solution.

Scheie's syndrome [MED] A hereditary disease transmitted as an autosomal recessive and characterized by high levels of chondroitin sulfate B in the urine, mild distortion of the facies, hypertrichosis, clouding of the cornea, and aortic valve disease.

schemochrome [ZOO] A feather color that originates within the feather structures, through refraction of light independent of pigments.

Schenck's disease *See* sporotrichosis.

Scheuermann's disease *See* osteochondrosis.

Schick test [IMMUNOL] A skin test for determining susceptibility to diphtheria performed by the intradermal injection of diluted diphtheria toxin; a positive reaction, showing edema and scaling after 5 to 7 days, indicates lack of immunity.

Schilder's disease [MED] **1.** A retrogressive disease of the white matter in the central nervous system characterized by diffuse loss of myelin. **2.** Any of the progressive degenerative diseases of the white matter in the central nervous system.

Schindleriidae [VERT ZOO] The single family of the order Schindlerioidei.

Schindlerioidei [VERT ZOO] A suborder of fishes in the order Perciformes composed of one monogeneric family comprising two tiny oceanic species that are transparent and neotenic.

schindylesis [ANAT] A synarthrosis in which a plate of one bone is fixed in a fissure of another.

Schistosoma [INV ZOO] A genus of blood flukes infecting man.

schistosome dermatitis [MED] A dermatitis caused by penetration of the skin by certain schistosome cercariae. Also known as swamp itch; swimmer's itch.

schistosomiasis [MED] A disease in which humans are parasitized by any of three species of blood flukes: *Schistosoma mansoni, S. haematobium,* and *S. japonicum;* adult worms inhabit the blood vessels. Also known as snail fever.

Schistostegiales [BOT] A monospecific order of mosses; the small, slender, glaucous plants are distinguished by the luminous protonema.

schizaxon [ANAT] An axon that divides, in its course, into equal or nearly equal branches.

schizocarp [BOT] A dry fruit that separates at maturity into single-seeded indehiscent carpels.

Schizocoela [INV ZOO] A group of animal phyla, including Bryozoa, Brachiopoda, Phoronida, Sipunculoidea, Echiuroidea, Priapuloidea, Mollusca, Annelida, and Arthropoda, all characterized by the appearance of the coelom as a space in the embryonic mesoderm.

schizodont [INV ZOO] A multinucleate trophozoite that segments into merozoites.

schizogamy [BIOL] A form of reproduction involving division of an organism into a sexual and an asexual individual.

schizogenesis [BIOL] Reproduction by fission.

schizogenous [BOT] Formed by a splitting or separation of adjacent cell walls.

schizognathous [VERT ZOO] Descriptive of birds having a palate in which the vomer is small and pointed, the maxillo-palatines are not united with each other or with the vomer, and the palatines articulate posteriorly with the rostrum.

Schizogoniales [BOT] A small order of the Chlorophyta containing algae that are either submicroscopic filaments or macroscopic ribbons or sheets a few centimeters wide and attached by rhizoids to rocks.

schizogony [INV ZOO] Asexual reproduction by multiple fission of a trophozoite; a characteristic of certain Sporozoa.

Schizomeridaceae [BOT] A family of green algae in the order Ulvales.

Schizomycetes [MICROBIO] A class of the division Protophyta which includes the bacteria.

schizont [INV ZOO] A multinucleate cell in certain members of the Sporozoa that is produced from a trophozoite in a cell of the host, and that segments into merozoites.

schizontocyte [INV ZOO] One of the cytomeres into which a schizont divides, and which then divides into a cluster of merozoites.

Schizopathidae [INV ZOO] A family of dimorphic zoantharians in the order Antipatharia.

schizopelmous [VERT ZOO] Having the two flexor tendons of the toes separate, as in certain birds.

Schizophora [INV ZOO] A series of the dipteran suborder Cyclorrhapha in which adults possess a frontal suture through which a distensible sac, or ptilinum, is pushed to help the organism escape from its pupal case.

schizophrenia [PSYCH] A group of mental disorders characterized by withdrawal from reality and by alterations in thinking, feeling, and concept formations. Also known as dementia praecox.

Schizophyceae [MICROBIO] The blue-green algae, a class of the division Protophyta.

Schizophyta [BOT] The prokaryotes, a division of the plant subkingdom Thallobionta; includes the bacteria and blue-green algae.

schizopod [INV ZOO] **1.** Having the limbs split so that each has an endopodite and an exopodite, as in certain crustaceans. **2.** A biramous appendage.

Schizopteridae [INV ZOO] A family of minute ground-inhabiting hemipterans in the group Dipsocoeoidea; individuals characteristically live in leaf mold.

schizorhinal [VERT ZOO] Having a deep cleft on the posterior margin of the osseous external nares, as in certain birds.

schizothecal [VERT ZOO] Having the horny envelope of the tarsus divided into scalelike plates; refers to most birds.

schizothoracic [INV ZOO] Having a prothorax that is large and loosely articulated with the thorax.

Schlemm's canal [ANAT] A space or series of spaces at the junction of the sclera and cornea in the eye; drains aqueous humor from the anterior chamber.

Schneiderian membrane [ANAT] The mucosa lining the nasal cavities and paranasal sinuses.

Schneider's index [MED] A test of general physical and circulatory efficiency, consisting of pulse and blood pressure observations under standard conditions of rest and exercise.

Schoenbiinae [INV ZOO] A subfamily of moths in the family Pyralididae, including the genus *Acentropus*, the most completely aquatic Lepidoptera.

Schubertellidae [PALEON] An extinct family of marine protozoans in the superfamily Fusulinacea.

Schultz-Charlton test [IMMUNOL] A skin test for the diagnosis of scarlet fever, performed by the intradermal injection of human scarlet fever immune serum; a positive reaction consists of blanching of the rash in the area surrounding the point of injection.

Schultz-Dale reaction [IMMUNOL] A method for demonstrating anaphylactic hypersensitivity outside the body by suspending an excised intestinal loop or uterine strip from a sensitized animal in an oxygenated, physiological salt solution; addition of the proper allergen causes contraction of the smooth muscle.

Schwagerinidae [PALEON] A family of fusulinacean protozoans that flourished during the Early and Middle Pennsylvanian and became extinct during the Late Permian.

Schwann cell [HISTOL] One of the cells that surround peripheral axons forming sheaths of the neurilemma.

schwannoma *See* neurilemmoma.

Sciaenidae [VERT ZOO] A family of perciform fishes in the suborder Percoidei, which includes the drums.

sciatica [MED] Neuralgic pain in the lower extremities, hips, and back caused by inflammation or injury to the sciatic nerve.

sciatic nerve [ANAT] Either of a pair of long nerves that originate in the lower spinal cord and send fibers to the upper thigh muscles and the joints, skin, and muscles of the leg.

science [SCI TECH] A branch of study in which facts are observed and classified, and, usually, quantitative laws are formulated and verified; involves the application of mathematical reasoning and data analysis to natural phenomena.

scientific method [SCI TECH] The systematic collection and classification of data and, usually, the formulation and testing of hypotheses based on the data.

Scincidae [VERT ZOO] The skinks, a family of the reptilian suborder Sauria which have reduced limbs and snakelike bodies.

Scinidae [INV ZOO] A family of bathypelagic, amphipod crustaceans in the suborder Hyperiidea.

Sciomyzidae [INV ZOO] A family of myodarian cyclorrhaphous dipteran insects in the subsection Acalypteratae.

scion [BOT] A section of a plant, usually a stem or bud, which is attached to the stock in grafting.

sciophilous [ECOL] Capable of thriving in shade.

sciophyte [BOT] A plant that thrives at lowered light intensity.

scirrhous carcinoma [MED] A hard, poorly differentiated adenocarcinoma in which the anaplastic cells are surrounded by dense bundles of collagenous fibers.

Scitaminales [BOT] An equivalent name for Zingiberales.

Scitamineae [BOT] An equivalent name for Zingiberales.

Sciuridae [VERT ZOO] A family of rodents including squirrels, chipmunks, marmots, and related forms.

Sciuromorpha [VERT ZOO] A suborder of Rodentia according to the classical arrangement of the order.

SCLERACTINIA

Solitary coral polyps, *Oulangia* species. *(After S. Hickson)*

sclera [ANAT] The hard outer coat of the eye, continuous with the cornea in front and with the sheath of the optic nerve behind.

Scleractinia [INV ZOO] An order of the subclass Zoantharia which comprises the true, or stony, corals; these are solitary or colonial anthozoans which attach to a firm substrate.

scleratogenous layer [VERT ZOO] A layer of fused sclerotomes along the neural tube which eventually surround the notochord.

Scleraxonia [INV ZOO] A suborder of coelenterates in the order Gorgonacea in which the axial skeleton has calcareous spicules.

sclereid [BOT] A thick-walled, lignified plant cell typically found in sclerenchyma.

sclerema neonatorum [MED] A disease of the newborn, particularly the premature or the undernourished, dehydrated, and debilitated, characterized by waxy-white hardening of subcutaneous tissue.

sclerenchyma [BOT] A supporting plant tissue composed principally of sclereids whose walls are often mineralized.

scleriasis *See* scleroderma.

sclerite [INV ZOO] One of the sclerotized plates of the integument of an arthropod.

sclerobasidium [MYCOL] A thick-walled resting body or encysted probasidium of rust and smut fungi.

scleroblast [INV ZOO] A spicule-secreting cell in Porifera.

scleroblastema [EMBRYO] Embryonic tissue from which bones are formed.

sclerocarp [BOT] The hard outer coat, usually the endocarp, of the seed of a succulent fruit.

sclerocaulous [BOT] Having a hard, dry stem because of exceptional development of sclerenchyma.

sclerochore [BOT] A plant that disperses disseminules without apparent morphological adaptations.

Sclerodactylidae [INV ZOO] A family of echinoderms in the order Dendrochirotida having a complex calcareous ring.

scleroderm [INV ZOO] An indurating integument composing the skeleton of corals.

scleroderma [MED] An abnormal increase in collagenous connective tissue in the skin. Also known as chorionitis; dermatosclerosis; scleriasis.

sclerodermatous [INV ZOO] Having a skeleton that is composed of scleroderm, as certain corals. [VERT ZOO] Having a hard outer covering, for example, hard plate or horny scale.

sclerodermite [INV ZOO] The hard outer integument of an arthropod segment.

sclerogen [BOT] The woody tissue in a plant cell.

Sclerogibbidae [INV ZOO] A monospecific family of the hymenopteran superfamily Bethyloidea.

sclerophyll [BOT] A stiff, evergreen leaf, usually covered with cutin.

sclerophyllous [BOT] Characterized by thick, hard foliage due to well-developed sclerenchymatous tissue.

scleroprotein [BIOCHEM] Any of a large class of fibrous proteins, such as collagen, keratin, and fibroin, which occur in tendons, bones, cartilages, ligaments, and other parts of the body. Also known as albuminoid.

scleroseptum [INV ZOO] A calcareous radial septum in certain corals.

sclerosing adenomatosis [MED] A form of mammary dysplasia in which ductular structures are encased in fibrous tissue, simulating invading cancerous ductular structures. Also known as fibrosing adenomatosis.

sclerosing hemangioma [MED] A type of benign histiocytoma marked by prominence of the capillary channels.

sclerosis [PATH] Hardening of a tissue, especially by proliferation of fibrous connective tissue.

sclerotic [ANAT] Pertaining to the sclera. [MED] **1.** Hard. **2.** Of or pertaining to sclerosis.

sclerotium [MICROBIO] The hardened, resting or encysted condition of the plasmodium of Myxomycetes. [MYCOL] A hardened, resting mass of hyphae, usually black on the outside, from which fructifications may develop.

sclerotome [EMBRYO] The part of a mesodermal somite which enters into the formation of the vertebrae. [MED] A knife used in sclerotomy. [VERT ZOO] The fibrous tissue separating successive myotomes in certain lower vertebrates.

sclerotomy [MED] Surgical incision of the sclera.

scobiform [BOT] Resembling sawdust.

scolecodont [PALEON] Any of the paired, pincerlike jaws occurring as fossils of annelid worms.

Scolecosporae [MYCOL] A spore group of Fungi Imperfecti characterized by filiform spores.

scolecospore [BOT] A threadlike spore.

scolex [INV ZOO] The head of certain tapeworms, typically having a muscular pad with hooks, and two pairs of lateral suckers.

Scoliidae [INV ZOO] A family of the Hymenoptera in the superfamily Scolioidea.

Scolioidea [INV ZOO] A superfamily of hymenopteran insects in the suborder Apocrita.

scoliosis [MED] Lateral curvature of the spine.

Scolopacidea [VERT ZOO] A large, cosmopolitan family of birds of the order Charadriiformes including snipes, sandpipers, curlews, and godwits.

Scolopendridae [INV ZOO] A family of centipedes in the order Scolopendromorpha which characteristically possess eyes.

Scolopendromorpha [INV ZOO] An order of the chilopod subclass Pleurostigmophora containing the dominant tropical forms, and also the largest of the centipedes.

scolophore *See* scolopophore.

scolopophore [INV ZOO] A spindle-shaped, bipolar nerve ending in the integument of insects, believed to be auditory in function. Also known as scolophore.

Scolytidae [INV ZOO] The bark beetles, a large family of coleopteran insects in the superfamily Curculionoidea characterized by a short beak and clubbed antennae.

Scombridae [VERT ZOO] A family of perciform fishes in the suborder Scombroidei including the mackerels and tunas.

Scombroidei [VERT ZOO] A suborder of fishes in the order Perciformes; all are moderate- to large-sized shore and oceanic fishes having fixed premaxillae.

scopa [INV ZOO] A brushlike arrangement of short stiff hairs on the body surface of certain insects.

Scopeumatidae [INV ZOO] The dung flies, a family of myodarian cyclorrhaphous dipteran insects in the subsection Calypteratae.

Scopidae [VERT ZOO] A family of birds in the order Ciconiiformes containing a single species, the hammerhead (*Scopus umbretta*) of tropical Africa.

scopolamine [PHARM] $C_{17}H_{21}O_4N$ An alkaloid derivative of several plants in the family Solanaceae, used as an anticholinergic drug; its hydrobromide salt is used as a sedative.

scopophilia [PSYCH] Sexual stimulation from looking at the unclad human figure; observed chiefly in normal adolescent and adult males where it takes the aim-inhibited form of "girl watching" or looking at nude or seminude females in magazines or as part of some stage performance, or where it may

SCOLECODONT

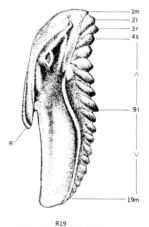

R19
1m − 2l − 3r − 4s < 9l > 19m

Dental formula for a large typical scolecodont. R = ramus, r = roder or largest denticle, l = large, s = small, m = minute. Teeth are numbered in sequence from anterior to posterior, and range in size from 1 to 2 millimeters.

SCOLEX

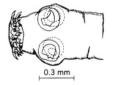

0.3 mm

Scolex of tapeworm showing the hooks and lateral suckers.

SCOLYTIDAE

Drawing of a bark beetle. *(From T. I. Storer and R. L. Usinger, General Zoology, 3d ed., McGraw-Hill, 1957)*

be sublimated as scientific curiosity; when present to a pathologic degree, it is deviant and called voyeurism.

scopula [ZOO] A tuft of hair, as on the feet and chelicerae of certain spiders.

scorching [PL PATH] Browning of plant tissues caused by heat or parasites; may also be symptomatic of disease.

Scorpaenidae [VERT ZOO] The scorpion fishes, a family of Perciformes in the suborder Cottoidei, including many tropical shorefishes, some of which are venomous.

Scorpaeniformes [VERT ZOO] An order of fishes coextensive with the perciform suborder Cottoidei in some systems of classification.

scorpioid cyme [BOT] A cyme with a curved axis and flowers arising two-ranked on alternate sides of the axis.

scorpion [INV ZOO] The common name for arachnids constituting the order Scorpionida.

Scorpionida [VERT ZOO] The scorpions, an order of arachnids characterized by a shieldlike carapace covering the cephalothorax and by large pedipalps armed with chelae.

Scotch pine [BOT] *Pinus sylvestris.* A hard pine of North America having two short, bluish needles in a cluster.

scotochromogen [MICROBIO] **1.** Any microorganism which produces pigment when grown without light as well as with light. **2.** A member of group II of the atypical mycobacteria.

scotodinia [MED] Dizziness and headache associated with the appearance of black spots before the eyes.

scotoma [MED] A blind spot or area of depressed vision in the visual field.

scotopic vision [PHYSIO] Vision that is due to the activity of the rods of the retina only; it is the type of vision that occurs at very low levels of illumination, and it can detect differences of brightness but not of hue.

scrapie [VET MED] A transmissible, usually fatal, virus disease of sheep, characterized by degeneration of the central nervous system.

Scraptidae [INV ZOO] An equivalent name for Melandryidae.

screen memory [PSYCH] A consciously tolerable but usually unimportant memory recalled in place of an associated important one, which would be painful and disturbing.

Scribner log rule [FOR] A method of scaling logs to derive board-foot calculations; uses a table showing expected log output in board feet that originated from diagrams of 1-inch boards drawn to scale within cylinders of various sizes.

scrod [VERT ZOO] A young fish, especially cod.

scrofula [MED] Tuberculosis of cervical lymph nodes.

Scrophulariaceae [BOT] A large family of dicotyledonous plants in the order Scrophulariales, characterized by a usually herbaceous habit, irregular flowers, axile placentation, and dry, dehiscent fruit.

Scrophulariales [BOT] An order of flowering plants in the subclass Asteridae distinguished by a usually superior ovary and, generally, either by an irregular corolla or by fewer stamens than corolla lobes, or commonly both.

scrotum [ANAT] The pouch containing the testes.

scrub [ECOL] A tract of land covered with a generally thick growth of dwarf or stunted trees and shrubs and a poor soil.

scrub typhus *See* tsutsugamushi disease.

scruple [PHARM] A unit of mass in apothecaries' measure, equal to 20 grains or to 1.2959782 grams.

sculpin [VERT ZOO] Any of several species of small fishes in the family Cottidae characterized by a large head that sometimes has spines, spiny fins, broad mouth, and smooth, scaleless skin.

SCORPIONIDA

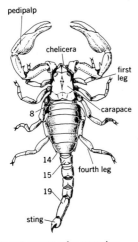

pedipalp

chelicera

first leg

8

carapace

14

15

fourth leg

19

sting

Dorsal aspect of a scorpion, *Chactas vanbenedein.* Some of the body segments have been numbered. *(From R. E. Snodgrass, A Textbook of Arthropod Anatomy, Cornell University Press, 1952)*

SCULPIN

The sculpin *Cottus cognatus*, with large mouth, winglike pectoral fins, and large anal and second dorsal fins.

scurf [MED] A branlike desquamation of the epidermis, especially from the scalp; dandruff.

scurvy [MED] An acute or chronic nutritional disorder due to vitamin C deficiency; characterized by weakness, subcutaneous hemorrhages, and alterations of any tissue containing collagen, ground substance, dentine, intercellular cement, or osteoid.

scute [INV ZOO] A cornified, epithelial, scalelike structure in lizards and snakes.

Scutechiniscidae [INV ZOO] A family of heterotardigrades in the suborder Echiniscoidea, with segmental and intersegmental thickenings of cuticle.

Scutelleridae [INV ZOO] The shield bugs, a family of Hemiptera in the superfamily Pentatomoidea.

scutellum [BOT] 1. A rounded apothecium with an elevated rim found in certain lichens. 2. The flattened cotyledon of a monocotyledonous plant embryo, such as a grass. [INV ZOO] The third of four pieces forming the upper part of the thoracic segment in certain insects. [VERT ZOO] One of the scales on the tarsi and toes of birds.

Scutigeromorpha [INV ZOO] The single order of notostigmophorous centipedes; members are distinguished by a dorsal respiratory opening, compound-type eyes, long flagellate multisegmental antennae, and long thin legs with multisegmental tarsi.

scutum [INV ZOO] 1. A bony, horny, or chitinous plate. 2. The second of four pieces forming the upper part of the thoracic segment in certain insects. 3. One or two lower opercular valves in certain barnacles.

Scydmaenidae [INV ZOO] The antlike stone beetles, a large cosmopolitan family of the Coleoptera in the superfamily Staphylinoidea.

Scyllaridae [INV ZOO] The Spanish, or shovel-nosed, lobsters, a family of the Scyllaridea.

Scyllaridea [INV ZOO] A superfamily of decapod crustaceans in the section Macrura including the heavily armored spiny lobsters and the Spanish lobsters, distinguished by the absence of a rostrum and chelae.

Scylliorhinidae [VERT ZOO] The catsharks, a family of the cacharinid group of galeoids; members exhibit the most exotic color patterns of all sharks.

scyphistoma [INV ZOO] A sessile, polyploid larva of many Scyphozoa which may produce either more scyphistomae or free-swimming medusae.

scyphomedusa [INV ZOO] A medusa of the scyphozoans.

Scyphomedusae [INV ZOO] A subclass of the class Scyphozoa characterized by reduced marginal tentacles, tetramerous medusae, and medusalike polyploids.

scyphopolyp [INV ZOO] A polyp of the scyphozoans.

Scyphozoa [INV ZOO] A class of the phylum Coelenterata; all members are marine and are characterized by large, well-developed medusae and by small, fairly well-organized polyps.

scyphus [BOT] 1. A funnel-shaped corolla. 2. A cup-shaped expansion of the podetium in some lichens.

Se *See* selenium.

sea anemone [INV ZOO] Any of the 1000 marine coelenterates that constitute the order Actiniaria; the adult is a cylindrical polyp or hydroid stage with the free end bearing tentacles that surround the mouth.

sea cucumber [INV ZOO] The common name for the echinoderms that make up the class Holothuroidea.

sea fan [INV ZOO] A form of horny coral that branches like a fan.

sea horse [INV ZOO] Any of about 50 species of tropical and

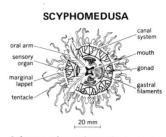

SCYPHOMEDUSA

oral arm
sensory organ
marginal lappet
tentacle
canal system
mouth
gonad
gastral filaments
20 mm

Schema of scyphomedusa, oral view.

SEA ANEMONE

Sea anemone *(Sagartia)*, a translucent, sessile species that has retractile tentacles.

SEA HORSE

The sea horse *Hippocampus hudsonius*, which has gill tufts and 11 trunk rings and grows to a length of 6 inches (15 centimeters).

SEAL

Alaska fur seal *(Callorhinus ursinus).*

SEA SQUIRT

Sea squirts; these saclike tunicates serve as scavengers and as food for higher forms.

SEA URCHIN

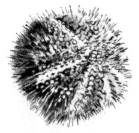

The sea urchin shell, which protects the soft internal organs, is covered with spines arranged in five broad areas that are separated by narrow unprotected areas.

subtropical marine fishes constituting the genus *Hippocampus* in the family Syngnathidae; the body is compressed, the head is bent ventrally and has a tubiform snout, and the tail is tapering and prehensile.

seal [VERT ZOO] Any of various carnivorous mammals of the suborder Pinnipedia, especially the families Phoridae, containing true seals, and Otariidae, containing the eared and fur seals.

sea lily [INV ZOO] The common name for those crinoids in which the body is flower-shaped and is carried at the tip of an anchored stem.

sea lion [VERT ZOO] Any of several large, eared seals of the Pacific Ocean; related to fur seals but lack a valuable coat.

sea otter [VERT ZOO] *Enhydra lutris.* A large marine otter found close to the shoreline in the North Pacific; these animals are diurnally active and live in herds.

sea pen [INV ZOO] The common name for coelenterates constituting the order Pennatulacea.

sea-run [VERT ZOO] Having the habit of ascending a river from the sea, especially to spawn, as salmon and brook trout.

seashell [INV ZOO] The shell of a marine invertebrate, especially a mollusk.

seasickness [MED] Motion sickness occurring at sea. Also known as pelagism.

sea slug [INV ZOO] The common name for the naked gastropods composing the suborder Nudibranchia.

sea spider [INV ZOO] The common name for arthropods in the subphylum Pycnogonida.

sea squirt [INV ZOO] A sessile, marine tunicate of the class Ascidiacea; it squirts water from two openings in the unattached end when touched or disturbed.

sea turtle [VERT ZOO] Any of various marine turtles, principally of the families Cheloniidae and Dermochelidae, having paddle-shaped feet.

seatworm *See* pinworm.

sea urchin [INV ZOO] A marine echinoderm of the class Echinoidea; the soft internal organs are enclosed in and protected by a test or shell consisting of a number of close-fitting plates beneath the skin.

seaweed [BOT] A marine plant, especially algae.

sebaceous gland [PHYSIO] A gland, arising in association with a hair follicle, which produces and liberates sebum.

Sebekidae [INV ZOO] A family of pentastomid arthropods in the suborder Porocephaloidea.

seborrheic dermatitis [MED] An acute inflammation of the skin, occurring usually on oily skin; characterized by scales, crusting yellowish patches, and itching.

sebum [PHYSIO] The secretion of sebaceous glands, composed of fat, cellular debris, and keratin.

secalose [BIOCHEM] A polysaccharide consisting of fructose units; occurs in green rye and oats, and in rye flour.

Secernentea [INV ZOO] A class of the phylum Nematoda in which the primary excretory system consists of intracellular tubular canals joined anteriorly and ventrally in an excretory sinus, into which two ventral excretory gland cells may also open.

secobarbital [PHARM] $C_{12}H_{18}N_2O_3$ 5-Allyl-5 (1-methylbutyl) barbituric acid, a short-acting barbiturate; white powder with bitterish taste; very soluble in alcohol and ether, slightly soluble in water; melts at 82°C; used as a sedative and hypnotic, frequently as the sodium derivative. Also known as quinalbarbitone.

secodont [VERT ZOO] Having teeth adapted for cutting.

Seconal [PHARM] A trade name for secobarbital.

secondary amyloidosis [MED] Amyloidosis that usually follows chronic suppurative, inflammatory diseases, such as tuberculosis, osteomyelitis, and bronchiectasis.

secondary cambium [BOT] One of the tissue layers formed after the initial cambial layers in certain plant roots, and that produce a ring of tissue.

secondary periderm [BOT] Any layer of the periderm except the first and outermost layer.

secondary phloem [BOT] Phloem produced by the cambium, consisting of two interpenetrating systems, the vertical or axial and the horizontal or ray.

secondary reinforcement [PSYCH] A stimulus or event that has been paired with a primary reinforcement.

secondary tympanic membrane [ANAT] The membrane closing the fenestra cochleae.

secondary wall [BOT] The portion of a plant cell wall produced internal to and following deposition of the primary wall; usually consists of several anisotropic layers, and often has prominent internal rings, spirals, bars, or reticulations.

secondary xylem [BOT] Xylem produced by cambium, composed of two interpenetrating systems, the horizontal (ray) and vertical (axial).

second crop [AGR] A crop succeeding one already harvested during a growing season; either a regrowth of the harvested crop, or a newly planted crop.

second-degree burn [MED] A burn that is more severe than a first-degree burn and is characterized by blistering as well as reddening of the skin, edema, and destruction of the superficial underlying tissues.

second generation *See* F_2.

second growth [FOR] New trees that naturally replace trees which have been removed from a forest by cutting or by fire.

second law of thermodynamics [THERMO] A general statement of the idea that there is a preferred direction for any process; there are many equivalent statements of the law, the best known being those of Clausius and of Kelvin.

second-order reaction [PHYS CHEM] A reaction whose rate of reaction is determined by the concentration of two chemical species.

secreta [PHYSIO] The secretions of a process.

secretin [BIOCHEM] A basic polypeptide hormone produced by the duodenum in response to the presence of acid; acts to excite the pancreas to activity.

secretion [PHYSIO] **1.** The act or process of producing a substance which is specialized to perform a certain function within the organism or is excreted from the body. **2.** The material produced by such a process.

secretor gene [GEN] A dominant autosomal gene in man which controls secretion of A and B antigenic material in saliva, urine, plasma, and other body fluids; it is not linked to the ABO genes.

secretory granules [CYTOL] Accumulations of material produced within a cell for secretion outside the cell.

secretory structure [BOT] Plant cells or organizations of plant cells which produce a variety of secretions.

sectional radiography [ELECTR] The technique of making radiographs of plane sections of a body or an object; its purpose is to show detail in a predetermined plane of the body, while blurring the images of structures in other planes. Also known as laminography; planigraphy; tomography.

secund [BOT] Having lateral members arranged on one side only.

SED *See* skin erythema dose.

sedation [MED] A state of lessened activity.

sedative [PHARM] An agent or drug that has a quieting effect on the central nervous system.

SECONDARY PHLOEM

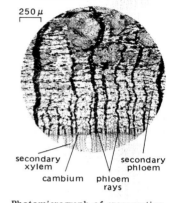

250 μ

secondary xylem

secondary phloem

cambium

phloem rays

Photomicrograph of cross section of the secondary phloem of paper birch *(Betula papyrifera).* *(Forest Products Laboratory, USDA)*

Sedentaria [INV ZOO] A group of families of polychaete annelids in which the anterior, or cephalic, region is more or less completely concealed by overhanging peristomial structures, or the body is divided into an anterior thoracic and a posterior abdominal region.

sedimentation coefficient [PHYS CHEM] In the sedimentation of molecules in an accelerating field, such as that of a centrifuge, the velocity of the boundary between the solution containing the molecules and the solvent divided by the accelerating field. (In the case of a centrifuge, the accelerating field equals the distance of the boundary from the axis of rotation multiplied by the square of the angular velocity in radians per second.)

sedimentation equilibrium [ANALY CHEM] The equilibrium between the forward movement of a sample's liquid-sediment boundary and reverse diffusion during centrifugation; used in molecular-weight determinations.

sedimentation rate [PATH] The rate at which red blood cells settle out of anticoagulated blood.

sedimentation velocity [ANALY CHEM] The rate of movement of the liquid-sediment boundary in the sample holder during centrifugation; used in molecular-weight determinations.

sedoheptulose [BIOCHEM] A seven-carbon ketose sugar widely distributed in plants of the Crassulaceae family; a significant intermediary compound in the cyclic regeneration of D-ribulose.

seed [BOT] A fertilized ovule containing an embryo which forms a new plant upon germination. [CHEM] A small, single crystal of a desired substance added to a solution to induce crystallization.

seed coat [BOT] The envelope which encloses the seed except for a tiny pore, the micropyle.

seed down [AGR] To sow seeds for grass or forage legumes.

seed fern [PALEOBOT] The common name for the extinct plants classified as Pteridospermae, characterized by naked seeds borne on large, fernlike fronds.

seeding [AGR] The planting of seed. [CHEM] The adding of a seed charge to a supersaturated solution, or a single crystal of a desired substance to a solution of the substance to induce crystallization.

seedling [BOT] **1.** A plant grown from seed. **2.** A tree younger and smaller than a sapling. **3.** A tree grown from a seed.

segmental reflex [PHYSIO] A reflex arc having afferent inputs by way of the spinal dorsal roots, and efferent outputs over spinal ventral roots of the same or adjacent segments.

segmentation *See* metamerism.

segmentation cavity *See* blastocoele.

segmentation nucleus [CYTOL] The body formed by fusion of the male and female pronuclei during fertilization.

segregation [GEN] The separation of alleles and homologous chromosomes during meiosis in the formation of gametes.

segregation distorter [GEN] An abnormality of meiosis which produces a distortion of the 1:1 segregation ratio in a heterozygote.

Seisonacea [INV ZOO] A monofamiliar order of the class Rotifera characterized by an elongated jointed body with a small head, a long slender neck region, a thick fusiform trunk, and an elongated foot terminating in a perforated disk.

Seisonidea [INV ZOO] The equivalent name for Seisonacea.

Seitz filter [MICROBIO] A bacterial filter made of asbestos and used to sterilize solutions without the use of heat.

seizure [MED] **1.** The sudden onset or recurrence of a disease or an attack. **2.** Specifically, an epileptic attack, fit, or convulsion.

Selachii [VERT ZOO] An order of elasmobranchs including all fossil sharks, except Cladoselachii and Pleuracanthodii.

Selaginellales [BOT] The plant order of small club mosses, containing one living genus, *Selaginella;* distinguished from other lycopods in being heterosporous and in having a ligule borne on the upper base of the leaf.

selection [GEN] Any natural or artificial process which favors the survival and propagation of individuals of a given genotype in a population.

selection bias [STAT] A bias built into an experiment by the method used to select the subjects which are to undergo treatment.

selection coefficient [GEN] A measure of the rate of transmission through successive generations of a given allele compared to the rate of transmission of another (usually the wild-type) allele.

selection pressure [EVOL] Those factors that influence the direction of natural selection.

selection rules [PHYS] Rules summarizing the changes that must take place in the quantum numbers of a quantum-mechanical system for a transition between two states to take place with appreciable probability; transitions that do not agree with the selection rules are called forbidden and have considerably lower probability.

selective breeding [BIOL] Breeding of animals or plants having desirable characters.

selective medium [MICROBIO] A bacterial culture medium containing an individual organic compound as the sole source of carbon, nitrogen, or sulfur for growth of an organism.

selective permeability [PHYS] The property of a membrane or other material that allows some substances to pass through it more easily than others.

selenium [CHEM] A highly toxic, nonmetallic element in group VI, symbol Se, atomic number 34; steel-gray color; soluble in carbon disulfide, insoluble in water and alcohol; melts at 217°C; and boils at 690°C; used in analytical chemistry, metallurgy, and photoelectric cells, and as a lube-oil stabilizer and chemicals intermediate.

selenodont [VERT ZOO] **1.** Being or pertaining to molars having crescentic ridges on the crown. **2.** A mammal with selenodont dentition.

selenosis [MED] Selenium poisoning.

self-analysis [PSYCH] The attempt to gain insight into one's own psychic state and behavior.

self-differentiation [PHYSIO] The differentiation of a tissue, even when isolated, solely as a result of intrinsic factors after determination.

self-incompatibility *See* self-sterility.

selfing [GEN] A plant breeding term usually meaning natural or artificial self-pollination.

self-pollination [BOT] Transfer of pollen from the anther to the stigma of the same flower or of another flower on the same plant.

self-sterility [BIOL] Inability to self-fertilize. [GEN] In higher plants and fungi, inability to fertilize an individual with the same allele or alleles in respect to particular self-compatibility genes. Also known as self-incompatibility.

Seliwanoff's test [ANALY CHEM] A color test helpful in the identification of ketoses, which develop a red color with resorcinol in hydrochloric acid.

sella turcica [ANAT] A depression in the upper surface of the sphenoid bone in which the pituitary gland rests in vertebrates.

selva *See* tropical rainforest.

Semaeostomeae [INV ZOO] An order of the class Scyphozoa

60 cm

Primitive selachian *Hybodus hauffianus*, a Mesozoic shark about 7½ feet (2.3 meters) long, showing claspers and the constricted base for paired fins that differentiates the Selachii from Cladoselachii. *(After Smith Woodward)*

SELENIUM

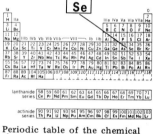

Periodic table of the chemical elements showing the position of selenium.

SEMAEOSTOMEAE

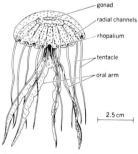

Pelagia, showing the tentacles arising between the lappets on the umbrella. *(From L. Hyman, The Invertebrates, vol. 1, McGraw-Hill, 1940)*

including most of the common medusae, characterized by a flat, domelike umbrella whose margin is divided into many lappets.

semeiography [MED] Description of the signs and symptoms of a disease.

semen [PHYSIO] The fluid that carries the male germ cells.

semicircular canal [ANAT] Any of three loop-shaped tubular structures of the vertebrate labyrinth; they are arranged in three different spatial planes at right angles to each other, and function in the maintenance of body equilibrium.

semiclasp [INV ZOO] Either of two apophyses which may join to form the clasper in certain male insects.

semicoma [MED] A mildly or partially comatose state in which the patient can be roused and responds to strong stimuli with some purposeful movements.

semiconservative replication [MOL BIO] Replication of deoxyribonucleic acid by longitudinal separation of the two complementary strands of the molecule, each being conserved and acting as a template for synthesis of a new complementary strand.

semidormancy [BOT] Decrease in plant growth rate; may be seasonal or associated with unfavorable environmental conditions.

semidouble [BOT] Pertaining to a flower that has more than the usual number of petals or disk florets while it retains some pollen-bearing stamens or some perfect disk florets.

semilate [BOT] Pertaining to a plant whose growing season is intermediate between midseason forms and late forms.

semilethal gene [GEN] A mutant causing the death of some of the individuals of the relevant genotype, but never 100%. Also known as sublethal gene.

semilunar cartilage [ANAT] One of the two interarticular knee cartilages.

semilunar ganglion *See* Gasserian ganglion.

semilunar valve [ANAT] Either of two tricuspid valves in the heart, one at the orifice of the pulmonary artery and the other at the orifice of the aorta.

semimembranosus [ANAT] One of the hamstring muscles, arising from the ischial tuber, and inserted into the tibia.

seminal receptacle *See* spermatheca.

seminal vesicle [ANAT] A saclike, glandular diverticulum on each ductus deferens in male vertebrates; it is united with the excretory duct and serves for temporary storage of semen.

seminiferous tubule [ANAT] Any of the tubercles of the testes which produce spermatozoa.

seminivorous [ZOO] Feeding on seeds.

seminoma [MED] A malignant tumor of the testes composed of large, uniform cells with clear cytoplasm.

Semionotiformes [VERT ZOO] An order of actinopterygian fishes represented by the single living genus *Lepisosteus,* the gars; the body is characteristically encased in a heavy armor of interlocking ganoid scales.

semioviparous [VERT ZOO] Producing young that are imperfectly developed when born, as marsupials.

semipalmate [VERT ZOO] Having a web halfway down the toes.

semiparasite *See* hemiparasite.

semipermeable membrane [PHYS] A membrane which allows a solvent to pass through it, but not certain dissolved or colloidal substances.

semiplacenta [VERT ZOO] A placenta that is not expelled at parturition.

semiplume [VERT ZOO] A feather having an ordinary shaft but a downy web.

semispinalis [ANAT] One of the deep longitudinal muscles of the back, attached to the vertebrae.

semitendinosus [ANAT] One of the hamstring muscles, arising from the ischium and inserted into the tibia.

Semper's larva [INV ZOO] A cylindrical larva in the life history of certain zoanthid corals, characterized by a hole at each end and an annular or longitudinal band of long cilia.

senescence [BIOL] The study of the biological changes related to aging, with special emphasis on plant, animal, and clinical observations which may apply to man.

senescent arthritis *See* degenerative joint disease.

senile [MED] Of, pertaining to, or caused by the aging process or by the infirmities of old age.

senile dementia [PSYCH] A chronic, organic brain syndrome associated with old age, characterized by intellectual deterioration, impairment of judgment, and gross emotional instability.

senile eczema [MED] A form of eczema associated with aging and caused by factors such as dryness of skin, soap sensitivity, or poor hygiene or diet.

senile emphysema [MED] Degenerative changes in the lungs and thoracic cage associated with aging.

senile gangrene [MED] A form of tissue death caused by deterioration of the blood supply to the extremities in the elderly.

senile psychosis [PSYCH] A severe form of senile dementia characterized by personality deterioration, progressive memory loss, eccentricity, irritability, and sometimes delusions and hallucinations.

senile vaginitis [MED] Inflammation of the vagina occurring in elderly women following chronic irritation of the thinned, atrophic mucosa.

Senn process [FOOD ENG] A butter-making process in which the cream is subjected to rapid agitation at 1500 revolutions per minute, decreasing to 20 revolutions per minute in the presence of 2–4 atmospheres ($2–4 \times 10^5$ newtons per square meter) of carbon dioxide.

sensation [PHYSIO] The subjective experience that results from the stimulation of a sense organ.

sense organ [PHYSIO] A structure which is a receptor for external or internal stimulation.

sensillum [ZOO] A simple, epithelial sense organ composed of one cell or of a few cells.

sensitivity [PHYSIO] The capacity for receiving sensory impressions from the environment. [SCI TECH] The ability of the output of a device, system, or organism to respond to an input stimulus.

sensorium [PHYSIO] **1.** A center, especially in the brain, for receiving and integrating sensations. **2.** The entire sensory apparatus of an individual.

sensory [PHYSIO] Pertaining to or conveying sensation.

sensory aphasia [MED] A form of aphasia in which the perception of sounds as language is partially preserved, but the patient is unable to comprehend the meaning of words, and in speaking, words are evoked with difficulty, are used incorrectly, and do not convey ideas correctly, resulting frequently in other forms of language impairment, particularly in agrammatism. Also known as receptive aphasia.

sensory area [PHYSIO] Any area of the cerebral cortex associated with the perception of sensations.

sensory cell [PHYSIO] **1.** A neuron having its terminal processes connected with sensory nerve endings. **2.** A modified epithelial or connective tissue cell adapted for the reception and transmission of sensations.

sensory learning [PSYCH] Learning situations in which a

SEMPER'S LARVA

3 cm

(a) (b)

Two types of pelagic Semper's larva. *(a) Zoanthina tentaculata. (b) Zoanthella galapagoensis. (After Senna)*

person or animal is trained to respond to changes in or differences between some aspects of a physical stimulus presented to one of the sense organs.

sensory nerve [PHYSIO] A nerve that conducts afferent impulses from the periphery to the central nervous system.

sepal [BOT] One of the leaves composing the calyx.

separation disk [BOT] A layer of gelatinous material between two adjacent negative cells in some blue-green algae; associated with hormogonium formation.

separation layer [BOT] A structurally distinct layer of the abscission zone of a plant containing abundant starch and dense cytoplasm.

Sepioidea [INV ZOO] An order of the molluscan subclass Coleoidea having a well-developed eye, an internal shell, fins separated posteriorly, and chromatophores in the dermis.

Sepsidae [INV ZOO] The spiny-legged flies, a family of myodarian cyclorrhaphous dipteran insects in the subsection Acalypteratae; development takes place in decaying organic matter.

sepsis [MED] **1.** Poisoning by products of putrefaction. **2.** The severe toxic, febrile state resulting from infection with pyogenic microorganisms, with or without associated septicemia.

septal neck [INV ZOO] In *Nautilus,* a hard tube that passes for some distance beyond each septum and supports the siphuncle.

septate [BIOL] Having a septum.

Septibranchia [INV ZOO] A small order of bivalve mollusks in which the anterior and posterior abductor muscles are about equal in size, the foot is long and slender, and the gills have been transformed into a muscular septum.

septic abortion [MED] An abortion complicated by acute infection of the endometrium.

septic embolus [MED] An embolus formed by bacteria.

septicemia [MED] A clinical syndrome in which infection is disseminated through the body in the bloodstream. Also known as blood poisoning.

septicidal [BOT] Dehiscing longitudinally through a septa, so that the carpels are separated.

septulum [ANAT] A small septum.

septum [BIOL] A partition or dividing wall between two cavities.

septum pellucidum [ANAT] A thin translucent septum forming the internal boundary of the lateral ventricles of the brain and enclosing between its two laminas the so-called fifth ventricle.

septum primum [EMBRYO] The first incomplete interatrial septum of the embryo.

septum secundum [EMBRYO] The second incomplete interatrial septum of the embryo, containing the foramen ovale; it develops to the right of the septum primum and fuses with it to form the adult interatrial septum.

septum transversum [ANAT] A ridge within the ampulla of a semicircular canal. [EMBRYO] The diaphragm of a fetus.

Sequanian [GEOL] Upper Lower Jurassic (Upper Lusitanian) geologic time. Also known as Astartian.

sequestrum [MED] A piece of dead or detached bone within a cavity, abscess, or wound.

Sequoia [BOT] A genus of conifers having overlapping, scalelike evergreen leaves and vertical grooves in the trunk; the giant sequoia (*Sequoia gigantea*) is the largest and oldest of all living things.

sere [ECOL] A temporary community which occurs during a successional sequence on a given site.

SEQUOIA

The giant sequoia tree, *Sequoia gigantea*, showing relative size of man at base.

Sergestidae [INV ZOO] A family of decapod crustaceans including several species of prawns.

serial homology [ZOO] The similarity between the members of a single series of structures, such as vertebrae, in an organism.

serial learning [PSYCH] The type of association in verbal learning involved in learning the alphabet; studied in the laboratory by giving the subject serial lists to learn, where each list would consist of a number of unrelated items.

sericeous [BOT] Of, pertaining to, or consisting of silk.

serine [BIOCHEM] $C_3H_7O_3N$ An amino acid obtained by hydrolysis of many proteins; a biosynthetic precursor of several metabolites, including cysteine, glycine, and choline.

serioscopy [NUCLEO] A radiographic technique enabling three-dimensional exploration by moving two of the three components of the system (tube, subject, film) in order to register the radiographic image of one plane in the object while images outside this slice have a continuous relative displacement and are blurred.

seroche *See* mountain sickness.

serodiagnosis [MED] Diagnosis based upon the reaction of blood serum of a patient.

serofibrinous [PHYSIO] Composed of serum and fibrin.

Serolidae [INV ZOO] A family of isopod crustaceans which contains greatly flattened forms that live partially buried on sandy bottoms.

serology [BIOL] The branch of science dealing with the properties and reactions of blood sera.

Seromycin [MICROBIO] A trade name for cycloserine.

seronegative [PATH] **1.** Having a negative serologic test for some condition. **2.** Specifically, having a negative serologic test for syphilis.

seropositive [PATH] **1.** Having a positive serologic test for some condition. **2.** Specifically, having a positive serologic test for syphilis.

seropurulent [MED] Composed of serum and pus, as a sero-purulent exudate.

seroresistance [PATH] Persistent positive serologic reaction for syphilis despite prolonged intensive treatment; the patient is said to be Wassermann-fast.

serosa [ANAT] The serous membrane lining the pleural, peritoneal, and pericardial cavities. [EMBRYO] The chorion of reptile and bird embryos.

serotherapy [MED] The treatment of disease by means of human or animal serum containing antibodies. Also known as immunotherapy.

serotonin [BIOCHEM] $C_{10}H_{12}ON_2$ A compound derived from tryptophan which functions as a local vasoconstrictor, plays a role in neurotransmission, and has pharmacologic properties. Also known as 5-hydroxytryptamine.

serotype [MICROBIO] A serological type of intimately related microorganisms, distinguished on the basis of antigenic composition.

serous [PHYSIO] **1.** Watery. **2.** Referring to serum.

serous cystadenoma [MED] A benign cystic tumor of the ovary, made up of cylindrical cells resembling those of the uterine tube; psammoma bodies often appear in the wall of the cyst. Also known as endosalpingioma; papillary adenoma of ovary; papillary cystadenoma of ovary; papillocystoma; psammomatus papilloma; serous cystoma.

serous cystoma *See* serous cystadenoma.

serous gland [PHYSIO] A structure that secretes a watery, albuminous fluid.

serous membrane [HISTOL] A delicate membrane covered with flat, mesothelial cells lining closed cavities of the body.

SERINE

$$CH_2OH$$
$$|$$
$$C$$
$$H_2N \quad \overset{|}{H} \quad COOH$$

Structural formula of serine.

SERPULIDAE

Serpula in right lateral view.

SESSOBLAST

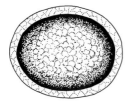

Sessoblast of *Stolella indica.*

serous plethora [MED] An increase in the watery part of the blood.

Serpentes [VERT ZOO] The snakes, a suborder of the Squamata characterized by the lack of limbs and pectoral girdle and external ear openings, immovable eyelids, and a braincase that is completely bony anteriorly.

serpentine locomotion [VERT ZOO] The wavelike or undulating movements characteristic of snakes.

Serpulidae [INV ZOO] A family of polychaete annelids belonging to the Sedentaria including many of the feather-duster worms which construct calcareous tubes in the earth, sometimes in such abundance as to clog drains and waterways.

serpulite [PALEON] The fossil tube of a polychaete.

Serranidae [VERT ZOO] A family of perciform fishes in the suborder Percoidei including the sea basses and groupers.

serrate [BIOL] Having a notched or toothed edge.

Serratieae [MICROBIO] A tribe of the Enterobacteriaceae containing the genus *Serratia,* with soil and water forms characterized by the production of a bright-orange to deep-red pigment, prodigiosin.

Serridentinae [PALEON] An extinct subfamily of elephantoids in the family Gomphotheriidae.

Serritermitidae [INV ZOO] A family of the Isoptera which contains the single monotypic genus *Serritermes.*

Serropalpidae [INV ZOO] An equivalent name for Melandryidae.

serrula [INV ZOO] A comblike ridge on the chelicerae of some Arachnida.

serrulate [BIOL] Finely serrate.

Sertoli cell [HISTOL] One of the sustentacular cells of the seminiferous tubules.

serum [PHYSIO] The liquid portion that remains when blood clots spontaneously and the formed and clotting elements are removed by centrifugation; it differs from plasma by the absence of fibrinogen.

serum accident [IMMUNOL] A serious allergic reaction which immediately follows the introduction of a foreign serum into a hypersensitive individual; dyspnea and flushing occur, soon followed by shock and occasionally by fatal termination.

serum albumin [BIOCHEM] The principal protein fraction of blood serum and serous fluids.

serum globulin [BIOCHEM] The globulin fraction of blood serum.

serum hepatitis [MED] A form of viral hepatitis transmitted by parenteral injection of human blood or blood products contaminated with the virus.

serum shock [MED] An anaphylactic reaction following the injection of foreign serum into a sensitized individual.

serum sickness [MED] A syndrome manifested in 8–12 days after the administration of serum by an urticarial rash, edema, enlargement of lymph nodes, arthralgia, and fever.

sesamoid bone [MED] A small bone developed in a tendon subjected to much pressure.

sessile [BOT] Attached directly to a branch or stem without an intervening stalk. [ZOO] Permanently attached to the substrate.

Sessilina [INV ZOO] A suborder of ciliates in the order Peritrichida.

sessoblast [INV ZOO] A statoblast that attaches to zooecial tubes or to the substratum.

seta [BIOL] A slender, usually rigid bristle or hair. Also known as chaeta.

setobranchia [INV ZOO] In certain decapods, a tuft of setae attached to the gills.

sex [BIOL] The state of condition of an organism which comes to expression in the production of germ cells.

sex chromatin *See* Barr body.

sex chromosome [GEN] Either member of a pair of chromosomes responsible for sex determination of an organism.

sex cords [EMBRYO] Cordlike masses of epithelial tissue that invaginate from germinal epithelium of the gonad and give rise to seminiferous tubules and rete testes in the male, and primary ovarian follicles and rete ovarii in the female.

sex determination [GEN] The mechanisms by which sex is determined in a species.

sexduction [GEN] A process in bacteria by which genetic material is carried from one bacterium to another by a sex factor, similar to transduction in which the carrier is a phage.

sex factor *See* fertility factor.

sex hormone [BIOCHEM] Any hormone secreted by a gonad, but also found in other tissues.

sex-influenced inheritance [GEN] That part of the inheritance pattern on which sex differences operate to promote character differences.

sex-limited inheritance [GEN] Expression of a phenotype in only one sex; may be due to either a sex-linked or autosomal gene.

sex-linked inheritance [GEN] The transmission to successive generations of differences that are due to genes located in the sex chromosomes.

sex organs [ANAT] The organs pertaining entirely to the sex of an individual, both physiologically and anatomically.

sex ratio [BIOL] The relative proportion of males and females in a population.

sexual cycle [PHYSIO] A cycle of physiological and structural changes associated with sex; examples are the estrous cycle and the menstrual cycle.

sexual dimorphism [BIOL] Diagnostic morphological differences between the sexes.

sexual isolation [EVOL] Differences in sexual, physiological, anatomical, or behavioral features that prevent or reduce interbreeding.

sexuality [BIOL] **1.** The sum of a person's sexual attributes, behavior, and tendencies. **2.** The psychological and physiological sexual impulses whose satisfaction affords pleasure. [PSYCH] The quality of being sexual, or the degree of a person's sexual attributes, attractiveness, and drives.

sexual reproduction [BIOL] Reproduction involving the paired union of special cells (gametes) from two individuals.

sexual spore [BIOL] A spore resulting from conjugation of gametes or nuclei of opposite sex.

Seymouriamorpha [PALEON] An extinct group of labyrinthodont Amphibia of the Upper Carboniferous and Permian in which the intercentra were reduced.

Sezary syndrome [MED] Exfoliative erythroderma with a cutaneous infiltrate of atypical mononuclear cells; similar cells are also present in the peripheral blood.

shaft [ANAT] The straight part of long bone. [HISTOL] The stem of a hair. [VERT ZOO] The stem of a feather.

shake culture [MICROBIO] **1.** A method for isolating anaerobic bacteria by shaking a deep liquid culture of an agar or gelatin to distribute the inoculum before solidification of the medium. **2.** A liquid medium in a flask that has been inoculated with an aerobic microorganism and placed on a shaking machine; action of the machine continually aerates the culture.

SHARK

The mackerel shark *(Lamnia nasus)* has a horizontal heellike tail and gill slits on both sides; claspers and prolonged pelvic fins, occur on males.

SHEEP

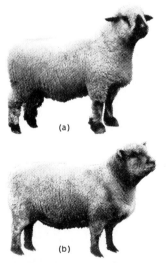

(a)

(b)

Examples of some prominent breeds of sheep. *(a)* Hampshire ram. *(b)* Southdown ram *(Southdown Association).*

shallot [BOT] *Allium ascalonicum.* A bulbous onionlike herb. Also known as scallion.

shape constancy [PHYSIO] Perception of the "true" shape of an object even when the image is distorted on the retina.

shark [VERT ZOO] Any of about 225 species of carnivorous elasmobranchs which occur principally in tropical and subtropical oceans; the body is fusiform with a heterocercal tail and a tough, usually gray, skin roughened by tubercles, and the snout extends beyond the mouth.

shearwater [VERT ZOO] Any of various species of oceanic birds of the genus *Puffinus* having tubular nostrils and long wings.

sheath [SCI TECH] A protective case or cover.

Sheehan's syndrome [MED] Hypopituitarism resulting from postpartum adenohypophyseal necrosis.

sheep [VERT ZOO] Any of various mammals of the genus *Ovis* in the family Bovidae characterized by a stocky build and horns, when present, which tend to curl in a spiral.

sheet composting [AGR] Addition of large amounts of organic residue to a soil; extra nitrogen is usually added for faster decomposition.

shell [ATOM PHYS] A set of orbital electron states that have the same principal quantum number and, therefore, have approximately the same energy level and average distance from the nucleus. [ZOO] **1.** A hard, usually calcareous, outer covering on an animal body, as of bivalves and turtles. **2.** The hard covering of an egg. **3.** Chitinous exoskeleton of certain arthropods.

shellfish [INV ZOO] An aquatic invertebrate, such as a mollusk or crustacean, that has a shell or exoskeleton.

shell gland [INV ZOO] An organ that secretes the embryonic shell in many mollusks. [VERT ZOO] A specialized structure attached to the oviduct in certain animals that secretes the eggshell material.

shell membrane [CYTOL] Either of a pair of membranes lining the inner surface of an egg shell; they allow free entry of oxygen but prevent rapid evaporation of moisture.

shelterbelt [ECOL] A natural or planned barrier of trees or shrubs to reduce erosion and provide shelter from wind and storm activity.

shelterwood method [FOR] A method for ensuring tree reproduction; older trees are removed by successive cuttings so that the amount of light reaching the seedlings is gradually increased.

shikimic acid [BIOCHEM] $C_7H_{10}O_5$ A crystalline acid that is a plant constituent, and an intermediate in the biochemical pathway from phosphoenolpyruvic acid to tyrosine.

shingles *See* herpes zoster.

shin splints [MED] An injury and an inflammation of the lower leg muscles and bones of the lower and middle third of the tibia and fibula, seen in athletes such as runners or basketball and football players.

shipping fever [VET MED] An acute, occasionally subacute, septicemic disease in cattle and sheep, probably caused by a combination of virus and *Pasteurella multocida* or *P. hemolytica.*

shipworm [INV ZOO] Any of several bivalve mollusk species belonging to the family Teredinidae and which superficially resemble earthworms because the two valves are reduced to a pair of plates at the anterior of the animal or are used for boring into wood.

shipyard eye *See* keratoconjunctivitis.

shistosoma [INV ZOO] A genus of blood flukes infecting man.

shock [MED] Clinical manifestations of circulatory insufficiency, including hypotension, weak pulse, tachycardia, pallor, and diminished urinary output.

shock organ [IMMUNOL] The organ or tissue that exhibits the most marked response to the antigen-antibody interaction in hypersensitivity, as the lungs in allergic asthma or the skin in allergic contact dermatitis.

shock therapy [PSYCH] The use of drugs, carbon dioxide, insulin, or electric current to induce coma in the treatment of psychiatric disorders.

shoot [BOT] **1.** The aerial portion of a plant, including stem, branches, and leaves. **2.** A new, immature growth on a plant.

Shope papilloma [VET MED] A transmissible, virus-induced papilloma occurring naturally on the skin of rabbits.

shop typhus *See* murine typhus.

shore bird [VERT ZOO] A general term applied to a large number of birds in 12 families of the suborder Charadrii which are always found near water, although the habitat and morphology is varied. Also known as wader.

short-day [BOT] Pertaining to plants whose flowering period is hastened by a relatively short photoperiod, generally less than 12 hours.

shoulder [ANAT] **1.** The area of union between the upper limb and the trunk in humans. **2.** The corresponding region in other vertebrates.

shoulder blade *See* scapula.

shoulder-elbow height [ANTHRO] A measure of the distance taken from the top of the acromion to the tip of the elbow, as the subject sits erect, with the upper arm vertical and the forearm horizontal.

shoulder girdle *See* pectoral girdle.

shoulder-hand syndrome [MED] A syndrome characterized by severe constant intractable pain in the shoulder and arm, limited joint motion, diffuse swelling of the distal part of the upper extremity, fibrosis and atrophy of muscles, and decalcification of underlying bones; the cause is not well understood; it is similar to, or may be a form of, causalgia. Also known as hand-shoulder syndrome.

shrew [VERT ZOO] Any of more than 250 species of insectivorous mammals of the family Soricidae; individuals are small with a moderately long tail, minute eyes, a sharp-pointed snout, and small ears.

shrimp [INV ZOO] The common name for a number of crustaceans, principally in the decapod suborder Natantia, characterized by having well-developed pleopods and by having the abdomen sharply bent in most species, producing a humped appearance.

shrub [BOT] A low woody plant with several stems.

Shwartzman phenomenon [IMMUNOL] A type of local tissue reactivity in the skin in which a preparatory injection of the endotoxin is followed by an intravenous injection of the same or another endotoxin 24 hours later, producing immediate neutropenia and thrombopenia with the development of leukocyte-platelet thrombi with subsequent hemorrhage.

Si *See* silicon.

sialadenitis [MED] Inflammation of a salivary gland.

sialagogue [PHARM] A drug producing a flow of saliva.

sialic acid [BIOCHEM] Any of a family of amino sugars, containing nine or more carbon atoms, that are nitrogen- and oxygen-substituted acyl derivatives of neuraminic acid; as components of lipids, polysaccharides, and mucoproteins, they are widely distributed in bacteria and in animal tissues.

sialogram [MED] Roentgenogram of a salivary duct system after the injection of an opaque medium.

sialography [MED] Radiographic examination of a salivary

SHREW

The Eurasian common shrew *(Sorex araneus)* has the soft, velvetlike fur characteristic of all shrews.

SHRIMP

Common shrimp *(Crangon vulgaris)*, with a laterally compressed body and curved abdomen.

gland following injection of an opaque substance into its duct.

sialolith [PATH] A salivary calculus.

sialolithiasis [MED] The presence of salivary calculi.

sialomucin [BIOCHEM] An acid mucopolysaccharide containing sialic acid as the acid component.

Siamese twins [MED] Viable conjoined twins.

Siberian tick typhus [MED] A relatively benign, rash- and eschar-producing spotted-fever-like disease in northern Asia, caused by *Rickettsia siberica;* transmitted by four species of *Dermacentor* and two of *Haemaphysalis.*

sibling rivalry [PSYCH] Competition between siblings for parental love, or for some other recognition.

Siboglinidae [INV ZOO] A family of pogonophores in the order Athecanephria.

sickle-cell anemia [MED] A chronic, hereditary hemolytic and thrombotic disorder in which hypoxia causes the erythrocyte to assume a sickle shape; occurs in individuals homozygous for sickle-cell hemoglobin trait.

sickle-cell hemoglobin [PATH] The hemoglobin found in sickle-cell anemia, differing in electrophoretic mobility and other physiochemical properties from normal adult hemoglobin. Also known as hemoglobin S.

sicula [INV ZOO] The cone-shaped chitinous skeleton of the first zooid of a graptolite colony.

Siderocapsaceae [MICROBIO] A family of aquatic gram-negative bacteria of the order Pseudomonadales, found in iron-bearing waters; all members possess the ability to deposit iron or manganese compounds around the cells.

siderocyte [CYTOL] An erythrocyte which contains granules staining blue with the Prussian blue reaction.

siderofibrosis [MED] Fibrosis associated with deposits of iron-bearing pigment.

sideropenic dysphagia *See* Plummer-Vinson syndrome.

siderosilicosis [MED] A pneumoconiosis resulting from prolonged inhalation of silica and iron dusts.

siderosis [MED] Pneumoconiosis due to prolonged inhalation of dust containing iron salt. Also known as arc-welder's disease. [PATH] The presence or concentration of stainable iron pigment in a tissue or organ.

Sierolomorphidae [INV ZOO] A small family of hymenopteran insects in the superfamily Scolioidea.

sieve area [BOT] An area in the wall of a sieve-tube element, sieve cell, or parenchyma cell characterized by clusters of pores through which strands of cytoplasm pass to adjoining cells.

sieve cell [BOT] A long, tapering cell characteristic of phloem in gymnosperms and lower vascular plants, in which all the sieve areas are of equal specialization.

sieve elements [BOT] The conducting parts of phloem, comprising sieve cells and sieve-tube cells.

sieve pit [BOT] A primary pit that gives rise to a sieve pore.

sieve plate [BOT] **1.** Part of the wall of a sieve cell; contains sieve areas. **2.** The perforated, thickened end of a sieve-tube cell.

sieve pore [BOT] A perforation in a sieve area or a sieve plate.

sieve tissue *See* phloem.

sieve tube [BOT] A phloem element consisting of a series of thin-walled cells arranged end to end, in which some sieve areas are more specialized than others.

Sigalionidae [INV ZOO] A family of scale-bearing polychaete annelids belonging to the Errantia.

Siganidae [VERT ZOO] A small family of herbivorous perciform fishes in the suborder Acanthuroidei having minute concealed scales embedded in the skin and strong, sharp fin spines.

Sigatoka [PL PATH] Leaf-spot disease of banana caused by the fungus *Mycosphaerella musicola.*

sight [PHYSIO] *See* vision.

sigma [INV ZOO] A C-shaped spicule.

sigmaspire [INV ZOO] An S-shaped sponge spicule.

sigmoid [BIOL] S-shaped.

sigmoid colon [ANAT] The S-shaped portion of the colon between the descending colon and the rectum.

sigmoiditis [MED] Inflammation of the sigmoid flexure of the colon.

sigmoidoscope [MED] An appliance for the inspection, by artificial light, of the sigmoid colon; it differs from the proctoscope in its greater length and diameter.

signal detection theory [PSYCH] A theory which characterizes not only the acuity of an individual's discrimination but also the psychological factors that bias his judgment.

signet-ring cell [HISTOL] A cell with a large fat- or carbohydrate-filled vacuole that pushes the nucleus against the cell membrane.

sign [MED] An objective abnormality indicative of disease.

sign test [STAT] A test which can be used whenever an experiment is conducted to compare a treatment with a control on a number of matched pairs, provided the two treatments are assigned to the members of each pair at random.

silage [AGR] Green or mature fodder that is fermented to retard spoilage and produce a succulent winter feed for livestock.

silicle [BOT] A many-seeded capsule formed from two united carpels, usually of equal length and width, and divided on the inside by a replum.

Silicoflagellata [BOT] A class of unicellular flagellates of the plant division Chrysophyta represented by a single living genus, *Dictyocha.*

Silicoflagellida [INV ZOO] An order of marine flagellates in the class Phytamastigophorea which have an internal, siliceous, tubular skeleton, numerous yellow chromatophores, and a single flagellum.

silicon [CHEM] A group IV nonmetallic element, symbol Si, with atomic number 14, atomic weight 28.086; dark-brown crystals that burn in air when ignited; soluble in hydrofluoric acid and alkalies; melts at 1410°C; used to make silicon-containing alloys, as an intermediate for silicon-containing compounds, and in rectifiers and transistors.

silicosiderosis [MED] Pneumoconiosis caused by inhalation of silicate- and iron-containing dust.

silicosis [MED] Pneumoconiosis due to the inhalation of silica (SiO_2) particles.

silique [BOT] A silicle-like capsule, but usually at least four times as long as it is wide, which opens by sutures at either margin and has parietal placentation.

silk [INV ZOO] A continuous protein fiber consisting principally of fibroin and secreted by various insects and arachnids, especially the silkworm, for use in spinning cocoons, webs, egg cases, and other structures.

silk cotton tree *See* kapok tree.

silk gland [INV ZOO] A gland in certain insects which secretes a viscous fluid in the form of filaments known as silk; it is a salivary gland in insects and an abdominal gland in spiders.

silkworm [INV ZOO] The larva of various moths, especially *Bombyx mori,* that produces a large amount of silk for building its cocoon.

SIGATOKA

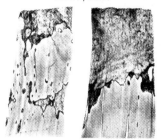

Sigatoka disease; banana with spotting, and necrosis of marginal areas on leaves of Gros Michel variety. *(Division of Tropical Research, United Brands Co.)*

SILICOFLAGELLATA

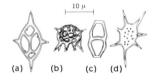

(a) (b) (c) (d)

Examples of fossil and modern Silicoflagellata. *(a) Dictyocha,* Cretaceous to Recent; *(b) Cannopilus,* Miocene; *(c) Naviculopsis,* Eocene to Miocene; and *(d) Vallacerta,* Upper Cretaceous.

SILICON

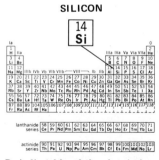

Periodic table of the chemical elements showing the position of silicon.

SILURIAN

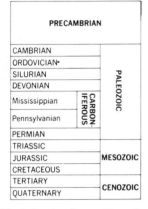

PRECAMBRIAN			
CAMBRIAN			
ORDOVICIAN•		PALEOZOIC	
SILURIAN			
DEVONIAN			
Mississippian	CARBONIFEROUS		
Pennsylvanian			
PERMIAN			
TRIASSIC			
JURASSIC		MESOZOIC	
CRETACEOUS			
TERTIARY		CENOZOIC	
QUATERNARY			

Chart showing the position of the Silurian in relation to the various periods of geologic time.

SILVER

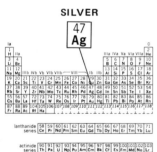

Periodic table of the chemical elements showing the position of silver.

SIMPLE MICROSCOPE

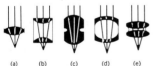

(a) (b) (c) (d) (e)

Types of lenses used in simple microscope. *(a)* Double convex. *(b)* Doublet. *(c)* Coddington. *(d)* Hastings triplet. *(e)* Achromat. *(From F. A. Jenkins and H. E. White, Fundamentals of Optics, 3d ed., McGraw-Hill, 1957)*

Silphidae [INV ZOO] The carrion beetles, a family of coleopteran insects in the superfamily Staphylinoidea.

Silurian [GEOL] **1.** A period of geologic time of the Paleozoic era, covering a time span of between 430-440 and 395 million years ago. **2.** The rock system of this period.

Siluridae [VERT ZOO] A family of European catfish in the suborder Siluroidei in which the adipose dorsal fin is rudimentary or lacking.

Siluriformes [VERT ZOO] The catfishes, a distinctive order of actinopterygian fishes in the superorder Ostariophysi, distinguished by a complex Weberian apparatus that involves the fifth vertebrae and one to four pair of barbels.

Siluroidei [VERT ZOO] A suborder of the Siluriformes.

Silvanidae [INV ZOO] An equivalent name for Cucujidae.

silver [CHEM] A white metallic element in group I, symbol Ag, with atomic number 47; soluble in acids and alkalies, insoluble in water; melts at 961°C, boils at 2212°C; used in photographic chemicals, alloys, conductors, and plating.

silverfish [INV ZOO] Any of over 350 species of insects of the order Thysanura; they are small, wingless forms with biting mouthparts.

silverline system [INV ZOO] A series of superficial argentophilic lines in many protozoans, especially ciliates.

silvicolous [ECOL] Inhabiting or growing in woodlands.

silviculture [FOR] The theory and practice of controlling the establishment, composition, and growth of stands of trees for any of the goods and benefits that they may be called upon to produce.

simian [VERT ZOO] Resembling or pertaining to anthropoid apes.

similarity coefficient [SYST] In numerical taxonomy, a factor S used to calculate the similarity between organisms, according to the formula $S = n_s/(n_s + n_d)$, where n_s represents the number of positive features shared by two strains, and n_d represents the number of features positive for one strain and negative for the other.

Simmonds' disease [MED] Hypopituitarism with marked insufficiency of the target glands and profound cachexia. Also known as hypophyseal cachexia; hypopituitary cachexia.

simple [BIOL] **1.** Made up of one piece. **2.** Unbranched. **3.** Consisting of identical units, as a simple tissue.

simple branched tubular gland [ANAT] A structure consisting of two or more unbranched, tubular secreting units joining a common outlet duct.

simple gland [ANAT] A gland having a single duct.

simple goiter [MED] Diffuse enlargement of the thyroid gland, usually not associated with constitutional features.

simple leaf [BOT] A leaf having one blade, or a lobed leaf in which the separate parts do not reach down to the midrib.

simple microscope [OPTICS] A diverging lens system, which can form an enlarged image of a small object. Also known as hand lens; magnifier; magnifying glass.

simple protein [BIOCHEM] One of a group of proteins which, upon hydrolysis, yield exclusively amino acids; included are globulins, glutelins, histones, prolamines, and protamines.

simple salt [CHEM] One of four classes of salts in a classification system that depends on the character of completeness of the ionization; examples are $NaCl$, $NaHCO_3$, and $Pb(OH)Cl$.

simple stomach [ANAT] A stomach consisting of a single dilation of the alimentary canal, as found in man, the dog, and many higher and lower vertebrates.

simple tubular gland [ANAT] A gland consisting of a single, tubular secreting unit.

simplex uterus [ANAT] A uterus consisting of a single cavity, representing the greatest degree of fusion of the Müllerian ducts; found in man and apes.

Simuliidae [INV ZOO] The black flies, a family of orthorrhaphous dipteran insects in the series Nematocera.

Sinemurian [GEOL] A European stage of geologic time; Lower Jurassic, above Hattangian and below Pliensbachian.

single-blind technique [BIOL] A method in which the subjects of a drug experiment do not know whether or not they have received the drug.

sinistrorse [BIOL] Twisting or coiling counterclockwise.

sinoatrial node [ANAT] A bundle of Purkinje fibers located near the junction of the superior vena cava with the right atrium which acts as a pacemaker for cardiac excitation. Abbreviated SA node. Also known as sinoauricular node.

sinuate [BOT] Having a wavy margin with strong indentations.

sinupalliate [INV ZOO] In mollusks, having a well-developed siphon, and therefore an indented pallial line.

sinus [BIOL] A cavity, recess, or depression in an organ, tissue, or other part of an animal body.

sinus gland [INV ZOO] An endocrine gland in higher crustaceans, lying in the eyestalk in most stalk-eyed species, which is the site of storage and release of a molt-inhibiting hormone.

sinus hairs *See* vibrissae.

sinusitis [MED] Inflammation of a paranasal sinus.

sinus of Morgagni [ANAT] The space between the upper border of the levator veli palatini muscle and the base of the skull.

sinusoid [ANAT] Any of the relatively large spaces comprising part of the venous circulation in certain organs, such as the liver.

sinus venosus [EMBRYO] The vessel in the transverse septum of the embryonic mammalian heart into which open the vitelline, allantoic, and common cardinal veins. [VERT ZOO] The chamber of the lower vertebrate heart to which the veins return blood from the body.

Siphinodentallidae [INV ZOO] A family of mollusks in the class Scaphopoda characterized by a subterminal epipodial ridge which is not slit dorsally and which terminates with a crenulated disk.

siphon [BOT] A tubular element in various algae. [INV ZOO] **1.** A tubular structure for intake or output of water in bivalves and other mollusks. **2.** The sucking-type of proboscis in many arthropods.

Siphonales [BOT] A large order of green algae (Chlorophyta) which are coenocytic, nonseptate, and mostly marine.

Siphonaptera [INV ZOO] The fleas, an order of insects characterized by a small, laterally compressed, oval body armed with spines and setae, three pairs of legs modified for jumping, and sucking mouthparts.

siphonium [VERT ZOO] A bony tube connecting the air passages of the tympanum with the air spaces in the articular piece of the mandible.

Siphonocladaceae [BOT] A family of green algae in the order Siphonocladales.

Siphonocladales [BOT] An order of green algae in the division Chlorophyta including marine, mostly tropical forms.

siphonogamous [BOT] In plants, especially seed plants, the accomplishment of fertilization by means of a pollen tube.

siphonoglyph [INV ZOO] A ciliated groove leading from the mouth to the gullet in certain anthozoans.

Siphonolaimidae [INV ZOO] A family of nematodes in the superfamily Monhysteroidea in which the stoma is modified into a narrow, elongate, hollow, spearlike apparatus.

Siphonophora [INV ZOO] An order of the coelenterate class

SIPHONAPTERA

Human flea, *Pulex irritans*. *(From E. O. Essig, College Entomology, Macmillan, 1942)*

SIPHONOCLADALES

Anadyomene, habit of thallus, showing expanded blades.

SIPHONOTRETACEA

Multispinula, external surface of pedicle valve. *(From R. C. Moore, ed., Treatis on Invertebrate Paleontology, pt. H, Geological Society of America, Inc. and University of Kansas Press, 1965)*

SIPUNCULIDA

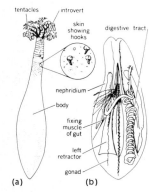

Two examples of Sipunculida. *(a) Dendrostoma petraeum. (b) Dendrostomum pyroides. (After Chamberlin)*

SISAL

Flower of the West Indian agave *(Agave sisalina)*.

Hydrozoa characterized by the complex organization of components which may be connected by a stemlike region or may be more closely united into a compact organism.

siphonoplax [INV ZOO] A calcareous plate connected with the siphon of certain mollusks.

siphonosome [INV ZOO] The lower part of a siphonophore colony, bearing the nutritive and reproductive zooids.

siphonostele [BOT] A type of stele consisting of pith surrounded by xylem and phloem.

siphonostomatous [INV ZOO] **1.** Having a tubular mouth. **2.** Having the anterior margin of the shell notched for passage of the siphon in certain mollusks.

Siphonotretacea [PALEON] A superfamily of extinct, inarticulate brachiopods in the suborder Acrotretidina of the order Acrotretida having an enlarged, tear-shaped, apical pedicle valve.

siphonozooid [INV ZOO] A zooid of certain alcyonarians that lacks tentacles and gonads.

siphuncle [INV ZOO] **1.** A honeydew-secreting tube (cornicle) in aphids. **2.** A tubular extension of the mantle extending through all the chambers to the apex of a shelled cephalopod.

Siphunculata [INV ZOO] The equivalent name for Anoplura.

Sipunculida [INV ZOO] A phylum of marine worms which dwell in burrows, secreted tubes, or adopted shells; the mouth and anus occur close together at one end of the elongated body, and the jawless mouth, surrounded by tentacles, is situated in an eversible proboscis.

Sipunculoidea [INV ZOO] An equivalent name for Sipunculida.

Sirenia [VERT ZOO] An order of aquatic placental mammals which include the living manatees and dugongs; these are nearly hairless, thick-skinned mammals without hindlimbs and with paddlelike forelimbs.

Siricidae [INV ZOO] The horntails, a family of the Hymenoptera in the superfamily Siricoidea; females use a stout, hornlike ovipositor to deposit eggs in wood.

Siricoidea [INV ZOO] A superfamily of wasps of the suborder Symphala in the order Hymenoptera.

sisal [BOT] *Agave sisalina.* An agave of the family Amaryllidaceae indigenous to Mexico and Central America; a coarse, stiff yellow fiber produced from the leaves is used for making twine and brush bristles.

sister chromatids [CYTOL] The two daughter strands of a chromosome after it has duplicated.

sitting height [ANTHRO] A measure of the vertical distance (taken along the back) from the table surface to the crest of the head as the subject sits erectly on the table, knees pressed against the table edge and head in the eye-ear horizontal plane.

situation therapy *See* milieu therapy.

situs inversus [MED] Reversed location or position.

Sjögren's syndrome [MED] A complex of symptoms including keratoconjunctivitis sicca, laryngopharyngitis sicca, rhinitis sicca, enlargement of the parotid gland, and polyarthritis. Also known as xerodermosteosis.

skate [VERT ZOO] Any of various batoid elasmobranchs in the family Rajidae which have flat bodies with winglike pectoral fins and a slender tail with two small dorsal fins.

skeletal muscle [ANAT] A striated, voluntary muscle attached to a bone and concerned with body movements.

skeletal system [ANAT] Structures composed of bone or cartilage or a combination of both which provide a framework for the vertebrate body and serve as attachment for muscles.

skelic index [ANTHRO] The ratio, multiplied by 100, of the length of the leg to that of the trunk.

skimmer [VERT ZOO] Any of various ternlike birds, members of the Rhynchopidae, that inhabit tropical waters around the world and are unique in having the knifelike lower mandible substantially longer than the wider upper mandible.

skin [ANAT] The external covering of the vertebrate body, consisting of two layers, the outer epidermis and the inner dermis. [BUILD] The exterior wall of a building.

skin erythema dose [NUCLEO] A unit of radioactive dose resulting from exposure to electromagnetic radiation, equal to the dose that slightly reddens or browns the skin of 80% of all persons within 3 weeks after exposure; it is approximately 1000 roentgens for gamma rays, 600 roentgens for x-rays. Abbreviated SED.

skin gills [INV ZOO] In Asteroidea, transparent contractile outgrowths from the integument which function in respiration.

skink [VERT ZOO] Any of numerous small- to medium-sized lizards comprising the family Scincidae with a cylindrical body; short, sometimes vestigial, legs; cores of bone in the body scales; and pleurodont dentition.

skin test [IMMUNOL] A procedure for evaluating immunity status involving the introduction of a reagent into or under the skin.

skipjack *See* bluefish.

skotoplankton [ECOL] Plankton found at depths below 500 meters.

skull [ANAT] The bones and cartilages of the vertebrate head which form the cranium and the face.

skunk [VERT ZOO] Any one of a group of carnivores in the family Mustelidae characterized by a glossy black and white coat and two musk glands at the base of the tail.

slant culture [MICROBIO] A method for maintaining bacteria in which the inoculum is streaked on the surface of agar that has solidified in inclined glass tubes.

slavery [INV ZOO] An interspecific association among ants in which members of one species bring pupae of another species to their nest, which, when adult, become slave workers in the colony.

sleep [PHYSIO] A state of rest in which consciousness and activity are diminished.

sleeping sickness *See* African sleeping sickness; encephalitis lethargica.

sleep paralysis [MED] Transient paralysis with spontaneous recovery occurring on falling asleep or on awakening.

sleep therapy *See* narcosis therapy.

slime bacteria [MICROBIO] The common name for bacteria in the order Myxobacterales, so named for the layer of slime deposited behind cells as they glide on a surface.

slime disease [PL PATH] Any of several diseases of plants characterized by slimy rot of the parts.

slime flux [PL PATH] The fluid or viscous outflow from the bark or wood of a deciduous tree that is indicative of injury or disease.

slime fungus *See* slime mold.

slime gland [ZOO] A glandular structure in many animals producing a mucous material.

slime mold [MYCOL] The common name for members of the Myxomycetes. Also known as slime fungus.

slipped disk *See* herniated disk.

slop culture [BOT] A method of growing plants in which surplus nutrient fluid is allowed to run through the sand or other medium in which the plants are growing.

sloth [VERT ZOO] Any of several edentate mammals in the

SKINK

The blue-tailed skink *(Eumeces fasciatus)*, a moderately stout species, is light green with four dorsal black stripes and one on each side between white ones.

SKULL

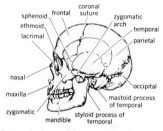

Lateral view of human skull. *(From W. T. Foster, Anatomy, Foster Art Service)*

SLOTH

The three-toed sloth *(Bradypus tridactylus)* in characteristic hanging position in which it spends most of its life.

family Bradypodidae found exclusively in Central and South America; all are slow-moving, arboreal species that cling to branches upside down by means of long, curved claws.

slough [MED] A necrotic mass of tissue in, or separating from, healthy tissue.

slow virus [VIROL] Any member of a group of animal viruses characterized by prolonged periods of incubation and an extended clinical course lasting months or years.

slow wave sleep *See* deep sleep.

sludged blood [MED] The intracapillary aggregation of erythrocytes associated with decreased blood flow in the involved capillary bed.

slug [INV ZOO] Any of a number of pulmonate gastropods which have a rudimentary shell and the body elevated toward the middle and front end where the mantle covers the lung region.

Sm *See* samarium.

Smalian's formula [FOR] A cubic volume formula used in log scaling, expressed cubic volume $= (B + b)/2L$, where $B =$ the cross-sectional area at the large end of the log, $b =$ the cross-sectional area at the small end of the log, and $L =$ log length.

small calorie *See* calorie.

small intestine [ANAT] The anterior portion of the intestine in man and other mammals; it is divided into three parts, the duodenum, the jejunum, and the ileum.

smallpox [MED] An acute, infectious, viral disease characterized by severe systemic involvement and a single crop of skin lesions which proceeds through macular, papular, vesicular, and pustular stages. Also known as variola.

smallpox vaccine [IMMUNOL] A vaccine prepared from a glycerinated suspension of the exudate from cowpox vesicles obtained from healthy vaccinated calves or sheep. Also known as antismallpox vaccine; glycerinated vaccine virus; Jennerian vaccine; virus vaccinium.

smear [BIOL] A preparation for microscopic examination made by spreading a drop of fluid, such as blood, across a slide and using the edge of another slide to leave a uniform film.

smegma [PHYSIO] The sebaceous secretion that accumulates around the glans penis and the clitoris.

smell [PHYSIO] To perceive by olfaction.

smelling salts [PHARM] A preparation containing ammonium carbonate and ammonia water, usually scented with aromatic substances.

smell prism [PSYCH] A diagram that attempts to systematize an odor classification system that has six main odor qualities: fruity, flowery, resinous, spicy, foul, and burnt.

Smilacaceae [BOT] A family of monocotyledonous plants in the order Liliales; members are usually climbing, leafy-stemmed plants with tendrils, trimerous flowers, and a superior ovary.

Sminthuridae [INV ZOO] A family of insects in the order Collembola which have simple tracheal systems.

Smith-Petersen nail [MED] A three-flanged nail used to fix fractures of the neck of the femur; it is inserted from just below the greater trochanter, through the neck, and into the head of the femur.

Smith-Recklinghausen's disease *See* neurofibromatosis.

smooth muscle [ANAT] The involuntary muscle tissue found in the walls of viscera and blood vessels, consisting of smooth muscle fibers.

smooth muscle fiber [HISTOL] Any of the elongated, nucleated, spindle-shaped cells comprising smooth muscles. Also known as involuntary fiber; nonstriated fiber; unstriated fiber.

SLUG

Field gray slug *(Deroceras agreste)* has a thin, white-bordered shell; body length is about $1\frac{1}{2}$ inches (4 centimeters). Eyes are borne at tips of the longer pair of feelers.

SMEAR

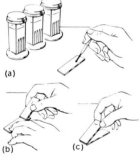

(a)

(b)   (c)

Making a smear preparation. *(a)* Place the drop about 1 inch (2.5 centimeters) from end of slide. *(b)* Apply a second slide just in front of the drop. *(c)* Push slide smoothly forward to spread the smear. *(From P. Gray, Handbook of Basic Microtechnique, 3d ed., McGraw-Hill, 1964)*

smudge [PL PATH] Any of several fungus diseases of cereals and other plants characterized by dark, sooty discolorations.

smut [MET] A reaction product left on the surface of a metal after pickling. [PL PATH] Any of various destructive fungus diseases of cereals and other plants characterized by large dusty masses of dark spores on the plant organs.

smut fungus [MYCOL] The common name for members of the Ustilaginales.

snail [INV ZOO] Any of a large number of gastropod mollusks distinguished by a spiral shell that encloses the body, a head, a foot, and a mantle.

snail fever *See* schistosomiasis.

snake [VERT ZOO] Any of about 3000 species of reptiles which belong to the 13 living families composing the suborder Serpentes in the order Squamata.

sneeze [PHYSIO] A sudden, noisy, spasmodic expiration through the mouth and nose.

Snellen test [MED] A test for visual acuity presenting letters, numbers, or letter E's in various positions, with the symbols varying in size; if the smallest are read at a distance of 20 feet (6 meters), vision is recorded as 20/20, or normal.

snout [VERT ZOO] The elongated nose of various mammals.

snow blindness [MED] A transient visual impairment and actinic keratoconjunctivitis caused by exposure of the eyes to ultraviolet rays reflected from snow. Also known as solar photophthalmia.

sobole [BOT] An underground creeping stem.

social animal [ZOO] An animal that exhibits social behavior.

social anthropology [ANTHRO] The study of social organization among nonliterate peoples.

social behavior [ZOO] Any behavior on the part of an organism stimulated by, or acting upon, another member of the same species.

social hierarchy [VERT ZOO] The establishment of a dominance-subordination relationship among higher animal societies.

socialization [PSYCH] The process whereby a child learns to get along with and to behave similarly to other people in the group, largely through imitation as well as group pressure.

social parasitism [VERT ZOO] An aberrant type of parasitism occurring in some birds, in which the female of one species lays her eggs in the nests of other species and permits the foster parents to raise the young.

social psychiatry [PSYCH] Psychiatry especially concerned with the study of social influences on the cause and dynamics of emotional and mental illness, the use of the social environment in treatment, and preventive community programs, as well as the application of psychiatry to social issues, industry, law, education, and other such activities and organizations.

social psychology [PSYCH] The study of the manner in which the attitudes, personality, and motivations of the individual influence, and are influenced by, the structure, dynamics, and behavior of the social group with which the individual interacts.

society [ECOL] A secondary or minor plant community forming part of a community.

sodium [CHEM] A metallic element of group I, symbol Na, with atomic number 11, atomic weight 22.9898; silver-white, soft, and malleable; oxidizes in air; melts at 97.6°C; used as a chemical intermediate and in pharmaceuticals, petroleum refining, and metallurgy; the source of the symbol Na is natrium.

sodium antimony bis (pyrocatechol-2,4-disulfonate) *See* stibophen.

sodium benzyl penicillinate *See* benzyl penicillin sodium.

SNAIL

The white-lipped sand snail *(Triodopsis albalabris)*; eyes are on the longer tentacles.

SNAKE

Garter snake *(Thamnophis ordinoides)*.

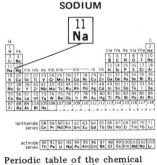

SODIUM

Periodic table of the chemical elements showing the position of sodium.

sodium chloride [INORG CHEM] NaCl Colorless or white crystals; soluble in water and glycerol, slightly soluble in alcohol; melts at 804°C; used in foods and as a chemical intermediate and an analytical reagent. Also known as common salt; rock salt; table salt.

sodium chloride solution *See* normal saline.

sodium citrate [ORG CHEM] $C_6H_5Na_3O_7 \cdot 2H_2O$ A white powder with the taste of salt; soluble in water, slightly soluble in alcohol; has an acid taste; loses water at 150°C; decomposes at red heat; used in medicine as an anticoagulant, in soft drinks, cheesemaking, and electroplating. Also known as disodium citrate; disodium hydrogen citrate.

sodium glucosulfone [PHARM] $C_{24}H_{34}N_2Na_2O_{18}S_3$ A leprostatic drug, and suppressant for dermatitis herpetiformis. Also known as glucosulfone sodium.

sodium *para*-diphenylaminoazobenzenesulfonate *See* tropeoline 00.

sodium pentobarbitone *See* pentobarbital sodium.

soft cataract [MED] A cataract, affecting the cortex of the lens of the eye, which is of soft consistency and has a milky appearance.

soft chancre *See* chancroid.

soft coral [INV ZOO] The common name for coelenterates composing the order Alcyonacea; the colony is supple and leathery.

soft palate [ANAT] The posterior part of the palate which consists of an aggregation of muscles, the tensor veli palatini, levator veli palatini, azygos uvulae, palatoglossus, and palatopharyngeus, and their covering mucous membrane.

soft rot [PL PATH] A mushy, watery, or slimy disintegration of plant parts caused by either fungi or bacteria.

soft-shell disease [INV ZOO] A disease of lobsters caused by a chitinous bacterium which extracts chitin from the exoskeleton.

soft wood [MATER] Wood from a coniferous tree.

soil conservation [ECOL] Management of soil to prevent or reduce soil erosion and depletion by wind and water.

soil microbiology [MICROBIO] A study of the microorganisms in soil, their functions, and the effect of their activities on the character of the soil and the growth and health of plant life.

soil rot [PL PATH] Plant rot caused by soil microorganisms.

Solanaceae [BOT] A family of dicotyledonous plants in the order Polemoniales having internal phloem, mostly numerous ovules and seeds on axile placentae, and mostly cellular endosperm.

solar dermatitis [MED] Any of various skin eruptions caused by exposure to the sun, excluding sunburn.

solar photophthalmia *See* snow blindness.

solar propagation [BOT] A method of rooting plant cuttings involving the use of a modified hotbed; bottom heat is provided by radiation of stored solar heat from bricks or stones in the bottom of the hotbed frame.

Solasteridae [INV ZOO] The sun stars, a family of asteroid echinoderms in the order Spinulosida.

Solemyidae [INV ZOO] A family of bivalve mollusks in the order Protobranchia.

Solenichthyes [VERT ZOO] An equivalent name for Gasterosteiformes.

solenium [INV ZOO] A diverticulum of the enteron in certain hydroids.

solenocyte [INV ZOO] Any of various hollow, flagellated cells in the nephridia of the larvae of certain annelids, mollusks, rotifers, and lancelets.

solenodon [VERT ZOO] Either of two species of insectivorous

SOLENODON

The solenodon uses its elongated snout to grub for food.

mammals comprising the family Solenodontidae; the almique (*Atopogale cubana*) is found only in Cuba, while the white agouta (*Solenodon paradoxus*) is confined to Haiti.

Solenodontidae [VERT ZOO] The solenodons, a family of insectivores belonging to the group Lipotyphla.

Solenogastres [INV ZOO] The equivalent name for Aplacophora.

Solenopora [PALEOBOT] A genus of extinct calcareous red algae in the family Solenoporaceae that appeared in the Late Cambrian and lasted until the Early Tertiary.

Solenoporaceae [PALEOBOT] A family of extinct red algae having compact tissue and the ability to deposit calcium carbonate within and between the cell walls.

soleus [ANAT] A flat muscle of the calf; origin is the fibula, popliteal fascia, and tibia, and insertion is the calcaneus; plantar-flexes the foot.

solid [PHYS] **1.** A substance that has a definite volume and shape and resists forces that tend to alter its volume or shape. **2.** A crystalline material, that is, one in which the constituent atoms are arranged in a three-dimensional lattice, periodic in three independent directions.

solid-liquid equilibrium [PHYS CHEM] **1.** The interrelation of a solid material and its melt at constant vapor pressure. **2.** The concentration relationship of a solid with a solvent liquid other than its melt. Also known as liquid-solid equilibrium.

solidus [PHYS CHEM] In a constitution or equilibrium diagram, the locus of points representing the temperature below which the various compositions finish freezing on cooling, or begin to melt on heating.

Solo man [PALEON] A relative but primitive form of fossil man from Java; this form had a small brain, heavy horizontal browridges, and a massive cranial base.

Solpugida [INV ZOO] The sun spiders, an order of nonvenomous, spiderlike, predatory arachnids having large chelicerae for holding and crushing prey.

solubility [PHYS CHEM] The ability of a substance to form a solution with another substance.

solubility curve [PHYS CHEM] A graph showing the concentration of a substance in its saturated solution in a solvent as a function of temperature.

solubility product constant [PHYS CHEM] A type of simplified equilibrium constant, K_{sp}, defined for and useful for equilibria between solids and their respective ions in solution; for example, the equilibrium

$$AgCl(s) \rightleftarrows Ag^+ + Cl^-, [Ag^+][Cl^-] \cong K_{sp},$$

where $[Ag^+]$ and $[Cl^-]$ are molar concentrations of silver ions and chloride ions.

solute [CHEM] The substance dissolved in a solvent.

solution [CHEM] A single, homogeneous liquid, solid, or gas phase that is a mixture in which the components (liquid, gas, solid, or combinations thereof) are uniformly distributed throughout the mixture.

solvation [CHEM] The process of swelling, gelling, or dissolving of a material by a solvent; for resins, the solvent can be a plasticizer.

solvent [CHEM] That part of a solution that is present in the largest amount, or the compound that is normally liquid in the pure state (as for solutions of solids or gases in liquids).

soma [BIOL] The whole of the body of an individual, excluding the germ tract.

Somasteroidea [INV ZOO] A subclass of Asterozoa comprising sea stars of generalized structure, the jaws often only partly developed, and the skeletal elements of the arm

SOLPUGIDA

Sun spider. *(From T. I Storer and R. L. Usinger, General Zoology, 3d ed., McGraw-Hill, 1957)*

arranged in a double series of transverse rows termed meta-pinnules.

somatic [BIOL] Pertaining solely to the bodily part of an animal or plant, as opposed to the germinal part.

somatic aneuploidy [CYTOL] An irregular variation in number of one or more individual chromosomes in the cells of a tissue.

somatic cell [BIOL] Any cell of the body of an organism except the germ cells.

somatic cell genetics [GEN] **1.** The genetics of somatic cells, especially in culture, of higher organisms. **2.** The genetics of higher organisms by means of their somatic cells, especially in culture.

somatic cell hybrid [GEN] A hybrid derived from two somatic cells of different types which undergo a process of cytoplasmic and nuclear fusion to give rise to a new cell, the nucleus of which contains genetic information from both parents.

somatic copulation [MYCOL] A form of reproduction in ascomycetes and basidiomycetes involving sexual fusion of undifferentiated vegetative cells.

somatic crossing-over [CYTOL] Crossing-over during mitosis in somatic or vegetative cells.

somatic death [BIOL] The cessation of characteristic life functions.

somatic mesoderm [EMBRYO] The external layer of the lateral mesoderm associated with the ectoderm after formation of the coelom.

somatic nervous system [PHYSIO] The portion of the nervous system concerned with the control of voluntary muscle and relating the organism with its environment.

somatic pairing [CYTOL] The pairing of homologous chromosomes at mitosis in somatic cells; occurs in Diptera.

somatic reflex system [PHYSIO] An involuntary control system characterized by a control loop which includes skeletal muscles.

somatization [PSYCH] A type of neurosis manifested in neurasthenias, hypochondriacal symptoms, and conversion hysterias.

somatoblast [INV ZOO] **1.** An undifferentiated cleavage cell that gives rise to somatic cells in annelids. **2.** The outer cell layer of the nematogen in Dicyemida.

somatochrome [CYTOL] A nerve cell possessing a well-defined body completely surrounding the nucleus on all sides, the cytoplasm having a distinct contour, and readily taking a stain.

somatocyst [INV ZOO] A cavity filled with air in the float of certain Siphonophora.

somatometry [ANAT] Measurement of the human body with the soft parts intact.

somatophyte [BOT] A plant composed of distinct somatic cells that develop especially into mature or adult tissue.

somatopleure [EMBRYO] A complex layer of tissue consisting of the somatic layer of the mesoblast together with the epiblast, forming the body wall in craniate vertebrates and the amnion and chorion in amniotes.

somatopsychic [MED] Pertaining to both the body and mind.

somatotonia [PSYCH] The temperamental trait associated with a mesomorphic somatotype, characterized by an active, aggressive, and risk-taking approach to life.

somatotropin [BIOCHEM] The growth hormone of the pituitary gland.

somatotype [PSYCH] A basic body type; three primary components are ectomorph, mesomorph, and endomorph.

somesthesis [PHYSIO] The general name for all systems of

SOMATIC COPULATION

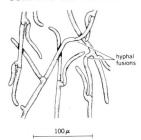

hyphal fusions

100μ

Somatic copulation by sexual fusion of vegetative cells of hyphae.

sensitivity present in the skin, muscles and their attachments, visceral organs, and nonauditory labyrinth of the ear.

somite *See* metamere.

sondage [ARCHEO] A trial excavation made prior to an archeological excavation.

Sonne dysentery [MED] An intestinal bacterial infection caused by *Shigella sonnei.*

sooty mold [MYCOL] Ascomycetous fungi of the family Capnodiaceae, with dark mycelium and conidia. [PL PATH] A plant disease, common on *Citrus* species, characterized by a dense velvety layer of a sooty mold on exposed parts of the plant.

soporific *See* hypnotic.

sorbin *See* sorbose.

sorbose [BIOCHEM] $C_6H_{12}O_6$ A carbohydrate prepared by fermentation; produced as water-soluble crystals that melt at 165°C; used in the production of vitamin C. Also known as sorbin; L-sorbose.

L-sorbose *See* sorbose.

soredium [BOT] A structure comprising algal cells wrapped in the hyphal tissue of lichens, as in certain Lecanorales; when separated from the thallus, it grows into a new thallus.

sore shin [PL PATH] A fungus disease of cowpea, cotton, tobacco, and other plants, beyond the seedling stage, marked by annular growth of the pathogen on the stem at the groundline.

sorghum [BOT] Any of a variety of widely cultivated grasses, especially *Sorghum bicolor* in the United States, grown for grain and herbage; growth habit and stem form are similar to Indian corn, but leaf margins are serrate and spikelets occur in pairs on a hairy rachis.

Soricidae [VERT ZOO] The shrews, a family of insectivorous mammals belonging to the Lipotyphla.

sorocarp [INV ZOO] The naked, simple fruit body of certain Myxomycetes.

sorogen [BOT] The cell or tissue which gives rise to a sorus.

sorophore [BOT] A stalk that bears a sorus.

sorosis [BOT] A composite fruit formed by union of the fleshy axis and the flowers, such as pineapple.

sorption [PHYS CHEM] A general term used to encompass the processes of adsorption, absorption, desorption, ion exchange, ion exclusion, ion retardation, chemisorption, and dialysis.

sorrel tree *See* sourwood.

sorus [BOT] **1.** A cluster of sporangia on the lower surface of a fertile fern leaf. **2.** A clump of reproductive bodies or spores in lower plants.

souma [VET MED] A disease caused by *Trypanosoma vivax* in domestic and wild animals; the insect vectors are the tsetse fly and the stable fly.

sourwood [BOT] *Oxydendrum arboreum.* A deciduous tree of the heath family (Ericaceae) indigenous along the Alleghenies and having long, simple, finely toothed, long-pointed leaves that have an acid taste, and white, urn-shaped flowers. Also known as sorrel tree.

South African tick-bite fever [MED] An infectious tick-borne rickettsial disease of humans which is similar to fièvre boutonneuse.

South American blastomycosis [MED] An infectious, yeast-like fungus disease of humans seen primarily in Brazil; caused by *Blastomyces brasiliensis* and characterized by massive enlargement of the cervical lymph nodes. Also known as paracoccidioidomycosis.

South American leishmaniasis *See* American mucocutaneous leishmaniasis.

SORUS

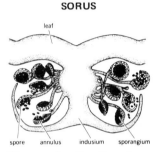

leaf

spore   annulus   indusium   sporangium

Diagram of section through a fern leaf, showing details of a sorus. *(From W. W. Robbins, T. E. Weier, and C. R. Stocking, Botany: An Introduction to Plant Science, 3d ed., Wiley, 1964)*

SOURWOOD

Sourwood *(Oxydendrum arboreum)* leaf, twig, and enlargement of twig showing axial bud and leaf scar.

South American trypanosomiasis *See* Chagas' disease.

South Australian faunal region [ECOL] A marine littoral region along the southwestern coast of Australia.

sow [VERT ZOO] An adult female swine.

soybean [BOT] *Glycine max.* An erect annual legume native to China and Manchuria and widely cultivated for forage and for its seed.

space biology [BIOL] A term for the various biological sciences and disciplines that are concerned with the study of living things in the space environment.

space medicine [MED] A branch of medicine that deals with the physiologic disturbances and diseases produced in man by high-velocity projection through and beyond the earth's atmosphere, flight through interplanetary space, and return to earth.

space perception [PHYSIO] The awareness of the spatial properties and relations of an object, or of one's own body, in space; especially, the sensory appreciation of position, size, form, distance, and direction of an object, or of the observer himself, in space.

spadix [BOT] A fleshy spike that is enclosed in a leaflike spathe and is the characteristic inflorescence of palms and arums. [INV ZOO] A cone-shaped structure in male Nautiloidea formed of four modified tentacles, and believed to be homologous with the hectocotylus in male squids.

Spanish collar *See* paraphimosis.

Sparganiaceae [BOT] A family of monocotyledonous plants in the order Typhales distinguished by the inflorescence of globose heads, a vestigial perianth, and achenes that are sessile or nearly sessile.

sparganosis [VET MED] An infection by the plerocercoid larva, or sparganum, of certain species of *Spirometra;* the adult form normally occurs in the intestine of dogs and cats.

sparganum [INV ZOO] The plerocercoid larva of a tapeworm.

Sparidae [VERT ZOO] A family of perciform fishes in the suborder Percoidei, including the porgies.

Sparnacean [GEOL] A European stage of geologic time; upper upper Paleocene, above Thanetian, below Ypresian of Eocene.

spasm [MED] **1.** A sudden muscular contraction. **2.** A convulsion or seizure.

spasmophilia [MED] A morbid tendency to convulsions, and to tonic spasms, such as those observed in tetany, infantile spasms, or spasmus nutans.

spastic colon *See* irritable colon.

spastic diplegia [MED] **1.** Spastic paralysis of the arms and legs caused by diffuse lesions of the cerebral cortex. **2.** A form of cerebral palsy, possibly due to prenatal or perinatal hypoxia or other injuries resulting in atrophic lobar sclerosis, or to congenital or developmental abnormalities.

spastic ileus [MED] A form of ileus in which temporary obstruction is due to segmental intestinal spasm. Also known as dynamic ileus; hyperdynamic ileus.

spastic paralysis [MED] A condition in which a group of muscles manifest increased tone, exaggerated tendon reflexes, depressed or absent superficial reflexes, and sometimes clonus, due to an upper motor neuron lesion.

spastic paraplegia [MED] Paralysis of the lower limbs with increased muscular tone and hyperactive tendon reflexes; commonly seen in diseases and injuries involving pyramidal tracts of the spinal cord.

spastic strabismus [MED] A squint resulting from the contraction of an ocular muscle.

spat [INV ZOO] The young of a bivalve mollusk.

Spatangoida [INV ZOO] An order of exocyclic Euechinoidea in which the posterior ambulacral plates form a shield-shaped area behind the mouth.

spatula [INV ZOO] A spoon-shaped process on the body of certain dipterous larvae.

spatulate [BIOL] Shaped like a spoon.

spawn [ZOO] **1.** The collection of eggs deposited by aquatic animals, such as fish. **2.** To produce or deposit eggs or discharge sperm; applied to aquatic animals.

spay [VET MED] To remove the ovaries.

spearmint [BOT] *Mentha spicata.* An aromatic plant of the mint family, Labiatae; the leaves are used as a flavoring in foods.

speciation [EVOL] The evolution of species.

species [SYST] A taxonomic category ranking immediately below a genus and including closely related, morphologically similar individuals which actually or potentially interbreed.

species concept [EVOL] The idea that the diversity of nature is divisible into a finite number of definable species.

species population [ECOL] A group of similar organisms residing in a defined space at a certain time.

specific locus test [GEN] A technique used to detect recessive induced mutations in diploid organisms; a strain which carries several known recessive mutants in a homozygous condition is crossed with a nonmutant strain treated to induce mutations in its germ cells; induced recessive mutations allelic with those of the test strain will be expressed in the progeny.

specific volume [MECH] The volume of a substance per unit mass; it is the reciprocal of the density. Abbreviated sp vol.

specimen [SCI TECH] **1.** An item representative of others in the same class or group. **2.** A sample selected for testing, examination, or display.

speck [PL PATH] A fungus or bacterial disease of rice characterized by speckled grains.

spectacle [ZOO] A colored marking in the form of rings around the eyes, as in certain birds, reptiles, and mammals (as the raccoon).

spectral sensitivity [PHYSIO] Visual sensitivity at different spectral wavelengths.

spectrograph [SPECT] A spectroscope provided with a photographic camera or other device for recording the spectrum.

spectrography [SPECT] The use of photography to record the electromagnetic spectrum displayed in a spectroscope.

spectrometer [SPECT] **1.** A spectroscope that is provided with a calibrated scale either for measurement of wavelength or for measurement of refractive indices of transparent prism materials. **2.** A spectroscope equipped with a photoelectric photometer to measure radiant intensities at various wavelengths.

spectrometry [SPECT] The use of spectrographic techniques for deriving the physical constants of materials.

spectrophotometer [SPECT] An instrument that measures transmission or apparent reflectance of visible light as a function of wavelength, permitting accurate analysis of color or accurate comparison of luminous intensities of two sources or specific wavelengths.

spectrophotometric titration [ANALY CHEM] An analytical method in which the radiant-energy absorption of a solution is measured spectrophotometrically after each increment of titrant is added.

spectrophotometry [SPECT] A procedure to measure photometrically the wavelength range of radiant energy absorbed by a sample under analysis; can be by visible light, ultraviolet light, or x-rays.

spectroscope [SPECT] An optical instrument consisting of a slit, collimator lens, prism or grating, and a telescope or objective lens which produces a spectrum for visual observation.

spectroscopy [PHYS] The branch of physics concerned with the production, measurement, and interpretation of electromagnetic spectra arising from either emission or absorption of radiant energy by various substances.

spectrum [PHYS] **1.** A display or plot of intensity of radiation (particles, photons, or acoustic radiation) as a function of mass, momentum, wavelength, frequency, or some related quantity. **2.** The set of frequencies, wavelengths, or related quantities, involved in some process; for example, each element has a characteristic discrete spectrum for emission and absorption of light. **3.** A range of frequencies within which radiation has some specified characteristic, such as audio-frequency spectrum, ultraviolet spectrum, or radio spectrum.

speculum [MED] A tubular instrument for inserting into a passage or cavity of the body to facilitate visual inspection or medication.

speech [PHYSIO] A set of audible sounds produced by disturbing the air through the integrated movements of certain groups of anatomical structures.

Spelaeogriphacea [INV ZOO] A peracaridan order of the Malacostraca comprised of the single species *Spelaeogriphus lepidops*, a small, blind, transparent, shrimplike crustacean with a short carapace that coalesces dorsally with the first thoracic somite.

sperm *See* spermatozoon.

spermangium [MYCOL] An organ that produces male spore-like cells in Ascomycetes.

spermatheca [ZOO] A sac in the female for receiving and storing sperm until fertilization; found in many invertebrates and certain vertebrates. Also known as seminal receptacle.

spermatic cord [ANAT] The cord consisting of the ductus deferens, epididymal and testicular nerves and blood vessels, and connective tissue that extends from the testis to the deep inguinal ring.

spermatid [HISTOL] A male germ cell immediately before assuming its final typical form.

spermatin [BIOCHEM] An albuminoid material occurring in semen.

spermatocele [MED] A cystic dilation of a duct in the head of the epididymis or in the rete testis.

spermatocyte [HISTOL] A cell of the last or next to the last generation of male germ cells which differentiates to form spermatozoa.

spermatogenesis [PHYSIO] The process by which spermatogonia undergo meiosis and transform into spermatozoa.

spermatogonium [HISTOL] A primitive male germ cell, the last generation of which gives rise to spermatocytes.

spermatophore [ZOO] A bundle or packet of sperm produced by certain animals, such as annelids, arthropods, and some vertebrates.

spermatophyte [BOT] A seed plant.

spermatoplasm [CYTOL] The protoplasm of a sperm cell.

spermatorrhea [MED] Involuntary discharge of semen without orgasm.

spermatozoon [HISTOL] A mature male germ cell. Also known as sperm.

spermaturia [MED] The presence of sperm in the urine.

spermidine [BIOCHEM] $H_2N(CH_2)_3NH(CH_2)_4NH_2$ The triamine found in semen and other animal tissues.

spermine [BIOCHEM] $C_{10}H_{26}N_4$ A tetramine found in semen, blood serum, and other body tissues.

SPELAEOGRIPHACEA

Spelaeogriphus lepidops Gordon, ovigerous female in dorsolateral aspect. *(British Museum of Natural History)*

spermiogenesis [CYTOL] Nuclear and cytoplasmic transformation of spermatids into spermatozoa.

sperm nucleus [BOT] One of the two nuclei in a pollen grain that function in double fertilization in seed plants.

spermoderm [BOT] The seed coat, consisting of the inner tegmen and the outer testa.

sperm whale [VERT ZOO] *Physeter catadon.* An aggressive toothed whale belonging to the group Odontoceti of the order Cetacea; it produces ambergris and contains a mixture of spermaceti and oil in a cavity of the nasal passage.

Sphaeractinoidea [PALEON] An extinct group of fossil marine hydrozoans distinguished in part by the relative prominence of either vertical or horizontal trabeculae and by the presence of long, tabulate tubes called autotubes.

Sphaeriales [MYCOL] An order of fungi in the subclass Euascomycetes characterized by hard, dark perithecia with definite ostioles.

Sphaeriidae [INV ZOO] The minute bog beetles, a small family of coleopteran insects in the suborder Myxophaga.

Sphaerioidaceae [MYCOL] A family of fungi of the order Sphaeropsidales in which the pycnidia are black or dark-colored and are flask-, cone-, or lens-shaped with thin walls and a round, relatively small pore.

Sphaeroceridae [INV ZOO] A family of myodarian cyclorrhaphous dipteran insects in the subsection Acalypteratae.

Sphaerodoridae [INV ZOO] A family of polychaete annelids belonging to the Errantia in which species are characterized by small bodies, and are usually papillated.

Sphaerolaimidae [INV ZOO] A family of free-living nematodes in the superfamily Monhysteroidea characterized by a spacious and deep stoma.

Sphaeromatidae [INV ZOO] A family of isopod crustaceans in the suborder Flabellifera in which the body is broad and oval and the inner branch of the uropod is immovable.

Sphaerophoraceae [BOT] A family of the Ascolichens in the order Caliciales which are fruticose with a solid thallus.

Sphaeropleineae [BOT] A suborder of green algae in the order Ulotrichales distinguished by long, coenocytic cells, numerous bandlike chloroplasts, and heterogametes produced in undifferentiated vegetative cells.

Sphaeropsidaceae [MYCOL] An equivalent name for Sphaerioidaceae.

Sphaeropsidales [MYCOL] An order of fungi of the class Fungi Imperfecti in which asexual spores are formed in pycnidia, which may be separate or joined to vegetative hyphae, conidiophores are short or absent, and conidia are always slime spores.

Sphagnaceae [BOT] The single monogeneric family of the order Sphagnales.

Sphagnales [BOT] The single order of mosses in the subclass Sphagnobrya containing the single family Sphagnaceae.

Sphagnobrya [BOT] A subclass of the Bryopsida; plants are grayish-green with numerous, spirally arranged branches and grow in deep tufts or mats, commonly in bogs and in other wet habitats.

sphagnum bog [ECOL] A bog composed principally of mosses of the genus *Sphagnum* (Sphagnales) but also of other plants, especially acid-tolerant species, which tend to form peat.

Sphecidae [INV ZOO] A large family of hymenopteran insects in the superfamily Sphecoidea.

Sphecoidea [INV ZOO] A superfamily of wasps belonging to the suborder Apocrita.

Sphenacodontia [PALEON] A suborder of extinct reptiles in

SPERMIOGENESIS

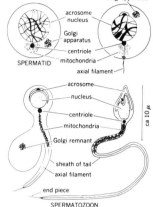

Diagram of spermiogenesis showing sequence of nuclear and cytoplasmic transformations of a spermatid into a spermatozoon; applies to the mammalian sperm.

SPERM WHALE

The sperm whale *(Physeter catodon)* has a head that measures 20 feet (6 meters), or one-third of the total body length.

SPHECIDAE

Typical member of the family Sphecidae.

the order Pelycosauria which were advanced, active carnivores.

sphenethmoid [VERT ZOO] A bone that surrounds the anterior portion of the brain in many amphibians.

Spheniscidae [VERT ZOO] The single family of the avian order Sphenisciformes.

Sphenisciformes [VERT ZOO] The penguins, an order of aquatic birds found only in the Southern Hemisphere and characterized by paddlelike wings, erect posture, and scalelike feathers.

Sphenodontidae [VERT ZOO] A family of lepidosaurian reptiles in the order Rhynchocephalia represented by a single living species, *Sphenodon punctatus*, a lizardlike form distinguished by lack of a penis.

sphenoid [CRYSTAL] An open crystal, occurring in monoclinic crystals of the sphenoidal class, and characterized by two nonparallel faces symmetrical with an axis of twofold symmetry. [SCI TECH] Wedge-shaped.

sphenoid bone [ANAT] The butterfly-shaped bone forming the anterior part of the base of the skull and portions of the cranial, orbital, and nasal cavities.

sphenoid sinus [ANAT] Either of a pair of paranasal sinuses located centrally between and behind the eyes, below the ethymoid sinus.

sphenopalatine [ANAT] Of or pertaining to the region of or surrounding the sphenoid and palatine bones.

sphenopalatine foramen [ANAT] The space between the sphenoid and orbital processes of the palatine bone; it opens into the nasal cavity and gives passage to branches from the pterygopalatine ganglion and the sphenopalatine branch of the maxillary artery.

sphenoparietal index [ANTHRO] The ratio, multiplied by 100, of the breadth of the skull from stenion to stenion to its greatest breadth.

Sphenophyllatae *See* Sphenophyllopsida.

Sphenophyllopsida [BOT] An extinct class of embryophytes in the division Equisetophyta. Formerly known as Sphenophyllatae.

Sphenopsida [BOT] A group of vascular cryptogams characterized by whorled, often very small leaves and by the absence of true leaf gaps in the stele; essentially equivalent to the division Equisetophyta.

spheraster [INV ZOO] A globular spicule having many rays.

spherocyte [PATH] A spherical red blood cell.

spherocytosis [MED] Preponderance of spherocytes in the blood.

sphincter [ANAT] A muscle that surrounds and functions to close an orifice.

sphincter of Oddi [ANAT] Sphincter of the hepatopancreatic ampulla.

Sphinctozoa [PALEON] A group of fossil sponges in the class Calcarea which have a skeleton of massive calcium carbonate organized in the form of hollow chambers.

Sphindidae [INV ZOO] The dry fungus beetles, a family of coleopteran insects in the superfamily Cucujoidea.

Sphingidae [INV ZOO] The single family of the lepidopteran superfamily Sphingoidea.

Sphingoidea [INV ZOO] A superfamily of Lepidoptera in the suborder Heteroneura consisting of the sphinx, hawk, or hummingbird moths; these are heavy-bodied forms with antennae that are thickened with a pointed apex, a well-developed proboscis, and narrow wings.

sphingolipid [BIOCHEM] Any lipid, such as a sphingomyelin, that yields sphingosine or one of its derivatives as a product of hydrolysis.

sphingolipidosis [MED] Any of a group of hereditary metabolic disorders characterized by excessive accumulations of certain glycolipids and phospholipids in various tissues of the body.

sphingomyelin [BIOCHEM] A phospholipid consisting of choline, sphingosine, phosphoric acid, and a fatty acid.

sphingosine [BIOCHEM] $C_{18}H_{37}O_2N$ A moiety of sphingomyelin, cerebrosides, and certain other phosphatides.

sphygmoid [BIOL] Pulsating.

sphygmomanometer [MED] An instrument for measuring the arterial blood pressure.

Sphyraenidae [VERT ZOO] A family of shore fishes in the suborder Mugiloidei of the order Perciformes comprising the barracudas.

Sphyriidae [INV ZOO] A family of ectoparasitic Crustacea belonging to the group Lernaeopodoida; the parasite embeds its head and part of its thorax into the host.

spice [FOOD ENG] An aromatic vegetable material used for food seasoning.

spicule [BOT] An empty diatom shell. [INV ZOO] A calcareous or siliceous, usually spikelike supporting structure in many invertebrates, particularly in sponges and alcyonarians.

spiculum [INV ZOO] A bristlelike copulatory organ in certain nematodes. Also known as copulatory spicule.

spider [INV ZOO] The common name for arachnids comprising the order Araneida.

spider nevus [MED] A type of telangiectasis characterized by a central, elevated, tiny red dot, pinhead in size, from which blood vessels radiate like strands of a spider's web. Also known as stellar nevus.

Spiegler's test [PATH] A test for the presence of protein in urine performed by overlaying clear acidulated urine on Spiegler's reagent (mercuric chloride, tartaric acid, glycerin, distilled water); opalescence at the fluid junction indicates protein.

spike [BOT] An indeterminate inflorescence with sessile flowers.

spikelet [BOT] The compound inflorescence of a grass consisting of one or several bracteate spikes.

spike-tooth harrow [AGR] An implement with steel spikes extending downward from a frame and pulled by a tractor to pulverize and smooth plowed soil.

spina bifida [MED] A congenital anomaly characterized by defective closure of the vertebral canal with herniation of the spinal cord meninges.

spina bifida occulta [MED] An asymptomatic congenital anomaly consisting of incomplete fusion of the posterior arch of the vertebral canal without hernial protrusion of the meninges.

spinacene *See* squalene.

spinach [BOT] *Spinacia oleracea.* An annual potherb of Asiatic origin belonging to the order Caryophyllales and grown for its edible foliage.

spinal anesthesia [MED] **1.** Anesthesia due to a lesion of the spinal cord. **2.** Anesthesia produced by the injection of an anesthetic into the spinal subarachnoid space.

spinal column *See* spine.

spinal cord [ANAT] The cordlike posterior portion of the central nervous system contained within the spinal canal of the vertebral column of all vertebrates.

spinal foramen [ANAT] Central canal of the spinal cord.

spinal ganglion [ANAT] Any one of the sensory ganglions, each associated with the dorsal root of a spinal nerve.

spinal nerve [ANAT] Any of the paired nerves arising from the spinal cord.

SPIKELET

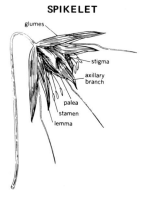

glumes

stigma

axillary branch

palea

stamen

lemma

Spikelet of wild oat *(Avena fatua).*

SPINE

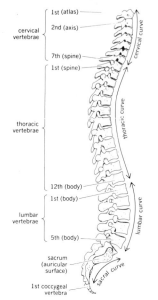

cervical vertebrae
- 1st (atlas)
- 2nd (axis)
- 7th (spine)

- 1st (spine)

thoracic vertebrae

- 12th (body)
- 1st (body)

lumbar vertebrae

- 5th (body)

sacrum (auricular surface)

1st coccygeal vertebra

cervical curve
thoracic curve
lumbar curve
sacral curve

The human spine in lateral view. (From W. J. Hamilton et al., Textbook of Human Anatomy, Macmillon, 1956)

SPINOBLAST

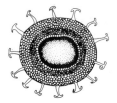

Spinoblast of *Pectinatella magnifica*.

SPINTHERIDAE

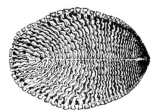

Spinther, dorsal view.

spinal reflex [PHYSIO] A reflex mediated through the spinal cord without the participation of the more cephalad structures of the brain or spinal cord.

spinasternum [INV ZOO] An intersegmental sclerite with an internal spine on the sternum of certain insects.

spindle [CYTOL] A structure formed of fiberlike elements just before metaphase that extends between the poles of the achromatic figure and is attached to the centromeric regions of the chromatid pairs.

spindle fiber [CYTOL] One of the fiberlike elements of the spindle; an aggregation of microtubules resulting from the polymerization of a series of small protein fibrils by primary $-S-S-$ linkages.

spindle tuber [PL PATH] A virus disease of the potato characterized by spindliness of the tops and tubers.

spine [ANAT] An articulated series of vertebrae forming the axial skeleton of the trunk and tail, and being a characteristic structure of vertebrates. Also known as backbone; spinal column; vertebral column. [BOT] A rigid sharp-pointed process in plants; many are modified leaves. [INV ZOO] One of the processes covering the surface of a sea urchin. [VERT ZOO] 1. One of the spiny rays supporting the fins of most fishes. 2. A sharp-pointed modified hair on certain mammals, such as the porcupine.

spinisternite [INV ZOO] A small sternite with a spiniform apodema, between the segments of the thorax in insects.

spinneret [INV ZOO] An organ that spins fiber from the secretion of silk glands.

spinning glands [INV ZOO] Glands which secrete threadlike material for webs in spiders and for cocoons in caterpillars.

spinoblast [INV ZOO] A statoblast having a float of air cells and barbs or hooks on the surface.

spinochrome [BIOCHEM] A type of echinochrome; an organic pigment that is known only from sea urchins and certain homopteran insects.

spinous process [ANAT] Any slender, sharp-pointed projection on a bone.

spin state [QUANT MECH] Condition of a particle in which its total spin, and the component of its spin along some specified axis, have definite values; more precisely, the particle's wave function is an eigenfunction of the operators corresponding to these quantities.

Spintheridae [INV ZOO] An amphinomorphan family of small polychaete annelids included in the Errantia.

Spinulosida [INV ZOO] An order of Asteroidea in which pedicellariae rarely occur, marginal plates bounding the arms and disk are small and inconspicuous, and spines occur in groups on the upper surface.

spiny-rayed fish [VERT ZOO] The common designation for actinopterygian fishes, so named for the presence of stiff, unbranched, pointed fin rays, known as spiny rays.

Spionidae [INV ZOO] A family of spioniform annelid worms belonging to the Sedentaria.

spioniform worm [INV ZOO] A polychaete annelid characterized by the presence of a pair of short to long, grooved palpi near the mouth.

spiracle [INV ZOO] An external breathing orifice of the tracheal system in insects and certain arachnids. [VERT ZOO] 1. The external respiratory orifice in cetaceous and amphibian larvae. 2. The first visceral cleft in fishes.

spiral cleavage [EMBRYO] A cleavage pattern characterized by formation of a cell mass showing spiral symmetry; occurs in mollusks.

spiral ganglion [ANAT] The ganglion of the cochlear part of the vestibulocochlear nerve embedded in the spiral canal of the modiolus.

spiral ligament [ANAT] The reticular connective tissue connecting the basilar membrane to the outer cochlear wall in the ear of mammals.

spiral organ *See* organ of Corti.

spiral valve [VERT ZOO] A spiral fold of mucous membrane in the small intestine of elasmobranchs and some primitive fishes which increases the surface area for absorption.

spiramycin [MICROBIO] A complex of related antibiotics, which resemble erythromycin structurally and in antibacterial spectrum, produced by *Streptomyces ambofaciens*.

spiraster [INV ZOO] A spiral spicule bearing rays in Porifera.

spire [BOT] A narrow, tapering blade or stalk. [INV ZOO] The overall sequence of whorls of a spiral shell.

spiricle [BOT] Any of the coiled threads in certain seed coats which uncoil when moistened.

Spiriferida [PALEON] An order of fossil articulate brachiopods distinguished by the spiralium, a pair of spirally coiled ribbons of calcite supported by the crura.

Spiriferidina [PALEON] A suborder of the extinct brachiopod order Spiriferida including mainly ribbed forms having laterally or ventrally directed spires, well-developed interareas, and a straight hinge line.

Spirillaceae [MICROBIO] A family of bacteria of the order Pseudomonadales comprising chemoheterotrophic, gram-negative, curved to spirally twisted rods, with rigid cell walls and polar flagella.

Spirillinacea [INV ZOO] A superfamily of foraminiferan protozoans in the suborder Rotaliina characterized by a planispiral or low conical test with a wall composed of radial calcite.

Spirobrachiidae [INV ZOO] A family of the Brachiata in the order Thecanephria.

Spirochaetaceae [MICROBIO] A family of the Spirochaetales including the large spirochetes, which are about 30 to 500 microns long.

Spirochaetales [MICROBIO] An order of bacteria characterized by elongate cells twisted three-dimensionally into a spiral shape.

spirochetal jaundice *See* Weil's disease.

spirochete [MICROBIO] A member of the Spirochaetales, distinguished by their spiral form and motility.

spirochetemia [MED] The presence of spirochetes in the blood.

spirocyst [INV ZOO] A thin-walled capsule that contains a long, unarmed, evertable, spirally coiled thread of uniform diameter; found in coelenterates.

spirograph [MED] An instrument for registering respiration.

spirometry [PHYSIO] The measurement, by a form of gas meter (spirometer), of volumes of air that can be moved in or out of the lungs.

spironeme [INV ZOO] The spiral thread in the stalk of an infusorian.

spironolactone [PHARM] $C_{24}H_{32}O_4S$ A steroid having a lactone ring attached at carbon-17; used as a diuretic.

Spirotrichia [INV ZOO] A subclass of the protozoan class Ciliatea which contains those ciliates characterized by conspicuous, compound ciliary structures, known as cirri, and buccal organelles.

Spirulidae [INV ZOO] A family of cephalopod mollusks containing several species of squids.

Spiruria [INV ZOO] A subclass of nematodes in the class Secernentea.

Spirurida [INV ZOO] An order of phasmid nematodes in the subclass Spiruria.

SPIRAL VALVE

ventral lobe of pancreas

pyloric end of stomach

blood vessels

spiral valve

large intestine

Small intestine of shark *(Squalus acanthias)* cut open to show spiral valve. *(From C. K. Weichert, Anatomy of the Chordates, 3d ed., McGraw-Hill, 1965)*

SPIRIFERIDINA

Dorsal view of *Spinocyrtia* valve.

Spiruroidea [INV ZOO] A superfamily of spirurid nematodes which are parasitic in the respiratory and digestive systems of vertebrates.

splanchnic mesoderm [EMBRYO] The internal layer of the lateral mesoderm that is associated with the entoderm after the formation of the coelom.

splanchnic nerve [ANAT] A nerve carrying nerve fibers from the lower thoracic paravertebral ganglions to the collateral ganglions.

splanchnocoel [EMBRYO] The body cavity of the embryo formed by splitting of the mesoderm.

splanchnocranium [ANAT] Portions of the skull derived from the primitive skeleton of the gill apparatus.

splanchnopleure [EMBRYO] The inner layer of the mesoblast from which part of the wall of the alimentary canal and portions of the visceral organs are derived in coelomates.

spleen [ANAT] A blood-forming lymphoid organ of the circulatory system, present in most vertebrates.

splenectomy [MED] Surgical removal of the spleen.

splenic fever *See* anthrax.

splenic flexure [ANAT] An abrupt turn of the colon beneath the lower end of the spleen, connecting the descending with the transverse colon.

splenium [ANAT] The rounded posterior extremity of the corpus callosum. [MED] A bandage.

splenomegaly [MED] Enlargement of the spleen.

split personality [PSYCH] The type of human being in whom there is a separation of various components of the normal personality unit, and each component functions as an independent entity.

spondylitis [MED] Inflammation of the vertebrae.

spondylolisthesis [MED] Forward displacement of a vertebra upon the one below as a result of a bilateral defect in the vertebral arch, or erosion of the articular surface of the posterior facets due to degenerative joint disease.

sponge [INV ZOO] The common name for members of the phylum Porifera.

spongicolous [ECOL] Living in sponges.

Spongiidae [INV ZOO] A family of sponges of the order Dictyoceratida; members are encrusting, massive, or branching in form and have small spherical flagellated chambers which characteristically join the exhalant canals by way of narrow channels.

Spongillidae [INV ZOO] A family of fresh- and brackish water sponges in the order Haplosclerida which are chiefly gray, brown, or white in color, and encrusting, massive, or branching in form.

spongin [BIOCHEM] A scleroprotein, occurring as the principal component of skeletal fibers in many sponges.

spongioblast [EMBRYO] A primordial cell arising from the ectoderm of the embryonic neural tube which differentiates to form the neuroglia, the ependymal cells, the neurolemma sheath cells, the satellite cells of ganglions, and Müller's fibers of the retina.

spongiocyte [HISTOL] **1.** A neuroglia cell. **2.** A cell of the adrenal cortex which has a spongy appearance due to the solution of lipids during tissue preparation for microscopical examination.

Spongiomorphida [PALEON] A small, extinct Mesozoic order of fossil colonial Hydrozoa in which the skeleton is a reticulum composed of perforate lamellae parallel to the upper surface and of regularly spaced vertical elements in the form of pillars.

Spongiomorphidae [PALEON] The single family of extinct hydrozoans comprising the order Spongiomorphida.

SPONGILLIDAE

An encrusting spongillid sponge growing on a twig.

spongocoel [INV ZOO] The branching, internal cavity of a sponge, connected to the outside by way of the osculum.

spongophyll [BOT] A leaf with spongy parenchyma, and no palisade tissue, between the upper and the lower epidermis, as in certain aquatic plants.

spongy mesophyll [BOT] A system of loosely and irregularly arranged parenchymal cells with numerous intercellular spaces found near the lower surface in well-differentiated broad leaves. Also known as spongy parenchyma.

spongy parenchyma *See* spongy mesophyll.

spontaneous abortion [MED] An unexpected, premature expulsion of the fetus.

spontaneous amputation [MED] **1.** Congenital amputation. **2.** Amputation not caused by external trauma or injury, as in ainhum.

spontaneous generation *See* abiogenesis.

spontaneous mutation [GEN] A mutation that occurs naturally.

spoon nail *See* koilonychia.

sporangiocarp [BOT] An enclosed cluster of sporangia.

sporangiophore [BOT] A stalk or filament on which sporangia are borne.

sporangiospore [BOT] A spore that forms in a sporangium.

sporangium [BOT] A case in which asexual spores are formed and borne.

spore [BIOL] A uni- or multicellular, asexual, reproductive or resting body that is resistant to unfavorable environmental conditions and produces a new vegetative individual when the environment is favorable.

spore mother cell [BOT] One of the cells of the archespore of a sporebearing plant from which a spore, but usually a tetrad of spores, is produced.

sporidium [MYCOL] A small spore, especially one formed on a promycelium.

sporine [INV ZOO] Pertaining to a sessile vegetative condition in protists in which cell division may take place.

sporoblast [INV ZOO] A sporozoan cell from which sporozoites arise.

Sporobolomycetaceae [MYCOL] The single family of the order Sporobolomycetales.

Sporobolomycetales [MYCOL] An order of yeastlike and moldlike fungi assigned to the class Basidiomycetes characterized by the formation of sterigmata, upon which the asexual ballistospores are formed.

sporocarp [BOT] Any multicellular structure in or on which spores are formed.

sporocyst [BOT] A unicellular resting body from which asexual spores arise. [INV ZOO] **1.** A resistant envelope containing an encysted sporozoan. **2.** An encysted sporozoan. **3.** The first reproductive form of a digenetic trematode in which rediae develop.

sporodesm [BOT] **1.** A compound spore, each of whose cells can germinate independently. **2.** A multilocular, septate, or pluricellular spore.

sporogenesis [BIOL] **1.** Reproduction by means of spores. **2.** Formation of spores.

sporogonium [BOT] In mosses, a structure formed from a fertilized oosphere of an archegonium; gives rise to asexual spores.

sporogony [BIOL] Reproduction by means of spores. [INV ZOO] Propagative reproduction involving formation, by sexual processes, and subsequent division of a zygote.

sporont [INV ZOO] A stage in the life history of sporozoans which gives rise to spores.

SPONGOCOEL

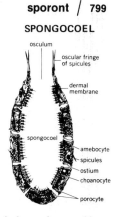

Morphology of asconoid calcareous sponge—longitudinal section. *(After Hyman, 1940)*

sporophore [MYCOL] A structure on the thallus of fungi which produces spores.

sporophyll [BOT] A modified leaf that develops sporangia.

sporophyte [BOT] **1.** An individual of the spore-bearing generation in plants exhibiting alternation of generation. **2.** The spore-producing generation. **3.** The diplophase in a plant life cycle.

sporopollenin [BIOCHEM] A substance related to suberin and cutin but more resistant to decay that is found in the exine of pollen grains.

sporosac [INV ZOO] A degenerate gonophore in certain hydroid coelenterates.

sporotrichosis [MED] A granulomatous fungus disease caused by *Sporotrichum schenckii*, with cutaneous lesions along the lymph channels and occasionally involving the internal organs. Also known as de Beurmann-Gougerot disease; Schenk's disease.

Sporozoa [INV ZOO] A subphylum of parasitic Protozoa, typically producing spores during the asexual stages of the life cycle.

sporozoid [BIOL] A motile spore.

sporozoite [INV ZOO] A motile, infective stage of certain sporozoans, which is the result of sexual reproduction and which gives rise to an asexual cycle in the new host.

sporulation [BIOL] The act and process of spore formation.

spot blight *See* grease spot.

spot blotch [PL PATH] A fungus disease of barley caused by *Helminthosporium sativum* and characterized by the appearance of dark, elongated spots on the foliage.

spot film [MED] A small, highly collimated radiograph of an anatomic part, usually obtained in conjunction with fluoroscopy.

spotted wilt [PL PATH] A virus disease of various crop and wild plants, especially tomato, characterized by bronzing and downward curling of the leaves.

spout hole [VERT ZOO] **1.** A blowhole of a cetacean mammal. **2.** A nostril of a walrus or seal.

sprain [MED] A wrenching of a joint, producing a stretching or laceration of the ligaments.

sprain fracture [MED] An injury in which a tendon or ligament, together with a shell of bone, is torn from its attachment; occurs most commonly at the ankle.

spread [STAT] The range within which the values of a variable quantity occur.

spreading factor *See* hyaluronidase.

Sprigginidae [INV ZOO] An extinct family of annelid worms distinguished by a horseshoe-shaped head.

springwood [BOT] The portion of an annual ring that is formed principally during the growing season; it is softer, more porous, and lighter than summerwood because of its higher proportion of large, thin-walled cells.

spruce [BOT] An evergreen tree belonging to the genus *Picea* characterized by single, four-sided needles borne on small peglike projections, pendulous cones, and resinous wood.

sprue [MED] A syndrome characterized by impaired absorption of food, water, and minerals by the small intestine; symptoms are the result of nutritional deficiencies.

Spumellina [INV ZOO] The equivalent name for Peripylina.

spur [BOT] **1.** A hollow process at the base of a petal or sepal. **2.** A short fruit-bearing tree branch. **3.** A short projecting root. [ZOO] A stiff, sharp outgrowth, as on the legs of certain birds and insects.

spur blight [PL PATH] A fungus disease of raspberries and blackberries caused by *Didymella applanata* which kills the fruit spurs and causes dark spotting of the cane.

SPRIGGINIDAE

Spriggina floundersi fossil showing characteristic horseshoe-shaped head.

sputum [PHYSIO] Material discharged from the surface of the respiratory passages, mouth, or throat; may contain saliva, mucus, pus, microorganisms, blood, or inhaled particulate matter in any combination.

squalene [BIOCHEM] $C_{30}H_{50}$ A liquid triterpene which is found in large quantities in shark liver oil, and which appears to play a role in the biosynthesis of sterols and polycyclic terpenes; used as a bactericide and as an intermediate in the synthesis of pharmaceuticals. Also known as spinacene.

Squalidae [VERT ZOO] The spiny dogfishes, a family of squaloid elasmobranchs recognized by their well-developed fin spines.

squama [ANAT] **1.** The vertical part of the frontal bone. **2.** That part of the occipital bone lying above and behind the foramen magnum. **3.** The upper, anterior part of the temporal bone. [BIOL] **1.** A scale. **2.** A scalelike part. [INV ZOO] **1.** A scale below the wing base in Diptera. **2.** A scalelike structure attached to the second podomere of an antenna of some Crustacea.

Squamata [VERT ZOO] An order of reptiles, composed of the lizards and snakes, distinguished by a highly modified skull that has only a single temporal opening, or none, by the lack of shells or secondary palates, and by possession of paired penes on the males.

squamosal bone [ANAT] The part of the temporal bone in man corresponding with the squamosal bone in lower vertebrates. [VERT ZOO] A membrane bone lying external and dorsal to the auditory capsule of many vertebrate skulls.

squamous [BIOL] Covered with or composed of scales.

squamous-cell carcinoma [MED] A carcinoma composed principally of anaplastic, squamous epithelial cells. Also known as epidermoid carcinoma.

squamous epithelium [HISTOL] A single-layered epithelium consisting of thin, flat cells.

squamulose [BIOL] Covered with or composed of minute scales.

square-root law [STAT] The standard deviation of the ratio of the number of successes to number of trials is inversely proportional to the square root of the number of trials.

squarrose [BOT] Having stiff divergent bracts, or other processes.

squarrulose [BOT] Mildly squarrose.

squash [BOT] Either of two plants of the genus *Cucurbita*, order Violales, cultivated for its fruit; some types are known as pumpkins.

Squatinidae [VERT ZOO] A group of squaloid elasmobranchs of uncertain affinity characterized by a greatly extended rostrum with enlarged denticles along the margins; maximum length is under 4 feet (1.2 meters).

squid [INV ZOO] Any of a number of marine cephalopod mollusks characterized by a reduced internal shell, ten tentacles, an ink sac, and chromatophores.

Squillidae [INV ZOO] The single family of the eumalacostracan order Stomatopoda, the mantis shrimp.

squint [MED] *See* strabismus.

squirrel [VERT ZOO] Any of over 200 species of arboreal rodents of the families Sciuridae and Anomaluridae having a bushy tail and long, strong hind limbs.

Sr *See* strontium.

S stage [CYTOL] The stage of the mitotic cycle, following G_1, in which DNA synthesis takes place.

stab culture [MICROBIO] A culture of anaerobic bacteria made by piercing a solid agar medium in a test tube with an inoculating needle covered with the bacterial inoculum.

stabilized feedback *See* negative feedback.

SQUAMOUS EPITHELIUM

Cellular arrangement in squamous epithelial tissue.

SQUASH

(a)

(b)

Two varieties of squash *(Cucurbita pepo)*. *(a)* Summer Crookneck. *(b)* Summer Bergen.

SQUID

Dorsal view of a squid *(Loligo)*, a cephalopod mollusk having worldwide distribution.

STAFFELLIDAE

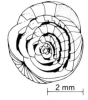

A representative of the fusulinacean family Staffellidae.

STAMEN

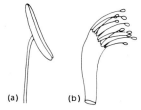

(a) (b)

Two types of stamens:
(a) with versatile anther;
(b) monodelphous, with filaments united into one set.

stabilizing selection [GEN] A type of selection within populations in which individuals closer to the mean for a character are favored over those at the extremes.

stable position effect [GEN] The production of a stable mutant phenotype as a consequence of a change in the position of a gene.

stachyosporous [BOT] Having the ovules attached terminally on branchlike structures (telomes), rather than on modified leaves.

Stader's splint [MED] A metal bar with pins affixed at right angles; the pins are driven into the fragments of a fracture, and the bar maintains the alignment.

Staffellidae [PALEON] An extinct family of marine protozoans (superfamily Fusulinacea) that persisted during the Pennsylvanian and Early Permian.

staggers [VET MED] Any of various diseases of livestock, sheep, and horses manifested by lack of coordination in movement and a staggering gait.

staghead *See* witches'-broom disease.

stain [MATER] **1.** A nonprotective coloring matter used on wood surfaces; imparts color without obscuring the wood grains. **2.** Any colored, organic compound used to stain tissues, cells, cell components, cell contents, or other biological substrates for microscopic examination.

stalked barnacle [INV ZOO] The common name for crustaceans composing the suborder Lepadomorpha.

stallion [VERT ZOO] **1.** A mature male equine mammal. **2.** A male horse not castrated.

stamen [BOT] The male reproductive structure of a flower, consisting of an anther and a filament.

staminate [BOT] Producing, or composed of, stamens.

stand [ECOL] A growth of plants, especially trees, on a given area. [OCEANOGR] The interval at high or low water when there is no appreciable change in the height of the tide. Also known as tidal stand.

standard conditions [PHYS] **1.** A temperature of 0°C and a pressure of 1 atmosphere (760 torr). Also known as standard temperature and pressure (STP). **2.** According to the American Gas Association, a temperature of 60°F (15 5/9°C) and a pressure of 762 millimeters (30 inches) of mercury. **3.** According to the Compressed Gas Institute, a temperature of 20°C (68°F) and a pressure of 1 atmosphere.

standard deviation [STAT] The positive square root of the expected value of the square of the difference between a random variable and its mean.

standard electrode potential [PHYS CHEM] The reversible or equilibrium potential of an electrode in an environment where reactants and products are at unit activity.

standard heat of formation [THERMO] The heat needed to produce one mole of a compound from its elements in their standard state.

standard state [PHYS] The stable and pure form of a substance at standard pressure and ordinary temperature.

stand fire [FOR] A forest fire igniting in the trunks of trees.

standing knee height [ANTHRO] The vertical distance taken from the top of the kneecap to the floor as the subject stands.

stand method [FOR] The practice of successively cutting trees of different ages so that ultimately the trees in the stand are new growth of a uniform age.

Stanford achievement test [PSYCH] A group test employing a primary (grades 2 and 3) and an advanced (grades 4 through 9) examination to measure achievement at various grade levels.

stapedius muscle [ANAT] The muscle which attaches to and controls the stapes in the inner ear.

stapes [ANAT] The stirrup-shaped middle-ear ossicle, articulating with the incus and the oval window. Also known as columella.

Staphylinidae [INV ZOO] The rove beetles, a very large family of coleopteran insects in the superfamily Staphylinoidea.

Staphylinoidea [INV ZOO] A superfamily of Coleoptera in the suborder Polyphaga.

staphylococcal pneumonia [MED] A severe form of pneumonia caused by *Staphylococcus aureus.*

staphylococcus toxoid [IMMUNOL] Formaldehyde-detoxified toxins of *Staphylococcus aureus* whose antigenicity is maintained; used prophylactically and therapeutically in the treatment of various staphylococcic pyodermas and localized pyogenic processes.

staphylomycin [MICROBIO] An antibiotic composed of three active components produced by a strain of *Actinomyces* that inhibits growth of gram-positive microorganisms and acid-fast bacilli.

staphylotoxin [BIOCHEM] Any of the various toxins elaborated by strains of *Staphylococcus aureus,* including hemolysins, enterotoxins, and leukocidin.

starch [BIOCHEM] Any one of a group of carbohydrates or polysaccharides, of the general composition $(C_6H_{10}O_5)_n$, occurring as organized or structural granules of varying size and markings in many plant cells; it hydrolyzes to several forms of dextrin and glucose; its chemical structure is not completely known, but the granules consist of concentric shells containing at least two fractions: an inner portion called amylose, and an outer portion called amylopectin.

starfish [INV ZOO] The common name for echinoderms belonging to the subclass Asteroidea.

startle response [PHYSIO] The complex, involuntary, usually spasmodic psychophysiological response movement of an organism to a sudden unexpected stimulus.

stasis [MED] A cessation of the normal flow of blood or other body fluids.

stasis dermatitis [MED] Chronic inflammation of the skin of the legs, resulting from poor circulation.

state [PHYS] The condition of a system which is specified as completely as possible by observations of a specified nature, for example, thermodynamic state, energy state.

stationary phase [MICROBIO] The period following termination of exponential growth in a bacterial culture when the number of viable microorganisms remains relatively constant for a time.

statistic [STAT] An estimate or piece of data, concerning some parameter, obtained from a sampling.

statistical distribution *See* distribution of a random variable.

statistical independence [STAT] Two events are statistically independent if the probability of their occurring jointly equals the product of their respective probabilities.

statistics [MATH] A discipline dealing with methods of obtaining data, analyzing and summarizing it, and drawing inferences from data samples by the use of probability theory.

statoblast [INV ZOO] A chitin-encapsulated body which serves as a special means of asexual reproduction in the Phylactolaemata.

statocone [INV ZOO] One of the minute calcareous granules found in the statocyst of certain animals.

statocyst [BOT] A cell containing statoliths in a fluid medium. Also known as statocyte. [INV ZOO] A sensory vesicle containing statoliths and which functions in the perception of the position of the body in space.

statocyte *See* statocyst.

statokinetic [PHYSIO] Pertaining to the balance and posture

STAPHYLINIDAE

A representative member of the coeleopteran superfamily Staphylinidae.

of the body or its parts during movement, as in walking.

statolith [BOT] A sand grain or other solid inclusion which moves readily in the fluid contents of a statocyst, comes to rest on the lower surface of the cell, and is believed to function in gravity perception. [INV ZOO] A secreted calcareous body, a sand grain, or other solid inclusion contained in a statocyst.

statoreceptor [PHYSIO] A sense organ concerned primarily with equilibrium.

statorhabd [INV ZOO] A short tentacular process bearing the statolith in Trachomedusae.

stature [ANTHRO] A measure of the distance from the floor to the vertex of the head, taken either front or back as the subject stands erectly with heels together.

status asthmaticus [MED] Intractable asthma lasting from a few days to a week or longer.

status epilepticus [MED] Occurrence of prolonged, generalized epileptic seizures in rapid succession with brief intervals of coma.

Stauromedusae [INV ZOO] An order of the class Scyphozoa in which the medusa is composed of a cuplike bell called a calyx and a stem that terminates in a pedal disk.

Staurosporae [MYCOL] A spore group of the Fungi Imperfecti characterized by star-shaped or forked spores.

steapsin [BIOCHEM] An enzyme in pancreatic juice that catalyzes the hydrolysis of fats. Also known as pancreatic lipase.

Steatornithidae [VERT ZOO] A family of birds in the order Caprimulgiformes which contains a single, South American species, the oilbird or guacharo (*Steatornis caripensis*).

steatorrhea [MED] **1.** Fatty stool. **2.** Increased flow of sebum.

Steganopodes [VERT ZOO] Formerly, an order of birds that included the totipalmate swimming birds.

steganopodous [VERT ZOO] Having fully webbed feet.

stegocarpous [BOT] Having a capsule with both an operculum and a peristome.

Stegodontinae [PALEON] An extinct subfamily of elephantoid proboscideans in the family Elephantidae.

Stegosauria [PALEON] A suborder of extinct reptiles of the order Ornithischia comprising the plated dinosaurs of the Jurassic which had tiny heads, great triangular plates arranged on the back in two alternating rows, and long spikes near the end of the tail.

Steinheim man [PALEON] A prehistoric man represented by a skull, without mandible, found near Stuttgart, Germany; the browridges are massive, the face is relatively small, and the braincase is similar in shape to that of *Homo sapiens*.

Stein-Leventhal syndrome [MED] A complex of symptoms characterized by amenorrhea or abnormal uterine bleeding or both, enlarged polycystic ovaries, frequently hirsutism, and occasionally retarded breast development.

Steinmann pin [MED] A surgical nail inserted in distal portions of such bones as the femur or tibia for skeletal tractions.

stele [BOT] The part of a plant stem including all tissues and regions of plants from the cortex inward, including the pericycle, phloem, cambium, xylem, and pith.

Stelenchopidae [INV ZOO] A family of polychaete annelids belonging to the Myzostomaria, represented by a single species from Crozet Island in the Antarctic Ocean.

stellar nevus *See* spider nevus.

stellate ganglion [ANAT] The ganglion formed by the fusion of the inferior cervical and the first thoracic sympathetic ganglions.

stellate reticulum [HISTOL] The part of the epithelial dental organ of a developing tooth which lies between the inner and the outer dental epithelium; composed of stellate cells with

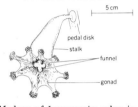

STAUROMEDUSAE

5 cm

pedal disk

stalk

funnel

gonad

Medusa of *Lucernaria*; calyx is eight-sided and has eight groups of short, capped tentacles on its margin.

long, anastomosing processes in a mucoid fluid in the interstitial spaces.

Stelleroidea [INV ZOO] The single class of echinoderms in the subphylum Asterozoa; characters coincide with those of the subphylum.

stem [BOT] The organ of vascular plants that usually develops branches and bears leaves and flowers.

stem blight [PL PATH] Any of various fungus blights that affect the plant stem.

stem break *See* browning.

stem canker [PL PATH] Any canker disease affecting the stem.

stem cell [EMBRYO] A formative cell. [HISTOL] *See* hemocytoblast.

stem-cell leukemia [MED] A form of leukemia in which the predominant cell type is so poorly differentiated that its series cannot be identified.

Stemonitaceae [MYCOL] The single family of the order Stemonitales.

Stemonitales [MYCOL] An order of fungi in the subclass Myxogastromycetidae of the class Myxomycetes.

stem rust [PL PATH] Any of several fungus diseases, especially of grasses, affecting the stem and marked by black or reddish-brown lesions.

Stenetrioidea [INV ZOO] A group of isopod crustaceans in the suborder Asellota consisting mostly of tropical marine forms in which the first pleopods are fused.

Stenocephalidae [INV ZOO] A family of Old World, neotropical Hemiptera included in the Pentatomorpha.

stenocephaly [MED] Unusual narrowness of the head.

Stenoglossa [INV ZOO] The equivalent name for Neogastropoda.

Stenolaemata [INV ZOO] A class of marine ectoproct bryozoans having lophophores which are circular in basal outline and zooecia which are long, slender, tubular or prismatic, and gradually tapering to their proximal ends.

Stenomasteridae [PALEON] An extinct family of Euchinoidea in the order Holasteroida comprising oval and heart-shaped forms with fully developed pore pairs.

stenopetalous [BOT] Having narrow petals.

stenophagous [BIOL] Feeding on a limited variety of food.

stenophyllous [BOT] Having narrow leaves.

Stenopodidea [INV ZOO] A section of decapod crustaceans in the suborder Natantia which includes shrimps having the third pereiopods chelate and much longer and stouter than the first pair.

stenopodium [INV ZOO] A crustacean limb with a protopodite that bears both an endopodite and an exopodite on the distal end.

stenosepalous [BOT] Having narrow sepals.

stenosis [MED] Constriction or narrowing, as of the heart or blood vessels.

Stenostomata [INV ZOO] The equivalent name for Cyclostomata.

Stenothoidae [INV ZOO] A family of amphipod crustaceans in the suborder Gammaridea containing semiparasitic and commensal species.

Stensen's duct *See* parotid duct.

Stensioellidae [PALEON] A family of Lower Devonian placoderms of the order Petalichthyida having large pectoral fins and a broad subterminal mouth.

Stenurida [PALEON] An order of Ophiuroidea, comprising the most primitive brittlestars, known only from Paleozoic sediments.

Stephanidae [INV ZOO] A small family of the Hymenoptera

in the superfamily Ichneumonoidea characterized by many-segmented filamentous antennae.

stepping reflex [PHYSIO] A reflex response of the newborn and young infant, characterized by alternating stepping movements with the legs, as in walking, elicited when the infant is held upright so that both soles touch a flat surface while the infant is moved forward to accompany any step taken.

stercobilin [BIOCHEM] Urobilin as a component of the brown fecal pigment.

stercobilinogen [BIOCHEM] A colorless reduction product of stercobilin found in feces.

Stercorariidae [VERT ZOO] A family of predatory birds of the order Charadriiformes including the skuas and jaegers.

Sterculiaceae [BOT] A family of dicotyledonous trees and shrubs of the order Malvales distinguished by imbricate or contorted petals, bilocular anthers, and ten to numerous stamens arranged in two or more whorls.

stereid [BOT] A lignified parenchyma cell having pit canals.

stereoblastula [EMBRYO] A blastula that lacks a cavity, making it unable to gastrulate.

stereochemistry [PHYS CHEM] The study of the spatial arrangement of atoms in molecules and the chemical and physical consequences of such arrangement.

stereocilia [CYTOL] **1.** Nonmotile tufts of secretory microvilli on the free surface of cells of the male reproductive tract. **2.** Homogeneous cilia within simple membrane coverings; found on the free-surface hair cells.

stereogastrula [EMBRYO] A gastrula that lacks a cavity.

stereognosis [PSYCH] The recognition or identification of objects by the sense of touch.

stereome [BOT] Rigid cellular tissue, such as sclerenchyma and collenchyma. [INV ZOO] Material composing the exoskeleton of an invertebrate.

stereopsis *See* stereoscopy.

stereoscopic vision *See* stereoscopy.

stereoscopy [PHYSIO] The phenomenon of simultaneous vision with two eyes in which there is a vivid perception of the distances of objects from the viewer; it is present because the two eyes view objects in space from two points, so that the retinal image patterns of the same object are slightly different in the two eyes. Also known as stereopsis; stereoscopic vision.

Stereospondyli [PALEON] A group of labyrinthodont amphibians from the Triassic characterized by a flat body without pleurocentra and with highly developed intercentra.

stereospondylous [VERT ZOO] Having the components of the vertebrae fused into one piece.

stereotaxis [BIOL] An orientation movement in response to stimulation by contact with a solid body. Also known as thigmotaxis.

stereotropism [BIOL] Growth or orientation of a sessile organism or part of an organism in response to the stimulus of a solid body. Also known as thigmotropism.

steric effect [PHYS CHEM] The influence of the spatial configuration of reacting substances upon the rate, nature, and extent of reaction.

sterigma [BOT] A peg-shaped structure to which needles are attached in certain conifers. [MYCOL] A slender stalk arising from the basidium of some fungi, from the top of which basidiospores are formed by abstriction.

sterility [PHYSIO] The inability to reproduce because of congenital or acquired reproductive system disorders involving lack of gamete production or production of abnormal gametes.

sterilization [MICROBIO] An act or process of destroying all forms of microbial life on and in an object.

sterilizer [ENG] An apparatus for sterilizing by dry heat, steam, or water.

Sternaspidae [INV ZOO] A monogeneric family of polychaete annelids belonging to the Sedentaria.

sternebra [VERT ZOO] A segment of the sternum in vertebrates.

sternellum [INV ZOO] A sclerite of the sternum of an insect.

Sterninae [VERT ZOO] A subfamily of birds in the family Laridae, including the Arctic tern.

sternite [INV ZOO] **1.** The ventral part of an arthropod somite. **2.** The chitinous plate on the ventral surface of an abdominal segment of an insect.

sternocleidomastoid [ANAT] **1.** Pertaining to the sternum, the clavicle, and the mastoid process. **2.** A neck muscle; functions in flexing the head.

sternocostal [ANAT] Pertaining to the sternum and the ribs.

sternohyoid [ANAT] A muscle arising from the manubrium of the sternum and inserted into the hyoid bone.

sternopleurite [INV ZOO] A thoracic sclerite in insects, formed by fusion of the episternum and the sternum.

Sternorrhyncha [INV ZOO] A series of the insect order Homoptera in which the beak appears to arise either between or behind the fore coxae, and the antennae are long and filamentous with no well-differentiated terminal setae.

sternothyroid [ANAT] Pertaining to the sternum and thyroid cartilage.

Sternoxia [INV ZOO] The equivalent name for Elateroidea.

sternum [ANAT] The bone, cartilage, or series of bony or cartilaginous segments in the median line of the anteroventral part of the body of vertebrates above fishes, connecting with the ribs or pectoral girdle.

sternum height [ANTHRO] Vertical distance taken from the lower tip of the sternum to the floor as the subject stands.

steroid [BIOCHEM] A compound composed of a series of four carbon rings joined together to form a structural unit called cyclopentanoperhydrophenanthrene.

sterol [BIOCHEM] Any of the natural products derived from the steroid nucleus; all are waxy, colorless solids soluble in most organic solvents but not in water, and contain one alcohol functional group.

sterrula [INV ZOO] The solid free-swimming larva of Alcyonaria; it precedes the planula.

stethoscope [MED] An instrument for indirect auscultation for the detection and study of sounds arising within the body; sounds are conveyed to the ears of the examiner through rubber tubing connected to a funnel or disk-shaped endpiece.

Sthenurinae [PALEON] An extinct subfamily of marsupials of the family Diprotodontidae, including the giant kangaroos.

stibophen [PHARM] $C_{12}H_4Na_5O_{16}S_4Sb \cdot 7H_2O$ A crystalline compound that is soluble in water, insoluble in organic solvents; used in medicine for protozoan infections. Also known as sodium antimony bis(pyrocatechol-2,4-disulfonate).

Stichaeidae [VERT ZOO] The pricklebacks, a family of perciform fishes in the suborder Blennioidei.

Stichocotylidae [INV ZOO] A family of trematodes in the subclass Aspidogastrea in which adults are elongate and have a single row of alveoli.

Stichopodidae [INV ZOO] A family of the echinoderm order Aspidochirotida characterized by tentacle ampullae and by left and right gonads.

Stickland reaction [BIOCHEM] An amino acid fermentation

STEROID

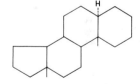

Shorthand formulation for steroid skeleton; lines attached to the rings represent methyl groups.

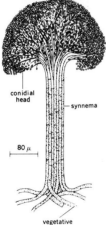

The three-spined stickleback. It is used for fish meal and oil in some Baltic countries.

STILBELLACEAE

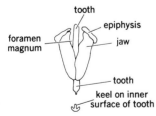

conidial head

synnema

80 μ

vegetative hyphae

Graphium ulmi, cause of Dutch elm disease. *(After C. Ferdinandsen and C. A. Jörgensen, 1938–1939)*

STIRODONT DENTITION

tooth
epiphysis
foramen magnum
jaw
tooth
keel on inner surface of tooth

Tooth and lantern structure in stirodont dentition.

involving the coupled decomposition of two or more substrates.

stickleback [VERT ZOO] Any fish which is a member of the family Gasterosteidae, so named for the variable number of free spines in front of the dorsal fin.

stigma [BOT] The rough or sticky apical surface of the pistil for reception of the pollen. [INV ZOO] **1.** The eyespot of certain protozoans, such as *Euglena.* **2.** The spiracle of an insect or arthropod. **3.** A colored spot on many lepidopteran wings.

stigmatism [PHYSIO] A condition of the refractive media of the eye in which rays of light from a point are accurately brought to a focus on the retina.

Stilbaceae [MYCOL] The equivalent name for Stilbellaceae.

Stilbellaceae [MYCOL] A family of fungi of the order Moniliales in which conidiophores are aggregated in long bundles or fascicles, forming synnemata or coremia, generally having the conidia in a head at the top.

stilbesterol *See* diethylstilbesterol.

stillbirth [MED] Birth of a dead infant.

Still's disease [MED] Juvenile rheumatoid arthritis in which involvement of the viscera is prominent. Also known as Chauffard-Still disease.

stimulation [BIOL] Excitation by external or internal influences evoking a characteristic physiologic activity.

stimulation deafness [MED] Deafness induced by noise; involves changes in the chemical interchange between the canals of the cochlea, as well as nerve destruction.

stimulus filtering [PSYCH] The apparent awareness of only a few of the great number of stimuli bombarding an animal.

stinger [ZOO] A sharp piercing organ, as of a bee, stingray, or wasp, usually connected with a poison gland.

stinging cell *See* cnidoblast.

stingray [VERT ZOO] Any of various rays having a whiplike tail armed with a long serrated spine, at the base of which is a poison gland.

stipe [BOT] **1.** The petiole of a fern frond. **2.** The stemlike portion of the thallus in certain algae. [MYCOL] The short stalk or stem of the fruit body of a fungus, such as a mushroom.

stipes [INV ZOO] **1.** The peduncle of a stalked eye. **2.** The distal part of a protopodite of the first maxilla of insects.

stipule [BOT] Either of a pair of appendages that are often present at the base of the petiole of a leaf.

Stirodonta [INV ZOO] Formerly, an order of Euechinoidea that included forms with stirodont dentition.

stirodont dentition [INV ZOO] In Echinoidea, the condition in which the teeth are keeled within and the foramen magnum is open.

stock [BOT] The stem receiving the bud or scion in grafting. [INV ZOO] An asexual zooid which produces unisexual zooids by gemmation, as in Polychaeta.

Stokes-Adams syndrome [MED] Syncopic or convulsive attacks occurring in patients with complete heart block.

Stokes' law [FL MECH] At low velocities, the frictional force on a spherical body moving through a fluid at constant velocity is equal to 6π times the product of the velocity, the fluid viscosity, and the radius of the sphere. [SPECT] The wavelength of luminescence excited by radiation is always greater than that of the exciting radiation.

stolon [BOT] *See* runner. [INV ZOO] An elongated projection of the body wall from which buds are formed giving rise to new zooids in Anthozoa, Hydrozoa, Bryozoa, and Ascidiacea. [MYCOL] A hypha produced above the surface and connecting a group of conidiophores.

Stolonifera [INV ZOO] An order of the Alcyonaria, lacking a coenenchyme; they form either simple (*Clavularia*) or rather complex colonies (*Tubipora*).

stoma [BIOL] A small opening or pore in a surface. [BOT] One of the minute openings in the epidermis of higher plants which are regulated by guard cells and through which gases and water vapor are exchanged between internal spaces and the external atmosphere.

stomach [ANAT] The tubular or saccular organ of the vertebrate digestive system located between the esophagus and the intestine and adapted for temporary food storage and for the preliminary stages of food breakdown.

stomatitis [MED] Inflammation of the soft tissues in the mouth.

stomatology [MED] The branch of medical science that concerns the anatomy, physiology, pathology, therapeutics, and hygiene of the oral cavity, of the tongue, teeth, and adjacent structures and tissues, and of the relationship of that field to the entire body.

Stomatopoda [INV ZOO] The single order of the Eumalacostraca in the superorder Hoplocarida distinguished by raptorial arms, especially the second pair of maxillipeds.

Stomiatoidei [VERT ZOO] A suborder of fishes of the order Salmoniformes including the lightfishes and allies, which are of small size and often grotesque form and are equipped with photophores.

stomion [INV ZOO] Any of the dermal pores or ostia which perforate the dermal membrane of a developing sponge.

stomium [BOT] **1.** A strip of thin-walled cells in fern sporangium where a mature capsule ruptures. **2.** The slit of a dehiscing anther.

stomocnidae nematocyst [INV ZOO] A nematocyst which has an open-ended thread.

stomodaeum [EMBRYO] The anterior part of the embryonic alimentary tract formed as an invagination of the ectoderm.

stone canal [INV ZOO] A canal in many echinoderms that has a more or less calcified wall and that leads from the madreporite to the ring vessel.

stone cell *See* brachysclereid.

stone fruit *See* drupe.

stonewort [BOT] The common name for algae comprising the class Charophyceae, so named because most·species are lime-encrusted.

stony coral [INV ZOO] Any coral characterized by a calcareous skeleton.

stork [VERT ZOO] Any of several species of long-legged wading birds in the family Ciconiidae.

strabismus [MED] Incoordinate action of the extrinsic ocular muscles resulting in failure of the visual axes to meet at the desired objective point. Also known as cast; heterotropia; squint.

straight sinus [ANAT] A sinus of the dura mater running from the inferior sagittal sinus along the junction of the falx cerebri and tentorium to the transverse sinus.

strangulated hernia [MED] A hernia involving the intestine in which circulation of the blood and fecal current are blocked.

strangulation [MED] **1.** Asphyxiation due to obstruction of the air passages, as by external pressure on the neck. **2.** Constriction of a part producing arrest of the circulation, as strangulation of a hernia.

stratification [ECOL] Vertical grouping within a community. [SCI TECH] Arrangement in layers.

stratified sampling [STAT] A random sample of specified size is drawn from each stratum of a population.

STOLONIFERA

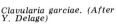

Clavularia garciae. (After Y. Delage)

STRATIFIED SQUAMOUS EPITHELIUM

Drawing of a section through stratified squamous epithelium showing arrangement of cells.

STRAUSS REACTION

Strauss reaction in the guinea pig.

STRAWBERRY

Flower and fruit of a strawberry plant.

stratified squamous epithelium [HISTOL] A multiple-layered epithelium composed of thin, flat superficial cells and cuboidal and columnar deeper cells.

Stratiomyidae [INV ZOO] The soldier flies, a family of orthorrhaphous dipteran insects in the series Brachycera.

stratum corneum [HISTOL] The outer layer of flattened keratinized cells of the epidermis.

stratum disjunctum [HISTOL] The outermost layer of desquamating keratinized cells of the stratum corneum.

stratum germinativum [HISTOL] The innermost germinative layer of the epidermis.

stratum granulosum [HISTOL] A layer of granular cells interposed between the stratum corneum and the stratum germinativum in the thick skin of the palms and soles.

stratum lucidum [HISTOL] A layer of irregular transparent epidermal cells with traces of nuclei interposed between the stratum corneum and stratum germinativum in the thick skin of the palms and soles.

Strauss reaction [IMMUNOL] The exudative swelling of the scrotum in male hamsters and guinea pigs upon subcutaneous or intraperitoneal inoculation of *Pseudomonas mallei,* the causative agent of glanders.

straw [AGR] Grain stalks after threshing and usually mixed with leaves and chaff. [BOT] A stem of grain, such as wheat or oats.

strawberry [BOT] A low-growing perennial of the genus *Fragaria,* order Rosales, that spreads by stolons; the juicy, usually red, edible fruit consists of a fleshy receptacle with numerous seeds in pits or nearly superficial on the receptacle.

strawberry hemangioma [MED] A vascular birthmark characterized by a soft, raised, bright-red, lobular appearance.

strawwalker [AGR] A set of reciprocating notched bars inside a thresher or combine that push the straw to the rear.

streak plate [MICROBIO] A method of culturing aerobic bacteria by streaking the surface of a solid medium in a petri dish with an inoculating wire or glass rod in a continuous movement so that most of the surface is covered; used to isolate majority members of a mixed population.

Streblidae [INV ZOO] The bat flies, a family of cyclorrhaphous dipteran insects in the section Pupipara; adults are ectoparasites on bats.

Strelitziaceae [BOT] A family of monocotyledonous plants in the order Zingiberales distinguished by perfect flowers with five functional stamens and without an evident hypanthium, penniveined leaves, and symmetrical guard cells.

strepogenin [BIOCHEM] A factor, possibly a peptide derivative of glutamic acid, reported to exist in certain proteins, acting as a growth stimulant in bacteria and mice in the presence of completely hydrolyzed protein. Also known as streptogenin.

Strepsiptera [INV ZOO] An order of the Coleoptera that is coextensive with the family Stylopidae.

streptobacillary fever *See* Haverhill fever.

streptobiosamine [BIOCHEM] $C_{13}H_{23}NO_9$ A nitrogen-containing disaccharide, obtained when streptomycin undergoes acid hydrolysis; in the streptomycin molecule it is glycosidally linked to streptidine.

Streptococceae [MICROBIO] A tribe of the family Lactobacillaceae including cocci that occur in pairs, short chains, or tetrads and which generally obtain energy by fermentation of carbohydrates or related compounds.

streptokinase [BIOCHEM] An enzyme occurring as a component of fibrinolysin in cultures of certain hemolytic streptococci.

streptolysin [BIOCHEM] Any of a group of hemolysins elaborated by *Streptococcus pyogenes.*

Streptomyces [MICROBIO] A genus of the family Streptomycetaceae which are aerobic, nonacid-fast, saprophytic soil-inhabiting organisms with branching filaments and conidia produced in aerial hyphae in chains.

Streptomycetaceae [MICROBIO] A family of soil-inhabiting bacteria which form nonfragmenting vegetative mycelia with conidia borne on sporophores.

streptomycin [MICROBIO] $C_{21}H_{39}O_{12}N_7$ Water-soluble antibiotic obtained from *Streptomyces griseus* that is used principally in the treatment of tuberculosis.

streptothricin [MICROBIO] $C_{19}H_{34}O_7N_8$ An antibiotic produced by *Streptomyces lavendulae*; active against various gram-negative and gram-positive microorganisms.

stress [BIOL] A stimulus or succession of stimuli of such magnitude as to tend to disrupt the homeostasis of the organism.

stretch reflex [PHYSIO] Contraction of a muscle in response to a sudden, brisk, longitudinal stretching of the same muscle. Also known as myostatic reflex.

stria [BIOL] A minute line, band, groove, or channel.

striated muscle [HISTOL] Muscle tissue consisting of muscle fibers having cross striations.

stridor [MED] A peculiar, harsh, vibrating sound produced during respiration.

stridulation [INV ZOO] Creaking and other audible sounds made by certain insects, produced by rubbing various parts of the body together.

Strigidae [VERT ZOO] A family of birds of the order Strigiformes containing the true owls.

Strigiformes [VERT ZOO] The order of birds containing the owls.

strigose [BIOL] Covered with stiff, pointed, hairlike scales or bristles.

Strigulaceae [BOT] A family of Ascolichenes in the order Pyrenulales comprising crustose species confined to tropical evergreens, and which form extensive crusts on or under the cuticle of leaves.

strip-cropping [AGR] Growing separate crops in adjacent strips that follow the contour of the land as a method of reducing soil erosion.

strip method [FOR] A lumbering method in which timbers are cleared from a forest in strips; new growth in the strip results from seeds sown in the adjoining forest.

strip survey [FOR] A survey of the value of a strip of forest; used to estimate the value of a larger area of the forest.

strobilation [INV ZOO] Asexual reproduction by segmentation of the body into zooids, proglottids, or separate individuals.

strobilocercus [INV ZOO] A larval tapeworm that has undergone strobilation.

strobilus [BIOL]·A chain of similar units, as of the proglottides of tapeworms, or the individuals produced by successive transverse division of a scyphozoan scyphistoma larva. [BOT] **1.** A conelike structure made up of sporophylls, or spore-bearing leaves, as in Equisetales. **2.** The cone of members of the Pinophyta.

stroma [ANAT] The supporting tissues of an organ, including connective and nervous tissues and blood vessels.

stromal endometriosis *See* interstitial endometrosis.

stromal myosis *See* interstitial endometrosis.

Stromateidae [INV ZOO] A family of perciform fishes in the suborder Stromateoidei containing the butterfishes.

Stromateoidei [VERT ZOO] A suborder of fishes of the order

STRONTIUM

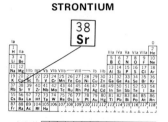

Periodic table of the chemical elements showing the position of strontium.

STROPHANTHUS

(a)   (b)

Strophanthus kombe seed
(a) in folicle, (b) out of folicle.

Perciformes in which most species have teeth in pockets behind the pharyngeal bone.

Stromatoporoidea [PALEON] An extinct order of fossil colonial organisms thought to belong to the class Hydrozoa; the skeleton is a coenosteum.

stromatosis *See* interstitial endometrosis.

Strombacea [PALEON] An extinct superfamily of gastropod mollusks in the order Prosobranchia.

Strombidae [INV ZOO] A family of gastopod mollusks comprising tropical conchs.

Strongyloidea [INV ZOO] The hookworms, an order or superfamily of roundworms which, as adults, are endoparasites of most vertebrates, including man.

strongyloidiasis [MED] An infestation of man with one of the roundworms of the genus *Strongyloides.*

strongylote [INV ZOO] Rounded at one end, referring to sponge spicules.

strontium [CHEM] A metallic element in group IIA, symbol Sr, with atomic number 38, atomic weight 87.62; flammable, soft, pale-yellow solid; soluble in alcohol and acids, decomposes in water; melts at 770°C, boils at 1380°C; chemistry is similar to that of calcium; used as electron-tube getter.

strophanthin [PHARM] A glycoside or mixture of glycosides extracted from the plant *Strophanthus kombe;* used as a cardioactive drug in the treatment of various heart ailments.

Strophanthus [BOT] A genus of woody climbers of the dogbane family (Apocynaceae), natives of tropical Asia and Africa; the source of arrow poisons and of strophanthin.

strophiole [BOT] A crestlike excrescence around the hilum in certain seeds.

Strophomenida [PALEON] A large diverse order of articulate brachiopods which first appeared in Lower Ordovician times and became extinct in the Late Triassic.

Strophomenidina [PALEON] A suborder of extinct, articulate brachiopods in the order Strophomenida characterized by a concavo-convex shell, the pseudodeltidium and socket plates disposed subparallel to the hinge.

structural gene *See* cistron.

struma lymphomatosa [MED] A type of chronic thyroiditis involving diffuse enlargement of the gland with retrogressive epithelial changes and lymphoid hyperplasia. Also known as Hashimoto's disease; Hashimoto's struma; lymphadenoid goiter.

struma ovarii [MED] An ovarian teratoma composed chiefly of thyroid tissue.

Struthionidae [VERT ZOO] The single family of the avian order Struthioniformes.

Struthioniformes [VERT ZOO] A monofamilial order of ratite birds containing the single living species of ostrich (*Struthio camelus*).

strychnine [ORG CHEM] $C_{21}H_{22}O_2N_2$ An alkaloid obtained primarily from the plant nux vomica, formerly used for therapeutic stimulation of the central nervous system.

strychninization [MED] The condition resulting from large doses of strychnine.

Stuart factor [BIOCHEM] A procoagulant in normal plasma but deficient in the blood of patients with a hereditary bleeding disorder; may be closely related to prothrombin since both are formed in the liver by action of vitamin K. Also known as factor X; Stuart-Power factor.

Stuart-Power factor *See* Stuart factor.

stubborn disease [PL PATH] A virus disease of citrus trees characterized by short internodes resulting in stiff brushy growth and chlorotic leaves.

Student's distribution [STAT] The probability distribution used to test the hypothesis that a random sample of *n* observations comes from a normal population with a given mean.

Student's t-statistic [STAT] A one-sample test statistic computed by $T = \sqrt{n}(\bar{X} - \mu_H)/S$, where $\bar{X}$ is the mean of a collection of *n* observations, *S* is the square root of the mean square deviation, and μ_H is the hypothesized mean.

Student's t-test [STAT] A test in a one-sample problem which uses Student's *t*-statistic.

stunt [PL PATH] Any of several plant diseases marked by reduction in size of the plant.

sturgeon [VERT ZOO] Any of 10 species of large bottom-living fish which comprise the family Acipenseridae; the body has five rows of bony plates, and the snout is elongate with four barbels on its lower surface.

stutter [MED] A speech disorder marked by repetition of words, syllables, or sounds, or by hesitations in manner, without apparent awareness by the speaker.

sty *See* hordeolum.

Styginae [INV ZOO] A subfamily of butterflies in the family Lycaenidae in which the prothoracic legs in the male are nonfunctional.

Stygocaridacea [INV ZOO] An order of crustaceans in the superorder Syncarida characterized by having a furca.

Stylasterina [INV ZOO] An order of the class Hydrozoa, including several brightly colored branching or encrusting coral-like coelenterates of warm seas.

style [BOT] The portion of a pistil connecting the stigma and ovary. [ZOO] A slender elongated process on an animal.

stylet [INV ZOO] A slender, rigid, elongated appendage. [MED] **1.** A slender probe used for surgery. **2.** A thin wire inserted in a catheter to provide support or in a hollow needle to clear the passage.

styloglossus [ANAT] A muscle arising from the styloid process of the temporal bone, and inserted into the tongue.

stylohyoid [ANAT] Pertaining to the styloid process of the temporal bone and the hyoid bone.

styloid [ZOO] Resembling a style.

stylomandibular [ANAT] Pertaining to the styloid process of the temporal bone and the mandible.

stylomastoid [ANAT] Relating to the styloid and the mastoid processes of the temporal bone.

Stylommatophora [INV ZOO] A large order of the molluscan subclass Pulmonata characterized by having two pairs of retractile tentacles with eyes located on the tips of the large tentacles.

stylopodium [BOT] A conical or disk-shaped enlargement at the base of the style in plants of the family Umbelliferae.

stylospore [MYCOL] A stalked spore.

stylostome [PATH] A tube in the skin formed by reaction of the host tissue to insertion of the chelicerae of a mite.

Stypocapitellidae [INV ZOO] A family of polychaete annelids belonging to the Sedentaria and consisting of a monotypic genus found in western Germany.

styramate [PHARM] $C_9H_{11}NO_3$ A compound that crystallizes from chloroform solution, and melts at 111–112°C; used in medicine as a muscle relaxant. Also known as carbamic acid β-hydroxyphenethyl ester.

subacute bacterial endocarditis *See* bacterial endocarditis.

subalpine [ECOL] Pertaining to the zone below the timber line.

subanconeus [ANAT] A variable muscle having its origin at the posterior distal surface of the humerus and its insertion in the posterior aspect of the capsule of the elbow joint.

STURGEON

Short-nosed sturgeon *(Acipenser brevirostrus)*, a species of sturgeon found in United States coastal waters.

subarachnoid hemorrhage [MED] Bleeding between the pia mater and the arachnoid of the brain.

subarachnoid space [ANAT] The space between the pia mater and the arachnoid of the brain.

subboreal [ECOL] A biogeographic zone whose climatic condition approaches that of the boreal.

subcardinal vein [VERT ZOO] Either of a pair of longitudinal veins of the mammalian embryo or the adult of some lower vertebrates which partly replace the postcardinals in the abdominal region, ventromedial to the mesonephros.

subcerebral plane [ANTHRO] The plane passing through a line traversing the lower angles of the parietal bones and the juncture of the superciliary ridge and the cheek bone.

subchela [INV ZOO] A prehensile claw whose last joint folds back on the preceding joint.

subclavian artery [ANAT] The proximal part of the principal artery in the arm or forelimb.

subclavian vein [ANAT] The proximal part of the principal vein in the arm or forelimb.

subclavius [ANAT] A small muscle attached to the clavicle and the first rib.

subclimax [ECOL] A community immediately preceding a climax in an ecological succession.

subcollateral [ANAT] Ventrad of the collateral sulcus of the brain.

subconscious [PSYCH] Pertaining to mental activity beyond the level of consciousness, including the preconscious and the unconscious.

subcrenate [BOT] Having rounded scallops on the margin, as certain leaves.

subcutaneous [BIOL] Beneath the cutis or skin.

subcutaneous connective tissue [HISTOL] The layer of loose connective tissue beneath the dermis.

subdural hematoma [MED] A mass of blood between the arachnoid and the dura mater.

subdural hemorrhage [MED] Bleeding between the dura mater and the arachnoid.

subendothelial layer [HISTOL] The middle layer of the tunica intima of veins and of medium and larger arteries, consisting of collagenous and elastic fibers and a few fibroblasts.

suber [BOT] Cork tissue.

suberin [BIOCHEM] A fatty substance found in many plant cell walls, especially cork.

suberization [BOT] Infiltration of plant cell walls by suberin resulting in the formation of corky tissue that is impervious to water.

suberose [BOT] Having a texture like cork due to or resembling that due to suberization.

subiculum [ANAT] **1.** A part of the hippocampus bordering the hippocampal fissure. **2.** A bony ridge bordering the oval opening in the wall of the middle ear. [MYCOL] A mass of hyphae forming a stratum which contains perithecia or pycnidia.

sublethal gene *See* semilethal gene.

subleukemic [MED] Less than leukemic; usually applied to states in which the peripheral blood manifestations of leukemia are temporarily suppressed.

sublimation [PSYCH] A defense mechanism whereby the energies of undesirable instinctual cravings and impulses are converted into socially acceptable activities. [THERMO] The process by which solids are transformed directly to the vapor state or vice versa without passing through the liquid phase.

subliminal [PHYSIO] Below the threshold of responsiveness, consciousness, or sensation to a stimulus.

sublingual [ANAT] **1.** Beneath the tongue. **2.** Pertaining to structures beneath the tongue.

sublingual gland [ANAT] A complex of salivary glands located in the sublingual fold on each side of the floor of the mouth.

sublittoral zone [OCEANOGR] The benthic region extending from mean low water (or 40–60 meters, according to some authorities) to a depth of about 100 fathoms (200 meters), or the edge of a continental shelf, beyond which most abundant attached plants do not grow.

submandibular duct [ANAT] The duct of the submandibular gland which empties into the mouth on the side of the frenulum of the tongue.

submandibular gland [ANAT] A large seromucous or mixed salivary gland located below the mandible on each side of the jaw. Also known as mandibular gland; submaxillary gland.

submaxillary gland *See* submandibular gland.

submentum [INV ZOO] The basal part of the labium of insects.

submerged culture [MICROBIO] A method for growing pure cultures of aerobic bacteria in which microorganisms are incubated in a liquid medium subjected to continuous, vigorous agitation.

submerged fermentation [MICROBIO] Industrial production of antibiotics, enzymes, and other substances by growing the microorganisms that produce the product in a submerged culture.

submucosa [HISTOL] The layer of fibrous connective tissue that attaches a mucous membrane to its subadjacent parts.

submucous plexus [ANAT] A visceral nerve network lying in the submucosa of the digestive tube. Also known as Meissner's plexus.

Suboscines [VERT ZOO] A major division of the order Passeriformes, usually divided into the suborders Eurylaimi, Tyranni, and Memirae.

subscapularis [ANAT] A muscle arising from the costal surface of the scapula and inserted on the lesser tubercle of the humerus.

subsere [ECOL] A secondary community that succeeds an interrupted climax.

subsidiary cell [BOT] One of the modified epidermal cells immediately adjacent to the guard cells on the lower surface of most leaves.

substrate [BIOCHEM] The substance upon which an enzyme acts. [BIOL] An inert material containing or receiving nutrients.

substratum [BIOL] The base to which a sessile organism is attached.

subsurface tillage [AGR] A method of stirring the soil with blades that leaves stubble on or just below the surface.

subtend [BOT] To lie adjacent to and below another structure, often enclosing it.

subtilin [MICROBIO] An antibiotic substance obtained from *Bacillus subtilis*, active against gram-positive bacteria.

Subtriquetridae [INV ZOO] A family of arthropods in the suborder Porocephaloidea.

subulate [BOT] Linear, delicate, and tapering to a sharp point.

Subulitacea [PALEON] An extinct superfamily of gastropod mollusks in the order Prosobranchia which possessed a basal fold but lacked an apertural sinus.

Subuluridae [INV ZOO] The equivalent name for Heterakidae.

subumbrella [INV ZOO] The concave undersurface of the body of a jellyfish.

SUCKER

A sucker, characterized by fleshy lips.

SUCTORIA

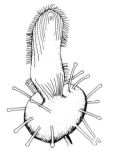

Endogenous budding in the suctorian *Podophrya*, a species which measures 10–28 micrometers.

succession [ECOL] A gradual process brought about by the change in the number of individuals of each species of a community and by the establishment of new species populations which may gradually replace the original inhabitants.

succession of crops [AGR] **1.** Growing a crop over a long season by either repeated sowings or a single sowing of varieties of the crop that mature at different rates. **2.** Growing two or more crops in a season on the same land by planting them in succession.

succinamide [BIOCHEM] $H_2NCOCH_2CONH_2$ The amide of succinic acid.

succinic acid [ORG CHEM] $CO_2H(CH_2)_2CO_2H$ Water-soluble, colorless crystals with an acid taste; melts at 185°C; used as a chemical intermediate, in medicine, and to make perfume esters. Also known as butanedioic acid.

succinic acid dehydrogenase [BIOCHEM] An enzyme that catalyzes the dehydrogenation of succinic acid to fumaric acid in the presence of a hydrogen acceptor. Also known as succinic dehydrogenase.

succinic dehydrogenase *See* succinic acid dehydrogenase.

succinoxidase [BIOCHEM] A complex enzyme system containing succinic dehydrogenase and cytochromes that catalyzes the conversion of succinate ion and molecular oxygen to fumarate ion.

succinylsulfathiazole [PHARM] $C_{13}H_{13}N_3O_5S_2$ A poorly absorbed sulfonamide used as an intestinal antibacterial agent in preoperative preparation of patients for abdominal surgery, and also postoperatively to maintain a low bacterial count.

succulent [BOT] Describing a plant having juicy, fleshy tissue.

succus [BOT] The fluid of a plant.

succus entericus [PHYSIO] The intestinal juice secreted by the glands of the intestinal mucous membrane; it is thin, opalescent, alkaline, and has a specific gravity of 1.011.

sucker [BOT] A shoot that develops rapidly from the lower portion of a plant, and usually at the expense of the plant. [ZOO] A disk-shaped organ in various animals for adhering to or holding onto an individual, usually of another species.

sucking disc [INV ZOO] A disc assisting in attachment by suction, as at the end of tube feet.

sucking louse [INV ZOO] The common name for insects of the order Anoplura, so named for the slender, tubular mouthparts.

sucrase *See* saccharase.

sucrose [ORG CHEM] $C_{12}H_{22}O_{11}$ Combustible, white crystals soluble in water, decomposes at 160 to 186°C; derived from sugarcane or sugarbeet; used as a sweetener in drinks and foods and to make syrups, preserves, and jams. Also known as saccharose; table sugar.

Suctoria [INV ZOO] A small subclass of the protozoan class Ciliatea, distinguished by having tentacles which serve as mouths.

suctorial [BIOL] **1.** Modified for sucking. **2.** Having suckers.

Suctorida [INV ZOO] The single order of the protozoan subclass Suctoria.

sudamen [MED] A skin disease in which sweat accumulates under the superficial horny layers of the epidermis to form small, clear, transparent vesicles.

sudatoria *See* hyperhidrosis.

sudden death syndrome *See* crib death.

sudomotor [PHYSIO] Pertaining to the efferent nerves that control the activity of sweat glands.

sudoriferous [PHYSIO] Conducting or secreting sweat.

suffrutescent [BOT] Of or pertaining to a stem intermediate between herbaceous and shrubby, becoming partly woody and perennial at the base.

suffruticose [BOT] Low stems which are woody, grading into herbaceous at the top.

sugar [BIOCHEM] A generic term for a class of carbohydrates usually crystalline, sweet, and water soluble; examples are glucose and fructose.

sugar alcohol [ORG CHEM] Any of the acyclic linear polyhydric alcohols; may be considered sugars in which the aldehydic group of the first carbon atom is reduced to a primary alcohol; classified according to the number of hydroxyl groups in the molecule; sorbitol (D-glucitol, sorbite) is one of the most widespread of all the naturally occurring sugar alcohols.

sugarbeet [BOT] *Beta vulgaris.* A beet characterized by a white root and cultivated for the high sugar content of the roots.

sugarcane [BOT] *Saccharum officinarum.* A stout, perennial grass plant characterized by two-ranked leaves, and a many-jointed stalk with a terminal inflorescence in the form of a silky panicle; the source of more than 50% of the world's annual sugar production.

sugarcane gummosis *See* Cobb's disease.

sugar maple [BOT] *Acer saccharum.* A commercially important species of maple tree recognized by its gray furrowed bark, sharp-pointed scaly winter buds, and symmetrical oval outline of the crown.

Suidae [VERT ZOO] A family of paleodont artiodactyls in the superfamily Suoidae including wild and domestic pigs.

sulcate [ZOO] Having furrows or grooves on the surface.

sulculus [ZOO] A small sulcus.

sulcus [ZOO] A furrow or groove, especially one on the surface of the cerebrum.

sulfadiazine [PHARM] $C_{10}H_{10}O_2N_4S$ An antibacterial sulfonamide used in the treatment of a variety of infections.

sulfa drug [PHARM] Any of a family of drugs of the sulfonamide type with marked bacteriostatic properties.

sulfaguanidine [PHARM] $C_7H_{10}N_4O_2S$ An intestinal antibacterial sulfonamide proposed for treatment of dysentery and for sterilization of the colon prior to gastrointestinal tract surgery.

sulfamerazine [PHARM] $C_{11}H_{12}N_4O_2S$ An antibacterial agent with uses similar to those of sulfadiazine, but generally used in combination with sulfadiazine and with sulfamethazine.

sulfanilamide [PHARM] $C_6H_8O_2N_2S$ White crystals slightly soluble in water, soluble in alcohol and most sulfa drugs, but less effective and more toxic than its derivatives. Also known as *para*-aminobenzenesulfonamide; sulfamidyl; sulfonamide P.

sulfapyridine [PHARM] $C_{11}H_{11}N_3O_2S$ A sulfonamide formerly used for the treatment of various infections but found to be too toxic for general use; now employed only as a suppressant for dermatitis herpetiformis.

sulfatase [BIOCHEM] Any of a group of esterases that catalyze the hydrolysis of sulfuric esters.

sulfathiazole [PHARM] $C_9H_9N_3O_2S_2$ A sulfa drug formerly widely used in the treatment of pneumococcal, staphylococcal, and urinary tract infections; it has been replaced by less toxic sulfonamides.

sulfatide lipidosis *See* metachromatic leukodystrophy.

sulfhemoglobin [BIOCHEM] A greenish substance derived from hemoglobin by the action of hydrogen sulfide; it may appear in the blood following the ingestion of sulfanilamide and other substances.

SUGARBEET

Root with leaves of a typical sugarbeet. *(USDA)*

SUGARCANE

Flowering sugarcane in Florida. *(USDA)*

sulfisoxazole [PHARM] $C_{11}H_{13}N_3O_3S$ A sulfonamide of general therapeutic utility; for parenteral administration the soluble salt sulfisoxazole diethanolamine is used; for pediatric use the tasteless derivative acetyl sulfisoxazole is given.

sulfonamide [ORG CHEM] One of a group of organosulfur compounds, RSO_2NH_2, prepared by the reaction of sulfonyl chloride and ammonia; used for sulfa drugs.

sulfur [CHEM] A nonmetallic element in group VIa, symbol S, atomic number 16, atomic weight 32.064, existing in a crystalline or amorphous form and in four stable isotopes; used as a chemical intermediate and fungicide, and in rubber vulcanization.

SULFUR

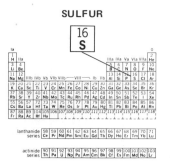

Periodic table of the chemical elements showing the position of sulfur.

SUNDEW

The sundew, *Drosera* species, a small, herbaceous, insectivorous plant. *(General Biological Supply House)*

SUNFISH

Black crappie *(Pomoxis nigromaculatus)*, a freshwater sunfish also known as the calico bass.

Sulidae [VERT ZOO] A family of aquatic birds in the order Pelecaniformes including the gannets and boobies.

summation [PHYSIO] The accumulation of effects, especially of those of muscular, sensory, or mental stimuli.

summerwood [BOT] The less porous, usually harder portion of an annual ring that forms in the latter part of the growing season.

sunburn [MED] Skin inflammation due to overexposure to sunlight.

sundew [BOT] Any plant of the genus *Drosera* of the family Droseraceae; the genus comprises small, herbaceous, insectivorous plants that grow on all continents, especially Australia.

sunfish [VERT ZOO] Any of several species of marine and freshwater fishes in the families Centrarchidae and Molidae characterized by brilliant metallic skin coloration.

sunflower [BOT] *Helianthus annuus.* An annual plant native to the United States characterized by broad, ovate leaves growing from a single, usually long (3–20 feet or 1–6 meters) stem, and large, composite flowers with yellow petals.

sunstroke [MED] Heat stroke resulting from prolonged exposure to the sun, characterized by extreme pyrexia, prostration, convulsion, and coma. Also known as thermic fever.

Suoidea [VERT ZOO] A superfamily of artiodactyls of the suborder Paleodonta which comprises the pigs and peccaries.

supercilium [ANAT] The eyebrow.

superego [PSYCH] The subdivision of the psyche that acts as the conscience of the unconscious; the components, derived from both the id and the ego, are associated with standards of behavior and self-criticism.

superfecundation [PHYSIO] Multiple, simultaneous fertilization by a number of sperm of many eggs released at ovulation.

superfemale [GEN] A female with three X chromosomes and two sets of autosomes resulting in sterility and generally early death.

superficial cleavage [EMBRYO] Meroblastic cleavage restricted to the peripheral cytoplasm, as in the centrolecithal insect ovum.

superficial palmar arch [ANAT] The arterial anastomosis formed by the ulnar artery in the palm with a branch from the radial artery. Also known as palmar arch.

superinfection [VIROL] An attack on a bacterial cell by several phages due to the introduction of large numbers of viruses into the bacterial culture.

superior [BOT] 1. Positioned above another organ or structure. 2. Referring to a calyx that is attached to the ovary. 3. Referring to an ovary that is above the insertion of the floral parts.

superior alveolar canals [ANAT] The alveolar canals of the maxilla.

superior ganglion [ANAT] 1. The upper sensory ganglion of the glossopharyngeal nerve, located in the upper part of the jugular foramen; it is inconstant. 2. The upper sensory ganglion of the vagus nerve, located in the jugular foramen.

superior mesenteric artery [ANAT] A major branch of the abdominal aorta with branches to the pancreas and intestine.

superior vena cava [ANAT] The principal vein collecting blood from the head, chest wall, and upper extremities and draining into the right atrium.

supermale [GEN] A male with one X chromosome and three or more sets of autosomes, resulting in sterility and generally early death.

supernumerary bud *See* accessory bud.

supernumerary chromosome [CYTOL] A chromosome present in addition to the normal chromosome complement. Also known as accessory chromosome.

superposed [BOT] **1.** Growing vertically over another part. **2.** Of or pertaining to floral parts that are opposite each other.

superposition eye [INV ZOO] A compound eye in which a given rhabdome receives light from a number of facets; visual acuity is reduced in this type of eye.

supination [ANAT] **1.** Turning the palm upward. **2.** Inversion of the foot.

suppository [PHARM] A medicated solid body of varying weight and shape, intended for introduction into different orifices of the body, as the rectum, urethra, or vagina; usually suppositories melt or are softened at body temperature, though in some instances release of medication is effected through use of a hydrophilic vehicle; typical vehicles or bases are theobroma oil (cocoa butter), glycerinated gelatin, sodium stearate, and propylene glycol monostearate.

suppressor gene [GEN] A mutant gene that masks the effect of another mutant, thus causing partial or total restoration of the normal phenotype; a suppressor can be intracistronic or nonallelic.

suppuration [MED] Pus formation.

supracardinal veins [VERT ZOO] Paired longitudinal veins in the mammalian embryo and various adult lower vertebrates in the thoracic and abdominal regions, dorsolateral to and on the sides of the descending aorta; they replace the postcardinal and subcardinal veins.

suprahyoid muscles [ANAT] The muscles attached to the upper margin of the hyoid bone.

supraliminal [PHYSIO] Above, or in excess of, a threshold.

supramastoid crest [ANAT] A bony ridge on the squamous part of the temporal bone behind the external acoustic meatus.

supranuclear [ANAT] In the nervous system, central to a nucleus.

supraoccipital [ANAT] Situated above the occipital bone.

supraoptic [ANAT] Situated above the optic tract.

suprarenal gland *See* adrenal gland.

suprascapula [ANAT] An anomalous bone sometimes found between the superior border of the scapula and the spines of the lower cervical or first thoracic vertebrae.

suprasegmental reflex [PHYSIO] A reflex employing complex multineuronal channels to integrate the body and limb musculature with fixed positions or movements of the head.

supraspinatus [ANAT] A muscle that originates above the spine of the scapula and has its insertion on the greater tubercle of the humerus.

suprasternal notch [ANAT] Jugular notch of the sternum.

suprasternal space [ANAT] The triangular space above the manubrium, enclosed by the layers of the deep cervical fascia which are attached to the front and back of this bone.

supravital [BIOL] Pertaining to the staining of living cells after removal from a living animal or of still living cells from a recently killed animal.

surculus [BOT] An underground shoot which ultimately becomes aerial and independent.

surface fire [FOR] A forest fire in which only surface litter and undergrowth burn.

surgery [MED] The branch of medicine that deals with conditions requiring operative procedures.

surgical needle [MED] Any sewing needle used in a surgical operation.

survival curve [NUCLEO] The curve obtained by plotting the number or percentage of organisms surviving at a given time against the dose of radiation, or the number surviving at different intervals after a particular dose of radiation.

survival ratio [BIOL] The number of organisms surviving irradiation by ionizing radiation divided by the number of organisms before irradiation.

suspension feeder [ZOO] An animal that feeds on small particles suspended in water; particles may be minute living plants or animals, or products of excretion or decay from these or larger organisms.

suspensor [BOT] A mass of cells in higher plants that pushes the embryo down into the embryo sac and into contact with the nutritive tissue. [MYCOL] A hypha which bears an apical gametangium in fungi of the Mucorales.

sustained yield [BIOL] In a biological resource such as timber or grain, the replacement of a harvest yield by growth or reproduction before another harvest occurs.

sustentacular cell [HISTOL] One of the supporting cells of an epithelium as contrasted with other cells with special function, as the nonnervous cells of the olfactory epithelium or the Sertoli cells of the seminiferous tubules.

suture [BIOL] A distinguishable line of union between two closely united parts. [MED] A fine thread used to close a wound or surgical incision.

svedberg [PHYS CHEM] A unit of sedimentation coefficient, equal to 10^{-13} second.

swamp [ECOL] A waterlogged land supporting a natural vegetation predominantly of shrubs and trees.

swamp fever *See* infectious anemia.

swamp itch *See* schistosome dermatitis.

swan [VERT ZOO] Any of several species of large waterfowl comprising the subfamily Anatinae; they are herbivorous stout-bodied forms with long necks and spatulate bills.

Swanscombe man [PALEON] A partial skull recovered in Swanscombe, Kent, England, which represents an early stage of *Homo sapiens* but differing in having a vertical temporal region and a rounded occipital profile.

swarm [BIOL] **1.** A large group of small, motile organisms considered collectively. **2.** A number of bees or termites active as a group.

swarmer cell [MICROBIO] The daughter cell which separates from the stalked mother cell in bacteria in the genus *Caulobacter*.

swayback [MED] Increased lumbar lordosis with compensatory increased thoracic kyphosis. [VET MED] Sinking of the back, or lordosis.

sweat [CHEM] Exudation of nitroglycerin from dynamite due to separation of nitroglycerin from its adsorbent. [PHYSIO] The secretion of the sweat glands. Also known as perspiration. [SCI TECH] Formation of moisture beads on a surface as a result of concentration.

sweat gland [PHYSIO] A coiled tubular gland of the skin which secretes sweat.

sweepstakes route [ECOL] A means that allows chance migration across a sea on natural rafts, so that oceanic islands can be colonized.

sweetgum [BOT] *Liquidambar styraciflua.* A deciduous tree of the order Hamamelidales found in the southeastern United States, and distinguished by its five-lobed, or star-shaped, leaves, and by the corky ridges developed on the twigs.

sweet oil *See* olive oil.

sweet orange oil [MATER] A sweet, yellow, mild essential oil expressed from the peel of an orange, *Citrus aurantium;* soluble in glacial acetic acid, somewhat in alcohol; used in flavors, perfumes, and medicine. Also known as orange oil.

sweet potato [BOT] *Ipomoea batatas.* A tropical vine having variously shaped leaves, purplish flowers, and a sweet, fleshy, edible tuberous root.

swim bladder [VERT ZOO] A gas-filled cavity found in the body cavities of most bony fishes; has various functions in different fishes, acting as a float, a lung, a hearing aid, and a sound-producing organ.

swimmeret [INV ZOO] Any of a series of paired biramous appendages under the abdomen of many crustaceans, used for swimming and egg carrying.

swimmer's conjunctivitis *See* inclusion conjunctivitis.

swimmer's itch *See* schistosome dermatitis.

swimming bird [VERT ZOO] Any bird belonging to the orders Charadriiformes and Pelacaniformes.

swimming funnel [INV ZOO] In dibranchiates, the tube through which water is expelled from the mantle cavity, providing a means of propulsion.

swimming plate [INV ZOO] In Ctenophora, any of eight ciliated comblike bands or ribs which serve to propel the organism.

swimming pool conjunctivitis *See* inclusion conjunctivitis.

swine [AGR] A domesticated member of the family Suidae. [VERT ZOO] Any of various species comprising the Suidae.

swine erysipelas [VET MED] An infectious bacterial disease of swine caused by *Erysipelothrix insidiosa* in which involvement of the skin is predominant.

swine influenza [VET MED] A disease of swine caused by the associated effects of a filterable virus and *Hemophilus suis,* characterized by inflammation of the upper respiratory tract.

swine plague [VET MED] Hemorrhagic septicemia of swine caused by *Pasteurella suiseptica,* characterized by pleuropneumonia.

swine pox [VET MED] A benign infection of young hogs characterized by pox lesions on the body and inner surfaces of the legs.

sycamore [BOT] **1.** Any of several species of deciduous trees of the genus *Platanus,* especially *P. occidentalis* of eastern and central North America, distinguished by simple, large, three- to five-lobed leaves and spherical fruit heads. **2.** The Eurasian maple (*Acer pseudoplatanus*).

Sycettida [INV ZOO] An order of calcareous sponges of the subclass Calcaronea in which choanocytes occur in flagellated chambers, and the spongocoel is not lined with these cells.

Sycettidae [INV ZOO] A family of sponges in the order Sycettida.

Sycidales [PALEON] A group of fossil aquatic plants assigned to the Charophyta, characterized by vertically ribbed gyrogonites.

sycon [INV ZOO] A canal system in sponges in which the flagellated layer is confined to outpocketings of the paragaster that are indirectly connected to the incurrent canals.

sycosis [MED] An inflammatory disease affecting the hair follicles, particularly of the beard, and characterized by papules, pustules, and tubercles, perforated by hairs, together with infiltration of the skin and crusting.

SWEET POTATO

The sweet, fleshy, tuberous roots of big-stem Jersey type sweet potatoes. *(USDA)*

SYCAMORE

Terminal bud, leaf with seed capsule, and twig of American sycamore *(Platanus occidentalis).*

SYLLIDAE

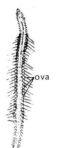

ova

Exogone, of the Syllidae
(Exogoninae), dorsal view
showing cirri and ova attached
to body segments.

SYMPHYLA

A symphylan, *Scutigerella
immaculata. (From R. E.
Snodgrass, A Textbook of
Arthropod Anatomy, Cornell
University Press, 1952)*

Sydenham's chorea   *See* St. Vitus dance.

Syllidae   [INV ZOO] A large family of polychaete annelids belonging to the Errantia; identified by their long, linear, translucent bodies with articulated cirri; size ranges from minute *Exogone* to *Trypanosyllis*, which may be 100 millimeters long.

Syllinae   [INV ZOO] A subfamily of polychaete annelids of the family Syllidae.

Sylonidae   [INV ZOO] A family of parasitic crustaceans in the order Rhizocephala.

Sylopidae   [INV ZOO] A family of coleopteran insects in the superfamily Meloidea in which the elytra in males are reduced to small leathery flaps while the hindwings are large and fan-shaped.

sylvatic plague   [VET MED] Plague occurring in rodents; may be transmitted to man. Also known as endemic rural plague.

Sylvicolidae   [INV ZOO] A family of orthorrhaphous dipteran insects in the series Nematocera.

symballophone   [ENG] A double stethoscope for the comparison and lateralization of sounds; permits the use of the acute function of the two ears to compare intensity and varying quality of sounds arising in the body or mechanical devices.

symbiont   [ECOL] A member of a symbiotic pair.

symbiosis   [ECOL] **1.** An interrelationship between two different species. **2.** An interrelationship between two different organisms in which the effects of that relationship is expressed as being harmful or beneficial. Also known as consortism.

symblepharon   [MED] Adhesion of the eyelids to the eyeball.

Symmetrodonta   [PALEON] An order of the extinct mammalian infraclass Pantotheria distinguished by the central high cusp, flanked by two smaller cusps and several low minor cusps, on the upper and lower molars.

symmetry   [BIOL] The disposition of organs and other constituent parts of the body of living organisms with respect to imaginary axes.

sympathetic nervous system   [ANAT] The portion of the autonomic nervous system, innervating smooth muscle and glands of the body, which upon stimulation produces a functional state of preparation for flight or combat.

sympathetic ophthalmia   [MED] A granulomatous inflammation of the uveal tract following ocular injury or intraocular surgery.

sympathicotropic cell   [HISTOL] Any of various cells possessing special affinity for the sympathetic nervous system.

sympathochromaffin cell   [HISTOL] One of the precursors of sympathetic and medullary cells in the adrenal medulla.

sympatholitic   [PHARM] Of or pertaining to an effect antagonistic to that of the sympathetic nervous system.

sympathomimetic   [PHARM] Having the ability to produce physiologic changes similar to those caused by action of the sympathetic nervous system.

sympatric   [ECOL] **1.** Of a species, occupying the same range as another species but maintaining identity by not interbreeding. **2.** Distributed in the same or overlapping geographical areas.

sympetalous   *See* gamopetalous.

symphile   [ECOL] An organism, usually a beetle, living as a guest in the nest of a social insect, such as an ant, where it is reared and bred in exchange for its exudates.

Symphyla   [INV ZOO] A class of the Myriapoda comprising tiny, pale, centipedelike creatures which inhabit humus or soil.

symphyllodium   [BOT] A compound structure formed by coalescence of the external coats of two or more ovules.

symphysis [ANAT] An immovable articulation of bones connected by fibrocartilaginous pads.

symphysis pubis *See* pubic symphysis.

Symphyta [INV ZOO] A suborder of the Hymenoptera including the sawflies and horntails characterized by a broad base attaching the abdomen to the thorax.

symplectic [VERT ZOO] In a fish skull, a bone between the quadrate and the hyomandibular.

sympodium [BOT] A branching system in trees in which the main axis is comprised of successive secondary branches, each representing the dominant fork of a dichotomy.

symptom [MED] A phenomenon of physical or mental disorder or disturbance which leads to complaints on the part of the patient.

symptomatology [MED] **1.** The science of symptoms. **2.** In common usage, the symptoms of disease taken together as a whole.

synacme [BOT] Simultaneous maturation of the stamens and the pistils.

Synallactidae [INV ZOO] A family of echinoderms of the order Aspidochirotida comprising mainly deep-sea forms which lack tentacle ampullae.

synandrous [BOT] Having several united stamens.

synangium [BOT] A compound sorus made up of united sporangia. [VERT ZOO] In lower vertebrates, a peripheral arterial trunk from which branches arise.

Synanthae [BOT] An equivalent name for Cyclanthales.

Synanthales [BOT] An equivalent name for Cyclanthales.

synantherous [BOT] Having the anthers united into a tube-like structure.

synanthous [BOT] **1.** Simultaneous appearance of the flowers and the leaves. **2.** Having the flowers united.

synapse [ANAT] A site where the axon of one neuron comes into contact with and influences the dendrites of another neuron or a cell body.

synapsid [VERT ZOO] Referring to skulls having the supra- and infratemporal fossae united in a single fossa.

synapsis [CYTOL] Pairing of homologous chromosomes during the zygotene stage of meiosis.

synaptic membrane [CYTOL] The single membrane at the synapse; separates the cytoplasm of the terminal enlargement of an axon from the cytoplasm of the neuron.

synaptic transmission [PHYSIO] The mechanisms by which a presynaptic neuron influences the activity of an anatomically adjacent postsynaptic neuron.

synapticulum [INV ZOO] A conical or cylindrical supporting process, as those extending between septa in some corals, or connecting gill bars in *Branchiostoma*.

Synaptidae [INV ZOO] A family of large sea cucumbers of the order Apodida lacking a respiratory tree and having a reduced water-vascular system.

synaptinemal complex [CYTOL] Ribbonlike structures that extend the length of synapsing chromosomes and are believed to function in exchange pairing.

synarthrosis [ANAT] An articulation in which the connecting material (fibrous connective tissue) is continuous, immovably binding the bones.

synascus [MYCOL] An ascogonium that contains several asci.

Synbranchiformes [VERT ZOO] An order of eellike actinopterygian fishes that, unlike true eels, have the premaxillae present as distinct bones.

Synbranchii [VERT ZOO] The equivalent name for Synbranchiformes.

Syncarida [INV ZOO] A superorder of crustaceans of the

SYMPODIUM

Sympodial branching in the American elm *(Ulmus americana)*. *(From E. W. Sinnott and K. S. Wilson, Botany: Principles and Problems, 5th ed., McGraw-Hill, 1955)*

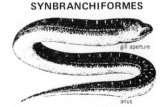

SYNBRANCHIFORMES

Rice eel *(Monopterus albus)*; length to 35 inches (89 centimeters). *(After M. Weber and L. F. de Beaufort, The Fishes of the Indo-Australian Archipelago, vol. 3, 1916)*

subclass Melacostraca lacking a carapace and oostegites and having exopodites on all thoracic limbs.

syncarp [BOT] A compound fleshy fruit.

syncarpous [BOT] Descriptive of a gynoecium having the carpels united in a compound ovary.

synchondrosis [ANAT] A type of synarthrosis in which the bone surfaces are connected by cartilage.

synchronous growth [MICROBIO] A population of bacteria in which all cells divide at approximately the same time.

syncope [MED] Swooning or fainting; temporary suspension of consciousness.

syncraniate [VERT ZOO] Having the vertebral elements fused with the skull.

synctial trophoblast *See* synctiotrophoblast.

syncytiotrophoblast [CYTOL] An irregular sheet or net of deeply staining cytoplasm in which nuclei are irregularly scattered. Also known as plasmoditrophoblast; synctial trophoblast.

syncytium [CYTOL] A mass of multinucleated cytoplasm without division into separate cells.

syndactyly [ANAT] The condition characterized by union of two or more digits, as in certain birds and mammals; it is a familial anomaly in man.

syndesmosis [ANAT] An articulation in which the bones are joined by collagen fibers.

syndrome [MED] A group of signs and symptoms which together characterize a disease. Also known as complex.

synecology [ECOL] The study of environmental relations of groups of organisms, such as communities.

Synentognathi [VERT ZOO] The equivalent name for Beloniformes.

synergid [BOT] Either of two small cells lying in the embryo sac in seed plants adjacent to the egg cell toward the micropylar end.

synergism [ECOL] An ecological association in which the physiological processes or behavior of an individual are enhanced by the nearby presence of another organism.

synesthesia [PHYSIO] The subjective association between stimulus of one type of sense receptor and concomitant response in another sense, as when an individual hearing music seems to see colors.

Syngamidae [INV ZOO] A family of roundworms belonging to the Strongyloidea and including parasites of birds and mammals.

syngamy [BIOL] Sexual reproduction involving union of gametes.

syngen [INV ZOO] A mating variety within species of protozoans; mating does not usually occur between syngens.

syngenesious [BOT] **1.** Having the anthers united in a cylindrical form. **2.** Having united anthers.

Syngnathidae [VERT ZOO] A family of fishes in the order Gasterosteiformes including the seahorses and pipefishes.

synkinesia [PHYSIO] Involuntary movement coincident with purposeful movements carried out by a distant part of the body, such as swinging the arms while walking. Also known as accessory movement; associated automatic movement.

synoecious [BOT] **1.** Having the antheridia and archegonia on the same receptacle. **2.** Having the stamens and pistils on the same flower. **3.** Having male and female flowers on the same capitulum.

synonym [SYST] A taxonomic name that is rejected as being incorrectly applied, or incorrect in form, or not representative of a natural genetic grouping.

synorchidism [MED] Partial or complete fusion of the two testes within the abdomen or scrotum.

synostosis [ANAT] A type of synarthrosis in which the bones are continuous. [MED] Union of originally separate bones into a single bone structure.

synotic tectum [VERT ZOO] A cartilaginous arch occurring between otic capsules in higher vertebrates.

synovia *See* synovial fluid.

synovial fluid [PHYSIO] A transparent viscid fluid secreted by synovial membranes. Also known as synovia.

synovial membrane [HISTOL] A layer of connective tissue which lines sheaths of tendons at freely moving articulations, ligamentous surfaces of articular capsules, and bursae.

synovioma [MED] Any tumor composed principally of cells similar to those covering the synovial membrane.

synovitis [MED] Inflammation of a synovial membrane.

synpelmous [VERT ZOO] Having the two main flexor tendons of the toes united beyond the branches that go to each digit, as in certain birds.

synsacrum [VERT ZOO] A fused-vertebrae structure supporting the pelvic girdle of birds.

synsepalous *See* gamosepalous.

synspermous [BOT] Having several seeds joined together.

Synteliidae [INV ZOO] The sap-flow beetles, a small family of coleopteran insects in the superfamily Histeroidea.

syntenic group [GEN] The loci on the same chromosome pair, irrespective of whether or not they are known to show linkage in heredity.

Syntexidae [INV ZOO] A family of the Hymenoptera in the superfamily Siricoidea.

synthetase *See* ligase.

Syntrophiidina [PALEON] A suborder of extinct articulate brachiopods of the order Pentamerida characterized by a strong dorsal median fold.

syntrophoblast [EMBRYO] The outer synctial layer of the trophoblast that forms the outermost fetal element of the placenta.

syntype [SYST] Any specimen of a series when no specimen is designated as the holotype. Also known as cotype.

synusia [ECOL] A structural unit of a community characterized by uniformity of life-form or of height.

Synxiphosura [PALEON] An extinct heteorgeneous order of arthropods in the subclass Merostomata possibly representing an explosive proliferation of aberrant, terminal, and apparently blind forms.

syphilis [MED] An infectious disease caused by the spirochete *Treponema pallidum*, transmitted principally by sexual intercourse.

syphilitic meningoencephalitis *See* general paresis.

syringe [MED] **1.** An apparatus commonly made of glass or plastic, fitted snugly onto a hollow needle, used to aspirate or inject fluids for diagnostic or therapeutic purposes. Also known as hypodermic syringe. **2.** A large glass barrel with a fitted rubber bulb at one end and a nozzle at the other, used primarily for irrigation purposes.

syringobulbia [MED] The presence of cavities in the medulla oblongata similar to those found in syringomyelia.

syringocystadenoma *See* syringoma.

syringocystoma *See* syringoma.

syringoma [MED] A multiple nevoid tumor of sweat glands. Also known as syringocystadenoma; syringocystoma.

syringomyelia [MED] A chronic disease characterized by the presence of cavities surrounded by gliosis near the canal of the spinal cord and often extending to the medulla.

Syringophyllidae [PALEON] A family of extinct corals in the order Tabulata.

syrinx [PALEON] A tube surrounding the pedicle in certain

SYNTROPHIIDINA

(a)   (b)

Pedicle valve of *Imbricata* in the Syntrophiidina; *(a)* exterior view, *(b)* interior view.

fossil brachiopods. [VERT ZOO] The vocal organ in birds.

Syrphidae [INV ZOO] The flower flies, a family of cyclorrhaphous dipteran insects in the series Aschiza.

Systellomatophora [INV ZOO] An order of the subclass Pulmonata in which the eyes are contractile but stalks are not retractile, the body is sluglike, oval, or lengthened, and the lung is posterior.

systematic error [STAT] An error which results from some bias in the measurement process and is not due to chance, in contrast to random error.

systematics [BIOL] The science of animal and plant classification.

systemic circulation [PHYSIO] The general circulation, as distinct from the pulmonary circulation.

systemic lupus erythematosus [MED] A fatal disease of unknown cause, characterized by fever, muscle and joint pains, anemia, leukopenia, and frequently by a skin eruption; primarily involved are the kidney, spleen, and endocardium.

systems ecology [ECOL] The combined approaches of systems analysis and the ecology of whole ecosystems and subsystems.

systole [PHYSIO] The contraction phase of the heart cycle.

syzygy [INV ZOO] End-to-end union of the sporonts of certain gregarine protozoans.

T

Ta   *See* tantalum.

Tabanidae   [INV ZOO] The deer and horse flies, a family of orthorrhaphous dipteran insects in the series Brachycera.

tabes dorsalis   [MED] A form of parenchymatous neurosyphilis in which there is demyelination and sclerosis of the posterior columns of the spinal cord. Also known as locomotor ataxia.

tabled whelk   [INV ZOO] *Neptunea tabulata.* A marine gastropod mollusk about 5 inches (13 centimeters) in length and 2 inches (5 centimeters) in diameter, found at depths of 150-200 feet (45-60 meters), off the west coast of Canada and the United States.

tabula   [PALEON] A transverse septum that closes off the lower part of the polyp cavity in certain extinct corals and hydroids.

tabulare   [VERT ZOO] A skull bone posterior to the parietal bone in some vertebrates.

Tabulata   [PALEON] An extinct Paleozoic order of corals of the subclass Zoantharia characterized by an exclusively colonial mode of growth and by secretion of a calcareous exoskeleton of slender tubes.

TAB vaccine   [IMMUNOL] A vaccine containing killed typhoid bacilli and the paratyphoid organisms (*Salmonella paratyphi* A and B) most frequently involved in paratyphoid fever.

tache noire   [MED] The primary painless lesion of the tickborne typhus fevers of Africa, manifested by a raised red area with a typical black necrotic center which appears at the site of the tick bite.

Tachinidae   [INV ZOO] The tachina flies, a family of bristly, grayish or black Diptera whose larvae are parasitic in caterpillars and other insects.

tachycardia   [MED] Excessive rapidity of the heart's action.

Tachyglossidae   [VERT ZOO] A family of monotreme mammals having relatively large brains with convoluted cerebral hemispheres; comprises the echidnas or spiny anteaters.

Tachyniscidae   [INV ZOO] A family of myodarian cyclorrhaphous dipteran insects in the subsection Acalypteratae.

tachyphylaxis   [IMMUNOL] Rapid desensitization against doses of organ extracts or serum by the previous inoculation of small subtoxic doses of the same preparation.

tachysterol   [BIOCHEM] The precursor of calciferol in the irradiation of ergosterol; an isomer of ergosterol.

tachytely   [EVOL] Evolution at a rapid rate resulting in differential selection and fixation of new types.

tactile   [PHYSIO] Pertaining to the sense of touch.

tactile agnosia   *See* astereognosis.

tactile hairs   *See* vibrissae.

tactile receptor   *See* tactoreceptor.

tactoid   [BIOCHEM] A particle that appears as a spindle-

TABULA

1 cm

Part of two corallites from a specimen of the species *Favosites gothlandica* Lam in the order Tabulata, from the Corniferous Limestone (Devonian) of Woodstock, Ontario. One of the corallites is cut away to show a series of transverse tabulae. *(After H. G. Nicholson)*

TABULATA

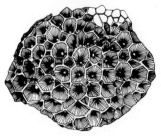

Specimen of *Michelinia convexa* D'Orb, a representative species of the Tabulata, seen from above, from the Carboniferous Limestone of Ontario. *(After Billings)*

TADPOLE

Tadpole of the common frog
(Rana temporaria).

TAENIA

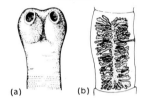

(a)   (b)

Morphological features of the
beef-inhabiting tapeworm *Taenia
saginata.* (a) Scolex. (b) Mature
proglottid. *(From T. I. Storer
and R. L. Usinger, General
Zoology, 3d ed., McGraw-Hill,
1957)*

TAENOGLOSSATE RADULA

Taenoglossate radula from
Lanistes, a fresh-water gastropod.
*(From R. R. Shrock and W. H.
Twenhofel, Principles of
Invertebrate Paleontology, 2d
ed., McGraw-Hill, 1953)*

TANAIDACEA

Female *Apseudes spinosus*
(M. Sars), a representative
species of the order Tanaidacea.
(After G. O. Sars)

shaped body under the polarizing microscope and occurs in
mosaic virus, fibrin, and myosin.

tactoreceptor [PHYSIO] A sense organ that responds to
touch. Also known as tactile receptor.

tadpole [VERT ZOO] The larva of a frog or toad; at hatching it
has a rounded body with a long fin-bordered tail, and the gills
are external but shortly become enclosed.

tadpole shrimp [INV ZOO] Any of the phyllopod crustaceans
that are members of the genus *Lepidurus.*

taenia [ANAT] A ribbon-shaped band of nerve fibers or muscle.

Taeniodidea [INV ZOO] An equivalent name for Cyclophyllidea.

Taeniodonta [PALEON] An order of extinct quadrupedal land
mammals, known from early Cenozoic deposits in North
America.

Taenioidea [INV ZOO] An equivalent name for Cyclophyllidea.

Taeniolabidoidea [PALEON] An advanced suborder of the
extinct mammalian order Multituberculata having incisors
that were self-sharpening in a limited way.

taenoglossate radula [INV ZOO] A long, narrow radula with
seven teeth in each transverse row, found in certain pectini-
branch bivalves.

tagua palm [BOT] *Phytelephas macrocarpa.* A palm tree of
tropical America; the endosperm of the seed is used as an
ivory substitute.

taiga [ECOL] A zone of forest vegetation encircling the
Northern Hemisphere between the arctic-subarctic tundras
in the north and the steppes, hardwood forests, and prairies in
the south.

tail [VERT ZOO] **1.** The usually slender appendage that arises
immediately above the anus in many vertebrates and contains
the caudal vertebrae. **2.** The uropygium, and its feathers, of a
bird. **3.** The caudal fin of a fish or aquatic mammal.

talbutal [PHARM] $C_{11}H_{16}N_2O_3$ A crystalline compound
that melts at 108–110°C, and is soluble in alcohol, chloroform, acetone, and ether; used in medicine as a short-acting
hypnotic and sedative. Also known as 5-allyl-5-*sec*-butylbar-
bituric acid.

talcosis [MED] A lung disease caused by inhalation of talc
dust and characterized by chronic induration and fibrosis.

talipes [MED] Any of several foot deformities, especially of
congenital origin.

talipes cavus [MED] A deformity of the foot marked by
exaggeration of the longitudinal arch and by dorsal contrac-
tures of the toes.

Talitridae [INV ZOO] A family of terrestrial amphipod crustaceans in the suborder Gammaridea.

talon [VERT ZOO] A sharply hooked claw on the foot of a bird
of prey.

Talpidae [VERT ZOO] The moles, a family of insectivoran
mammals; distinguished by the forelimbs which are adapted
for digging, having powerful muscles and a spadelike bony
structure.

talus *See* astragalus.

tamarack [BOT] *Larix laricina.* A larch and a member of the
pine family; it has an erect narrowly pyramidal habit, and
grows in wet and moist soils in the northeastern United
States, west to the Lake States and across Canada to Alaska;
used for railroad ties, posts, sills, and boats. Also known as
hackmarack.

tampon [MED] A plug of absorbent material, such as cotton
or sponge, inserted into a cavity as packing.

Tanaidacea [INV ZOO] An order of eumalacostracans of the crustacean superorder Peracarida; the body is linear, more or less cylindrical or dorsoventrally depressed, and the first and second thoracic segments are fused with the head, forming a carapace.

Tanaostigmatidae [INV ZOO] A small family of hymenopteran insects in the superfamily Chalcidoidea.

tandem duplication [CYTOL] The occurrence of two identical sequences, one following the other, in a chromosome segment.

tangelo [BOT] A tree that is hybrid between a tangerine or other mandarin and a grapefruit or shaddock; produces an edible fruit.

tangerine [BOT] Any of several trees of the species *Citrus reticulata*; the fruit is a loose-skinned mandarin with a deep-orange or scarlet rind.

tangoreceptor [PHYSIO] A sense organ in the skin that responds to touch and pressure.

tannase [BIOCHEM] An enzyme that catalyzes the hydrolysis of tannic acid to gallic acid; found in cultures of *Aspergillus* and *Penicillium*.

tan rot [PL PATH] A fungus disease of strawberries caused by *Pezizella lythri* and characterized by the appearance of tan depressions on the fruit.

tantalum [CHEM] Metallic element in group V, symbol Ta, atomic number 73, atomic weight 180.948; black powder or steel-blue solid soluble in fused alkalies, insoluble in acids (except hydrofluoric and fuming sulfuric); melts about 3000°C.

Tanyderidae [INV ZOO] The primitive crane flies, a family of orthorrhaphous dipteran insects in the series Nematocera.

Tanypezidae [INV ZOO] A family of myodarian cyclorrhaphous dipteran insects in the subsection Acalypteratae.

tapesium [MYCOL] A dense outer mycelium that bears hyphae producing asci.

tapetum [ANAT] **1.** A reflecting layer in the choroid coat behind the neural retina, chiefly in the eyes of nocturnal mammals. **2.** A tract of nerve fibers forming part of the roof of each lateral ventricle in the vertebrate brain. [BOT] A layer of nutritive cells surrounding the spore mother cells in the sporangium in higher plants; it is broken down to provide nourishment for developing spores.

tapeworm [INV ZOO] Any member of the class Cestoidea; all are vertebrate endoparasites, characterized by a ribbonlike body divided into proglottids, and the anterior end modified into a holdfast organ.

tapir [VERT ZOO] Any of several large odd-toed ungulates of the family Tapiridae that have a heavy, sparsely hairy body, stout legs, a prehensile muzzle, a short tail, and small eyes.

Tapiridae [VERT ZOO] The tapirs, a family of perissodactyl mammals in the superfamily Tapiroidea.

Tapiroidea [VERT ZOO] A superfamily of the mammalian order Perissodactyla in the suborder Ceratomorpha.

taproot [BOT] A root system in which the primary root forms a dominant central axis that penetrates vertically and rather deeply into the soil; it is generally larger in diameter than its branches.

tarantula [INV ZOO] **1.** Any of various large hairy spiders of the araneid suborder Mygalomorphae. **2.** Any of the wolf spiders comprising the family Lycosidae.

Tardigrada [INV ZOO] A class of microscopic, bilaterally symmetrical invertebrates in the subphylum Malacopoda; the body consists of an anterior prostomium and five segments surrounded by a soft, nonchitinous cuticle, with four pairs of ventrolateral legs.

TANTALUM

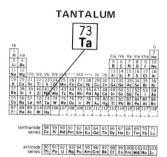

Periodic table of the chemical elements showing the position of tantalum.

TAPIR

Brazilian tapir *(Tapirus terrestris)*, the most common South American species of tapir.

TAPROOT

Taproot system of a dandelion.

TARPON

The tarpon *(Megalops atlantica)*, largest of the herringlike fishes, reaching a length of 8 feet (2 meters).

TARSIER

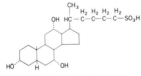

The tarsier has adhesive pads on the expanded ends of the fingers and toes.

TAUROCHOLIC ACID

Structural formula of taurocholic acid.

target spot [PL PATH] Any plant disease characterized by lesions in the form of concentric markings.

tarpon [VERT ZOO] *Megalops atlantica.* A herringlike fish of the family Elopidae weighing up to 300 pounds and reaching a length of 8 feet; it has a single soft, rayed dorsal fin, strong jaws, a bony plate under the mouth, numerous small teeth, and coarse, bony flesh covered with large scales.

tarsal gland [ANAT] Any of the sebaceous glands in the tarsal plates of the eyelids. Also known as Meibomian gland.

tarsier [VERT ZOO] Any of several species of primates comprising the genus *Tarsius* of the family Tarsiidae characterized by a round skull, a flattened face, and large eyes that are separated from the temporal fossae in the orbital depression, and by adhesive pads on the expanded ends of the fingers and toes.

Tarsiidae [VERT ZOO] The tarsiers, a family of prosimian primates distinguished by incomplete postorbital closure and a greatly elongated ankle region.

tarsomere [INV ZOO] Either of the two parts (basitarsus and telotarsus) of the dactylopodite in spiders.

tarsometatarsus [VERT ZOO] In birds, a short, straight leg bone formed by fusion of the distal row of tarsals with the second to fifth metatarsals.

Tarsonemidae [INV ZOO] A small family of phytophagous mites in the suborder Trombidiformes.

tarsus [ANAT] **1.** The instep of the foot consisting of the calcaneus, talus, cuboid, navicular, medial, intermediate, and lateral cuneiform bones. **2.** The dense connective tissues supporting an eyelid.

tartar *See* dental calculi.

tartary buckwheat [BOT] One of three buckwheat species grown commercially; the leaves are narrower than the other two species and arrow-shaped, and the flowers are smaller with inconspicuous greenish-white sepals. Also known as duck wheat; hulless buckwheat; rye buckwheat.

tassel [BOT] The male inflorescence of corn and certain other plants.

taste [PHYSIO] A chemical sense by which flavors are perceived depending on taste, tactile, and warm and cold receptors in the mouth, as well as smell receptors in the nose.

taste bud [ANAT] An end organ consisting of goblet-shaped clusters of elongate cells with microvilli on the distal end to mediate the sense of taste.

TAT *See* thematic apperception test.

Tauber test [ANALY CHEM] A color test for identification of pentose sugars; the sugars produce a cherry-red color when heated with a solution of benzidine in glacial acetic acid.

taurocholic acid [BIOCHEM] $C_{26}H_{45}NO_7S$ A common bile acid with a five-carbon chain; it is the product of the conjugation of taurine with cholic acid; crystallizes from an alcohol ether solution, and decomposes at about 125°C. Also known as cholaic acid; cholytaurine.

taurodont [ANAT] Of teeth, having a large pulp cavity and reduced roots.

tautomerism [CHEM] The reversible interconversion of structural isomers of organic chemical compounds; such interconversions usually involve transfer of a proton.

Taxales [BOT] A small order of gymnosperms in the class Pinatae; members are trees or shrubs with evergreen, often needlelike leaves, with a well-developed fleshy covering surrounding the individual seeds, which are terminal or subterminal on short shoots.

taxeopodous [VERT ZOO] Having the proximal and distal tarsal bones oriented in straight lines parallel to the limb axis.

taxis [PHYSIO] A mechanism of orientation by means of

which an animal moves in a direction related to a source of stimulation.

Taxocrinida [PALEON] An order of flexible crinoids distributed from Ordovician to Mississippian.

Taxodonta [INV ZOO] A subclass of pelecypod mollusks in which the hinge is of the taxodontal type, that is, the dentition is a series of similar alternating teeth and sockets along the hinge margin.

taxon [SYST] A taxonomic group or entity.

taxonomic category [SYST] One of a hierarchy of levels in the biological classification of organisms; the seven major categories are kingdom, phylum, class, order, family, genus, species.

taxonomy [SYST] A study aimed at producing a hierarchical system of classification of organisms which best reflects the totality of similarities and differences.

Tayassuidae [VERT ZOO] The peccaries, a family of artiodactyl mammals in the superfamily Suoidae.

Tay-Sachs disease [MED] A form of sphingolipidosis, transmitted as an autosomal recessive, in which there is an accumulation in neuronal cells of the neuraminic fraction of gangliosides; manifested clinically within the first few months of life by hypotonia progressing to spasticity, convulsions, and visual loss accompanied by the appearance of a cherry-red spot at the macula lutea. Also known as infantile amaurotic familial idiocy.

Tb *See* terbium.

Tc *See* technetium.

T cell [IMMUNOL] One of a heterogeneous population of thymus-derived lymphocytes which participates in the immune responses.

Te *See* tellurium.

tea [BOT] *Thea sinensis.* A small tree of the family Theaceae having lanceolate leaves and fragrant white flowers; a caffeine beverage is made from the leaves of the plant.

teakwood [MATER] The strong, durable, yellowish-brown wood obtained from the teak tree, *Tectona grandis.*

tear gland *See* lacrimal gland.

technetium [CHEM] A member of group VII, symbol Tc, atomic number 43; derived from uranium and plutonium fission products; chemically similar to rhenium and manganese; isotope Tc^{99} has a half-life of 2×10^5 years; used to absorb slow neutrons in reactor technology.

Tectibranchia [INV ZOO] An order of mollusks in the subclass Opisthobranchia containing the sea hares and the bubble shells; the shell may be present, rudimentary, or absent.

tectorial membrane [ANAT] **1.** A jellylike membrane covering the organ of Corti in the ear. **2.** A strong sheet of connective tissue running from the basilar part of the occipital bone to the dorsal surface of the bodies of the axis and third cervical vertebra.

tectospondylic [VERT ZOO] Having several concentric rings of calcification in the vertebrae, as in some elasmobranchs.

tectum [ANAT] A rooflike structure of the body, especially the roof of the midbrain including the corpora quadrigemina.

teflon shakes *See* metal fume fever.

tegmen [BIOL] An integument or covering. [BOT] The inner layer of a seed coat. [INV ZOO] A thickened forewing of Orthoptera, Coleoptera, and certain other insects.

tegmentum [ANAT] A mass of white fibers with gray matter in the cerebral peduncles of higher vertebrates. [BOT] The outer layer, or scales, of a leaf bud. [INV ZOO] The upper layer of a shell plate in Amphineura.

tegula [INV ZOO] **1.** A small sclerite on the mesothorax which overhangs the wing articulation in Lepidoptera and Hymen-

TECHNETIUM

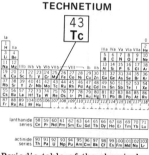

Periodic table of the chemical elements showing the position of technetium.

TECTIBRANCHIA

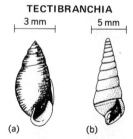

Shells of two genera of Tectibranchia. *(a) Acteon. (b) Pyramidella. (From A. M. Keen and J. C. Pearson, Illustrated Key to West North American Gastropod Genera, Stanford University Press, 1958)*

optera. **2.** A small lobe at the base of the wing of Diptera.

teichoic acid [BIOCHEM] A polymer of ribitol or glycerol phosphate with additional compounds such as glucose linked to the backbone of the polymer; found in the cell walls of some bacteria.

Teiidae [VERT ZOO] The tegus lizards, a diverse family of the suborder Sauria that is especially abundant and widespread in South America.

telamon [INV ZOO] A curved chitinous outgrowth of the cloacal wall in various male nematodes.

telangiectasis [MED] Localized dilation of capillaries forming dark-red, wartlike elevations varying in size from about 1 to 7 millimeters.

Telegeusidae [INV ZOO] The long-lipped beetles, a small family of colepteran insects in the superfamily Cantharoidea confined to the western United States.

teleianthous [BOT] Pertaining to a flower having bóth a gynoecium and an androecium.

teleiochrysalis [INV ZOO] A nymph of the resting stage that precedes the adult form of certain mites.

telencephalon [EMBRYO] The anterior subdivision of the forebrain in a vertebrate embryo; gives rise to the olfactory lobes, cerebral cortex, and corpora striata.

teleology [SCI TECH] The doctrine that explanations of phenomena are to be sought in terms of final causes, purpose, or design in nature.

Teleosauridae [PALEON] A family of Jurassic reptiles in the order Crocodilia characterized by a long snout and heavy armor.

Teleostei [VERT ZOO] An infraclass of the subclass Actinopterygii, or rayfin fishes; distinguished by paired bracing bones in the supporting skeleton of the caudal fin, a homocercal caudal fin, thin cycloid scales, and a swim bladder with a hydrostatic function.

Telestacea [INV ZOO] An order of the subclass Alcyonaria comprised of individuals which form erect branching colonies by lateral budding from the body wall of an elongated axial polyp.

teleutospore *See* teliospore.

teliospore [MYCOL] A thick-walled spore of the terminal stage of Uredinales and Ustilaginales which is a probosidium or a group of probosidia. Also known as teleutospore.

tellurium [CHEM] A member of group VI, symbol Te, atomic number 52, atomic weight 127.60; dark-gray crystals, insoluble in water, soluble in nitric and sulfuric acids and potassium hydroxide; melts at 452°C, boils at 1390°C; used in alloys (with lead or steel), glass, and ceramics.

telocentric [CYTOL] Pertaining to a chromosome with a terminal centromere.

telocoel [EMBRYO] A cavity of the telencephalon.

telodendrion [ANAT] The terminal branching of an axon. Also known as telodendron.

telodendron *See* telodendrion.

telofemur [INV ZOO] The distal segment of the femur in certain Acarina.

telogen [PHYSIO] A quiescent phase in the cycle of hair growth when the hair is retained in the hair follicle as a dead or "club" hair.

telolecithal [CYTOL] Of an ovum, having a large, evenly dispersed volume of yolk and a small amount of cytoplasm concentrated at one pole.

telome [BOT] A small terminal branchlet of a dichotomously branching stem having a sporangium attached to the distal end.

TELEOSAURIDAE

Skull of typical Teleosauridae, with long snout.

TELLURIUM

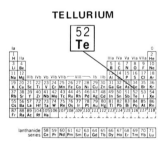

Periodic table of the chemical elements showing the position of tellurium.

TELOLECITHAL

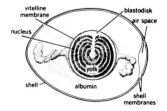

Telolecithal ovum of a hen.

telomere [CYTOL] A centromere in the terminal position on a chromosome.

telophase [CYTOL] The phase of meiosis or mitosis at which the chromosomes, having reached the poles, reorganize into interphase nuclei with the disappearance of the spindle and the reappearance of the nuclear membrane; in many organisms telophase does not occur at the end of the first meiotic division.

Telosporea [INV ZOO] A class of the protozoan subphylum Sporozoa in which the spores lack a polar capsule and develop from an oocyst.

telotarsus [INV ZOO] The distal part, or tarsus, of a dactylopodite of spiders.

telotaxis [BIOL] Tactic movement of an organism by the orientation of one or the other of two bilaterally symmetrical receptors toward the stimulus source.

telotroch [INV ZOO] A preanal tuft of cilia in a trochophore larva.

telson [INV ZOO] The postabdominal segment in lobsters, amphipods, and certain other invertebrates.

telum [INV ZOO] The last segment of an insect abdomen.

Temnocephalida [INV ZOO] A group of rhabdocoeles sometimes considered a distinct order but usually classified under the Neorhabdocoela; members are characterized by the possession of tentacles and adhesive organs.

Temnochilidae [INV ZOO] The equivalent name for Ostomidae.

Temnopleuridae [INV ZOO] A family of echinoderms in the order Temnopleuroida whose tubercles are imperforate, though usually crenulate.

Temnopleuroida [INV ZOO] An order of echinoderms in the superorder Echinacea with a camarodont lantern, smooth or sculptured test, imperforate or perforate tubercles, and bronchial slits which are usually shallow.

Temnospondyli [PALEON] An order of extinct amphibians in the subclass Labyrinthodontia having vertebrae with reduced pleurocentra and large intercentra.

temnospondylous [VERT ZOO] Having vertebrae in articulated pieces rather than fused.

temperate and cold savannah [ECOL] A regional vegetation zone, very extensively represented in North America and in Eurasia at high altitudes; consists of scattered or clumped trees (very often conifers and mostly needle-leaved evergreens) and a shrub layer of varying coverage; mosses and, even more abundantly, lichens form an almost continuous carpet.

temperate and cold scrub [ECOL] Regional vegetation zone whose density and periodicity vary a good deal; requires a considerable amount of moisture in the soil, whether from mist, seasonal downpour, or snowmelt; shrubs may be evergreen or deciduous; and undergrowth of ferns and other large-leaved herbs are quite frequent, especially at subalpine level; wind shearing and very cold winters prevent tree growth. Also known as bosque; fourré; heath.

temperate mixed forest [ECOL] A forest of the North Temperate Zone containing a high proportion of conifers with a few broad-leafed species.

temperate phage [VIROL] A deoxyribonucleic acid phage, the genome (DNA) of which can under certain circumstances become integrated with the genome of the host.

temperate rainforest [ECOL] A vegetation class in temperate areas of high and evenly distributed rainfall characterized by comparatively few species with large populations of each species; evergreens are somewhat short with small leaves, and there is an abundance of large tree ferns. Also known as

TEMPERATE RAINFOREST

Temperate rainforest in Olympic National Park, Washington, showing sheath of moss on maple tree. *(Photograph by J. O. Sumner, from National Audubon Society)*

cloud forest; laurel forest; laurisilva; moss forest; subtropical forest.

temperate woodland [ECOL] A vegetation class similar to tropical woodland in spacing, height, and stratification, but it can be either deciduous or evergreen, broad-leaved or needle-leaved. Also known as parkland; woodland.

temperature [THERMO] A property of an object which determines the direction of heat flow when the object is placed in thermal contact with another object: heat flows from a region of higher temperature to one of lower temperature; it is measured either by an empirical temperature scale, based on some convenient property of a material or instrument, or by a scale of absolute temperature, for example, the Kelvin scale.

temperature scale [THERMO] An assignment of numbers to temperatures in a continuous manner, such that the resulting function is single valued; it is either an empirical temperature scale, based on some convenient property of a substance or object, or it measures the absolute temperature.

temperature-sensitive mutant [GEN] A mutant gene that is functional at high (low) temperature but is inactivated by lowering (elevating) the temperature.

temporal bone [ANAT] The bone forming a portion of the lateral aspect of the skull and part of the base of the cranium in vertebrates.

temporalis [ANAT] A muscle of mastication; originates at the temporal fossa and is inserted into the coronoid process of the mandible. Also known as temporal muscle.

temporal lobe [ANAT] The portion of the cerebral cortex below the lateral fissure. [VERT ZOO] That part of the vertebrate cerebrum containing centers for speech and hearing.

temporal-lobe epilepsy [MED] Recurrent seizures originating in lesions of the temporal lobe of the brain.

temporal muscle *See* temporalis.

temporary plankton *See* meroplankton.

tenaculum [BOT] The holdfast of algae. [INV ZOO] In Collembola, a structure on the third segment of the abdomen which holds the furcula in place. [VERT ZOO] The claspers of a shark.

Tendipedidae [INV ZOO] The midges, a family of orthorrhaphous dipteran insects in the series Nematocera whose larvae occupy intertidal wave-swept rocks on the seacoasts.

tendon [ANAT] A white, glistening, fibrous cord which joins a muscle to some movable structure such as a bone or cartilage; tendons permit concentration of muscle force into a small area and allow the muscle to act at a distance.

tendonitis [MED] Inflammation of a tendon.

tendon reflex [PHYSIO] Contraction of a muscle in response to sudden stretching of the muscle by a brisk tap against its tendon.

tendon sheath [ANAT] The synovial membrane surrounding a tendon.

tendril [BOT] A stem modification in the form of a slender coiling structure capable of twining about a support to which the plant is then attached.

Tenebrionidae [INV ZOO] The darkling beetles, a large cosmopolitan family of coleopteran insects in the superfamily Tenebrionoidea; members are common pests of grains, dried fruits, beans, and other food products.

Tenebrionoidea [INV ZOO] A superfamily of the Coleoptera in the suborder Polyphaga.

teniae coli [HISTOL] The three bands comprising the longitudinal layer of the tunica muscularis of the colon: the tenia libera, tenia mesocolica, and tenia omentalis.

TENDRIL

Portion of a grape plant stem with leaf and tendril.

TENEBRIONIDAE

Representative species of family Tenebrionidae. (*From T. I. Storer and R. L. Usinger, General Zoology, 3d ed., McGraw-Hill, 1957*)

tenosynonitis [MED] Inflammation of a tendon and its sheath.

tenrec [VERT ZOO] Any of about 30 species of unspecialized, insectivorous mammals indigenous to Madagascar, and which have poor vision and clawed digits.

Tenrecidae [VERT ZOO] The tenrecs, a family of insectivores in the group Lipotyphla.

tensor muscle [PHYSIO] A muscle that stretches a part or makes it tense.

tentacle [INV ZOO] Any of various elongate, flexible processes with tactile, prehensile, and sometimes other functions, and which are borne on the head or about the mouth of many animals.

Tentaculata [INV ZOO] A class of the phylum Ctenophora whose members are characterized by having variously modified tentacles.

tentaculocyst [INV ZOO] A sense organ located at the margin of the umbrella in some coelenterate medusoids, consisting of a modified tentacle with a cavity that often contains lithites.

tentaculozoid [INV ZOO] A slender tentacular individual of a hydrozoan colony.

tented arch [ANAT] A fingerprint pattern which possesses either an angle, an upthrust, or two of the three basic characteristics of a loop.

tenthmeter *See* angstrom.

Tenthredinidae [INV ZOO] A family of hymenopteran insects in the superfamily Tenthredinoidea including economically important species whose larvae are plant pests.

Tenthredinoidea [INV ZOO] A superfamily of Hymenoptera in the suborder Symphyla.

tentillum [INV ZOO] A contractile branch of a tentacle containing many nematocysts in certain siphonophores.

tentorium [ANAT] A transverse fold of the dura mater between the cerebellum and the occipital lobes of brain; it is ossified in some mammals. [INV ZOO] A chitinous framework supporting the brain of insects.

tenuinucellate [BOT] Having the nucellus composed of a single layer of cells.

Tenuipalpidae [INV ZOO] A small family of mites in the suborder Trombidiformes.

tepal [BOT] 1. A sepal or a petal. 2. Any of the modified leaves composing the perianth.

Tephritidae [INV ZOO] The fruit flies, a family of myodarian cyclorrhaphous dipteran insects in the subsection Acalypteratae.

teratocarcinoma [MED] A teratoma with carcinomatous elements.

teratogen [MED] An agent causing formation of a congenital anomaly or monstrosity.

teratogenesis [MED] The formation of a fetal monstrosity.

teratoid adenocystoma *See* mesonephroma.

teratology [MED] The science of fetal malformations and monstrosities.

teratoma [MED] A true neoplasm composed of bizarre and chaotically arranged tissues that are foreign embryologically as well as histologically to the area in which the tumor is found.

Teratornithidae [PALEON] An extinct family of vulturelike birds of the Pleistocene of western North America included in the order Falconiformes.

terbium [CHEM] A rare-earth element, symbol Tb, in the yttrium subgroup of group III, atomic number 65, atomic weight 158.924.

tercine [BOT] 1. The third coat of an ovule. 2. A layer of the second coat of an ovule.

TENREC

Tenrec, showing characteristic features.

TENTED ARCH

Fingerprint having tented arch pattern with upthrust. *(Federal Bureau of Investigation)*

TENTHREDINIDAE

Representative species of family Tenthredinidae. *(From T. I. Storer and R. L. Usinger, General Zoology, 3d ed., McGraw-Hill, 1957)*

TERBIUM

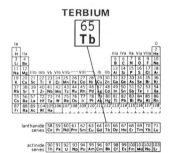

Periodic table of the chemical elements showing the position of terbium.

TEREBELLIDAE

Left-lateral view of the anterior end of *Artacamella* species in the subfamily Artacaminae of the Terebillidae.

TEREBRATELLIDINA

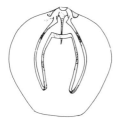

Representative species of suborder Terebratellidina, in genus *Magellania*, showing characteristic loop. *(From R. C. Moore, ed., Treatise on Invertebrate Paleontology, pt. H, Geological Society of America and University of Kansas Press, 1965)*

TERMITE

Winged female termite of the reproductive caste.

Terebellidae [INV ZOO] A family of polychaete annelids belonging to the Sedentaria which are chiefly large, thick-bodied, tubicolous forms with the anterior end covered by a matted mass of tentacular cirri.

terebra [INV ZOO] A modified ovipositor used for boring, sawing, or stinging, as in certain Hymenoptera.

Terebratellidina [PALEON] An extinct suborder of articulate brachiopods in the order Terebratulida in which the loop is long and offers substantial support to the side arms of the lophophore.

Terebratulida [INV ZOO] An order of articulate brachiopods that has a punctate shell structure and is characterized by the possession of a loop extending anteriorly from the crural bases, providing some degree of support for the lophophore.

Terebratulidina [INV ZOO] A suborder of articulate brachiopods in the order Terebratulida distinguished by a short V- or W-shaped loop.

Teredinidae [INV ZOO] The pileworms or shipworms, a family of bivalve mollusks in the subclass Eulamellibranchia distinguished by having the two valves reduced to a pair of small plates at the anterior end of the animal.

teres [ANAT] A cylindrical muscle.

terete [BOT] Of a stem, cylindrical in section, but tapering at both ends.

tergite [INV ZOO] The dorsal plate covering a somite in arthropods and certain other articulate animals.

tergum [INV ZOO] A dorsal plate of the operculum in barnacles.

terminal bar [CYTOL] One of the structures formed in certain epithelial cells by the combination of local modifications of contiguous surfaces and intervening intercellular substances; they become visible with the light microscope after suitable staining and appear to close the spaces between the epithelial cells of the intestine at their free surfaces.

terminal bud [BOT] A bud that develops at the apex of a stem. Also known as apical bud.

terminal endocarditis *See* verrucous endocarditis.

terminal hair [ANAT] One of three types of hair in man based on hair size, time of appearance, and structural variations; the larger, coarser hair in the adult that replaces the vellus hair.

terminal nerve [ANAT] Either of a pair of small cranial nerves that run from the nasal area to the forebrain, present in most vertebrates; the function is not known.

terminal sinus [EMBRYO] The vascular sinus bounding the area vasculosa of the blastoderm of a meroblastic ovum. Also known as marginal sinus.

Termitaphididae [INV ZOO] The termite bugs, a small family of Hemiptera in the superfamily Aradoidea.

termitarium [INV ZOO] A termites' nest.

termite [INV ZOO] A soft-bodied insect of the order Isoptera; individuals feed on cellulose and live in colonies with a caste system comprising three types of functional individuals: sterile workers and soldiers, and the reproductives. Also known as white ant.

termiticole [ECOL] An organism that lives in a termites' nest.

Termitidae [INV ZOO] A large family of the order Isoptera which contains the higher termites, representing 80% of the species.

termitophile [ECOL] An organism that lives in a termites' nest in a symbiotic association with the termites.

Termopsidae [INV ZOO] A family of insects in the order Isoptera composed of damp wood-dwelling forms.

ternate [BOT] Composed of three subdivisions, as a leaf with three leaflets.

Ternifine man [PALEON] The name for a fossil human type, represented by three lower jaws and a parietal bone discovered in France and thought to be from the upper part of the middle Pleistocene.

Terramycin [MICROBIO] Trade name for the antibiotic oxytetracycline.

terrapin [VERT ZOO] Any of several North American tortoises in the family Testudinidae, especially the diamondback terrapin.

terricolous [ECOL] Inhabiting soil.

territoriality [ZOO] A pattern of behavior in which one or more animals occupy and defend a definite area or territory.

tertian malaria *See* vivax malaria.

Tertiary [GEOL] The older major subdivision (period) of the Cenozoic era, extending from the end of the Cretaceous to the beginning of the Quaternary, from 70,000,000 to 2,000,000 years ago.

tertiary structure [BIOCHEM] The characteristic three-dimensional folding of the polypeptide chains in a protein molecule.

Tessaratomidae [INV ZOO] A family of large tropical Hemiptera in the superfamily Pentatomoidea.

tessellate [BOT] Marked by a pattern of small squares resembling a tiled pavement.

test [INV ZOO] A hard external covering or shell that is calcareous, siliceous, chitinous, fibrous, or membranous.

testa [BOT] A seed coat. Also known as episperm.

Testacellidae [INV ZOO] A family of pulmonate gastropods that includes some species of slugs.

testicular hormone [BIOCHEM] Any of various hormones secreted by the testes.

testis [ANAT] One of a pair of male reproductive glands in vertebrates; after sexual maturity, the source of sperm and hormones.

test of hypothesis [STAT] A rule for rejecting or accepting a hypothesis concerning a population which is based upon a given sample of data.

test of significance [STAT] A test of a hypothetical population property against a sample property where an acceptance interval is used as the rule for rejection.

testosterone [BIOCHEM] $C_{19}H_{28}O_2$ The principal androgenic hormone released by the human testis; may be synthesized from cholesterol and certain other sterols.

Testudinata [VERT ZOO] The equivalent name for Chelonia.

Testudinellidae [INV ZOO] A family of free-swimming rotifers in the suborder Flosculariacea.

Testudinidae [VERT ZOO] A family of tortoises in the suborder Cryptodira; there are about 30 species found on all continents except Australia.

tetanospasmin [BIOCHEM] A neurotoxin elaborated by the bacterium *Clostridium tetani* and which is responsible for the manifestations of tetanus.

tetanus [MED] An infectious disease of humans and animals caused by the toxin of *Clostridium tetani* and characterized by convulsive tonic contractions of voluntary muscles; infection commonly follows dirt contamination of deep wounds or other injured tissue. Also known as lockjaw.

tetanus antitoxin [IMMUNOL] A serum containing antibodies that neutralize tetanus toxin.

tetanus toxoid [IMMUNOL] Detoxified tetanus toxin used to produce active immunity against tetanus.

tetany [MED] A state of increased neuromuscular irritability caused by a decrease of serum calcium, manifested by intermittent numbness and cramps or twitchings of the extremities, laryngospasm, bizarre behavior, loss of consciousness, and convulsions.

TERRAPIN

Diamondback terrapin
(Malaclemys terrapin).

TERTIARY

PRECAMBRIAN		
CAMBRIAN	PALEOZOIC	
ORDOVICIAN		
SILURIAN		
DEVONIAN		
Mississippian	CARBON-IFEROUS	
Pennsylvanian		
PERMIAN		
TRIASSIC	MESOZOIC	
JURASSIC		
CRETACEOUS		
TERTIARY	CENOZOIC	
QUATERNARY		

Chart showing the position of the Tertiary period in relation to the other periods and the eras of geologic time.

TETRACYCLINE

	R_1	R_2	R_3
Tetracycline	H	H	CH_3
Chlortetracycline	Cl	H	CH_3
Oxytetracycline	H	OH	CH_3
6-Demethyl tetracycline	H	H	H
7-Chloro-6-demethyl tetracycline	Cl	H	H

Chemical structural relationships of the tetracyclines (def. 1).

TETRAODONTIFORMES

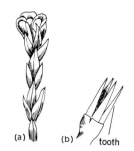

Gray triggerfish *(Balistes capriscus)*, a representative species of the order Tetraodontiformes. *(After G. B. Goode, Fishery Industries of the United States, sect. 1, 1884)*

TETRAPHIDALES

(a) 　　 *(b)* / tooth

Tetraphis pellucida, a representative species of the order Tetraphidales. *(a)* Gemmiferous branch. *(b)* Enlarged peristome. *(From W. H. Welch, Mosses of Indiana, Ind. Dep. Conserv., 1957)*

tetany of the newborn [MED] A temporary increase of neuromuscular irritability during the first two months of life, especially in infants who are premature, are delivered by cesarean section, or receive an exchange transfusion, or in twins.

Tethinidae [INV ZOO] A family of myodarian cyclorrhaphous dipteran insects in the subsection Acalypteratae.

Tetrabranchia [INV ZOO] A subclass of primitive mollusks of the class Cephalopoda; *Nautilus* is the only living form and is characterized by having four gills.

Tetracentraceae [BOT] A family of dicotyledonous trees in the order Trochodendrales distinguished by possession of a perianth, four stamens, palmately veined leaves, and secretory idioblasts.

Tetracorallia [PALEON] The equivalent name for Rugosa.

Tetractinomorpha [INV ZOO] A heterogeneous subclass of Porifera in the class Demospongiae.

tetracycline [MICROBIO] **1.** Any of a group of broad-spectrum antibiotics produced biosynthetically by fermentation with a strain of *Streptomyces aureofaciens* and certain other species or chemically by hydrogenolysis of chlortetracycline. **2.** ($C_{22}H_{24}O_8N_2$) A broad-spectrum antibiotic belonging to the tetracycline group of antibiotics; useful because of broad antimicrobial action, with low toxicity, in the therapy of infections caused by gram-positive and gram-negative bacteria as well as rickettsiae and large viruses such as psittacosis-lymphogranuloma viruses.

tetrad [BIOL] A group of four. [BOT] Four spores or pollen grains forming a coherent group. [CYTOL] A group of four chromatids lying parallel to each other as a result of the longitudinal division of each of a pair of homologous chromosomes during the pachytene and later stages of the prophase of meiosis.

tetradactylous [VERT ZOO] Having four digits on a limb.

tetrad of Fallot *See* tetralogy of Fallot.

tetradynamous [BOT] Having four long and two short stamens.

Tetrahymenina [INV ZOO] A suborder of ciliated protozoans in the order Hymenostomatida.

tetralogy of Fallot [MED] A congenital abnormality of the heart consisting of pulmonary stenosis, defect of the interventricular septum, hypertrophy of the right ventricle, and overriding or dextroposition of the aorta. Also known as tetrad of Fallot.

tetralophodont [VERT ZOO] Having four ridges on the molar teeth.

Tetralophodontinae [PALEON] An extinct subfamily of proboscidean mammals in the family Gomphotheridae.

tetramerous [BIOL] Characterized by or having four parts.

Tetranychidae [INV ZOO] The spider mites, a family of phytophagous trombidiform mites.

Tetraodontiformes [VERT ZOO] An order of specialized teleost fishes that includes the trigger fishes, puffers, trunkfishes, and ocean sunfishes.

Tetraonidae [VERT ZOO] The ptarmigans and grouse, a family of upland game birds in the order Galliformes characterized by rounded tails and wings and feathered nostrils.

Tetraphidaceae [BOT] The single family of the plant order Tetraphidales.

Tetraphidales [BOT] A monofamilial order of mosses distinguished by scalelike protonema and the peristomes of four rigid, nonsegmented teeth.

Tetraphyllidea [INV ZOO] An order of small tapeworms of the subclass Cestoda characterized by the variation in the structure of the scolex; all species are intestinal parasites of elasmobranch fishes.

tetraploidy [CYTOL] The occurrence of related forms possessing in the somatic cells chromosome numbers four times the haploid number.

tetrapod [VERT ZOO] A four-footed animal.

Tetrapoda [VERT ZOO] The superclass of the subphylum Vertebrata whose members typically possess four limbs; includes all forms above fishes.

tetrapyrrol [ORG CHEM] A molecule composed of four pyrrol units, each unit being a five-membered ring of carbon and nitrogen; the pyrrol units may be arranged linearly or in a ring.

Tetrarhynchoidea [INV ZOO] The equivalent name for Trypanorhyncha.

tetraselenodont [VERT ZOO] Having four crescentic ridges on the molars.

Tetrasporales [BOT] A heterogeneous and artificial assemblage of colonial fresh-water and marine algae in the division Chlorophyta.

tetraspore [BOT] One of the haploid asexual spores of the red algae formed in groups of four.

tetrasporic embryo sac [BOT] An embryo sac that arises from four megaspores.

tetrasternum [INV ZOO] In Acarina, the sternite of the fourth segment of the prosoma or the second segment of the podosoma.

tetraterpene [ORG CHEM] A class of terpene compounds that contain isoprene units; best known are the carotenoid pigments from plants and animals, such as lycopene, the red coloring matter in tomatoes.

tetraxon [INV ZOO] A type of sponge spicule with four axes.

Tetrigidae [INV ZOO] The grouse locusts or pygmy grasshoppers in the order Orthoptera in which the front wings are reduced to small scalelike structures.

tetrodotoxin [BIOCHEM] $C_{11}H_{17}N_3O_8$ A toxin that blocks the action potential in the nerve impulse.

tetrose [BIOCHEM] Any of a group of monosaccharides that have a four-carbon chain; an example is erythrose, $CH_2OH - \cdot(CHOH)_2 \cdot CHO$.

Tettigoniidae [INV ZOO] A family of insects in the order Orthoptera which have long antennae, hindlegs fitted for jumping, and an elongate, vertically flattened ovipositor; consists of the longhorn or green grasshopper.

Teuthidae [VERT ZOO] The rabbitfishes, a family of perciform fishes in the suborder Acanthuroidei.

Teuthoidea [INV ZOO] An order of the molluscan subclass Coleoidea in which the rostrum is not developed, the proostracum is represented by the elongated pen or gladus, and ten arms are present.

Texas fever [VET MED] A tick-borne infectious disease of cattle caused by the parasite *Babesia annulatus* which invades erythrocytes; characterized by fever, hemoglobinuria, and splenomegaly.

textile microbiology [MICROBIO] That branch of microbiology concerned with textile materials; deals with microorganisms that are harmful either to the fibers or to the consumer, and microorganisms that are useful, as in the retting process.

Textulariina [INV ZOO] A suborder of foraminiferan protozoans characterized by an agglutinated wall.

Th *See* thorium.

thalamus [ANAT] Either one of two masses of gray matter located on the sides of the third ventricle and forming part of the lateral wall of that cavity.

thalassemia [MED] A hereditary form of hemolytic anemia

TETRASPORALES

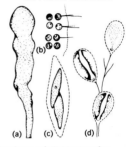

Members of Tetrasporales.
(a) Tetraspora, habit of a gelatinous thallus;
(b) arrangement of cells with pseudocilia. *(c) Elakatothrix*, simple colony. *(d) Chlorangium*, an attached, dendroid colony.

resulting from a defective synthesis of hemoglobin: thalassemia major is the homozygous state accompanied by clinical illness, and thalassemia minor is the heterozygous state and may not have evident clinical manifestations. Also known as Mediterranean anemia.

Thalassinidea [INV ZOO] The mud shrimps, a group of thin-shelled, burrowing decapod crustaceans belonging to the Macrura; individuals have large chelate or subchelate first pereiopods, and no chelae on the third pereiopods.

Thalattosauria [PALEON] A suborder of extinct reptiles in the order Eosuchia from the Middle Triassic.

Thaliacea [INV ZOO] A small class of pelagic Tunicata in which oral and atrial apertures occur at opposite ends of the body.

thalidomide [PHARM] $C_{13}H_{10}N_2O_4$ A drug used as a sedative and hypnotic; may produce teratogenic effects when administered during pregnancy.

thallium [CHEM] A metallic element in group III, symbol Tl, atomic number 81, atomic weight 204.37; insoluble in water, soluble in nitric and sulfuric acids, melts at 302°C, boils at 1457°C.

Thallobionta [BOT] One of the two subkingdoms of plants, characterized by the absence of specialized tissues or organs and multicellular sex organs.

thalloid [BOT] Pertaining to, resembling, or composed of a thallus.

Thallophyta [BOT] The equivalent name for Thallobionta.

thallophyte [BOT] A plant that is not differentiated into stem and root, as algae, fungi, and lichens.

thallospore [BOT] A spore that develops by budding of hyphal cells.

thallotoxicosis [MED] Poisoning due to ingestion of thallium or its derivatives.

thallus [BOT] A plant body that is not differentiated into special tissue systems or organs and may vary from a single cell to a complex, branching multicellular structure.

thanatology [MED] The study of the phenomenon of somatic death.

Thaumaleidae [INV ZOO] A family of orthorrhaphous dipteran insects in the series Nematocera.

Thaumastellidae [INV ZOO] A monospecific family of the Hemiptera assigned to the Pentatomorpha found only in Ethiopia.

Thaumastocoridae [INV ZOO] The single family of the hemipteran superfamily Thaumastocoroidea.

Thaumastocoroidea [INV ZOO] A monofamilial superfamily of the Hemiptera in the subdivision Geocorisae which occurs in Australia and the New World tropics.

Thaumatoxenidae [INV ZOO] A family of cyclorrhaphous dipteran insects in the series Aschiza.

Theaceae [BOT] A family of dicotyledonous erect trees or shrubs in the order Theales characterized by alternate, exstipulate leaves, usually five petals, and mostly numerous stamens.

Theales [BOT] An order of dicotyledonous mostly woody plants in the subclass Dilleniidae with simple or occasionally compound leaves, petals usually separate, numerous stamens, and the calyx arranged in a tight spiral.

theca [ANAT] The sheath of dura mater covering the spinal cord. [BOT] **1.** A moss capsule. **2.** A pollen sac. [HISTOL] The layer of stroma surrounding a Graafian follicle. [INV ZOO] The test of a testate protozoan or a rotifer.

theca folliculi [HISTOL] The capsule surrounding a developing or mature Graafian follicle; consists of two layers, theca interna and theca externa.

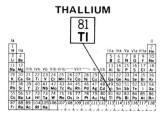

THALLIUM

Periodic table of the chemical elements showing the position of thallium.

Thecanephria [INV ZOO] An order of the phylum Brachiata containing a group of elongate, tube-dwelling tentaculate, deep-sea animals of bizarre structure.

thecate [BIOL] Having a theca.

Thecideidina [PALEON] An extinct suborder of articulate brachiopods doubtfully included in the order Terebratulida.

thecodont [VERT ZOO] Having the teeth in sockets.

Thecodontia [PALEON] An order of archosaurian reptiles, confined to the Triassic and distinguished by the absence of a supratemporal bone, parietal foramen, and palatal teeth, and by the presence of an antorbital fenestra.

Thelastomidae [INV ZOO] A family of nematode worms in the superfamily Oxyuroidea.

thematic apperception test [PSYCH] A projective psychological test using a set of pictures suggesting life situations from which the subject constructs a story; designed to reveal to the trained interpreter some of the dominant drives, emotions, sentiments, complexes, and conflicts of personality. Abbreviated TAT.

thenar [ANAT] The ball of the thumb.

Theophrastaceae [BOT] A family of tropical and subtropical dicotyledonous woody plants in the order Primulales characterized by flowers having staminodes alternate with the corolla lobes.

theory of antecedent conflicts [PSYCH] A theory that the effects of an emotionally disturbing event in adult life may be greatly multiplied through the conditioning or sensitizing effects of the vicissitudes of early life.

therapeutic abortion [MED] Abortion performed when pregnancy jeopardizes the health or life of the mother.

Therapsida [PALEON] An order of mammallike reptiles of the subclass Synapsida which first appeared in mid-Permian times and persisted until the end of the Triassic.

Therevidae [INV ZOO] The stiletto flies, a family of orthorrhaphous dipteran insects in the series Brachycera.

Theria [VERT ZOO] A subclass of the class Mammalia including all living mammals except the monotremes.

Theridiidae [INV ZOO] The comb-footed spiders, a family of the suborder Dipneumonomorphae.

thermal diffusion [PHYS CHEM] A phenomenon in which a temperature gradient in a mixture of fluids gives rise to a flow of one constituent relative to the mixture as a whole. Also known as thermodiffusion.

thermic fever *See* sunstroke.

thermocoagulation [MED] Destruction of tissue by means of electrocautery or a high-frequency current.

thermodiffusion *See* thermal diffusion.

thermoduric bacteria [MICROBIO] Bacteria which survive pasteurization, but do not grow at temperatures used in a pasteurizing process.

thermodynamics [PHYS] The branch of physics which seeks to derive, from a few basic postulates, relationships between properties of matter, especially those which are affected by changes in temperature, and a description of the conversion of energy from one form to another.

thermometer bird [VERT ZOO] The name applied to the brush turkey, native to Australia, because it lays its eggs in holes in mounds of earth and vegetation, with the heat from the decaying vegetation serving to incubate the eggs.

thermometric analysis [PHYS CHEM] A method for determination of the transformations a substance undergoes while being heated or cooled at an essentially constant rate, for example, freezing-point determinations.

thermoperiodicity [BOT] The totality of responses of a plant to appropriately fluctuating temperatures.

THETA RHYTHM

alpha rhythms

beta rhythms

theta rhythms

delta rhythms |—— 1 sec ——| |̄ 50 μv

Electroencephalogram of brain
rhythms taken from scalp leads
on human subject.

THIGMOTRICHIDA

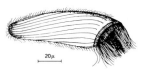

20μ

Drawing of *Boveria*, an example
of a thigmotrichid, showing
ciliature.

thermophile [BIOL] An organism that thrives at high temperatures.

thermoreception [PHYSIO] The process by which environmental temperature affects specialized sense organs (thermo-
receptors).

thermoreceptor [PHYSIO] A sense receptor that responds to
stimulation by heat and cold.

thermoregulation [PHYSIO] A mechanism by which mammals and birds attempt to balance heat gain and heat loss in
order to maintain a constant body temperature when exposed
to variations in cooling power of the external medium.

Thermosbaenacea [INV ZOO] An order of small crustaceans
in the superorder Pancarida.

Thermosbaenidae [INV ZOO] A family of the crustacean
order Thermosbaenacea.

thermotaxis [BIOL] Orientation movement of a motile organism in response to the stimulus of a temperature gradient.

thermotherapy [MED] The treatment of disease by heat of any
kind; involves the local or general application of heat to the
body.

therophyte [ECOL] An annual plant whose seed is the only
overwintering structure.

Theropoda [PALEON] A suborder of carnivorous bipedal saurischian reptiles which first appeared in the Upper Triassic
and culminated in the uppermost Cretaceous.

Theropsida [PALEON] An order of extinct mammallike reptiles in the subclass Synapsida.

Thesium [BOT] The hymenium of the apothecium in lichens.

thesocyte [INV ZOO] An amebocyte in Porifera containing
ergastic cytoplasmic inclusions.

theta antigen [IMMUNOL] A cell membrane constituent
which distinguishes T cells from other lymphocytes.

theta rhythm [PSYCH] A brain rhythm having a frequency of
about 4–7 hertz, and somewhat greater voltage than the alpha
rhythm; thought to originate in the hippocampus.

thiamine [BIOCHEM] $C_{12}H_{17}ClN_4OS$ A member of the vitamin B complex that occurs in many natural sources, frequently in the form of cocarboxylase. Also known as aneurine; vitamin B_1.

thiamine pyrophosphate [BIOCHEM] The coenzyme or prosthetic component of carboxylase; catalyzes decarboxylation
of various α-keto acids. Also known as cocarboxylase.

Thiaridae [INV ZOO] A family of freshwater gastropod mollusks in the order Pectinibranchia.

thicket *See* tropical scrub.

thigh [ANAT] The upper part of the leg, from the pelvis to the
knee.

thigh circumference [ANTHRO] The measurement around
the thigh of the left leg midway between the crotch and the
knee when the subject is in a standing position.

thigmotaxis *See* stereotaxis.

Thigmotrichida [INV ZOO] An order of ciliated protozoans in
the subclass Holotrichia.

thigmotropism *See* stereotropism.

thimerosal [ORG CHEM] $C_9H_9HgNaO_2S$ Sodium ethyl-
mercurithiosalicylate, an organomercurial antiseptic used
topically, and also as a preservative of certain biological
products.

thin-layer chromatography [ANALY CHEM] Chromatograph-
ing on thin layers of adsorbents rather than in columns;
adsorbent can be alumina, silica gel, silicates, charcoals, or
cellulose.

Thinocoridae [VERT ZOO] The seed snipes, family of South
American birds in the order Charadriiformes.

thinophyte [ECOL] A plant that lives on dunes.

Thiobacteriaceae [MICROBIO] A family of nonfilamentous, gram-negative bacteria of the suborder Pseudomonadineae which oxidize hydrogen sulfide, free sulfur, and inorganic sulfur compounds to sulfuric acid.

thiobarbiturate [PHARM] A derivative of thiobarbituric acid that differs from the barbiturates in the replacement of one oxygen atom by sulfur but resembles the barbiturates in its effects.

thiobarbituric acid [ORG CHEM] $C_6H_4N_2O_2S$ Malonyl thiourea, the parent compound of the thiobarbiturates; represents barbituric acid in which the oxygen atom of the urea component has been replaced by sulfur.

thiocarbamazine [PHARM] $C_{21}H_{17}AsN_2O_5S_2$ A white crystalline powder, freely soluble in dilute alkali; used in medicine as an amebicide. Also known as (*para*-ureidophenylarsylenedithio)di-*o*-benzoic acid; thiocarbamisin.

thiocarbamisin *See* thiocarbamazine.

thiocarbarsone [PHARM] $C_{11}H_{13}AsN_2O_5S_2$ A white crystalline powder, freely soluble in dilute alkali; used in medicine as an amebicide. Also known as (*para*-ureidophenylarsylenedithio)diacetic acid.

thioglycolic acid [ORG CHEM] $HSCH_2COOH$ A liquid with a strong unpleasant odor; used as a reagent for metals such as iron, molybdenum, silver, and tin, and in bacteriology. Also known as mercaptoacetic acid.

thiophil [MICROBIO] An organism that thrives in the presence of sulfur compounds, as certain bacteria.

Thiorhodaceae [MICROBIO] A family of bacteria in the suborder Rhodobacteriineae composed of the purple, red, orange, and brown sulfur bacteria; nearly all are strict anaerobes, oxidize hydrogen sulfide, and store sulfur globules internally.

thiouracil [PHARM] $C_4H_4N_2OS$ An antithyroid drug that acts by interfering with thyroxine synthesis.

thiourea [PHARM] CSN_2H_4 Thiocarbamide, an antithyroid drug used in the treatment of hyperthyroidism.

thirst [PHYSIO] A sensation, as of dryness in the mouth and throat, resulting from water deprivation.

thistle [BOT] Any of the various prickly plants comprising the family Compositae.

Thlipsuridae [PALEON] A Paleozoic family of ostracod crustaceans in the suborder Platycopa.

Thoracica [INV ZOO] An order of the subclass Cirripedia; individuals are permanently attached in the adult stage, the mantle is usually protected by calcareous plates, and six pairs of biramous thoracic appendages are present.

thoracic cavity *See* thorax.

thoracic duct [ANAT] The common lymph trunk beginning in the crura of the diaphragm at the level of the last thoracic vertebra, passing upward, and emptying into the left subclavian vein at its junction with the left internal jugular vein.

thoracic vertebrae [ANAT] The vertebrae associated with the chest and ribs in vertebrates; there are 12 in humans.

thoracoabdominal breathing [PHYSIO] The process of air breathing in reptiles, birds, and mammals that depends upon aspiration or sucking inspiration, and involves trunk musculature to supply pulmonary ventilation.

thoracopod [INV ZOO] A thoracic leg of malacostracans.

Thoracostomopsidae [INV ZOO] A family of marine nematodes in the superfamily Enoploidea, which have the stomatal armature modified to form a hollow tube.

thorax [ANAT] The chest; the cavity of the mammalian body between the neck and the diaphragm, containing the heart, lungs, and mediastinal structures. Also known as thoracic cavity. [INV ZOO] The middle of three principal divisions of the body of certain classes of arthropods.

THORIUM

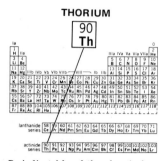

Periodic table of the chemical elements showing the position of thorium.

Thorazine [PHARM] Trademark for chlorpromazine, a tranquilizer and antiemetic drug, used as the hydrochloride salt.

Thorictidae [INV ZOO] The ant blood beetles, a family of coleopteran insects in the superfamily Dermestoidea.

thorium [CHEM] An element of the actinium series, symbol Th, atomic number 90, atomic weight 232; soft, radioactive, insoluble in water and alkalies, soluble in acids, melts at 1750°C, boils at 4500°C.

thorn [BOT] A short, sharp, rigid, leafless branch on a plant. [ZOO] Any of various sharp spinose structures on an animal.

thornback [VERT ZOO] *Raja clavata.* A ray found in European waters and characterized by spines on its back.

thornbush [ECOL] A vegetation class that is dominated by tall succulents and profusely branching smooth-barked deciduous hardwoods which vary in density from mesquite bush in the Caribbean to the open spurge thicket in Central Africa; the climate is that of a warm desert, except for a rather short intense rainy season. Also known as Dorngeholz; Dorngestrauch; dornveld; savane armée; savane épineuse; thorn scrub.

thorn forest [ECOL] A type of forest formation, mostly tropical and subtropical, intermediate between desert and steppe; dominated by small trees and shrubs, many armed with thorns and spines; leaves are absent, succulent, or deciduous during long dry periods, which may also be cool; an example is the caatinga of northeastern Brazil.

thorn scrub *See* thornbush.

thread blight [PL PATH] A fungus disease of a number of tropical and semitropical woody plants, including cocoa and tea, caused by species of *Pellicularia* and *Marasmius* which form filamentous mycelia on the surface of twigs and leaves.

threadfin [VERT ZOO] Common name for any of the fishes in the family Polynemidae.

three-day fever *See* phlebotomus fever.

threonine [BIOCHEM] $CH_3CHOHCH(NH_2)COOH$ A crystalline α-amino acid considered essential for normal growth of animals; it is biosynthesized from aspartic acid and is a precursor of isoleucine in microorganisms.

thresher [AGR] A machine that separates grain or seeds from straw. Also known as threshing machine.

thresher shark [VERT ZOO] Common name for fishes in the family Alopiidae; pelagic predacious sharks of generally wide distribution that have an extremely long, whiplike tail with which they thrash the water, destroying schools of small fishes.

threshing machine *See* thresher.

threshold [PHYS] The minimum level of some input quantity needed for some process to take place, such as a threshold energy for a reaction, or the minimum level of pumping at which a laser can go into self-excited oscillation. [PHYSIO] The minimum level of a stimulus that will evoke a response in an irritable tissue.

threshold of audibility [PHYSIO] The minimum effective sound pressure of a specified signal that is capable of evoking an auditory sensation in a specified fraction of the trials; the threshold may be expressed in decibels relative to 0.0002 microbar or 1 microbar. Also known as threshold of detectability; threshold of hearing.

threshold of detectability *See* threshold of audibility.

threshold of hearing *See* threshold of audibility.

Threskiornithidae [VERT ZOO] The ibises, a family of long-legged birds in the order Ciconiiformes.

thrip [INV ZOO] A small, slender-bodied phytophagous insect of the order Thysanoptera with suctorial mouthparts, a stout

THRIP

Flower thrip *(Frankliniella tritici)*, a common and widely distributed species of thrip, about 1 millimeter long, showing two pairs of long featherlike wings.

proboscis, a vestigial right mandible, and a fully developed left mandible, while wings may be present or absent.

Thripidae [INV ZOO] A large family of thrips, order Thysanoptera, which includes the most common species.

throat [ANAT] The region of the vertebrate body that includes the pharynx, the larynx, and related structures. [BOT] The upper, spreading part of the tube of a gamopetalous calyx or corolla.

thrombin [BIOCHEM] An enzyme elaborated from prothrombin in shed blood which induces clotting by converting fibrinogen to fibrin.

thromboangitis obliterans [MED] Thrombosis with organization and a variable degree of associated inflammation in the arteries and veins of the extremities, occasionally of the viscera, progressing to fibrosis about these structures and associated nerves, and complicated by ischemic changes in the parts supplied. Also known as Buerger's disease.

thrombocyte [HISTOL] One of the minute protoplasmic disks found in vertebrate blood; thought to be fragments of megakaryocytes. Also known as blood platelet; platelet.

thrombocythemia See thrombocytosis.

thrombocytopenia [MED] A condition characterized by a decrease in the absolute number of thrombocytes in the circulation.

thrombocytopenic purpura [MED] Hemorrhages in the skin, mucous membranes, and elsewhere associated with a decreased number of thrombocytes per unit volume of blood.

thrombocytosis [MED] A condition characterized by an increase in the absolute number of thrombocytes in the circulation. Also known as piastrenemia; thrombocythemia.

thromboembolectomy [MED] Surgical removal of an embolus that stems from a dislodged thrombus or part of a thrombus.

thromboembolism [MED] An embolism resulting from a dislodged thrombus or part of a thrombus.

thrombokinase [BIOCHEM] A proteolytic enzyme in blood plasma that, together with thromboplastin, calcium, and factor V, converts prothrombin to thrombin.

thrombophlebitis [MED] Inflammation of a vein associated with thrombosis.

thromboplastin [BIOCHEM] Any of a group of lipid and protein complexes in blood that accelerate the conversion of prothrombin to thrombin. Also known as factor III; plasma thromboplastin component (PTC).

thromboplastinogen See antihemophilic factor.

thrombosis [MED] Formation of a thrombus.

thrombotic thrombocytopenic purpura [MED] Thrombi in blood vessels associated with deposits of hyaline substances in the walls and with thrombocytopenia. Also known as Moschcowitz's disease.

thrombus [MED] A blood clot occurring on the wall of a blood vessel where the endothelium is damaged.

Throscidae [INV ZOO] The false metallic wood-boring beetles, a cosmopolitan family of the Coleoptera in the superfamily Elateroidea.

thrush [MED] A form of candidiasis due to infection by *Candida albicans* and characterized by small whitish spots on the tip and sides of the tongue and the mucous membranes of the buccal cavity. Also known as mycotic stomatitis; parasitic stomatitis. [VET MED] A disease of the frog of a horse's foot accompanied by a fetid discharge.

thulium [CHEM] A rare-earth element, symbol Tm, group IIIB, of the lanthanide group, atomic number 69, atomic weight 168.934; reacts slowly with water, soluble in dilute

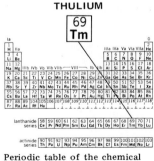

THULIUM

Periodic table of the chemical elements showing the position of thulium.

acids, melts at 1550°C, boils at 1727°C; the dust is a fire hazard; used as x-ray source and to make ferrites.

Thunburg technique [BIOCHEM] A technique used to study oxidation of a substrate occurring by dehydrogenation reactions; methylene blue, a reversibly oxidizable indicator, substitutes for molecular oxygen as the ultimate hydrogen acceptor (oxidant), becoming reduced to the colorless leuco form.

Thunnidae [VERT ZOO] The tunas, a family of perciform fishes; there are no scales on the posterior part of the body, and those on the anterior are fused to form an armored covering, the body is streamlined, and the tail is crescent-shaped.

Thurniaceae [BOT] A small family of monocotyledonous plants in the order Juncales distinguished by an inflorescence of one or more dense heads, vascular bundles of the leaf in vertical pairs, and silica bodies in the leaf epidermis.

Thylacinidae [VERT ZOO] A family of Australian carnivorous marsupials in the superfamily Dasyuroidea.

Thylacoleonidae [PALEON] An extinct family of carnivorous marsupials in the superfamily Phalangeroidea.

thyme [BOT] A perennial mint plant of the genus *Thymus*; pungent aromatic herb is made from the leaves.

thyme camphor *See* thymol.

Thymelaeceae [BOT] A family of dicotyledonous woody plants in the order Myrtales characterized by a superior ovary with a solitary ovule, and petals, if present, are scalelike.

thymic aplasia [MED] Congenital absence of the thymus and of the parathyroids with deficient cellular immunity. Also known as Di George's syndrome.

thymic corpuscle [HISTOL] A characteristic, rounded, acidophil body in the medulla of the thymus; composed of hyalinized cells concentrically arranged about a core which is occasionally calcified. Also known as Hassal's body.

thymidine [BIOCHEM] $C_{10}H_{14}N_2O_5$ A nucleoside derived from deoxyribonucleic acid; essential growth factor for certain microorganisms in mediums lacking vitamin B_{12} and folic acid.

thymidylic acid [BIOCHEM] The phosphate ester of thymidine; nucleotide of thymine.

thymine [BIOCHEM] $C_5H_6N_2O_2$ A pyrimidine component of nucleic acid, first isolated from the thymus.

thymocyte [HISTOL] A lymphocyte formed in the thymus.

thymol [ORG CHEM] $C_{10}H_{14}O$ A naturally occurring crystalline phenol obtained from thyme or thyme oil, melting at 515°C; used to kill parasites in herbaria, to preserve anatomical specimens, and in medicine as a topical antifungal agent. Also known as isopropyl-*meta*-cresol; thyme camphor.

thymoma [MED] A usually benign primary tumor of the thymus composed principally of lymphocytic and epithelial cells in varying proportions.

thymopharyngeal duct [EMBRYO] The third pharyngo-branchial duct; it may elongate between the pharynx and thymus.

thymus gland [ANAT] A lymphoid organ in the neck or upper thorax of all vertebrates; it is most prominent in early life and is essential for normal development of the circulating pool of lymphocytes.

thyreothecium [MYCOL] A shieldlike fruit body of certain ectoparasitic fungi.

Thyrididae [INV ZOO] The window-winged moths, a small tropical family of lepidopteran insects in the suborder Heteroneura.

thyridium [INV ZOO] A hairless, whitish area on the wings of certain insects.

thyrocalcitonin *See* calcitonin.

thyroglobulin [BIOCHEM] An iodinated protein found as the storage form of the iodinated hormones in the thyroid follicular lumen and epithelial cells.

thyroglossal cyst [MED] A cyst formed from the remnants of the thyroglossal duct.

thyroglossal duct [EMBRYO] A narrow temporary channel connecting the anlage of the thyroid with the surface of the tongue.

thyrohyoid [ANAT] Pertaining to the thyroid cartilage and the hyoid bone.

thyroid [PHARM] Dried and powdered thyroid gland which contains about 0.2% iodine in combination, especially as thyroxine, and is used therapeutically in the treatment of thyroid deficiencies.

thyroid cartilage [ANAT] The largest of the laryngeal cartilages in humans and most other mammals, located anterior to the cricoid; in man, forms the Adam's apple.

thyroidectomy [MED] Surgical removal of the thyroid gland.

thyroid gland [ANAT] An endocrine gland found in all vertebrates that produces, stores, and secretes the thyroid hormones.

thyroid hormone [BIOCHEM] Commonly, thyroxine or triiodothyronine, or both; a metabolically active compound formed and stored in the thyroid gland which functions to regulate the rate of metabolism.

thyroiditis [MED] Inflammation of the thyroid gland.

thyroid-stimulating hormone *See* thyrotropic hormone.

thyroprotein [BIOCHEM] A protein secreted in the thyroid gland, such as thyroxine.

Thyropteridae [VERT ZOO] The New-World disk-winged bats, a family of the Chiroptera found in Central and South America, characterized by a stalked sucking disk and a well-developed claw on the thumb.

thyrotoxicosis *See* hyperthyroidism.

thyrotropic hormone [BIOCHEM] A hormone produced by the adenohypophysis which regulates thyroid gland function. Also known as thyroid-stimulating hormone (TSH).

thyrotropin [BIOCHEM] A thyroid-stimulating hormone produced by the adenohypophysis.

thyroxine [BIOCHEM] $C_{15}H_{11}I_4NO_4$ The active physiologic principle of the thyroid gland; used in the form of the sodium salt for replacement therapy in states of hypothyroidism or absent thyroid function.

thyrse [BOT] An inflorescence with a racemose primary axis and cymose secondary and later axes.

Thysanidae [INV ZOO] A family of hymenopteran insects in the superfamily Chalcidoidea.

Thysanoptera [INV ZOO] The thrips, an order of small, slender insects having exopterygote development, sucking mouthparts, and exceptionally narrow wings with few or no veins and bordered by long hairs.

Thysanura [INV ZOO] The silverfish, machilids, and allies, an order of primarily wingless insects with soft, fusiform bodies.

Ti *See* titanium.

tibia [ANAT] The larger of the two leg bones, articulating with the femur, fibula, and talus.

tibialis [ANAT] **1.** A muscle of the leg arising from the proximal end of the tibia and inserted into the first cuneiform and first metatarsal bones. **2.** A deep muscle of the leg arising proximally from the tibia and fibula and inserted into the navicular and first cuneiform bones.

tibiofibula [VERT ZOO] A bone formed by fusion of the tibia and the fibula.

THYSANURA

Firebrat, *Thermobia domestica* Packard, a species of Thysanura found in all areas of temperate climate.

TIMOTHY

Drawing of timothy *(Phleum pratense)* showing leafy stems and cylindrical inflorescence.

TIN

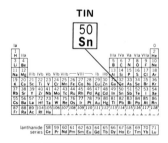

Periodic table of the chemical elements showing the position of tin.

tibiotarsus [VERT ZOO] In birds, a tibial bone fused with the proximal tarsals.

tic douloureux *See* trigeminal neuralgia.

tick [INV ZOO] Any arachnid comprising Ixodoidea; a blood-sucking parasite and important vector of various infectious diseases of humans and lower animals.

tick-bite paralysis [VET MED] A flaccid paralysis in animals, and occasionally in humans, caused by a feeding tick attached to the body.

tick-borne typhus fever of Africa [MED] Any of several infections caused by *Rickettsia conori,* transmitted by ixodid ticks, and occurring in Africa and adjacent areas; includes boutonneuse fever, Marseilles fever, Kenya tick typhus fever, and South African tick-bite fever.

tick fever *See* Rocky Mountain spotted fever.

tickle [PHYSIO] A tingling sensation of the skin or a mucous membrane following light, tactile stimulation.

tick typhus *See* Rocky Mountain spotted fever.

tic polonga [VERT ZOO] *Vipera russellii.* A member of the Viperidae; one of the most deadly and most common snakes in India; it may reach a length of 5 feet (1.5 meters), is nocturnal in its habits, and pursues rodents into houses. Also known as Russell's viper.

Tiedenmann's body [INV ZOO] One of the small glands opening into the ring vessel in many echinoderms in which amebocytes are produced.

tigellum [BOT] The central embryonic axis of the plant embryo; consists of the radicle and the plumule.

tiger [VERT ZOO] *Felis tigris.* An Asiatic carnivorous mammal in the family Felidae characterized by a tawny coat with transverse black stripes and white underparts.

tiger beetle [INV ZOO] The common name for any of the bright-colored beetles in the family Cicindelidae; there are about 1300 species distributed all over the world.

tiger salamander [VERT ZOO] *Ambystoma tigrinum.* A salamander in the family Ambystomatidae, found in a variety of subspecific forms from Canada to Mexico and over most of the United States; lives in arid and humid regions, and is the only salamander in much of the Great Plains and Rocky Mountains.

tiger shark *See* sand shark.

Tilletiaceae [MYCOL] A family of fungi in the order Ustilaginales in which basidiospores form at the tip of the apibasidium.

Tillodontia [PALEON] An order of extinct quadrupedal land mammals known from early Cenozoic deposits in the Northern Hemisphere and distinguished by large, rodentlike incisors, blunt-cuspid cheek teeth, and five clawed toes.

timbal [INV ZOO] The sound-producing organ in cicadas.

timberline [ECOL] The elevation or latitudinal limits for arboreal growth. Also known as tree line.

timothy [BOT] *Phleum pratense.* A perennial hay grass of the order Cyperales characterized by moderately leafy stems and a dense cylindrical inflorescence.

tin [CHEM] Metallic element in group IV, symbol Sn, atomic number 50, atomic weight 118.69; insoluble in water, soluble in acids and hot potassium hydroxide solution; melts at 232°C, boils at 2260°C.

Tinamidae [VERT ZOO] The single family of the avian order Tinamiformes.

Tinamiformes [VERT ZOO] The tinamous, an order of South and Central American birds which are superficially fowllike but have fully developed wings and are weak fliers.

tincture of iodine [PHARM] A medicinal preparation used as an anti-infective containing 20 grams iodine and 24 grams sodium iodide in 1000 milliliters of alcohol. Also known as iodine solution; iodine tincture.

tinea *See* ringworm.

tinea favosa *See* favus.

Tineidae [INV ZOO] A family of small moths in the superfamily Tineoidea distinguished by an erect, bristling vestiture on the head.

Tineoidea [INV ZOO] A superfamily of heteroneuran Lepidoptera which includes small moths that usually have well-developed maxillary palpi.

Tingidae [INV ZOO] The lace bugs, the single family of the hemipteran superfamily Tingoidea.

Tingoidea [INV ZOO] A superfamily of the Hemiptera in the subdivision Geocorisae characterized by the wings with many lacelike areolae.

tinnitus [MED] A ringing, roaring, or hissing sound in one or both ears.

Tintinnida [INV ZOO] An order of ciliated protozoans in the subclass Spirotrichia whose members are conical or trumpet-shaped pelagic forms bearing shells.

tipburn [PL PATH] A disease of certain cultivated plants, such as potato and lettuce, characterized by browning of the leaf margins due to excessive loss of water.

Tiphiidae [INV ZOO] A family of the Hymenoptera in the superfamily Scolioidea.

Tipulidae [INV ZOO] The crane flies, a family of orthorrhaphous dipteran insects in the series Nematocera.

tissue [HISTOL] An aggregation of cells more or less similar morphologically and functionally.

tissue culture [CYTOL] Growth of tissue cells in artificial media.

titanium [CHEM] A metallic element in group IV, symbol Ti, atomic number 22, atomic weight 47.90; ninth most abundant element in the earth's crust; insoluble in water, melts at 1660°C, boils above 3000°C.

Titanoideidae [PALEON] A family of extinct land mammals in the order Pantodonta.

titanothere [PALEON] Any member of the family Brontotheriidae.

titer [CHEM] **1.** The concentration in a solution of a dissolved substance as shown by titration. **2.** The least amount or volume needed to give a desired result in titration. **3.** The solidification point of hydrolyzed fatty acids.

Tithonian [GEOL] Southern European equivalent of the Portlandian stage (uppermost Jurassic) of geologic time.

titration [ANALY CHEM] A method of analyzing the composition of a solution by adding known amounts of a standardized solution until a given reaction (color change, precipitation, or conductivity change) is produced.

Tl *See* thallium.

Tm *See* thulium.

T method [BOT] A budding method in which a T-shaped cut is made through the bark at the internode of the stock, the bark of the scion is separated from the xylem along the cambium and removed, and the scion is forced into the incision on the stock.

toad [VERT ZOO] Any of several species of the amphibian order Anura, especially in the family Bufonidae; glandular structures in the skin secrete acrid, irritating substances of varying toxicity.

toadstool [MYCOL] Any of various fleshy, poisonous or inedible fungi with a large umbrella-shaped fruiting body.

Toarcian [GEOL] A European stage of geologic time; Lower Jurassic (above Pliensbachian, below Bajocian).

tobacco [BOT] **1.** Any plant of the genus *Nicotinia* cultivated

TINTINNIDA

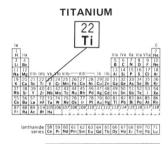

A living specimen of a representative species of the order Tintinnida in the genus *Tintinnopsis*, shown protruding from its lorica, or shell.

TITANIUM

Periodic table of the chemical elements showing the position of titanium.

TOAD

A toad of the genus *Bufo*, covered with warty structures that secrete poisons for purposes of defense.

TOMATO

Drawing of tomato
(*Lycopersicon esculentum*)
showing fruit and usual
arrangement of leaves.

for its leaves, which contain 1–3% of the alkaloid nicotine. **2.** The dried leaves of the plant.

tobacco mosaic [PL PATH] Any of a complex of virus diseases of tobacco and other solanaceous plants in which the leaves are mottled with light- and dark-green patches, sometimes interspersed with yellow.

tocopherol [ORG CHEM] $C_{29}H_{50}O_2$ Any of several substances having vitamin E activity that occur naturally in certain oils; alpha tocopherol is the most potent.

Todidae [VERT ZOO] The todies, a family of birds in the order Coraciiformes found in the West Indies.

toe [ANAT] One of the digits on the foot of man and other vertebrates.

tokocytes [INV ZOO] Reproductive cells in Porifera.

toleragen [IMMUNOL] A substance which, in appropriate dosages, produces a state of specific immunological tolerance in humans or animals.

tolerance [PHARM] **1.** The ability of enduring or being less responsive to the influence of a drug or poison, particularly when acquired by continued use of the substance. **2.** The allowable deviation from a standard, as the range of variation permitted for the content of a drug in one of its dosage forms.

tomato [BOT] A plant of the genus *Lycopersicon*, especially *L. esculentum*, in the family Solanaceae cultivated for its fleshy edible fruit, which is red, pink, orange, yellow, white, or green, with fleshy placentas containing many small, oval seeds with short hairs and covered with a gelatinous matrix.

tomentose [BOT] Covered with densely matted hairs.

tomentum [ANAT] The deep layer of the pia mater composed principally of minute blood vessels. [BIOL] Pubescence consisting of densely matted wooly hairs.

tomite [INV ZOO] A motile, nonfeeding stage that follows the protomite stage during the life cycle of the Holotricha.

tomium [VERT ZOO] The cutting edge of a bird's beak.

tomography *See* sectional radiography.

Tomopteridae [INV ZOO] The glass worms, a family of pelagic polychaete annelids belonging to the group Errantia.

Tongrian [GEOL] A European stage of geologic time; lower Oligocene (above Ludian of Eocene, below Rupelian). Also known as Lattorfian.

tongue [ANAT] A muscular organ located on the floor of the mouth in man and most vertebrates which may serve various functions, such as taking and swallowing food or tasting or as a tactile organ or sometimes a prehensile organ.

tongue worm *See* acorn worm.

tonic convulsion *See* tonic postural epilepsy; tonic spasm.

tonic labyrinthine reflexes [PHYSIO] Rotation or deviation of the head causes extension of the limbs on the same side as the chin, and flexion of the opposite extremities: dorsiflexion of the head produces increased extensor tonus of the upper extremities and relaxation of the lower limbs, while ventroflexion of the head produces the reverse; seen in the young infant and patients with a lesion at the midbrain level or above.

tonic neck reflexes [PHYSIO] Reflexes in which rotation or deviation of the head causes extension of the limbs on the same side as the chin, and flexion of the opposite extremities; dorsiflexion of the head produces increased extension tonus of the upper extremities and relaxation of the lower limbs, and ventroflexion of the head, the reverse; seen normally in incomplete forms in the very young infant, and thereafter in patients with a lesion at the midbrain level or above.

tonic postural epilepsy [MED] A form of epilepsy in which seizures are characterized by a rigid posture with the arms

TONGUE

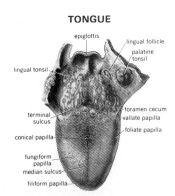

epiglottis
lingual follicle
palatine tonsil
lingual tonsil
foramen cecum
vallate papilla
foliate papilla
terminal sulcus
conical papilla
fungiform papilla
median sulcus
filiform papilla

The human tongue, dorsal view.
(*From M. W. Woerdeman, Atlas
of Human Anatomy, vol. 2,
McGraw-Hill, 1950*)

and legs extended, hands pronated, and feet held in plantar flexion. Also known as tonic convulsion.

tonic spasm [MED] A spasm which persists for some time without relaxation. Also known as tonic convulsion.

tonofibril [CYTOL] Any of the fibrils converging on desmosomes in epithelial cells.

tonometer [MED] An electronic instrument that measures hydrostatic pressure within the eye: when placed in position, a tiny movable plate is pressed against the eye, flattening a circular section of the cornea (no eyeball anesthesia is required); a current is then sent through a small electromagnet, of such value that it will just pull the plate away from the eye; the value of the current is proportional to eye pressure; a measurement can be made in about 1 second; used in diagnosis of glaucoma. Also known as electronic tonometer.

tonoplast [BOT] The membrane surrounding a plant-cell vacuole.

tonsil [ANAT] **1.** Localized aggregation of diffuse and nodular lymphoid tissue found in the throat where the nasal and oral cavities open into the pharynx. **2.** *See* palatine tonsil.

tonsillectomy [MED] Surgical removal of the palatine tonsil.

tonsillitis [MED] Inflammation of the tonsils.

tonus [PHYSIO] The degree of muscular contraction when not undergoing shortening.

tooth [ANAT] One of the hard bony structures supported by the jaws in mammals and by other bones of the mouth and pharynx in lower vertebrates serving principally for prehension and mastication. [INV ZOO] Any of various sharp, horny, chitinous, or calcareous processes on or about any part of an invertebrate that functions like or resembles vertebrate jaws.

toothache [MED] Pain in or about a tooth. Also known as odontalgia.

tooth decay [MED] Caries of the teeth.

tooth shell [INV ZOO] A mollusk of the class Scaphopoda characterized by the elongate, tube-shaped, or cylindrical shell which is open at both ends and slightly curved.

top grafting [BOT] Grafting a scion of one variety of tree onto the main branch of another.

tophus [MED] A localized swelling principally in cartilage and connective tissues in or adjacent to the small joints of the hands and feet; occurs specifically in gout.

topocline [ECOL] A graded series of characters exhibited by a species or other closely related organisms along a geographical axis.

topogenesis *See* morphogenesis.

topographic anatomy [ANAT] The use of bony and soft tissue landmarks on the surface of the body to indicate the known location of deeper structures.

topographic climax [ECOL] A climax plant community under a uniform macroclimate over which minor topographic features such as hills, rivers, valleys, or undrained depressions exert a controlling influence.

topotaxis *See* tropism.

topotype [SYST] A specimen of a species not of the original type series collected at the type locality.

top shell [INV ZOO] Any of the marine snails of the family Trochidae characterized by a spiral conical shell with a flat base.

torma [INV ZOO] A thickening at the point where the labrum and the clypeus join.

tornaria [INV ZOO] The larva of some acorn worms (Enteropneusta) which is large and marked by complex bands of cilia.

tornote [INV ZOO] A monaxon spicule in certain Porifera having both ends terminating abruptly in points.

TOOTH SHELL

Tube-shaped tooth shell.

torose [INV ZOO] **1.** Having knobby prominences on the surface. **2.** Cylindrical with alternate swellings and contractions.

Torpedinidae [VERT ZOO] The electric rays or torpedoes, a family of batoid sharks.

torpor [PHYSIO] The condition in hibernating poikilotherms during winter when body temperature drops in a parallel relation to ambient environmental temperatures.

Torridincolidae [INV ZOO] A small family of coleopteran insects in the suborder Myxophaga found only in Africa and Brazil.

torticollis [MED] A deformity of the neck resulting from contraction of the cervical muscles or fascia. Also known as wryneck.

tortoise [VERT ZOO] Any of various large terrestrial reptiles in the order Chelonia, especially the family Testudinidae.

Tortonian [GEOL] A European stage of geologic time: Miocene (above Helvetian, below Sarmatian).

Tortricidae [INV ZOO] A family of phytophagous moths in the superfamily Tortricoidea which have a stout body, lightly fringed wings, and threadlike antennae.

Tortricoidea [INV ZOO] A superfamily of small wide-winged moths in the suborder Heteroneura.

Torulopsidales [MYCOL] The equivalent name for Cryptococcales.

torulosis See cryptococcosis.

torulus [INV ZOO] The socket for insertion of the antenna in insects.

torus [ANAT] A rounded protruberance on a body part. [BOT] The thickened membrane closing a bordered pit.

Torymidae [INV ZOO] A family of hymenopteran insects in the superfamily Chalcidoidea.

total heat See enthalpy.

total heat of solution See heat of solution.

total span [ANTHRO] An anthropometric determination of the distance between the tips of the middle fingers at maximum arm stretch without straining.

totipalmate [VERT ZOO] Having all four toes connected by webs, as in the Pelecaniformes.

totipotence [EMBRYO] Capacity of a blastomere to develop into a fully formed embryo.

toucan [VERT ZOO] Any of numerous fruit-eating birds, of the family Ramphastidae, noted for their large and colorful bills.

touch [PHYSIO] The array of sensations arising from pressure sensitivity of the skin.

Tournaisian [GEOL] European stage of lowermost Carboniferous time.

tourniquet [MED] An apparatus for controlling hemorrhage from, or circulation in, a limb or part of the body, where pressure can be brought upon the blood vessels by means of straps, cords, rubber tubes, or pads.

toxa [INV ZOO] A curved sponge spicule.

toxaspire [INV ZOO] A spiral spicule having more than one revolution.

Toxasteridae [PALEON] A family of Cretaceous echinoderms in the order Spatangoida which lacked fascioles and petals.

toxemia [MED] A condition in which the blood contains toxic substances, either of microbial origin or as by-products of abnormal protein metabolism.

toxemia of pregnancy See preeclampsia.

toxic [MED] Relating to a harmful effect by a poisonous substance on the human body by physical contact, ingestion, or inhalation.

toxic amaurosis [MED] Blindness following the introduction

TORTOISE

Desert tortoise *(Gopherus agassizi).*

of toxic substances into the body, such as ethyl and methyl alcohol, tobacco, lead, and metabolites of uremia and diabetes.

toxic goiter *See* hyperthyroidism.

toxic hepatitis [MED] Inflammation of the liver caused by chemical agents ingested or inhaled into the body, such as chlorinated hydrocarbons and some alkaloids.

toxicity [PHARM] **1.** The quality of being toxic. **2.** The kind and amount of poison or toxin produced by a microorganism, or possessed by a chemical substance not of biological origin.

toxicology [PHARM] The study of poisons, including their nature, effects, and detection, and methods of treatment.

toxic psychosis [MED] A brain disorder due to a toxic agent such as lead or alcohol.

toxicyst [INV ZOO] A type of trichocyst in Protozoa which may, upon contact, induce paralysis or lysis of the prey.

toxin [BIOCHEM] Any of various poisonous substances produced by certain plant and animal cells, including bacterial toxins, phytotoxins, and zootoxins.

Toxodontia [PALEON] An extinct suborder of mammals representing a central stock of the order Notoungulata.

Toxoglossa [INV ZOO] A group of carnivorous marine gastropod mollusks distinguished by a highly modified radula (toxoglossate).

toxoglossate radula [INV ZOO] A radula in certain carnivorous gastropods having elongated, spearlike teeth often perforated by the ducts of large poison glands.

toxoid [IMMUNOL] Detoxified toxin, but with antigenic properties intact; toxoids of tetanus and diphtheria are used for immunization.

Toxoplasmea [INV ZOO] A class of the protozoan subphylum Sporozoa composed of small, crescent-shaped organisms that move by body flexion or gliding and are characterized by a two-layered pellicle with underlying microtubules, micropyle, paired organelles, and micronemes.

Toxoplasmida [INV ZOO] An order of the class Toxoplasmea; members are parasites of vertebrates.

toxoplasmin [BIOCHEM] The *Toxoplasma* antigen; used in a skin test to demonstrate delayed hypersensitivity to toxoplasmosis.

toxoplasmosis [MED] Infection by the protozoan *Toxoplasma gondii*, manifested clinically in severe cases by jaundice, hepatomegaly, and splenomegaly.

Toxopneustidae [INV ZOO] A family of Tertiary and extant echinoderms of the order Temnopleuroida where the branchial slits are deep and the test tends to be absent.

Toxotidae [VERT ZOO] The archerfishes, a family of small fresh-water forms in the order Perciformes.

T phage [VIROL] Any of a series (T1–T7) of deoxyribonucleic acid phages which lyse strains of the gram-negative bacterium *Escherichia coli* and its relatives.

TPI test *See* Treponema pallidum immobilization test.

TPN *See* triphosphopyridine nucleotide.

trabecula [ANAT] A band of fibrous or muscular tissue extending from the capsule or wall into the interior of an organ.

trace fossil [GEOL] A trail, track, or burrow made by an animal and found in ancient sediments such as sandstone, shale, or limestone. Also known as ichnofossil.

tracer [CHEM] A foreign substance, usually radioactive, that is mixed with or attached to a given substance so the distribution or location of the latter can later be determined; used to trace chemical behavior of a natural element in an organism. Also known as tracer element.

tracer element *See* tracer.

trachea [ANAT] The cartilaginous and membranous tube by

TOXOGLOSSATE RADULA

(a)   (b)

Toxoglossate radula from marine gastropods: *(a) Conus* species; *(b) Terebra* species. *(From R. R. Shrock and W. H. Twenhofel, Principles of Invertebrate Paleontology, 2d ed., McGraw-Hill, 1953)*

TRACE FOSSIL

Feeding trail, a trace fossil resulting from the grazing of animals along the surface of a sediment, in ichnogenus *Cosmorhaphe*, from Tertiary sediments at Pologne. *(From R. C. Moore, ed., Treatise on Invertebrate Paleontology, pt. W, University of Kansas Press, 1962)*

TRACHEID

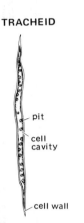

— pit

— cell
cavity

— cell wall

Drawing of a tracheid, showing
characteristic features. *(From
H. J. Fuller and O. Tippo,
College Botany, rev. ed., Holt,
1954)*

which air passes to and from the lungs in man and many
vertebrates. [BOT] A xylem vessel resembling the trachea of
vertebrates. [INV ZOO] One of the anastomosing air-convey-
ing tubules composing the respiratory system in most insects.

tracheid [BOT] An elongate, spindle-shaped xylem cell, lacking protoplasm at maturity, and having secondary walls laid
in various thicknesses and patterns over the primary wall.

tracheole [INV ZOO] A terminal branch of the tracheal system in insects.

Tracheophyta [BOT] A large group of plants characterized by
the presence of specialized conducting tissues (xylem and
phloem) in the roots, stems, and leaves.

trachoma [MED] An infectious disease of the conjunctiva and
cornea caused by *Chlamydia trachomatis* producing photophobia, pain, and excessive lacrimation.

trachomatous conjunctivitis [MED] Inflammation of the conjunctiva associated with trachoma, characterized by a subepi-
thelial cellular infiltration with a follicular distribution.

trachyglossate [VERT ZOO] Having a rasping or toothed
tongue.

Trachylina [INV ZOO] An order of moderate-sized jellyfish of
the class Hydrozoa distinguished by having balancing organs
and either a small polyp stage or none.

Trachymedusae [INV ZOO] A group of marine jellyfish, recognized as a separate order or as belonging to the order
Trachylina whose tentacles have a solid core consisting of a
single row of endodermal cells.

Trachypsammiacea [INV ZOO] An order of colonial anthozoan coelenterates characterized by a dendroid skeleton.

Trachystomata [VERT ZOO] The name given to the Meantes
when the group is considered to be an order.

tractellum [INV ZOO] A flagellum on the anterior end of
Mastigophora, or of zoospores, characterized by a circum-
ductory motion.

traction diverticulum [MED] A circumscribed sacculation,
usually of the esophagus, with bulging of the full thickness of
the wall; caused by the pull of adhesions arising from adjacent
organs.

tragacanth [MATER] The gummy exudate produced by certain Asiatic species of *Astragalus*; consists of a soluble portion
containing uronic acid and arabinose, and an insoluble
portion that absorbs water and swells to make a stiff opalescent mucilage.

tragion–nasal root [ANTHRO] A measure of the distance from
the tragion to the deepest concavity of the nasal root.

Tragulidae [VERT ZOO] The chevrotains, a family of pecoran
ruminants in the superfamily Traguloidea.

Traguloidea [VERT ZOO] A superfamily of pecoran ruminants, comprised of the most primitive forms with large
canines; the chevrotain is the only extant member.

tragus [ANAT] 1. The prominence in front of the opening of
the external ear. 2. One of the hairs in the external ear canal.

trama [MYCOL] The loosely woven hyphal tissue between
adjacent hymenia in basidiomycetes.

tranquilizer [PHARM] 1. Any agent that brings about a state
of relief from anxiety, or peace of mind. 2. Any agent that
produces a calming or sedative effect without inducing sleep.
3. Any drug, such as chlorpromazine, used primarily for its
calming and antispsychotic effects, or such as meprobamate,
used for symptomatic treatment of common psychoneuroses
and as an adjunct in somatic disorders complicated by anxiety
and tension.

transaminase [BIOCHEM] One of a group of enzymes that
catalyze the transfer of the amino group of an amino acid to

a keto acid to form another amino acid. Also known as aminotransferase.

transamination [CHEM] **1.** The transfer of one or more amino groups from one compound to another. **2.** The transposition of an amino group within a single compound.

trans arrangement *See* repulsion linkage.

transcapsidation [VIROL] Change in the capsid of PARA (particle aiding replication of adenovirus) from one type of adenovirus to another.

transduction [MICROBIO] Transfer of genetic material between bacterial cells by bacteriophages.

transect [SCI TECH] To cut across, or to cut transversely.

transeptate [BIOL] Having transverse septae.

transferase [BIOCHEM] Any of various enzymes that catalyze the transfer of a chemical group from one molecule to another.

transference [PSYCH] The unconscious transfer of the patient's feelings and reactions originally associated with important persons in the patient's life, usually father, mother, or siblings, toward others and in the analytic situation, toward the analyst.

transfer ribonucleic acid [MOL BIO] The smallest ribonucleic acid molecule found in cells; its structure is complementary to messenger ribonucleic acid and it functions by transferring amino acids from the free state to the polymeric form of growing polypeptide chains. Abbreviated t-RNA.

transferrin [BIOCHEM] Any of various beta globulins in blood serum which bind and transport iron to the bone marrow and storage areas.

transformation [GEN] A change in the genome of a bacterium as the result of exposure to the DNA of a bacterium with a different genome.

transfusion [MED] The administration of blood, or one of its components, as a part of treatment.

transient polymorphism [GEN] A form of polymorphism in which two forms coexist while one is in the process of replacing the other.

transient situational disturbance [PSYCH] A form of personality disorder, more or less transient, and generally an acute symptom response to a specific situation, without persistent personality disturbance.

transitional epithelium [HISTOL] A form of stratified epithelium found in the urinary bladder; cells vary between squamous, when the tissue is stretched, and columnar, when not stretched.

transketolase [BIOCHEM] An enzyme that cleaves a substrate at the position of the carbonyl carbon and transports a two-carbon fragment to an acceptor compound to form a new compound.

translation [MOL BIO] The process by which the linear sequence of nucleotides in a molecule of messenger ribonucleic acid directs the specific linear sequence of amino acids, as during protein synthesis.

translocation [BOT] Movement of water, mineral salts, and organic substances from one part of a plant to another. [CYTOL] The transfer of a chromosome segment from its usual position to a new position in the same or in a different chromosome.

translocation heterozygote [GEN] An individual heterozygous for an exchange of segments between two nonhomologous chromosomes; it produces gametes of which a proportion are imbalanced, having duplications and deficiencies in respect to the exchanged segments.

transmethylase [BIOCHEM] A transferase enzyme involved in

TRANSITIONAL EPITHELIUM

Cellular arrangement in transitional epithelium.

TRANSKETOLASE

$$
\begin{array}{cccc}
\mathrm{CH_2OH} & \mathrm{CHO} & & \mathrm{CH_2OH} \\
| & | & & | \\
\mathrm{CO} & \mathrm{HCOH} & \mathrm{CHO} & \mathrm{CO} \\
| & | & | & | \\
\mathrm{HOCH} & +\mathrm{HCOH} \rightleftharpoons \mathrm{HCOH} & +\mathrm{HOCH} \\
| & | & | & | \\
\mathrm{HCOH} & \mathrm{HCOH} & \mathrm{CH_2OPO_3H_2} & \mathrm{HCOH} \\
| & | & & | \\
\mathrm{CH_2OPO_3H_2} & \mathrm{CH_2OPO_3H_2} & & \mathrm{HCOH} \\
& & & | \\
& & & \mathrm{HCOH} \\
& & & | \\
& & & \mathrm{CH_2OPO_3H_2} \\
\mathbf{I} & \mathbf{II} & \mathbf{III} & \mathbf{IV}
\end{array}
$$

Example of a reaction catalyzed by transketolase. I is xylulose-5-phosphate. II is ribulose-5-phosphate, III is glyceraldehyde-3-phosphate, and IV is sedoheptulose-7-phosphate.

TRANSLOCATION

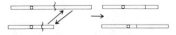

Translocation in which nonhomologous chromosomes become broken, switching their broken ends in the process of rejoining.

catalyzing chemical reactions in which methyl groups are transferred from a substrate to a new compound.

transmethylation [BIOCHEM] A metabolic reaction in which a methyl group is transferred from one compound to another; methionine and choline are important donors of methyl groups.

transmission electron microscope [ELECTR] A type of electron microscope in which the specimen transmits an electron beam focused on it, image contrasts are formed by the scattering of electrons out of the beam, and various magnetic lenses perform functions analogous to those of ordinary lenses in a light microscope.

transorbital lobotomy [MED] A lobotomy performed through the roof of the orbit.

transpalatine [VERT ZOO] A cranial bone connecting the pterygoid with the jugal and maxilla in crocodiles.

trans phase *See* repulsion linkage.

transpinalis [ANAT] A muscle that connects the transverse processes of vertebrae.

transpiration [BIOL] The passage of a gas or liquid (in the form of vapor) through the skin, a membrane, or other tissue.

transplantation [BIOL] **1.** The artificial removal of part of an organism and its replacement in the body of the same or of a different individual. **2.** To remove a plant from one location and replant it in another place.

transplantation antigen [IMMUNOL] An antigen in a cell which induces a histocompatibility reaction when the cell is transplanted into an organism not having that antigen.

transplanter [AGR] A special kind of equipment designed for the planting of cuttings or small plants; it transports one or more men who assist the action of the machine in placing plants in a furrow and covering them; it commonly supplies a small quantity of water to each plant.

transsexual [PSYCH] An individual whose chromosomes, gonads, and body habitus mark that individual as a member of one sex, but who feels psychically to be of the other sex, with an overwhelming desire for sex reassignment through surgical and hormonal intervention.

transversion [MOL BIO] A mutation resulting from the substitution in deoxyribonucleic acid or ribonucleic acid of a purine for a pyrimidine or a pyrimidine for a purine.

transversum [VERT ZOO] A cranial bone extending from the pterygoid to the maxilla in most reptiles.

trapeziform [BIOL] Having the form of a trapezium.

trapezium [ANAT] The first bone of the second row of carpal bones.

trapezius [ANAT] A back muscle having its origin at the occipital bone, the nuchal ligament, and the spines of the thoracic vertebrae, and its insertion in the clavicle, acromion, and spine of the scapula.

trauma [MED] An injury caused by a mechanical or physical agent. [PSYCH] A severe psychic injury.

traumatic pneumonosis [MED] The acute, noninflammatory pathologic pulmonary changes produced by a large momentary deceleration.

traumatotropism [BIOL] Orientation response of an organ of a sessile organism in response to a wound.

Trebidae [INV ZOO] A family of copepod crustaceans of the order Caligoida which are external parasites on selachians.

tree [BOT] A perennial woody plant at least 20 feet (6 meters) in height at maturity, having an erect stem or trunk and a well-developed crown or leaf canopy.

tree fern [BOT] The common name for plants belonging to the families Cyatheaceae and Dicksoniaceae; all are ferns that exhibit an arborescent habit.

tree frog [VERT ZOO] Any of the arboreal frogs comprising the family Hylidae characterized by expanded digital adhesive disks.

tree line *See* timberline.

Trematoda [INV ZOO] A loose grouping of acoelomate, parasitic flatworms of the phylum Platyhelminthes; they exhibit cephalization, bilateral symmetry, and well-developed holdfast structures.

trematodiasis [MED] Infection caused by a member of the Trematoda (trematode).

Trematosauria [PALEON] A group of Triassic amphibians in the order Temnospondyli.

trembling ill *See* louping ill.

Tremellales [MYCOL] An order of basidiomycetous fungi in the subclass Heterobasidiomycetidae in which basidia have longitudinal walls.

tremor [MED] Involuntary, rhythmic trembling of voluntary muscles resulting from alternate contraction and relaxation of opposing muscle groups.

trench fever [MED] A louse-borne infection that is caused by *Rickettsia quintana* and is characterized by headache, chills, rash, pain in the legs and back, and often by a relapsing fever.

trench mouth [MED] A common form of Vincent's angina; characterized by redness, congestion, and edema of the gums, with involvement of the entire oral cavity in severe cases.

trend [STAT] The general drift, tendency, or bent of a set of statistical data as related to time or another related set of statistical data.

Trentepohliaceae [BOT] A family of green algae belonging to the Ulotrichales having thick walls, bandlike or reticulate chloroplasts, and zoospores or isogametes produced in enlarged, specialized cells.

Trentonian [GEOL] A North American stage of geologic time; Middle Ordovician (above Wilderness, below Edenian); equivalent to the upper Mohawkian.

Treponema pallidum immobilization test [IMMUNOL] A serologic test for syphilis in which suspensions of *Treponema pallidum* are immobilized in the presence of syphilitic serum and complement. Abbreviated TPI test.

Treponemataceae [MICROBIO] A family of the bacterial order Spirochaetales including the spirochetes less than 20 micrometers long and less than 5 micrometers in diameter; most species are parasitic.

treponematosis [MED] Infection caused by any species of the genus *Treponema*. Also known as treponemiasis.

treponemiasis *See* treponematosis.

Trepostomata [PALEON] An extinct order of ectoproct bryozoans in the class Stenolaemata characterized by delicate to massive colonies composed of tightly packed zooecia with solid calcareous zooecial walls.

Treroninae [VERT ZOO] The fruit pigeons, a subfamily of the avian family Columbidae distinguished by the gaudy coloration of the feathers.

Tretothoracidae [INV ZOO] A family of the Coleoptera in the superfamily Tenebrionoidea which contains a single species found in Queensland, Australia.

triacetyloleandomycin [MICROBIO] An antibiotic produced by *Streptomyces antibioticus* and used clinically in the treatment of pneumonia, osteomyelitis, furuncles, and carbuncles.

triaene [INV ZOO] An elongated spicule in certain Porifera with three rays diverging from one end.

triadelphous [BOT] Having the stamens united into three bundles by their filaments.

triage [MED] The process of determining which casualties (as from an accident, disaster, military battle, or explosion of nuclear weapons) need urgent treatment, which ones are well enough to go untreated, and which ones are beyond hope of benefit from treatment.

Triakidae [VERT ZOO] A family of galeoid sharks in the carcharinid line.

triandrous [BOT] Possessing three stamens.

triangular ligament *See* urogenital diaphragm.

triaperturate [BIOL] Having three apertures.

triarch [BOT] Having three united xylem bundles forming the woody tissue of the root.

Triassic [GEOL] The first period of the Mesozoic era, lying above Permian and below Jurassic, 180–225 million years ago.

Triatominae [INV ZOO] The kissing bugs, a subfamily of hemipteran insects in the family Reduviidae, distinguished by a long, slender rostrum.

triaxon [INV ZOO] A spicule in Porifera having three axes which cross each other at right angles.

tribe [SYST] A subdivision of a family.

tribromoethanol [PHARM] $C_2H_3Br_3O$ A white crystalline compound, melting at 79–82°C; used in medicine in anesthesia. Also known as tribromoethyl alcohol.

tribromoethyl alcohol *See* tribromoethanol.

trica [BOT] An apothecium of a lichen characterized by a ridged spherical surface.

tricarboxylic acid cycle *See* Krebs cycle.

triceps [ANAT] A muscle with three heads; an example is the triceps brachii muscle of the upper arm.

Trichechidae [VERT ZOO] The manatees, a family of nocturnal, solitary sirenian mammals in the suborder Trichechiformes.

Trichechiformes [VERT ZOO] A suborder of mammals in the order Sirenia which contains the manatees and dugongids.

Trichiaceae [MYCOL] A family of slime molds in the order Trichiales.

Trichiales [MYCOL] An order of Myxomycetes in the subclass Myxogastromycetidae.

trichilium [BOT] A mass of matted hairs at the base of certain leaf petioles.

trichinosis [MED] Infection by the nematode *Trichinella spiralis* following ingestion of encysted larvae in raw or partially cooked pork; characterized by eosinophilia, nausea, fever, diarrhea, stiffness and painful swelling of muscles, and facial edema.

trichite [INV ZOO] **1.** Any of the fine rhabdoid structures in the oral basket of certain infusorians. **2.** A siliceous spicule in certain Porifera.

Trichiuridae [VERT ZOO] The cutlass-fishes, a family of the suborder Scombroidei.

trichloromethane *See* chloroform.

trichobezoar [MED] A ball of hair or similar concretion in the stomach or intestine.

trichoblast [BOT] An epidermal cell which develops into a root hair.

trichobothrium [INV ZOO] An erect, bristlelike sensory hair found on certain arthropods, insects, and other invertebrates.

trichobranchiate gill [INV ZOO] A gill with filamentous branches arranged in several series around the axis; found in some decapod crustaceans.

Trichobranchidae [INV ZOO] A family of polychaete annelids belonging to the Sedentaria; most members are rare and live at great ocean depths.

trichocarpous [BOT] Having hairy fruit

trichocercous cercaria [INV ZOO] A trematode larva distinguished by a spiny tail.

TRIASSIC

PRECAMBRIAN		
CAMBRIAN		
ORDOVICIAN		
SILURIAN		PALEOZOIC
DEVONIAN		
Mississippian	CARBONIFEROUS	
Pennsylvanian		
PERMIAN		
TRIASSIC		
JURASSIC	MESOZOIC	
CRETACEOUS		
TERTIARY	CENOZOIC	
QUATERNARY		

Chart showing position of the Triassic period in relation to the other periods and to the eras of geologic time.

Trichocomaceae [MYCOL] A small tropical family of asco-mycetous fungi in the order Eurotiales with ascocarps from which a tuft of capillitial threads extrudes, releasing the ascospores after dissolution of the asci.

trichocyst [INV ZOO] A minute structure in the cortex of certain protozoans that releases filamentous or fibrillar threads when discharged.

Trichodactylidae [INV ZOO] A family of fresh-water crabs in the section Brachyura, found mainly in tropical regions.

trichodragmata [INV ZOO] Straight, hairlike spicules, ar-ranged in bundles.

trichoepithelioma [MED] A benign tumor characterized by small, round, yellow, or flesh-colored papules, chiefly on the center of the face.

trichogen [INV ZOO] A cell that produces a hair or a bristle in insects.

Trichogrammatidae [INV ZOO] A family of the Hymenoptera in the superfamily Chalcidoidea whose larvae are parasitic in the eggs of other insects.

trichogyne [BOT] A terminal portion of a procarp or archi-carp which receives a spermatium.

trichome [BOT] An appendage derived from the protoderm in plants, including hairs and scales. [INV ZOO] A brightly colored tuft of hairs on the body of a myrmecophile that releases an aromatic substance attractive to ants.

Trichomonadida [INV ZOO] An order of the protozoan class Zoomastigophorea which contains four families of uninucle-ate species.

Trichomonadidae [INV ZOO] A family of flagellate protozo-ans in the order Trichomonadida.

Trichomonas [INV ZOO] A genus of flagellate, parasitic pro-tozoans, belonging to the Mastigophora, characterized by three to five flagella and an undulating membrane.

Trichomonas vaginalis [INV ZOO] A species of Protozoa that causes vaginitis.

trichomoniasis [MED] An infection caused by a species of the genus *Trichomonas*.

Trichomycetes [MYCOL] A class of true fungi, division Fun-gi.

trichomycin [MICROBIO] An antibiotic produced by *Strepto-myces hachijoensis* and *S. abikoensis;* a water-soluble yellow powder that inhibits yeasts and fungi.

Trichoniscidae [INV ZOO] A primitive family of isopod crus-taceans in the suborder Oniscoidea found in damp littoral, halophilic, or riparian habitats.

Trichophilopteridae [INV ZOO] A family of lice in the order Mallophaga adapted to life upon the lemurs of Madagascar.

trichophytin [IMMUNOL] A group antigen obtained from fil-trates of *Trichophyton mentagrophytes;* used in a skin test to ascertain past or present infection with the dermatophytes.

Trichoptera [INV ZOO] The caddis flies, an aquatic order of the class Insecta; larvae are wormlike and adults have two pairs of well-veined hairy wings, long antennae, and mouth-parts capable of lapping only liquids.

Trichopterygidae [INV ZOO] The equivalent name for Pti-liidae.

Trichostomatida [INV ZOO] An order of ciliated protozoans in the subclass Holotrichia in which no true buccal ciliature is present but there is a vestibulum.

Trichostrongylidae [INV ZOO] A family of parasitic round-worms belonging to the Strongyloidea; hosts are cattle, sheep, goats, swine, and cats.

trichothallic [BOT] Having a filmentous thallus, as certain algae.

Trichuroidea [INV ZOO] A group of nematodes parasitic in

TRICHOPTERA

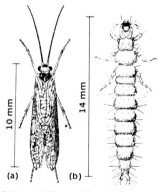

Rhyacophila, a widespread genus of Trichoptera. *(a)* Adult, showing well-veined wings and long antennae. *(b)* Free-living larva. *(Illinois Natural History Survey)*

TRICONODONT

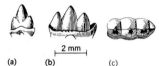

2 mm

(a)   (b)   (c)

Triconodont molars, showing
three main cusps in logitudinal
series. *(a)* External view of a
lower molar of *Amphilestes*,
an amphilestine triconodont.
(b) Internal and *(c)* occlusal
views of a lower molar of
Priacodon, a triconodontine
triconodont. *(After G. G.
Simpson, 1929)*

TRIFOLIATE

Trifoliate leaf, with three leaflets.

various vertebrates and characterized by a slender body
sometimes having a thickened posterior portion.

Tricladida [INV ZOO] The planarians, an order of the Turbellaria distinguished by diverticulated intestines with a single
anterior branch and two posterior branches separated by the
pharynx.

tricoccous [BOT] Pertaining to a three-carpel fruit.

triconodont [VERT ZOO] **1.** A tooth with three main conical
cusps. **2.** Having such teeth.

Triconodonta [PALEON] An extinct mammalian order of
small flesh-eating creatures of the Mesozoic era having no
angle or a pseudoangle on the lower jaw and triconodont
molars.

Trictenotomidae [INV ZOO] A small family of Indian and
Malaysian beetles in the superfamily Tenebrionoidea.

tricuspid [ANAT] Having three cusps or points, as a tooth.

tricuspid valve [ANAT] A valve consisting of three flaps
located between the right atrium and right ventricle of the
heart.

Tridacnidae [INV ZOO] A family of bivalve mollusks in the
subclass Eulamellibranchia which contains the giant clams of
the tropical Pacific.

Tridactylidae [INV ZOO] The pygmy mole crickets, a family
of insects in the order Orthoptera, highly specialized for
fossorial existence.

tridynamous [BOT] Having three long and three short stamens.

trifid [BIOL] Divided into three lobes separated by narrow
sinuses partway to the base.

trifoliate [BOT] Having three leaves or leaflets.

trifoliosis [VET MED] An acute photosensitization characterized by superficial necrosis of white or light-skinned animals
feeding on certain leguminous plants.

trigamous [BOT] Pertaining to a flower head bearing staminate, pistillate, and hermaphrodite flowers.

trigeminal nerve [ANAT] The fifth cranial nerve in vertebrates; either of a pair of composite nerves rising from the side
of the medulla, and with three great branches: the ophthalmic, maxillary, and mandibular nerves.

trigeminal neuralgia [MED] Sudden severe pains of unknown
cause along the path of one or more branches of the
trigeminal nerve. Also known as tic douloureux.

Triglidae [VERT ZOO] The searobins, a family of perciform
fishes in the suborder Cottoidei.

trigon [VERT ZOO] Triangular arrangement of cusps on the
molar teeth of the upper jaw.

Trigonalidae [INV ZOO] A small family of hymenopteran
insects in the superfamily Proctotrupoidea.

trigone [ANAT] A triangular area inside the bladder limited
by the openings of the ureters and urethra.   [BOT] A thickening of plant cell walls formed when three or more cells adjoin.

Trigonostylopoidea [PALEON] A suborder of Paleocene-Eo-
cene ungulate mammals in the order Astrapotheria.

trigonous [BIOL] **1.** Having three corners. **2.** Having a triangular cross section.

trilacunar node [BOT] A node having three leaf gaps.

Trilobita [PALEON] The trilobites, a class of extinct Cam-
brian-Permian arthropods characterized by an exoskeleton
covering the dorsal surface, delicate biramous appendages,
body segments divided by furrows on the dorsal surface, and
a pygidium composed of fused segments.

Trilobitoidea [PALEON] A class of Cambrian arthropods that
are closely related to the Trilobita.

Trilobitomorpha [INV ZOO] A subphylum of the Arthropoda
including Trilobita.

trilocular [BIOL] Having three cavities or cells.

trilophodont [VERT ZOO] Having three-crested teeth.

trilophous [INV ZOO] Pertaining to a spicule having three branched or ridged rays.

Trimenoponidae [INV ZOO] A family of lice in the order Mallophaga occurring as parasites on South American rodents.

Trimerellacea [PALEON] A superfamily of extinct inarticulate brachiopods in the order Lingulida; they have valves, usually consisting of calcium carbonate.

Trimerophytatae *See* Trimerophytopsida.

Trimerophytopsida [PALEOBOT] A group of extinct land vascular plants with leafless, dichotomously branched stems that bear terminal sporangia. Formerly known as Trimerophytatae.

trimerous [BOT] Having parts in sets of three. [INV ZOO] In insects, having the tarsus divided or apparently divided into three segments.

trimonoecious [BOT] Having male, female, and hermaphrodite flowers on a single plant.

trimorphous [BIOL] Characterized by occurring in three distinct forms, as an organ or whole organism.

trinomial [SYST] A nomenclatural designation for an organism composed of three terms: genus, species, and subspecies or variety.

trioecious [BOT] Having the male, female, and hermaphrodite flowers on different individuals.

Trionychidae [VERT ZOO] The soft-shelled turtles, a family of reptiles in the order Chelonia.

triose [BIOCHEM] A group of monosaccharide compounds that have a three-carbon chain length.

tripetalous [BOT] Having three petals.

triphosphopyridine dinucleotide *See* triphosphopyridine nucleotide.

triphosphopyridine nucleotide [BIOCHEM] $C_{12}H_{28}N_7O_{17}P_3$ A grayish-white powder, soluble in methanol and in water; a coenzyme and an important component of enzymatic systems concerned with biological oxidation-reduction systems. Abbreviated TPN. Also known as codehydrogenase II; coenzyme II; triphosphopyridine dinucleotide.

tripinnate [BIOL] Being bipinnate and having each division pinnate.

triple point [PHYS CHEM] A particular temperature and pressure at which three different phases of one substance can coexist in equilibrium.

triple response [PHYSIO] The three stages of vasomotor reaction consisting of reddening, flushing of adjacent skin, and development of wheals, when a pointed instrument is drawn heavily across the skin.

Triploblastica [ZOO] Animals that develop from three germ layers.

Tripylina [INV ZOO] A subdivision of the protozoan order Oculosida in which the major opening (astropyle) usually contains a perforated plate.

triquetrum [ANAT] The third carpal bone from the radial side in the proximal row of carpals.

trisaccate pollen [BOT] A three-pored pollen grain, often having a triangular outline in cross section.

trisaccharide [BIOCHEM] A carbohydrate which, on hydrolysis, yields three molecules of monosaccharides.

trisepalous [BOT] Having three sepals.

trisomic syndrome [MED] Any pathological condition characterized by the presence in triplicate of one of the chromosomes of a complement.

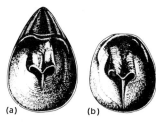

TRIMERELLACEA

Internal views of Silurian trimerellacean *Trimerella:* (a) pedicle valve; (b) brachial valve. (From C. D. Walcott, *Cambrian Brachiopoda, USGS Monogr. no. 51, 1912*)

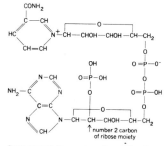

TRIPHOSPHOPYRIDINE NUCLEOTIDE

Structural formula of oxidized form of triphosphopyridine nucleotide (TPN).

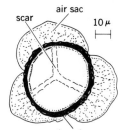

TRISACCATE POLLEN

Trisaccate pollen (young grain of *Podocarpus dacrydioides*) with scar on cap.

trisomy [CYTOL] The presence in triplicate of one of the chromosomes of the complement.

trisomy 13–15 *See* D₁ trisomy.

trisomy 18 syndrome [MED] A congenital disorder due to trisomy of all or a large part of chromosome 18, characterized by severe mental deficiency, hypertonicity with clenched hands, and anomalies of the hands, sternum, pelvis, and facies; most infants so afflicted fail to thrive. Also known as Edwards' syndrome; E trisomy.

trisomy 21 syndrome *See* Down's syndrome.

tritanopia [MED] A defect in a third constituent essential for color vision, as in violet blindness.

tritonymph [INV ZOO] The third stage of development in certain acarids.

tritosternum [INV ZOO] The sternite of the third prosomal segment or the first podosomal segment in Acarina.

Triuridaceae [BOT] A family of monocotyledonous plants in the order Triuridales distinguished by unisexual flowers and several carpels with one seed per carpel.

Triuridales [BOT] A small order of terrestrial, mycotrophic monocots in the subclass Alismatidae without chlorophyll, and with separate carpels, trinucleate pollen, and a well-developed endosperm.

trivalent [CYTOL] An association of three, usually homologous chromosomes at meiosis in a polyploid.

trivial name [ORG CHEM] In systematic nomenclature, being the name of a chemical compound derived from the names of the natural source of the compound at the time of its isolation and before anything is known about its molecular structure.

trivium [INV ZOO] The three rays opposite the madreporite in starfish.

t-RNA *See* transfer ribonucleic acid.

Trochacea [PALEON] A recent subfamily of primitive gastropod mollusks in the order Aspidobranchia.

trochal disk [INV ZOO] A flat or funnel-shaped ciliated disk at the anterior end of a rotifer that functions in locomotion and food ingestion.

trochantellus [INV ZOO] A leg segment between the trochanter and the femur in some insects.

trochanter [ANAT] A process on the proximal end of the femur in many vertebrates, which serves for muscle attachment and, in birds, for articulation with the ilium. [INV ZOO] The second segment of an insect leg, counting from the base.

trochantin [INV ZOO] **1.** A small sclerite at the base of the coxa of the leg of an insect. **2.** A sclerite for articulation of the mandible in Orthoptera.

Trochidae [INV ZOO] A family of gastropod mollusks in the order Aspidobranchia, including many of the top shells.

Trochili [VERT ZOO] A suborder of the avian order Apodiformes.

Trochilidae [VERT ZOO] The hummingbirds, a tropical New World family of the suborder Trochili with tubular tongues modified for nectar feeding; slender bills and the ability to hover are further feeding adaptations.

Trochiliscales [PALEOBOT] A group of extinct plants belonging to the Charophyta in which the gyrogonites are dextrally spiraled.

trochlea [ANAT] A pulleylike anatomical structure.

trochlear nerve [ANAT] The fourth cranial nerve; either of a pair of somatic motor nerves which innervate the superior oblique muscle of the eyeball.

trochoblast [INV ZOO] A cell bearing cilia on a trochophore.

Trochodendraceae [BOT] A family of dicotyledonous trees in the order Trochodendrales distinguished by the absence of a

TROCHILISCALES

1 mm

Dextrally spiraled gyrogonite of Trochiliscales.

perianth and stipules, numerous stamens, and pinnately veined leaves.

Trochodendrales [BOT] An order of dicotyledonous trees in the subclass Hamamelidae characterized by primitively vesselless wood and unique, elongate, often branched idioblasts in the leaves.

trochoid *See* pivot joint.

trochophore [INV ZOO] A generalized but distinct free-swimming larva found in several invertebrate groups, having a pear-shaped form with an external circlet of cilia, apical ciliary tufts, a complete functional digestive tract, and paired nephridia with excretory tubules. Also known as trochosphere.

trochosphere *See* trochophore.

trochus [INV ZOO] The inner band of cilia on a trochal disk.

Troglodytidae [VERT ZOO] The wrens, a family of songbirds in the order Passeriformes.

Trogonidae [VERT ZOO] The trogons, the single, pantropical family of the avian order Trogoniformes.

Trogoniformes [VERT ZOO] An order of brightly colored, slow-moving birds characterized by a unique foot structure with the first and second toes directed backward.

Trombiculidae [INV ZOO] The chiggers, or red bugs, a family of mites in the suborder Trombidiformes whose larvae are parasites of vertebrates.

Trombidiformes [INV ZOO] The trombidiform mites, a suborder of the Acarina distinguished by the presence of a respiratory system opening at or near the base of the chelicerae.

Tropaeolaceae [BOT] A family of dicotyledonous plants in the order Geraniales characterized by strongly irregular flowers, simple peltate leaves, eight stamens, and schizocarpous fruit.

tropeoline 00 [ORG CHEM] $NaSO_3C_6H_4NNC_6H_4NHC_6H_5$ An acid-base indicator with a pH range of 1.4–3.0, color change (from acid to base) red to yellow; used as a biological stain. Also known as sodium *para*-diphenylaminoazobenzene sulfonate.

trophallaxis [ECOL] Exchange of food between organisms, not only of the same species but between different species, especially among social insects.

trophic [BIOL] Pertaining to or functioning in nutrition.

trophic level [ECOL] Any of the feeding levels through which the passage of energy through an ecosystem proceeds; examples are photosynthetic plants, herbivorous animals, and microorganisms of decay.

trophidium [INV ZOO] In some ant species, the first larval stage.

trophifer [INV ZOO] The region of an insect head which articulates with the mouthparts.

trophobiosis [ECOL] A nutritional relationship associated only with certain species of ants in which alien insects supply food to the ants and are milked by the ants for their secretions.

trophoblast [EMBRYO] A layer of ectodermal epithelium covering the outer surface of the chorion and chorionic villi of many mammals.

trophocyte [INV ZOO] A nutritive cell of the ovary or testis of an insect.

trophodisc [INV ZOO] A female gonophore.

trophogenic [ECOL] Originating from nutritional differences rather than resulting from genetic determinants, such as various castes of social insects.

trophogone [MYCOL] A modified antheridium that functions as a nutritive organ in Ascomycetes.

tropholytic [ECOL] Pertaining to the deep zone in a lake where dissimilation of organic matter predominates.

TROCHOPHORE

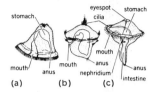

(a) (b) (c)

Some trochophore larvae, showing pear-shaped form, circlet of cilia, digestive tract and nephridium. *(a)* Bryozoan. *(b) Patella*, a mollusk. *(c) Polygardius*, an annelid. *(From T. I. Storer and R. L. Usinger, General Zoology, 3d ed., McGraw-Hill, 1957)*

TROMBIDIFORMES

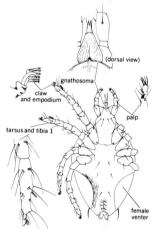

A trombidiform mite, showing characteristic features. *(Acarology Laboratory, Ohio State University)*

trophonemata [VERT ZOO] Villi or hairlike projections of the uterus which provide nourishment to the embryo of elasmobranchs.

trophosome [INV ZOO] The nutritional zooids of a hydroid colony.

trophospongia [VERT ZOO] A spongy, vascular mucous membrane between the uterine wall and the trophoblast in many mammals.

trophotaeniae [VERT ZOO] Vascular rectal processes which establish placental relationships with the ovarian tissue in live-bearing fishes.

trophothylax [INV ZOO] A food pocket on the first abdominal segment of certain ant larvae.

trophozoite [INV ZOO] A vegetative protozoan; used especially of a parasite.

trophozooid [INV ZOO] A nutritive zooid occurring in free-swimming tunicate colonies.

trophus [INV ZOO] Masticatory apparatus in Rotifera.

tropical life zone [ECOL] A subdivision of the eastern division of Merriam's life zones; an example is southern Florida, where the vegetation is the broadleaf evergreen forest; typical and important plants are palms and mangroves; typical and important animals are the armadillo and alligator; typical and important crops are citrus fruits, avocado, and banana.

tropical rainforest [ECOL] A vegetation class consisting of tall, close-growing trees, their columnar trunks more or less unbranched in the lower two-thirds, and forming a spreading and frequently flat crown; occurs in areas of high temperature and high rainfall. Also known as hylaea; selva.

tropical savanna *See* tropical woodland.

tropical scrub [ECOL] A class of vegetation composed of low woody plants (shrubs), sometimes growing quite close together, but more often separated by large patches of bare ground, with clumps of herbs scattered throughout; an example is the Ghanaian evergreen coastal thicket. Also known as bush; brush; fourré; mallee; thicket.

tropical woodland [ECOL] A vegetation class similar to a forest but with wider spacing between trees and sparse lower strata characterized by evergreen shrubs and seasonal graminoids; the climate is warm and moist. Also known as parkland; savanna-woodland; tropical savanna.

Tropiometridae [INV ZOO] A family of feather stars in the class Crinoidea which are bottom crawlers.

tropism [BIOL] Orientation movement of a sessile organism in response to a stimulus. Also known as topotaxis.

tropocollagen [BIOCHEM] The fundamental units of collagen fibrils.

tropomyosin [BIOCHEM] A muscle protein similar to myosin and implicated as being part of the structure of the Z bands separating sarcomeres from each other.

troponin [BIOCHEM] A protein species located at specific stations every 36.5 nanometers on the actin helix in muscle sarcomere.

tropophytia [BOT] Plants that thrive in a climate that undergoes marked periodic changes.

trout [VERT ZOO] Any of various edible fresh-water fishes in the order Salmoniformes that are generally much smaller than the salmon.

Trucherognathidae [PALEON] A family of conodonts in the order Conodontophorida in which the attachment scar permits the conodont to rest on the jaw ramus.

truffle [BOT] The edible underground fruiting body of various European fungi in the family Tuberaceae, especially the genus *Tuber.*

trumpeter [VERT ZOO] A bird belonging to the Psophiidae, a

TROPOMYOSIN

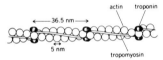

Thin filament of muscle tissue, made up primarily of protein actin, showing position of tropomyosin, and also of troponin. *(From S. Ebashi, M. Endo, and I. Ohtsuki, Control of muscle contraction, Quart. Rev. Biophys., 2:351–384, 1969)*

TRUCHEROGNATHIDAE

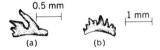

Typical examples of Trucherognathidae; (a) Polycaulodus, and (b) Curtognathus. (After illustration in R. R. Shrock and W. H. Twenhofel, Principles of Invertebrate Paleontology, McGraw-Hill, 2d ed., 1953)

family with three South American species; the common trumpeter (*Psophia crepitans*) is the size of a pheasant and resembles a long-legged guinea fowl.

truncate [BIOL] Abbreviated at an end, as if cut off.

truncus arteriosis [EMBRYO] The embryonic arterial trunk between the bulbous arteriosis and the ventral aorta in anamniotes and early stages of amniotes.

trunk [ANAT] The main mass of the human body, exclusive of the head, neck, and extremities; it is divided into thorax, abdomen, and pelvis. [BOT] The main stem of a tree.

trunk height [ANTHRO] A vertical measurement (taken in front) of the distance from the table top to the upper edge of the sternum, when the individual is seated as in a sitting-height measurement.

trunk legs [INV ZOO] Locomotory legs on the thorax of decapods.

Tryblidiidae [PALEON] An extinct family of Paleozoic mollusks.

tryma [BOT] A nutlike drupe having a rind that is separable from the two-valved endocarp, as walnut fruit.

trypan blue [MATER] An acid disazo dye of the benzo-purpurine series used as a vital stain.

Trypanorhyncha [INV ZOO] An order of tapeworms of the subclass Cestoda; all are parasites in the intestine of elasmobranch fishes.

Trypanosoma [INV ZOO] A genus of slender, polymorphic, elongate protozoans belonging to the Mastigophora, characterized by a central nucleus, posterior blepharoplast, and an undulating membrane; responsible for such infections as trypanosomiasis and Chagas' disease in man.

Trypanosomatidae [INV ZOO] A family of Protozoa, order Kinetoplastida, containing flagellated parasites which exhibit polymorphism during their life cycle.

trypanosome [INV ZOO] A flagellated protozoan of the genus *Trypanosoma.*

trypanosomiasis [MED] Any of many diseases of man and animals caused by infection with species of *Trypanosoma* and transmitted by tsetse flies and other insects.

trypsin [BIOCHEM] A proteolytic enzyme which catalyzes the hydrolysis of peptide linkages in proteins and partially hydrolyzed proteins; derived from trypsinogen by the action of enterokinase in intestinal juice.

trypsinogen [BIOCHEM] The zymogen of trypsin, secreted in the pancreatic juice. Also known as protrypsin.

tryptophan [BIOCHEM] $C_{11}H_{12}O_2N_2$ An amino acid obtained from casein, fibrin, and certain other proteins; it is a precursor of indoleacetic acid, serotonin, and nicotinic acid.

tsetse fly [INV ZOO] Any of various South African muscoid flies of the genus *Glossina;* medically important as vectors of sleeping sickness or trypanosomiasis.

TSH *See* thyrotropic hormone.

tsutsugamushi disease [MED] A rickettsial disease of man caused by *Rickettsia tsutsugamushi,* transmitted by larval mites, and characterized by headache, high fever, and a rash. Also known as scrub typhus.

t-test [STAT] A statistical test involving means of normal populations with unknown standard deviations; small samples are used, based on a variable *t* equal to the difference between the mean of the sample and the mean of the population divided by a result obtained by dividing the standard deviation of the sample by the square root of the number of individuals in the sample.

tubal bladder [VERT ZOO] A urine reservoir organ that is an enlargement of the mesonephric ducts in most fish; there are

TRUNK

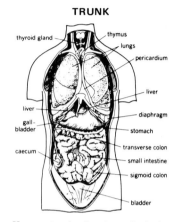

Human trunk, showing principal organs. *(From Franz Frohse et al., Atlas of Human Anatomy, rev. ed., Barnes and Noble, 1957)*

TUBER

Tuber of a potato.

TUBERCULARIACEAE

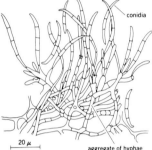

conidia

20 μ

aggregate of hyphae
forming a sporodochium

Sporodochium of *Fusarium lini*,
a representative species of
Tuberculariaceae, showing sickle-
shaped, multicelled conidia.
(After H. L. Bolley, 1901)

TUBULIDENTATA

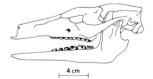

4 cm

Skull and jaw of *Orycteropus
gaudryi*, a Pliocene tubulidentate
from Samos, Greece. *(After E. H.
Colbert, 1941)*

four types: duplex, bilobed, simplex with ureters tied, and simplex with separate ureters.

tubal ligation [MED] Surgical tying of the uterine tubes to prevent conception.

tube foot [INV ZOO] One of the tentaclelike outpushings of the radial vessels of the water-vascular system in echinoderms; may be suctorial, or serve as stiltlike limbs or tentacles.

tuber [BOT] The enlarged end of a rhizome in which food accumulates, as in the potato.

tuber cinereum [ANAT] An area of gray matter extending from the optic chiasma to the mammillary bodies and forming part of the floor of the third ventricle.

tubercle [BIOL] A small knoblike prominence.

tubercle bacillus *See* Mycobacterium tuberculosis.

Tuberculariaceae [MYCOL] A family of fungi of the order Moniliales having short conidia that form cushion-shaped, often waxy or gelatinous aggregates (sporodochia).

tuberculate [BIOL] Having or characterized by knoblike processes.

tuberculin [IMMUNOL] A preparation containing tuberculoproteins derived from *Mycobacterium tuberculosis* used in the tuberculin test to determine sensitization to tubercle bacilli.

tuberculin test [IMMUNOL] A test for past or present infection with tubercle bacilli based on a delayed hypersensitivity reaction at the site where tuberculin or purified protein derivative was introduced.

tuberculosis [MED] A chronic infectious disease of humans and animals primarily involving the lungs caused by the tubercle bacillus, *Mycobacterium tuberculosis*, or by *M. bovis*. Also known as consumption; phthisis.

tuberosity [ANAT] A large or obtuse prominence, especially as on bone for muscle attachment.

tuberous sclerosis [MED] A familial neurocutaneous syndrome characterized in its complete form by epilepsy, adenoma sebaceum, and mental deficiency, and pathologically by nodular sclerosis of the cerebral cortex. Also known as Bourneville's disease.

Tubicola [INV ZOO] An order of sedentary polychaete annelids that surround themselves with a calcareous tube or one which is composed of agglutinated foreign particles.

tubicolous [ECOL] Living in a tube.

Tubulanidae [INV ZOO] A family of the order Palaeonemertini.

tubular gland [ANAT] A secreting structure whose secretory endpieces are tubelike or cylindrical in shape.

tubule [ANAT] A slender, elongated microscopic tube in an anatomical structure.

Tubulidentata [VERT ZOO] An order of mammals which contains a single living genus, the aardvark (*Orycteropus*) of Africa.

tubuloacinous gland *See* tubuloalveolar gland.

tubuloalveolar gland [ANAT] A secreting structure having both tubular and alveolar secretory endpieces. Also known as acinotubular gland; tubuloacinous gland.

tularemia [VET MED] A bacterial infection of wild rodents caused by *Pasteurella tularensis;* it may be generalized, or it may be localized in the eyes, skin, or lymph nodes, or in the respiratory tract or gastrointestinal tract; may be transmitted to humans and to some domesticated animals.

tulip [BOT] Any of various plants with showy flowers constituting the genus *Tulipa* in the family Liliaceae; characterized by coated bulbs, lanceolate leaves, and a single flower with six equal perianth segments and six stamens.

tulip poplar *See* tulip tree.

tulip tree [BOT] *Liriodendron tulipifera.* A tree belonging to the magnolia family (Magnoliaceae) distinguished by leaves which are squarish at the tip, true terminal buds, cone-shaped fruit, and large greenish-yellow and orange-colored flowers. Also known as tulip poplar.

tumbleweed [BOT] Any of various plants that break loose from their roots in autumn and are driven by the wind in rolling masses over the ground.

tumid [BIOL] Marked by swelling or inflation.

tumor [MED] Any abnormal mass of cells resulting from excessive cellular multiplication.

tumorigenic [MED] Tumor-forming.

tuna [VERT ZOO] Any of the large, pelagic, cosmopolitan marine fishes which form the family Thunnidae including species that rank among the most valuable of food and game fish.

tundra [ECOL] An area supporting some vegetation between the northern upper limit of trees and the lower limit of perennial snow on mountains, and on the fringes of the Antarctic continent and its neighboring islands.

tung nut [BOT] The seed of the tung tree (*Aleurites fordii*), which is the source of tung oil.

tungsten [CHEM] Also known as wolfram. A metallic element in group VI, symbol W, atomic number 74, atomic weight 183.85; soluble in mixed nitric and hydrofluoric acids; melts at 3400°C.

tung tree [BOT] *Aleurites fordii.* A plant of the spurge family in the order Euphorbiales, native to central and western China and grown in the southern United States.

tunica [BIOL] A membrane or layer of tissue that covers or envelops an organ or other anatomical structure.

tunica adventitia *See* adventitia.

tunica intima *See* intima.

tunica mucosa *See* mucous membrane.

Tunicata [INV ZOO] A subphylum of the Chordata characterized by restriction of the notochord to the tail and posterior body of the larva, absence of mesodermal segmentation, and secretion of an outer covering or tunic about the body.

tunicate [BIOL] Having a tunic or test.

Tupaiidae [VERT ZOO] The tree shrews, a family of mammals in the order Insectivora.

tupelo [BOT] Any of various trees belonging to the genus *Nyssa* of the sour gum family, Nyssaceae, distinguished by small, obovate, shiny leaves, a small blue-black drupaceous fruit, and branches growing at a wide angle from the axis.

Turbellaria [INV ZOO] A class of the phylum Platyhelminthes having bodies that are elongate and flat to oval or circular in cross section.

turbidostat [MICROBIO] A device in which a bacterial culture is maintained at a constant volume and cell density (turbidity) by adjusting the flow rate of fresh medium into the growth tube by means of a photocell and appropriate electrical connections.

turbinate [BOT] Shaped like an inverted cone. [INV ZOO] Spiral with rapidly decreasing whorls from base to apex.

turbinate bone *See* concha.

Turbinidae [INV ZOO] A family of gastropod mollusks including species of top shells.

Turdidae [VERT ZOO] The thrushes, a family of passeriform birds in the suborder Oscines.

turgor [BOT] Distension of a plant cell wall and membrane by the fluid contents.

turgor movement [BOT] A reversible change in the position of plant parts due to a change in turgor pressure in certain

TUNA

Bluefin tuna *(Thunnus thynnus),* a species of tuna which occurs in all warm and temperate seas and serves as a food fish.

TUNGSTEN

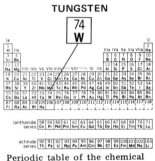

Periodic table of the chemical elements showing the position of tungsten.

TURBELLARIA

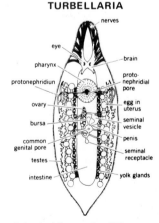

A typical hermaphroditic turbellarian, *Mesostoma ehrenbergii wardii,* showing principal organs. *(Modified from Ruebush, 1940)*

TURNIP

Turnip, showing enlarged root and foliage.

specialized cells; movement of *Mimosa* leaves when touched is an example.

turgor pressure [BOT] The actual pressure developed by the fluid content of a turgid plant cell.

turion [BOT] A scaly shoot, such as asparagus, developed from an underground bud.

turkey [VERT ZOO] Either of two species of wild birds, and any of various derived domestic breeds, in the family Meleagrididae characterized by a bare head and neck, and in the male a large pendant wattle which hangs on one side from the base of the bill.

turmeric [BOT] *Curcuma longa.* An East Indian perennial of the ginger family (Zingiberaceae) with a short stem, tufted leaves, and short thick rhizomes; a spice with a pungent, bitter taste and a musky odor is derived from the rhizome.

Turner's syndrome [MED] A sex aberration in humans in which the chromosome complement includes only one sex chromosome, an X.

Turnicidae [VERT ZOO] The button quails, a family of Old World birds in the order Gruiformes.

turnip [BOT] *Brassica rapa* or *B. campestris* var. *rapa.* An annual crucifer of Asiatic origin belonging to the family Brassiaceae in the order Capparales and grown for its foliage and edible root.

turnover [MOL BIO] The number of substrate molecules transformed by a single molecule of enzyme per minute, when the enzyme is operating at maximum rate.

turnover number [BIOCHEM] The number of molecules of a substrate acted upon in a period of 1 minute by a single enzyme molecule, with the enzyme working at a maximum rate.

Turonian [GEOL] A European stage of geologic time: Upper or Middle Cretaceous (above Cenomanian, below Coniacian).

turtle [VERT ZOO] Any of about 240 species of reptiles which constitute the order Chelonia distinguished by the two bony shells enclosing the body.

twin [BIOL] One of two individuals born at the same time.

twiner [BOT] A climbing stem that winds about its support, as pole beans or many tropical lianas.

twin spots [GEN] Two areas of somatic or colonial growth in which, as a result of somatic crossing-over, the cells carry the complementary products of crossing-over.

two-sided test [STAT] A test which rejects the null hypothesis when the test statistic T is either less than or equal to c or greater than or equal to d, where c and d are critical values.

two-stage design [STAT] The design of an experiment which employs a pilot study in order to decide how to design the main experiment.

two-stage experiment [STAT] An experiment in two parts, the outcome of the first part deciding the procedure for the second.

two-stage sampling [STAT] Sampling from a population whose members are themselves sets of objects and then sampling from the sets selected in the first sampling; for example, to first draw a sample of states and then to draw a sample of representatives to Congress from each state selected.

Tylenchida [INV ZOO] An order of soil-dwelling or phytoparasitic nematodes in the subclass Rhabdita.

Tylenchoidea [INV ZOO] A superfamily of mainly soil and insect-associated nematodes in the order Tylenchida with a stylet for piercing live cells and sucking the juices.

Tylopoda [VERT ZOO] An infraorder of artiodactyls in the

TURTLE

Box turtle, *Terrapene carolina*, a common United States turtle. (*From J. J. Shomon, ed., Virginia Wildlife, 15(6):27, 1954*)

suborder Ruminantia that contains the camels and extinct related forms.

tylostyle [INV ZOO] A uniradiate spicule in Porifera with a point at one end and a knob at the other end.

tylote [INV ZOO] A slender sponge spicule with a knob at each end.

tylus [INV ZOO] A medial projection on the head of certain Hemiptera.

tympanic cavity [ANAT] The irregular, air-containing, mucous-membrane-lined space of the middle ear; contains the three auditory ossicles and communicates with the nasopharynx through the auditory tube.

tympanic membrane [ANAT] The membrane separating the external from the middle ear. Also known as eardrum; tympanum.

tympanum [ANAT] See tympanic membrane. [INV ZOO] A thin membrane covering an organ of hearing in insects.

type [SYST] A specimen on which a species or subspecies is based.

type A encephalitis See lethargic encephalitis.

type I of Cori See von Gierke's disease.

Typhaceae [BOT] A family of monocotyledonous plants in the order Typhales characterized by an inflorescence of dense, cylindrical spikes and absence of a perianth.

Typhales [BOT] An order of marsh or aquatic monocotyledons in the subclass Commelinidae with emergent or floating stems and leaves and reduced, unisexual flowers having a single ovule in an ovary composed of a single carpel.

Typhlopidae [VERT ZOO] A family of small, burrowing circumtropical snakes, suborder Serpentes, with vestigial eyes and toothless jaws.

Typhloscolecidae [INV ZOO] A family of pelagic polychaete annelids belonging to the Errantia.

typhlosole [INV ZOO] A dorsal longitudinal invagination of the intestinal wall in certain invertebrates serving to increase the absorptive surface.

typhoid fever [MED] A highly infectious, septicemic disease of humans caused by *Salmonella typhi* which enters the body by the oral route through ingestion of food or water contaminated by contact with fecal matter.

typhoid vaccine [IMMUNOL] A type of killed vaccine used for active immunity production; made from killed typhoid bacillus (*Salmonella typhi*).

typhus fever [MED] Any of three louse-borne human diseases caused by *Rickettsia prowazakii* characterized by fever, stupor, headaches, and a dark-red rash.

Typotheria [PALEON] A suborder of extinct rodentlike herbivores in the order Notoungulata.

Tyranni [VERT ZOO] A suborder of suboscine Passeriformes containing birds with limited song power and having the tendon of the hind toe separate and the intrinsic muscles of the syrinx reduced to one pair.

Tyrannidae [VERT ZOO] The tyrant flycatchers, a family of passeriform birds in the suborder Tyranni confined to the Americas.

Tyrannoidea [VERT ZOO] The flycatchers, a superfamily of suboscine birds in the suborder Tyranni.

tyrocidine [MICROBIO] A peptide antibiotic produced by *Bacillus brevis*; used to control fungi, bacteria, and protozoa.

tyrosinase [BIOCHEM] An enzyme found in plants, molds, crustaceans, mollusks, and some bacteria which, in the presence of oxygen, catalyzes the oxidation of monophenols and polyphenols with the introduction of −OH groups and the formation of quinones.

tyrosine [BIOCHEM] $C_9H_{11}NO_3$ A phenolic alpha amino

TYROSINE

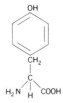

Structural formula of tyrosine.

acid found in many proteins; a precursor of the hormones epinephrine, norepinephrine, thyroxine, and triiodothyronine, and of the black pigment melanin.

tyrosinemia [MED] An inborn metabolic disorder in which there is a deficiency of the enzyme p-hydroxyphenylpyruvic acid oxidase with abnormally high blood levels of tyrosine and sometimes methionine.

tyrosinosis [MED] Excretion of excessive amounts of tyrosine and its first oxidation products in the urine.

tyrothricin [MICROBIO] A polypeptide mixture produced by *Bacillus brevis* and consisting of the antibiotic substances gramicidin and tyrocidine; effective as an antibacterial applied locally in infections due to germ-positive organisms.

Tyson's gland [ANAT] A small scent gland in the human male which secretes the smegma. Also known as preputial gland.

Tytonidae [VERT ZOO] The barn owls, a family of birds in the order Strigiformes distinguished by an unnotched sternum which is fused to large clavicles.

tyvelose [BIOCHEM] A dideoxy sugar found in bacterial lipopolysaccharides.

U

U   *See* uranium.

udder   [VERT ZOO] A pendulous organ consisting of several mammary glands enclosed in a single envelope; each gland has its own nipple; found in some mammals, such as the cow and goat.

UDP   *See* uridine diphosphate.

UDPG   *See* uridine diphosphoglucose.

Uintatheriidae   [PALEON] The single family of the extinct mammalian order Dinocerata.

Uintatheriinae   [PALEON] A subfamily of extinct herbivores in the family Uintatheriidae including all horned forms.

ulcer   [MED] Localized interruption of the continuity of an epithelial surface, with an inflamed base.

ulcerative colitis   [MED] An idiopathic inflammatory disease of the mucosa and submucosa of the colon manifested clinically by pain, diarrhea, and rectal bleeding.

Ullrich-Turner syndrome   [MED] A complex of symptoms including webbing of the neck, short stature, cubitus valgus, and hypogonadism in the male. Also known as male Turner's syndrome.

Ulmaceae   [BOT] A family of dicotyledonous trees in the order Urticales distinguished by alternate stipulate leaves, two styles, a pendulous ovule, and lack of a latex system.

ulna   [ANAT] The larger of the two bones of the forearm or forelimb in vertebrates; articulates proximally with the humerus and distally with the radius.

Ulotrichaceae   [BOT] A family of green algae in the suborder Ulotrichineae; contains both attached and floating filamentous species with cells having parietal, platelike or bandlike chloroplasts.

Ulotrichales   [BOT] A large, artificial order of the Chlorophyta composed mostly of fresh-water, branched or unbranched filamentous species with mostly cylindrical, uninucleate cells having cellulose, but often mucilaginous walls.

Ulotrichineae   [BOT] A suborder of the Ulotrichales characterized by short cylindrical cells.

ultimobranchial body   [EMBRYO] One of the small, endocrine structures which originate as terminal outpocketings from each side of the embryonic vertebrate pharynx; can produce the hormone calcitonin.

ultracentrifuge   [ENG] A laboratory centrifuge that develops centrifugal fields of more than 100,000 times gravity.

ultramicrotome   [ENG] A microtome which uses a glass or diamond knife, allowing sections of cells to be cut 300 nanometers in thickness.

ultrasonic therapy   *See* ultrasound diathermy.

ultrasound diathermy   [MED] The application of high-frequency sound waves (0.7 to 1.0 megahertz) and the conversion of this mechanical energy into heat for local thermotherapy. Also known as ultrasonic therapy.

UINTATHERIINAE

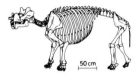

Skeleton of *Uintatherium*, a middle Eocene member of the subfamily Uintatheriinae in the order Dinocerata, showing the five-toed feet, a characteristic of the order. *(After Flerov)*

ULOTRICHACEAE

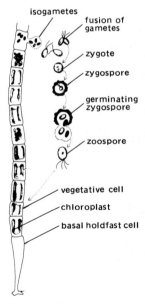

Ulothrix, an attached form of Ulotrichaceae, showing cylindrical cells and other features.

ULVALES

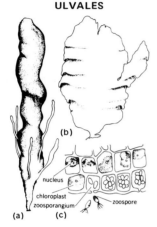

Ulvales. *(a) Enteromorpha,* a tubular form. *(b) Ulva,* the sea lettuce, an expanded thallus. *(c) Ulva,* a portion of the marginal cells of the thallus showing chloroplasts and zoospore formation.

UMBEL

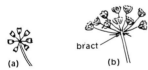

Two types of umbel inflorescence: *(a)* simple; *(b)* compound.

ultrastructure [MOL BIO] The ultimate physiochemical organization of protoplasm.

ultraviolet absorption spectrophotometry [SPECT] The study of the spectra produced by the absorption of ultraviolet radiant energy during the transformation of an electron from the ground state to an excited state as a function of the wavelength causing the transformation.

ultraviolet microscope [OPTICS] A special type of microscope which uses electromagnetic radiation in the range 180–400 nanometers; it requires reflecting optics or special quartz and crystal objectives.

ultraviolet radiation [ELECTROMAG] Electromagnetic radiation in the wavelength range 4–400 nanometers; this range begins at the short-wavelength limit of visible light and overlaps the wavelengths of long x-rays (some scientists place the lower limit at higher values, up to 40 nanometers). Also known as ultraviolet light.

ultraviolet spectrometer [SPECT] A device which produces a spectrum of ultraviolet light and is provided with a calibrated scale for measurement of wavelength.

ultraviolet spectrophotometry [SPECT] Determination of the spectra of ultraviolet absorption by specific molecules in gases or liquids (for example, Cl_2, SO_2, NO_2, CS_2, ozone, mercury vapor, and various unsaturated compounds).

ultraviolet spectroscopy [SPECT] Absorption spectroscopy involving electromagnetic wavelengths in the range 4–400 nanometers.

ultraviolet spectrum [ELECTROMAG] **1.** The range of wavelengths of ultraviolet radiation, covering 4–400 nanometers. **2.** A display or graph of the intensity of ultraviolet radiation emitted or absorbed by a material as a function of wavelength or some related parameter.

Ulvaceae [BOT] A large family of green algae in the order Ulvales.

Ulvales [BOT] An order of algae in the division Chlorophyta in which the thalli are macroscopic, attached tubes or sheets.

umbel [BOT] An indeterminate inflorescence with the pedicels all arising at the top of the peduncle and radiating like umbrella ribs; there are two types, simple and compound.

Umbellales [BOT] An order of dicotyledonous herbs or woody plants in the subclass Rosidae with mostly compound or conspicuously lobed or dissected leaves, well-developed schizogenous secretory canals, separate petals, and an inferior ovary.

Umbelliferae [BOT] A large family of aromatic dicotyledonous herbs in the order Umbellales; flowers have an ovary of two carpels, ripening to form a dry fruit that splits into two halves, each containing a single seed.

umbilical artery [EMBRYO] Either of a pair of arteries passing through the umbilical cord to carry impure blood from the mammalian fetus to the placenta.

umbilical cord [EMBRYO] The long, cylindrical structure containing the umbilical arteries and vein, and connecting the fetus with the placenta.

umbilical duct *See* vitelline duct.

umbilical hernia [MED] Herniation through the umbilical ring. Also known as annular hernia.

umbilical vein [EMBRYO] A vein passing through the umbilical cord and conveying purified, nutrient-rich blood from placenta to fetus.

Umbilicariaceae [BOT] The rock tripes, a family of Ascolichenes in the order Lecanorales having a large, circular, umbilicate thallus.

umbilicus [ANAT] The navel; the round, depressed cicatrix in the median line of the abdomen, marking the site of the aperture through which passed the fetal umbilical vessels.

umbo [ANAT] A rounded elevation of the surface of the tympanic membrane. [INV ZOO] A prominence above the hinge of a bivalve mollusk shell.

umbonate [BIOL] Having or forming an umbo.

umbrella [INV ZOO] **1.** The contractile discoid body of a jellyfish. **2.** The web between the arms of certain Octopoda.

UMP *See* uridylic acid.

uncinate trophus [INV ZOO] A trophus in rotifers characterized by a hooked or curved uncus.

uncompetitive enzyme inhibition [BIOCHEM] The prevention of an enzymic process as a result of the interaction of an inhibitor with the enzyme-substrate complex or a subsequent intermediate form of the enzyme, but not with the free enzyme.

unconditioned reflex [PSYCH] A response evoked by an unconditioned stimulus.

unconditioned stimulus [PSYCH] A stimulus which always elicits a response.

unconscious [MED] Insensible; in a state lacking conscious awareness, with reflexes abolished. [PSYCH] **1.** Pertaining to behavior or experience not controlled by the ego. **2.** The part of the mind, mental functioning, or personality not in the immediate field of awareness.

underground stem [BOT] Any of the stems that grow underground and are often mistaken for roots; principal kinds are rhizomes, tubers, corms, bulbs, and rhizomorphic droppers.

underwing [INV ZOO] Either of a pair of posterior wings on certain insects, as the moth.

undulant fever *See* brucellosis.

undulating membrane [INV ZOO] **1.** A membrane composed of fused cilia, for moving food to the mouth of ciliates. **2.** A protoplasmic membrane extending between the body and the basal part of the flagellum in flagellates.

ungula [VERT ZOO] A nail, hoof, or claw.

ungulate [VERT ZOO] Referring to an animal that has hoofs.

ungulicutate [VERT ZOO] Having claws or nails.

unguligrade [VERT ZOO] Walking on hoofs.

uniaperturate [BIOL] Having a single aperture.

unicellular [BIOL] Composed of a single cell.

unicuspid [ANAT] Having one cusp, as certain teeth.

uniform distribution [STAT] The distribution of a random variable in which each value has the same probability of occurrence. Also known as rectangular distribution.

unilacunar node [BOT] A node with a single leaf gap.

unilateral hermaphroditism [ZOO] A type of hermaphroditism in which there is a combination of ovatestis on one side of the body with an ovary or testis on the other side.

unilocular [BIOL] Having a single cavity.

uninhibited bladder [MED] An abnormal urinary bladder that shows only a variable loss of cerebral inhibition over reflex bladder contractions, representing, of all neurogenic bladders, the least variance from normal.

Unionidae [INV ZOO] The fresh-water mussels, a family of bivalve mollusks in the subclass Eulamellibranchia; the larvae, known as glochidia, are parasitic on fish.

Unipolarina [INV ZOO] A suborder of the protozoan order Myxosporida characterized by spores with one to six (never five) polar capsules located at the anterior end.

unisexual [BOT] Having either an androecium or a gynoecium, but not both.

unitegmic [BOT] Referring to an ovule having a single integument.

universal donor [IMMUNOL] An individual of O blood group; can give blood to persons of all blood types.

UNCINATE TROPHUS

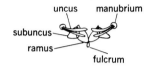

Uncinate trophus of a predacious rotifer, *Stephanoceros.*

universal recipient [IMMUNOL] An individual of AB blood group; can receive a blood transfusion of all blood types, A, AB, B, or O.

unmyelinated [HISTOL] Lacking myelin, either as a normal condition or as the result of a disease.

unrelated frequencies [STAT] The long run frequency of any result in one part of an experiment is approximately equal to the long run conditional frequency of that result, given that any specified result has occurred in the other part of the experiment.

unresolved pneumonia *See* organizing pneumonia.

uns-, unsym- [ORG CHEM] A chemical prefix denoting that the substituents of an organic compound are structurally unsymmetrical with respect to the carbon skeleton, or with respect to a function group (for example, double or triple bond).

unsaturated hydrocarbon [ORG CHEM] One of a class of hydrocarbons that have at least one double or triple carbon-to-carbon bond that is not in an aromatic ring; examples are ethylene, propadiene, and acetylene.

unstable colon *See* irritable colon.

unstriated fibers *See* smooth muscle fibers.

upper-arm circumference [ANTHRO] A measure of the horizontal circumference at the largest part of the biceps muscle.

Upper Cambrian [GEOL] The latest epoch of the Cambrian period of geologic time, beginning approximately 510 million years ago.

Upper Carboniferous [GEOL] The European epoch of geologic time equivalent to the Pennsylvanian of North America.

Upper Cretaceous [GEOL] The late epoch of the Cretaceous period of geologic time, beginning about 90 million years ago.

Upper Devonian [GEOL] The latest epoch of the Devonian period of geologic time, beginning about 365 million years ago.

upper face height [ANTHRO] A measure of the distance from the nasion to the lower gum edge between the two central upper teeth.

Upper Huronian *See* Animikean.

Upper Jurassic [GEOL] The latest epoch of the Jurassic period of geologic time, beginning approximately 155 million years ago.

Upper Mississippian [GEOL] The latest epoch of the Mississippian period of geologic time.

Upper Ordovician [GEOL] The latest epoch of the Ordovician period of geologic time, beginning approximately 440 million years ago.

Upper Pennsylvanian [GEOL] The latest epoch of the Pennsylvanian period of geologic time.

Upper Permian [GEOL] The latest epoch of the Permian period of geologic time, beginning about 245 million years ago.

Upper Silurian [GEOL] The latest epoch of the Silurian period of geologic time.

Upper Triassic [GEOL] The latest epoch of the Triassic period of geologic time, beginning about 200 million years ago.

Upupidae [VERT ZOO] The Old World hoopoes, a family of birds in the order Coraciiformes whose young are hatched with sparse down.

urachus [EMBRYO] A cord or tube of epithelium connecting the apex of the urinary bladder with the allantois; its connective tissue forms the median umbilical ligament.

uracil [BIOCHEM] $C_4H_4N_2O_2$ A pyrimidine base important as a component of ribonucleic acid.

Uralean [GEOL] A stage of geologic time in Russia: upper-

URACIL

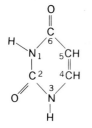

Structural formula of uracil.

most Carboniferous (above Gzhelian, below Sakmarian of Permian).

Uraniidae [INV ZOO] A tropical family of moths in the superfamily Geometroidea including some slender-bodied, brilliantly colored diurnal insects which lack a frenulum and are often mistaken for butterflies.

uranium [CHEM] A metallic element in the actinide series, symbol U, atomic number 92, atomic weight 238.03; highly toxic and radioactive; ignites spontaneously in air and reacts with nearly all nonmetals; melts at 1132°C, boils at 3818°C; used in nuclear fuel and as the source of U^{235} and plutonium.

uranium-lead dating [GEOL] A method for calculating the geologic age of a material in years based on the radioactive decay rate of uranium-238 to lead-206 and of uranium-235 to lead-207.

urate [BIOCHEM] A salt of uric acid.

urate calculi [PATH] Kidney stones composed of uric acid salts and found particularly in people suffering from gout.

urban typhus See murine typhus.

urceolate [BIOL] Shaped like an urn.

urea [BIOCHEM] $CO(NH_2)_2$ Carbamide, a product of protein metabolism; used therapeutically as a diuretic.

urease [BIOCHEM] An enzyme that catalyzes the degradation of urea to ammonia and carbon dioxide; obtained from the seed of jack bean.

Urechinidae [INV ZOO] A family of echinoderms in the order Holasteroida which have an ovoid test lacking a marginal fasciole.

Uredinales [MYCOL] An order of parasitic fungi of the subclass Heterobasidiomycetidae characterized by the teleutospore, a spore with one or more cells, each of which is a modified hypobasidium; members cause plant diseases known as rusts.

uredinium [MYCOL] The aggregation of sporebearing hyphae and urediospores of a rust fungus that forms beneath the cuticle or epidermis of a host plant.

urediospore [MYCOL] A thin-walled spore produced by rust fungi; gives rise to a vegetative mycelium which may produce more urediospores.

(*para*-ureidophenylarsylenedithio)diacetic acid See thiocarbarsone.

(*para*-ureidophenylarsylenedithio)di-o-benzoic acid See thiocarbamazine.

uremia [MED] A condition resulting from kidney failure and characterized by azotemia, chronic acidosis, anemia, and a variety of systemic signs and symptoms.

ureotelic [PHYSIO] Excreting nitrogen in the form of urea, referring specifically to mammals.

ureter [ANAT] A long tube conveying urine from the renal pelvis to the urinary bladder or cloaca in vertebrates.

urethra [ANAT] The canal in most mammals through which urine is discharged from the urinary bladder to the outside.

urethral gland [ANAT] One of the small, branched tubular mucous glands in the mucosa lining the urethra.

urethritis [MED] Inflammation of the urethra.

uric acid [BIOCHEM] $C_5H_4N_4O_3$ A white, crystalline compound, the excretory end product in amino acid metabolism by uricotelic species. Also known as 8-hydroxyxanthine.

uricase [BIOCHEM] An enzyme present in the liver, spleen, and kidney of most mammals except man; converts uric acid to allantoin in the presence of gaseous oxygen.

uricotelism [PHYSIO] An adaptation of terrestrial reptiles and birds which effectively provides for detoxification of ammonia and also for efficient conservation of water due to a relatively low rate of glomerular filtration and active secre-

URANIUM

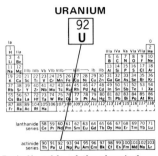

Periodic table of the chemical elements showing the position of uranium.

URIC ACID

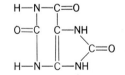

Structural formula of uric acid.

tion of uric acid by the tubules to form a urine practically saturated with urate.

uridine [BIOCHEM] $C_9H_{12}N_2O_6$ A crystalline nucleoside composed of one molecule of uracil and one molecule of D-ribose; a component of ribonucleic acid.

uridine diphosphate [BIOCHEM] The chief transferring coenzyme in carbohydrate metabolism. Abbreviated UDP.

uridine diphosphoglucose [BIOCHEM] A compound in which α-glucopyranose is esterified, at carbon atom 1, with the terminal phosphate group of uridine-5'-pyrophosphate (that is, uridine diphosphate); occurs in animal, plant, and microbial cells; functions as a key in the transformation of glucose to other sugars. Abbreviated UDPG.

uridine monophosphate See uridylic acid.

uridine phosphoric acid See uridylic acid.

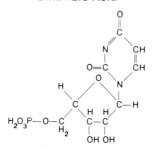
uridylic acid [BIOCHEM] $C_9H_{13}N_2O_9P$ Water- and alcohol-soluble crystals, melting at 202°C; used in biochemical research. Also known as uridine monophosphate (UMP); uridine phosphoric acid.

urinalysis [PATH] Analysis of the urine, involving chemical, physical, and microscopic tests.

urinary bladder [ANAT] A hollow organ which serves as a reservoir for urine.

urinary system [ANAT] The system which functions in the elaboration and excretion of urine in vertebrates; in man and most mammals, consists of the kidneys, ureters, urinary bladder, and urethra.

urination [PHYSIO] The discharge of urine from the bladder. Also known as micturition.

urine [PHYSIO] The fluid excreted by the kidneys.

uriniferous tubule [ANAT] One of the numerous winding tubules of the kidney. Also known as nephric tubule.

urite [INV ZOO] **1.** A segment of the abdomen in arthropods. **2.** The anal cirrus in polychaetes.

urn [BOT] The theca of a moss.

urobilin [BIOCHEM] A bile pigment produced by reduction of bilirubin by intestinal bacteria and excreted by the kidneys or removed by the liver.

urobilinogen [BIOCHEM] A chromogen, formed in feces and present in urine, from which urobilin is formed by oxidation.

urocanic acid [BIOCHEM] $C_6H_6N_2O_2$ A crystalline compound formed as an intermediate in the degradative pathway of histidine.

urochord [VERT ZOO] The notochord when limited to the caudal region, as in tunicates.

Urochordata [INV ZOO] The equivalent name for Tunicata.

Urodela [VERT ZOO] The tailed amphibians or salamanders, an order of the class Amphibia distinguished superficially from frogs and toads by the possession of a tail, and from caecilians by the possession of limbs.

urodeum [VERT ZOO] The cloacal chamber into which the ureters and the genital ducts open.

urogenital diaphragm [ANAT] The sheet of tissue stretching across the pubic arch, formed by the deep transverse perineal and the sphincter urethrae muscles. Also known as triangular ligament.

urogenital ridge [EMBRYO] Either of the paired ridges from which the urinary and genital systems develop.

urogenital system [ANAT] The combined urinary and genital system in vertebrates, which are intimately related embryologically and anatomically. Also known as genitourinary system.

urokinase [BIOCHEM] An enzyme, present in human urine, that catalyzes the conversion of plasminogen to plasmin.

urolithiasis [MED] **1.** Condition associated with the presence of urinary calculi. **2.** Formation or presence of urinary calculi.

urology [MED] The scientific study of urine and the diseases and abnormalities of the urinary and urogenital tracts.

uropatagium [VERT ZOO] In bats the membrane which stretches from one femur to the other and usually includes the tail. [INV ZOO] One of the plates on either side of the anus in insects.

uropepsin [BIOCHEM] The end product of the secretion of pepsinogen into the blood by gastric cells; occurs in urine.

uropod [INV ZOO] One of the flattened abdominal appendages of various crustaceans that with the telson forms the tail fan.

uropore [INV ZOO] The opening of the excretory duct in Acarina.

uroporphyrin [BIOCHEM] Any of several isomeric, metal-free porphyrins, occurring in small quantities in normal urine and feces; molecule has four acetic acid ($-CH_2COOH$) and four propionic acid ($-CH_2CH_2COOH$) groups.

Uropygi [INV ZOO] The tailed whip scorpions, an order of arachnids characterized by an elongate, flattened body which bears in front a pair of thickened, raptorial pedipalps set with sharp spines and used to hold and crush insect prey.

uropygial gland [VERT ZOO] A relatively large, compact, bilobed, secretory organ located at the base of the tail (uropygium) of most birds having a keeled sternum. Also known as oil gland; preen gland.

uropygium [VERT ZOO] The hump containing the caudal vertebrae at the end of a bird's trunk; it supports the tail feathers.

urosternite [INV ZOO] The ventral plate of an abdominal segment in arthropods.

urostyle [VERT ZOO] An unsegmented bone representing several fused vertebrae and forming the posterior part of the vertebral column in Anura.

Urostylidae [INV ZOO] A family of hemipteran insects in the superfamily Pentatomoidea.

Ursidae [VERT ZOO] A family of mammals in the order Carnivora including the bears and their allies.

ursolic acid [BIOCHEM] $C_{30}H_{48}O_3$ A pentacyclic terpene that crystallizes from absolute alcohol solution, found in leaves and berries of plants; used in pharmaceutical and food industries as an emulsifying agent.

Urticaceae [BOT] A family of dicotyledonous herbs in the order Urticales characterized by a single unbranched style, a straight embryo, and the lack of milky juice (latex).

Urticales [BOT] An order of dicotyledons in the subclass Hamamelidae; woody plants or herbs with simple, usually stipulate leaves, and reduced clustered flowers that usually have a vestigial perianth.

urticaria [MED] Hives or nettle rash; a skin condition characterized by the appearance of intensely itching wheals or welts with elevated, usually white centers and a surrounding area of erythema. Also known as hives.

use-dilution test [MICROBIO] A bioassay method for testing disinfectants for use on surfaces where a substantial reduction of bacterial contamination is not achieved by prior cleaning; the test organisms *Salmonella choleraesuis* and *Staphylococcus aureus* are deposited in stainless steel cylinders which are then exposed to the action of the test disinfectant.

Usneaceae [BOT] The beard lichens, a family of Ascolichenes in the order Lecanorales distinguished by their conspicuous fruticose growth form.

usnic acid [BIOCHEM] $C_{18}H_{16}O_7$ Yellow crystals, insoluble in water, slightly soluble in alcohol and ether, melts about

198°C; found in lichens; used as an antibiotic. Also known as usninic acid.

usninic acid   *See* usnic acid.

Ustilaginaceae   [MYCOL] A family of fungi in the order Ustilaginales in which basidiospores bud from the sides of the septate epibasidium.

Ustilaginales   [MYCOL] An order of the subclass Heterobasidiomycetidae comprising the smut fungi which parasitize plants and cause diseases known as smut or bunt.

uterus   [ANAT] The organ of gestation in mammals which receives and retains the fertilized ovum, holds the fetus during development, and becomes the principal agent of its expulsion at term.

uterus bicornis   [ANAT] A uterus divided into two horns or compartments; an abnormal condition in humans but normal in many mammals, such as carnivores.

utricle   *See* utriculus.

utriculus   [ANAT] **1.** That part of the membranous labyrinth of the ear into which the semicircular canals open. **2.** A small, blind pouch extending from the urethra into the prostate. Also known as utricle.

U-tube manometer   [ENG] A manometer consisting of a U-shaped glass tube partly filled with a liquid of known specific gravity; when the legs of the manometer are connected to separate sources of pressure, the liquid rises in one leg and drops in the other; the difference between the levels is proportional to the difference in pressures and inversely proportional to the liquid's specific gravity. Also known as liquid-column gage.

uva   [BOT] A pulpy indehiscent fruit having a central placenta, such as the grape.

uvea   [ANAT] The pigmented, vascular layer of the eye: the iris, ciliary body, and choroid.

uveitis   *See* iridocyclochoroiditis.

uvula   [ANAT] **1.** A fingerlike projection in the midline of the posterior border of the soft palate. **2.** A lobe of the vermiform process of the lower surface of the cerebellum.

UTERUS

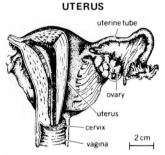

Human uterus and associated structures. *(From L. B. Arey, Developmental Anatomy, 7th ed., Saunders, 1965)*

V

V *See* vanadium.

vaccination [IMMUNOL] Inoculation of viral or bacterial organisms or antigens to produce immunity in the recipient.

vaccination encephalitis [MED] Encephalitis caused by vaccination with rabies vaccine.

vaccine [IMMUNOL] A suspension of killed or attenuated bacteria or viruses or fractions thereof, injected to produce active immunity.

vaccinia [VET MED] A contagious disease of cows which is characterized by vesicopustular lesions of the skin that are prone to appear on the teats and udder, and which is transmissible to humans by handling infected cows and by vaccination; confers immunity against smallpox. Also known as cowpox.

vacuole [CYTOL] A membrane-bound cavity within a cell; may function in digestion, storage, secretion, or excretion.

vacuum [PHYS] **1.** Theoretically, a space in which there is no matter. **2.** Practically, a space in which the pressure is far below normal atmospheric pressure so that the remaining gases do not affect processes being carried on in the space.

vacuum behavior [PSYCH] The carrying out of a series of model action patterns in apparent absence of any obviously appropriate behavior.

vagility [ECOL] The ability of organisms to disseminate.

vagina [ANAT] The canal from the vulvar opening to the cervix uteri.

vaginal protocele *See* rectocele.

vaginate [BIOL] Invested in a sheath.

vaginismus [MED] Painful vaginal spasm.

vaginitis [MED] **1.** Inflammation of the vagina. **2.** Inflammation of a tendon sheath.

vagus [ANAT] The tenth cranial nerve; either of a pair of sensory and motor nerves forming an important part of the parasympathetic system in vertebrates.

valence [BIOCHEM] The relative ability of a biological substance to react or combine. [CHEM] A positive number that characterizes the combining power of an element for other elements, as measured by the number of bonds to other atoms which one atom of the given element forms upon chemical combination; hydrogen is assigned valence 1, and the valence is the number of hydrogen atoms, or their equivalent, with which an atom of the given element combines.

valence bond [PHYS CHEM] The bond formed between the electrons of two or more atoms.

valence-bond method [PHYS CHEM] A method of calculating binding energies and other parameters of molecules by taking linear combinations of electronic wave functions, some of which represent covalent structures, others ionic structures; the coefficients in the linear combination are calculated by

VACUOLE

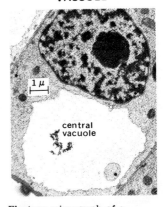

1 μ

central
vacuole

Electron micrograph of a
partially mature cell from a
developing root tip showing the
central vacuole. The osmotic
pressure in this vacuole is
responsible for the turgor and
provides the force to expand
the cell to its mature size.

VALINE

Structural formula of valine.

VALONIACEAE

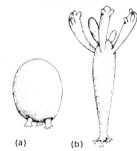

(a)   (b)

Valonia. (a) Spherical form.
(b) Clavate form.

VALVATINA

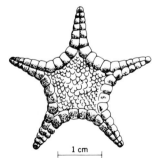

├─ 1 cm ─┤

A representative valvate starfish,
*Iconaster perierctus. (After
A. G. Fisher, 1919)*

the variational method. Also known as valence-bond resonance method.

valence-bond resonance method   *See* valence-bond method.

valence-bond theory   [CHEM] A theory of the structure of
chemical compounds according to which the principal requirements for the formation of a covalent bond are a pair of
electrons and suitably oriented electron orbitals on each of
the atoms being bonded; the geometry of the atoms in the
resulting coordination polyhedron is coordinated with the
orientation of the orbitals on the central atom.

valence electron   [ATOM PHYS] An electron that belongs to
the outermost shell of an atom.

valence number   [CHEM] A number that is equal to the
valence of an atom or ion multiplied by $+1$ or -1, depending
on whether the ion is positive or negative, or equivalently on
whether the atom in the molecule under consideration has
lost or gained electrons from its free state.

valence shell   [ATOM PHYS] The electrons that form the outermost shell of an atom.

valid   [SYST] Describing a taxon classified on the basis of
distinctive characters of accepted importance.

valine   [BIOCHEM] $C_5H_{11}NO_2$ An amino acid considered essential for normal growth of animals, and biosynthesized
from pyruvic acid. Also known as 2-aminoisovaleric acid; α-
aminoisovaleric acid; 2-amino-3-methylbutyric acid.

Valium   [PHARM] Trademark for diazepam, a tranquilizer.

vallate papilla   [ANAT] One of the large, flat papillae, each
surrounded by a trench, in a group anterior to the sulcus
terminalis of the tongue. Also known as circumvallate papilla.

Valoniaceae   [BOT] A family of green algae in the order
Siphonocladales consisting of plants that are essentially
unicellular, coenocytic vesicles, spherical or clavate, and up
to 6 centimeters in diameter.

Valvatacea   [PALEON] A superfamily of extinct gastropod
mollusks in the order Prosobranchia.

valvate   [BOT] Having valvelike parts, as those which meet
edge to edge or which open as if by valves.

Valvatida   [INV ZOO] An order of echinoderms in the subclass
Asteroidea.

Valvatina   [INV ZOO] A suborder of echinoderms in the order
Phanerozonida in which the upper marginals lie directly
over, and not alternate with, the corresponding lower marginals.

valve   [ANAT] A flat of tissue, as in the veins or between the
chambers in the heart, which permits movement of fluid in
one direction only.   [BOT] **1.** A segment of a dehiscing
capsule or legume. **2.** The lidlike portion of certain anthers.
[INV ZOO] **1.** One of the distinct, articulated pieces composing
the shell of certain animals, such as barnacles and brachiopods. **2.** One of two shells encasing the body of a bivalve
mollusk or a diatom.

valvifer   [INV ZOO] One of the basal plates of the ovipositor in
certain insects.

Valvifera   [INV ZOO] A suborder of isopod crustaceans distinguished by having a pair of flat, valvelike uropods which
hinge laterally and fold inward beneath the rear part of the
body.

valvula   [BIOL] A small valve.   [INV ZOO] One of the small
processes forming a sheath for the ovipositor in certain
insects.

valvulate   [BIOL] Having valvules.

vampire   [VERT ZOO] The common name for bats making up
the family Desmodontidae which have teeth specialized for
cutting and which subsist on a blood diet.

Vampyrellidae [INV ZOO] A family of protozoans in the order Proteomyxida including species which invade filamentous algae and sometimes higher plants.

Vampyromorpha [INV ZOO] An order of dibranchiate cephalopod mollusks represented by *Vampyroteuthis infernalis,* an inhabitant of the deeper waters of tropical and temperate seas.

vanadium [CHEM] A metal in group Vb, symbol V, atomic number 23; soluble in strong acids and alkalies; melts at 1900°C, boils about 3000°C; used as a catalyst.

vancomycin [MICROBIO] A complex antibiotic substance produced by *Streptomyces orientalis;* useful for treatment of severe staphylococcic infections.

van Crevald–von Gierke's disease *See* von Gierke's disease.

Van den Bergh reaction [PATH] A liver function test in which diazotized serum or plasma is compared with a standard solution of diazotized bilirubin.

Vaneyellidae [INV ZOO] A family of holothurian echinoderms in the order Dactylochirotida.

Vanhorniidae [INV ZOO] A monospecific family of the Hymenoptera in the superfamily Proctotrupoidea.

vannus [INV ZOO] The fanlike posterior lobe of the hindwing in some insects.

vapor [THERMO] A gas at a temperature below the critical temperature, so that it can be liquefied by compression, without lowering the temperature.

vapor-liquid equilibrium *See* liquid-vapor equilibrium.

Varanidae [VERT ZOO] The monitors, a family of reptiles in the suborder Sauria found in the hot regions of Africa, Asia, Australia, and Malaya.

variance [STAT] The square of the standard deviation.

variance ratio test [STAT] A technique for comparing the spreads or variabilities of two sets of figures to determine whether the two sets of figures were drawn from the same population. Also known as F test.

variate difference method [STAT] A technique for estimating the correlation between the random parts of two given time series.

varicella *See* chickenpox.

varicocele [MED] Dilatation of the veins of the pampiniform plexus of the spermatic cord, forming a soft, elastic, often uncomfortable swelling.

varicose [MED] Pertaining to blood vessels that are dilated, knotted, and tortuous.

varicose vein [ANAT] An enlarged tortuous blood vessel that occurs chiefly in the superficial veins and their tributaries in the lower extremities. Also known as varicosity.

varicosity *See* varicose vein.

variegate [BIOL] Having irregular patches of diverse colors.

variegated position effect [GEN] A phenomenon observed in some cases when a chromosome aberration causes a wild-type gene from the euchromatin to be relocated adjacent to heterochromatin; the phenotypic expression of the wild-type allele will be unstable, producing patches of phenotypically mutant tissue that differ from the surrounding wild-type tissue.

variety [SYST] A taxonomic group or category inferior in rank to a subspecies.

variola *See* smallpox.

varix [INV ZOO] A conspicuous ridge across each whorl of certain univalves marking the ancestral position of the outer lip of the aperture. [MED] A dilated and tortuous vein, artery, or lymphatic vessel.

varnish tree [BOT] *Rhus vernicifera.* A member of the sumac family (Anacardiaceae) cultivated in Japan; the cut bark exudes a juicy milk which darkens and thickens on exposure

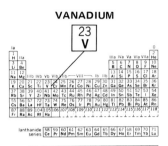

VANADIUM

Periodic table of the chemical elements showing the position of vanadium.

VASCULAR BUNDLE

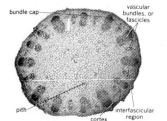

Cross section of sunflower
(Helianthus) stem showing
vascular bundles.

and is applied as a thin film to become a varnish of extreme hardness. Also known as lacquer tree.

vas [ANAT] A vessel.

vasa vasorum [ANAT] The blood vessels supplying the walls of arteries and veins.

vascular bundle [BOT] A strandlike part of the plant vascular system containing xylem and phloem.

vascular cambium [BOT] The lateral meristem which produces secondary xylem and phloem.

vascular ray [BOT] A ray derived from cambium and found in the stele of some vascular plants, often separating vascular bundles.

vascular retinopathy [MED] Pathological changes in the retina associated with diseases such as arterial hypertension, chronic nephritis, eclampsia, and advanced arteriosclerosis. Also known as retinal retinitis.

vascular tissue [BOT] The conducting tissue found in higher plants, consisting principally of xylem and phloem.

vas deferens [ANAT] The portion of the excretory duct system of the testis which runs from the epididymal duct to the ejaculatory duct. Also known as ductus deferens.

vasectomy [MED] Cutting, or removing a section from, the ductus deferens.

vasicentric [BOT] Concentrated, as a sheath, around a vessel in wood.

vasoconstrictor [PHYSIO] A nerve or an agent that causes blood vessel constriction.

vasodilator [PHYSIO] A nerve or an agent that causes blood vessel dilation.

vasogenic shock [MED] Failure of peripheral circulation due to vasodilation of arterioles and capillaries.

vasomotor [PHYSIO] Referring to any nerve that controls muscles in the wall of a blood vessel to regulate the diameter of the vessel; the nerves have both vasoconstrictor and vasodilator functions.

vasomotor center [PHYSIO] A large, diffuse area in the reticular formation of the lower brainstem; stimulation of different portions of this center causes either a rise in blood pressure and tachycardia (pressor area) or a fall in blood pressure and bradycardia (depressor area).

vasopressin [BIOCHEM] A peptide hormone which is elaborated by the posterior pituitary and which has a pressor effect; used medicinally as an antidiuretic. Also known as antidiuretic hormone (ADH).

vasotocin [BIOCHEM] A hormone from the neurosecretory cells of the posterior pituitary of lower vertebrates; increases permeability to water in amphibian skin and in bladder.

vault [BIOL] An anatomical structure that is arched or dome-shaped.

v-body *See* nucleosome.

VD *See* venereal disease.

vector [MATH] An element of a vector space. [MED] An agent, such as an insect, capable of mechanically or biologically transferring a pathogen from one organism to another.

vectorcardiogram [PHYSIO] The part of the pathway of instantaneous vectors during one cardiac cycle. Also known as monocardiogram.

vectorcardiography [PHYSIO] A method of recording the magnitude and direction of the instantaneous cardiac vectors.

vegetable [AGR] The edible portion of a usually herbaceous plant; customarily served with the main course of a meal. [BOT] Resembling or relating to plants.

vegetable diastase *See* diastase.

vegetal pole [CYTOL] The portion of a telolecithal egg which divides more slowly due to yolk content. [EMBRYO] That portion of a blastula composed of megameres.

vegetation [BOT] The total mass of plant life that occupies a given area.

vegetational plant geography [ECOL] A field of study concerned with the mapping of vegetation regions and the interpretation of these in terms of environmental or ecological influences.

vegetation and ecosystem mapping [BOT] An art and a science concerned with the drawing of maps which locate different kinds of plant cover in a geographic area.

vegetation management [ECOL] The art and practice of manipulating vegetation such as timber, forage, crops, or wild life, so as to produce a desired part or aspect of that material in higher quantity or quality.

vegetation zone [ECOL] **1.** An extensive, even transcontinental, band of physiognomically similar vegetation on the earth's surface. **2.** Plant communities assembled into regional patterns by the area's physiography, geological parent material, and history.

vegetative [BIOL] Having nutritive or growth functions, as opposed to reproductive.

vegetative propagation [BOT] Production of a new plant from a portion of another plant, such as a stem or branch.

vein [ANAT] A relatively thin-walled blood vessel that carries blood from capillaries to the heart in vertebrates. [BOT] One of the vascular bundles in a leaf. [INV ZOO] **1.** One of the thick, stiff ribs providing support for the wing of an insect. **2.** A venous sinus in invertebrates.

velamen [BOT] The corky epidermis covering the aerial roots of an epiphytic orchid.

velarium [INV ZOO] The velum of certain scyphozoans and cubomedusans distinguished by the presence of canals lined with endoderm.

veldt [ECOL] Grasslands of eastern and southern Africa that are usually level and mixed with trees and shrubs. Also spelled veld.

veliger [INV ZOO] A mollusk larval stage following the trochophore, distinguished by an enlarged girdle of ciliated cells (velum).

Veliidae [INV ZOO] A family of the Hemiptera in the subdivision Amphibicorisae composed of small water striders which have short legs and a longitudinal groove between the eyes.

vellus [ANAT] Fine body hair that is present until puberty.

Velocipedidae [INV ZOO] A tropical family of hemipteran insects in the superfamily Cimicoidea.

velum [BIOL] A veil- or curtainlike membrane. [INV ZOO] A swimming organ on the larva of certain marine gastropod mollusks that develops as a contractile ciliated collar-shaped ridge.

velvet [VERT ZOO] Soft vascular tissue covering deer antlers during growth.

vena cava [ANAT] One of two large veins which in air-breathing vertebrates conveys blood from the systemic circulation to the right atrium.

venation [BOT] The system or pattern of veins in the tissues of a leaf. [INV ZOO] The arrangement of veins in an insect wing.

venereal bubo *See* lymphogranuloma venereum.

venereal disease [MED] Any of several contagious diseases generally acquired during sexual intercourse; includes gonorrhea, syphilis, chancroid, granuloma inguinale, and lymphogranuloma venereum. Abbreviated VD.

venereal wart [MED] A warty growth of the penis, frequent in

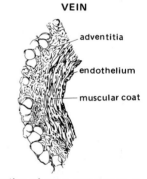

VEIN

adventitia

endothelium

muscular coat

Portion of cross section through common digital vein of a human. *(From A. A. Maximow and W. Bloom, A Textbook of Histology, 6th ed., Saunders, 1953)*

some parts of the world, and probably acquired during sexual intercourse.

venom [PHYSIO] Any of various poisonous materials secreted by certain animals, such as snakes or bees.

venous pressure [PHYSIO] Tension of the blood within the veins.

vent [ZOO] The external opening of the cloaca or rectum, especially in fish, birds, and amphibians.

venter [ANAT] The abdomen, or other body cavity containing organs. [BOT] The thickened basal portion of an archegonium. [INV ZOO] **1.** The undersurface of an arthropod's abdomen. **2.** The outer, convex part of a curved or coiled gastropod or cephalopod shell.

ventral [BOT] On the lower surface of a dorsiventral plant structure, such as a leaf. [ZOO] On or belonging to the lower or anterior surface of an animal, that is, on the side opposite the back.

ventral aorta [VERT ZOO] The arterial trunk or trunks between the heart and the first aortic arch in embryos or lower vertebrates.

ventral hernia [MED] A hernia of the abdominal wall not involving the umbilical, femoral, or inguinal openings. Also known as abdominal hernia.

ventral light reflex [INV ZOO] A basic means of orientation in aquatic invertebrates, such as shrimp, which swim belly up toward the light.

ventral rib [VERT ZOO] Any of the ribs which lie in the septa dividing the trunk musculature into segments in fish. Also known as pleural rib.

ventricle [ANAT] **1.** A chamber, or one of two chambers, in the vertebrate heart which receives blood from the atrium and forces it into the arteries by contraction of the muscular wall. **2.** One of the interconnecting, fluid-filled chambers of the vertebrate brain that are continuous with the canal of the spinal cord. [ZOO] A cavity in a body part or organ.

ventricose [BIOL] Swollen or distended, especially on one side.

ventricular depolarization complex *See* QRS complex.

ventricular septum *See* interventricular septum.

ventriculus [ZOO] A ventricle that performs digestive functions, such as a stomach or a gizzard.

ventromedial nucleus [ANAT] A central nervous system nucleus in the hypothalamus that appears to be the satiation center; bilateral surgical damage to this nucleus results in overeating.

venule [ANAT] A small vein.

Venus' flytrap [BOT] *Dionaea muscipula.* An insectivorous plant (order Sarraceniales) of North and South Carolina; the two halves of a leaf blade can swing upward and inward as though hinged, thus trapping insects between the closing halves of the leaf blade.

verbal learning [PSYCH] A field of experimental psychology which studies the formation of certain verbal associations; deals with acquisition of the associations.

Verbeekinidae [PALEON] A family of extinct marine protozoans in the superfamily Fusulinacea.

Verbenaceae [BOT] A family of variously woody or herbaceous dicotyledons in the order Lamiales characterized by opposite or whorled leaves and regular or irregular flowers, usually with four or two functional stamens.

Vermes [INV ZOO] An artificial taxon considered to be a phylum in some systems of classification, but variously defined as including all invertebrates except arthropods, or including all vermiform invertebrates.

vermiform [BIOL] Wormlike; resembling a worm.

VENUS' FLYTRAP

Venus' flytrap *(Dionaea muscipula).* The leaves capture insects.

VERBEEKINIDAE

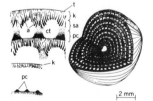

Cutaway diagram of representative species of Verbeekinidae; t = tectum, k = keriotheca, sa = septula, pc = parachromata, a = alveoli, and ct = chamberlets.

vermiform appendix [ANAT] A small, blind sac projecting from the cecum. Also known as appendix.

Vermilingua [VERT ZOO] An infraorder of the mammalian order Edentata distinguished by lack of teeth and in having a vermiform tongue; includes the South American true anteaters.

vermis [ANAT] The median lobe of the cerebellum.

vernalization [BOT] The induction in plants of the competence or ripeness to flower by the influence of cold, that is, at temperatures below the optimal temperature for growth.

vernation [BOT] The characteristic arrangement of young leaves within the bud. Also known as prefoliation.

vernix caseosa [EMBRYO] A cheesy deposit on the surface of the fetus derived from the stratum corneum, sebaceous secretion, and remnants of the epitrichium.

Veronal [PHARM] A trade name for barbital.

verruca [BIOL] A wartlike elevation on the surface of a plant or animal.

verruca peruana *See* verruca peruviana.

verruca peruviana [MED] A benign eruptive form of bartonellosis with chronic cutaneous lesions. Also known as verruca peruana.

Verrucariaceae [BOT] A family of crustose lichens in the order Pyrenulales typically found on rocks, especially in intertidal or salt-spray zones along rocky coastlines.

Verrucomorpha [INV ZOO] A suborder of the crustacean order Thoracica composed of sessile, asymmetrical barnacles.

verrucose [BIOL] Having the surface covered with wartlike protuberances.

verrucous endocarditis [MED] Small thrombotic, nonbacterial, wartlike lesions on the heart valves and endocardium, occurring frequently in systemic lupus erythematosus. Also known as terminal endocarditis.

versatile anther [BOT] An anther whose attachment is near its middle, thus enabling it to swing freely.

vertebra [ANAT] One of the bones that make up the spine in vertebrates.

vertebral arch [ANAT] An arch formed by the paired pedicles and laminas of a vertebra; the posterior part of a vertebra which together with the anterior part, the body, encloses the vertebral foramen in which the spinal cord is lodged in vertebrates. Also known as neural arch.

vertebral column *See* spine.

Vertebrata [VERT ZOO] The major subphylum of the phylum Chordata including all animals with backbones, from fish to man.

vertebrate zoology [ZOO] That branch of zoology concerned with the study of members of the Vertebrata.

verticil [BOT] An arrangement of flowers, inflorescences, or other structures around a common point on the axis.

verticillaster [BOT] A condensed cyme having the appearance of a whorl, but arising in the axils of opposite leaves.

verticillate [BOT] Whorled, in an arrangement resembling the spokes of a wheel.

vertigo [MED] The sensation that the outer world is revolving about the patient (objective vertigo) or that the patient is moving in space (subjective vertigo).

vesicant [PHARM] An agent that causes blistering.

vesication [MED] **1.** A blister. **2.** Formation of a blister.

vesicle [BIOL] A small, thin-walled bladderlike cavity, usually filled with fluid.

Vespertilionidae [VERT ZOO] The common bats, a large cosmopolitan family of the Chiroptera characterized by a long tail, extending to the edge of the uropatagium; almost all members are insect-eating.

VERRUCOMORPHA

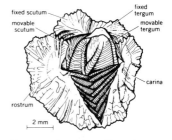

Verruca stroemia, a Recent species of Verrucomorpha. (*O. Müller*)

vespertine [VERT ZOO] Active in the evening.

Vespidae [INV ZOO] A widely distributed family of Hymenoptera in the superfamily Vespoidea including hornets, yellow jackets, and potter wasps.

Vespoidea [INV ZOO] A superfamily of wasps in the suborder Apocrita.

vessel [BOT] A xylem tube formed from several modified tracheids oriented end to end.

vestibular apparatus [ANAT] The anatomical structures concerned with the vestibular portion of the eighth cranial nerve; includes the saccule, utricle, semicircular canals, vestibular nerve, and vestibular nuclei of the ear.

vestibular membrane of Reissner *See* Reissner's membrane.

vestibular nerve [ANAT] A somatic sensory branch of the auditory nerve, which is distributed about the ampullae of the semicircular canals, macula sacculi, and macula utriculi.

vestibular reflexes [PHYSIO] The responses of the vestibular apparatus to strong stimulation; responses include pallor, nausea, vomiting, and postural changes.

vestibule [ANAT] **1.** The central cavity of the bony labyrinth of the ear. **2.** The parts of the membranous labyrinth within the cavity of the bony labyrinth. **3.** The space between the labia minora. **4.** *See* buccal cavity.

vestibulocochlear nerve *See* auditory nerve.

vestibulospinal tract [ANAT] A tract of nerve fibers that originates principally from the lateral vestibular nucleus and descends in the anterior funiculus of the spinal cord.

vestige [BIOL] A degenerate anatomical structure or organ that remains from one more fully developed and functional in an earlier phylogenetic form of the individual.

vestigial [BIOL] Of, being, or resembling a vestige.

VETCH

vetch [BOT] Any of a group of mostly annual legumes, especially of the genus *Vicia*, with weak viny stems terminating in tendrils and having compound leaves; some varieties are grown for their edible seed.

veterinary medicine [MED] The branch of medical practice which treats of the diseases and injuries of animals.

vexillum [VERT ZOO] The vane of a feather.

Vianaidae [INV ZOO] A small family of South American Hemiptera in the superfamily Tingordea.

vibraculum [INV ZOO] A specially modified bryozoan zooid with a bristlelike seta that sweeps debris from the surface of the colony.

vibriosis [VET MED] An infectious bacterial disease, primarily of cattle, sheep, and goats, caused by *Vibrio fetus* and characterized by abortion, retained placenta, and metritis.

vibrissae [VERT ZOO] Hairs with specialized erectile tissue; found on all mammals except man. Also known as sinus hairs; tactile hairs; whiskers.

vicuna [VERT ZOO] *Lama vicugna.* A rare, wild ruminant found in the Andes mountains; the fiber of the vicuna is strong, resilient, and elastic but is the softest and most delicate of animal fibers.

villous adenoma [MED] A slow-growing, potentially malignant neoplasm of the mucosa of the rectum; manifested by bleeding and mucoid diarrhea.

villous placenta *See* epitheliochorial placenta.

villous tenosynovitis [MED] A chronic inflammatory reaction of a tendon sheath producing hypertrophy of the lining, with the formation of redundant folds and villi.

villus [ANAT] A fingerlike projection from the surface of a membrane.

vinblastine [PHARM] $C_{46}H_{58}O_9N_4$ An alkaloid obtained from the periwinkle plant (*Vinca rosea*) and used, as the sulfate salt, as an antineoplastic drug.

Purple vetch *(Vicia bengalensis)*, a representative species of vetch, showing characteristic features. *(USDA)*

Vincent's angina [MED] Vincent's infection involving tissues of the pharynx and tonsils.

Vincent's infection [MED] A noncontagious bacterial infection of the oral mucosa characterized by ulceration and formation of a gray pseudomembrane; caused by certain fusiform bacteria and spirochetes.

vincristine [PHARM] $C_{46}H_{56}O_{10}N_4$ An alkaloid extracted from the periwinkle plant (*Vinca rosea*) and used, as the sulfate salt, as an antineoplastic drug. Also known as leurocristine.

Vindobonian [GEOL] A European stage of geologic time, middle Miocene.

vine [BOT] A plant having a stem that is too flexible or weak to support itself.

vinegar [MATER] The product of the incomplete oxidation to acetic acid of ethyl alcohol produced by a primary fermentation of vegetable materials; contains not less than 4 grams of acetic acid per gallon; used in preparation of pickled fruits and vegetables and in salad dressing.

vinegar eel [INV ZOO] *Turbatrix aceti.* A very small nematode often found in large numbers in vinegar fermentation. Also known as vinegar worm.

vinegar worm *See* vinegar eel.

Viocin [MICROBIO] A trademark for viomycin.

Violaceae [BOT] A family of dicolyledonous plants in the order Violales characterized by polypetalous, mostly perfect, hypogynous flowers with a single style and five stamens.

Violales [BOT] A heterogeneous order of dicotyledons in the subclass Dilleniidae distinguished by a unilocular, compound ovary and mostly parietal placentation.

viomycin [MICROBIO] A polypeptide antibiotic or mixture of antibiotic substances produced by strains of *Streptomyces griseus* var. *purpureus* (*Streptomyces puniceus*); the sulfate salt is administered intramuscularly for treatment of tuberculosis resistant to other therapy.

viper [VERT ZOO] The common name for reptiles of the family Viperidae; thick-bodied poisonous snakes having a pair of long fangs, present on the anterior part of the upper jaw, which fold against the roof of the mouth when the jaws are closed.

Viperidae [VERT ZOO] A family of reptiles in the suborder Serpentes found in Eurasia and Africa; all species are proglyphodont.

viral encephalomyelitides [MED] A group of several encephalitis diseases caused by various viruses; includes epidemic encephalitis, equine encephalitides, and Japanese B encephalitis.

viral gastroenteritis [MED] An acute infectious gastroenteritis thought to be caused by various viruses and characterized by diarrhea, nausea, vomiting, and variable systemic symptoms.

viral hepatitis [MED] Either of two forms of hepatitis caused by a virus, serum hepatitis or infectious hepatitis.

viral pneumonia [MED] A form of pneumonia caused by a virus of various types, in which the inflammatory reaction predominates in the septa, and the alveoli contain fibrin, edema fluid, and some inflammatory cells.

viremia [MED] Presence of viral particles in the blood.

Vireonidae [VERT ZOO] The vireos, a family of New World passeriform birds in the suborder Oscines.

virgate [BOT] Banded.

virgate trophus [INV ZOO] A piercing type of trophus in rotifers that is thin and slightly toothed.

virgo-forcipate trophus [INV ZOO] A type of muscular chamber in rotifers containing jaws of a cuticular material interme-

VIPER

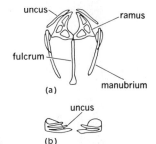

The fer-de-lance *(Bothrops atrox)* is from 5 to 6 feet (1.5 to 1.8 meters) in length; ranges from southern Mexico to northern South America and is found in the West Indies.

VIRGATE TROPHUS

uncus — ramus

fulcrum

(a) manubrium

uncus

(b)

Virgate trophus of *Notommata.* (a) Ventral view and (b) upper view, showing thin, slightly toothed structure. *(After Meyers)*

diate between a piercing and grasping type of structure.

viridans streptococci [MICROBIO] A group of pathogenic and saprophytic streptococci including strains not causing beta hemolysis, although many cause alpha hemolysis, and none which elaborate a C substance.

virilism [MED] *See* gynandry. [PSYCH] **1.** Masculinity. **2.** Manifestation of male behavioral patterns in the female.

virion [VIROL] The complete, mature virus particle.

virology [MICROBIO] The study of submicroscopic organisms known as viruses.

virulence [MICROBIO] The disease-producing power of a microorganism; infectiousness.

virus [VIROL] A large group of infectious agents ranging from 10 to 250 nanometers in diameter, composed of a protein sheath surrounding a nucleic acid core and capable of infecting all animals, plants, and bacteria; characterized by total dependence on living cells for reproduction and by lack of independent metabolism.

virus hepatitis *See* infectious hepatitis.

virus interference [MICROBIO] A phenomenon which may be defined as protection of host cells against one virus, conferred as a result of prior infection with a different virus.

viscera [ANAT] The organs within the cavities of the body of an organism.

visceral arch [ANAT] One of the series of mesodermal ridges covered by epithelium bounding the lateral wall of the oral and pharyngeal regions of vertebrates; embryonic in higher forms, they contribute to formation of the face and neck. [VERT ZOO] One of the first two arches of the series in gill-bearing forms.

visceral cleft [VERT ZOO] A furrow between successive visceral arches.

visceral leishmaniasis [MED] A severe, generalized, and often fatal infection, caused by any of three pathogenic hemoflagellates of the genus *Leishmania*, affecting organs rich in endothelial cells; accompanied by fever, spleen and liver enlargement, anemia, leukopenia, skin pigmentation, and changes in plasma protein.

visceral peritoneum [ANAT] That portion of the peritoneum covering the organs of the abdominal cavity.

visceral pouch *See* pharyngeal pouch.

visceratonia [PSYCH] The behavioral type assigned to the endomorphic somatotype, manifested by a desire for assimilation and the conservation of energy through social interactions, relaxation, and love of food.

visceromotor [PHYSIO] Referring to a nerve transmitting motor impulses to the viscera.

visceroptosis [MED] Prolapse of a viscus, especially the intestine; downward displacement of the intestine in the abdominal cavity. Also known as enteroptosis.

viscid [BOT] Having a sticky surface, as certain leaves.

viscoelasticity [MECH] Property of a material which is viscous but which also exhibits certain elastic properties such as the ability to store energy of deformation, and in which the application of a stress gives rise to a strain that approaches its equilibrium value slowly.

viscosity [FL MECH] Energy dissipation and generation of stresses in a fluid by the distortion of fluid elements; quantitatively, when otherwise qualified, the absolute viscosity. Also known as flow resistance; internal friction.

viscous flow [FL MECH] **1.** The flow of a viscous fluid. **2.** The flow of a fluid through a duct under conditions such that the mean free path is small in comparison with the smallest, transverse section of the duct.

viscous fluid [FL MECH] A fluid whose viscosity is sufficiently large to make the viscous forces a significant part of the total force field in the fluid.

viscus [ANAT] Singular of viscera.

visible spectrum [SPECT] **1.** The range of wavelengths of visible radiation. **2.** A display or graph of the intensity of visible radiation emitted or absorbed by a material as a function of wavelength or some related parameter.

vision [PHYSIO] The sense which perceives the form, color, size, movement, and distance of objects. Also known as sight.

visual acuity [PHYSIO] The ability to see fine details of an object; specifically, the ability to see an object whose angle subtended at the eye is 1 minute of arc.

visual learning [PSYCH] A type of sensory learning that is controlled by the cortical visual areas of the brain.

visual pigment [BIOCHEM] Any of various photosensitive pigments of vertebrate and invertebrate photoreceptors.

visual projection area [PHYSIO] The receptive center for visual images in the cortex of the brain, located in the walls and margins of the calcarine sulcus of the occipital lobe. Also known as Brodmann's area 17.

visual purple *See* rhodopsin.

visual-righting reflex [PHYSIO] A reflex mechanism whereby righting of the head and body is caused by visual stimuli. Also known as optical-righting reflex.

visual seizure [MED] A form of epileptic seizure in which the patient experiences visual sensations in the form of light flashes, sometimes of varied colors.

visual yellow [BIOCHEM] An intermediary substance formed from rhodopsin in the retina after exposure to light; it is ultimately broken down to retinene and vitamin A.

Vitaceae [BOT] A family of dicotyledonous plants in the order Rhamnales; mostly tendril-bearing climbers with compound or lobed leaves, as in grapes (*Vitis*).

vital capacity [PHYSIO] The volume of air that can be forcibly expelled from the lungs after the deepest inspiration.

vitalism [BIOL] The theory that the activities of a living organism are under the guidance of an agency which has none of the attributes of matter or energy.

vitamin [BIOCHEM] An organic compound present in variable, minute quantities in natural foodstuffs and essential for the normal processes of growth and maintenance of the body; vitamins do not furnish energy, but are essential for energy transformation and regulation of metabolism.

vitamin A [BIOCHEM] $C_{20}H_{29}OH$ A pale-yellow alcohol that is soluble in fat and insoluble in water; found in liver oils and carotenoids, and produced synthetically; it is a component of visual pigments and is essential for normal growth and maintenance of epithelial tissue. Also known as antiinfective vitamin; antixerophthalmic vitamin; retinol.

vitamin B$_1$ *See* thiamine.

vitamin B$_2$ *See* riboflavin.

vitamin B$_3$ *See* pantothenic acid.

vitamin B$_6$ [BIOCHEM] A vitamin which exists as three chemically related and water-soluble forms found in food: pyridoxine, pyridoxal, and pyridoxamine; dietary requirements and physiological activities are uncertain.

vitamin B$_{12}$ [BIOCHEM] A group of closely related polypyrrole compounds containing trivalent cobalt; the antipernicious anemia factor, essential for normal hemopoiesis. Also known as cobalamin; cyanocobalamin; extrinsic factor.

vitamin B complex [BIOCHEM] A group of water-soluble vitamins that include thiamine, riboflavin, nicotinic acid, pyridoxine, panthothenic acid, inositol, *p*-aminobenzoic acid, biotin, adenylic acid, folic acid, and vitamin B$_{12}$.

vitamin B$_6$ hydrochloride *See* pyridoxine hydrochloride.

VITAMIN A

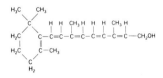

Structural formula for vitamin A (retinol).

VITAMIN B$_6$

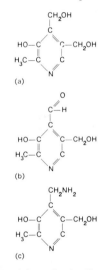

Structural formulas for the three naturally occurring forms of vitamin B$_6$. *(a)* Pyridoxine (pyridoxol). *(b)* Pyridoxal. *(c)* Pyridoxamine.

VITAMIN D

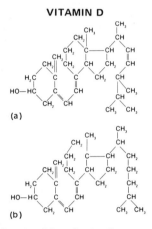

(a)

(b)

Structural fomulas for the
two vitamin D compounds.
(a) Vitamin D₂ (calciferol).
(b) Vitamin D₃ (cholecalciferol).

vitamin B₂ phosphate   *See* riboflavin 5′-phosphate.

vitamin C   *See* ascorbic acid.

vitamin D   [BIOCHEM] Either of two fat-soluble, sterol-like compounds, calciferol or ergocalciferol (vitamin D₂) and cholecalciferol (vitamin D₃); occurs in fish liver oils and is essential for normal calcium and phosphorus deposition in bones and teeth. Also known as antirachitic vitamin.

vitamin E   [BIOCHEM] $C_{29}H_{50}O_2$ Any of a series of eight related compounds called tocopherols, α-tocopherol having the highest biological activity; occurs in wheat germ and other oils and is believed to be needed in certain human physiological processes.

vitamin G   *See* riboflavin.

vitamin K   [BIOCHEM] Any of three yellowish oils which are fat-soluble, nonsteroid, and nonsaponifiable; it is essential for formation of prothrombin. Also known as antihemorrhagic vitamin; prothrombin factor.

vitamin P   [BIOCHEM] A substance, such as citrin or one or more of its components, believed to be concerned with maintenance of the normal state of the walls of small blood vessels.

vitamin P complex   *See* bioflavanoid.

vitellarium   [INV ZOO] The part of the ovary in certain rotifers and flatworms that produces nutritive cells filled with yolk. Also known as yolk larva.

vitelline artery   [EMBRYO] An artery passing from the yolk sac to the primitive aorta in young vertebrate embryos. Also known as omphalomesenteric artery.

vitelline duct   [EMBRYO] The constricted part of the yolk sac opening into the midgut region of the future ileum. Also known as omphalomesenteric duct; umbilical duct.

vitelline membrane   [CYTOL] The cytoplasmic membrane on the surface of the mammalian ovum.

vitelline vein   [EMBRYO] Any of the embryonic veins in vertebrates uniting the yolk sac and the sinus venosus; their proximal fused ends form the portal vein. Also known as omphalomesenteric vein.

vitellogenesis   [PHYSIO] The process by which yolk is formed in the ooplasm of an oocyte.

viticulture   [AGR] That division of horticulture concerned with grape growing, studies of grape varieties, methods of culture, and insect and disease control.

vitiligo   [MED] A skin disease characterized by an acquired ochromia in areas of various sizes and shapes.

Vitreoscillaceae   [MICROBIO] A family of bacteria in the order Beggiatoales; organisms have a filamentous habit and move by gliding, but never store sulfur, and rely on organic nutrients in their metabolism.

vitreous body   *See* vitreous humor.

vitreous chamber   [ANAT] A cavity of the eye posterior to the crystalline lens and anterior to the retina, which is filled with vitreous humor.

vitreous humor   [PHYSIO] The transparent gel-like substance filling the greater part of the globe of the eye, the vitreous chamber. Also known as vitreous body.

vitta   [BOT] **1.** An oil receptacle in the pericarp of Umbelliferae. **2.** A longitudinal ridge in diatoms.

vittate   [BOT] **1.** Having longitudinal stripes. **2.** Bearing specialized oil tubes (vittae).

vivax malaria   [MED] Malaria caused by *Plasmodium vivax* and characterized by typical paroxysms occurring every few days, commonly every 2 days. Also known as benign tertian malaria; tertian malaria.

Viverridae   [VERT ZOO] A family of carnivorous mammals in the superfamily Feloidea composed of the civets, genets, and mongooses.

Viviparidae [INV ZOO] A family of fresh-water gastropod mollusks in the order Pectinibranchia.

viviparous [PHYSIO] Bringing forth live young.

vocal cord *See* vocal fold.

vocal fold [ANAT] Either of a pair of folds of tissue covered by mucous membrane in the larynx. Also known as vocal cord.

vocal sac [VERT ZOO] An expansible pocket of skin beneath the chin or behind the jaws of certain frogs; may be inflated to a great volume and serves as a resonator.

Vochysiaceae [BOT] A family of dicotyledonous plants in the order Polygalales characterized by mostly three carpels, usually stipulate leaves, one fertile stamen, and capsular fruit.

Voges-Proskauer test [MICROBIO] One of the four tests of the IMVIC test; a qualitative test for the formation of acetyl methylcarbinol from glucose, in which solutions of α-naphthol, potassium hydroxide, and creatinine are added to an incubated culture of test organisms in a glucose broth, and a pink to rose color indicates a positive reaction.

volar [ANAT] Pertaining to, or on the same side as, the palm of the hand or the sole of the foot.

volatile [CHEM] Readily passing off by evaporation.

vole [VERT ZOO] Any of about 79 species of rodent in the tribe Microtini of the family Cricetidae; individuals have a stout body with short legs, small ears, and a blunt nose.

voltametry [PHYS CHEM] Any electrochemical technique in which a faradaic current passing through the electrolysis solution is measured while an appropriate potential is applied to the polarizable or indicator electrode; for example, polarography.

volumetric pipet [ANALY CHEM] A graduated glass tubing used to measure quantities of a solution; the tube is open at the top and bottom, and a slight vacuum (suction) at the top pulls liquid into the calibrated section; breaking the vacuum allows liquid to leave the tube.

voluntary muscle [PHYSIO] A muscle directly under the control of the will of the organism.

Volutidae [INV ZOO] A family of gastropod mollusks in the order Neogastropoda.

volutin [BIOCHEM] A basophilic substance, thought to be a nucleic acid, occurring as granules in the cytoplasm and vacuoles of algae and other microorganisms.

volva [MYCOL] A cuplike membrane surrounding the base of the stipe in certain gill fungi.

volvent nematocyst [INV ZOO] A nematocyst in the form of an unarmed, coiled tube that is closed at the end.

Volvocales [BOT] An order of one-celled or colonial green algae in the division Chlorophyta; individuals are motile with two, four, or rarely eight whiplike flagella.

Volvocida [INV ZOO] An order of the protozoan class Phytamastigophorea; individuals are grass-green or colorless, have one, two, four, or eight flagella, and thick cell walls of cellulose.

volvulus [MED] A twisting of the bowel upon itself so as to occlude the lumen and, in severe cases, compromise its circulation.

Vombatidae [VERT ZOO] A family of marsupial mammals in the order Diprotodonta in some classification systems.

vomer [ANAT] A skull bone below the ethmoid region constituting part of the nasal septum in most vertebrates.

vomeronasal cartilage [ANAT] A strip of hyaline cartilage extending from the anterior nasal spine upward and backward on either side of the septal cartilage of the nose and

VOCAL SAC

Toad of the genus *Bufo* giving mating call with vocal sac expanded. *(American Museum of Natural History photograph)*

VOLVENT NEMATOCYST

Volvent nematocyst of *Hydra*. *(After Schulze, from T. I. Storer and R. L. Usinger, General Zoology, 3d ed., McGraw-Hill, 1957)*

VOLVOCALES

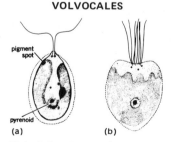

pigment spot

pyrenoid

(a)   (b)

Unicellular algae of the order Volvocales. *(a) Chlamydomonas,* a unicellular organism with two flagella. *(b) Polyblepharides,* showing massive chloroplast and eight flagella.

attached to the anterior margin of the vomer. Also known as Jacobson's cartilage.

von Gierke's disease [MED] A form of glycogenosis characterized by marked diminution in or absence of hepatic glucose-6-phosphatase, resulting in hepatic glycogenosis, hypoglycemia, and acidosis. Also known as glycogen storage disease; hepatic glycogenosis; type I of Cori; van Crevald–von Gierke's disease.

vulture [VERT ZOO] The common name for any of various birds of prey in the families Cathartidae and Accipitridae of the order Falconiformes; the head of these birds is usually naked.

vulva [ANAT] The external genital organs of women.

vulval gland [ANAT] A scent gland in the vulval tissues of the human female.

vulvovaginitis [MED] Simultaneous inflammation of the vulva and the vagina.

wader *See* shore bird.

wading bird [VERT ZOO] Any of the long-legged, long-necked birds composing the order Ciconiiformes, including storks, herons, egrets, and ibises.

Waldeyer's ring [ANAT] A circular arrangement of the lymphatic tissues formed by the palatine and pharyngeal tonsils and the lymphatic follicles at the base of the tongue and behind the posterior pillars of the fauces.

walking bird [VERT ZOO] Any bird of the order Columbiformes, including the pigeons, doves, and sandgrouse.

walnut [BOT] The common name for about a dozen species of deciduous trees in the genus *Juglans* characterized by pinnately compound, aromatic leaves and chambered or laminate pith; the edible nut of the tree is distinguished by a deeply furrowed or sculptured shell.

walrus [VERT ZOO] *Odobenus rosmarus.* The single species of the pinniped family Odobenidae distinguished by the upper canines in both sexes being prolonged as tusks.

wart [MED] A papillomatous growth which occurs singly or in groups on the skin surface; thought to be caused by a viral agent.

wasp [INV ZOO] The common name for insects belonging to the order Hymenoptera; all are important as parasites or predators of injurious pests.

Wasserman test [IMMUNOL] A complement-fixation test for syphilis using sensitized lipid extracts of beef heart as antigen.

water [CHEM] H_2O Clear, odorless, tasteless liquid that is essential for most animal and plant life and is an excellent solvent for many substances; melting point 0°C (32°F), boiling point 100°C (212°F); the chemical compound may be termed hydrogen oxide.

waterborne disease [MED] Disease transmitted by drinking water or by contact with potable or bathing water.

water bug [INV ZOO] Any insect which lives in an aquatic habitat during all phases of its life history.

water conservation [ECOL] The protection, development, and efficient management of water resources for beneficial purposes.

watercress [BOT] *Nasturtium officinale.* A perennial cress generally grown in flooded soil beds and used for salads and food garnishing.

waterfowl [VERT ZOO] Aquatic birds which constitute the order Anseriformes, including the swans, ducks, geese, and screamers.

watermelon [BOT] *Citrullus vulgaris.* An annual trailing vine with light-yellow flowers and leaves having five to seven deep lobes; the edible, oblong or roundish fruit has a smooth, hard, green rind filled with sweet, tender, juicy, pink to red tissue containing many seeds.

water microbiology [MICROBIO] An aspect of microbiology

WALRUS

Walrus *(Odobenus rosmarus),* showing tusks.

WASP

Paper wasp *(Polistes),* a type of wasp that grows to 2.5 centimeters or more. It is common on flowers and near buildings where open-comb paper nests are constructed.

WATER

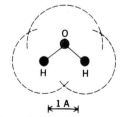

Water molecule, showing positions of atoms. Dotted circles show effective sizes of isolated atoms.

that deals with the normal and adventitious microflora of natural and artificial water bodies.

water mocassin [VERT ZOO] *Agkistrodon piscivorus.* A semi-aquatic venomous pit viper; skin is brownish or olive on the dorsal aspect, paler on the sides, and has indistinct black bars. Also known as cottonmouth.

water-vascular system [INV ZOO] An internal closed system of reservoirs and ducts containing a watery fluid, found only in echinoderms.

wattle [VERT ZOO] A fleshy pendulous process beneath the throat of a cock or turkey and of certain reptiles.

wavelength [PHYS] The distance between two points having the same phase in two consecutive cycles of a periodic wave, along a line in the direction of propagation.

wax gland *See* ceruminous gland.

weasel [VERT ZOO] The common name for at least 12 species of small, slim carnivores which belong to the family Mustelidae and which have a reddish-brown coat with whitish underparts; species in the northern regions have white fur during the winter and are called ermine.

web [VERT ZOO] The membrane between digits in many birds and amphibians.

Weberian apparatus [VERT ZOO] A series of bony ossicles which form a chain connecting the swim bladder with the inner ear in fishes of the superorder Ostariophysi.

Weberian ossicle [VERT ZOO] One of a chain of three or four small bones that make up the Weberian apparatus.

Weber's law [PHYSIO] The law that the stimulus increment which can barely be detected (the just noticeable difference) is a constant fraction of the initial magnitude of the stimulus; this is only an approximate rule of thumb.

weed [BOT] A plant that is useless or of low economic value, especially one growing on cultivated land to the detriment of the crop.

weevil [INV ZOO] Any of various snout beetles whose larvae destroy crops by eating the interior of the fruit or grain, or bore through the bark into the pith of many trees.

Wegener's granulomatosis [MED] A rare disease of unknown causation characterized by necrotizing granulomas in the air passages, necrotizing vasculitis, and glomerulitis.

weighted average [STAT] The number obtained by adding the product of α_i times the ith number in a set of N numbers for $i = 1,2, \ldots, N$, where α_i are numbers (weights) such that $\alpha_1 + \alpha_2 + \ldots + \alpha_N = 1$.

Weil-Felix test [IMMUNOL] An agglutination test for various rickettsial infections based on production of nonspecific agglutinins in the blood of infected patients, and using various strains of *Proteus vulgaris* as antigen.

Weil's disease [MED] A severe form of leptospirosis characterized by jaundice, oliguria, circulatory collapse, and tendency to hemorrhage. Also known as icterohemorrhagic fever; leptospirosis icterohemorrhagia; spirochetal jaundice.

Welwitschiales [BOT] An order of gymnosperms in the subdivision Geneticae represented by the single species *Welwitschia mirabilis* of southwestern Africa; distinguished by only two leaves and short, unbranched, cushion- or saucer-shaped woody main stem which tapers to a very long taproot.

Wenlockian [GEOL] A European stage of geologic time: Middle Silurian (above Tarannon, below Ludlovian).

Wentworth classification [GEOL] A logarithmic grade for size classification of sediment particles starting at 1 millimeter and using the ratio of ½ in one direction (and 2 in the other), providing diameter limits to the size classes of 1, ½, ¼, etc. and 1, 2, 4, etc.

WEASEL

The common or European weasel *(Mustela nivalis).*

Werdnig-Hoffman disease [MED] Infantile spinal muscular atrophy.

Werner's syndrome [MED] A complex of symptoms, thought to be inherited as an autosomal recessive, including premature senescence, dwarfism, cataracts, sclerodermalike changes of the skin, osteoporosis, and multiglandular dysfunction.

Wernicke's encephalopathy [MED] A disease due to thiamine deficiency, characterized by vomiting, ophthalmoplegia, ptosis, nystagmus, ataxia, weakness, dementia, and hemorrhaging of neurons around the third ventricle, cerebral aqueduct, and mammillary bodies.

West Nile fever [MED] An acute, usually mild, mosquito-borne virus disease occurring in summer, chiefly in Egypt, Israel, Africa, India, and Korea; signs are fever and lymphadenopathy, sometimes with a rash.

Westphal-Pilcz reflex *See* pupillary reflex.

Westphal's pupillary reflex *See* pupillary reflex.

wetwood [PL PATH] Wood having a water-soaked appearance because of a high water content; may be caused by bacteria or by physiological factors.

whale [VERT ZOO] A large marine mammal of the order Cetacea; the body is streamlined, the broad flat tail is used for propulsion, and the limbs are balancing structures.

whalebone *See* baleen.

Wharton's duct *See* submandibular duct.

Wharton's jelly [EMBRYO] A gelatinous layer, surrounding the umbilical cord vessels and made up of connective tissue.

wheat [BOT] A food grain crop of the genus *Triticum*; plants are self-pollinating; the inflorescence is a spike bearing sessile spikelets arranged alternately on a zigzag rachis.

wheat germ [BOT] The embryo of a wheat grain.

wheel organ [INV ZOO] A ciliated ring or trochal disc functioning in locomotion of Rotifera.

whelk [INV ZOO] A gastropod mollusk belonging to the order Neogastropoda; species are carnivorous but also scavenge.

whip grafting [BOT] A method of grafting by fitting a small tongue and notch cut in the base of the scion into corresponding cuts in the stock.

Whipple's disease [MED] A disease characterized by infiltration of the intestinal wall and lymphatics by macrophages filled with glycoprotein. Also known as intestinal lipodystrophy.

whipworm disease [MED] A chronic, wasting diarrhea produced by heavy parasitization of the large intestine by the nematode *Trichuris trichiura*, particularly in undernourished children in the tropics.

whiskers *See* vibrissae.

white ant *See* termite.

white blood cell *See* leukocyte.

white corpuscle *See* leukocyte.

white diarrhea *See* pullorum disease.

whitefish [VERT ZOO] Any of various food fishes in the family Salmonidae, especially of the genus *Coregonus*, characterized by an adipose dorsal fin and nearly toothless mouth.

white infarct [MED] An infarct in which hemorrhage is slight, or that has been decolorized by removal of blood or its pigments.

white matter [HISTOL] The part of the central nervous system composed of nerve fibers covered with myelin.

white potato *See* potato.

whooping cough *See* pertussis.

whooping crane [VERT ZOO] *Grus americana.* A member of a rare North American migratory species of wading birds; the entire species forms a single population.

WHALE

Sperm whale *(Physeter catodon),* a species of whale having worldwide distribution. Head measures 20 feet (6 meters), or one-third of total body length.

WHEAT

(a) (b)　(c)　　(d)　(e) (f)

Spikes of some representative varieties of wheat. *(a)* Wild and *(b)* cultivated forms of einkorn *(Triticum monococcum);* *(c)* wild emmer *(T. dicocciodes);* *(d)* emmer *(T. dicoccum);* *(e)* durum *(T. durum);* *(f)* poulard *(T. turgidum).* *(Courtesy of H. Kihara)*

WHOOPING CRANE

Whooping cranes *(Grus americana),* a rare North American species in danger of extinction. *(Bureau of Sport Fisheries and Wildlife)*

WHORL

Plain whorl fingerprint pattern.
(Federal Bureau of Investigation)

WILLOW

Twig, terminal bud, and leaf
of the Babylon weeping willow
(Salix babylonia).

whorl [ANAT] A fingerprint pattern in which at least two deltas are present with a recurve in front of each. [BOT] An arrangement of several identical anatomical parts, such as petals, in a circle around the same point.

Whytt's reflex *See* pupillary reflex.

Widal test [IMMUNOL] A macroscopic or microscopic agglutination test for the diagnosis of typhoid fever and other *Salmonella* infections by using killed or preserved bacteria as the antigen.

wild boar [VERT ZOO] *Sus scrofa.* A wild hog with coarse, grizzled hair and enlarged tusks or canines on both jaws. Also known as boar.

wild cinnamon *See* bayberry.

wildfire [PL PATH] A bacterial disease of tobacco caused by *Pseudomonas tabaci* and characterized by the appearance of brown spots surrounded by yellow rings, which turn dark, rot, and fall out.

wild type [GEN] The natural or unmutated organism or character.

Williamsoniaceae [PALEOBOT] A family of extinct plants in the order Cycadeoidales distinguished by profuse branching.

willow [BOT] A deciduous tree and shrub of the genus *Salix*, order Salicales; twigs are often yellow-green and bear alternate leaves which are characteristically long, narrow, and pointed, usually with fine teeth along the margins.

Wilms' tumor [MED] A malignant renal tumor composed principally of mesodermal tissues. Also known as nephroblastoma.

Wilson's disease [MED] A hereditary disease of ceruloplasmin formation transmitted as an autosomal recessive and characterized by decreased serum ceruloplasmin and copper values, and increased excretion of copper in the urine. Also known as hepatolenticular degeneration.

wilt [PL PATH] Any of various plant diseases characterized by drooping and shriveling, following loss of turgidity.

windburn [BOT] Injury to plant foliage, caused by strong, hot, dry winds. [MED] A superficial inflammation of the skin, analogous to sunburn, caused by exposure to wind, especially a hot dry wind, inducing a dilation of the surface blood vessels.

wing [ZOO] Any of the paired appendages serving as organs of flight on many animals.

wing pad [INV ZOO] The undeveloped wing of an insect pupa.

Winteraceae [BOT] A family of dicotyledonous plants in the order Magnoliales distinguished by hypogynous flowers, exstipulate leaves, air vessels absent, and stamens usually laminar.

winter bud [BOT] A dormant bud that is protected by hard scales during the winter.

Wirsung's duct [ANAT] The adult pancreatic duct in man, sheep, ganoid fish, teleost fish, and frog.

Wisconsin [GEOL] Pertaining to the fourth, and last, glacial stage of the Pleistocene epoch in North America; followed the Sangamon interglacial, beginning about $85,000 \pm 15,000$ years ago and ending 7000 years ago.

Wisconsin false blossom *See* false blossom.

witches'-broom disease [PL PATH] An abnormal cluster of small branches or twigs that grow on a tree or shrub as a result of attack by fungi, viruses, dwarf mistletoes, or insect injury. Also known as hexenbesen; staghead.

withertip [PL PATH] A blighting of the terminal shoots or the tips of leaves associated with certain plant diseases, such as anthracnose of citrus plants.

Wolbachiae [MICROBIO] A tribe of rickettsial organisms containing symbiotic forms that inhabit arthropods.

wolf [VERT ZOO] Any of several wild species of the genus *Canis* in the family Canidae which are fierce and rapacious, sometimes attacking man; includes the red wolf, gray wolf, and coyote.

Wolfcampian [GEOL] A North American provincial series of geologic time; lowermost Permian (below Leonardian, above Virgilian of Pennsylvania).

Wolffian duct *See* mesonephric duct.

wolverine [VERT ZOO] *Gulo gulo.* A carnivorous mammal which is the largest and most vicious member of the family Mustelidae.

wood [BOT] The hard fibrous substance that makes up the trunks and large branches of trees beneath the bark.   [ECOL] A dense growth of trees, more extensive than a grove and smaller than a forest.

woodland *See* forest; temperate woodland.

woodpecker [VERT ZOO] A bird of the family Picidae characterized by stiff tail feathers and zygodactyl feet which enable them to cling to a tree trunk while drilling into the bark for insects.

wood sugar *See* xylose.

wool [VERT ZOO] The soft undercoat of various animals such as sheep, angora, goat, camel, alpaca, llama, and vicuna.

wool-sorter's disease *See* anthrax.

worker [INV ZOO] One of the neuter, usually sterile individuals making up a caste of social insects, such as ants, termites, or bees, which labor for the colony.

worm [INV ZOO] **1.** The common name for members of the Annelida. **2.** Any of various elongated, naked, soft-bodied animals resembling an earthworm.

wren [VERT ZOO] Any of the various small brown singing birds in the family Troglodytidae; they are insectivorous and tend to inhabit dense, low vegetation.

wrist [ANAT] The part joining forearm and hand.

wrist breadth [ANTHRO] A measurement from the outside projection of the distal part of the ulna to the radius at the wrist joint.

wrist thickness [ANTHRO] The measurement transverse to the wrist breadth.

wryneck *See* torticollis.

Wuchereria [INV ZOO] A genus of filarial worms parasitic in man in all worm regions of the world.

wuchereriasis [MED] Infection with worms of the genus *Wuchereria*. Also known as Bancroft's filariasis.

Würm [GEOL] **1.** A European stage of geologic time: uppermost Pleistocene (above Riss, below Holocene). **2.** Pertaining to the fourth glaciation of the Pleistocene epoch in the Alps, equivalent to the Wisconsin glaciation in North America, following the Riss-Würm interglacial.

Wynyardiidae [PALEON] An extinct family of herbivorous marsupial mammals in the order Diprotodonta.

WOODPECKER

Red-headed woodpecker
(Melanerpes erythrocephalus),
a typical woodpecker with
clawed feet and stiff tail feathers.

X

xanthelasma [MED] Raised yellow plaques occurring around the eyelids, resulting from lipid-filled cells in the dermis.

Xanthidae [INV ZOO] The mud crabs, a family of decapod crustaceans in the section Brachyura.

xanthine oxidase [BIOCHEM] A flavoprotein enzyme catalyzing the oxidation of certain purines.

xanthoma [MED] A yellowish mass of lipid-filled histocytes occurring in subcutaneous tissue, often around tendons.

xanthomatosis [MED] A condition marked by the deposit of a yellowish or orange lipoid material in the reticuloendothelial cells, the skin, and the internal organs.

xanthomycin [MICROBIO] An antibiotic produced by a strain of *Streptomyces* and composed of two varieties, A and B; active in low concentrations against a number of gram-positive microorganisms.

xanthophore [CYTOL] A yellow chromatophore.

Xanthophyceae [BOT] A class of yellow-green to green flagellate organisms of the division Chrysophyta; zoologists classify these organisms in the order Heterochlorida.

xanthophyll [BIOCHEM] $C_{40}H_{56}O_2$ Any of a group of yellow, alcohol-soluble carotenoid pigments that are oxygen derivatives of the carotenes, and are found in certain flowers, fruits, and leaves. Also known as lutein.

Xantusiidae [VERT ZOO] The night lizards, a family of reptiles in the suborder Sauria. ·

X chromosome [GEN] The sex chromosome occurring in double dose in the homogametic sex and in single dose in the heterogametic sex.

Xe *See* xenon.

Xenarthra [VERT ZOO] A suborder of mammals in the order Edentata including sloths, anteaters, and related forms; posterior vertebrae have extra articular facets and vertebrae in the hip, and shoulder regions tend to be fused.

xenogamy [BOT] Cross-fertilization between flowers on different plants.

xenon [CHEM] An element, symbol Xe, member of the noble gas family, group O, atomic number 54, atomic weight 131.30; colorless, boiling point −108°C (1 atm), noncombustible, nontoxic, and nonreactive; used in photographic flash lamps, luminescent tubes, and lasers, and as an anesthetic.

Xenophyophorida [INV ZOO] An order of Protozoa in the subclass Granuloreticulosia; includes deep-sea forms that develop as discoid to fan-shaped branching forms which are multinucleate at maturity.

Xenopneusta [INV ZOO] A small order of wormlike animals belonging to the Echiurida.

Xenopterygii [VERT ZOO] The equivalent name for Gobiesociformes.

Xenosauridae [VERT ZOO] A family of four rare species of lizards in the suborder Sauria; composed of the Chinese lizard

XENON

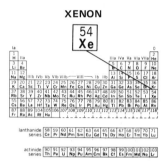

Periodic table of the chemical elements showing the position of xenon.

XENOPHYOPHORIDA

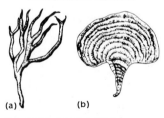

(a)   (b)

Representative fan-shaped branching forms of xenophyophorids. *(a) Stanomma dendroides (after Schulze). (b) Stannophyllum zonarium (after Schulze). (Doflein and Reichenon, Lehrbuch der Protozoenkunde, 1929)*

(*Shinisaurus crocodilurus*) and three Central American species of the genus *Xenosaurus*.

Xenungulata [PALEON] An order of large, digitigrade, extinct, tapirlike mammals with relatively short, slender limbs and five-toed feet with broad, flat phalanges; restricted to the Paleocene deposits of Brazil and Argentina.

xeric [ECOL] **1.** Of or pertaining to a habitat having a low or inadequate supply of moisture. **2.** Of or pertaining to an organism living in such an environment.

xerochasy [BOT] Dehiscence induced by arid conditions.

xeroderma pigmentosum [MED] A genodermatosis characterized by premature degenerative changes in the form of keratoses, malignant epitheliomatosis, and hyper- and hypopigmentation.

xerodermosteosis *See* Sjögren's syndrome.

xeromorphic [BOT] Modified structurally to retard transpiration, referring specifically to xerophytes. [ECOL] Thriving under dry conditions.

xerophilous [ECOL] Able to withstand arid conditions, referring to plants adapted to a limited water supply.

xerophthalmia [MED] Dryness and thickening of the conjunctiva, sometimes following chronic conjunctivitis, disease of the lacrimal apparatus, or vitamin A deficiency.

xerophyte [ECOL] A plant adapted to life in areas where the water supply is limited.

xerosere [ECOL] A temporary community in an ecological succession on dry, sterile ground such as rock, sand, or clay.

xerotherm [ECOL] A plant that thrives in conditions of aridity and high temperature.

xerothermic [CLIMATOL] Characterized by dryness and heat.

xiphidio cercaria [INV ZOO] A digenetic trematode larva having a stylet in the oral sucker.

Xiphiidae [VERT ZOO] The swordfishes, a family of perciform fishes in the suborder Scombroidei characterized by a tremendously produced bill.

xiphiplastron [VERT ZOO] The fourth lateral plate of the plastron of Chelonia.

xiphisternum [ANAT] The elongated posterior portion of the sternum.

Xiphodontidae [PALEON] A family of primitive tylopod ruminants in the superfamily Anaplotherioidea from the late Eocene to the middle Oligocene of Europe.

Xiphosura [INV ZOO] The equivalent name for Xiphosurida.

Xiphosurida [INV ZOO] A subclass of primitive arthropods in the class Merostomata characterized by cephalothoracic appendages, ocelli, book lungs, a somewhat trilobed body, and freely articulating styliform telson.

Xiphydriidae [INV ZOO] A family of the Hymenoptera in the superfamily Siricoidea.

X organ [INV ZOO] A cluster of neurosecretory cells of the medulla terminales, a portion of the brain lying in the eyestalk in stalk-eyed crustaceans.

x-radiation *See* x-rays.

x-ray diffraction [PHYS] The scattering of x-rays by matter, especially crystals, with accompanying variation in intensity due to interference effects. Also known as x-ray microdiffraction.

x-ray fluorescence analysis [ANALY CHEM] A nondestructive physical method used for chemical analyses of solids and liquids; the specimen is irradiated by an intense x-ray beam and the lines in the spectrum of the resulting x-ray fluorescence are diffracted at various angles by a crystal with known lattice spacing; the elements in the specimen are identified by the wavelengths of their spectral lines, and their concentra-

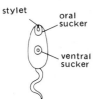

XIPHIDIO CERCARIA

stylet   oral sucker

ventral sucker

Xiphidio cercaria showing stylet in oral sucker. *(From R. M. Cable, An Illustrated Laboratory Manual of Parasitology, Burgess, 1958)*

XIPHOSURIDA

2 mm

(a)   (b)

Drawing of *Paleomerus* species of the subclass Xiphosurida. *(a)* Side view. *(b)* Dorsal view.

tions are determined by the intensities of these lines. Also known as x-ray fluorimetry.

x-ray fluorescent emission spectrometer [SPECT] An x-ray crystal spectrometer used to measure wavelengths of x-ray fluorescence; in order to concentrate beams of low intensity, it has bent reflecting or transmitting crystals arranged so that the theoretical curvature required can be varied with the diffraction angle of a spectrum line.

x-ray fluorimetry *See* x-ray fluorescence analysis.

x-ray microdiffraction *See* x-ray diffraction.

x-ray microprobe *See* microprobe.

x-ray microscope [ENG] **1.** A device in which an ultra-fine-focus x-ray tube or electron gun produces an electron beam focused to an extremely small image on a transmission-type x-ray target that serves as a vacuum seal; the magnification is by projection; specimens being examined can thus be in air, as also can the photographic film that records the magnified image. **2.** Any of several instruments which utilize x-radiation for chemical analysis and for magnification of 100–1000 diameters; it is based on contact or projection microradiography, reflection x-ray microscopy, or x-ray image spectrography.

x-rays [PHYS] A penetrating electromagnetic radiation, usually generated by accelerating electrons to high velocity and suddenly stopping them by collision with a solid body, or by inner-shell transitions of atoms with atomic number greater than 10; their wavelengths range from about 10^{-5} angstrom to 10^3 angstroms, the average wavelength used in research being about 1 angstrom. Also known as roentgen rays; x-radiation.

x-ray spectrograph [SPECT] An x-ray spectrometer equipped with photographic or other recording apparatus; one application is fluorescence analysis.

x-ray spectrometer [SPECT] An instrument for producing the x-ray spectrum of a material and measuring the wavelengths of the various components.

x-ray spectrometry [SPECT] The measure of wavelengths of x-rays by observing their diffraction by crystals of known lattice spacing. Also known as roentgen spectrometry; x-ray spectroscopy.

x-ray spectroscopy *See* x-ray spectrometry.

x-ray spectrum [SPECT] A display or graph of the intensity of x-rays, produced when electrons strike a solid object, as a function of wavelengths or some related parameter; it consists of a continuous bremsstrahlung spectrum on which are superimposed groups of sharp lines characteristic of the elements in the target.

x-ray therapy [MED] Medical treatment by controlled application of x-rays; a type of radiotherapy.

X test [STAT] A one-sample test which rejects the hypothesis $\mu = \mu_H$ in favor of the alternative $\mu > \mu_H$ if $X - \mu_H \geq c$ where c is an appropriate critical value, X is the arithmetic mean of observations, μ_H is a given number, and μ is the (unknown) expected value of the random variable X.

Xyelidae [INV ZOO] A family of hymenopteran insects in the superfamily Megalodontoidea.

xylem [BOT] The principal water-conducting tissue and the chief supporting tissue of higher plants; composed of tracheids, vessel members, fibers, and parenchyma.

xylocarp [BOT] A fruit that is hard and woody.

Xylocopidae [INV ZOO] A family of hairy tropical bees in the superfamily Apoidea.

xyloma [MYCOL] A hard mass of mycelia which gives rise to spore-bearing structures in certain fungi. [PL PATH] A tree tumor.

X-RAY SPECTROGRAPH

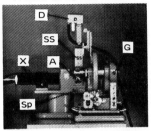

Modern x-ray spectrograph for fluorescence analysis. X is x-ray tube; Sp, specimen; A, crystal analyzer; SS, Soller (parallel) slits; D, counter tube detector; G, goniometer. *(Philips Electronic Instruments, Inc.)*

XYLEM

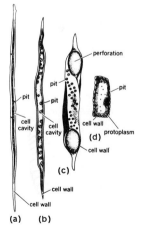

Xylem cell types. *(a)* Wood (xylem) fiber. *(b)* Tracheid. *(c)* Vessel member. *(d)* Xylem parenchyma cell. All can vary widely in structure. *(From H. J. Fuller and O. Tippo, College Botany, rev. ed., Holt, 1954)*

Xylomyiidae [INV ZOO] A family of orthorrhaphous dipteran insects in the series Brachycera.

xylophagous [BIOL] Feeding on wood.

xylose [BIOCHEM] $C_5H_{10}O_5$ A pentose sugar found in many woody materials; combustible, white crystals with a sweet taste; soluble in water and alcohol; melts about 148°C; used as a nonnutritive sweetener and in dyeing and tanning. Also known as wood sugar.

Xyridaceae [BOT] A family of terrestrial monocotyledonous plants in the order Commelinales characterized by an open leaf sheath, three stamens, and a simple racemose head for the inflorescence.

Y

Y *See* yttrium.

yak [VERT ZOO] *Poephagus grunniens.* A heavily built, long-haired mammal of the order Artiodactyla, with a shoulder hump; related to the bison, and resembles it in having 14 pairs of ribs.

yam [BOT] An erroneous name for the Puerto Rico variety of sweet potato; the edible, starchy tuberous root of the plant.

yaws [MED] An infectious tropical disease of humans caused by the spirochete *Treponema pertenue;* manifested by a primary cutaneous lesion followed by a granulomatous skin eruption.

Yb *See* ytterbium.

Y chromosome [GEN] The sex chromosome found only in the heterogametic sex.

yeast [MYCOL] A collective name for those fungi which possess, under normal conditions of growth, a vegetative body (thallus) consisting, at least in part, of simple, individual cells.

yellow cartilage [HISTOL] Elastic cartilage.

yellow dwarf [PL PATH] Any of several plant viral diseases characterized by yellowing of the foliage and stunting of the plant.

yellow fat cell [HISTOL] A large, generally spherical fat cell with a thin shell of protoplasm and a single enlarged fat droplet which appears yellowish.

yellow fever [MED] An acute, febrile, mosquito-borne viral disease characterized in severe cases by jaundice, albuminuria, and hemorrhage.

yellow-green algae [BOT] The common name for members of the class Xanthophyceae.

yellow leaf blotch [PL PATH] A fungus disease of alfalfa caused by *Pyrenopeziza medicaginis* characterized by the appearance of yellow or orange blotches with small black dots on the foliage.

yellows [PL PATH] Any of various fungus diseases of plants characterized by yellowing of the leaves which later turn brown, become brittle, and die; affects cabbage, lettuce, cauliflower, peach, sugarbeet, and other plants.

yew [BOT] A genus of evergreen trees and shrubs, *Taxus,* with the fruit, an aril, containing a single seed surrounded by a scarlet, fleshy, cuplike envelope; the leaves are flat and acicular.

Y ligament *See* iliofemoral ligament.

yolk [BIOCHEM] **1.** Nutritive material stored in an ovum. **2.** The yellow spherical mass of food material that makes up the central portion of the egg of a bird or reptile.

yolk larva *See* vitellarium.

yolk plug [VERT ZOO] A mass of yolk cells which fills up the blastopore, as in the frog.

yolk sac [EMBRYO] A distended extraembryonic extension,

YAK

Yak *(Poephagus grunniens)* showing characteristic features.

YAM

Puerto Rico sweet potato known as the yam. *(USDA)*

YOUNGINIFORMES

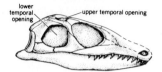

Lateral view of skull of *Youngina*, a representative of Younginiformes, Upper Permian to Lower Triassic. *(From A. S. Romer, Vertebrate Paleontology, 3d ed., University of Chicago Press, 1966)*

YTTERBIUM

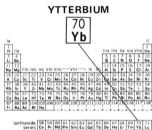

Periodic table of the chemical elements showing the position of ytterbium.

YTTRIUM

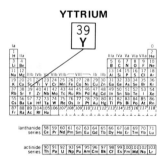

Periodic table of the chemical elements showing the position of yttrium.

heavy-laden with yolk, through the umbilicus of the midgut of the vertebrate embryo.

Y organ [INV ZOO] Either of a pair of nonneural structures found in the anterior portion of the crustacean body; source of the molting hormone, ecdysone.

Young-Helmholtz theory [PHYSIO] A theory of color vision according to which there are three types of color receptors that respond to short, medium, and long waves respectively; primary colors are those that stimulate most successfully the three types of receptors. Also known as Helmholtz theory.

Younginiformes [PALEON] A suborder of extinct small lizardlike reptiles in the order Eosuchia, ranging from the Middle Permian to the Lower Triassic in South Africa.

Yponomeutidae [INV ZOO] A heterogeneous family of small, often brightly colored moths in the superfamily Tineoidea; the head is usually smooth with reduced or absent ocelli.

Ypsilothuriidae [INV ZOO] A family of echinoderms in the order Dactylochirotida having 8-10 tentacles, a permanent spire on the plates of the test, and the body fusiform or U-shaped.

ytterbium [CHEM] A rare-earth metal of the yttrium subgroup, symbol Yb, atomic number 70, atomic weight 173.04; lustrous, malleable, soluble in dilute acids and liquid ammonia, reacts slowly with water; melts at 824°C, boils at 1427°C; used in chemical research, lasers, garnet doping, and x-ray tubes.

yttrium [CHEM] A rare-earth metal, symbol Y, atomic number 39, atomic weight 88.905; dark-gray, flammable (as powder), soluble in dilute acids and potassium hydroxide solution, and decomposes in water; melts at 1500°C, boils at 2927°C; used in alloys and nuclear technology and as a metal deoxidizer.

Z

Zalambdalestidae [PALEON] A family of extinct insectivorous mammals belonging to the group Proteutherea; they occur in the Late Cretaceous of Mongolia.

Zanclidae [VERT ZOO] The Moorish idols, a family of Indo-Pacific perciform fishes in the suborder Acanthuroidei.

Zapodidae [VERT ZOO] The Northern Hemisphere jumping mice, a family of the order Rodentia with long legs and large feet adapted for jumping.

zebra [VERT ZOO] Any of three species of African mammals belonging to the family Equidae distinguished by a coat of black and white stripes.

zebu [VERT ZOO] A domestic breed of cattle, indigenous to India, belonging to the family Bovidae, distinguished by long drooping ears, a dorsal hump between the shoulders, and a dewlap under the neck; known as the Brahman in the United States.

Zechstein [GEOL] A European series of geologic time, especially in Germany: Upper Permian (above Rothliegende).

Zeiformes [VERT ZOO] The dories, a small order of teleost fishes, distinguished by the absence of an orbitosphenoid bone, a spinous dorsal fin, and a pelvic fin with a spine and five to nine soft rays.

zenodiagnosis [MED] A procedure of using a suitable arthropod to transfer an infectious agent from a patient to a susceptible laboratory animal.

Zeoidea [VERT ZOO] An equivalent name for Zeiformes.

Zeomorphi [VERT ZOO] An equivalent name for Zeiformes.

Ziehl-Neelsen stain [MICROBIO] A procedure for acid-fast staining of tubercle bacilli with carbol fuchsin.

zinc [CHEM] A metal of group IIb, symbol Zn, atomic number 30, atomic weight 65.37; explosive as powder; soluble in acids and alkalies, insoluble in water; strongly electropositive; melts at 419°C, boils at 907°C.

zinc chills See metal fume fever.

Zingiberaceae [BOT] A family of aromatic monocotyledonous plants in the order Zingiberales characterized by one functional stamen with two pollen sacs, distichously arranged leaves and bracts, and abundant oil cells.

Zingiberales [BOT] An order of monocotyledonous herbs or scarcely branched shrubs in the subclass Commelinidae characterized by pinnately veined leaves and irregular flowers that have well-differentiated sepals and petals, an inferior ovary, and either one or five functional stamens.

zirconium [CHEM] A metallic element of group IVb, symbol Zr, atomic number 40, atomic weight 91.22; occurs as crystals, flammable as powder; insoluble in water, soluble in hot, concentrated acids; melts at 1850°C, boils at 4377°C.

Z line [HISTOL] The line formed by attachment of the actin filaments between two sarcomeres.

Zn See zinc.

ZEIFORMES

John dory *(Zenopsis ocellata)*, a representative of Zeiformes, showing characteristic features. *(After G. B. Goode, Fishery Industries of the United States, sect. 1, 1884)*

ZINC

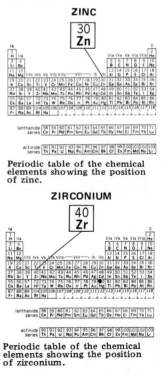

Periodic table of the chemical elements showing the position of zinc.

ZIRCONIUM

Periodic table of the chemical elements showing the position of zirconium.

ZOANTHIDEA

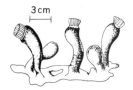

3 cm

Colony of *Zoanthina tentaculata*, a representative species of Zoanthidea. *(After Y. Dalage)*

ZOEA

0.5 mm

Zoea larva of crab, showing characteristic features. *(Smithsonian Institution)*

Zoantharia [INV ZOO] A subclass of the class Anthozoa; individuals are monomorphic and most have retractile, simple, tubular tentacles.

Zoanthidea [INV ZOO] An order of anthozoans in the subclass Zoantharia; these are mostly colonial, sedentary, skeletonless, anemonelike animals that live in warm, shallow waters and coral reefs.

Zoarcidae [VERT ZOO] The eelpouts, a family of actinopterygian fishes in the order Gadiformes which inhabit cold northern and far southern seas.

zoea [INV ZOO] An early larval stage of decapod crustaceans distinguished by a relatively large cephalothorax, conspicuous eyes, and large, fringed antennae.

Zollinger-Ellison syndrome [MED] Gastric hypersecretion and hyperacidity, fulminating intractable atypical peptic ulceration, and hyperplasia of the islet cells of the pancreas.

zona fasciculata [HISTOL] The central part of the adrenal cortex in which the cellular cords are radially disposed.

zona glomerulosa [HISTOL] The outer part of the adrenal cortex in which cells are grouped in rounded masses.

zonal centrifuge [BIOL] A centrifuge that uses a rotating chamber of large capacity in which to separate cell organelles by density-gradient centrifugation.

zona orbicularis [ANAT] A thickening of the capsular ligament around the acetabulum.

zona pellucida [HISTOL] The thick, solid, elastic envelope of the ovum. Also known as oolemma.

zonation [ECOL] Arrangement of organisms in biogeographic zones.

zone of optimal proportion [IMMUNOL] One of three zones considered to appear when antigen and antibody are mixed; it is that zone in which there is no uncombined antigen or antibody. Also known as equivalence zone.

zonite [INV ZOO] A body segment in Diplopoda.

zoochlorellae [BIOL] Unicellular green algae which live as symbionts in the cytoplasm of certain protozoans, sponges, and other invertebrates.

zoochory [BOT] Dispersal of plant disseminules by animals.

zooecium [INV ZOO] The exoskeleton of a feeding zooid in bryozoans.

zoogeographic region [ECOL] A major unit of the earth's surface characterized by faunal homogeneity.

zoogeography [BIOL] The science that attempts to describe and explain the distribution of animals in space and time.

zoogloea [MICROBIO] A gelatinous or mucilaginous mass characteristic of certain bacteria grown in organic-rich fluid media.

zooid [INV ZOO] A more or less independent individual of colonial animals such as bryozoans and coral.

zoology [BIOL] The science that deals with knowledge of animal life.

Zoomastigina [INV ZOO] The equivalent name for Zoomastigophorea.

Zoomastigophorea [INV ZOO] A class of flagellate protozoans in the subphylum Sarcomastigophora; some are simple, some are specialized, and all are colorless.

zoonoses [BIOL] Diseases which are biologically adapted to and normally found in lower animals but which under some conditions also infect man.

zooplankton [ECOL] Microscopic animals which move passively in aquatic ecosystems.

zoosphere [ECOL] The world community of animals.

zoosporangium [BOT] A spore case bearing zoospores.

zoospore [BIOL] An independently motile spore.

Zoraptera [INV ZOO] An order of insects, related to termites and psocids, which live in decaying wood, sheltered from light; most individuals are wingless, pale in color, and blind.

Zoroasteridae [INV ZOO] A family of deep-water asteroid echinoderms in the order Forcipulatida.

Zorotypidae [INV ZOO] The single family, containing one genus, *Zorotypus*, in the order Zoraptera.

zoster *See* herpes zoster.

Zosteraceae [BOT] A family of monocotyledonous plants in the order Najadales; the group is unique among flowering plants in that they grow submerged in shallow ocean waters near the shore.

Zosterophyllatae [PALEOBOT] *See* Zosterophyllopsida.

Zosterophyllopsida [PALEOBOT] A group of early land vascular plants ranging from the Lower to the Upper Devonian; individuals were leafless and rootless.

Zr *See* zirconium.

zwitterion *See* dipolar ion.

Zygaenidae [INV ZOO] A diverse family of small, often brightly colored African moths in the superfamily Zygaenoidea.

Zygaenoidea [INV ZOO] A superfamily of moths in the suborder Heteroneura characterized by complete venation, rudimentary palpi, and usually a rudimentary proboscis.

Zygnemataceae [BOT] A family of filamentous plants in the order Conjugales; they are differentiated into genera by chloroplast morphology, which may be spiral, bandlike, or cushionlike.

zygodactyl [VERT ZOO] Of birds, having a toe arrangement of two in front and two behind.

zygomatic bone [ANAT] A bone of the side of the face below the eye; forms part of the zygomatic arch and part of the orbit in mammals. Also known as malar bone.

zygomorphic [BIOL] Bilaterally symmetrical.

Zygomycetes [MYCOL] A class of fungi in the division Eumycetes.

zygopophysis [ANAT] One of the articular processes of the neural arch of a vertebra.

Zygoptera [INV ZOO] The damsel flies, a suborder of insects in the order Odonata; individuals are slender, dainty creatures, often with bright-blue or orange coloring and usually with clear or transparent wings.

zygote [EMBRYO] **1.** An organism produced by the union of two gametes. **2.** The fertilized ovum before cleavage.

zygotene [CYTOL] The stage of meiotic prophase during which homologous chromosomes synapse; visible bodies in the nucleus are now bivalents. Also known as amphitene.

zygotic induction [VIROL] Phage induction following conjugation of a lysogenic bacterium with a nonlysogenic one.

zymase [BIOCHEM] A complex of enzymes that catalyze glycosis.

zymogen [BIOCHEM] The inactive precursor of an enzyme; liberates an active enzyme on reaction with an appropriate kinose. Also known as proenzyme.

zymogen granules [BIOCHEM] Granules of zymogen in gland cells, particularly those of the pancreatic acini and of the gastric chief cells.

zymogenic [MICROBIO] Obtaining energy by amylolitic processes.

zymophore [BIOCHEM] The active portion of an enzyme.

zymosterol [BIOCHEM] $C_{27}H_{43}OH$ An unsaturated sterol obtained from yeast fat; yields cholesterol on hydrogenation.

Zythiaceae [MYCOL] A family of fungi of the order Sphaeropsidales which contains many plant and insect pathogens.

ZOSTEROPHYLLATAE

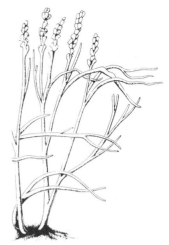

Reconstruction of *Zosterophyllum*, a typical genus of Zosterophyllatae, showing characteristic features. (*Modified from Kräusel and Weyland, 1935*)

ZYGOPTERA

Two adult damsel flies of the suborder Zygoptera. Note mites attached.

Appendix

U.S. Customary System and the metric system

To date, scientists and engineers have used two major systems of units in measurement. These are commonly called the U.S. Customary System (inherited from the British Imperial System) and the metric system.

In the U.S. Customary System the units yard and pound with their divisions, such as the inch, and multiples, such as the ton, are basic. The metric system was evolved during the 18th century and has been adopted for general use by most countries. Nearly everywhere it is used for precise measurements in science. The meter and kilogram with their multiples, such as the kilometer, and fractions, such as the gram, are basic to the metric system.

In the U.S. Customary System, units of the same kind are related almost at random. For example, there are the units of length, the inch, yard, and mile. In the metric system the relationships between units of the same kind are strictly decimal (millimeter, meter, and kilometer).

However, to complicate matters in scientific writing, there is no uniformity within each of these two systems as to the choice of units for the same quantities. For example, the hour or the second, the foot or the inch, and the centimeter or the millimeter could be chosen by a scientist as the unit of measurement for the quantities time and length.

The International System, or SI

To simplify matters and to make communication more understandable, an internationally accepted system of units is coming into use. This is termed the International System of Units, which is abbreviated SI in all languages.

Fundamentally the system is metric with the base units derived from scientific formulas or natural constants. For example, the meter in the SI is defined as the length equal to 1 650 763.73 wavelengths in vacuum of the radiation corresponding to the transition between the electronic energy levels $2p_{10}$ and $5d_5$ of the krypton-86 atom. Previously, in the metric system, the meter was defined as the distance between two marks on a specific metal bar.

In a similar way the second in the SI is defined as the duration of 9 192 631 770 periods of the radiation corresponding to the transition between two hyperfine levels of the ground state of the cesium-133 atom.

Interestingly, the kilogram, the SI unit of mass, is still the mass of the kilogram kept at Sèvres, France. However, it is possible that eventually the unit will be redefined in terms of atomic mass.

Although the SI is increasing in usage by scientists and engineers, there are some units in everyday use which will probably remain, for example, minute, hour, day, degree (angle), and liter. The point should be made, however, that these terms will not be employed in a scientific context if the SI is fully adopted.

Because of their extremely common use among scientists, several units are still permitted in conjunction with SI units, for example, the electron volt, gauss, barn, and curie. In time their usage might be phased out.

One further point is that in October, 1967, the Thirteenth General Conference of Weights and Measures decided to name the SI unit of thermodynamic temperature "kelvin" (symbol K) instead of "degree Kelvin" (symbol °K). For example, the notation is 273 K and not 273°K.

The basic units and derived units of the SI are shown in **Table 1** and **Table 2.** Some common units defined in terms of SI units are given in **Table 3** (the definitions in the fourth column are exact).

In the SI the prefixes differ from a unit in steps of 10^3. A list of prefix terms, symbols, and their factors is given in **Table 4.** Prefixes are used as follows:

$$
\begin{aligned}
1000 \text{ m} &= 1 \text{ kilometer} &&= 1 \text{ km} \\
1000 \text{ V} &= 1 \text{ kilovolt} &&= 1 \text{ kV} \\
1\,000\,000 \ \Omega &= 1 \text{ megohm} &&= 1 \text{ M}\Omega \\
0.000\,000\,001 \text{ s} &= 1 \text{ nanosecond} &&= 1 \text{ ns}
\end{aligned}
$$

Only one prefix is to be employed for a unit. For example:

$$
\begin{aligned}
1000 \text{ kg} &= 1 \text{ Mg} &&\text{not } 1 \text{ kkg} \\
10^{-9} \text{ s} &= 1 \text{ ns} &&\text{not } 1 \text{ m}\mu\text{s} \\
1\,000\,000 \text{ m} &= 1 \text{ Mm} &&\text{not } 1 \text{ kkm}
\end{aligned}
$$

Also, when a unit is raised to a power, the power applies to the whole unit including the prefix. For example:

$$
\text{km}^2 = (\text{km})^2 = (1000 \text{ m})^2 = 10^6 \text{ m}^2 \qquad \text{not } 1000 \text{ m}^2
$$

Table 1. Basic units of the International System

Quantity	Name of unit	Unit symbol
length	meter	m
mass	kilogram	kg
time	second	s
electric current	ampere	A
temperature	kelvin	K
luminous intensity	candela	cd
amount of substance	mole	mol

Table 2. Derived units of the International System

Quantity	Name of unit	Unit symbol or abbreviation, where differing from base form	Unit expressed in terms of base or supplementary units*
area	square meter		m^2
volume	cubic meter		m^3
frequency	hertz	Hz	s^{-1}
density	kilogram per cubic meter		kg/m^3
velocity	meter per second		m/s
angular velocity	radian per second		rad/s
acceleration	meter per second squared		m/s^2
angular acceleration	radian per second squared		rad/s^2
volumetric flow rate	cubic meter per second		m^3/s
force	newton	N	$kg \cdot m/s^2$
surface tension	newton per meter, joule per square meter	N/m, J/m^2	kg/s^2
pressure	newton per square meter, pascal	N/m^2, Pa	$kg/m \cdot s^2$
viscosity, dynamic	newton-second per square meter, pascal-second	$N \cdot s/m^2$, Pa·s	$kg/m \cdot s$
viscosity, kinematic	meter squared per second		m^2/s
work, torque, energy, quantity of heat	joule, newton-meter, watt-second	J, N·m, W·s	$kg \cdot m^2/s^2$
power, heat flux	watt, joule per second	W, J/s	$kg \cdot m^2/s^3$
heat flux density	watt per square meter	W/m^2	kg/s^3

Table 2. Derived units of the International System (cont.)

Quantity	Name of unit	Unit symbol or abbreviation, where differing from base form	Unit expressed in terms of base or supplementary units*
volumetric heat release rate	watt per cubic meter	W/m^3	$kg/m \cdot s^3$
heat transfer coefficient	watt per square meter kelvin	$W/m^2 \cdot K$	$kg/s^3 \cdot K$
heat capacity (specific)	joule per kilogram kelvin	$J/kg \cdot K$	$m^2/s^2 \cdot K$
capacity rate	watt per kelvin	W/K	$kg \cdot m^2/s^3 \cdot K$
thermal conductivity	watt per meter kelvin	$W/m \cdot K$, $\dfrac{J \cdot m}{s \cdot m^2 \cdot K}$	$kg \cdot m/s^3 \cdot K$
quantity of electricity	coulomb	C	$A \cdot s$
electromotive force	volt	V, W/A	$kg \cdot m^2/A \cdot s^3$
electric field strength	volt per meter	V/m	V/m
electric resistance	ohm	Ω, V/A	$kg \cdot m^2/A^2 \cdot s^3$
electric conductivity	ampere per volt meter	$A/V \cdot m$	$A^2 \cdot s^3/kg \cdot m^3$
electric capacitance	farad	F, $A \cdot s/V$	$A^3 \cdot s^4/kg \cdot m^2$
magnetic flux	weber	Wb, $V \cdot s$	$kg \cdot m^2/A \cdot s^2$
inductance	henry	H, $V \cdot s/A$	$kg \cdot m^2/A^2 \cdot s^2$
magnetic permeability	henry per meter	H/m	$kg \cdot m/A^2 \cdot s^2$
magnetic flux density	tesla, weber per square meter	T, Wb/m^2	$kg/A \cdot s^2$
magnetic field strength	ampere per meter		A/m
magnetomotive force	ampere		A
luminous flux	lumen	lm	$cd \cdot sr$
luminance	candela per square meter		cd/m^2
illumination	lux, lumen per square meter	lx, lm/m^2	$cd \cdot sr/m^2$

*Supplementary units are: plane angle, radian (rad), solid angle, steradian (sr).

Table 3. Some common units defined in terms of SI units

Quantity	Name of unit	Unit symbol	Definition of unit
length	inch	in.	2.54×10^{-2} m
mass	pound (avoirdupois)	lb	0.45359237 kg
force	kilogram-force	kgf	9.80665 N
pressure	atmosphere	atm	101325 N·m^{-2}
pressure	torr	Torr	$(101325/760)$ N·m^{-2}
pressure	conventional millimeter of mercury*	mmHg	13.5951×980.665 $\times 10^{-2}$ N·m^{-2}
energy	kilowatt-hour	kWh	3.6×10^6 J
energy	thermochemical calorie	cal	4.184 J
energy	international steam table calorie	cal$_{IT}$	4.1868 J
thermodynamic temperature (T)	degree Rankine	°R	$(5/9)$ K
customary temperature (t)	degree Celsius	°C	$t(°C) = T(K) - 273.16$
customary temperature (t)	degree Fahrenheit	°F	$t(°F) = T(°R) - 459.68$
radioactivity	curie	Ci	3.7×10^{10} s^{-1}
energy†	electron volt	eV	$eV \approx 1.60219 \times 10^{-19}$ J
mass†	unified atomic mass unit	u	$u \approx 1.66057 \times 10^{-27}$ kg

*The conventional millimeter of mercury, symbol mmHg (not mm Hg), is the pressure exerted by a column exactly 1 mm high of a fluid of density exactly 13.5951 g·cm^{-3} in a place where the gravitational acceleration is exactly 980.665 cm·s^{-2}. The mmHg differs from the Torr by less than 2×10^{-7} Torr.
†These units defined in terms of the best available experimental values of certain physical constants may be converted to SI units. The factors for conversion of these units are subject to change in the light of new experimental measurements of the constants involved.

Table 4. Prefixes for units in the International System

Prefix	Symbol	Power	Example
tera	T	10^{12}	
giga	G	10^{9}	
mega	M	10^{6}	megahertz (MHz)
kilo	k	10^{3}	kilometer (km)
hecto	h	10^{2}	
deca	da	10^{1}	
deci	d	10^{-1}	
centi	c	10^{-2}	
milli	m	10^{-3}	milligram (mg)
micro	μ	10^{-6}	microgram (μg)
nano	n	10^{-9}	nanosecond (ns)
pico	p	10^{-12}	picofarad (pf)
femto	f	10^{-15}	
atto	a	10^{-18}	

Conversion factors for the measurement systems

Because it will take some years for all scientists and engineers to convert to the SI, this dictionary has often retained the U.S. Customary units, followed by parenthetical SI equivalents. Conversion factors are given in **Table 5** for some prevalent units; in each of the subtables the user proceeds as follows:

To convert a quantity expressed in a unit in the left-hand column to the equivalent in a unit in the top row of a subtable, multiply the quantity by the factor common to both units.

The factors have been carried out to seven significant figures, as derived from the fundamental constants and the definitions of the units. However, this does not mean that the factors are always known to that accuracy. Numbers followed by ellipses are to be continued indefinitely with repetition of the same pattern of digits. Factors written with fewer than seven significant digits are exact values. Numbers followed by an asterisk are definitions of the relation between the two units. In "G. UNITS OF ENERGY," the electrical units are those in terms of which certification of standard cells, standard resistances, and so forth, is made by the National Bureau of Standards; unless otherwise indicated, all electrical units are absolute.

Table 5. Conversion factors for the U.S. Customary System, metric system, and International System

A. UNITS OF LENGTH

Units	cm	m	in.	ft	yd	mile
1 cm =	1	0.01*	0.3937008	0.03280840	0.01093613	6.213712×10^{-6}
1 m =	100.	1	39.37008	3.280840	1.093613	6.213712×10^{-4}
1 in. =	2.54*	0.0254	1	0.08333333...	0.02777777...	1.578283×10^{-5}
1 ft =	30.48	0.3048	12.*	1	0.3333333...	$1.893939 \times ...10^{-4}$
1 yd =	91.44	0.9144	36.	3.*	1	$5.681818 \times ...10^{-4}$
1 mile =	1.609344×10^{5}	1.609344×10^{3}	6.336×10^{4}	5280.*	1760.	1

B. UNITS OF AREA

Units	cm²	m²	in.²	ft²	yd²	mile²
1 cm² =	1	10^{-4}*	0.1550003	1.076391×10^{-3}	1.195990×10^{-4}	3.861022×10^{-11}
1 m² =	10^{4}	1	1550.003	10.76391	1.195990	3.861022×10^{-7}
1 in.² =	6.4516*	6.4516×10^{-4}	1	$6.944444 \times 10^{-3}...$	7.716049×10^{-4}	2.490977×10^{-10}
1 ft² =	929.0304	0.09290304	144.*	1	0.1111111...	3.587007×10^{-8}
1 yd² =	8361.273	0.8361273	1296.	9.*	1	3.228306×10^{-7}
1 mile² =	2.589988×10^{10}	2.589988×10^{6}	4.014490×10^{9}	2.78784×10^{7}*	3.0976×10^{6}	1

Table 5. Conversion factors for the U.S. Customary System, metric system, and International System (cont.)

C. UNITS OF VOLUME

Units	cm³	liter	in.³	ft³	qt	gal
1 cm³	= 1	10^{-3}	0.06102374	3.531467×10^{-5}	1.056688×10^{-3}	2.641721×10^{-4}
1 liter	= 1000.*	1	61.02374	0.03531467	1.056688	0.2641721
1 in.³	= 16.38706*	0.01638706	1	5.787037×10^{-4}	0.01731602	4.329004×10^{-3}
1 ft³	= 28316.85	28.31685	1728.*	1	2.992208	7.480520
1 qt	= 946.353	0.946353	57.75	0.0342014	1	0.25
1 gal (U.S.)	= 3785.412	3.785412	231.*	0.1336806	4.*	1

D. UNITS OF MASS

Units	g	kg	oz	lb	metric ton	ton
1 g	= 1	10^{-3}	0.03527396	2.204623×10^{-3}	10^{-6}	1.102311×10^{-6}
1 kg	= 1000.	1	35.27396	2.204623	10^{-3}	1.102311×10^{-3}
1 oz (avdp)	= 28.34952	0.02834952	1	0.0625	2.834952×10^{-5}	$5. \times 10^{-4}$
1 lb (avdp)	= 453.5924	0.4535924	16.*	1	4.535924×10^{-4}	0.0005
1 metric ton	= 10^{6}	1000.*	35273.96	2204.623	1	1.102311
1 ton	= 907184.7	907.1847	32000.	2000.*	0.9071847	1

continued

Table 5. Conversion factors for the U.S. Customary System, metric system, and International System (cont.)

E. UNITS OF DENSITY

Units	g cm^{-3}	g l.$^{-1}$	oz in.$^{-3}$	lb in.$^{-3}$	lb ft^{-3}	lb gal^{-1}
1 g cm^{-3} = 1	1000.	0.5780365	0.03612728	62.42795	8.345403
1 g l.$^{-1}$ = 10^{-3}	1	5.780365 × 10^{-4}	3.612728 × 10^{-5}	0.06242795	8.345403 × 10^{-3}
1 oz in.$^{-3}$ = 1.729994	1729.994	1	0.0625	108.	14.4375
1 lb in.$^{-3}$ = 27.67991	27679.91	16.	1	1728.	231.
1 lb ft^{-3} = 0.01601847	16.01847	9.259259 × 10^{-3}	5.7870370 × 10^{-4}	1	0.1336806
1 lb gal^{-1} = 0.1198264	119.8264	4.749536 × 10^{-3}	4.3290043 × 10^{-3}	7.480519	1

F. UNITS OF PRESSURE

Units	dyn cm^{-2}	bar	atm	kg (wt) cm^{-2}	mmHg (Torr)	in. Hg	lb (wt) in.$^{-2}$
1 dyn cm^{-2} = 1	10^{-6}	9.869233 × 10^{-7}	1.019716 × 10^{-6}	7.500617 × 10^{-4}	2.952999 × 10^{-5}	1.450377 × 10^{-5}
1 bar = 10^{6}*	1	0.9869233	1.019716	750.0617	29.52999	14.50377
1 atm = 1013250.*	1.013250	1	1.033227	760.	29.92126	14.69595
1 kg (wt) cm^{-2} = 980665.	0.980665	0.9678411	1	735.5592	28.95903	14.22334
1 mmHg (Torr) = 1333.224	1.333224 × 10^{-3}	1.3157895 × 10^{-3}	1.3595099 × 10^{-3}	1	0.03937008	0.01933678
1 in. Hg = 33863.88	0.03386388	0.03342105	0.03453155	25.4	1	0.4911541
1 lb (wt) in.$^{-2}$ = 68947.57	0.06894757	0.06804596	0.07030696	51.71493	2.036021	1

Table 5. Conversion factors for the U.S. Customary System, metric system, and International System (cont.)

G. UNITS OF ENERGY

Units	g mass (energy equiv)	J	int J	cal	cal$_{IT}$	Btu$_{IT}$	kW hr	hp hr	ft-lb (wt)	cu ft-lb (wt) in.$^{-2}$	l.-atm
1 g mass (energy equiv) =	1	8.987554×10^{13}	8.986071×10^{13}	2.148077×10^{13}	2.146640×10^{13}	8.518558×10^{10}	2.496543×10^{7}	3.347919×10^{7}	6.628880×10^{13}	4.603399×10^{11}	8.870026×10^{11}
1 J =	1.112650×10^{-14}	1	0.999835	0.2390057	0.2388459	9.478172×10^{-4}	$2.777777... \times 10^{-7}$	3.725062×10^{-7}	0.7375622	5.121960×10^{-3}	9.869233×10^{-3}
1 int J =	1.112833×10^{-14}	1.000165	1	0.2390452	0.2388853	9.479735×10^{-4}	2.778236×10^{-7}	3.725676×10^{-7}	0.7376839	5.122805×10^{-3}	9.870862×10^{-3}
1 cal =	4.655327×10^{-14}	4.184*	4.183310	1	0.9993312	3.965667×10^{-3}	$1.162222... \times 10^{-6}$	1.558562×10^{-6}	3.085960	2.143028×10^{-2}	0.04129287
1 cal$_{IT}$ =	4.658442×10^{-14}	4.1868*	4.186109	1.000669	1	3.968321×10^{-3}	1.163000×10^{-6}	1.559609×10^{-6}	3.088025	2.144462×10^{-2}	0.04132050
1 Btu$_{IT}$ =	1.173908×10^{-11}	1055.056	1054.882	252.1644	251.9958*	1	2.930711×10^{-4}	3.930148×10^{-4}	778.1693	5.403953	10.41259
1 kW hr =	4.005539×10^{-8}	3600000.*	3599406.	860420.7	859845.2	3412.142	1	1.341022	2655224.	18439.06	35529.24
1 hp hr =	2.986930×10^{-8}	2684519.	2684077.	641615.6	641186.5	2544.33	0.7456998	1	1980000.*	13750.	26494.15
1 ft-lb (wt) =	1.508550×10^{-14}	1.355818	1.355594	0.3240483	0.3233315	1.285067×10^{-3}	3.766161×10^{-7}	$5.050505... \times 10^{-7}$	1	$6.944444... \times 10^{-3}$	0.01338088
1 cu ft-lb (wt) in.$^{-2}$ =	2.172313×10^{-12}	195.2378	195.2056	46.66295	46.63174	0.1850497	5.423272×10^{-5}	$7.272727... \times 10^{-5}$	144.*	1	1.926847
1 l.-atm =	1.127392×10^{-12}	101.3250	101.3083	24.21726	24.20106	0.09603757	2.814583×10^{-5}	3.774419×10^{-5}	74.73349	0.5189825	1

Symbols and atomic numbers for the chemical elements*

Name	Symbol	At. no.	Name	Symbol	At. no.	Name	Symbol	At. no.	Name	Symbol	At. no.
Actinium	Ac	89	Erbium	Er	68	Mercury	Hg	80	Samarium	Sm	62
Aluminum	Al	13	Europium	Eu	63	Molybdenum	Mo	42	Scandium	Sc	21
Americium	Am	95	Fermium	Fm	100	Neodymium	Nd	60	Selenium	Se	34
Antimony	Sb	51	Fluorine	F	9	Neon	Ne	10	Silicon	Si	14
Argon	Ar	18	Francium	Fr	87	Neptunium	Np	93	Silver	Ag	47
Arsenic	As	33	Gadolinium	Gd	64	Nickel	Ni	28	Sodium	Na	11
Astatine	At	85	Gallium	Ga	31	Niobium	Nb	41	Strontium	Sr	38
Barium	Ba	56	Germanium	Ge	32	Nitrogen	N	7	Sulfur	S	16
Berkelium	Bk	97	Gold	Au	79	Nobelium	No	102	Tantalum	Ta	73
Beryllium	Be	4	Hafnium	Hf	72	Osmium	Os	76	Technetium	Tc	43
Bismuth	Bi	83	Helium	He	2	Oxygen	O	8	Tellurium	Te	52
Boron	B	5	Holmium	Ho	67	Palladium	Pd	46	Terbium	Tb	65
Bromine	Br	35	Hydrogen	H	1	Phosphorus	P	15	Thallium	Tl	81
Cadmium	Cd	48	Indium	In	49	Platinum	Pt	78	Thorium	Th	90
Calcium	Ca	20	Iodine	I	53	Plutonium	Pu	94	Thulium	Tm	69
Californium	Cf	98	Iridium	Ir	77	Polonium	Po	84	Tin	Sn	50
Carbon	C	6	Iron	Fe	26	Potassium	K	19	Titanium	Ti	22
Cerium	Ce	58	Krypton	Kr	36	Praseodymium	Pr	59	Tungsten	W	74
Cesium	Cs	55	Lanthanum	La	57	Promethium	Pm	61	Uranium	U	92
Chlorine	Cl	17	Lawrencium	Lr (Lw)	103	Protactinium	Pa	91	Vanadium	V	23
Chromium	Cr	24	Lead	Pb	82	Radium	Ra	88	Xenon	Xe	54
Cobalt	Co	27	Lithium	Li	3	Radon	Rn	86	Ytterbium	Yb	70
Copper	Cu	29	Lutetium	Lu	71	Rhenium	Re	75	Yttrium	Y	39
Curium	Cm	96	Magnesium	Mg	12	Rhodium	Rh	45	Zinc	Zn	30
Dysprosium	Dy	66	Manganese	Mn	25	Rubidium	Rb	37	Zirconium	Zr	40
Einsteinium	Es	99	Mendelevium	Md	101	Ruthenium	Ru	44			

*Elements 104, 105, and 106 have been reported, but no official names or symbols have yet been assigned.

Fundamental constants

Compiled by E. R. Cohen and B. N. Taylor under the auspices of the CODATA Task Group on Fundamental Constants. This set has been officially adopted by CODATA and is taken from J. Phys. Chem. Ref. Data, Vol. 2, No. 4, p. 663 (1973) and CODATA Bulletin No. 11 (December 1973).

Quantity	Symbol	Numerical Value [*]	Uncert. (ppm)	SI [†] ← Units →	cgs [‡]
Speed of light in vacuum	c	299792458(1.2)	0.004	$m \cdot s^{-1}$	$10^2\ cm \cdot s^{-1}$
Permeability of vacuum	μ_0	4π		$10^{-7}\ H \cdot m^{-1}$	
		$=12.5663706144$		$10^{-7}\ H \cdot m^{-1}$	
Permittivity of vacuum, $1/\mu_0 c^2$	ϵ_0	8.854187818(71)	0.008	$10^{-12}\ F \cdot m^{-1}$	
Fine-structure constant, $[\mu_0 c^2/4\pi](e^2\hbar c)$	α	7.2973506(60)	0.82	10^{-3}	10^{-3}
	α^{-1}	137.03604(11)	0.82		
Elementary charge	e	1.6021892(46)	2.9	$10^{-19}\ C$	$10^{-20}\ emu$
		4.803242(14)	2.9		$10^{-10}\ esu$
Planck constant	h	6.626176(36)	5.4	$10^{-34}\ J \cdot s$	$10^{-27}\ erg \cdot s$
	$\hbar = h/2\pi$	1.0545887(57)	5.4	$10^{-34}\ J \cdot s$	$10^{-27}\ erg \cdot s$
Avogadro constant	N_A	6.022045(31)	5.1	$10^{23}\ mol^{-1}$	$10^{23}\ mol^{-1}$
Atomic mass unit, $10^{-3} kg \cdot mol^{-1} N_A^{-1}$	u	1.6605655(86)	5.1	$10^{-27}\ kg$	$10^{-24}\ g$
Electron rest mass	m_e	9.109534(47)	5.1	$10^{-31}\ kg$	$10^{-28}\ g$
		5.4858026(21)	0.38	$10^{-4}\ u$	$10^{-4}\ u$
Proton rest mass	m_p	1.6726485(86)	5.1	$10^{-27}\ kg$	$10^{-24}\ g$
		1.007276470(11)	0.011	u	u
Ratio of proton mass to electron mass	m_p/m_e	1836.15152(70)	0.38		
Neutron rest mass	m_n	1.6749543(86)	5.1	$10^{-27}\ kg$	$10^{-24}\ g$
		1.008665012(37)	0.037	u	u
Electron charge to mass ratio	e/m_e	1.7588047(49)	2.8	$10^{11}\ C \cdot kg^{-1}$	$10^7\ emu \cdot g^{-1}$
		5.272764(15)	2.8		$10^{17}\ esu \cdot g^{-1}$
Magnetic flux quantum, $[c]^{-1}(hc/2e)$	Φ_0	2.0678506(54)	2.6	$10^{-15}\ Wb$	$10^{-7}\ G \cdot cm^2$
	h/e	4.135701(11)	2.6	$10^{-15}\ J \cdot s \cdot C^{-1}$	$10^{-7}\ erg \cdot s \cdot emu^{-1}$
		1.3795215(36)	2.6		$10^{-17}\ erg \cdot s \cdot esu^{-1}$
Josephson frequency-voltage ratio	$2e/h$	4.835939(13)	2.6	$10^{14}\ Hz \cdot V^{-1}$	
Quantum of circulation	$h/2m_e$	3.6369455(60)	1.6	$10^{-4}\ J \cdot s \cdot kg^{-1}$	$erg \cdot s \cdot g^{-1}$
	h/m_e	7.273891(12)	1.6	$10^{-4}\ J \cdot s \cdot kg^{-1}$	$erg \cdot s \cdot g^{-1}$
Faraday constant, $N_A e$	F	9.648456(27)	2.8	$10^4\ C \cdot mol^{-1}$	$10^3\ emu \cdot mol^{-1}$
		2.8925342(82)	2.8		$10^{14}\ esu \cdot mol^{-1}$
Rydberg constant, $[\mu_0 c^2/4\pi]^2(m_e e^4/4\pi\hbar^3 c)$	R_∞	1.097373177(83)	0.075	$10^7\ m^{-1}$	$10^5\ cm^{-1}$
Bohr radius, $[\mu_0 c^2/4\pi]^{-1}(\hbar^2/m_e e^2)=\alpha/4\pi R_\infty$	a_0	5.2917706(44)	0.82	$10^{-11}\ m$	$10^{-9}\ cm$
Classical electron radius, $[\mu_0 c^2/4\pi](e^2/m_e c^2)=\alpha^3/4\pi R_\infty$	$r_e=\alpha \lambdabar_C$	2.8179380(70)	2.5	$10^{-15}\ m$	$10^{-13}\ cm$
Thomson cross section, $(8/3)\pi r_e^2$	σ_e	0.6652448(33)	4.9	$10^{-28}\ m^2$	$10^{-24}\ cm^2$
Free electron g-factor, or electron magnetic moment in Bohr magnetons	$g_e/2=\mu_e/\mu_B$	1.0011596567(35)	0.0035		
Free muon g-factor, or muon magnetic moment in units of $[c](e\hbar/2m_\mu c)$	$g_\mu/2$	1.00116616(31)	0.31		
Bohr magneton, $[c](e\hbar/2m_e c)$	μ_B	9.274078(36)	3.9	$10^{-24}\ J \cdot T^{-1}$	$10^{-21}\ erg \cdot G^{-1}$
Electron magnetic moment	μ_e	9.284832(36)	3.9	$10^{-24}\ J \cdot T^{-1}$	$10^{-21}\ erg \cdot G^{-1}$

*For footnotes, see page A15.

continued

Fundamental constants (cont.)

Quantity	Symbol	Numerical Value *	Uncert. (ppm)	SI †	← Units →	cgs ‡
Gyromagnetic ratio of protons in H_2O	γ'_p	2.6751301(75)	2.8	10^8 $s^{-1} \cdot T^{-1}$		10^4 $s^{-1} \cdot G^{-1}$
	$\gamma'_p / 2\pi$	4.257602(12)	2.8	10^7 $Hz \cdot T^{-1}$		10^3 $Hz \cdot G^{-1}$
γ'_p corrected for diamagnetism of H_2O	γ_p	2.6751987(75)	2.8	10^8 $s^{-1} \cdot T^{-1}$		10^4 $s^{-1} \cdot G^{-1}$
	$\gamma_p / 2\pi$	4.257711(12)	2.8	10^7 $Hz \cdot T^{-1}$		10^3 $Hz \cdot G^{-1}$
Magnetic moment of protons in H_2O in Bohr magnetons	μ'_p / μ_B	1.52099322(10)	0.066	10^{-3}		10^{-3}
Proton magnetic moment in Bohr magnetons	μ_p / μ_B	1.521032209(16)	0.011	10^{-3}		10^{-3}
Ratio of electron and proton magnetic moments	μ_e / μ_p	658.2106880(66)	0.010			
Proton magnetic moment	μ_p	1.4106171(55)	3.9	10^{-26} $J \cdot T^{-1}$		10^{-23} $erg \cdot G^{-1}$
Magnetic moment of protons in H_2O in nuclear magnetons	μ'_p / μ_N	2.7927740(11)	0.38			
μ'_p / μ_N corrected for diamagnetism of H_2O	μ_p / μ_N	2.7928456(11)	0.38			
Nuclear magneton, [c]$(e\hbar/2m_pc)$	μ_N	5.050824(20)	3.9	10^{-27} $J \cdot T^{-1}$		10^{-24} $erg \cdot G^{-1}$
Ratio of muon and proton magnetic moments	μ_μ / μ_p	3.1833402(72)	2.3			
Muon magnetic moment	μ_μ	4.490474(18)	3.9	10^{-26} $J \cdot T^{-1}$		10^{-23} $erg \cdot G^{-1}$
Ratio of muon mass to electron mass	m_μ / m_e	206.76865(47)	2.3			
Muon rest mass	m_μ	1.883566(11)	5.6	10^{-28} kg		10^{-25} g
		0.11342920(26)	2.3	u		u
Compton wavelength of the electron, $h/m_ec = \alpha^2/2R_\infty$	λ_C	2.4263089(40)	1.6	10^{-12} m		10^{-10} cm
	$\lambdabar_C = \lambda_C/2\pi = \alpha a_0$	3.8615905(64)	1.6	10^{-13} m		10^{-11} cm
Compton wavelength of the proton, h/m_pc	$\lambda_{C,p}$	1.3214099(22)	1.7	10^{-15} m		10^{-13} cm
	$\lambdabar_{C,p} = \lambda_{C,p}/2\pi$	2.1030892(36)	1.7	10^{-16} m		10^{-14} cm
Compton wavelength of the neutron, h/m_nc	$\lambda_{C,n}$	1.3195909(22)	1.7	10^{-15} m		10^{-13} cm
	$\lambdabar_{C,n} = \lambda_{C,n}/2\pi$	2.1001941(35)	1.7	10^{-16} m		10^{-14} cm
Molar volume of ideal gas at s.t.p.	V_m	22.41383(70)	31	10^{-3} $m^3 \cdot mol^{-1}$		10^3 $cm^3 \cdot mol^{-1}$
Molar gas constant, V_mp_0/T_0 ($T_0 \equiv 273.15$ K; $p_0 \equiv 101325$ Pa $\equiv 1$atm)	R	8.31441(26)	31	$J \cdot mol^{-1} \cdot K^{-1}$		10^7 $erg \cdot mol^{-1} \cdot K^{-1}$
		8.20568(26)	31	10^{-5} $m^3 \cdot atm \cdot mol^{-1} \cdot K^{-1}$		10 $cm^3 \cdot atm \cdot mol^{-1} \cdot K^{-1}$
Boltzmann constant, R/N_A	k	1.380662(44)	32	10^{-23} $J \cdot K^{-1}$		10^{-16} $erg \cdot K^{-1}$
Stefan-Boltzmann constant, $\pi^2 k^4 / 60\hbar^3 c^2$	σ	5.67032(71)	125	10^{-8} $W \cdot m^{-2} \cdot K^{-4}$		10^{-5} $erg \cdot s^{-1} \cdot cm^{-2} \cdot K^{-4}$
First radiation constant, $2\pi hc^2$	c_1	3.741832(20)	5.4	10^{-16} $W \cdot m^2$		10^{-5} $erg \cdot cm^2 \cdot s^{-1}$
Second radiation constant, hc/k	c_2	1.438786(45)	31	10^{-2} $m \cdot K$		$cm \cdot K$
Gravitational constant	G	6.6720(41)	615	10^{-11} $m^3 \cdot s^{-2} \cdot kg^{-1}$		10^{-8} $cm^3 \cdot s^{-2} \cdot g^{-1}$
Ratio, kx-unit to ångström, $\Lambda = \lambda(\text{Å})/\lambda(\text{kxu})$; $\lambda(\text{Cu}K\alpha_1) \equiv 1.537400$ kxu	Λ	1.0020772(54)	5.3			
Ratio, Å* to ångström, $\Lambda^* = \lambda(\text{Å})/\lambda(\text{Å}^*)$; $\lambda(\text{W}K\alpha_1) \equiv 0.2090100$ Å*	Λ^*	1.0000205(56)	5.6			

*For footnotes, see page A15.

Fundamental constants (cont.): Energy conversion factors and equivalents

Quantity	Symbol	Numerical Value *	Units	Uncert. (ppm)
1 kilogram ($kg \cdot c^2$)		8.987551786(72)	10^{16} J	0.008
		5.609545(16)	10^{29} MeV	2.9
1 Atomic mass unit ($u \cdot c^2$)		1.4924418(77)	10^{-10} J	5.1
		931.5016(26)	MeV	2.8
1 Electron mass $m_{\nu} \cdot c^2$)		8.187241(42)	10^{-14} J	5.1
		0.5110034(14)	MeV	2.8
1 Muon mass ($m_{\mu} \cdot c^2$)		1.6928648(96)	10^{-11} J	5.6
		105.65948(35)	MeV	3.3
1 Proton mass ($m_{p} \cdot c^2$)		1.5033015(77)	10^{-10} J	5.1
		938.2796(27)	MeV	2.8
1 Neutron mass ($m_{n} \cdot c^2$)		1.5053738(78)	10^{-10} J	5.1
		939.5731(27)	MeV	2.8
1 Electron volt		1.6021892(46)	10^{-19} J	2.9
			10^{-12} erg	2.9
	1 eV/h	2.4179696(63)	10^{14} Hz	2.6
	1 eV/hc	8.065479(21)	10^{5} m^{-1}	2.6
			10^{3} cm^{-1}	2.6
	1 eV/k	1.160450(36)	10^{4} K	31
Voltage-wavelength conversion, hc		1.986478(11)	10^{-25} J$\cdot$m	5.4
		1.2398520(32)	10^{-6} eV$\cdot$m	2.6
			10^{-4} eV$\cdot$cm	2.6
Rydberg constant	$R_x hc$	2.179907(12)	10^{-18} J	5.4
			10^{-11} erg	5.4
		13.605804(36)	eV	2.6
	$R_x c$	3.28984200(25)	10^{15} Hz	0.075
	$R_x hc/k$	1.578885(49)	10^{5} K	31
Bohr magneton	μ_B	9.274078(36)	10^{-24} J$\cdot$T^{-1}	3.9
		5.7883785(95)	10^{-5} eV$\cdot$T^{-1}	1.6
	μ_B/h	1.3996123(39)	10^{10} Hz$\cdot$T^{-1}	2.8
	μ_B/hc	46.68604(13)	m$^{-1} \cdot$T^{-1}	2.8
			10^{-2} cm$^{-1} \cdot$T^{-1}	2.8
	μ_B/k	0.671712(21)	K$\cdot$T^{-1}	31
Nuclear magneton	μ_N	5.505824(20)	10^{-27} J$\cdot$T^{-1}	3.9
		3.1524515(53)	10^{-8} eV$\cdot$T^{-1}	1.7
	μ_N/h	7.622532(22)	10^{6} Hz$\cdot$T^{-1}	2.8
	μ_N/hc	2.5426030(72)	10^{-2} m$^{-1} \cdot$T^{-1}	2.8
			10^{-4} cm$^{-1} \cdot$T^{-1}	2.8
	μ_N/k	3.65826(12)	10^{-4} K$\cdot$T^{-1}	31

* Note that the numbers in parentheses are the one standard-deviation uncertainties in the last digits of the quoted value computed on the basis of internal consistency, that the unified atomic mass scale $^{12}C \triangleq 12$ has been used throughout, that u=atomic mass unit, C=coulomb, F=farad, G=gauss, H=henry, Hz=hertz=cycle/s, J=joule, K=kelvin (degree Kelvin), Pa=pascal=N$\cdot$m^{-2}, T=tesla (10^4 G), V=volt, Wb=weber= T$\cdot$m^2, and W=watt. In cases where formulas for constants are given (e.g., R_x), the relations are written as the product of two factors. The first factor, in parentheses, is the expression to be used when all quantities are expressed in cgs units, with the electron charge in electrostatic units. The first factor, in brackets, is to be included only if all quantities are expressed in SI units. We remind the reader that with the exception of the auxiliary constants which have been taken to be exact, the uncertainties of these constants are correlated, and therefore the general law of error propagation must be used in calculating additional quantities requiring two or more of these constants.

† Quantities given in u and atm are for the convenience of the reader; these units are not part of the International System of Units (SI).

‡ In order to avoid separate columns for "electromagnetic" and "electrostatic" units, both are given under the single heading "cgs Units." When using these units, the elementary charge e in the second column should be understood to be replaced by e_m or e_s, respectively.

Spectrum of activity of antibiotics and other antimicrobial agents

			Diplococcus pneumoniae	Streptococcus hemolyticus	Streptococcus viridans	Staphylococcus aureus	Corynebacterium diphtheriae	Clostridium tetani
	Infecting microorganism							
			Disease produced					
Generic name	**Trade name**	**Produced by**	Pneumonia Otitis media etc.	Scarlet fever Rheumatic fever Erysipelas Tonsillitis Septic sore throat, etc.	Subacute bacterial endocarditis	Pneumonia Osteomyelitis Furuncles Carbuncles, etc.	Diphtheria	Tetanus
Amphotericin	Fungizone	*Streptomyces nodosus*						
Bacitracin		*Bacillus subtilis* *Bacillus licheniformis*	+	+	+	+	+	+
Cephalothin	Keflin	*Cephalosporium* sp. and chemical modification	⊕	⊕	⊕	⊕	+	
Chloramphenicol	Chloromycetin	*Streptomyces venezuelae*	+	+	+	⊕	+	+
Cycloserine	Seromycin	*Streptomyces orchidaceus*						
Cycloheximide	Actidione	*Streptomyces griseus* *Streptomyces noursei*						
Erythromycin	Ilotycin, Erythrocin	*Streptomyces erythreus*	⊕	⊕	⊕	⊕	+	+
Furazolidone	Furoxone	Chemical synthesis						
Gentamicin	Garamycin	*Micromonospora purpurea*	+	+		⊕		
Griseofulvin	Fulvicin, Grifulvin, Grisactin	*Penicillium griseofulvum*						
Isonicotinic acid hydrazide	Cotinazin, Rimifon, Nydrazid	Chemical synthesis						
Kanamycin	Kantrex	*Streptomyces kanamyceticus*				+		
Lincomycin	Lincocin	*Streptomyces lincolnensis*	⊕	⊕	⊕	⊕	+	+
Methenamine mandelate	Mandelamine	Chemical synthesis						
Neomycin	Mycifradin	*Streptomyces fradiae* and other *Streptomyces* species	+	+		+	+	
Novobiocin	Albamycin, Cathomycin	*Streptomyces niveus* *Streptomyces spheroides*	+		+	⊕		
Nystatin	Mycostatin	*Streptomyces noursei*						
Nitrofurantoin	Furadantin	Chemical synthesis		+		+		
Oleandomycin	Romicil, Matromycin	*Streptomyces antibioticus*	+	+		+		
Triacetyloleandomycin	Cyclamycin, TAO	*Streptomyces antibioticus*	+	+		⊕		
Polymyxin B	Aerosporin	*Bacillus polymyxa*						
Penicillin		*Penicillium notatum* *Penicillium chrysogenum*	⊕	⊕	⊕	⊕	⊕	⊕
Ampicillin	Polycillin	*Penicillium chrysogenum* and chemical modification	⊕	⊕	⊕	⊕	⊕	⊕
Puromycin	Stylomycin	*Streptomyces albo-niger*						
p-Aminosalicylic acid		Chemical synthesis						
Ristocetin	Spontin	*Nocardia lurida*	+	+				+
Streptomycin		*Streptomyces griseus* *Streptomyces bikiniensis* *Streptomyces mashuemsis*	+	+	⊕	+	+	

+ Signifies test-tube activity. ⊕ Signifies clinical usefulness.

Spectrum of activity of antibiotics and other antimicrobial agents (cont.)

Disease produced

Bacillus anthracis	Mycobacterium tuberculosis	Hemophilus species	Neisseria species	Brucella species	Escherichia coli	Aerobacter aerogenes	Klebsiella pneumoniae	Salmonella species	Shigella species	Proteus species	Pseudomonas species	Pasteurella species	Erysipelothrix rhusiopathiae	Leptospira icterohemorrhagiae	Treponema species	Vibrio species	Fungi	Endamoeba histolytica	Actinomyces species
Anthrax	Tuberculosis	Meningitis / Whooping cough	Gonorrhea / Meningitis	Brucellosis (undulant fever)	Urinary tract infections, etc.	Urinary tract infections, etc.	Pneumonia	Typhoid fever / Enteritis, etc.	Dysentery	Urinary tract infections, etc.	Urinary tract infections / Otitis media, etc.	Tularemia / Plague		Leptospirosis	Syphilis / Yaws		Coccidiomycosis / Histoplasmosis etc.	Amebiasis	Actinomycosis
		+	+														⊕		
		+	⊕		⊕	⊕	⊕	⊕	+	⊕								+	
+		⊕	+	⊕	⊕	⊕	⊕	⊕	⊕	⊕	⊕	⊕					+		+
	⊕																		
																	+		
⊕		+	+	+								+		+			+		
	+	+			+	+	+	+		+	⊕								
	⊕																		
	+				+	+					⊕								
					⊕	+					+								
	+	+			+	+		+	+	+		+							
																	⊕		
					+	+					⊕								
		+	+	+	+	+	+	+	+	+		⊕							
⊕			⊕								+			+	+		⊕	+	⊕
⊕		⊕	⊕		⊕	⊕	+	⊕	+										
																		+	
	⊕																		
	+																		+
+	⊕	⊕	+	⊕	⊕	⊕	⊕	+	+	+	+	+					+		

continued

Spectrum of activity of antibiotics and other antimicrobial agents (cont.)

Generic name	Trade name	Produced by	*Diplococcus pneumoniae* Pneumonia Otitis media etc.	*Streptococcus hemolyticus* Scarlet fever Rheumatic fever Erysipelas Tonsillitis Septic sore throat, etc.	*Streptococcus viridans* Subacute bacterial endocarditis	*Staphylococcus aureus* Pneumonia Osteomyelitis Furuncles Carbuncles, etc.	*Corynebacterium diphtheriae* Diphtheria	*Clostridium tetani* Tetanus
Dihydrostreptomycin		See streptomycin	+	+	+	+	+	
Spiramycin	Rovamycin	*Streptomyces ambofaciens*	+	+		+	+	+
Subtilin		*Bacillus subtilis*	+	+				
Sulfa drugs		Chemical synthesis	+	+		+		
Tyrothricin		*Bacillus brevis*	+	+		+	+	+
Tetracycline	Tetracyn, Polycycline, Achromycin, Steclin, Panmycin	*Streptomyces* species	⊕	⊕	+	⊕	⊕	⊕
Chlortetracycline	Aureomycin	*Streptomyces aureofaciens*	⊕	⊕	+	⊕	⊕	⊕
Oxytetracycline	Terramycin	*Streptomyces rimosus*	⊕	⊕	+	⊕	⊕	⊕
Vancomycin	Vanococin	*Streptomyces orientalis*	+	+		+		+
Viomycin	Viocin, Vinactane	*Streptomyces floridae Streptomyces puniceus*						

+ Signifies test-tube activity.   ⊕ Signifies clinical usefulness.

Spectrum of activity of antibiotics and other antimicrobial agents (cont.)

Disease produced

Bacillus anthracis / Anthrax	Mycobacterium tuberculosis / Tuberculosis	Hemophilus species / Meningitis Whooping cough	Neisseria species / Gonorrhea Meningitis	Brucella species / Brucellosis (undulant fever)	Escherichia coli / Urinary tract infections, etc.	Aerobacter aerogenes / Urinary tract infections, etc.	Klebsiella pneumoniae / Pneumonia	Salmonella species / Typhoid fever Enteritis, etc.	Shigella species / Dysentery	Proteus species / Urinary tract infections, etc.	Pseudomonas species / Urinary tract infections Otitis media, etc.	Pasteurella species / Tularemia Plague	Erysipelothrix rhusiopathiae	Leptospira icterohemorrhagiae / Leptospirosis	Treponema species / Syphilis Yaws	Vibrio species	Fungi / Coccidiomycosis Histoplasmosis etc.	Endamoeba histolytica / Amebiasis	Actinomyces species / Actinomycosis
+	⊕	+	+	+	⊕	⊕	⊕	+	+	+	+	+					+		
			+																
	+		+																
	+		+														+		
⊕	+	⊕	⊕	⊕	⊕	⊕	⊕	+	⊕	⊕	+	⊕			⊕	⊕	⊕		
⊕	+	⊕	⊕	⊕	⊕	⊕	⊕	+	⊕	⊕	+	⊕			⊕	⊕	⊕		
⊕	+	⊕	⊕	⊕	⊕	⊕	⊕	+	⊕	+	+	+			⊕	+	⊕		
	⊕																		

Table 1. Procedures and normal values for urinalysis

Procedures	Normal values
Physical characteristics	
Volume	1000–1400 ml/day
Color	Straw-amber
Transparency	Clear
Odor	Aromatic
Specific gravity	1.003–1.024
Routine chemical characteristics	
Reaction (pH)	4.5–7.5
Protein	0–50 mg/day
Sugar	None
Acetone	None
Bile	None
Microscopic examination	
Cells	
Erythrocytes	0–2/hpf*
Leukocytes	0–5/hpf
Epithelial	Variable/hpf
Casts	O-occasional Hyalin/lpf†
Crystals	Variable/hpf

*High-power field.   †Low-power field.

Table 2. Normal values in a complete blood count

Procedure	Normal values
Erythrocyte count:	
Males	4,500,000–5,500,000/mm^3
Females	4,000,000–5,000,000/mm^3
Hemoglobin:	
Males	12–17 g/100 ml
Females	11–15 g/100 ml
Hematocrit:	
Males	46 vol%
Females	42 vol%
Sedimentation rate (Wintrobe method)	
Males	4–9 mm in 1 hr
Females	4–20 mm in 1 hr
Color index	0.9–1.1
Differential leukocyte count	
Segmented neutrophilic leukocytes	50–70% (2000–7000/mm^3)
Eosinophils	1–4% (50–400/mm^3)
Basophils	0.25–0.5% (15–50/mm^3)
Lymphocytes	25–33% (1200–3000/mm^3)
Monocytes	2–6% (100–600/mm^3)

Table 3. Factors in blood coagulation

Factor	Activity
I (fibrinogen)	The protein in liquid blood which is converted into fibrin by thrombin
II (prothrombin)	The precursor of thrombin
III (thromboplastin)	A lipid derived from tissue juices or platelets after injury; its release initiates clotting
IV (calcium)	Must be available to permit the various reactions to occur
V (labile factor)	Needed to activate prothrombin (factor II) and is so named because of rapid loss on storage
VI (no longer recognized)	Once used to designate the active form of factor V
VII (stable factor)	Required to activate thromboplastin; it is the first factor inhibited by treatment with anticoagulant drugs
VIII (antihemophilic globulin, or PTF-A factor)	Lacking in the common form of hemophilia
IX (Christmas, or PTC, factor)	Named after first patient in which it was found to be absent; its absence produces a hemophilioid disease
X (Stuart-Prower factor)	Named after the first two patients in which it was found to be absent; it is inhibited by anticoagulant drugs
XI (plasma thromboplastin antecedent, or PTA)	Absence is rare and produces a hemophilioid condition
XII (Hageman factor)	Named after first known patient in which it was deficient; it helps activate factor XI, but patients with this defect are asymptomatic
XIII (fibrin-stabilizing factor)	Helps produce chemical cross-linkages in fibrin in the clot so that it does not break up prematurely

Table 4. Blood chemical constituents

Constituent	Normal values
Albumin (see Proteins)	
Amylase (see Enzymes)	
Bilirubin	
Total	0.2 – 1.0 mg/dl
Direct	0 – 0.2 mg/dl
Indirect	0.2 – 0.8 mg/dl
Copper	100 – 200 μg/dl
Enzymes	
Amylase	60 – 160 Dy Amyl-L units
Creatine phosphokinase (CPK)	0 – 50 mU/ml
Glutamic-oxaloacetic transaminase (GOT)	0 – 40 mU/ml
Glutamic pyruvic transaminase (GPT)	0 – 45 mU/ml
Leucine aminopeptidase (LAP)	70 – 200 mU/ml
Lactic dehydrogenase (LDH), total	0 – 225 mU/ml
LDH, isoenzymes	Normal progression = 2,1,3,4,5
Lipase	0.2 – 1.5 units/dl
Phosphatase	
Acid	0 – 3.5 mU/ml
Alkaline	
Adults	35 – 85 mU/ml
Children	38 – 138 mU/ml
Gamma glutamyl transpeptidase (GGT)	
Male	6 – 28 mU/ml
Female	4 – 18 mU/ml
Electrolytes	
Calcium (serum)	9 – 11.5 mg/dl, 4.5 – 5.5 meq/liter
Potassium (serum)	16 – 22 mg/dl, 4.1 – 5.6 meq/liter
Chloride (serum)	570 – 620 mg/dl, 96 – 105 meq/liter
Sodium (serum)	300 – 330 mg/dl, 135 – 152 meq/liter

Gases	*Arterial blood*	*Venous blood*
Carbon dioxide		
tension (pCO$_2$)	34 – 46 mm Hg	42 – 55 mm Hg
Oxygen saturation	97 – 100%	55 – 71%
O$_2$ tension	80 – 100 mm Hg	80 – 100 mm Hg

Constituent	Normal values
Glucose	
Absolute value	60 – 90 mg/dl
Total reducing substances	80 – 120 mg/dl
Iron	
Total	60 – 120 μg/dl
Binding capacity	100 – 300 μg/dl
Lipids	
Cholesterol	
Adults	150 – 250 mg/dl
Children	80 – 170 mg/dl
Fatty acids	190 – 420 mg/dl
Phospholipids	175 – 250 mg/dl
Total lipids	900 mg/dl
Triglycerides	37 – 134 mg/dl
Magnesium	1.5 – 2.1 meq/liter
Nonprotein nitrogen	
Total	25 – 35 mg/dl
Creatinine	1 – 2 mg/dl
Urea nitrogen	10 – 20 mg/dl
Uric acid nitrogen	
Adult males	3.5 – 7.3 mg/dl
Adult females	3.5 – 5.5 mg/dl
Children	2.1 – 3.9 mg/dl

Table 4. Blood chemical constitutents (cont.)

Constituent	Normal values
Osmolality (serum)	275−295 mOsm/liter
Phenylalanine	0−3.5 mg/dl
Proteins	
Total	6−8 g/dl
Albumin	3.5−5.5 g/dl
Globulin	1.3−3.3 g/dl

Proteins by paper electrophoresis	Relative %	Absolute g/dl
Albumin	50−60	3.3−4.5
Alpha-1 globulin	4− 8	0.2−0.4
Alpha-2 globulin	8−10	0.4−0.7
Beta globulin	12−14	0.7−1.0
Gamma globulin	12−18	0.7−1.2

Vitamins	
Ascorbic acid (vitamin C)	0.1−1.5 mg/dl
Vitamin B_{12}	330−1025 pg/ml
Vitamin A	100−250 IU
Vitamin K	Prothrombin time (12−13 sec)
Folate	5−21 µg/ml

Table 5. Normal cellular composition of bone marrow

Cell types	Distribution
Myeloblasts	0−1%
Promyelocytes	1−5%
Neutrophilic myelocytes	2−10%
Neutrophilic metamyelocytes	5−15%
Neutrophilic band (stab) forms	10−40%
Neutrophilic segmented forms	5−30%
Esoinophils	0− 3%
Basophils	0− 1%
Lymphocytes	5−15%
Monocytes	0− 2%
Plasmocytes	0− 1%
Rubriblasts	0− 1%
Prorubricytes	1− 4%
Rubricytes	5−10%
Metarubricytes	10−20%
Megakaryocytes	1−4 cells/hpf*
Platelets	Innumerable

*High-power field.

Table 6. Agglutinogens and agglutinins of the major blood groups

Blood group (agglutinogens)	Agglutinins (serum)	Incidence in population, %
AB	o	5
A	b	10
B	a	40
O	ab	50

Animal kingdom

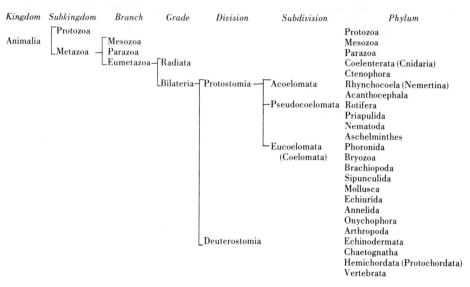

Kingdom	Subkingdom	Branch	Grade	Division	Subdivision	Phylum
Animalia	Protozoa					Protozoa
	Metazoa	Mesozoa				Mesozoa
		Parazoa				Parazoa
		Eumetazoa	Radiata			Coelenterata (Cnidaria)
						Ctenophora
			Bilateria	Protostomia	Acoelomata	Rhynchocoela (Nemertina)
						Acanthocephala
					Pseudocoelomata	Rotifera
						Priapulida
						Nematoda
						Aschelminthes
					Eucoelomata (Coelomata)	Phoronida
						Bryozoa
						Brachiopoda
						Sipunculida
						Mollusca
						Echiurida
						Annelida
						Onychophora
						Arthropoda
				Deuterostomia		Echinodermata
						Chaetognatha
						Hemichordata (Protochordata)
						Vertebrata

Animal taxonomy

Phylum Protozoa
  Subphylum Sarcomastigophora
    Superclass Mastigophora
      Class Phytamastigophorea
        Order Chrysomonadida
        Order Silicoflagellida
        Order Coccolithophorida
        Order Heterochlorida
        Order Cryptomonadida
        Order Dinoflagellida
        Order Ebriida
        Order Euglenida
        Order Chloromonadida
        Order Volvocida
          (Phytomonadida)
      Class Zoomastigophorea
        Order Choanoflagellida
        Order Bicosoecida
        Order Rhizomastigida
        Order Kinetoplastida
          (Protomastigida)
        Order Retortomonadida
        Order Diplomonadida
        Order Oxymonadida

        Order Trichomonadida
        Order Hypermastigida
    Superclass Sarcodina
      Class Rhizopodea
        Subclass Lobosia
          Order Amoebida
          Order Arcellinida
        Subclass Filosia
          Order Aconchulinida
          Order Gromiida
        Subclass Granuloreticulosia
          Order Athalamida
          Order Foraminiferida
          Order Xenophyophorida
        Subclass Mycetozoia
          Order Acrasida
          Order Eumycetozoida
          Order Plasmodiophorida
        Subclass Labyrinthulia
      Class Piroplasmea
        Actinopodea
        Subclass Radiolaria
          Order Porulosida
          Order Oculosida

Subclass Acantharia
   Order Acanthometrida
   Order Acanthophractida
Subclass Heliozoia
   Order Actinophryida
   Order Centrohelida
   Order Desmothoracida
Subclass Proteomyxidia
Subphylum Sporozoa
  Class Telosporea
  Subclass Gregarinia
   Order Archigregarinida
   Order Eugregarinida
   Order Neogregarinida
  Subclass Coccidia
   Order Protococcida
   Order Eucoccida
  Class Toxoplasmea
   Order Toxoplasmida
  Class Haplosporea
   Order Haplosporida
   Order Sarcosporida
Subphylum Cnidospora
  Class Myxosporidea
   Order Myxosporida
   Order Actinomyxida
   Order Helicosporida
  Class Microsporidea
Subphylum Ciliophora
  Class Ciliatea
  Subclass Holotrichia
   Order Gymnostomatida
   Order Trichostomatida
   Order Chonotrichida
   Order Apostomatida
   Order Astomatida
   Order Hymenostomatida
   Order Thigmotrichida
  Subclass Peritrichia
  Subclass Suctoria
   Order Suctorida
  Subclass Spirotrichia
   Order Heterotrichida
   Order Oligotrichida
   Order Tintinnida
   Order Entodiniomorphida
   Order Odontostomatida
   Order Hypotrichida
Phylum Mesozoa
   Order Dicyemida
   Order Orthonectida
Phylum Parazoa (Porifera)
  Class Hexactinellida
  Subclass Amphidiscophora
   Order Amphidiscosa
   Order Hemidiscosa
  Subclass Hexasterophora
   Order Hexactinosa
   Order Lychniscosa
   Order Lyssacinosa
   Order Reticulosa

Class Calcarea
  Subclass Calcinea
   Order Clathrinida
   Order Leucettida
  Subclass Calcaronea
   Order Leucosoleniida
   Order Sycettida
  Subclass Pharetronidia
Class Demospongiae
  Subclass Tetractinomorpha
   Order Homosclerophorida
   Order Choristida
   Order Spirophorida
   Order Hadromerida
   Order Axinellida
  Subclass Ceractinomorpha
   Order Dendroceratida
   Order Dictyoceratida
   Order Haplosclerida
   Order Poecilosclerida
   Order Halichondrida
Class Sclerospongiae
Phylum Collenterata
  Class Hydrozoa
   Order Hydroida
   Order Milleporina
   Order Stylasterina
   Order Trachylina
   Order Siphonophora
   Order Spongiomorphida
   Order Stromatoporoidea
  Class Scyphozoa
  Subclass Scyphomedusae
   Order Stauromedusae
   Order Cubomedusae
   Order Coronatae
   Order Semaeostomeae
   Order Rhizostomeae
  Subclass Conulta
   Order Conularida
  Class Anthozoa
  Subclass Alcyonaria
   Order Stolonifera
   Order Telestacea
   Order Alcyonacea
   Order Coenothecalia
   Order Gorgonacea
   Order Pennatulacea
   Order Trachypsammiacea
  Subclass Zoantharia
   Order Actiniaria
   Order Ptychodactiaria
   Order Corallimorpharia
   Order Rugosa
   Order Heterocorallia
   Order Seleractinia
   Order Zoanthidea
   Order Tabulata
  Subclass Cerianthipatharia
   Order Antipatharia
   Order Ceriantharia

Phylum Ctenophora
  Class Tentaculata
    Order Cydippidea
    Order Lobata
    Order Cestida
    Order Platyctenea
  Class Nuda
    Order Beroida
Phylum Rhynchocoela
  Class Anopla
    Order Paleonemertini
    Order Heteronemertini
  Class Enopla
    Order Hoplonemetrini
    Order Bdellomorpha
      (Bdellonemertini)
Phylum Acanthocephala
    Order Archiacantho-
      cephala
    Order Palaeacantho-
      cephala
    Order Eocanthocephala
Phylum Rotifera
    Order Seisonacea
    Order Bdelloidea
    Order Monogonta
Phylum Priapulida
Phylum Nematoda (Nemata,
      Nematoidea)
  Class Secernentea (Phasmidia,
      Phasmidea)
    Subclass Rhabditia
      Order Rhabditida
      Order Strongylida
      Order Ascaridida
      Order Tylenchida
    Subclass Spiruria
      Order Spirurida
      Order Camallanida
  Class Adenophorea (Aphasmidia,
      Aphasmidea)
    Subclass Chromadoria
      Order Chromadorida
      Order Monhysterida
    Subclass Enoplia
      Order Enoplida
      Order Dorylaimida
      Order Dioctophymatida
Phylum Aschilminthes
  Class Rotifera
  Class Gastrotricha
  Class Kinorhynca (Echinodera)
  Class Nematoda
  Class Nematomorpha
Phylum Phoronida
Phylum Bryozoa
  Subphylum Entoprocta
  Class Calyssozoa
    Order Pedicellinida
  Subphylum Ectoprocta
  Class Phylactolaemata

    Order Plumatellina
  Class Gymnolaemata
    Order Ctenostomata
    Order Cheilostomata
  Class Stenolaemata
    Order Fenestrata
    Order Cryptostomata
    Order Rhabdomesonata
    Order Trepostomata
    Order Cystoporata
    Order Cyclostomata
Phylum Brachiopoda
  Class Inarticulata
    Order Lingulida
    Order Acrotretida
    Order Obolellida
    Order Paterinida
  Class Incertae Sedis
    Order Kutorginida
  Class Articulata
    Order Orthida
    Order Strophomenida
    Order Pentamerida
    Order Rhynchonellida
    Order Spiriferida
    Order Terebratulida
Phylum Sipunculida
Phylum Mollusca
  Class Monoplacophora
  Class Amphineura
    Subclass Aplacophora
      (Solenogastres)
    Subclass Polyplacophora
      (Loricata)
  Class Scaphopoda
  Class Gastropoda
    Subclass Prosobranchia
      Order Archaeogastropoda
        (Aspidobranchia)
      Order Mesogastropoda
        (Pectinibranchia)
      Order Neogastropoda
    Subclass Opisthobranchia
      Order Nudibranchia
      Order Tectibranchia
    Subclass Pulmonata
      Order Stylommatophora
      Order Basommatophora
  Class Bivalvia (Pelecypoda)
    Subclass Protobranchia
    Subclass Taxodonta
    Subclass Anisomyaria
    Subclass Eulamellibranchia
    Subclass Septibranchia
  Class Cephalopoda
    Subclass Nautiloidea
    Subclass Ammonoidea
    Subclass Coleoidea
Phylum Echiurida
    Order Echiuroinea
    Order Xenopneusta

Order Heteromyota
Phylum Annelida
  Class Polychaeta
  Class Errantia
  Class Sedentaria
  Class Oligochaeta
  Class Hirudinea
      Order Arhynchobdellae
      Order Rhynchobdellae
  Class Archiannelida
  Class Mysostomaria
Phylum Onychophora
Phylum Arthropoda
  Subphylum Trilobitomorpha
  Subphylum Chelicerata
    Class Merostomata
      Subclass Xiphosura
        Order Aglaspida
        Order Xiphosurida
      Subclass Eurypterida
    Class Pycongonida
      Order Pantopoda
      Order Palaeonpantopoda
    Class Arachnida
      Order Scorpionida
      Order Amblypygi
      Order Uropygi
      Order Palpigradida
      Order Araneida
      Order Solpugida
      Order Pseudoscor-
        pionida
      Order Ricinuleida
      Order Phalangida
      Order Acarina
  Subphylum Mandibulata
    Class Crustacea
      Subclass Cephalocarida
      Subclass Branchiopoda
        Order Anostraca
        Order Lipostraca
        Order Notostraca
        Order Conchostraca
        Order Cladocera
      Subclass Ostracoda
        Order Myodocopa
        Order Cladocopa
        Order Podocopa
        Order Platycopa
      Subclass Mystacocarida
      Subclass Copepoda
        Order Calanoida
        Order Harpacticoida
        Order Cyclopoida
        Order Notodelphyoida
        Order Monstrilloida
        Order Caligoida
        Order Lernaepodoida
      Subclass Branchiura
        Order Arguloida
      Subclass Cirripedia

        Order Thoracica
        Order Acrothoracica
        Order Ascothoracica
        Order Apoda
        Order Rhizocephala
      Subclass Malacostraca
        Superorder Leptostraca
          Order Nebaliacea
        Superorder Hoplocarida
          Order Stomatopoda
        Superorder Syncarida
          Order Anaspidacea
          Order Bathynellacea
        Superorder Pancarida
          Order Thermosbaenacea
        Superorder Peracarida
          Order Spelaeogriphacca
          Order Mysidacea
          Order Cumacea
          Order Tanaidacea
          Order Isopoda
          Order Amphipoda
        Superorder Eucarida
          Order Euphausiacea
          Order Decapoda
    Class Pauropoda
    Class Symphyla
    Class Diplopoda
    Class Chilopoda
      Subclass Notostigmorphora
        Order Scutigeromorpha
      Subclass Pleurostigmophora
        Order Lithobiomorpha
        Order Craterostigmo-
          morpha
        Order Scolopendromor-
          pha
        Order Geophilomorpha
    Class Insecta (Hexapoda)
      Subclass Apterygota
        Order Protura
          (Myrienomata)
        Order Thysanura
        Order Collembola
      Subclass Pterygota
      Exopterygota
        Order Ephemeroptera
          (Ephemerida,
          Plectoptera)
        Order Odonata
        Order Orthoptera
        Order Isoptera
        Order Plecoptera
        Order Dermaptera
          (Euplexoptera)
        Order Embioptera
          (Embiidina)
        Order Psocoptera
          (Corrodentia)
        Order Zoraptera
        Order Mallophaga

Order Anoplura
(Siphunculata)
Order Thysanoptera
(Physopoda)
Order Hemiptera
(Heteroptera)
Order Homoptera
Endopterygota
Order Neuroptera
Order Coleoptera
Order Mecoptera
Order Trichoptera
Order Lepidoptera
Order Diptera
Order Siphonaptera
Order Hymenoptera
Phylum Echinodermata
Subphylum Homalozoa
Subphylum Crinozoa
Class Cystoidea
Order Diploporita
Order Rhombifera
Class Blastoidea
Class Eocrinoidea
Class Paracrinoidea
Class Crinoidea
Subclass Inadunata
Subclass Camerata
Subclass Flexibilia
Subclass Articulata
Subphylum Echinozoa
Class Edrioasteroidea
Class Helicoplacoidea
Class Ophiocistioidea
Class Cyclocystoidea
Class Holothuroidea
Subclass Dendrochirotacea
Order Dactylochirotida
Order Dendrochirotida
Subclass Aspidochirotacea
Order Aspidochirotida
Order Elasipodida
Subclass Apodacea
Order Molpadida
Order Apodida
Class Echinoidea
Subclass Perischoechinoidea
Order Bothriocidaroida
Order Echinocystitoida
Order Palaechinoida
Order Cidaroida
Subclass Euchinoidea
Superorder Diadematacea
Order Diadematoida
Order Pedinoida
Order Echinothurioida
Order Pygasteroida
Superorder Echinacea
Order Salenioida
Order Hemicidaroida
Order Phymosomatoida

Order Arbacioida
Order Temnopleuroida
Order Echinoida
Superorder Gnathostomata
Order Holectypoida
Order Clypeasteroida
Order Cassiduloida
Superorder Atelostomata
Order Holasteroida
Order Spatangoida
Subphylum Asterozoa
Class Stelleroidea
Subclass Somasteroidea
Subclass Asteroidea
Order Platyasterida
Order Hemizonida
Order Phanerozonida
Order Spinulosida
Order Euclasterida
Order Forcipulatida
Subclass Ophiuroidea
Order Stenurida
Order Oegophiurida
Order Phrynophiurida
Order Ophiurida
Phylum Chaetognatha
Phylum Hemichordata
Class Enteropneusta
Class Pterobronchia
Class Graptolithina (Graptozoa)
Phylum Chordata
Subphylum Tunicata (Urochordata)
Class Ascidiacea
Class Thaliaceae
Class Larvaceae
Subphylum Cephalochordate
(Leptocardii)
Subphylum Craniata (Vertebrata)
Class Agnatha
Subclass Diplorhina
Order Heterostraci
Order Coelolepida
Subclass Monorhina
Superorder Hyperotreti
Order Myxinoidea
Superorder Hyperoartii
Order Petromyzonida
Order Osteostraci
Order Anaspida
Subclass Cyclostomata
Class Placodermi
Order Arthrodira
Order Ptyctodontia
Order Phyllolepida
Order Petalichthyida
Order Rhenanida
Order Antiarchi
Class Chondrichthyes
Subclass Elasmobranchii
Order Cladoselachii
Order Pleurocanthodii

Order Selachii
Order Batoidea
Subclass Holocephali
Order Chimaeriformes
Class Osteichthyes
Subclass Actinopterygii
Infraclass Chondrostei
Order Palaeonisciformes
Order Polypteriformes
(Cladistia)
Order Acipenseriformes
Infraclass Holostei
Order Semionotiformes
(Protospondyli,
Ginglymodi, and
Lepisostei)
Order Pycnodontiformes
Order Amiiformes
(Halecomorphi)
Order Aspidorhynchi-
formes
Order Pholidophori-
formes (Halecostomi)
Infraclass Teleostei
Order Leptolepiformes
Order Elopiformes
(Isospondyli in part)
Order Anguilliformes
(Apodes and
Saccopharyngiformes
or Lyomeri)
Order Notacanthiformes
(Lyopomi and
Heteromi)
Order Clupeiformes
Order Osteoglossiformes
(Isospondyli in part
and Mormyriformes)
Order Salmoniformes
(Myctophiformes or
Iniomi, and Haplomi)
Order Cetomimiformes
(Cetunculi)
Order Ctenothrissiformes
Order Gonorynchiformes
Order Cypriniformes
(Heterognathi,
Eventognathi, and
Gymnonoti)
Order Siluriformes
(Nematognathi)
Order Percopsiformes
(Salmopercae and
Amblyopsiformes)
Order Batrachoidiformes
(Haplodoci)
Order Gobiesociformes
(Xenopterygii)
Order Lophiiformes
(Pediculati)
Order Gadiformes

(Anacanthini)
Order Atheriniformes
(Beloniformes or
Synentognathi,
Cyprinodontiformes or
Microcyprini, and
Phallostethiformes)
Order Beryciformes
(Berycomorphi)
Order Zeiformes
(Zeomorphi)
Order Lampridiformes
(Allotriognathi)
Order Gasterosteiformes
(Thoracostei and
Solenichthyes)
Order Pegasiformes
(Hypostomides)
Order Synbranchiformes
(Synbranchii)
Order Perciformes
(Acanthopterygii or
Percomorphi)
Order Pleuronectiformes
(Heterosomata)
Order Tetraodontiformes
(Plectognathi)
Subclass Crossopterygii
Order Osteolepiformes
(Rhipidistia)
Order Coelacanthiformes
(Coelacanthini)
Subclass Dipnoi (Dipneusti)
Order Dipteriformes
Class Amphibia
Subclass Labyrinthodontia
Order Ichthyostegalia
Order Temnospondyli
Order Anthracosauria
Subclass Lepospondyli
Order Nectridea
Order Aistopoda
Order Microsauria
Subclass Lissamphibia
Order Anura
Order Urodela
Order Apoda
Class Reptilia
Subclass Anapsida
Order Cotylosauria
Order Mesosauria
Order Chelonia
Subclass Ichthyopterygia
Order Ichthyosauria
Subclass Euryapsida
Order Sauroptergia
Order Placodontia
Order Araeoscelidia
Subclass Lepidosauria
Order Eosuchia
Order Rhynchocephalia

Order Squamata
Subclass Archosauria
    Order Thecodontia
    Order Crocodilia
    Order Pterosauria
    Order Saurischia
    Order Ornithischia
Subclass Synapsida
    Order Pelycosauria
    Order Theropsida
    Order Ictidosauria
Class Aves
Subclass Archaeornithes
    Order Archaeopterygi-
    formes
Subclass Neornithes
    Superorder Odontognathae
    Order Hesperornithiformes
    Superorder Neognathae
    Order Struthioniformes
    Order Aepyornithiformes
    Order Dinornithiformes
    Order Casuariiformes
    Order Apterygiformes
    Order Rheiformes
    Order Tinamiformes
    Order Gaviiformes
    Order Podicipediformes
    Order Procellariiformes
    Order Sphenisciformes
    Order Pelecaniformes
    Order Ciconiiformes
    Order Anseriformes
    Order Falconiformes
    Order Galliformes
    Order Gruiformes
    Order Diatrymiformes
    Order Charadriiformes
    Order Ichthyornithiformes
    Order Columbiformes
    Order Psittaciformes
    Order Cuculiformes
    Order Strigiformes
    Order Caprimulgiformes
    Order Apodiformes
    Order Coliiformes
    Order Trogoniformes
    Order Coraciiformes

Order Piciformes
Order Passeriformes
Class Mammalia
Subclass Prototheria
    Order Monotremata
Subclass Allotheria
    Order Multituberculata
Subclass Theria
Infraclass Triconodonta
    Order Triconodonta
    Order Docondonta
Infraclass Pantotheria
    Order Pantotheria
    Order Symmetrodonta
Infraclass Metatheria
    Order Marsupialia
Infraclass Eutheria
    Order Insectivora
    Order Deltatheridia
    Order Dermoptera
    Order Tillodontia
    Order Taeniodonta
    Order Chiroptera
    Order Macroscelidea
    Order Primates
    Order Carnivora
    Order Condylarthra
    Order Pantodonta
    Order Dinocerata
    Order Pyrotheria
    Order Proboscidea
    Order Sirenia
    Order Desmostylia
    Order Hyracoidea
    Order Embrithopoda
    Order Notoungulata
    Order Astrapotheria
    Order Xenungulata
    Order Litopterna
    Order Perissodactyla
    Order Artiodactyla
    Order Edentata
    Order Pholidota
    Order Tubulidentata
    Order Cetacea
    Order Rodentia
    Order Lagomorpha

Bacterial taxonomy

Bergey's Manual of Determinative Bacteriology is the standard reference work for taxonomic organization of bacteria. This dictionary observes the taxonomic organization presented in the seventh edition (R. S. Breed, E. G. D. Murray, N. R. Smith, eds.; Baltimore: Williams and Wilkins, 1957). However, the eighth edition (R. E. Buchanan and N. E. Gibbons, eds.; 1974) shows a much modified organization. For comparison, the taxonomic schemes for both editions are outlined in the following pages.

This change in the microbiologists' view of identification and classification of bacteria has evolved slowly from the application of three sets of experimental-observational approaches: (1) the use of comparative cytology with the light microscope and staining methods to observe the behavior and form of that part of the nucleus that contains DNA; (2) the use of the electron microscope for the extension of comparative cytology to the level of the ultrastructure of the bacterial cell; and (3) the use of the techniques of biochemistry and biophysics to further define those features of cellular organization unique to bacteria.

The seventh edition of *Bergey's Manual* classifies bacteria in the form of a complete hierarchy, while the eighth abandons this approach and groups bacteria into 19 sections based on a few readily determined criteria, including structure, genetic data, and biochemical, nutritional, physiological, staining, and ecological characteristics. Many families, orders, and classes, have been eliminated and a large number of genera are considered to be of uncertain affiliation.

Outline from Bergey's eighth edition

The Phototrophic Bacteria
Order I. Rhodospirillales
Suborder: Rhodospirillineae
Family I. Rhodospirillaceae
Genus   I. *Rhodospirillum*
Genus II. *Rhodopseudomonas*
Genus III. *Rhodomicrobium*
Family II. Chromatiaceae
Genus   I. *Chromatium*
Genus   II. *Thiocystis*
Genus III. *Thiosarcina*
Genus IV. *Thiospirillum*
Genus   V. *Thiocapsa*
Genus   VI. *Lamprocystis*
Genus VII. *Thiodictyon*
Genus VIII. *Thiopedia*
Genus   IX. *Amoebobacter*
Genus   X. *Ectothiorhodospira*
Suborder: Chlorobiineae
Family III. Chlorobiaceae
Genus   I. *Chlorobium*
Genus II. *Prosthecochloris*
Genus III. *Chloropseudomonas*
Genus IV. *Pelodictyon*

Genus V. *Clathrochloris*
Addenda
    Genus: *Chlorochromatium*
    Genus: *Cylindrogloea*
    Genus: *Chlorobacterium*
The Gliding Bacteria
  Order I. Mycobacterales
    Family I. Myxococcaceae
      Genus I. *Mycococcus*
    Family II. Archangiaceae
      Genus I. *Archangium*
    Family III. Cystobacteraceae
      Genus I. *Cystobacter*
      Genus II. *Melittangium*
      Genus III. *Stigmatella*
    Family IV. Polyangiaceae
      Genus I. *Polyangium*
      Genus II. *Nannocystis*
      Genus III. *Chondromyces*
  Order II. Cytophagales
    Family I. Cytophagaceae
      Genus I. *Cytophaga*
      Genus II. *Flexibacter*
      Genus III. *Herpetosiphon*
      Genus IV. *Flexithrix*
      Genus V. *Saprospira*
      Genus VI. *Sporocytophaga*
    Family II. Beggiatoaceae
      Genus I. *Beggiatoa*
      Genus II. *Vitreoscilla*
      Genus III. *Thioploca*
    Family III. Simonsiellaceae
      Genus I. *Simonsiella*
      Genus II. *Alysiella*
    Family IV. Leucotrichaceae
      Genus I. *Leucothrix*
      Genus II. *Thiothrix*
    Incertae sedis
      Genus: *Toxothrix*
    Familiae incertae sedis
    Achromatiaceae
      Genus: *Achromatium*
    Pelonemataceae
      Genus I. *Pelonema*
      Genus II. *Achroonema*
      Genus III. *Peloploca*
      Genus IV. *Desmanthos*
The Sheathed Bacteria
    Genus: *Sphaerotilus*
    Genus: *Leptothrix*
    Genus: *Streptothrix*
    Genus: *Lieskeela*
    Genus: *Phragmidiothrix*
    Genus: *Chrenothrix*
    Genus: *Clonothrix*
Budding and/or Appendaged Bacteria
    Genus: *Hyphomicrobium*
    Genus: *Hyphomonas*
    Genus: *Pedomicrobium*
    Genus: *Caulobacter*
    Genus: *Asticcacaulis*

    Genus: *Ancalomicrobium*
    Genus: *Prosthecomicrobium*
    Genus: *Thiodendron*
    Genus: *Pasteuria*
    Genus: *Blastobacter*
    Genus: *Seliberia*
    Genus: *Gallionella*
    Genus: *Nevskia*
    Genus: *Planctomyces*
    Genus: *Metallogenium*
    Genus: *Caulococcus*
    Genus: *Kusnezonia*
The Spirochetes
  Order I. Spirochaetales
    Family I. Spirochaetaceae
      Genus I. *Spirochaeta*
      Genus II. *Cristispira*
      Genus III. *Treponema*
      Genus IV. *Borrelia*
      Genus V. *Leptospira*
Spiral and Curved Bacteria
    Family I. Spirillaceae
      Genus I. *Spirillum*
      Genus II. *Compylobacter*
    Incertae sedis
      Genus: *Bdellovibrio*
      Genus: *Microcyclus*
      Genus: *Pelosigma*
      Genus: *Brachyarcus*
Gram-negative Aerobic Rods and Cocci
    Family I. Pseudomonadaceae
      Genus I. *Pseudomonas*
      Genus II. *Xanthomonas*
      Genus III. *Zoogloea*
      Genus IV. *Glauconobacter*
    Family II. Azotobacteraceae
      Genus I. *Azotobacter*
      Genus II. *Azomonas*
      Genus III. *Beijerinckia*
      Genus IV. *Derxia*
    Family III. Rhizobiaceae
      Genus I. *Rhizobium*
      Genus II. *Agrobacterium*
    Family IV. Methylomonadaceae
      Genus I. *Methylomonas*
      Genus II. *Methylococcus*
    Family V. Halobacteriaceae
      Genus I. *Halobacterium*
      Genus II. *Halococcus*
    Incertae sedis
      Genus: *Alcaligenes*
      Genus: *Acetobacter*
      Genus: *Brucella*
      Genus: *Bordetella*
      Genus: *Francisella*
      Genus: *Thermus*
Gram-negative Facultatively Anaerobic Rods
    Family I. Enterobacteriaceae
      Genus I. *Escherichia*
      Genus II. *Edwardsiella*
      Genus III. *Citrobacter*

Genus   IV. *Salmonella*
Genus   V. *Shigella*
Genus   VI. *Klebsiella*
Genus   VII. *Enterobacter*
Genus VIII. *Hafnia*
Genus   IX. *Serratia*
Genus   X. *Proteus*
Genus   XI. *Yersinia*
Genus   XII. *Erwinia*
Family II. Vibrionaceae
Genus   I. *Vibrio*
Genus   II. *Aeromonas*
Genus III. *Plesiomonas*
Genus IV. *Photobacterium*
Genus   V. *Lucibacterium*
Incertae sedis
Genus: *Zymomonas*
Genus: *Chromobacterium*
Genus: *Flavobacterium*
Genus: *Haemophilus*
Genus: *Pasteurella*
Genus: *Actinobacillus*
Genus: *Cardiobacterium*
Genus: *Streptobacillus*
Genus: *Calymmatobacterium*
Gram-negative Anaerobic Bacteria
Family I. Bacteroidaceae
Genus   I. *Bacteroides*
Genus   II. *Fusobacterium*
Genus III. *Leptotrichia*
Incertae sedis
Genus: *Desulfovibrio*
Genus: *Butyrivibrio*
Genus: *Succinivibrio*
Genus: *Succinimonas*
Genus: *Lachnospira*
Genus: *Selenomonas*
Gram-negative Cocci and Coccobacilli
Family I. Neisseriaceae
Genus   I. *Neisseria*
Genus   II. *Branhamella*
Genus III. *Moraxella*
Genus IV. *Acinetobacter*
Incertae sedis
Genus: *Paracoccus*
Genus: *Lampropedia*
Gram-negative Anaerobic Cocci
Family I. Veillonellaceae
Genus   I. *Veillonella*
Genus   II. *Acidaminococcus*
Genus III. *Megasphaera*
Gram-negative Chemolithotrophic Bacteria
Family I. Nitrobacteraceae
Genus   I. *Nitrobacter*
Genus   II. *Nitrospina*
Genus III. *Nitrococcus*
Genus IV. *Nitrosomonas*
Genus   V. *Nitrospira*
Genus VI. *Nitrosococcus*
Genus VII. *Nitrosolobus*
Organisms Metabolizing Sulfur

Genus 1. *Thiobacillus*
Genus 2. *Sulfolobus*
Genus 3. *Thiobacterium*
Genus 4. *Macromonas*
Genus 5. *Thiovulum*
Genus 6. *Thiospira*
Family II. Siderocapsaceae
Genus   I. *Siderocapsa*
Genus   II. *Ochrobium*
Genus III. *Siderococcus*
Methane-producing Bacteria
Family I. Methanobacteriaceae
Genus   I. *Methanobacterium*
Genus   II. *Methanosarcina*
Genus III. *Methanococcus*
Gram-positive Cocci
Family I. Micrococcaceae
Genus   I. *Micrococcus*
Genus   II. *Staphylococcus*
Genus III. *Planococcus*
Family II. Streptococcaceae
Genus   I. *Streptococcus*
Genus   II. *Leuconostoc*
Genus III. *Pediococcus*
Genus IV. *Aerococcus*
Genus   V. *Gemella*
Family III. Peptococcaceae
Genus   I. *Peptococcus*
Genus   II. *Peptostreptococcus*
Genus III. *Ruminococcus*
Genus IV. *Sarcina*
Endospore-forming Rods and Cocci
Family I. Bacillaceae
Genus   I. *Bacillus*
Genus   II. *Sporolactobacillus*
Genus III. *Clostridium*
Genus IV. *Desulfotomaculum*
Genus   V. *Sporosarcina*
Incertae sedis
Genus: *Oscillospira*
Gram-positive, Asporogenous Rod-shaped Bacteria
Family I. Lactobacillaceae
Genus I. *Lactobacillus*
Incertae sedis
Genus: *Listeria*
Genus: *Erysipelothrix*
Genus: *Caryophanon*
Actinomycetes and Related Organisms
Coryneform Group of Bacteria
Genus   I. *Corynebacterium*
Genus   II. *Arthrobacter*
Incertae sedis
Genus   A. *Brevibacterium*
Genus   B. *Microbacterium*
Genus III. *Cellulomonas*
Genus IV. *Kurthia*
Family I. Propionibacteriaceae
Genus   I. *Propionibacterium*
Genus   II. *Eubacterium*
Order I. Actinomycetales
Family I. Actinomycetaceae

Genus   I. *Actinomyces*
Genus   II. *Arachnia*
Genus III. *Bifidobacterium*
Genus IV. *Bacterionema*
Genus   V. *Rothia*
Family II. Mycobacteriaceae
Genus I. *Mycobacterium*
Family III. Frankiaceae
Genus I. *Frankia*
Family IV. Actinoplanaceae
Genus   I. *Actinoplanes*
Genus   II. *Spirillospora*
Genus   III. *Streptosporangium*
Genus   IV. *Amorphosporangium*
Genus   V. *Ampullariella*
Genus   VI. *Pilemelia*
Genus VII. *Planomonospora*
Genus VIII. *Planobispora*
Genus   IX. *Dactylosporangium*
Genus   X. *Kitastoa*
Family V. Dermatophilaceae
Genus I. *Dermatophilus*
Genus II. *Geodermatophilus*
Family VI. Nocardiaceae
Genus I. *Nocardia*
Genus II. *Pseudonocardia*
Family VII. Streptomycetaceae
Genus   I. *Streptomyces*
Genus   II. *Streptoverticillium*
Genus III. *Sporichthya*
Genus IV. *Microellobosporia*
Family VIII. Micromonosporaceae
Genus   I. *Micromonospora*
Genus   II. *Thermoactinomyces*
Genus III. *Actinobifida*
Genus IV. *Thermomonospora*
Genus   V. *Microbispora*
Genus VI. *Micropolyspora*
The Rickettsias

Order I. Rickettsiales
Family: Rickettsiaceae
Tribe I. Rickettsieae
Genus   I. *Rickettsia*
Genus   II. *Rochalimaea*
Genus   III. *Coxiella*
Tribe II. Ehrlichieae
Genus   IV. *Ehrlichia*
Genus   V. *Cowdria*
Genus   VI. *Neorickettsia*
Tribe III. Wolbachieae
Genus VII. *Wolbachia*
Genus VIII. *Symbiotes*
Genus   IX. *Blattabacterium*
Genus   X. *Rickettsiella*
Family: Bartonellaceae
Genus I. *Bartonella*
Genus II. *Grahamella*
Family: Anaplasmataceae
Genus   I. *Anaplasma*
Genus II. *Paranaplasma*
Genus III. *Aegyptianella*
Genus IV. *Haemobartonella*
Genus V. *Eperythrozoon*
Order II. Chlamydiales
Family I. Chlamydiaceae
Genus I. *Chlamydia*
The Mycoplasmas
Class: Mollicutes
Order I. Mycoplasmatales
Family I. Mycoplasmataceae
Genus I. *Mycoplasma*
Family II. Acholeplasmataceae
Genus I. *Acholeplasma*
Incertae sedis
Genus: *Thermoplasma*
Incertae sedis
Genus: *Spiroplasma*

Outline from Bergey's seventh edition

Division I. Protophyta
Class I. Schizophyceae
Class II. Schizomycetes
Order I. Pseudomonadales
Suborder I. Rhodobacteriineae
Family I. Thiorhodaceae
Genus I. *Thiosarcina*
Genus   II. *Thiopedia*
Genus   III. *Thiocapsa*
Genus   IV. *Thiodictyon*
Genus   V. *Thiothece*
Genus   VI. *Thiocystis*
Genus VII. *Lamprocystis*
Genus VIII. *Amoebobacter*
Genus   IX. *Thiopolycoccus*
Genus   X. *Thiospirillum*
Genus   XI. *Rhabdomonas*
Genus XII. *Rhodothece*

Genus XIII. *Chromatium*
Family II. Athiorhodaceae
Genus I. *Rhodopseudomonas*
Genus II. *Rhodospirillum*
Family III. Chlorobacteriaceae
Genus   I. *Chlorobium*
Genus   II. *Pelodictyon*
Genus III. *Clathrochloris*
Genus   IV. *Chlorobacterium*
Genus   V. *Chlorochromatium*
Genus   VI. *Cylindrogloea*
Suborder II. Pseudomonadineae
Family I. Nitrobacteraceae
Genus   I. *Nitrosomonas*
Genus   II. *Nitrosococcus*
Genus III. *Nitrosospira*
Genus   IV. *Nitrosocystis*
Genus   V. *Nitrosogloea*

Genus VI. *Nitrobacter*
Genus VII. *Nitrocystis*
Family II. Methanomonadaceae
  Genus I. *Methanomonas*
  Genus II. *Hydrogenomonas*
  Genus III. *Carboxydomonas*
Family III. Thiobacteriaceae
  Genus I. *Thiobacterium*
  Genus II. *Macromonas*
  Genus III. *Thiovulum*
  Genus IV. *Thiospira*
  Genus V. *Thiobacillus*
Family IV. Pseudomonadaceae
  Genus I. *Pseudomonas*
  Genus II. *Xanthomonas*
  Genus III. *Acetobacter*
  Genus IV. *Aeromonas*
  Genus V. *Photobacterium*
  Genus VI. *Azotomonas*
  Genus VII. *Zymomonas*
  Genus VIII. *Protaminobacter*
  Genus IX. *Alginomonas*
  Genus X. *Mycoplana*
  Genus XI. *Zoogloea*
  Genus XII. *Halobacterium*
Family V. Caulobacteraceae
  Genus I. *Caulobacter*
  Genus II. *Gallionella*
  Genus III. *Siderophacus*
  Genus IV. *Nevskia*
Family VI. Siderocapsaceae
  Genus I. *Siderocapsa*
  Genus II. *Siderosphaera*
  Genus III: *Sideronema*
  Genus IV. *Ferribacterium*
  Genus V. *Sideromonas*
  Genus VI. *Naumanniella*
  Genus VII. *Ochrobium*
  Genus VIII. *Siderococcus*
  Genus IX. *Siderobacter*
  Genus X. *Ferrobacillus*
Family VII. Spirillaceae
  Genus I. *Vibrio*
  Genus II. *Desulfovibrio*
  Genus III. *Methanobacterium*
  Genus IV. *Cellvibrio*
  Genus V. *Cellfalcicula*
  Genus VI. *Microcyclus*
  Genus VII. *Spirillum*
  Genus VIII. *Paraspirillum*
  Genus IX. *Selenomonas*
  Genus X. *Myconostoc*
Order II. Chlamydobacteriales
  Family I. Chlamydobacteriaceae
    Genus I. *Sphaerotilus*
    Genus II. *Leptothrix*
    Genus III. *Toxothrix*
  Family II. Peloplocaceae
    Genus I. *Peloploca*
    Genus II. *Pelonema*
  Family III. Crenotrichaceae

Genus I. *Crenothrix*
Genus II. *Phragmidiothrix*
Genus III. *Clonothrix*
Order III. Hyphomicrobiales
  Family I. Hyphomicrobiaceae
    Genus I. *Hyphomicrobium*
    Genus II. *Rhodomicrobium*
  Family II. Pasteuriaceae
    Genus I. *Pasteuria*
    Genus II. *Blastocaulis*
Order IV. Eubacteriales
  Family I. Azotobacteraceae
    Genus I. *Azotobacter*
  Family II. Rhizobiaceae
    Genus I. *Rhizobium*
    Genus II. *Agrobacterium*
    Genus III. *Chromobacterium*
  Family III. Achromobacteraceae
    Genus I. *Alcaligenes*
    Genus II. *Achromobacter*
    Genus III. *Flavobacterium*
    Genus IV. *Agarbacterium*
    Genus V. *Beneckea*
  Family IV. Enterobacteriaceae
    Tribe I. Escherichieae
      Genus I. *Escherichia*
      Genus II. *Aerobacter*
      Genus III. *Klebsiella*
      Genus IV. *Paracolobactrum*
      Genus V. *Alginobacter*
    Tribe II. Erwinieae
      Genus VI. *Erwinia*
    Tribe III. Serrateae
      Genus VII. *Serratia*
    Tribe IV. Proteeae
      Genus VIII. *Proteus*
    Tribe V. Salmonelleae
      Genus IX. *Salmonella*
      Genus X. *Shigella*
  Family V. Brucellaceae
    Genus I. *Pasturella*
    Genus II. *Bordetella*
    Genus III. *Brucella*
    Genus IV. *Haemophilus*
    Genus V. *Actinobacillus*
    Genus VI. *Calymmato-bacterium*
    Genus VII. *Morazella*
    Genus VIII. *Noguchia*
  Family VI. Bacteroidaceae
    Genus I. *Bacteroides*
    Genus II. *Fusobacterium*
    Genus III. *Dialister*
    Genus IV. *Sphaerophorus*
    Genus V. *Streptobacillus*
  Family VII. Micrococcaceae
    Genus I. *Micrococcus*
    Genus II. *Staphylococcus*
    Genus III. *Gaffkya*
    Genus IV. *Sarcina*
      Subgenus I. *Zymosarcina*

Subgenus II. *Methanosarcina*
Subgenus III. *Sarcinococcus*
Subgenus IV. *Urosarcina*
Genus V. *Methanococcus*
Genus VI. *Peptococcus*
Familly VIII. Neisseriaceae
Genus I. *Neisseria*
Genus II. *Veillonella*
Family IX. Brevibacteriaceae
Genus I. *Brevibacterium*
Genus II. *Kurthia*
Family X. Lactobacillaceae
Tribe I. Streptococceae
Genus I. *Diplococcus*
Genus II. *Streptococcus*
Genus III. *Pediococcus*
Genus IV. *Leuconostoc*
Genus V. *Peptostreptococcus*
Tribe II. Lactobacillae
Genus I. *Lactobacillus*
Subgenus I. *Lactobacillus*
Subgenus II. *Saccharobacillus*
Genus II. *Eubacterium*
Genus III. *Catenabacterium*
Genus IV. *Ramibacterium*
Genus V. *Cillobacterium*
Family XI. Propionibacteriaceae
Genus I. *Propionibacterium*
Genus II. *Butyribacterium*
Genus III. *Zymobacterium*
Family XII. Corynebacteriaceae
Genus I. *Corynebacterium*
Genus II. *Listeria*
Genus III. *Erysipelothrix*
Genus IV. *Microbacterium*
Genus V. *Cellulomonas*
Genus VI. *Arthrobacter*
Family XIII. Bacillaceae
Genus I. *Bacillus*
Genus II. *Clostridium*
Order V. Caryophanales
Family I. Caryophanaceae
Genus I. *Caryophanon*
Genus II. *Lineola*
Genus III. *Simonsiella*
Family II. Oscillospiraceae
Genus I. *Oscillospira*
Family III. Arthromitaceae
Genus I. *Arthromitus*
Genus II. *Coleomitus*
Order VI. Actinomycetales
Family I. Mycobacteriaceae
Genus I. *Mycobacterium*
Genus II. *Mycococcus*
Family II. Actinomycetaceae
Genus I. *Nocardia*
Genus II. *Actinomyces*
Family III. Streptomycetaceae
Genus I. *Streptomyces*
Genus II. *Micromonospora*
Genus III. *Thermoactinomyces*

Family IV. Actinoplanaceae
Genus I. *Actinoplanes*
Genus II. *Streptosporangium*
Order VII. Beggiatoales
Family I. Beggiatoaceae
Genus I. *Beggiatoa*
Genus II. *Thiospirillopsis*
Genus III. *Thioploca*
Genus IV. *Thiothrix*
Family II. Vitreoscillaceae
Genus I. *Vitreoscilla*
Genus II. *Bactoscilla*
Genus III. *Microscilla*
Family III. Leucotrichaceae
Genus I. *Leucothrix*
Family IV. Achromatiaceae
Genus I. *Achromatium*
Order VIII. Myxobacterales
Family I. Cytophagaceae
Genus I. *Cytophaga*
Family II. Archangiaceae
Genus I. *Archangium*
Genus II. *Stelangium*
Family III. Sorangiaceae
Genus I. *Sorangium*
Family IV. Polyangiaceae
Genus I. *Polyangium*
Genus II. *Synangium*
Genus III. *Podangium*
Genus IV. *Chondromyces*
Family V. Myxococcaceae
Genus I. *Myxococcus*
Genus II. *Chondrococcus*
Genus III. *Angiococcus*
Genus IV. *Sporocytophaga*
Order IX. Spirochaetales
Family I. Spirochaetaceae
Genus I. *Spirochaeta*
Genus II. *Saprospira*
Genus III. *Cristispira*
Family II. Treponemataceae
Genus I. *Borrelia*
Genus II. *Treponema*
Genus III. *Leptospira*
Order X. Mycoplasmatales
Family I. Mycoplasmataceae
Genus I. *Mycoplasma*
Addendum to Class II: Schizomycetes
Class III. Microtatobiotes
Order I. Rickettsiales
Family I. Rickettsiaceae
Tribe I. Rickettsieae
Genus I. *Rickettsia*
Subgenus A. *Rickettsia*
Subgenus B. *Zinssera*
Subgenus C. *Dermacentroxenus*
Subgenus D. *Rochalimaea*
Genus II. *Coxiella*
Tribe II. Ehrlichieae
Genus III. *Ehrlichia*
Genus IV. *Cowdria*

Genus V. *Neorickettsia*
Tribe III. Wolbachieae
    Genus VI. *Wolbachia*
    Genus VII. *Symbiotes*
    Genus VIII. *Rickettsiella*
Family II. Chlamydiaceae
    Genus I. *Chlamydia*
    Genus II. *Colesiota*
    Genus III. *Ricolesia*
    Genus IV. *Colettsia*

Genus V. *Miyagawanella*
Family III. Bartonellaceae
    Genus I. *Bartonella*
    Genus II. *Grahamella*
    Genus III. *Haemobartonella*
    Genus IV. *Eperythrozoon*
Family IV. Anaplasmataceae
    Genus I. *Anaplasma*
Order II. Virales

Plant taxonomy

Subkingdom Thallobionta (thallophytes)
  Division Schizophyta (prokaryotes)
    Class Schizomycetes (bacteria)
    Class Cyanophyceae (blue-green algae)
  Division Rhodophyta (red algae)
    Class Bangiophyceae
    Class Rhodophyceae
  Division Chlorophyta (green algae)
    Class Chlorophyceae
    Class Charophyceae
  Division Euglenophyta (englenoids)
    Class Euglenophyceae
  Division Pyrrhophyta
    Class Cryptophyceae (cryptomonads)
    Class Desmophyceae
    Class Dinophyceae (dinoflagellates)
  Division Chrysophyta
    Class Chloromonadophyceae (chloromonads)
    Class Xanthophyceae (yellow-green algae)
    Class Chrysophyceae (golden algae)
    Class Bacillariophyceae (diatoms)
  Division Phaeophyta (brown algae)
    Class Laminariophyceae
    Class Fucophyceae
  Division Fungi
    Class Myxomycetes (slime molds)
    Class Oomycetes
    Class Chytridiomycetes (chytrids)
    Class Hyphochrytridiomycetes
    Class Zygomycetes
    Class Ascomycetes (sac fungi)
    Class Basidiomycetes (club fungi)
Subkingdom Embryobionta (embryophytes)
  Division Rhyniophyta
    Class Rhyniatae
  Division Bryophyta
    Class Anthocerotae (horned liverworts)
    Class Marchantiatae (liverworts)
    Class Bryatae (mosses)
  Division Psilotophyta
    Class Psieotatae
  Division Lycopodiophyta
    Class Lycopodiatae (club mosses)

    Class Isoetatae
  Division Equisetophyta
    Class Hyeniatae
    Class Sphenophyllatae
    Class Equisetatae (horsetails)
  Division Polypodiophyta (ferns)
    Class Polypodiatae
      Order Protopteridales
      Order Archaeopteridales
      Order Ophioglossales
      Order Noeggerathiales
      Order Marattiales
      Order Polypodiales
      Order Marsileales
      Order Salviniales
  Division Pinophyta (gymnosperms)
    Subdivision Cycadicae
      Class Lyginopteridatae
        Order Lyginopteridales (seed ferns)
        Order Caytoniales
      Class Cycadatae (cycads)
        Order Cycadales
      Class Bennettitatae (cycadeoids)
        Order Bennettitales
    Subdivision Pinicae
      Class Ginkgoatae (maidenhair tree)
        Order Ginkgoales
      Class Pinatae
        Order Pinales (conifers)
        Order Taxales
    Subdivision Gneticae
      Class Gnetatae
        Order Ephedrales
        Order Welwitschiales
        Order Gnetales
  Division Magnoliophyta (angiosperms; flowering plants)
    Class Magnoliatae (dicotyledons)
      Subclass Magnoliidae
        Order Magnoliales
        Order Piperales
        Order Aristolochiales
        Order Nymphaeales
        Order Ranunculales

Order Papaverales
Subclass Hamamelidae
  Order Trochodendrales
  Order Hamamelidales
  Order Eucommiales
  Order Urticales
  Order Leitneriales
  Order Juglandales
  Order Myricales
  Order Fagales
  Order Casuarinales
Subclass Caryophyllidae
  Order Caryophyllales
  Order Batales
  Order Polygonales
  Order Plumbaginales
Subclass Dilleniidae
  Order Dilleniales
  Order Theales
  Order Malvales
  Order Lecythidales
  Order Sarraceniales
  Order Violales
  Order Salicales
  Order Capparales
  Order Ericales
  Order Diapensiales
  Order Ebenales
  Order Primulales
Subclass Rosidae
  Order Rosales
  Order Podostemales
  Order Haloragales
  Order Myrtales
  Order Proteales
  Order Cornales
  Order Santalales
  Order Rafflesiales
  Order Celastrales
  Order Euphorbiales

Order Rhamnales
Order Sapindales
Order Geraniales
Order Linales
Order Polygalales
Order Umbellales
Subclass Asteridae
  Order Gentianales
  Order Polemoniales
  Order Lamiales
  Order Plantaginales
  Order Scrophulariales
  Order Campanulales
  Order Rubiales
  Order Dipsacales
  Order Asterales
Class Liliatae (monocotyledons)
Subclass Alismatidae
  Order Alismatales
  Order Hydrocharitales
  Order Najadales
  Order Triuridales
Subclass Commelinidae
  Order Commelinales
  Order Eriocaulales
  Order Restionales
  Order Juncales
  Order Cyperales
  Order Typhales
  Order Bromeliales
  Other Zingiberales
Subclass Arecidae
  Order Arecales
  Ordery Cyclanthales
  Order Pandanales
  Order Arales
Subclass Liliidae
  Order Liliales
  Order Orchiadales